Earth

TWELFTH EDITION

Earth
An Introduction to Physical Geology

EDWARD J. TARBUCK

FREDERICK K. LUTGENS

Illustrated by
DENNIS TASA

PEARSON

Boston   Columbus   Indianapolis   New York   San Francisco   Hoboken
Amsterdam   Cape Town   Dubai   London   Madrid   Milan   Munich   Paris   Montréal   Toronto
Delhi   Mexico City   São Paulo   Sydney   Hong Kong   Seoul   Singapore   Taipei   Tokyo

Editor-in-Chief: Beth Wilbur
Senior Marketing Manager: Mary Salzman
Senior Acquisitions Editor: Andrew Dunaway
Executive Marketing Manager: Neena Bali
Senior Project Manager: Crissy Dudonis/Nicole Antonio
Director of Development: Jennifer Hart
Development Editor: Veronica Jurgena
Senior Content Producer: Timothy Hainley
Program Manager: Sarah Shefveland
Editorial Assistant: Michelle Koski
Marketing Assistant: Ami Sampat
Team Lead, Project Management: David Zielonka
Team Lead, Program Management: Kristen Flathman

Project Manager, Instructor Media: Kyle Doctor
Full Service/Composition: Cenveo® Publisher Services
Full Service Project Manager: Heidi Allgair/Cenveo®
    Publisher Services
Design Manager: Derek Bacchus
Cover and Interior Design: Jeff Puda Design
Photo and Illustration Support: International Mapping
Photo Manager: Rachel Youdelman
Photo Researcher: Kristin Piljay
Text Permissions Manager: Timothy Nicholls
Text and Photo Permissions Researchers: Cordes Hoffman
    and Erica Gordon, QBS Learning
Procurement Specialist: Maura Zaldivar-Garcia

Cover Image Credit: © Michael Collier

Credits and acknowledgments for materials borrowed from other sources and reproduced, with permission, in this textbook appear on the appropriate page within the book.

Library of Congress Cataloging-in-Publication Data

Tarbuck, Edward J.
  Earth: an introduction to physical geology/Edward J. Tarbuck, Frederick K. Lutgens; illustrated by Dennis Tasa.--Twelfth edition.
     pages cm
Includes index.
ISBN 978-0-13-407425-2–ISBN 0-13-407425-4
1. Physical geology–Textbooks. I. Lutgens, Frederick K. II. Title.
  QE28.2.T37 2016
  550-dc23

          2015011695

3 18

Student Edition
ISBN-10: 0-134-07425-4
ISBN-13: 978-0-134-07425-2

Instructor's Review Copy
ISBN-10: 0-134-25411-2
ISBN-13: 978-0-134-25411-1

www.pearsonhighered.com

To Our Grandchildren

Shannon, Amy, Andy, Ali, and Michael

Allison and Lauren

Each is a bright promise for the future.

Brief Contents

Table of Contents

Page 36

Page 72

4
Magma, Igneous Rocks, and Intrusive Activity 106

Page 106

5
Volcanoes and Volcanic Hazards 140

Page 180

Page 272

Page 240

Page 362

Page 418

Page 442

17
Groundwater 500

18
Glaciers and Glaciation 532

19
Deserts and Wind 570

Page 532

Page 594

Page 702

SmartFigures

Using a cutting-edge technology called augmented reality, Pearson's BouncePages app launches engaging, interactive videos and animations that bring textbook pages to life. Use your mobile device to scan a SmartFigure identified by the BouncePages icon, and an animation/video/interactive illustrating the SmartFigure's concept launches immediately. No slow websites or hard-to-remember logins required. BouncePages' augmented reality technology transforms textbooks into convenient digital platforms, breathes life into your learning experience, and helps you grasp difficult academic concepts. Learning geology from a textbook will never be the same.

Preface

Earth is a very small part of a vast universe, but it is our home. It provides the resources that support our modern society and the ingredients necessary to maintain life. Knowledge of our physical environment is critical to our well-being and vital to our survival. A basic geology course can help a person gain such an understanding, and can also take advantage of the interest and curiosity many of us have about our planet—its landscapes and the processes that create and alter our physical environment.

This 12th edition of *Earth: An Introduction to Physical Geology*, like its predecessors, is a college-level text that is intended to be a meaningful, nontechnical survey for students taking their first course in geology. In addition to being informative and up-to-date, a major goal of *Earth* is to meet the need for a readable and user-friendly text, a book that is a highly usable tool for students learning the basic principles and concepts of geology.

Although many topical issues are examined in the 12th edition of *Earth*, it should be emphasized that the main focus of this new edition remains the same as the focus of earlier editions: to promote student understanding of basic principles. As much as possible, we have attempted to provide the reader with a sense of the observational techniques and reasoning processes that constitute the science of geology.

New and Important Features

The 12th edition represents an extensive and thorough revision of *Earth* that integrates improved textbook resources with new online features to enhance the learning experience,

- **Significant updating and revision of content.** A basic function of a college science textbook is to present material in a clear, understandable way that is accurate, engaging, and up-to-date. In the long history of this textbook, our number-one goal has always been to keep *Earth* current, relevant, and highly readable for beginning students. To that end, every part of this text has been examined carefully. Many discussions, case studies, examples, and illustrations have been updated and revised.

- **SmartFigures that make *Earth* much more than a traditional textbook.** Through its many editions, an important strength of *Earth* has always been clear, logically organized, and well-illustrated explanations. Now, complementing and reinforcing this strength are a series of SmartFigures. Simply by scanning a SmartFigure with a mobile device and **Pearson's BouncePages Augmented Reality app** (FREE and available for iOS and Android),

students can link to hundreds of unique and innovative digital learning opportunities that will increase their insight and understanding of important ideas and concepts. We have also placed short URLs in the caption for every SmartFigure. This will ensure that students who may not have a smart phone, will have the ability to access these videos easily. SmartFigures are truly art that teaches! This 12th edition of *Earth* has more than 200 SmartFigures, of five different types:

1. **SmartFigure Tutorials.** Each of these 2- to 4-minute tutorials, prepared and narrated by Professor Callan Bentley, is a mini-lesson that examines and explains the concepts illustrated by the figure.
2. **SmartFigure Mobile Field Trips.** Scattered throughout this new edition are 24 video field trips that explore classic geologic sites from Iceland to Hawaii. On each trip you will accompany geologist-pilot-photographer Michael Collier in the air and on the ground to see and learn about landscapes that relate to discussions in the chapter.
3. **SmartFigure Condor Videos.** The 10 *Condor* videos take you to sites in the American West. By coupling aerial footage acquired by a quadcopter aircraft with ground-level views, effective narratives, and helpful animations, these videos will engage you in real-life case studies.
4. **SmartFigure Animations.** Scanning the many figures with this designation brings art to life. These animations and accompanying narrations illustrate and explain many difficult-to-visualize topics and ideas more effectively than static art alone.
5. **SmartFigure Videos.** Rather than providing a single image to illustrate an idea, these figures include short video clips that help illustrate such diverse subjects as mineral properties and the structure of ice sheets.

As mentioned, please visit your iOS or Android App Stores to download the FREE Pearson BouncePages App. By scanning one of the two QR codes to the right, you will be taken directly to the BouncePages App.

Android Store   iTunes App Store

- **Enhanced Modular, learning objective-driven, active learning path.** *Earth* is designed for learning. Every chapter begins with *Focus on Concepts*. Each numbered learning objective corresponds to a major section in the chapter. The statements identify the knowledge and skills students should master by the end of the chapter and help students prioritize key concepts. Within the chapter, each major section concludes with *Concept Checks* that allow students to check their understanding and comprehension of important ideas and terms before moving on to the next section. Two end-of-chapter features complete the learning path. *Concepts in Review* coordinates with the *Focus on Concepts* at the start of the chapter and with the numbered sections within the chapter. It is a concise overview of key ideas, with photos, diagrams, and questions that help students focus on important ideas and test their understanding of key concepts. Chapters conclude with *Give It Some Thought* questions that challenge learners by involving them in activities that require higher-order thinking skills, such as application, analysis, and synthesis of chapter material.

- **An unparalleled visual program.** In addition to more than 100 new, high-quality photos and satellite images, dozens of figures are new or have been redrawn by the gifted and highly respected geoscience illustrator Dennis Tasa. Maps and diagrams are frequently paired with photographs for greater effectiveness. Further, many new and revised figures have additional labels that narrate the process being illustrated and guide students as they examine the figures. *Earth's* visual program is clear and easy to understand.

- **MasteringGeology.** MasteringGeology delivers engaging, dynamic learning opportunities—focused on course objectives and responsive to each student's progress—that are proven to help students learn course material and understand difficult concepts. Assignable activities in MasteringGeology include SmartFigure (Tutorial, Condor, Animation, Mobile Field Trip, Video) activities, GigaPan activities, Encounter Earth activities using Google Earth, GeoTutor activities, Geoscience Animation activities, GEODe tutorials, and more. MasteringGeology also includes all instructor resources and a robust Study Area with resources for students.

The Teaching and Learning Package

MasteringGeology™ with Pearson eText

Used by over 1 million science students, the Mastering platform is the most effective and widely used online tutorial, homework, and assessment system for the sciences. Now available with *Earth*, 12th edition, **MasteringGeology™** offers tools for use before, during, and after class:

- **Before class:** Assign adaptive Dynamic Study Modules and reading assignments from the eText with Reading Quizzes to ensure that students come prepared to class, having done the reading.

- **During class:** Learning Catalytics, a "bring your own device" student engagement, assessment, and classroom intelligence system, allows students to use a smartphone, tablet, or laptop to respond to questions in class. With Learning Catalytics, you can assess students in real-time, using open-ended question formats to uncover student misconceptions and adjust lectures accordingly.

- **After class:** Assign an array of assessment resources such as Mobile Field Trips, Project Condor tutorials, Interactive Simulations, GeoDrone activities, Google Earth Encounter Activities, and much more. Students receive wrong-answer feedback personalized to their answers, which will help them get back on track.

MasteringGeology Student Study Area also provides students with self-study materials like geoscience animations, *GEODe: Earth* activities, *In the News* RSS feeds, Self Study Quizzes, Web Links, Glossary, and Flashcards.

For more information or access to MasteringGeology, please visit www.masteringgeology.com.

Instructor's Resource Materials (Download Only)

The authors and publisher have been pleased to work with a number of talented people who have produced an excellent supplements package.

Instructor's Resource Materials (IRM)

The IRM puts all your lecture resources in one easy-to-reach place:

- *All* of the line art, tables, and photos from the text in .jpg files

- PowerPoint presentations

  - The IRM provides three PowerPoint files for each chapter. Cut down on your preparation time, no matter what your lecture needs, by taking advantage of these components of the PowerPoint files:

- **Exclusive art.** All of the photos, art, and tables from the text, in order, loaded into PowerPoint slides.

- **Lecture outlines.** This set averages 50 slides per chapter and includes customizable lecture outlines with supporting art.

- **Classroom Response System (CRS) questions.** Authored for use in conjunction with classroom response systems, these PowerPoints allow you to electronically poll your class for responses to questions, pop quizzes, attendance, and more.

Instructor Manual (Download Only)

The Instructor Manual has been designed to help seasoned and new professors alike, offering the following for each chapter: an introduction to the chapter, an outline, and learning objectives/Focus on Concepts; teaching strategies; teacher resources; and answers to *Concept Checks*, *Eye on Earth*, and *Give It Some Thought* questions from the textbook.

TestGen Computerized Test Bank (Download Only)

TestGen is a computerized test generator that lets instructors view and edit Test Bank questions, transfer questions to tests, and print tests in a variety of customized formats. The Test Bank includes more than 2,000 multiple-choice, matching, and essay questions. Questions are correlated to Bloom's Taxonomy, each chapter's learning objectives, the Earth Science Learning Objectives, and the Pearson Science Global Outcomes to help instructors better map the assessments against both broad and specific teaching and learning objectives. The Test Bank is also available in Microsoft Word and can be imported into Blackboard.

Blackboard

Already have your own website set up? We will provide a Test Bank in Blackboard or formats for importation upon request. Additional course resources are available on the IRC and are available for use, with permission.

Acknowledgments

Writing a college textbook requires the talents and cooperation of many people. It is truly a team effort, and the authors are fortunate to be part of an extraordinary team at Pearson Education. In addition to being great people to work with, all of them are committed to producing the best textbooks possible. Special thanks to our geology editor, Andy Dunaway, who invested a great deal of time, energy, and effort in this project. We appreciate his enthusiasm, hard work, and quest for excellence. We also appreciate our conscientious project managers, Crissy Dudonis and Nicole Antonio, whose job it was to keep track of all that was going on—and a lot was going on. As always, our marketing managers, Neena Bali and Mary Salzman, who talk with faculty daily, provide us with helpful advice and many good ideas. The 12th edition of *Earth* was certainly improved by the talents of our developmental editor, Veronica Jurgena. Many thanks. It was the job of the production team, led by Heidi Allgair at Cenveo® Publisher Services, to turn our manuscript into a finished product. The team also included copyeditor Kitty Wilson, compositor Annamarie Boley, proofreader Heather Mann, and photo researcher Kristin Piljay. We think these talented people did great work. All are true professionals, with whom we are very fortunate to be associated.

The authors owe special thanks to three people who were very important contributors to this project:

- Working with Dennis Tasa, who is responsible for all of the text's outstanding illustrations and several of its animations, is always special for us. He has been part of our team for more than 30 years. We not only value his artistic talents, hard work, patience, and imagination but his friendship as well.

- As you read this text, you will see dozens of extraordinary photographs by Michael Collier. Most are aerial shots taken from his nearly 60-year-old Cessna 180. Michael was also responsible for preparing the 24 remarkable Mobile Field Trips that are scattered through the text. Among his many awards is the American Geological Institute Award for Outstanding Contribution to the Public Understanding of Geosciences. We think that Michael's photographs and field trips are the next best thing to being there. We were very fortunate to have had Michael's assistance on *Earth*, 12th edition. Thanks, Michael.

- Callan Bentley has been an important addition to the *Earth* team. Callan is an assistant professor of geology at Northern Virginia Community College in Annandale, where he has been honored many times as an outstanding teacher. He is a frequent contributor to *EARTH* magazine and is author of the popular geology blog *Mountain Beltway*. Callan was responsible for preparing the SmartFigure Tutorials that appear throughout the text. As you take advantage of these outstanding learning aids, you will hear his voice explaining the ideas. We appreciate Callan's contributions to this new edition of *Earth*.

Great thanks also go to our colleagues who prepared in-depth reviews. Their critical comments and thoughtful input helped guide our work and clearly strengthened the text. Special thanks to:

Richard Barca, Jackson College
Lisa Barlow, University of Colorado
Marianne Caldwell, Hillsborough Community College–Dale Mabry Campus
Samuel Castonguay, University of Wisconsin–Eau Claire
Alvin Coleman, Cape Fear Community College
Carol Edson, Las Positas College
Anne Egger, Central Washington University
Tom Evans, Western Washington University
Bill Garcia, University of North Carolina–Charlotte
Nicholas Gioppo, Mohawk Valley Community College
Melinda Gutierrez, Missouri State University–Springfield
Emily Hamecher, California State University–Fullerton
Brennan Jordan, University of South Dakota
Amanda Julson, Blinn College–Bryan
Deborah Leslie, Ohio State University
Wendy Nelson, University of Houston
Richard Orndorff, Eastern Washington University
David Peate, University of Iowa
Tiffany Roberts, Louisiana State University
Jeffrey Ryan, University of South Florida
Paul Scrivner, Glendale Community College
Halie Sims, University of Iowa
Henry Turner, University of Arkansas

Last but certainly not least, we gratefully acknowledge the support and encouragement of our wives, Joanne Bannon and Nancy Lutgens. Preparation of this edition of *Earth* would have been far more difficult without their patience and understanding.

Ed Tarbuck
Fred Lutgens

Augmented Reality Enhances the Reading Experience, Bringing the Textbook to Life

Using a cutting-edge technology called augmented reality, Pearson's BouncePages app launches engaging, interactive videos and animations that bring textbook pages to life. Use your mobile device to scan a SmartFigure identified by the BouncePages icon, and an animation or video illustrating the SmartFigure's concept launches immediately. No slow websites or hard-to-remember logins required.

BouncePages' augmented reality technology transforms textbooks into convenient digital platforms, breathes life into your learning experience, and helps you grasp difficult academic concepts. Learning geology from a textbook will never be the same.

Download the FREE BP App for Android

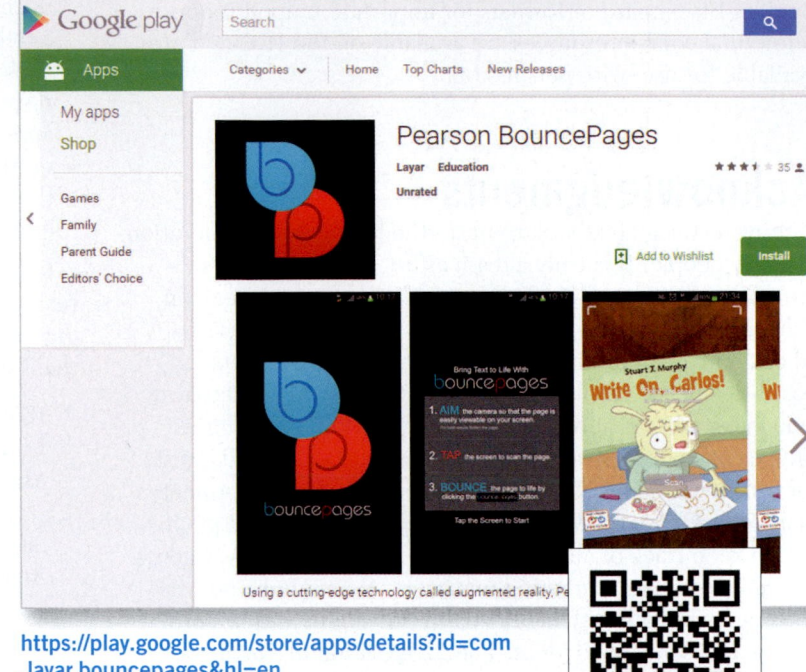

https://play.google.com/store/apps/details?id=com.layar.bouncepages&hl=en

Download the FREE BP App for iOS

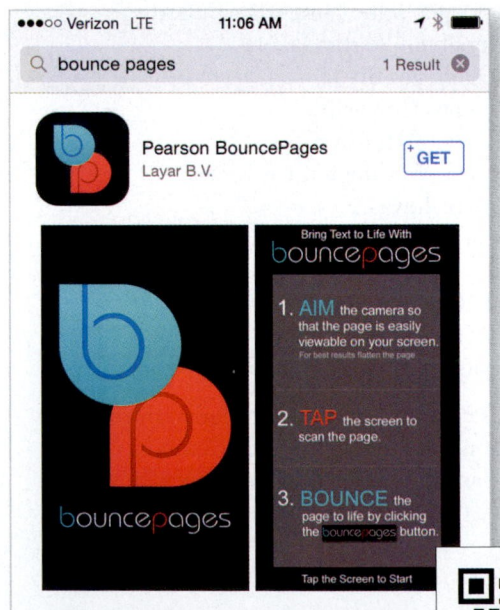

https://itunes.apple.com/us/app/pearson-bouncepages/id659370955?mt=8

By scanning figures associated with the BouncePages icon, students will be immediately connected to the digital world and will deepen their learning experience with the printed text.

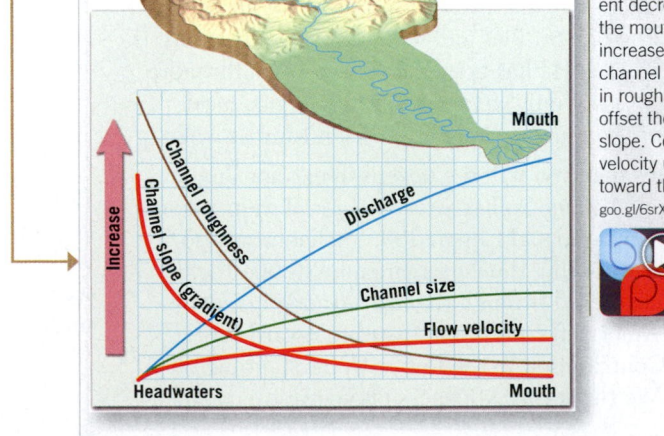

SmartFigure 16.14
Channel changes from head to mouth Although the gradient decreases toward the mouth of a stream, increases in discharge and channel size and decreases in roughness more than offset the decrease in slope. Consequently, flow velocity usually increases toward the mouth. (https://goo.gl/6srX2s)

▶ Tutorial

NEW! SmartFigure: Condor Videos. Bringing Physical Geology to life for GenEd students, three geologists, using a quadcopter with a GoPro camera mounted to it, have ventured out into the field to film 10 key geologic locations. These process-oriented videos, accessed through BouncePages technology, are designed to bring the field to the classroom or dorm room and enhance the learning experience in our texts.

NEW! SmartFigure: Mobile Field Trips. Scattered throughout this new edition of Earth are **24 video field trips**. On each trip, you will accompany geologist-pilot-photographer Michael Collier in the air and on the ground to see and learn about iconic landscapes that relate to discussions in the chapter. These extraordinary field trips are accessed by using the BouncePages app to scan the figure in the chapter—usually one of Michael's outstanding photos.

Visualize Processes and Tough Topics

Dip-slip fault   Dip-slip fault   Dip-slip fault

NEW! SmartFigure: Animations are brief videos, many created by text illustrator Dennis Tasa, that animate a process or concept depicted in the textbook's figures. This technology allows students to view moving figures rather than static art to depict how a geologic process actually changes through time. The videos can be accessed using Pearson's BouncePages app for use on mobile devices, and will also be available via MasteringGeology.

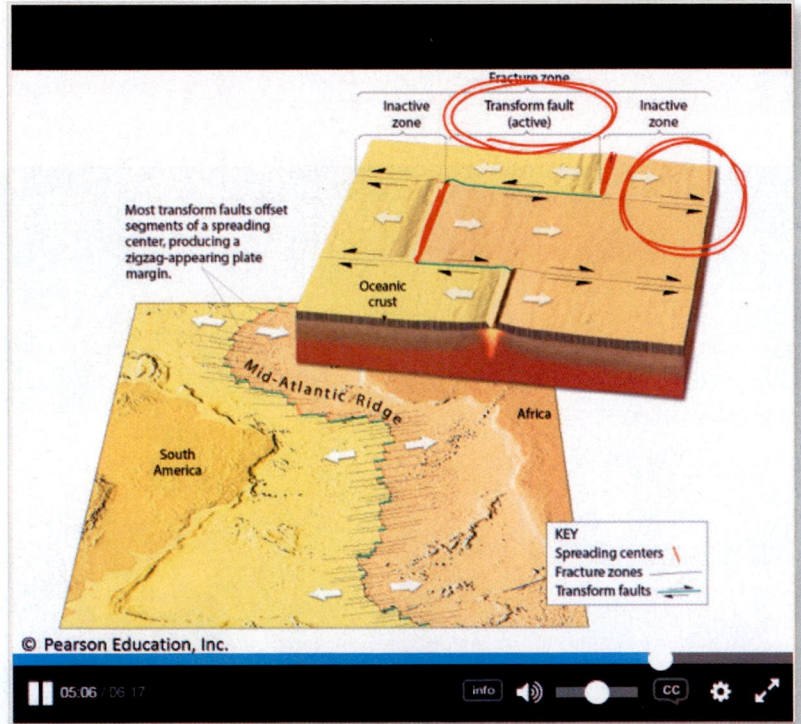

Callan Bentley, SmartFigure Tutorial author, is a Chancellor's Commonwealth Professor of Geology at Northern Virginia Community College (NOVA) in Annandale, Virginia. Trained as a structural geologist, Callan teaches introductory level geology at NOVA, including field-based and hybrid courses. Callan writes a popular geology blog called *Mountain Beltway*, contributes cartoons, travel articles, and book reviews to *EARTH* magazine, and is a digital education leader in the two-year college geoscience community.

SmartFigure: Tutorials bring key chapter illustrations to life! Found throughout the book, these Tutorials are sophisticated, annotated illustrations that are also narrated videos. They are accessible on mobile devices via scannable BouncePages printed in the text and through the Study Area in MasteringGeology.

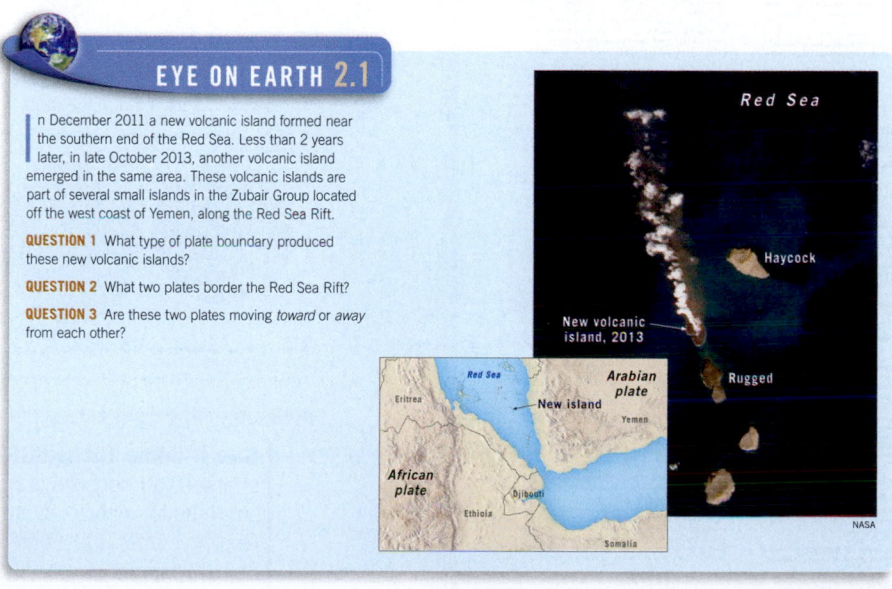

GEOGRAPHICS 15.2

Landslide Risks: United States and Worldwide

According to the U. S. Geological Survey, each year in the United States, landslides cost nearly $4 billion (2010 dollars) in damage repair and cause between 25 and 50 deaths. All states experience rapid mass-wasting processes, but not all areas have the same landslide potential. What's the risk where you live?

U.S. LANDSLIDE POTENTIAL

KEY
- VERY HIGH POTENTIAL
- HIGH POTENTIAL
- MODERATE POTENTIAL
- LOW POTENTIAL

1. In parts of the **Seattle area,** volcanic mudflows called lahars are a potential threat.
2. In the mountainous parts of the **Pacific Northwest,** heavy rains and melting snow often trigger rapid forms of mass wasting.
3. **Coastal California's** steep slopes have a high landslide potential often triggered by winter storms or ground shaking associated with earthquakes.
4. Strong wave activity undercuts and oversteepens coastal cliffs.
5. In the **center of the country,** the plains states are relatively flat, so landslide potential is mostly low-to-moderate.
6. High potential occurs along steep bluffs that flank river valleys.
7. **Florida** and the adjacent Atlantic and Gulf coastal plains have some of the lowest potential because steep slopes are largely absent.
8. In the East, landslides are most common in the Appalachian Mountains.

GLOBAL LANDSLIDE RISKS

? Question: What do areas with the highest landslide potential have in common?

NASA scientists compiled this risk map based on topographic data, land cover classifications and soil types.

Purple and dark red indicate areas at highest risk.

Black dots identify locations of major landslides over a four-year span (2003-2006)

LANDSLIDE RISK
SLIGHT — MODERATE — SEVERE

GEOgraphics use contemporary, compelling visual representations to illustrate complex concepts, enhancing students' ability to synthesize and recall information and important data.

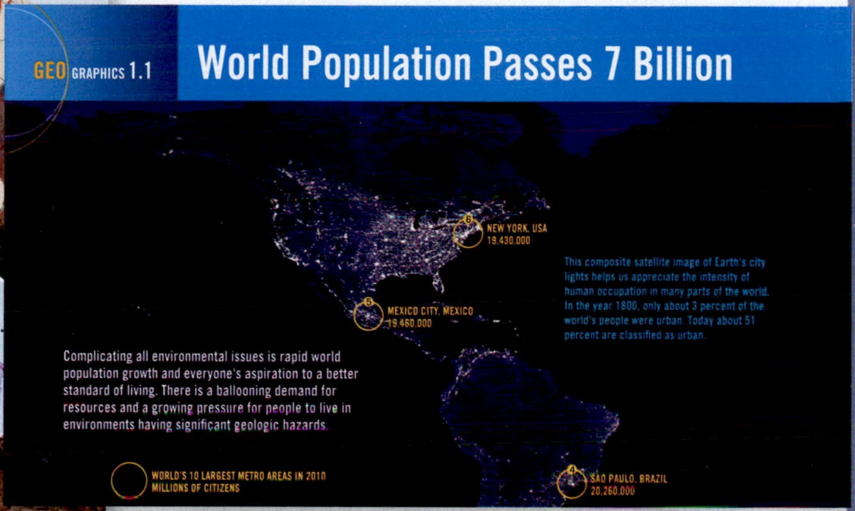

GEOGRAPHICS 1.1

World Population Passes 7 Billion

This composite satellite image of Earth's city lights helps us appreciate the intensity of human occupation in many parts of the world. In the year 1800, only about 3 percent of the world's people were urban. Today about 51 percent are classified as urban.

Complicating all environmental issues is rapid world population growth and everyone's aspiration to a better standard of living. There is a ballooning demand for resources and a growing pressure for people to live in environments having significant geologic hazards.

NEW YORK, USA 19,430,000

MEXICO CITY, MEXICO 19,460,000

SÃO PAULO, BRAZIL 20,260,000

⬤ WORLD'S 10 LARGEST METRO AREAS IN 2010 MILLIONS OF CITIZENS

Eye on Earth features engage students in active learning, asking them to perform critical thinking and visual analysis tasks to evaluate data and make predictions.

EYE ON EARTH 2.1

In December 2011 a new volcanic island formed near the southern end of the Red Sea. Less than 2 years later, in late October 2013, another volcanic island emerged in the same area. These volcanic islands are part of several small islands in the Zubair Group located off the west coast of Yemen, along the Red Sea Rift.

QUESTION 1 What type of plate boundary produced these new volcanic islands?

QUESTION 2 What two plates border the Red Sea Rift?

QUESTION 3 Are these two plates moving *toward* or *away* from each other?

Red Sea

Haycock

New volcanic island, 2013

Rugged

Red Sea
Eritrea
New island
Arabian plate
Yemen
African plate
Ethiopia
Djibouti
Somalia

NASA

Modular Approach Driven by Learning Objectives

The new edition is designed to support a four-part learning path, an innovative structure which facilitates active learning and allows students to focus on important ideas as they pause to assess their progress at frequent intervals.

The chapter-opening **Focus on Concepts** lists the learning objectives for each chapter. Each section of the chapter is tied to a specific learning objective, providing students with a clear learning path to the chapter content.

Concepts in Review, a fresh approach to the typical end-of-chapter material, provides students with a structured and highly visual review of each chapter. Consistent with the Focus on Concepts and Concept Checks, the **Concepts in Review** is structured around the section title and the corresponding learning objective for each section.

Each chapter section concludes with **Concept Checks,** a feature that lists questions tied to the section's learning objective, allowing students to monitor their grasp of significant facts and ideas.

10.5   Concept Checks

1. Distinguish between the two measurements used to establish the orientation of deformed strata.

2. Briefly describe the method geologists use to infer the orientation of rock structures that lie mainly below Earth's surface.

Give It Some Thought (GIST) is found at the end of each chapter and consists of questions and problems asking students to analyze, synthesize, and think critically about Geology. GIST questions relate back to the chapter's learning objectives, and can easily be assigned using MasteringGeology.

Continuous Learning Before, During, and After Class with MasteringGeology™

MasteringGeology delivers engaging, dynamic learning opportunities—focusing on course objectives responsive to each student's progress—that are proven to help students learn geology course material and understand challenging concepts.

Before Class

Dynamic Study Modules and eText 2.0 provide students with a preview of what's to come.

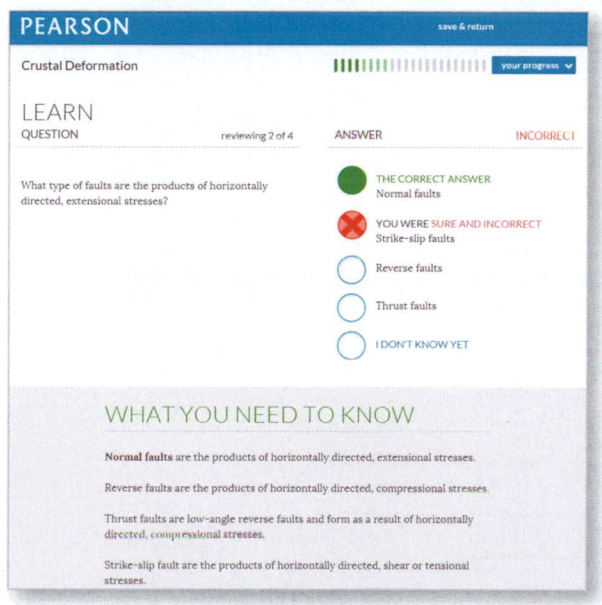

Dynamic Study Modules enable students to study effectively on their own in an adaptive format. Students receive an initial set of questions with a unique answer format asking them to indicate their confidence.

Once completed, Dynamic Study Modules include explanations using material taken directly from the text.

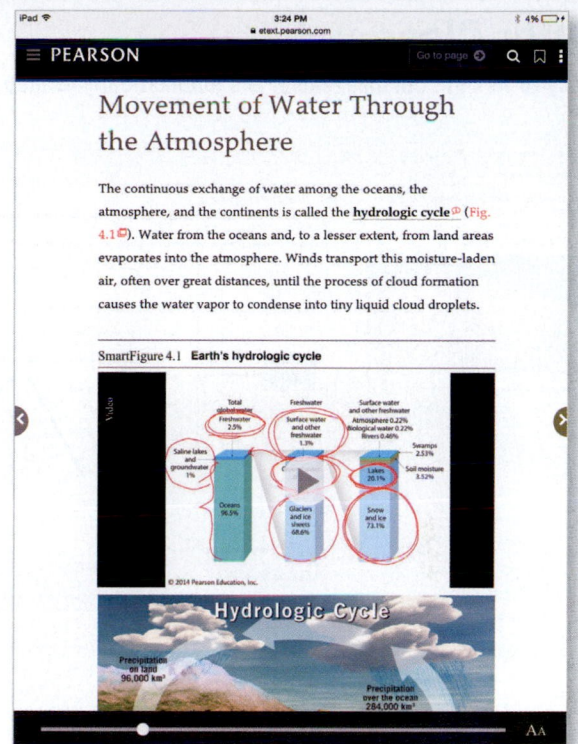

NEW! Interactive eText 2.0 complete with embedded media. eText 2.0 is mobile friendly and ADA accessible.

- Now available on smartphones and tablets.
- Seamlessly integrated videos and other rich media.
- Accessible (screen-reader ready).
- Configurable reading settings, including resizable type and night reading mode.
- Instructor and student note-taking, highlighting, bookmarking, and search.

During Class

Engage Students with Learning Catalytics

Learning Catalytics, a "bring your own device" student engagement, assessment, and classroom intelligence system, allows students to use their smartphone, tablet, or laptop to respond to questions in class.

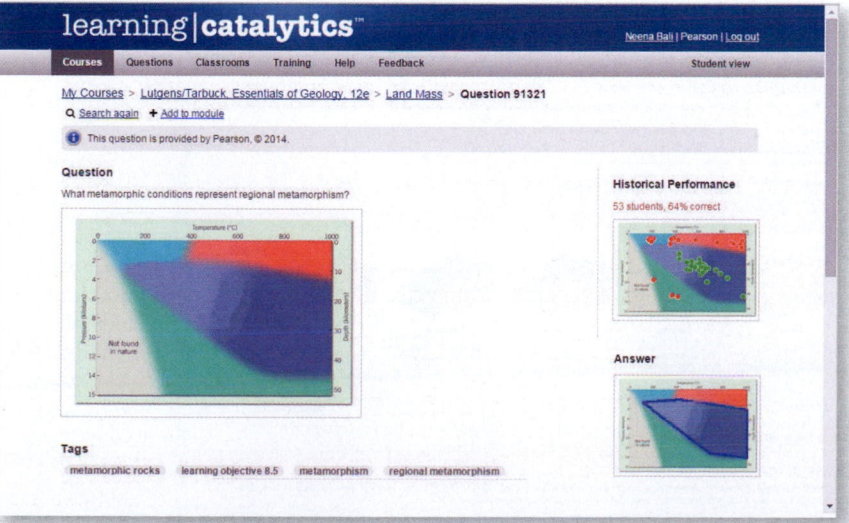

MasteringGeology™

After Class
Easy-to-Assign, Customizeable, and Automatically Graded Assignments

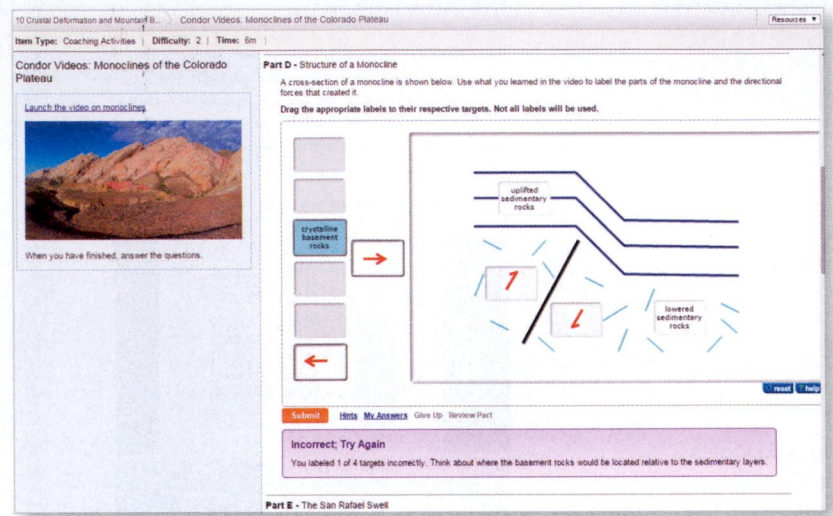

NEW! Project Condor Videos capture stunning footage of the Mountain West region with a quadcopter and a GoPro camera. A series of videos have been created with annotations, sketching, and narration to improve the way students learn about faults and folds, streams, volcanoes, and so much more. In Mastering, these videos are accompanied by questions designed to assess students on the main takeaways from each video.

NEW! 24 Mobile Field Trips take students to classic geologic locations as they accompany geologist–pilot–photographer–author Michael Collier in the air and on the ground to see and learn about landscapes that relate to concepts in the chapter. In Mastering, these videos will be accompanied by auto-gradable assessments that will track what students have learned.

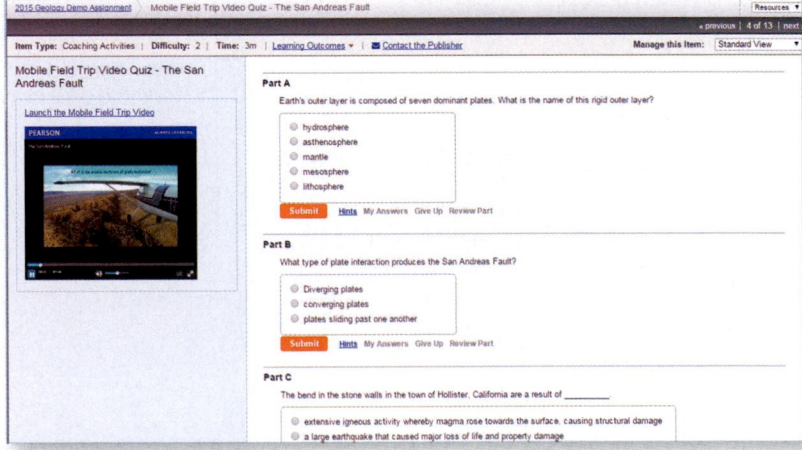

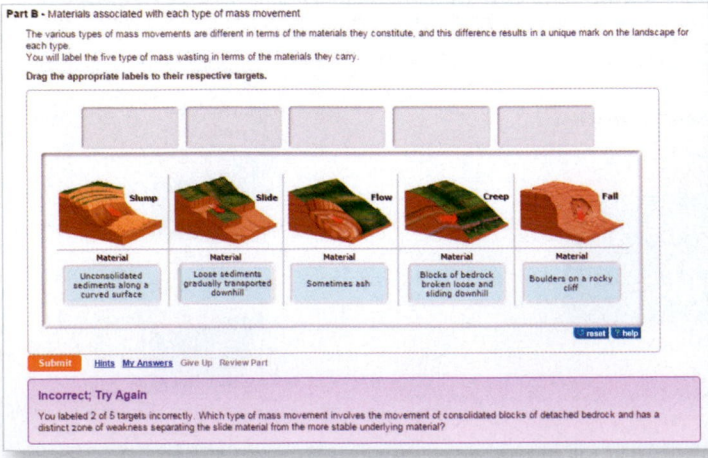

GeoTutor coaching activities help students master important geologic concepts with highly visual, kinesthetic activities focused on critical thinking and application of core geoscience concepts.

MasteringGeology™

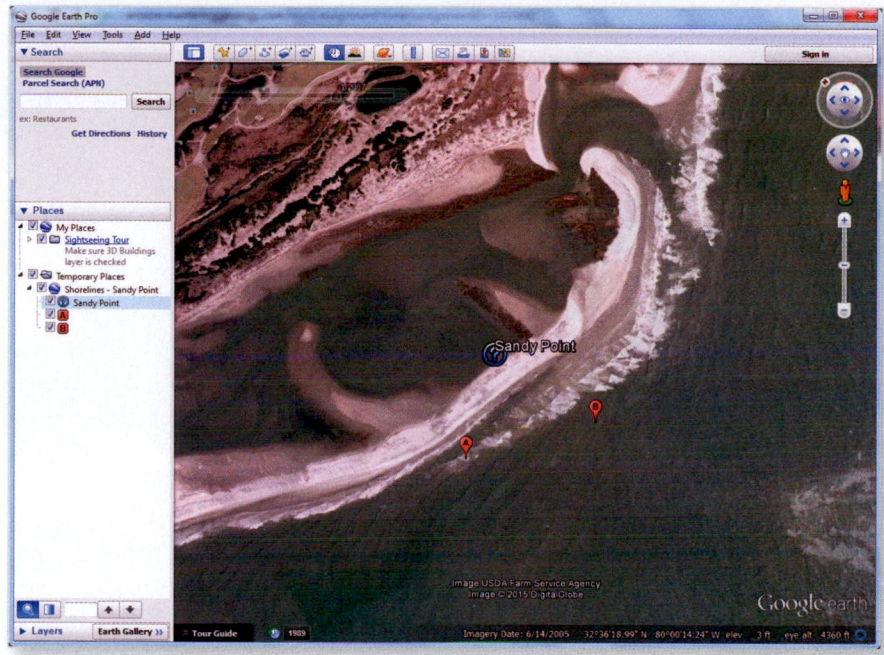

Encounter Activities provide rich, interactive explorations of geology and earth science concepts using the dynamic features of Google Earth™ to visualize and explore earth's physical landscape. Dynamic assessment includes questions related to core geology concepts. All explorations include corresponding Google Earth KMZ media files, and questions include hints and specific wrong-answer feedback to help coach students towards mastery of the concepts while improving students geospatial skills.

NEW! GigaPan Activities allow students to take advantage of a virtual field experience with high-resolution picture technology that has been developed by Carnegie Mellon University in conjunction with NASA.

Part D - Making Observations

After exploring the Gigapan field site, arrange the following observations/inferences by their respective rock unit. These observations/inferences describe the material, appearance and weathering pattern of the respective rock units.

Drag the appropriate items into their respective bins. Each item may be used only once.

Rock Unit 1
- Red and white in color
- Appears to be made up of many thin layers
- Weathers in small irregular shapes
- Weathers in large blocks
- Appears to be massive (no layers)
- Sediments too small to see

Rock Unit 2
- Black and dark gray in color
- Crystals too small to see

reset help

Submit Hints My Answers Give Up Review Part

Incorrect; Try Again

You sorted 2 out of 8 items incorrectly. Compare the weathering pattern of rock unit #2 to the weathering pattern of rock unit #1. Which rock unit produces large blocks?

Additional MasteringGeology assignments available:

- SmartFigures
- Interactive Animations
- Give It Some Thought Activities
- Reading Quizzes
- MapMaster Interactive Maps

1

An Introduction to Geology

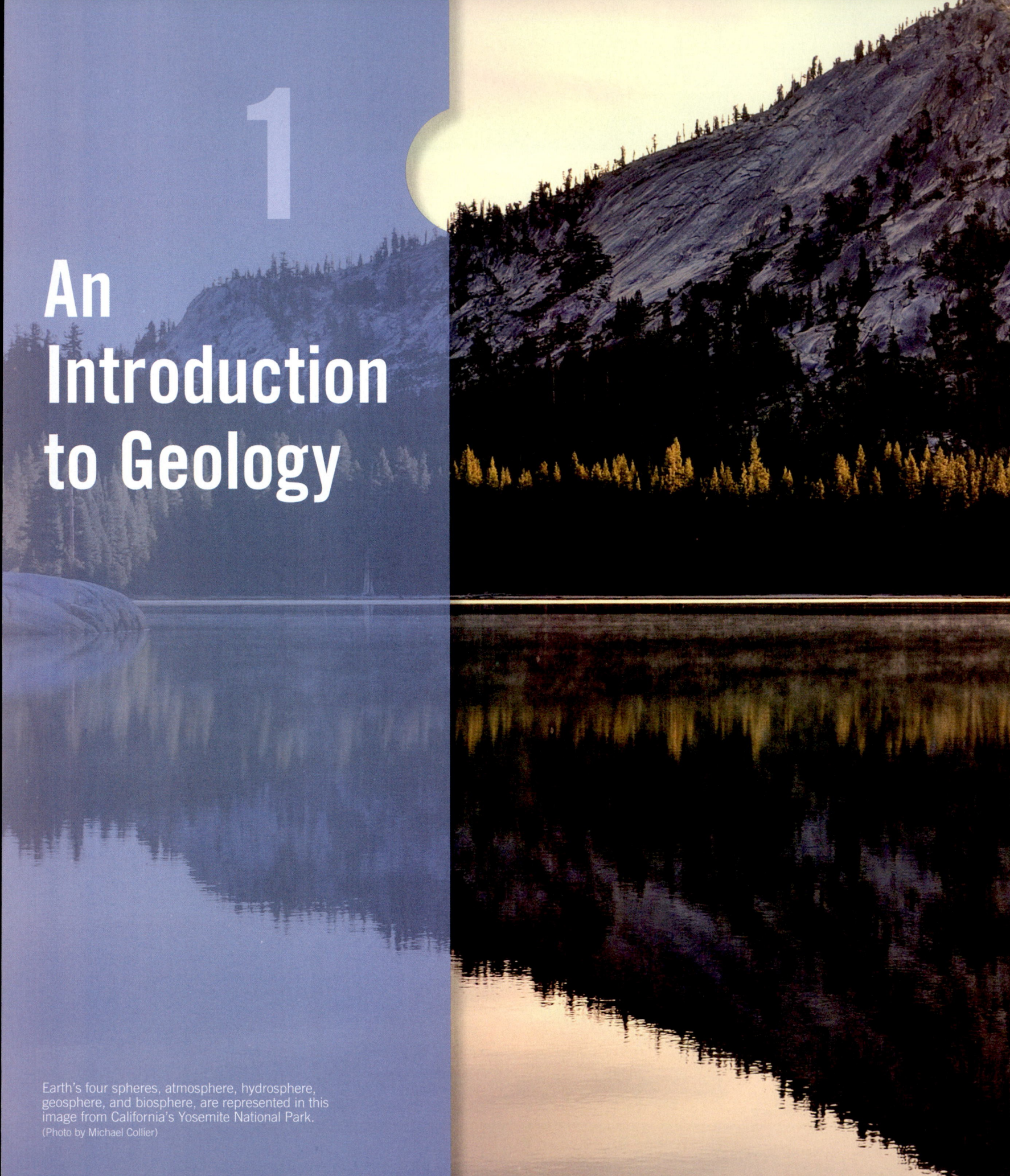

Earth's four spheres, atmosphere, hydrosphere, geosphere, and biosphere, are represented in this image from California's Yosemite National Park. (Photo by Michael Collier)

Each statement represents the primary **LEARNING OBJECTIVE** for the corresponding major heading within the chapter. After you complete the chapter, you should be able to:

1.1 Distinguish between physical and historical geology and describe the connections between people and geology.

1.2 Summarize early and modern views on how change occurs on Earth and relate them to the prevailing ideas about the age of Earth.

1.3 Discuss the nature of scientific inquiry, including the construction of hypotheses and the development of theories.

1.4 List and describe Earth's four major spheres. Define *system* and explain why Earth is considered to be a system.

1.5 Outline the stages in the formation of our solar system.

1.6 Sketch Earth's internal structure and label and describe the main subdivisions.

1.7 Sketch, label, and explain the rock cycle.

1.8 List and describe the major features of the continents and ocean basins.

The spectacular eruption of a volcano, the terror brought by an earthquake, the magnificent scenery of a mountain range, and the destruction created by a landslide or flood are all subjects for a geologist. The study of geology deals with many fascinating and practical questions about our physical environment. What forces produce mountains? When will the next major earthquake occur in California? What was the Ice Age like, and will there be another? How are ore deposits formed? Where should we search for water? Will we find plentiful oil if we drill a well in a particular location? Geologists seek to answer these and many other questions about Earth, its history, and its resources.

1.1 Geology: The Science of Earth

Distinguish between physical and historical geology and describe the connections between people and geology.

The subject of this text is **geology**, from the Greek *geo* (Earth) and *logos* (discourse). Geology is the science that pursues an understanding of planet Earth. Understanding Earth is challenging because our planet is a dynamic body with many interacting parts and a complex history. Throughout its long existence, Earth has been changing. In fact, it is changing as you read this page and will continue to do so. Sometimes the changes are rapid and violent, as when landslides or volcanic eruptions occur. Just as often, change takes place so slowly that it goes unnoticed during a lifetime. Scales of size and space also vary greatly among the phenomena that geologists study. Sometimes geologists must focus on phenomena that are microscopic, such as the crystalline structure of minerals, and at other times they must deal with processes that are continental or global in scale, such as the formation of major mountain ranges.

Figure 1.1
Internal and external processes The processes that operate beneath and upon Earth's surface are an important focus of physical geology. (River photo by Michael Collier; volcano photo by AM Design/Alamy Live News/Alamy Images)

Physical and Historical Geology

Geology is traditionally divided into two broad areas— physical and historical. **Physical geology**, which is the primary focus of this book, examines the materials composing Earth and seeks to understand the many processes that operate beneath and upon its surface (**Figure 1.1**). The aim of **historical geology**, on the other hand, is to understand Earth's origins and its development through time. Thus, it strives to establish an orderly chronological arrangement of the multitude of physical and biological changes that have occurred in the geologic past. The study of physical geology logically precedes the study of Earth history because we must first understand how Earth works before we attempt to unravel its past. It should also be pointed out that physical and historical

A.

External processes, such as landslides, rivers, and glaciers, erode and sculpt surface features. The Colorado River played a major role in creating the Grand Canyon.

B.

Internal processes are those that occur beneath Earth's surface. Sometimes they lead to the formation of major features at the surface, such as Italy's Mt. Etna.

geology are divided into many areas of specialization. Every chapter of this book represents one or more areas of specialization in geology.

Geology is perceived as a science that is done outdoors—and rightly so. A great deal of geology is based on observations, measurements, and experiments conducted in the field. But geologists also work in the laboratory, where, for example, their analysis of minerals and rocks provides insights into many basic processes and the microscopic study of fossils unlocks clues to past environments (**Figure 1.2**). Geologists must also understand and apply knowledge and principles from physics, chemistry, and biology. Geology is a science that seeks to expand our knowledge of the natural world and our place in it.

Geology, People, and the Environment

The primary focus of this book is to develop your understanding of basic geologic principles, but along

Figure 1.2
In the field and in the lab Geology involves not only outdoor fieldwork but work in the laboratory as well. **A.** This research team is gathering data at Mount Nyiragongo, an active volcano in the Democratic Republic of the Congo. (Carsten Peter/National Geographic Image Collection/Alamy) **B.** This researcher is using a petrographic microscope to study the mineral composition of rock samples. (Photo by Jon Wilson/Science Source)

the way, we explore numerous important relationships between people and the natural environment. Many of the problems and issues that geologists address are of practical value.

Natural hazards are a part of living on Earth. Every day they adversely affect millions of people worldwide and are responsible for staggering damages. Among the hazardous Earth processes that geologists study are volcanoes, floods, tsunamis, earthquakes, and landslides. Of course, geologic hazards are *natural* processes. They become hazards only when people try to live where these processes occur (**Figure 1.3**).

Figure 1.3
Tsunami destruction Undersea earthquakes sometimes create large, fast-moving ocean waves that can cause significant death and destruction in coastal areas. This tsunami struck densely populated Fukushima, Japan, in 2011, causing major damage to a nuclear power plant. Geologic hazards are natural processes. They become hazards only when people try to live where these processes occur. (Photo by Mainichi Newspaper/Aflo /Newscom)

World Population Passes 7 Billion

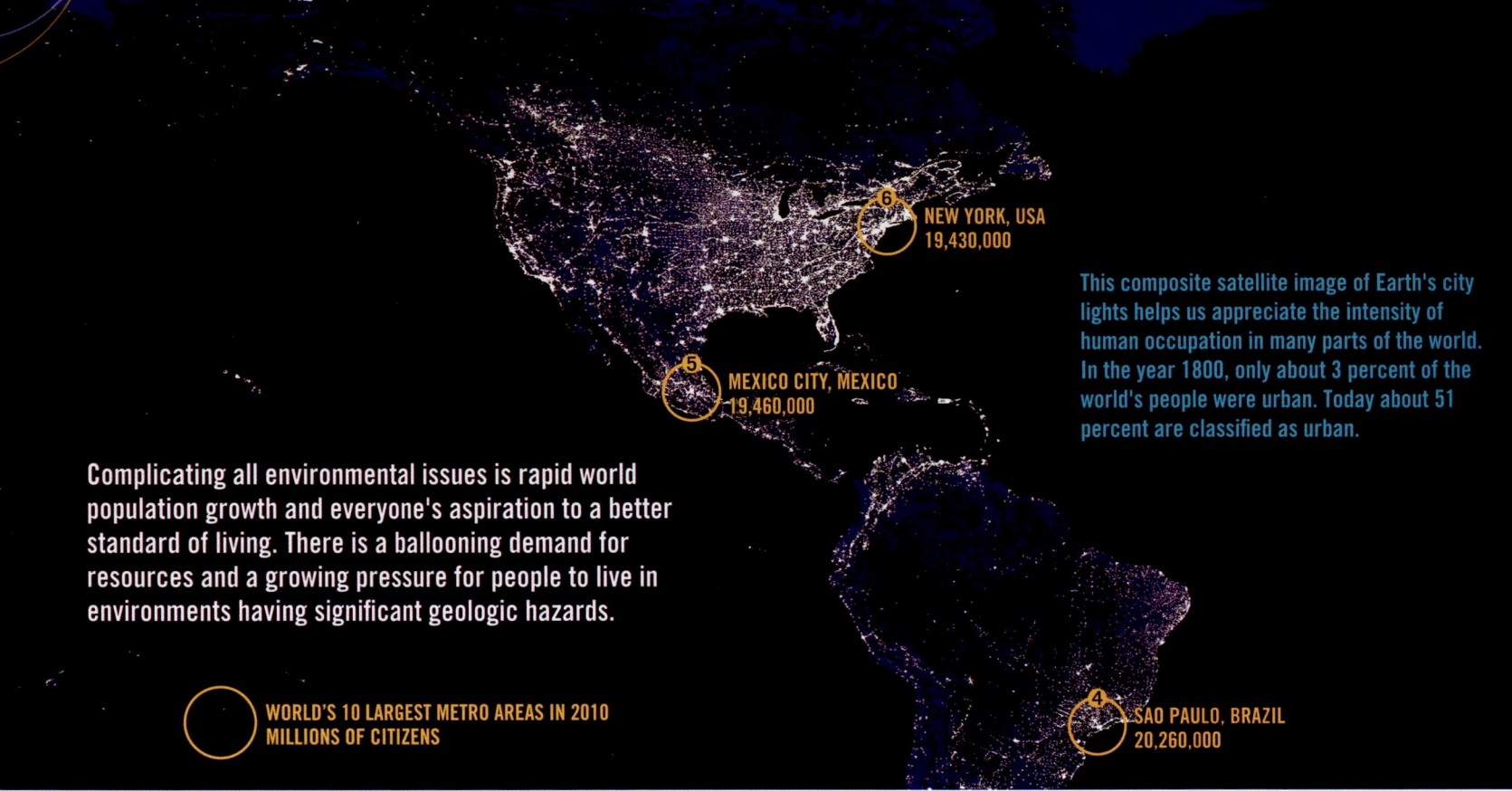

NEW YORK, USA
19,430,000

This composite satellite image of Earth's city lights helps us appreciate the intensity of human occupation in many parts of the world. In the year 1800, only about 3 percent of the world's people were urban. Today about 51 percent are classified as urban.

MEXICO CITY, MEXICO
19,460,000

Complicating all environmental issues is rapid world population growth and everyone's aspiration to a better standard of living. There is a ballooning demand for resources and a growing pressure for people to live in environments having significant geologic hazards.

WORLD'S 10 LARGEST METRO AREAS IN 2010
MILLIONS OF CITIZENS

SAO PAULO, BRAZIL
20,260,000

According to the United Nations, more people live in cities than in rural areas (see GEOgraphics 1.1). This global trend toward urbanization concentrates millions of people into megacities, many of which are vulnerable to natural hazards. Coastal sites are especially vulnerable because development often destroys natural defenses such as wetlands and sand dunes. In addition, threats associated with human influences on the Earth system are increasing; one example is sea-level rise linked to global climate change. Some megacities are exposed to seismic (earthquake) and volcanic hazards because inappropriate land use and poor construction practices, coupled with rapid population growth, increase the risk of death and damage.

Resources are another important focus of geology that is of great practical value to people. They include water and soil, a great variety of metallic and nonmetallic minerals, and energy (**Figure 1.4**). Together these form the very foundation of modern civilization. Geology deals not only with how and where these vital resources form but also with maintaining supplies and with the environmental impacts of their extraction and use.

Geologic processes clearly have an impact on people. Conversely, we humans can dramatically influence geologic processes. For example, landslides and river flooding occur naturally, but the magnitude and frequency of these processes can be affected significantly by human activities such as clearing forests, building cities, and constructing roads and dams. Unfortunately, natural systems do not always adjust to artificial changes in ways that we can anticipate. Thus, an alteration to the environment that was intended to benefit society sometimes has the opposite effect.

At appropriate places throughout this book, you will examine different aspects of our relationship with the physical environment. Nearly every chapter addresses some aspect of natural hazards, resources, and the environmental issues associated with each. Significant parts of some chapters provide the basic geologic knowledge and principles needed to understand environmental problems.

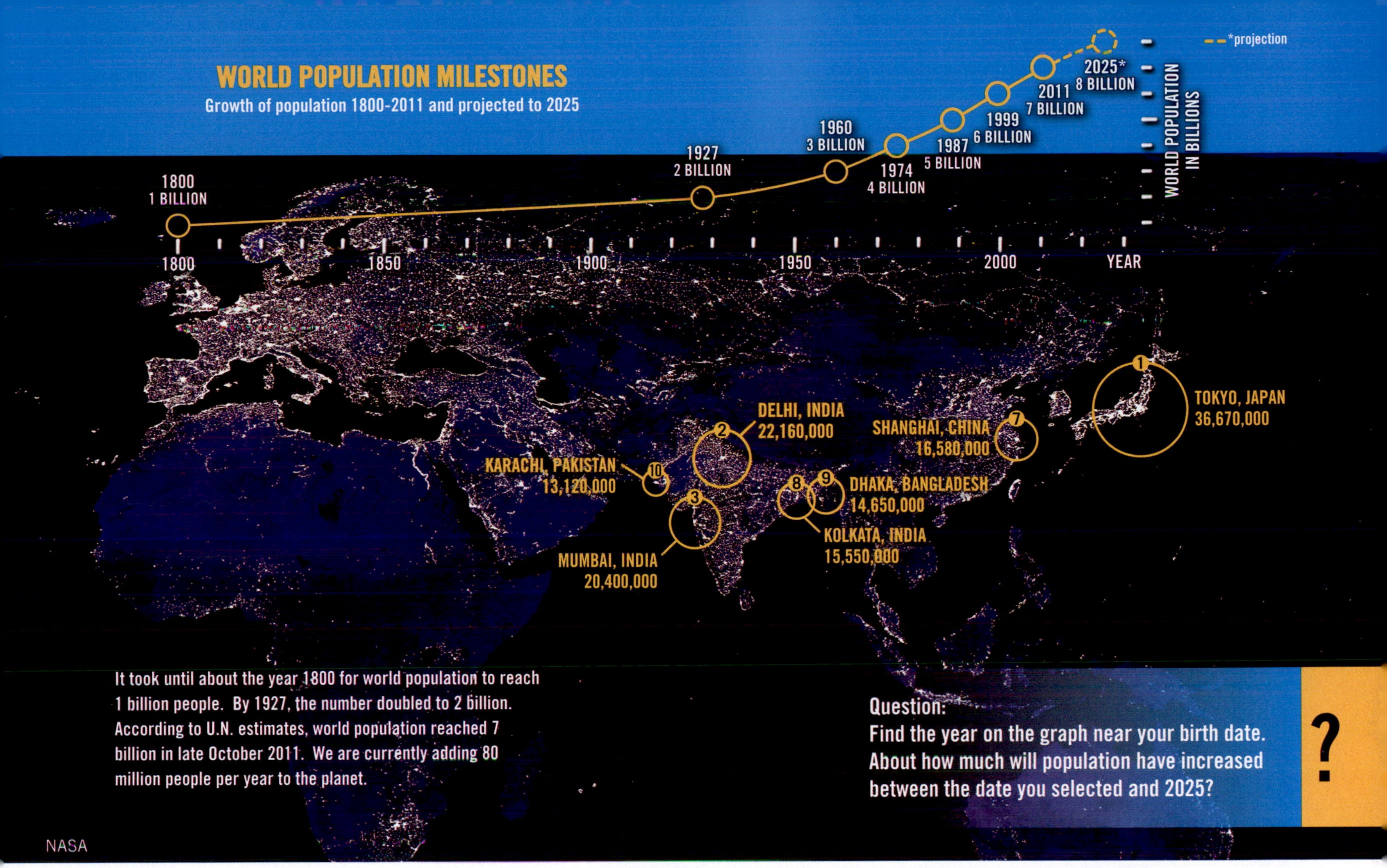

WORLD POPULATION MILESTONES
Growth of population 1800-2011 and projected to 2025

- - *projection

2025*
8 BILLION

2011
7 BILLION

1999
6 BILLION

1987
5 BILLION

1974
4 BILLION

1960
3 BILLION

1927
2 BILLION

1800
1 BILLION

WORLD POPULATION IN BILLIONS

1800   1850   1900   1950   2000   YEAR

TOKYO, JAPAN
36,670,000

DELHI, INDIA
22,160,000

SHANGHAI, CHINA
16,580,000

KARACHI, PAKISTAN
13,120,000

DHAKA, BANGLADESH
14,650,000

KOLKATA, INDIA
15,550,000

MUMBAI, INDIA
20,400,000

It took until about the year 1800 for world population to reach 1 billion people. By 1927, the number doubled to 2 billion. According to U.N. estimates, world population reached 7 billion in late October 2011. We are currently adding 80 million people per year to the planet.

Question:
Find the year on the graph near your birth date. About how much will population have increased between the date you selected and 2025?

?

Figure 1.4
Mineral resources represent an important link between people and geology

Each year an average American requires huge quantities of Earth materials—more than 6 tons of stone, 4.5 tons of sand and gravel, nearly a half ton of cement, almost 400 pounds of salt, 360 pounds of phosphate, and about a half ton of other nonmetals. In addition, per capita consumption of metals such as iron, aluminum, and copper exceeds 700 pounds. This open pit copper mine is in southern Arizona. (Ball Miwako/Alamy)

1.1 Concept Checks

1. Name and distinguish between the two broad subdivisions of geology.

2. List at least three different geologic hazards.

3. Aside from geologic hazards, describe another important connection between people and geology.

1.2 | The Development of Geology

Summarize early and modern views on how change occurs on Earth and relate them to the prevailing ideas about the age of Earth.

The nature of our Earth—its materials and processes—has been a focus of study for centuries. Writings about such topics as fossils, gems, earthquakes, and volcanoes date back to the early Greeks, more than 2300 years ago.

Certainly the most influential Greek philosopher was Aristotle. Unfortunately, Aristotle's explanations about the natural world were not based on keen observations and experiments. He arbitrarily stated that rocks were created under the "influence" of the stars and that earthquakes occurred when air crowded into the ground, was heated by central fires, and escaped explosively. When confronted with a fossil fish, he explained that "a great many fishes live in the earth motionless and are found when excavations are made." Although Aristotle's explanations may have been adequate for his day, they continued to be viewed as authoritative for many centuries, thus inhibiting the acceptance of more up-to-date ideas. After the Renaissance of the 1500s, however, more people became interested in finding answers to questions about Earth.

Catastrophism

In the mid-1600s James Ussher, Archbishop of Armagh, Primate of all Ireland, published a major work that had immediate and profound influences. A respected biblical scholar, Ussher constructed a chronology of human and Earth history in which he calculated that Earth was only a few thousand years old, having been created in 4004 B.C.E. Ussher's treatise was widely accepted by Europe's scientific and religious leaders, and his chronology was soon printed in the margins of the Bible itself.

During the seventeenth and eighteenth centuries, Western thought about Earth's features and processes was strongly influenced by Ussher's calculation. The result was a guiding doctrine called **catastrophism**. Catastrophists believed that Earth's landscapes were shaped primarily by great catastrophes. Features such as mountains and canyons, which today we know take great spans of time to form, were explained as resulting from sudden, often worldwide disasters produced by unknowable causes that no longer operate. This philosophy was an attempt to fit the rates of Earth processes to then-current ideas about the age of Earth.

The Birth of Modern Geology

Against the backdrop of Aristotle's views and a conception of an Earth created in 4004 B.C.E., a Scottish physician and gentleman farmer, James Hutton, published *Theory of the Earth* in 1795. In this work, Hutton put forth a fundamental principle that is a pillar of geology today: **uniformitarianism**. It states that the *physical, chemical, and biological processes that operate today have also operated in the geologic past*. This means

SmartFigure 1.5
Earth history—Written in the rocks The Grand Canyon of the Colorado River in northern Arizona. (Photo by Dennis Tasa)
(http://goo.gl/7KwQLk)

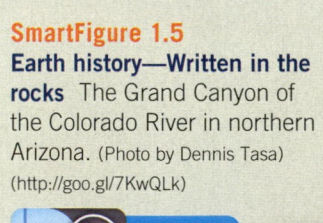

Grand Canyon rocks span more than 1.5 billion years of Earth history.

The uppermost layer, the Kaibab Formation, is about 270 million years old.

Rocks at the bottom are nearly 2 billion years old.

that the forces that we observe presently shaping our planet have been at work for a very long time. Thus, to understand ancient rocks, we must first understand present-day processes and their results. This idea is commonly stated as *the present is the key to the past.*

Prior to Hutton's *Theory of the Earth,* no one had effectively demonstrated that geologic processes occur over extremely long periods of time. However, Hutton persuasively argued that seemingly small forces can, over long spans of time, produce effects that are just as great as those resulting from sudden catastrophic events. Unlike his predecessors, Hutton carefully cited verifiable observations to support his ideas.

For example, when Hutton argued that mountains are sculpted and ultimately destroyed by weathering and the work of running water, and that their waste materials are carried to the oceans by observable processes, he said, "We have a chain of facts which clearly demonstrate . . . that the materials of the wasted mountains have traveled through the rivers"; and further, "There is not one step in all this progress . . . that is not to be actually perceived." He then summarized this thought by asking a question and immediately providing the answer: "What more can we require? Nothing but time."

Geology Today

Today the basic tenets of uniformitarianism are just as viable as in Hutton's day. Indeed, we realize more strongly than ever before that the present gives us insight into the past and that the physical, chemical, and biological laws that govern geologic processes remain unchanging through time. However, we also understand that the doctrine should not be taken too literally. To say that geologic processes in the past were the same as those occurring today is not to suggest that they have always had the same relative importance or that they have operated at precisely the same rate. Moreover, some important geologic processes are not currently observable, but there is well-established evidence that they occur. For example, we know that Earth has experienced impacts from large meteorites even though we have no human witnesses to those impacts. Nevertheless, such events have altered Earth's crust, modified its climate, and strongly influenced life on the planet.

Acceptance of uniformitarianism meant the acceptance of a very long history for Earth. Although Earth processes vary in intensity, they still take a very long time to create or destroy major landscape features. The Grand Canyon provides a good example (**Figure 1.5**).

The rock record contains evidence showing that Earth has experienced many cycles of mountain building and erosion (**Figure 1.6**). Concerning the ever-changing nature of Earth through great expanses of geologic time, Hutton famously stated in 1788: "The results, therefore,

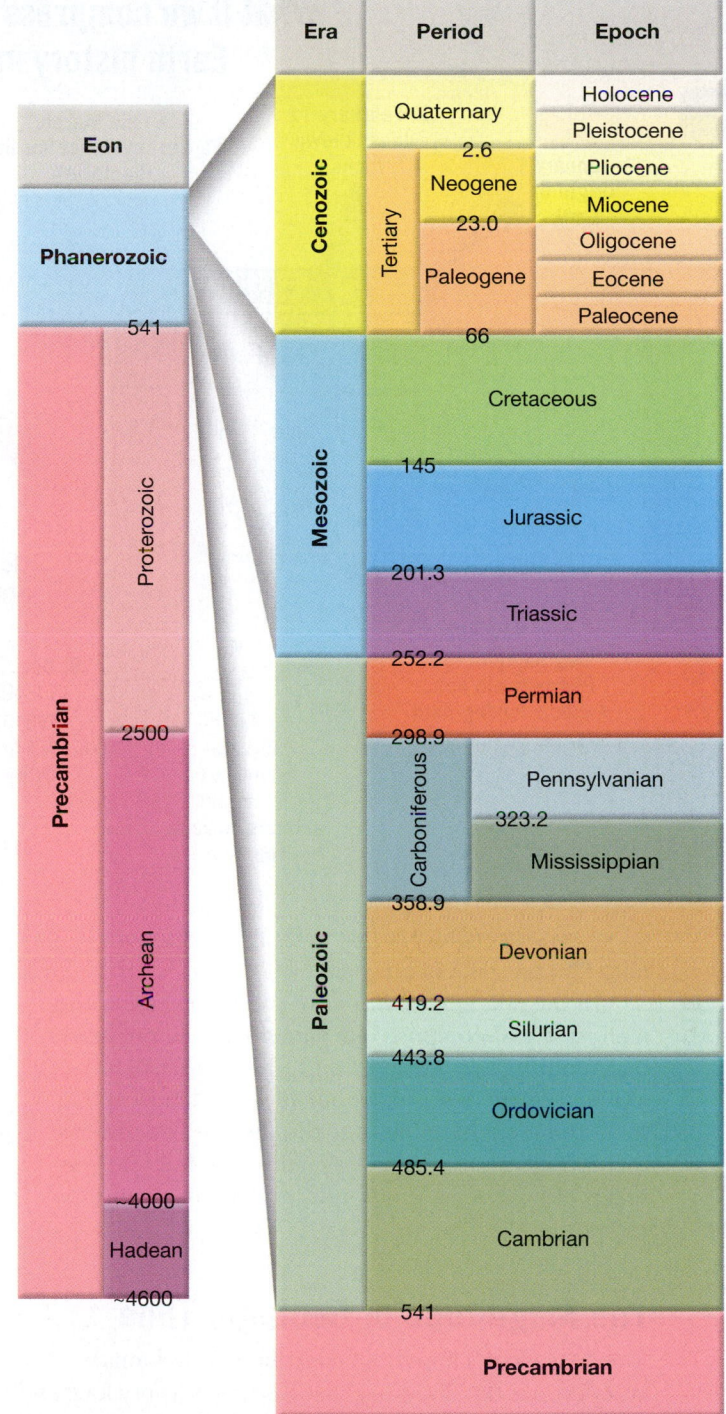

Figure 1.6
Geologic time scale: A basic reference The time scale divides the vast 4.6-billion-year history of Earth into eons, eras, periods, and epochs. Numbers on the time scale represent time in millions of years before the present. The Precambrian accounts for more than 88 percent of geologic time. The geologic time scale is a dynamic tool that is periodically updated. Numerical ages appearing on this time scale are those that were currently accepted by the International Commission on Stratigraphy (ICS) in 2014. The color scheme used on this chart was selected because it is similar to that used by the ICS. The ICS is responsible for establishing global standards for the time scale.

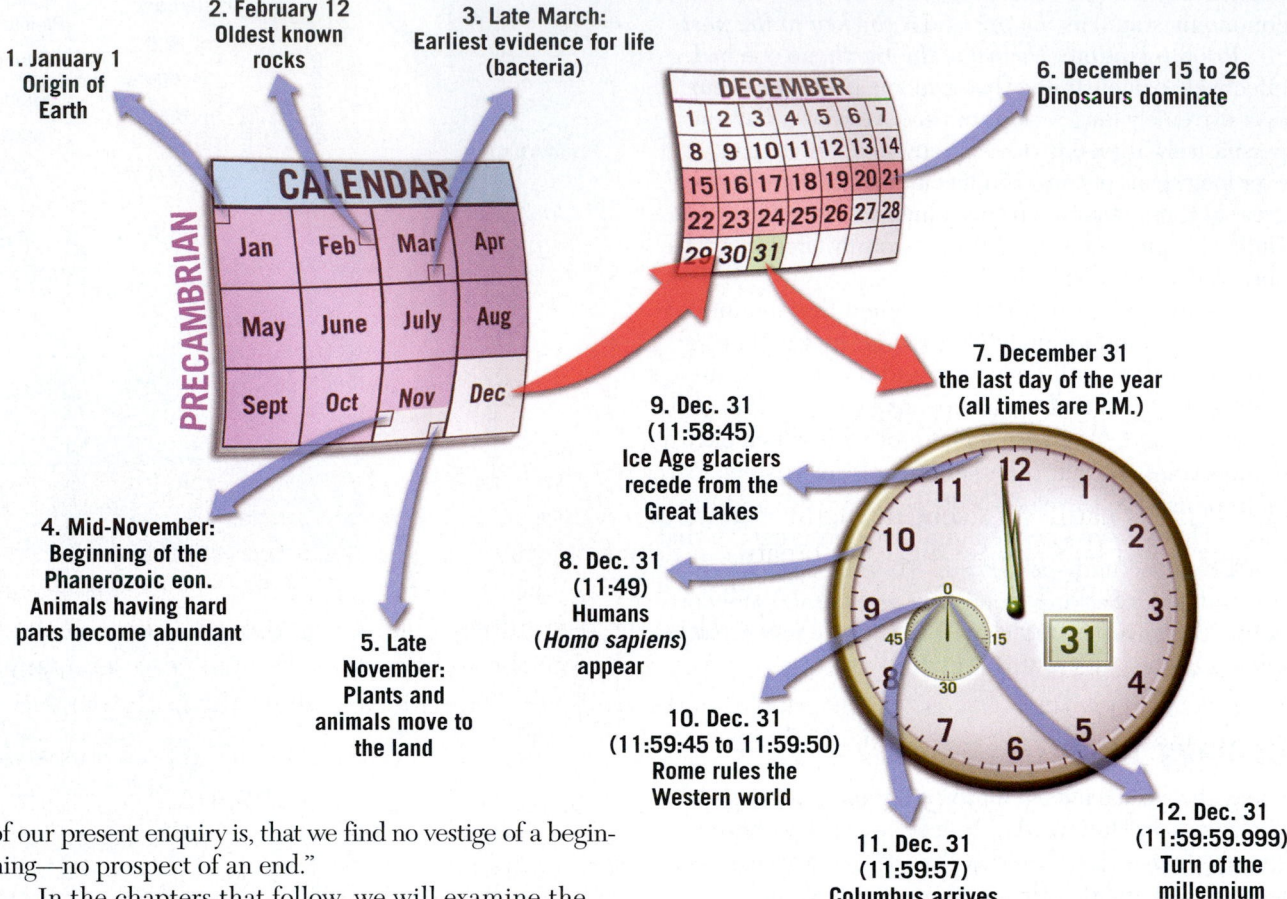

What if we compress the 4.6 billion years of Earth history into a single year?

1. January 1 Origin of Earth

2. February 12 Oldest known rocks

3. Late March: Earliest evidence for life (bacteria)

6. December 15 to 26 Dinosaurs dominate

PRECAMBRIAN

CALENDAR
Jan	Feb	Mar	Apr
May	June	July	Aug
Sept	Oct	Nov	Dec

DECEMBER

4. Mid-November: Beginning of the Phanerozoic eon. Animals having hard parts become abundant

5. Late November: Plants and animals move to the land

9. Dec. 31 (11:58:45) Ice Age glaciers recede from the Great Lakes

8. Dec. 31 (11:49) Humans (*Homo sapiens*) appear

7. December 31 the last day of the year (all times are P.M.)

10. Dec. 31 (11:59:45 to 11:59:50) Rome rules the Western world

11. Dec. 31 (11:59:57) Columbus arrives in the New World

12. Dec. 31 (11:59:59.999) Turn of the millennium

of our present enquiry is, that we find no vestige of a beginning—no prospect of an end."

In the chapters that follow, we will examine the materials that compose our planet and the processes that modify it. It is important to remember that, although many features of our physical landscape may seem to be unchanging over the decades we observe them, they are nevertheless changing—but on time scales of hundreds, thousands, or even many millions of years.

The Magnitude of Geologic Time

Among geology's important contributions to human knowledge is the discovery that Earth has a very long and complex history. Although Hutton and others recognized that geologic time is exceedingly long, they had no methods to accurately determine the age of Earth. Early time scales simply placed the events of Earth history in the proper sequence or order, without knowing how long ago in years they occurred.

Today our understanding of radioactivity—and the fact that rocks and minerals contain certain radioactive isotopes having decay rates ranging from decades to billions of years—allows us to accurately determine numerical dates for rocks that represent important events in Earth's distant past (Figure 1.6). For example,

we know that the dinosaurs died out about 65 million years ago. Today geologists put the age of Earth at about 4.6 billion years. Chapter 9 is devoted to a much more complete discussion of geologic time and the geologic time scale.

The concept of geologic time is new to many nongeologists. People are accustomed to dealing with increments of time measured in hours, days, weeks, and years. History books often examine events over spans of centuries, but even a century is difficult to appreciate fully. For most of us, someone or something that is 90 years old is *very old*, and a 1000-year-old artifact is *ancient*.

By contrast, geologists must routinely deal with vast time periods—millions or billions (thousands of millions) of years. When viewed in the context of Earth's 4.6-billion-year history, a geologic event that occurred 100 million years ago may be characterized as "recent" by a geologist, and a rock sample that has been dated at 10 million years may be called "young." An appreciation for the magnitude of geologic time is important in the study of geology because many processes are so gradual that vast spans of time are needed before significant changes occur. How long is 4.6 billion years? If you were to begin counting at the rate of one number per second and continued 24 hours a day, 7 days a week and never stopped, it would take about two lifetimes (150 years) to reach 4.6 billion! **Figure 1.7** provides another interesting way of viewing the expanse of geologic time. Although helpful in conveying the magnitude of geologic time, this figure and other analogies, no matter how clever, only begin to help us comprehend the vast expanse of Earth history.

1.2	**Concept Checks**

1. Describe Aristotle's influence on geology.
2. Contrast catastrophism and uniformitarianism. How did each view the age of Earth?
3. How old is Earth?
4. Refer to Figure 1.6 and list the eon, era, period, and epoch in which we live.
5. Why is an understanding of the magnitude of geologic time important for a geologist?

1.3 | The Nature of Scientific Inquiry

Discuss the nature of scientific inquiry, including the construction of hypotheses and the development of theories.

In our modern society, we are constantly reminded of the benefits derived from science. But what exactly is the nature of scientific inquiry? Science is a process of producing knowledge, based on making careful observations and on creating explanations that make sense of the observations. Developing an understanding of how science is done and how scientists work is an important theme appearing throughout this book. You will explore the difficulties in gathering data and some of the ingenious methods that have been developed to overcome these difficulties. You will also see many examples of how hypotheses are formulated and tested, and you will learn about the evolution and development of some major scientific theories.

All science is based on the assumption that the natural world behaves in a consistent and predictable manner that is comprehensible through careful, systematic study. The overall goal of science is to discover the underlying patterns in nature and then use that knowledge to make predictions about what should or should not be expected, given certain facts or circumstances. For example, by knowing how oil deposits form, geologists can predict the most favorable sites for exploration and, perhaps as importantly, avoid regions that have little or no potential.

EYE ON EARTH 1.1

These rock layers consist of sediments such as sand, mud, and gravel that were deposited by rivers, waves, wind, and glaciers. The material was buried and eventually compacted and cemented into solid rock. Later, erosion uncovered the layers.

QUESTION 1 Can you establish a relative time scale for these rocks? That is, can you determine which one of the layers shown here is likely oldest and which is probably youngest?

QUESTION 2 Explain the logic you used.

(Photo by David Carriere/AGE Fotostock)

Figure 1.8
**Observation and measure-
ment** Scientific data are
gathered in many ways.
(Satellite image by NASA)

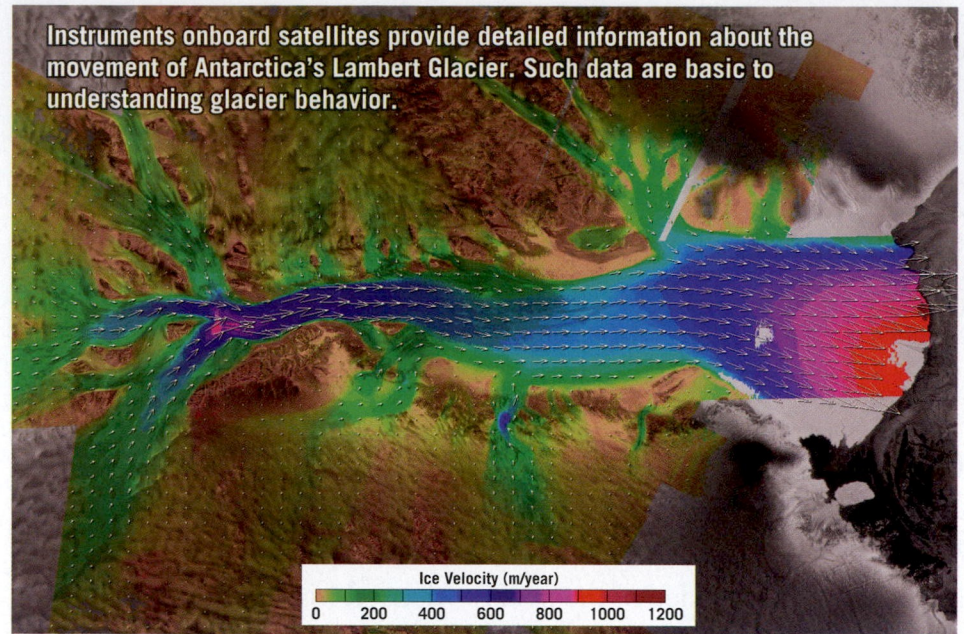

Instruments onboard satellites provide detailed information about the movement of Antarctica's Lambert Glacier. Such data are basic to understanding glacier behavior.

Ice Velocity (m/year)
0   200   400   600   800   1000   1200

fit observations other than those used to formulate them in the first place. Hypotheses that fail rigorous testing are ultimately discarded. The history of science is littered with discarded hypotheses. One of the best known is the Earth-centered model of the universe—a proposal that was supported by the apparent daily motion of the Sun, Moon, and stars around Earth. As mathematician Jacob Bronowski so ably stated, "Science is a great many things, but in the end they all return to this: Science is the acceptance of what works and the rejection of what does not."

The development of new scientific knowledge involves basic logical processes that are universally accepted. To determine what is occurring in the natural world, scientists collect data through observation and measurement (**Figure 1.8**). The data collected often help answer well-defined questions about the natural world. Because some error is inevitable, the accuracy of a particular measurement or observation is always open to question. Nevertheless, these data are essential to science and serve as a springboard for the development of scientific theories.

Hypothesis

Once data have been gathered and principles have been formulated to describe a natural phenomenon, investigators try to explain how or why things happen in the manner observed. They often do this by constructing a tentative (or untested) explanation, which is called a scientific **hypothesis**. It is best if an investigator can formulate more than one hypothesis to explain a given set of observations. If an individual scientist cannot devise multiple hypotheses, others in the scientific community will almost always develop alternative explanations. A spirited debate frequently ensues. As a result, proponents of opposing hypotheses conduct extensive research, and scientific journals make the results available to the wider scientific community.

Before a hypothesis can become an accepted part of scientific knowledge, it must pass objective testing and analysis. If a hypothesis cannot be tested, it is not scientifically useful, no matter how interesting it might seem. The verification process requires that *predictions* be made, based on the hypothesis being considered, and that the predictions be tested through comparison with objective observations. Put another way, hypotheses must

Theory

When a hypothesis has survived extensive scrutiny and when competing hypotheses have been eliminated, it may be elevated to the status of a scientific **theory**. In everyday speech, we often hear that something is "only a theory." But a scientific theory is a well-tested and widely accepted view that the scientific community agrees best explains certain observable facts. Some theories that are extensively documented and extremely well supported are comprehensive in scope. For example, the theory of plate tectonics provides a framework for understanding the origins of mountains, earthquakes, and volcanic activity. Plate tectonics also explains the evolution of the continents and the ocean basins through time—ideas that are explored in detail in Chapters 2, 13, and 14.

Scientific Methods

The process just described, in which researchers gather data through observations and formulate scientific hypotheses and theories, is called the **scientific method**. Contrary to popular belief, the scientific method is not a standard recipe that scientists apply in a routine manner to unravel the secrets of our natural world; rather, it is an endeavor that involves creativity and insight. Rutherford and Ahlgren put it this way: "Inventing hypotheses or theories to imagine how the world works and then figuring out how they can be put to the test of reality is as creative as writing poetry, composing music, or designing skyscrapers."*

*F. James Rutherford and Andrew Ahlgren, *Science for All Americans* (New York: Oxford University Press, 1990), p. 7.

There is no fixed path that scientists always follow that leads unerringly to scientific knowledge. However, many scientific investigations involve the steps outlined in **Figure 1.9**. In addition, some scientific discoveries result from purely theoretical ideas that stand up to extensive examination. Some researchers use high-speed computers to create models that simulate what is happening in the "real" world. These models are useful when dealing with natural processes that occur on very long time scales or take place in extreme or inaccessible locations. Still other scientific advancements are made when a totally unexpected happening occurs during an experiment. These serendipitous discoveries are more than pure luck, for as the nineteenth-century French scientist Louis Pasteur said, "In the field of observation, chance favors only the prepared mind."

Scientific knowledge is acquired through several avenues, so it might be best to describe the nature of scientific inquiry as the methods of science rather than as the scientific method. In addition, we should always remember that even the most compelling scientific theories are still simplified explanations of the natural world.

Plate Tectonics and Scientific Inquiry

This book offers many opportunities to develop and reinforce your understanding of how science works and, in particular, how the science of geology works. You will learn about data-gathering methods and the observational techniques and reasoning processes used by geologists.

Chapter 2 provides an excellent example. During the past several decades, we have learned a great deal about the workings of our dynamic planet. This period has seen an unequaled revolution in our understanding of Earth. The revolution began in the early part of the twentieth century, with the radical proposal of *continental drift*—the idea that the continents move about the face of the planet. This hypothesis contradicted the established view that the continents and ocean basins are permanent and stationary features on the face of Earth. For that reason, the notion of drifting continents was received with great skepticism and even ridicule. More than 50 years passed before enough data were gathered to transform this controversial hypothesis into a sound theory that wove

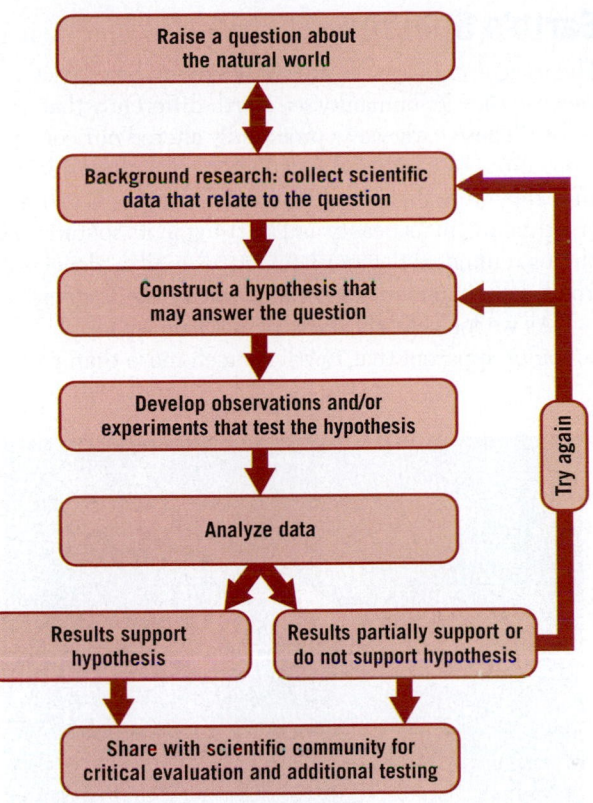

Figure 1.9
Steps frequently followed in scientific investigations
The diagram depicts the steps involved in the process many refer to as the *scientific method*.

together the basic processes known to operate on Earth. The theory that finally emerged, called the *theory of plate tectonics*, provided geologists with the first comprehensive model of Earth's internal workings.

In Chapter 2, you will not only gain insights into the workings of our planet, you will also see an excellent example of the way geologic "truths" are uncovered and reworked.

1.3 Concept Checks

1. How is a scientific hypothesis different from a scientific theory?

2. Summarize the basic steps followed in many scientific investigations.

1.4 | Earth as a System

List and describe Earth's four major spheres. Define *system* and explain why Earth is considered to be a system.

Anyone who studies Earth soon learns that our planet is a dynamic body with many separate but interacting parts, or *spheres*. The hydrosphere, atmosphere, biosphere, and geosphere and all of their components can be studied separately. However, the parts are *not* isolated. Each is related in some way to the others, producing a complex and continuously interacting whole that we call the *Earth system*.

Earth's Spheres

The images in **Figure 1.10** are considered to be classics because they let humanity see Earth differently than ever before. These early views profoundly altered our conceptualizations of Earth and remain powerful images decades after they were first viewed. Seen from space, Earth is breathtaking in its beauty and startling in its solitude. The photos remind us that our home is, after all, a planet— small, self-contained, and in some ways even fragile.

As we look closely at our planet from space, it becomes apparent that Earth is much more than rock and soil. In fact, among the most conspicuous features in both views of Earth in Figure 1.10 are swirling clouds suspended above the surface of the vast global ocean. These features emphasize the importance of water on our planet.

The closer view of Earth from space shown in Figure 1.10 helps us appreciate why the physical environment is traditionally divided into three major parts: the water portion of our planet, the *hydrosphere*; Earth's gaseous envelope, the *atmosphere*; and, of course, the solid Earth, or *geosphere*. It needs to be emphasized that our environment is highly integrated and not dominated by rock, water, or air alone. Rather, it is characterized by continuous interactions as air comes in contact with rock, rock with water, and water with air. Moreover, the *biosphere*, which is the totality of all plant and animal life on our planet, interacts with each of the three physical realms and is an equally integral part of the planet. Thus, Earth can be thought of as consisting of four major spheres: the hydrosphere, atmosphere, geosphere, and biosphere.

The interactions among Earth's spheres are incalculable. **Figure 1.11** provides us with one easy-to-visualize example. The shoreline is an obvious meeting place for rock, water, and air. In this scene, ocean waves created by the drag of air moving across the water are breaking against the rocky shore.

Figure 1.10
Two classic views of Earth from space (NASA)

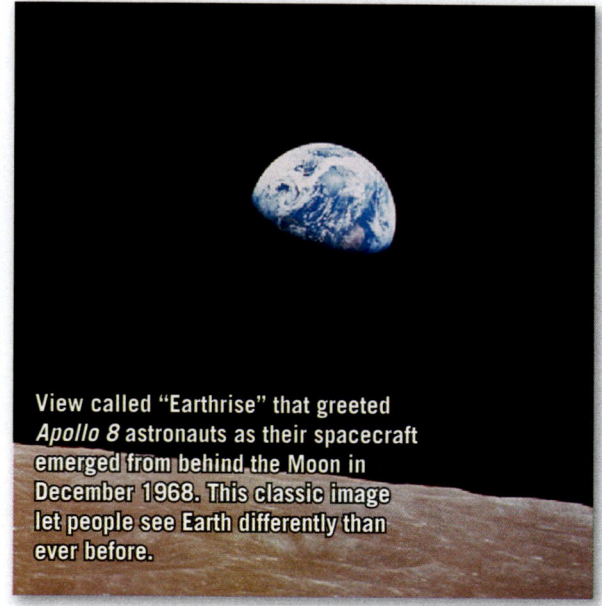

View called "Earthrise" that greeted *Apollo 8* astronauts as their spacecraft emerged from behind the Moon in December 1968. This classic image let people see Earth differently than ever before.

This image taken from *Apollo 17* in December 1972 is perhaps the first to be called "The Blue Marble." The dark blue ocean and swirling cloud patterns remind us of the importance of the oceans and atmosphere.

Hydrosphere   Earth is sometimes called the *blue planet*. Water, more than anything else, makes Earth unique. The **hydrosphere** is a dynamic mass of water that is continually on the move, evaporating from the oceans to the atmosphere, precipitating to the land, and running back to the ocean again. The global ocean is certainly the most prominent feature of the hydrosphere, blanketing nearly 71 percent of Earth's surface to an average depth of about 3800 meters (12,500 feet). It accounts for about 97 percent of Earth's water (**Figure 1.12**). However, the hydrosphere also includes the freshwater found underground and in streams, lakes, and glaciers. Moreover, water is an important component of all living things.

Although these latter sources constitute just a tiny fraction of the total, they are much more important than their meager percentages indicates. In addition to providing the freshwater that is so vital to life on land, streams, glaciers, and groundwater are responsible for sculpting and creating many of our planet's varied landforms.

Atmosphere   Earth is surrounded by a life-giving gaseous envelope called the **atmosphere** (**Figure 1.13**). When we watch a high-flying jet plane cross the sky, it seems that the atmosphere extends upward for a great distance. However, when compared to the thickness (radius) of the solid Earth (about 6400 kilometers [4000 miles]), the atmosphere is a very shallow layer. Despite its modest dimensions, this thin blanket of air is an integral part of the planet. It not only provides the air we breathe but also protects us from the Sun's intense heat and dangerous ultraviolet radiation. The energy exchanges that continually occur between the atmosphere and Earth's surface and between the atmosphere and space produce the effects we call *weather* and *climate*. Climate has a strong influence on the nature and intensity of Earth's external processes. When climate changes, these processes respond.

If, like the Moon, Earth had no atmosphere, our planet would be lifeless, and many of the processes and interactions that make the surface such a dynamic place could not operate. Without weathering and erosion, the face of our planet might more closely resemble the lunar surface, which has not changed appreciably in nearly 3 billion years.

Biosphere   The **biosphere** includes all life on Earth (**Figure 1.14**). Ocean life is concentrated in the sunlit surface waters of the sea. Most life on land is also concentrated near the surface, with tree roots and burrowing animals tunneling a few meters underground and flying insects and birds reaching a kilometer or so into the atmosphere. A surprising variety of life-forms are also adapted to extreme environments. For example, on the ocean floor, where pressures are extreme and no light penetrates, there are places where vents spew hot, mineral-rich fluids that support communities of exotic life-forms. On land, some bacteria thrive in rocks as deep as 4 kilometers (2.5 miles) and in boiling hot springs. Moreover, air currents can carry microorganisms many kilometers into the atmosphere. But even when we consider these extremes, life still must be thought of as being confined to a narrow band very near Earth's surface.

Plants and animals depend on the physical environment for the basics of life. However, organisms do not just respond to their physical environment. Through countless interactions, life-forms help maintain and alter the physical environment. Without life, the makeup and nature of the geosphere, hydrosphere, and atmosphere would be very different.

Figure 1.11
Interactions among Earth's spheres The shoreline is one obvious interface—a common boundary where different parts of a system interact. In this scene at California's Montano del Oro State Park, ocean waves (hydrosphere) that were created by the force of moving air (atmosphere) break against a rocky shore (geosphere). The force of the water can be powerful, and the erosional work that is accomplished can be great. (Photo by Michael Collier)

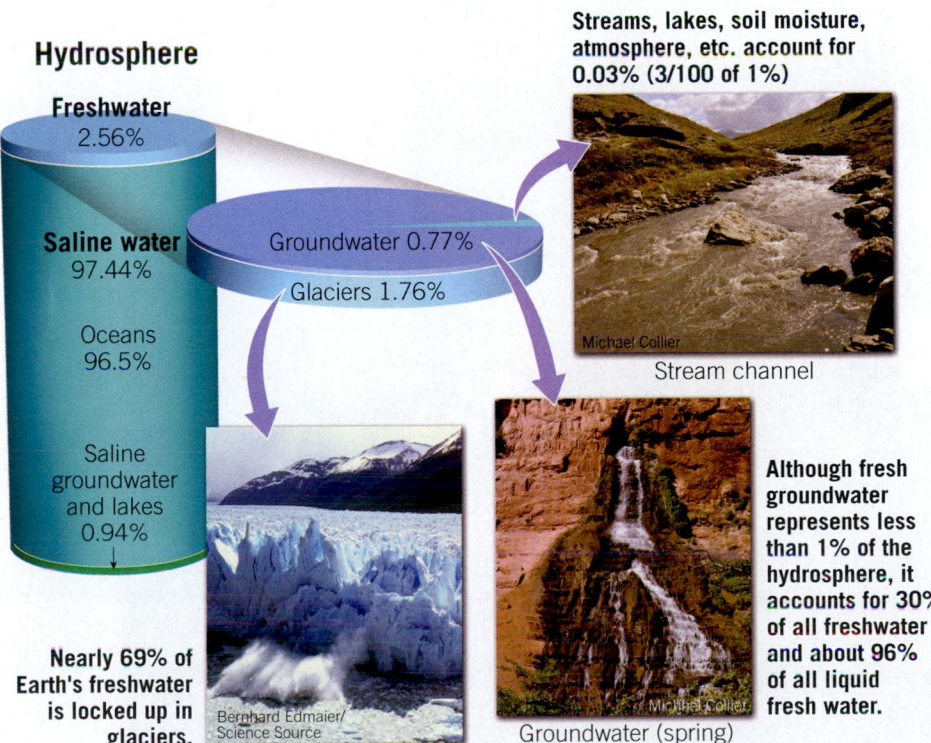

Hydrosphere

Freshwater 2.56%
Saline water 97.44%
Oceans 96.5%
Saline groundwater and lakes 0.94%

Groundwater 0.77%
Glaciers 1.76%

Streams, lakes, soil moisture, atmosphere, etc. account for 0.03% (3/100 of 1%)

Michael Collier
Stream channel

Nearly 69% of Earth's freshwater is locked up in glaciers.

Bernhard Edmaier/Science Source
Glaciers

Michael Collier
Groundwater (spring)

Although fresh groundwater represents less than 1% of the hydrosphere, it accounts for 30% of all freshwater and about 96% of all liquid fresh water.

Figure 1.12
The water planet Distribution of water in the hydrosphere.

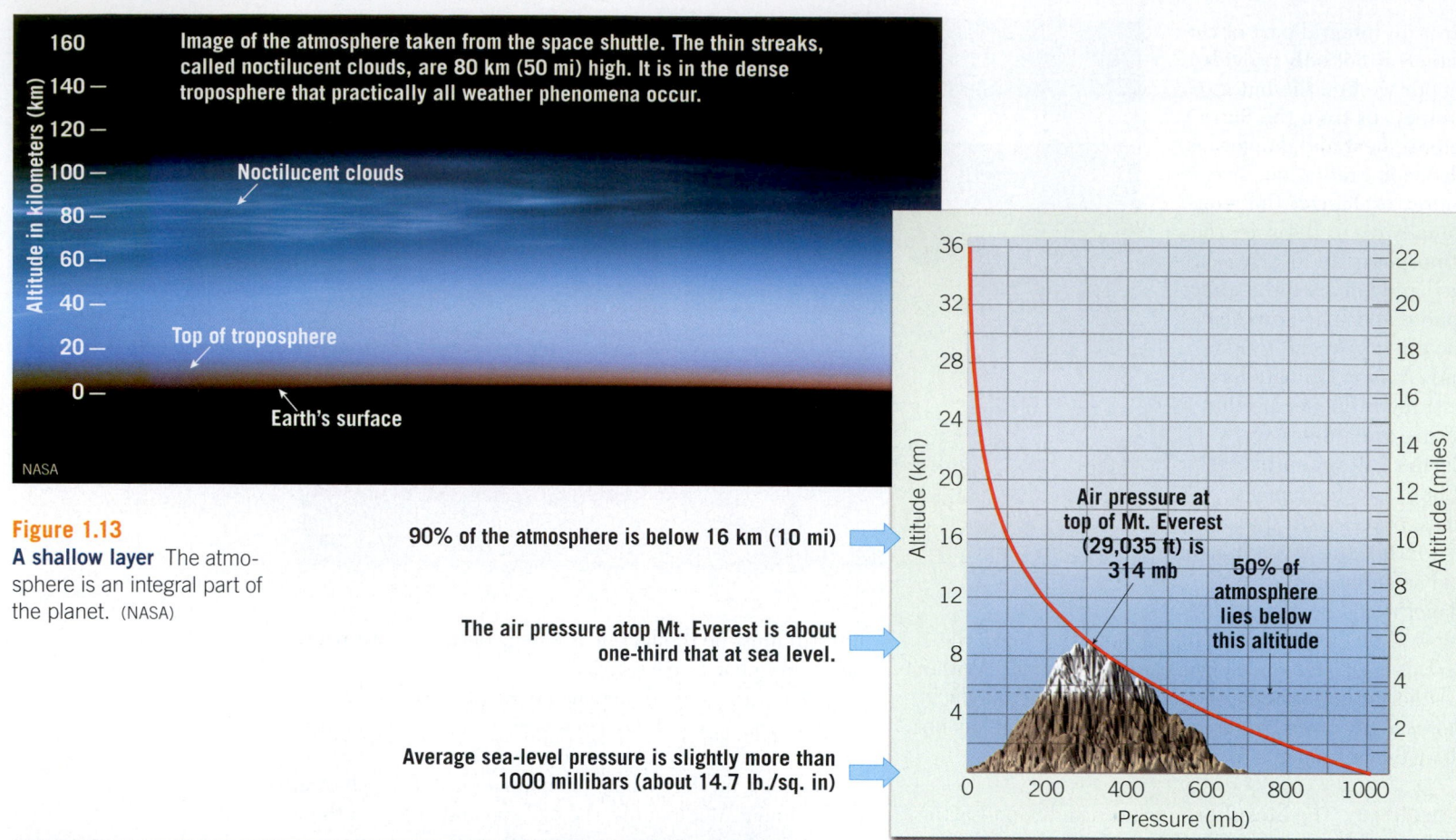

Image of the atmosphere taken from the space shuttle. The thin streaks, called noctilucent clouds, are 80 km (50 mi) high. It is in the dense troposphere that practically all weather phenomena occur.

Noctilucent clouds

Top of troposphere

Earth's surface

NASA

Figure 1.13
A shallow layer The atmosphere is an integral part of the planet. (NASA)

90% of the atmosphere is below 16 km (10 mi)

The air pressure atop Mt. Everest is about one-third that at sea level.

Average sea-level pressure is slightly more than 1000 millibars (about 14.7 lb./sq. in)

Air pressure at top of Mt. Everest (29,035 ft) is 314 mb

50% of atmosphere lies below this altitude

Altitude (km)

Altitude (miles)

Pressure (mb)

Geosphere   Beneath the atmosphere and the oceans is the solid Earth, or **geosphere**. The geosphere extends from the surface to the center of the planet, a depth of nearly 6400 kilometers (nearly 4000 miles), making it by far the largest of Earth's four spheres. Much of our study of the solid Earth focuses on the more accessible surface features. Fortunately, many of these features represent the outward expressions of the dynamic behavior of Earth's interior. Examining the most prominent surface features and their global extent gives us clues to the dynamic processes that have shaped our planet. A first look at the structure of Earth's interior and at the major surface features of the geosphere comes later in the chapter.

Soil, the thin veneer of material at Earth's surface that supports the growth of plants, may be thought of as part of all four spheres. The solid portion is a mixture of weathered

The ocean contains a significant portion of Earth's biosphere. Modern coral reefs are unique and complex examples and are home to about 25% of all marine species. Because of this diversity, they are sometimes referred to as the ocean equivalent of a rain forest.

Figure 1.14
The biosphere The biosphere, one of Earth's four spheres, includes all life. (Coral reef photo by Darryl Leniuk /AGE Fotostock; rain forest photo by AGE Fotostock/SuperStock)

Tropical rain forests are characterized by hundreds of different species per square kilometer.

rock debris (geosphere) and organic matter from decayed plant and animal life (biosphere). The decomposed and disintegrated rock debris is the product of weathering processes that require air (atmosphere) and water (hydrosphere). Air and water also occupy the open spaces between the solid particles.

Earth System Science

A simple example of the interactions among different parts of the Earth system occurs every winter and spring, as moisture evaporates from the Pacific Ocean and subsequently falls as rain in coastal hills and mountains, triggering destructive debris flows. The processes that move water from the hydrosphere to the atmosphere and then to the solid Earth have a profound impact on the plants and animals (including humans) that inhabit the affected regions (**Figure 1.15**).

Scientists have recognized that in order to more fully understand our planet, they must learn how its individual components (land, water, air, and life-forms) are interconnected. This endeavor, called **Earth system science**, aims to study Earth as a *system* composed of numerous interacting parts, or *subsystems*. Rather than look through the limited lens of only one of the traditional sciences—geology, atmospheric science, chemistry, biology, and so on—Earth system science attempts to integrate the knowledge of several academic fields. Using an interdisciplinary approach, those engaged in Earth system science attempt to achieve the level of understanding necessary to comprehend and solve many of our global environmental problems.

A **system** is a group of interacting, or interdependent, parts that form a complex whole. Most of us hear and use the term *system* frequently. We may service our car's cooling *system*, make use of the city's transportation *system*, and participate in our political *system*. A news report might inform us of an approaching weather *system*. Further, we know that Earth is just a small part of a larger system known as the *solar system*, which in turn is a subsystem of an even larger system, the Milky Way Galaxy.

The Earth System

The Earth system has a nearly endless array of subsystems in which matter is recycled over and over. One familiar

Figure 1.15
Deadly debris flow This image provides an example of interactions among different parts of the Earth system. Extraordinary rains triggered this debris flow (popularly called a mudslide) on March 22, 2014, near Oso, Washington. The mass of mud and debris blocked the North Fork of the Stillaguamish River and engulfed an area of about 2.6 square kilometers (1 square mile). Forty-three people perished. (Photo by Michael Collier)

Figure 1.16
Change is a geologic constant When Mount St. Helens erupted in May 1980, the area shown here was buried by a volcanic mudflow. Now plants are reestablished, and new soil is forming. (Photo by Terry Donnelly/Alamy Images)

loop or subsystem is the *hydrologic cycle*. It represents the unending circulation of Earth's water among the hydrosphere, atmosphere, biosphere, and geosphere. Water enters the atmosphere through evaporation from Earth's surface and transpiration from plants. Water vapor condenses in the atmosphere to form clouds, which in turn produce precipitation that falls back to Earth's surface. Some of the rain that falls onto the land infiltrates (soaks in) to be taken up by plants or become groundwater, and some flows across the surface toward the ocean.

Viewed over long time spans, the rocks of the geosphere are constantly forming, changing, and re-forming. The loop that involves the processes by which one rock changes to another is called the *rock cycle* and will be discussed at some length later in the chapter. The cycles of the Earth system are not independent of one another; to the contrary, these cycles come in contact and interact in many places.

The parts of the Earth system are linked so that a change in one part can produce changes in any or all of the other parts. For example, when a volcano erupts, lava from Earth's interior may flow out at the surface and block a nearby valley. This new obstruction influences the region's drainage system by creating a lake or causing streams to change course. The large quantities of volcanic ash and gases that can be emitted during an eruption might be blown high into the atmosphere and influence the amount of solar energy that can reach Earth's surface. The result could be a drop in air temperatures over the entire hemisphere.

Where the surface is covered by lava flows or a thick layer of volcanic ash, existing soils are buried. This causes soil-forming processes to begin anew to transform the new surface material into soil (**Figure 1.16**). The soil that eventually forms will reflect the interactions among many parts of the Earth system—the volcanic parent material, the climate, and the impact of biological activity. Of course, there would also be significant changes in the biosphere. Some organisms and their habitats would be eliminated by the lava and ash, whereas new settings for life, such as a lake formed by a lava dam, would be created. The potential climate change could also impact sensitive life-forms.

The Earth system is characterized by processes that vary on spatial scales from fractions of millimeters to thousands of kilometers. Time scales for Earth's processes range from seconds to billions of years. As we learn about Earth, it becomes increasingly clear that despite significant separations in distance or time, many processes are connected, and a change in one component can influence the entire system.

The Earth system is powered by energy from two sources. The Sun drives external processes that occur in the atmosphere, in the hydrosphere, and at Earth's surface. Weather and climate, ocean circulation, and erosional processes are driven by energy from the Sun. Earth's interior is the second source of energy. Heat remaining from when our planet formed and heat that is continuously generated by radioactive decay power the internal processes that produce volcanoes, earthquakes, and mountains.

Humans are *part of* the Earth system, a system in which the living and nonliving components are entwined and interconnected. Therefore, our actions produce changes in all the other parts. When we burn gasoline and coal, dispose of our wastes, and clear the land, we cause other parts of the system to respond, often in unforeseen ways. Throughout this book you will learn about many of Earth's subsystems, including the hydrologic system, the tectonic (mountain-building) system, the rock cycle, and the climate system. Remember that these components *and we humans* are all part of the complex interacting whole we call the Earth system.

1.4 Concept Checks

1. List and briefly describe the four spheres that constitute the Earth system.

2. Compare the height of the atmosphere to the thickness of the geosphere.

3. How much of Earth's surface do oceans cover? What percentage of Earth's water supply do oceans represent?

4. What is a system? List three examples.

5. What are the two sources of energy for the Earth system?

1.5 Origin and Early Evolution of Earth
Outline the stages in the formation of our solar system.

Recent earthquakes caused by displacements of Earth's crust and lavas spewed from active volcanoes represent only the latest in a long line of events by which our planet has attained its present form and structure. The geologic processes operating in Earth's interior can be best understood when viewed in the context of much earlier events in Earth history.

Origin of Our Solar System

This section describes the most widely accepted views on the origin of our solar system. The theory described here represents the most consistent set of ideas we have to explain what we know about our solar system today. GEOgraphics 1.2 provides a useful perspective on size and scale in our solar system. In addition, the origins of Earth and other bodies of our solar system are discussed in more detail in Chapters 2 and 24.

The Universe Begins   Our scenario begins about 13.7 billion years ago with the *Big Bang*, an incomprehensibly large explosion that sent all matter of the universe flying outward at incredible speeds. In time, the debris from this explosion, which was almost entirely hydrogen and helium, began to cool and condense into the first stars and galaxies. It was in one of these galaxies, the Milky Way, that our solar system and planet Earth took form.

The Solar System Forms   Earth is one of eight planets that, along with several dozen moons and numerous smaller bodies, revolve around the Sun. The orderly nature of our solar system leads researchers to conclude that Earth and the other planets formed at essentially the same time and from the same primordial material as the Sun. The **nebular theory** proposes that the bodies of our solar system evolved from an enormous rotating cloud called the **solar nebula** (Figure 1.17). Besides the hydrogen and helium atoms generated during the Big Bang, the solar nebula consisted of microscopic dust grains and the ejected matter of long-dead stars. (Nuclear fusion in stars converts hydrogen and helium into the other elements found in the universe.)

Nearly 5 billion years ago, this huge cloud of gases and minute grains of heavier elements began to slowly contract due to the gravitational interactions among its particles. Some external influence, such as a shock wave traveling from a catastrophic explosion (*supernova*), may have triggered the collapse. As this slowly spiraling nebula contracted, it rotated faster and faster for the same reason ice skaters do when they draw their arms toward their bodies. Eventually the inward pull of gravity came into balance with the outward force caused by the rotational motion of the nebula (see Figure 1.17). By this time, the once-vast cloud had assumed a flat disk shape with a large concentration of material at its center called the *protosun* (pre-Sun). (Astronomers are fairly confident that the nebular cloud formed a disk because similar structures have been detected around other stars.)

During the collapse, gravitational energy was converted to thermal (heat) energy, causing the temperature of the inner portion of the nebula to dramatically rise. At these high temperatures, the dust grains broke up into molecules and extremely energetic atomic particles. However, at distances beyond the orbit of Mars, temperatures probably remained quite low. At −200°C (−328°F), the tiny particles in the outer portion of the nebula were likely covered with a thick layer of frozen water, carbon dioxide, ammonia, and methane. The disk-shaped cloud also contained appreciable amounts of the lighter gases hydrogen and helium.

The birth of our solar system began as dust and gases (nebula) started to gravitationally collapse.

The nebula contracted into a flattened, rotating disk that was heated by the conversion of gravitational energy into thermal energy.

Cooling of the nebular cloud caused rocks and metallic material to condense into tiny particles.

Repeated collisions caused the dust-size particles to gradually coalesce into asteroid-size bodies that accreted into planets within a few million years.

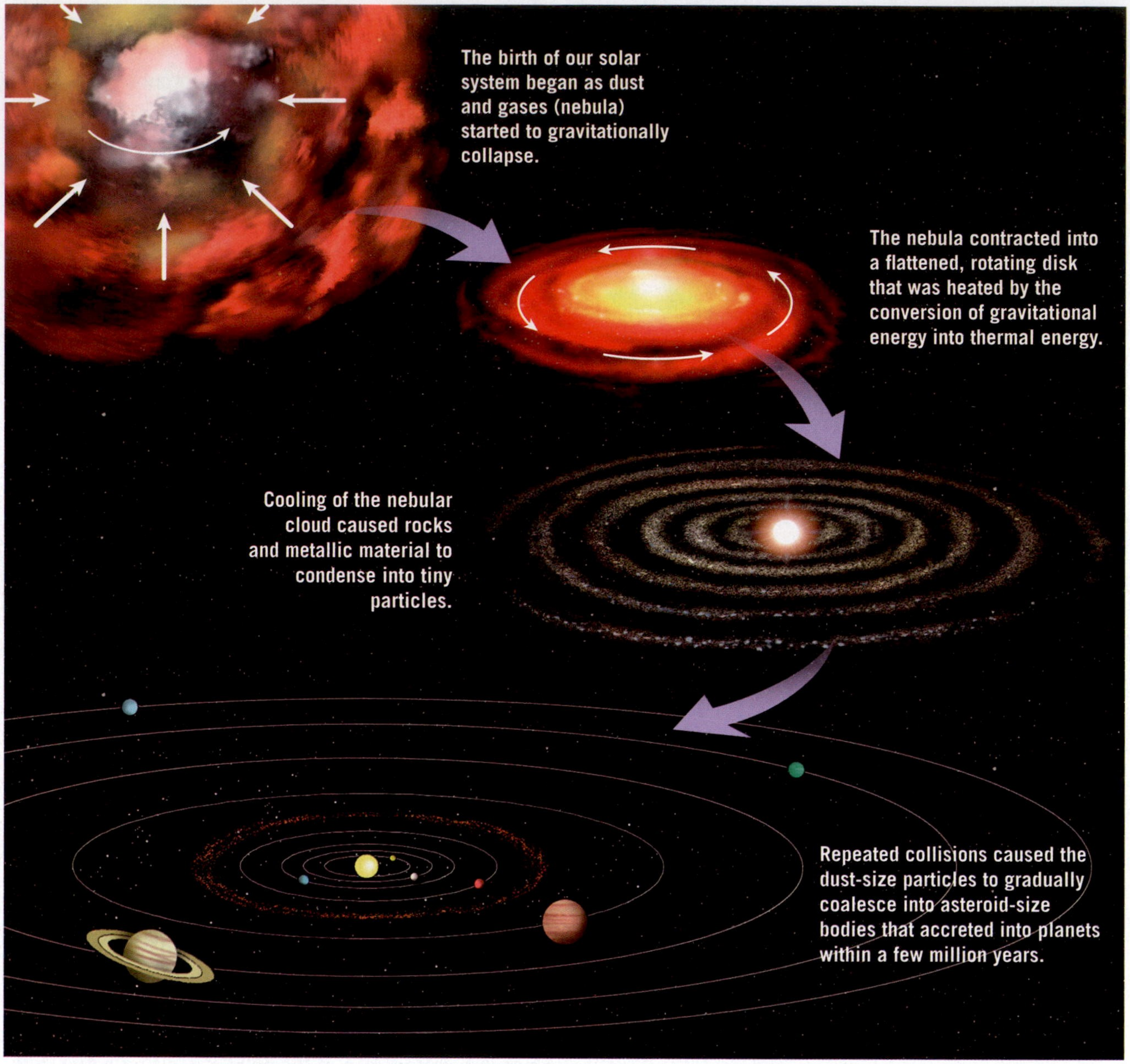

The Inner Planets Form   The formation of the Sun marked the end of the period of contraction and thus the end of gravitational heating. Temperatures in the region where the inner planets now reside began to decline. This decrease in temperature caused those substances with high melting points to condense into tiny particles that began to coalesce (join together). Materials such as iron and nickel and the elements of which the rock-forming minerals are composed—silicon, calcium, sodium, and so forth—formed metallic and rocky clumps that orbited the Sun (see Figure 1.17). Repeated collisions caused these masses to coalesce into larger asteroid-size bodies, called *planetesimals*, which in a few tens of millions of years accreted into the four inner planets we call Mercury, Venus, Earth, and Mars (**Figure 1.18**). Not all of these clumps of matter were incorporated into the planetesimals. Those rocky and metallic pieces that remained in orbit are called *meteorites* when they survive an impact with Earth.

As more and more material was swept up by the planets, the high-velocity impact of nebular debris caused the temperatures of these bodies to rise. Because of their relatively high temperatures and weak gravitational fields, the inner planets were unable to accumulate much of the lighter components of the nebular cloud. The lightest of these, hydrogen and helium, were eventually whisked from the inner solar system by the solar wind.

The Outer Planets Develop   At the same time that the inner planets were forming, the larger outer planets (Jupiter, Saturn, Uranus, and Neptune), along with their

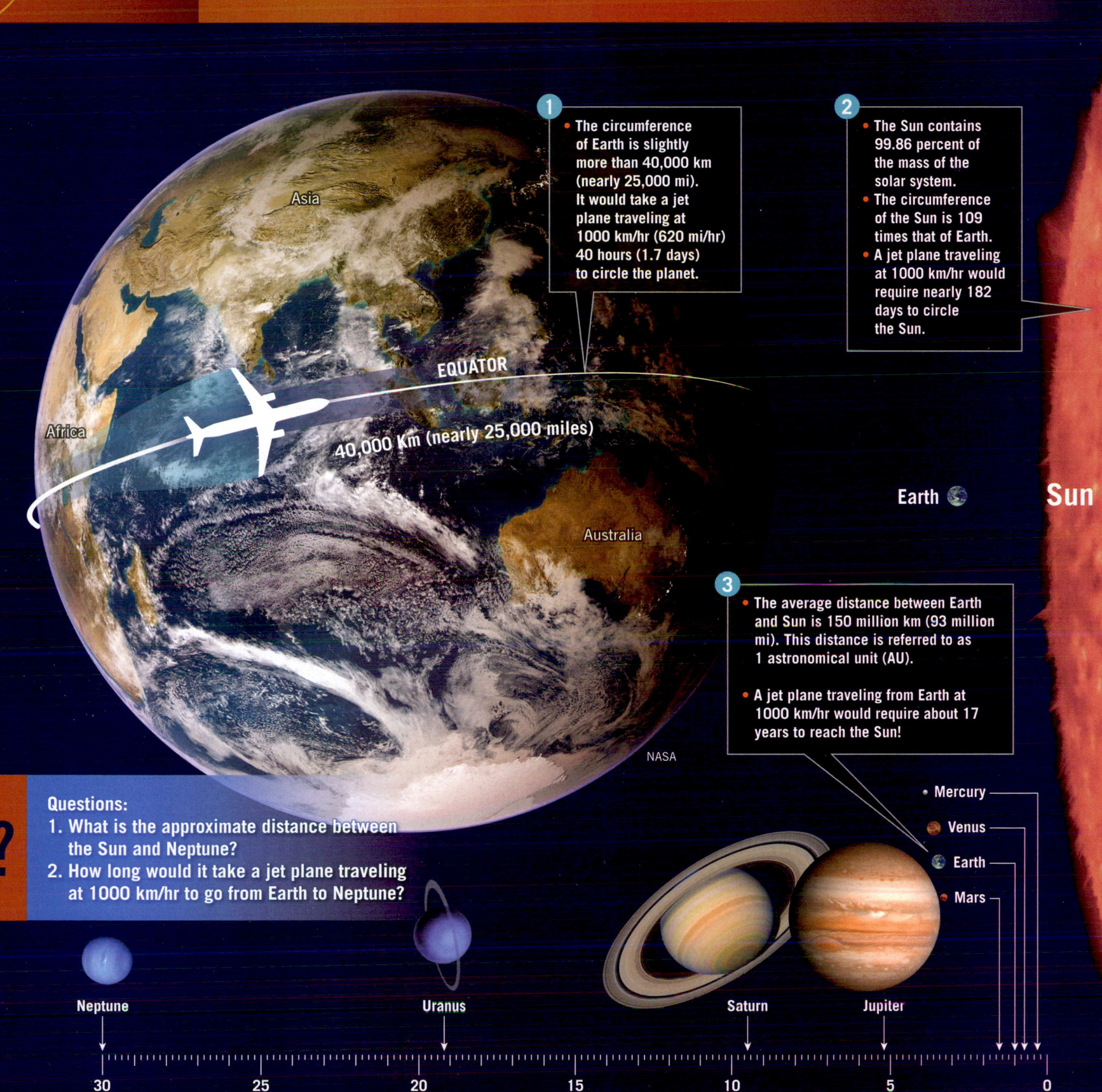

Solar System: Size and Scale

The Sun is the center of a revolving system trillions of miles across, consisting of 8 planets, their satellites, and numerous dwarf planets, asteroids, comets, and meteoroids.

1
- The circumference of Earth is slightly more than 40,000 km (nearly 25,000 mi). It would take a jet plane traveling at 1000 km/hr (620 mi/hr) 40 hours (1.7 days) to circle the planet.

2
- The Sun contains 99.86 percent of the mass of the solar system.
- The circumference of the Sun is 109 times that of Earth.
- A jet plane traveling at 1000 km/hr would require nearly 182 days to circle the Sun.

Asia

EQUATOR

Africa

40,000 Km (nearly 25,000 miles)

Australia

Earth 🌍   Sun

3
- The average distance between Earth and Sun is 150 million km (93 million mi). This distance is referred to as 1 astronomical unit (AU).

- A jet plane traveling from Earth at 1000 km/hr would require about 17 years to reach the Sun!

NASA

Questions:
1. What is the approximate distance between the Sun and Neptune?
2. How long would it take a jet plane traveling at 1000 km/hr to go from Earth to Neptune?

• Mercury

• Venus

🌍 Earth

• Mars

Neptune   Uranus   Saturn   Jupiter

30   25   20   15   10   5   0

Figure 1.18
A remnant planetesimal This image of Asteroid 21 Lutetia was obtained by special cameras aboard the *Rosetta* spacecraft on July 10, 2010. Spacecraft instruments showed that Lutetia is a primitive body (planetesimal) left over from when the solar system formed. (NASA)

extensive satellite systems, were also developing. Because of low temperatures far from the Sun, the material from which these planets formed contained a high percentage of ices—frozen water, carbon dioxide, ammonia, and methane—as well as rocky and metallic debris. The accumulation of ices partly accounts for the large size and low density of the outer planets. The two most massive planets, Jupiter and Saturn, had a surface gravity sufficient to attract and hold large quantities of even the lightest elements—hydrogen and helium.

Formation of Earth's Layered Structure

As material accumulated to form Earth (and for a short period afterward), the high-velocity impact of nebular debris and the decay of radioactive elements caused the temperature of our planet to steadily increase. During this time of intense heating, Earth became hot enough that iron and nickel began to melt. Melting produced liquid blobs of dense metal that sank toward the center of the planet. This process occurred rapidly on the scale of geologic time and produced Earth's dense, iron-rich core.

Chemical Differentiation and Earth's Layers The early period of heating resulted in another process of chemical differentiation, whereby melting formed buoyant masses of molten rock that rose toward the surface and solidified to produce a primitive crust. These rocky materials were enriched in oxygen and "oxygen-seeking" elements, particularly silicon and aluminum, along with lesser amounts of calcium, sodium, potassium, iron, and magnesium. In addition, some heavy metals such as gold, lead, and uranium, which have low melting points or were highly soluble in the ascending molten masses, were scavenged from Earth's interior and concentrated in the developing crust. This early period of chemical differentiation established the three basic divisions of Earth's interior: the iron-rich *core*; the thin *primitive crust*; and Earth's largest layer, called the *mantle*, which is located between the core and crust.

An Atmosphere Develops An important consequence of the early period of chemical differentiation is that large quantities of gaseous materials were allowed to escape from Earth's interior, as happens today during volcanic eruptions. By this process, a primitive atmosphere gradually evolved. It is on this planet, with this atmosphere, that life as we know it came into existence.

Continents and Ocean Basins Evolve Following the events that established Earth's basic structure, the primitive crust was lost to erosion and other geologic processes, so we have no direct record of its makeup. When and exactly how the continental crust—and thus Earth's first landmasses—came into existence is a matter of ongoing research. Nevertheless, there is general agreement that the continental crust formed gradually over the past 4 billion years. (The oldest rocks yet discovered are isolated fragments found in the Northwest Territories of Canada that have radiometric dates of about 4 billion years.) In addition, as you will see in subsequent chapters, Earth is an evolving planet whose continents and ocean basins have continually changed shape and even location during much of this period.

1.5   Concept Checks

1. Name and briefly outline the theory that describes the formation of our solar system.

2. List the inner planets and outer planets. Describe basic differences in size and composition.

3. Explain why density and buoyancy were important in the development of Earth's layered structure.

1.6 | Earth's Internal Structure

Sketch Earth's internal structure and label and describe the main subdivisions.

The preceding section outlined how differentiation of material early in Earth's history resulted in the formation of three major layers defined by their chemical composition: the crust, mantle, and core. In addition to these compositionally distinct layers, Earth is divided into layers based on physical properties. The physical properties used to define such zones include whether the layer is solid or liquid and how weak or strong it is. Important examples include the lithosphere, asthenosphere, outer core, and inner core. Knowledge of both chemical and physical layers is important to our understanding of many geologic processes, including volcanism, earthquakes, and mountain building. **Figure 1.19** shows different views of Earth's layered structure.

How did we learn about the composition and structure of Earth's interior? The nature of Earth's interior is primarily determined by analyzing seismic waves from earthquakes. As these waves of energy penetrate the planet, they change speed and are bent and reflected as they move through zones that have different properties. Monitoring stations around the world detect and record this energy. With the aid of computers, these data are analyzed and used to build a detailed picture of Earth's interior. There is more about this in Chapter 12.

SmartFigure 1.19
Earth's layers Structure of Earth's interior based on chemical composition (right side of diagram) and physical properties (left side of diagram). (https://goo.gl/70au1N)

Tutorial

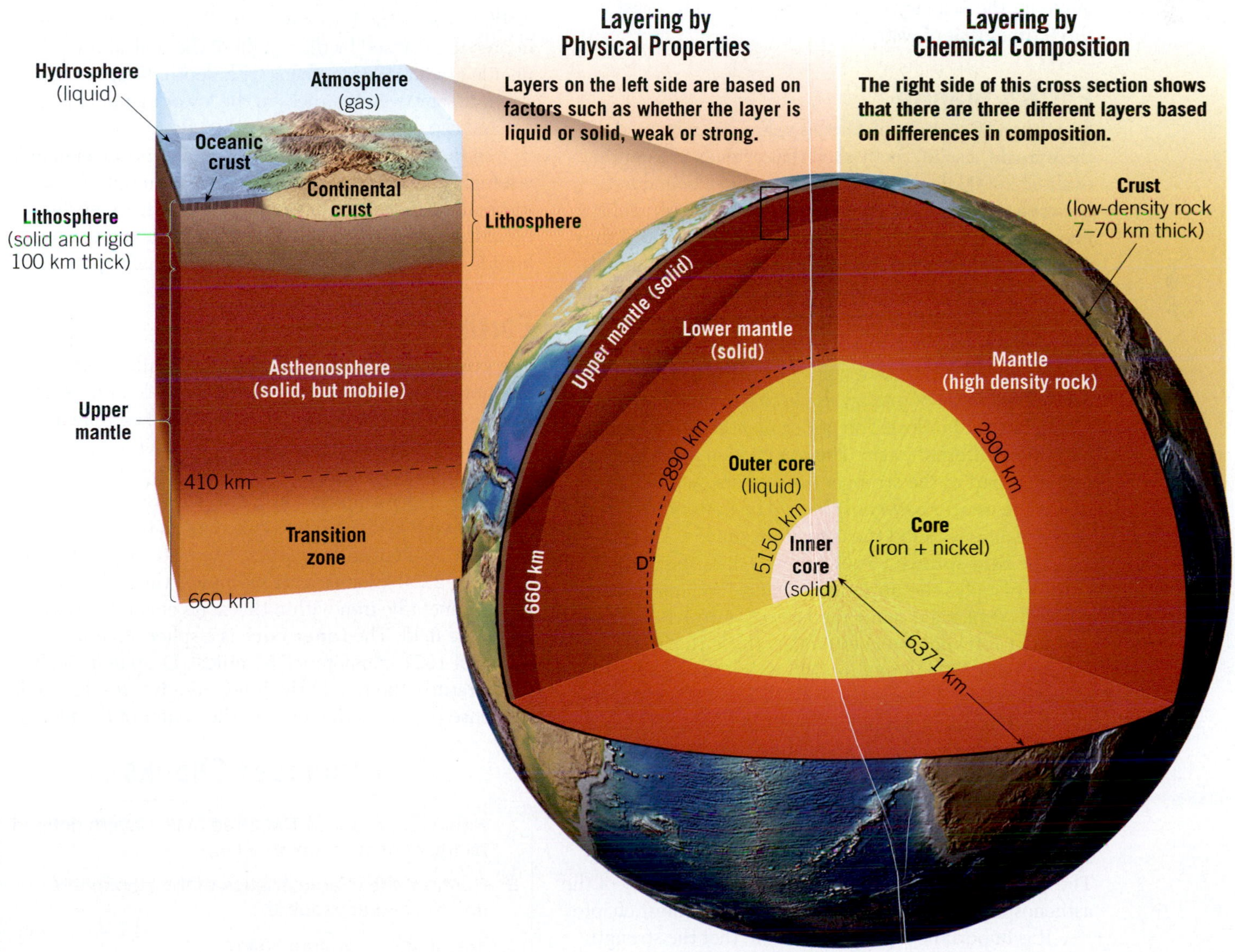

Layering by Physical Properties

Layers on the left side are based on factors such as whether the layer is liquid or solid, weak or strong.

Layering by Chemical Composition

The right side of this cross section shows that there are three different layers based on differences in composition.

Hydrosphere (liquid)

Atmosphere (gas)

Oceanic crust

Continental crust

Lithosphere (solid and rigid 100 km thick)

Lithosphere

Upper mantle (solid)

Lower mantle (solid)

Crust (low-density rock 7–70 km thick)

Mantle (high density rock)

Asthenosphere (solid, but mobile)

Upper mantle

410 km

Transition zone

660 km

660 km

2890 km

D"

Outer core (liquid)

5150 km

Inner core (solid)

Core (iron + nickel)

2900 km

6371 km

Earth's Crust

The **crust**, Earth's relatively thin, rocky outer skin, is of two different types—continental crust and oceanic crust. Both share the word *crust*, but the similarity ends there. The oceanic crust is roughly 7 kilometers (4.5 miles) thick and composed of the dark igneous rock *basalt*. By contrast, the continental crust averages about 35 kilometers (22 miles) thick but may exceed about 70 kilometers (45 miles) in some mountainous regions such as the Rockies and Himalayas. Unlike the oceanic crust, which has a relatively homogeneous chemical composition, the continental crust consists of many rock types. Although the upper crust has an average composition of a *granitic rock* called *granodiorite*, it varies considerably from place to place.

Continental rocks have an average density of about 2.7 g/cm^3, and some have been discovered that are more than 4 billion years old. The rocks of the oceanic crust are younger (180 million years or less) and denser (about 3.0 g/cm^3) than continental rocks. For comparison, liquid water has a density of 1 g/cm^3; therefore, the density of basalt, the primary rock composing oceanic crust, is three times that of water.

Earth's Mantle

More than 82 percent of Earth's volume is contained in the **mantle**, a solid, rocky shell that extends to a depth of about 2900 kilometers (1800 miles). The boundary between the crust and mantle represents a marked change in chemical composition. The dominant rock type in the uppermost mantle is *peridotite*, which contains minerals richer in the metals magnesium and iron compared to the minerals found in either the continental or oceanic crust.

The Upper Mantle   The upper mantle extends from the crust–mantle boundary down to a depth of about 660 kilometers (410 miles). The upper mantle can be divided into three different parts. The top portion of the upper mantle is part of the stronger *lithosphere*, and beneath that is the weaker *asthenosphere*. The bottom part of the upper mantle is called the *transition zone*.

The **lithosphere** ("rock sphere") consists of the entire crust plus the uppermost mantle and forms Earth's relatively cool, rigid outer shell (see Figure 1.19). Averaging about 100 kilometers (60 miles) thick, the lithosphere is more than 250 kilometers (155 miles) thick below the oldest portions of the continents. Beneath this stiff layer to a depth of about 410 kilometers (255 miles) lies a soft, comparatively weak layer known as the **asthenosphere** ("weak sphere"). The top portion of the asthenosphere has a temperature/pressure regime that results in a small amount of melting. Within this very weak zone, the lithosphere is mechanically detached from the layer below. The lithosphere thus is able to move independently of the asthenosphere, a fact we will consider in the next chapter.

It is important to emphasize here that the strength of various Earth materials is a function of both their composition and the temperature and pressure of their environment. You should not get the idea that the entire lithosphere behaves like a rigid or brittle solid similar to rocks found on the surface. Rather, the rocks of the lithosphere get progressively hotter and weaker (more easily deformed) with increasing depth. At the depth of the uppermost asthenosphere, the rocks are close enough to their melting temperature (some melting may actually occur) that they are very easily deformed. Thus, the uppermost asthenosphere is weak because it is near its melting point, just as hot wax is weaker than cold wax.

From about 410 kilometers (255 miles) to about 660 kilometers (410 miles) in depth is the part of the upper mantle called the **transition zone** (Figure 1.19). The top of the transition zone is identified by a sudden increase in density from about 3.5 to 3.7 g/cm^3. This change occurs because minerals in the rock peridotite respond to the increase in pressure by forming new minerals with closely packed atomic structures.

The Lower Mantle   From a depth of 660 kilometers (410 miles) to the top of the core, at a depth of 2900 kilometers (1800 miles), is the **lower mantle**. Because of an increase in pressure (caused by the weight of the rock above), the mantle gradually strengthens with depth. Despite their strength, however, the rocks in the lower mantle are very hot and capable of extremely gradual flow.

In the bottom few hundred kilometers of the mantle is a highly variable and unusual layer called the D″ layer (pronounced "dee double-prime"). The nature of this boundary layer between the rocky mantle and the hot liquid iron outer core will be examined in Chapter 12.

Earth's Core

The **core** is composed of an iron–nickel alloy with minor amounts of oxygen, silicon, and sulfur—elements that readily form compounds with iron. At the extreme pressure found in the core, this iron-rich material has an average density of nearly 11 g/cm^3 and approaches 14 times the density of water at Earth's center.

The core is divided into two regions that exhibit very different mechanical strengths. The **outer core** is a *liquid layer* 2250 kilometers (1395 miles) thick. The movement of metallic iron within this zone generates Earth's magnetic field. The **inner core** is a sphere that has a radius of 1221 kilometers (757 miles). Despite its higher temperature, the iron in the inner core is *solid* due to the immense pressures that exist in the center of the planet.

1.6   Concept Checks

1. Name and describe the three major layers defined by their chemical composition.

2. Contrast the characteristics of the lithosphere and the asthenosphere.

3. Why is the inner core solid?

1.7 | Rocks and the Rock Cycle

Sketch, label, and explain the rock cycle.

Rock is the most common and abundant material on Earth. To a curious traveler, the variety seems nearly endless. When a rock is examined closely, we find that it usually consists of smaller crystals called minerals. *Minerals* are chemical compounds (or sometimes single elements), each with its own composition and physical properties. The grains or crystals may be microscopically small or easily seen with the unaided eye.

The minerals that compose a rock strongly influence its nature and appearance. In addition, a rock's *texture*—the size, shape, and/or arrangement of its constituent minerals—also has a significant effect on its appearance. A rock's mineral composition and texture, in turn, reflect the geologic processes that created it (**Figure 1.20**). Such analyses are critical to understanding our planet. This understanding also has many practical applications, including finding energy and mineral resources and solving environmental problems.

Geologists divide rocks into three major groups: igneous, sedimentary, and metamorphic. **Figure 1.21** provides some examples. As you will learn, each group is linked to the others by the processes that act upon and within the planet.

A. The large crystals of light-colored minerals in granite result from the slow cooling of molten rock deep beneath the surface. Granite is abundant in the continental crust.

B. Basalt is rich in dark minerals. Rapid cooling of molten rock at Earth's surface is responsible for the rock's microscopically small crystals. Oceanic crust is composed mainly of basalt.

Figure 1.20

Two basic rock characteristics Texture and mineral composition are basic rock features. These two samples are the common igneous rocks granite (**A**) and basalt (**B**). (Photo A by geoz/alamy Images; photo B by Tyler Boyes/Shutterstock)

Earlier in this chapter, you learned that Earth is a system. This means that our planet consists of many interacting parts that form a complex whole. Nowhere is this idea better illustrated than when we examine the rock cycle (**Figure 1.22**). The **rock cycle** allows us to view many of the interrelationships among different parts of the Earth system. It helps us understand the origin of igneous, sedimentary, and metamorphic rocks and to see that each type is linked to the others by processes that act upon and within the planet. Consider the rock cycle to be a simplified but useful overview of physical geology. Learn the rock cycle well; you will be examining its interrelationships in greater detail throughout this book.

The Basic Cycle

Magma is molten rock that forms deep beneath Earth's surface. Over time, magma cools and solidifies. This process, called *crystallization*, may occur either beneath the surface or, following a volcanic eruption, at the surface. In either situation, the resulting rocks are called **igneous rocks**.

If igneous rocks are exposed at the surface, they undergo *weathering*, in which the day-in and day-out influences of the atmosphere, hydrosphere, and biosphere slowly disintegrate and decompose rocks. The materials that result are often moved downslope by gravity before being picked up and transported by any of a number of erosional agents, such as running water, glaciers, wind, or waves. Eventually these particles and dissolved substances, called **sediment**, are deposited. Although most sediment ultimately comes to rest in the ocean, other sites of deposition include river floodplains, desert basins, swamps, and sand dunes.

Next, the sediments undergo *lithification*, a term meaning "conversion into rock." Sediment is usually lithified into **sedimentary rock** when compacted by the weight of overlying layers or when cemented as percolating groundwater fills the pores with mineral matter.

If the resulting sedimentary rock is buried deep within Earth and involved in the dynamics of mountain building or intruded by a mass of magma, it is subjected to great pressures and/or intense heat. The sedimentary rock reacts to the changing environment and turns into the third rock type, **metamorphic rock**. When metamorphic rock is subjected to additional pressure changes or to still higher temperatures, it melts, creating magma, which eventually crystallizes into igneous rock, starting the cycle all over again.

Three rock groups Geologists divide rocks into three groups: igneous, sedimentary, and metamorphic.

Igneous rocks form when molten rock solidifies at the surface (extrusive) or beneath the surface (intrusive). The lava flow in the foreground is the fine-grained rock basalt and came from SP Crater in northern Arizona.

Sedimentary rocks consist of particles derived from the weathering of other rocks. This layer consists of durable sand-size grains of the glassy mineral quartz that are cemented into a solid rock. The grains were once a part of extensive dunes. This rock layer, called the Navajo Sandstone, is prominent in southern Utah.

The metamorphic rock pictured here, known as the Vishnu Schist, is exposed in the inner gorge of the Grand Canyon. Its formation is associated with environments deep below Earth's surface where temperatures and pressures are high and with the forces associated with ancient mountain-building processes that occurred in Precambrian time.

Where does the energy that drives Earth's rock cycle come from? Processes driven by heat from Earth's interior are responsible for creating igneous and metamorphic rocks. Weathering and erosion, external processes powered by energy from the Sun, produce the sediment from which sedimentary rocks form.

Alternative Paths

The paths shown in the basic cycle are not the only ones possible. To the contrary, other paths are just as likely to be followed as those described in the preceding section. These alternatives are indicated by the light blue arrows in Figure 1.22.

Rather than being exposed to weathering and erosion at Earth's surface, igneous rocks may remain deeply buried. Eventually these masses may be subjected to the strong compressional forces and high temperatures associated with mountain building. When this occurs, they are transformed directly into metamorphic rocks.

Metamorphic and sedimentary rocks, as well as sediment, do not always remain buried. Rather, overlying layers may be stripped away, exposing the once-buried rock. This exposed material is attacked by weathering processes and turned into new raw materials for sedimentary rocks.

Although rocks may seem to be unchanging masses, the rock cycle shows that they are not. The changes,

ROCK CYCLE

Viewed over long time spans, rocks are constantly forming, changing, and re-forming.

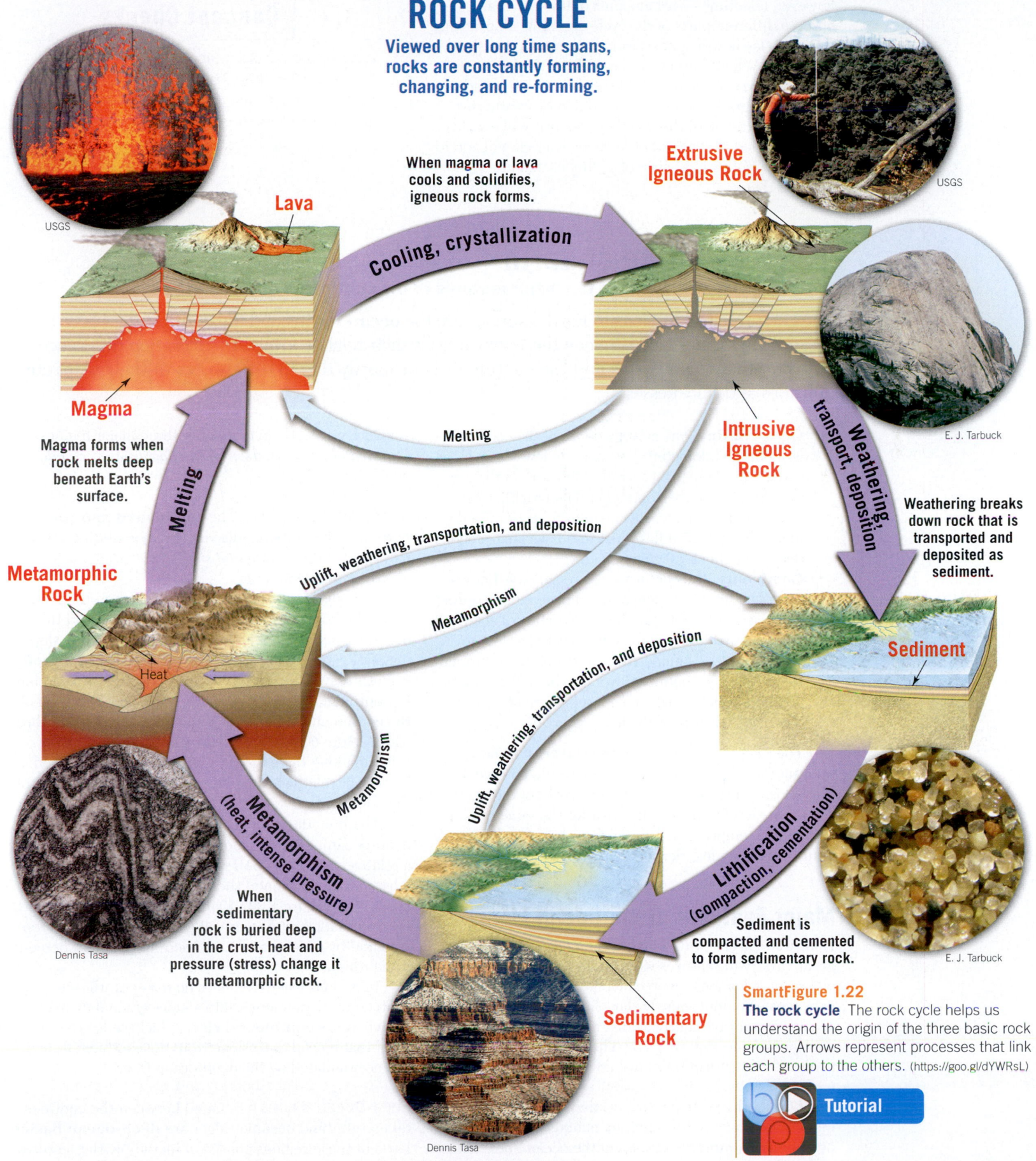

When magma or lava cools and solidifies, igneous rock forms.

Lava

Extrusive Igneous Rock

USGS

USGS

Cooling, crystallization

Magma

Melting

Magma forms when rock melts deep beneath Earth's surface.

Intrusive Igneous Rock

E. J. Tarbuck

Weathering, transport, deposition

Weathering breaks down rock that is transported and deposited as sediment.

Melting

Metamorphic Rock

Uplift, weathering, transportation, and deposition

Metamorphism

Heat

Sediment

Metamorphism

Sediment

Uplift, weathering, transportation, and deposition

Metamorphism (heat, intense pressure)

When sedimentary rock is buried deep in the crust, heat and pressure (stress) change it to metamorphic rock.

Dennis Tasa

Lithification (compaction, cementation)

Sediment is compacted and cemented to form sedimentary rock.

E. J. Tarbuck

Sedimentary Rock

Dennis Tasa

SmartFigure 1.22

The rock cycle The rock cycle helps us understand the origin of the three basic rock groups. Arrows represent processes that link each group to the others. (https://goo.gl/dYWRsL)

Tutorial

however, take time—vast amounts of time. We can observe different parts of the cycle operating all over the world. Today new magma is forming beneath the island of Hawaii. When it erupts at the surface, the lava flows add to the size of the island. Meanwhile, the Colorado Rockies are gradually being worn down by weathering and erosion. Some of this weathered debris will eventually be carried to the Gulf of Mexico, where it will add to the already substantial mass of sediment that has accumulated there.

1.7   Concept Checks

1. List two rock characteristics that are used to determine the processes that created a rock.
2. Sketch and label a basic rock cycle. Make sure to include alternate paths.

1.8 | The Face of Earth

List and describe the major features of the continents and ocean basins.

The two principal divisions of Earth's surface are the **ocean basins** and the **continents** (**Figure 1.23**). A significant difference between these two areas is their relative levels. The elevation difference between the ocean basins and the continents is primarily due to differences in their respective densities and thicknesses:

- **Ocean basins.** The average depth of the ocean floor is about 3.8 kilometers (2.4 miles) below sea level, or about 4.5 kilometers (2.8 miles) lower than the average elevation of the continents. The basaltic rocks that comprise the oceanic crust average only 7 kilometers (about 4.5 miles) thick and have an average density of about 3.0 g/cm³.
- **Continents.** The continents are remarkably flat features that have the appearance of plateaus protruding above sea level. With an average elevation of about 0.8 kilometer (0.5 mile), continental blocks lie close to sea level, except for limited areas of mountainous terrain. Recall that the continents average about 35 kilometers (22 miles) thick and are composed of granitic rocks that have a density of about 2.7 g/cm³.

The thicker, less dense continental crust is more buoyant than the oceanic crust. As a result, continental crust floats on top of the deformable rocks of the mantle at a higher level than oceanic crust for the same reason that a large, empty (less dense) cargo ship rides higher than a small, loaded (denser) one.

Major Features of the Ocean Floor

If all water were drained from the ocean basins, a great variety of features would be seen, including chains of volcanoes, deep canyons, plateaus, and large expanses of monotonously flat plains. In fact, the scenery would be nearly as diverse as that on the continents (see Figure 1.23). These features and the processes that form them are covered in detail in Chapter 13.

During the past 65 years, oceanographers have used modern depth-sounding equipment and satellite technology to map significant portions of the ocean floor. These studies have led them to identify three major regions: *continental margins*, *deep-ocean basins*, and *oceanic (mid-ocean) ridges*.

Continental Margins   The **continental margin** is the portion of the seafloor adjacent to major landmasses. It may include the *continental shelf*, the *continental slope*, and the *continental rise*.

Although land and sea meet at the shoreline, this is *not* the boundary between the continents and the ocean basins. Rather, along most coasts, a gently sloping platform of material, called the **continental shelf**, extends seaward from the shore. Because it is underlain by continental crust, it is clearly a flooded extension of the continents. A glance at Figure 1.23 shows that the width of the continental shelf varies. For example, it is broad along the east and Gulf coasts of the United States but relatively narrow along the Pacific margin of the continent.

The boundary between the continents and the deep-ocean basins lies along the **continental slope**, a relatively steep dropoff that extends from the outer edge of the continental shelf to the floor of the deep ocean (see Figure 1.23). Using this as the dividing line, we find that about 60 percent of Earth's surface is represented by ocean basins and the remaining 40 percent by continents.

In regions where trenches do not exist, the steep continental slope merges into a more gradual incline known as the **continental rise**, a thick wedge of sediment that moved downslope from the continental shelf and accumulated on the deep-ocean floor.

Deep-Ocean Basins   Situated between the continental margins and oceanic ridges are **deep-ocean basins**. Parts of these regions consist of incredibly flat features

EYE ON EARTH 1.3

This is a shoreline scene along the east coast of the United States. The Atlantic Ocean is on the right.

QUESTION 1 Does the shoreline, the line where the water meets the land, mark the outer edge of the North American continent?

QUESTION 2 Explain your answer to Question 1.

Michael Collier

called **abyssal plains**. The ocean floor also contains extremely deep depressions, some more than 11,000 meters (36,000 feet) deep. Although these **deep-ocean trenches** are relatively narrow and represent only a small fraction of the ocean floor, they are nevertheless very significant features. Some trenches are located adjacent to young mountains that flank the continents. For example, in Figure 1.23 the Peru–Chile trench off the west coast of South America parallels the Andes Mountains. Other trenches parallel island chains called *volcanic island arcs.*

Dotting the ocean floor are submerged volcanic structures called **seamounts**, which sometimes form long, narrow chains. Volcanic activity has also produced several large *lava plateaus*, such as the Ontong Java Plateau located northeast of New Guinea. In addition, some submerged plateaus are composed of continental-type crust. Examples include the Campbell Plateau southeast of New Zealand and the Seychelles Bank northeast of Madagascar.

Oceanic Ridges   The most prominent feature on the ocean floor is the **oceanic ridge**, or **mid-ocean ridge**. As shown in Figure 1.23, the Mid-Atlantic Ridge and the East Pacific Rise are parts of this system. This broad elevated feature forms a continuous belt winding more than 70,000 kilometers (43,500 miles) around the globe, in a manner similar to the seam of a baseball. Unlike most continental mountains that consist of highly deformed rock, the oceanic ridge system consists of layer upon layer of igneous rock that has been fractured and uplifted.

Major Features of the Continents

The major features of the continents can be grouped into two distinct categories: uplifted regions of deformed rocks that make up present-day mountain belts and extensive flat, stable areas that have been eroded nearly to sea level. Notice in **Figure 1.24** that the young mountain belts tend to be long, narrow features at the margins of continents and that the flat, stable areas are typically located in the interiors of the continents. Mountain building is discussed in more detail in Chapter 14.

Mountain Belts   The most prominent continental features are mountains. Although their distribution appears to be random, this is not the case. The youngest mountains (those less than 100 million years old) are located principally in two major zones. The circum-Pacific belt (the region surrounding the Pacific Ocean) includes the mountains of the western Americas and continues into the western Pacific in the form of volcanic island arcs (see Figure 1.23). Island arcs are active mountainous regions composed largely of volcanic rocks and deformed sedimentary rocks. Examples include the Aleutian Islands, Japan, the Philippines, and New Guinea.

The other major **mountain belt** extends eastward from the Alps through Iran and the Himalayas and then dips southward into Indonesia. Careful examination of mountainous terrains reveals that most are places where thick sequences of rocks have been squeezed and highly deformed, as if placed in a gigantic vise. Older mountains are also found on the continents. Examples include the Appalachians in the eastern United States and the Urals in Russia. Their once lofty peaks are now worn low, the result of millions of years of weathering and erosion.

The Stable Interior   Unlike the young mountain belts that have formed within the past 100 million years, the interiors of the continents, called **cratons**, have been relatively stable (undisturbed) for the past 600 million years or even longer.

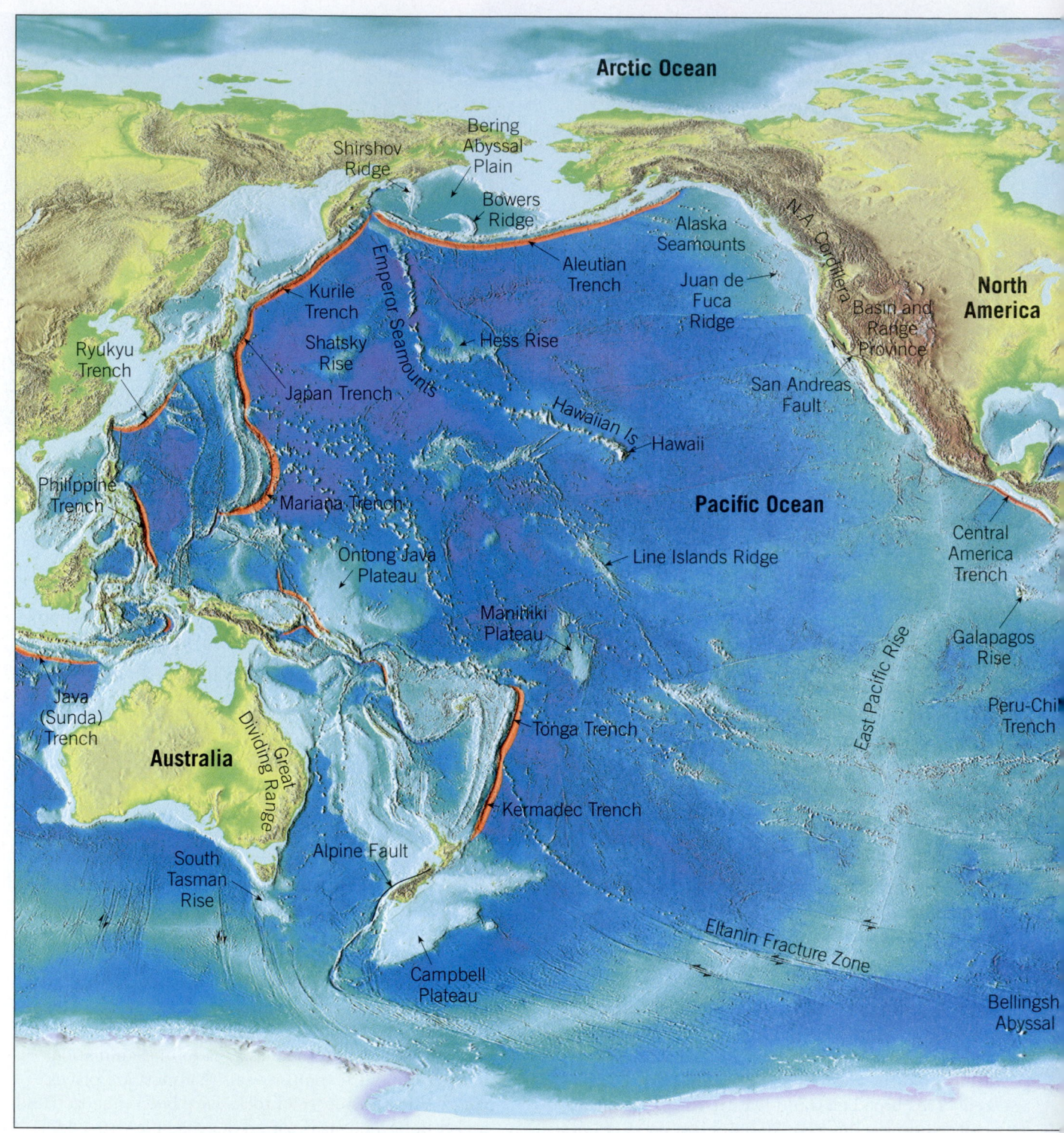

Figure 1.23
The face of Earth Major surface features of the geosphere.

Typically these regions were involved in mountain-building episodes much earlier in Earth's history.

Within the stable interiors are areas known as **shields**—expansive, flat regions composed largely of deformed igneous and metamorphic rocks. Notice in Figure 1.24 that the Canadian Shield is exposed in much of the northeastern part of North America. Radiometric dating of shields indicates that they are truly ancient regions. All contain Precambrian-age rocks more than 1 billion years old, with some samples approaching 4 billion years in age. Even these oldest-known rocks exhibit evidence of enormous forces that have folded, faulted, and metamorphosed them.

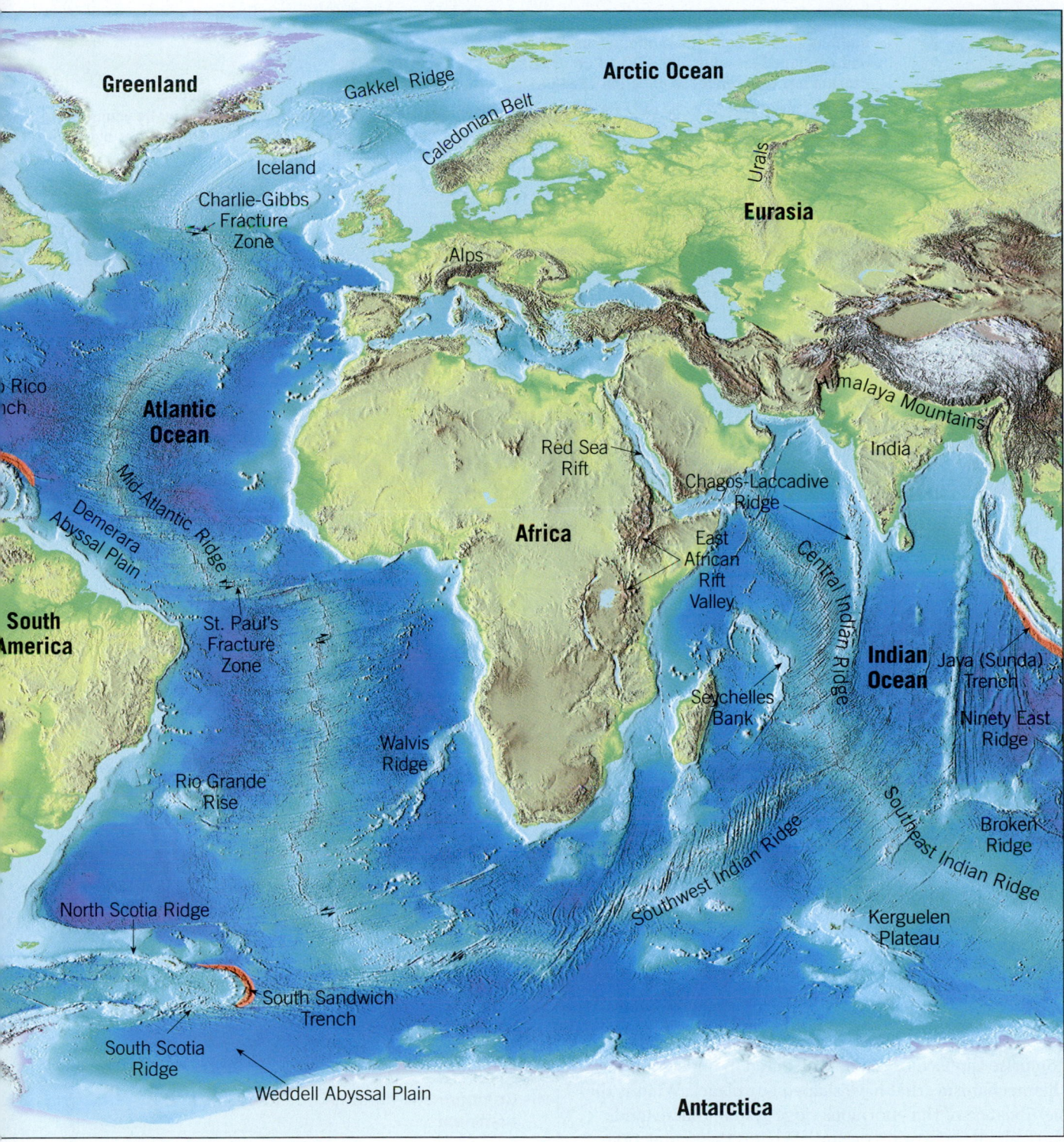

Thus, we conclude that these rocks were once part of an ancient mountain system that has since been eroded away to produce these expansive, flat regions.

Other flat areas of the craton exist, where highly deformed rocks, like those found in the shields, are covered by a relatively thin veneer of sedimentary rocks.

These areas are called **stable platforms**. The sedimentary rocks in stable platforms are nearly horizontal, except where they have been warped to form large basins or domes. In North America a major portion of the stable platform is located between the Canadian Shield and the Rocky Mountains.

The Canadian Shield is an expansive region of ancient Precambrian rocks, some more than 4 billion years old. It was recently scoured by Ice Age glaciers.

The Appalachians are old mountains. Mountain building began about 480 million years ago and continued for more than 200 million years. Erosion has lowered these once lofty peaks.

The rugged Himalayas are the highest mountains on Earth and are geologically young. They began forming about 50 million years ago and uplift continues today.

Superstock

Michael Collier

Alamy Images

Greenland shield

N.A. Cordillera

Canadian shield

Appalachians

Caledonian Belt

Baltic shield

Urals

Angara shield

Alps

Himalaya Mountains

Indian shield

African shield

Orinoco shield

Brazilian shield

Andes Mountains

Australian shield

Great Dividing Range

Key

Young mountain belts (less than 100 million years old)

Old mountain belts

Shields

Stable platforms (shields covered by sedimentary rock)

SmartFigure 1.24
The continents Distribution of mountain belts, stable platforms, and shields. (https://goo.gl/z2Sbxd)

Tutorial

Being familiar with the topographic features that comprise the face of Earth is essential to understanding the mechanisms that have shaped our planet. What is the significance of the enormous ridge system that extends through all the world's oceans? What is the connection, if any, between young, active mountain belts and oceanic trenches? What forces crumple rocks to produce majestic mountain ranges? These are a few of the questions that will be addressed beginning in the next chapter, as we investigate the dynamic processes that shaped our planet in the geologic past and will continue to shape it in the future.

1.8   **Concept Checks**

1. Compare and contrast ocean basins and continents.

2. Name the three major regions of the ocean floor. What are some features associated with each?

3. Describe the general distribution of Earth's youngest mountains.

4. What is the difference between shields and stable platforms?

1.1 Geology: The Science of Earth

Distinguish between physical and historical geology and describe the connections between people and geology.

KEY TERMS geology, physical geology, historical geology

- Geologists study Earth. Physical geologists focus on the processes by which Earth operates and the materials that result from those processes. Historical geologists apply an understanding of Earth materials and processes to reconstruct the history of our planet.

- People have a relationship with planet Earth that can be positive and negative. Earth processes and products sustain us every day, but they can also harm us. Similarly, people have the ability to alter or harm natural systems, including those that sustain civilization.

Q Consider the question of when a given volcano is likely to erupt, and also the question of whether volcanic eruptions played a part in the extinction of the dinosaurs. Which is an issue that a physical geologist would address? Which question would a historical geologist focus on?

1.2 The Development of Geology

Summarize early and modern views on how change occurs on Earth and relate them to the prevailing ideas about the age of Earth.

KEY TERMS catastrophism, uniformitarianism

- Early ideas about the nature of Earth were based on religious traditions and notions of great catastrophes. In 1795, James Hutton emphasized that the same slow processes have acted over great spans of time and are responsible for Earth's rocks, mountains, and landforms. This similarity of process over vast spans of time led to this principle being dubbed "uniformitarianism."

- Based on the rate of radioactive decay of certain elements, the age of Earth has been calculated to be about 4,600,000,000 (4.6 billion) years. That is an incredibly vast amount of time.

Q In what eon, era, period, and epoch do we live?

1.3 The Nature of Scientific Inquiry

Discuss the nature of scientific inquiry, including the construction of hypotheses and the development of theories.

KEY TERMS hypothesis, theory, scientific method

- Geologists make observations, construct tentative explanations for those observations (hypotheses), and then test those hypotheses with field investigations and laboratory work. In science, a theory is a well-tested and widely accepted view that the scientific community agrees best explains certain observable facts.

- As flawed hypotheses are discarded, scientific knowledge moves closer to a correct understanding, but we can never be fully confident that we know all the answers. Scientists must always be open to new information that forces changes in our model of the world.

1.4 Earth as a System

List and describe Earth's four major spheres. Define *system* and explain why Earth is considered to be a system.

KEY TERMS hydrosphere, atmosphere, biosphere, geosphere, Earth system science, system

- Earth's physical environment is traditionally divided into three major parts: the solid Earth, called the geosphere; the water portion of our planet, called the hydrosphere; and Earth's gaseous envelope, called the atmosphere.

- A fourth Earth sphere is the biosphere, the totality of life on Earth. It is concentrated in a relatively narrow zone that extends a few kilometers into the hydrosphere and geosphere and a few kilometers up into the atmosphere.

- Of all the water on Earth, more than 96 percent is in the oceans, which cover nearly 71 percent of the planet's surface.

- Although each of Earth's four spheres can be studied separately, they are all related in a complex and continuously interacting whole that is called the Earth system.

- Earth system science uses an interdisciplinary approach to integrate the knowledge of several academic fields in the study of our planet and its global environmental problems.

- The two sources of energy that power the Earth system are (1) the Sun, which drives the external processes that occur in the atmosphere, hydrosphere, and at Earth's surface, and (2) heat from Earth's interior that powers the internal processes that produce volcanoes, earthquakes, and mountains.

Michael Collier

Q Is glacial ice part of the geosphere, or does it belong to the hydrosphere? Explain your answer.

1.5 Origin and Early Evolution of Earth

Outline the stages in the formation of our solar system.

KEY TERMS nebular theory, solar nebula

- The nebular theory describes the formation of the solar system. The planets and Sun began forming about 4.6 billion years ago from a large cloud of dust and gases.
- As the cloud contracted, it began to rotate and assume a disk shape. Material that was gravitationally pulled toward the center became the protosun. Within the rotating disk, small centers, called planetesimals, swept up more and more of the cloud's debris.
- Because of their high temperatures and weak gravitational fields, the inner planets were unable to accumulate and retain many of the lighter components. Because of the very cold temperatures existing far from the Sun, the large outer planets consist of huge amounts of lighter materials. These gaseous substances account for the comparatively large sizes and low densities of the outer planets.

Q Earth is about 4.6 billion years old. If all of the planets in our solar system formed at about the same time, how old would you expect Mars to be? Jupiter? The Sun?

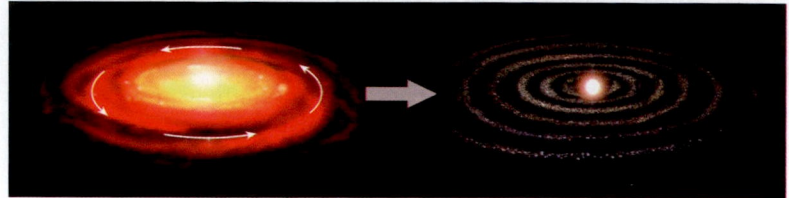

1.6 Earth's Internal Structure

Sketch Earth's internal structure and label and describe the main subdivisions.

KEY TERMS crust, mantle, lithosphere, asthenosphere, transition zone, lower mantle, core, outer core, inner core

- Compositionally, the solid Earth has three layers: core, mantle, and crust. The core is most dense, and the crust is least dense.
- Earth's interior can also be divided into layers based on physical properties. The crust and upper mantle make a two-part layer called the lithosphere, which is broken into the plates of plate tectonics. Beneath that is the "weak" asthenosphere. The lower mantle is stronger than the asthenosphere and overlies the molten outer core. This liquid is made of the same iron–nickel alloy as the inner core, but the extremely high pressure of Earth's center compacts the inner core into a solid form.

Q The diagram represents Earth's layered structure. Does it show layering based on physical properties or layering based on composition? Identify the lettered layers.

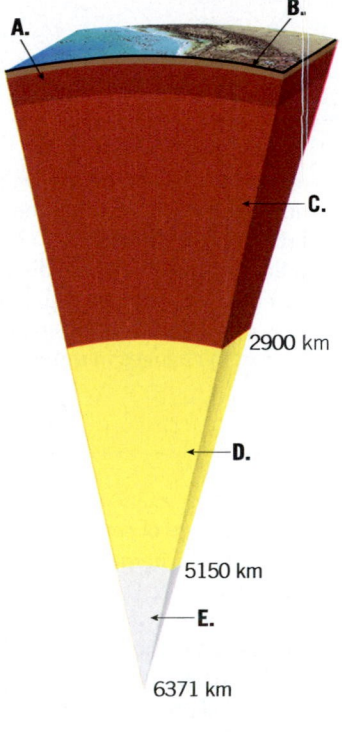

A.
B.
C.
2900 km
D.
5150 km
E.
6371 km

1.7 Rocks and the Rock Cycle

Sketch, label, and explain the rock cycle.

KEY TERMS rock cycle, igneous rock, sediment, sedimentary rock, metamorphic rock

- The rock cycle is a good model for thinking about the transformation of one rock to another due to Earth processes. All igneous rocks are made from molten rock. All sedimentary rocks are made from weathered products of other rocks. All metamorphic rocks are the products of pre-existing rocks that are transformed at high temperatures or pressures. Given the right conditions, any kind of rock can be transformed into any other kind of rock.

Q Name the processes that are represented by each of the letters in this simplified rock cycle diagram.

A.
Dennis Tasa
B.
USGS
E.
Rock cycle
Shutterstock
C.
Dennis Tasa
D.
Dennis Tasa

1.8 The Face of Earth

List and describe the major features of the continents and ocean basins.

KEY TERMS ocean basin, continent, continental margin, continental shelf, continental slope, continental rise, deep-ocean basin, abyssal plain, deep-ocean trench, seamount, oceanic ridge (mid-ocean ridge), mountain belt, craton, shield, stable platform

- Two principal divisions of Earth's surface are the ocean basins and the continents. A significant difference is their relative levels. The elevation differences between ocean basins and continents is primarily due to differences in their respective densities and thicknesses.

- There are shallow portions of the oceans that are essentially flooded margins of the continents, and there are deeper portions that include vast abyssal plains and deep-ocean trenches. Seamounts and lava plateaus interrupt the abyssal plain in some places.
- Continents consist of relatively flat, stable areas called cratons. Where a craton is blanketed by a relatively thin layer of sediment or sedimentary rock, it is called a stable platform. Where a craton is exposed at the surface, it is known as a shield. Wrapping around the edges of some cratons are mountain belts, linear zones of intense deformation and metamorphism.

Q Put these features of the ocean floor in order from shallowest to deepest: continental slope, deep-ocean trench, continental shelf, abyssal plain, continental rise.

Give It Some **Thought**

1. The length of recorded history for humankind is about 5000 years. Clearly, most people view this span as being very long. How does it compare to the length of geologic time? Calculate the percentage or fraction of geologic time that is represented by recorded history. To make calculations easier, round the age of Earth to the nearest billion.

2. Refer to the graph in Figure 1.13 to answer the following questions.
   a. If you were to climb to the top of Mount Everest, how many breaths of air would you have to take at that altitude to equal one breath at sea level?
   b. If you are flying in a commercial jet at an altitude of 12 kilometers (about 39,000 feet), about what percentage of the atmosphere's mass is below you?

3. Making accurate measurements and observations is a basic part of scientific inquiry. Identify two images in this chapter that illustrate ways in which scientific data are gathered. Suggest an advantage that might be associated with the examples you select.

4. The accompanying photo provides an example of interactions among different parts of the Earth system. It is a view of a mudflow that was triggered by extraordinary rains. Which of Earth's four spheres were involved in this natural disaster that buried a small town on the Philippine island of Leyte? Describe how each contributed to or was influenced by the event.

AP Photo/PatRoque

5. Refer to Figure 1.22. How does the rock cycle diagram, particularly the process arrows, support the fact that sedimentary rock is the most abundant rock type on the surface of Earth?

6. This photo shows the picturesque coastal bluffs and rocky shoreline along a portion of the California coast south of San Simeon State Park. This area, like other shorelines, is described as an *interface*. What does this mean? Describe another interface in the Earth system.

Michael Collier

7. After entering a dark room, you turn on a wall switch, but the light does not come on. Suggest at least three hypotheses that might explain this observation. Once you have formulated your hypotheses, what is the next logical step?

2

Plate Tectonics: A Scientific Revolution Unfolds

Hikers crossing a crevasse in Khumbu glacier, Mount Everest, Nepal. (Photo by Christian Kober/Robert Harding)

FOCUS ON CONCEPTS

Each statement represents the primary **LEARNING OBJECTIVE** for the corresponding major heading within the chapter. After you complete the chapter, you should be able to:

2.1 Summarize the view that most geologists held prior to the 1960s regarding the geographic positions of the ocean basins and continents.

2.2 List and explain the evidence Wegener presented to support his continental drift hypothesis.

2.3 Summarize the two main objections to the continental drift hypothesis.

2.4 List the major differences between Earth's lithosphere and asthenosphere and explain the importance of each in the plate tectonics theory.

2.5 Sketch and describe the movement along a divergent plate boundary that results in the formation of new oceanic lithosphere.

2.6 Compare and contrast the three types of convergent plate boundaries and name a location where each type can be found.

2.7 Describe the relative motion along a transform fault boundary and be able to locate several examples on a plate boundary map.

2.8 Explain why plates such as the African and Antarctic plates are increasing in size, while the Pacific plate is decreasing in size.

2.9 List and explain the evidence used to support the plate tectonics theory.

2.10 Describe two methods researchers use to measure relative plate motion.

2.11 Describe plate–mantle convection and explain two of the primary driving forces of plate motion.

P late tectonics is the first theory to provide a comprehensive view of the processes that produced Earth's major surface features, including the continents and ocean basins. Within the framework of this theory, geologists have found explanations for the basic causes and distribution of earthquakes, volcanoes, and mountain belts. Further, we are now better able to explain the distribution of plants and animals in the geologic past, as well as the distribution of economically significant mineral deposits.

2.1 | From Continental Drift to Plate Tectonics

Summarize the view that most geologists held prior to the 1960s regarding the geographic positions of the ocean basins and continents.

Prior to the late 1960s most geologists held the view that the ocean basins and continents had fixed geographic positions and were of great antiquity. Scientists came to realize that Earth's continents are not static; instead, they gradually migrate across the globe. These movements cause blocks of continental material to collide, deforming the intervening crust and thereby creating Earth's great mountain chains (**Figure 2.1**). Furthermore, landmasses occasionally split apart. As continental blocks separate, a new ocean basin emerges between them. Meanwhile, other portions of the seafloor plunge into the mantle. In short, a dramatically different model of Earth's tectonic processes emerged. Tectonic processes deform Earth's crust to create major structural features, such as mountains, continents, and ocean basins.

Figure 2.1
The Himalayan mountains where created when the subcontinent of India collided with southeastern Asia. (Photo by Hartmut Postges/ Robert Harding)

This profound reversal in scientific thought has been appropriately called a *scientific revolution*. The revolution began early in the twentieth century as a relatively straightforward proposal termed *continental drift*. For more than 50 years, the scientific community categorically rejected the idea that continents are capable of

movement. North American geologists in particular had difficulty accepting continental drift, perhaps because much of the supporting evidence had been gathered from Africa, South America, and Australia, continents with which most North American geologists were unfamiliar.

After World War II, modern instruments replaced rock hammers as the tools of choice for many researchers. Armed with more advanced tools, geologists and a new breed of researchers, including *geophysicists* and *geochemists*, made several surprising discoveries that rekindled interest in the drift hypothesis. By 1968 these developments had led to the unfolding of a far more encompassing explanation known as the *theory of plate tectonics*.

In this chapter, we will examine the events that led to this dramatic reversal of scientific opinion. We will also briefly trace the development of the *continental drift hypothesis*, examine why it was initially rejected, and consider the evidence that finally led to the acceptance of its direct descendant—the theory of plate tectonics.

2.1 Concept Checks

1. Briefly describe the view held by most geologists regarding the ocean basins and continents prior to the 1960s.

2. What group of geologists were the least receptive to the continental drift hypothesis? Why?

2.2 Continental Drift: An Idea Before Its Time

List and explain the evidence Wegener presented to support his continental drift hypothesis.

The idea that continents, particularly South America and Africa, fit together like pieces of a jigsaw puzzle came about during the 1600s, as better world maps became available. However, little significance was given to this notion until 1915, when Alfred Wegener (1880–1930), a German meteorologist and geophysicist, wrote *The Origin of Continents and Oceans*. This book outlined Wegener's hypothesis called **continental drift**, which dared to challenge the long-held assumption that the continents and ocean basins had fixed geographic positions.

Wegener suggested that a single **supercontinent** consisting of all Earth's landmasses once existed.[*] He named this giant landmass **Pangaea** (pronounced "Pan-jee-ah," meaning "all lands") (**Figure 2.2**). Wegener further hypothesized that about 200 million years ago, during the early part of the Mesozoic era, this supercontinent began to fragment into smaller landmasses. These continental blocks then "drifted" to their present positions over a span of millions of years.

Wegener and others who advocated the continental drift hypothesis collected substantial evidence to support their point of view. The fit of South America and Africa and the geographic distribution of fossils and ancient climates all seemed to buttress the idea that these now separate landmasses were once joined. Let us examine some of this evidence.

Evidence: The Continental Jigsaw Puzzle

Like a few others before him, Wegener suspected that the continents might once have been joined when he

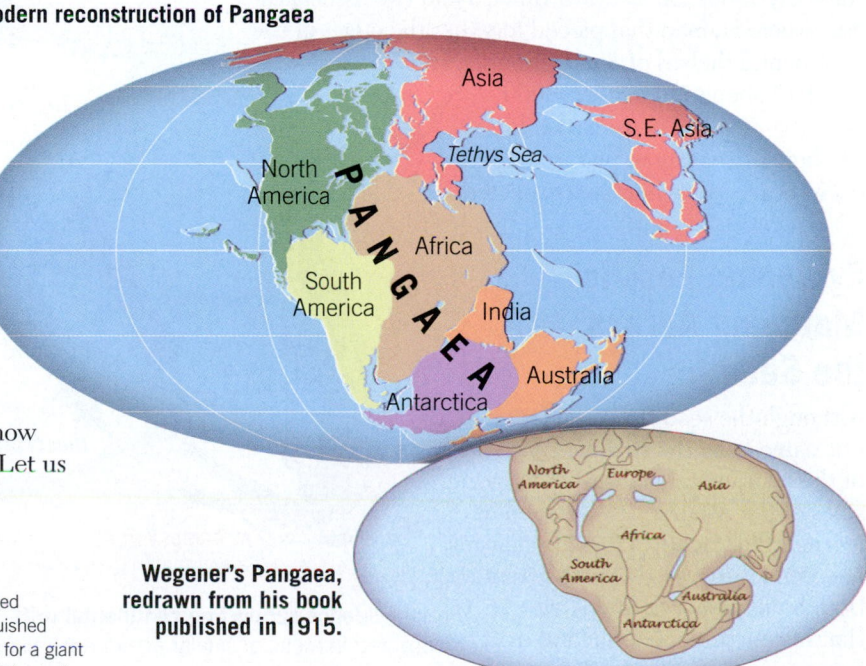

Modern reconstruction of Pangaea

Asia

S.E. Asia

Tethys Sea

North America

PANGAEA

Africa

South America

India

Australia

Antarctica

Wegener's Pangaea, redrawn from his book published in 1915.

North America Europe Asia

Africa

South America

Australia

Antarctica

SmartFigure 2.2 Reconstructions of Pangaea The supercontinent of Pangaea, as it is thought to have formed in the late Paleozoic and early Mesozoic eras, more than 200 million years ago. (https://goo.gl/eOttu9)

 **Tutorial**

---

[*]Wegener was not the first to conceive of a long-vanished supercontinent. Edward Suess (1831–1914), a distinguished nineteenth-century geologist, pieced together evidence for a giant landmass comprising South America, Africa, India, and Australia.

Figure 2.3
Two of the puzzle pieces
The best fit of South America and Africa along the continental slope at a depth of 500 fathoms (about 900 meters [3000 feet]). (Based on A. G. Smith, "Continental Drift," in Understanding the Earth, edited by I. G. Gass, Artemis Press.)

noticed the remarkable similarity between the coastlines on opposite sides of the Atlantic Ocean. However, other Earth scientists challenged Wegener's use of present-day shorelines to "fit" these continents together. These opponents correctly argued that wave erosion and depositional processes continually modify shorelines. Even if continental displacement had taken place, a good fit today would be unlikely. Because Wegener's original jigsaw fit of the continents was crude, it is assumed that he was aware of this problem (see Figure 2.2).

Scientists later determined that a much better approximation of the outer boundary of a continent is the seaward edge of its continental shelf, which lies submerged a few hundred meters below sea level. In the early 1960s, Sir Edward Bullard and two associates constructed a map that pieced together the edges of the continental shelves of South America and Africa at a depth of about 900 meters (3000 feet) (**Figure 2.3**). The remarkable fit obtained was more precise than even these researchers had expected.

Evidence: Fossils Matching Across the Seas

Although the seed for Wegener's hypothesis came from the striking similarities of the continental margins on opposite sides of the Atlantic, it was when he learned that identical fossil organisms had been discovered in rocks from both South America and Africa that his pursuit of continental drift became more focused. Wegener learned that most paleontologists

(scientists who study the fossilized remains of ancient organisms) agreed that some type of land connection was needed to explain the existence of similar Mesozoic age life-forms on widely separated landmasses. Just as modern life-forms native to North America are not the same as those of Africa and Australia, during the Mesozoic era, organisms on widely separated continents should have been distinctly different.

Mesosaurus   To add credibility to his argument, Wegener documented cases of several fossil organisms found on different landmasses, even though their living forms were unlikely to have crossed the vast ocean presently separating them (**Figure 2.4**). A classic example is *Mesosaurus*, a small aquatic freshwater reptile whose fossil remains are limited to black shales of the Permian period (about 260 million years ago) in eastern South America and southwestern Africa. If *Mesosaurus* had been able to make the long journey across the South Atlantic, its remains should be more widely distributed. As this is not the case, Wegener asserted that South America and Africa must have been joined during that period of Earth history.

How did opponents of continental drift explain the existence of identical fossil organisms in places separated by thousands of kilometers of open ocean? Rafting, transoceanic land bridges (isthmian links), and island stepping stones were the most widely invoked explanations for these migrations (**Figure 2.5**). We know, for example, that during the Ice Age that ended about 8000 years ago, the lowering of sea level allowed mammals (including humans) to cross the narrow Bering Strait that separates Russia and Alaska. Was it possible that land bridges once

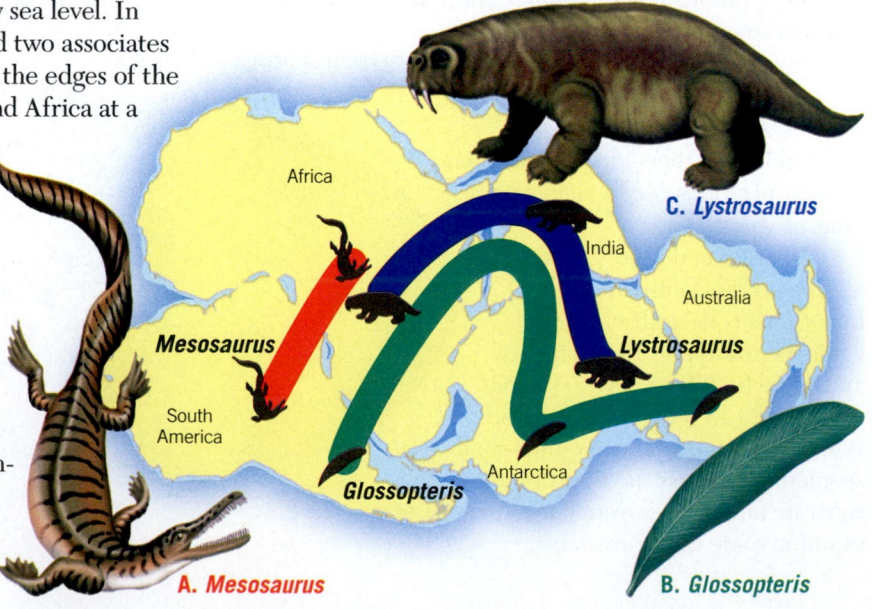

Figure 2.4
Fossil evidence supporting continental drift Fossils of identical organisms have been discovered in rocks of similar age in Australia, Africa, South America, Antarctica, and India—continents that are currently widely separated by ocean barriers. Wegener accounted for these occurrences by placing these continents in their pre-drift locations.

connected Africa and South America but later subsided below sea level? Modern maps of the seafloor substantiate Wegener's contention that if land bridges of this magnitude existed, their remnants would still lie below sea level.

Glossopteris   Wegener also cited the distribution of the fossil "seed fern" *Glossopteris* as evidence for Pangaea's existence (see Figure 2.4). With tongue-shaped leaves and seeds too large to be carried by the wind, this plant was known to be widely dispersed among Africa, Australia, India, and South America. Later, fossil remains of *Glossopteris* were also discovered in Antarctica.* Wegener also learned that these seed ferns and associated flora grew only in cool climates—similar to central Alaska. Therefore, he concluded that when these landmasses were joined, they were located much closer to the South Pole.

RAFTING

ISTHMIAN LINKS

ISLAND STEPPING STONES

CONTINENTAL DRIFT

Figure 2.5
How do land animals cross vast oceans? These sketches illustrate various proposals to explain the occurrence of similar species on landmasses now separated by vast oceans. (Used by permission of John C. Holden)

Evidence: Rock Types and Geologic Features

You know that successfully completing a jigsaw puzzle requires fitting the pieces together while maintaining the continuity of the picture. The "picture" that must match

* In 1912 Captain Robert Scott and two companions froze to death lying beside 35 pounds (16 kilograms) of rock on their return from a failed attempt to be the first to reach the South Pole. These samples, collected on Beardmore Glacier, contained fossil remains of *Glossopteris*.

in the "continental drift puzzle" is one of rock types and geologic features such as mountain belts. If the continents were once together, the rocks found in a particular region on one continent should closely match in age and type those found in adjacent positions on the adjoining continent. Wegener found evidence of highly deformed igneous rocks in Brazil that closely resembled similar rocks in Africa.

Similar evidence can be found in mountain belts that terminate at one coastline and reappear on landmasses across the ocean. For instance, the mountain belt that includes the Appalachians trends northeastward through the eastern United States and disappears off the coast of Newfoundland (**Figure 2.6A**). Mountains of comparable age and structure are found in the British Isles and Scandinavia. When these landmasses are positioned as they were about 200 million years ago, as shown in **Figure 2.6B**, the mountain chains form a nearly continuous belt.

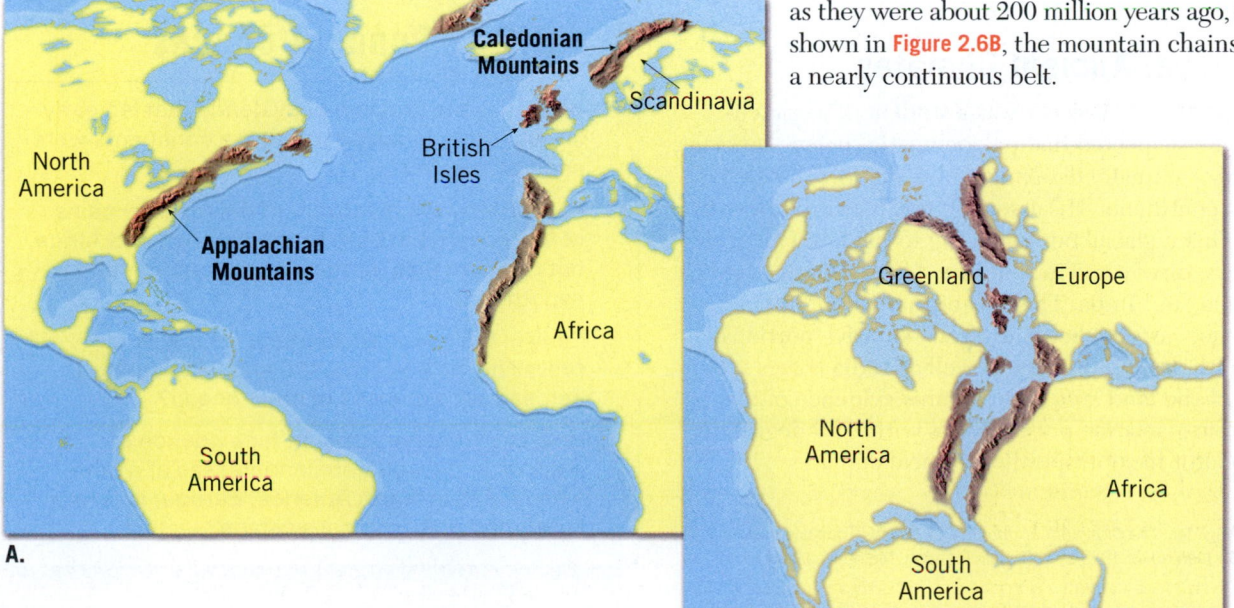

Figure 2.6
Matching mountain ranges across the North Atlantic

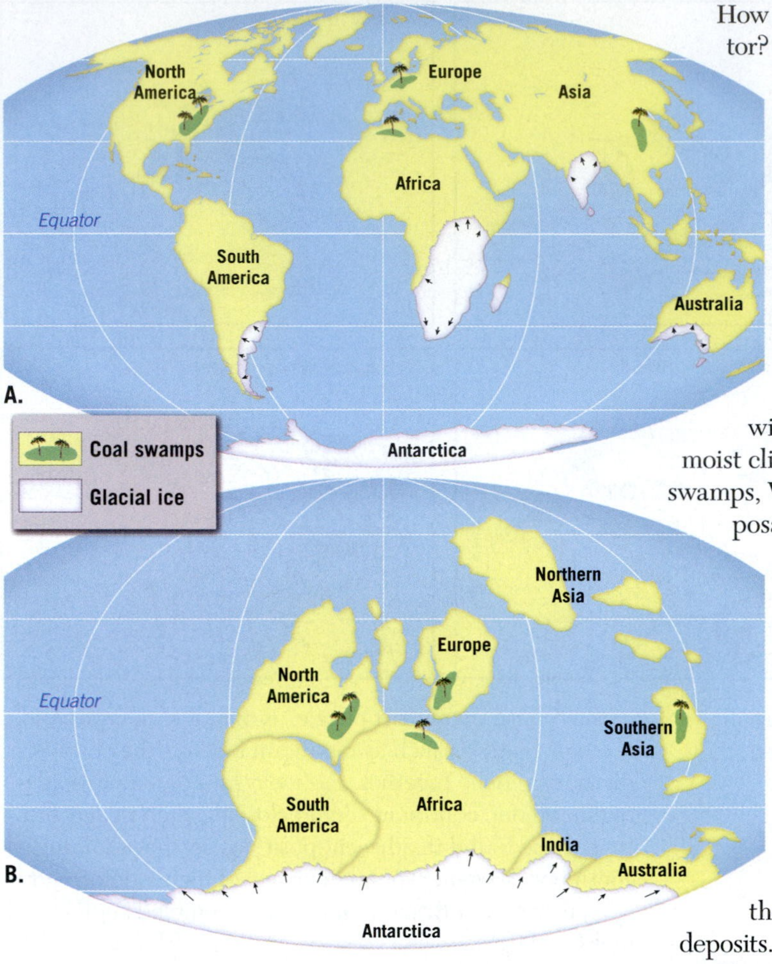

Figure 2.7
Paleoclimatic evidence for continental drift
A. About 300 million years ago, ice sheets covered extensive areas of the Southern Hemisphere and India. Arrows show the direction of ice movement that can be inferred from the pattern of glacial scratches and grooves found in the bedrock.
B. The continents restored to their pre-drift positions account for tropical coal swamps that existed in areas presently located in temperate climates.

Wegener described how the similarities in geologic features on both sides of the Atlantic linked these landmasses: "It is just as if we were to refit the torn pieces of a newspaper by matching their edges and then check whether the lines of print run smoothly across. If they do, there is nothing left but to conclude that the pieces were in fact joined in this way."†

Evidence: Ancient Climates

Because Alfred Wegener was a student of world climates, he suspected that paleoclimatic (*paleo* = ancient, *climatic* = climate) data might also support the idea of mobile continents. His assertion was bolstered by evidence that a glacial period dating to the late Paleozoic had been discovered in southern Africa, South America, Australia, and India. This meant that about 300 million years ago, vast ice sheets covered extensive portions of the Southern Hemisphere as well as India (**Figure 2.7A**). Much of the land area that contains evidence of this Paleozoic glaciation presently lies within 30 degrees of the equator in subtropical or tropical climates.

†Alfred Wegener, *The Origin of Continents and Oceans*, translated from the 4th revised German ed. of 1929 by J. Birman (London: Methuen, 1966).

How could extensive ice sheets form near the equator? One proposal suggested that our planet experienced a period of extreme global cooling. Wegener rejected this explanation because during the same span of geologic time, large tropical swamps existed in several locations in the Northern Hemisphere. The lush vegetation in these swamps was eventually buried and converted to coal (**Figure 2.7B**). Today these deposits comprise major coal fields in the eastern United States and Northern Europe. Many of the fossils found in these coal-bearing rocks were produced by tree ferns with large fronds—a fact consistent with warm, moist climates.°° The existence of these large tropical swamps, Wegener argued, was inconsistent with the proposal that extreme global cooling caused glaciers to form in areas that are currently tropical.

Wegener suggested a more plausible explanation for the late Paleozoic glaciation: the supercontinent of Pangaea. The southern continents being joined together and located near the South Pole would account for the polar conditions necessary to generate extensive expanses of glacial ice over much of these landmasses (Figure 2.7B). At the same time, this geography places today's northern continents nearer the equator and accounts for the tropical swamps that generated the vast coal deposits.

How does a glacier develop in hot, arid central Australia? How do land animals migrate across wide expanses of the ocean? As compelling as this evidence may have been, 50 years passed before most of the scientific community accepted the concept of continental drift and the logical conclusions to which it led.

** It is important to note that coal can form in a variety of climates, provided that large quantities of plant life are buried.

2.2   Concept Checks

1. What was the first line of evidence that led early investigators to suspect that the continents were once connected?

2. Explain why the discovery of the fossil remains of *Mesosaurus* in both South America and Africa, but nowhere else, supports the continental drift hypothesis.

3. Early in the twentieth century, what was the prevailing view of how land animals migrated across vast expanses of open ocean?

4. How did Wegener account for evidence of glaciers in the southern landmasses at a time when areas in North America, Europe, and Asia supported lush tropical swamps?

2.3 | The Great Debate

Summarize the two main objections to the continental drift hypothesis.

From 1924, when Wegener's book was translated into English, French, Spanish, and Russian, until his death in 1930, his proposed drift hypothesis encountered a great deal of hostile criticism. The respected American geologist R. T. Chamberlain stated, "Wegener's hypothesis in general is of the foot-loose type, in that it takes considerable liberty with our globe, and is less bound by restrictions or tied down by awkward, ugly facts than most of its rival theories."

Rejection of the Drift Hypothesis

One of the main objections to Wegener's hypothesis stemmed from his inability to identify a credible mechanism for continental drift. Wegener proposed that gravitational forces of the Moon and Sun that produce Earth's tides were also capable of gradually moving the continents across the globe. However, the prominent physicist Harold Jeffreys correctly argued that tidal forces strong enough to move Earth's continents would have resulted in halting our planet's rotation, which, of course, has not happened.

Wegener also incorrectly suggested that the larger and sturdier continents broke through thinner oceanic crust, much as icebreakers cut through ice. However, no evidence existed to suggest that the ocean floor was weak enough to permit passage of the continents without the continents being appreciably deformed in the process.

In 1930, Wegener made his fourth and final trip to the Greenland ice sheet (**Figure 2.8**). Although the primary focus of this expedition was to study this great ice cap and its climate, Wegener continued to test his continental drift hypothesis. While returning from Eismitte, an experimental station located in the center of Greenland, Wegener perished along with his Greenland companion. His intriguing idea, however, did not die.

Why was Wegener unable to overturn the established scientific views of his day? Foremost was the fact that, although the central theme of Wegener's drift hypothesis was correct, some details were incorrect. For example, continents do not break through the ocean floor, and tidal energy is much too weak to move continents. Moreover, for any comprehensive scientific theory to gain wide acceptance, it must withstand critical testing from all areas of science. Despite Wegener's great contribution to our understanding of Earth, not *all* of the evidence supported the continental drift hypothesis as he had proposed it.

Although many of Wegener's contemporaries opposed and even ridiculed his views, some considered his ideas plausible. For those geologists who continued the search, the exciting concept of continents adrift held their interest. Others viewed continental drift as a solution to previously unexplainable observations such as the cause of earthquakes. Nevertheless, most of the scientific community, particularly in North America, either categorically rejected continental drift or treated it with considerable skepticism.

2.3 | Concept Checks

1. Describe two aspects of Wegener's continental drift hypothesis that were objectionable to most Earth scientists.

2. What analogy did Wegener use to describe how the continents move through the ocean floor?

Figure 2.8
Alfred Wegener, during an expedition to Greenland
(Photo courtesy of Archive of Alfred Wegener Institute)

Alfred Wegener shown waiting out the 1912–1913 Arctic winter during an expedition to Greenland, where he made a 1200-kilometer traverse across the widest part of the island's ice sheet.

2.4 | The Theory of Plate Tectonics

List the major differences between Earth's lithosphere and asthenosphere and explain the importance of each in the plate tectonics theory.

Following World War II, oceanographers equipped with new marine tools and ample funding from the U.S. Office of Naval Research embarked on an unprecedented period of oceanographic exploration. Over the next two decades, a much better picture of large expanses of the seafloor slowly and painstakingly began to emerge. From this work came the discovery of a global oceanic ridge system that winds through all the major oceans.

In other parts of the ocean, more new discoveries were being made. Studies conducted in the western Pacific demonstrated that earthquakes were occurring at great depths beneath deep-ocean trenches. Of equal importance was the fact that dredging of the seafloor did not bring up any oceanic crust that was older than 180 million years. Further, sediment accumulations in the deep-ocean basins were found to be thin, not the thousands of meters that were predicted. By 1968 these developments, among others, had led to the unfolding of a far more encompassing theory than continental drift, known as the **theory of plate tectonics** (*tekto* = to build).

Rigid Lithosphere Overlies Weak Asthenosphere

According to the plate tectonics model, the crust and the uppermost, and therefore coolest, part of the mantle constitute Earth's strong outer layer, the **lithosphere** (*lithos* = stone). The lithosphere varies in both thickness and density, depending on whether it is oceanic or continental lithosphere (**Figure 2.9**). Oceanic lithosphere is about 100 kilometers (60 miles) thick in the deep-ocean basins but is considerably thinner along the crest of the oceanic ridge system—a topic we will consider later. By contrast, continental lithosphere averages about 150 kilometers (90 miles) thick but may extend to depths of 200 kilometers

(125 miles) or more beneath the stable interiors of the continents. Further, the composition of both the oceanic and continental crusts affects their respective densities. Oceanic crust is composed of basalt, rich in dense iron and magnesium, whereas continental crust is composed largely of less dense granitic rocks. Because of these differences, the overall density of oceanic lithosphere (crust and upper mantle) is greater than the overall density of continental lithosphere. This important difference will be considered in greater detail later in this chapter.

The **asthenosphere** (*asthenos* = weak) is a hotter, weaker region in the mantle that lies below the lithosphere (see Figure 2.9). The temperatures and pressures in the upper asthenosphere (100 to 200 kilometers [60 to 125 miles] in depth) are such that rocks at this depth are very near their melting temperatures and, hence, respond to forces by *flowing*, similar to the way a thick liquid would flow. By contrast, the relatively cool and rigid lithosphere tends to respond to forces acting on it by *bending or breaking but not flowing*. Because of these differences, Earth's rigid outer shell is effectively detached from the asthenosphere, which allows these layers to move independently.

Earth's Major Plates

The lithosphere is broken into about two dozen segments of irregular size and shape called **lithospheric plates**, or simply **plates**, that are in constant motion with respect to one another (**Figure 2.10**). Seven major lithospheric plates are recognized and account for 94 percent of Earth's surface area: the *North American*, *South American*, *Pacific*, *African*, *Eurasian*, *Australian-Indian*, and *Antarctic plates*. The largest is the Pacific plate, which encompasses a significant portion of the Pacific basin. Each of the six other large plates consists of an entire continent, as well as a significant amount of oceanic crust. Notice in **Figure 2.11** that the South American plate encompasses almost all of South America and about one-half of the floor of the South Atlantic. Note also that none of the plates are defined entirely by the margins of a single continent. This is a major departure from Wegener's continental drift hypothesis, which proposed that the continents move through the ocean floor, not with it.

Intermediate-sized plates include the *Caribbean*, *Nazca*, *Philippine*, *Arabian*, *Cocos*, *Scotia*, and *Juan de Fuca plates*. These plates, with the exception of the

SmartFigure 2.9
Rigid lithosphere overlies the weak asthenosphere
(https://goo.gl/KH1iAR)

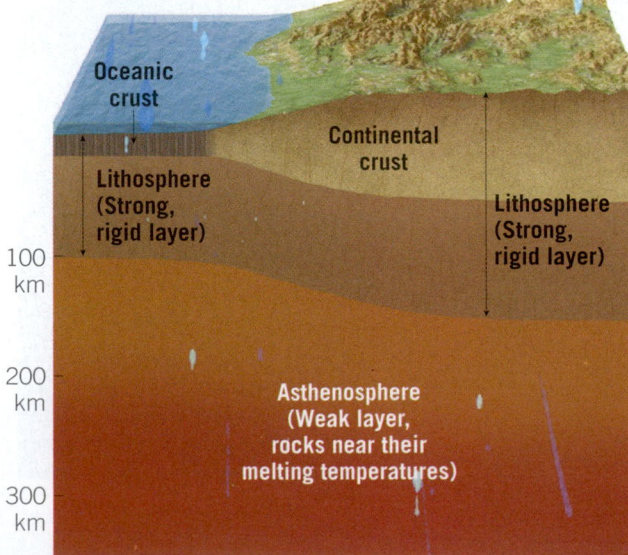

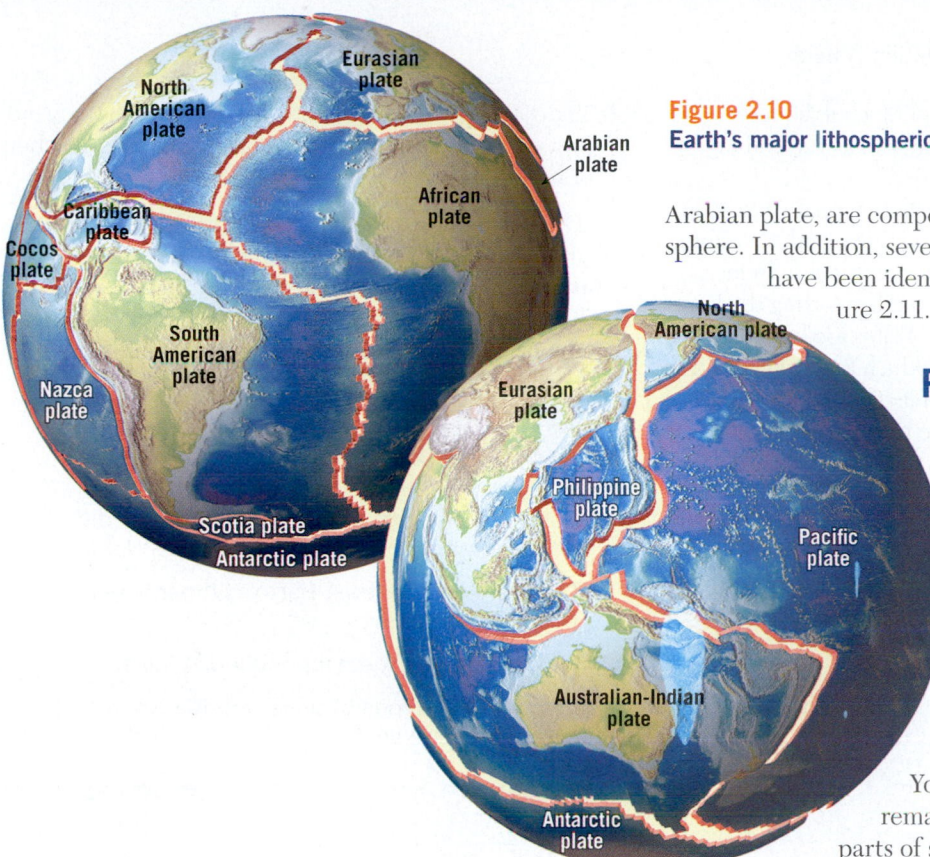

Figure 2.10
Earth's major lithospheric plates

Arabian plate, are composed mostly of oceanic lithosphere. In addition, several smaller plates (*microplates*) have been identified but are not shown in Figure 2.11.

Plate Movement

One of the main tenets of the plate tectonics theory is that plates move as somewhat rigid units relative to all other plates. As plates move, the distance between two locations on different plates, such as New York and London, gradually changes, whereas the distance between sites on the same plate—New York and Denver, for example—remains relatively constant. However, parts of some plates are comparatively

Figure 2.11
Divergent, convergent, and transform plate boundaries
(Based on W. B. Hamilton, U.S. Geological Survey)

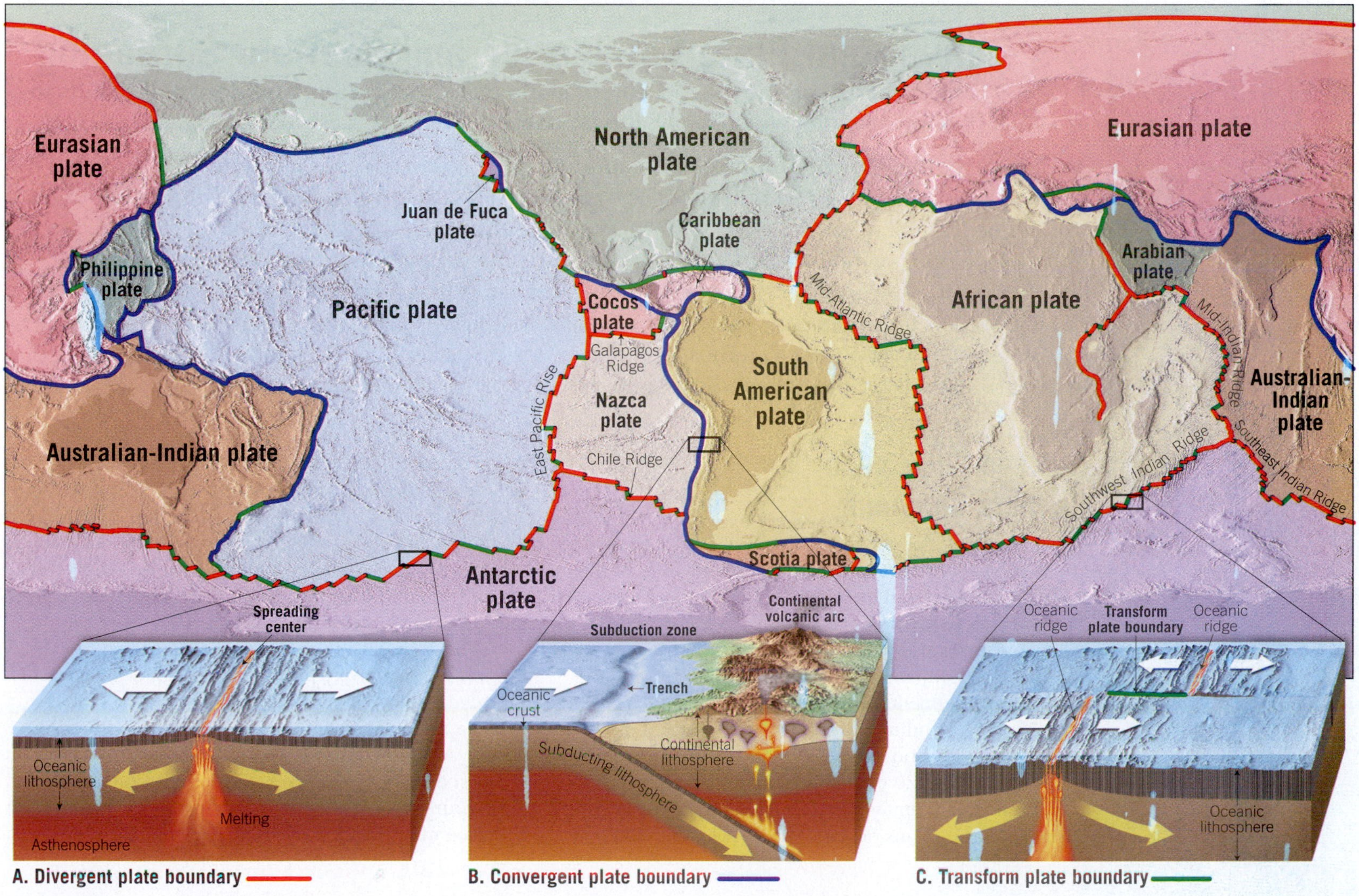

A. Divergent plate boundary ——

B. Convergent plate boundary ——

C. Transform plate boundary ——

"weak and can become fragmented" such as southern China, which is literally being squeezed as the Indian subcontinent rams into Asia.

Because plates are in constant motion relative to each other, most major interactions among them (and, therefore, most deformation) occur along their *boundaries*. In fact, plate boundaries were first established by plotting the locations of earthquakes and volcanoes. Plates are bounded by three distinct types of boundaries, which are differentiated by the type of movement they exhibit. These boundaries are depicted in Figure 2.11 and are briefly described here:

- Divergent plate boundaries—where two plates move apart, resulting in upwelling and partial melting of hot material from the mantle to create new seafloor (see Figure 2.11A)
- Convergent plate boundaries—where two plates move together, resulting either in oceanic lithosphere descending beneath an overriding plate, eventually to be reabsorbed into the mantle, or possibly in the collision of two continental blocks to create a mountain belt (see Figure 2.11B)

- Transform plate boundaries—where two plates grind past each other without the production or destruction of lithosphere (see Figure 2.11C)

Divergent and convergent plate boundaries each account for about 40 percent of all plate boundaries. Transform boundaries, or faults account for the remaining 20 percent. In the following sections we will summarize the nature of the three types of plate boundaries.

2.4 Concept Checks

1. What new findings about the ocean floor did oceanographers discover after World War II?

2. Compare and contrast Earth's lithosphere and asthenosphere.

3. List the seven largest lithospheric plates.

4. List the three types of plate boundaries and describe the relative motion along each.

2.5 | Divergent Plate Boundaries and Seafloor Spreading

Sketch and describe the movement along a divergent plate boundary that results in the formation of new oceanic lithosphere.

Most **divergent plate boundaries** (*di* = apart, *vergere* = to move) are located along the crests of oceanic ridges and can be thought of as *constructive plate margins* because this is where new ocean floor is generated (**Figure 2.12**). Here, two adjacent plates move away from each other, producing long, narrow fractures in the ocean crust. As a result, hot molten rock from the mantle below migrates upward to fill the voids left as the crust is being ripped apart. This molten material gradually cools to produce new slivers of seafloor. In a slow yet unending manner, adjacent plates spread apart, and new oceanic lithosphere forms between them. For this reason, divergent plate boundaries are also called **spreading centers**.

Oceanic Ridges and Seafloor Spreading

The majority of, but not all, divergent plate boundaries are associated with *oceanic ridges*: elevated areas of the seafloor characterized by high heat flow and volcanism. The global **oceanic ridge system** is the longest topographic feature on Earth's surface, exceeding 70,000 kilometers (43,000 miles) in length. As shown in Figure 2.11, various segments of the global ridge system have been named, including the Mid-Atlantic Ridge, East Pacific Rise, and Mid-Indian Ridge.

Representing 20 percent of Earth's surface, the oceanic ridge system winds through all major ocean basins, like the seams on a baseball. Although the crest

of the oceanic ridge is commonly 2 to 3 kilometers (1 to 2 miles) higher than the adjacent ocean basins, the term *ridge* may be misleading because it implies "narrow" when, in fact, ridges vary in width from 1000 kilometers (600 miles) to more than 4000 (2500 miles) kilometers. Further, along the crest of some ridge segments is a deep canyonlike structure called a **rift valley** (**Figure 2.13**). This structure is evidence that tensional (opposing) forces are actively pulling the ocean crust apart at the ridge crest.

The mechanism that operates along the oceanic ridge system to create new seafloor is appropriately called **seafloor spreading**. Spreading typically averages around 5 centimeters (2 inches) per year, roughly the same rate at which human fingernails grow.

Figure 2.12
Seafloor spreading Most divergent plate boundaries are situated along the crests of oceanic ridges—the sites of seafloor spreading.

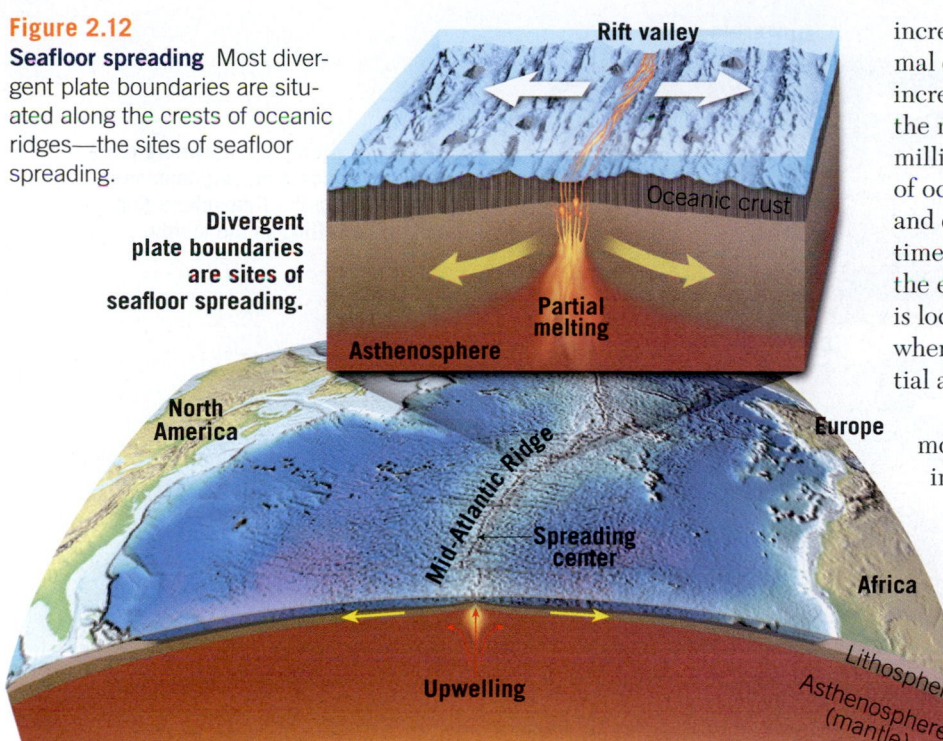

increasing in density. This thermal contraction accounts for the increase in ocean depths away from the ridge crest. It takes about 80 million years for the temperature of oceanic lithosphere to stabilize and contraction to cease. By this time, rock that was once part of the elevated oceanic ridge system is located in the deep-ocean basin, where it may be buried by substantial accumulations of sediment.

In addition, as the plate moves away from the ridge, cooling of the underlying asthenosphere causes it to become increasingly more rigid. Thus, oceanic lithosphere is generated by cooling of the asthenosphere from the top down. Stated another way, the thickness of oceanic lithosphere is

Comparatively slow spreading rates of 2 centimeters per year are found along the Mid-Atlantic Ridge, whereas spreading rates exceeding 15 centimeters (6 inches) per year have been measured along sections of the East Pacific Rise. Although these rates of seafloor production are slow on a human time scale, they are nevertheless rapid enough to have generated all of Earth's ocean basins within the past 200 million years.

The primary reason for the elevated position of the oceanic ridge is that newly created oceanic lithosphere is hot and, therefore, less dense than cooler rocks found away from the ridge axis. (Geologists use the term *axis* to refer to a line that follows the general trend of the ridge crest.) As soon as new lithosphere forms, it is slowly yet continually displaced away from the zone of upwelling. Thus, it begins to cool and contract, thereby

SmartFigure 2.13
Rift valley Thingvellir National Park, Iceland, is located on the western margin of a rift valley roughly 30 kilometers (20 miles) wide. This rift valley is connected to a similar feature that extends along the crest of the Mid-Atlantic Ridge. The cliff in the left half of the image approximates the eastern edge of the North American plate. (Photo by Ragnar Sigurdsson/Arctic/Alamy) (http://goo.gl/RsbHWM)

Mobile Field Trip

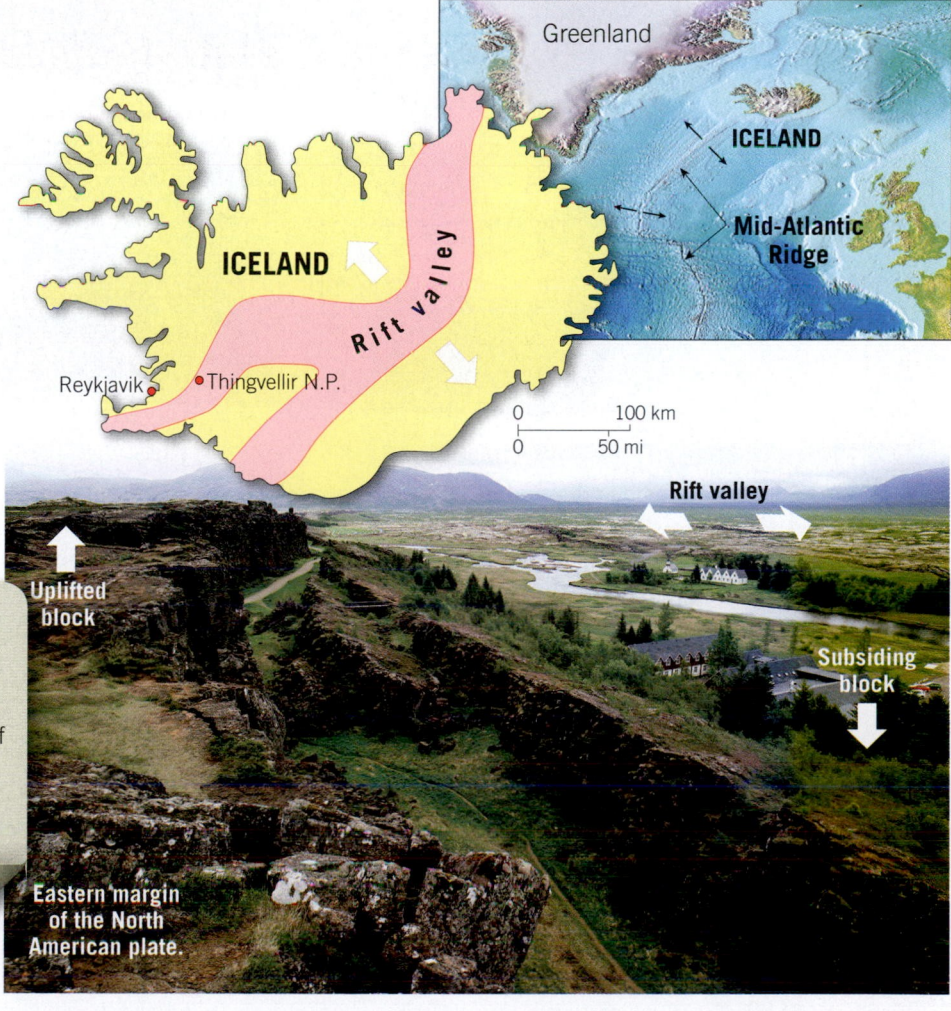

Upwarping

Continental crust

Lithosphere

Upwelling

Asthenosphere

A.

Continental rifting occurs where plate motions produce opposing tensional forces that thin the lithosphere and promote upwelling in the mantle.

Continental rift

Lithosphere

Upwelling

Asthenosphere

B.

Stretching causes the brittle crust to break into large blocks that sink, generating a rift valley.

T I M E

Linear sea

Lithosphere

Upwelling

Asthenosphere

C.

Continued spreading generates a long narrow sea similar to the present-day Red Sea.

|← **Mid-ocean ridge** →|

Rift valley

Continental lithosphere

Oceanic lithosphere

Upwelling

Asthenosphere

D.

Eventually, an expansive deep-ocean basin containing a centrally located oceanic ridge is formed by continued seafloor spreading.

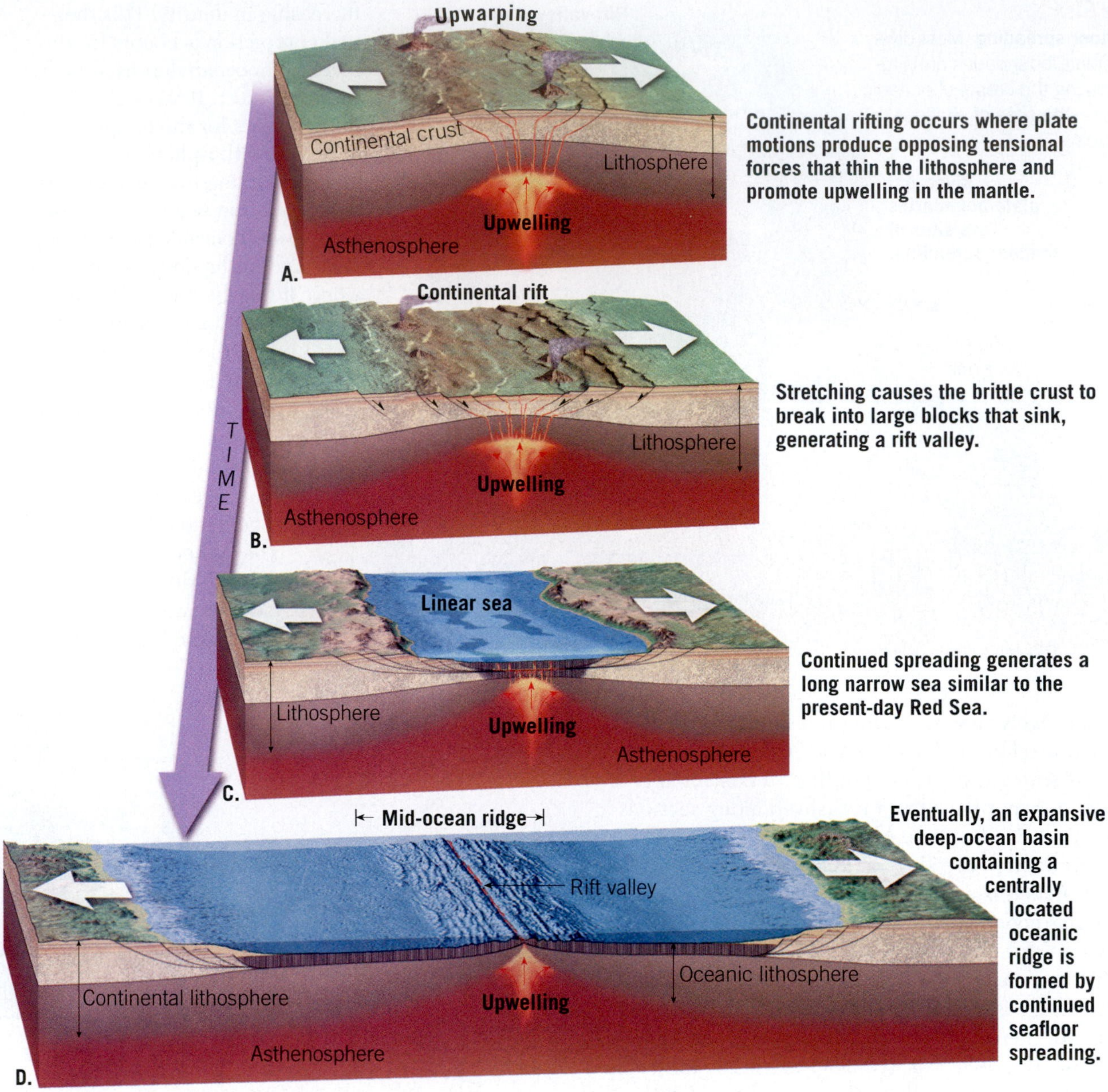

age dependent. The older (cooler) it is, the greater its thickness. Oceanic lithosphere that exceeds 80 million years in age is about 100 kilometers (60 miles) thick—approximately its maximum thickness.

Continental Rifting

Divergent boundaries can develop within a continent, in which case the landmass may split into two or more smaller segments separated by an ocean basin. Continental rifting begins when plate motions produce tensional forces that pull and stretch the lithosphere. This stretching, in turn, promotes mantle upwelling and broad upwarping of the overlying lithosphere (**Figure 2.14A**). During this process, the lithosphere is thinned, while the brittle crustal rocks break into large blocks. As

the tectonic forces continue to pull apart the crust, the broken crustal fragments sink, generating an elongated depression called a **continental rift**, which can widen to form a narrow sea (**Figure 2.14B,C**) and eventually a new ocean basin (**Figure 2.14D**).

An example of an active continental rift is the East African Rift (**Figure 2.15**). Whether this rift will eventually result in the breakup of Africa is a topic of ongoing research. Nevertheless, the East African Rift is an excellent model of the initial stage in the breakup of a continent. Here, tensional forces have stretched and thinned the lithosphere, allowing molten rock to ascend from the mantle. Evidence for this upwelling includes several large volcanic mountains, including Mount Kilimanjaro and Mount Kenya, the tallest peaks in Africa. Research suggests that if rifting continues, the rift valley will lengthen and deepen (see Figure 2.14C). At some point, the rift valley will become a narrow sea with an outlet to the ocean. The Red Sea, formed when the Arabian Peninsula split from Africa, is a modern example of such a feature and provides us with a view of how the Atlantic Ocean may have looked in its infancy (see Figure 2.14D).

2.5 Concept Checks

1. Sketch or describe how two plates move in relation to each other along divergent plate boundaries.

2. What is the average rate of seafloor spreading in modern oceans?

3. List four features that characterize the oceanic ridge system.

4. Briefly describe the process of continental rifting. Where is it occurring today?

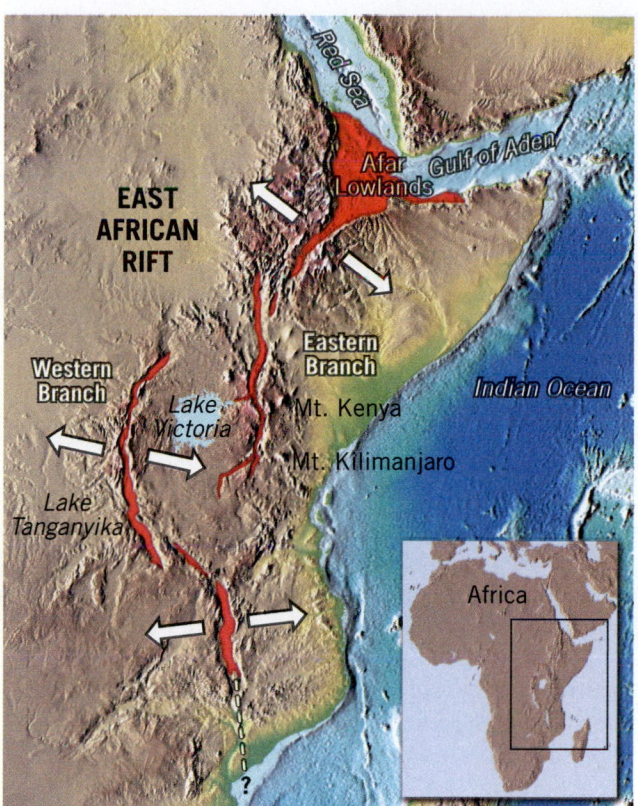

SmartFigure 2.15 **East African Rift valley** The East African Rift valley represents the early stage in the breakup of a continent. Areas shown in red consist of lithosphere that has been stretched and thinned, allowing magma to well up from the mantle. (https://goo.gl/Gp4pje)

Condor Video

EYE ON EARTH 2.1

In December 2011 a new volcanic island formed near the southern end of the Red Sea. Less than 2 years later, in late October 2013, another volcanic island emerged in the same area. These volcanic islands are part of several small islands in the Zubair Group located off the west coast of Yemen, along the Red Sea Rift.

QUESTION 1 What type of plate boundary produced these new volcanic islands?

QUESTION 2 What two plates border the Red Sea Rift?

QUESTION 3 Are these two plates moving *toward* or *away* from each other?

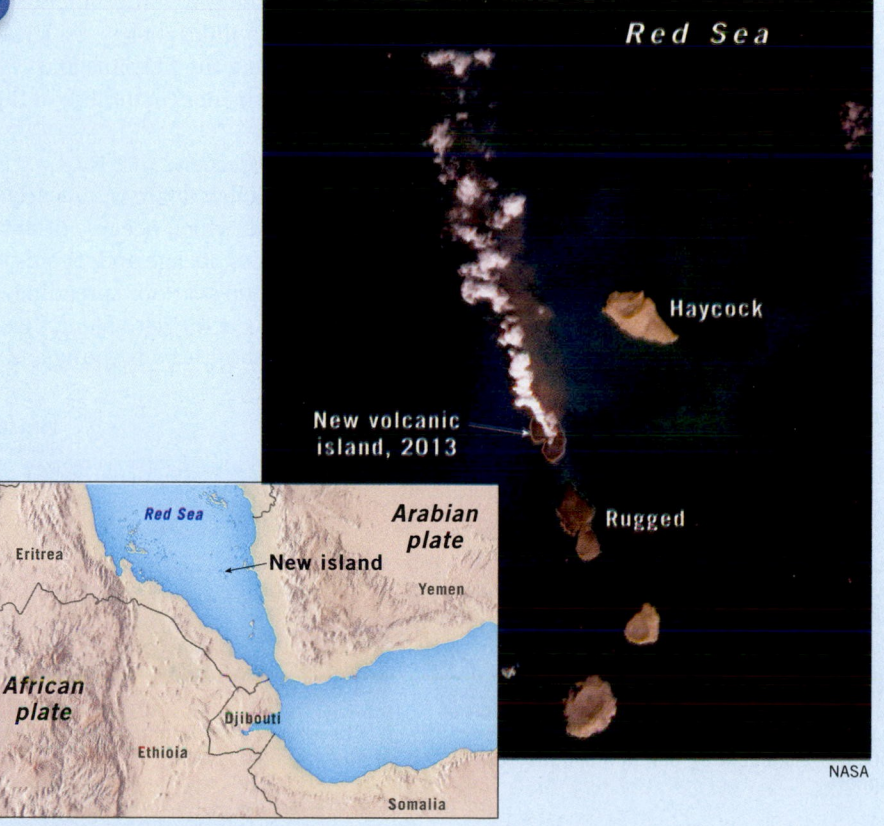

NASA

2.6 | Convergent Plate Boundaries and Subduction

Compare and contrast the three types of convergent plate boundaries and name a location where each type can be found.

New lithosphere is constantly being produced at the oceanic ridges. However, our planet is not growing larger; its total surface area remains constant. A balance is maintained because older, denser portions of oceanic lithosphere descend into the mantle at a rate equal to seafloor production. This activity occurs along **convergent plate boundaries**, where two plates move toward each other and the leading edge of one is bent downward as it slides beneath the other.

Convergent boundaries are also called **subduction zones** because they are sites where lithosphere is descending (being subducted) into the mantle. Subduction occurs because the density of the descending lithospheric plate is greater than the density of the underlying asthenosphere. Recall that oceanic crust is more dense than continental crust. In general, old oceanic lithosphere is about 2 percent more dense than the underlying asthenosphere, which causes it to subduct. Continental lithosphere, in contrast, is less dense than the underlying asthenosphere and resists subduction. As a consequence, only oceanic lithosphere will subduct to great depths.

Deep-ocean trenches are the surface manifestations produced as oceanic lithosphere descends into the mantle (see Figure 1.23, page 30). These large linear depressions are remarkably long and deep. The Peru–Chile trench along the west coast of South America is more than 4500 kilometers (2800 miles) long, and its floor is as much as 8 kilometers (5 miles) below sea level. Western Pacific trenches, including the Mariana and Tonga trenches, tend to be even deeper than those of the eastern Pacific.

Slabs of oceanic lithosphere descend into the mantle at angles that vary from a few degrees to nearly vertical (90 degrees). The angle at which oceanic lithosphere subducts depends largely on its age and, therefore, its density. For example, when seafloor spreading occurs near a subduction zone, as is the case along the coast of Chile, the subducting lithosphere is young and

buoyant, which results in a low angle of descent. As the two plates converge, the overriding plate scrapes over the top of the subducting plate below—a type of forced subduction. Consequently, the region around the Peru–Chile trench experiences great earthquakes, including the 2010 Chilean earthquake—one of the 10 largest on record.

As oceanic lithosphere ages (gets farther from the spreading center), it gradually cools, which causes it to thicken and increase in density. In parts of the western Pacific, some oceanic lithosphere is 180 million years old—the thickest and densest in today's oceans. The very dense slabs in this region typically plunge into the mantle at angles approaching 90 degrees. This largely explains why most trenches in the western Pacific are deeper than trenches in the eastern Pacific.

Although all convergent zones have the same basic characteristics, they vary considerably depending on the type of crustal material involved and the tectonic setting. Convergent boundaries can form *between one oceanic plate and one continental plate, between two oceanic plates, or between two continental plates* (**Figure 2.16**).

Oceanic–Continental Convergence

When the leading edge of a plate capped with continental crust converges with a slab of oceanic lithosphere, the buoyant continental block remains "floating," while the denser oceanic slab sinks into the mantle (see Figure

SmartFigure 2.16
Three types of convergent plate boundaries
(https://goo.gl/Zlylbf)

▶ Tutorial

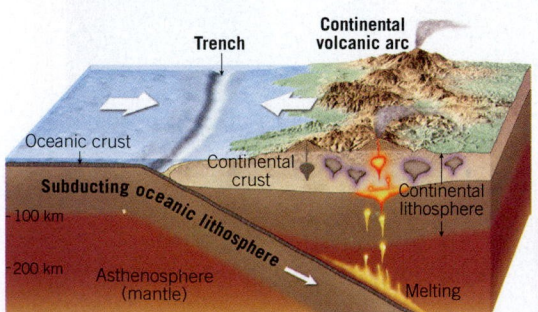

A. Convergent plate boundary where oceanic lithosphere is subducting beneath continental lithosphere.

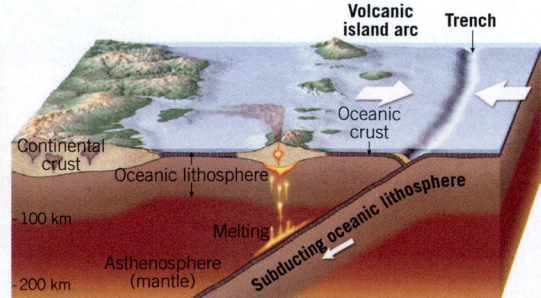

B. Convergent plate boundary involving two slabs of oceanic lithosphere.

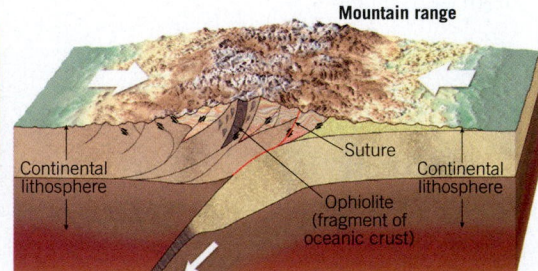

C. Continental collisions occur along convergent plate boundaries when both plates are capped with continental crust.

2.16A). When a descending oceanic slab reaches a depth of about 100 kilometers (60 miles), melting is triggered within the wedge of hot asthenosphere that lies above it. But how does the subduction of a cool slab of oceanic lithosphere cause mantle rock to melt? The answer lies in the fact that water contained in the descending plates acts the way salt does to melt ice. That is, "wet" rock in a high-pressure environment melts at substantially lower temperatures than does "dry" rock of the same composition.

Sediments and oceanic crust contain large amounts of water, which is carried to great depths by a subducting plate. As the plate plunges downward, heat and pressure drive water from the hydrated (water-rich) minerals in the subducting slab. At a depth of roughly 100 kilometers (60 miles), the wedge of mantle rock is sufficiently hot that the introduction of water from the slab below leads to some melting. This process, called **partial melting**, is thought to generate some molten material, which is mixed with unmelted mantle rock. Being less dense than the surrounding mantle, this hot mobile material gradually rises toward the surface. Depending on the environment, these mantle-derived masses of molten rock may ascend through the crust and give rise to a volcanic eruption. However, much of this material never reaches the surface but solidifies at depth—a process that thickens the crust.

The volcanoes of the towering Andes are the product of molten rock generated by the subduction of the Nazca plate beneath the South American continent (see Figure 2.11). Mountain systems like the Andes, which are produced in part by volcanic activity associated with the subduction of oceanic lithosphere, are called **continental volcanic arcs**. The Cascade Range in Washington, Oregon, and California is another mountain system consisting of several well-known volcanoes, including Mount Rainier, Mount Shasta, Mount St. Helens, and Mount Hood (**Figure 2.17**). This active volcanic arc also extends into Canada, where it includes Mount Garibaldi and Mount Silverthrone.

Oceanic–Oceanic Convergence

An *oceanic–oceanic convergent boundary* has many features in common with oceanic–continental plate margins (see Figure 2.16A,B). Where two oceanic slabs converge, one descends beneath the other, initiating volcanic activity by the same mechanism that operates at all subduction zones (see Figure 2.11). Water driven from the subducting slab of oceanic lithosphere triggers melting in the hot wedge of mantle rock above. In this setting, volcanoes grow up from the ocean floor rather than on a continental platform. When subduction is sustained, it will eventually build a chain of volcanic structures large enough to emerge as islands. The newly

Figure 2.17
Oceanic–continental convergent plate boundary Mount Hood, Oregon, is one of more than a dozen large composite volcanoes in the Cascade Range, a continental volcanic arc.

Wallace Garrison/Getty Images

Mt. Hood, Oregon

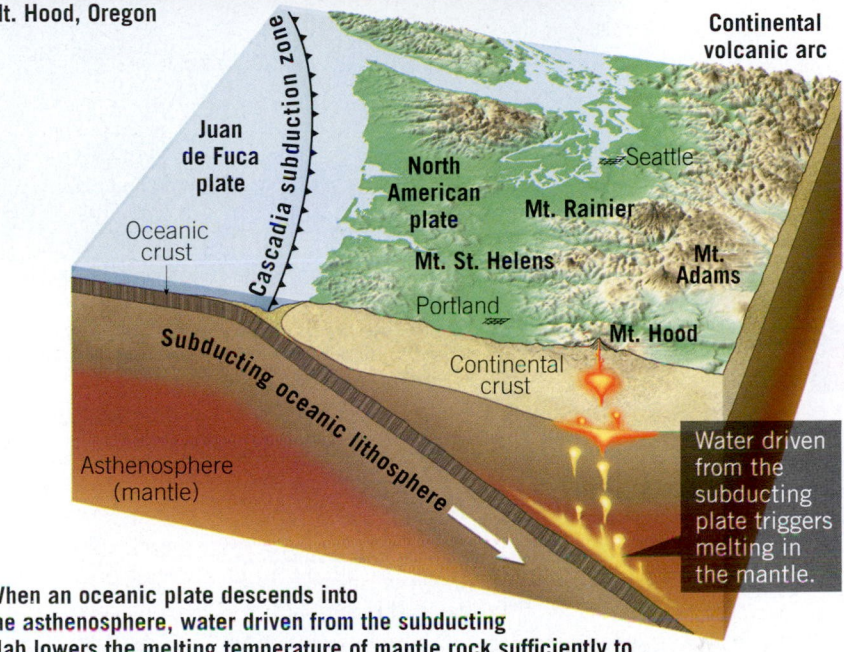

When an oceanic plate descends into the asthenosphere, water driven from the subducting slab lowers the melting temperature of mantle rock sufficiently to generate magma. The Cascade Range is a continental volcanic arc formed by the subduction of the Juan de Fuca plate under the North American plate.

formed land consisting of an arc-shaped chain of volcanic islands is called a **volcanic island arc**, or simply an **island arc** (**Figure 2.18**).

The Aleutian, Mariana, and Tonga Islands are examples of relatively young volcanic island arcs. Island arcs are generally located 120 to 360 kilometers (75 to 225 miles) from a deep-ocean trench. Located adjacent to the island arcs just mentioned are the Aleutian trench, the Mariana trench, and the Tonga trench.

Most volcanic island arcs are located in the western Pacific. Only two are located in the Atlantic—the Lesser Antilles arc, on the eastern margin of the Caribbean Sea, and the Sandwich Islands, located off the tip of South America. The Lesser Antilles are a product of the subduction of the Atlantic seafloor beneath the Caribbean plate. Located within this volcanic arc are the Virgin Islands of the United States and Britain as well as the island of Martinique, where Mount Pelée erupted in 1902, destroying the town of

Figure 2.18
Volcanoes of the Aleutian Islands The Aleutian Islands are a volcanic island arc produced by the subduction of the Pacific plate beneath the North American plate.

Active volcanoes of the Aleutian chain, Alaska

Alaska
Redoubt
Augustine
Katmai
Aniakchak
Pavlof
Shishaldin
Cleveland
Garelol
Kanaga
Great Sitkin

St. Pierre and killing an estimated 28,000 people. This chain of islands also includes Montserrat, where volcanic activity has occurred as recently as 2010.

Island arcs are typically simple structures made of numerous volcanic cones underlain by oceanic crust that is generally less than 20 kilometers (12 miles) thick. By contrast, some island arcs are more complex and are underlain by highly deformed crust that may reach 35 kilometers (22 miles) in thickness. Examples include Japan, Indonesia, and the Alaskan Peninsula. These island arcs are built on material generated by earlier episodes of subduction or on small slivers of continental crust that have rafted away from the mainland.

SmartFigure 2.19
The collision of India and Eurasia formed the Himalayas
(https://goo.gl/9IDLvo)

Animation

Continental–Continental Convergence

The third type of convergent boundary results when one landmass moves toward the margin of another because of subduction of the intervening seafloor (**Figure 2.19A**). Whereas oceanic lithosphere tends to be dense and sinks into the mantle, the buoyancy of continental material inhibits it from being subducted. Consequently, a collision between two converging continental fragments ensues (**Figure 2.19B**). This event folds and deforms the accumulation of sediments and sedimentary rocks along the continental margins as if they

The ongoing collision of the subcontinent of India with Eurasia began about 50 million years ago and produced the majestic Himalayas.

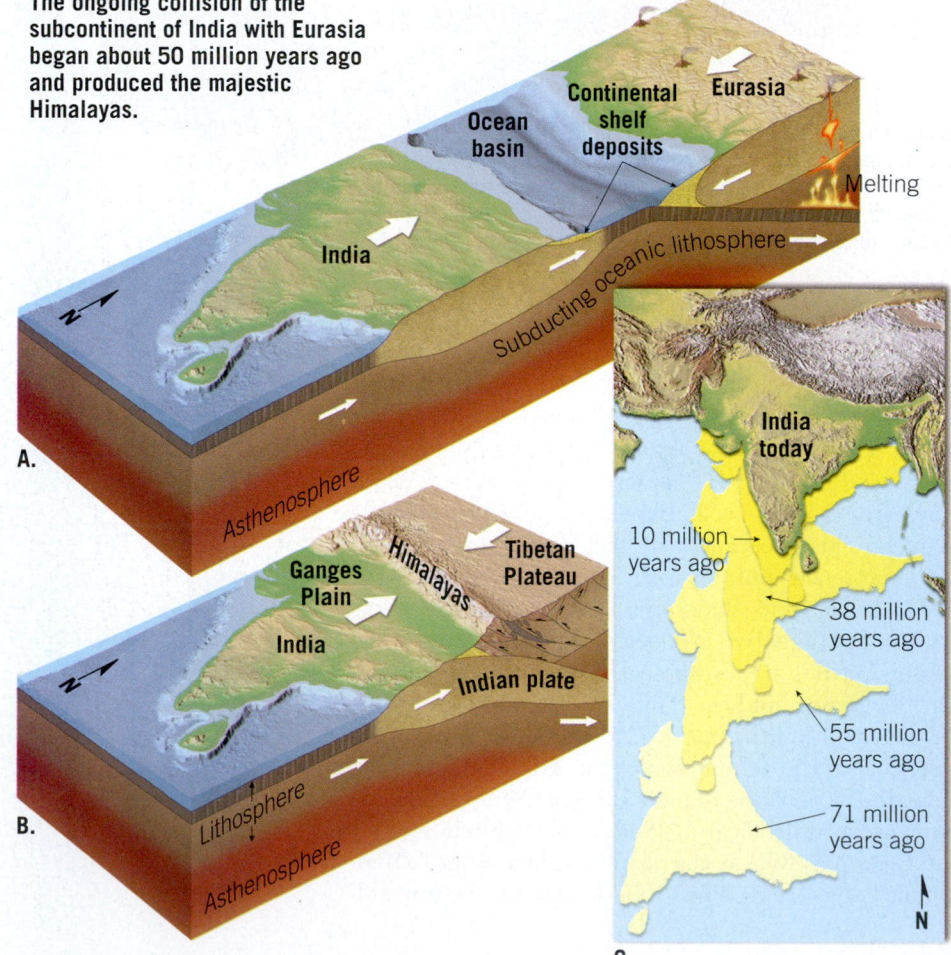

Continental volcanic arc
Eurasia
Continental shelf deposits
Ocean basin
Melting
India
Subducting oceanic lithosphere
A.
Asthenosphere

Himalayas
Tibetan Plateau
Ganges Plain
India
Indian plate
B.
Lithosphere
Asthenosphere

India today
10 million years ago
38 million years ago
55 million years ago
71 million years ago
N
C.

D.

(Photo by Peter Giovannini/Robert Harding)

had been placed in a gigantic vise. The result is the formation of a new mountain belt composed of deformed sedimentary and metamorphic rocks that often contain slivers of oceanic lithosphere.

Such a collision began about 50 million years ago, when the subcontinent of India rammed into Asia, producing the Himalayas—the most spectacular mountain range on Earth (**Figure 2.19C**). During this collision, the continental crust buckled and fractured and was generally shortened horizontally and thickened vertically. In addition to the Himalayas, several other major mountain systems, including the Alps, Appalachians, and Urals, formed as continental fragments collided. This topic will be considered further in Chapter 14.

2.6 | Concept Checks

1. Explain why the rate of lithosphere production is roughly equal to the rate of lithosphere destruction.

2. Why does oceanic lithosphere subduct, while continental lithosphere does not?

3. What characteristic of a slab of oceanic lithosphere leads to the formation of deep-ocean trenches instead of shallow trenches?

4. What distinguishes a continental volcanic arc from a volcanic island arc?

5. Briefly describe how mountain belts such as the Himalayas form.

2.7 | Transform Plate Boundaries

Describe the relative motion along a transform fault boundary and be able to locate several examples on a plate boundary map.

Along a **transform plate boundary**, also called a **transform fault**, plates slide horizontally past one another, without the production or destruction of lithosphere. The nature of transform faults was discovered in 1965 by Canadian geologist J. Tuzo Wilson, who proposed that these large faults connected two spreading centers (divergent boundaries) or, less commonly, two trenches (convergent boundaries). Most transform faults are found on the ocean floor, where they offset segments of the oceanic ridge system, producing a steplike plate margin (**Figure 2.20A**). Notice that the zigzag shape of the Mid-Atlantic Ridge in Figure 2.11 roughly reflects the shape of the original rifting that caused the breakup of the supercontinent of Pangaea. (Compare the shapes of the continental margins of the landmasses on both sides of the Atlantic with the shape of the Mid-Atlantic Ridge.)

SmartFigure 2.20 Transform plate boundaries
(https://goo.gl/SaoJ2o)

 **Tutorial**

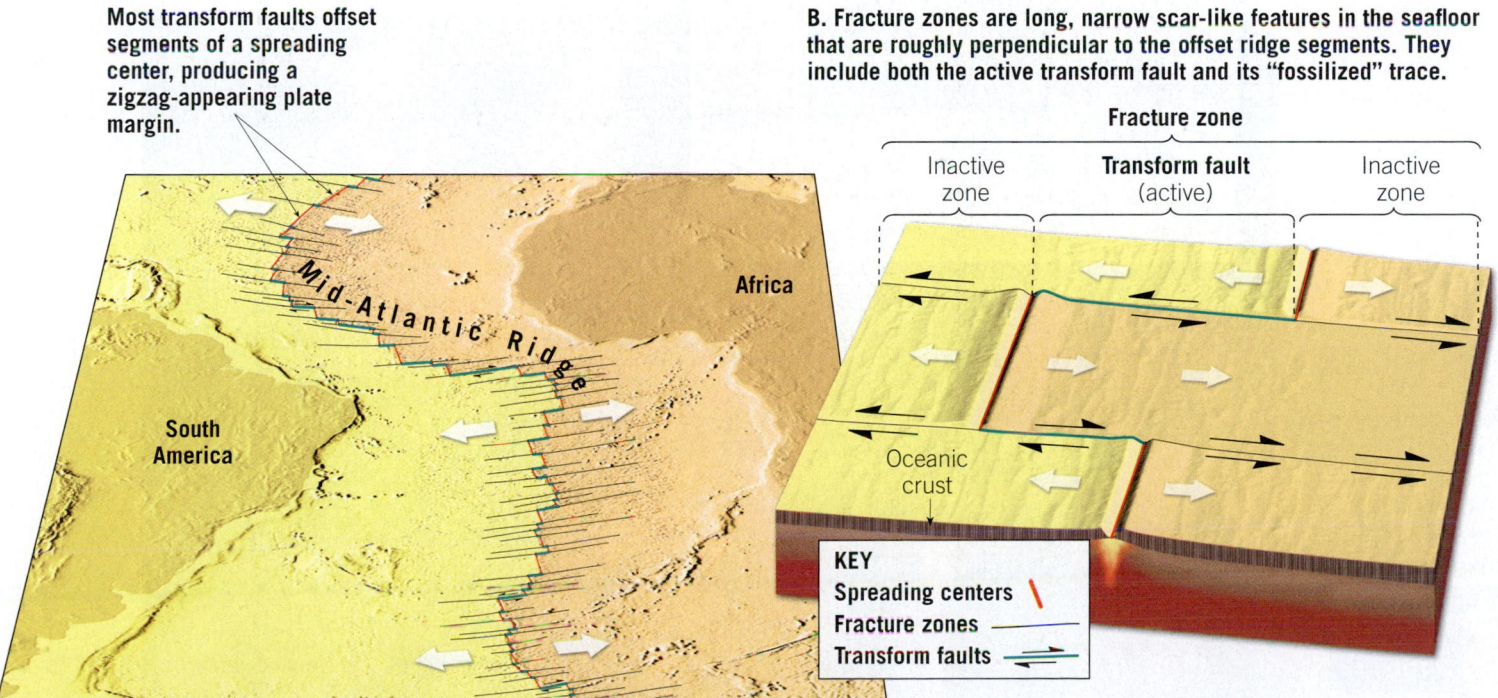

Most transform faults offset segments of a spreading center, producing a zigzag-appearing plate margin.

B. Fracture zones are long, narrow scar-like features in the seafloor that are roughly perpendicular to the offset ridge segments. They include both the active transform fault and its "fossilized" trace.

Fracture zone

Inactive zone | Transform fault (active) | Inactive zone

Africa

Mid-Atlantic Ridge

South America

Oceanic crust

KEY
Spreading centers
Fracture zones
Transform faults

A. The Mid-Atlantic Ridge, with its zigzag pattern, roughly reflects the shape of the rifting zone that resulted in the breakup of Pangaea.

Figure 2.21
Transform faults facilitate plate motion Seafloor generated along the Juan de Fuca Ridge moves southeastward, past the Pacific plate. Eventually it subducts beneath the North American plate. Thus, this transform fault connects a spreading center (divergent boundary) to a subduction zone (convergent boundary). Also shown is the San Andreas Fault, a transform fault that connects a spreading center located in the Gulf of California and the Mendocino Fault.

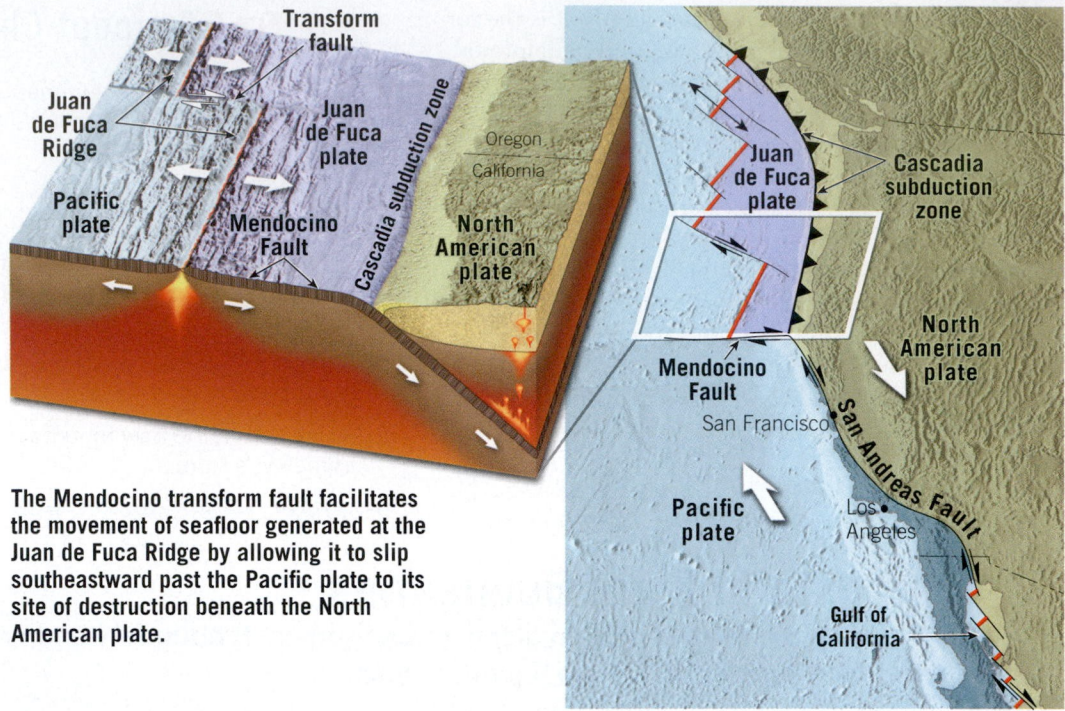

The Mendocino transform fault facilitates the movement of seafloor generated at the Juan de Fuca Ridge by allowing it to slip southeastward past the Pacific plate to its site of destruction beneath the North American plate.

Typically, transform faults are part of prominent linear breaks in the seafloor known as **fracture zones**, which include both active transform faults and their inactive extensions into the plate interior (**Figure 2.20B**). Active transform faults lie *only between* the two offset ridge segments and are generally defined by weak, shallow earthquakes. Here seafloor produced at one ridge axis moves in the opposite direction of seafloor produced at an opposing ridge segment. Thus, between the ridge segments, these adjacent slabs of oceanic crust are grinding past each other along a transform fault. Beyond the ridge crests are inactive zones, where the fractures are

Geologist's Sketch

SmartFigure 2.22
Movement along the San Andreas Fault This aerial view shows the offset in the dry channel of Wallace Creek near Taft, California. (Photo by Michael Collier) (http://goo.gl/tKTXky)

Mobile Field Trip

preserved as linear topographic depressions. The trend of these fracture zones roughly parallels the direction of plate motion at the time of their formation. Thus, these structures help geologists map the direction of plate motion in the geologic past.

Transform faults also provide the means by which the oceanic crust created at ridge crests can be transported to a site of destruction—the deep-ocean trenches. **Figure 2.21** illustrates this situation. Notice that the Juan de Fuca plate moves in a southeasterly direction, eventually being subducted under the west coast of the United States and Canada. The southern end of this plate is bounded by a transform fault called the Mendocino Fault. This transform boundary connects the Juan de Fuca Ridge to the Cascadia subduction zone. Therefore, it facilitates the movement of the crustal material created at the Juan de Fuca Ridge to its destination beneath the North American continent.

Like the Mendocino Fault, most other transform fault boundaries are located within the ocean basins; however, a few cut through continental crust. Two examples are the earthquake-prone San Andreas Fault

of California and New Zealand's Alpine Fault. Notice in Figure 2.21 that the San Andreas Fault connects a spreading center located in the Gulf of California to the Cascadia subduction zone and the Mendocino Fault. Along the San Andreas Fault, the Pacific plate is moving toward the northwest, past the North American plate (**Figure 2.22**). If this movement continues, the part of California west of the fault zone, including Mexico's Baja Peninsula, will become an island off the west coast of the United States and Canada. However, a more immediate concern is the earthquake activity triggered by movements along this fault system.

2.7 Concept Checks

1. Sketch or describe how two plates move in relation to each other along a transform plate boundary.

2. List two characteristics that differentiate transform faults from the two other types of plate boundaries.

EYE ON EARTH 2.2

Baja California is separated from mainland Mexico by a long narrow sea called the Gulf of California (also known to local residents as the Sea of Cortez). The Gulf of California contains many islands that were created by volcanic activity. (Photo courtesy of NASA)

QUESTION 1 What type of plate boundary is responsible for opening the Gulf of California?

QUESTION 2 What major U.S. river originates in the Colorado Rockies and created a large delta at the northern end of the Gulf of California?

QUESTION 3 If the material carried by the river in Question 2 had *not* been deposited, the Gulf of California would extend northward to include the inland sea shown in this satellite image. What is the name of this inland sea?

2.8 | How Do Plates and Plate Boundaries Change?

Explain why plates such as the African and Antarctic plates are increasing in size, while the Pacific plate is decreasing in size.

Although Earth's total surface area does not change, the size and shape of individual plates are constantly changing. For example, the African and Antarctic plates, which are mainly bounded by divergent boundaries—sites of seafloor production—are continually growing in size as new lithosphere is added to their margins. By contrast, the Pacific plate is being consumed into the mantle along much of its flanks faster that it is growing along the East Pacific Rise and thus is diminishing in size.

Another result of plate motion is that boundaries also migrate. For example, the position of the Peru–Chile trench, which is the result of the Nazca plate being bent downward as it descends beneath the South American plate, has changed over time (see Figure 2.11). Because of the westward drift of the South American plate relative to the Nazca plate, the Peru–Chile trench has migrated in a westerly direction as well.

Plate boundaries can also be created or destroyed in response to changes in the forces acting on the lithosphere. Plates carrying continental crust are presently moving toward one another. For example, in the South Pacific, Australia is moving northward toward southern Asia. If Australia continues its northward migration, the boundary separating it from Asia will disappear as these plates become one. Other plates are moving apart. Recall that the Red Sea is the site of a relatively new spreading center that came into existence less than 20 million years ago, when the Arabian Peninsula began to break apart from Africa. The breakup of Pangaea is a classic example of how plate boundaries change through geologic time.

The Breakup of Pangaea

Wegener used evidence from fossils, rock types, and ancient climates to create a jigsaw-puzzle fit of the continents, thereby creating his supercontinent of Pangaea. By employing modern tools not available to Wegener, geologists have re-created the steps in the breakup of this supercontinent, an event that began about 180 million years ago. From this work, the dates when individual crustal fragments separated from one another and their relative motions have been well established (Figure 2.23).

An important consequence of Pangaea's breakup was the creation of a "new" ocean basin: the Atlantic. As you can see in Figure 2.23, splitting of the supercontinent did not occur simultaneously along the margins of the Atlantic. The first split developed between North America and Africa. Here, the continental crust was highly fractured, providing pathways for huge quantities of fluid lavas to reach

Figure 2.23
The breakup of Pangaea

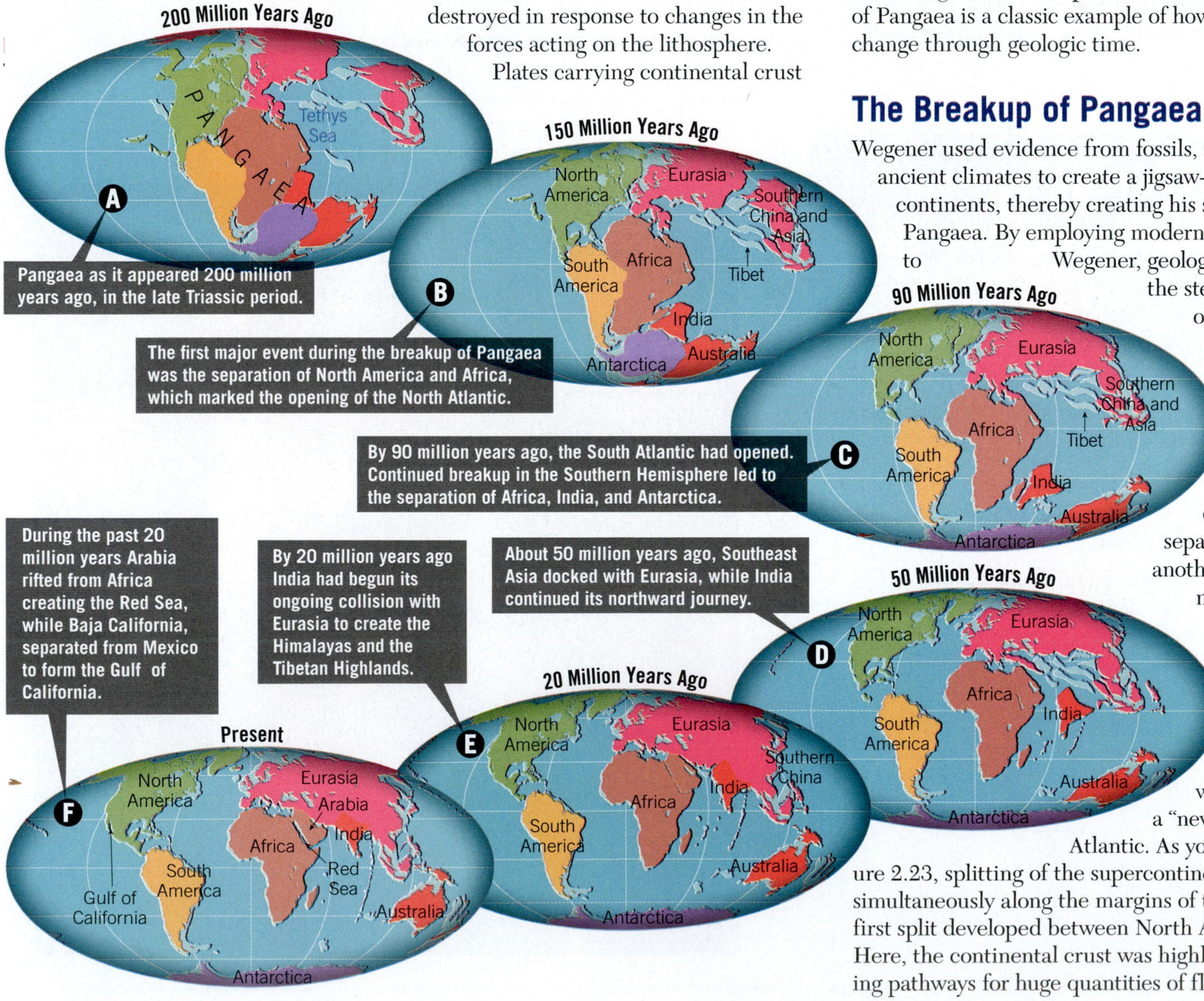

200 Million Years Ago

A — Pangaea as it appeared 200 million years ago, in the late Triassic period.

150 Million Years Ago

B — The first major event during the breakup of Pangaea was the separation of North America and Africa, which marked the opening of the North Atlantic.

90 Million Years Ago

C — By 90 million years ago, the South Atlantic had opened. Continued breakup in the Southern Hemisphere led to the separation of Africa, India, and Antarctica.

50 Million Years Ago

D — About 50 million years ago, Southeast Asia docked with Eurasia, while India continued its northward journey.

20 Million Years Ago

E — By 20 million years ago India had begun its ongoing collision with Eurasia to create the Himalayas and the Tibetan Highlands.

Present

F — During the past 20 million years Arabia rifted from Africa creating the Red Sea, while Baja California, separated from Mexico to form the Gulf of California.

the surface. Today, these lavas are represented by weathered igneous rocks found along the eastern seaboard of the United States—primarily buried beneath the sedimentary rocks that form the continental shelf. Radiometric dating of these solidified lavas indicates that rifting began between 200 million and 190 million years ago. This time span represents the "birth date" for this section of the North Atlantic.

By 130 million years ago, the South Atlantic began to open near the tip of what is now South Africa. As this zone of rifting migrated northward, it gradually opened the South Atlantic (**Figures 2.23B,C**). Continued breakup of the southern landmass led to the separation of Africa and Antarctica and sent India on a northward journey. By the early Cenozoic, about 50 million years ago, Australia had separated from Antarctica, and the South Atlantic had emerged as a full-fledged ocean (**Figure 2.23D**).

India eventually collided with Asia (**Figure 2.23E**), an event that began about 50 million years ago and created the Himalayas and the Tibetan Highlands. About the same time, the separation of Greenland from Eurasia completed the breakup of the northern landmass. During the past 20 million years or so of Earth's history, Arabia has rifted from Africa to form the Red Sea, and Baja California has separated from Mexico to form the Gulf of California (**Figure 2.23F**). Meanwhile, the Panama Arc joined North America and South America to produce our globe's familiar modern appearance.

Plate Tectonics in the Future

Geologists have extrapolated present-day plate movements into the future. **Figure 2.24** illustrates where Earth's landmasses may be 50 million years from now if present plate movements persist during this time span.

In North America we see that the Baja Peninsula and the portion of southern California that lies west of the San Andreas Fault will have slid past the North American plate. If this northward migration takes place, Los Angeles and San Francisco will pass each other in about 10 million years, and in about 60 million years the Baja Peninsula will begin to collide with the Aleutian Islands.

If Africa continues on a northward path, it will continue to collide with Eurasia. The result will be the closing of the

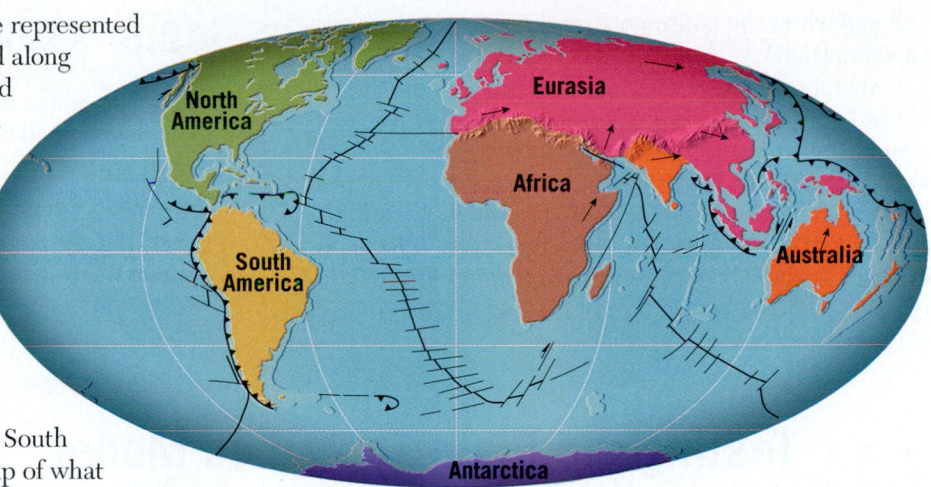

Figure 2.24
The world as it may look 50 million years from now This reconstruction is highly idealized and based on the assumption that the processes that caused the breakup of Pangaea will continue to operate.
(Based on Robert S. Dietz, John C. Holden, C. Scotese, and others)

Mediterranean, the last remnant of a once-vast ocean called the Tethys Ocean, and the initiation of another major mountain-building episode (see Figure 2.24). Australia will be astride the equator and, along with New Guinea, will be on a collision course with Asia. Meanwhile, North and South America will begin to separate, while the Atlantic and Indian Oceans will continue to grow, at the expense of the Pacific Ocean.

A few geologists have even speculated on the nature of the globe 250 million years in the future. In this scenario the Atlantic seafloor will eventually become old and dense enough to form subduction zones around much of its margins, not unlike the present-day Pacific basin. Continued subduction of the floor of the Atlantic Ocean will result in the closing of the Atlantic basin and the collision of the Americas with the Eurasian–African landmass to form the next supercontinent, shown in **Figure 2.25**. Support for the possible closing of the Atlantic comes from a similar event, when an ocean predating the Atlantic closed during Pangaea's formation. Australia is also projected to collide with Southeast Asia by that time. If this scenario is accurate, the dispersal of Pangaea

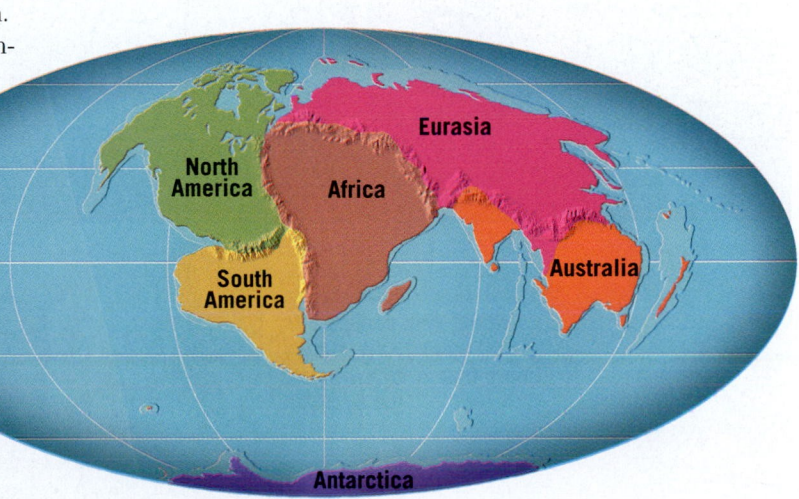

Figure 2.25
Earth as it may appear 250 million years from now

will end when the continents reorganize into the next supercontinent.

Such projections, although interesting, must be viewed with considerable skepticism because many assumptions must be correct for these events to unfold as just described. Nevertheless, changes in the shapes and positions of continents that are equally profound will undoubtedly occur for many hundreds of millions of years to come. Only after much more of Earth's internal heat has been lost will the engine that drives plate motions cease.

2.8 | Concept Checks

1. Name two plates that are growing in size. Which plate is shrinking in size?

2. What new ocean basin was created by the breakup of Pangaea?

3. Briefly describe changes in the positions of the continents if we assume that the plate motions we see today continue 50 million years into the future.

2.9 | Testing the Plate Tectonics Model

List and explain the evidence used to support the plate tectonics theory.

Some of the evidence supporting continental drift was presented in Section 2.3. With the development of plate tectonics theory, researchers began testing this new model of how Earth works. Although new supporting data were obtained, it has often been new interpretations of already existing data that have swayed the tide of opinion.

Evidence: Ocean Drilling

Some of the most convincing evidence for seafloor spreading came from the Deep Sea Drilling Project, which operated from 1966 until 1983. One of the early goals of the project was to gather samples of the ocean floor in order to establish its age. To accomplish this, the *Glomar Challenger*, a drilling ship capable of working in water thousands of meters deep, was built. Hundreds of holes were drilled through the layers of sediments that blanket the ocean crust, as well as into the basaltic rocks below. Rather than use radiometric dating, which can be unreliable on oceanic rocks because of the alteration of basalt by seawater, researchers used the fossil remains of microorganisms found in the sediments resting directly on the crust to date the seafloor at each site.

When the oldest sediment from each drill site was plotted against its distance from the ridge crest, the plot showed that the sediments increased in age with increasing distance from the ridge (**Figure 2.26A**). This finding supported the seafloor-spreading hypothesis, which predicted that the youngest oceanic crust would be found at the ridge crest, the site of seafloor production, and the oldest oceanic crust would be located adjacent to the continents.

The distribution and thickness of ocean-floor sediments provided additional verification of seafloor spreading. Drill cores from the *Glomar Challenger* revealed that sediments are almost entirely absent on the ridge crest and that sediment thickness increases with increasing distance from the ridge (see Figure 2.26A). This pattern of

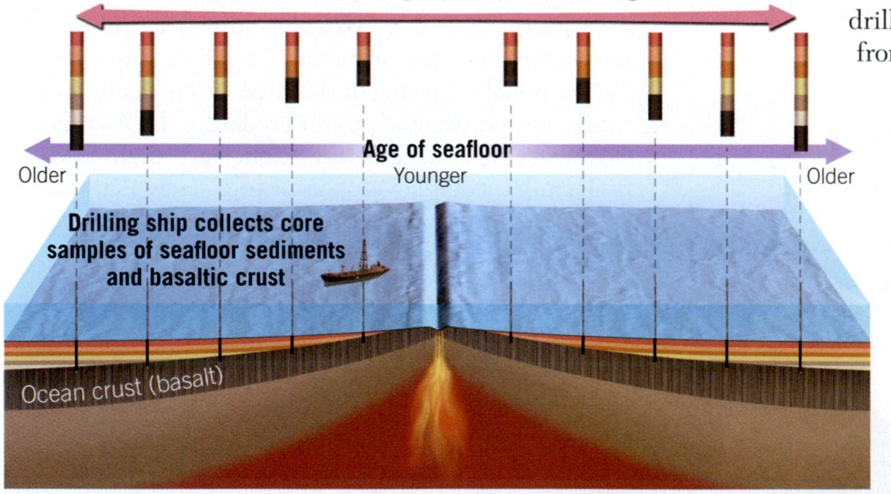

Figure 2.26
Deep-sea drilling A. Data collected through deep-sea drilling have shown that the ocean floor is indeed youngest at the ridge axis. **B.** The Japanese deep-sea drilling ship *Chikyu* became operational in 2007. (Photo by Katsumi Kasahara/Getty)

Core samples show that the thickness of sediments increases with increasing distance from the ridge crest.

Age of seafloor
Older   Younger   Older

Drilling ship collects core samples of seafloor sediments and basaltic crust

Ocean crust (basalt)

A.

Chikyu is a state-of-the-art drilling ship designed to drill up to 7,000 meters (more than 4 miles) below the seafloor.

B.

Katsumi Kasahara/Getty

sediment distribution should be expected if the seafloor-spreading hypothesis is correct.

The data collected by the Deep Sea Drilling Project also reinforced the idea that the ocean basins are geologically young because no seafloor older than 180 million years was found. By comparison, most continental crust exceeds several hundred million years in age, and some has been located that exceeds 4 billion years in age.

In 1983, a new ocean-drilling program was launched by the Joint Oceanographic Institutions for Deep Earth Sampling (JOIDES). Now the International Ocean Discovery Program (IODP), this ongoing international effort uses multiple vessels for exploration, including the massive 210-meter-long (nearly 690-foot-long) *Chikyu* ("planet Earth" in Japanese), which began operations in 2007 (**Figure 2.26B**). One of the goals of the IODP is to recover a complete section of the ocean crust, from top to bottom.

Evidence: Mantle Plumes and Hot Spots

Mapping volcanic islands and *seamounts* (submarine volcanoes) in the Pacific Ocean revealed several linear chains of volcanic structures. One of the most-studied chains consists of at least 129 volcanoes that extend from the Hawaiian Islands to Midway Island and continue northwestward toward the Aleutian trench (**Figure 2.27**). Radiometric dating of this linear structure, called the Hawaiian Island–Emperor Seamount chain, showed that

the volcanoes increase in age with increasing distance from the Big Island of Hawaii. The youngest volcanic island in the chain (Hawaii) rose from the ocean floor less than 1 million years ago, whereas Midway Island is 27 million years old, and Detroit Seamount, near the Aleutian trench, is about 80 million years old (see Figure 2.27).

Most researchers agree that a cylindrically shaped upwelling of hot rock, called a **mantle plume**, is located beneath the island of Hawaii. As the hot, rocky plume ascends through the mantle, the confining pressure drops, which triggers partial melting. (This process, called *decompression melting*, is discussed in Chapter 4.) The surface manifestation of this activity is a **hot spot**, an area of volcanism, high heat flow, and crustal uplifting that is a few hundred kilometers across. As the Pacific plate moved over a hot spot, a chain of volcanic structures known as a **hot-spot track** was built. As shown in Figure 2.27, the age of each volcano indicates how much time has elapsed since it was situated over the mantle plume.

A closer look at the five largest Hawaiian Islands reveals a similar pattern of ages, from the volcanically active island of Hawaii to the inactive volcanoes that make up the oldest island, Kauai (see Figure 2.27). Five million years ago, when Kauai was positioned over the

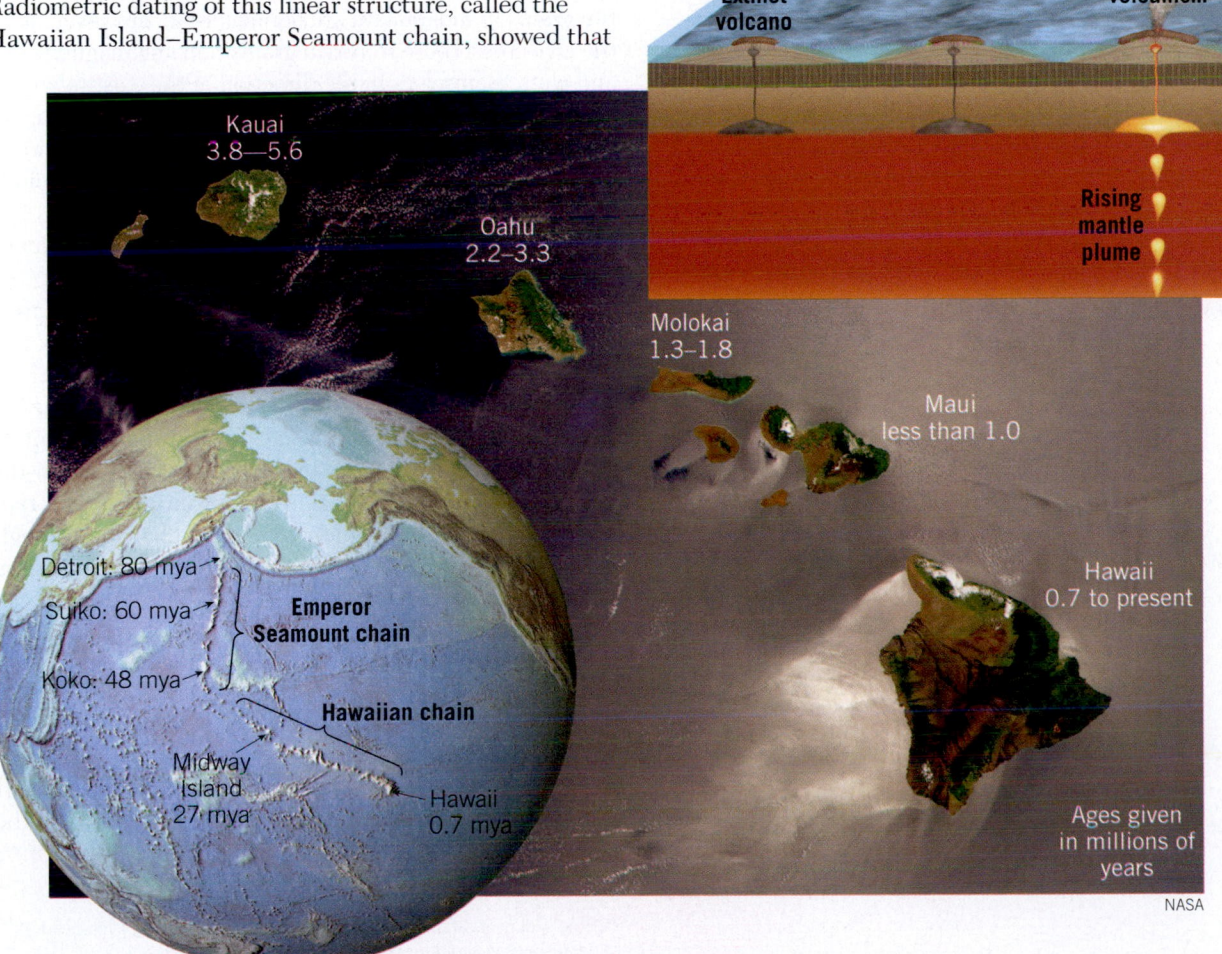

Figure 2.27
Hot-spot volcanism and the formation of the Hawaiian chain
Radiometric dating of the Hawaiian Islands shows that volcanic activity increases in age moving away from the Big Island of Hawaii.

Figure 2.28
Hot spots and hot-spot tracks Locations of some well-documented hot spots and hot-spot tracks thought to be associated with mantle plumes.

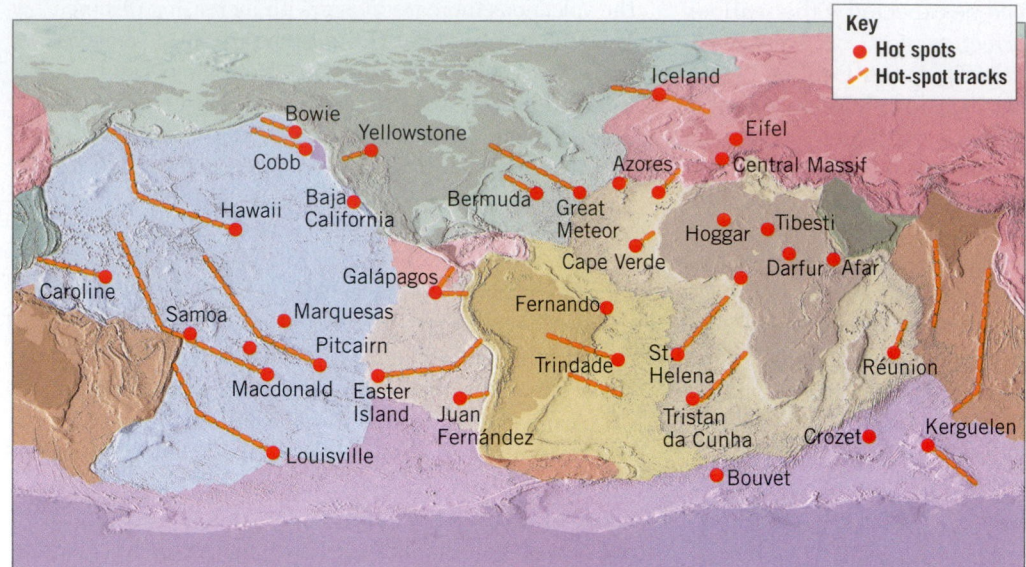

hot spot, it was the *only* modern Hawaiian Island in existence. Kauai's age is evident by examining its extinct volcanoes, which have been eroded into jagged peaks and vast canyons. By contrast, the relatively young island of Hawaii exhibits many fresh lava flows, and one of its five major volcanoes, Kilauea, remains active today.

Of approximately 40 hot spots that are thought to have formed because of upwelling of hot mantle plumes, most, but not all have hot-spot tracks (**Figure 2.28**). Also notice in Figure 2.28 that nearby hot spots have nearly parallel tracks, confirming large-scale plate motion.

Evidence: Paleomagnetism

You are probably familiar with how a compass operates and know that Earth's magnetic field has north and south magnetic poles. Today these magnetic poles roughly align with the geographic poles that are located where Earth's rotational axis intersects the surface. Earth's magnetic

Figure 2.29
Earth's magnetic field Earth's magnetic field consists of lines of force much like those a giant bar magnet would produce if placed at the center of Earth.

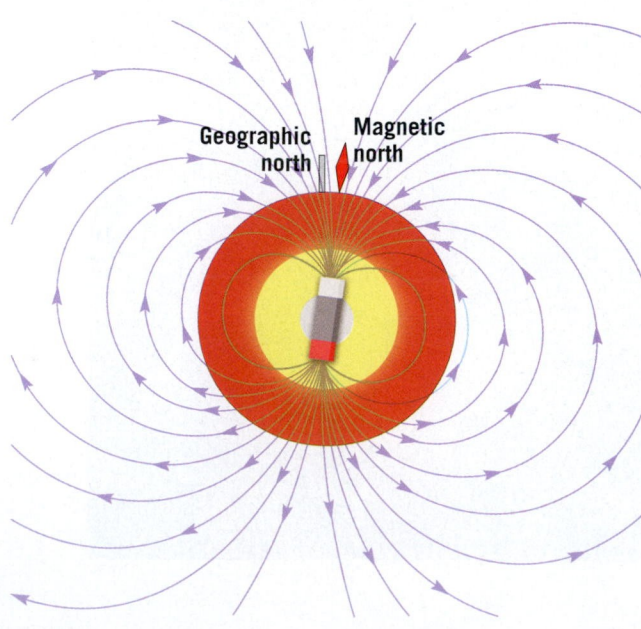

field is similar to that produced by a simple bar magnet. Invisible lines of force pass through the planet and extend from one magnetic pole to the other (**Figure 2.29**). A compass needle, itself a small magnet free to rotate on an axis, becomes aligned with the magnetic lines of force and points to the magnetic poles.

Earth's magnetic field is less obvious than the pull of gravity because humans cannot feel it. Movement of a compass needle, however, confirms its presence. In addition, some naturally occurring minerals are magnetic and are influenced by Earth's magnetic field. One of the most common is the iron-rich mineral *magnetite*, which is abundant in lava flows of basaltic composition.[*] Basaltic lavas erupt at the surface at temperatures greater than 1000°C (1800°F), exceeding a threshold temperature for magnetism known as the **Curie point** (about 585°C [1085°F]). Consequently, the magnetite grains in molten lava are nonmagnetic. However, as the lava cools, these iron-rich grains become magnetized and align themselves in the direction of the existing magnetic lines of force. Once the minerals solidify, the magnetism they possess usually remains "frozen" in this position. Thus, they act like a compass needle because they "point" toward the position of the magnetic poles at the time of their formation. Rocks that formed thousands or millions of years ago and contain a "record" of the direction of the magnetic poles at the time of their formation are said to possess **paleomagnetism**, or **fossil magnetism**.

Apparent Polar Wandering A study of paleomagnetism in ancient lava flows throughout Europe led to an interesting discovery. The magnetic alignment of iron-rich minerals in lava flows of different ages indicated that the position of the paleomagnetic poles had changed through time. A plot of the location of the magnetic north pole, as measured from Europe, revealed that during the past 500 million years, the pole had gradually "wandered" from a location near Hawaii northeastward to its present location over the Arctic Ocean (**Figure 2.30**). This was strong evidence that either the magnetic north pole had migrated, an idea known as *polar wandering*, or that the poles had remained in place and the continents had drifted beneath them—in other words, Europe had drifted relative to the magnetic north pole.

---

[*] Some sediments and sedimentary rocks also contain enough iron-bearing mineral grains to acquire a measurable amount of magnetization.

Although the magnetic poles are known to move in a somewhat erratic path, studies of paleomagnetism from numerous locations show that the positions of the magnetic poles, averaged over thousands of years, correspond closely to the positions of the geographic poles. Therefore, a more acceptable explanation for the apparent polar wander was provided by Wegener's hypothesis: If the magnetic poles remain stationary, their *apparent movement* is produced by continental drift.

Further evidence for continental drift came a few years later, when a polar-wandering path was constructed for North America (see Figure 2.30A). For the first 300 million years or so, the paths for North America and Europe were found to be similar in direction—but separated by about 5000 kilometers (3000 miles). Then, during the middle of the Mesozoic era (180 million years ago), they began to converge on the present North Pole. The explanation for these curves is that North America and Europe were joined until the Mesozoic, when the Atlantic began to open. From this time forward, these continents continuously moved apart. When North America and Europe are moved back to their pre-drift positions, as shown in Figure 2.30B, these paths of apparent polar wandering coincide. This is evidence that North America and Europe were once joined and moved relative to the poles as part of the same continent.

Magnetic Reversals and Seafloor Spreading
More evidence emerged when geophysicists learned that over periods of hundreds of thousands of years, Earth's magnetic field periodically reverses polarity. During a **magnetic reversal**, the magnetic north pole becomes the

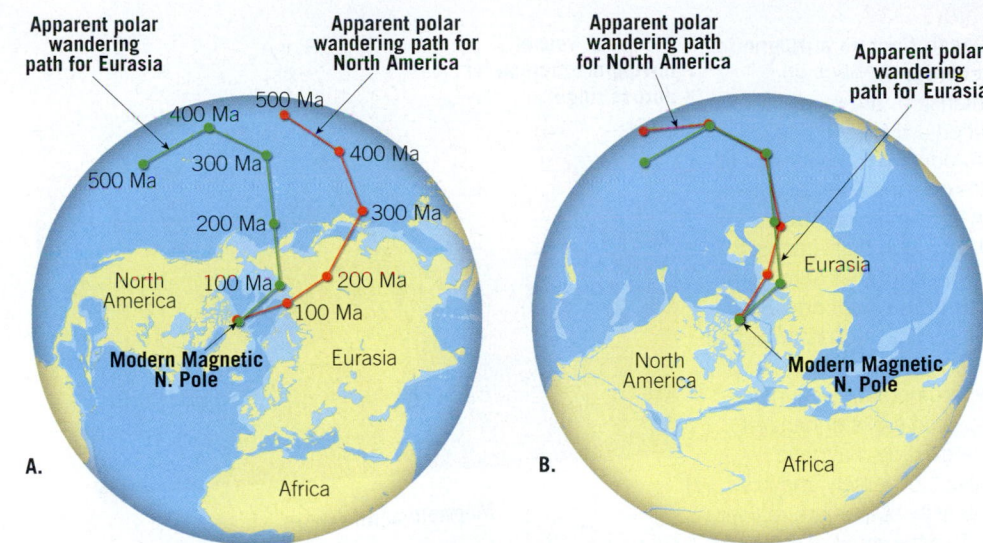

Figure 2.30   Apparent polar-wandering paths A. The more westerly path determined from North American data is thought to have been caused by the westward drift of North America by about 24 degrees from Eurasia. **B.** The positions of the wandering paths when the landmasses are reassembled in their pre-drift locations.

magnetic south pole and vice versa. Lava solidifying during a period of reverse polarity will be magnetized with the polarity opposite that of volcanic rocks being formed today. When rocks exhibit the same magnetism as the present magnetic field, they are said to possess **normal polarity**, whereas rocks exhibiting the opposite magnetism are said to have **reverse polarity**.

Once the concept of magnetic reversals was confirmed, researchers set out to establish a time scale for these occurrences. The task was to measure the magnetic polarity of hundreds of lava flows and use radiometric dating techniques to establish the age of each flow. **Figure 2.31** shows the **magnetic time scale** established using

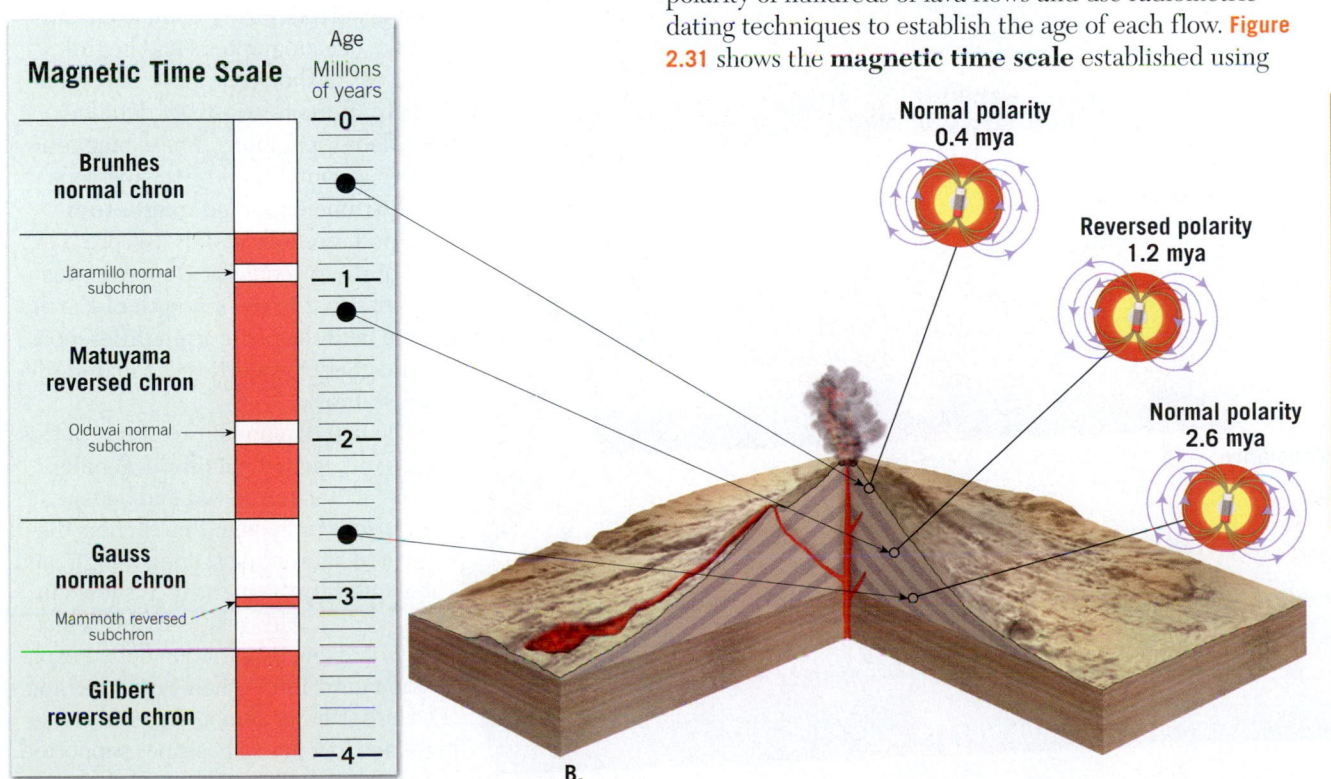

A.   B.

SmartFigure 2.31
Time scale of magnetic reversals A. Time scale of Earth's magnetic reversals for the past 4 million years. **B.** This time scale was developed by establishing the magnetic polarity for lava flows of known age. (Data from Allen Cox and G. B. Dalrymple) (https://goo.gl/e9Qwe2)

Tutorial

Figure 2.32
Ocean floor as a magnetic recorder A. Magnetic intensities are recorded when a magnetometer is towed across a segment of the oceanic floor. **B.** Notice the symmetrical stripes of low- and high-intensity magnetism that parallel the axis of the Juan de Fuca Ridge. The stripes of high-intensity magnetism occur where normally magnetized oceanic rocks enhanced the existing magnetic field. Conversely, the low-intensity stripes are regions where the crust is polarized in the reverse direction, which weakens the existing magnetic field.

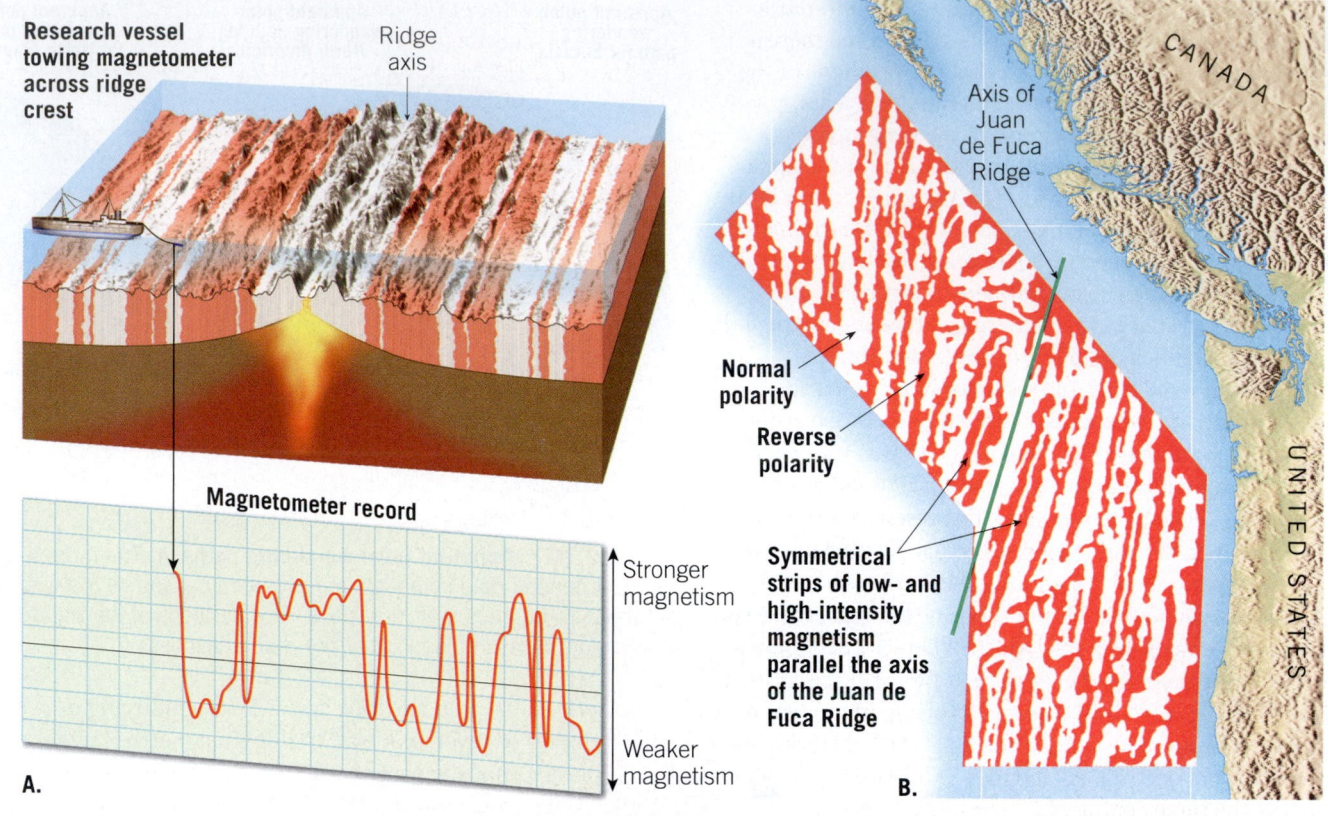

Research vessel towing magnetometer across ridge crest

Ridge axis

Magnetometer record

Stronger magnetism

Weaker magnetism

A.

Normal polarity

Reverse polarity

Symmetrical strips of low- and high-intensity magnetism parallel the axis of the Juan de Fuca Ridge

Axis of Juan de Fuca Ridge

CANADA

UNITED STATES

B.

SmartFigure 2.33
Magnetic reversals and seafloor spreading When new basaltic rocks form at mid-ocean ridges, they magnetize according to Earth's existing magnetic field. Hence, oceanic crust provides a permanent record of each reversal of our planet's magnetic field over the past 200 million years.
(https://goo.gl/5gKsdz)

Animation

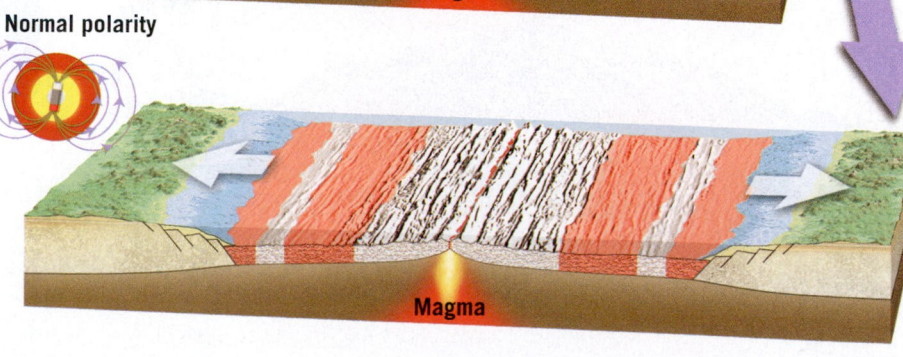

A. Normal polarity

Magma

B. Reverse polarity

Magma

C. Normal polarity

Magma

TIME

this technique for the past few million years. The major divisions of the magnetic time scale are called *chrons* and last for roughly 1 million years. As more measurements became available, researchers realized that several short-lived reversals (less than 200,000 years long) often occurred during a single chron.

Meanwhile, oceanographers had begun magnetic surveys of the ocean floor in conjunction with their efforts to construct detailed maps of seafloor topography. These magnetic surveys were accomplished by towing very sensitive instruments, called **magnetometers**, behind research vessels (**Figure 2.32A**). The goal of these geophysical surveys was to map variations in the strength of Earth's magnetic field that arise from differences in the magnetic properties of the underlying crustal rocks.

The first comprehensive study of this type was carried out off the Pacific coast of North America and had an unexpected outcome. Researchers discovered alternating stripes of high- and low-intensity magnetism, as shown in **Figure 2.32B**. This relatively simple pattern of magnetic variation defied explanation until 1963, when Fred Vine and D. H. Matthews demonstrated that the high- and low-intensity stripes supported the concept of seafloor spreading. Vine and Matthews suggested that the stripes of high-intensity magnetism are regions where the paleomagnetism of the ocean crust exhibits normal polarity (see Figure 2.32). Consequently, these rocks *enhance* (reinforce) Earth's magnetic field. Conversely, the low-intensity stripes are regions where the ocean crust is polarized in the reverse direction and therefore *weaken* the existing magnetic field. But how do parallel stripes of normally and reversely magnetized rock become distributed across the ocean floor?

Vine and Matthews reasoned that as magma solidifies along narrow rifts at the crest of an oceanic ridge, it is magnetized with the polarity of the existing magnetic field (**Figure 2.33**). Because of seafloor spreading, this strip of magnetized crust would gradually increase in width. When Earth's magnetic field reverses polarity, any newly formed seafloor having the opposite polarity would form in the middle of the old strip. Gradually, the two halves of the old strip would be carried in opposite directions, away from the ridge crest. Subsequent reversals would build a pattern of normal and reverse magnetic stripes, as shown in Figure 2.33. Because new rock is added in equal amounts to both trailing edges of the spreading ocean floor, we should

expect the pattern of stripes (width and polarity) found on one side of an oceanic ridge to be a mirror image of those on the other side. In fact, a survey across the Mid-Atlantic Ridge just south of Iceland reveals a pattern of magnetic stripes exhibiting a remarkable degree of symmetry in relation to the ridge axis.

2.9 | Concept Checks

1. What is the age of the oldest sediments recovered using deep-ocean drilling? How do the ages of these sediments compare to the ages of the oldest continental rocks?

2. How do sedimentary cores from the ocean floor support the concept of seafloor spreading?

3. Assuming that hot spots remain fixed, in what direction was the Pacific plate moving while the Hawaiian Islands were forming?

4. Describe how Fred Vine and D. H. Matthews related the seafloor-spreading hypothesis to magnetic reversals.

2.10 | How Is Plate Motion Measured?

Describe two methods researchers use to measure relative plate motion.

A number of methods are used to establish the direction and rate of plate motion. Some of these techniques not only confirm that lithospheric plates move but allow us to trace those movements back in geologic time.

Geologic Measurement of Plate Motion

Using ocean-drilling ships, researchers have obtained dates for hundreds of locations on the ocean floor. By knowing the age of a sample and its distance from the ridge axis where it was generated, an average rate of plate motion can be calculated.

Scientists used these data, combined with their knowledge of paleomagnetism stored in hardened lavas on the ocean floor and seafloor topography, to create maps that show the age of the ocean floor. The reddish-orange colored bands shown in **Figure 2.34** range in age from the present to about 30 million years ago. The width of the bands indicates how much crust formed during that time period. For example, the reddish-orange band along the East Pacific Rise is more than three times wider than the same-color band along the Mid-Atlantic Ridge. Therefore, the rate of seafloor spreading has been approximately three times faster in the Pacific basin than in the Atlantic.

Maps of this type also provide clues to the current direction of plate movement. Notice the offsets in the ridges; these are transform faults that connect the spreading centers. Recall that transform faults are

aligned parallel to the direction of spreading. When measured carefully, transform faults reveal the direction of plate movement.

To establish the direction of plate motion in the past, geologists can examine the long fracture zones that extend for hundreds or even thousands of kilometers from ridge crests. Fracture zones are inactive extensions of transform faults and therefore preserve a record of past directions of plate motion. Unfortunately, most of the ocean floor is less than 180 million years old, so to look deeper into the past, researchers must rely on paleomagnetic evidence provided by continental rocks.

Measuring Plate Motion from Space

You are likely familiar with the Global Positioning System (GPS) used to locate one's position in order to provide directions to some other location. The Global Positioning System employs satellites that send radio signals that are intercepted by GPS receivers located at Earth's surface. The exact position of a site is determined by simultaneously establishing the distance from the receiver to four or more satellites. Researchers use specially designed equipment to locate a point on Earth to within a few

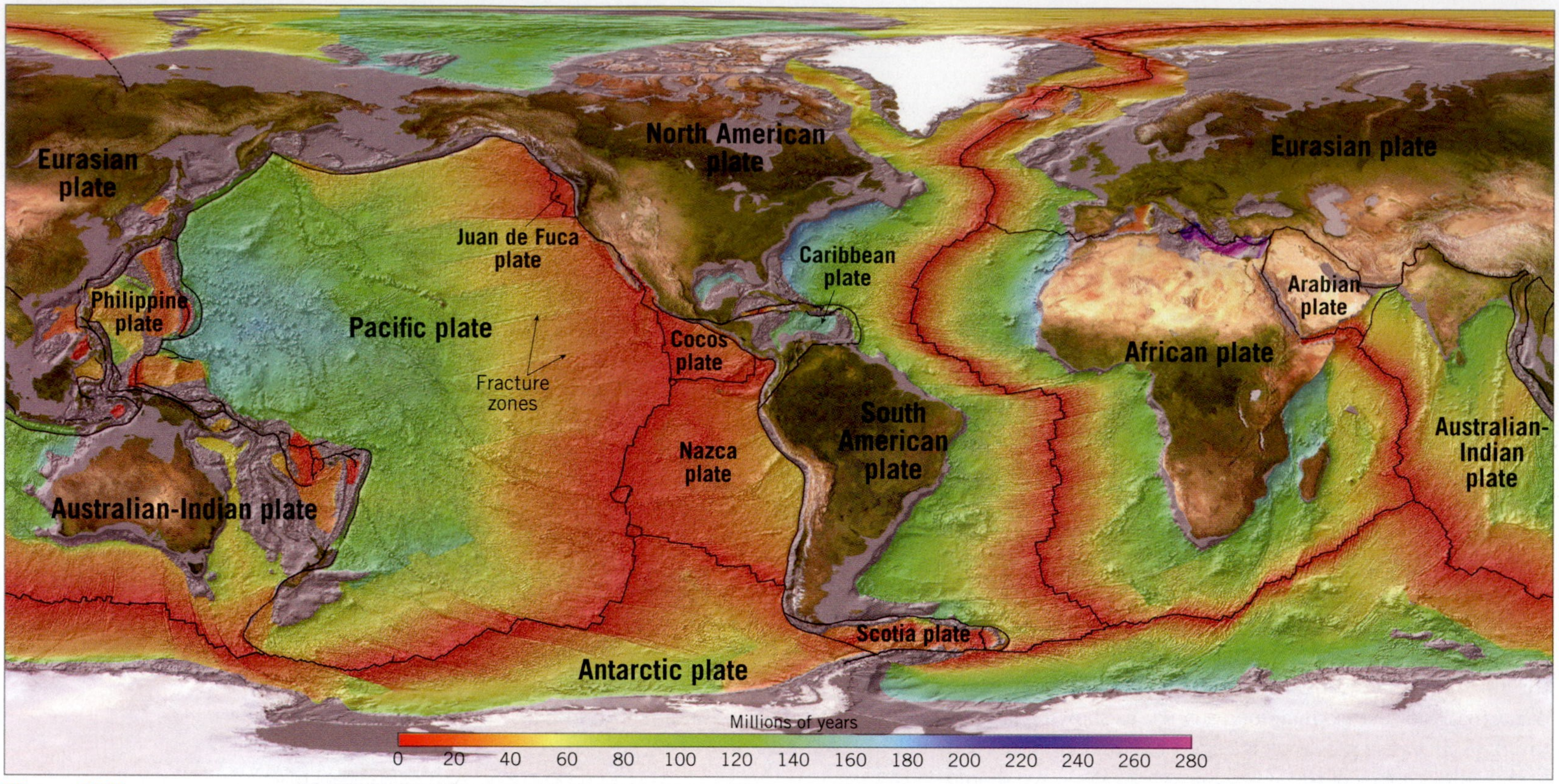

Figure 2.34
Age of the ocean floor

Millions of years

0 20 40 60 80 100 120 140 160 180 200 220 240 260 280

millimeters (about the diameter of a small pea). To establish plate motions, GPS data are collected at numerous sites repeatedly over a number of years.

Data obtained from these and other techniques are shown in **Figure 2.35**. Calculations show that Hawaii is moving in a northwesterly direction and approaching Japan at 8.3 centimeters per year. A location in Maryland is retreating from a location in England at a speed of 1.7 centimeters per year—a value close to the 2.0-centimeters-per-year spreading rate established from paleomagnetic evidence obtained for the North Atlantic. Techniques involving GPS devices have also been useful in

confirming small-scale crustal movements such as those that occur along faults in regions known to be tectonically active (for example, the San Andreas Fault).

2.10 | Concept Checks

1. What do transform faults that connect spreading centers indicate about plate motion?

2. Refer to Figure 2.35 and determine which three plates appear to exhibit the highest rates of motion.

Figure 2.35
Rates of plate motion
The red arrows show plate motion at selected locations based on GPS data. Longer arrows represent faster spreading rates. The small black arrows and labels show seafloor spreading velocities based mainly on paleomagnetic data. (Seafloor data from DeMets and others; GPS data from Jet Propulsion Laboratory)

Directions and rates of plate motions measured in centimeters per year

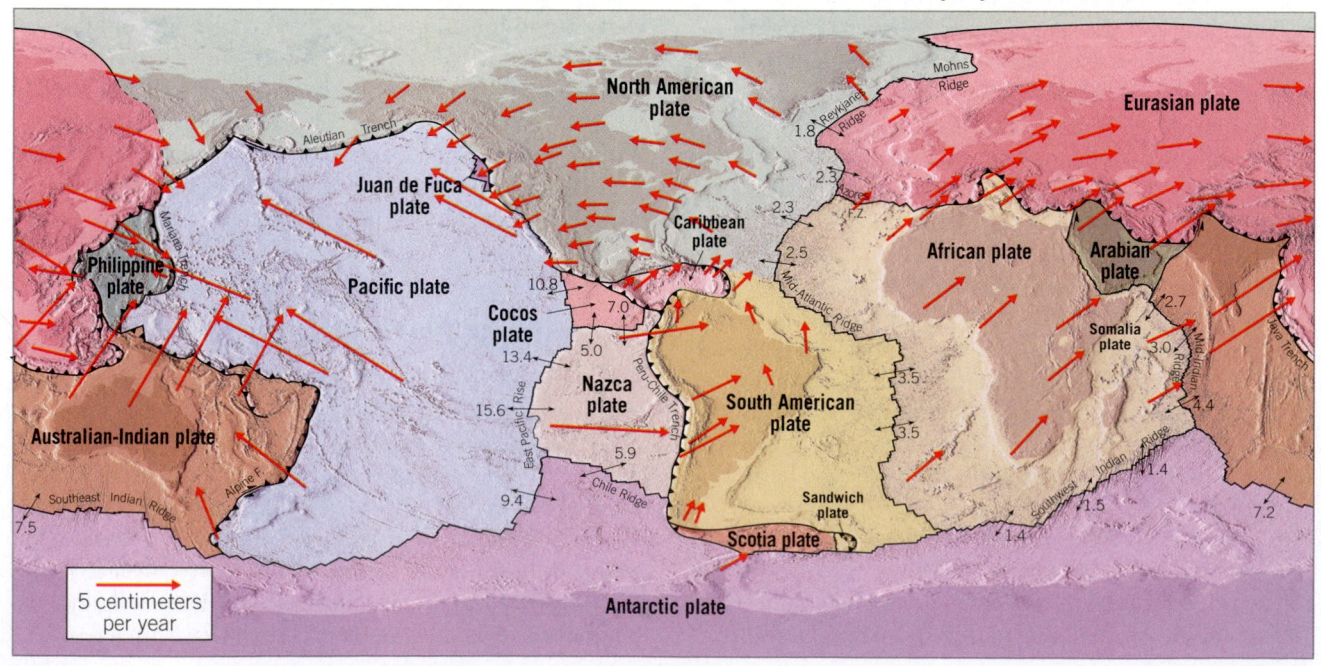

2.11 | What Drives Plate Motions?

Describe plate–mantle convection and explain two of the primary driving forces of plate motion.

Researchers are in general agreement that some type of *convection*—where hot mantle rocks rise and cold, dense oceanic lithosphere sinks—is the ultimate driver of plate tectonics. Many of the details of this convective flow, however, remain topics of debate in the scientific community.

Forces That Drive Plate Motion

Geophysical evidence confirms that although the mantle consists almost entirely of solid rock, it is hot and weak enough to exhibit a slow, fluid-like convective flow. The simplest type of **convection** is analogous to heating a pot of water on a stove (**Figure 2.36**). Heating the base of a pot causes the water to become less dense (more buoyant), causing it to rise in relatively thin sheets or blobs that spread out at the surface. As the surface layer cools, its density increases, and the cooler water sinks back to the bottom of the pot, where it is reheated until it achieves enough buoyancy to rise again. Mantle convection is similar to, but considerably more complex than, the model just described.

Geologists generally agree that subduction of cold, dense slabs of oceanic lithosphere is a major driving force of plate motion (**Figure 2.37**). This phenomenon, called **slab pull**, occurs because cold slabs of oceanic lithosphere are more dense than the underlying warm asthenosphere and hence "sink like a rock"—meaning that they are pulled down into the mantle by gravity.

Another important driving force is **ridge push** (see Figure 2.37). This gravity-driven mechanism results from the elevated position of the oceanic ridge, which causes slabs of lithosphere to "slide" down the flanks of the ridge. Ridge push appears to contribute far less to plate motions than slab pull. The primary evidence for this is that the fastest-moving plates—Pacific, Nazca, and Cocos plates—have extensive subduction zones along their margins. By contrast, the spreading rate in the North Atlantic basin, which is nearly devoid of subduction zones, is one of the lowest, at about 2.5 centimeters (1 inch) per year.

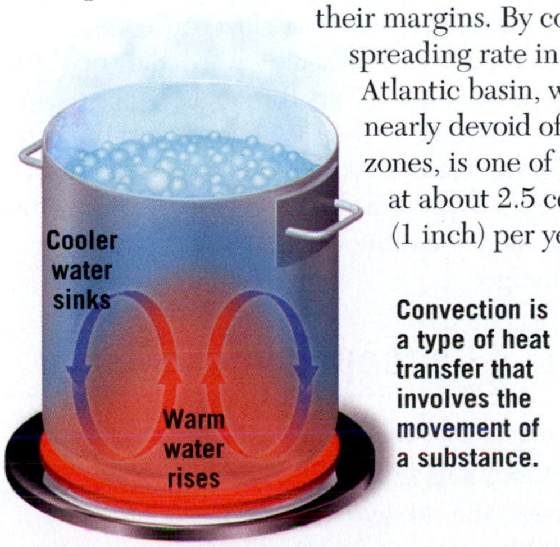

Cooler water sinks

Warm water rises

Convection is a type of heat transfer that involves the movement of a substance.

Figure 2.36 **Convection in a cooking pot** As a stove warms the water in the bottom of a cooking pot, the heated water expands, becomes less dense (more buoyant), and rises. Simultaneously, the cooler, denser water near the top sinks.

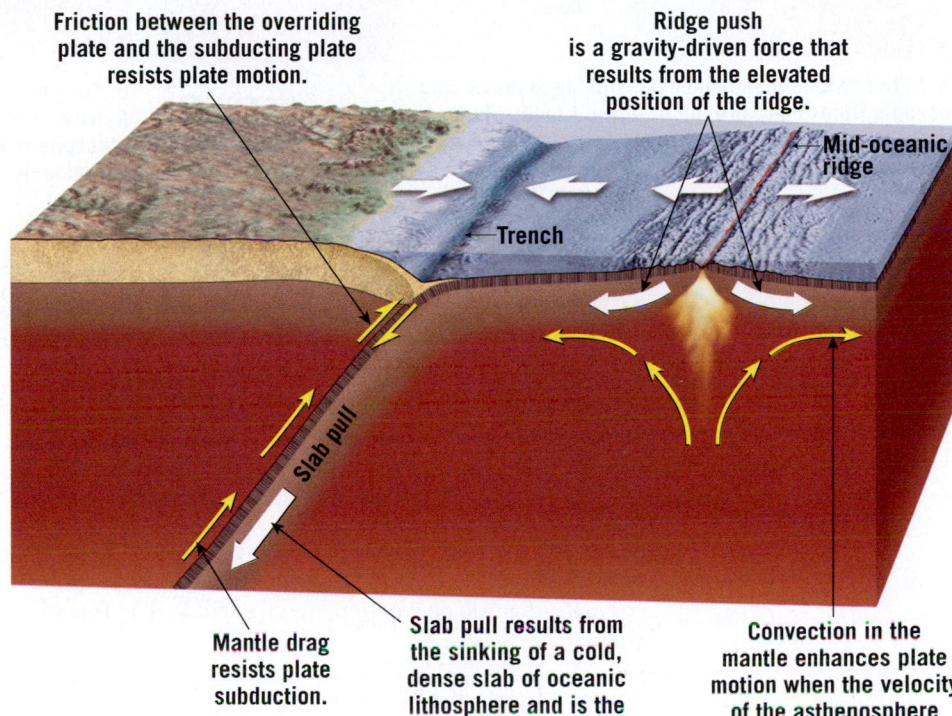

Friction between the overriding plate and the subducting plate resists plate motion.

Ridge push is a gravity-driven force that results from the elevated position of the ridge.

Mid-oceanic ridge

Trench

Slab pull

Mantle drag resists plate subduction.

Slab pull results from the sinking of a cold, dense slab of oceanic lithosphere and is the major driving force of plate motion.

Convection in the mantle enhances plate motion when the velocity of the asthenosphere exceeds that of the overlying plate.

Figure 2.37 **Forces that act on lithospheric plates**

Although the subduction of cold, dense lithospheric plates appears to be the dominant force acting on plates, other factors are also at work. Flow in the mantle, perhaps best described as "mantle drag," is also thought to affect plate motion (see Figure 2.37). When flow in the asthenosphere is moving at a velocity that exceeds that of the plate, mantle drag enhances plate motion. However, if the asthenosphere is moving more slowly than the plate, or if it is moving in the opposite direction, this force tends to resist plate motion. Another type of resistance to plate motion occurs along some subduction zones, where friction between the overriding plate and the descending slab generates significant earthquake activity.

Models of Plate–Mantle Convection

Although *convection* in the mantle has yet to be fully understood, researchers generally agree on the following:

- Convective flow in the rocky 2900-kilometer- (1800 mile-) thick mantle—in which warm, buoyant rock rises and cooler, denser material sinks—is the underlying driving force for plate movement.

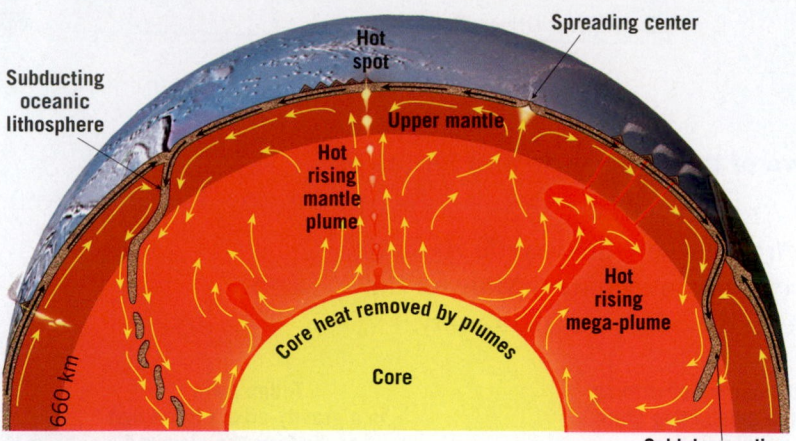

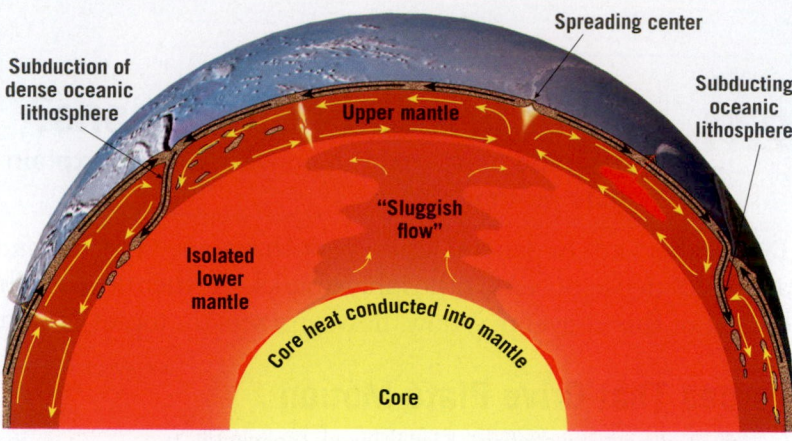

A. In the "whole-mantle model," sinking slabs of cold oceanic lithosphere are the downward limbs of convection cells, while rising mantle plumes carry hot material from the core–mantle boundary toward the surface.

B. The "layer cake model" has two largely disconnected convective layers. A dynamic upper layer driven by descending slabs of cold oceanic lithosphere and a sluggish lower layer that carries heat upward without appreciably mixing with the layer above.

Figure 2.38
Models of mantle convection

- Mantle convection and plate tectonics are part of the same system. Subducting oceanic plates drive the cold downward-moving portion of convective flow, while shallow upwelling of hot rock along the oceanic ridge and buoyant mantle plumes are the upward-flowing arms of the convective mechanism.
- Convective flow in the mantle is a major mechanism for transporting heat away from Earth's interior to the surface, where it is eventually radiated into space.

What is not known with certainty is the exact structure of this convective flow. Several models have been proposed for plate–mantle convection, and we will look at two of them.

Whole-Mantle Convection   One group of researchers favor some type of *whole-mantle convection* model, also called the *plume model*, in which cold oceanic lithosphere sinks to great depths and stirs the entire mantle (**Figure 2.38A**). The whole-mantle model suggests that the ultimate burial ground for these subducting lithospheric slabs is the core–mantle boundary. The downward flow of these subducting slabs is balanced by buoyantly rising mantle plumes that transport hot mantle rock toward the surface.

Two kinds of plumes have been proposed—narrow tube-like plumes and giant upwellings, often referred to as *mega-plumes*. The long, narrow plumes are thought to originate from the core–mantle boundary and produce hot-spot volcanism of the type associated with the Hawaiian Islands, Iceland, and Yellowstone. Scientists believe that areas of large mega-plumes, as shown in Figure 2.38A, occur beneath the Pacific basin and southern Africa. The latter structure is thought to explain why southern Africa has an elevation much higher than would be predicted for a stable continental landmass. In the whole-mantle convection model, heat for both types of plumes is thought to arise mainly from Earth's core, while the deep mantle provides a source for chemically distinct magmas. However, some researchers have questioned that idea and instead propose that the source of magma for most hot spot volcanism is found in the upper mantle (asthenosphere).

Layer Cake Model   Some researchers argue that the mantle resembles a "layer cake" divided at a depth of perhaps 660 kilometers (410 miles) but no deeper than 1000 kilometers (620 miles). As shown in **Figure 2.38B**, this layered model has two zones of convection—a thin, dynamic layer in the upper mantle and a thick, larger, sluggish one located below. As with the whole-mantle model, the downward convective flow is driven by the subduction of cold, dense oceanic lithosphere. However, rather than reach the lower mantle, these subducting slabs penetrate to depths of no more than 1000 kilometers (620 miles). Notice in Figure 2.38B that the upper layer in the layer cake model is littered with recycled oceanic lithosphere of various ages. Melting of these fragments is thought to be the source of magma for some of the volcanism that occurs away from plate boundaries, such as the hot-spot volcanism of Hawaii.

In contrast to the active upper mantle, the lower mantle is sluggish and does not provide material to support volcanism at the surface. Very slow convection within this layer likely carries heat upward, but very little mixing between these two layers is thought to occur.

Geologists continue to debate the nature of the convective flow in the mantle. As they investigate the possibilities, perhaps a hypothesis that combines features from the layer cake model and the whole-mantle convection model will emerge.

2.11   Concept Checks

1. Describe slab pull and ridge push. Which of these forces appears to contribute more to plate motion?

2. Briefly describe the two models of mantle–plate convection.

3. What geologic processes are associated with the upward and downward circulation in the mantle?

2.1 From Continental Drift to Plate Tectonics

Summarize the view that most geologists held prior to the 1960s regarding the geographic positions of the ocean basins and continents.

- Fifty years ago, most geologists thought that ocean basins were very old and that continents were fixed in place. Those ideas were discarded with a scientific revolution that revitalized geology: the theory of plate tectonics. Supported by multiple kinds of evidence, plate tectonics is the foundation of modern Earth science.

2.2 Continental Drift: An Idea Before Its Time

List and explain the evidence Wegener presented to support his continental drift hypothesis.

KEY TERMS continental drift, supercontinent, Pangaea

- German meteorologist Alfred Wegener formulated the idea of continental drift in 1912. He suggested that Earth's continents are not fixed in place but moved slowly over geologic time.
- Wegener proposed a supercontinent called Pangaea that existed about 200 million years ago, during the late Paleozoic and early Mesozoic.
- Wegener's evidence that Pangaea existed but later broke into pieces that drifted apart included (1) the shape of the continents, (2) continental fossil organisms that matched across oceans, (3) matching rock types and modern mountain belts, and (4) sedimentary rocks that recorded ancient climates, including glaciers on the southern portion of Pangaea.

Q Why did Wegener choose organisms such as *Glossopteris* and *Mesosaurus* as evidence for continental drift, as opposed to other fossil organisms such as sharks or jellyfish?

0   10 cm

0   5 inches

John Cancalosi

2.3 The Great Debate

Summarize the two main objections to the continental drift hypothesis.

- Wegener's hypothesis suffered from two flaws: It proposed tidal forces as the mechanism for the motion of continents, and it implied that the continents would have plowed their way through weaker oceanic crust, like a boat cutting through a thin layer of sea ice. Geologists rejected the idea of continental drift when Wegener proposed it, and it wasn't resurrected for another 50 years.

Q Today, we know that the early twentieth-century scientists who rejected the idea of continental drift were wrong. Would you consider them bad scientists? Why or why not?

2.4 The Theory of Plate Tectonics

List the major differences between Earth's lithosphere and asthenosphere and explain the importance of each in the plate tectonics theory.

KEY TERMS theory of plate tectonics, lithosphere, asthenosphere, lithospheric plate (plate)

- Research conducted after World War II led to new insights that helped revive Wegener's hypothesis of continental drift. Exploration of the seafloor revealed previously unknown features, including an extremely long mid-ocean ridge system. Sampling of the oceanic crust revealed that it was quite young relative to the continents.
- The lithosphere, Earth's outermost rocky layer, is broken into the plates of plate tectonics. It is relatively stiff and deforms by breaking or bending. Beneath the lithosphere is the asthenosphere, a relatively weak layer that deforms by flowing. The lithosphere consists both of crust (either oceanic or continental) and underlying upper mantle. Plates are usually a combination of both continental and oceanic lithosphere.
- There are seven large plates, another seven intermediate-size plates, and numerous relatively small microplates. Plates meet along boundaries that may either be divergent (moving apart from each other), convergent (moving toward each other), or transform (moving laterally past each other).

Q Compare and contrast Earth's lithosphere and asthenosphere.

2.5 Divergent Plate Boundaries and Seafloor Spreading

Sketch and describe the movement along a divergent plate boundary that results in the formation of new oceanic lithosphere.

KEY TERMS divergent plate boundary (spreading center), oceanic ridge system, rift valley, seafloor spreading, continental rift

- Seafloor spreading leads to the generation of new oceanic lithosphere at mid-ocean ridge systems. As two plates move apart from one another, tensional forces open cracks in the plates, allowing magma to well up and generate new slivers of seafloor. This process generates new oceanic lithosphere at a rate of 2 to 15 centimeters (1 to 6 inches) each year.
- As it ages, oceanic lithosphere cools and becomes denser. It therefore subsides as it is transported away from the mid-ocean ridge. At the same time, new material is added to its underside, causing the plate to grow thicker.
- Divergent boundaries are not limited to the seafloor. Continents can break apart, too, starting with a continental rift (like modern-day east Africa) and eventually leading to a new ocean basin opening between the two sides of the rift.

2.6 Convergent Plate Boundaries and Subduction

Compare and contrast the three types of convergent plate boundaries and name a location where each type can be found.

KEY TERMS convergent plate boundary (subduction zone), deep-ocean trench, partial melting, continental volcanic arc, volcanic island arc (island arc)

- When plates move toward one another, oceanic lithosphere is subducted into the mantle, where it is recycled. Subduction manifests itself on the ocean floor as a deep linear trench. The subducting slab of oceanic lithosphere can descend at a variety of angles, from nearly horizontal to nearly vertical.
- Aided by the presence of water, the subducted oceanic lithosphere triggers melting in the mantle, which produces magma. The magma is less dense than the surrounding rock and will rise. It may cool at depth, thickening the crust, or it may make it all the way to Earth's surface, where it erupts as a volcano.
- A line of volcanoes that emerge through continental crust is termed a continental volcanic arc, while a line of volcanoes that emerge through an overriding plate of oceanic lithosphere is a volcanic island arc.
- Continental crust resists subduction due to its relatively low density, and so when an intervening ocean basin is completely destroyed through subduction, the continents on either side collide, generating a new mountain range.

Q Sketch a typical continental volcanic arc and label the key parts. Then repeat the drawing with an overriding plate made of oceanic lithosphere.

2.7 Transform Plate Boundaries

Describe the relative motion along a transform fault boundary and be able to locate several examples on a plate boundary map.

KEY TERMS transform plate boundary (transform fault), fracture zone

- At a transform boundary, lithospheric plates slide horizontally past one another. No new lithosphere is generated, and no old lithosphere is consumed. Shallow earthquakes signal the movement of these slabs of rock as they grind past their neighbors.
- The San Andreas Fault in California is an example of a transform boundary in continental crust, while the fracture zones between segments of the Mid-Atlantic Ridge are transform faults in oceanic crust.

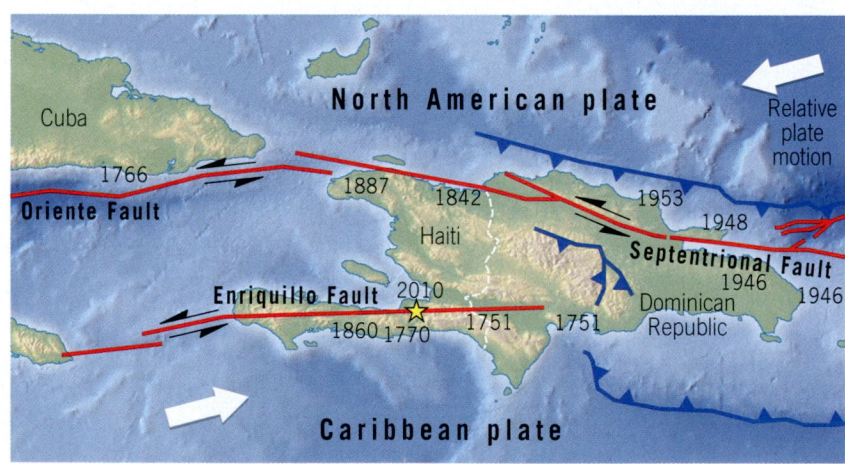

Q On the accompanying tectonic map of the Caribbean, find the Enriquillo fault. (The location of the 2010 Haiti earthquake is shown as a yellow star.) What kind of plate boundary is shown here? Are there any other faults in the area that show the same type of motion?

2.8 How Do Plates and Plate Boundaries Change?

Explain why plates such as the African and Antarctic plates are increasing in size, while the Pacific plate is decreasing in size.

- Although the total surface area of Earth does not change, the shape and size of individual plates are constantly changing as a result of subduction and seafloor spreading. Plate boundaries can also be created or destroyed in response to changes in the forces acting on the lithosphere.
- The breakup of Pangaea and the collision of India with Eurasia are two examples of how plates change through geologic time.

2.9 Testing the Plate Tectonics Model

List and explain the evidence used to support the plate tectonics theory.

KEY TERMS mantle plume, hot spot, hot-spot track, Curie point, paleomagnetism (fossil magnetism), magnetic reversal, normal polarity, reverse polarity, magnetic time scale, magnetometer

- Multiple lines of evidence have verified the plate tectonics model. For instance, the Deep Sea Drilling Project found that the age of the seafloor increases with distance from a mid-ocean ridge. The thickness of sediment atop this seafloor is also proportional to distance from the ridge: Older lithosphere has had more time to accumulate sediment.
- Overall, oceanic lithosphere is quite young, with none older than about 180 million years.

- A hot spot is an area of volcanic activity where a mantle plume reaches Earth's surface. Volcanic rocks generated by hot-spot volcanism provide evidence of both the direction and rate of plate movement over time.
- Magnetic minerals such as magnetite align themselves with Earth's magnetic field as rock forms. These fossil magnets act as recorders of the ancient orientation of Earth's magnetic field. This is useful to geologists in two ways: (1) It allows a given stack of rock layers to be interpreted in terms of changing position relative to the magnetic poles through time, and (2) reversals in the orientation of the magnetic field are preserved as "stripes" of normal and reversed polarity in the oceanic crust. Magnetometers reveal this signature of seafloor spreading as a symmetrical pattern of magnetic stripes parallel to the axis of the mid-ocean ridge.

2.10 How Is Plate Motion Measured?

Describe two methods researchers use to measure relative plate motion.

- Data collected from the ocean floor has established the direction and rate of motion of lithospheric plates. Transform faults point in the direction a plate is moving. Sediments with diagnostic fossils and radiometric dates of igneous rocks both help to calibrate the rate of motion.

- GPS satellites can be used to accurately measure the motion of special receivers to within a few millimeters. These "real-time" data support the inferences made from seafloor observations. Plates move at about the same rate human fingernails grow: an average of about 5 centimeters (2 inches) per year.

2.11 What Drives Plate Motions?

Describe plate–mantle convection and explain two of the primary driving forces of plate motion.

KEY TERMS convection, slab pull, ridge push

- Some kind of convection (upward movement of less dense material and downward movement of more dense material) appears to drive the motion of plates.
- Slabs of oceanic lithosphere sink at subduction zones because the subducted slab is denser than the underlying asthenosphere. In this process, called slab pull, Earth's gravity tugs at the slab, drawing the rest of the plate toward the subduction zone. As lithosphere slides down the mid-ocean ridge, it exerts a small additional force, the outward-directed ridge push. Frictional drag exerted on the underside of plates by flowing mantle is another force that acts on plates and influences their direction and rate of movement.
- The exact patterns of mantle convection are not fully understood. Convection may occur throughout the entire mantle, as suggested by the whole-mantle model. Or it may occur in two layers within the mantle—the active upper mantle and the sluggish lower mantle—as proposed in the layer cake model.

Steve Bower/Shutterstock

Q Compare and contrast mantle convection with the operation of a lava lamp.

Give It Some Thought

1. After referring to the section in Chapter 1 titled "The Nature of Scientific Inquiry," answer the following:

   a. What observation led Alfred Wegener to develop his continental drift hypothesis?

   b. Why did the majority of the scientific community reject the continental drift hypothesis?

   c. Do you think Wegener followed the basic principles of scientific inquiry? Support your answer.

2. Referring to the accompanying diagrams that illustrate the three types of convergent plate boundaries, complete the following:

   a. Identify each type of convergent boundary.

   b. On what type of crust do volcanic island arcs develop?

   c. Why are volcanoes largely absent where two continental blocks collide?

   d. Describe two ways that oceanic–oceanic convergent boundaries are different from oceanic–continental boundaries. How are they similar?

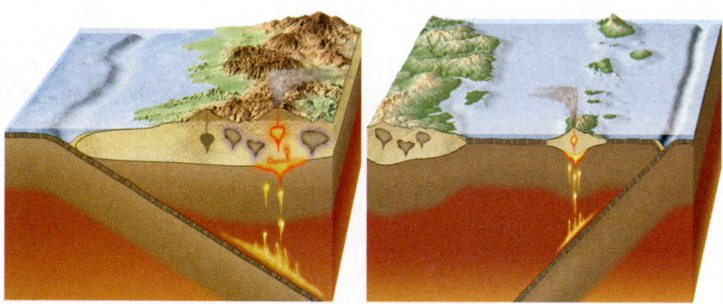

A.   B.

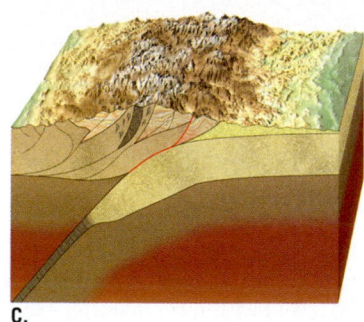

C.

3. Some predict that California will sink into the ocean. Is this idea consistent with the theory of plate tectonics? Explain.

4. Refer to the accompanying hypothetical plate map to answer the following questions:

   a. How many portions of plates are shown?

   b. Are continents A, B, and C moving toward or away from each other? How did you determine your answer?

   c. Explain why active volcanoes are more likely to be found on continents A and B than on continent C.

   d. Provide one scenario in which volcanic activity might be triggered on continent C.

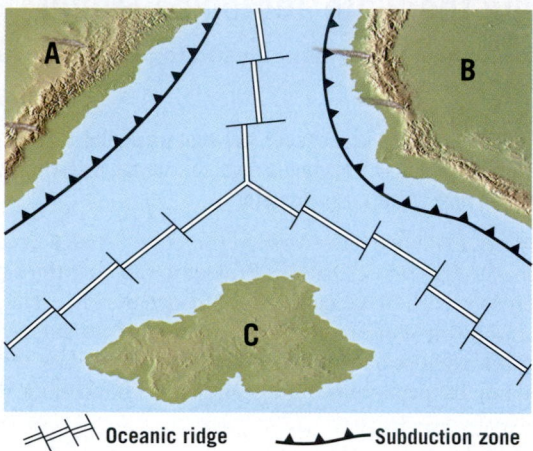

⊬⊬ Oceanic ridge   ▲▲▲ Subduction zone

5. Volcanoes, such as the Hawaiian chain, that form over mantle plumes are some of the largest shield volcanoes on Earth. However, several shield volcanoes on Mars are gigantic compared to those on Earth. What does this difference tell us about the role of plate motion in shaping the Martian surface?

6. Imagine that you are studying seafloor spreading along two different oceanic ridges. Using data from a magnetometer, you produced the two accompanying maps. From these maps, what can you determine about the relative rates of seafloor spreading along these two ridges? Explain.

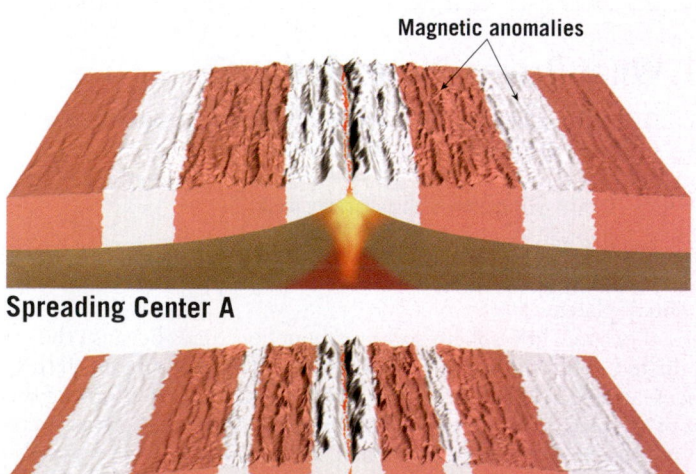

Magnetic anomalies

Spreading Center A

Spreading Center B

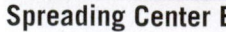

7. Australian marsupials (kangaroos, koala bears, etc.) have direct fossil links to marsupial opossums found in the Americas. Yet the modern marsupials in Australia are markedly different from their American relatives. How does the breakup of Pangaea help to explain these differences (see Figure 2.23)?

8. Density is a key component in the behavior of Earth materials and is especially important in understanding key aspects of plate tectonics. Describe three different ways that density and/or density differences play a role in plate tectonics.

9. Explain how the processes that create hot-spot volcanic chains differ from the processes that generate volcanic island arcs.

10. Refer to the accompanying map and the pairs of cities below to complete the following:

(Boston, Denver), (London, Boston), (Honolulu, Beijing)

a. Which pair of cities is moving apart as a result of plate motion?

b. Which pair of cities is moving closer as a result of plate motion?

c. Which pair of cities is not presently moving relative to each other?

Plate motion measured in centimeters per year

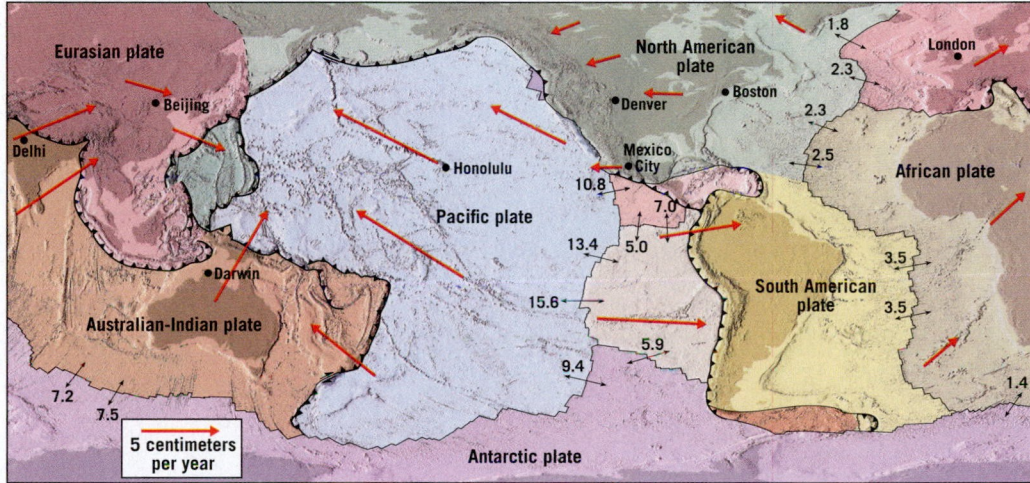

MasteringGeology™

www.masteringgeology.com

Looking for additional review and test prep materials? With individualized coaching on the toughest topics of the course, MasteringGeology offers a wide variety of ways for you to move beyond memorization to begin thinking like a geologist. Visit the Study Area in www.masteringgeology.com to find practice quizzes, study tools, and multimedia that will improve your understanding of this chapter's content. Sign in today to enjoy the following features: **Self Study Quizzes, SmartFigures: Tutorials/Animations/Condor Videos/Mobile Field Trips, Geoscience Animation Library, GEODe, RSS Feeds, Digital Study Modules,** and an optional **Pearson eText.**

3
Matter and Minerals

The Cave of Crystals, Chihuahua, Mexico, contains giant gypsum crystals, some of the largest natural crystals ever found. (Photo by Carsten Peter/Speleoresearch & Films/National Geographic Stock/Getty Images)

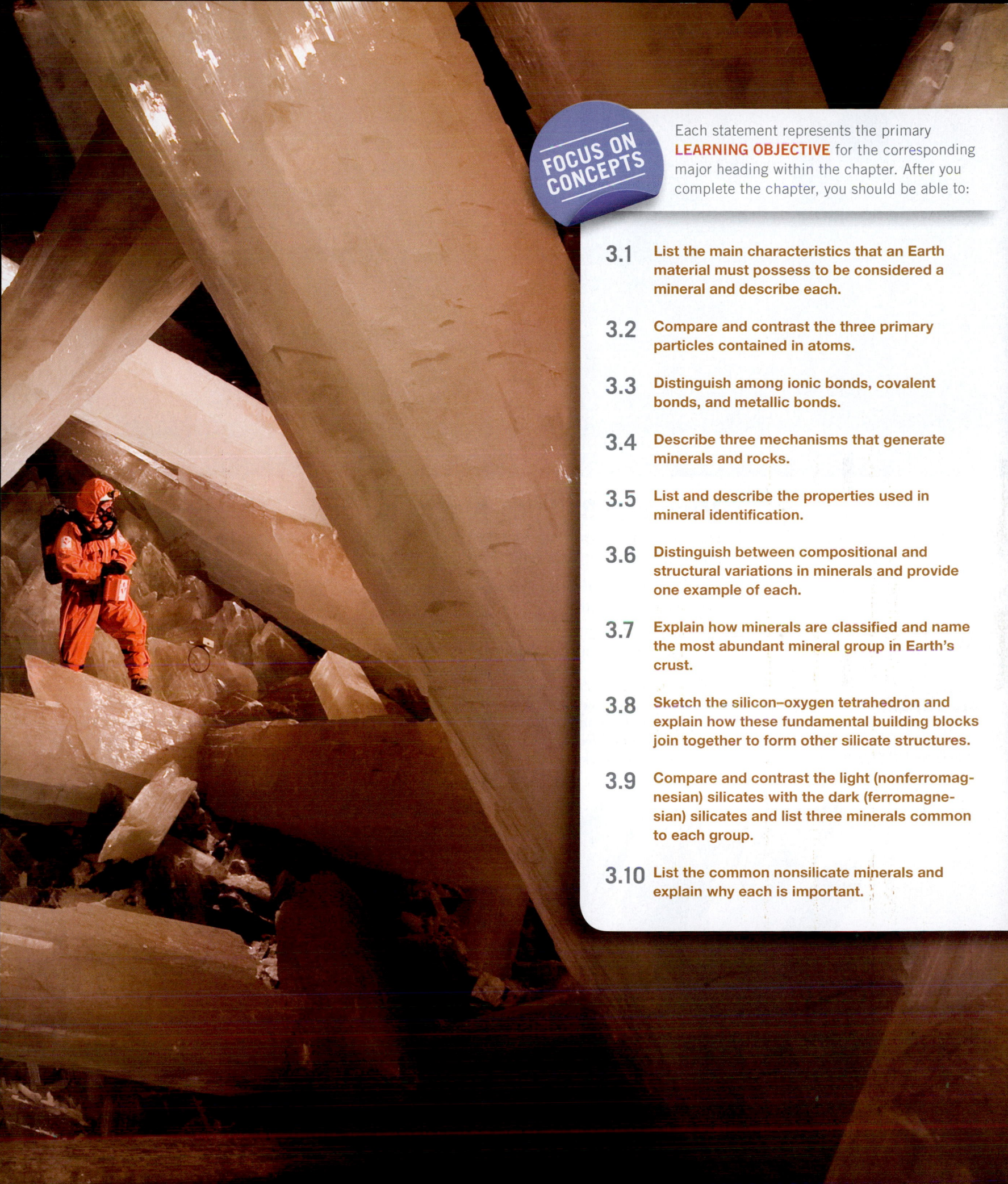

FOCUS ON CONCEPTS

Each statement represents the primary **LEARNING OBJECTIVE** for the corresponding major heading within the chapter. After you complete the chapter, you should be able to:

3.1 List the main characteristics that an Earth material must possess to be considered a mineral and describe each.

3.2 Compare and contrast the three primary particles contained in atoms.

3.3 Distinguish among ionic bonds, covalent bonds, and metallic bonds.

3.4 Describe three mechanisms that generate minerals and rocks.

3.5 List and describe the properties used in mineral identification.

3.6 Distinguish between compositional and structural variations in minerals and provide one example of each.

3.7 Explain how minerals are classified and name the most abundant mineral group in Earth's crust.

3.8 Sketch the silicon–oxygen tetrahedron and explain how these fundamental building blocks join together to form other silicate structures.

3.9 Compare and contrast the light (nonferromagnesian) silicates with the dark (ferromagnesian) silicates and list three minerals common to each group.

3.10 List the common nonsilicate minerals and explain why each is important.

Earth's crust and oceans are home to a wide variety of useful and essential minerals. Most people are familiar with the common uses of many basic metals, including aluminum in beverage cans, copper in electrical wiring, and gold and silver in jewelry. But some people are not aware that pencil "lead" contains the greasy-feeling mineral graphite and that bath powders and many cosmetics contain the mineral talc. Moreover, many do not know that dentists use drill bits impregnated with diamonds to drill through tooth enamel. In fact, practically every manufactured product contains materials obtained from minerals.

In addition to the economic uses of rocks and minerals, every process that geologists study in some way depends on the properties of these basic Earth materials. Events such as volcanic eruptions, mountain building, weathering and erosion, and even earthquakes involve rocks and minerals. Consequently, a basic knowledge of Earth materials is essential to understanding all geologic phenomena.

3.1 | Minerals: Building Blocks of Rocks

List the main characteristics that an Earth material must possess to be considered a mineral and describe each.

We begin our discussion of Earth materials with an overview of **mineralogy** (*mineral* = mineral, *ology* = study of) because minerals are the building blocks of rocks. In addition, humans have used minerals for both practical and decorative purposes for thousands of years. For example, the common mineral quartz is the source of silicon for computer chips (**Figure 3.1**).

Figure 3.1
Quartz crystals
A collection of well-developed quartz crystals found near Hot Springs, Arkansas. (Photo by Jeff Scovil)

The first minerals mined were flint and chert, which people fashioned into weapons and cutting tools. As early as 3700 B.C.E., Egyptians began mining gold, silver, and copper. By 2200 B.C.E., humans had discovered how to combine copper with tin to make bronze—a strong, hard alloy. Later, humans developed a process to extract iron from minerals such as hematite—a discovery that marked the decline of the Bronze Age. During the Middle Ages, mining of a variety of minerals became common, and the impetus for the formal study of minerals was in place.

The term *mineral* is used in several different ways. For example, those concerned with health and fitness extol the benefits of vitamins and minerals. The mining industry typically uses the word *mineral* to refer to anything extracted from Earth, such as coal, iron ore, or sand and gravel. The guessing game *Twenty Questions* usually begins with the question *Is it animal, vegetable, or mineral?* What criteria do geologists use to determine whether something is a mineral?

Defining a Mineral

Geologists define **mineral** as *any naturally occurring inorganic solid that possesses an orderly crystalline structure and a definite chemical composition that allows for some variation.* Thus, Earth materials that are classified as minerals exhibit the following characteristics:

A.   **B.**

Figure 3.2 Ordered versus unordered arrangement of atoms Minerals **A.** have a repetitive, orderly arrangement of atoms, whereas substances such a glass **B.** have an unordered atomic structure.

1. **Naturally occurring.** Minerals form by natural geologic processes. Synthetic materials, meaning those produced in a laboratory or by human intervention, are not considered minerals.

2. **Generally inorganic.** Inorganic crystalline solids, such as ordinary table salt (halite), that are found naturally in the ground are considered minerals. (Organic compounds, on the other hand, are generally not. Sugar, a crystalline solid like salt but extracted from sugarcane or sugar beets, is a common example of such an organic compound.) Many marine animals secrete inorganic compounds, such as calcium carbonate (calcite), in the form of shells and coral reefs. If these materials are buried and become part of the rock record, geologists consider them minerals.

3. **Solid substance.** Only solid crystalline substances are considered minerals. Ice (frozen water) fits this criterion and is considered a mineral, whereas liquid water and water vapor do not. The exception is mercury, which is found in its liquid form in nature.

4. **Orderly crystalline structure.** Minerals are crystalline substances, made up of atoms (or ions) that are arranged in an orderly, repetitive manner (**Figure 3.2**). This orderly packing of atoms is reflected in the regularly shaped objects called *crystals*. Some naturally occurring solids, such as volcanic glass (obsidian), lack a repetitive atomic structure and are not considered minerals.

5. **Definite chemical composition that allows for some variation.** Minerals are chemical compounds having compositions that can be expressed by a chemical formula. For example, the common mineral quartz has the formula SiO_2, which indicates that quartz consists of silicon (Si) and oxygen (O) atoms in a 1:2 ratio. This proportion of silicon to oxygen is true for any sample of pure quartz, regardless of its origin. However, the compositions of some minerals vary *within specific, well-defined limits*. This occurs because certain elements can substitute for others of similar size without changing the mineral's internal structure.

What Is a Rock?

In contrast to minerals, rocks are more loosely defined. Simply, a **rock** is any solid mass of mineral, or mineral-like, matter that occurs naturally as part of our planet. Most rocks, like the sample of granite shown in **Figure 3.3**, occur as aggregates of several different minerals. The term *aggregate* implies that the minerals are joined in such a way that their individual properties are retained. Note that the different minerals that make up granite can be easily identified. However, some rocks are composed almost entirely of one mineral. A common example is the sedimentary rock *limestone*, which is an impure mass of the mineral calcite.

In addition, some rocks are composed of nonmineral matter. These include the volcanic rocks *obsidian* and *pumice*, which are noncrystalline glassy substances, and *coal*, which consists of solid organic debris.

Although this chapter deals primarily with the nature of minerals, keep in mind that most rocks are simply aggregates of minerals. Because the properties of rocks are determined largely by the chemical composition and crystalline structure of the minerals contained within them, we will first consider these Earth materials.

3.1 | Concept Checks

1. List five characteristics of a mineral.

2. Based on the definition of a mineral, which of the following—gold, water, synthetic diamonds, ice, and wood materials—are *not* classified as minerals?

3. Define the term *rock*. How do rocks differ from minerals?

Granite (Rock)

Quartz (Mineral)

Hornblende (Mineral)

Feldspar (Mineral)

SmartFigure 3.3 Most rocks are aggregates of minerals Shown here is a hand sample of the igneous rock granite and three of its major constituent minerals. (Photos by E. J. Tarbuck) (https://goo.gl/7cZXyr)

Tutorial

EYE ON EARTH 3.1

The accompanying image is of the world's largest open pit gold mine, located near Kalgoorlie, Australia. Known as the Super Pit, it originally consisted of a number of small underground mines that were consolidated into a single, open pit mine. Each year, about 28 metric tons of gold are extracted from the 15 million tons of rock shattered by blasting and then transported to the surface.

QUESTION 1 What is one environmental advantage that underground mining has over open pit mining?

QUESTION 2 If you had been employed at this mine, what change in working conditions would have occurred as it evolved from an underground mine to an open pit mine?

McPHOTO/AGE Fotostock

3.2 | Atoms: Building Blocks of Minerals
Compare and contrast the three primary particles contained in atoms.

When minerals are carefully examined, even under optical microscopes, the innumerable tiny particles of their internal structures are not visible. Nevertheless, scientists have discovered that all matter, including minerals, is composed of minute building blocks called **atoms**—the smallest particles that cannot be chemically split. Atoms, in turn, contain even smaller particles: *protons* and *neutrons* located in a central **nucleus** that is surrounded by *electrons* (**Figure 3.4**).

Properties of Protons, Neutrons, and Electrons

Protons and **neutrons** are very dense particles with almost identical masses. By contrast, **electrons** have a negligible mass, about 1/2000 that of a proton. To illustrate this difference, imagine a scale where a proton has the mass of a baseball and an electron has the mass of a single grain of rice.

Both protons and electrons share a fundamental property, called *electrical charge*. Protons have an electrical charge of +1, and electrons have a charge of –1. Neutrons, as the name suggests, have no charge. The

Figure 3.4 Two models of an atom A. Simplified view of an atom consisting of a central nucleus composed of protons and neutrons encircled by high-speed electrons. **B.** This model of an atom shows spherically shaped electron clouds (shells) surrounding a central nucleus. The nucleus contains virtually all of the mass of the atom. The remainder of the atom is the space occupied by negatively charged electrons. (The relative sizes of the nuclei shown are greatly exaggerated.)

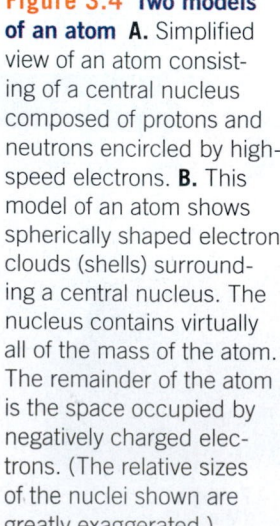

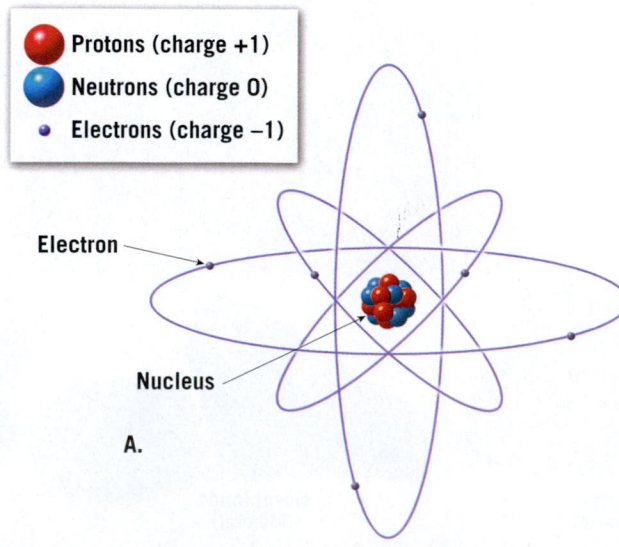

- Protons (charge +1)
- Neutrons (charge 0)
- Electrons (charge –1)

Electron

Nucleus

A.

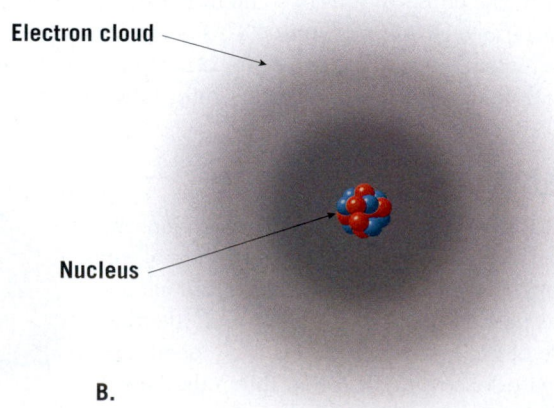

Electron cloud

Nucleus

B.

charges of protons and electrons are equal in magnitude but opposite in polarity, so when these two particles are paired, the charges cancel each other out. Since matter typically contains equal numbers of positively charged protons and negatively charged electrons, most substances are electrically neutral.

Illustrations sometimes show electrons orbiting the nucleus in a manner that resembles the planets of our solar system orbiting the Sun (see Figure 3.4A). However, electrons do not actually behave this way. A more realistic depiction is to show electrons as a cloud of negative charges surrounding the nucleus (see Figure 3.4B). Studies of the arrangements of electrons show that they move about the nucleus in regions called *principal shells*, each with an associated energy level. In addition, each shell can hold a specific number of electrons, with the outermost shell containing **valence electrons**. These electrons can be transferred to or shared with other atoms to form chemical bonds.

Most of the atoms in the universe (except hydrogen and helium) were created inside massive stars by nuclear fusion and released into interstellar space during hot, fiery supernova explosions. As this ejected material cooled, the newly formed nuclei attracted electrons to complete their atomic structure. At

the temperatures found at Earth's surface, all free atoms (those not bonded to other atoms) have a full complement of electrons—one for each proton in the nucleus.

Elements: Defined by Their Number of Protons

The simplest atoms have only 1 proton in their nuclei, whereas others have more than 100. The number of protons in the nucleus of an atom, called the **atomic number**, determines its chemical nature. All atoms with the same number of protons have the same chemical and physical properties. Together, a group of the same kind of atoms is called an **element**. There are about 90 naturally occurring elements and several more that have been synthesized in the laboratory. You are probably familiar with the names of many elements, including carbon, nitrogen, and oxygen. All carbon atoms have six protons, whereas all nitrogen atoms have seven protons, and all oxygen atoms have eight.

Scientists have organized elements so those with similar properties line up in columns, referred to as groups. This arrangement, called the **periodic table**, is shown in **Figure 3.5**. Each element is assigned a one- or

Figure 3.5
Periodic table of the elements

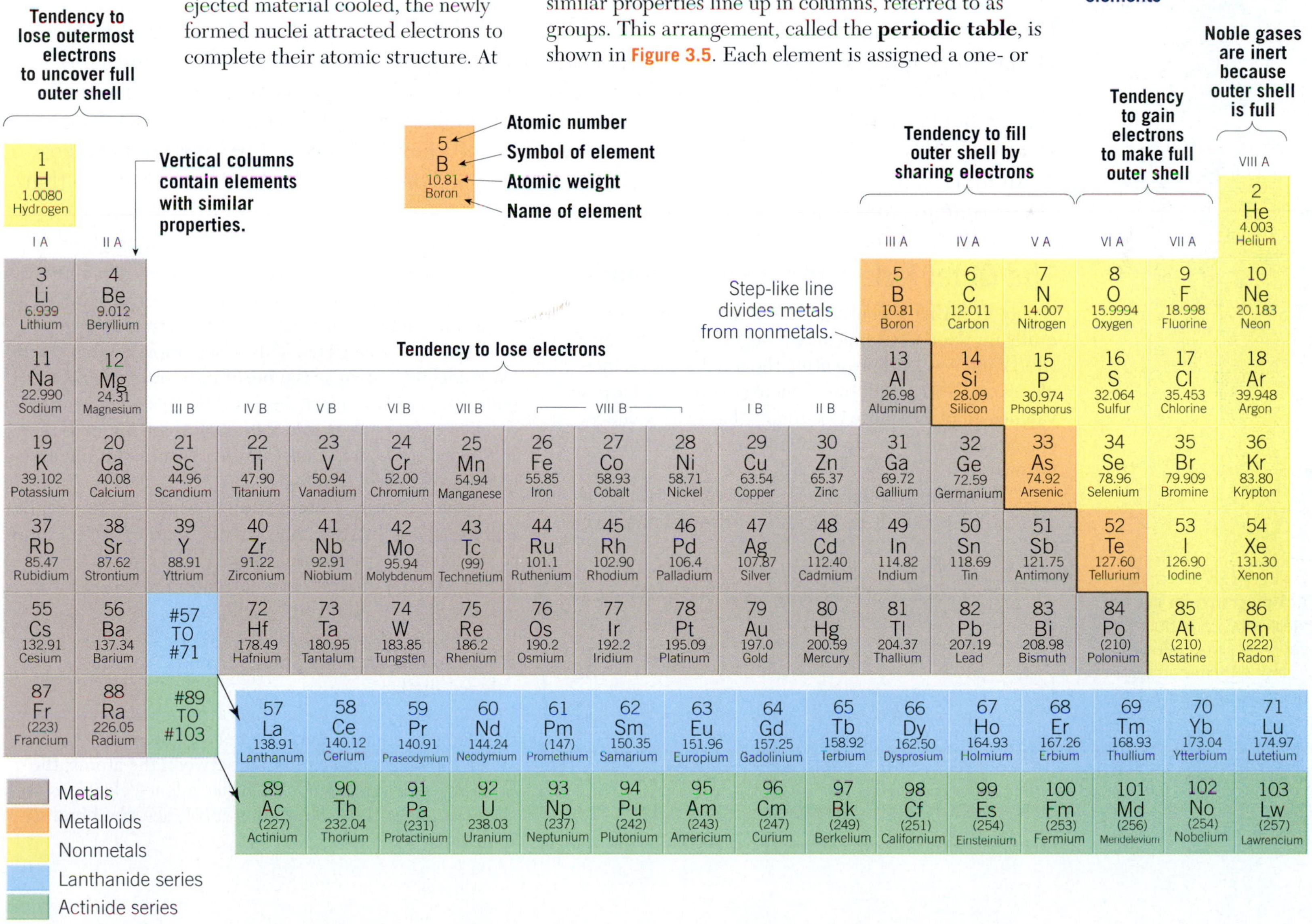

A. Gold on quartz

B. Sulfur

C. Copper

These include the minerals quartz (SiO_2), halite (NaCl), and calcite ($CaCO_3$). However, a few minerals, such as native (occurring in pure form in nature) copper, diamonds, sulfur, and gold, are made entirely of atoms of only one element (**Figure 3.6**).

two-letter symbol. The atomic numbers and masses for each element are also shown on the periodic table.

Atoms of the naturally occurring elements are the basic building blocks of Earth's minerals. Most elements join with atoms of other elements to form **chemical compounds**. Therefore, most minerals are chemical compounds composed of atoms of two or more elements.

3.2 | Concept Checks

1. Make a simple sketch of an atom and label its three main particles. Explain how these particles differ from one another.

2. What is the significance of valence electrons?

3.3 | Why Atoms Bond

Distinguish among ionic bonds, covalent bonds, and metallic bonds.

Except for a group of elements known as the noble gases, atoms bond to one another under the temperature and pressure conditions that occur on Earth. Some atoms bond to form *ionic compounds*, some form *molecules*, and still others form *metallic substances*. Why does this happen? Experiments show that electrical forces hold atoms together and bond them to each other. These electrical attractions lower the total energy of the bonded atoms, which, in turn, generally makes them more stable. Consequently, atoms that are bonded in compounds tend to be more stable than atoms that are free (not bonded).

The Octet Rule and Chemical Bonds

As noted earlier, valence (outer shell) electrons are generally involved in chemical bonding. **Figure 3.7** shows a shorthand way of representing the number of valence electrons. Notice that the elements in Group I have one valence electron, those in Group II have two valence electrons, and so on, up to eight valence electrons in Group VIII.

The noble gases have very stable electron arrangements with eight valence electrons (except helium which

has two) and, therefore, tend to lack chemical reactivity. Many other atoms gain, lose, or share electrons during chemical reactions, ending up with electron arrangements of the noble gases. This observation led to a chemical guideline known as the **octet rule**: *Atoms tend to gain, lose, or share electrons until they are surrounded by eight valence electrons.* Although there are exceptions to the octet rule, it is a useful rule of thumb for understanding chemical bonding.

When an atom's outer shell does not contain eight electrons, it is likely to chemically bond to other atoms to fill its shell. A **chemical bond** is a transfer or sharing of electrons that allows each atom to attain a full valence shell of electrons. Some atoms do this by transferring all their valence electrons to other atoms, so that an inner shell becomes the full valence shell.

When the valence electrons are transferred between the elements to form ions, the bond is an *ionic bond*. When the electrons are shared between the atoms, the bond is a *covalent bond*. When the valence electrons are shared among all the atoms in a substance, the bonding is *metallic*.

Electron Dot Diagrams for Some Representative Elements

I	II	III	IV	V	VI	VII	VIII
H·							He:
Li·	·Be·	·B·	·C·	·N·	:O·	:F·	:Ne:
Na·	·Mg·	·Al·	·Si·	·P·	:S·	:Cl·	:Ar:
K·	·Ca·	·Ga·	·Ge·	·As·	:Se·	:Br·	:Kr:

Ionic Bonds: Electrons Transferred

Perhaps the easiest type of bond to visualize is the **ionic bond**, in which one atom gives up one or more valence electrons to another atom to form **ions**—*positively and negatively charged atoms*. The atom that loses electrons becomes a positive ion, and the atom that gains electrons becomes a negative ion. Oppositely charged ions are strongly attracted to one another and join to form ionic compounds.

Consider the ionic bonding that occurs between sodium (Na) and chlorine (Cl) to produce the solid ionic compound sodium chloride—the mineral halite (common table salt). Notice in **Figure 3.8A** that a sodium atom gives up its single valence electron to chlorine and, as a result, becomes a positively charged sodium ion. Chlorine, on the other hand, gains one electron and becomes a negatively charged chloride ion. We know that ions having unlike charges attract. Thus, an ionic bond is the attraction of oppositely charged ions to one another, and it produces an electrically neutral ionic compound.

Figure 3.8B illustrates the arrangement of sodium and chlorine ions in ordinary table salt. Notice that salt consists of alternating sodium and chlorine ions, positioned in such a manner that each positive ion is attracted to and surrounded on all sides by negative ions and vice versa. This arrangement maximizes the attraction between ions with opposite charges while minimizing the repulsion between ions with identical charges. Thus, ionic compounds consist of an orderly arrangement of oppositely charged ions assembled in a definite ratio that provides overall electrical neutrality.

The properties of a chemical compound are dramatically different from the properties of the various elements comprising it. For example, sodium is a soft silvery metal that is extremely reactive and poisonous. If you were to consume even a small amount of elemental sodium, you would need immediate medical attention. Chlorine, a green poisonous gas, is so toxic that it was used as a chemical weapon during World War I. Together, however, these elements produce sodium chloride, a harmless flavor enhancer that we call table salt. Thus, when elements combine to form compounds, their properties change significantly.

Covalent Bonds: Electron Sharing

Sometimes the forces that hold atoms together cannot be understood on the basis of the attraction of oppositely charged ions. One example is the hydrogen molecule (H_2), in which the two hydrogen atoms are held together tightly and no ions are present. The strong attractive force that holds two hydrogen atoms together results from a **covalent bond**, a chemical bond formed by the *sharing* of one or more valence electrons between a pair of atoms.

Imagine two hydrogen atoms (each with one proton and one electron) approaching one another, as shown in **Figure 3.9**. Once they meet, the electron configuration will change so that both electrons will primarily occupy the space between the atoms. In other words, the two electrons are shared by both hydrogen atoms and are attracted simultaneously by the positive

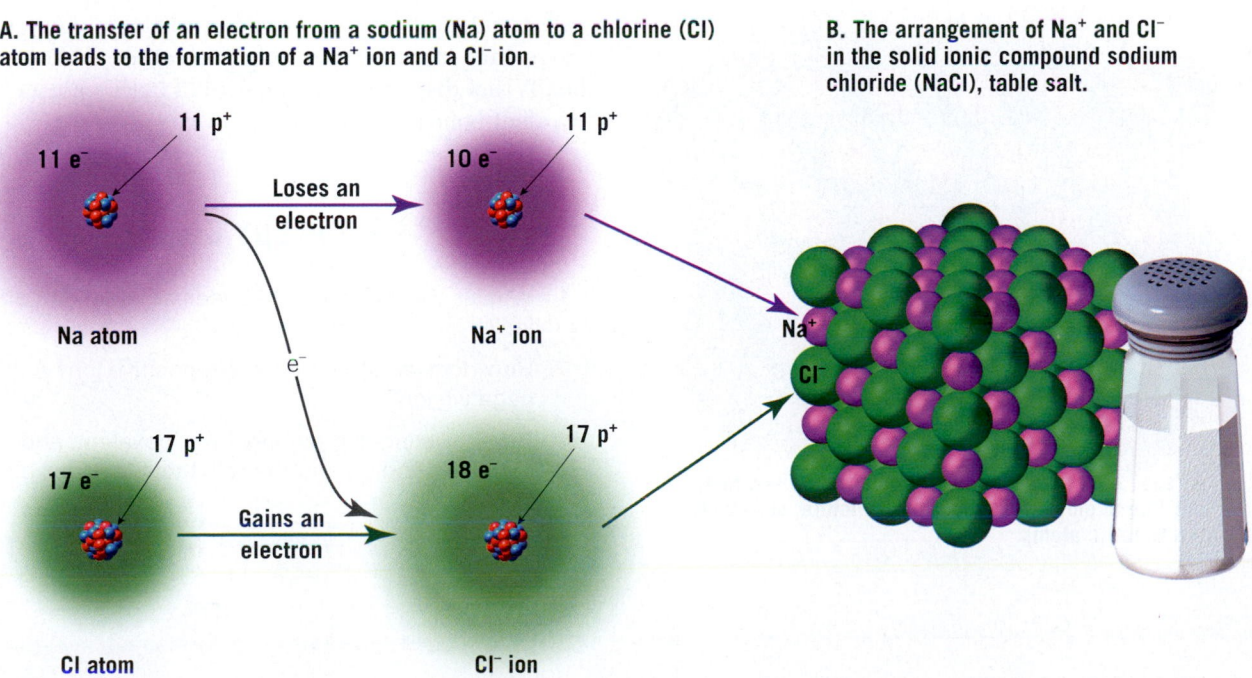

A. The transfer of an electron from a sodium (Na) atom to a chlorine (Cl) atom leads to the formation of a Na⁺ ion and a Cl⁻ ion.

11 p⁺
11 e⁻
Na atom

Loses an electron

11 p⁺
10 e⁻
Na⁺ ion

e⁻

17 p⁺
17 e⁻
Cl atom

Gains an electron

17 p⁺
18 e⁻
Cl⁻ ion

B. The arrangement of Na⁺ and Cl⁻ in the solid ionic compound sodium chloride (NaCl), table salt.

Na⁺
Cl⁻

Figure 3.8
Formation of the ionic compound sodium chloride

Figure 3.9
Formation of a covalent bond When hydrogen atoms bond, the negatively charged electrons are shared by both hydrogen atoms and attracted simultaneously by the positive charge of the proton in the nucleus of each atom.

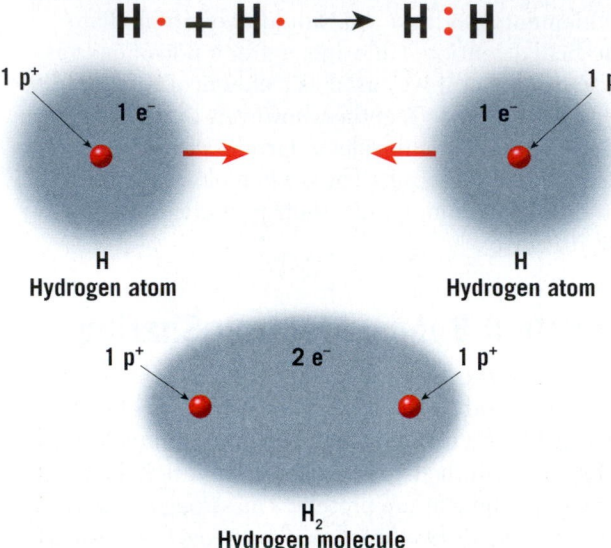

Two hydrogen atoms combine to form a hydrogen molecule, held together by the attraction of oppositely charged particles—positively charged protons in each nucleus and negatively charged electrons that surround these nuclei.

H· + H· → H:H

1 p⁺

1 e⁻

H
Hydrogen atom

1 p⁺

1 e⁻

H
Hydrogen atom

1 p⁺   2 e⁻   1 p⁺

H₂
Hydrogen molecule

charge of the proton in the nucleus of each atom. In this situation the hydrogen atoms do not form ions, rather the force that holds these atoms together arises from the attraction of oppositely charged particles—positively charged protons in the nuclei and negatively charged electrons that surround these nuclei.

Figure 3.10
Metallic bonding
Metallic bonding is the result of each atom contributing its valence electrons to a common pool of electrons that are free to move throughout the entire metallic structure. The attraction between the "sea" of negatively charged electrons and the positive ions produces the metallic bonds that give metals their unique properties.

The central core of each metallic atom, which has an overall positive charge, consists of the nucleus and inner electrons.

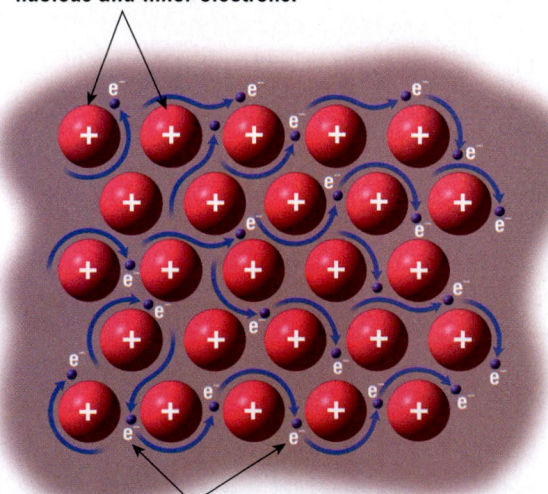

A "sea" of negatively charged outer electrons, that are free to move throughout the structure, surrounds the metallic atoms.

Metallic Bonds: Electrons Free to Move

A few minerals, such as native gold, silver, and copper, are made entirely of metal atoms packed tightly together in an orderly way. The bonding that holds these atoms together results from each atom contributing its valence electrons to a common pool of electrons, which freely move throughout the entire metallic structure. The contribution of one or more valence electrons leaves an array of positive ions immersed in a "sea" of valence electrons, as shown in **Figure 3.10**.

The attraction between this "sea" of negatively charged electrons and the positive ions produces the **metallic bonds** that give metals their unique properties. Metals are good conductors of electricity because the valence electrons are free to move from one atom to another. Metals are also *malleable*, which means they can be hammered into thin sheets, and *ductile*, which means they can be drawn into thin wires. By contrast, ionic and covalent solids tend to be *brittle*, and fracture when stress is applied. Consider the difference between dropping a metal frying pan and a ceramic plate onto a concrete floor.

Hybrid Bonds

We have described the extremes in chemical bonding—complete transfer of electrons and perfect sharing of electrons. As you may suspect, many chemical bonds are actually hybrids that exhibit some degree of electron sharing and some degree of electron transfer. Bonds can be found with almost every possible combination of covalent and ionic character. For example, silicate minerals are composed of silicon and oxygen atoms that are joined together with other elements by bonds that display characteristics of both ionic and covalent bonds.

3.3   **Concept Checks**

1. What is the difference between an atom and an ion?

2. How does an atom become a positive ion? A negative ion?

3. Briefly distinguish among ionic, covalent, and metallic bonding and the role that electrons play in each.

3.4 | How Do Minerals Form?

Describe three mechanisms that generate minerals and rocks.

Minerals form through a wide variety of processes and in many different environments. We will examine three ways minerals can form: *precipitation* of mineral matter from a solution, *crystallization* of molten rock by cooling, and mineral matter *deposition* as a result of biological processes. Because most rocks are aggregates of minerals, they also form by these same processes. In Chapter 8 we will examine the ways existing minerals can be altered to form new minerals through the process of metamorphism.

Precipitation of Mineral Matter

Perhaps the most familiar way in which minerals grow is from an aqueous (water) solution containing dissolved material matter (ions). As long as the solution is not saturated, the motion of the dissolved ions keeps them from joining together. Two factors—a drop in temperature and water lost through evaporation—cause the solution to become closer to reaching saturation. Once saturation is reached, the ions begin to bond, forming crystalline solids (called salts) that precipitate from (settle out of) the solution.

The Great Salt Lake in Utah and Bolivia's great salt flat, Salar de Uyuni, provide good examples of this process at work (**Figure 3.11**). Because they are located in arid regions with high evaporation rates, these water bodies regularly precipitate the minerals halite, sylvite, and gypsum, as well as other soluble salts, called *evaporite deposits*. Worldwide, extensive evaporite deposits, some exceeding hundreds of meters in thickness, provide evidence of ancient seas that have long since evaporated (see Figure 3.42, page 100).

Minerals can also precipitate from slowly moving groundwater filling in fractures and voids in rocks and sediments. One interesting example, called a *geode*, is a somewhat spherically shaped object with inward-projecting crystals that were gradually deposited by groundwater (**Figure 3.12**). Geodes sometimes contain spectacular crystals of quartz, calcite, or other less common minerals.

Crystallization of Molten Rock

The crystallization of minerals from molten rock, although more complicated, is a process similar to water freezing (**Figure 3.13**). When magma is hot, the atoms are very mobile, but as the molten material cools, the atoms slow and begin to chemically combine. Crystallization of a molten mass generates

Mining salt in Bolivia's Salar de Uyuni, the largest known reserve of lithium chloride in the world. Lithium is in great demand because it is used to manufacture batteries for cell phones, computers, and electric and hybrid automobiles.

igneous rocks that consist of a mosaic of intergrown crystals that tend to lack well-developed planar surfaces, or faces (see Figure 3.3). This process is discussed in greater detail in Chapter 4.

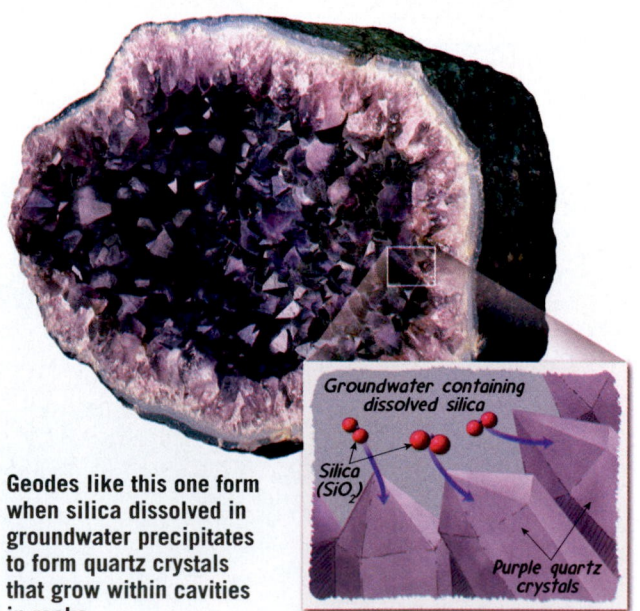

Geodes like this one form when silica dissolved in groundwater precipitates to form quartz crystals that grow within cavities in rocks.

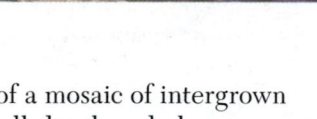

Groundwater containing dissolved silica

Silica (SiO_2)

Purple quartz crystals

Geologist's Sketch

Figure 3.11

Salt being mined in Bolivia's Salar de Uyuni
The Salar de Uyuni is a salt flat that overlies a lake of brine up to 20 meters (66 feet) deep. It is the world's largest salt flat—25 times larger than the Bonneville Salt Flats in Utah. This region is completely surrounded by mountains and is quite arid, so water that enters the valley leaves via evaporation—which results in deposition of salts, including sodium chloride, magnesium chloride, and lithium chloride. (Photo by imagebrocker.net/SuperStock)

Figure 3.12

Geode partially filled with amethyst Geodes form in cavities in rocks, such as limestone and volcanic rocks. Slowly moving groundwater deposits dissolved mineral matter in these voids. (Photo by Jeff Scovil)

Gold

Shutterstock

Gold has been treasured since long before recorded history for its beauty.

How valuable is gold? → **$41,280**

In early 2015, the value of one troy ounce of gold was about US$1,290. Based on that value, a 1000-gram (32-ounce) bar of gold, like the one shown, was worth $41,280. In 1970, the price of gold was less than $40 per troy ounce!

Where is the world's gold produced?

In 1970, South Africa dominated global gold production—accounting for 79% of production throughout the world.

Since 2011 China has become the world's largest producer of gold.

Although its nickname is the "Silver State," Nevada accounts for about 74% of the U.S. gold production, and more than all other countries, except for China, Australia, and Russia.

Metric Tons Per Year (y-axis: 0 to 1100)

Legend:
1. China
2. Australia
3. United States
4. Russia
5. South Africa

(x-axis: 1960, 1970, 1980, 1990, 2000, 2010, 2020)

Figure 3.13
Minerals can form when molten rock solidifies
(Photo by Arctic-Images/Getty Images)

Deposition as a Result of Biological Processes

Water-dwelling organisms are responsible for transforming substantial quantities of dissolved material into mineral matter. For example, corals are organisms capable of creating large quantities of marine limestones, rocks composed of the mineral calcite. These relatively simple invertebrate animals use calcium (Ca) ions from seawater and in turn secrete external skeletons composed of calcium carbonate ($CaCO_3$). Over time, these small organisms are capable of creating massive limestone structures called reefs.

Mollusks (such as clams) and other marine invertebrates also secrete shells composed of the carbonate

Native gold

Because gold does not easily react with other elements, it often occurs as a native element in nuggets found in stream deposits or as grains in igneous rocks.

Shutterstock

Shutterstock

Shutterstock

What are the uses of gold?

About 50% of gold is used in jewelry. Another 40% is used for currency and investment, and about 10% is used in industry, including electronic devices such as cell phones and televisions. Gold is also used in gourmet foods and cocktails as a decorative ingredient. Because metallic gold is one of the least reactive materials, it has no taste, provides no nutritional value, and leaves the human body unaltered.

Shutterstock

Questions:
1. What is the chemical symbol for gold?
2. What is the term for the property of tenacity, which allows gold to be easily hammered into different shapes?

?

---

minerals calcite and aragonite. When the remains of these shells are buried, they become the major component of the sedimentary rock limestone. Limestone is a very common rock at Earth's surface and is visible across the entire United States.

Although most water-dwelling organisms produce hard parts made of calcium carbonate, some, such as diatoms and radiolarians, produce glasslike silica skeletons. During burial, this material forms microscopic silicon dioxide (quartz) crystals that are the main constituents of rocks such as chert and flint.

3.4 Concept Checks

1. Describe three ways minerals and rocks can form.

2. Crystallization of molten rock produces which one of the three types of rock?

3. What is the chemical composition of the mineral matter secreted by most organisms? What is the mineral and rock resulting from this process?

3.5 | Properties of Minerals

List and describe the properties used in mineral identification.

Minerals have definite crystalline structures and chemical compositions that give them unique sets of physical and chemical properties shared by all specimens of that mineral, regardless of when or where they formed. For example, two samples of the mineral quartz will be equally hard and equally dense, and they will break in a similar manner. However, the physical properties of individual samples may vary within specific limits due to ionic substitutions, inclusions of foreign elements (impurities), and defects in the crystalline structure.

Figure 3.14
Metallic versus submetallic luster (Photo courtesy of E. J. Tarbuck)

A. This freshly broken sample of galena displays a metallic luster.

B. This sample of galena is tarnished and has a submetallic luster.

Metallic

Submetallic

Some mineral properties, called **diagnostic properties**, are particularly useful in identifying an unknown mineral. The mineral halite, for example has a salty taste. Because so few minerals share this property, a salty taste is considered a diagnostic property of halite. Other properties of certain minerals vary among different specimens of the same mineral. These properties are referred to as **ambiguous properties**.

Optical Properties

Of the many optical properties of minerals, their luster, their ability to transmit light, their color, and their streak are most frequently used for mineral identification.

SmartFigure 3.15
Color variations in minerals Some minerals, such as quartz, occur in a variety of colors. (Photo by E. J. Tarbuck)
(https://goo.gl/wznOpk)

Tutorial

Quartz

Luster The appearance or quality of light reflected from the surface of a mineral is known as **luster**. Minerals that have the appearance of a metal, regardless of color, are said to have a *metallic luster* (**Figure 3.14A**). Some metallic minerals, such as native copper and galena, develop a dull coating or tarnish when exposed to the atmosphere. Because they are not as shiny as samples with freshly broken surfaces, these samples are often said to exhibit a *submetallic luster* (**Figure 3.14B**).

Most minerals have a *nonmetallic luster* and are described using various adjectives, such as *vitreous* or *glassy*. Other nonmetallic minerals are described as having a *dull*, or *earthy*, *luster* (a dull appearance like soil) or a *pearly luster* (such as a pearl or the inside of a clamshell). Still others exhibit a *silky luster* (like satin cloth) or a *greasy luster* (as though coated in oil).

Ability to Transmit Light An optical property used to identify minerals is the ability to transmit light. When no light is transmitted, the mineral is described as *opaque*; when light, but not an image, is transmitted through a mineral sample, the mineral is said to be *translucent*. When both light and an image are visible through the sample, the mineral is described as *transparent*.

Color Although **color** is generally the most conspicuous characteristic of any mineral, it is considered a diagnostic property of only a few minerals. Slight impurities in the common mineral quartz, for example, give it a variety of tints, including pink, purple, yellow, white, gray, and even black (**Figure 3.15**). Other minerals, such as tourmaline, also exhibit a variety of hues, with multiple colors sometimes occurring in the same sample. Thus, the use of color as a means of identification is often ambiguous or even misleading.

Streak The color of a mineral in powdered form, called **streak**, is often useful in identification. A mineral's streak is obtained by rubbing it across a *streak plate* (a piece of unglazed porcelain) and observing the color of the mark it leaves (**Figure 3.16**). Although a mineral's color may vary from sample to sample, its streak is usually consistent in color. (Note that not all minerals produce

Mineral
(Pyrite)

(Photo by Dennis Tasa)

Color
(Brass yellow)

Streak
(Black)

SmartFigure 3.16
Streak (https://goo.gl/ULVLM4)

Video

Although the color of a mineral is not always helpful in identification, the streak, which is the color of the powdered mineral, can be very useful.

Figure 3.17
Common crystal shapes of pyrite

Although most minerals exhibit only one common crystal shape, some, such as pyrite, have two or more characteristic habits.

Dennis Tasa

a streak when rubbed across a streak plate. Quartz, for example, is harder than a porcelain streak plate and therefore leaves no streak.)

Streak can also help distinguish between minerals with metallic luster and those with nonmetallic luster. Metallic minerals generally have a dense, dark streak, whereas minerals with nonmetallic luster typically have a light-colored streak.

Crystal Shape, or Habit

Mineralogists use the term **crystal shape**, or **habit**, to refer to the common or characteristic shape of individual crystals or aggregates of crystals. Some minerals tend to grow equally in all three dimensions, whereas others tend to be elongated in one direction or flattened if growth in one dimension is suppressed. A few minerals have crystals that exhibit regular polygons that are helpful in their identification. For example, magnetite crystals sometimes occur as octahedrons, garnets often form dodecahedrons, and halite and fluorite crystals tend to grow as cubes or near-cubes. Minerals tend to have one common crystal shape, but a few, such as the pyrite samples shown in **Figure 3.17**, have two or more characteristic crystal shapes.

In addition, some mineral samples consist of numerous intergrown crystals that exhibit characteristic shapes that are useful for identification. Terms commonly used to describe these and other crystal habits include *equant* (equidimensional), *bladed*, *fibrous*, *tabular*, *prismatic*, *platy*, and *blocky*. Some of these habits are pictured in **Figure 3.18**.

Mineral Strength

How easily minerals break or deform under stress is determined by the type and strength of the chemical bonds that hold the crystals together. Mineralogists use

E.J. Tarbuck

A. Fibrous

Dennis Tasa

B. Bladed

Dennis Tasa

C. Banded

Dennis Tasa

D. Cubic crystals

SmartFigure 3.18
Some common crystal habits A. Thin, rounded crystals that break into fibers. **B.** Elongated crystals that are flattened in one direction. **C.** Minerals that have stripes or bands of different color or texture. **D.** Groups of crystals that are cube shaped. (https://goo.gl/IBw4OJ)

Tutorial

SmartFigure 3.19
Hardness scales A. The Mohs scale of hardness, with the hardnesses of some common objects. **B.** Relationship between the Mohs relative hardness scale and an absolute hardness scale. (https://goo.gl/ZN0DAG)

Tutorial

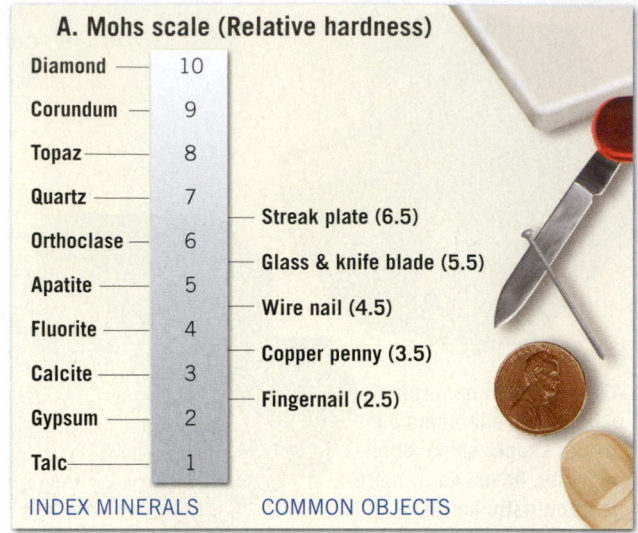

A. Mohs scale (Relative hardness)

INDEX MINERALS		COMMON OBJECTS
Diamond	10	
Corundum	9	
Topaz	8	
Quartz	7	Streak plate (6.5)
Orthoclase	6	Glass & knife blade (5.5)
Apatite	5	Wire nail (4.5)
Fluorite	4	Copper penny (3.5)
Calcite	3	Fingernail (2.5)
Gypsum	2	
Talc	1	

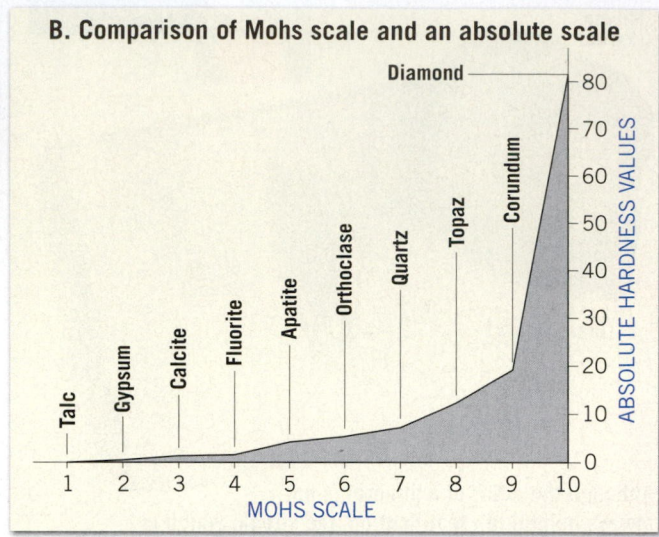

B. Comparison of Mohs scale and an absolute scale

SmartFigure 3.20
Micas exhibit perfect cleavage The thin sheets shown here exhibit one plane of cleavage. (Photo by Chip Clark/Fundamental Photos) (https://goo.gl/JYCISi)

Animation

terms including *hardness, cleavage, fracture,* and *tenacity* to describe mineral strength and how minerals break when stress is applied.

Hardness   One of the most useful diagnostic properties is **hardness**, a measure of the resistance of a mineral to abrasion or scratching. This property is determined by rubbing a mineral of unknown hardness against one of known hardness or vice versa. A numerical value of hardness can be obtained by using the **Mohs scale** of hardness, which consists of 10 minerals arranged in order from 1 (softest) to 10 (hardest), as shown in **Figure 3.19A**. It should be noted that the Mohs scale is a relative ranking and does not imply that a mineral with a hardness of 2, such as gypsum, is twice as hard as a mineral with a hardness of 1, like talc. In fact, gypsum is only slightly harder than talc, as **Figure 3.19B** indicates.

In the laboratory, common objects used to determine the hardness of a mineral can include a human fingernail, which has a hardness of about 2.5, a copper penny (3.5), and a piece of glass (5.5). The mineral gypsum, which has a hardness of 2, can be easily scratched with a fingernail. On the other hand, the mineral calcite, which has a hardness of 3, will scratch a fingernail but will not scratch glass. Quartz, one of the hardest common minerals, will easily scratch glass. Diamonds, hardest of all, scratch anything, including other diamonds.

Cleavage   In the crystal structure of many minerals, some atomic bonds are weaker than others. It is along these weak bonds that minerals tend to break when they are stressed. **Cleavage** (*kleiben* = carve) is the tendency of a mineral to break (cleave) along planes of weak bonding. Not all minerals have cleavage, but those that do can be identified by the relatively smooth, flat surfaces that are produced when the mineral is broken.

The simplest type of cleavage is exhibited by the micas (**Figure 3.20**). Because these minerals have very weak bonds in one direction, they cleave to form thin, flat sheets. Some minerals have excellent cleavage in one, two, three, or more directions, whereas others exhibit fair or poor cleavage, and still others have no cleavage at all. When minerals break evenly in more than one direction, cleavage is described by *the number of cleavage directions and the angle(s) at which they meet* (**Figure 3.21**).

Each cleavage surface that has a different orientation is counted as a different direction of cleavage. For example, some minerals cleave to form six-sided cubes. Because cubes are defined by three different sets of parallel planes that intersect at 90-degree angles, cleavage is described as *three directions of cleavage that meet at 90 degrees.*

Do not confuse cleavage with crystal shape. When a mineral exhibits cleavage, it will break into pieces that all have the same geometry. By contrast, the smooth-sided quartz crystals shown in Figure 3.1 do not have cleavage. If broken, they fracture into shapes that do not resemble one another or the original crystals.

Fracture Minerals having chemical bonds that are equally, or nearly equally, strong in all directions exhibit a property called **fracture**. When minerals fracture, most produce uneven surfaces and are described as exhibiting *irregular fracture* (**Figure 3.22A**). However, some minerals, including quartz, can break into smooth, curved surfaces resembling broken glass. Such breaks are called *conchoidal fractures* (**Figure 3.22B**). Still other minerals exhibit fractures that produce splinters or fibers, referred to as *splintery fracture* and *fibrous fracture*, respectively.

Tenacity The term **tenacity** describes a mineral's resistance to breaking, bending, cutting, or other forms of deformation. As mentioned earlier, nonmetallic minerals such as quartz and halite tend to be *brittle* and fracture or exhibit cleavage when struck. Minerals that are ionically bonded, such as fluorite and halite, tend to be *brittle* and shatter into small pieces when struck. By contrast, native metals, such as copper and gold, are *malleable*, which means they can be hammered without breaking. In addition, minerals that can be cut into thin shavings, including gypsum and talc, are described as *sectile*. Still others, notably the micas, are *elastic* and will bend and snap back to their original shape after stress is released.

Density and Specific Gravity

Density, an important property of matter, is defined as mass per unit volume. Mineralogists often use a related measure called **specific gravity** to describe the density of minerals. Specific gravity is a number representing the ratio of a mineral's weight to the weight of an equal volume of water.

Most common minerals have a specific gravity between 2 and 3. For example, quartz has a specific gravity of 2.65. By contrast, some metallic minerals, such as pyrite, native copper, and magnetite, are more than twice as dense and thus have more than twice the specific gravity of quartz. Galena, an ore from which lead is extracted, has a specific gravity of roughly 7.5, whereas 24-karat gold has a specific gravity of approximately 20.

With a little practice, you can estimate the specific gravity of a mineral by hefting it in your hand. Does

SmartFigure 3.21
Cleavage directions exhibited by minerals (Photos by E. J. Tarbuck and Dennis Tasa) (https://goo.gl/5lkkd6)
Tutorial

A. Cleavage in one direction.
Example: Muscovite

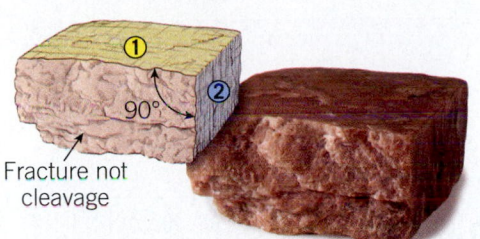

B. Cleavage in two directions at 90° angles.
Example: Feldspar

C. Cleavage in two directions not at 90° angles. Example: Hornblende

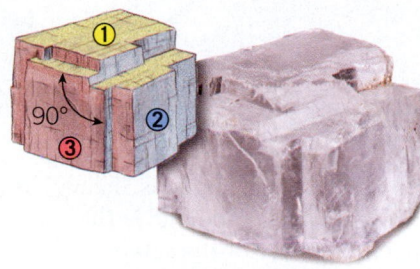

D. Cleavage in three directions at 90° angles. Example: Halite

E. Cleavage in three directions not at 90° angles.
Example: Calcite

F. Cleavage in four directions.
Example: Fluorite

Figure 3.22
Irregular versus conchoidal fracture (Photos by E. J. Tarbuck)

A. Irregular fracture (Quartz)

B. Conchoidal fracture (Quartz)

Figure 3.23
Double refraction
This sample of calcite exhibits double refraction. (Photo by Chip Clark/Fundamental Photos)

can be picked up with a magnet, while some varieties (such as lodestone) are themselves natural magnets and will pick up small iron-based objects such as pins and paper clips (see Figure 3.41F, page 99).

Moreover, some minerals exhibit special optical properties. For example, when a transparent piece of calcite is placed over printed text, the letters appear twice. This optical property is known as *double refraction* (**Figure 3.23**).

One very simple chemical test to detect carbonate minerals involves placing a drop of dilute hydrochloric acid onto a freshly broken mineral surface. Samples containing carbonates will effervesce (fizz) as carbon dioxide gas is released (**Figure 3.24**). This test is especially useful in identifying calcite, a common carbonate mineral.

this mineral feel about as "heavy" as similarly sized rocks you have handled? If the answer is "yes," the specific gravity of the sample will likely be between 2.5 and 3.

SmartFigure 3.24
Calcite reacting with a weak acid (Photo by Chip Clark/Fundamental Photos) (https://goo.gl/pnGkML)

Video

Other Properties of Minerals

In addition to the properties discussed thus far, some minerals can be recognized by other distinctive properties. For example, halite is ordinary salt, so it can be quickly identified through taste. Talc and graphite both have distinctive feels: Talc feels soapy, and graphite feels greasy. Further, the streaks of many sulfur-bearing minerals smell like rotten eggs. A few minerals, such as magnetite, have a high iron content and

3.6 | Mineral Structures and Compositions

Distinguish between compositional and structural variations in minerals and provide one example of each.

Many people associate the word *crystal* with delicate wine goblets or glassy objects with smooth sides and gem-like shapes. In geology, the term **crystal** or **crystalline** refers to *any natural solid with an orderly, repeating internal structure*. Therefore, all mineral samples are crystals or crystalline solids, even if they lack smooth-sided faces. The specimen shown in Figure 3.1, for example, exhibits the characteristic crystal form associated with quartz—a six-sided prismatic shape with pyramidal ends. However, the quartz crystals in the sample of granite shown in Figure 3.3 do not display well-defined faces. Both quartz samples are nonetheless crystalline.

Mineral Structures

The smooth faces and symmetry possessed by well-developed crystals are surface manifestations of the orderly packing of the atoms or ions that constitute a mineral's internal structure. This highly ordered atomic arrangement within minerals can be illustrated by using spherically shaped atoms held together by ionic, covalent, or metallic bonds. The simplest crystal structures are those of native metals, such as gold and silver, which are composed of only one element. These materials consist of atoms packed together in a rather simple three-dimensional network that minimizes voids. Imagine a group of cannon balls stacked in layers such that the spheres in one layer nestle in the hollows between spheres in the adjacent layers.

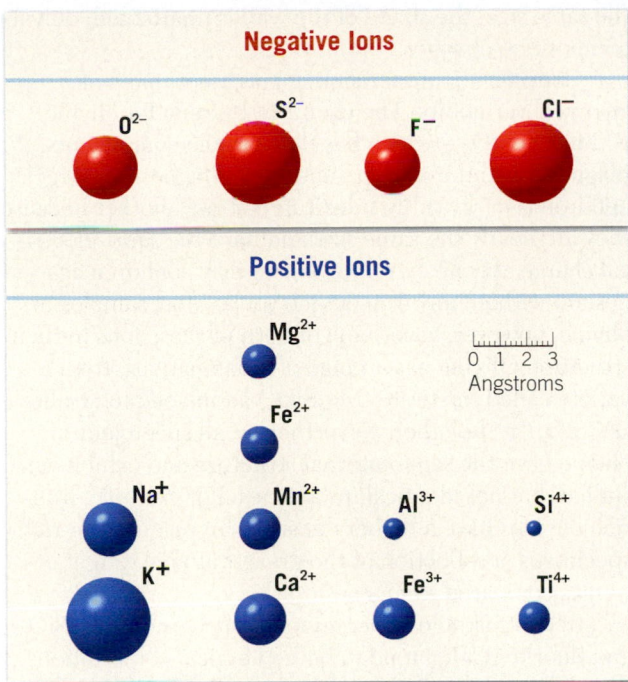

Figure 3.25
Relative sizes and charges of selected ions Ionic radii are usually expressed in angstroms (1 angstrom equals 10^{-8} cm).

The atomic structure of most minerals consists of at least two different ions (often of very different sizes). **Figure 3.25** illustrates the relative sizes of some of the most common ions found in minerals. Notice that the negative ions, which are atoms that gained electrons, tend to be larger than the positive ions, which lost electrons.

Crystal structures can be considered three-dimensional stacks of larger spheres (negative ions) with smaller spheres (positive ions) located in the spaces between them, so that the positive and negative charges cancel each other out. Consider the mineral halite (NaCl), which has a relatively simple framework composed of an equal number of positively charged sodium ions and negatively charged chlorine ions. Because ions of similar charge repel, they are spaced as far apart from each other as possible. Consequently, in halite, each sodium ion (Na^+) is surrounded on all sides by chlorine ions and vice versa (**Figure 3.26**). This particular arrangement forms basic building blocks, called **unit cells**, that have cubic shapes. As shown in Figure 3.26C, these cubic unit cells combine to form cube-shaped halite crystals, including those that come out of salt shakers.

The shape and symmetry of these building blocks relate to the shape and symmetry of the entire crystal. It is important to note, however, that two minerals can be constructed of geometrically similar building blocks yet exhibit different external forms. For example, fluorite, magnetite, and garnet are minerals constructed of cubic unit cells, but these unit cells can join to produce crystals of many shapes. Typically, fluorite crystals are cubes, whereas magnetite crystals are octahedrons, and garnets form dodecahedrons built up of many small cubes, as shown in **Figure 3.27**. Because the building blocks are so small, the resulting crystal faces are smooth and flat.

Despite the fact that natural crystals are rarely perfect, the angles between equivalent crystal faces of the same mineral are remarkably consistent. This observation was first made by Nicolas Steno in 1669. Steno found that the angles between adjacent prism faces of quartz crystals are 120 degrees, regardless of sample size, the size of the crystal faces, or where the crystals were collected (**Figure 3.28**). This observation is commonly called **Steno's Law**, or the **Law of Constancy of Interfacial Angles**, because it applies to all minerals. Because Steno's Law holds for all minerals, crystal shape is frequently a valuable tool in mineral identification.

A. Cube (fluorite)

B. Octahedron (magnetite)

C. Dodecahedron (garnet)

Figure 3.27
Cubic unit cells These cells stack together in different ways to produce crystals that exhibit different shapes. Fluorite **A.** tends to display cubic crystals, whereas magnetite crystals **B.** are typically octahedrons, and garnets **C.** usually occur as dodecahedrons. (Photos by Dennis Tasa)

A. Sodium and chlorine ions.

B. Basic building block of the mineral halite.

C. Collection of basic building blocks (crystal).

D. Crystals of the mineral halite.

Figure 3.26
Arrangement of sodium and chloride ions in the mineral halite The arrangement of atoms into basic building blocks that have a cubic shape results in regularly shaped cubic crystals. (Photo by Dennis Tasa)

Figure 3.28
Steno's Law Because some faces of a crystal may grow larger than others, two crystals of the same mineral may *not* have identical shapes. Nevertheless, the angles between equivalent faces are remarkably consistent.

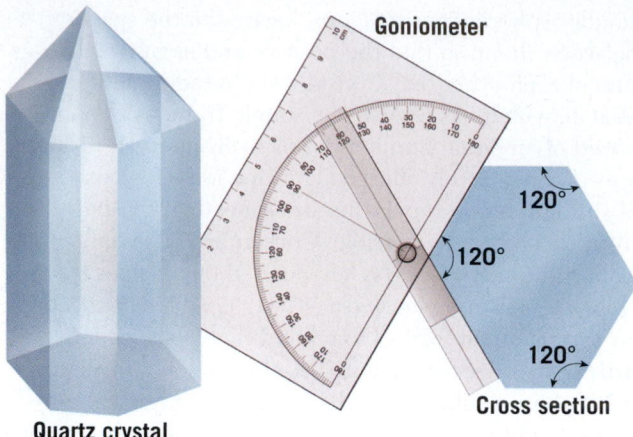

Goniometer

Quartz crystal

Cross section

Compositional Variations in Minerals

Mineralogists have determined that the chemical composition of some minerals varies substantially from sample to sample. These compositional variations are possible because ions of similar size can readily substitute for one another without disrupting a mineral's internal framework. This is analogous to a wall made of bricks of different colors and materials. As long as the bricks are roughly

Figure 3.29
Diamond versus graphite Both diamond and graphite are natural substances with the same chemical composition: carbon atoms. Nevertheless, their internal structures and physical properties reflect the fact that each formed in a very different environment. (Photo A Marcel Clemens/Shutterstock; photo B by E. J. Tarbuck)

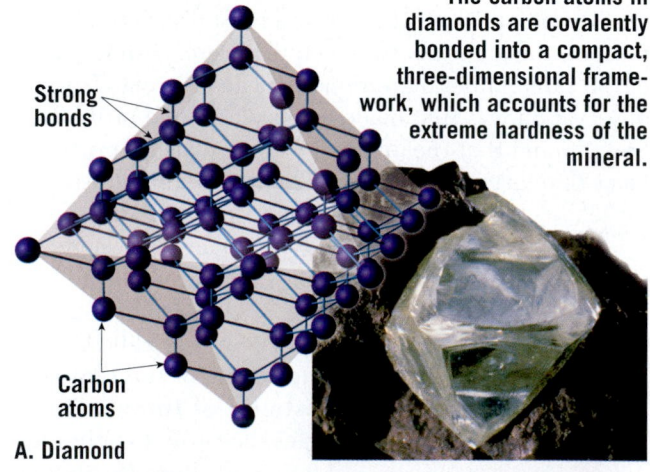

Strong bonds

The carbon atoms in diamonds are covalently bonded into a compact, three-dimensional framework, which accounts for the extreme hardness of the mineral.

Carbon atoms

A. Diamond

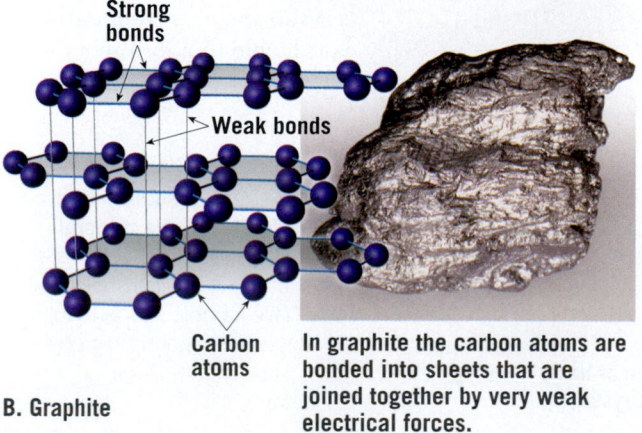

Strong bonds

Weak bonds

Carbon atoms

In graphite the carbon atoms are bonded into sheets that are joined together by very weak electrical forces.

B. Graphite

the same size, the shape of the wall is unaffected; only its composition changes.

Consider the mineral olivine as an example of chemical variability. The chemical formula for olivine is $(Mg,Fe)_2SiO_4$—which has the variable components magnesium and iron in parentheses. Magnesium (Mg^{2+}) and iron (Fe^{2+}) readily substitute for one another because they are nearly the same size and have the same electrical charge. At one extreme, olivine may contain magnesium without any iron or vice versa. Most samples of olivine, however, have some of both of these ions in their structure. Olivine has a range of combinations, from a variety called forsterite (Mg_2SiO_4) at one end to fayalite (Fe_2SiO_4) at the other. Nevertheless, all specimens of olivine have the same internal structure and exhibit very similar, but not identical, properties. For example, iron-rich olivines have a higher density than magnesium-rich specimens, a reflection of the greater atomic weight of iron as compared to magnesium.

In contrast to olivine, minerals such as quartz (SiO_2) and fluorite (CaF_2) tend to have chemical compositions that differ very little from their chemical formulas. However, even these minerals often contain tiny amounts of other less common elements, referred to as *trace elements*. Although trace elements have little effect on most mineral properties, they can significantly influence a mineral's color.

Structural Variations in Minerals

It is possible for two minerals with exactly the same chemical composition to have different internal structures and, hence, different external forms. Minerals of this type are called **polymorphs** (*poly* = many, *morph* = form). Graphite and diamond are particularly good examples of polymorphism because, when pure, they are both made up exclusively of carbon atoms. Graphite is the soft gray mineral from which pencil "lead" is made, whereas diamond is the hardest-known mineral. The differences between these minerals can be attributed to the conditions under which they form. Diamonds form at depths that may exceed 200 kilometers (nearly 125 miles), where extreme pressures and temperatures produce the compact structure shown in **Figure 3.29A**. Graphite, on the other hand, forms under comparatively low pressures and consists of sheets of carbon atoms that are widely spaced and weakly bonded (**Figure 3.29B**). Because the carbon sheets in graphite easily slide past one another, graphite has a greasy feel and makes an excellent lubricant.

Scientists have learned that by heating graphite under high confining pressures, they can generate synthetic diamonds. Because human-made diamonds often contain flaws, they are generally not gem quality, but due to their hardness, they have many industrial uses. Further, because diamonds form in environments of extreme pressure and temperature, they are somewhat unstable at Earth's surface. Fortunately for jewelers, "diamonds are forever" because the rate at which diamonds change to their more stable form, graphite, is infinitesimally slow.

The transformation of one polymorph to another is an example of a *phase change*. In nature, certain minerals go through phase changes as they move from one environment to another. For example, when a slab of ocean crust composed of olivine-rich basalt is carried to great depths by a subducting plate, olivine changes to a more compact, denser polymorph with the same structure as the mineral *spinel*.

Recall that oceanic lithosphere sinks because it is colder and more dense than the underlying mantle. It follows, therefore, that during subduction, the

transformation of olivine from its low- to high-density form would contribute to plate subduction. Stated another way, this phase change causes an increase in the overall density of the slab, thereby enhancing its rate of descent.

3.6 Concept Checks

1. Explain Steno's Law in your own words.

2. Define *polymorph* and give an example.

3.7 | Mineral Groups

Explain how minerals are classified and name the most abundant mineral group in Earth's crust.

More than 4000 minerals have been named, and several new ones are identified each year. Fortunately for students who are beginning to study minerals, no more than a few dozen are abundant. Collectively, these few make up most of the rocks of Earth's crust and they are therefore often referred to as the **rock-forming minerals**.

Although less abundant, many other minerals are used extensively in the manufacture of products and are called **economic minerals**. However, rock-forming minerals and economic minerals are not mutually exclusive groups. When found in large deposits, some rock-forming minerals are economically significant. One example is calcite, the primary component of the sedimentary rock limestone. Calcite has many uses, including the production of concrete. *Chapter 23 provides an overview of the importance of Earth's vast natural resources.*

Classifying Minerals

Minerals are placed into categories in much the same way that plants and animals are classified. Mineralogists use the term *mineral species* for a collection of specimens that exhibit similar internal structures and chemical compositions. Some common mineral species are quartz, calcite, galena, and pyrite. However, just as individual plants and animals within a species differ somewhat from one another, so do most specimens of the same mineral.

Some mineral species are further subdivided into mineral *varieties*. For example, pure quartz (SiO_2) is colorless and transparent. However, when small amounts of aluminum are incorporated into its atomic structure, quartz appears quite dark, in a variety called *smoky quartz*. *Amethyst*, another variety of quartz, owes its violet color to the presence of trace amounts of iron.

Mineral species are assigned to mineral *classes*. Some important mineral classes are the silicates (SiO_4^{4-}), carbonates (CO_3^{2-}), halides (Cl^{1-}, F^{1-}, Br^{1-}), and sulfates (SO_4^{2-}). Minerals within each class tend to have similar

internal structures and, hence, similar properties. For example, minerals belonging to the carbonate class react chemically with acid—albeit to varying degrees—and many exhibit similar cleavage. Furthermore, minerals within the same class are often found together in the same rock. For example, halite (NaCl) and silvite (KCl) belong to the halide class and commonly occur together in evaporite deposits.

Silicate Versus Nonsilicate Minerals

It is worth noting that only *eight elements* make up the vast majority of the rock-forming minerals and represent more than 98 percent (by weight) of the continental crust (**Figure 3.30**). These elements, in order of most to least

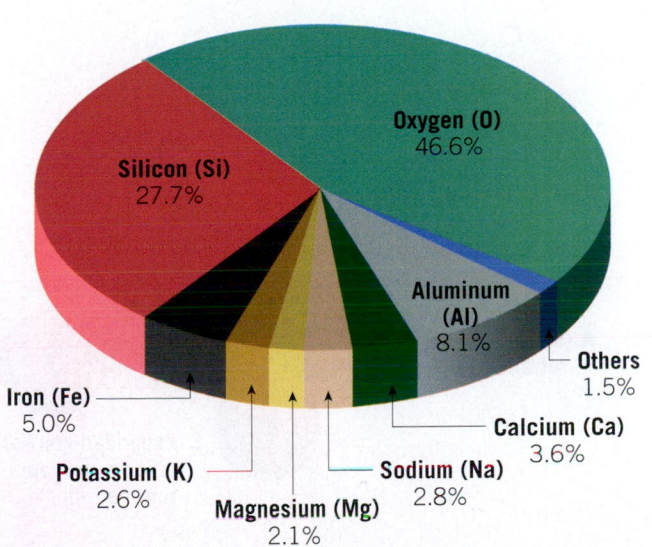

Figure 3.30
The eight most abundant elements in the continental crust

Oxygen (O) 46.6%

Silicon (Si) 27.7%

Aluminum (Al) 8.1%

Others 1.5%

Iron (Fe) 5.0%

Calcium (Ca) 3.6%

Potassium (K) 2.6%

Sodium (Na) 2.8%

Magnesium (Mg) 2.1%

abundant, are oxygen (O), silicon (Si), aluminum (Al), iron (Fe), calcium (Ca), sodium (Na), potassium (K), and magnesium (Mg). As shown in Figure 3.30, oxygen and silicon are by far the most common elements in Earth's crust. Furthermore, these two elements readily combine to form the basic "building block" for the most common mineral group, the **silicates**. More than 800 silicate minerals are known, and they account for more than 90 percent of Earth's crust.

Because other mineral groups are far less abundant in Earth's crust than the silicates, they are often grouped together under the heading **nonsilicates**. Although not as common as silicates, some nonsilicate minerals are very important economically. They provide us with iron and aluminum to build automobiles, gypsum for plaster and drywall for home construction, and copper wire

that carries electricity and connects us to the Internet. Common nonsilicate mineral groups include the carbonates, sulfates, and halides. In addition to their economic importance, these groups include minerals that are major constituents in sediments and sedimentary rocks.

3.8 | The Silicates

Sketch the silicon–oxygen tetrahedron and explain how these fundamental building blocks join together to form other silicate structures.

Every silicate mineral contains the two most abundant elements in Earth's crust, oxygen and silicon. Further, most contain one or more of the other common elements. Together, these elements give rise to hundreds of silicate minerals with a wide variety of properties, including hard quartz, soft talc, sheetlike mica, fibrous asbestos, green olivine, and blood-red garnet.

Silicate Structures

All silicate minerals have the same fundamental building block, the **silicon–oxygen tetrahedron** (SiO_4^{4-}). This structure consists of four oxygen ions that are covalently bonded to one comparatively small silicon ion, forming a *tetrahedron*—a pyramid shape with four identical faces (**Figure 3.31**). These tetrahedra are not chemical compounds but rather complex ions (SiO_4^{4-}) having a net charge of −4. To become electrically balanced, these

complex ions bond to positively charged metal ions. Specifically, each O^{2-} has one of its valence electrons bonding with the Si^{4+} located at the center of the tetrahedron. The remaining $1-$ charge on each oxygen is available to bond with another positive ion, or with the silicon ion in an adjacent tetrahedron. It is important to note that the silicon–oxygen tetrahedron is the fundamental building block that all silicate minerals have in common.

Minerals with Independent Tetrahedra One of the simplest silicate structures consists of independent tetrahedra that have their four oxygen ions bonded to positive ions, such as Mg^{2+}, Fe^{2+}, and Ca^{2+}. The mineral olivine, with the formula $(Mg,Fe)_2SiO_4$, is a good example. In olivine, magnesium (Mg^{2+}) and/or iron (Fe^{2+}) ions pack between comparatively large independent SiO_4 tetrahedra, forming a dense three-dimensional structure (**Figure 3.32**). Garnet, another common silicate, is also composed of independent tetrahedra that are ionically bonded by positive ions. Both olivine and garnet form dense, hard, equidimensional crystals that lack cleavage.

Minerals with Chain or Sheet Structures One reason for the great variety of silicate minerals is the ability of SiO_4 tetrahedra to link to one another in a variety of configurations. This important phenomenon, called

Figure 3.31
Two representations of the silicon–oxygen tetrahedron

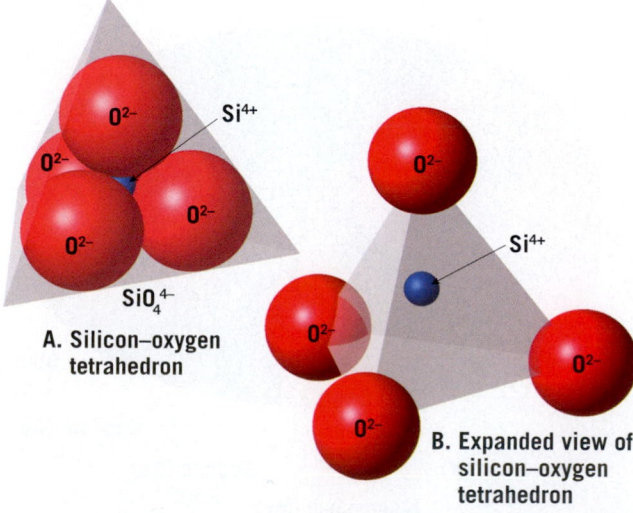

A. Silicon–oxygen tetrahedron

SiO_4^{4-}

B. Expanded view of silicon–oxygen tetrahedron

polymerization, is achieved by the sharing of one, two, three, or all four of the oxygen atoms with adjacent tetrahedra. Vast numbers of tetrahedra join together to form single chains, double chains, sheet structures, or three-dimensional frameworks, as shown in **Figure 3.33**.

To see how oxygen atoms are shared between adjacent tetrahedra, select one of the silicon ions (small blue spheres) near the middle of the single chain shown in **Figure 3.33B**. Notice that this silicon ion is completely surrounded by four larger oxygen ions. Also notice that two of the four oxygen atoms are bonded to two silicon atoms, whereas the other two are not shared in this manner. It is the linkage across the shared oxygen ions that joins the tetrahedra into a chain structure. Now examine a silicon ion near the middle of the sheet structure (**Figure 3.33D**) and count the number of shared and unshared oxygen ions surrounding it. As you likely observed, the sheet structure is the result of three of the four oxygen atoms being shared by adjacent tetrahedra.

Minerals with Three-Dimensional Frameworks

In the most common silicate structure, all four oxygen ions are shared, producing a complex three-dimensional framework (**Figure 3.33E**). Quartz and the most common mineral group, the feldspars, have structures in which all of the oxygens are shared.

The ratio of oxygen ions to silicon ions differs in each type of silicate structure. In independent tetrahedra (SiO_4) there are four oxygen ions for every silicon ion. In single chains, the oxygen-to-silicon ratio is 3:1 (SiO_3), and in three-dimensional frameworks, as found in quartz, the ratio is 2:1 (SiO_2). As more oxygen ions are shared, the percentage of silicon in the structure increases. Silicate minerals are therefore described as having a low or high silicon content, based on their ratio of oxygen to silicon.

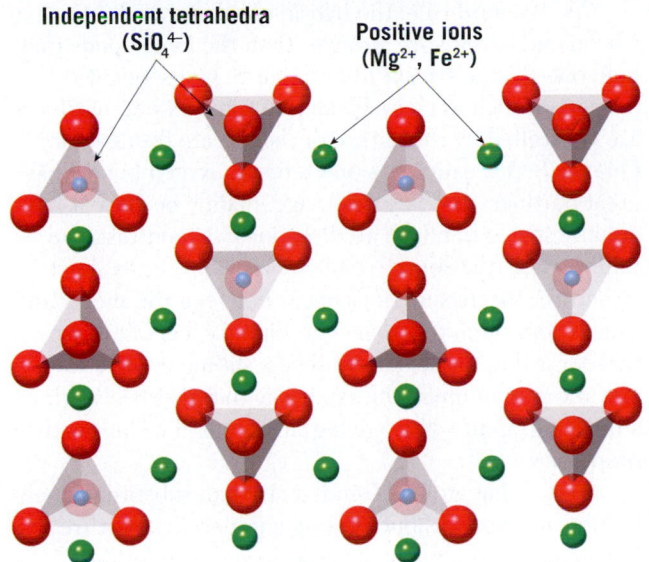

Independent tetrahedra, which have an overall negative charge, form bonds (not shown) with positively charged metallic ions (green) to form a stable chemical compound.

Independent tetrahedra (SiO_4^{4-})   Positive ions (Mg^{2+}, Fe^{2+})

Silicate minerals with three-dimensional structures have the highest silicon content, while those composed of independent tetrahedra have the lowest.

Figure 3.32
How independent tetrahedra bond with metallic ions The internal structure of the mineral olivine, which is composed of independent tetrahedra that are bonded together, with metallic ions (iron and magnesium) shown in green. This is an expanded view that does not show the bonds between the metallic ions and the tetrahedra.

Joining Silicate Structures

Except for quartz (SiO_2), the basic structure (chains, sheets, or three-dimensional frameworks) of most silicate minerals has a net negative charge. Therefore, metal ions are required to bring the overall charge into balance and to serve as the "mortar" that holds these structures together. The positive ions that most often link silicate structures are iron (Fe^{2+}), magnesium (Mg^{2+}), potassium (K^{1+}), sodium (Na^{1+}), aluminum (Al^{3+}), and calcium

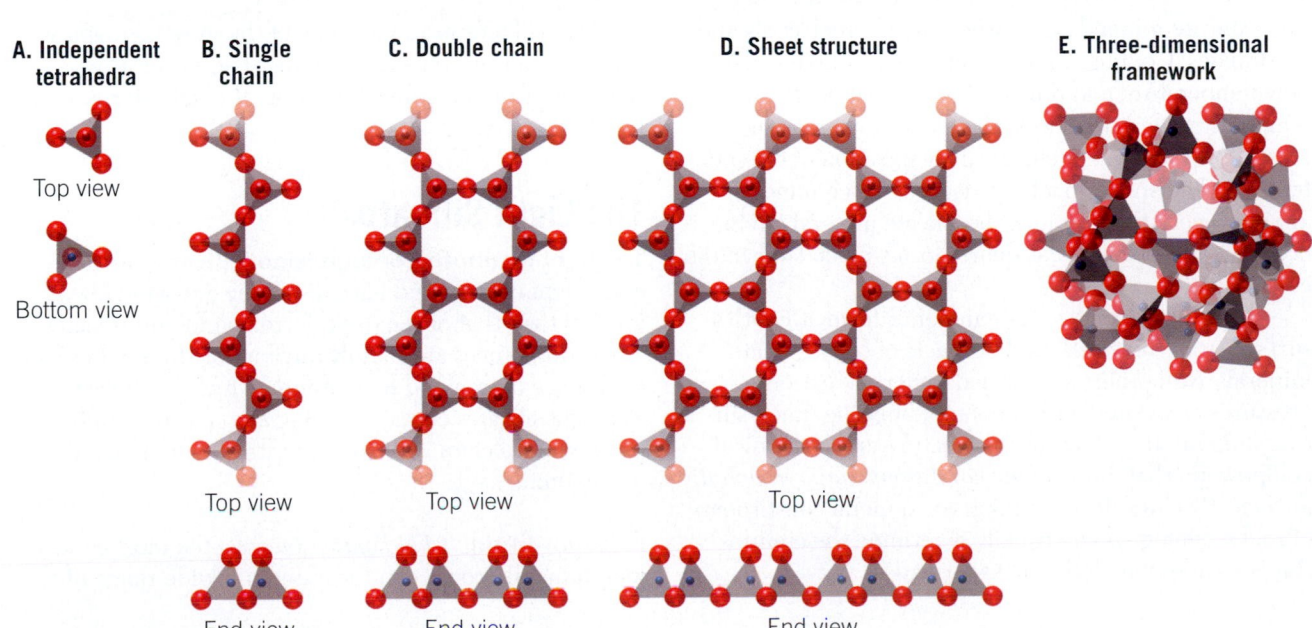

A. Independent tetrahedra — Top view, Bottom view. B. Single chain — Top view, End view. C. Double chain — Top view, End view. D. Sheet structure — Top view, End view. E. Three-dimensional framework

SmartFigure 3.33
Five basic silicate structures A. Independent tetrahedra. **B.** Single chains. **C.** Double chains. **D.** Sheet structures. **E.** Three-dimensional framework.
(https://goo.gl/xpaEPC)

 **Tutorial**

(Ca^{2+}). These positively charged ions bond with the unshared oxygen ions that occupy the corners of the silicate tetrahedra.

As a general rule, the hybrid covalent bonds between silicon and oxygen are stronger than the ionic bonds that hold one silicate structure to the next. Consequently, properties such as cleavage and, to some extent, hardness are controlled by the nature of the silicate framework. Quartz (SiO_2), which has only silicon–oxygen bonds, has great hardness and lacks cleavage, mainly because it has equally strong bonds in all directions. By contrast, the mineral talc (the source of talcum powder) has a sheet structure. Magnesium ions occur between the sheets and weakly join them together. The slippery feel of talcum powder is due to the silicate sheets sliding relative to one another, in much the same way that sheets of carbon atoms in graphite slide, giving this mineral its lubricating properties.

Recall that atoms of similar size can substitute freely for one another without altering a mineral's structure. For example, in olivine, iron (Fe^{2+}) and magnesium (Mg^{2+}) substitute for each other. This also holds true for the third-most-common element in Earth's crust,

aluminum (Al^{3+}), which often substitutes for silicon (Si^{4+}) in the center of silicon–oxygen tetrahedra.

Because most silicate structures will readily accommodate two or more different positive ions at a given bonding site, individual specimens of a particular mineral may contain varying amounts of certain elements. As a result, many silicate minerals form a mineral group that exhibits a range of compositions between two end members. Examples include the olivines, pyroxenes, amphiboles, micas, and feldspars.

3.8   Concept Checks

1. Sketch the silicon–oxygen tetrahedron and label its parts.

2. What is the ratio of oxygen to silicon found in single tetrahedrons? How about framework structures? Which has the highest silicon content?

3. What differences in their silicate structures account for the slipperiness of talc and the hardness of quartz?

3.9 | Common Silicate Minerals

Compare and contrast the light (nonferromagnesian) silicates with the dark (ferromagnesian) silicates and list three minerals common to each group.

The major groups of silicate minerals and common examples are given in **Figure 3.34**. The feldspars are by far the most plentiful silicate group, comprising more than 50 percent of Earth's crust. Quartz, the second-most-abundant mineral in the continental crust, is the only common mineral made completely of silicon and oxygen.

Most silicate minerals form when molten rock cools and crystallizes. Cooling can occur at or near Earth's surface (low temperature and pressure) or at great depths (high temperature and pressure). The environment during crystallization and the chemical composition of the molten rock determine, to a large degree, which minerals are produced. For example, the silicate mineral olivine crystallizes early, whereas quartz forms much later in the crystallization process.

In addition, some silicate minerals form at Earth's surface from the weathered products of other silicate minerals. Still others are formed under the extreme pressures associated with mountain building. Each silicate mineral, therefore, has a structure and a chemical composition that *indicate the conditions under which it formed*. By carefully examining the mineral constituents of rocks, geologists can usually determine the circumstances under which the rocks formed.

We will now examine some of the most common silicate minerals, which we divide into two major groups on the basis of their chemical makeup: the light silicates and the dark silicates.

The Light Silicates

The **light** (or **nonferromagnesian**) **silicates** are generally light in color and have a specific gravity of about 2.7, less than that of the dark (ferromagnesian) silicates. These differences are mainly attributable to the presence or absence of iron and magnesium, which are "heavy." The light silicates contain varying amounts of aluminum, potassium, calcium, and sodium rather than iron and magnesium.

Feldspar Group   *Feldspar minerals*, the most common mineral group, can form under a wide range of

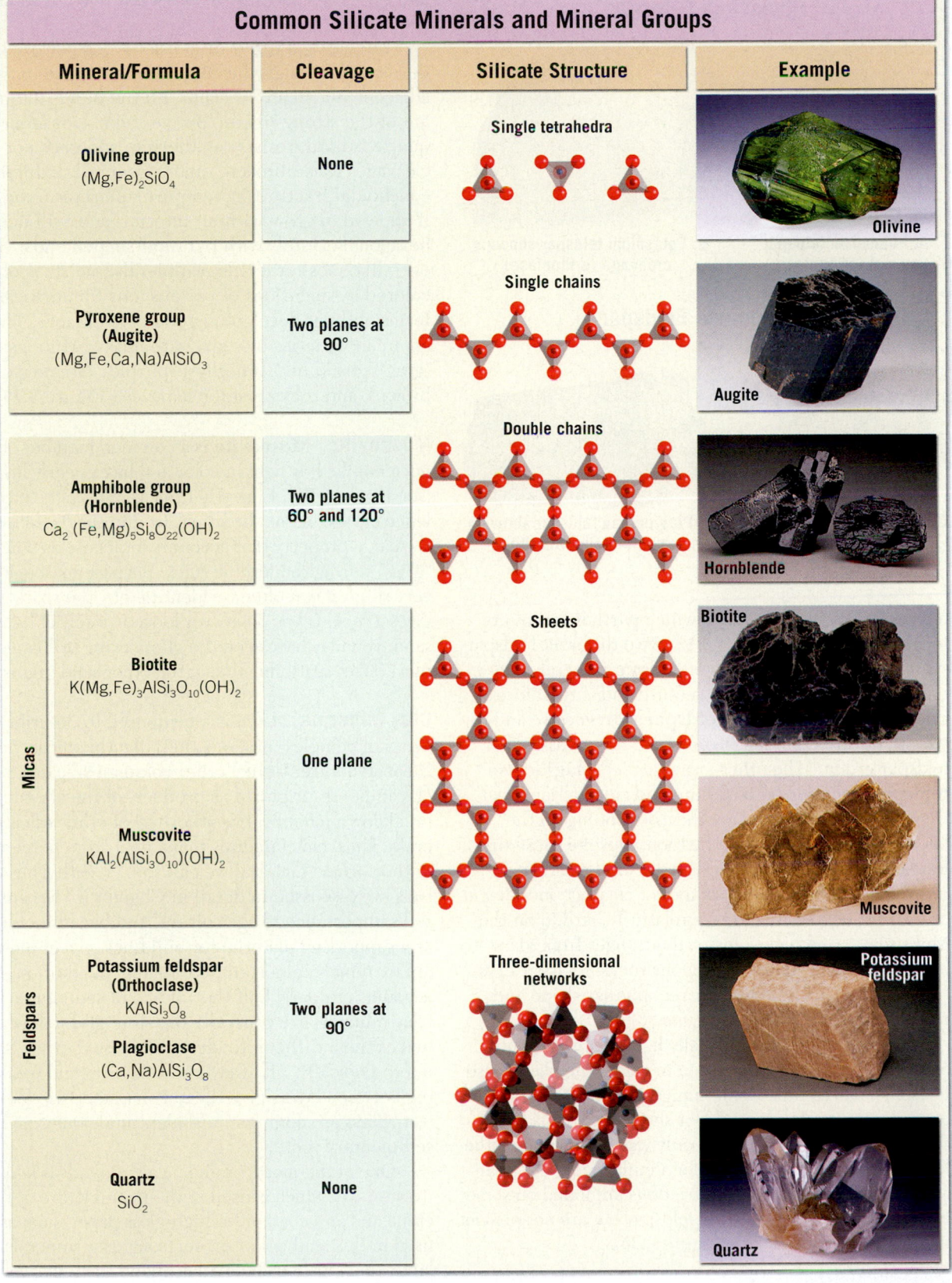

Common Silicate Minerals and Mineral Groups

Mineral/Formula	Cleavage	Silicate Structure	Example
Olivine group $(Mg,Fe)_2SiO_4$	None	Single tetrahedra	Olivine
Pyroxene group (Augite) $(Mg,Fe,Ca,Na)AlSiO_3$	Two planes at 90°	Single chains	Augite
Amphibole group (Hornblende) $Ca_2(Fe,Mg)_5Si_8O_{22}(OH)_2$	Two planes at 60° and 120°	Double chains	Hornblende
Micas — **Biotite** $K(Mg,Fe)_3AlSi_3O_{10}(OH)_2$ / **Muscovite** $KAl_2(AlSi_3O_{10})(OH)_2$	One plane	Sheets	Biotite / Muscovite
Feldspars — **Potassium feldspar (Orthoclase)** $KAlSi_3O_8$ / **Plagioclase** $(Ca,Na)AlSi_3O_8$	Two planes at 90°	Three-dimensional networks	Potassium feldspar
Quartz SiO_2	None		Quartz

Figure 3.34
Common silicate minerals Note that the complexity of the silicate structure increases from top to bottom. (Photos by Dennis Tasa and E. J. Tarbuck)

Figure 3.35
Feldspar minerals
A. Characteristic crystal form of potassium feldspar. **B.** Most salmon-colored feldspar belongs to the potassium feldspar subgroup. **C.** Sodium-rich plagioclase feldspar tends to be light colored and has a pearly luster. **D.** Calcium-rich plagioclase feldspar tends to be gray, blue-gray, or black in color. Labradorite, the variety shown here, exhibits striations on one of its crystal faces. (Photos by Dennis Tasa and E. J. Tarbuck)

Potassium Feldspar

A. Potassium feldspar crystal (orthoclase)

B. Potassium feldspar showing cleavage (orthoclase)

Plagioclase Feldspar

C. Sodium-rich plagioclase feldspar (albite)

D. Plagioclase feldspar showing striations (labradorite)

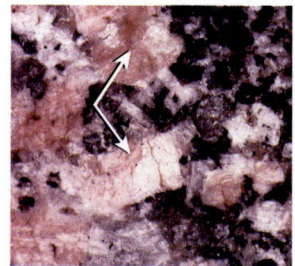

A. Potassium feldspar

B. Plagioclase feldspar

Figure 3.36
Feldspar crystals in igneous rocks When found in igneous rocks, feldspar crystals tend to be rectangular in shape and exhibit smooth, shiny faces. (Photos by E. J. Tarbuck)

temperatures and pressures, which partially accounts for their abundance (**Figure 3.35**). Two different feldspar structures exist. One group of feldspar minerals contains potassium ions in its structure and is therefore referred to as **potassium feldspar**. (*Orthoclase* and *microcline* are common members of the potassium feldspar group.) The other group, called **plagioclase feldspar**, contains both sodium and calcium ions that freely substitute for one another, depending on the environment during crystallization. Despite these differences, all feldspar minerals have similar physical properties. They have two planes of cleavage meeting at or near 90-degree angles, are relatively hard (6 on the Mohs scale), and have a luster that ranges from glassy to pearly. As a component in igneous rocks, feldspar crystals can be identified by their rectangular shape and rather smooth, shiny faces (**Figure 3.36**).

Potassium feldspar is usually light cream, salmon pink, or occasionally blue-green in color. The plagioclase feldspars, on the other hand, range in color from white to medium gray. However, color should not be used to distinguish these groups. The only way to distinguish the feldspars physically is to look for a multitude of fine parallel lines, called *striations*. Striations are found on some cleavage planes of plagioclase feldspar but are not present on potassium feldspar (see Figure 3.35).

Quartz   **Quartz** (SiO_2) is the only common silicate mineral that consists entirely of silicon and oxygen. As such, the term *silica* is applied to quartz. Because quartz contains a ratio of two oxygen ions (O^{2-}) for every silicon

ion (Si^{4+}), no other positive ions are needed to attain neutrality.

In quartz, a three-dimensional framework is developed through the complete sharing of oxygen by adjacent silicon atoms. Thus, all the bonds in quartz are of the strong silicon–oxygen type. Consequently, quartz is hard, resists weathering, and does not have cleavage. When broken, quartz generally exhibits conchoidal fracture. When pure, quartz is clear and, if allowed to grow without interference, will develop hexagonal crystals with pyramid-shaped ends. However, like most other clear minerals, quartz is often colored by inclusions of various ions (impurities) and forms without developing good crystal faces. The most common varieties of quartz are milky (white), smoky (gray), rose (pink), amethyst (purple), citrine (yellow to brown), and rock crystal (clear) (see Figure 3.15).

Muscovite   **Muscovite** is a common member of the mica family. It is light in color and has a pearly luster (see Figure 3.20). Like other micas, muscovite has excellent cleavage in one direction. In thin sheets, muscovite is clear, a property that accounts for its use as window "glass" during the Middle Ages. Because muscovite is very shiny, it can often be identified by the sparkle it gives a rock. If you have ever looked closely at beach sand, you may have seen the glimmering brilliance of the mica flakes scattered among the other sand grains.

Clay Minerals   *Clay* is a term used to describe a category of complex minerals that, like the micas, have a sheet structure. Unlike other common silicates, most clay minerals originate as products of the chemical breakdown (chemical weathering) of other silicate minerals. Thus, clay minerals make up a large percentage of the surface material we call soil. (Weathering and soils are discussed in detail in Chapter 6.) Because of soil's importance to agriculture, and because of its role as a supporting material for buildings, clay minerals are extremely important to humans. In addition, clays account for nearly half the volume of sedimentary rocks. Clay minerals are generally very fine grained, which makes them difficult to identify unless they are studied microscopically. Their layered structure and weak bonding between layers give them a characteristic feel when wet. Clays are common in shales, mudstones, and other sedimentary rocks.

One of the most common clay minerals is *kaolinite* (**Figure 3.37**), which is used in the manufacture of fine china and as a coating for high-gloss paper, such as that used in this textbook. Further, some clay minerals absorb large amounts of water, which allows them to swell to several times their normal size. These clays have been used commercially in a variety of ingenious ways, including as an additive to thicken milkshakes in fast-food restaurants.

Figure 3.37
Kaolinite Kaolinite is a common clay mineral formed by weathering of feldspar minerals.

The Dark Silicates

The **dark** (or **ferromagnesian**) **silicates** are minerals containing ions of iron (*ferro* = iron) and/or magnesium in their structure. Because of their iron content, ferromagnesian silicates are dark in color and have a greater specific gravity, between 3.2 and 3.6, than nonferromagnesian silicates. The most common dark silicate minerals are olivine, the pyroxenes, the amphiboles, dark mica (biotite), and garnet.

Olivine Group   **Olivine**, a family of high-temperature silicate minerals, is black to olive green in color and has a glassy luster and a conchoidal fracture (see Figure 3.34). Transparent olivine is occasionally used as a gemstone called peridot. Rather than develop large crystals, olivine commonly forms small, rounded crystals that give olivine-rich rocks a granular appearance (**Figure 3.38**). Olivine and related forms are typically found in basalt, a common igneous rock of the oceanic crust and volcanic areas on the continents, and are thought to constitute up to 50 percent of Earth's upper mantle.

Pyroxene Group   The *pyroxenes* are a group of diverse minerals that are important components in dark-colored igneous rocks. The most common member, **augite**, is a black, opaque mineral with two directions of cleavage that meet at nearly a 90-degree angle. Augite is one of the dominant minerals in basalt (**Figure 3.39A**).

Amphibole Group   The most common member of a chemically complex group of minerals called *amphiboles* (**Figure 3.39B**) is **hornblende**. Hornblende is usually dark green to black in color, and except for its cleavage angles, which are about 60 degrees and 120 degrees, it is very similar in appearance to augite. In a rock, hornblende often forms elongated crystals. This helps distinguish it from pyroxene, which forms rather blocky crystals. Hornblende is found in igneous rocks, where it often makes up the dark portion of an otherwise light-colored rock (see Figure 3.3).

Biotite   **Biotite** is a dark, iron-rich member of the mica family (see Figure 3.34). Like other micas, biotite possesses a sheet structure that gives it excellent cleavage in one direction. Biotite also has a shiny black appearance that helps distinguish it from the other dark ferromagnesian minerals. Like hornblende, biotite is a common constituent of igneous rocks, including the rock granite.

Garnet   *Garnet* is similar to olivine in that its structure is composed of individual tetrahedra linked by metallic ions. Also like olivine, garnet has a glassy luster, lacks cleavage, and exhibits conchoidal fracture. Although the colors of garnet are varied, this mineral is most often brown to deep red. Well-developed garnet crystals have 12 diamond-shaped faces and are most commonly found in metamorphic rocks (**Figure 3.40**). Transparent garnets are prized as semiprecious gemstones.

Olivine-rich peridotite (variety dunite)

Figure 3.38
Olivine Commonly black to olive green in color, olivine has a glassy luster and is often granular in appearance.

A. Augite

B. Hornblende

Figure 3.39   Augite and hornblende These dark-colored silicate minerals are common constituents of a variety of igneous rocks. (Photos courtesy of E. J. Tarbuck)

← 2 cm →

Figure 3.40
Well-formed garnet crystal Garnets come in a variety of colors and are commonly found in mica-rich metamorphic rocks. (Photo by E. J. Tarbuck)

3.9   Concept Checks

1. Apart from their difference in color, what is one main distinction between light and dark silicates? What accounts for this difference?

2. Based on the chart in Figure 3.34, what do muscovite and biotite have in common? How do they differ?

3. Is color a good way to distinguish between orthoclase and plagioclase feldspar? If not, what is a more effective means of distinguishing them?

EYE ON EARTH 3.2

Glass bottles, like most other manufactured products, contain substances obtained from minerals extracted from Earth's crust and oceans. The primary ingredient in commercially produced glass bottles is the mineral quartz. Glass also contains lesser amounts of the mineral calcite. (Photo by Chris Brignell/Shutterstock)

QUESTION 1 In what mineral group does quartz belong?

QUESTION 2 Glass beer bottles are usually clear, green, or brown. Based on what you know about how the mineral quartz is colored, what do glass manufacturers do to make bottles green and brown?

QUESTION 3 Why did some brewers start using brown bottles instead of the green bottles that were popular until the 1930s?

3.10 | Important Nonsilicate Minerals

List the common nonsilicate minerals and explain why each is important.

Nonsilicate minerals are typically divided into groups, based on the negatively charged ion or complex ion that the members have in common. For example, the *oxides* contain the negative oxygen ions (O^{2-}), which bond to one or more kinds of positive ions (**Figure 3.41**). Thus, within each mineral group, the basic structure and type of bonding is similar. As a result, the minerals in each group have similar physical properties that are useful in mineral identification.

Although the nonsilicates make up only about 8 percent of Earth's crust, some minerals, such as gypsum, calcite, and halite, occur as constituents in sedimentary rocks in significant amounts. Furthermore, many others are important economically. Figure 3.41 lists some of the major nonsilicate mineral groups and gives a few examples of each.

Some of the most common nonsilicate minerals belong to one of three classes of minerals: the carbonates (CO_3^{2-}), the sulfates (SO_4^{2-}), and the halides (Cl^{1-}, F^{1-}, Br^{1-}). The carbonate minerals are much simpler structurally than the silicates. This mineral group is composed of the carbonate ion (CO_3^{2-}) and one or more kinds of positive ions. The two most common carbonate minerals are *calcite*, $CaCO_3$ (calcium carbonate), and *dolomite*, $CaMg(CO_3)_2$ (calcium/magnesium carbonate) (see Figure 3.41A,B). Calcite and dolomite are usually found together as the primary constituents in the sedimentary rocks limestone and dolostone. When calcite is the dominant

mineral, the rock is called *limestone*, whereas *dolostone* results from a predominance of dolomite. Limestone has many uses, including as road aggregate, as building stone, and as the main ingredient in Portland cement.

Two other nonsilicate minerals frequently found in sedimentary rocks are *halite* and *gypsum* (see Figure 3.41C,I). Both minerals are commonly found in thick layers that are the last vestiges of ancient seas that have long since evaporated (**Figure 3.42**). Like limestone, both halite and gypsum are important nonmetallic resources. Halite is the mineral name for common table salt ($NaCl$). Gypsum ($CaSO_4 \cdot 2\,H_2O$), which is calcium sulfate with water bound into the structure, is the mineral from which plaster and other similar building materials are formed.

Most nonsilicate mineral classes contain members prized for their economic value. This includes the oxides, whose members *hematite* and *magnetite* are important ores of iron (see Figure 3.41E,F). Also

Common Nonsilicate Mineral Groups

Mineral Group (key ion(s) or element(s))	Mineral Name	Chemical Formula	Economic Use	Examples
Carbonates (CO_3^{2-})	Calcite Dolomite	$CaCO_3$ $CaMg(CO_3)_2$	Portland cement, lime Portland cement, lime	 B. Dolomite A. Calcite
Halides (Cl^{1-}, F^{1-}, Br^{1-})	Halite Fluorite Sylvite	$NaCl$ CaF_2 KCl	Common salt Used in steelmaking Used as fertilizer	 C. Halite D. Fluorite
Oxides (O^{2-})	Hematite Magnetite Corundum Ice	Fe_2O_3 Fe_3O_4 Al_2O_3 H_2O	Ore of iron, pigment Ore of iron Gemstone, abrasive Solid form of water	 E. Hematite F. Magnetite
Sulfides (S^{2-})	Galena Sphalerite Pyrite Chalcopyrite Cinnabar	PbS ZnS FeS_2 $CuFeS_2$ HgS	Ore of lead Ore of zinc Sulfuric acid production Ore of copper Ore of mercury	 H. Chalcopyrite G. Galena
Sulfates (SO_4^{2-})	Gypsum Anhydrite Barite	$CaSO_4 \cdot 2H_2O$ $CaSO_4$ $BaSO_4$	Plaster Plaster Drilling mud	 J. Anhydrite I. Gypsum
Native elements (single elements)	Gold Copper Diamond Graphite Sulfur Silver	Au Cu C C S Ag	Trade, jewelry Electrical conductor Gemstone, abrasive Pencil lead Sulfadrugs, chemicals Jewelry, photography	 K. Copper L. Sulfur

Figure 3.41
Important nonsilicate minerals (Photos by Dennis Tasa and E. J. Tarbuck)

Figure 3.42
Thick bed of halite exposed in an underground mine Halite (salt) mine in Grand Saline, Texas. Note the person for scale. (Photo by Tom Bochsler)

significant are the sulfides, which are basically compounds of sulfur (S) and one or more metals. Important sulfide minerals include galena (lead), sphalerite (zinc), and chalcopyrite (copper). In addition, native elements—including gold, silver, and carbon (diamonds)—plus a host of other nonsilicate minerals—fluorite (flux in making steel), corundum (gemstone, abrasive), and uraninite (a uranium source)—are economically important (see GEOgraphics 3.1, p.82 and 3.2, p.101).

3.10 | Concept Checks

1. List six common nonsilicate mineral groups. What key ions or elements define each group?

2. What is the most common carbonate mineral?

3. List eight common nonsilicate minerals and their economic uses.

Gemstones

Precious stones have been prized since antiquity. Although most gemstones are varieties of a particular mineral, misinformation abounds regarding gems and their mineral makeup.

Important Gemstones

Gemstones are classified in one of two categories: precious or semiprecious. Precious gems are rare and generally have hardnesses that exceed 9 on the Mohs scale. Therefore, they are more valuable and thus more expensive than semiprecious gems.

GEM	MINERAL NAME		PRIZED HUES
PRECIOUS			
Diamond	Diamond	●●●	Colorless, pinks, blues
Emerald	Beryl	●	Greens
Ruby	Corundum	●	Reds
Sapphire	Corundum	●	Blues
Opal	Opal	●	Brilliant hues
SEMIPRECIOUS			
Alexandrite	Chrysoberyl		Variable
Amethyst	Quartz	●	Purples
Cat's-eye	Chrysoberyl	●	Yellows
Chalcedony	Quartz (agate)	●	Banded
Citrine	Quartz	●	Yellows
Garnet	Garnet	●●	Red, greens
Jade	Jadeite or nephrite	●	Greens
Moonstone	Feldspar	●	Transparent blues
Peridot	Olivine	●	Olive greens
Smoky quartz	Quartz	●	Browns
Spinel	Spinel	●	Reds
Topaz	Topaz	●●	Purples, reds
Tourmaline	Tourmaline	●●●	Reds, blue-greens
Turquoise	Turquoise	●	Blues
Zircon	Zircon	●	Reds

Reuters

The Famous Hope Diamond

The deep-blue Hope Diamond is a 45.52-carat gem that is thought to have been cut from a much larger 115-carat stone discovered in India in the mid-1600s. The original 115-carrat stone was cut into a smaller gem that became part of the crown jewels of France and was in the possession of King Louis XVI and Marie Antoinette before they attempted to escape France. Stolen during the French Revolution in 1792, the gem is thought to have been recut to its present size and shape. In the 1800s, it became part of the collection of Henry Hope (hence its name) and is on display at the Smithsonian in Washington, DC.

What Constitutes a Gemstone?

When found in their natural state, most gemstones are dull and would be passed over by most people as "just another rock." Gems must be cut and polished by experienced professionals before their true beauty is displayed. Cutting and polishing is accomplished using abrasive material, most often tiny fragments of diamonds that are embedded in a metal disk.

Charles D. Winters/ Photo Researchers, Inc.

Naming Gemstones

Most precious stones are given names that differ from their parent mineral. For example, sapphire is one of two gems that are varieties of the same mineral, corundum. Trace elements can produce vivid sapphires of nearly every color. Tiny amounts of titanium and iron in corundum produce the most prized blue sapphires. When the mineral corundum contains a sufficient quantity of chromium, it exhibits a brilliant red color. This variety of corundum is called ruby.

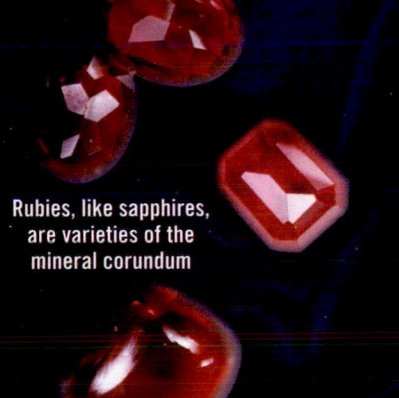

Rubies, like sapphires, are varieties of the mineral corundum

Sapphires, showing variations of colors

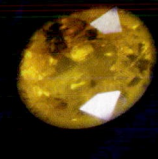

Dorling Kindersley

Greg C. Grace/ Alamy

? **Question:**
Why are diamonds used as an abrasive material to cut and polish gemstones?

3.1 Minerals: Building Blocks of Rocks

List the main characteristics that an Earth material must possess to be considered a mineral and describe each.

KEY TERMS mineralogy, mineral, rock

- In Earth science, the word *mineral* refers to naturally occurring inorganic solids that possess an orderly crystalline structure and a characteristic chemical composition. The study of minerals is called mineralogy.

- Minerals are the building blocks of rocks. Rocks are naturally occurring masses of minerals or mineral-like matter, such as glass or organic material.

3.2 Atoms: Building Blocks of Minerals

Compare and contrast the three primary particles contained in atoms.

KEY TERMS atom, nucleus, proton, neutron, electron, valence electron, atomic number, element, periodic table, chemical compound

- Minerals are composed of atoms of one or more elements. The atoms of any element consist of the same three basic ingredients: protons, neutrons, and electrons.

- The atomic number is the number of protons in an atom. For example, an oxygen atom has eight protons, so its atomic number is eight. Protons and neutrons are approximately the same size and mass, but while protons are positively charged, neutrons have no charge.

- Electrons are much smaller than both protons and neutrons, weighing about 2000 times less. Each electron has a negative charge, equal in magnitude to the positive charge of a proton. Electrons swarm around an atom's nucleus at a distance, in several distinctive energy levels called principal shells. The electrons in the outermost principal shell, called valence electrons, are important when the atom bonds with other atoms to form chemical compounds.

- Elements with similar numbers of valence electrons tend to behave in similar ways. The periodic table displays these similarities in its graphical arrangement of the elements.

Q Use the periodic table (Figure 3.5) to identify these geologically important elements by their number of protons: (A) 14, (B) 6, (C) 13, (D) 17, and (E) 26.

3.3 Why Atoms Bond

Distinguish among ionic bonds, covalent bonds, and metallic bonds.

KEY TERMS octet rule, chemical bond, ionic bond, ion, covalent bond, metallic bond

- When atoms are attracted to other atoms, they can form chemical bonds, which generally involve the transfer or sharing of valence electrons. The most stable arrangement for most atoms is to have eight electrons in the outermost principal shell. This idea is called the octet rule.

- Ionic bonds involve atoms of one element giving up electrons to atoms of another element, forming positively and negatively charged atoms called ions. Positively charged ions bond with negative ions to form ionic bonds.

- Covalent bonds involve the sharing of one or more electrons between two adjacent atoms.

- In metallic bonds, the sharing is more extensive: Electrons can freely move from one atom to the next throughout the entire mass.

Q Match the diagrams labeled A–C with the type of bonding each illustrates: ionic, covalent, or metallic. What are the distinguishing characteristics of each type of bonding?

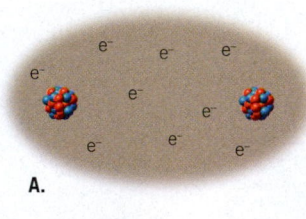

A.

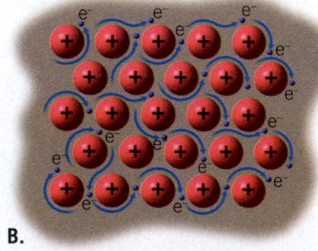

B.

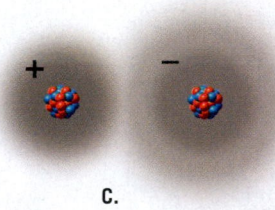

C.

3.4 How Do Minerals Form?

Describe three mechanisms that generate minerals and rocks.

- Mineral crystals can form in several ways. For elements dissolved in solutions of water, precipitation of minerals may be triggered by a drop in temperature or when loss of water by evaporation causes the concentration of ions to increase.

- In molten rock, free-moving atoms form bonds with other atoms as the liquid cools, and that "nucleus" of a mineral crystal will add additional atoms on its outer edges, growing larger as more atoms enter the lower-energy solid state.

- Marine organisms can extract ions from the surrounding seawater and then biochemically secrete skeletal material made of either calcium carbonate or silica.

1. Crystallization from an aqueous solution

National Park Service

Givaga/Shutterstock

2. Crystallization by living organisms

USGS

3. Crystallization of molten rock

3.5 Physical Properties of Minerals

List and describe the properties used in mineral identification.

KEY TERMS diagnostic properties, ambiguous properties, luster, color, streak, crystal shape (habit), hardness, Mohs scale, cleavage, fracture, tenacity, density, specific gravity

- The composition and internal crystalline structure of a mineral's crystal lattice gives it specific physical properties. These properties can be used to identify and distinguish minerals and make minerals useful to humans.

Quartz

Dennis Tasa

Calcite

Dennis Tasa

- Luster is a mineral's ability to reflect light. The terms *transparent*, *translucent*, and *opaque* describe the degree to which a mineral can transmit light. Color can be unreliable for mineral identification, as impurities can "stain" minerals with misleading colors. A more trustworthy identifier is streak, the color of the powder generated by scraping a mineral against a porcelain streak plate.

- The shape a crystal assumes as it grows is often very useful for identification. We call this its habit.

- Variations in the strength of chemical bonds give minerals properties such as hardness (resistance to being scratched) and tenacity (whether a mineral breaks in a brittle fashion or bends when stressed). Cleavage, the preferential breakage of a mineral along planes of weakly bonded atoms, is very useful in identifying minerals.

- The amount of matter packed into a given volume determines a mineral's density. To compare the densities of minerals, mineralogists use a related property, specific gravity, which is the ratio between a mineral's density and the density of water.

- Other properties are diagnostic for certain minerals but rare in most others. A distinctive smell, taste, feel, reaction to hydrochloric acid, magnetism, and double refraction are examples.

Q Research the minerals *quartz* and *calcite*. List five physical characteristics that may be used to distinguish one from the other.

3.6 Mineral Structures and Compositions

Distinguish between compositional and structural variations in minerals and provide one example of each.

KEY TERMS crystal, crystalline, unit cell, Steno's Law (Law of Constancy of Interfacial Angles), polymorph

- Crystals are naturally occurring solids with orderly, repeating internal structures. All minerals occur in crystal form, even if those crystals are not stereotypically large, perfectly formed prisms. The atoms forming a mineral are packed together in some geometrically regular fashion. The smallest expression of that arrangement is called a unit cell; many unit cells repeating in three dimensions yield a mineral crystal.

- Steno's Law, or the Law of Constancy of Interfacial Angles, says that no matter how big a crystal of a given mineral may be, the angles between its faces will always be the same.

- The composition of minerals may vary within certain parameters. Certain portions of the chemical formula of a mineral may be fixed and invariable, while other parts have flexibility in terms of which atoms are present. Generally, atoms of similar size and charge can "swap out" for one another.

- The same chemical compound can grow into different mineral crystals, with different arrangements of atoms. Calcite and aragonite are examples of this "polymorph" relationship: Though they are different minerals, they both have the same chemical composition—$CaCO_3$.

Q A crystal of the mineral corundum measuring 1 millimeter long has an angle between two crystal faces that measures 60 degrees. Assume the crystal continues to grow until it has tripled in size. What is the angle between the crystal faces now?

3.7 Mineral Groups

Explain how minerals are classified and name the most abundant mineral group in Earth's crust.

KEY TERMS rock-forming mineral, economic mineral, silicate, nonsilicate

- More than 4000 different minerals have been identified, but only a few dozen are especially common in Earth's crust: These are the rock-forming minerals. Many minerals have economic value.

- Minerals are placed into classes on the basis of similar crystal structures and compositions. Minerals of the same class tend to have similar properties and are found in similar geologic settings.

- Silicon and oxygen are the most common elements in Earth's crust, and so the most common minerals in the crust are silicate minerals. In comparison, nonsilicate minerals make up only about 8 percent of the crust.

3.8 The Silicates

Sketch the silicon–oxygen tetrahedron and explain how these fundamental building blocks join together to form other silicate structures.

KEY TERMS silicon–oxygen tetrahedron, polymerization

- Silicate minerals have a basic building block in common: a small pyramid-shaped structure consisting of one silicon atom surrounded by four oxygen atoms. Because this structure has four sides, it is called

the *silicon–oxygen tetrahedron*. Individual tetrahedra can be bonded to other elements, such as aluminum, iron, or potassium. Neighboring tetrahedra can share some of their oxygen atoms, causing them to develop long chains. This is the process of polymerization.

- Polymerization can produce silicate mineral structures with high or low degrees of oxygen sharing. The more sharing there is, the higher the ratio of silicon to oxygen. Polymerization can produce unit cells that are single "chains" of tetrahedra, double chains, sheets of shared tetrahedra, or even complicated three-dimensional networks of tetrahedra that share all the oxygen atoms in the mineral.

3.9 Common Silicate Minerals

Compare and contrast the light (nonferromagnesian) silicates with the dark (ferromagnesian) silicates and list three minerals common to each group.

KEY TERMS light (nonferromagnesian) silicate, potassium feldspar, plagioclase feldspar, quartz, muscovite, dark (ferromagnesian) silicate, olivine, augite, hornblende, biotite

- Silicate minerals are the most common mineral class on Earth. They are subdivided into minerals that contain iron and/or magnesium (dark, or ferromagnesian, minerals) and those that do not (light, or nonferromagnesian, minerals).

- Nonferromagnesian minerals are generally light in color and generally have relatively low specific gravity. Feldspar, quartz, muscovite, and clays are all examples.

- Ferromagnesian minerals are generally dark in color and relatively dense. Olivine, pyroxene, hornblende, biotite, and garnet are all examples.

Dennis Tasa

Smoky quartz

Q In general, nonferromagnesian minerals are light in color—shades of peach, tan, clear, or white. What could account for the fact that some nonferromagnesian minerals are dark colored, like the smoky quartz in this photo?

3.10 Important Nonsilicate Minerals

List the common nonsilicate minerals and explain why each is important.

- Nonsilicate mineral groups don't have the silicon–oxygen tetrahedron as the fundamental unit of their structures. Instead, these minerals are made of other elements in other chemical arrangements.

- Oxides are dominated by the bonding of other elements (usually metals) to oxygen ions. Carbonates have CO_3 as a critical part of their crystal structure. Sulfates have SO_4 as their basic building block. Halides have

a nonmetal ion such as chlorine, bromine, or fluorine bonded to a metal ion such as sodium or calcium.

- Nonsilicate minerals are often economic minerals. Hematite is an important source of industrial iron, while calcite is a critical component of cement. Halite makes popcorn taste good.

Q Examine Figure 3.41. What would be the most abundant elements on a planet composed of mostly halide minerals instead of silicate minerals? What about a carbonate planet?

Give It Some Thought

1. Using the geologic definition of *mineral* as your guide, determine which of the items on this list are minerals and which are not. If something in this list is not a mineral, explain.

- **a.** Gold nugget
- **b.** Seawater
- **c.** Quartz
- **d.** Cubic zirconia
- **e.** Obsidian
- **f.** Ruby
- **g.** Glacial ice
- **h.** Amber

Refer to the periodic table of the elements (Figure 3.5) to help answer Question 2.

2. Assume that the number of protons in a neutral atom is 92, and its mass number is 238.

   a. What is the name of the element?

   b. How many electrons does it have?

   c. How many neutrons does it have?

3. Referring to the accompanying photos of five minerals, determine which of these specimens exhibit a metallic luster and which have a nonmetallic luster.

A. B. C.
D. E.
(Photos by Dennis Tasa)

4. Gold has a specific gravity of almost 20. A 5-gallon bucket of water weighs 40 pounds. How much would a 5-gallon bucket of gold weigh?

5. Examine the accompanying photo of a mineral that has several smooth, flat surfaces that resulted when the specimen was broken.

Dennis Tasa
Cleaved sample

   a. How many flat surfaces are present on this specimen?

   b. How many different directions of cleavage does this specimen have?

   c. Do the cleavage directions meet at 90-degree angles?

6. Each of the following statements describes a silicate mineral or mineral group. In each case, provide the appropriate name:

   a. The most common member of the amphibole group

   b. The most common light-colored member of the mica family

   c. The only common silicate mineral made entirely of silicon and oxygen

   d. A silicate mineral with a name based on its color

   e. A silicate mineral characterized by striations

   f. A silicate mineral that originates as a product of chemical weathering

7. What mineral property is illustrated in the accompanying photo?

Dennis Tasa

8. Do an Internet search to determine which minerals are used to manufacture the following products.

   a. Stainless steel utensils

   b. Cat litter

   c. Tums brand antacid tablets

   d. Lithium batteries

   e. Aluminum beverage cans

9. The accompanying diagram shows one of several different arrangements for bonding silicon–oxygen tetrahedra. List each arrangement described in this chapter and give an example of a mineral for each.

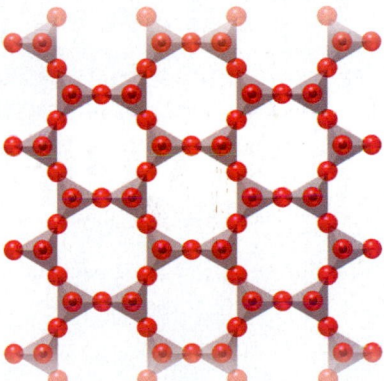

Minerals such as muscovite (in the mica mineral group) exhibit a sheet-like structure with the sheets consisting of shared silicon-oxygen tetrahedra.

10. Most states have designated a state mineral, rock, or gemstone to promote interest in that state's natural resources. Describe your state mineral, rock, or gemstone and explain why it was selected. If your state does not have a state mineral, rock, or gemstone, describe one from a state adjacent to yours.

4

Magma, Igneous Rocks, and Intrusive Activity

Aerial view of Yosemite valley which makes up a portion of the Sierra Nevada batholith, California. (Photo by Michael Collier)

Each statement represents the primary **LEARNING OBJECTIVE** for the corresponding major heading within the chapter. After you complete the chapter, you should be able to:

4.1 List and describe the three major components of magma.

4.2 Compare and contrast the four basic igneous compositions: basaltic (mafic), granitic (felsic), andesitic (intermediate), and ultramafic.

4.3 Identify and describe the six major igneous textures.

4.4 Distinguish among the common igneous rocks based on texture and mineral composition.

4.5 Summarize the major processes that generate magma from solid rock.

4.6 Describe how magmatic differentiation can generate a magma body that has a mineralogy (chemical composition) that is different from its parent magma.

4.7 Describe how partial melting of the mantle rock peridotite can generate a basaltic (mafic) magma.

4.8 Compare and contrast these intrusive igneous structures: dikes, sills, batholiths, stocks, and laccoliths.

Understanding the structure, composition, and internal workings of our planet requires a basic knowledge of igneous rocks. Igneous rocks and metamorphic rocks derived from igneous "parents" make up most of Earth's crust and mantle. Thus, Earth can be described as a huge mass of igneous and metamorphic rocks that is covered with a thin veneer of sedimentary rock and has a relatively small iron-rich core.

Many prominent landforms are composed of igneous rocks, including volcanoes such as Mount Rainier and the large igneous bodies that make up the Sierra Nevada, the Black Hills, and the high peaks of the Adirondacks. Igneous rocks also make excellent building stones and are widely used as decorative materials, such as for monuments and household countertops.

4.1 | Magma: Parent Material of Igneous Rock

List and describe the three major components of magma.

Our discussion of the rock cycle in Chapter 1 explained that **igneous rocks** (*ignis* = fire) form as molten rock cools and solidifies. Considerable evidence supports the idea that the parent material for igneous rocks, called **magma**, is formed by partial melting that occurs at various levels within Earth's crust and upper mantle to depths of about 250 kilometers (about 150 miles). Once formed, a magma body buoyantly rises toward the surface because it is less dense than the surrounding rocks. (When rock melts, it takes up more space and, hence, it becomes less dense than the surrounding solid rock.) Occasionally, molten rock reaches Earth's surface, where it is called **lava** (**Figure 4.1**). Sometimes lava is emitted as fountains that are produced when escaping gases propel it from a magma chamber. On other occasions, lava is explosively ejected, producing dramatic steam and ash eruptions. However, not all eruptions are violent; many volcanoes emit quiet outpourings of fluid lava.

The Nature of Magma

Magma is completely or partly molten rock, which when cooled solidifies to form igneous rocks composed mainly of silicate minerals. Most magmas consist of three materials: a *liquid* component, a *solid* component, and a *gaseous* component.

The liquid portion, called **melt**, is composed mainly of mobile ions of the eight most common elements found in Earth's crust—silicon and oxygen, along with lesser amounts of aluminum, potassium, calcium, sodium, iron, and magnesium.

The solid components (if any) in magma are crystals of silicate minerals. As a magma body cools, the size and number of crystals increase. During the last stage of cooling, a magma body is like a "crystalline mush" (resembling a bowl of very thick oatmeal) that contains only small amounts of melt.

The gaseous components of magma, called **volatiles**, are materials that vaporize (form a gas) at surface pressures. The most common volatiles found in magma are water vapor (H_2O), carbon dioxide

Figure 4.1
Fluid lava emitted from Bardarbunga volcano, Iceland Iceland is home to some of the most active volcanoes on the planet.
(Lukas Gawenda/Alamy)

(CO_2), and sulfur dioxide (SO_2), which are confined by the immense pressure exerted by the overlying rocks. These gases tend to separate from the melt as it moves toward the surface (from a high- to a low-pressure environment). As the gases build up, they may eventually propel magma from the vent. When deeply buried magma bodies crystallize, the remaining volatiles collect as hot, water-rich fluids that migrate through the surrounding rocks. These hot fluids play an important role in metamorphism and will be considered in Chapter 8.

From Magma to Crystalline Rock

To better understand how magma crystallizes, let us consider how a simple crystalline solid melts. Recall that, in any crystalline solid, the ions are arranged in a closely packed regular pattern. However, they are not without some motion; they exhibit a restricted vibration about fixed points. As the temperature rises, ions vibrate more rapidly and consequently collide with ever-increasing vigor with their neighbors. Thus, heating causes the ions to occupy more space, which in turn causes the solid to expand. When the ions are vibrating rapidly enough to overcome the force of their chemical bonds, melting occurs. At this stage, the ions are able to slide past one another, and the orderly crystalline structure disintegrates. Thus, melting converts a solid consisting of tight, uniformly packed ions into a liquid composed of unordered ions moving randomly about.

In the process called **crystallization**, cooling reverses the events of melting. As the temperature of the liquid drops, ions pack more closely together as their rate of movement slows. When they are cooled sufficiently, the forces of the chemical bonds will again confine the ions to an orderly crystalline arrangement.

When a magma body cools, the silicon and oxygen atoms link together first to form silicon–oxygen tetrahedra, the basic building blocks of the silicate minerals (see

Potassium feldspar (pink)   **Amphibole** (black)

Quartz (gray, glassy)   **Plagioclase feldspar** (white)

Dennis Tasa

Figure 4.2
Igneous rock composed of interlocking crystals The largest crystals are about 1 centimeter in length.

Figure 3.31, page 92). As magma continues to lose heat to its surroundings, the tetrahedra join with each other and with other ions to form embryonic crystal nuclei. Slowly each nucleus grows as ions lose their mobility and join the crystalline network.

The minerals that form the earliest have space to grow and tend to have better-developed crystal faces than do the ones that form later and occupy the remaining spaces. Eventually all of the melt is transformed into a solid mass of interlocking silicate minerals that we call an *igneous rock* (**Figure 4.2**).

Igneous Processes

Igneous rocks form in two basic settings. Molten rock may crystallize within Earth's crust over a range of depths, or it may solidify at Earth's surface (**Figure 4.3**).

SmartFigure 4.3
Intrusive versus extrusive igneous rocks (https://goo.gl/wnDbqk)

Tutorial

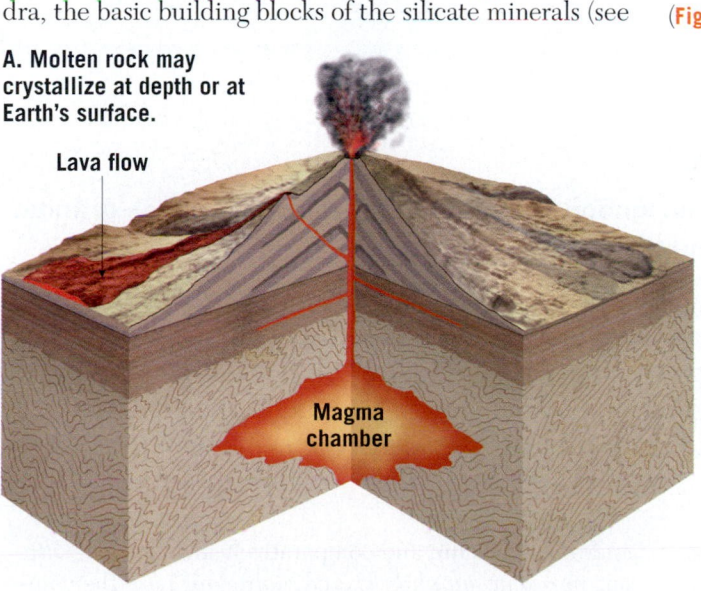

A. Molten rock may crystallize at depth or at Earth's surface.

Lava flow

Magma chamber

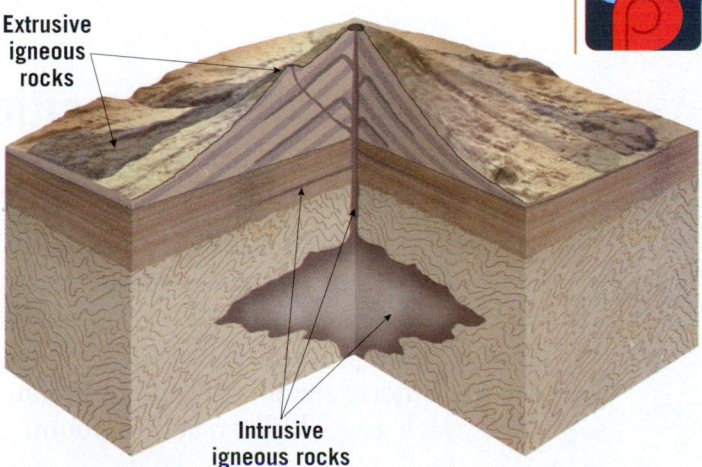

Extrusive igneous rocks

Intrusive igneous rocks

B. When magma crystallizes at depth, intrusive igneous rocks form. When magma solidifies on Earth's surface, extrusive igneous rocks form.

Figure 4.4
Mount Rushmore National Memorial This memorial, located in the Black Hills of South Dakota, is carved from intrusive igneous rocks. (Photo by Barbara A. Harvey/Shutterstock)

When magma crystallizes *at depth*, it forms **intrusive igneous rocks**, also known as **plutonic rocks**—after Pluto, the god of the underworld in classical mythology. These rocks are observed at the surface in locations where uplifting and erosion have stripped away the overlying rocks. Exposures of intrusive igneous rocks occur in many places, including the White Mountains, New Hampshire; Stone Mountain, Georgia; Mount Rushmore in the Black Hills of South Dakota; and Yosemite National Park, California (**Figure 4.4**).

Igneous rocks that form when molten rock solidifies *at the surface* are classified as **extrusive igneous rocks**. They are also called **volcanic rocks**—after Vulcan, the Roman fire god. Extrusive igneous rocks form when lava solidifies or when volcanic debris falls to Earth's surface. Extrusive igneous rocks are abundant in western portions of the Americas, where they make up the volcanic peaks of the Cascade Range

and the Andes Mountains. In addition, many oceanic islands, including the Hawaiian chain and Alaska's Aleutian Islands, are composed almost entirely of extrusive igneous rocks. The nature of volcanic activity will be addressed in more detail in Chapter 5.

> ### 4.1 | Concept Checks
>
> 1. What is magma? How does magma differ from lava?
> 2. List and describe the three components of magma.
> 3. Describe the process of crystallization.
> 4. Compare and contrast extrusive and intrusive igneous rocks.

4.2 | Igneous Compositions

Compare and contrast the four basic igneous compositions: basaltic (mafic), granitic (felsic), andesitic (intermediate), and ultramafic.

Igneous rocks are composed mainly of silicate minerals. Chemical analyses show that silicon and oxygen are by far the most abundant constituents of igneous rocks. These two elements, plus ions of aluminum (Al), calcium (Ca), sodium (Na), potassium (K), magnesium (Mg), and iron (Fe), make up roughly 98 percent, by weight, of most magmas. In addition, magma contains small amounts of many other elements, including titanium and manganese, and trace amounts of much rarer elements, such as gold, silver, and uranium.

As magma cools and solidifies, these elements combine to form two major groups of silicate minerals (**Figure 4.5**). The *dark* (or *ferromagnesian*) *silicates* are rich in iron

and/or magnesium and comparatively low in silica. *Olivine, pyroxene, amphibole,* and *biotite mica* are the common dark silicate minerals of Earth's crust. By contrast,

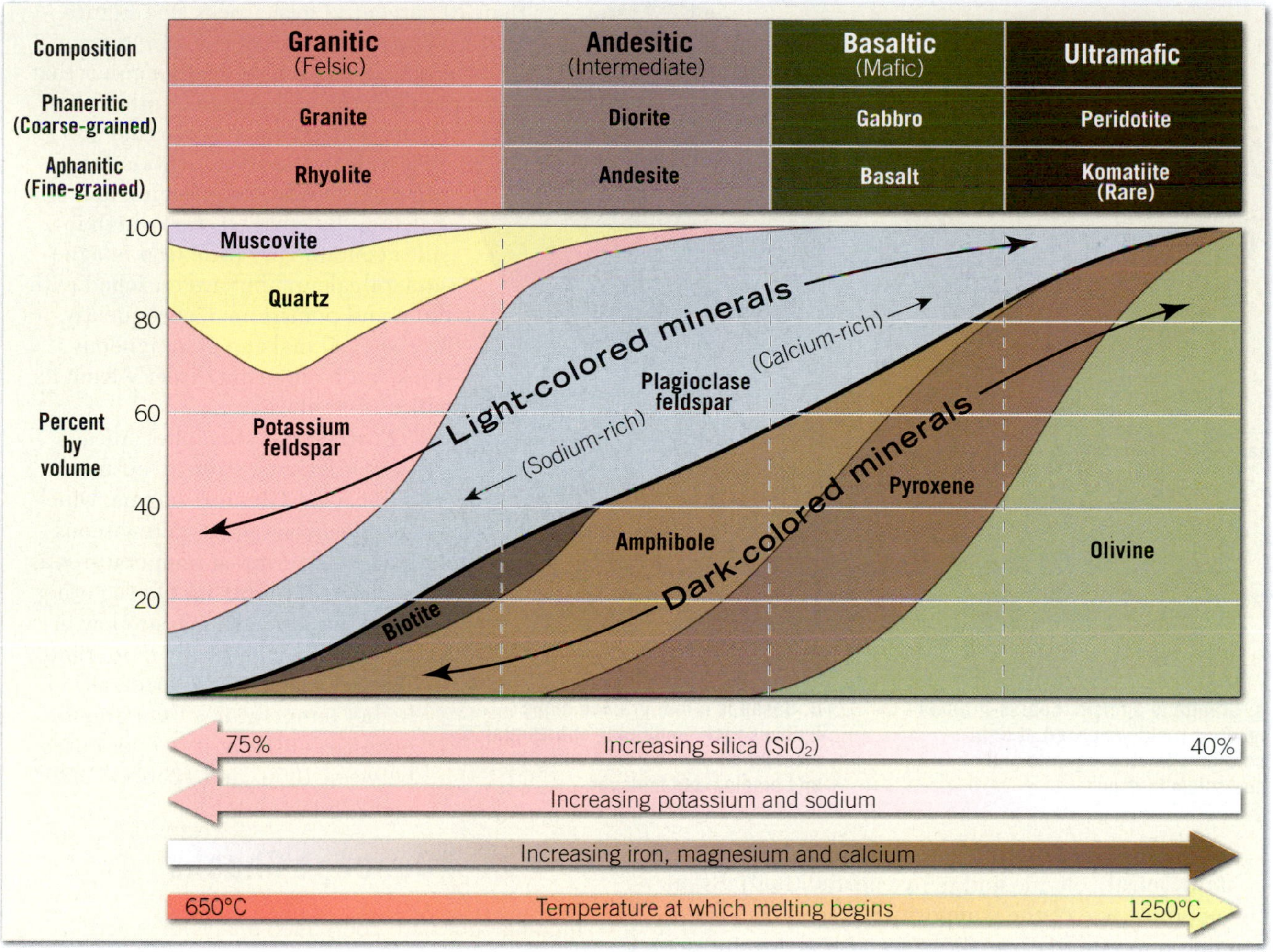

SmartFigure 4.5
Mineralogy of common igneous rocks (Based on Dietrich, Daily, and Larson.) (https://goo.gl/vST3z8)

Tutorial

Composition	Granitic (Felsic)	Andesitic (Intermediate)	Basaltic (Mafic)	Ultramafic
Phaneritic (Coarse-grained)	Granite	Diorite	Gabbro	Peridotite
Aphanitic (Fine-grained)	Rhyolite	Andesite	Basalt	Komatiite (Rare)

Figure 4.5 — diagram with Percent by volume (vertical axis 0–100) showing mineral fields: Muscovite, Quartz, Potassium feldspar, Biotite, Plagioclase feldspar (Sodium-rich / Calcium-rich), Amphibole, Pyroxene, Olivine. Labels: Light-colored minerals, Dark-colored minerals.

Increasing silica (SiO₂) 75% ← → 40%
Increasing potassium and sodium
Increasing iron, magnesium and calcium
Temperature at which melting begins 650°C — 1250°C

---

the *light* (or *nonferromagnesian*) *silicates* contain greater amounts of potassium, sodium, and calcium rather than iron and magnesium. The light silicate minerals, including *quartz*, *muscovite mica*, and the most abundant mineral group, the *feldspars*, are richer in silica than the dark silicates.

Granitic (Felsic) Versus Basaltic (Mafic) Compositions

Despite their great compositional diversity, igneous rocks (and the magmas from which they form) can be divided into broad groups according to their proportions of light and dark minerals (see Figure 4.5). Near one end of the continuum are rocks composed almost entirely of light-colored silicates—quartz and potassium feldspar. Igneous rocks in which these are the dominant minerals have a **granitic composition** (**Figure 4.6A**). Geologists also refer to granitic rocks as being **felsic**, a term derived from *fel*dspar and *si*lica (quartz). In addition to quartz and feldspar, most granitic rocks contain about 10 percent dark silicate minerals, usually biotite mica and amphibole. Granitic rocks are rich in silica (about 70 percent) and are major constituents of the continental crust.

Rocks that contain at least 45 percent dark silicate minerals and calcium-rich plagioclase feldspar (but no quartz) are said to have a **basaltic composition** (**Figure 4.6B**). Basaltic rocks contain a high percentage of ferromagnesian minerals, so geologists also refer to them as **mafic** (from *ma*gnesium and *f*errum, the Latin name for iron). Because of their iron content, mafic rocks are typically darker and denser than granitic rocks. Basaltic rocks make up the ocean floor as well as many of the volcanic islands located within the ocean basins. Basalt also forms extensive lava flows on the continents.

Other Compositional Groups

As you can see in Figure 4.5, rocks with a composition between granitic and basaltic rocks are said to have an **andesitic composition**, or **intermediate composition**, after the common volcanic rock *andesite*. Intermediate rocks contain at least 25 percent dark silicate minerals, mainly amphibole, pyroxene, and biotite mica, with the other dominant mineral being plagioclase feldspar. This important category of igneous rocks is often associated with volcanic activity that is typically confined to the seaward margins of the continents and on volcanic island arcs such as the Aleutian chain.

Figure 4.6
Granitic (felsic) versus basaltic (mafic) compositions Inset images are photomicrographs that show the interlocking crystals that make up granite and basalt, respectively. (Photos by E. J. Tarbuck)

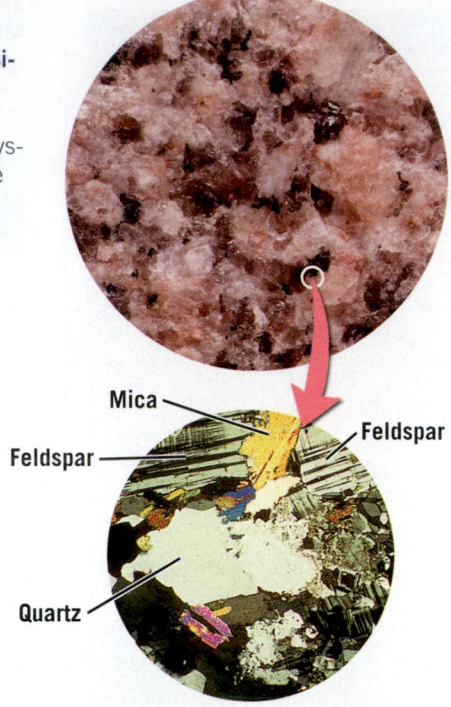

Mica
Feldspar
Feldspar
Quartz

Feldspar (white)
Pyroxene (black)

A. Granite is a felsic, coarse-grained igneous rock composed of light-colored silicates—quartz and potassium feldspar.

B. Basalt is a fine-grained mafic igneous rock containing substantial amounts of dark colored silicates and plagioclase feldspar.

than 70 percent in felsic rocks (see Figure 4.5). The percentage of silica in igneous rocks varies in a systematic manner that parallels the abundance of other elements. For example, rocks that are relatively low in silica contain large amounts of iron, magnesium, and calcium. By contrast, rocks high in silica contain very little iron, magnesium, or calcium but are enriched with sodium and potassium. Consequently, the chemical makeup of an igneous rock can be inferred directly from its silica content.

Further, the amount of silica present in magma strongly influences its behavior. Granitic magma, which has a high silica content, is quite viscous ("thick") and may erupt at temperatures as low as 650°C (1200°F). On the other hand, basaltic magmas are low in silica and are generally more fluid. Basaltic magmas also erupt at higher temperatures than granitic magmas—usually at temperatures between 1050° and 1250°C (1920° and 2280°F).

Another important igneous rock, **peridotite**, contains mostly olivine and pyroxene and thus falls on the opposite side of the compositional spectrum from granitic rocks (see Figure 4.5). Because peridotite is composed almost entirely of ferromagnesian minerals, its chemical composition is referred to as **ultramafic**. Although ultramafic rocks are rare at Earth's surface, peridotite is the main constituent of the upper mantle.

Silica Content as an Indicator of Composition

An important aspect of the chemical composition of igneous rocks is silica (SiO_2) content. Typically, the silica content of crustal rocks ranges from a low of about 40 percent in ultramafic rocks to a high of more

4.2   Concept Checks

1. Igneous rocks are composed mainly of which group of minerals?

2. How do light-colored igneous rocks differ in composition from dark-colored igneous rocks?

3. List the four basic compositional groups of igneous rocks, in order from the group with the highest silica content to the group with the lowest silica content.

4. Name two minerals typically found in rocks with high silica content and two minerals found in rocks with relatively low silica content.

4.3 | Igneous Textures: What Can They Tell Us?

Identify and describe the six major igneous textures.

The term **texture** is used to describe the overall appearance of a rock based on the size, shape, and arrangement of its mineral grains—not how it feels to touch. Texture is an important property because it reveals a great deal about the environment in which the rock formed (**Figure 4.7**). Geologists can make inferences about a rock's origin based on careful observations of grain size and other characteristics of the rock.

A. Glassy texture
Composed of unordered atoms and resembles dark manufactured glass. (Obsidian is a natural glass that usually forms when highly silica-rich magmas solidify.)

D. Vesicular texture
Extrusive rock containing voids left by gas bubbles that escape as lava solidifies. (Pumice is a frothy volcanic glass that displays a vesicular texture.)

B. Porphyritic texture
Composed of two distinctly different crystal sizes.

E. Pyroclastic (fragmental) texture
Produced by the consolidation of fragments that may include ash, once molten blobs, or large angular blocks that were ejected during an explosive volcanic eruption.

C. Phaneritic (coarse-grained) texture
Composed of mineral grains that are large enough to be identified without a microscope.

F. Aphanitic (fine-grained) texture
Composed of crystals that are too small for the individual minerals to be identified without a microscope.

Three factors influence the textures of igneous rocks:

- The rate at which molten rock cools
- The amount of silica present in the magma
- The amount of dissolved gases in the magma

Among these, the rate of cooling tends to be the dominant factor. A very large magma body located many kilometers beneath Earth's surface remains insulated from lower surface temperatures by the surrounding rock and thus will cool very slowly over a period of perhaps tens of thousands to millions of years. Initially, relatively few crystal nuclei form. Slow cooling permits ions to migrate freely until they eventually join one of the existing crystals. Consequently, slow cooling promotes the growth of fewer but larger crystals.

On the other hand, when cooling occurs rapidly—for example, in a thin lava flow—the ions quickly lose their mobility and readily combine to form crystals. This results in the development of numerous embryonic nuclei, all of which compete for the available ions. The result is a solid mass of many tiny intergrown crystals.

Types of Igneous Textures

The effect of cooling on rock textures is fairly straightforward: Slow cooling promotes the growth of large crystals, whereas rapid cooling tends to generate small crystals. However, a magma body may migrate to a new location or erupt at the surface before it completely solidifies. As a result, several types of igneous textures exist, including

SmartFigure 4.7
Igneous rock textures
(Photos by E. J. Tarbuck)
(https://goo.gl/U00Ix8)

Tutorial

Figure 4.8
Porphyritic texture The large crystals in porphyritic rocks are called *phenocrysts*, and the matrix of smaller crystals is called *groundmass*. (Photo by Dennis Tasa)

Groundmass

Phenocryst

1 cm

aphanitic (fine-grained), phaneritic (coarse-grained), porphyritic, vesicular, glassy, pyroclastic (fragmented), and pegmatitic.

Aphanitic (Fine-Grained) Texture

Igneous rocks that form at the surface or as small intrusive masses within the upper crust where cooling is relatively rapid exhibit a **fine-grained texture** termed **aphanitic**

(*a* = not, *phaner* = visible). By definition, the crystals that make up aphanitic rocks are so small that individual minerals can be distinguished only with the aid of a polarizing microscope or other sophisticated techniques (see Figure 4.6B). Therefore, we commonly characterize fine-grained rocks as being light, intermediate, or dark in color. Using this system of grouping, light-colored aphanitic rocks are those containing primarily light-colored nonferromagnesian silicate minerals.

Phaneritic (Coarse-Grained) Texture

When large masses of magma slowly crystallize at great depth, they form igneous rocks that exhibit a **coarse-grained texture** described as **phaneritic** (*phaner* = visible). Coarse-grained rocks consist of a mass of intergrown crystals that are roughly equal in size and large enough to distinguish the individual minerals without the aid of a microscope (see Figure 4.6A). Geologists often use a small magnifying lens to aid in identifying minerals in a phaneritic rock.

Porphyritic Texture

A large mass of magma may require thousands or even millions of years to solidify. Because different minerals crystallize under different environmental conditions (temperatures and pressure), it is possible for crystals of one mineral to become quite large before others even begin to form. If molten rock containing some large crystals moves to a different environment—for example, by erupting at the surface—the remaining liquid portion of the lava cools more quickly. The resulting rock, which has large crystals embedded in a matrix of smaller crystals, is said to have a **porphyritic texture** (**Figure 4.8**). The large crystals in porphyritic rocks are referred to as **phenocrysts** (*pheno* = show, *cryst* = crystal), whereas the matrix of smaller crystals is called **groundmass**. A rock with a *porphyritic* texture is termed a **porphyry**.

Vesicular Texture

Common features of many extrusive rocks are the voids left by gas bubbles that escape as lava solidifies. These nearly spherical openings are called *vesicles*, and the rocks that contain them are said to have a **vesicular texture**. Rocks that exhibit a vesicular texture often form in the upper zone of a lava flow, where cooling occurs rapidly enough to preserve the openings produced by the expanding gas bubbles (**Figure 4.9**). Another common vesicular rock, called *pumice*, forms when silica-rich lava is ejected during an explosive eruption (see Figure 4.7D).

Glassy Texture

During some volcanic eruptions, molten rock is ejected into the atmosphere, where it is quenched (very quickly cooled) to become a solid. Rapid

Lava flow

Vesicular texture

Figure 4.9
Vesicular texture The larger image shows a lava flow on Hawaii's Kilauea volcano. The inset photo is a close-up showing the vesicular texture of hardened lava. Vesicles are small holes left by escaping gas bubbles. (Inset photo by E. J. Tarbuck)

USGS

cooling of this type may generate rocks having a **glassy texture**. Glass results when unordered ions are "frozen in place" before they are able to unite into an orderly crystalline structure.

Obsidian, a common type of natural glass, is similar in appearance to dark chunks of manufactured glass. Because of its excellent conchoidal fracture and ability to hold a sharp, hard edge, obsidian was a prized material from which Native Americans chipped arrowheads and cutting tools (**Figure 4.10**).

Obsidian flows, typically a few hundred feet thick, provide evidence that rapid cooling is not the only mechanism that produces a glassy texture. Magmas with high silica content tend to form long, chainlike structures (polymers) before crystallization is complete. These structures, in turn, impede ionic transport and increase the magma's viscosity. (*Viscosity* is a measure of a fluid's resistance to flow.) So, granitic magma, which is rich in silica, may be extruded as an extremely viscous mass that eventually solidifies to form obsidian.

By contrast, basaltic magma, which is low in silica, forms very fluid lavas that, upon cooling, usually generate fine-grained crystalline rocks. However, when a basaltic lava flow enters the ocean, its surface is quenched rapidly enough to form a thin, glassy skin.

Pyroclastic (Fragmental) Texture

Another group of igneous rocks is formed from the consolidation of individual rock fragments ejected during explosive volcanic eruptions. The ejected particles might be very fine ash, molten blobs, or large angular blocks torn from the walls of a vent during an eruption (**Figure 4.11**). Igneous rocks composed of these rock fragments are said to have a **pyroclastic texture**, or **fragmental texture** (see Figure 4.7E).

A common type of pyroclastic rock, called *welded tuff*, is composed of fine fragments of glass that remained hot enough to fuse together. Other pyroclastic rocks are composed of fragments that solidified before impact and became cemented together at some later time. Because pyroclastic rocks are made of individual particles or fragments rather than interlocking crystals, their textures often resemble those exhibited by sedimentary rocks rather than those associated with igneous rocks.

Pegmatitic Texture

Under special conditions, exceptionally coarse-grained igneous rocks, called

Figure 4.10
Obsidian arrowhead Native Americans made arrowheads and cutting tools from obsidian, a natural glass. (Photo by Jeffrey Scovil)

Figure 4.11
Pyroclastic rocks are the product of explosive eruptions These volcanic fragments may eventually consolidate to become rocks displaying a pyroclastic texture. (Photo by Richard Roscoe/Getty Images)

pegmatites, may form. Rocks of this type, in which most of the crystals are larger than 1 centimeter in diameter, are described as having a **pegmatitic texture** (**Figure 4.12**). Most pegmatites occur as small masses or thin veins within or around the margins of large intrusive igneous bodies.

Pegmatites form late in the crystallization of a magma, when water and other materials, such as carbon dioxide, chlorine, and fluorine, make up an unusually high percentage of the melt. Because ion

Feldspar

Quartz

Figure 4.12
Pegmatitic texture This granite pegmatite, found in the inner gorge of the Grand Canyon, is composed mainly of quartz and feldspar. (Photo by Joanne Bannon/E. J. Tarbuck)

EYE ON EARTH 4.1

These two types of hardened lava, called Pele's hair and Pele's tears, are commonly generated during lava fountaining that occurs at Kilauea volcano. They are named after Pele, the Hawaiian goddess of fire and volcanoes. Pele's hair consists of goldish strands that are created when tiny blobs of lava are stretched by strong winds. When lava droplets cool quickly, they form Pele's tears, which are sometimes connected to strands of Pele's hair.

QUESTION 1 What is the texture of Pele's hair and Pele's tears?

QUESTION 2 What kind of igneous rock is Pele's tears?

Pele's hair   Marli Miller

Pele's tears   USGS

migration is enhanced in these fluid-rich environments, the crystals that form are abnormally large. Thus, the large crystals in pegmatites do not result from inordinately long cooling histories; rather, they are the consequence of a fluid-rich environment that enhances crystallization.

The composition of most pegmatites is similar to that of granite. Thus, pegmatites contain large crystals of quartz, feldspar, and muscovite. However, some contain significant quantities of relatively rare and hence valuable elements—gold, tungsten, beryllium, and the rare earth elements that are used in modern high-technology devices, including cell phones and hybrid autos.

4.3   Concept Checks

1. Define *texture*.

2. How does the rate of cooling influence crystal size? What other factors influence the texture of igneous rocks?

3. List the six major igneous rock textures.

4. Explain how the crystals in pegmatites are able to grow so large.

5. What does a porphyritic texture indicate about the cooling history of an igneous rock?

4.4 Naming Igneous Rocks

Distinguish among the common igneous rocks based on texture and mineral composition.

Geologists classify igneous rocks on the basis of their texture and mineral composition (**Figure 4.13**). The various igneous textures described in the previous section result mainly from different cooling histories, whereas the mineral composition of an igneous rock depends on the chemical makeup of its parent magma. Because igneous rocks are classified on the basis of both mineral composition and texture, some rocks having similar mineral constituents but exhibiting different textures are given different names.

Granitic (Felsic) Igneous Rocks

Granite   Of all the igneous rocks, **granite** is perhaps the best known (see GEOgraphics 4.1). This is because of its natural beauty, which is enhanced when it is polished, and its abundance in the continental crust. Slabs of polished granite are commonly used for tombstones and monuments and as building stones. Well-known areas in the United States where granite is quarried

include Barre, Vermont; Mount Airy, North Carolina; and St. Cloud, Minnesota.

Granite is a coarse-grained rock composed of about 10 to 20 percent quartz and roughly 50 percent feldspar. When examined close up, the quartz grains appear somewhat rounded in shape, glassy, and clear to gray in color. By contrast, feldspar crystals are generally white, gray, or salmon pink in color, and they are blocky or rectangular in shape. Other minor constituents of granite include

IGNEOUS ROCK CLASSIFICATION CHART

		MINERAL COMPOSITION			
		Granitic (Felsic)	**Andesitic** (Intermediate)	**Basaltic** (Mafic)	**Ultramafic**
	Dominant Minerals	Quartz Potassium feldspar	Amphibole Plagioclase feldspar	Pyroxene Plagioclase feldspar	Olivine Pyroxene
	Accessory Minerals	Plagioclase feldspar Amphibole Muscovite Biotite	Pyroxene Biotite	Amphibole Olivine	Plagioclase feldspar
TEXTURE	**Phaneritic** (coarse-grained)	Granite	Diorite	Gabbro	Peridotite
	Aphanitic (fine-grained)	Rhyolite	Andesite	Basalt	Komatiite (rare)
	Porphyritic (two distinct grain sizes)	Granite porphyry	Andesite porphyry	Basalt porphyry	Uncommon
	Glassy	Obsidian	Less common	Less common	Uncommon
	Vesicular (contains voids)	Pumice (also glassy)		Scoria	Uncommon
	Pyroclastic (fragmental)	Most fragments < 4mm — Tuff or welded tuff		Most fragments > 4mm — Volcanic breccia	Uncommon
	Rock Color (based on % of dark minerals)	0% to 25%	25% to 45%	45% to 85%	85% to 100%

SmartFigure 4.13
Classification of igneous rocks Igneous rocks are classified based on mineral composition and texture.
(Photos by Dennis Tasa and E. J. Tarbuck) (https://goo.gl/WiyTul)

▶ Tutorial

SmartFigure 4.14
Rocks contain information about the processes that produced them This massive granitic monolith (El Capitan) located in Yosemite National Park, California, was once a molten mass deep within Earth. (http://goo.gl/z8XJ1i)

Cory Rich/Getty Images

Mobile Field Trip

Michael Collier

Granite

Dennis Tasa

small amounts of dark silicates, particularly biotite and amphibole, and sometimes muscovite. Although the dark components generally make up less than 10 percent of most granites, dark minerals appear to be more prominent than their percentage indicates.

At a distance, most granitic rocks appear gray in color (**Figure 4.14**). However, when composed of dark pink feldspar grains, granite can exhibit a reddish color. In addition, some granites have a porphyritic texture. These specimens contain elongated feldspar crystals a few centimeters in length that are scattered among smaller crystals of quartz and amphibole (see Figure 4.13).

Rhyolite   **Rhyolite** is the fine-grained equivalent of granite and, like granite, is composed essentially of the light-colored silicates (see Figure 4.13). This fact accounts for its color, which is usually buff to pink or occasionally light gray. Rhyolite is fine-grained and frequently contains glass fragments and voids, indicating that it cooled rapidly in a surface, or near-surface, environment. In contrast to granite, which is widely distributed as large intrusive masses, rhyolite deposits are less common and generally less voluminous. The thick rhyolite lava flows and extensive ash deposits in and around Yellowstone National Park are well-known exceptions to this generalization.

Obsidian   **Obsidian** is a dark-colored glassy rock that usually forms when highly silica-rich lava cools quickly at Earth's surface (see Figure 4.13). In contrast to the orderly arrangement of ions characteristic of minerals, the arrangement of ions in glass is unordered. Consequently, glassy rocks such as obsidian are not composed of minerals in the same sense as most other rocks.

Although usually black or reddish-brown in color, obsidian has a chemical composition that is roughly equivalent to that of the light-colored igneous rock granite, not like dark rocks such as basalt. Obsidian's dark color results from small amounts of metallic ions in an otherwise relatively clear, glassy substance. If you examine a thin edge, obsidian will appear nearly transparent (see Figure 4.7A).

Pumice   **Pumice** is a glassy volcanic rock with a vesicular texture that forms when large amounts of gas escape through silica-rich lava to generate a gray, frothy mass. In some samples, the voids are quite noticeable, whereas in others the pumice resembles fine shards of intertwined glass. Because of the large percentage of voids, many samples of pumice float when placed in water (**Figure 4.15**). Oftentimes, flow lines are visible in pumice, indicating that some movement occurred before solidification was complete. Moreover, pumice and obsidian can often be

Figure 4.16
Basaltic lava flowing from Kilauea volcano, Hawaii (Photo by David Reggie/ Getty Images)

Figure 4.15
Pumice, a vesicular (and also glassy) igneous rock Most samples of pumice will float in water because they contain numerous vesicles. (Left photo by E. J. Tarbuck; right photo by Chip Clark/ Fundamental Photos)

found in the same rock mass, where they exist in alternating layers.

Andesitic (Intermediate) Igneous Rocks

Andesite **Andesite** is a medium-gray, fine-grained rock typically of volcanic origin. Its name comes from South America's Andes Mountains, where numerous volcanoes are composed of this rock type. In addition, the volcanoes of the Cascade Range and many of the volcanic structures occupying the continental margins that surround the Pacific Ocean are of andesitic composition. Andesite commonly exhibits a porphyritic texture (see Figure 4.13). When this is the case, the phenocrysts are often light, rectangular crystals of plagioclase feldspar or black, elongated amphibole crystals. Andesite may also resemble rhyolite, so its identification usually requires microscopic examination to verify mineral makeup.

Diorite **Diorite** is the intrusive equivalent of andesite. It is a coarse-grained rock that looks somewhat like gray granite. However, it can be distinguished from granite because it contains little or no visible quartz crystals and has a higher percentage of dark silicate minerals. The mineral makeup of diorite is primarily plagioclase feldspar and amphibole. Because the light-colored feldspar grains and dark amphibole crystals appear to be roughly equal in abundance, diorite has a salt-and-pepper appearance (see Figure 4.13).

Basaltic (Mafic) Igneous Rocks

Basalt **Basalt** is a very dark green to black, fine-grained rock composed primarily of pyroxene and calcium-rich plagioclase feldspar, with lesser amounts of olivine and amphibole (see Figure 4.13). When it

is porphyritic, basalt commonly contains small, light-colored feldspar phenocrysts or green, glassy-appearing olivine grains embedded in a dark groundmass.

Basalt is the most common extrusive igneous rock. Many volcanic islands, such as the Hawaiian Islands and Iceland, are composed mainly of basalt (**Figure 4.16**). Further, the upper layers of the oceanic crust consist of basalt. In the United States, large portions of central Oregon and Washington were the sites of extensive basaltic outpourings (discussed in detail in Chapter 5). At some locations, these once-fluid basaltic flows have accumulated to a combined thicknesses approaching 3 kilometers (nearly 2 miles).

Gabbro **Gabbro** is the intrusive equivalent of basalt (see Figure 4.13). Like basalt, it tends to be dark green to black in color and composed primarily of pyroxene and calcium-rich plagioclase feldspar. Although gabbro is uncommon in the continental crust, it makes up a significant percentage of oceanic crust.

Pyroclastic Rocks

Pyroclastic rocks are composed of fragments ejected during a volcanic eruption. One of the most common pyroclastic rocks, called *tuff*, is composed mainly of tiny,

Granite: An Intrusive Igneous Rock

To geologists, granite is an intrusive igneous rock with a rather precise mineral composition. However, the term granite is commonly used to describe a group of related rocks that are rich in quartz and feldspar. We will follow the latter convention here.

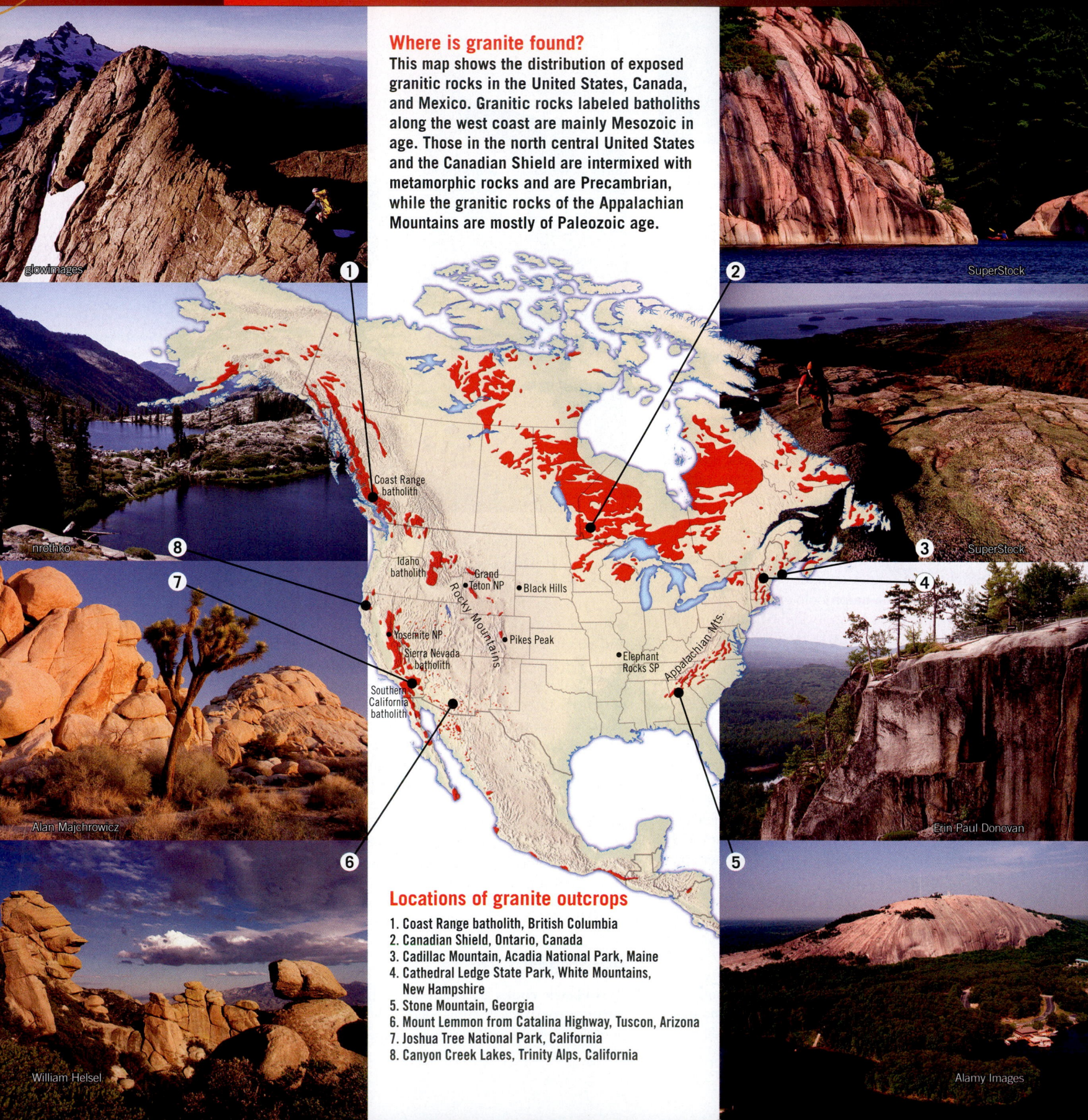

Where is granite found?

This map shows the distribution of exposed granitic rocks in the United States, Canada, and Mexico. Granitic rocks labeled batholiths along the west coast are mainly Mesozoic in age. Those in the north central United States and the Canadian Shield are intermixed with metamorphic rocks and are Precambrian, while the granitic rocks of the Appalachian Mountains are mostly of Paleozoic age.

glowimages

SuperStock

nrothko

SuperStock

Erin Paul Donovan

Alan Majchrowicz

Coast Range batholith

Idaho batholith

Grand Teton NP

Black Hills

Rocky Mountains

Yosemite NP

Pikes Peak

Sierra Nevada batholith

Southern California batholith

Elephant Rocks SP

Appalachian Mts.

Locations of granite outcrops

1. Coast Range batholith, British Columbia
2. Canadian Shield, Ontario, Canada
3. Cadillac Mountain, Acadia National Park, Maine
4. Cathedral Ledge State Park, White Mountains, New Hampshire
5. Stone Mountain, Georgia
6. Mount Lemmon from Catalina Highway, Tuscon, Arizona
7. Joshua Tree National Park, California
8. Canyon Creek Lakes, Trinity Alps, California

William Helsel

Alamy Images

What are the uses of granite?

Granite is widely used as a building material for countertops, floor tiles, building stones, and memorials.

Shutterstock

Polished granite is popular for countertops because it is durable as well as aesthetically pleasing. However, in the stone trade the name "granite" is applied to almost every coarse-grained igneous rock. "Black granite", for example, is usually the rock gabbro.

Shutterstock

Granite is used in large construction projects as a structural element and decorative facing or veneer. The Arlington Memorial Bridge, constructed mainly of concrete, is covered by a decorative granite veneer.

John Greim

Granite is commonly used for memorials and gravestones because of its resistance to chemical weathering. The Trenton Battle Monument commemorates the December 26, 1776 battle in which the Continental Army defeated the British forces in Trenton, New Jersey.

"Rock of Ages" granite quarry, Barre, Vermont

Granite was originally quarried using handsaws and explosives to blast away slabs of rock. Today, saws with diamond chips imbedded in steel wires and saw blades are used.

? **Question:**
List at least four different uses for granite.

Raymond Forbes

Alamy Images

Figure 4.17
Welded tuff, a pyroclastic igneous rock Outcrop of welded tuff from Valles Caldera near Los Alamos, New Mexico. Tuff is composed mainly of ash-sized particles and may contain larger fragments of pumice or other volcanic rocks. (Photo by Marli Miller)

Close-up

hundreds of feet thick and extend for more than 100 kilometers (60 miles) from their source. Most formed millions of years ago as volcanic ash spewed from large volcanic structures (calderas), sometimes spreading laterally at speeds approaching 100 kilometers per hour. Early investigators of these deposits incorrectly classified them as rhyolite lava flows. Today, we know that silica-rich lava is too viscous (thick) to flow more than a few miles from a vent.

Pyroclastic rocks composed mainly of particles larger than ash are called *volcanic breccia.* The particles in volcanic breccia may consist of streamlined lava blobs that solidified in air, blocks broken from the walls of the vent, ash, and glass fragments.

Unlike most igneous rock names, such as granite and basalt, the terms *tuff* and *volcanic breccia* do not imply mineral composition. Instead, they are frequently identified with a modifier; for example, *rhyolite tuff* indicates a rock composed of ash-size particles having a felsic composition.

ash-size fragments that were later cemented together. In situations where the ash particles remained hot enough to fuse, the rock is called *welded tuff.* Although welded tuff consists mostly of tiny glass shards, it may contain walnut-size pieces of pumice and other rock fragments.

Welded tuff deposits cover vast portions of previously volcanically active areas of the western United States (**Figure 4.17**). Some of these tuff deposits are

---

4.4 Concept Checks

1. List the two criteria by which igneous rocks are classified.

2. How are granite and rhyolite different? In what way are they similar?

3. Classify each of the following rocks by their mineral composition (felsic, intermediate, or mafic): gabbro, obsidian, granite, and andesite.

4. Describe each of the following in terms of composition and texture: diorite, rhyolite, and basalt porphyry.

5. In what way does tuff and volcanic breccia differ from other igneous rocks such as granite and basalt?

EYE ON EARTH 4.2

This photo shows rock exposed at Hole in the Wall, California, which formed as a result of an explosive ash eruption about 18.5 million years ago.

QUESTION 1 Most of this rock is composed of tiny ash fragments. What name is given to rocks of this type?

QUESTION 2 What rock name is given to the dark fragment in the upper part of the photo?

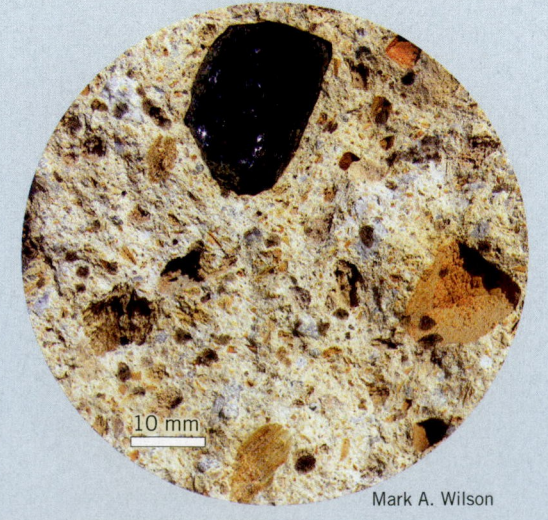

10 mm

Mark A. Wilson

4.5 | Origin of Magma

Summarize the major processes that generate magma from solid rock.

Based on evidence from the study of earthquake waves, we know that *Earth's crust and mantle are composed primarily of solid, not molten, rock.* Although the outer core is fluid, this iron-rich material is very dense and remains deep within Earth. So where does magma come from?

Most magma originates in Earth's uppermost mantle. The greatest quantities are produced at divergent plate boundaries, in association with seafloor spreading, with lesser amounts forming at subduction zones, where oceanic lithosphere descends into the mantle. Magma also can be generated when crustal rocks are heated sufficiently to melt.

Generating Magma from Solid Rock

Workers in underground mines know that temperatures increase as they descend deeper below Earth's surface. Although the rate of temperature change varies considerably from place to place, it *averages* about 25°C (75°F) per kilometer in the *upper* crust. This increase in temperature with depth is known as the **geothermal gradient**. As shown in **Figure 4.18**, when a typical geothermal gradient is compared to the melting point curve for the mantle rock peridotite, the temperature at which peridotite melts is higher than the geothermal gradient. Thus, under normal conditions, the mantle is solid. However, tectonic processes trigger melting though various means, including reducing the melting point (temperature) of mantle rock.

Decrease in Pressure: Decompression Melting

If temperature were the only factor that determined whether rock melts, our planet would be a molten ball covered with a thin, solid outer shell. This is not the case because pressure, which also increases with depth, influences the melting temperatures of rocks.

Melting, which is accompanied by an increase in volume, *occurs at progressively higher temperatures with increased depth*. This is the result of the steady increase in confining pressure exerted by the weight of overlying rocks. Conversely, reducing confining pressure lowers a rock's

melting temperature. When confining pressure drops sufficiently, **decompression melting** is triggered. Decompression melting occurs wherever hot, solid mantle rock ascends, thereby moving into regions of lower pressure.

Recall from Chapter 2 that tensional forces along spreading centers promote upwelling where plates diverge. This process is responsible for generating magma along oceanic ridges (divergent plate boundaries) where plates are rifting apart (**Figure 4.19**). Below the ridge crest, hot mantle rock rises and melts, generating a magma that replaces the material that shifted horizontally away from the ridge axis.

Decompression melting also occurs when ascending mantle plumes reach the uppermost mantle. If this rising magma reaches the surface, it will trigger an episode of hot-spot volcanism.

Addition of Water   An important factor affecting the melting temperature of rock is its water content. Water and other volatiles act as salt does to melt ice. That is,

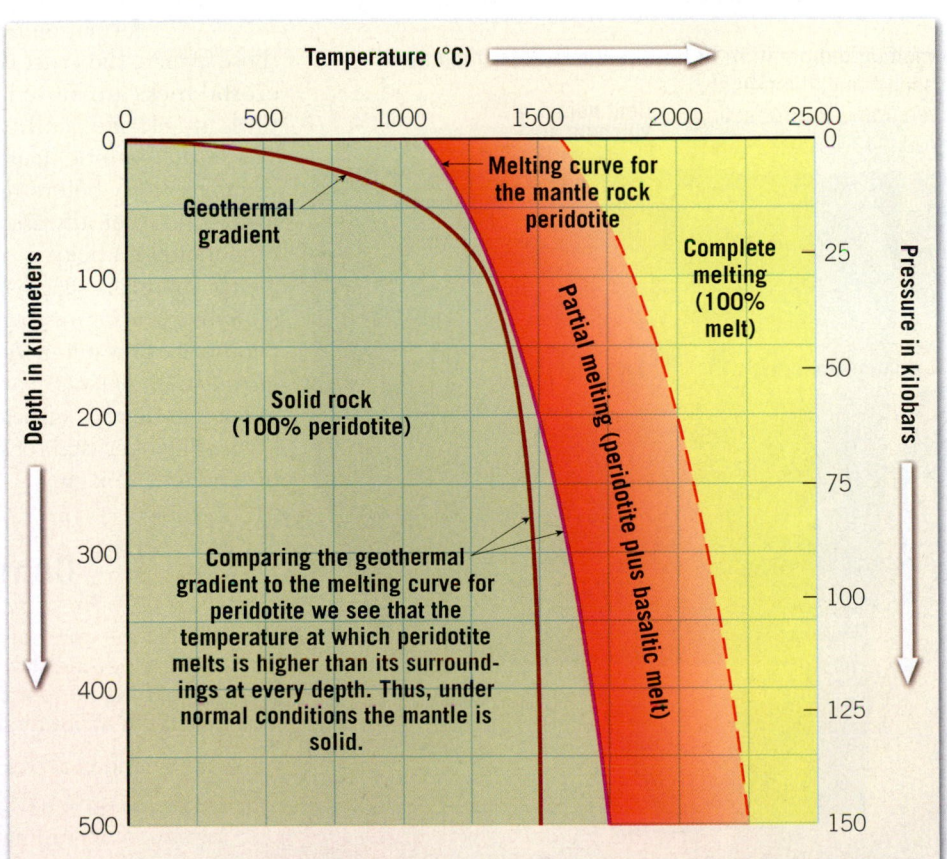

Temperature (°C)

| | 0 | 500 | 1000 | 1500 | 2000 | 2500 |

Melting curve for the mantle rock peridotite

Geothermal gradient

Complete melting (100% melt)

Solid rock (100% peridotite)

partial melting (peridotite plus basaltic melt)

Comparing the geothermal gradient to the melting curve for peridotite we see that the temperature at which peridotite melts is higher than its surroundings at every depth. Thus, under normal conditions the mantle is solid.

Depth in kilometers: 0, 100, 200, 300, 400, 500

Pressure in kilobars: 0, 25, 50, 75, 100, 125, 150

SmartFigure 4.18
Why the mantle is mainly solid This diagram shows the geothermal gradient (the increase in temperature with depth) for the crust and upper mantle. Also illustrated is the melting point curve for the mantle rock peridotite. (https://goo.gl/1ZogEy)

Animation

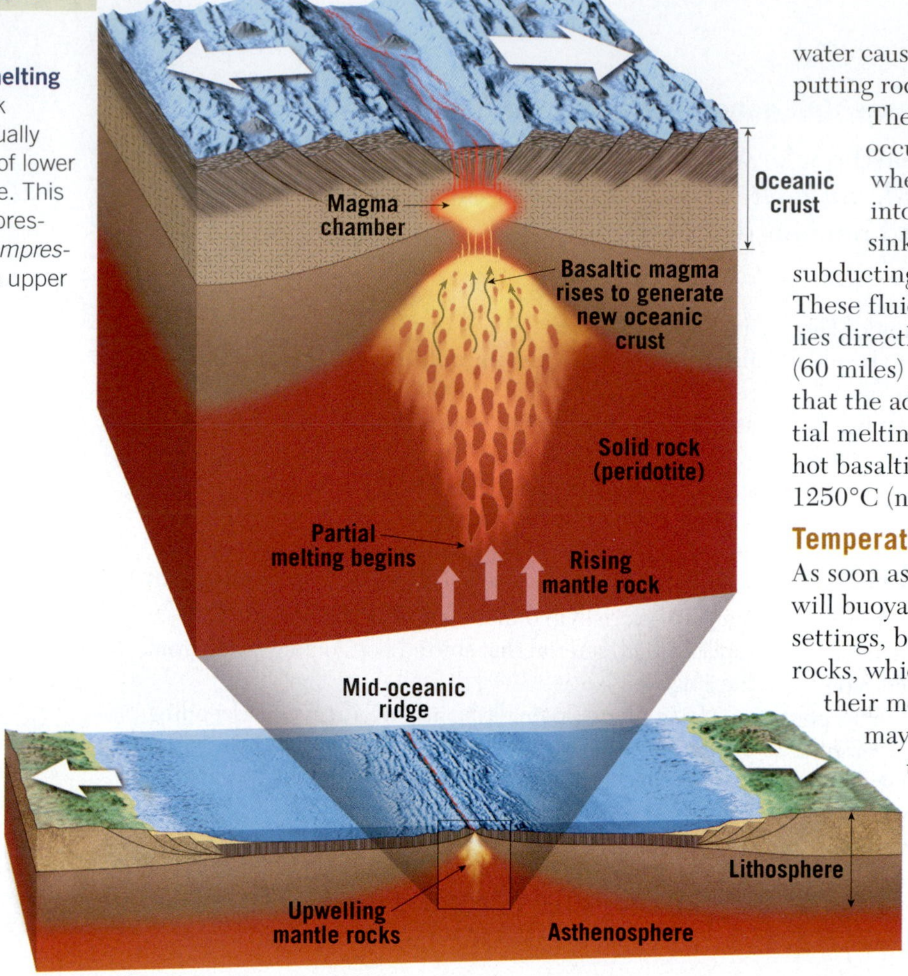

Figure 4.19
Decompression melting
As hot mantle rock ascends, it continually moves into zones of lower and lower pressure. This drop in confining pressure initiates *decompression melting* in the upper mantle.

water causes rock to melt at lower temperatures, just as putting rock salt on an icy sidewalk induces melting.

The introduction of water to generate magma occurs mainly at convergent plate boundaries, where cool slabs of oceanic lithosphere descend into the mantle (**Figure 4.20**). As an oceanic plate sinks, heat and pressure drive water from the subducting oceanic crust and overlying sediments. These fluids migrate into the wedge of hot mantle that lies directly above. At a depth of about 100 kilometers (60 miles) the wedge of mantle rock is sufficiently hot that the addition of water leads to some melting. Partial melting of the mantle rock peridotite generates hot basaltic magma whose temperatures may exceed 1250°C (nearly 2300°F).

Temperature Increase: Melting Crustal Rocks

As soon as mantle-derived basaltic magma forms, it will buoyantly rise toward the surface. In continental settings, basaltic magma often "ponds" beneath crustal rocks, which have a lower density and are already near their melting temperature. The hot basaltic magma may heat the overlying crustal rocks sufficiently to generate a secondary, silica-rich magma. If these low-density, silica-rich magmas reach the surface, they tend to produce explosive eruptions that we associate with convergent plate boundaries.

Crustal rocks can also melt during continental collisions that result in the formation of a large mountain belt. During these events, the crust is greatly thickened, and some crustal rocks are buried to depths where the temperatures are elevated sufficiently to cause partial melting. The felsic (granitic) magmas produced in this manner usually solidify before reaching the surface, so volcanism is not typically associated with these collision-type mountain belts.

In summary, magma can be generated three ways: (1) A *decrease in pressure* (without an increase in temperature) can result in *decompression melting*; (2) the *introduction of water* can lower the melting temperature of hot mantle rock sufficiently to generate magma; and (3) *heating* of crustal rocks above their melting temperature produces magma.

Figure 4.20 Water lowers the melting temperature of hot mantle rock to trigger partial melting As an oceanic plate descends into the mantle, water and other volatiles are driven from the subducting crustal rocks into the mantle above.

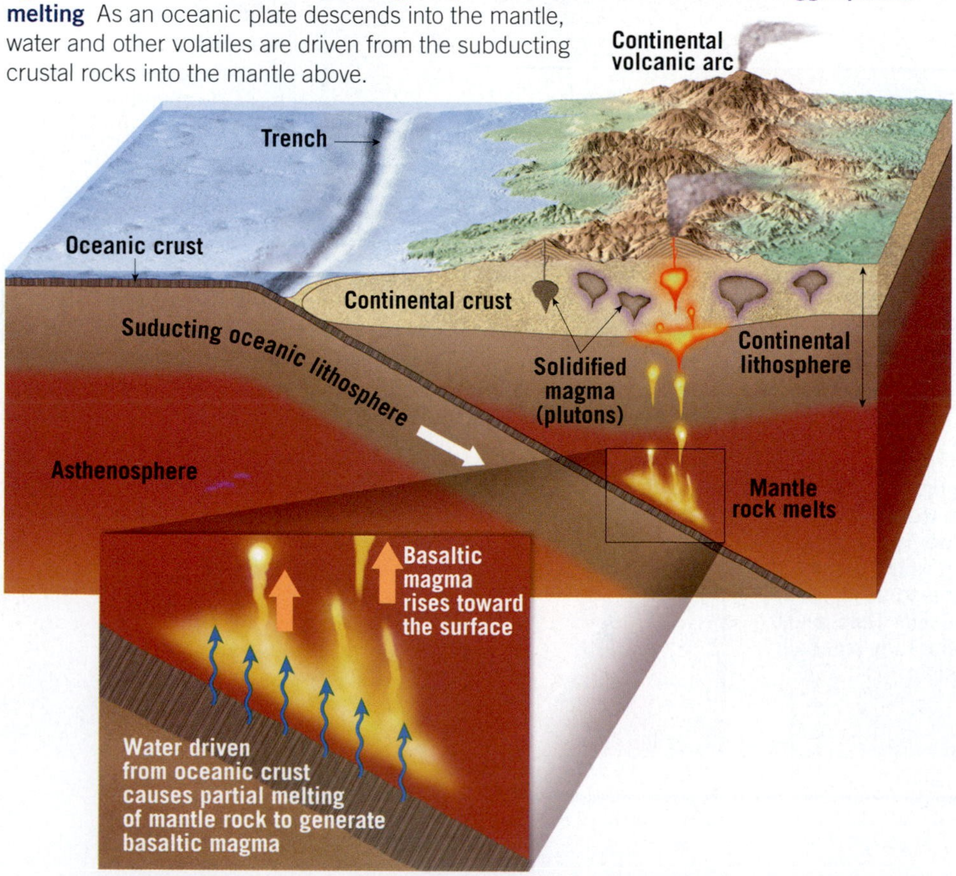

4.5 Concept Checks

1. What is the geothermal gradient? Describe how the geothermal gradient compares with the melting temperatures of the mantle rock peridotite at various depths.

2. Explain the process of decompression melting.

3. What role do water and other volatiles play in the formation of magma?

4. Name two plate tectonic settings in which you would expect magma to be generated.

4.6 | How Magmas Evolve

Describe how magmatic differentiation can generate a magma body that has a mineralogy (chemical composition) that is different from its parent magma.

Geologists have observed that, over time, a volcano may extrude lavas that vary in composition. Data of this type led them to examine the possibility that magma might change (evolve), and thus one magma body could become the parent to a variety of igneous rocks. To explore this idea, N. L. Bowen carried out a pioneering investigation into the crystallization of magma early in the twentieth century.

Bowen's Reaction Series and the Composition of Igneous Rocks

Recall that ice freezes at a specific temperature, whereas basaltic magma crystallizes over a range of at least 200°C of cooling (about 1200° to 1000°C). In a laboratory setting, Bowen and his coworkers demonstrated that as a basaltic magma cools, minerals tend to crystallize in a systematic fashion, based on their melting temperatures. As shown in **Figure 4.21**, the first mineral to crystallize is the ferromagnesian mineral olivine. Further cooling generates calcium-rich plagioclase feldspar as well as pyroxene, and so forth down the diagram.

During this crystallization process, the composition of the remaining liquid portion of the magma also continually changes. For example, at the stage when about one-third of the magma has solidified, the remaining molten material will be nearly depleted in iron, magnesium, and calcium because these elements are major constituents of the minerals that form earliest in the process.

The absence of these elements causes the melt to become enriched in sodium and potassium. Further, because the original basaltic magma contained about 50 percent silica (SiO_2), the crystallization of the earliest-formed mineral, olivine, which is only about 40 percent silica, leaves the remaining melt richer in SiO_2. Thus, the silica component of the remaining melt becomes enriched as the magma evolves.

Bowen also demonstrated that if the solid components in a magma remain in contact with the remaining melt, they will chemically react and change mineralogy (chemical composition), as shown in Figure 4.21. For this reason, this order of mineral formation became known as **Bowen's reaction series**. However, in nature, the earliest-formed minerals can separate from the melt, thus halting any further chemical reaction.

The diagram of Bowen's reaction series in Figure 4.21 depicts the sequence in which minerals crystallize from a magma of basaltic composition under laboratory conditions. Evidence that this highly idealized crystallization

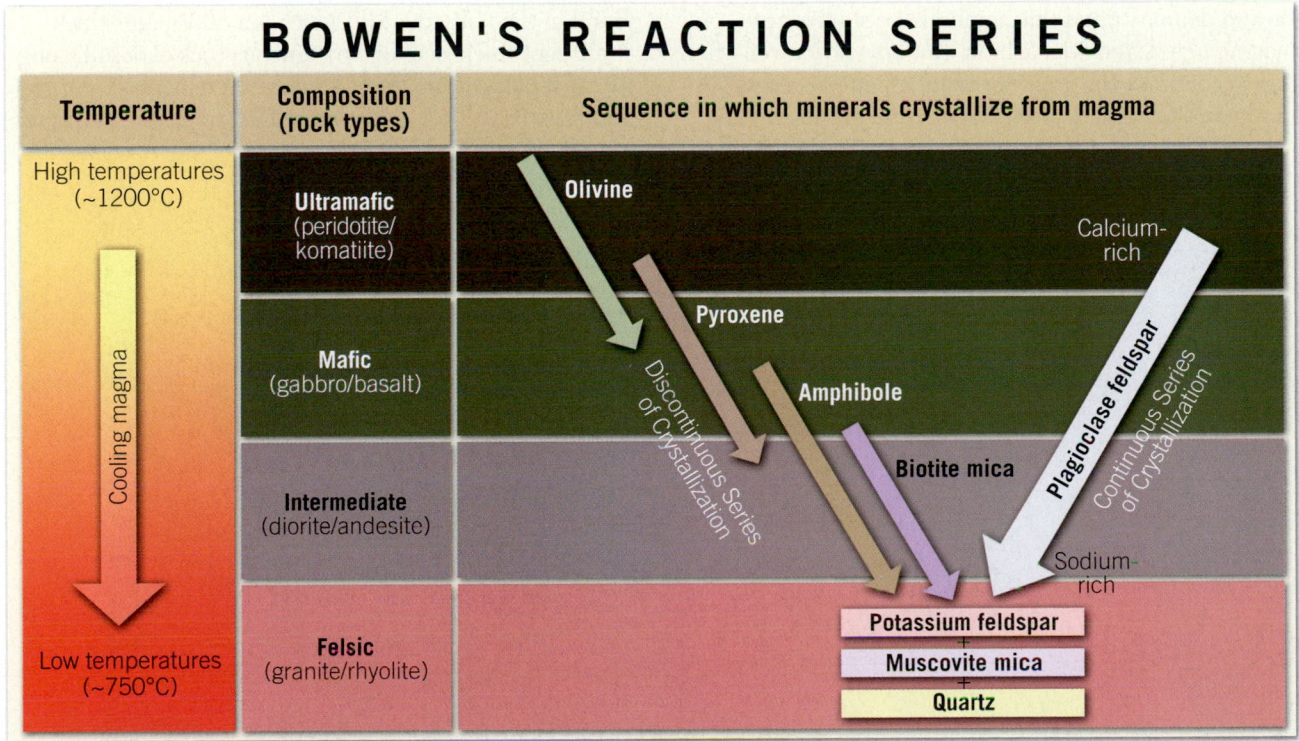

Figure 4.21
Bowen's reaction series
This diagram shows the sequence in which minerals crystallize from a mafic magma. Compare this figure to the mineral composition of the rock groups in Figure 4.13. Note that each rock group consists of minerals that crystallize in the same temperature range.

Figure 4.22
Crystal settling results in a change in the composition of the remaining melt A magma evolves as the earliest-formed minerals (those richer in iron, magnesium, and calcium) crystallize and settle to the bottom of the magma chamber, leaving the remaining melt richer in sodium, potassium, and silica (SiO_2).

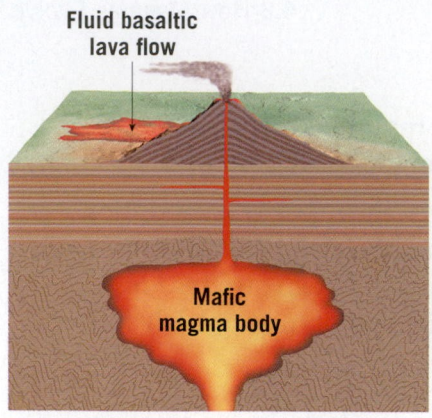

A. Magma having a mafic composition erupts fluid basaltic lavas.

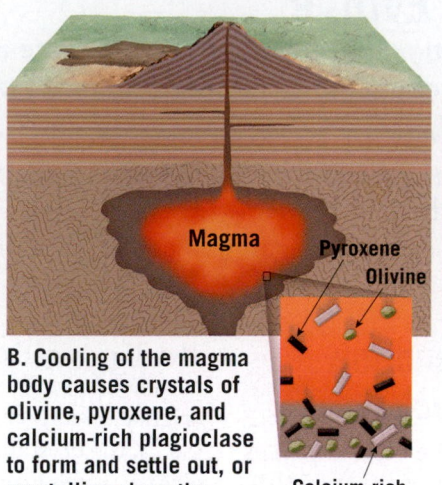

B. Cooling of the magma body causes crystals of olivine, pyroxene, and calcium-rich plagioclase to form and settle out, or crystallize along the magma body's cool margins.

C. The remaining melt will be enriched with silica, and should a subsequent eruption occur, the rocks generated will be more silica-rich and closer to the felsic end of the compositional range than the initial magma.

model approximates what can happen in nature comes from analysis of igneous rocks. In particular, scientists know that minerals that form in the same general temperature regime depicted in Bowen's reaction series are found together in the same igneous rocks. For example, notice in Figure 4.21 that the minerals quartz, potassium feldspar, and muscovite, which are located in the same region of Bowen's diagram, are typically found together as major constituents of the intrusive igneous rock granite.

Magmatic Differentiation and Crystal Settling

Bowen demonstrated that minerals crystallize from magma in a systematic fashion. But how do Bowen's findings account for the great diversity of igneous rocks? It has been shown that, at one or more stages during the crystallization of magma, a separation of various components can occur. One mechanism that causes this to happen is called **crystal settling**. This process occurs when the earlier-formed minerals are denser (heavier) than the liquid portion and sink toward the bottom of the magma chamber, as shown in **Figure 4.22**. When the remaining melt solidifies—either in place or in another location, if it migrates into fractures in the surrounding rocks—it will form a rock with a mineral composition that is different than the parent magma. The formation of a magma body having a mineralogy or chemical composition that is different than the parent magma is called **magmatic differentiation**.

A classic example of magmatic differentiation is found in the Palisades Sill, which is a 300-meter-thick (1000-foot-thick) slab of dark igneous rock exposed along the west bank of the lower Hudson River across from New York City (**Figure 4.23**). Because of its great thickness and

Figure 4.23
The Palisades Sill, as seen from New York City The Palisades form impressive cliffs along the west side of the Hudson River for more than 80 kilometers (50 miles). This structure, which is visible from Manhattan, formed when magma was injected between layers of sandstone and shale. (Terese Loeb Kreuzerl/Alamy)

subsequent slow rate of solidification, crystals of olivine (the first mineral to form) sank and make up about 25 percent of the lower portion of the Palisades Sill. By contrast, near the top of this igneous body, where the last melt crystallized, olivine represents only 1 percent of the rock mass.[*]

Assimilation and Magma Mixing

Bowen successfully demonstrated that through magmatic differentiation, a single parent magma can generate several mineralogically different igneous rocks. However, more recent work indicates that magmatic differentiation involving crystal settling cannot, by itself, account for the entire compositional spectrum of igneous rocks.

Once a magma body forms, the incorporation of foreign material can also change its composition. For example, in near-surface environments where rocks are brittle, the magma pushing upward can cause numerous fractures in the overlying rock. The force of the injected magma is often sufficient to dislodge and incorporate the surrounding host rock (**Figure 4.24**). Melting of these

[*]Recent studies indicate that the Palisades Sill was produced by multiple injections of magma and does not represent a simple case of crystal settling. However, it is nonetheless an instructional example of that process.

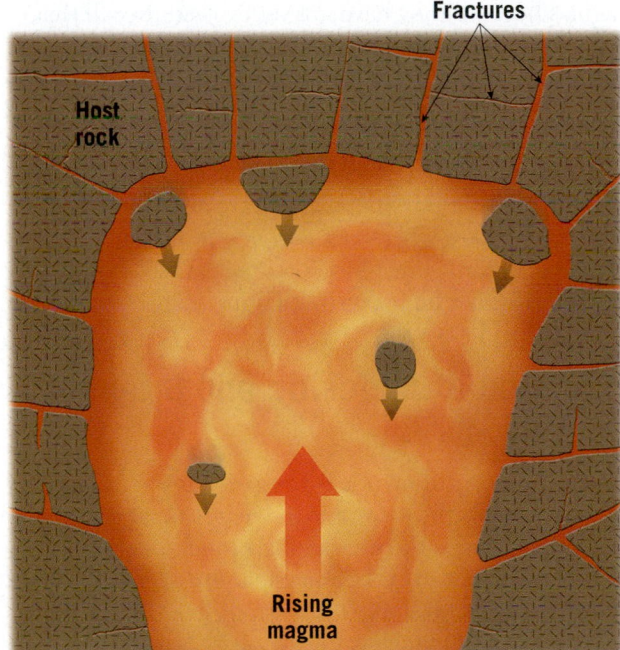

As magma rises through Earth's brittle upper crust, it may dislodge and incorporate the surrounding host rocks. Melting of these blocks, a process called *assimilation*, changes the overall composition of the rising magma body.

Figure 4.24
Assimilation of the host rock by a magma body The composition of magma changes when the molten mass incorporates pieces of surrounding host rock, in a process called assimilation.

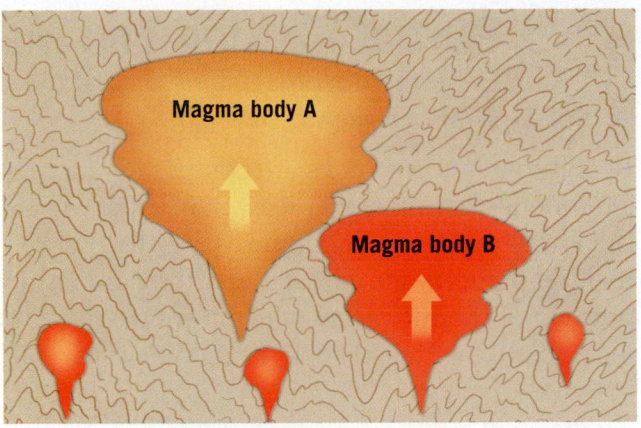

A. During the ascent of two chemically distinct magma bodies, the more buoyant mass may overtake the slower rising body.

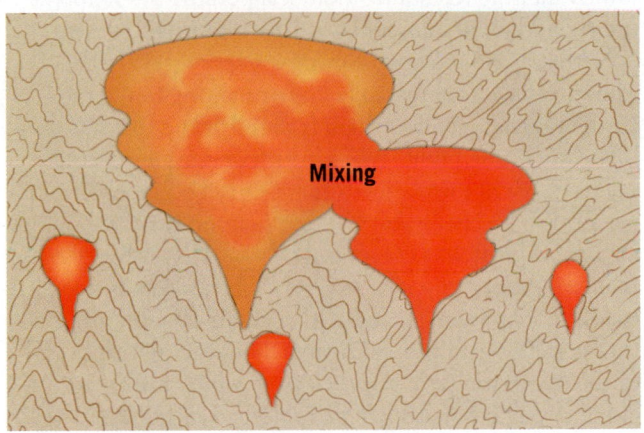

B. Once joined, convective flow will mix the two magmas, generating a mass that is a blend of the two magma bodies.

blocks, a process called **assimilation**, changes the overall chemical composition of the magma body.

Another means by which the composition of magma can be altered is called **magma mixing**. Magma mixing may occur during the ascent of two chemically distinct magma bodies as the more buoyant mass overtakes the more slowly rising body (**Figure 4.25**). Once they are joined, convective flow stirs the two magmas, generating a single mass that has an intermediate composition.

Figure 4.25
Magma mixing This is one way the composition of a magma body can change.

4.6 Concept Checks

1. Define Bowen's reaction series.

2. How does the crystallization and settling of the earliest formed minerals affect the composition of the remaining magma?

3. Compare the processes of assimilation and magma mixing.

4.7 | Partial Melting and Magma Composition

Describe how partial melting of the mantle rock peridotite can generate a basaltic (mafic) magma.

Recall that igneous rocks are composed of a mixture of minerals and, therefore, tend to melt over a temperature range of at least 200°C (nearly 400°F). As rock begins to melt, the minerals with the lowest melting temperatures are the first to melt. If melting continues, minerals with higher melting points begin to melt, and the composition of the melt steadily approaches the overall composition of the rock from which it was derived. Most often, however, melting is not complete. The incomplete melting of rocks is known as **partial melting**, a process that produces most magma.

Recall from Bowen's reaction series that rocks with a granitic composition are composed of minerals with the lowest melting (crystallization) temperatures—namely, quartz and potassium feldspar (see Figure 4.21). Also note that as we move up Bowen's reaction series, the minerals have progressively higher melting temperatures and that olivine, which is found at the top, has the highest melting point. When a rock undergoes partial melting, it will form a melt that is enriched in ions from minerals with the lowest melting temperatures, while the unmelted portion is composed of minerals with higher melting temperatures (**Figure 4.26**). Separation of these two fractions yields a melt with a chemical composition that is richer in silica and nearer the felsic (granitic) end of the spectrum than the rock from which it formed. In general, partial melting of *ultramafic* rocks tends to yield *mafic (basaltic) magmas*, partial melting of *mafic* rocks generally yields *intermediate (andesitic) magmas*, and partial melting of *intermediate* rocks can generate *felsic (granitic) magmas*.

Formation of Basaltic Magma

Most magma that erupts at Earth's surface is basaltic in composition and falls in a temperature range of 1000° to 1250°C. Experiments show that under the high-pressure conditions calculated for the upper mantle, partial melting of the ultramafic rock peridotite can generate a magma of basaltic composition. Further evidence that many basaltic magmas have a mantle source are the inclusions of the rock peridotite, which basaltic magmas often carry up to Earth's surface from the mantle.

Basaltic (mafic) magmas that originate from partial melting of mantle rocks are called *primary* or *primitive* magmas because they have not yet evolved. Recall that partial melting that produces mantle-derived magmas may be triggered by a reduction in confining pressure during the process of decompression melting. This can occur, for example, where hot mantle rock ascends as part of slow-moving convective flow at mid-ocean ridges (see Figure 4.19). Basaltic magmas are also generated at subduction zones, where water driven from the descending slab of oceanic crust promotes partial melting of the mantle rocks that lie above (see Figure 4.20).

Formation of Andesitic and Granitic Magmas

If partial melting of mantle rocks generates most basaltic magmas, what is the source of the magma that crystallizes to form andesitic (intermediate) and granitic (felsic) rocks? Recall that silica-rich magmas erupt mainly along the continental margins. This is strong evidence that continental crust, which is thicker and has a lower density than oceanic crust, must play a role in generating these more highly evolved magmas.

One way andesitic magma can form is when a rising mantle-derived basaltic magma undergoes magmatic differentiation as it slowly makes its way through the continental crust. Recall from our discussion of Bowen's reaction series that as basaltic magma solidifies, the silica-poor ferromagnesian minerals crystallize first. If these iron-rich components are separated from the

SmartFigure 4.26
Partial melting Partial melting generates a magma that is nearer the felsic (granitic) end of the compositional spectrum than the parent rock from which it was derived. (https://goo.gl/s5Vigs)

Tutorial

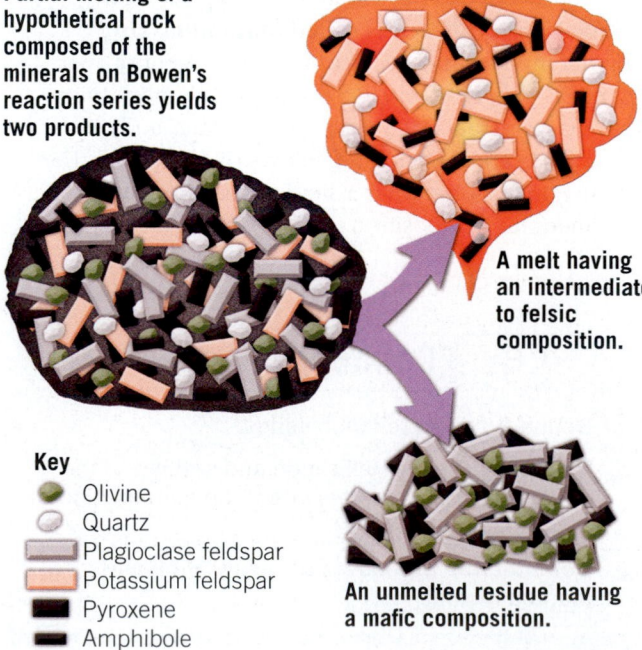

Partial melting of a hypothetical rock composed of the minerals on Bowen's reaction series yields two products.

A melt having an intermediate to felsic composition.

An unmelted residue having a mafic composition.

Key
- Olivine
- Quartz
- Plagioclase feldspar
- Potassium feldspar
- Pyroxene
- Amphibole

liquid by crystal settling, the remaining melt will have an andesitic composition (see Figure 4.22).

Andesitic magmas can also form when rising basaltic magmas assimilate crustal rocks that tend to be silica rich. Partial melting of basaltic rocks is yet another way in which at least some andesitic magmas are thought to be produced.

Although granitic magmas can be formed through magmatic differentiation of andesitic magmas, most granitic magmas probably form when hot basaltic magma ponds (becomes trapped because of its greater density) below continental crust (**Figure 4.27**). When the heat from the hot basaltic magma partially melts the overlying crustal rocks, which are silica rich and have a much lower melting temperature, the result can be the production of large quantities of granitic magmas. This process is thought to have been responsible for the volcanic activity in and around Yellowstone National Park in the distant past.

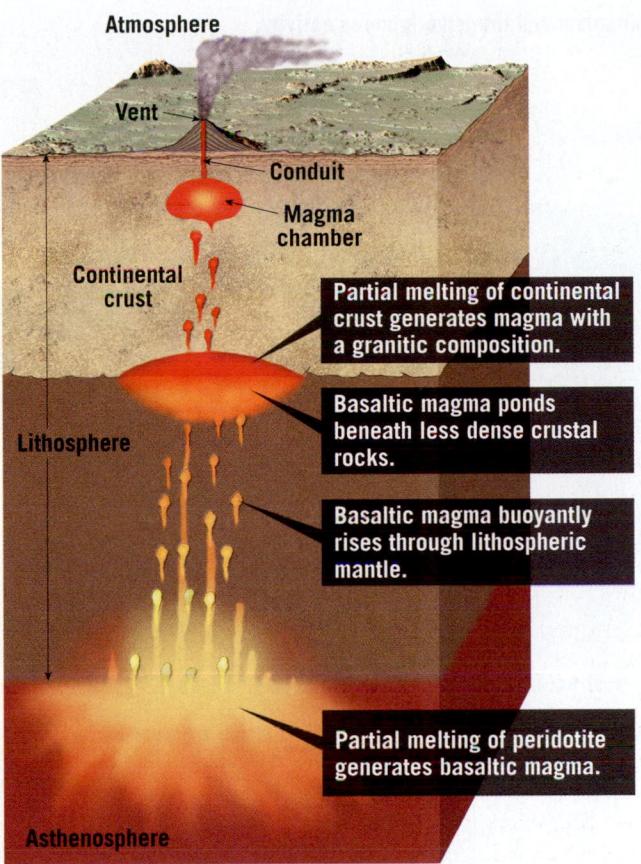

SmartFigure 4.27
Formation of granitic magma Granitic magmas are generated by the partial melting of continental crust. (https://goo.gl/P4Bffs)

4.7 | Concept Checks

1. Briefly describe why partial melting results in a magma having a composition different from the rock from which it was derived.

2. How are most basaltic magmas thought to have formed?

3. What is the process that is thought to generate most granitic magmas?

4.8 | Intrusive Igneous Activity

Compare and contrast these intrusive igneous structures: dikes, sills, batholiths, stocks, and laccoliths.

Although volcanic eruptions can be violent and spectacular events, most magma is emplaced and crystallizes at depth, without fanfare. Therefore, understanding the igneous processes that occur deep underground is as important to geologists as studying volcanic events.

Nature of Intrusive Bodies

When magma rises through the crust, it forcefully displaces preexisting crustal rocks, termed **host rock**, or **country rock**. The structures that result from the emplacement of magma into preexisting rocks are called **intrusions**, or **plutons**. Because all intrusions form far below Earth's surface, they are studied primarily after uplifting and erosion (covered in later chapters) have exposed them. The challenge lies in reconstructing the events that generated these structures in vastly different conditions deep underground, millions of years ago.

Intrusions are known to occur in a great variety of sizes and shapes. Some of the most common types are illustrated in **Figure 4.28**. Notice that some plutons have a **tabular** (*tabula* = table) shape, whereas others are best described as **massive** (blob shaped). Also, observe that some of these bodies cut across existing structures, such as sedimentary strata, whereas others form when magma is injected between sedimentary layers. Because of these differences, intrusive igneous bodies are generally classified according to their shape, as either tabular or massive, and by their orientation with respect to the host rock. Igneous bodies are said to be **discordant** (*discordare* = to disagree) if they cut across existing structures and **concordant** (*concordare* = to agree) if they inject parallel to features such as sedimentary strata.

A. Interrelationship between volcanism and intrusive igneous activity.

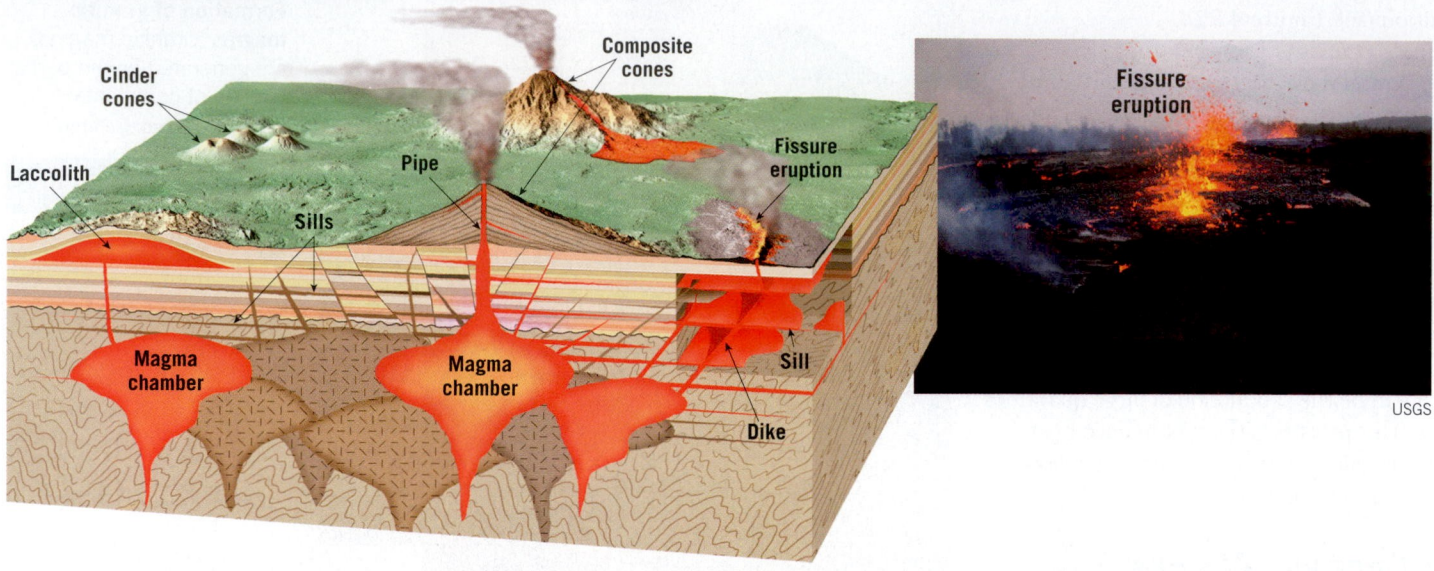

USGS

B. Basic intrusive structures, some of which have been exposed by erosion.

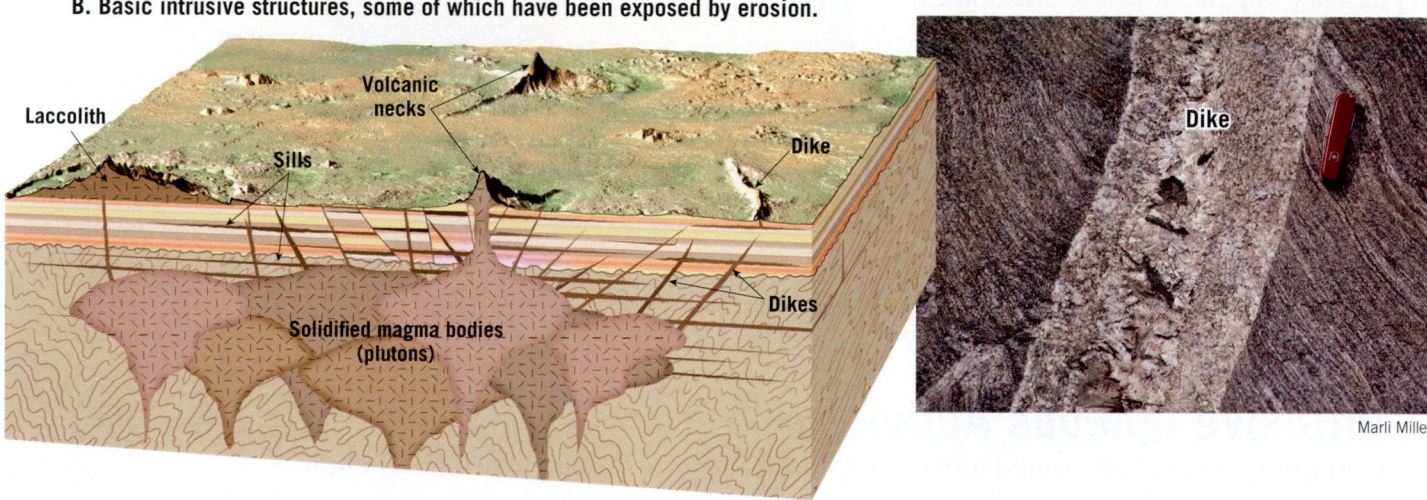

Marli Miller

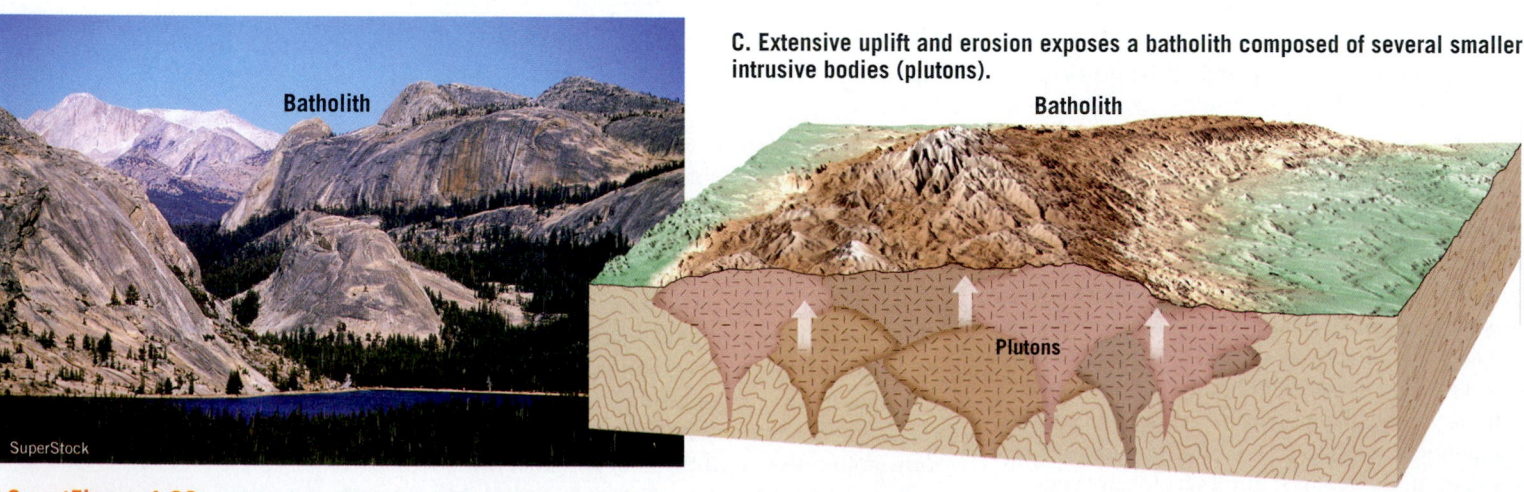

C. Extensive uplift and erosion exposes a batholith composed of several smaller intrusive bodies (plutons).

SuperStock

SmartFigure 4.28
Intrusive igneous structures (https://goo.gl/3sS9U5)

Animation

Tabular Intrusive Bodies: Dikes and Sills

Dikes and Sills Tabular intrusive bodies are produced when magma is forcibly injected into a fracture or zone of weakness, such as a bedding surface (see Figure 4.28). **Dikes** are discordant bodies that form when magma is forcibly injected into fractures and cut across bedding surfaces and other structures in the host rock. By contrast, **sills** are nearly horizontal, concordant bodies that form when magma exploits weaknesses between sedimentary beds or other rock structures (**Figure 4.29**).

SmartFigure 4.29
Sill exposed in Sinbad Country, Utah The dark, essentially horizontal band is a sill of basaltic composition that intruded horizontal layers of sedimentary rock. (Photo by Michael Collier) (http://goo.gl/4MZelh)

Mobile Field Trip

In general, dikes serve as tabular conduits that transport magma upward, whereas sills tend to accumulate magma and increase in thickness.

Dikes and sills are typically shallow features, occurring where the country rocks are sufficiently brittle to fracture. They can range in thickness from less than 1 millimeter to more than 1 kilometer.

While dikes and sills can occur as solitary bodies, dikes tend to form in roughly parallel groups called *dike swarms*. These multiple structures reflect the tendency for fractures to form in sets when tensional forces pull apart brittle country rock. Dikes can also radiate from an eroded volcanic neck, like spokes on a wheel. In these situations, the active ascent of magma generated fissures in the volcanic cone out of which lava flowed. Dikes frequently are more resistant and thus weather more slowly than the surrounding rock. Consequently, when exposed by erosion, dikes tend to have a wall-like appearance, as shown in **Figure 4.30**.

Because dikes and sills are relatively uniform in thickness and can extend for many kilometers, they are assumed to be the product of very fluid, and therefore mobile, magmas. One of the largest and most studied of all sills in the United States is the Palisades Sill (see Figure 4.23). Exposed for 80 kilometers (50 miles) along the west bank of the Hudson River in southeastern New York and northeastern New Jersey, this sill is about 300 meters (1000 feet) thick. Because it is resistant to erosion, the Palisades Sill forms an imposing cliff that can be easily seen from the opposite side of the Hudson.

Columnar Jointing In many respects, sills closely resemble buried lava flows. Both are tabular and can extend over a wide area, and both may exhibit columnar jointing. **Columnar jointing** occurs when igneous rocks cool and develop shrinkage fractures that produce elongated, pillar-like columns that most often have six sides

SmartFigure 4.30
Dike exposed in the Spanish Peaks, Colorado This wall-like dike is composed of igneous rock that is more resistant to weathering than the surrounding material. (Photo by Michael Collier) (https://goo.gl/MJHkem)

Condor Video

Figure 4.31
Columnar jointing
Columnar jointing on Akun Island located in the Aleutian Islands, Alaska. (Photo by Steve Hillebrand, U.S. Fish and Wildlife Service)

Columnar jointing

Rapid cooling from the outside causes shrinkage cracks

Columnar jointing tends to produce 6-sided columns

Geologist's Sketch

(**Figure 4.31**). Further, because sills and dikes generally form in near-surface environments and may be only a few meters thick, the emplaced magma often cools quickly enough to generate a fine-grained texture. (Recall that most intrusive igneous bodies have a coarse-grained texture.)

Massive Intrusive Bodies: Batholiths, Stocks, and Laccoliths

Batholiths   By far the largest intrusive igneous bodies are **batholiths** (*bathos* = depth, *lithos* = stone). Batholiths occur as mammoth linear structures several hundred kilometers long and up to 100 kilometers (60 miles) wide (**Figure 4.32**). The Sierra Nevada batholith, for example, is a continuous granitic structure that forms much of the "backbone" of the Sierra Nevada in California. An even larger batholith extends for over 1800 kilometers (1100 miles) along the Coast Mountains of western Canada and into southern Alaska. Although batholiths can cover a large area, recent geophysical studies indicate that most are less than 10 kilometers (6 miles) thick. Some are even thinner; the coastal batholith of Peru, for example, is essentially a flat slab with an average thickness of only 2 to 3 kilometers (1 to 2 miles). Batholiths are typically composed of felsic (granitic) and intermediate rock types and are often called "granite batholiths."

Early investigators thought the Sierra Nevada batholith was a huge single body of intrusive igneous

rock. Today we know that large batholiths are produced by hundreds of discrete injection of magma that form smaller intrusive bodies (plutons) that intimately crowd against or penetrate one another. These bulbous masses are emplaced over spans of millions of years. The intrusive activity that created the Sierra Nevada batholith, for example, occurred nearly continuously over a 130-million-year period that ended about 80 million years ago (see Figure 4.32).

By definition, a plutonic body must have a surface exposure greater than 100 square kilometers (40 square miles) to be considered a batholith. Smaller plutons are termed **stocks**. However, many stocks appear to be portions of much larger intrusive bodies that would be classified as batholiths if they were fully exposed.

Emplacement of Batholiths   How did massive granitic batholiths come to reside within sedimentary and metamorphic rocks that are only moderately deformed? What happened to the rock that was displaced by these huge igneous masses? (Geologists call this the "room problem.") How do magma bodies make their way through several kilometers of solid rock? Geologists continue to study and debate these questions.

We know that magma rises because it is less dense than the surrounding rock, much as a cork held at the bottom of a container of water will rise when it is released. In the upper mantle and lower crust, where temperatures and pressures are high, rock is ductile (able to flow). In this setting, buoyant magma bodies some of the host rock that was displaced will fill in the space left by the magma body as it passes.[*]

As a magma body nears the surface, it encounters relatively cool, brittle country rock that is not easily pushed aside. Further upward movement may be accomplished by a process called *stoping*, in which blocks of the roof overlying a hot, rising mass become dislodged and sink through the magma (see Figure 4.24). Evidence supporting stoping is found in plutons that contain suspended blocks of country rock called **xenoliths** (*xenos* = a stranger, *lithos* = stone) (**Figure 4.33**).

Magma may also *melt* and *assimilate* some of the overlying host rock. However, this process is greatly limited by the available thermal energy contained in the magma body. When plutons are emplaced near the surface, the room problem may be solved by "lifting the roof" that overlies the intrusive body.

Laccoliths   A nineteenth-century study by G. K. Gilbert of the U.S. Geological Survey in the Henry

---

[*] An analogous situation occurs when a can of oil-based paint is left in storage. The oily component of the paint is less dense than the pigments used for coloration; thus, oil collects into drops that slowly migrate upward, while the heavier pigments settle to the bottom.

Figure 4.32
Granitic batholiths along the western margin of North America These gigantic, elongated bodies consist of numerous plutons that were emplaced beginning about 150 million years ago.

are assumed to rise in the form of *diapirs*, inverted-teardrop-shaped masses with rounded heads and tapered tails. However, in the upper crust, large dike-like structures may provide conduits for the ascent of magma.

Depending on the tectonic environment, several mechanisms have been proposed to solve the room problem. At great depths, where rock is ductile, a mass of buoyant, rising magma can forcibly make room for itself by pushing aside the overlying rock—a process called *shouldering*. As the magma continues to move upward,

Figure 4.33
Xenolith Xenoliths are inclusions of host rock contained within igneous bodies. This unmelted chunk of dark (mafic) rock was incorporated into a granitic magma, eastern Sierra Nevada, Rock Creek Canyon, California. (Photo by Mark A. Wilson)

Figure 4.34
Laccoliths Mount Ellen in Utah's Henry Mountains is one of five peaks that make up this small mountain range. Although the main intrusions in the Henry Mountains are stocks, numerous laccoliths formed as offshoots of these structures. (Photo by Michael DeFreitas North America/ Alamy)

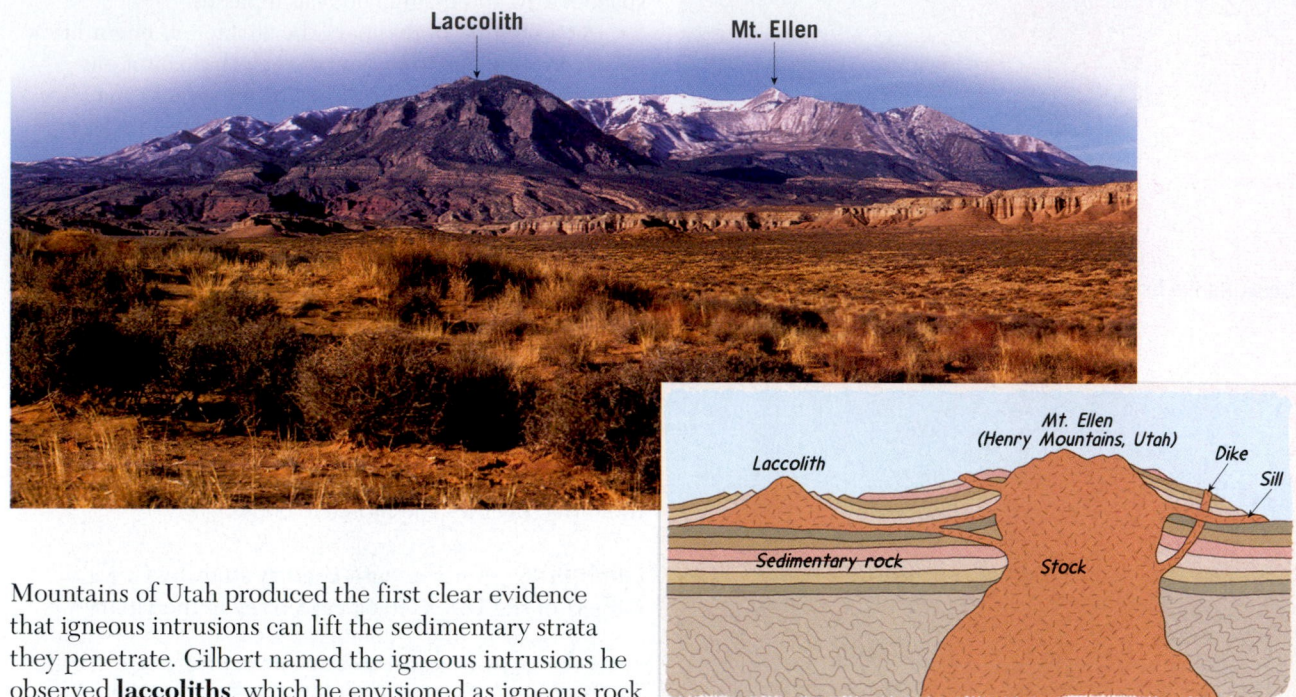

Laccolith   Mt. Ellen

Geologist's Sketch

Mountains of Utah produced the first clear evidence that igneous intrusions can lift the sedimentary strata they penetrate. Gilbert named the igneous intrusions he observed **laccoliths**, which he envisioned as igneous rock forcibly injected between sedimentary strata, so as to arch the beds above while leaving those below relatively flat. It is now known that the five major peaks of the Henry Mountains are not laccoliths but stocks. However, these central magma bodies are the source material for branching offshoots that are true laccoliths, as Gilbert defined them (**Figure 4.34**).

Numerous other granitic laccoliths have since been identified in Utah. The largest is a part of the Pine Valley Mountains located north of St. George, Utah. Others are found in the La Sal Mountains near Arches National Park and in the Abajo Mountains directly to the south.

4.8   Concept Checks

1. What is meant by the term *country rock*?

2. Describe *dikes* and *sills*, using the appropriate terms from the following list: massive, discordant, tabular, and concordant.

3. Distinguish among batholiths, stocks, and laccoliths in terms of size and shape.

EYE ON EARTH 4.3

Shiprock, New Mexico, is an igneous structure that rises more than 510 meters (1700 feet) above the surrounding desert in northwestern New Mexico. It consists of rock that accumulated in the vent of a volcano that has since been eliminated by erosion. (Photo by Michael Collier)

QUESTION 1 What type of landform is Shiprock?

QUESTION 2 What type of structure is the long, narrow ridge extending away from Shiprock?

4.1 Magma: Parent Material of Igneous Rock

List and describe the three major components of magma.

KEY TERMS igneous rock, magma, lava, melt, volatile, crystallization, intrusive igneous rock (plutonic rock), extrusive igneous rock (volcanic rock)

- Completely or partly molten rock is called magma if it is below Earth's surface and lava if it has erupted onto the surface. It consists of a liquid melt with possible additions of solids (mineral crystals) and gases (volatiles), such as water vapor or carbon dioxide.

- As magma cools, silicate minerals begin to form from the "cocktail" of mobile ions in the melt. These tiny crystals grow through the addition of ions to their outer surface. As cooling proceeds, crystallization gradually transforms the magma into a solid mass of interlocking crystals—an igneous rock.

- Magmas that cool below the surface produce intrusive igneous rocks, whereas those that erupt onto Earth's surface produce extrusive igneous rocks.

4.2 Igneous Compositions

Compare and contrast the four basic igneous compositions: basaltic (mafic), granitic (felsic), andesitic (intermediate), and ultramafic.

KEY TERMS granitic (felsic) composition, basaltic (mafic) composition, andesitic (intermediate) composition, peridotite, ultramafic

- Igneous rocks are composed mostly of silicate minerals. If a given igneous rock is mostly composed of nonferromagnesian minerals, it is described as felsic. If the rock has a greater proportion of ferromagnesian minerals, it may be classified as mafic. Mafic rocks are generally darker in color and of greater density than their felsic counterparts. Broadly, continental crust is felsic in composition, and oceanic crust is mafic.

- Intermediate rocks in which plagioclase feldspar predominates are compositionally between felsic and mafic. They are typical of continental volcanic arcs. Ultramafic rocks, which are rich in the minerals olivine and pyroxene, dominate in the upper mantle.

- The amount of silica (SiO_2) in an igneous rock is an indication of its overall composition. Rocks that contain a lot of silica (up to 70 percent, or more) are felsic, while rocks that are poor in silica (as low as 40 percent) fall on the ultramafic end of the spectrum. The amount of silica present in a magma determines both its viscosity and its crystallization temperature.

Q Describe igneous rocks having the compositions of samples A and D, using terms such as mafic, felsic, etc. Would you ever expect to find quartz and olivine in the same rock? Why or why not?

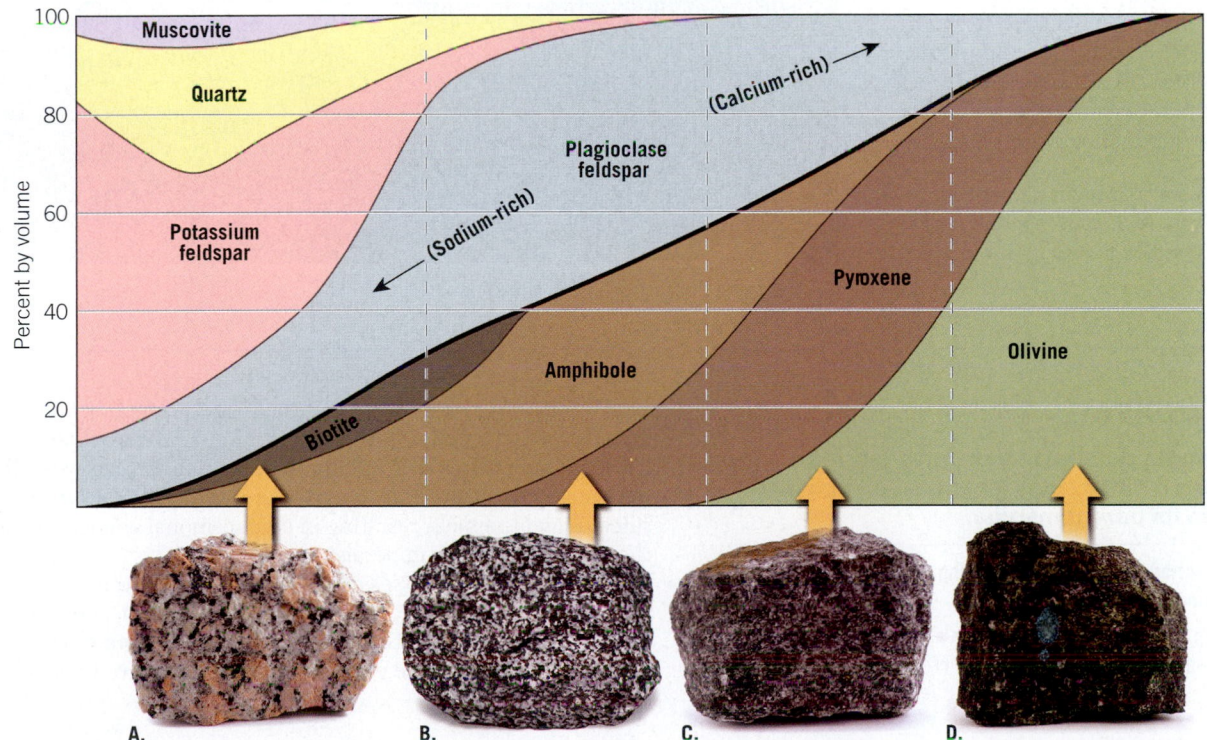

4.3 Igneous Textures: What Can They Tell Us?

Identify and describe the six major igneous textures.

KEY TERMS texture, aphanitic (fine-grained texture), phaneritic (coarse-grained texture), porphyritic texture, phenocryst, groundmass, porphyry, vesicular texture, glassy texture, pyroclastic (fragmental) texture, pegmatite, pegmatitic texture

- To geologists, "texture" is a description of the size, shape, and arrangement of mineral grains in a rock. Careful observation of the texture of igneous rocks can tell us about the conditions under which they formed. The rate at which magma or lava cools is an important factor in the rock's final texture.

- The cooling rate is quick for lava at or close to the surface, so crystallization is rapid and results in a large number of very small crystals. The result is a fine-grained texture. Magma cooling at depth loses heat more slowly. This allows sufficient time for the magma's ions to be organized into larger crystals, resulting in a rock with a coarse-grained texture. If crystals begin to form at depth and then the magma moves to a shallow depth or erupts at the surface, it will have a two-stage cooling history. The result is a rock with a porphyritic texture.

- Volcanic rocks may exhibit additional textures: vesicular if they had a high gas content, glassy if they were high in silica, pyroclastic if they erupted explosively. The large crystals that characterize pegmatitic textures result from the crystallization of magmas with high water content.

4.4 Naming Igneous Rocks

Distinguish among the common igneous rocks based on texture and mineral composition.

KEY TERMS granite, rhyolite, obsidian, pumice, andesite, diorite, basalt, gabbro, pyroclastic rocks

- Igneous rocks are classified on the basis of their textures and their compositions. Figure 4.13 summarizes the naming system based on

these two criteria. Two magmas with the same composition can cool at different rates, resulting in different final textures. On the other hand, two magmas that have different compositions may attain similar textures if they cool under similar circumstances.

Q Could granite ever be transformed into a rhyolite? If so, what processes would have to be involved?

4.5 Origin of Magma

Summarize the major processes that generate magma from solid rock.

KEY TERMS geothermal gradient, decompression melting

- Solid rock may melt under three geologic circumstances: when heat is added to the rock, raising its temperature; when already hot rock experiences lower pressures (decompression, as seen at mid-ocean ridges); and when water is added (as occurs at subduction zones).

Q Different processes produce magma in different tectonic settings. Consider situations A, B, and C in the diagram and describe the processes that would be most likely to trigger melting in each.

4.6 How Magmas Evolve

Describe how magmatic differentiation can generate a magma body that has a mineralogy (chemical composition) that is different from its parent magma.

KEY TERMS Bowen's reaction series, crystal settling, magmatic differentiation, assimilation, magma mixing

- Pioneering experimentation by N. L. Bowen revealed that in a cooling magma, minerals crystallize in a specific order. Ferromagnesian silicates

such as olivine crystallize first, at the highest temperatures (1250°C), and nonferromagnesian silicates such as quartz crystallize last, at the lowest temperatures (650°C). Bowen found that in between these temperatures, chemical reactions take place between the crystallized silicates and the melt, resulting in compositional changes to each and the formation of new minerals.

- Various physical processes can cause changes in the composition of magma. For instance, if crystallized silicates are denser than the remaining magma, they will sink to the bottom of the magma chamber. Because these early-formed minerals are likely to be ferromagnesians, the magma has now differentiated toward a more felsic composition.

- As they migrate, magmas may react with their "host" rocks or with other magma bodies. Assimilation of host rock or magma mixing will alter the magma's composition.

Q Consider the accompanying diagram, which shows a cross-sectional view of a hypothetical magma chamber. Using your understanding of Bowen's reaction series and magma evolution, interpret the layered structure by explaining how crystallization occurred.

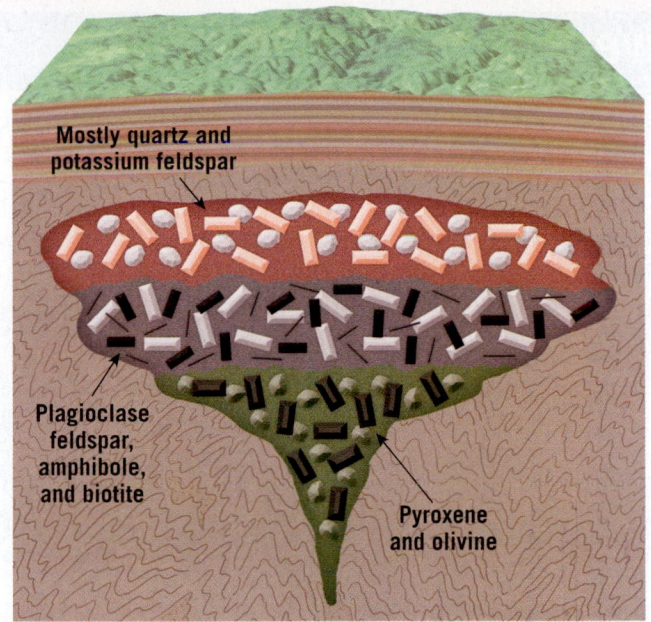

Mostly quartz and potassium feldspar

Plagioclase feldspar, amphibole, and biotite

Pyroxene and olivine

4.7 Partial Melting and Magma Composition

Describe how partial melting of the mantle rock peridotite can generate a basaltic (mafic) magma.

KEY TERM partial melting

- In most circumstances, when rocks melt, they do not melt completely. Different minerals have different temperatures at which they change state from solid to liquid (or liquid to solid). As rocks melt, minerals with the lowest melting temperatures melt first.
- Partial melting of the ultramafic mantle yields mafic oceanic crust. Partial melting of the lower continental crust at subduction zones produces magmas that have intermediate or felsic compositions.

Q If all silicate minerals melted at exactly the same temperature, would magmas of different compositions exist? How is partial melting important for generating the different kinds of rocks on Earth?

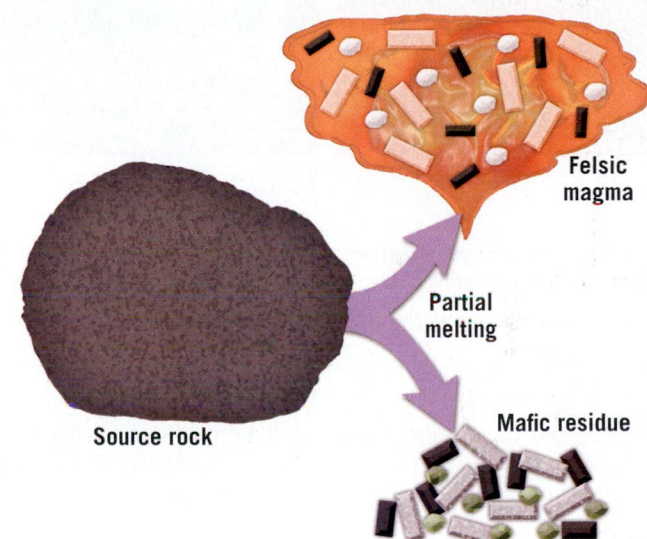

Felsic magma

Partial melting

Source rock

Mafic residue

4.8 Intrusive Igneous Activity

Compare and contrast these intrusive igneous structures: dikes, sills, batholiths, stocks, and laccoliths.

KEY TERMS host (country) rock, intrusion (pluton), tabular, massive, discordant, concordant, dike, sill, columnar jointing, batholith, stock, xenolith, laccolith

- When magma intrudes other rocks, it may cool and crystallize before reaching the surface to produce intrusions called plutons. Plutons come in many shapes. They may cut across the host rocks without regard for preexisting structures, or the magma may flow along weak zones in the host rock, such as between the horizontal layers of sedimentary bedding.

- Tabular intrusions may be concordant (sills) or discordant (dikes). Massive plutons may be small (stocks) or very large (batholiths). Blister-like intrusions also exist (laccoliths). As solid igneous rock cools, its volume decreases. Contraction can produce a distinctive fracture pattern called columnar jointing.

- Several processes contribute to magma's intrusion into host rocks. Rising diapirs are one possibility, and another is shouldering aside of host rocks. Stoping of xenoliths from the host rock can open up more room, or the magma can melt and assimilate some of the host rock.

Q Create a sketch, including labels, showing a series of igneous intrusions into sedimentary rocks. Include a dike, a sill, a stock, a batholith, a laccolith, some xenoliths, and a diapir.

Give It Some Thought

1. Would you expect all the crystals in an intrusive igneous rock to be the same size? Explain why or why not.

2. Apply your understanding of igneous rock textures to describe the cooling history of each of the igneous rocks pictured here. (Photos by E. J. Tarbuck)

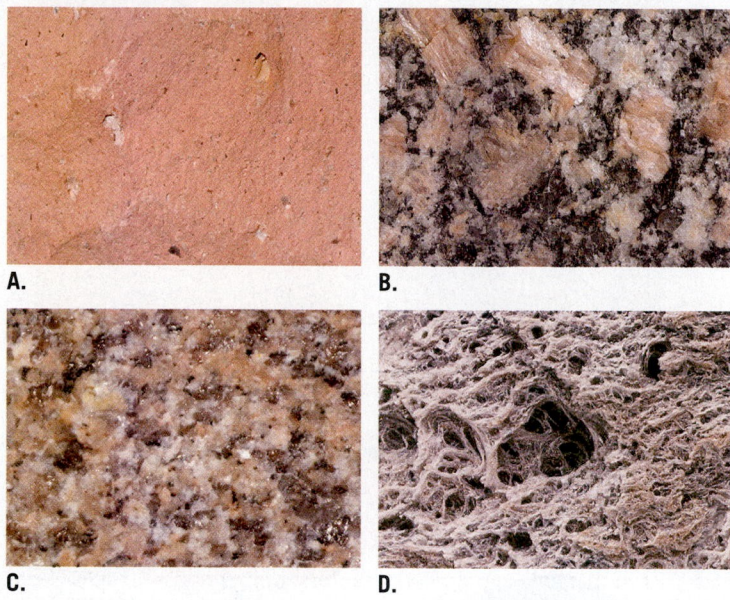

A.

B.

C.

D.

3. Is it possible for two igneous rocks to have the same mineral composition but be different rocks? Use an example to support your answer.

4. Use Figure 4.5 to classify the following igneous rocks.

   a. An aphanitic rock containing about 30 percent calcium-rich plagioclase feldspar, 55 percent pyroxene, and 15 percent olivine

   b. A phaneritic rock containing about 20 percent quartz, 40 percent potassium feldspar, 20 percent sodium-rich plagioclase feldspar, a few percent muscovite, and the remainder dark-colored silicate

   c. An aphanitic rock containing about 50 percent plagioclase feldspar, 35 percent amphibole, 10 percent pyroxene, and minor amounts of other light-colored silicates

   d. A phaneritic rock made mainly of olivine and pyroxene, with lesser amounts of calcium-rich plagioclase feldspar

5. Identify the igneous rock textures described by each of the following statements.

   a. Openings produced by escaping gases

   b. The texture of obsidian

   c. A matrix of fine crystals surrounding phenocrysts

   d. Consists of crystals that are too small to be seen without a microscope

   e. A texture characterized by rock fragments welded together

   f. Coarse grained, with crystals of roughly equal size

   g. Exceptionally large crystals, most exceeding 1 centimeter in diameter

6. During a hike, you pick up the igneous rock shown in the accompanying photo.

   a. What is the mineral name of the small, rounded, glassy green crystals?

   b. Did the magma from which this rock formed likely originate in the mantle or in the crust? Explain.

   c. Was the magma likely a high-temperature magma or a low-temperature magma? Explain.

   d. Describe the texture of this rock.

Unclesam/Fotolia

7. A common misconception about Earth's upper mantle is that it is a thick shell of molten rock. Explain why Earth's mantle is actually solid under most conditions.

8. Describe two mechanisms by which mantle rock can melt without an increase in temperature. How do these magma-generating mechanisms relate to plate tectonics?

9. Use your understanding of Bowen's reaction series (Figure 4.21) to explain how partial melting can generate magmas that have different compositions.

10. During a field trip with your geology class, you visit an exposure of rock layers similar to the one sketched here. A fellow student suggests that the layer of basalt is a sill. You disagree. Why do you think the other student is incorrect? What is a more likely explanation for the basalt layer?

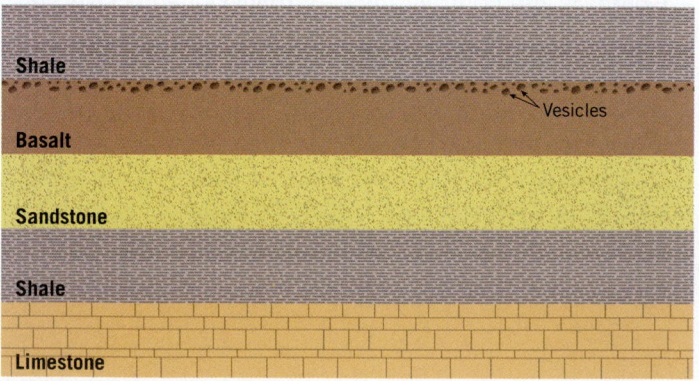

Shale

Vesicles

Basalt

Sandstone

Shale

Limestone

11. Each of the following statements describes how an intrusive feature appears when exposed at Earth's surface due to erosion. Name each feature.

   a. A dome-shaped mountainous structure flanked by upturned layers of sedimentary rocks

   b. A vertical wall-like feature a few meters wide and hundreds of meters long

   c. A huge expanse of granitic rock forming a mountainous terrain tens of kilometers wide

   d. A relatively thin layer of basalt sandwiched between horizontal layers of sedimentary rocks exposed along the walls of a river valley

12. Mount Whitney, the highest summit (4421 meters [14,505 feet]) in the contiguous United States, is located in the Sierra Nevada batholith. Based on its location, is Mount Whitney likely composed mainly of granitic, andesitic, or basaltic rocks?

Mount Whitney →

Photo by John Greim/Getty Images

MasteringGeology™

www.masteringgeology.com

Looking for additional review and test prep materials? With individualized coaching on the toughest topics of the course, MasteringGeology offers a wide variety of ways for you to move beyond memorization to begin thinking like a geologist. Visit the Study Area in www.masteringgeology.com to find practice quizzes, study tools, and multimedia that will improve your understanding of this chapter's content. Sign in today to enjoy the following features: **Self Study Quizzes, SmartFigures: Tutorials/Animations/Condor Videos/Mobile Field Trips, Geoscience Animation Library, GEODe, RSS Feeds, Digital Study Modules,** and an optional **Pearson eText.**

5

Volcanoes and Volcanic Hazards

Villarrica volcano ejecting incandescent lava and volcanic bombs, Chile, March 3, 2015. (Photo by REUTERS/S Aditya)

Each statement represents the primary **LEARNING OBJECTIVE** for the corresponding major heading within the chapter. After you complete the chapter, you should be able to:

5.1 Explain why some volcanic eruptions are explosive and others are quiescent.

5.2 List and describe the three categories of materials extruded during volcanic eruptions.

5.3 Label a diagram that illustrates the basic features of a typical volcanic cone.

5.4 Summarize the characteristics of shield volcanoes and provide one example.

5.5 Describe the formation, size, and composition of cinder cones.

5.6 List the characteristics of composite volcanoes and describe how they form.

5.7 Discuss the major geologic hazards associated with volcanoes.

5.8 List volcanic landforms other than shield, cinder, and composite volcanoes and describe their formation.

5.9 Explain how the global distribution of volcanic activity is related to plate tectonics.

5.10 List and describe the techniques used to monitor potentially dangerous volcanoes.

The significance of igneous activity may not be obvious at first glance. However, because volcanoes extrude molten rock that formed at great depth, they provide our only means of directly observing processes that occur many kilometers below Earth's surface. Furthermore, Earth's atmosphere and oceans have evolved from gases emitted during volcanic eruptions. Either of these facts is reason enough for igneous activity to warrant our attention.

5.1 | The Nature of Volcanic Eruptions

Explain why some volcanic eruptions are explosive and others are quiescent.

Volcanic activity is commonly perceived as a process that produces a picturesque, cone-shaped structure that periodically erupts in a violent manner (**Figure 5.1**). However, many eruptions are not explosive. What determines the manner in which volcanoes erupt?

Factors Affecting Viscosity

The source material for volcanic eruptions is **magma**, molten rock that usually contains some solid crystalline material and varying amounts of dissolved gas. Erupted magma is called **lava**. The primary factors that affect the behavior of magma are its *temperature* and *composition* and, to a lesser extent, the amount of *dissolved gases* it contains. To varying degrees, these factors determine a magma's mobility, or **viscosity** (*viscos* = sticky). The more viscous the material, the greater its resistance to flow. For example, syrup is more viscous, and thus more resistant to flow, than water.

Temperature The effect of temperature on viscosity is easily seen. Just as heating syrup makes it more fluid (less viscous), temperature also strongly influences the mobility of lava. As lava cools and begins to congeal, its viscosity increases, and eventually the flow halts.

Composition Another significant factor influencing volcanic behavior is the chemical composition of the magma. Recall that a major difference among various igneous rocks is their silica (SiO_2) content (**Table 5.1**). Magmas that produce mafic rocks such as basalt contain about

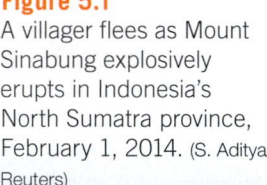

Figure 5.1
A villager flees as Mount Sinabung explosively erupts in Indonesia's North Sumatra province, February 1, 2014. (S. Aditya/Reuters)

TABLE 5.1   Different Compositions of Magmas Cause Properties to Vary

Composition	Silica Content	Gas Content	Eruptive Temperatures	Viscosity	Tendency to Form Pyroclastics	Volcanic Landform
Basaltic (mafic)	Least (~50%)	Least (1–2%)	1000–1250°C	Least	Least	Shield volcanoes, basalt plateaus, cinder cones
Andesitic (intermediate)	Intermediate (~60%)	Intermediate (3–4%)	800–1050°C	Intermediate	Intermediate	Composite cones
Rhyolitic (felsic)	Most (~70%)	Most (4–6%)	650–900°C	Greatest	Greatest	Pyroclastic flow deposits, lava domes

50 percent silica, whereas magmas that produce felsic rocks (granite and its extrusive equivalent, rhyolite) contain more than 70 percent silica. Intermediate rock types—andesite and diorite—contain about 60 percent silica. A magma's viscosity is directly related to its silica content: *The more silica in magma, the greater its viscosity.* Silica impedes the flow of magma because silicate structures begin to link together into long chains early in the crystallization process.

Consequently, felsic (rhyolitic) lavas are very viscous and tend to form comparatively short, thick flows. By contrast, mafic (basaltic) lavas, which contain less silica, are relatively fluid and have been known to travel 300 kilometers (180 miles) or more before congealing.

Dissolved Gases

The gaseous components in magma, mainly dissolved water vapor, also affect the mobility of magma. Other factors being equal, water vapor dissolved in magma tends to increase fluidity because it reduces formation of long silicate chains by breaking silicon–oxygen bonds. It follows, therefore, that the loss of gases renders magma (lava) more viscous. Gases also give magmas their explosive character.

Quiescent Versus Explosive Eruptions

You learned in Chapter 4 that most magma is generated by partial melting of the rock peridotite in Earth's upper mantle, and it tends to have a basaltic composition. Newly formed basaltic magma, which is less dense than the surrounding rock, slowly rises toward the surface. In some settings, high-temperature basaltic magmas reach Earth's surface, where they produce highly fluid lavas. This most commonly occurs on the ocean floor, in association with seafloor spreading. In continental settings, however, the density of crustal rocks is less than that of the ascending material, causing the magma to pond at the crust–mantle boundary. Heat from the hot magma is often sufficient to partially melt the overlying crustal rocks, generating a less-dense, silica-rich magma, which then continues the journey toward Earth's surface.

Quiescent Hawaiian-Type Eruptions

Eruptions that involve very fluid basaltic lavas, such as the eruptions of Kilauea on Hawaii's Big Island, are often triggered by the arrival of a new batch of molten rock rising into a near-surface magma chamber. Geologists can often detect such an event because the summit of the volcano begins to inflate

and rise months or even years before an eruption. The injection of a fresh supply of hot molten rock heats and remobilizes the semi-liquid magma chamber. In addition, swelling of the magma chamber fractures the rock above, allowing the fluid magma to move upward along the newly formed openings, often generating outpourings of lava for weeks, months, or possibly years. The eruption of Kilauea that began in 1983 has been ongoing ever since.

Triggering Explosive Eruptions

All magmas contain some water vapor and other gases that are kept in solution by the immense pressure of the overlying rock. As magma rises (or the rocks confining the magma fail), the confining pressure drops, causing the dissolved gases to separate from the melt and form large numbers of tiny bubbles. This is analogous to opening a can of soda, where the carbon dioxide dissolved in the soda quickly forms bubbles that rise and escape.

When fluid basaltic magmas erupt, the pressurized gases readily escape. At temperatures that often exceed 1100°C (2000°F), these gases can quickly expand to occupy hundreds of times their original volumes. Occasionally, these expanding gases propel incandescent lava hundreds of meters into the air, producing lava fountains (**Figure 5.2**). Although spectacular, these fountains are

Figure 5.2
Lava fountain produced by gases escaping fluid basaltic lava Lava erupting from Mount Etna, Italy. (Photo by D. Szczepanski terras/AGE Fotostock)

Gases readily escape hot fluid basaltic flows, producing lava fountains. Although often spectacular, these features generally do not cause great loss of life or property.

usually harmless and generally not associated with major explosive events that cause great loss of life and property.

At the other extreme, highly viscous magmas expel fragmented lava at nearly supersonic speeds, creating buoyant plumes consisting mainly of volcanic ash and gases called **eruption columns**. Eruption columns can rise perhaps 40 kilometers (25 miles) into the atmosphere. Therefore, volcanoes associated with highly viscous magmas are the most destructive to property and human life (**Figure 5.3**).

Because silica-rich magmas are sticky (viscous), a significant portion of the gaseous materials remain dissolved until the magma nears Earth's surface, at which time tiny bubbles begin to form and grow. When the pressure exerted by the expanding magma body exceeds the strength of the overlying rock, fracturing occurs. As magma moves up the fractures, a further drop in confining pressure causes even more gas bubbles to form and grow. This chain reaction may generate an explosive event in which magma is literally blown into fragments (ash and pumice) that are carried to great heights by the hot gases. As exemplified by the 1980 eruption of Mount St. Helens, the collapse of a volcano's flank can greatly reduce the pressure on the magma below, triggering an explosive eruption (see GEOgraphics 5.1).

When magma in the uppermost portion of the magma chamber is forcefully ejected by the escaping gases, the confining pressure on the molten rock directly below also drops suddenly. Thus, rather than a single "bang," volcanic eruptions are really a series of explosions that can last for days. Following explosive eruptions, degassed lava may slowly ooze out of the vent to form rhyolitic flows or dome-shaped lava bodies that grow over the vent.

5.1 | Concept Checks

1. Define *viscosity* and list three factors that influence the viscosity of magma.

2. List these three magmas in order, from the *most* silica-rich to the *least* silica-rich, based on their composition: basaltic magma, rhyolitic magma, andesitic magma.

3. What type of magma must erupt to produce an eruption column?

4. Are volcanoes fed by highly viscous magma *more likely* or *less likely* to be a greater threat to life and property than volcanoes supplied with very fluid magma?

SmartFigure 5.3
Eruption column generated by viscous, silica-rich magma Steam and ash eruption column from Mount Sinabung, Indonesia, February 7, 2014. A deadly cloud of fiery ash can be seen racing down the volcanoes slope in the foreground.
(Photo by REUTERS/Beawiharta)
(https://goo.gl/Gd11FL)

Video

Eruptions of highly viscous lavas may produce explosive clouds of hot ash and gases called eruption columns.

5.2 | Materials Extruded During an Eruption

List and describe the three categories of materials extruded during volcanic eruptions.

Volcanoes erupt lava, large volumes of gas, and pyroclastic materials (broken rock, lava "bombs," and ash). In this section we will examine each of these materials.

Lava Flows

The vast majority of lava on Earth—more than 90 percent of the total volume—is estimated to be basaltic (mafic) in composition. Andesitic lavas and other lavas of intermediate composition account for most of the rest, while rhyolitic (felsic) flows make up as little as 1 percent of the total.

Hot basaltic lavas, which are usually very fluid, generally flow in thin, broad sheets or streamlike ribbons. On the island of Hawaii, these lavas have been clocked at 30 kilometers (19 miles) per hour down steep slopes. However, flow rates of 10 to 300 meters (30 to 1000 feet) per hour are more common. By contrast, the movement of silica-rich rhyolitic lava may be too slow to perceive. Furthermore, most rhyolitic lavas seldom travel more than a few kilometers from their vents. As you might expect, andesitic lavas, which are intermediate in composition, exhibit flow characteristics between these extremes.

Aa and Pahoehoe Flows

Fluid basaltic magmas tend to generate two types of lava flows known by Hawaiian names. The first, called **aa** (pronounced "ah-ah") **flows**, have surfaces of rough jagged blocks with dangerously sharp edges and spiny projections (**Figure 5.4A**). Crossing a hardened aa flow can be a trying and miserable experience. The second type, **pahoehoe** (pronounced "pah-hoy-hoy") **flows**, exhibit smooth surfaces that sometimes resemble twisted braids of ropes (**Figure 5.4B**). Pahoehoe means "on which one can walk."

Although both lava types can erupt from the same volcano, pahoehoe lavas are hotter and more fluid than aa flows. In addition, pahoehoe lavas can change into aa lava flows, although the reverse (aa to pahoehoe) does not occur.

Cooling that occurs as the flow moves away from the vent is one factor that facilitates the change from pahoehoe to aa. The lower temperature increases viscosity and promotes bubble formation. Escaping gas bubbles produce numerous voids (vesicles) and sharp spines in the surface of the congealing lava. As the molten interior advances, the outer crust is broken, transforming a relatively smooth surface of a pahoehoe flow into an aa flow made up of an advancing mass of rough, sharp, broken lava blocks.

Pahoehoe flows often develop cave-like tunnels called **lava tubes** that were previously conduits for carrying lava from an active vent to the flow's leading edge (**Figure 5.5**). Lava tubes form in the interior of a lava flow, where the temperature remains high long after the exposed surface cools and hardens. Because they serve as insulated pathways that facilitate the advance of lava

Figure 5.4
Lava flows A. A slow-moving, basaltic aa flow advancing over hardened pahoehoe lava. **B.** A typical fluid pahoehoe (ropy) lava. Both of these lava flows erupted from a rift on the flank of Hawaii's Kilauea volcano. (Photos courtesy of U.S. Geological Survey)

A. Active aa flow overriding an older pahoehoe flow.

Aa flow

Pahoehoe flow

B. Pahoehoe flow displaying the characteristic ropy appearance.

Eruption of Mount St. Helens

On Sunday, May 18, 1980, the most destructive volcanic eruption to occur in North America in recorded history transformed a picturesque volcano into a decapitated remnant. Despite the fact that most residents of the area heeded evacuation warnings the event claimed 57 human lives.

Spirit Lake

USGS

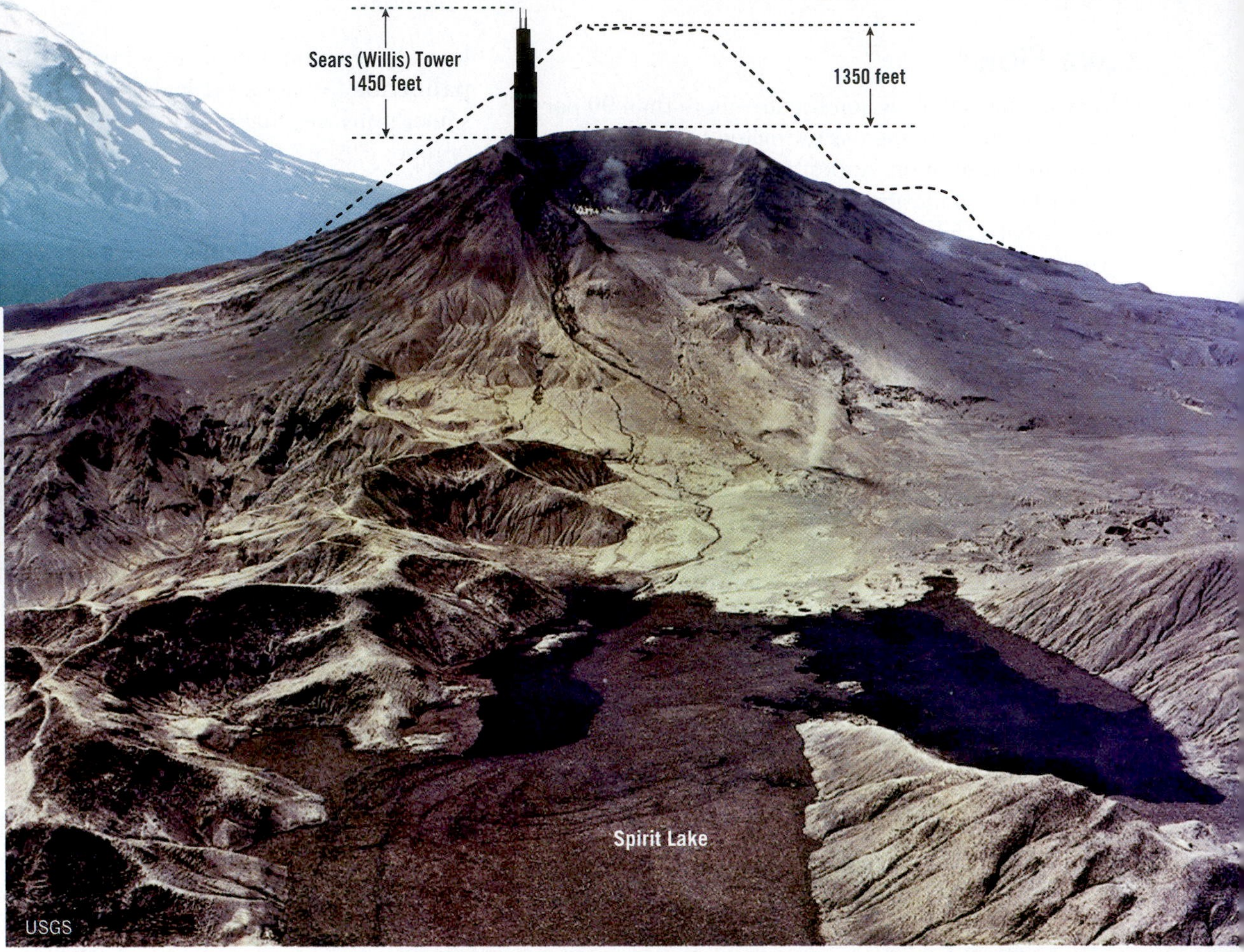

Sears (Willis) Tower 1450 feet

1350 feet

Spirit Lake

USGS

The blast blew out the entire north flank of Mount St. Helen's, leaving a gaping hole. In a brief moment, a prominent volcano was lowered by 1350 feet.

The initial eruption devastated 400 square kilometers of timber-rich land. Trees lay flattened and intertwined, stripped of their branches, and looking like toothpicks strewn about. (Inset image by John Burnley/Photo Researchers, Inc.)

Volcanologist David Johnston, monitoring the volcano from 6 miles away on the morning of May 18, 1980, transmitted the message "Vancouver! Vancouver! This is it!" before being killed by the blast. In commemoration, the USGS office in Vancouver, Washington, was renamed the David A. Johnston Cascades Volcano Observatory.

USGS

USGS

Mount St. Helens May 18, 1980
Ash Fallout Distribution within the U.S.

Ritzville
Spokane
Mount St. Helens
Yakima

2 to 5 inches
1/2 to 2 inches
Trace to 1/2 inch

USGS

During the days following the historic eruption, fine volcanic ash was carried around Earth by strong upper-air winds.

Mt. St. Helens Eruption Stages

1 Accumulation of magma over several months caused a large bulge to form on the volcano's north flank.

2 An earthquake caused the collapse of the north flank of the volcano, which triggered a lateral blast that destroyed a 400 square-kilometer area.

3 Within seconds of the lateral blast, an enormous vertical eruption sent a column of volcanic ash to an altitude of 18 kilometers (11 miles).

Question:
Briefly describe the sequence of events that culminated in the May 18, 1980 eruption of Mount St. Helens.

?

A. Lava tubes are cave-like tunnels that once served as conduits carrying lava from an active vent to the flow's leading edge.

Valentine Cave, a lava tube at Lava Beds National Monument, California.

B. Skylights develop where the roofs of lava tubes collapse and reveal the hot lava flowing through the tube.

Figure 5.5
Lava tubes A. A lava flow may develop a solid upper crust, while the molten lava below continues to advance in a conduit called a lava tube. Some lava tubes exhibit extraordinary dimensions. Kazumura Cave, located on the southeastern slope of Hawaii's Mauna Loa volcano, is a lava tube extending more than 60 kilometers (40 miles). (Photo by Dave Bunell) **B.** The collapsed section of the roof of a lava tunnel results in a skylight. (Photo courtesy of U.S. Geological Survey)

great distances from its source, lava tubes are important features of fluid lava flows.

Block Lavas

In contrast to fluid basaltic magmas that can travel many kilometers, viscous andesitic and rhyolitic magmas tend to generate relatively short prominent flows—a few hundred meters to a few kilometers long. Their upper surface consists largely of massive, detached blocks—hence the name **block lava**. Although similar to aa flows, these lavas consist of blocks with slightly curved, smooth surfaces rather than the rough, sharp, spiny surfaces typical of aa flows.

Figure 5.6
Pillow lava Pillow lava that formed off the coast of Hawaii. (Photo courtesy of U.S. Geological Survey)

Pillow lavas form on the ocean floor and have elongated shapes, resembling toothpaste coming out of a tube.

Pillow Lavas Recall that much of Earth's volcanic output occurs along oceanic ridges (divergent plate boundaries). When outpourings of lava occur on the ocean floor, the flow's outer skin quickly freezes (solidifies) to form basaltic glass. However, the interior lava is able to move forward by breaking through the hardened surface. This process occurs over and over, as molten basalt is extruded—like toothpaste from a tightly squeezed tube. The result is a lava flow composed of numerous tube-like structures called **pillow lavas**, stacked one atop the other (**Figure 5.6**). Pillow lavas are useful when reconstructing geologic history because their presence indicates that the lava flow formed below the surface of a water body.

Gases

Magmas contain varying amounts of dissolved gases, called **volatiles**, held in the molten rock by confining pressure, just as carbon dioxide is held in cans and bottles of soft drinks. As with soft drinks, as soon as the pressure is reduced, the gases begin to escape. Obtaining gas samples from an erupting volcano is difficult and dangerous, so geologists usually must estimate the amount of gas originally contained in the magma.

The gaseous portion of most magmas makes up 1 to 6 percent of the total weight, with most of this in the form of water vapor. Although the percentage may be small, the actual quantity of emitted gas can exceed thousands of tons per day. Occasionally, eruptions emit colossal amounts of volcanic gases that rise high into the atmosphere, where they may reside for several years. Some of these eruptions may have an impact on Earth's climate, a topic we consider later in this chapter.

The composition of volcanic gases is important because these gases contribute significantly to our planet's atmosphere. Analyses of samples taken during Hawaiian eruptions indicate that the gas component is about 70 percent water vapor, 15 percent carbon dioxide, 5 percent nitrogen, and 5 percent sulfur dioxide, with lesser amounts of chlorine, hydrogen, and argon. (The relative proportion of each gas varies significantly from one volcanic region to another.) Sulfur compounds are easily recognized by their pungent odor. Volcanoes are also natural sources of air pollution; some emit large quantities of sulfur dioxide (SO_2), which readily combines with atmospheric gases to form toxic sulfuric acid and other sulfate compounds.

Pyroclastic Materials

When volcanoes erupt energetically, they eject pulverized rock, lava, and glass fragments from the vent. The particles produced are called **pyroclastic materials** (*pyro* = fire, *clast* = fragment) and are also referred to as **tephra**. These fragments range in size from very fine dust and sand-sized volcanic ash (less than 2 millimeters) to pieces that weigh several tons (**Figure 5.7**).

Ash and *dust* particles are produced when gas-rich viscous magma erupts explosively. As magma moves up in the vent, the gases rapidly expand, generating a melt that resembles the froth that flows from a bottle of champagne. As the hot gases expand explosively, the froth is blown into very fine glassy fragments. When the hot ash falls, the glassy shards often fuse to form a rock

called *welded tuff*. Sheets of this material, as well as ash deposits that later consolidate, cover vast portions of the western United States.

Somewhat larger pyroclasts that range in size from small beads to walnuts are known as *lapilli* ("little stones"). These ejecta are commonly called *cinders* (2–64 millimeters [0.08–2.5 inches]). Particles larger than 64 millimeters (2.5 inches) in diameter are called *blocks* when they are made of hardened lava and *bombs* when they are ejected as incandescent lava (see Figure 5.7). Because bombs are semi-molten

Particle name	Particle size	Image
Volcanic Ash*	Less than 2 mm (0.08 inch)	
Lapilli (Cinders)	Between 2 mm and 64 mm (0.08–2.5 inches)	
Volcanic Bombs	More than 64 mm (2.5 inches)	
Volcanic Blocks		

Pyroclastic Materials (Tephra)

*The term volcanic dust is used for fine volcanic ash less than 0.063 mm (0.0025 inch).

Figure 5.7
Types of pyroclastic materials Pyroclastic materials are also commonly referred to as tephra.

A. Scoria is a vesicular rock commonly having a basaltic or andesitic composition. Pea-to-basketball size scoria fragments make up a large portion of most cinder cones (also called *scoria cones*).

B. Pumice is a low density vesicular rock that forms during explosive eruptions of viscous magma having an andesitic to rhyolitic composition.

from the vent during an eruption of the Japanese volcano Asama.

So far we have distinguished various pyroclastic materials based largely on the sizes of the fragments. Some materials are also identified by their texture and composition. In particular, **scoria** is the name applied to vesicular ejecta produced from basaltic magma (**Figure 5.8A**). These black to reddish-brown fragments are generally found in the size range of lapilli and resemble cinders and clinkers produced by furnaces used to smelt iron. When magmas with intermediate (andesitic) or felsic (rhyolitic) compositions erupt explosively, they emit ash and the vesicular rock pumice (**Figure 5.8B**). Pumice is usually lighter in color and less dense than scoria, and many pumice fragments have so many vesicles that they are light enough to float.

upon ejection, they often take on a streamlined shape as they hurtle through the air. Because of their size, bombs and blocks usually fall near the vent; however, they are occasionally propelled great distances. For instance, bombs 6 meters (20 feet) long and weighing about 200 tons were blown 600 meters (2000 feet)

5.2 | Concept Checks

1. Contrast pahoehoe and aa lava flows.

2. How do lava tubes form?

3. List the main gases released during a volcanic eruption. What role do gases play in eruptions?

4. How do volcanic bombs differ from blocks of pyroclastic debris?

5. What is scoria? How is scoria different from pumice?

5.3 | Anatomy of a Volcano
Label a diagram that illustrates the basic features of a typical volcanic cone.

The popular image of a volcano is a solitary, graceful, snowcapped cone, such as Mount Hood in Oregon or Japan's Fujiyama. These picturesque, conical mountains are produced by volcanic activity that occurred intermittently over thousands, or even hundreds of thousands, of years. However, many volcanoes do not fit this image. Cinder cones are quite small and form during a single eruptive phase that lasts a few days to a few years. Alaska's Valley of Ten Thousand Smokes is a flat-topped ash deposit that blanketed a river valley to a depth of 200 meters (650 feet). This event lasted less than 60 hours yet emitted more than 20 times the volcanic material of the 1980 Mount St. Helens eruption.

Volcanic landforms come in a wide variety of shapes and sizes, and each volcano has a unique eruptive history. Nevertheless, volcanologists have been able to classify volcanic landforms and determine their eruptive patterns. In this section we will consider the general anatomy of an idealized volcanic cone. We will follow this discussion by exploring the three major types of volcanic cones—shield volcanoes, cinder cones, and composite volcanoes—as well as their associated hazards.

Volcanic activity frequently begins when a **fissure** (crack) develops in Earth's crust as magma moves forcefully toward the surface. As the gas-rich magma moves up through a fissure, its path is usually localized into a somewhat circular **conduit** that terminates at a surface opening called a **vent** (**Figure 5.9**). The cone-shaped structure we call a **volcanic cone** is often created by successive eruptions of lava, pyroclastic material, or frequently a combination of both, often separated by long periods of inactivity.

Located at the summit of most volcanic cones is a somewhat funnel-shaped depression, called a **crater** (*crater* = a bowl). Volcanoes built primarily of pyroclastic materials typically have craters that form by gradual accumulation of volcanic debris on the surrounding rim. Other craters form during explosive eruptions, as the rapidly ejected particles erode the crater walls. Craters also form when the summit area of a volcano collapses following an eruption. Some volcanoes have very large circular depressions, called **calderas**, that have diameters greater than 1 kilometer (0.6 mile) and in rare cases can exceed 50 kilometers (30 miles). The formation of various types of calderas will be considered later in this chapter.

During early stages of growth, most volcanic discharges come from a central summit vent. As a volcano matures, material also tends to be emitted from fissures that develop along the flanks (sides) or at the base of the volcano. Continued activity from a flank eruption may produce one or more small **parasitic cones** (*parasitus* = one who eats at the table of another). Italy's Mount Etna, for example, has more than 200 secondary vents, some of which have built parasitic cones. Many of these vents, however, emit only gases and are appropriately called **fumaroles** (*fumus* = smoke).

SmartFigure 5.9
Anatomy of a volcano
Compare the structure of a "typical" composite cone to that of a shield volcano (Figure 5.10) and a cinder cone (Figure 5.13).
(https://goo.gl/Nl9iNq)

Tutorial

Labels on figure: Bombs, Lava, Crater, Vent, Parasitic cone, Pyroclastic material, Conduit, Magma chamber

5.3 Concept Checks

1. Distinguish among a conduit, a vent, and a crater.
2. How is a crater different from a caldera?
3. What is a parasitic cone, and where does it form?
4. What is emitted from a fumarole?

5.4 | Shield Volcanoes

Summarize the characteristics of shield volcanoes and provide one example.

A **shield volcano** is produced by the accumulation of fluid basaltic lavas and exhibits the shape of a broad, slightly domed structure that resembles a warrior's shield (**Figure 5.10**). Most shield volcanoes begin on the ocean floor as *seamounts* (submarine volcanoes), a few of which grow large enough to form volcanic islands. In fact, many small- to modest-sized oceanic islands are either a single shield volcano or, more often, the coalescence of two or more shields built on massive amounts of pillow lavas. Examples include the Hawaiian Islands, the Canary Islands, Iceland, the Galapagos Islands, and Easter Island. Although less common, some shield volcanoes form on continental crust. Included in this group are Nyamuragira, Africa's most active volcano, and Newberry volcano in Oregon.

Shield Volcanoes of Hawaii

Extensive study of the Hawaiian Islands has revealed that they are constructed of a myriad of thin basaltic lava flows averaging a few meters thick intermixed with relatively minor amounts of ejected pyroclastic material. Mauna Loa is the largest of five overlapping shield volcanoes that comprise the Big Island of Hawaii (see Figure 5.10). From its base on the floor of the Pacific Ocean

Figure 5.10
Volcanoes of Hawaii
Mauna Loa, Earth's largest volcano, is one of five shield volcanoes that collectively make up the Big Island of Hawaii. Shield volcanoes are built primarily of fluid basaltic lava flows and contain only a small percentage of pyroclastic materials.

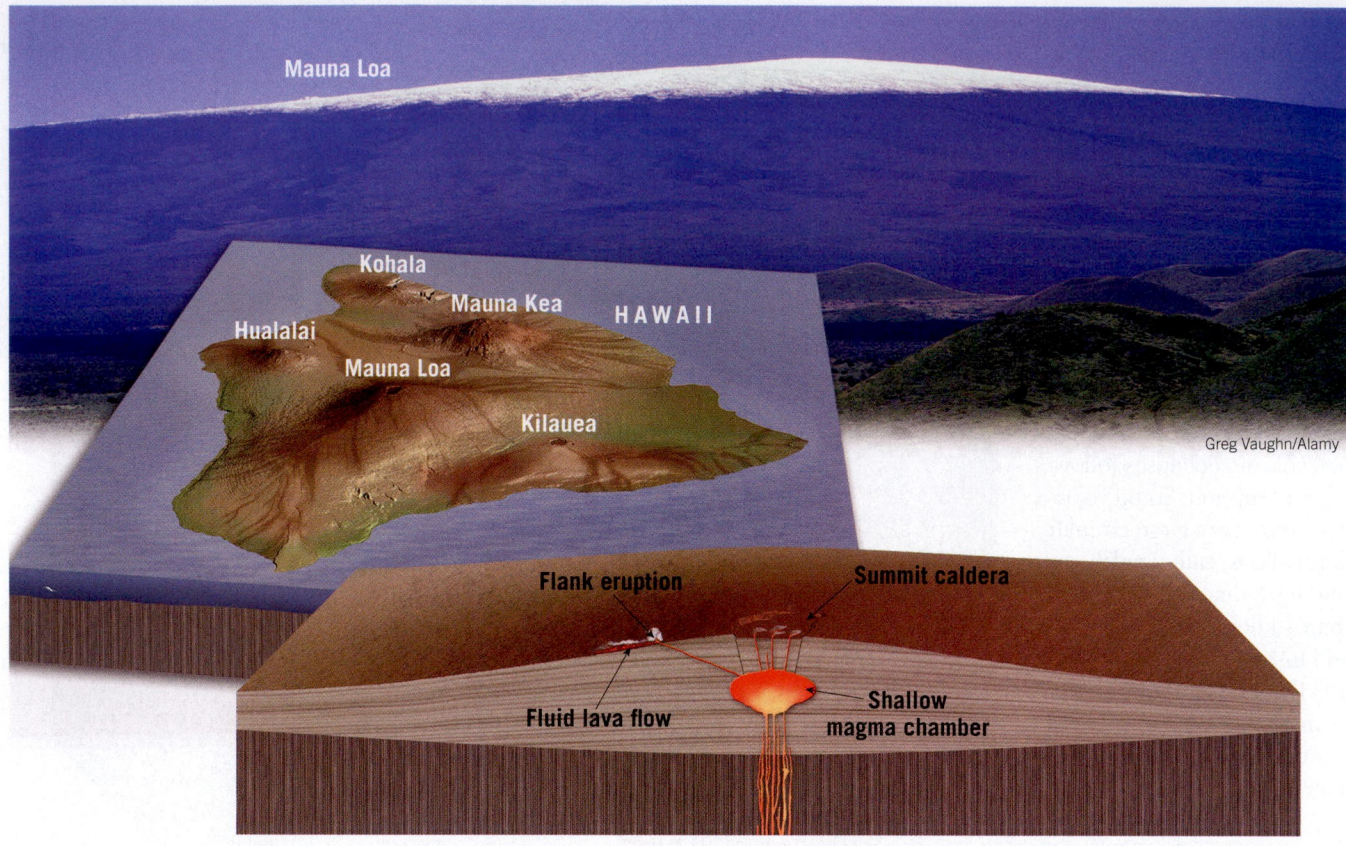

Greg Vaughn/Alamy

SmartFigure 5.11
Comparing scales of different types of volcanoes
A. Profile of Mauna Loa, Hawaii, the largest shield volcano in the Hawaiian chain. Note size comparison with Mount Rainier, Washington, a large composite cone.
B. Profile of Mount Rainier, Washington. Note how it dwarfs a typical cinder cone.
C. Profile of Sunset Crater, Arizona, a typical steep-sided cinder cone.
(http://goo.gl/VXEa4X)

Animation

to its summit, Mauna Loa is over 9 kilometers (6 miles) high, exceeding the elevation of Mount Everest. The volume of material composing Mauna Loa is roughly 200 times greater than that of a large composite cone such as Mount Rainier (**Figure 5.11**).

As with Hawaii's other shield volcanoes, the flanks of Mauna Loa have gentle slopes of only a few degrees. This low angle is due to the very hot, fluid lava that traveled "fast and far" from the vent. In addition, most of the lava (perhaps 80 percent) flowed through a well-developed system of lava tubes. Another feature common to active shield volcanoes is one or more large, steep-walled calderas that occupy the summit (see Figure 5.22, page 164). Calderas on shield volcanoes usually form when the roof above the magma chamber collapses. This occurs after the magma reservoir empties, either following a large eruption or as magma migrates to the flank of a volcano to feed a fissure eruption.

In their final stage of growth, shield volcanoes erupt more sporadically, and pyroclastic ejections are more common. The lava emitted later tends to be more viscous, resulting in thicker, shorter flows. These eruptions steepen the slope of the summit area, which often becomes capped with clusters of cinder cones. This explains why Mauna Kea, a more mature volcano that has not erupted in historic times, has a steeper summit than Mauna Loa, which erupted as recently as 1984. Astronomers are so certain that Mauna Kea is "over the hill" that they built an elaborate astronomical observatory on its summit to house some of the world's most advanced and expensive telescopes.

The zone of greatest volcanic activity on Hawaii began in what is presently the northwestern flanks of the island and has gradually migrated in a southeasterly direction. Its current location is Kilauea volcano, one of the most active and intensely studied

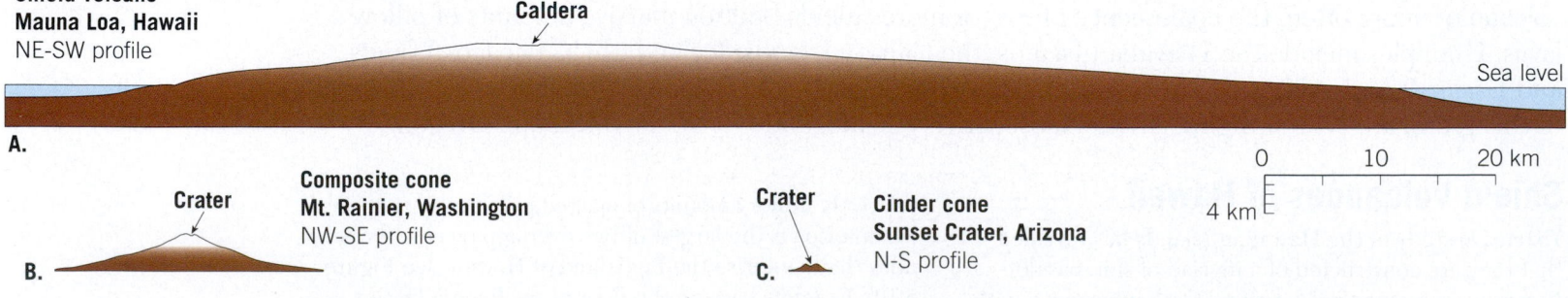

Shield volcano
Mauna Loa, Hawaii
NE-SW profile

Caldera

Sea level

A.

0   10   20 km

Composite cone
Mt. Rainier, Washington
NW-SE profile

Crater

4 km

Crater

Cinder cone
Sunset Crater, Arizona
N-S profile

B.

C.

shield volcanoes in the world. Kilauea, located in the shadow of Mauna Loa, has experienced more than 50 eruptions since record keeping began in 1823. Several months before each eruptive phase, Kilauea inflates as magma gradually migrates upward and accumulates in a central reservoir located a few kilometers below the summit. For up to 24 hours before an eruption, swarms of small earthquakes warn of the impending activity. Most of the recent activity on Kilauea has occurred along the flanks of the volcano, in a region called the East Rift Zone (**Figure 5.12**). The longest and largest rift eruption ever recorded on Kilauea began in 1983 and continues to this day, with no signs of abating (see GEOgraphics 5.2).

SmartFigure 5.12
A lava flow from Kilauea volcano advances toward the village of Pahoa, Hawaii, October 28, 2014. (Photo by Andrew Hara/Getty Images) (http://goo.gl/UHGrvC)

Mobile Field Trip

Evolution of Volcanic Islands

Although the shield volcanoes of Hawaii are commonly regarded as "typical" shield volcanoes, other ocean island volcanoes exhibit significant differences. The basaltic shields that comprise the Canary Islands, for example, have been active for as long as 20 million years, compared to most Hawaiian volcanoes that tend to have life spans of about 1 million years. Further, Canary Islands volcanoes tend to eject more pyroclastic material than do Hawaiian eruptions.

Despite their differences, most large oceanic shields have similar origins. They form above a long-lived, rising plume of hot mantle rock, called a *mantle plume*, described in more detail in Section 5.9. As the hot plume ascends through the mantle, the confining pressure drops, triggering partial melting and basaltic magma. This partial melting results in volcanic activity on the ocean floor, producing piles of pillow lavas that evolve into seamounts. Eventually, a few of these structures grow large enough to emerge as volcanic islands.

Volcanic islands continue to grow as long as they are located over a mantle plume, which provides them with a magma source. However, as the lithospheric plate moves away from the zone of melting, volcanic activity ceases. As the lithosphere cools and contracts, the overlying island gradually subsides. Most of the smallest volcanic islands ultimately sink below sea level as plate motion carries them further from the magma source.

5.4 Concept Checks

1. Describe the composition and viscosity of the lava associated with shield volcanoes.

2. Are pyroclastic materials a significant component of shield volcanoes?

3. Where do most shield volcanoes form—on the ocean floor or on the continents?

4. Where are the best-known shield volcanoes in the United States? Name some examples in other parts of the world.

5.5 | Cinder Cones
Describe the formation, size, and composition of cinder cones.

As the name suggests, **cinder cones** (also called **scoria cones**) are built from ejected lava fragments that begin to harden in flight to produce the vesicular rock scoria (**Figure 5.13**). These pyroclastic fragments range in size from fine ash to bombs that may exceed 1 meter (3 feet) in diameter. However, most of the volume of a cinder cone consists of pea- to walnut-sized fragments that are markedly vesicular and have a black to reddish-brown color. In addition, this pyroclastic material tends to have basaltic composition.

Kilauea's East Rift Zone Eruption

Kilauea, one of the most active volcanoes in the world, is located on the island of Hawaii in the shadow of Mauna Loa. Most of the recent activity on Kilauea has occurred along the flanks of the volcano in a region called the East Rift Zone. The longest and largest eruption ever recorded on Kilauea began in 1983, and continues with no signs of abating.

1 Kilauea's most recent eruptive phase began along a 6-kilometer (4 mile) fissure where a 100-meter (300-foot) high "curtain of fire" formed as red-hot basaltic lava was ejected skyward.

2 One of many fluid pahoehoe flows that have moved down the flanks of Kilauea since 1983.

3 The activity became localized at a single vent and a series of 44 short-lived episodes of lava fountaining built a cinder and spatter cone—given the Hawaiian name *Puu Oo.*

4 By the summer of 1986 a new vent opened along the rift. Pahoehoe lava flowing from this vent cut off the coastal highway and destroyed more than 180 structures including the National Park Visitor Center.

Greg Vaughn/Alamy

David Reggie/ Getty Images

Michael Coll

USGS

1983 to Present

Kilauea summit caldera

Halemaumau Crater

Pahoa

2014–2015

East rift zone

11

130

Kupaianaha

Puu Oo

Royal Gardens

Kalapana

N

0 1 2 3 4 miles
0   3   6 kilometers

Pacific Ocean

1983–1986

1986–1992

2007–2011

1992–2007

Royal Gardens

Kalapana

N

At night Halemaumau crater, continues to thrill visitors with the vivid glow that illuminates the plume of gases rising from its molten churning lava lake.

USGS

ERUPTION SUMMARY

1983-1986

The most recent eruption of Kilauea began on January 3, 1983 as a fissure eruption along the East Rift Zone in an area southeast of the summit caldera. These lava fountains quickly built a cinder-and-spatter cone called Puu Oo that produced abundant lava that flowed down the volcano's slope toward the sea.

1986-1992

In 1986 the eruption shifted eastward to new vent that built a small shield volcano called Kupainanaha. Lava tubes fed flows that extended about 12 kilometers (7 miles) to the sea. These flows buried most of the homes in the village of Kalapana under 15-25 meters (50-80 feet) of lava. It also destroyed a section of the Chain of Craters Road, buried Royal Gardens subdivision, and engulfed the famous Black Sand Beach at Kaimu.

1992-2012

Volcanic activity returned to the flanks of Puu Oo in 1992 when lava flowed nearly continuously to the ocean. By 2011, the eruption had added over 500 acres of land to the island, destroyed over 200 structures and buried 14 kilometers (9 miles) of the only major highway through that part of the island.

2014-2015

Kilauea volcano continues its activity in two locations. An eruption that began in 2008 in Halemaumau crater, which resides within the summit caldera, remains active. At night, the molten lava lake churns energetically and appears to be spewing fire. Lava continues to flow from the flanks of Puu Oo. The most recent flow began on June 27, 2014 and is slowly advancing toward the town of Pahoa and has the potential to engulf both the town and Highway 130.

Question:
What is the name of the area on Kilauea where the 1983 eruption began?

?

SmartFigure 5.13
Cinder cones Cinder cones are built from ejected lava fragments (mostly cinders and bombs) and are relatively small—usually less than 300 meters (1000 feet)—in height.
(http://goo.gl/hM5KVK)

Mobile Field Trip

Lava flow

Crater

Pyroclastic material

Central vent filled with rock fragments

SP Crater is a classic cinder cone located north of Flagstaff, Arizona.

Michael Collier

Although cinder cones are composed mostly of loose scoria fragments, some produce extensive lava fields. These lava flows generally form in the final stages of the volcano's life span, when the magma body has lost most of its gas content. Because cinder cones are composed of loose fragments rather than solid rock, the lava usually flows out from the unconsolidated base of the cone rather than from the crater.

A cinder cone has a very simple, distinct shape that is determined by the slope the loose pyroclastic material maintains as it comes to rest (Figure 5.13). Because cinders have a high angle of repose (the steepest angle at which material remains stable), cinder cones are steep-sided, having slopes between 30 and 40 degrees. In addition, a cinder cone has quite a large, deep crater in relation to the overall size of the structure. Although relatively symmetrical, some cinder cones are elongated and higher on the side that was downwind during the final eruptive phase.

Most cinder cones are produced by a single, short-lived eruptive event. One study found that half of all cinder cones examined were constructed in less than 1 month, and 95 percent formed in less than 1 year. Once the event ceases, the magma in the "plumbing" connecting the vent to the magma source solidifies, and the volcano usually does not erupt again. (One exception is Cerro Negro, a cinder cone in Nicaragua, which has erupted more than 20 times since it formed in 1850.) As a consequence of this typically short life span, cinder cones are small, usually between 30 and 300 meters (100 and 1000 feet) tall. A few rare examples exceed 700 meters (2300 feet) in height.

Cinder cones number in the thousands around the globe. Some occur in groups, such as the volcanic field near Flagstaff, Arizona, which consists of about 600 cones. Others are parasitic cones that are found on the flanks or within the calderas of larger volcanic structures.

Parícutin: Life of a Garden-Variety Cinder Cone

One of the very few volcanoes studied by geologists from its very beginning is a cinder cone called Parícutin, located about 320 kilometers (200 miles) west of Mexico City. In 1943, its eruptive phase began in a cornfield owned by Dionisio Pulido, who witnessed the event.

For 2 weeks prior to the first eruption, numerous tremors caused apprehension in the nearby village of Parícutin. Then, on February 20, sulfurous gases began billowing from a small depression that had been in the cornfield for as long as local residents could remember. During the night, hot, glowing rock fragments were ejected from the vent, producing a spectacular fireworks display. Explosive discharges continued, throwing hot fragments and ash occasionally as high as 6000 meters (20,000 feet) into the air. Larger fragments fell near the crater, some remaining incandescent as they rolled down the slope. These built an aesthetically pleasing cone, while finer ash fell over a much larger area, burning and eventually covering the village of Parícutin. In the first day, the cone grew to 40 meters (130 feet), and by the fifth day it was more than 100 meters (330 feet) high.

The first lava flow came from a fissure that opened just north of the cone, but after a few months, flows began to emerge from the base of the cone. In June 1944, a clinkery aa flow 10 meters (30 feet) thick

Parícutin, a cinder cone located in Mexico, erupted for nine years.

An aa flow emanating from the base of the cone buried much of the village of San Juan Parangaricutiro, leaving only remnants of the village's church.

SmartFigure 5.14
Parícutin, a well-known cinder cone The village of San Juan Parangaricutiro was engulfed by aa lava from Parícutin. The church steeple remains. (Photos by Michael Collier) (https://goo.gl/PQuC6A)

Condor Video

moved over much of the village of San Juan Parangaricutiro, leaving only the church steeple exposed (**Figure 5.14**). After 9 years of intermittent pyroclastic explosions and nearly continuous discharge of lava from vents at its base, the activity ceased almost as quickly as it had begun. Today, Parícutin is just another one of the scores of cinder cones dotting the landscape in this region of Mexico. Like the others, it will not erupt again.

5.5   Concept Checks

1. Describe the composition of a cinder cone.

2. How do cinder cones compare to shield volcanoes in terms of size and steepness of their flanks?

3. Over what time span does a typical cinder cone form?

5.6 | Composite Volcanoes

List the characteristics of composite volcanoes and describe how they form.

Earth's most picturesque yet potentially dangerous volcanoes are **composite volcanoes**, also known as **stratovolcanoes**. Most are located in a relatively narrow zone that rims the Pacific Ocean, appropriately called the *Ring of Fire* and discussed in detail in Section 5.9 (see Figure 5.29, page 169). This active zone consists of a chain of continental volcanoes distributed along the west coast of the Americas, including the large cones of the Andes in South America and the Cascade Range of the western United States and Canada.

Classic composite cones are large, nearly symmetrical structures consisting of alternating layers of explosively erupted cinders and ash interbedded with lava flows. A few composite cones, notably Italy's Etna and Stromboli, display very persistent eruption activity, and molten lava has been observed in their summit craters for decades.

Stromboli is so well known for eruptions that eject incandescent blobs of lava that it has been called the "Lighthouse of the Mediterranean." Mount Etna has erupted, on average, once every 2 years since 1979.

Just as shield volcanoes owe their shape to fluid basaltic lavas, composite cones reflect the viscous nature of the material from which they are made. In general, composite cones are the product of silica-rich magma having an andesitic composition. However, many composite cones also emit various amounts of fluid basaltic lava and, occasionally, pyroclastic material having a felsic (rhyolitic) composition. The silica-rich magmas typical of composite cones generate thick, viscous lavas that travel less than a few kilometers. Composite cones are also noted for generating explosive eruptions that eject huge quantities of pyroclastic material.

A conical shape with a steep summit area and gradually sloping flanks is typical of most large composite cones. This classic profile, which adorns calendars and postcards, is partially a result of the way viscous lavas and pyroclastic ejected materials contribute to the cone's growth. Coarse fragments ejected from the summit crater tend to accumulate near their source and contribute to the steep slopes around the summit. Finer ejected materials, on the other hand, are deposited as a thin layer over a large area that acts to flatten the flank of the cone. In addition, during the early stages of growth, lavas tend to be more abundant and flow greater distances from the vent than they do later in the volcano's history, which contributes to the cone's broad base. As a composite volcano matures, the shorter flows that come from the central vent serve to armor and strengthen the summit area. Consequently, steep slopes exceeding 40 degrees are possible. Two of the most perfect cones—Mount Mayon in the Philippines and Fujiyama in Japan—exhibit the classic form we expect of composite cones, with steep summits and gently sloping flanks (**Figure 5.15**).

Despite the symmetrical forms of many composite cones, most have complex histories. Many composite volcanoes have secondary vents on their flanks that have produced cinder cones or even much larger volcanic structures. Huge mounds of volcanic debris surrounding these structures provide evidence that large sections of these volcanoes slid downslope as massive landslides. Some develop amphitheater-shaped depressions at their summits as a result of explosive lateral eruptions—as occurred during the 1980 eruption of Mount St. Helens. Often, so much rebuilding has occurred since these eruptions that no trace of these amphitheater-shaped scars remain. Others, such as Crater Lake, have been truncated by the collapse of their summit (see Figure 5.21).

5.6   Concept Checks

1. What name is given to the region having the greatest concentration of composite volcanoes?

2. Describe the materials that compose composite volcanoes.

3. How do the composition and viscosity of lava flows differ between composite volcanoes and shield volcanoes?

Figure 5.15
Mount Fujiyama, a classic composite volcano Japan's Fujiyama exhibits the classic form of a composite cone—a steep summit and gently sloping flanks. (Photo by Koji Nakano/ Getty Images, Inc-Liaison)

5.7 | Volcanic Hazards

Discuss the major geologic hazards associated with volcanoes.

Roughly 1500 of Earth's known volcanoes have erupted at least once, and some several times, in the past 10,000 years. Based on historical records and studies of active volcanoes, 70 volcanic eruptions can be expected each year, and 1 large-volume eruption is likely to occur every decade. Large-magnitude eruptions account for the vast majority of human fatalities. For example, the 1902 eruption of Mount Pelée killed 28,000 people—nearly the entire population of the nearby city of St. Pierre.

Today, an estimated 500 million people from Japan and Indonesia, and from Italy to Oregon, live near active volcanoes. They face a number of volcanic hazards, such as destructive pyroclastic flows, molten lava flows, mudflows called lahars, and falling ash and volcanic bombs.

Pyroclastic Flow: A Deadly Force of Nature

Some of the most destructive forces of nature are **pyroclastic flows**, which consist of hot gases infused with incandescent ash and larger lava fragments. Also referred to as **nuée ardentes** (*glowing avalanches*), these fiery flows can race down steep volcanic slopes at speeds exceeding 100 kilometers (60 miles) per hour (**Figure 5.16**). Pyroclastic flows have two components: a low-density cloud of hot expanding gases containing fine ash particles and a ground-hugging portion composed of pumice and other vesicular pyroclastic material.

Driven by Gravity Pyroclastic flows are propelled by the force of gravity and tend to move in a manner similar to snow avalanches. They are mobilized by expanding volcanic gases released from the lava fragments and by the expansion of heated air that is overtaken and trapped in the moving front. These gases reduce friction between ash and pumice fragments, which gravity propels downslope in a nearly frictionless environment. This is why some pyroclastic flow deposits are found many miles from their source.

Occasionally, powerful hot blasts that carry small amounts of ash separate from the main body of a pyroclastic flow. These low-density clouds, called *surges*, can be deadly but seldom have sufficient force to destroy buildings in their paths. Nevertheless, in 2014, a hot ash cloud from Japan's Mount Ontake killed 47 hikers and injured 69 more.

Pyroclastic flows may originate in a variety of volcanic settings. Some occur when a powerful eruption blasts pyroclastic material out of the side of a volcano. More frequently, however, pyroclastic flows are generated by the collapse of tall eruption columns during an explosive event. When gravity eventually overcomes the initial upward thrust provided by the escaping gases, the ejected materials begin to fall, sending massive amounts of incandescent blocks, ash, and pumice cascading downslope.

The Destruction of St. Pierre In 1902, an infamous pyroclastic flow and associated surge from Mount Pelée, a small volcano on the Caribbean island of Martinique, destroyed the port town of St. Pierre. Although the main pyroclastic flow was largely confined to the valley of Riviere Blanche, a low-density fiery surge spread south of the river and quickly engulfed the entire city. The destruction happened in moments and was so devastating that nearly all of St. Pierre's 28,000 inhabitants were killed. Only 1 person on the outskirts of town—a

Figure 5.16
Pyroclastic flows Pyroclastic flows are composed of hot ash and pumice and/or blocky lava fragments that race down the slopes of volcanoes. (Photo by Ulet Ifansasti/Getty Images)

Figure 5.17
Destruction of St. Pierre A. St. Pierre as it appeared shortly after the eruption of Mount Pelée in 1902. (Reproduced from the collection of the Library of Congress) **B.** St. Pierre before the eruption. Many vessels were anchored offshore when this photo was taken, as was the case on the day of the eruption. (Photo by UPPA/Photoshot)

A. St. Pierre following the eruption of Mount Pelée.

B. St. Pierre before the 1902 eruption.

prisoner protected in a dungeon—and a few people on ships in the harbor were spared (**Figure 5.17**).

Scientists arrived on the scene within days and found that although St. Pierre was mantled by only a thin layer of volcanic debris, masonry walls nearly 1 meter (3 feet) thick had been knocked over like dominoes, large trees had been uprooted, and cannons had been torn from their mounts.

The Destruction of Pompeii

One well-documented event of historic proportions was the C.E. 79 eruption of the Italian volcano we now call Mount Vesuvius. For centuries prior to this eruption, Vesuvius had been dormant, with vineyards adorning its sunny slopes. Yet in less than 24 hours, the entire city of Pompeii (near Naples) and a few thousand of its residents were entombed beneath a layer of volcanic ash and pumice. The city and the victims of the eruption remained buried for nearly 17 centuries. The excavation of Pompeii gave archaeologists a superb picture of ancient Roman life (**Figure 5.18A**).

By reconciling historic records with detailed scientific studies of the region, volcanologists reconstructed the sequence of events. During the first day of the eruption, a rain of ash and pumice accumulated at a rate of 12 to 15 centimeters (5 to 6 inches) per hour, causing most of the roofs in Pompeii to eventually give way. Then, suddenly, a surge of searing hot ash and gas swept rapidly down the flanks of Vesuvius. This deadly pyroclastic flow killed those who had somehow managed to survive the initial ash and pumice fall. Their remains were quickly buried by falling ash, and subsequent rainfall caused the ash to harden. Over the centuries, the remains decomposed, creating cavities that were discovered by nineteenth-century excavators. Casts were then produced by pouring plaster of Paris into the voids (**Figure 5.18B**). Mount Vesuvius has had more than two dozen explosive eruptions since C.E. 79, the most recent occurring in 1944. Today, Vesuvius towers over the Naples skyline, a region occupied by roughly 3 million people. Such an image should prompt us to consider how volcanic crises might be managed in the future.

Lahars: Mudflows on Active and Inactive Cones

In addition to violent eruptions, large composite cones may generate a type of fluid mudflow known by its Indonesian name, **lahar**. These destructive flows occur when volcanic debris becomes saturated with water and rapidly moves down steep volcanic slopes, generally following stream valleys. Some lahars are triggered when magma nears the surface of a glacially clad volcano, causing large volumes of ice and snow to melt. Others are generated when heavy rains saturate weathered volcanic deposits. Thus, lahars may occur even when a volcano is *not* erupting.

Figure 5.18
Eruption of Mount Vesuvius, AD 79 In less than 24 hours, Pompeii and all of its residents were buried under a layer of volcanic ash and pumice that fell like rain. Today roughly 3 million people inhabit the area around this potentially hazardous volcano.

A. Pompeii was excavated nearly 17 centuries after the eruption of AD 79.

B. Plaster casts of some of the victims of the eruption of Mount Vesuvius.

Lahars are mudflows that originate on volcanic slopes.

This lahar was generated by the 1980 eruption of Mount St. Helens and carried this bridge down the valley of the Toutle River.

USGS

A.

Close-up view of bridge.

USGS

Lahar deposits invade a storage shed.

B. USGS

Figure 5.19
Lahars, mudflows that originate on volcanic slopes A. This lahar raced down the Toutle River, which flows westward from Mount St. Helens, following the May 18, 1980, eruption. **B.** Note the muddy consistency of lahar deposits.

When Mount St. Helens erupted in 1980, several lahars were generated. These flows and accompanying floodwaters raced down nearby river valleys at speeds exceeding 30 kilometers (20 miles) per hour. These raging rivers of mud destroyed or severely damaged nearly all the homes and bridges along their paths (**Figure 5.19**). Fortunately, the area was not densely populated.

In 1985, deadly lahars were produced during a small eruption of Nevado del Ruiz, a 5300-meter (17,400-foot) volcano in the Andes Mountains of Colombia. Hot pyroclastic material melted ice and snow that capped the mountain (*nevado* means "snow" in Spanish) and sent torrents of ash and debris down three major river valleys that flank the volcano. Reaching speeds of 100 kilometers (60 miles) per hour, these mudflows tragically claimed 25,000 lives.

Many consider Mount Rainier, Washington, to be America's most dangerous volcano because, like Nevado del Ruiz, it has a thick, year-round mantle of snow and glacial ice. Adding to the risk, more than 100,000 people live in the valleys around Rainier, and many homes are built on deposits left by lahars that flowed down the volcano hundreds or thousands of years ago. A future eruption, or perhaps just a period of heavier-than-average rainfall, may produce lahars that could be similarly destructive.

Other Volcanic Hazards

Volcanoes can be hazardous to human health and property in other ways. Ash and other pyroclastic material can collapse the roofs of buildings or can be drawn into the lungs of humans and other animals or into aircraft engines (**Figure 5.20**). Volcanic gases, most notably sulfur

Figure 5.20
Volcanic hazards In addition to generating destructive pyroclastic flows and lahars, volcanoes can be hazardous to human health and property in many other ways.

Ash and other pyroclastic materials can collapse roofs, or completely cover buildings.

AFLO/Nippon News/Corbis

Lava flows can destroy homes, roads, and other structures in their paths.

USGS

Prevailing wind

Eruption cloud

Ash fall

Acid rain

Eruption column

Bombs

Pyroclastic flow

Collapse of flank

Lava dome collapse

Emission of sulfur dioxide gases

Lava flow

Lahar (mudflow)

dioxide, pollute the air—and when mixed with rainwater, can destroy vegetation and reduce the quality of ground-water. Despite the known risks, millions of people live in close proximity to active volcanoes.

Volcano-Related Tsunamis One hazard associated with volcanoes is their ability to generate **tsunamis**. Although usually caused by strong earthquakes, these destructive sea waves sometimes result from powerful volcanic explosions or the sudden collapse of flanks of volcanoes into the ocean. In 1883, a volcanic eruption on the Indonesian island Krakatau literally tore the island apart and generated a tsunami that claimed an estimated 36,000 lives.

Volcanic Ash and Aviation During the past 15 years, at least 80 commercial jets have been damaged by inadvertently flying into clouds of volcanic ash. For example, in 1989, a Boeing 747 carrying more than 300 passengers encountered an ash cloud from Alaska's Redoubt Volcano. All four of the engines stalled when they became clogged with ash. Fortunately, the pilots were able to restart the engines and safely land the aircraft in Anchorage.

More recently, the 2010 eruption of Iceland's Eyjafjallajökull volcano sent ash high into the atmosphere. The thick plume of ash drifted over Europe, causing airlines to cancel thousands of flights, leaving hundreds of thousands of travelers stranded. Several weeks passed before air travel resumed its normal schedule.

Volcanic Gases and Respiratory Health One of the most destructive volcanic events, called the Laki eruption, began along a large fissure in southern Iceland in 1783. An estimated 14 cubic kilometers (3.4 cubic miles) of fluid basaltic lavas were released along with 130 million tons of sulfur dioxide and other poisonous gases. When sulfur dioxide is inhaled, it reacts with moisture in the lungs to produce sulfuric acid, a deadly toxin. More than half of Iceland's livestock died, and the ensuing famine killed 25 percent of the island's human population.

This huge eruption also endangered people and property all across Europe. Crop failure occurred in parts of Western Europe, and thousands of residents perished from lung-related diseases. A recent report estimated that a similar eruption today would cause more than 140,000 cardiopulmonary fatalities in Europe, as well as thousands of deaths elsewhere in the world.

Effects of Volcanic Ash and Gases on Weather and Climate Volcanic eruptions can eject dust-sized particles of volcanic ash and sulfur dioxide gas high into the atmosphere. The ash particles reflect sunlight back to space, producing temporary atmospheric cooling. The 1783 Laki eruption in Iceland appears to have affected atmospheric circulation around the globe. Drought conditions prevailed in the Nile River Valley and India, while the winter of 1784 brought the longest period of below-zero temperatures in New England's history.

Other eruptions that have produced significant effects on climate worldwide include the eruption of Indonesia's Mount Tambora in 1815, which produced the "year without a summer" (1816), and the eruption of El Chichón in Mexico in 1982. El Chichón's eruption, although small, emitted an unusually large quantity of sulfur dioxide that reacted with water vapor in the atmosphere to produce a dense cloud of tiny sulfuric acid droplets. These particles, called aerosols, take several years to settle out of the atmosphere. Like fine ash, these aerosols lower the mean temperature of the atmosphere by reflecting solar radiation back to space.

5.7   **Concept Checks**

1. **Describe pyroclastic flows and explain why they are capable of traveling great distances.**

2. **What is a lahar?**

3. **List at least three volcanic hazards besides pyroclastic flows and lahars.**

5.8 | Other Volcanic Landforms

List volcanic landforms other than shield, cinder, and composite volcanoes and describe their formation.

The most widely recognized volcanic structures are the cone-shaped edifices of composite volcanoes that dot Earth's surface. However, volcanic activity produces other distinctive and important landforms.

Calderas

Recall that *calderas* are large steep-sided depressions with diameters exceeding 1 kilometer (0.6 miles) that have a somewhat circular form. Similar depressions less than 1 kilometer across are called *collapse pits*, or *craters*. Most calderas are formed by one of the following processes: (1) the collapse of the summit of a large composite volcano following an explosive eruption of silica-rich pumice and ash fragments (*Crater Lake–type calderas*); (2) the collapse of the top of a shield volcano caused by subterranean drainage from a central magma chamber (*Hawaiian-type calderas*); and (3) the collapse of a large area, caused by the discharge of colossal volumes of silica-rich pumice and ash along ring fractures (*Yellowstone-type calderas*).

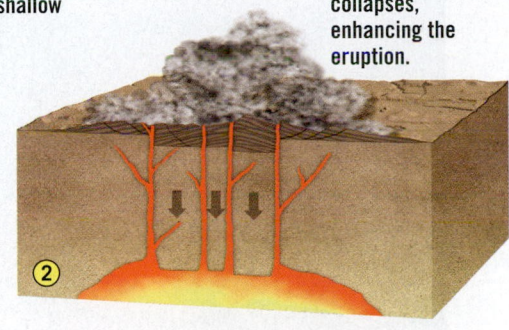

An explosive eruption partially empties a shallow magma chamber.

① Magma chamber

Summit of volcano collapses, enhancing the eruption.

②

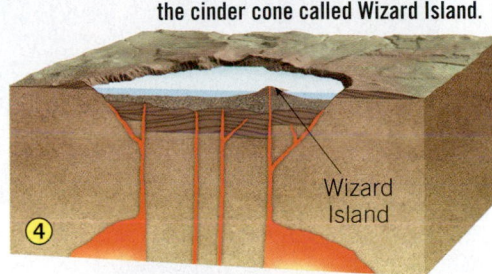

Newly formed caldera fills with rain and groundwater.

③

Subsequent eruptions produce the cinder cone called Wizard Island.

④

Wizard Island

SmartFigure 5.21
Formation of Crater Lake–type calderas About 7000 years ago, a violent eruption partly emptied the magma chamber of former Mount Mazama, causing its summit to collapse. Rainfall and groundwater contributed to forming Crater Lake, the deepest lake in the United States—594 meters (1949 feet) deep—and the ninth deepest in the world. (Inset photo courtesy of USGS) (https://goo.gl/FGvNQJ)

 **Animation**

Crater Lake–Type Calderas

Crater Lake, Oregon, is situated in a caldera approximately 10 kilometers (6 miles) wide and 1175 meters (more than 3800 feet) deep. This caldera formed about 7000 years ago, when a composite cone named Mount Mazama violently extruded 50 to 70 cubic kilometers of pyroclastic material (**Figure 5.21**). With the loss of support, 1500 meters (nearly 1 mile) of the summit of this once-prominent cone collapsed, producing a caldera that eventually filled with rainwater. Later, volcanic activity built a small cinder cone in the caldera. Today this cone, called Wizard Island, provides a mute reminder of past activity.

Crater Lake

Wizard Island

Michael Collier

Close-up view of Wizard Island.

Hawaiian-Type Calderas

Unlike Crater Lake–type calderas, many calderas form gradually because of the loss of lava from a shallow magma chamber underlying the volcano's summit. For example, Hawaii's active shield volcanoes, Mauna Loa and Kilauea, both have large calderas at their summits. Kilauea's measures 3.3 by 4.4 kilometers (about 2 by 3 miles) and is 150 meters (500 feet) deep (**Figure 5.22**). The walls are almost vertical, and as a result, the caldera looks like a vast, nearly flat-bottomed pit. Kilauea's caldera formed by gradual subsidence as magma slowly drained laterally from the underlying magma chamber, leaving the summit unsupported.

Yellowstone-Type Calderas

Historic and destructive eruptions such as Mount St. Helens and Vesuvius pale in comparison to what happened 630,000 years ago in the region now occupied by Yellowstone National Park, when approximately 1000 cubic kilometers of pyroclastic material erupted. This catastrophic eruption sent showers of ash as far as the Gulf of Mexico and formed a caldera 70 kilometers (43 miles) across (**Figure 5.23A**). Vestiges of this event are the many hot springs and geysers in the Yellowstone region.

Based on the extraordinary volume of erupted material, researchers have determined that the magma chambers associated with Yellowstone-type calderas must be similarly monstrous. As more and

SmartFigure 5.22
Kilauea's Summit Caldera
Kilauea's caldera measures 3.3 by 4.4 kilometers (about 2 by 3 miles) and is 150 meters (500 feet) deep. More recent flows are dark, while the older flows are pale because the iron in the lava oxidizes into "rust." Halemáumáu, the pit crater in the southwestern part of the caldera, was active as recently as 2012.
(https://goo.gl/OcQxed)

 **Video**

Lava lake in Halema'uma'u crater during a period of recent activity.

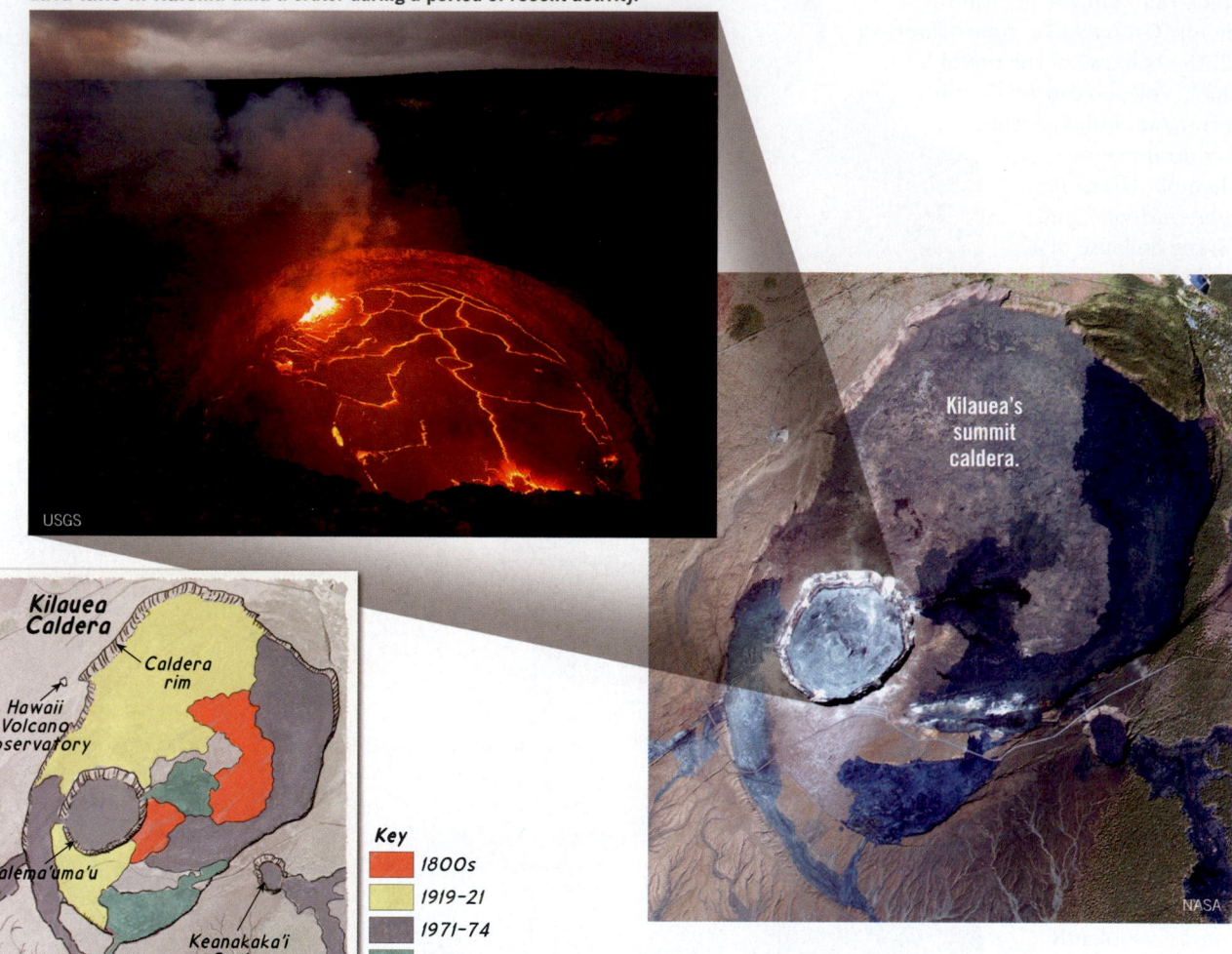

Kilauea's summit caldera.

Kilauea Caldera

Caldera rim

Hawaii Volcano Observatory

Haléma'uma'u

Keanakaka'i Crater

Key
- 1800s
- 1919–21
- 1971–74
- 1982

Geologist's Sketch

more magma accumulates, the pressure within the magma chamber begins to exceed the pressure exerted by the weight of the overlying rocks (**Figure 5.24**). An eruption occurs when the gas-rich magma raises the overlying crust enough to create vertical fractures extending to the surface. Magma surges upward along these cracks, forming a ring-shaped eruption. With a loss of support, the roof of the magma chamber collapses.

Caldera-forming eruptions are of colossal proportions, ejecting huge volumes of pyroclastic materials, mainly in the form of ash and pumice fragments. Typically, these materials form pyroclastic flows that sweep across the

EYE ON EARTH 5.1

This image was obtained during the 1991 eruption of Mount Pinatubo in the Philippines. This was the largest eruption to affect a densely populated area in recent times. Timely forecasts of the event by scientists were credited with saving at least 5,000 lives. (Alberto Garcia/CORBIS)

QUESTION 1 What name is given to the ash- and pumice-laden cloud that is racing toward the photographer?

QUESTION 2 At what speeds can these fiery clouds move down steep mountain slopes?

Jeep

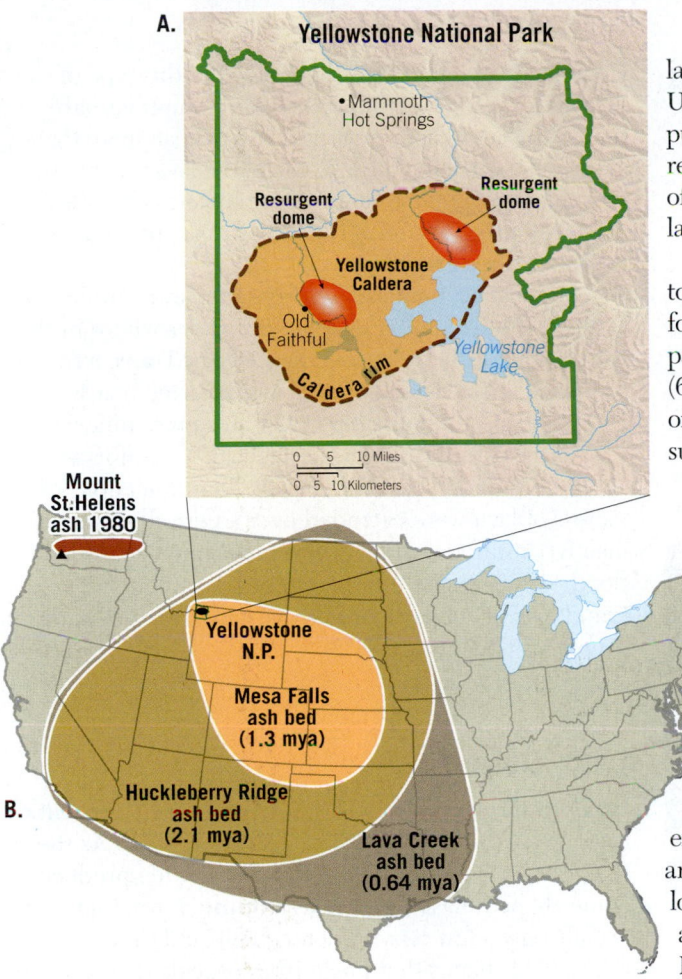

A. Yellowstone National Park

- Mammoth Hot Springs
- Resurgent dome
- Resurgent dome
- Yellowstone Caldera
- Old Faithful
- Caldera rim
- Yellowstone Lake

0   5   10 Miles
0 5 10 Kilometers

B.

- Mount St. Helens ash 1980
- Yellowstone N.P.
- Mesa Falls ash bed (1.3 mya)
- Huckleberry Ridge ash bed (2.1 mya)
- Lava Creek ash bed (0.64 mya)

Figure 5.23

Super-eruptions at Yellowstone **A.** This map shows Yellowstone National Park and the location and size of the Yellowstone caldera. **B.** Three huge eruptions, separated by relatively regular intervals of about 700,000 years, were responsible for the ash layers shown. The largest of these eruptions was 6000 times greater than the 1980 eruption of Mount St. Helens.

landscape, destroying most living things in their paths. Upon coming to rest, the hot fragments of ash and pumice fuse together, forming a welded tuff that closely resembles a solidified lava flow. Despite the immense size of these calderas, the eruptions that cause them are brief, lasting hours to perhaps a few days.

Large calderas tend to exhibit a complex eruptive history. In the Yellowstone region, for example, three caldera-forming episodes are known to have occurred over the past 2.1 million years (**Figure 5.23B**). The most recent event (630,000 years ago) was followed by episodic outpourings of degassed rhyolitic and basaltic lavas. Geologic evidence suggests that a magma reservoir still exists beneath Yellowstone; thus, another caldera-forming eruption is likely—but not necessarily imminent.

A distinctive characteristic of large caldera-forming eruptions is a slow upheaval of the floor of the caldera that produces a central elevated region called a *resurgent dome* (see Figure 5.23A). Unlike calderas associated with shield volcanoes or composite cones, Yellowstone-type depressions are so large and poorly defined that many remained undetected until high-quality aerial and satellite images became available. Other examples of large calderas located in the United States are California's Long Valley caldera; LaGarita caldera, located in the San Juan Mountains of southern Colorado; and the Valles caldera, west of Los Alamos, New Mexico. These and similar calderas found elsewhere around the globe are among the largest volcanic structures on Earth, hence the name "supervolcanoes." Volcanologists compare their destructive force to that of the impact of a small asteroid. Fortunately, no Yellowstone-type eruptions have occurred in historic times.

1. The emplacement of silica-rich magma causes the layers of rock above to bulge and crack, producing a set of *ring fractures*.

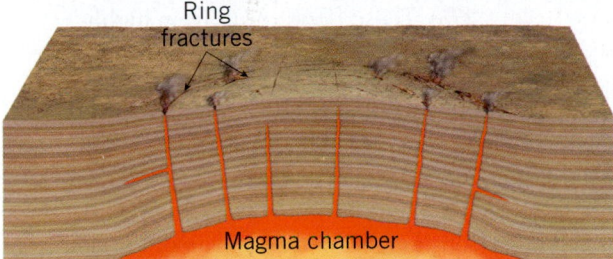

Ring fractures

Magma chamber

2. Massive amounts of magma erupt, producing fiery clouds of ash and gases called *pyroclastic flows* that devastate the surrounding landscape.

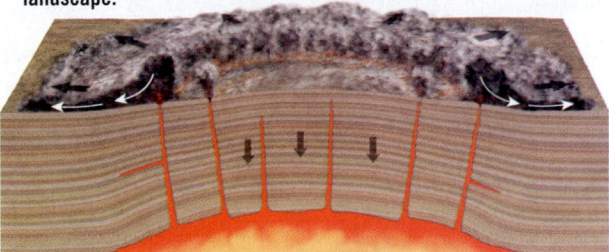

3. With loss of support, the roof of the magma chamber collapses forming a large caldera.

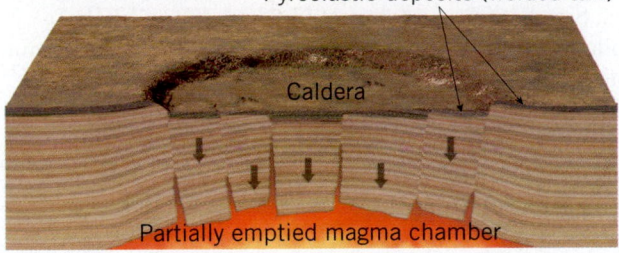

Pyroclastic deposits (welded tuff)

Caldera

Partially emptied magma chamber

4. Following the eruption, the caldera floor experiences a long period of slow upheaval that produces an elevated central region, called a *resurgent dome.*

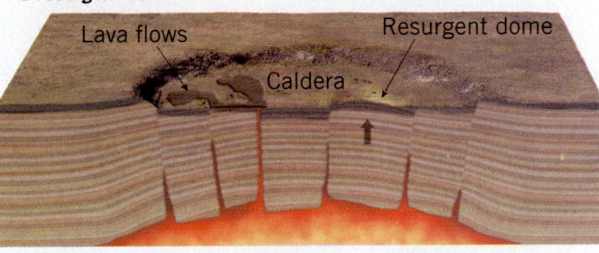

Lava flows

Resurgent dome

Caldera

SmartFigure 5.24
Formation of a Yellowstone-type caldera
Several Yellowstone-type calderas exist around the globe. These large volcanic structures are capable of rendering a large area of Earth's surface lifeless in perhaps only a few days, hence the name "supervolcanoes."
(https://goo.gl/6al3WM)

Tutorial

Fissure Eruptions and Basalt Plateaus

The greatest volume of volcanic material is extruded from fractures in Earth's crust called **fissures** (*fissura* = to split). Rather than build cones, fissure eruptions usually emit fluid basaltic lavas that blanket wide areas (**Figure 5.25**). In some locations, extraordinary amounts of lava have been extruded along fissures in a relatively short time, geographically speaking. These voluminous accumulations are commonly referred to as **basalt plateaus** because most have a basaltic composition and tend to be rather flat and broad. The Columbia Plateau in the northwestern United States, which consists of the

Columbia River basalts, is the product of this type of activity (**Figure 5.26**). Numerous fissure eruptions have buried the landscape, creating a lava plateau more than 1500 meters (1 mile) thick. Some of the lava remained molten long enough to flow 300 kilometers (180 miles) from its source. The term **flood basalts** appropriately describes these extrusions.

Massive accumulations of basaltic lava, similar to those of the Columbia Plateau, occur elsewhere in the world. One of the largest is the Deccan Traps, a thick sequence of flat-lying basalt flows covering nearly 500,000 square kilometers (195,000 square miles) of west-central India. When the Deccan Traps formed about 66 million years ago, nearly 2 million cubic kilometers of lava were extruded over a period of approximately 1 million years. Several other huge deposits of flood basalts, including the Ontong Java Plateau, are found on the ocean floor and are called *large igneous provinces* (see Figure 5.32).

Figure 5.25
Basaltic fissure eruptions Lava fountaining from a fissure and formation of fluid lava flows called *flood basalts*. The lower photo shows flood basalt flows near Idaho Falls.

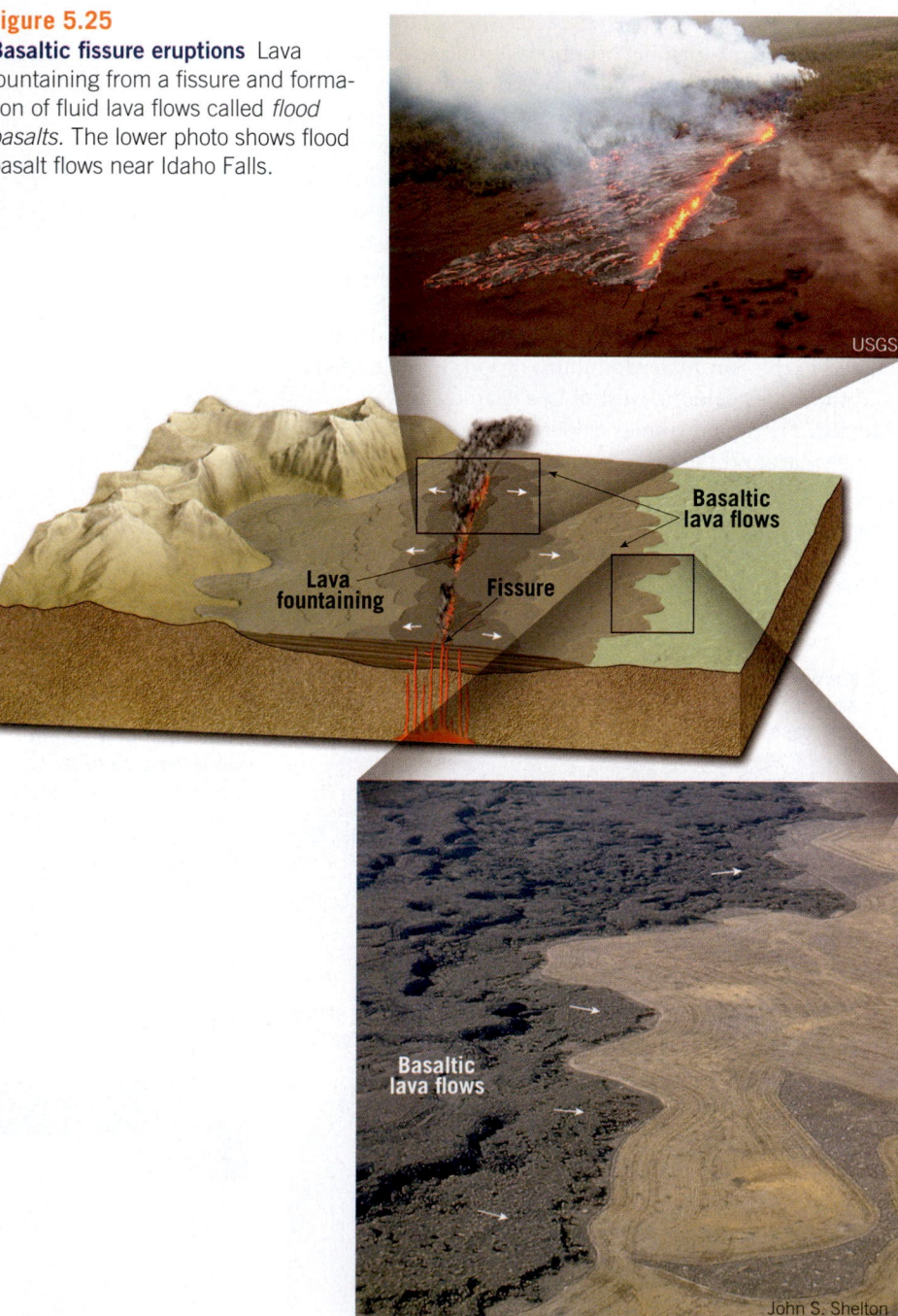

USGS

Basaltic lava flows

Lava fountaining   Fissure

Basaltic lava flows

Basaltic lava flows

John S. Shelton

Lava Domes

In contrast to hot basaltic lavas, cool silica-rich rhyolitic lavas are so viscous that they hardly flow at all. As the thick lava is "squeezed" out of a vent, it often produces a dome-shaped mass called a **lava dome**. Lava domes are usually only a few tens of meters high, and they come in a variety of shapes that range from pancake-like flows to steep-sided plugs that were pushed upward like pistons. Most lava domes grow over a period of several years, following an explosive eruption of silica-rich magma. A recent example is the dome that began to grow in the crater of Mount St. Helens immediately following the 1980 eruption (**Figure 5.27A**).

The collapse of lava domes, particularly those that form on the summit or along the steep flanks of composite cones, often produce powerful pyroclastic flows (**Figure 5.27B**). These flows result from highly viscous magma slowly entering the dome, causing it to expand and steepen its flanks. Over time, the cooler outer layer of the dome may start to crumble, producing relatively small pyroclastic flows consisting of dense blocks of lava. Occasionally, the rapid removal of the outer layer causes a significant decrease in pressure on the hot gaseous magma in the interior of the dome. Explosive degassing of the interior magma then triggers a fiery pyroclastic flow that races down the flanks of the volcano (see Figure 5.27B).

Since 1995, pyroclastic flows generated by the collapse of several lava domes on Soufriére Hills volcano have rendered more than half of the Caribbean island of Montserrat uninhabitable. The capital city, Plymouth, was destroyed, and two-thirds of the population has evacuated. A few years earlier, a collapsed lava dome at the summit of Japan's Mount Unzen produced a pyroclastic flow that claimed 42 lives. Many of the

Figure 5.26
Columbia River basalts A. The Columbia River basalts cover an area of nearly 164,000 square kilometers (63,000 square miles) that is commonly called the Columbia Plateau. Activity here began about 17 million years ago, as lava began to pour out of large fissures, eventually producing a basalt plateau with an average thickness of more than 1.5 kilometers. **B.** Columbia River basalt flows exposed in the Palouse River Canyon in southwestern Washington State. (Photo by Williamborg)

The Palouse River in Washington State has cut a canyon about 300 meters (1000 feet) deep into the flood basalts of the Columbia Plateau.

KEY

▮ Columbia River Basalts

▮ Other basaltic rocks

▲ Large Cascade volcanoes

A.

B.

A.

Lava domes are produced when highly viscous magma slowly rises over a period of months or years.

When a growing lava dome becomes over-steepened, it may collapse producing a blocky pyroclastic flow.

Blocky pyroclastic flow

Decompression of the interior magma may produce an explosive eruption and pyroclastic flow.

Fiery pyroclastic flow

B.

Figure 5.27
Lava domes can generate pyroclastic flows A. This lava dome began to develop in the vent of Mount St. Helens following the May 1980 eruption. **B.** The collapse of a lava dome often results in a powerful pyroclastic flow. (Photo by Lyn Topinka/U.S. Geological Survey)

SmartFigure 5.28
Volcanic neck Shiprock, New Mexico, is a volcanic neck that stands about 520 meters (1700 feet) high. It consists of igneous rock that crystallized in the vent of a volcano that has long since been eroded.

(https://goo.gl/XSRxzc)

Tutorial

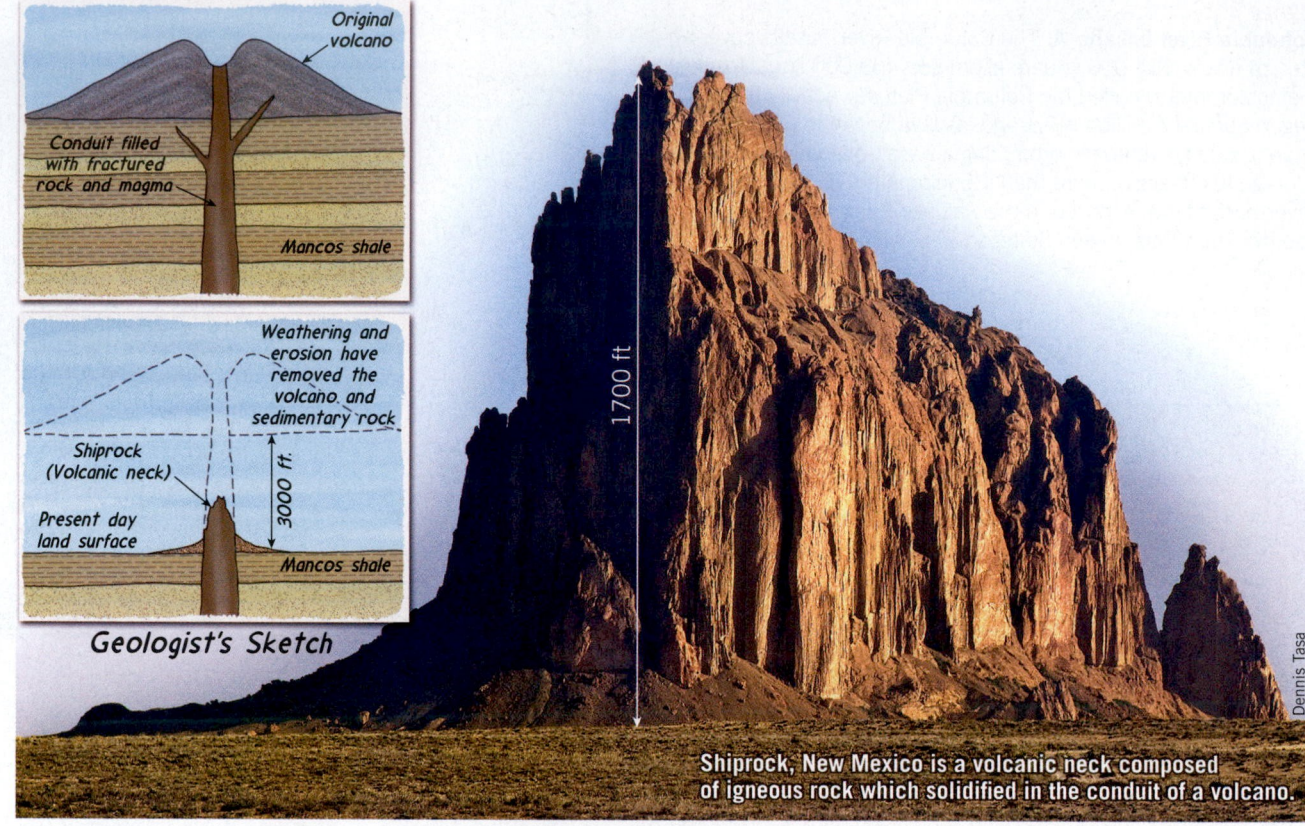

Geologist's Sketch

Shiprock, New Mexico is a volcanic neck composed of igneous rock which solidified in the conduit of a volcano.

Dennis Tasa

victims were journalists and film makers who ventured too close to the volcano in order to obtain photographs and document the event.

Volcanic Necks and Pipes

In most volcanic eruptions, lava is fed through short conduits that connect shallow magma chambers to vents located at the surface. When a volcano becomes inactive, congealed magma often becomes preserved in the feeding conduit of the volcano as a crudely cylindrical mass. However, all volcanoes succumb to forces of weathering and erosion. As erosion progresses, the rock occupying the volcanic conduit, which is highly resistant to weathering, may remain standing above the surrounding terrain long after the cone has been worn away. Shiprock, New Mexico, is a widely recognized and spectacular example of these structures, which geologists call **volcanic necks** (or **plugs**) (**Figure 5.28**). More than 510 meters (1700 feet) high, Shiprock is taller than most skyscrapers and is one of many such landforms that protrude conspicuously from the red desert landscapes of the American Southwest.

One rare type of conduit, called a **pipe**, carries magma that originated in the mantle at depths that may exceed 150 kilometers (93 miles). The gas-laden magmas that migrate through pipes travel rapidly enough to remain essentially unaltered during their ascent.

The best-known volcanic pipes are the diamond-bearing kimberlite pipes of South Africa. The rocks filling these pipes originated at great depths, where pressure is sufficiently high to generate diamonds and other minerals formed under high pressure. The process of transporting essentially unaltered magma (along with diamond inclusions) through 150 kilometers (100 miles) of solid rock is exceptional—and accounts for the scarcity of natural diamonds. Geologists consider pipes to be "windows" into Earth that allow us to view rocks that normally form only at great depths.

5.8   Concept Checks

1. Describe the formation of Crater Lake. Compare it to the calderas found on shield volcanoes such as Kilauea.

2. In addition to composite volcanoes, what other volcanic landform can generate a pyroclastic flow?

3. How do the eruptions that created the Columbia Plateau differ from eruptions that create large composite volcanoes?

4. Contrast the composition of a typical lava dome and a typical fissure eruption.

5. What type of volcanic structure is Shiprock, New Mexico, and how did it form?

5.9 | Plate Tectonics and Volcanic Activity

Explain how the global distribution of volcanic activity is related to plate tectonics.

Geologists have known for decades that the global distribution of most of Earth's volcanoes is not random. Most active volcanoes on land are located along the margins of the ocean basins—notably within the circum-Pacific belt known as the **Ring of Fire** (**Figure 5.29**). These volcanoes consist mainly of composite cones that emit volatile-rich magma that has an andesitic (intermediate) composition and occasionally produce awe-inspiring eruptions.

Another group of volcanoes includes the innumerable seamounts that form along the crest of the mid-ocean ridges. At these depths (1 to 3 kilometers below sea level), the pressures are so intense that the gases emitted quickly dissolve in the seawater and never reach the surface.

There are some volcanic structures, however, that appear to be somewhat randomly distributed around the globe. These volcanic structures comprise most of the islands of the deep-ocean basins, including the Hawaiian Islands, the Galapagos Islands, and Easter Island.

Prior to the development of the theory of plate tectonics, geologists did not have an acceptable explanation for the distribution of Earth's volcanoes. Recall that most magma originates in rocks from Earth's upper mantle that is essentially solid, *not molten*. The basic connection between plate tectonics and volcanism is that *plate motions provide the mechanisms by which mantle rocks undergo partial melting to generate magma.*

Volcanism at Convergent Plate Boundaries

Recall from the previous chapter that along certain convergent plate boundaries, two plates move toward each other, and a slab of oceanic lithosphere descends into the mantle, generating a deep-ocean trench. As the slab sinks deeper into the mantle, the increases in temperature and pressure drive water and carbon dioxide from the subducting oceanic crust. These mobile fluids migrate upward and reduce the melting point of hot mantle rock sufficiently to trigger some melting (**Figure 5.30A**). Recall that this partial melting of mantle rock (peridotite) generates magma with a basaltic composition. After a sufficient quantity of the rock has melted, blobs of buoyant magma slowly migrate upward.

Volcanism at a convergent plate margin results in the development of a slightly curved chain of volcanoes called a *volcanic arc*. These volcanic chains develop roughly parallel to the associated trench—at distances of 200 to 300 kilometers (120 to 180 miles). Volcanic arcs can be constructed on oceanic as well as continental lithosphere. Those that develop within the ocean and grow large enough for their tops to rise above the surface are labeled *archipelagos* in most atlases. Geologists prefer the more descriptive term **volcanic island arcs**, or simply **island arcs** (Figure 5.30A). Several young volcanic island arcs border the western Pacific basin, including the Aleutians, the Tongas, and the Marianas.

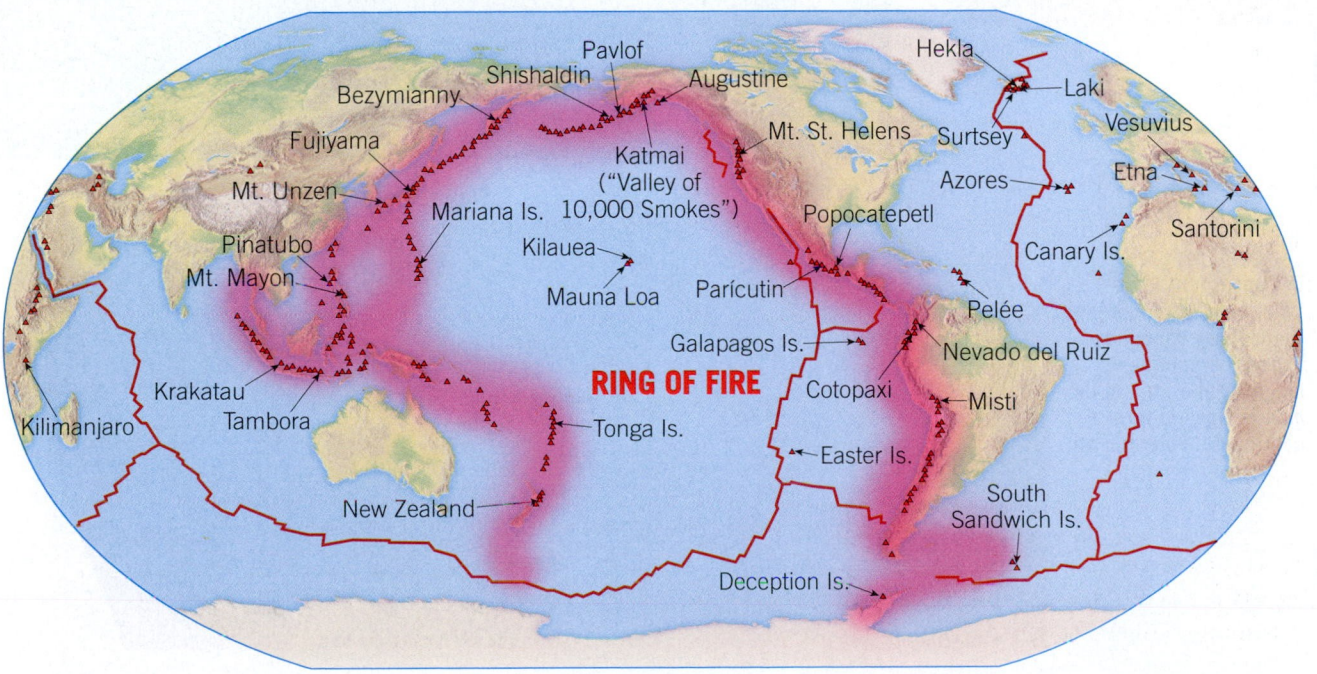

Figure 5.29

Ring of Fire Most of Earth's major volcanoes are located in a zone around the Pacific called the Ring of Fire. Another large group of active volcanoes lie undiscovered along the mid-ocean ridge system.

A. Convergent Plate Volcanism When an oceanic plate subducts, melting in the mantle produces magma that gives rise to a volcanic island arc on the overlying oceanic crust.

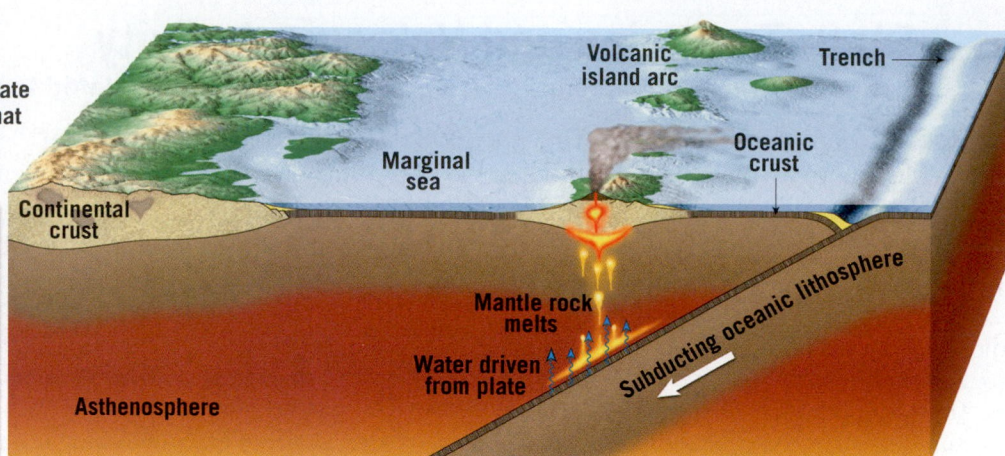

Cleveland Volcano, Aleutian Islands (USGS)

C. Intraplate Volcanism
When an oceanic plate moves over a hot spot, a chain of volcanic structures such as the Hawaiian Islands is created.

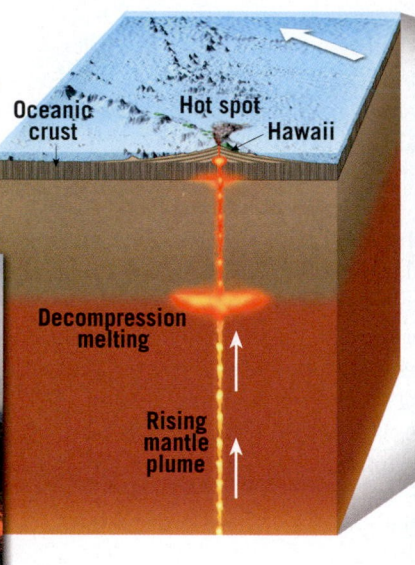

Kilauea, Hawaii (USGS)

E. Convergent Plate Volcanism When oceanic lithosphere descends beneath a continent, magma generated in the mantle rises to form a continental volcanic arc.

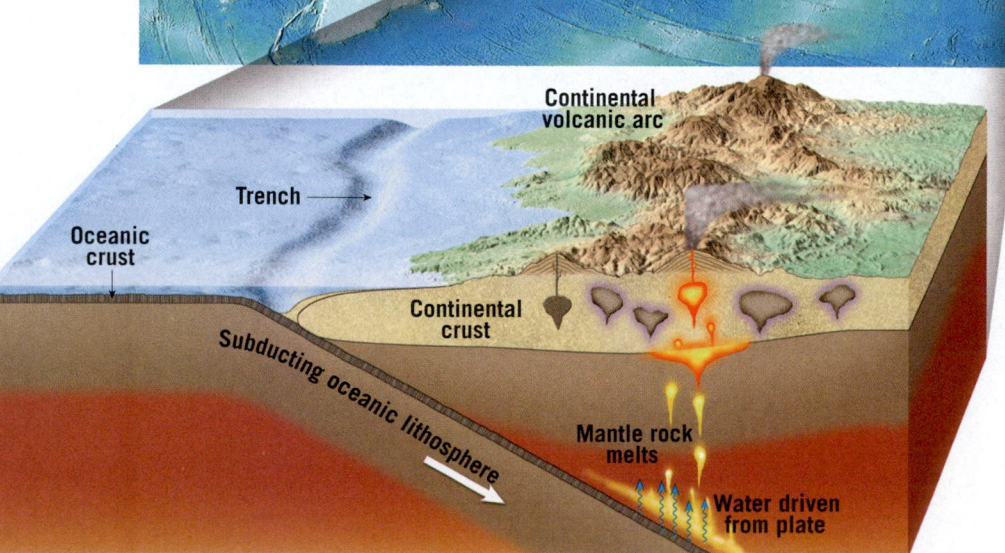

SmartFigure 5.30
Earth's zones of volcanism
(https://goo.gl/MoqLrr)

Tutorial

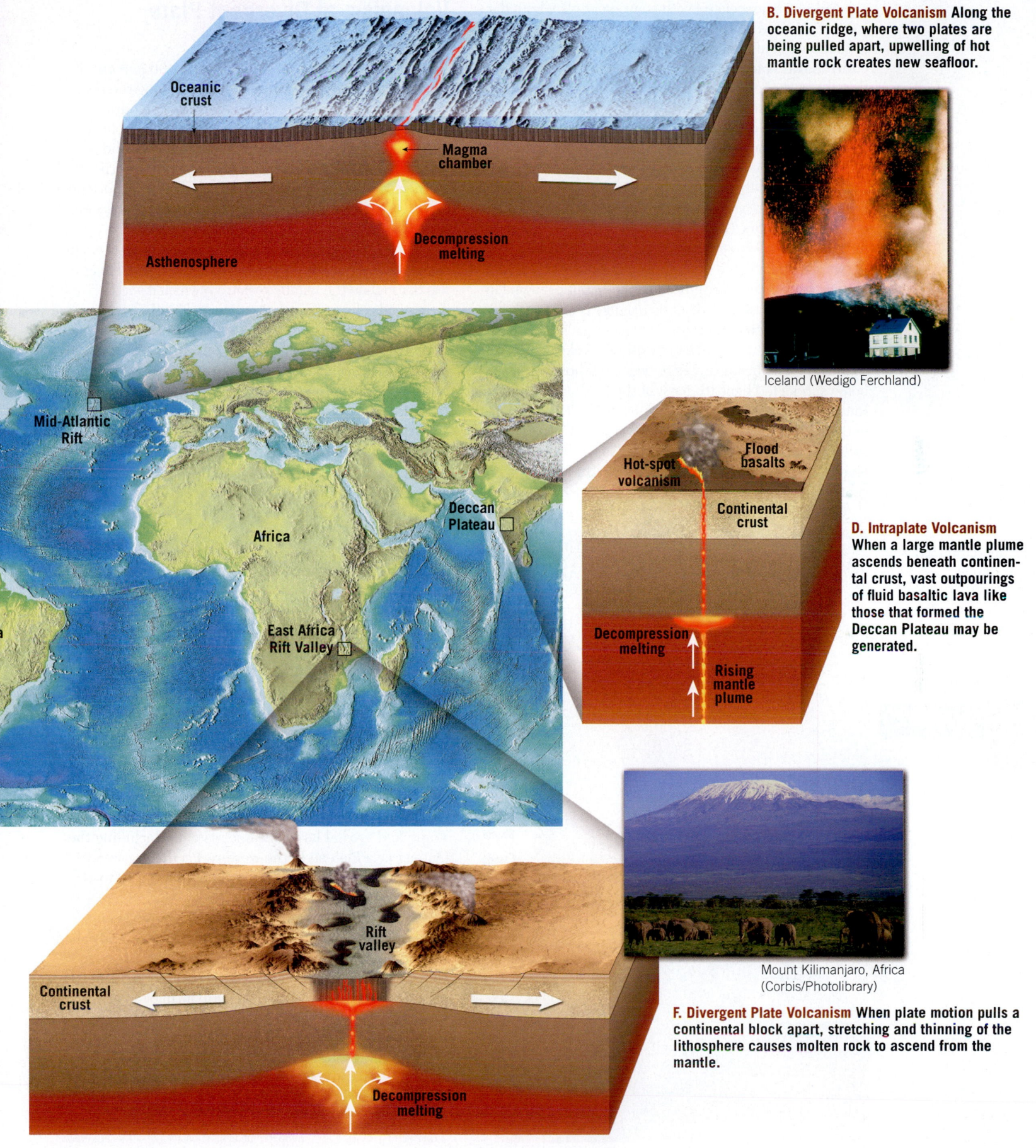

B. Divergent Plate Volcanism Along the oceanic ridge, where two plates are being pulled apart, upwelling of hot mantle rock creates new seafloor.

Oceanic crust

Magma chamber

Asthenosphere

Decompression melting

Iceland (Wedigo Ferchland)

Mid-Atlantic Rift

Deccan Plateau

Africa

East Africa Rift Valley

D. Intraplate Volcanism
When a large mantle plume ascends beneath continental crust, vast outpourings of fluid basaltic lava like those that formed the Deccan Plateau may be generated.

Hot-spot volcanism

Flood basalts

Continental crust

Decompression melting

Rising mantle plume

Rift valley

Continental crust

Decompression melting

Mount Kilimanjaro, Africa (Corbis/Photolibrary)

F. Divergent Plate Volcanism When plate motion pulls a continental block apart, stretching and thinning of the lithosphere causes molten rock to ascend from the mantle.

Volcanism associated with convergent plate boundaries may also develop where slabs of oceanic lithosphere are subducted under continental lithosphere to produce a **continental volcanic arc** (Figure 5.30E). The mechanisms that generate these mantle-derived magmas are essentially the same as those that create volcanic island arcs. The most significant difference is that continental crust is much thicker and is composed of rocks having higher silica content than oceanic crust. Hence, a mantle-derived magma changes chemically as it rises by assimilating silica-rich crustal rocks. At the same time, extensive magmatic differentiation occurs. (Recall from Chapter 4 that magmatic differentiation is the formation of secondary magmas from parent magma.) Stated another way, the magma generated in the mantle may change from a fluid basaltic magma to a silica-rich andesitic or rhyolitic magma as it moves up through the continental crust.

The Ring of Fire, a belt of explosive volcanoes, surrounds the Pacific basin where oceanic lithosphere is being subducted beneath most of the landmasses that surround the Pacific Ocean (see Figure 5.29). The volcanoes of the Cascade Range in the northwestern United States, including Mount Hood, Mount Rainier, and Mount Shasta, are examples of volcanoes generated at a convergent plate boundary (Figure 5.31).

SmartFigure 5.31
Subduction-produced Cascade Range volcanoes Subduction of the Juan de Fuca plate along the Cascadia subduction zone produced the Cascade volcanoes.
(https://goo.gl/4To5ak)

Tutorial

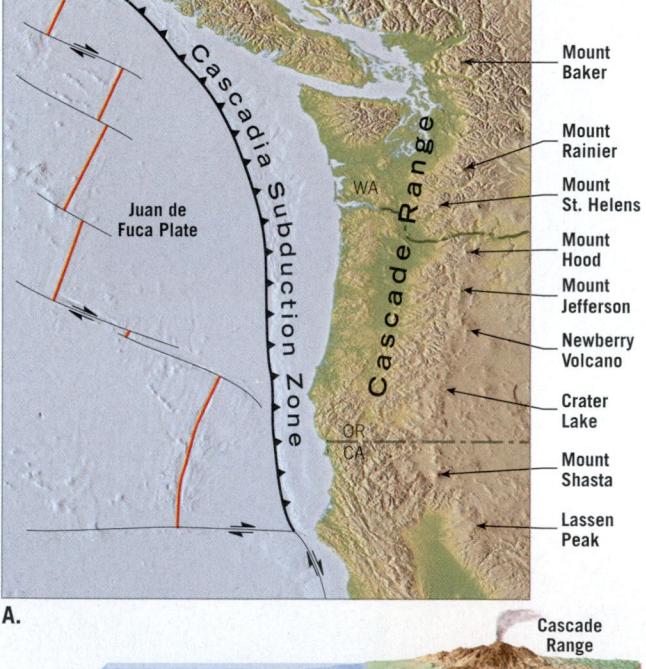

A.

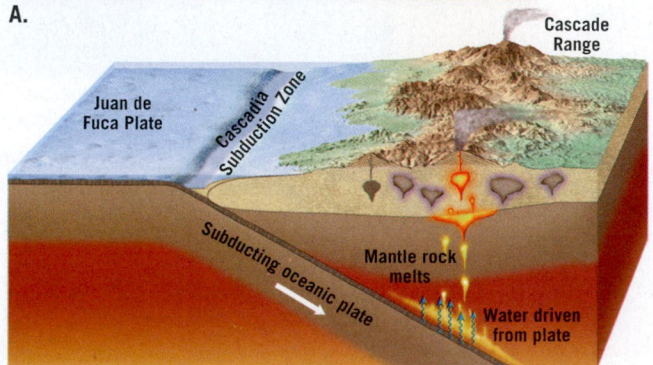

B.

Volcanism at Divergent Plate Boundaries

The greatest volume of magma (perhaps 60 percent of Earth's total yearly output) is produced along the oceanic ridge system in association with seafloor spreading (Figure 5.30B). Below the ridge axis where lithospheric plates are continually being pulled apart, the solid yet mobile mantle responds by ascending to fill the rift. You learned in Chapter 4 that as hot rock rises, it experiences a decrease in confining pressure and undergoes *decompression melting* without the addition of heat.

Partial melting of mantle rock at spreading centers produces basaltic magma. Because this newly formed magma is less dense than the mantle rock from which it was derived, it rises and collects in reservoirs located just below the ridge crest. This activity continuously adds new basaltic rock to plate margins, temporarily welding them together, only to have them break again as spreading continues. Along some ridge segments, outpourings of pillow lavas build numerous volcanic structures, the largest of which is Iceland.

Although most spreading centers are located along the axis of an oceanic ridge, some are not. In particular, the East African Rift is a site where continental lithosphere is being pulled apart (see Figure 5.30F and Figure 2.15, page 49). Vast outpourings of fluid basaltic lavas as well as several active composite volcanoes are found in this region of the globe.

Intraplate Volcanism

We know why igneous activity is initiated along plate boundaries, but why do eruptions occur in the interiors of plates? Hawaii's Kilauea, considered one of the world's most active volcanoes, is situated thousands of kilometers from the nearest plate boundary, in the middle of the vast Pacific plate (Figure 5.30C).

Sites of **intraplate volcanism** (meaning "within the plate") include the large outpourings of fluid basaltic lavas such as those that compose the Columbia River basalts, the Siberian Traps in Russia, India's Deccan Traps, and several large oceanic plateaus, including the Ontong Java Plateau in the western Pacific (Figure 5.32). These massive structures are estimated to be 10 to 40 kilometers (6 to 25 miles) thick.

It is thought that most intraplate volcanism occurs where a mass of hotter-than-normal mantle material called a **mantle plume** ascends toward the surface (Figure 5.33).[*] Although the depth at which mantle plumes originate is a topic of debate, some are thought to form deep within Earth, at the core–mantle boundary. These plumes of solid yet mobile rock rise toward the surface in a manner similar to the blobs that form within a lava lamp, which contain two immiscible liquids in a glass container. As the base of the lamp is heated, the denser liquid at the bottom

[*]Recent research on the nature of mantle plumes has caused some geologists to question their role, if any, in the formation of at least some of the large basalt plateaus.

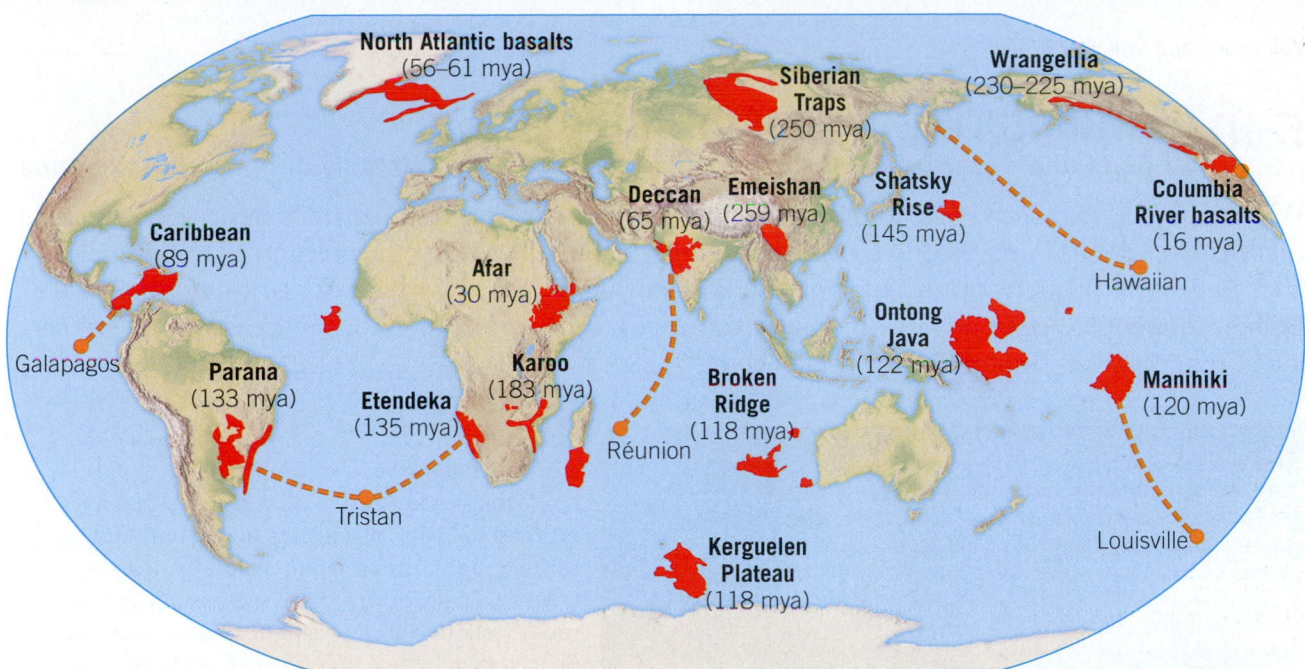

becomes buoyant and forms blobs that rise to the top. Like the blobs in a lava lamp, a mantle plume has a bulbous head that draws out a narrow stalk beneath it as it rises. The surface manifestation of this activity is called a *hot spot*, an area of volcanism and high heat flow.

Large mantle plumes, dubbed *superplumes*, are thought to be responsible for the vast outpourings of basaltic lava that created the large basalt plateaus. When the head of the plume reaches the base of the lithosphere, decompression melting progresses rapidly. This causes the burst of volcanism that emits voluminous outpourings of lava over a period of 1 million or so years (see **Figure 5.33B**). Due to the extreme nature of the eruptions required to produce the large basalt plateaus, some researchers believe these eruptions may have contributed to the extinction of many of Earth's life-forms.

The comparatively short initial eruptive phase is often followed by millions of years of less voluminous activity, as the plume tail slowly rises to the surface. Extending away from some large flood basalt plateaus is a chain of volcanic structures, similar to the Hawaiian chain (**Figure 5.33C**).

Intraplate volcanism that involves mantle plumes is also thought to be responsible for some massive eruptions of felsic (rhyolitic) pyroclastic material that occurred in a continental setting. The best know of these hot-spot eruptions are the three caldera-forming eruptions that occurred in the Yellowstone region.

5.9 Concept Checks

1. Are volcanoes in the Ring of Fire generally described as quiescent or violent? Name an example that supports your answer.

2. How is magma generated along convergent plate boundaries?

3. Volcanism at divergent plate boundaries is most often associated with which rock type? What causes rocks to melt in these settings?

4. What is the source of magma for most intraplate volcanism?

5. Which type of plate boundary generates the greatest quantity of magma?

**Figure 5.33
Mantle plumes and large basalt provinces** Model of hot-spot volcanism thought to explain the formation of large basalt provinces and the chains of volcanic islands associated with these features.

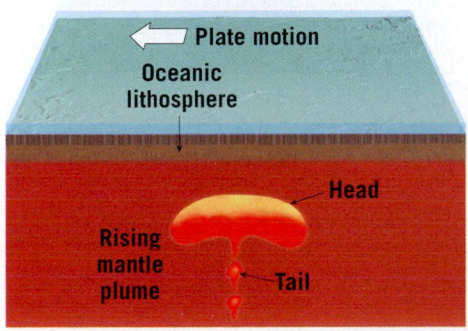

A rising mantle plume with a large bulbous head is thought to generate Earth's large basalt provinces.

A.

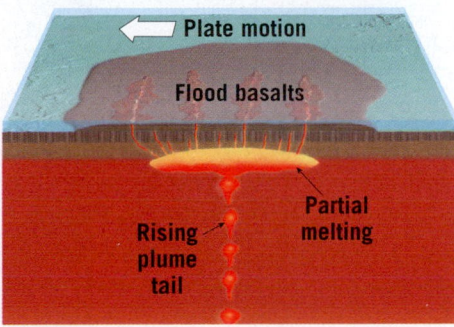

Rapid decompression melting of the plume head produces extensive outpourings of flood basalts over a relatively short time span.

B.

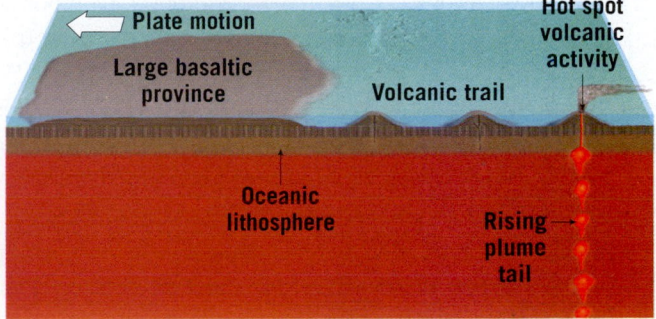

Because of plate movement, volcanic activity from the rising tail of the plume generates a linear chain of smaller volcanic structures.

C.

5.10 | Monitoring Volcanic Activity

List and describe the techniques used to monitor potentially dangerous volcanoes.

Why monitor volcanoes? Volcano monitoring has two primary goals. It provides basic scientific data that helps geologists understand the structure and dynamics of a specific volcano as well as volcanoes in general. Equally important, volcanic monitoring is critical for hazard assessment because millions of people live and work on or near volcanoes. In this role, monitoring addresses questions like, "When will an impending eruption occur?"

Figure 5.34

Monitoring volcanoes using infrared cameras Remote-sensing cameras on satellites have greatly enhanced our ability to monitor volcanoes. Infrared (heat) sensors produce false-color images of volcanoes that show hot objects in warm colors and cold objects in cool colors. This image shows an eruption of Mount St. Augustine volcano, Alaska, that occurred on March 27, 1986.

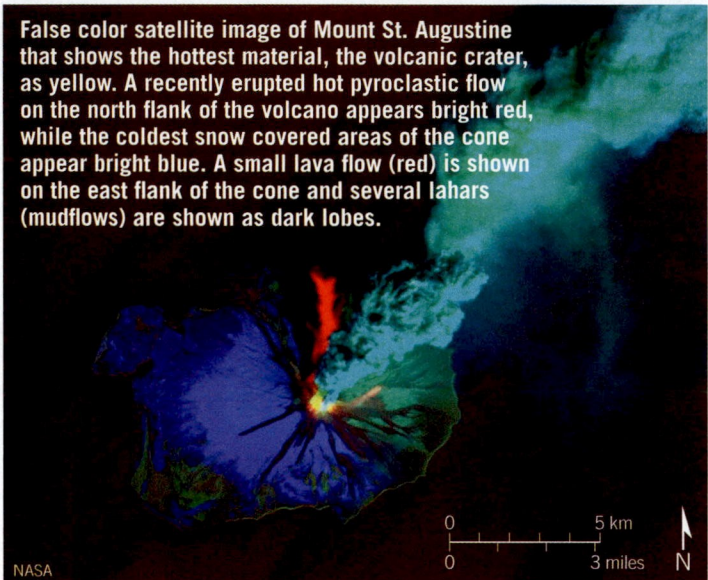

False color satellite image of Mount St. Augustine that shows the hottest material, the volcanic crater, as yellow. A recently erupted hot pyroclastic flow on the north flank of the volcano appears bright red, while the coldest snow covered areas of the cone appear bright blue. A small lava flow (red) is shown on the east flank of the cone and several lahars (mudflows) are shown as dark lobes.

0   5 km
0   3 miles   N

NASA

Volcanologists utilize several volcano monitoring techniques, with most of the efforts aimed at detecting the movement of magma from a subterranean reservoir (typically several kilometers deep) toward the surface. The three most noticeable changes in a volcanic landscape caused by the migration of magma are (1) changes in the pattern of earthquakes produced by the movement of magma; (2) magma entering a near-surface magma chamber, which leads to inflation of the volcano; and (3) changes in the amount and/or composition of gases released from a volcano.

Almost one-third of all volcanoes that have erupted in historic times are now monitored using seismographs, instruments that detect earthquake tremors. In general, a sharp increase in seismic unrest followed by a period of relative quiet has been shown to be a precursor of many volcanic eruptions. However, some large volcanic structures have exhibited lengthy periods of seismic unrest. For example, in 1981, a strong increase in seismicity was recorded for Rabaul caldera in New Guinea. The activity lasted 13 years and culminated with an eruption in 1994. Occasionally, a large earthquake triggers a volcanic eruption, or at least disturbs the volcano's plumbing. Hawaii's Kilauea, for example, began to erupt after the strong Kalapana earthquake (magnitude 7.2) of 1975.

The development of remote-sensing devices has greatly enhanced our ability to monitor volcanoes. These instruments and techniques are particularly useful for monitoring eruptions in progress. Photographic images and infrared (heat) sensors can detect lava flows and volcanic columns rising from a volcano (**Figure 5.34**). Furthermore, satellites can detect ground deformation as well as monitor sulfur dioxide emissions.

Because accessibility to many volcanoes is limited, remote-sensing devices, including lasers, Global

EYE ON EARTH 5.2

This image shows the Buddhist monastery Taung Kalat, located in central Myanmar (Burma). The monastery sits high on a sheer-sided rock made mainly of magmas that solidified in the conduit of an ancient volcano. The volcano has since been worn away. (Photo by Dreamstime)

QUESTION 1 Based on this information, what volcanic structure do you think is shown in this photo?

QUESTION 2 Would this volcanic structure most likely have been associated with a composite volcano or cinder cone? Explain how you arrived at your answer.

Mount Adams

GPS devices are used to monitor the inflation or deflation of a volcano as magma moves within. Inflation often precedes an eruption phase by a few months, or longer.

A.

An increase in sulfur dioxide (SO_2) gas emissions has been shown to precede an eruption.

B.

Figure 5.35
Monitoring volcanoes A. A high-precision Global Positioning System (GPS) antenna used to monitor horizontal and vertical ground motions on Mount St. Helens. **B.** Geologist monitoring gas emissions from a fumarole. The analysis of volcanic gases is used in the long-term study of volcanic systems. (Photos courtesy of U.S. Geological Society)

Positioning System (GPS) devices, and Earth-orbiting satellites, are often used to determine potential swelling of a volcano (**Figure 5.35A**). For example, swelling occurs when the roof of a volcano rises as new magma accumulates in its interior—a phenomenon that precedes many volcanic eruptions. The discovery of ground doming at Oregon's Three Sisters volcanoes in 2000 was first detected using images obtained from satellites.

Volcanologists also frequently monitor volcanoes in an effort to detect changes in the quantity and/or composition of the gases being released (**Figure 5.35B**). Some volcanoes show an increase in sulfur dioxide emissions months or years prior to an eruption.

The overriding goal of all monitoring is to discover precursors that may warn of an impending or imminent eruption. Volcanologists accomplish this by first diagnosing the current condition of a volcano and using the baseline data to predict its future behavior. A volcanologist must observe a volcano over an extended period of time in order to recognize significant changes from its "resting state."

Unfortunately, accurately predicting the timing and the potential hazard of a volcanic eruption still eludes scientists. This is perhaps best demonstrated by the February 2015 eruption of Sinabung volcano in Sumatra, Indonesia. This deadly eruption claimed at least 16 lives and occurred just days after authorities gave the "all-clear" for residents to return to their homes on Sinabung's slopes.

5.10 | Concept Checks

1. What three factors do volcanologists monitor in order to determine whether magma is migrating toward Earth's surface?

2. What is the overriding goal of volcano monitoring?

5.1 The Nature of Volcanic Eruptions

Explain why some volcanic eruptions are explosive and others are quiescent.

KEY TERMS magma, lava, viscosity, eruption column

- An important characteristic that differentiates various lavas is their viscosity (resistance to flow). In general, the more silica in the lava, the more viscous it is; lavas with a lower silica content are more fluid. Temperature also influences viscosity. Hot lavas are more fluid, while cool lavas are more viscous.

- High-silica, low-temperature lavas are the most viscous and allow the greatest amount of pressure to build up before they "let go" in an eruption. In contrast, hot, low-silica lavas are the most fluid and least explosive. Basaltic lavas result in relatively gentle eruptions, while volcanoes that erupt felsic lavas (rhyolitic and andesitic) tend to be more explosive.

Q Although Kilauea mostly erupts in a gentle manner, what risks might you encounter if you chose to live nearby?

USGS

5.2 Materials Extruded During an Eruption

List and describe the three categories of materials extruded during volcanic eruptions.

KEY TERMS aa flow, pahoehoe flow, lava tube, block lava, pillow lava, volatile, pyroclastic material, tephra, scoria

- Volcanoes bring liquid lava, gases, and solid materials to Earth's surface.
- Low-viscosity basaltic lava flows can extend great distances from a volcano, where they travel over the surface as pahoehoe or aa flows. Sometimes the surface of the flow congeals, but lava continues to flow below in tunnels called lava tubes. When lava erupts underwater, the outer surface is chilled instantly, while the inside continues to flow, producing pillow lavas.
- The gases most commonly emitted by volcanoes are water vapor and carbon dioxide. Upon reaching the surface, these gases rapidly expand, leading to explosive eruptions that can generate a mass of lava fragments called pyroclastic materials.
- Pyroclastic materials come in several sizes. From smallest to largest, they are ash, lapilli, and blocks or bombs, depending on whether the material left the volcano as solid fragments or as liquid blobs.

- If bubbles of gas in lava don't pop before the lava solidifies, they are preserved as voids called vesicles. Especially frothy, silica-rich lava can cool to make lightweight pumice, while basaltic lava with lots of bubbles cools to make scoria.

Q This photo shows layers of volcanic material ejected by a violent eruption and deposited in roughly horizontal layer. What term is used to describe this type of volcanic material?

Erik Klemetti

5.3 Anatomy of a Volcano

Label a diagram that illustrates the basic features of a typical volcanic cone.

KEY TERMS fissure, conduit, vent, volcanic cone, crater, caldera, parasitic cone, fumarole

- Volcanoes vary in form but share a few common features. Most are roughly conical piles of extruded material that collect around a central vent. The vent is usually within a summit crater or caldera. On the flanks of the volcano, there may be smaller vents marked by small parasitic cones, or there may be fumaroles, spots where gas is expelled.

Q Label the diagram using the following terms: conduit, vent, lava, parasitic cone, bomb, pyroclastic material.

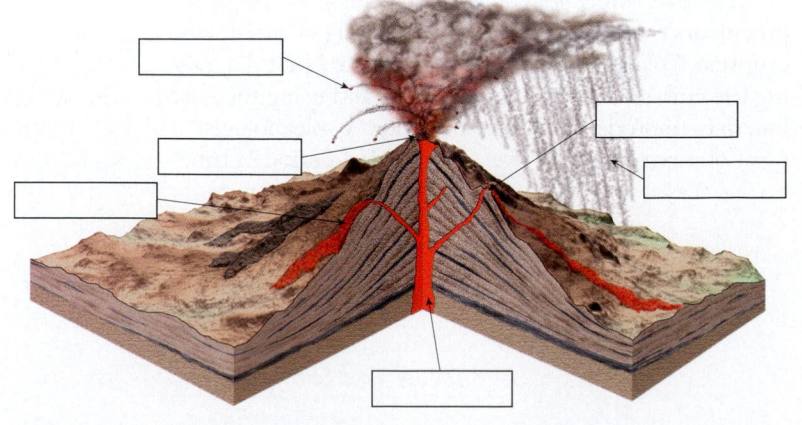

5.4 Shield Volcanoes

Summarize the characteristics of shield volcanoes and provide one example.

KEY TERM shield volcano

- Shield volcanoes consist of many successive lava flows of low-viscosity basaltic lava but lack significant amounts of pyroclastic debris. Lava tubes help transport lava far from the main vent, resulting in very gentle, shield-like profiles.

- Most shield volcanoes are associated with hot-spot volcanism. Mauna Loa, Mauna Kea, and Kilauea in Hawaii are classic examples of the low, wide form characteristic of shield volcanoes.

5.5 Cinder Cones

Describe the formation, size, and composition of cinder cones.

KEY TERM cinder cone (scoria cone)

- Cinder cones are steep-sided structures composed mainly of pyroclastic debris, typically having a basaltic composition. Lava flows sometimes emerge from the base of a cinder cone but typically do not flow out of the crater.

- Cinder cones are small relative to the other major kinds of volcanoes, reflecting the fact that they form quickly, as single eruptive events. Because they are unconsolidated, cinder cones easily succumb to weathering and erosion.

5.6 Composite Volcanoes

List the characteristics of composite volcanoes and describe how they form.

KEY TERM composite volcano (stratovolcano)

- Composite volcanoes are called "composite" because they consist of both pyroclastic material and lava flows. They typically erupt silica-rich lavas that cool to produce andesite or rhyolite. They are much larger than cinder cones and form from multiple eruptions over millions of years.

- Because andesitic and rhyolitic lavas are more viscous than basaltic lava, they accumulate at a steeper angle than does the lava from shield volcanoes. Over time, a composite volcano's combination of lava and cinders produces a towering volcano with a classic symmetrical shape.

- Mount Rainier and the other volcanoes of the Cascade Range in the northwestern United States are good examples of composite volcanoes, as are the other volcanoes of the Pacific Ocean's Ring of Fire.

Q If your family had to live next to a volcano, would you rather it be a shield volcano, a cinder cone, or a composite volcano? Explain

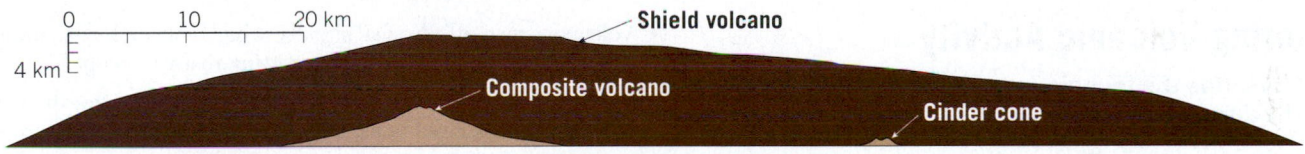

5.7 Volcanic Hazards

Discuss the major geologic hazards associated with volcanoes.

KEY TERMS pyroclastic flow (nuée ardente), lahar, tsunami

- The greatest volcanic hazard to human life is the pyroclastic flow, or nuée ardente. This dense mix of hot gas and pyroclastic fragments races downhill at great speed and incinerates everything in its path. A pyroclastic flow can travel many miles from its source volcano. Because pyroclastic flows are hot, their deposits frequently "weld" together into a solid rock called welded tuff.

- Lahars are volcanic mudflows. These rapidly moving slurries of ash and debris suspended in water can occur even when a volcano isn't actively erupting. Lahars tend to follow stream valleys and can result in loss of life and/or significant damage to structures in their path.

- Volcanic ash in the atmosphere can be a risk to air travel if it is sucked into airplane engines. Volcanoes at sea level can generate tsunamis when they erupt or when their flanks collapse into the ocean. Those that spew large amounts of gas such as sulfur dioxide can cause human respiratory problems. If volcanic gases reach the stratosphere, they screen out a portion of incoming solar radiation and can trigger short-term cooling at Earth's surface.

Q What do lahars and pyroclastic flows have in common?

5.8 Other Volcanic Landforms

List volcanic landforms other than shield, cinder, and composite volcanoes and describe their formation.

KEY TERMS fissure, basalt plateau, flood basalt, lava dome, volcanic neck (plug), pipe

- Calderas, among the largest volcanic structures, form when the rigid, cold rock above a magma chamber cannot be supported and collapses, creating a broad, bowl-like depression. On shield volcanoes, calderas form slowly as lava is drained from the magma chamber beneath the volcano. On a composite volcano, caldera collapse often follows an explosive eruption that can result in significant loss of life and destruction of property.

- Fissure eruptions produce massive floods of low-viscosity, silica-poor lava from large cracks in the crust. Layer upon layer of these flood basalts may build up to significant thicknesses, as in the Columbia Plateau. The defining feature of a flood basalt is the broad area it covers.
- Lava domes are thick masses of high-viscosity, silica-rich lava that accumulate in the summit crater or caldera of a composite volcano.

When they collapse, lava domes can produce extensive pyroclastic flows.
- An example of a volcanic neck is preserved at Shiprock, New Mexico. The lava in the "throat" of this ancient volcano crystallized to form a "plug" of solid rock that weathered more slowly than the conical volcano in which it formed. After the pyroclastic debris has been eroded away, the resistant neck is a distinctive landform.

5.9 Plate Tectonics and Volcanic Activity

Explain how the global distribution of volcanic activity is related to plate tectonics.

KEY TERMS Ring of Fire, volcanic island arc (island arc), continental volcanic arc, intraplate volcanism, mantle plume

- Volcanoes occur at both convergent and divergent plate boundaries, as well as in intraplate settings.
- Convergent plate boundaries that involve the subduction of oceanic crust are the sites where the explosive volcanoes of the Pacific Ring of Fire are most prevalent. Here, release of water from the subducted plate triggers melting in the overlying mantle. The resulting magma interacts with the lower crust of the overlying plate during its ascent and can form a volcanic arc at the surface.
- At divergent boundaries, decompression melting is the dominant generator of magma. As warm rock rises, it can begin to melt without the addition of heat. The result is a rift valley if the overlying crust is continental or a mid-ocean ridge if it is oceanic.

- In an intraplate setting, the source of magma is a mantle plume: a column of warm, rising solid rock in the mantle.

Q The accompanying diagram shows one of the tectonic settings where volcanism is a dominant process. Name the tectonic setting and briefly explain how magma is generated in this setting.

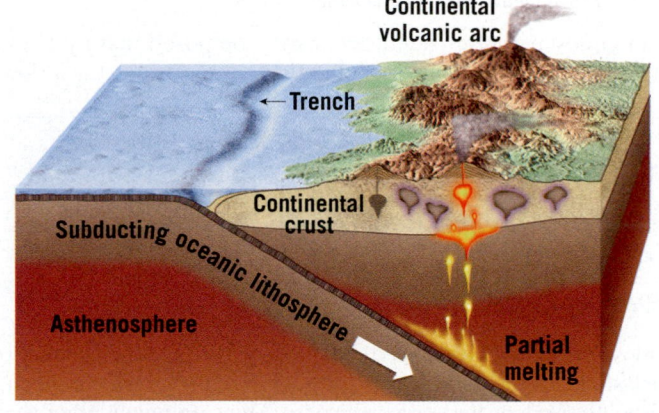

5.10 Monitoring Volcanic Activity

List and describe the techniques used to monitor potentially dangerous volcanoes.

- Volcanoes give off physical signals, which volcanologists monitor to determine whether a volcano is getting ready to erupt.
- Volcano monitoring involves observing changes in the shape of a volcano, earthquakes beneath a volcano that could signal magma movement, and the composition and quantity of gas output.

Give It Some Thought

1. Match each of these volcanic regions with one of the three zones of volcanism (convergent plate boundaries, divergent plate boundaries, or intraplate volcanism):
    a. Crater Lake
    b. Hawaii's Kilauea
    c. Mount St. Helens
    d. East African Rift
    e. Yellowstone
    f. Mount Pelée
    g. Deccan Plateau
    h. Mount Fujiyama

2. Examine the accompanying photo and complete the following:
    a. What type of volcano is shown here? What features helped you classify it as such?
    b. What is the eruptive style of such volcanoes? Describe the likely composition and viscosity of its magma.
    c. Which type of plate boundary is the likely setting for this volcano?
    d. Name a city that is vulnerable to the effects of a volcano of this type.

3. Answer the following questions about divergent boundaries, such as the Mid-Atlantic Ridge, and their associated lavas:

   a. Divergent boundaries are characterized by outpourings of what type of lava: andesitic, basaltic lava, or rhyolitic?

   b. What is the source of these lavas?

   c. What causes the source rocks to melt?

4. For each of the accompanying four sketches, identify the geologic setting (zone of volcanism). Which of these settings will most likely generate explosive eruptions? Which will produce outpouring of fluid basaltic lavas?

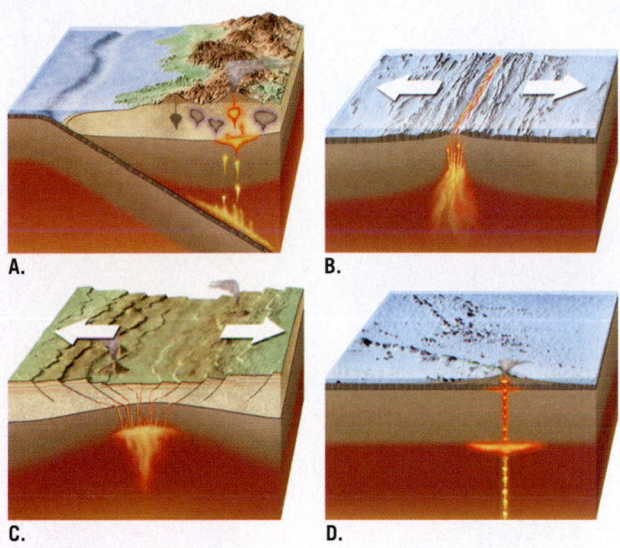

A.   B.

C.   D.

5. Examine the accompanying photo and identify the volcanic landform indicated by the white arrow.

USGS

6. Explain why an eruption of Mount Rainier would be considerably more destructive than the similar eruption of Mount St. Helens that occurred in 1980.

7. The accompanying image shows a geologist at the end of an unconsolidated flow consisting of rounded lava blocks that rapidly descended down the flank of Mount Augustine.

   a. What term best describes this type of flow: an aa flow, a pahoehoe flow, or a pyroclastic flow?

   b. What do volcanologists call the domed structure composed of broken lava blocks (shown with an arrow) visible near the summit of this volcano?

USGS

8. Assume that you are monitoring a volcano that has erupted several times in the recent past but appears to be quiet now. How might you determine whether magma is actually moving through the crust beneath the volcano? Suggest at least two phenomena you would observe or measure.

9. The formula for the volume of a cone is $V = 1/3\pi r^2 h$ (where V = volume, π = 3.14 r = radius, and h = height). If Mauna Loa is 9 kilometers high and has a radius of roughly 85 kilometers, what is its total volume?

10. Imagine that you are a geologist charged with the task of choosing three sites where state-of-the-art volcano-monitoring systems will be deployed. The sites can be anywhere in the world, but the budget available to oversee the operations is limited. What criteria would you use to select sites where monitoring might be most beneficial? List some potential choices and your reasons for considering them.

MasteringGeology™

6

Weathering and Soils

Weathering processes helped shape the rock formations in California's Joshua Tree National Park. (Photo by Alan Majchrowicz/AGE Fotostock)

Each statement represents the primary **LEARNING OBJECTIVE** for the corresponding major heading within the chapter. After you complete the chapter, you should be able to:

6.1 Define *weathering* and distinguish between the two main categories of weathering.

6.2 List and describe four examples of mechanical weathering.

6.3 Discuss the role of water in each of three chemical weathering processes.

6.4 Summarize the factors that influence the type and rate of rock weathering.

6.5 Define *soil* and explain why soil is referred to as an *interface*.

6.6 List and briefly discuss five controls of soil formation.

6.7 Sketch, label, and describe an idealized soil profile. Explain the need for classifying soils.

6.8 Explain the detrimental impact of human activities on soil and list several ways to combat soil erosion.

Earth's surface is constantly changing. Rock is disintegrated and decomposed, moved to lower elevations by gravity, and carried away by water, wind, or ice. In this manner, Earth's physical landscape is sculpted. This chapter focuses on the first step of this never-ending process—weathering. It looks at what causes solid rock to crumble and why the type and rate of weathering vary from place to place. Soil, an important product of the weathering process and a vital resource, is also examined.

6.1 | Weathering

Define *weathering* and distinguish between the two main categories of weathering.

Weathering involves the physical breakdown (disintegration) and chemical alteration (decomposition) of rock at or near Earth's surface. Weathering goes on all around us, but it seems like such a slow and subtle process that it is easy to underestimate its importance. Yet weathering is a basic part of the rock cycle and thus a key process in the Earth system. Weathering is also important to humans—even to those of us who do not study geology. For example, many of the life-sustaining minerals and elements found in soil, and ultimately in the food we eat, were freed from solid rock by weathering processes. As the chapter-opening photo, **Figure 6.1**, and many other images throughout this book illustrate, weathering also contributes to the formation of some of Earth's most spectacular scenery. Of course, these same processes are also responsible for causing the deterioration of many of the structures we build.

SmartFigure 6.1
Arches National Park Both mechanical and chemical weathering contributed greatly to the creation of the arches and other rock formations in Utah's Arches National Park. (Photo by Whit Richardson/Aurora Open/SuperStock) (https://goo.gl/dT0BdL)

Animation

There are two basic categories of weathering. **Mechanical weathering** is accomplished by physical forces that break rock into smaller and smaller pieces without changing the rock's mineral composition. **Chemical weathering** involves a chemical transformation of rock into one or more new compounds. These two concepts can be illustrated by a large log. The log disintegrates when it is split into smaller and smaller pieces, whereas decomposition occurs when the log is set afire and burned. GEOgraphics 6.1 provides other examples of weathering.

Why does rock weather? Simply, weathering is the response of Earth materials to a changing environment. For instance, after millions of years of uplift and erosion (the removal and transport of weathered rock material by water, wind, or ice), the rocks overlying a large, intrusive igneous body may be removed, exposing it at the surface. This mass of crystalline rock—formed deep below ground, where temperatures and pressures are high—is now subjected to a very different and comparatively hostile surface environment. In response, this rock mass will gradually change. This transformation of rock is what we call *weathering*.

In the following sections we will examine the various types of mechanical and chemical weathering. Although we will consider these two categories separately, keep in mind that mechanical and chemical weathering processes usually work simultaneously in nature and reinforce each other.

6.1 | Concept Checks

1. What are the two basic categories of weathering?

2. How do the products of each category of weathering differ?

6.2 | Mechanical Weathering

List and describe four examples of mechanical weathering.

When a rock undergoes *mechanical weathering*, it is broken into smaller and smaller pieces, each retaining the characteristics of the original material. The end result is many small pieces from a single large one. **Figure 6.2** shows that breaking a rock into smaller pieces increases the surface area available for chemical attack. An analogous situation occurs when sugar is added to a liquid. A sugar cube dissolves much more slowly than an equal volume of sugar granules because the cube has much less surface area available for dissolution. Hence, by breaking rocks into smaller pieces, mechanical weathering increases the amount of surface area available for chemical weathering.

In nature, four important physical processes lead to the fragmentation of rock: frost wedging, salt crystal growth, sheeting, and biological activity. In addition, although the work of erosional agents such as wind, waves, glacial ice, and running water is usually considered separately from mechanical weathering, this work is nevertheless related. As these mobile agents (discussed in detail in later chapters) transport rock debris, particles continue to be broken and abraded.

Frost Wedging

If you leave a glass bottle of water in the freezer too long, you will find the bottle fractured, as in **Figure 6.3**. The bottle breaks because liquid water has the unique property of expanding about 9 percent upon freezing. This is also the reason that poorly insulated or exposed water pipes rupture during frigid weather. You might also expect this same process to fracture rocks in nature. This is, in fact, the basis for the traditional explanation of **frost wedging**. After water works its way into the cracks in rock, the freezing water enlarges the cracks, and angular fragments eventually break off (**Figure 6.4** and GEOgraphics 6.2).

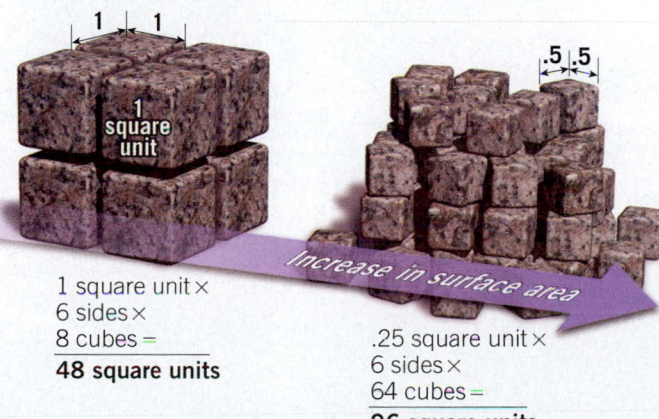

As mechanical weathering breaks rock into smaller pieces, more surface area is exposed to chemical weathering.

2 | 2

4 square units

4 square units ×
6 sides ×
1 cube =
24 square units

1 | 1

1 square unit

1 square unit ×
6 sides ×
8 cubes =
48 square units

.5 | .5

Increase in surface area

.25 square unit ×
6 sides ×
64 cubes =
96 square units

SmartFigure 6.2
Mechanical weathering increases surface area Mechanical weathering adds to the effectiveness of chemical weathering because chemical weathering can occur only on exposed surfaces. (https://goo.gl/XkBfd2)

Tutorial

Figure 6.3

Ice breaks bottle The bottle broke because water expands about 9 percent when it freezes. (Photo by Martyn F. Chillmaid/Science Source)

For many years, the conventional wisdom was that most frost wedging occurred in this way. However, research has shown that frost wedging can also occur in a different way.[*] It has long been known that when moist soils freeze, they expand, or *frost heave*, due to the growth of ice lenses. These masses of ice grow larger because they are supplied with water migrating from unfrozen areas as thin liquid films. As more water accumulates and freezes, the soil is heaved upward. A similar process occurs within the cracks and pore spaces of rocks. Lenses of ice grow larger as they attract liquid water from surrounding pores. The growth of these ice masses gradually weakens the rock, causing it to fracture.

[*]Bernard Hallet, "Why Do Freezing Rocks Break?" *Science* 314(17): 1092–1093, November 2006.

SmartFigure 6.4

Ice breaks rock In mountainous areas, frost wedging creates angular rock fragments that accumulate to form piles of debris called talus slopes. (Photo by Marli Miller) (https://goo.gl/5uqnS1)

Tutorial

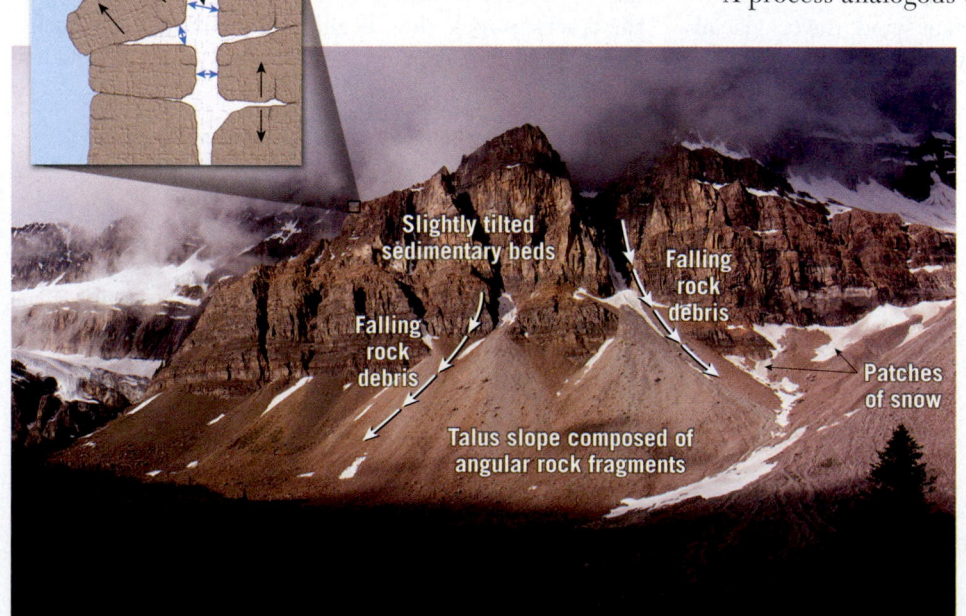

Frost wedging

Slightly tilted sedimentary beds

Falling rock debris

Falling rock debris

Patches of snow

Talus slope composed of angular rock fragments

Salt Crystal Growth

Another expansive force that can split rocks is created by the growth of salt crystals. Rocky shorelines and arid regions are common settings for this process. It begins when sea spray from breaking waves or salty groundwater penetrates crevices and pore spaces in rock. As this water evaporates, salt crystals form. As these crystals gradually grow larger, they weaken the rock by pushing apart the surrounding grains or enlarging tiny cracks.

This same process can also contribute to the crumbling of roadways where salt is spread to melt snow and ice in winter. The salt dissolves in water and seeps into cracks that quite likely originated from frost action. When the water evaporates, the growth of salt crystals further breaks the pavement.

Sheeting

When large masses of igneous rock, particularly granite, are exposed at Earth's surface by erosion, concentric slabs begin to break loose. The process that generates these onion-like layers is called **sheeting**. It takes place, at least in part, due to the great reduction in pressure that occurs as the overlying rock is eroded away, a process called *unloading*. **Figure 6.5** illustrates what happens: As the overburden is removed, the outer parts of the granitic mass expand more than the rock below and separate from the rock body. Continued weathering eventually causes the slabs to separate and peel off, creating an **exfoliation dome** (*ex* = off, *folium* = leaf). Excellent examples of exfoliation domes are Stone Mountain, Georgia, and Half Dome and Liberty Cap in Yosemite National Park.

A process analogous to sheeting can also occur when human activities reduce the confining pressure, similar to what occurs during unloading. For example, in deep mines, large rock slabs have been known to explode off the walls of newly cut tunnels. In quarries, fractures occur parallel to the floor when large blocks of rock are removed.

Although many fractures are created by expansion, others are produced by contraction during the crystallization of magma, and still others are produced by tectonic forces during mountain building.

Some Everyday Examples of Weathering

Weathering processes are responsible for the deterioration of objects and materials at Earth's surface. Day to day changes are not obvious, but over time the effects are significant.

Weathering gradually deteriorated the paint protecting this wood siding.

Question:
List one or more additional examples of weathering that you have seen.

?

Even the most "solid" stone structure that people erect gradually succumbs to weathering processes.

Michelle Milano/Dreamstime

William Britten/iStockphoto

Plant roots can break concrete and rock.

Potholes are especially common in places that experience wintry conditions. The freeze-thaw cycle coupled with water and road salt contribute to pavement destruction.

The chemical weathering process called oxidation is responsible for the rust on this car.

EuroStyle Graphics/Alamy

Christian Delbert/Dreamstime

Ondrejgara/Dreamstime

The Old Man of the Mountain

The Old Man of the Mountain in New Hampshire's Franconia State Park was one of the most famous rock faces. From forehead to chin his profile measured 12 meters (40 feet). On May 3, 2003, this iconic feature, weakened by weathering, collapsed.

The Old Man of the Mountain as it appeared **before** May 3, 2003. The state emblem featured the famous rock face.

Jim Cole/AP Photo

The famous granite outcrop **after** it collapsed on May 3, 2003. The collapse ended decades of efforts to protect and reinforce the state symbol from the same natural processes that created it in the first place. Ultimately frost wedging and other weathering processes prevailed.

AP Photo

Nina Shannon/iStockphoto

The New Hampshire quarter featured the state emblem with the Old Man.

Question:
Was there a specific geologic event on May 3, 2003, that triggered the collapse of the Old Man?

SmartFigure 6.5
Unloading leads to sheeting Sheeting leads to the formation of an exfoliation dome. (Photo by Gary Moon/AGE Fotostock America, Inc.) (https://goo.gl/k4N2o3)

Tutorial

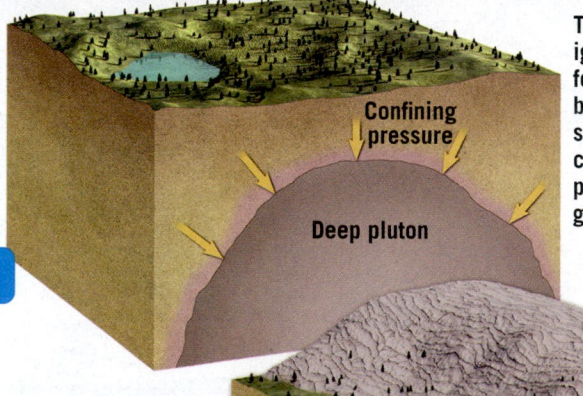

Confining pressure

Deep pluton

This large igneous mass formed deep beneath the surface, where confining pressure is great.

Joints

Expansion and sheeting

Uplift

As erosion removes the overlying bedrock (unloading), the outer parts of the igneous mass expand. Joints form parallel to the surface. Continued weathering causes thin slabs to separate and fall off.

The summit of Half Dome in California's Yosemite National Park is an exfoliation dome and illustrates the onion-like layers created by sheeting.

Figure 6.6
Joints aid weathering Aerial view of nearly parallel joints near Moab, Utah. (Photo by Michael Collier)

Fractures produced by these activities generally form a definite pattern and are called **joints** (**Figure 6.6**). Joints are important rock structures that allow water to penetrate to depth and start the process of weathering long before the rock is exposed.

Biological Activity

Weathering can be accomplished by the activities of organisms, including plants, burrowing animals, and humans. Plant roots in search of nutrients and water grow into fractures, and as the roots grow, they wedge apart the rock (**Figure 6.7**). Burrowing animals further break down rock by moving fresh material to the surface, where physical and chemical processes can more effectively attack it. Decaying organisms also produce acids that contribute to chemical weathering. Where rock has been blasted in search of minerals or for road construction, the impact of humans is particularly noticeable.

Figure 6.7
Plants can break rock Root wedging near Boulder, Colorado. (Photo by Kristin Piljay)

6.2 | Concept Checks

1. When a rock is mechanically weathered, how does its surface area change? How does this influence chemical weathering?

2. Explain how water can cause mechanical weathering.

3. Describe how an exfoliation dome forms.

4. How do joints promote weathering?

5. How does biological activity contribute to weathering?

6.3 | Chemical Weathering

Discuss the role of water in each of three chemical weathering processes.

In the preceding discussion of mechanical weathering, you learned that breaking rock into smaller pieces aids chemical weathering by increasing the surface area available for chemical attack. It should also be pointed out that chemical weathering contributes to mechanical weathering by weakening the outer portions of some rocks, which, in turn, makes them more susceptible to being broken by mechanical weathering processes.

Chemical weathering involves the complex processes that break down rock components and internal structures of minerals. Such processes convert the constituents to new minerals or release them to the surrounding environment. During this transformation, the original rock decomposes into substances that are stable in the surface environment. Consequently, the products of chemical weathering will remain essentially unchanged as long as they remain in an environment similar to the one in which they formed.

Water is by far the most important agent of chemical weathering. Pure water alone is a good solvent, and small amounts of dissolved materials result in increased chemical activity for weathering solutions. The major processes of chemical weathering are dissolution, oxidation, and hydrolysis. Water plays a leading role in each.

Dissolution

Perhaps the easiest type of decomposition to envision is the process of **dissolution**. Just as sugar dissolves in water, so too do certain minerals. One of the most water-soluble minerals is halite (common salt), which,

as you may recall, is composed of sodium and chloride ions. Halite readily dissolves in water because, although this compound maintains overall electrical neutrality, the individual ions retain their respective charges.

Moreover, the surrounding water molecules are polar—that is, the oxygen end of the molecule has a small residual negative charge; the hydrogen end has a small positive charge. As the water molecules come in contact with halite, their negative ends approach sodium ions, and their positive ends cluster about chloride ions. This disrupts the attractive forces in the halite crystal and releases the ions to the water solution (**Figure 6.8**).

Although most minerals are, for all practical purposes, insoluble in pure water, the presence of even a small amount of acid dramatically increases the corrosive force of water—thus causing dissolution. (An acidic solution contains the reactive hydrogen ion, H^+.) In nature, acids are produced by a number of processes. For example, carbonic acid is created when carbon dioxide in the atmosphere dissolves in raindrops. As acidic rainwater soaks into the ground, carbon dioxide in the soil may increase the acidity of the weathering solution. Various organic acids are also released into the soil as organisms

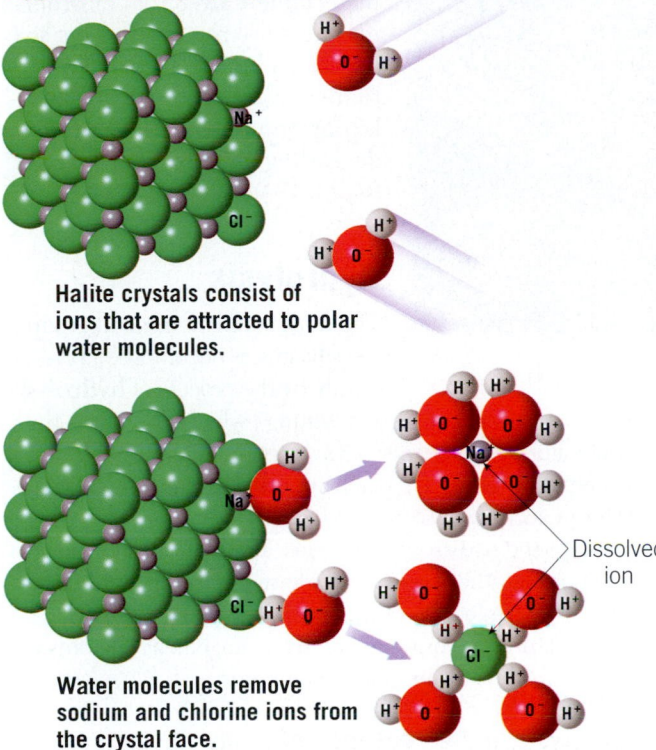

Water molecules are polar because both hydrogen atoms bond to the same side of an oxygen atom. Thus, the hydrogen side of the molecule is slightly positive, and the oxygen side is slightly negative.

Halite crystals consist of ions that are attracted to polar water molecules.

Water molecules remove sodium and chlorine ions from the crystal face.

Dissolved ion

Figure 6.8

Water can dissolve some rocks Dissolution is chemical weathering in which minerals dissolve in water. Halite readily dissolves in pure water. Other minerals, such as calcite, readily dissolve only in water that is acidic.

decay, and sulfuric acid is produced by the weathering of pyrite and other sulfide minerals.

Regardless of the source of the acid, this highly reactive substance readily decomposes most rocks and produces certain products that are water soluble. For example, the mineral calcite, $CaCO_3$, which composes the common building stones marble and limestone, is easily attacked by even a weakly acidic solution.

The overall reaction by which calcite dissolves in water containing carbon dioxide is:

$$CaCO_3 + (H^+ + HCO_3^-) \longrightarrow Ca^{2+} + 2HCO_3^-$$

calcite   carbonic acid   calcium ion   bicarbonate ion

During this process, the insoluble calcium carbonate is transformed into soluble products. In nature, over periods of thousands of years, large quantities of limestone are dissolved and carried away by underground water. This activity is largely responsible for the formation of limestone caverns (**Figure 6.9**). Monuments and buildings made of limestone or marble are also subjected to the corrosive work of acids, particularly in urban and industrial areas that have smoggy, polluted air.

The soluble ions from reactions of this type are retained in our underground water supply. These dissolved ions are responsible for the so-called hard water found in many locales. Hard water is considered undesirable because the active ions react with soap to produce an insoluble material that renders soap nearly useless in removing dirt. To solve this problem, a water softener can be used to remove these ions, generally by replacing them with others that do not chemically react with soap.

Oxidation

Everyone has seen iron and steel objects that have rusted when exposed to water. The same thing can happen to iron-rich minerals. The process of rusting occurs when oxygen combines with iron to form iron oxide, as follows:

$$4Fe + 3O_2 \longrightarrow 2Fe_2O_3$$

iron   oxygen   iron oxide (hematite)

This type of chemical reaction, called **oxidation**,[*] occurs when electrons are lost from one element during the reaction. In this case, we say that iron was oxidized because it lost electrons to oxygen. Although the oxidation of iron progresses very slowly in a dry environment, the addition of water greatly speeds the reaction.

Oxidation is important in decomposing such ferromagnesian minerals as olivine, pyroxene, hornblende, and biotite. Oxygen readily combines with the iron in these minerals to form the reddish-brown iron oxide called *hematite* (Fe_2O_3), or in other cases a yellowish-colored rust called *limonite* [FeO(OH)]. These products are responsible

[*]Note that *oxidation* is a term that refers to any chemical reaction in which a compound or radical loses electrons. The element oxygen is not necessarily present.

Figure 6.9

Acidic waters create caves The dissolving power of carbonic acid plays an important role in creating limestone caverns. This image shows the Chinese Theater area in New Mexico's Carlsbad Caverns. (Photo by Dennis MacDonald/AGE Fotostock)

Figure 6.10

Iron oxides add color Many sedimentary rocks are very colorful. The most important "pigments" are small amounts of iron oxide. Just as iron oxide colors the rusty barrels in **A**, this product of chemical weathering is also responsible for the reds and oranges seen in the rocks composing the Supai Formation in the Grand Canyon in **B**. (Photo A by Vladimir Melnik/Shutterstock; photo B by Cedric Weber/Shutterstock)

B.

A.

for the rusty color on the surfaces of dark igneous rocks, such as basalt, as they begin to weather. Hematite and limonite are also important cementing and coloring agents in many sedimentary rocks (**Figure 6.10**). However, oxidation can occur only after iron is freed from the silicate structure by another process, called *hydrolysis*.

Another important oxidation reaction occurs when sulfide minerals such as pyrite decompose. Sulfide minerals are major constituents of many metallic ores, and pyrite is frequently associated with coal deposits as well. In a moist environment, chemical weathering of pyrite (FeS_2) yields sulfuric acid (H_2SO_4) and iron oxide [$FeO(OH)$]. In many mining locales, this weathering process creates a serious environmental

hazard, particularly in humid areas where abundant rainfall infiltrates spoil banks (waste material left after coal or other minerals are removed). This so-called acid mine drainage eventually makes its way to streams, killing aquatic organisms and degrading aquatic habitats (**Figure 6.11**).

Hydrolysis

The most common mineral group, the silicates, is decomposed primarily by the process of **hydrolysis** (*hydro* = water, *lysis* = a loosening), which basically is the reaction of any substance with water. Ideally, the hydrolysis of a mineral could take place in pure water as some of the water molecules dissociate to form the very reactive hydrogen (H^+) and hydroxyl (OH^-) ions. The hydrogen ion attacks and replaces other positive ions found in the crystal lattice. With the introduction of hydrogen ions into the crystalline structure, the original orderly arrangement of atoms is destroyed, and the mineral decomposes.

Hydrolysis in the Presence of Acids In nature, water usually contains other substances that contribute additional hydrogen ions, thereby greatly accelerating hydrolysis. The most common of these substances is carbon dioxide, CO_2, which dissolves in water to form carbonic acid, H_2CO_3. Rain dissolves some carbon dioxide in the atmosphere, and additional amounts, released by decaying organic matter, are acquired as the water percolates down through the soil.

In water, carbonic acid dissociates to form hydrogen ions (H^+) and bicarbonate ions (HCO_3^-). To illustrate how a rock undergoes hydrolysis in the presence of carbonic acid, let's examine the chemical weathering of granite, a common continental rock. Recall that granite consists mainly of quartz and potassium feldspar. The weathering of the potassium feldspar component of granite is as follows:

$$2KAlSi_3O_8 + 2(H^+ + HCO_3^-) + H_2O \rightarrow$$

$$\underset{\text{potassium}}{} \quad \underset{\text{carbonic acid}}{} \quad \underset{\text{water}}{}$$
feldspar

$$Al_2Si_2O_5(OH)_4 + 2K^+ + 2HCO_3^- + 4SiO_2$$

kaolinite   potassium bicarbonate   silica
(residual clay)   ion   ion
                                    in solution

In this reaction, the hydrogen ions (H^+) attack and replace potassium ions (K^+) in the feldspar structure, thereby disrupting the crystalline network. Once removed, the potassium is available as a nutrient for plants or becomes the soluble salt potassium bicarbonate ($KHCO_3$), which may be incorporated into other minerals or carried to the ocean.

Figure 6.11

Acid mine drainage The water seeping from an abandoned mine in Clarion County, Pennsylvania, is an example of *acid mine drainage*. Acid mine drainage is water with a high concentration of sulfuric acid (H_2SO_4) produced by the oxidation of sulfide minerals such as pyrite. When such acid-rich water migrates from its source, it may pollute surface waters and groundwater and cause significant ecological damage. (Photo by Joel Bosch)

Acid mine drainage

Products of Silicate-Mineral Weathering The most abundant product of the chemical breakdown of potassium feldspar is the clay mineral *kaolinite*. Clay minerals are the end products of weathering and are very stable under surface conditions. Consequently, clay minerals make up a high percentage of the inorganic material in soils. Moreover, the most abundant sedimentary rock, shale, contains a high proportion of clay minerals.

In addition to the formation of clay minerals during the weathering of potassium feldspar, some silica is removed from the feldspar structure and carried away by groundwater. This dissolved silica will eventually precipitate to produce nodules of chert or flint, or it will fill in the pore spaces between grains of sediment, or it will be carried to the ocean, where microscopic animals will remove it from the water to build hard silica shells. To summarize, the weathering of potassium feldspar generates a residual clay mineral, a soluble salt (potassium bicarbonate), and some silica, which enters into solution.

Quartz, the other main component of granite, is *very* resistant to chemical weathering and remains substantially unaltered when attacked by weak acidic solutions. As a result, when granite weathers, the feldspar crystals dull and slowly turn to clay, releasing the once-interlocked quartz grains, which still retain their fresh, glassy appearance. Although some of the quartz remains in the soil, much is eventually transported to the sea or to other sites of deposition, where it becomes the main constituent of such features as sandy beaches and sand dunes. In time these quartz grains may be lithified to form the sedimentary rock sandstone.

Table 6.1 lists the weathered products of some of the most common silicate minerals. Remember that silicate

TABLE 6.1	Products of Chemical Weathering	
Mineral	**Residual Products**	**Material in Solution**
Quartz	Quartz grains	Silica
Feldspars	Clay minerals	Silica, K^+, Na^+, Ca^{2+}
Amphibole (hornblende)	Clay minerals	Silica, Ca^{2+}, Mg^{2+}
	Limonite	
	Hematite	
Olivine	Limonite	Silica
	Hematite	Mg^{2+}

minerals make up most of Earth's crust and that these minerals are essentially composed of only eight elements. When chemically weathered, these silicate minerals yield sodium, calcium, potassium, and magnesium ions that form soluble products, which may be removed from groundwater. The element iron combines with oxygen, producing relatively insoluble iron oxides, most notably hematite and limonite, which give soil a reddish-brown or yellowish color. Under most conditions, the three remaining elements—aluminum, silicon, and oxygen—join with water to produce residual clay minerals. However, even the highly insoluble clay minerals are very slowly removed by subsurface water.

Spheroidal Weathering

Many rock outcrops have a rounded appearance. This occurs because chemical weathering works inward from exposed surfaces. Figure 6.12 illustrates how angular masses of jointed rock change through time. The process is aptly called **spheroidal weathering**.

SmartFigure 6.12
The formation of rounded boulders Spheroidal weathering of extensively jointed rock. (Photo by E. J. Tarbuck) (https://goo.gl/GkxWQd)

Tutorial

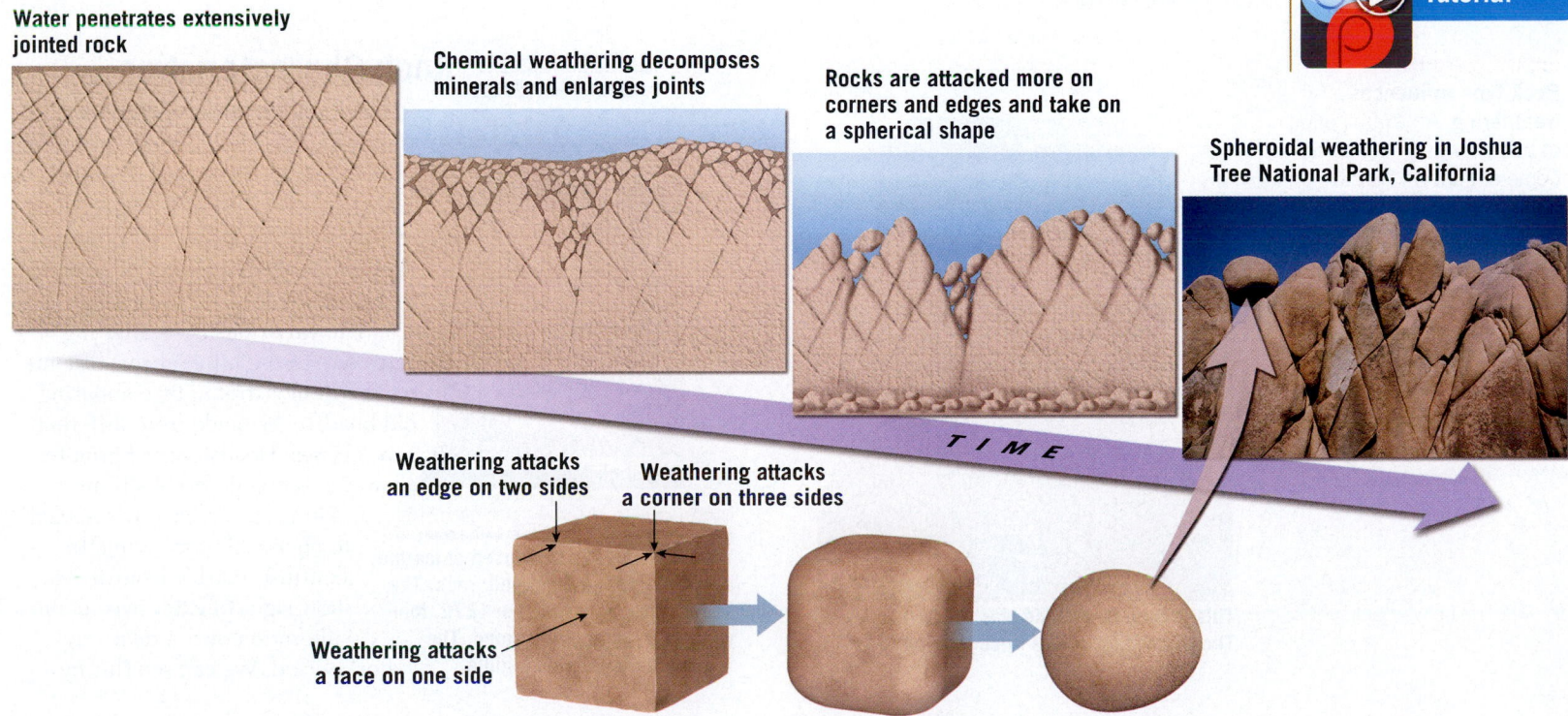

Water penetrates extensively jointed rock

Chemical weathering decomposes minerals and enlarges joints

Rocks are attacked more on corners and edges and take on a spherical shape

Spheroidal weathering in Joshua Tree National Park, California

TIME

Weathering attacks an edge on two sides

Weathering attacks a corner on three sides

Weathering attacks a face on one side

EYE ON EARTH 6.2

This sample of granite is rich in potassium feldspar and quartz. There are minor quantities of biotite and hornblende.

Dennis Tasa

QUESTION 1 If this rock were to undergo chemical weathering, how would its minerals change? Describe the products you would expect to result from each of the minerals in the sample.

QUESTION 2 Would all of the minerals decompose? If not, which mineral would likely be most resistant and remain relatively intact?

Because weathering attacks edges from two sides and corners from three sides, these areas wear down faster than a single flat surface. Gradually, sharp edges and corners become smooth and rounded. Eventually an angular block may evolve into a nearly spherical boulder. Once this occurs, the boulder's shape does not change, but the spherical mass continues to get smaller.

6.3 Concept Checks

1. How is carbonic acid formed in nature?

2. What occurs when carbonic acid reacts with calcite-rich rocks such as limestone?

3. What products result when carbonic acid reacts with potassium feldspar?

4. List several minerals that are especially susceptible to oxidation and list two common products of oxidation.

5. Explain how angular masses of rock often become spherical boulders.

6.4 | Rates of Weathering

Summarize the factors that influence the type and rate of rock weathering.

We have already seen how mechanical weathering affects the rate of weathering. When rock is broken into smaller pieces, the amount of surface area exposed to chemical weathering increases. Other important factors that influence the type and rate of rock weathering include rock characteristics and climate.

SmartFigure 6.13
Rock type influences weathering An examination of headstones in the same cemetery shows that the rate of chemical weathering is influenced by rock type. (Photos by E. J. Tarbuck) (https://goo.gl/NkVbSs)

 **Tutorial**

PETER SWEAT
1801 1868

SWEAT

This granite headstone was erected in 1868. The inscription is still fresh looking.

This monument is composed of marble, a calcite-rich metamorphic rock. The headstone was erected in 1872, four years after the granite stone. The inscription is nearly illegible.

Rock Characteristics

Rock characteristics encompass all of the chemical traits of rocks, including mineral composition and solubility. In addition, any physical features, such as joints, can be important because they influence the ability of water to penetrate rock.

The variations in weathering rates due to the mineral constituents can be demonstrated by comparing old headstones made from different rock types. Headstones of granite, which is composed of silicate minerals, are relatively resistant to chemical weathering. In contrast, marble headstones show signs of extensive chemical alteration over a relatively short period. We can see this by examining the inscriptions on the headstones shown in **Figure 6.13**. Marble is composed of calcite (calcium carbonate), which readily dissolves even in a weakly acidic solution.

The silicates, the most abundant mineral group, chemically weather in essentially the same order in which they crystallize. By examining Bowen's reaction series (see Figure 4.21 page 125), you can see that olivine crystallizes first and is therefore least resistant to chemical weathering, whereas quartz, which crystallizes last, is the most resistant.

Climate

Climatic factors, particularly temperature and precipitation, are crucial to the rate of rock weathering. For example, the frequency of freeze–thaw cycles greatly affects the amount of frost wedging, an important type of mechanical weathering. Temperature and moisture also exert a strong influence on rates of chemical weathering and determine the kind and amount of vegetation present. Regions with lush vegetation often have a thick mantle of soil rich in decayed organic matter from which chemically active fluids such as carbonic acid and humic acids are derived.

The optimum environment for chemical weathering is a combination of warm temperatures and abundant moisture. In polar regions, chemical weathering is ineffective because frigid temperatures keep the available moisture locked up as ice, whereas in arid regions, there is insufficient moisture to promote rapid chemical weathering.

Human activities often produce pollutants that alter the composition of the atmosphere. Such changes can, in turn, influence the rate of chemical weathering. One well-known example is acid rain (**Figure 6.14**).

Differential Weathering

Masses of rock do not weather uniformly. Take a moment to look back at

Figure 6.14
Acid rain accelerates the chemical weathering of stone monuments and structures As a result of burning large quantities of coal and petroleum, more than 20 million tons of sulfur and nitrogen oxides are released into the atmosphere each year in the United States. Through a series of complex chemical reactions, some of these pollutants are converted into acids that then fall to Earth's surface as rain or snow. This decomposing building facade is in Leipzig, Germany. (Photo by Doug Plummer/Science Source)

the photo of Shiprock, New Mexico, in Figure 5.28 (page 168). The durable volcanic neck protrudes high above the surrounding terrain. A glance at the chapter-opening photo of Joshua Tree National Park shows an additional example of this phenomenon, called **differential weathering**. The results vary in scale from the rough, uneven surface of the marble headstone in Figure 6.13 to the boldly sculpted exposures of bedrock in New Mexico's Bisti Badlands (**Figure 6.15**).

Differential weathering and subsequent erosion are responsible for creating many unusual, often spectacular rock formations and landforms. Many factors influence

SmartFigure 6.15
Monuments to weathering This example of differential weathering is in New Mexico's Bisti Badlands. When weathering accentuates differences in rocks, spectacular landforms are sometimes created. (Photo by Michael Collier) (http://goo.gl/MGqs7b)

Mobile Field Trip

the different rates of rock weathering. Among the most important are variations in rock composition. More resistant rock protrudes as ridges or pinnacles, or as steeper cliffs on an irregular hillside (see Figure 7.5, page 216). The number and spacing of joints can also be a significant factor (see Figure 6.12).

6.4   **Concept Checks**

1. Explain why the headstones in Figure 6.13 have weathered so differently.

2. How does climate influence weathering?

6.5 | Soil

Define *soil* and explain why soil is referred to as an *interface*.

Weathering is a key process in the formation of soil. Along with air and water, soil is one of our most indispensable resources. Also like air and water, soil is often taken for granted. The following quote helps put this vital layer in perspective:

> Science, in recent years, has focused more and more on the Earth as a planet, one that for all we know is unique—where a thin blanket of air, a thinner film of water, and the thinnest veneer of soil combine to support a web of life of wondrous diversity in continuous change.*

Soil has accurately been called "the bridge between life and the inanimate world." All life—the entire biosphere—owes its existence to a dozen or so elements that must ultimately come from Earth's crust. Once weathering and other processes create soil, plants carry out the intermediary role of assimilating the necessary elements and making them available to animals, including humans.

An Interface in the Earth System

When Earth is viewed as a system, as discussed in Chapter 1, soil is considered an *interface*—a common boundary where different parts of a system interact. This is an appropriate designation because soil forms

where the geosphere, the atmosphere, the hydrosphere, and the biosphere meet. Soil is a material that develops in response to complex environmental interactions among different parts of the Earth system. Over time, soil gradually evolves to a state of equilibrium, or balance, with the environment. Soil is dynamic and sensitive to almost every aspect of its surroundings. Thus, when environmental changes occur, such as changes in climate, vegetative cover, and animal (including human) activity, the soil responds. Any such change gradually alters soil characteristics until a new balance is reached. Although thinly distributed over the land surface, soil functions as a fundamental interface, providing an excellent example of the integration among many parts of the Earth system.

What Is Soil?

With few exceptions, Earth's land surface is covered by **regolith**, a layer of rock and mineral fragments produced by weathering. Some would call this material soil, but soil is more than an accumulation of weathered debris. **Soil** is a combination of mineral and organic matter, water, and air—the portion of the regolith that supports the growth of plants. Although the proportions of the major

*Jack Eddy, "A Fragile Seam of Dark Blue Light," in *Proceedings of the Global Change Research Forum*. U.S. Geological Survey Circular 1086, 1993, p. 15.

EYE ON EARTH 6.3

The rounded boulders in this image gradually formed in place from a rock mass that had many fractures. Initially the rocks had sharp corners and edges.
(Photo by Mike Brine/Alamy Images)

QUESTION 1 Explain the process that transformed angular blocks of bedrock into rounded boulders.

QUESTION 2 What term is applied to this process?

components in soil vary, the same four components are always present to some extent (**Figure 6.16**). About one-half of the total volume of good-quality surface soil is a mixture of disintegrated and decomposed rock (mineral matter) and **humus**, the decayed remains of animal and plant life (organic matter). The remaining half consists of pore spaces among the solid particles where air and water circulate.

Although the mineral portion of the soil is usually much greater than the organic portion, humus is an essential component. In addition to being an important source of plant nutrients, humus enhances the soil's ability to retain water. Because plants require air and water to live and grow, the portion of the soil consisting of pore spaces that allow these fluids to circulate is as vital as the solid soil constituents.

Soil water is far from "pure" water; instead, it is a complex solution that contains many soluble nutrients. Soil water not only provides the necessary moisture for the chemical reactions that sustain life, it also supplies plants with nutrients in a form they can use. The pore spaces that are not filled with water contain air. This air is the source of necessary oxygen and carbon dioxide for most microorganisms and plants that live in the soil.

Soil Texture and Structure

Most soils are far from uniform and contain particles of different sizes. **Soil texture** refers to the proportions of different particle sizes. Texture is a basic soil property because it strongly influences the soil's ability to retain and transmit water and air, both of which are essential to plant growth. Sandy soils may drain too rapidly and dry out quickly. At the opposite extreme, the pore spaces of clay-rich soils may be so small that they inhibit drainage, and long-lasting puddles result. Moreover, when the clay and silt content is very high, plant roots may have difficulty penetrating the soil.

Because soils rarely consist of particles of only one size, *textural categories* have been established based on the varying proportions of clay, silt, and sand. The standard system of classes used by the U.S. Department of Agriculture is shown in **Figure 6.17**. For example, point A on this triangular diagram (left center) represents a soil composed of 10 percent silt, 40 percent clay, and 50 percent sand. Such a soil is called a *sandy clay*. The soils called *loam*, which occupy the central portion of the diagram, are those in which no single particle size predominates over the other two. Loam soils are best suited to support plant life because they generally hold moisture and nutrients better than do soils composed predominantly of clay or coarse sand.

Soil particles are seldom completely independent of one another. Rather, they usually form clumps called *peds* that give soils a particular structure. Four basic soil structures are recognized: platy, prismatic, blocky,

Figure 6.16
What is soil? The pie chart depicts the composition (by volume) of a soil in good condition for plant growth. Although percentages vary, each soil is composed of mineral and organic matter, water, and air. (Photo by i love images/gardening/Alamy Images)

25% air
45% mineral matter
25% water
5% organic matter

and spheroidal. Soil structure is important because it influences how easily a soil can be cultivated as well as how susceptible a soil is to erosion. Soil structure also affects a soil's porosity and permeability (the ease with which water can penetrate). This in turn influences the movement of nutrients to plant roots. Prismatic and blocky peds usually allow for moderate water infiltration, whereas platy and spheroidal structures are characterized by slower infiltration rates.

6.5 Concept Checks

1. Explain why soil is considered an interface in the Earth system.

2. How is regolith different from soil?

3. Why is texture an important soil property?

4. Using the soil texture diagram in Figure 6.17, name the soil that consists of 60 percent sand, 30 percent silt, and 10 percent clay.

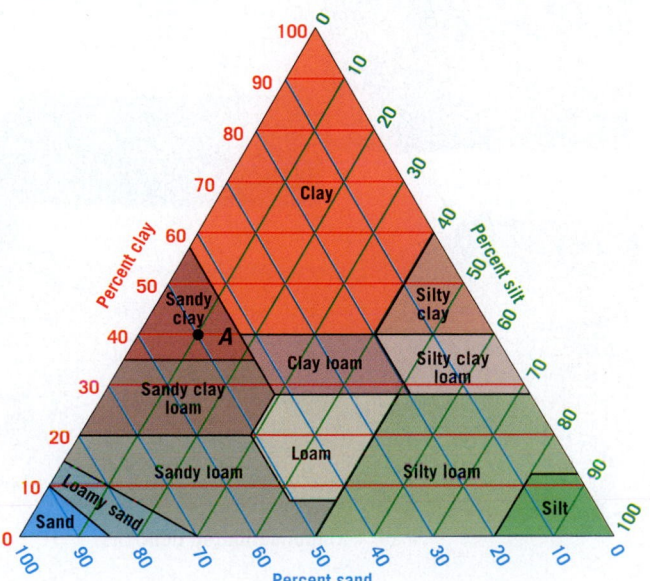

Figure 6.17
Soil texture diagram The texture of any soil can be represented by a point on this diagram. Soil texture is one of the factors used to estimate agricultural potential and engineering characteristics. (U.S. Department of Agriculture)

6.6 | Controls of Soil Formation

List and briefly discuss five controls of soil formation.

Soil is the product of the complex interplay of several factors, including parent material, climate, plants and animals, time, and topography. Although all these factors are interdependent, their roles will be examined separately.

Parent Material

The source of the weathered mineral matter from which soils develop is called the **parent material** and is a major factor influencing newly forming soil. Gradually this weathered matter undergoes physical and chemical changes as soil formation progresses. Parent material can either be the underlying bedrock or a layer of unconsolidated deposits. When the parent material is bedrock, the soils are termed *residual soils*, while soils developed on unconsolidated sediment are called *transported soils* (**Figure 6.18**). It should be pointed out that transported soils form *in place* on parent materials that have been carried from elsewhere and deposited by gravity, water, wind, or ice.

Parent material influences soils in two ways. First, the type of parent material influences the rate of weathering and thus the rate of soil formation. Also, because unconsolidated deposits are already partly weathered, soil development on such material will likely progress more rapidly than when bedrock is the parent material. Second, the chemical makeup of the parent material will affect the soil's fertility. This influences the character of the natural vegetation the soil can support.

At one time, the parent material was thought to be the primary factor causing differences among soils.

However, soil scientists have come to understand that other factors, especially climate, are more important. In fact, similar soils often develop from different parent materials, and dissimilar soils can develop from the same parent material. Such discoveries reinforce the importance of other soil-forming factors.

Climate

Climate is considered to be the most influential control of soil formation. Temperature and precipitation are the elements that exert the strongest impact. As noted earlier in this chapter, variations in temperature and precipitation determine whether chemical or mechanical weathering will predominate and also greatly influence the rate and depth of weathering. For instance, a hot, wet climate may produce a thick layer of chemically weathered soil in the same amount of time that a cold, dry climate produces a thin mantle of mechanically weathered debris. Also, the amount of precipitation influences the degree to which various materials are removed from the soil by percolating water (a process called *leaching*), thereby affecting soil fertility. Finally, climatic conditions are an important control on the types and numbers of plant and animal life present.

Plants and Animals

Plants and animals play a vital role in soil formation. The types and abundance of organisms strongly influence the physical and chemical properties of a soil (**Figure 6.19**). In fact, for well-developed soils in many regions, the significance of natural vegetation on soil type is frequently implied in the names used by soil scientists, such as *prairie soil*, *forest soil*, and *tundra soil*.

Plants and animals furnish organic matter to the soil. Certain bog soils are composed almost entirely of organic matter, whereas desert soils might contain as little as a small fraction of 1 percent. Although the quantity of organic matter varies substantially among soils, it is a rare soil that completely lacks it.

Figure 6.18
Slopes and soil development The parent material for residual soils is the underlying bedrock. Transported soils form on unconsolidated deposits. Also note that as slopes become steeper, soil becomes thinner. (Left and center photos by E. J. Tarbuck; right photo by Lucarelli Temistocle/ Shutterstock)

No soil development because of very steep slope

Transported soil developed on unconsolidated stream deposits

Residual soil is developed on bedrock

Thicker soil develops on flat terrain

Bedrock

Unconsolidated deposits

Thinner soil on steep slope because of erosion

Meager desert rainfall means reduced rates of weathering and relatively meager vegetation. Desert soils are typically thin and lack much organic matter.

In the northern coniferous forest, the organic litter is high in acid resin, which contributes to an accumulation of acid in the soil. As a result, acid leaching is an important soil-forming process.

Soils that develop in well-drained prairie regions typically have a humus-rich surface horizon that is rich in calcium and magnesium. Fertility is usually excellent.

Figure 6.19
Plants influence soil The nature of the vegetation in an area can have a significant influence on soil formation. (Photos by Bill Brooks/Alamy Images, Nickolay Stanev/Shutterstock, and Elizabeth C. Doemer/Shutterstock)

The primary source of organic matter in soil is plants, although animals and an infinite number of microorganisms also contribute. Decomposed organic matter supplies important nutrients to plants, as well as to animals and microorganisms living in the soil. Consequently, soil fertility is in part related to the amount of organic matter present. Furthermore, the decay of plant and animal remains causes the formation of various organic acids. These complex acids hasten the weathering process. Organic matter also has a high water-holding ability and thus aids water retention in a soil.

Microorganisms, including fungi, bacteria, and single-celled protozoa, play an active role in the decay of plant and animal remains. The end product is humus, a material that no longer resembles the plants and animals from which it is formed. In addition, certain microorganisms aid soil fertility by converting atmospheric nitrogen into soil nitrogen.

Earthworms and other burrowing animals mix the mineral and organic portions of a soil. Earthworms, for example, feed on organic matter and thoroughly mix soils in which they live, often moving and enriching many tons per acre each year. Burrows and holes also aid the passage of water and air through the soil.

Time

Time is an important component of *every* geologic process, including soil formation. The nature of soil is strongly influenced by the length of time processes have been operating. If weathering has been going on for a comparatively short time, the character of the parent material strongly influences soil characteristics. As weathering processes continue, the influence of parent material on soil is overshadowed by other soil-forming factors, especially climate. The time required for various soils to evolve cannot be listed because the soil-forming processes act at varying rates under different circumstances. However, as a rule, the longer a soil has been forming, the thicker it becomes and the less it resembles the parent material.

Topography

The lay of the land can vary greatly over short distances. Variations in topography can lead to the development of a variety of localized soil types. Many of the differences exist because the length and steepness of slopes significantly affect the amount of erosion and the water content of soil.

On steep slopes, soils are often poorly developed. Due to rapid runoff, the quantity of water soaking in is slight; as a result, the soil's moisture content may not be sufficient for vigorous plant growth. Further, because of accelerated erosion on steep slopes, the soils are thin or in some cases nonexistent (see Figure 6.18).

In contrast, poorly drained and waterlogged soils found in bottomlands have a much different character. Such soils are usually thick and dark. The dark color results from the large quantity of organic matter that accumulates because saturated conditions retard the decay of vegetation. The optimum terrain for soil development is a flat-to-undulating upland surface. Here we find good drainage, minimum erosion, and sufficient infiltration of water into the soil.

Slope orientation, or the direction a slope is facing, is another consideration. In the midlatitudes of the Northern Hemisphere, a south-facing slope receives a great deal more sunlight than a north-facing slope. In fact, a steep north-facing slope may receive no direct sunlight at all. The difference in the amount of solar radiation received causes differences in soil temperature and moisture, which in turn influence the nature of the vegetation and the character of the soil.

Although this section deals separately with each of the soil-forming factors, remember that all of them work together to form soil. No single factor is responsible for a soil's character; rather, it is the combined influence of parent material, climate, plants and animals, time, and topography that determines this character.

6.6   Concept Checks

1. **List the five basic controls of soil formation. Which factor is most influential in soil formation?**

2. **How might the direction a slope is facing influence soil formation?**

6.7   Describing and Classifying Soils

Sketch, label, and describe an idealized soil profile. Explain the need for classifying soils.

The factors controlling soil formation vary greatly from place to place and from time to time, leading to an amazing variety of soil types.

The Soil Profile

Because soil-forming processes operate from the surface downward, soil composition, texture, structure, and color gradually evolve differently at varying depths.

SmartFigure 6.20
Soil horizons Idealized soil profile from a humid climate in the middle latitudes.
(https://goo.gl/dgNbWj)

Tutorial

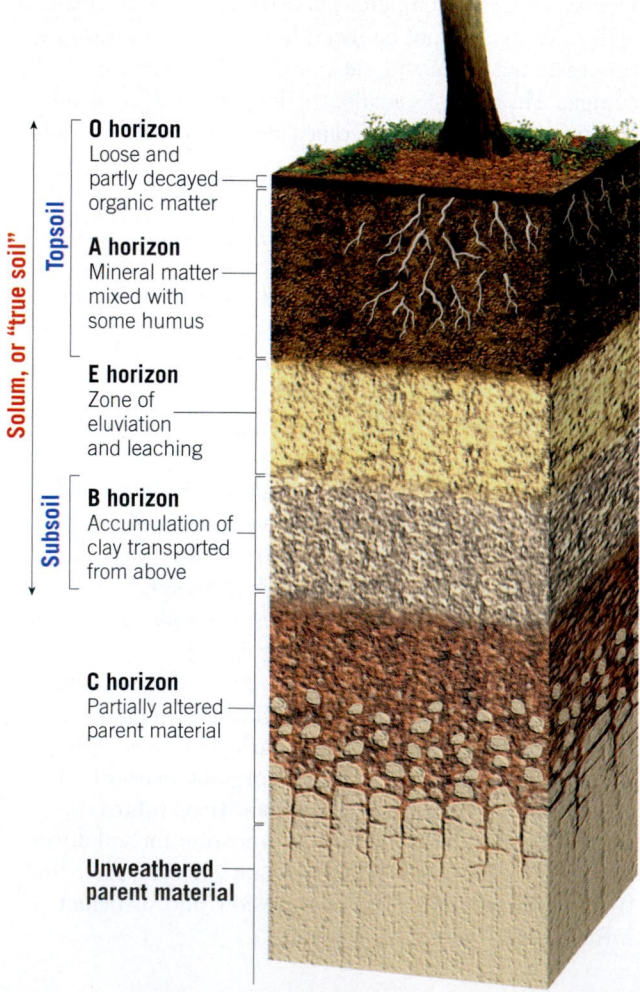

Solum, or "true soil"

Topsoil

O horizon
Loose and partly decayed organic matter

A horizon
Mineral matter mixed with some humus

E horizon
Zone of eluviation and leaching

Subsoil

B horizon
Accumulation of clay transported from above

C horizon
Partially altered parent material

Unweathered parent material

These vertical differences, which usually become more pronounced as time passes, divide the soil into zones or layers known as **horizons**. If you were to dig a pit in soil, you would see that its walls are layered. Such a vertical section through all of the soil horizons constitutes the **soil profile**.

Figure 6.20 presents an idealized view of a well-developed soil profile in which five horizons are identified. From the surface downward, they are designated as *O, A, E, B,* and *C*. These five horizons are common to soils in temperate regions; not all soils have these five layers. The characteristics and extent of horizon development vary in different environments. Thus, different localities exhibit soil profiles that can contrast greatly with one another:

- The *O soil horizon* consists largely of organic material, in contrast to the layers beneath it, which consist mainly of mineral matter. The upper portion of the *O* horizon is primarily plant litter, such as loose leaves and other organic debris that are still recognizable. By contrast, the lower portion of the *O* horizon is made up of partly decomposed organic matter (humus) in which plant structures can no longer be identified. In addition to plants, the *O* horizon is teeming with microscopic life, including bacteria, fungi, algae, and insects. All these organisms contribute oxygen, carbon dioxide, and organic acids to the developing soil.

- The *A horizon* is largely mineral matter, yet biological activity is high, and humus is generally present—up to 30 percent in some instances. Together the *O* and *A* horizons make up what is commonly called the *topsoil*.

- The *E horizon* is a light-colored layer that contains little organic material. As water percolates downward through this zone, finer particles are carried away.

This washing out of fine soil components is termed **eluviation**. Water percolating downward also dissolves soluble inorganic soil components and carries them to deeper zones. This depletion of soluble materials from the upper soil is termed **leaching**.

- The *B horizon*, or *subsoil*, is where much of the material removed from the *E* horizon by eluviation is deposited. Thus, the *B* horizon is often referred to as the *zone of accumulation*. The accumulation of the fine clay particles enhances this horizon's ability to hold water. In extreme cases, clay accumulation can form a very compact, impermeable layer called *hardpan*.

- The *O*, *A*, *E*, and *B* horizons together constitute the **solum**, or "true soil." It is in the solum that soil-forming processes are active and that living roots and other plant and animal life are largely confined.

- The *C horizon* is characterized by partially altered parent material. Whereas the parent material is difficult to see in the *O*, *A*, *E*, and *B* horizons, it is easily identifiable in the *C* horizon. Although this material is undergoing changes that will eventually transform it into soil, it has not yet crossed the threshold that separates regolith from soil.

The characteristics and extent of development can vary greatly among soils in different environments (**Figure 6.21**). The boundaries between soil horizons may be sharp, or the horizons may blend gradually from one to another. Consequently, a well-developed soil profile indicates that environmental conditions have been relatively stable over an extended time span and that the soil is *mature*. By contrast, some soils lack horizons altogether. Such soils are called *immature* because soil building has been going on for only a short time. Immature soils are also characteristic of steep slopes, where erosion continually strips away the soil, preventing full development.

Horizons are indistinct in this soil in Puerto Rico, giving it a relatively uniform appearance.

This profile shows a soil in southeastern South Dakota with well-developed horizons.

Figure 6.21
Contrasting soil profiles Soil characteristics and development vary greatly in different environments. (Left photo by USDA; right photo courtesy of E. J. Tarbuck)

Classifying Soils

The great variety of soils on Earth makes it essential to devise some means of classifying the vast array of soil data. Establishing categories of items having certain important characteristics in common introduces order and simplicity, which not only aids comprehension and understanding but also facilitates analysis and explanation.

Soil scientists in the United States have devised a system for classifying soils known as the **Soil Taxonomy**. It emphasizes the physical and chemical properties of the soil profile and is organized on the basis of observable

EYE ON EARTH 6.4

This thick red soil is exposed somewhere in the United States and is either a gelisol, a mollisol, or an oxisol. (Photo by Sandra A. Dunlap/Shutterstock)

QUESTION 1 Refer to the descriptions in Table 6.2 and determine the likely soil order shown in this image. Explain your choice.

QUESTION 2 Which one of these states is the most likely location of the soil—Alaska, Illinois, or Hawaii?

TABLE 6.2	Basic Soil Orders	
Soil Order	**Description**	**Percentage***
Alfisol	Moderately weathered soils formed under boreal forests or broadleaf deciduous forests, rich in iron and aluminum. Clay particles accumulate in a subsurface layer due to leaching in moist environments. Fertile, productive soils because they are neither too wet nor too dry.	9.65
Andisol	Young soils in which the parent material is volcanic ash and cinders, deposited by recent volcanic activity.	0.7
Aridosol	Soils that develop in dry places with insufficient water to remove soluble minerals; may have calcium carbonate, gypsum, or salt accumulation in subsoil; low organic content.	12.02
Entisol	Young soils with limited development and exhibiting properties of the parent material. Productivity ranges from very high for some forming on recent river deposits to very low for those forming on shifting sand or rocky slopes.	16.16
Gelisol	Young soils with little profile development, found in regions with permafrost. Low temperatures and frozen conditions for much of the year; slow soil-forming processes.	8.61
Histosol	Organic soils found in any climate where organic debris accumulates to form a bog soil. Dark, partially decomposed organic material commonly referred to as *peat*.	1.17
Inceptisol	Weakly developed young soils showing the beginning (inception) of profile development. Most common in humid climates but found from the arctic to the tropics. Native vegetation is most often forest.	9.81
Mollisol	Dark, soft soils developed under grass vegetation, generally found in prairie areas. Humus-rich surface horizon that is rich in calcium and magnesium; excellent fertility. Also found in hardwood forests with significant earthworm activity. Climatic range is boreal or alpine to tropical. Dry seasons are normal.	6.89
Oxisol	Soils formed on old land surfaces unless parent materials were strongly weathered before they were deposited. Generally found in the tropics and subtropical regions. Rich in iron and aluminum oxides, oxisols are heavily leached and hence are poor soils for cultivation.	7.5
Spodosol	Soils found only in humid regions on sandy material. Common in northern coniferous forests and cool humid forests. Beneath the dark upper horizon of weathered organic material lies a light-colored leached horizon, the distinctive property of this soil.	2.56
Ultisol	Soils representing the products of long periods of weathering. Percolating water concentrates clay particles in the lower horizons. Restricted to humid climates in the temperate regions and the tropics, where the growing season is long. Abundant water and a long frost-free period contribute to extensive leaching and poor fertility.	8.45
Vertisol	Soils containing large amounts of clay, which shrink when dry and swell with the addition of water. Found in subhumid to arid climates if sufficient water is available to saturate the soil after periods of drought. Soil expansion and contraction exert stresses on human structures.	2.24

*Percentages refer to the world's ice-free surface.

Figure 6.22

Global soil regions Worldwide distribution of the Soil Taxonomy's 12 soil orders. Points A and B are references for a *Give It Some Thought* item at the end of the chapter. (Natural Resources Conservation Service/USDA)

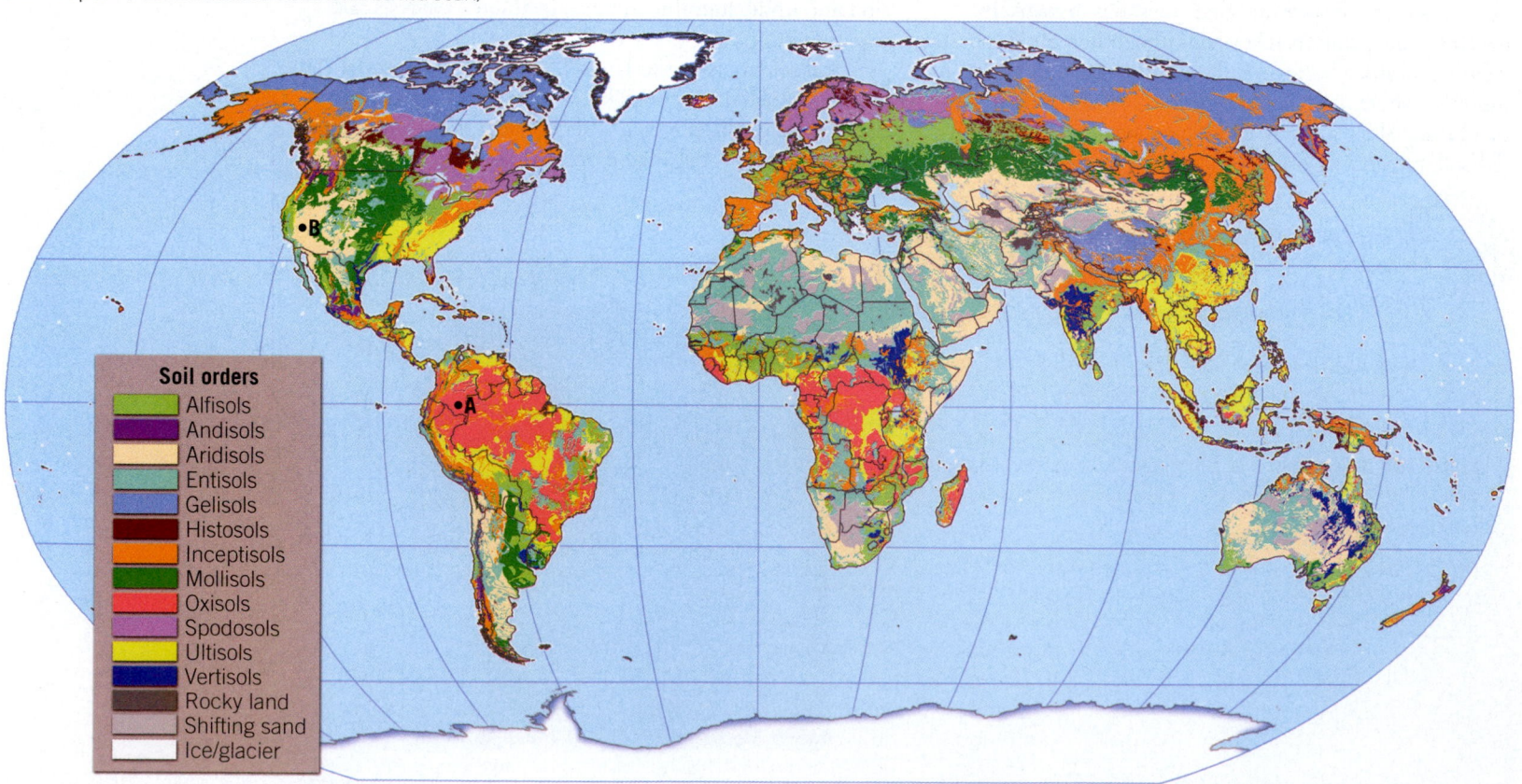

soil characteristics. There are 6 hierarchical categories of classification, ranging from *order*, the broadest category, to *series*, the most specific category. The system recognizes 12 soil orders and more than 19,000 soil series.

The names of the classification units are mostly combinations of Latin or Greek descriptive terms. For example, soils of the order aridosol (from the Latin *aridus* = dry and *solum* = soil) are characteristically dry soils in arid regions. Soils in the order inceptisol (from Latin *inceptum* = beginning and *solum* = soil) are soils with only the beginning, or inception, of profile development.

Brief descriptions of the 12 basic soil orders are provided in **Table 6.2**. **Figure 6.22** shows the complex worldwide distribution pattern of the Soil Taxonomy's 12 soil orders. Like many other classification systems, the Soil Taxonomy is not suitable for every purpose. It is especially useful for agricultural and related land-use purposes, but it is not a useful system for engineers who are preparing evaluations of potential construction sites.

6.7	**Concept Checks**

1. Sketch and label the main soil horizons in a well-developed soil profile.

2. Describe the following features or processes: eluviation, leaching, zone of accumulation, and hardpan.

3. Why are soils classified?

4. Refer to Figure 6.22 and identify three particularly extensive soil orders that occur in the contiguous 48 United States. Describe two soil orders in Alaska.

6.8 | The Impact of Human Activities on Soil

Explain the detrimental impact of human activities on soil and list several ways to combat soil erosion.

Soils are just a tiny fraction of all Earth materials, yet they are vital. Soils are necessary for the growth of rooted plants and thus are a basic foundation of the human life-support system. Because soil forms very slowly, it must be considered a nonrenewable resource. Just as human ingenuity can increase the agricultural productivity of soils through fertilization and irrigation, soils can be damaged or destroyed by careless activities. Despite their role in providing food, fiber, and other basic materials, soils are among our most abused resources.

Clearing the Tropical Rain Forest: A Case Study of Human Impact on Soil

Over the past few decades, the destruction of tropical forests has become a serious environmental issue. Each year millions of acres are cleared for agriculture and logging (**Figure 6.23**). This clearing results in soil degradation, loss of biodiversity, and climate change.

Thick red-orange soils (oxisols) are common in the wet tropics and subtropics (see Figure 6.22). They are the end product of extreme chemical weathering. Because lush tropical rain forests are associated with these soils, many people assume that they are fertile and have great potential for agriculture. However, just the opposite is true: Oxisols are among the poorest soils for farming. How can this be?

Rain forest soils develop under conditions of high temperature and heavy rainfall and are therefore severely leached. Not only does leaching remove the soluble materials such as calcium carbonate, but the great quantities of percolating rainwater also remove much of the silica, and as a result, insoluble oxides of iron and aluminum become concentrated in the soil. Iron oxides give the soil its distinctive color. Because bacterial activity is high in the wet tropics, organic matter quickly breaks down, and rain forest soils contain very little humus. Moreover, leaching destroys fertility because most plant nutrients in the soil are removed by the large volume of downward-percolating water. Despite the dense and luxuriant rain forest vegetation, the soil itself contains few available nutrients.

Most nutrients that support the rain forest are locked up in the trees themselves. As vegetation dies and decomposes, the roots of the rain forest trees quickly absorb the nutrients before they are leached from the soil. The nutrients are continuously recycled as trees die and decompose. Therefore, when forests are cleared to provide land for farming or to harvest timber, most of the nutrients are removed as well. What remains is a soil that contains little to nourish planted crops.

Rain forest clearing not only removes plant nutrients but also accelerates soil erosion. The roots of rain forest vegetation anchor the soil, and leaves and branches provide a canopy that protects the ground by deflecting the full force of the frequent heavy rains. When the protective vegetation is gone, soil erosion increases.

Figure 6.23
Tropical deforestation Clearing the Amazon rain forest in Surinam. The thick soils (oxisols) are highly leached. Clearing of the tropical rain forest is a serious environmental problem. (Photo by Wesley Bocxe/Science Source)

The removal of vegetation also exposes the ground to strong direct sunlight. When baked by the Sun, these tropical soils can harden to a bricklike consistency and become practically impenetrable to water and crop roots. In just a few years, a freshly cleared area may no longer be cultivable.

Soil Erosion: Losing a Vital Resource

Many people do not realize that soil erosion—the removal of topsoil—is a serious environmental problem. Perhaps this is the case because a substantial amount of soil seems to remain even where soil erosion is serious.

Figure 6.24
Raindrop impact Soil dislodged by raindrop impact is more easily moved by sheet erosion.
(Photo courtesy U.S. Department of the Navy/Soil Conservation Service/USDA)

Raindrops may strike the surface at velocities approaching 35 km per hour. When a drop strikes an exposed surface, soil particles may splash as high as one meter and land more than a meter away from the point of raindrop impact.

Nevertheless, although the loss of fertile topsoil may not be obvious to the untrained eye, it is a significant and growing problem as human activities expand and disturb more and more of Earth's surface.

Soil erosion is a natural process; it is part of the constant recycling of Earth materials that we call the *rock cycle*. Once soil forms, erosional forces, especially water and wind, move soil components from one place to another. Every time it rains, raindrops strike the land with surprising force (**Figure 6.24**). Each drop acts like a tiny bomb, blasting movable soil particles out of their positions in the soil mass. Then, water flowing across the surface carries away the dislodged soil particles. Because the soil is moved by thin sheets of water, this process is termed *sheet erosion*.

After the water flows as a thin, unconfined sheet for a relatively short distance, threads of current typically develop, and tiny channels called *rills* begin to form. Still deeper cuts in the soil, known as *gullies*, are created as rills enlarge (**Figure 6.25**). When normal farm cultivation cannot eliminate the channels, we know the rills have grown large enough to be called gullies. Although most dislodged soil particles move only a short distance during each rainfall, substantial quantities eventually leave the fields and make their way downslope to a stream. Once in the stream channel, these soil particles, which can now be called *sediment*, are transported downstream and eventually deposited.

Rates of Erosion We know that soil erosion is the ultimate fate of practically all soils. In the past, erosion

Severe sheet and rill erosion on an Iowa farm following heavy rains.

A.

Gully erosion on unprotected soil on a Wisconsin farm. Just one millimeter of soil from a single acre amounts to about 5 tons.

B.

Figure 6.25
Soil erosion on unprotected soils A. Sheetflow and rills. **B.** Rills can grow into deep gullies. (Photo A by Lynn Betts/NRCS; photo B by D. P. Burnside/Science Source)

occurred at slower rates than it does today because more of the land surface was covered and protected by trees, shrubs, grasses, and other plants. However, human activities such as farming, logging, and construction, which remove or disrupt the natural vegetation, have greatly accelerated the rate of soil erosion. Without the stabilizing effect of plants, the soil is more easily swept away by the wind or carried downslope by sheet wash (see GEOgraphics 6.3).

Natural rates of soil erosion vary greatly from one place to another and depend on soil characteristics as well as factors such as climate, slope, and type of vegetation. Over a broad area, erosion caused by surface runoff may be estimated by determining how much sediment is carried by the streams that drain the region. Studies of this kind made on a global scale indicate that prior to the appearance of humans, sediment transport by rivers to the ocean amounted to just over 9 billion metric tons per year. In contrast, the amount of material currently transported to the sea by rivers is about 24 billion metric tons per year, or more than two and a half times the earlier rate.

It is estimated that flowing water is responsible for about two-thirds of the soil erosion in the United States. Much of the remainder is caused by wind. When dry conditions prevail, strong winds can remove large quantities of soil from unprotected fields (**Figure 6.26**). At present, it is estimated that topsoil is

eroding faster than it forms on more than one-third of the world's croplands. The results are lower productivity, poorer crop quality, reduced agricultural income, and an ominous future.

Controlling Soil Erosion   On every continent, unnecessary soil loss is occurring because appropriate conservation measures are not being taken. Although we recognize that soil erosion can never be completely eliminated, soil conservation programs can substantially reduce the loss of this basic resource.

Steepness of slope is an important factor in soil erosion. The steeper the slope, the faster the water runs off and the greater the erosion. It is best to leave steep slopes undisturbed, but when such slopes are farmed, terraces can be constructed. These nearly flat, steplike surfaces slow runoff and thus decrease soil loss while allowing more water to soak into the ground.

The man is pointing to where the ground surface was when the grasses began to grow. Wind erosion lowered the land surface to the level of his feet.

Clumps of anchored soil

Unanchored soil

Sand dune

1.2 meters

Figure 6.26
Wind erosion   When the land is dry and largely unprotected by anchoring vegetation, soil erosion by wind can be significant. (Photo courtesy Natural Resources Conservation Service/USDA)

The 1930s Dust Bowl
An Environmental Disaster

During a span of dry years in the 1930s, large dust storms plagued the Great Plains. Topsoil was stripped from millions of acres. Because of the size and severity of these storms, the region came to be called the Dust Bowl and the time period the Dirty Thirties.

In places, dust drifted like snow, covering farm buildings, fences, and fields. Crop failure and economic hardship resulted in many farms being abandoned.

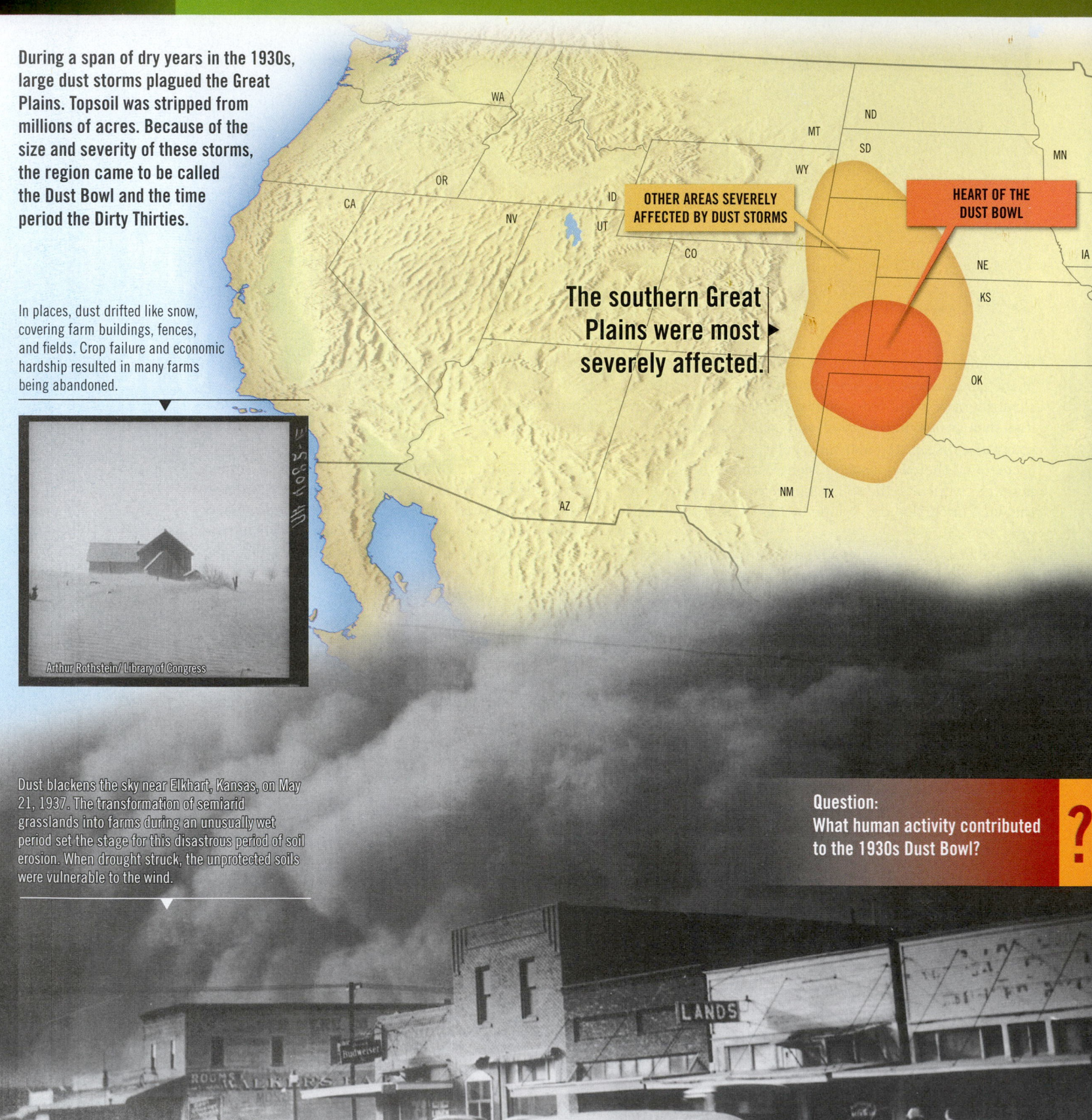

Arthur Rothstein/ Library of Congress

OTHER AREAS SEVERELY AFFECTED BY DUST STORMS

HEART OF THE DUST BOWL

The southern Great Plains were most severely affected. ▶

WA · ND · MT · SD · MN · WY · OR · ID · NE · IA · CA · NV · UT · CO · KS · AZ · NM · TX · OK

Dust blackens the sky near Elkhart, Kansas, on May 21, 1937. The transformation of semiarid grasslands into farms during an unusually wet period set the stage for this disastrous period of soil erosion. When drought struck, the unprotected soils were vulnerable to the wind.

Question:
What human activity contributed to the 1930s Dust Bowl? **?**

Library of Congress

Crops planted in strips
along contours of hillside

Corn

Grass planted
along drainage

Hay

Corn and hay have been planted in strips that follow the contours of the hillside.
This pattern reduces soil loss because it slows the rate of water runoff.

Figure 6.27
Soil conservation Crops on a farm in northeastern Iowa are planted to decrease water erosion. (Photo courtesy of Erwin C. Cole/USDA/NRCS)

Soil erosion by water also occurs on gentle slopes. **Figure 6.27** illustrates one conservation method in which crops are planted parallel to the contours of the slope. This pattern reduces soil loss by slowing runoff. Strips of grass or cover crops such as hay slow runoff even more and act to promote water infiltration and trap sediment.

Creating grassed waterways is another common practice (**Figure 6.28**). Natural drainageways are shaped to form smooth, shallow channels and then planted with grass. The grass prevents the formation of gullies and traps soil washed from cropland. Frequently crop residues are also left on fields. This debris protects the surface from both water and wind erosion. To protect fields from excessive wind erosion, rows of trees and shrubs are planted as windbreaks to slow the wind and deflect it upward (**Figure 6.29**).

6.8   Concept Checks

1. Why are soils in tropical rain forests not well suited for farming?

2. Place these phenomena related to soil erosion in the proper sequence: sheet erosion, gullies, raindrop impact, rills, stream.

3. Explain how human activities have affected the rate of soil erosion.

4. What are two detrimental effects of soil erosion, aside from the loss of topsoil?

5. Briefly describe three ways to control soil erosion.

The grassed waterway prevents the formation of gullies and traps soil washed from cropland.

Figure 6.28
Reducing erosion by water Grassed waterway on a Pennsylvania farm. (Photo courtesy Bob Nichols/NRCS/USDA)

These flat expanses are susceptible to wind erosion, especially when the fields are bare. The rows of trees slow and deflect the wind, which decreases the loss of top soil.

Figure 6.29
Reducing wind erosion Windbreaks protect wheat fields in North Dakota. (Photo courtesy Natural Resources Conservation Service/USDA)

6.1 Weathering

Define *weathering* and distinguish between the two main categories of weathering.

KEY TERMS weathering, mechanical weathering, chemical weathering

- Weathering is the disintegration and decomposition of rocks on the surface of Earth. Rocks might break into many smaller pieces through physical processes called mechanical weathering. Rocks can also decompose as minerals react with environmental agents such as oxygen and water to produce new substances that are stable at Earth's surface. This is called chemical weathering.

Q Would a shattered window be an example of mechanical weathering or chemical weathering? What about a rusty bicycle?

Bare Essence Photography/Alamy

6.2 Mechanical Weathering

List and describe four examples of mechanical weathering.

KEY TERMS frost wedging, sheeting, exfoliation dome, joint

- Mechanical weathering forces include the expansion of ice, the crystallization of salt, and the growth of plant roots. All work to pry apart grains and enlarge fractures.

- Rocks that form under lots of pressure deep in Earth expand when they are exposed at the surface, and sometimes this expansion is great enough to cause the rock to break into onion-like layers. This sheeting can generate broad dome-shaped exposures of rock called exfoliation domes.

6.3 Chemical Weathering

Discuss the role of water in each of three chemical weathering processes.

KEY TERMS dissolution, oxidation, hydrolysis, spheroidal weathering

- Water plays an important role in the chemical reactions that take place at the surface of Earth. It can liberate and transport ions from some minerals through dissolution. Water can also facilitate reactions such as rusting, an example of oxidation. Acid mine drainage is an environmental consequence of the oxidation of pyrite in old coal mines.

- Water may also directly react with exposed minerals, producing new minerals that are stable at Earth's surface. The hydrolysis of feldspar to form kaolinite clay is an example. Clays are stable minerals at Earth's surface conditions, and they are profusely generated by the hydrolysis of silicate minerals. As a result, clay is a common constituent of soil and sedimentary rocks.

- Spheroidal weathering results when sharp edges and corners of rocks are chemically weathered more rapidly than flat rock faces. The higher proportion of surface area for a given volume of rock at the edges and corners means there is more mineral material exposed to chemical attack. Faster weathering at the corners produces weathered rocks that become increasingly sphere shaped over time.

Q Which one of these minerals—potassium feldspar, halite, or pyrite—is most likely to chemically weather via dissolution? Oxidation? Hydrolysis?

6.4 Rates of Weathering

Summarize the factors that influence the type and rate of rock weathering.

KEY TERM differential weathering

- Some types of rocks are more stable at Earth's surface than others, due to the minerals they contain. Different minerals break down at different rates under the same conditions. Quartz is the most stable silicate mineral, while minerals that crystallize early in Bowen's reaction series such as olivine tend to decompose more rapidly.

- Rock weathers most rapidly in an environment with lots of heat to drive reactions and water to facilitate those reactions. Consequently, rocks decompose relatively quickly in hot, wet climates and slowly in cold, dry conditions.

- Frequently, rocks exposed at Earth's surface do not weather at the same rate. This differential weathering of rocks is influenced by factors such as mineral composition and degree of jointing. In addition, if a rock mass is protected from weathering by another, more resistant rock, then it will weather at a slower rate than a fully exposed equivalent rock. Differential weathering produces many of our most spectacular landforms.

Q Imagine that we broke an unweathered sample of granite into two equal pieces. We put one piece in the Dry Valleys of Antarctica and the other in the Amazon rain forest. Which one will weather more rapidly, and why? How would the products of weathering differ between the two places?

Alamy Images

6.5 Soil

Define *soil* and explain why soil is referred to as an *interface*.

KEY TERMS regolith, soil, humus, soil texture

- Soils are vital combinations of organic and inorganic components found at the interface where the geosphere, atmosphere, hydrosphere, and biosphere meet. This dynamic zone is the overlap between different parts of the Earth system. It includes the regolith's rocky debris, mixed with humus, water, and air.

- Soil texture refers to the proportions of different particle sizes (clay, silt, and sand) found in soil. Soil particles often form clumps called peds that give soils a particular structure.

Q Label the four components of a soil on this pie chart.

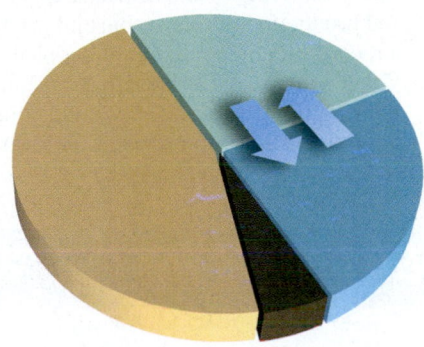

6.6 Controls of Soil Formation

List and briefly discuss five controls of soil formation.

KEY TERM parent material

- Residual soils form in place due to the weathering of bedrock, whereas transported soils develop on unconsolidated sediment.

- Soils formed in different climates are different in part due to temperature and moisture differences but also due to the organisms that live in those different environments. These organisms can add organic matter or chemical compounds to the developing soil or can help mix the soil through their growth and movement.

- It takes time for soil to form. Soils that have developed for a longer period of time will have different characteristics than young soils. In addition, some minerals break down more readily than others. Soils produced from the weathering of different parent rocks are produced at different rates.

- The steepness of the slope on which a soil is forming is a key variable, with shallow slopes retaining their soils and steeper slopes losing them to accumulate elsewhere.

Q How might the addition of acids from coniferous trees affect soil formation?

6.7 Describing and Classifying Soils

Sketch, label, and describe an idealized soil profile. Explain the need for classifying soils.

KEY TERMS horizon, soil profile, eluviation, leaching, solum, Soil Taxonomy

- Despite the great diversity of soils around the world, there are some broad patterns to the vertical anatomy of soil layers. Organic material, called humus, is added at the top (*O* horizon), mainly from plant sources. There, it mixes with mineral matter (*A* horizon). At the bottom, bedrock breaks down and contributes mineral matter (*C* horizon). In between, some materials are leached out or eluviated from higher levels (*E* horizon) and transported to lower levels (*B* horizon), where they may form an impermeable layer called hardpan.

- The need to bring order to huge quantities of data motivated the establishment of a classification scheme for the world's soils. This Soil Taxonomy features 12 broad orders.

Q Which soil order would likely contain a higher proportion of humus: inceptisol or histosol? Which soil order would be more likely found in Brazil: gelisol or oxisol?

6.8 The Impact of Human Activities on Soil

Explain the detrimental impact of human activities on soil and list several ways to combat soil erosion.

- The clearing of tropical rain forests is an issue of concern. Most of the nutrients in the tropical rain forest ecosystem are not in the soil but in the trees themselves. When the trees are removed, most of the nutrients are removed. The loss of vegetation also makes the soils highly susceptible to erosion. Once cleared of vegetation, soils may also be baked by the Sun into a bricklike consistency.
- Soil erosion is a natural process, part of the constant recycling of Earth materials that we call the rock cycle. But human activities have increased soil erosion rates over the past several hundred years. Because natural soil production rates are constant, there is a net loss of soil at a time when a record-breaking number of people live on the planet.
- Using windbreaks and terraces, installing grassed waterways, and plowing the land along horizontal contour lines are all practices that have been shown to reduce soil erosion.

Q Why was this row of evergreens planted on an Indiana farm?

Edwin C. Cole/NCRS

Give It Some Thought

1. How are the two main categories of weathering represented in this image that shows human-made objects? (Photo by Michael Collier)

Michael Collier

2. Describe how plants contribute to mechanical and chemical weathering but inhibit erosion.

3. Granite and basalt are exposed at Earth's surface in a hot, wet region. Will mechanical weathering or chemical weathering predominate? Which rock will weather more rapidly? Why?

4. The accompanying photo shows Shiprock, a well-known landmark in the northwestern corner of New Mexico. It is a mass of igneous rock that represents the "plumbing" of a now-vanished volcanic feature. Extending toward the upper left is a related wall-like igneous structure known as a dike. The igneous features are surrounded by sedimentary rocks. Explain why these once deeply buried igneous features now stand high above the surrounding terrain. What term in Section 6.4, "Rates of Weathering," applies to this situation?

Michael Collier

5. Due to burning of fossil fuels such as coal and petroleum, the level of carbon dioxide (CO_2) in the atmosphere has been increasing for more than 150 years. Should this increase tend to accelerate or slow down the rate of chemical weathering of Earth's surface rocks? Explain how you arrived at your conclusion.

6. In Chapter 4, you learned that feldspars are very common minerals in igneous rocks. When you learn about the common minerals that compose sedimentary rocks in Chapter 7, you will find that feldspars are relatively rare. Applying what you have learned about chemical weathering, explain why this is true. Based on this explanation, what mineral might you expect to be common in sedimentary rocks that is not found in igneous rocks?

7. The accompanying photo shows a footprint on the Moon left by an *Apollo* astronaut in material popularly called *lunar soil*. Does this material satisfy the definition we use for soil on Earth? Explain why or why not. You may want to refer to Figure 6.16.

NASA

8. What might cause different soils to develop from the same kind of parent material or similar soils to form from different parent materials?

9. Using the map of global soil regions in Figure 6.22, identify the main soil order in the region adjacent to South America's Amazon River (point A on the map) and the predominant soil order in the American Southwest (point B). Briefly contrast these soils. Do they have anything in common? Referring to Table 6.2 might be helpful.

10. This soil sample is from a farm in the Midwest. From which horizon was the sample most likely taken—*A*, *E*, *B*, or *C*? Explain.

Lynn Betts/NRCS

7

Sedimentary Rocks

These eroded sedimentary rocks are exposed in South Dakota's Badlands National Park. (Photo by Dennis MacDonald/ Fotostock)

FOCUS ON CONCEPTS

Each statement represents the primary **LEARNING OBJECTIVE** for the corresponding major heading within the chapter. After you complete the chapter, you should be able to:

7.1 Explain the importance of sedimentary rocks and summarize the part of the rock cycle that pertains to sediments and sedimentary rocks. List the three categories of sedimentary rocks.

7.2 Discuss the primary basis for distinguishing among detrital rocks and describe how the origin and history of such rocks might be determined.

7.3 Explain the processes involved in the formation of chemical sedimentary rocks and describe several examples.

7.4 Outline the successive stages in the formation of coal.

7.5 Describe the processes that convert sediment into sedimentary rock and other changes associated with burial.

7.6 Summarize the criteria used to classify sedimentary rocks.

7.7 Distinguish among three broad categories of sedimentary environments and provide an example of each. List several sedimentary structures and explain why these features are useful to geologists.

7.8 Relate weathering processes and sedimentary rocks to the carbon cycle.

hapter 6 provides the background you need to understand the origin of sedimentary rocks. Recall that weathering of existing rocks begins the process. Next, gravity and agents of erosion such as running water, wind, and glacial ice remove the products of weathering and carry them to a new location, where they are deposited. Usually the particles are broken down further during this transport phase. Following deposition, this material, which is now called *sediment*, becomes *lithified* (turned to rock). It is from sedimentary rocks that geologists reconstruct many details of Earth's history. Because sediments are deposited in a variety of settings at the surface, the rock layers that they eventually form hold many clues about past surface environments. A layer may represent a desert sand dune, the muddy floor of a swamp, or a tropical coral reef. There are many possibilities. Many sedimentary rocks are associated with important energy and mineral resources and are therefore important economically as well.

7.1 | An Introduction to Sedimentary Rocks

Explain the importance of sedimentary rocks and summarize the part of the rock cycle that pertains to sediments and sedimentary rocks. List the three categories of sedimentary rocks.

Most of the solid Earth consists of igneous and metamorphic rocks. Geologists estimate that these two categories represent 90 to 95 percent of the outer 16 kilometers (10 miles) of the crust. Nevertheless, most of Earth's solid surface consists of either sediment or sedimentary rock.

Importance

About 75 percent of land areas are covered by sediments and sedimentary rocks. Across the ocean floor, which represents about 70 percent of Earth's solid surface, virtually everything is covered by sediment. Igneous rocks are exposed only at the crests of mid-ocean ridges and in some volcanic areas. Thus, while sediment and sedimentary rocks make up only a small percentage of Earth's crust, they are concentrated at or near the surface—the interface among the geosphere, hydrosphere, atmosphere, and biosphere. Because of this unique position, sediments and the rock layers that they eventually form contain evidence of past conditions and events at the surface. Based on the compositions, textures, structures, and fossils in sedimentary rocks, experienced geologists can decipher clues that provide insights into past climates, ecosystems, and ocean environments. Furthermore, by studying sedimentary rocks, geologists can reconstruct the configuration of ancient landmasses and the locations and compositions of long-vanished mountain systems. In short, this group of rocks provides geologists with much of the basic information needed to reconstruct the details of Earth history (**Figure 7.1**).

The study of sedimentary rocks has economic significance as well. Coal, which provides a significant portion of our electrical energy, is classified as a sedimentary rock. Moreover, other major energy sources—including oil, natural gas,

Figure 7.1
Sedimentary rocks record change Because they contain fossils and other clues about the geologic past, sedimentary rocks are important in the study of Earth history. Vertical changes in rock types represent environmental changes through time. These strata are exposed at Karijini National Park, Western Australia. (Photo by S. Sailer/A. Sailer/AGE Fotostock)

E. J. Tarbuck

Bob Gibbons/Alamy Images

Premium Stock Photography GmbH/Alamy

Gravity moves solid particles downslope.

Glaciers, rivers, and wind transport sediment.

Deposition of solid particles produces many different features—glacial ridges, dunes, floodplains, deltas. Ultimately much sediment reaches the ocean floor.

Landslide

Glacier

Wind

Dunes

River

Ocean

Lake

Chemical and mechanical weathering decompose and disintegrate rock.

Soluble products of chemical weathering become dissolved in groundwater and streams.

Reef

When material dissolved in water precipitates, it is the source of such features as reefs and deposits rich in shells.

As sediments are buried, they become compacted and cemented into solid rock.

SmartFigure 7.2
The big picture This is an outline of the portion of the rock cycle that pertains to the formation of sedimentary rocks. (https://goo.gl/2296bj)

and uranium—are derived from sedimentary rocks. So are major sources of iron, aluminum, manganese, and phosphate fertilizer, plus numerous materials that are essential to the construction industry, such as cement and aggregate. Sediments and sedimentary rocks are also the primary reservoir of groundwater. Thus, having an understanding of this group of rocks and the processes that form and modify them is basic to locating and maintaining supplies of many important resources.

Origins

Like other rocks, the **sedimentary rocks** that we see around us and use in so many different ways have their origin in the rock cycle. **Figure 7.2** illustrates the portion of the rock cycle that occurs near Earth's surface—the part that pertains to sediments and sedimentary rocks. A brief overview of these processes provides a useful perspective:

- Weathering begins the process. It involves the physical disintegration and chemical decomposition of preexisting igneous, metamorphic, and sedimentary rocks. Weathering generates a variety of products subject to erosion, including various solid particles and ions in solution. These are the raw materials for sedimentary rocks.

- Soluble constituents are dissolved and carried away by runoff and groundwater. Solid particles are frequently moved downslope by gravity, a process termed *mass wasting*, before running water, groundwater, wave activity, wind, and glacial ice remove them. These agents of transport, covered in detail in later chapters, move these materials from the sites where they originated to locations where they accumulate. The transport of sediment is usually intermittent. For example, during a flood, a rapidly moving river moves large quantities of sand and gravel. As the floodwaters recede, particles are temporarily deposited, only to be moved again by a subsequent flood.

- Deposition of solid particles occurs when wind and water currents slow down and as glacial ice melts. The word *sedimentary* actually refers to this process. It is derived from the Latin *sedimentum*, which means "to settle," a reference to solid material settling out of a fluid (water or air). The mud on the floor of a lake, a delta at the mouth of a river, a gravel bar in a stream bed, the particles in a desert sand dune, and even household dust are examples.

- The deposition of material dissolved in water is not related to the strength of water currents. Rather, ions in solution are removed when chemical or temperature changes cause material to crystallize and precipitate (solidify out of a liquid solution) or when organisms remove dissolved material to build hard parts such as shells.

- As deposition continues, older sediments are buried beneath younger layers and gradually converted to sedimentary rock (lithified) by compaction and cementation. This and other changes are referred to as *diagenesis* (*dia* = change; *genesis* = origin), a collective term for all the changes (short of metamorphism, discussed in Chapter 8) that take place in texture, composition, and other physical properties after sediments are deposited.

Because there are a variety of ways that the products of weathering are transported, deposited, and transformed into solid rock, geologists recognize three categories of sedimentary rocks. As this overview reminds us, sediment has two principal sources. First, it may be an accumulation of material that originates and is transported as solid particles derived from both mechanical and chemical weathering. Deposits of this type are termed *detrital*, and the sedimentary rocks they form are called **detrital sedimentary rocks**.

The second major source of sediment is soluble material produced largely by chemical weathering. When these ions in solution are precipitated by either inorganic or biological processes, the material is known as *chemical sediment*, and the rocks formed from it are called **chemical sedimentary rocks**.

The third category is **organic sedimentary rocks**, which form from the carbon-rich remains of organisms. The primary example is coal. This black combustible rock consists of organic carbon from the remains of plants that died and accumulated on the floor of a swamp. The bits and pieces of undecayed plant material that constitute the "sediments" in coal are quite unlike the weathering products that make up detrital and chemical sedimentary rocks.

7.1   Concept Checks

1. How does the volume of sedimentary rocks in Earth's crust compare to the volume of igneous and metamorphic rocks?

2. List two ways in which sedimentary rocks are important.

3. Outline the steps that would transform an exposure of granite in the mountains into various sedimentary rocks.

4. List and briefly describe the differences among the three basic sedimentary rock categories.

7.2 | Detrital Sedimentary Rocks

Discuss the primary basis for distinguishing among detrital rocks and describe how the origin and history of such rocks might be determined.

Though a wide variety of minerals and rock fragments (*clasts*) may be found in detrital rocks, clay minerals and quartz are the chief constituents of most sedimentary rocks in this category. Recall from Chapter 6 that clay minerals are the most abundant product of the chemical weathering of silicate minerals, especially the feldspars. Clays are fine-grained minerals with sheetlike crystalline structures similar to the micas. The other common mineral, quartz, is abundant because it is extremely durable and very resistant to chemical weathering. Thus, when igneous rocks such as granite are attacked by weathering processes, individual quartz grains are freed.

Other common minerals in detrital rocks are feldspars and micas. Because chemical weathering rapidly transforms these minerals into new substances, their presence in sedimentary rocks indicates that erosion and deposition occurred fast enough to preserve some of the primary minerals from the source rock before they could be decomposed.

Particle size is the primary basis for distinguishing among various detrital sedimentary rocks. **Figure 7.3** presents the size categories for particles making up detrital rocks. Particle size is not only a convenient method of dividing detrital rocks; the sizes of the component grains also provide useful information about environments of deposition. Currents of water or air sort the particles by size; the stronger the current, the larger the particle size that can be carried. Gravels, for example, are moved by swiftly flowing rivers as well as by landslides and glaciers. Less energy is required to transport sand; thus, it is common to such features as windblown dunes and some river deposits and beaches. Very little energy is needed to transport clay, so it settles very slowly. Accumulation of these tiny particles is generally associated with the quiet water of a lake, lagoon, swamp, or certain marine environments.

In order of increasing particle size, common detrital sedimentary rocks include shale, sandstone, and conglomerate or breccia. We will now look at each type and how it forms.

Shale

Shale is a sedimentary rock consisting of silt- and clay-size particles (**Figure 7.4**). These fine-grained detrital rocks account for well over half of all sedimentary rocks. The particles in these rocks are so small that they cannot be readily identified without great magnification, and this makes shale more difficult to study and analyze than most other sedimentary rocks.

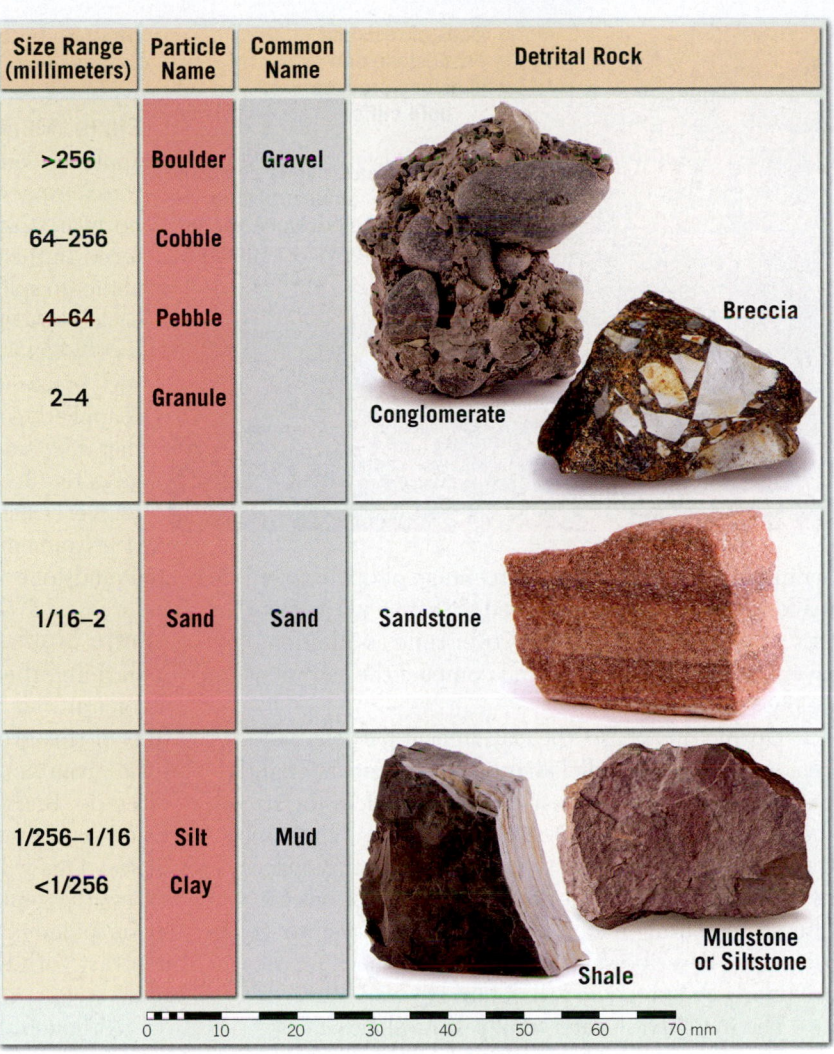

Size Range (millimeters)	Particle Name	Common Name	Detrital Rock
>256	Boulder	Gravel	
64–256	Cobble		Conglomerate / Breccia
4–64	Pebble		
2–4	Granule		
1/16–2	Sand	Sand	Sandstone
1/256–1/16	Silt	Mud	Shale / Mudstone or Siltstone
<1/256	Clay		

0 10 20 30 40 50 60 70 mm

Figure 7.3
Particle size categories Particle size is the primary basis for distinguishing among various detrital sedimentary rocks. (Breccia photo by E. J. Tarbuck; all other photos by Dennis Tasa)

How Does Shale Form? Much of what can be learned about the process that forms shale is related to particle size. The tiny grains in shale indicate that deposition occurs as a result of gradual settling from relatively quiet, nonturbulent currents. Such environments include lakes, river floodplains, lagoons, and portions of the deep-ocean basins. Even in these "quiet" environments, there is usually enough turbulence to keep clay-size particles suspended almost indefinitely. Consequently, much of the clay is deposited only after the individual particles coalesce to form larger aggregates.

Sometimes the chemical composition of the rock provides additional information. One example is black shale, which is black because it contains abundant organic matter (carbon). When such a rock is found, it strongly implies that deposition occurred in an oxygen-poor environment such as a swamp, where organic materials do not readily oxidize and decay.

Thin Layers As silt and clay accumulate, they tend to form thin layers, which are commonly referred to as *laminae* (*lamin* = thin sheet). Initially the particles in the laminae are oriented randomly. This disordered

Figure 7.4
Shale—the most abundant sedimentary rock Dark shale containing fossilized plant remains is relatively common. (Photo by E. J. Tarbuck)

Figure 7.5
Shale crumbles easily
This image was taken in the Grand Canyon. Hikers soon notice that the trail is usually gentler when layers of shale are encountered. (Photo by Dennis Tasa)

Beds of resistant sandstone and limestone produce bold cliffs

Weak, poorly cemented shale crumbles and produces gentler slopes of weathered debris

arrangement leaves a high percentage of open space (called *pore space*) that is filled with water. However, this situation usually changes over time as additional layers of sediment pile up and compact the sediment below.

During this phase, the clay and silt particles take on a more nearly parallel alignment and become tightly packed. This rearrangement of grains reduces the size of the pore spaces and forces out much of the water. Once the grains are pressed closely together, the tiny spaces between particles do not readily permit solutions containing cementing material to circulate. Therefore, geologists often describe shales as being weak because they are poorly cemented and therefore not well lithified.

The inability of water to penetrate shale's microscopic pore spaces explains why this rock often forms barriers to the subsurface movement of water and petroleum. Indeed, rock layers that contain groundwater are commonly underlain by shale beds that block further downward movement. The opposite is true for underground reservoirs of petroleum. They are often capped by shale beds that effectively prevent oil and gas from escaping to the surface.°

Shale, Mudstone, or Siltstone?   It is common to apply the term *shale* to all fine-grained sedimentary rocks, especially in a nontechnical context. However, be aware that there is a more restricted use of the term. In this narrower usage, shale must exhibit the ability to split into thin layers along well-developed, closely spaced planes. This property is termed **fissility** (*fissilis* = that which can be cleft or split). If the rock breaks into chunks or blocks, the name *mudstone* is applied. Another fine-grained sedimentary rock that, like mudstone, is often grouped with shale but lacks fissility is *siltstone* (see Figure 7.3). As its name implies, siltstone is composed largely of silt-size particles and contains less clay-size material than shale and mudstone.

Gentle Slopes   Although shale is far more common than other sedimentary rocks, it does not usually attract as much notice as other, less abundant, members of this group. The reason is that shale does not form prominent outcrops, as sandstone and limestone often do. Rather, shale crumbles easily and usually forms a cover of soil that hides the unweathered rock below. This is illustrated nicely in the Grand Canyon, where the gentler slopes of weathered shale are quite inconspicuous and overgrown with vegetation, in sharp contrast with the bold cliffs produced by more durable rocks (**Figure 7.5**).

Although shale beds may not form striking cliffs and prominent outcrops, some deposits have economic value. Certain shales are quarried to obtain raw material for pottery, brick, tile, and china. Moreover, when mixed with limestone, shale is used to make Portland cement. In the future, one type of shale, called oil shale, may become a valuable energy resource. Oil shale potential is discussed in Chapter 23.

Sandstone

Sandstone is the name given to rocks in which sand-size grains predominate (**Figure 7.6**). After shale, sandstone is the next most abundant sedimentary rock, accounting for approximately 20 percent of the entire group. Sandstones form in a variety of environments and often contain significant clues about their origin, including sorting, particle shape, and composition.

Figure 7.6
Quartz sandstone   After shale, sandstone is the next most abundant sedimentary rock. (Photos by Dennis Tasa)

Close up

Sorting   All the particles in sandstone are not necessarily identical in size. **Sorting** refers to the degree of similarity in particle size in a sedimentary rock. For example, if all the grains in a sample of sandstone are

---

* The relationship between impermeable beds and the occurrence and movement of groundwater is examined in Chapter 17. Shale beds can be cap rocks in oil traps and are discussed in Chapter 23.

about the same size, the sand is considered *well sorted*. Conversely, if the rock contains mixed large and small particles, the sand is said to be *poorly sorted* (Figure 7.7). By studying the degree of sorting, we can learn much about the depositing current. Deposits of wind-blown sand are usually better sorted than deposits sorted by wave activity (Figure 7.8). Particles washed by waves are commonly better sorted than materials deposited by streams. Sediment accumulations that exhibit poor sorting usually result when particles are transported for only a relatively short time and then rapidly deposited. For example, when a turbulent stream reaches the gentler slopes at the base of a steep mountain, its velocity is quickly reduced, and poorly sorted sands and gravels are deposited.

Sorting

Very poorly sorted   **Poorly sorted**   **Well sorted**   **Very well sorted**

Sediments are "very poorly sorted" when there is a wide range of different sizes.

Rocks with particles that are nearly all the same size are "well sorted."

Angularity and Sphericity

Angular   **Subangular**   **Subrounded**   **Rounded**

High sphericity

Low sphericity

Transportation reduces the size and angularity of particles but does not change their general shape.

SmartFigure 7.7
Sorting and particle shape *Sorting* refers to the range of particle sizes present in a rock. Geologists describe a particle's shape in terms of its *angularity* (the degree to which edges and corners are rounded) and *sphericity* (how close the shape is to a sphere).
(https://goo.gl/hcKMT3)

Tutorial

Particle Shape The shapes of sand grains can also help decipher the history of a sandstone (see Figure 7.7). When streams, winds, or waves move sand and other larger sedimentary particles, the grains lose their sharp edges and corners and become more rounded as they collide with other particles during transport. Thus, rounded grains likely have been airborne or waterborne. Further, the degree of rounding indicates the distance or time involved in the transportation of sediment by currents of air or water. Highly rounded grains indicate that a great deal of abrasion and hence a great deal of transport has occurred.

Very angular grains, on the other hand, imply two things: that the rock materials were transported only a short distance before they were deposited or that some other medium may have transported them. For example, when glaciers move sediment, the particles are usually made more irregular by the crushing and grinding action of the ice.

Transport Affects Mineral Composition In addition to affecting the degree of rounding and the amount of sorting that particles undergo, the length of transport by turbulent air and water currents also influences the mineral composition of a sedimentary deposit. Substantial weathering and long transport lead to the gradual destruction of weaker and less stable minerals, including

Figure 7.8
Sand dunes consist of well-sorted sediment
A. The Navajo Sandstone represents a vast area of ancient sand dunes that once covered an area the size of California. **B.** These modern dunes are among the highest in North America. (Photo A by Dennis Tasa; photo B by George H. H. Huey/Alamy Images)

A. The orange and yellow cliffs of Utah's Zion National Park expose thousands of feet of Jurassic-age Navajo Sandstone.

B. The quartz grains composing the Navajo Sandstone were deposited by wind as dunes similar to these in Colorado's Great Sand Dunes National Park. The sand is well sorted because all of the particles are practically the same size.

Figure 7.9
Conglomerate The gravel-size particles in this rock are rounded. (Photo by E. J. Tarbuck)

the feldspars and ferromagnesians. Because quartz is very durable, it is usually the mineral that survives a long trip in a turbulent environment.

To summarize, the origin and history of sandstone can often be deduced by examining the sorting, roundness, and mineral composition of its constituent grains. Knowing this information allows us to infer that a well-sorted, quartz-rich sandstone consisting of highly rounded grains must be the result of a great deal of transport. Such a rock, in fact, may represent several cycles of weathering, transport, and deposition. We may also conclude that a sandstone containing significant amounts of feldspar and angular grains of ferromagnesian minerals underwent little chemical

Figure 7.10
Poorly sorted sediments Gravel deposits along Carbon Creek in Grand Canyon National Park are poorly sorted. (Photo by Michael Collier)

weathering and transport and was probably deposited close to the source area of the rock particles.

Varieties of Sandstone   Due to its durability, quartz is the predominant mineral in most sandstones. Such rock is often simply called *quartz sandstone* (see Figure 7.6). When a sandstone contains appreciable quantities of feldspar (25 percent or more), the rock is called *arkose*. In addition to feldspar, arkose usually contains quartz and sparkling bits of mica. The mineral composition of arkose indicates that the grains were derived from granitic source rocks. The particles are generally poorly sorted and angular, which suggests short-distance transport, minimal chemical weathering in a relatively dry climate, and rapid deposition and burial.

A third variety of sandstone is known as *graywacke*. Along with quartz and feldspar, this dark-colored rock contains abundant rock fragments and matrix—finer-grained material in which the fragments are embedded. More than 15 percent of graywacke's volume is matrix. The poor sorting and angular grains characteristic of graywacke suggest that the particles were transported only a relatively short distance from their source area and were then rapidly deposited. Before the sediment could be reworked and sorted further, it was buried by additional layers of material. Graywacke is frequently associated with submarine deposits made by dense sediment-choked torrents called *turbidity currents*.

Conglomerate and Breccia

Conglomerate consists largely of rounded gravel-size particles (**Figure 7.9**). As Figure 7.3 indicates, these particles can range in size from large boulders to particles as small as peas. The particles are often large enough to be identified as distinctive rock types; thus, they can be valuable in identifying the source areas of sediments. More often than not, conglomerates are poorly sorted because the openings between the large gravel particles contain sand or mud (**Figure 7.10**).

Gravels accumulate in a variety of environments and usually indicate the existence of steep slopes or very turbulent currents. The coarse particles in a conglomerate

EYE ON EARTH 7.1

This detrital rock consists of angular grains and is rich in potassium feldspar and quartz. (Photo by E. J. Tarbuck)

QUESTION 1 What do the angular grains indicate about the distance the sediment was transported?

QUESTION 2 The source of the sediment in this rock was an igneous mass. Name the likely rock type.

QUESTION 3 Did the sediment in this sample undergo a great deal of chemical weathering? Explain.

may reflect the action of energetic mountain streams or result from strong wave activity along a rapidly eroding coast. Some glacial and landslide deposits also contain plentiful gravel.

If the large particles are angular rather than rounded, the rock is called **breccia** (**Figure 7.11**). Because

Figure 7.11
Breccia The gravel-size sediments in this rock are sharp and angular. (Photo by E. J. Tarbuck)

large particles abrade and become rounded very rapidly during transport, the pebbles and cobbles in a breccia indicate that they did not travel far from their source area before they were deposited. Thus, as with many other sedimentary rocks, conglomerates and breccias contain clues to their history. Their particle sizes reveal the strength of the currents that transported them, whereas the degree of rounding indicates how far the particles traveled. The fragments within a sample identify the source rocks that supplied them.

7.2 | Concept Checks

1. What minerals are most abundant in detrital sedimentary rocks? In which rocks do these minerals predominate?

2. What is the primary basis for distinguishing among detrital rocks?

3. Describe how sediments become sorted. What would cause sediments to be poorly sorted?

4. Distinguish between conglomerate and breccia.

7.3 | Chemical Sedimentary Rocks

Explain the processes involved in the formation of chemical sedimentary rocks and describe several examples.

In contrast to detrital rocks, which form from the solid products of weathering, chemical sediments derive from ions that are carried *in solution* to lakes and seas. This material does not remain dissolved in the water indefinitely, however. Some of it precipitates to form chemical sediments. These become rocks such as limestone, chert, and rock salt.

This precipitation of material occurs in two ways. *Inorganic* (*in* = not, *organicus* = life) processes such as evaporation and chemical activity can produce chemical sediments. *Organic* (life) processes of water-dwelling organisms also form chemical sediments, said to be of **biochemical** origin.

One example of a deposit resulting from inorganic chemical processes is the dripstone that decorates many caves (**Figure 7.12**). Another is the salt left behind as a body of seawater evaporates. In contrast, many water-dwelling animals and plants extract dissolved mineral matter to form shells and other hard parts. After the organisms die, their skeletons collect by the millions on the floor of a lake or an ocean as biochemical sediment (**Figure 7.13**).

Limestone

Representing about 10 percent of the total volume of all sedimentary rocks, **limestone** is the most abundant chemical sedimentary rock, and it has economic significance as well (see GEOgraphics 7.1). It is composed

Delicate calcite crystals forming in a drop of water at the tip of soda straw stalactite. The formation of crystals is triggered when some carbon dioxide escapes from the water drop.

Figure 7.12
Cave deposits An example of a chemical sedimentary rock with an inorganic origin. (Photo by Guillen Photography/Alamy Images; inset photo by Dante Fenolio/Science Source)

Limestone
An Important and Versatile Commodity

Limestone, as defined by the minerals industry, refers to any rock composed mostly of the carbonate minerals calcite and dolomite. The U.S. Geological Survey characterizes limestone as an *"essential mineral commodity of national importance."*

Why is limestone important?

Limestone that has undergone metamorphism is called marble. It is used as floor tile, table and countertops, and as a building stone.

Purified limestone is added to bread and cereal as a source of calcium and is an ingredient in antacid tablets and calcium supplements. It is even used to neutralize acids in wine and beer making!

Limestone is used as a filler and white pigment in many products including paper, plastics, paint, and even toothpaste.

Limestone is a key ingredient in making Portland cement, an essential product to the building industry.

White roofing granules.

USGS

Question:
List at least five different commercial uses for limestone.

?

Huge quantities of limestone are crushed and used as aggregate—the solid base of many roads and an ingredient in concrete. It is the raw material for making lime (CaO), which is used to treat soils, purify water, and smelt copper among many uses.

Limestone has been used as a building stone for centuries—from Egypt's ancient pyramids and Europe's medieval castles, to modern buildings such as this.

Martin Bond/Photo Researchers, Inc.   B. Christopher/Alamy Images

UNITED STATES

FEDERAL TRADE COMMISSION BUILDING

chiefly of the mineral calcite ($CaCO_3$) and forms either by inorganic means or as a result of biochemical processes. Regardless of origin, the mineral composition of all limestone is similar, yet many different types exist. This is true because limestones are produced under a variety of conditions. Forms that have a marine biochemical origin are by far the most common.

Carbonate Reefs Corals are one important example of organisms that are capable of creating large quantities of marine limestone. These relatively simple invertebrate animals secrete a calcareous (calcium carbonate) external skeleton. Although they are small, corals are capable of creating massive structures called *reefs* (**Figure 7.14**). Reefs consist of coral colonies made up of great numbers of individuals that live side by side on a calcite structure secreted by the animals. In addition, calcium carbonate–secreting algae live with the corals and help cement the entire structure into a solid mass. A wide variety of other organisms also live in and near the reefs.

Certainly the best-known modern reef is Australia's 2600-kilometer-long (1600-mile-long) Great Barrier Reef, but many lesser reefs also exist. They develop in the shallow, warm waters of the tropics and subtropics equatorward of about 30 degrees latitude. Striking examples exist in The Bahamas, Hawaii, and the Florida Keys.

Modern corals were not the first reef builders. Earth's first reef-building organisms were photosynthesizing bacteria that lived during Precambrian time, more than 2 billion years ago. From fossil remains, it is known that a variety of organisms have constructed reefs, including bivalves (clams and oysters), bryozoans (coral-like animals), and sponges. Corals have been found in fossil reefs as ancient as 500 million years old, but corals

Close up

Figure 7.13
Coquina This variety of limestone consists of shell fragments; therefore, it has a biochemical origin. (Rock sample photo by E. J. Tarbuck; beach photo by Donald R. Frazier Photolibrary, Inc./Alamy Images)

Aerial view showing a small portion of Australia's Great Barrier Reef. Located off the coast of Queensland, it extends for 2600 kilometers and consists of more than 2900 individual reefs.

This image in Guadalupe Mountains National Park, Texas, shows a small portion of an ancient (Permian-age) reef complex that once formed a 600-kilometer loop around the margin of the Delaware Basin.

Figure 7.14
Carbonate reefs Large quantities of biochemical limestone are created by reef-building organisms.
(Photos by JC Photo/Shutterstock and Michael Collier)

The massive White Chalk Cliffs. Chalk is a biochemical limestone made up almost entirely of the tiny hard parts of microscopic marine organisms, mainly plankton.

View of a group of plankton called *coccolithophores* from a scanning electron microscope. Individual plates shaped like hubcaps are only three one-thousandths of a millimeter in diameter; so tiny they could pass through the eye of a needle.

that formed during the Permian period (251 to 299 million years ago) is strikingly exposed in Guadalupe Mountains National Park (see Figure 7.14).

Coquina and Chalk Although a great deal of limestone is produced by biological processes, this origin is not always evident because shells and skeletons may undergo considerable change before becoming lithified into rock. However, one easily identified biochemical limestone is *coquina*, a coarse rock composed of poorly cemented shells and shell fragments (see Figure 7.13). Another less obvious but nevertheless familiar example is *chalk*, a soft, porous rock made up almost entirely of the hard parts of microscopic marine organisms. Among the most famous chalk deposits are those exposed along the southeast coast of England (**Figure 7.15**).

Inorganic Limestones Limestones having an inorganic origin form when chemical changes or high water temperatures increase the water's concentration of calcium carbonate to the point that it precipitates. *Travertine*, the type of limestone commonly seen in caves, is an example (see Figure 7.12). When travertine is deposited in caves, groundwater is the source of the calcium carbonate. As water droplets become exposed to the air in a cavern, some of the carbon dioxide dissolved in the water escapes, causing calcium carbonate to precipitate.

similar to the modern colonial varieties have constructed reefs only during the past 60 million years.

In the United States, reefs of Silurian age (416 to 444 million years ago) are prominent features in Wisconsin, Illinois, and Indiana. In western Texas and adjacent southeastern New Mexico, a massive reef complex

Small spherical grains called *ooids* are formed by chemical precipitation of calcium carbonate around a tiny nucleus and are the raw material for oolitic limestone.

Another variety of inorganic limestone is *oolitic limestone*, a rock composed of small spherical grains called *ooids*. Ooids form in shallow marine waters as tiny "seed" particles (commonly small shell fragments) are moved back and forth with currents. As the grains are rolled about in the warm water, which is supersaturated with calcium carbonate, they become coated with layer upon layer of the chemical precipitate (**Figure 7.16**).

Dolostone

Closely related to limestone is **dolostone**, a rock composed of the calcium-magnesium carbonate mineral dolomite [$CaMg(CO_3)_2$]. Although dolostone and limestone sometimes closely resemble one another, they can be easily distinguished by observing their reaction to dilute hydrochloric acid. When a drop of acid is placed on limestone, the reaction (fizzing) is obvious. However, unless dolostone is powdered, it does not visibly react to the acid.

Flint

Jasper

Chert arrowhead

Petrified wood

Figure 7.17

Colorful chert Chert is the name applied to a number of dense, hard chemical sedimentary rocks made of microcrystalline quartz. (Petrified wood photo by gracious_tiger/Shutterstock; flint and jasper photos by E. J. Tarbuck; arrowhead photo by Daniel Sambraus/Science Source)

Dolostone's origins remain a subject of discussion among geologists. No marine organisms produce hard parts of dolomite, and the chemical precipitation of dolomite from seawater occurs only under conditions of unusual water chemistry in certain near-shore sites. Yet dolostone is abundant in many ancient sedimentary rock successions. It appears that significant quantities of dolostone are produced when magnesium-rich waters circulate through limestone and convert calcite to dolomite when some calcium ions are replaced by magnesium ions (a process called *dolomitization*). However, not all dolostones appear to be formed by such a process, and their origin remains uncertain.

Chert

Chert is a name used for a number of very compact and hard rocks made of microcrystalline quartz (SiO_2). It can vary in color from nearly white to black; chert often appears gray, brown, or brownish gray and can even appear red or greenish. Color variations are related to trace elements present in the rock. **Figure 7.17** shows some varieties. One well-known form is *flint*, whose dark color results from the organic matter it contains. *Jasper*, a red variety, gets its bright color from iron oxide. *Petrified wood* is chert that is made when silica-rich material such as volcanic ash buries trees. Groundwater rich in dissolved silica from the ash penetrates the wood. As the dissolved silica precipitates, it gradually replaces the wood. The shape and structures such as growth rings are often preserved. Like glass, most chert has a conchoidal fracture. Its hardness, ease of chipping, and ability to hold a sharp edge made chert a favorite of Native Americans for fashioning points for spears and arrows. Because of chert's durability and extensive use, arrowheads are found in many parts of North America.

Chert deposits are commonly found in one of two situations: as layered deposits referred to as *bedded cherts* and as somewhat spherical masses called *nodules*, which vary in diameter from a few millimeters (pea size) to a few centimeters. Most water-dwelling organisms that produce hard parts make them of calcium carbonate. But some, such as diatoms and radiolarians, produce glasslike silica skeletons. These tiny organisms are able to extract silica even though seawater contains only tiny quantities. It is from their remains that most bedded cherts are believed to originate. Some bedded cherts occur in association with lava flows and layers of volcanic

EYE ON EARTH 7.2

This is a mass of chemical sedimentary rock in Yellowstone National Park. It was formed by the following process: Rainwater became acidic when it absorbed carbon dioxide in the air. As the water seeped beneath the surface, it dissolved calcite in the limestone bedrock. Eventually, Yellowstone's underground plumbing returned the water, now saturated with calcium carbonate, to the surface as a hot spring. When the water emerged, some carbon dioxide escaped into the air, triggering the deposition of the rock seen here. (Photo by Ross Davidson/Alamy Images)

QUESTION 1 Did this rock have a biochemical origin or an inorganic origin?

QUESTION 2 Is the rock most likely chert or limestone? Explain.

QUESTION 3 Name the particular variety of this rock. What figure in this chapter provides another example?

ash. For these occurrences, it is probable that the silica was derived from the decomposition of the volcanic ash and not from biochemical sources. Chert nodules are sometimes referred to as *secondary cherts* or *replacement cherts*, and they most often occur within beds of limestone. They form when silica, originally deposited in one place, dissolves, migrates, and then chemically precipitates elsewhere, replacing older material.

Evaporites

In the geologic past, many areas that are now dry land were basins, submerged under shallow arms of a sea that had only narrow connections to the open ocean. Under these conditions, seawater continually moved into the bay to replace water lost by evaporation. Eventually the waters of the bay became saturated, and salt deposition began. Such deposits are called **evaporites**.

Minerals commonly precipitated in this fashion include halite (sodium chloride, NaCl), which is the chief component of *rock salt*, and gypsum (hydrous calcium sulfate, $CaSO_4 \cdot 2H_2O$), which is the main ingredient in *rock gypsum* (**Figure 7.18**). Both have significant importance. Halite is familiar to everyone as the common salt used in cooking and seasoning foods. Of course it has many other uses, from melting ice on roads to making

hydrochloric acid, and it has been considered important enough that people have sought, traded, and fought over it for much of human history. Gypsum is the basic ingredient in plaster of Paris. This material is used most extensively in the construction industry for wallboard (drywall) and interior plaster.

When a body of seawater evaporates, the minerals that precipitate do so in a sequence that is determined by their solubility. Less soluble minerals precipitate first, and more soluble minerals precipitate later, as salinity increases. For example, gypsum precipitates when about 80 percent of the seawater has evaporated, and halite crystallizes when 90 percent of the water has been removed. During the last stages of this process, potassium and magnesium salts precipitate. One of these last-formed salts, the mineral *sylvite*, is mined as a significant source of potassium ("potash") for fertilizer.

On a smaller scale, evaporite deposits can be seen in such places as Death Valley, California, and Utah's Bonneville Salt Flats. Following rains or periods of snowmelt in the mountains, streams flow from the surrounding mountains into an enclosed basin. As the water evaporates, **salt flats** form when dissolved materials are precipitated as a white crust on the ground (**Figure 7.19**).

7.3   Concept Checks

1. Explain how the formation of biochemical sediments differs from the formation of sediments by inorganic processes. Use examples as part of your explanation.

2. Distinguish among limestone, dolostone, and chert. Describe several varieties of each.

3. How do evaporites form? What are some examples?

 Tutorial

This extensive evaporite deposit is a 30,000-acre expanse of hard white salt that in places is nearly 2 meters thick.

7.4 | Coal: An Organic Sedimentary Rock
Outline the successive stages in the formation of coal.

Coal is different from other sedimentary rocks. Unlike limestone and chert, which are calcite and silica rich, coal is made of organic matter. Close examination of coal under a magnifying glass often reveals plant structures such as leaves, bark, and wood that have been chemically altered but are still identifiable. This supports the conclusion that coal is the end product of large amounts of plant material, buried for millions of years (**Figure 7.20**). The formation of coal involves these stages:

1. **Accumulation of plant remains.** The initial stage in coal formation is the accumulation of large quantities of plant remains. Such accumulations result from special conditions because dead plants readily decompose when exposed to the atmosphere or other oxygen-rich environments. One important environment that allows for the buildup of plant material is a swamp. Stagnant swamp water is oxygen deficient, so complete decay (oxidation) of the plant material is not possible. Instead, the plants are attacked by certain bacteria that partly decompose the organic material and liberate oxygen and hydrogen. As these elements escape, the percentage of carbon in the plant matter gradually increases. The bacteria are not able to finish the job of decomposition because their growth is impeded by acids liberated from the plants.

2. **Formation of peat and lignite.** The partial decomposition of plant remains in an oxygen-poor swamp creates a layer of *peat*, a soft brown material in which plant structures are still easily recognized. With shallow burial, peat slowly changes to *lignite*, a soft brown coal. Burial increases the temperature of sediments as well as the pressure on them.

3. **Formation of bituminous coal.** The higher temperatures bring about chemical reactions within the plant materials and yield water and organic gases (volatiles). As the load increases from more sediment on top of the developing coal, the water and volatiles are pressed out, and the proportion of *fixed carbon* (the remaining solid combustible material) increases. The greater the carbon content, the greater the coal's energy ranking as a fuel. During burial, the coal also becomes increasingly compact. For example, deeper burial transforms lignite into a harder, more compacted black rock called *bituminous* coal. A bed of *bituminous* coal may be only one-tenth as thick as the peat bed from which it formed.

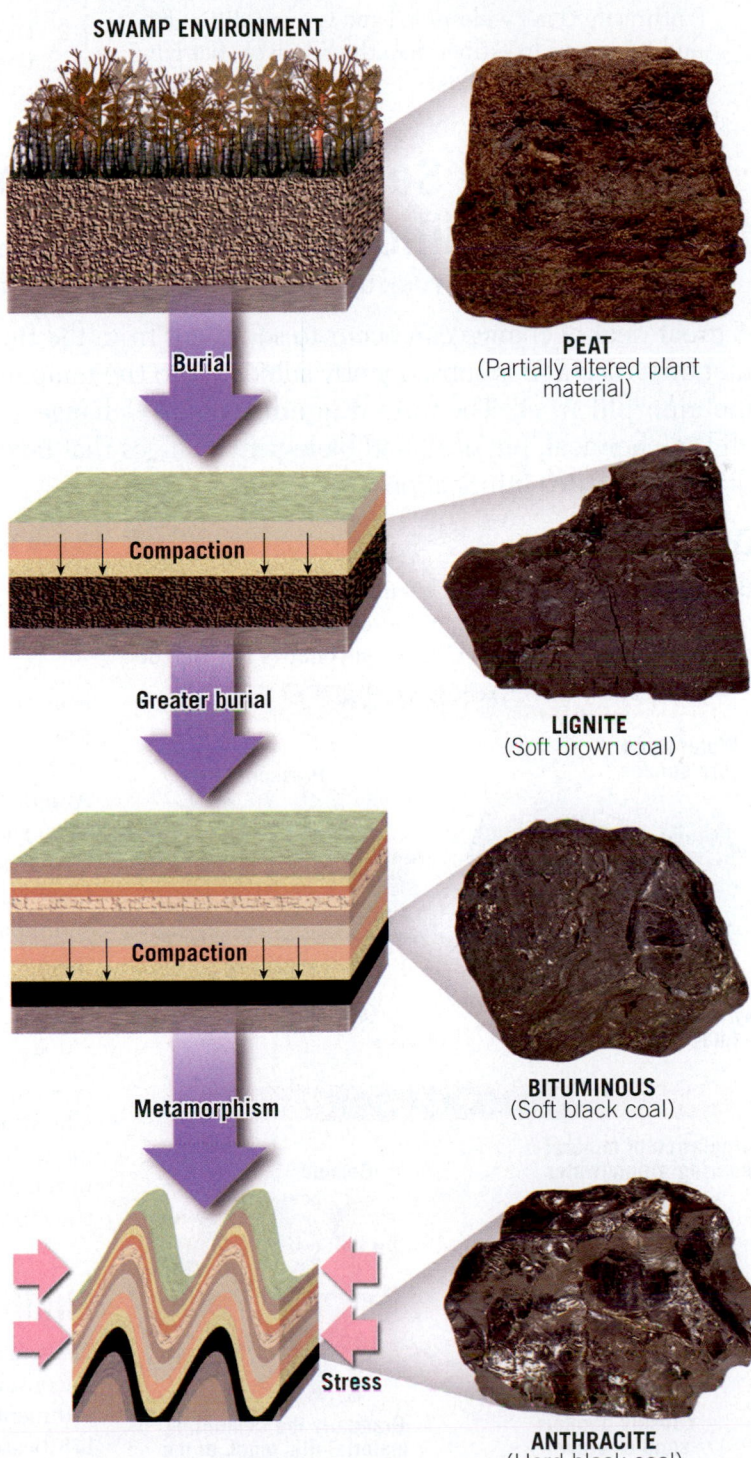

SWAMP ENVIRONMENT

Burial

PEAT
(Partially altered plant material)

Compaction

Greater burial

LIGNITE
(Soft brown coal)

Compaction

Metamorphism

BITUMINOUS
(Soft black coal)

Stress

ANTHRACITE
(Hard black coal)

SmartFigure 7.20
From plants to coal
Successive stages in the formation of coal. (Photos by E. J. Tarbuck) (https://goo.gl/EHKyKU)

Tutorial

4. **Formation of anthracite coal.** Lignite and bituminous coals are sedimentary rocks. However, when sedimentary layers are subjected to the folding and deformation associated with mountain building, the heat and pressure cause a further loss of volatiles and water, thus increasing the concentration of fixed carbon. This metamorphoses bituminous coal into *anthracite*, a very hard, shiny, black *metamorphic* rock (see Chapter 8). Although anthracite is a clean-burning fuel, only a relatively small amount is mined. Anthracite is not widespread and is more difficult and expensive to extract than the relatively flat-lying layers of bituminous coal.

Coal is a major energy resource. Its role as a fuel and some of the problems associated with burning coal are discussed in Chapter 23.

> ### 7.4 | Concept Checks
>
> 1. What is the "raw material" for coal? Under what circumstances does it accumulate?
> 2. Outline the successive stages in the formation of coal.

7.5 | Turning Sediment into Sedimentary Rock: Diagenesis and Lithification
Describe the processes that convert sediment into sedimentary rock and other changes associated with burial.

A great deal of change can occur to sediment from the time it is deposited until it becomes a sedimentary rock and is subsequently subjected to the temperatures and pressures that convert it to metamorphic rock. The term **diagenesis** (*dia* = change, *genesis* = origin) is a collective term for all the chemical, physical, and biological changes that take place after sediments are deposited and during and after lithification.

Diagenesis

Burial promotes diagenesis because as sediments are buried, they are subjected to increasingly higher temperatures and pressures. Diagenesis occurs within the upper few kilometers of Earth's crust at temperatures that are generally less than 150° to 200°C (300° to 400°F). Beyond this somewhat arbitrary temperature threshold, metamorphism is said to occur.

One example of diagenetic change is *recrystallization*, the development of more stable minerals from less stable ones. For example, the mineral aragonite is the less stable form of calcium carbonate ($CaCO_3$). Aragonite is secreted by many marine organisms to form shells and other hard parts, such as the skeletal structures produced by corals. In some environments, large quantities of these solid materials accumulate as sediment. As burial takes place, aragonite recrystallizes to the more stable form of calcium carbonate, calcite, the main constituent in the sedimentary rock limestone.

Another example of diagenesis is provided in the preceding discussion of coal. It involves the chemical alteration of organic matter in an oxygen-poor environment. Instead of completely decaying, as would occur in the presence of oxygen, the organic matter is slowly transformed into solid carbon.

Lithification

Diagenesis includes **lithification**, the processes by which unconsolidated sediments are transformed into solid sedimentary rocks (*lithos* = stone, *fic* = making). Basic lithification processes include compaction and cementation (**Figure 7.21**).

Figure 7.21
Compaction and cementation

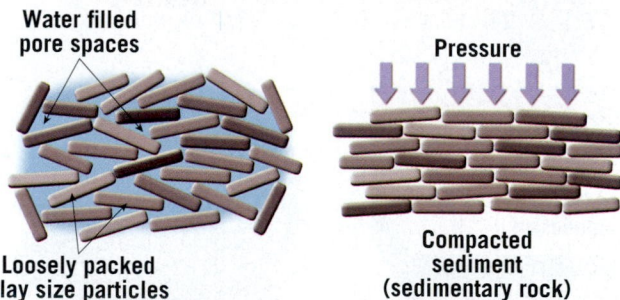

COMPACTION

Water filled pore spaces

Loosely packed clay size particles (magnified)

Pressure

Compacted sediment (sedimentary rock)

CEMENTATION

Circulation of mineral-bearing groundwater

Loosely packed sand or gravel size particles (magnified)

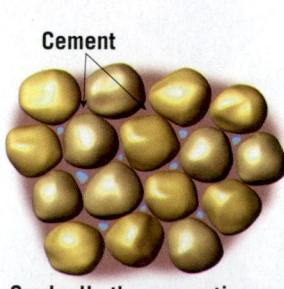

Cement

Gradually the cementing material fills much of the pore space and "glues" the grains together

Compaction   The most common physical diagenetic change is **compaction**. As described earlier, as sediment accumulates, the weight of overlying material compresses the deeper sediments. The deeper a sediment is buried, the more it is compacted and the firmer it becomes. As the grains are pressed closer and closer, there is considerable reduction in pore space. For example, when clays are buried beneath several thousand meters of material, the volume of the clay layer may be reduced by as much as 40 percent. As pore space decreases, much of the water that was trapped in the sediments is driven out. Because sands and other coarse sediments are less compressible, compaction is most significant as a lithification process in fine-grained sedimentary rocks like shale.

Cementation   The most important process by which sediments are converted to sedimentary rock is **cementation**. This change involves the crystallization of minerals among the individual sediment grains. Groundwater carries ions in solution. Gradually, the crystallization of new minerals from these ions takes place in the pore spaces, cementing the clasts together. Just as the amount of pore space is reduced during compaction, the addition of cement to a sedimentary deposit reduces its porosity as well.

Calcite, silica, and iron oxide are the most common cements. It is often a relatively simple matter to identify the cementing material. Calcite cement will effervesce in contact with dilute hydrochloric acid. Silica is the hardest cement and thus produces the hardest sedimentary rocks. An orange or dark-red color in a sedimentary rock means that iron oxide is present.

Most sedimentary rocks are lithified by means of compaction and/or cementation. However, some, like the evaporites, initially form as solid masses of intergrown crystals rather than beginning as accumulations of separate particles that later become solid. Other crystalline sedimentary rocks do not begin that way but are transformed into masses of interlocking crystals sometime after the sediment is deposited.

For example, with time and burial, loose sediment consisting of delicate calcium carbonate–rich skeletal debris may be recrystallized into a relatively dense crystalline limestone. Because crystals grow until they fill all the available space, crystalline sedimentary rocks often lack pore spaces. Unless the rocks later develop joints and fractures, they will be relatively impermeable to fluids such as water and oil.

7.5 | Concept Checks

1. What is diagenesis?

2. Compaction is most important as a lithification process with which sediment size?

3. List three common cements. How might each be identified?

7.6 | Classification of Sedimentary Rocks

Summarize the criteria used to classify sedimentary rocks.

The classification scheme in **Figure 7.22** divides sedimentary rocks into major groups: detrital on the left side and chemical/organic on the right. Further, we can see that the main criterion for subdividing the detrital rocks is particle size, whereas the primary basis for distinguishing among different rocks in the chemical group is their mineral composition.

As is the case with many (perhaps most) classifications of natural phenomena, the categories presented in Figure 7.22 are more rigid than the actual state of nature. In reality, many of the sedimentary rocks classified into the chemical group also contain at least small quantities of detrital sediment. Many limestones, for example, contain varying amounts of mud or sand, giving them a "sandy" or "shaly" quality. Conversely, because practically all detrital rocks are cemented with material that was originally dissolved in water, they too are far from being "pure."

As was the case with the igneous rocks examined in Chapter 5, *texture* is a part of sedimentary rock classification. There are two major textures used in the

classification of sedimentary rocks: clastic and nonclastic. The term **clastic** is taken from a Greek word meaning "broken." Rocks that display a clastic texture consist of discrete fragments and particles that are cemented and compacted together. Although cement is present in the spaces between particles, these openings are rarely filled completely. All detrital rocks have a clastic texture. In addition, some chemical sedimentary rocks exhibit this texture. For example, coquina, the limestone composed of shells and shell fragments, is obviously as clastic as a conglomerate or sandstone. The same applies for some varieties of oolitic limestone.

Some chemical sedimentary rocks have a **nonclastic**, or **crystalline** texture, in which the minerals

Figure 7.22
Identification of sedimentary rocks The main criterion for naming detrital rocks is particle size. The primary basis for naming chemical and organic sedimentary rocks is their composition.

Detrital Sedimentary Rocks				
Clastic Texture (particle size)		**Sediment Name**	**Rock Name**	
Coarse (over 2 mm)		Gravel (Rounded particles)	Conglomerate	
		Gravel (Angular particles)	Breccia	
Medium (1/16 to 2 mm)		Sand	Sandstone (Arkose)*	
Fine (1/16 to 1/256 mm)		Mud	Siltstone	
Very fine (less than 1/256 mm)		Mud	Shale or Mudstone	

*If abundant feldspar is present the rock is called Arkose.

Chemical and Organic Sedimentary Rocks			
Composition	**Texture**	**Rock Name**	
Calcite, CaCO₃	Nonclastic: Fine to coarse crystalline	Crystalline Limestone	
		Travertine	
	Clastic: Visible shells and shell fragments loosely cemented	Coquina	Biochemical Limestone
	Clastic: Various size shells and shell fragments cemented with calcite cement	Fossiliferous Limestone	
	Clastic: Microscopic shells and clay	Chalk	
Quartz, SiO₂	Nonclastic: Very fine crystalline	Chert (light colored) Flint (dark colored) Jasper (red) Agate (banded)	
Gypsum CaSO₄•2H₂O	Nonclastic: Fine to coarse crystalline	Rock Gypsum	
Halite, NaCl	Nonclastic: Fine to coarse crystalline	Rock Salt	
Altered plant fragments	Nonclastic: Fine-grained organic matter	Bituminous Coal	

form a pattern of interlocking crystals. The crystals may be microscopically small or large enough to be visible without magnification. Common examples of rocks with nonclastic textures are those deposited when saline water evaporates (**Figure 7.23**). The materials that make up many other nonclastic rocks may actually have originated as detrital deposits. In these instances, the particles probably consisted of shell fragments and other hard parts rich in calcium carbonate or silica. The clastic nature of the grains was subsequently obliterated or obscured because the particles recrystallized when they were consolidated into limestone or chert.

Nonclastic rocks consist of intergrown crystals, and some may resemble igneous rocks, which are also crystalline. The two rock types are usually easy to distinguish because the minerals contained in nonclastic sedimentary rocks are quite unlike those found in most igneous rocks. For example, rock salt, rock gypsum, and some forms of limestone consist of intergrown crystals, but the minerals within these rocks (halite, gypsum, and calcite) are seldom associated with igneous rocks.

Figure 7.23
Rock salt Like other evaporites, rock salt has a nonclastic texture because it is composed of intergrown crystals. (Photos by E. J. Tarbuck)

Close up

7.6 Concept Checks

1. What is the primary basis for distinguishing (naming) different chemical sedimentary rocks? How is the naming of detrital rocks different?

2. Distinguish between clastic and nonclastic. Which texture is associated with all detrital rocks?

7.7 | Sedimentary Rocks Represent Past Environments

Distinguish among three broad categories of sedimentary environments and provide an example of each. List several sedimentary structures and explain why these features are useful to geologists.

Sedimentary rocks are important to interpreting Earth's history (**Figure 7.24**). By understanding the conditions under which sedimentary rocks form, geologists can often deduce the history of a rock, including information about the origin of its component particles, the method of sediment transport, and the nature of the place where the grains eventually came to rest—that is, the environment of deposition.

An **environment of deposition**, or **sedimentary environment**, is a geographic setting where sediment is accumulating. Each site is characterized by a particular combination of geologic processes and environmental conditions. Some sediments, such as the chemical sediments that precipitate in water bodies, are solely products of their sedimentary environment. That is, their component minerals originated and were deposited in the same place. Other sediments originate far from the site where they accumulate. These materials are transported great distances from their source by some combination of gravity, water, wind, and ice.

At any given time, the geographic setting and environmental conditions of a sedimentary environment determine the nature of the sediments that accumulate. Geologists carefully study the sediments in present-day depositional environments because the features they find can also be observed in ancient sedimentary rocks.

Geologists apply a thorough knowledge of present-day conditions to reconstruct the ancient environments and geographic relationships of an area at the time a particular set of sedimentary layers were deposited. This process is an excellent example of applying a fundamental principle of modern geology: "The present is the key to the past."* Such analyses often lead to the creation of maps depicting the past geographic distribution of land and sea, mountains and river valleys, deserts and glaciers, and other environments of deposition.

Types of Sedimentary Environments

Sedimentary environments are commonly placed into one of three broad categories: continental, marine, or transitional (shoreline). Each category includes many specific subenvironments. **Figure 7.25** is an idealized diagram illustrating a number of important sedimentary environments associated with each category. Realize that this is just a sampling of the great diversity of depositional environments. Chapters 16 through 20 examine many of these environments in greater detail. Each of these three categories is an

SmartFigure 7.24
Utah's Capitol Reef National Park These tilted sedimentary strata are part of the Waterpocket Fold at Halls Creek. The strata here record changing environments during the Mesozoic era. (Photo by Michael Collier) (http://goo.gl/eeryMM)

Mobile Field Trip

area where sediment accumulates and where organisms live and die. Each produces a characteristic sedimentary rock or assemblage that reflects prevailing conditions.

Continental Environments Continental environments are dominated by the erosion and deposition associated with streams. In some cold regions, moving masses of glacial ice replace running water as the dominant process. In arid regions (as well as some coastal settings), wind takes on greater importance. Clearly, the nature of the sediments deposited in continental environments is strongly influenced by climate.

Streams are the dominant agent of landscape alteration, eroding more land and transporting and depositing more sediment than any other process. In addition to channel deposits, large quantities of sediment are dropped when floodwaters periodically inundate broad, flat valley floors (called *floodplains*). Where rapid streams emerge from a mountainous area onto a flatter surface, a distinctive cone-shaped accumulation of sediment known as an *alluvial fan* forms.

In frigid, high-latitude or high-altitude settings, glaciers pick up and transport huge volumes of sediment. Materials deposited directly from ice are typically unsorted mixtures of particles that range in size from fine clay to huge boulders. Water from melting glaciers transports and redeposits some of this glacial sediment, creating stratified, sorted accumulations.

*For more on this idea, see Section 1.2 in Chapter 1.

Beaches that form where wave activity is strong, consist mainly of pebbles and cobbles.

Michael Collier

Caves that develop in limestone are sites where calcium carbonate is deposited as dripstone.

Dennis Tasa

Sand dunes consist of well-sorted sand grains deposited by the wind.

Michael Collier

PixAchi/Shutterstock

Beaches, **bars**, and **spits** along low-lying coasts and in sheltered coves are typically composed of well-sorted sand and/or shell fragments.

Salt flat

Lake

Estuary

Deep-sea fans

Turbidity current

Barrett & MacKay/Glow Images

Tidal flats and **lagoons** are areas where fine clay particles or carbonate-rich muds accumulate.

Figure 7.25
Sedimentary environments Each environment is characterized by certain physical, chemical, and biological conditions. Because sediments contain clues about the environment in which they were deposited, sedimentary rocks are important in the interpretation of Earth history. A number of important examples are represented in this idealized diagram.

Marli Miller

Deep marine environments adjacent to the continental slope often contain material that was transported by dense underwater currents of suspended sediment. Each layer has coarser particles at the bottom and finer material on top.

E.J. Tarbuck

Shallow marine environments are sites where sand, clay, and carbonate-rich muds are often deposited. Ripple marks caused by wave activity may be present.

Inland seas and **lakes** in arid environments where evaporation exceeds precipitation, produce evaporite deposits such as rock salt and gypsum.

Alluvial fans consist of coarse sediments that are deposited when mountain streams reach flat lowlands.

Glacial deposits often consist of a poorly-sorted mixture of many different sediment sizes ranging from clay to boulders.

Swamps and **bogs** are quiet-water environments where mud and decayed plant material accumulate.

Streams in mountainous areas erode and deposit a wide variety of sediment, while those in lowlands transport and deposit mostly mud (silt and clay) and sand.

Landslides produce an unsorted jumble of many sediment sizes.

Coral reefs are massive limestone structures that form in warm, shallow clear seas and consist of material secreted by corals and other marine life.

The work of wind and its resulting deposits are referred to as *eolian*, after Aeolus, the Greek god of wind. Unlike glacial deposits, eolian sediments are well sorted. Wind can lift fine dust high into the atmosphere and transport it great distances. Where winds are strong and the surface is not anchored by vegetation, sand is transported closer to the ground, where it accumulates in *dunes*. Deserts and coasts are common sites for this type of deposition.

In addition to being areas where dunes sometimes develop, desert basins are sites where shallow *playa lakes* occasionally form following heavy rains or periods of snowmelt in adjacent mountains. They rapidly dry up, sometimes leaving behind evaporites and other characteristic deposits. Figures 19.9 and 19.10 on page 581 illustrate such an environment. In humid regions, lakes are more enduring features, and their quiet waters are excellent sediment traps. Small deltas, beaches, and bars form along the lakeshore, with finer sediments coming to rest on the lake floor.

Marine Environments   Marine depositional environments are divided according to depth. The *shallow marine* environment reaches to depths of about 200 meters (nearly 700 feet) and extends from the shore to the outer edge of the continental shelf. The *deep marine* environment lies seaward of the continental shelf in waters deeper than 200 meters.

The shallow marine environment borders all of the world's continents. Its width varies greatly, from practically nonexistent in some places to broad expanses extending as far as 1500 kilometers (more than 900 miles) in other locations. On average this zone is about 80 kilometers (50 miles) wide. The kind of sediment deposited here depends on several factors, including distance from shore, elevation of the adjacent land area, water depth, water temperature, and climate.

Due to the ongoing erosion of the adjacent continent, the shallow marine environment receives huge quantities of land-derived sediment. Where the influx of such sediment is small and the seas are relatively warm, carbonate-rich muds may be the predominant sediment. Most of this material consists of the skeletal debris of carbonate-secreting organisms mixed with inorganic precipitates. Coral reefs are also associated with warm, shallow marine environments. In hot regions where the sea occupies a basin with restricted circulation, evaporation triggers the precipitation of soluble materials and the formation of marine evaporite deposits.

Deep marine environments include all the floors of the deep ocean. Far from landmasses, tiny particles from many sources remain adrift for long spans. Gradually these small grains "rain" down on the ocean floor, where they accumulate very slowly. Significant exceptions are thick deposits of relatively coarse sediment that occur at the base of the continental slope. These materials move down from the continental shelf as turbidity currents—dense gravity-driven masses of sediment and water (see Figure 7.29).

Transitional Environments   The shoreline is the transition zone between marine and continental

EYE ON EARTH 7.3

This is an aerial view of North Carolina's Hatteras Island (looking toward the south). To the left (east) of the island is the Atlantic Ocean. The sheltered waters of Pamlico Sound are on the right, the side of the island facing the mainland. Assume that the sediments accumulating in the area are primarily detrital. (Photo by Michael Collier)

QUESTION 1 If you were to sample the sediments at points A and B, which would likely have the coarser particles?

QUESTION 2 Explain why the sediments at these two sites would probably be different.

QUESTION 3 Which one of the three broad categories of sedimentary environments is represented in this image?

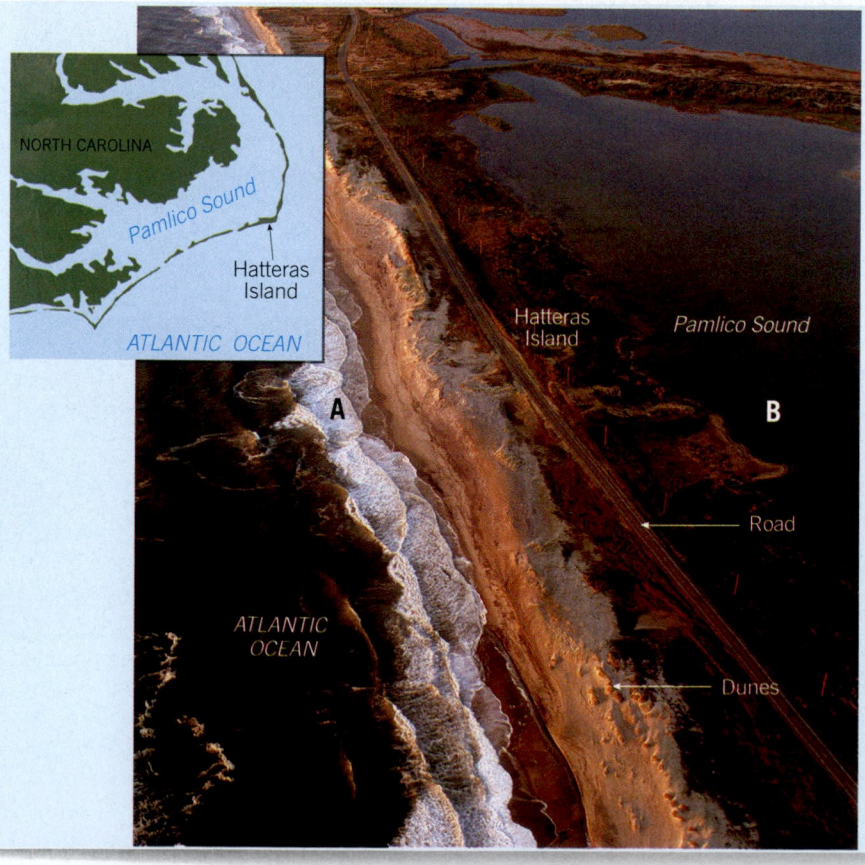

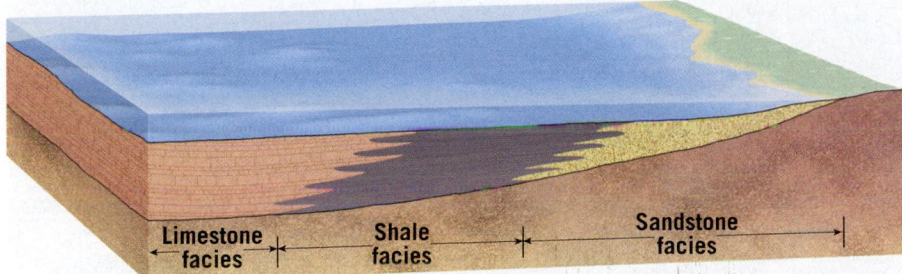

SmartFigure 7.26
Lateral change When a sedimentary layer is traced laterally, we may find that it is made up of several different rock types. This occurs because many sedimentary environments can exist at the same time over a broad area. The term *facies* is used to describe such sets of sedimentary rocks. Each facies grades laterally into another that formed at the same time but in a different environment. (https://goo.gl/HOJsQ8)

environments. Here we find the familiar deposits of sand or gravel called *beaches*. Mud-covered *tidal flats* are alternately covered with shallow sheets of water and then exposed to air as tides rise and fall. Along and near the shore, the work of waves and currents distributes sand, creating *spits*, *bars*, and *barrier islands*. Offshore bars and reefs create *lagoons*. The quieter waters in these sheltered areas are another site of deposition in the transition zone.

Deltas are among the most significant deposits associated with transitional environments. The complex accumulations of sediment build outward into the sea when rivers experience an abrupt loss of velocity and deposit their load of detrital material.

Sedimentary Facies

When we study a series of sedimentary layers, we can see the successive changes in environmental conditions that occurred at a particular place with the passage of time. Changes in past environments may also be seen when a single layer of sedimentary rock is traced laterally. This is true because at any one time, many different depositional environments can exist over a broad area. For example, when sand is accumulating in a beach environment, finer muds are often being deposited in quieter offshore waters. Still farther out, perhaps in a zone where biological activity is high and land-derived sediments are scarce, the deposits consist largely of the calcite-rich remains of small organisms. In this example, different sediments are accumulating adjacent to one another at the same time. Different parts of each layer possess a distinctive set of characteristics that reflect the conditions in a particular environment. The term **facies** is used to describe such sets of sediments. When a sedimentary layer is examined in cross section from one end to the other, each facies grades laterally into another that formed at the same time but that exhibits different characteristics (**Figure 7.26**). The merging of adjacent facies tends to be a gradual transition rather than a sharp boundary, but abrupt changes do sometimes occur.

Sedimentary Structures

In addition to variations in grain size, mineral composition, and texture, sediments exhibit a variety of structures that are often preserved when they change to sedimentary rock. Some, such as graded beds, are created when

sediments are accumulating and are a reflection of the transporting medium. Others, such as *mud cracks*, form after the materials have been deposited and result from processes occurring in the environment. When present, sedimentary structures provide additional information that can be useful in interpreting Earth's history.

Sedimentary rocks form as layer upon layer of sediment accumulates in various depositional environments. These layers, called **strata**, or **beds**, are probably *the single most common and characteristic feature of sedimentary rocks*. Each stratum is unique. It may be a coarse sandstone, a fossil-rich limestone, a black shale, and so on. When you look at **Figure 7.27**, or look back at the chapter-opening photo and at Figures 7.1 and 7.5,

Figure 7.27
Layers are called strata This outcrop of sedimentary strata illustrates the characteristic layering of this group of rocks. This exposure of sedimentary strata is in New York's Shawangunk Mountains.
(Photo by Colin D. Young/ Shutterstock)

Figure 7.28
Cross-bedding Sand dunes typically exhibit thin layers inclined at an angle to the main bedding.

John S. Shelton/ University of Washington Libraries

The cutaway section of this sand dune shows the characteristic cross-bedding

The cross-bedding in this sandstone indicates it was once a sand dune

Weathered bedding planes within unit

Principal bedding surfaces

Cross-bedding unit

Steep truncated cross-beds

0 1 2 3
meters

Dennis Tasa

you see many such layers, each different from the others. The variations in texture, composition, and thickness reflect the different conditions under which each layer was deposited.

Beds range from microscopically thin to tens of meters thick. Separating these strata are **bedding planes**, relatively flat surfaces along which rocks tend to separate or break. Changes in the grain size or in the composition of the sediment being deposited can create bedding planes. Pauses in deposition can also lead to layering because chances are slight that newly deposited

material will be exactly the same as previously deposited sediment. Generally, each bedding plane marks the end of one episode of sedimentation and the beginning of another.

Because sediments usually accumulate as particles that settle from a fluid, most strata are originally deposited as horizontal layers. There are circumstances, however, when sediments do not accumulate in horizontal beds. Sometimes when a bed of sedimentary rock is examined, we see layers within it that are inclined to the horizontal. This is called **cross-bedding**, and it is most

Figure 7.29
Graded bedding A graded bed is characterized by a decrease in sediment size from bottom to top. Graded beds are associated with submarine currents known as *turbidity currents*. (Photo by Marli Miller)

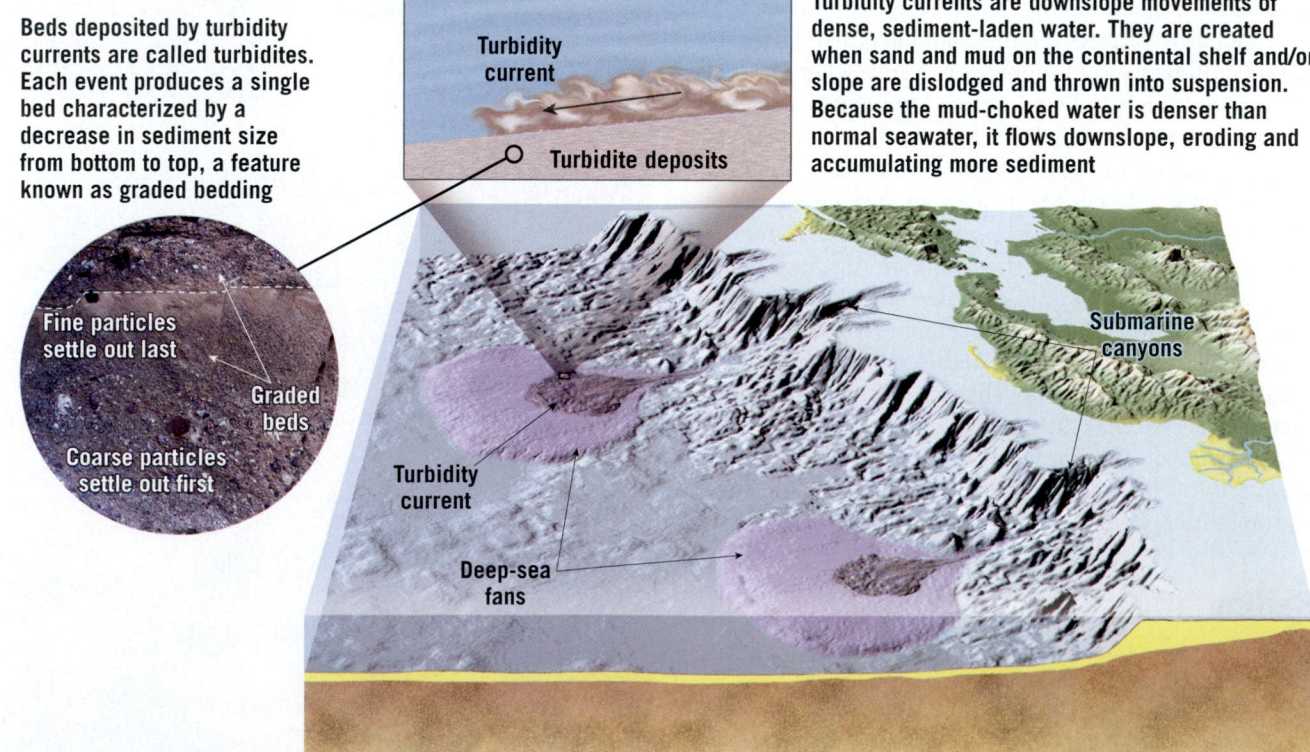

Beds deposited by turbidity currents are called turbidites. Each event produces a single bed characterized by a decrease in sediment size from bottom to top, a feature known as graded bedding

Turbidity currents are downslope movements of dense, sediment-laden water. They are created when sand and mud on the continental shelf and/or slope are dislodged and thrown into suspension. Because the mud-choked water is denser than normal seawater, it flows downslope, eroding and accumulating more sediment

Turbidity current

Turbidite deposits

Fine particles settle out last

Graded beds

Coarse particles settle out first

Submarine canyons

Turbidity current

Deep-sea fans

characteristic of sand dunes, river deltas, and certain stream channel deposits (see Figure 7.8 and **Figure 7.28**).

    Graded beds are another special type of bedding. In this case, the particles within a single sedimentary layer gradually change from coarse at the bottom to fine at the top. Graded beds are most characteristic of rapid deposition from water containing sediment of varying sizes. When a current experiences a rapid energy loss, the largest particles settle first, followed by successively smaller grains. The deposition of a graded bed is most often associated with a turbidity current, a mass of sediment-choked water that is denser than clear water and that moves downslope along the bottom of a lake or an ocean (**Figure 7.29**).

    As geologists examine sedimentary rocks, they can deduce quite a bit. A conglomerate, for example, may indicate a high-energy environment such as a surf zone or rushing stream, where only coarse materials remain and finer particles are kept suspended. An arkose may signify a dry climate, where little chemical alteration of feldspar is possible. Carbonaceous shale is a sign of a low-energy, organic-rich environment, such as a swamp or lagoon.

    Other features found in some sedimentary rocks also give clues to past environments. Ripple marks are such a feature. **Ripple marks** are small waves of sand that develop on the surface of a sediment layer through the action of moving water or air (**Figure 7.30A**). The ridges form at right angles to the direction of motion. If the ripple marks were formed by air or water moving in essentially one direction, their form will be asymmetrical. These *current ripple marks* will have steeper sides in the downcurrent direction and more gradual slopes on the upcurrent side. Ripple marks produced by a stream flowing across a sandy channel and by wind blowing over a sand dune are two common examples of current ripples. When present in solid rock, they may be used to determine the direction of movement of ancient wind or water currents. Other ripple marks have a symmetrical form. These features, called *oscillation ripple marks*, result from the back-and-forth movement of surface waves in a shallow near-shore environment.

    Mud cracks (**Figure 7.30B**) indicate that the sediment in which they were formed was alternately wet and dry. When exposed to air, wet mud dries out and shrinks, producing cracks. Mud cracks are associated with environments such as tidal flats, shallow lakes, and desert basins.

A.

B.

Figure 7.30
Frozen in stone A. These current ripple marks formed in sandy sediment and are now preserved in rock. (Photo by Tim Graham/Alamy Images)
B. When muddy sediments dry out, they shrink and create cracks. When the sediment is turned into rock, the mud cracks are preserved. (Photo by Marli Miller)

    Fossils, the remains or traces of prehistoric life, are important inclusions in sediment and sedimentary rocks. They are important tools for interpreting the geologic past. Knowing the nature of the life-forms that existed at a particular time helps researchers understand past environmental conditions. In addition, fossils are important time indicators and play a key role in correlating rocks of similar ages that are from different places. Fossils are discussed further in Chapter 9.

7.7   Concept Checks

1. What are the three broad categories of sedimentary environments? List a specific example associated with each category (see Figure 7.25).

2. Why might a single layer exhibit different types of sedimentary rock? What term applies to the different parts of such a layer?

3. What is the single most characteristic feature of sedimentary rocks?

4. What is the difference between cross-bedding and graded bedding?

5. How might mud cracks and ripple marks be useful clues about the geologic past?

7.8 | The Carbon Cycle and Sedimentary Rocks

Relate weathering processes and sedimentary rocks to the carbon cycle.

To illustrate the movement of material and energy among the spheres of the Earth system, let us take a brief look at the **carbon cycle** (**Figure 7.31**). Most carbon is bonded chemically to other elements to form compounds such as carbon dioxide, calcium carbonate, and the hydrocarbons found in coal and petroleum. Carbon is also the basic building block of life, as it readily combines with hydrogen and oxygen to form the fundamental organic compounds that compose living things.

Certainly one of the most active parts of the carbon cycle is the movement of carbon from the atmosphere to the biosphere and back again. In the atmosphere, carbon is found mainly as carbon dioxide (CO_2). Atmospheric carbon dioxide is significant because it is a greenhouse gas, which means it is an efficient absorber of energy emitted by Earth and thus influences the heating of the atmosphere, a process described in more detail in Chapter 21. Because many of the processes that operate on Earth involve carbon dioxide, this gas is constantly moving into and out of the atmosphere. Through the process of photosynthesis, plants absorb carbon dioxide from the atmosphere to produce the essential organic compounds needed for growth. Animals that consume these plants (or consume other animals that eat plants) use these organic compounds as a source of energy and, through the process of respiration, return carbon dioxide to the atmosphere. (Plants also return some CO_2 to the atmosphere via respiration.) Further, when plants die and decay or are burned, this biomass is oxidized, and carbon dioxide is returned to the atmosphere.

Not all dead plant material decays immediately back to carbon dioxide. A small percentage is deposited as sediment. Over long spans of geologic time, considerable biomass is buried with sediment. Under the right conditions, some of these carbon-rich deposits are converted to fossil fuels—coal, petroleum, or natural gas. Some of the fuels are eventually recovered (mined or pumped from a well) and burned to generate electricity and fuel our transportation system. One result of fossil-fuel combustion is the release of huge quantities of carbon dioxide back into the atmosphere. Carbon also moves from the geosphere and hydrosphere to the atmosphere and back again. For example, volcanic activity early in Earth's history is thought to be the source of much of our atmospheric carbon dioxide. One way that carbon dioxide makes its way back to the hydrosphere and then to the solid Earth is by first combining with water to form carbonic acid (H_2CO_3), which then attacks the rocks that compose Earth's crust. One product of this chemical weathering of solid rock is the soluble bicarbonate ion (HCO_3^-) which is carried by groundwater and streams to the ocean. Water-dwelling organisms extract this dissolved material to produce hard parts of calcium carbonate ($CaCO_3$). When the organisms die, these skeletal remains settle to the ocean floor as biochemical sediment and become sedimentary rock. In fact, the crust is by far Earth's largest depository of carbon, where it is a constituent of a variety of rocks, the most abundant being limestone. Eventually the limestone may be exposed at Earth's surface, where chemical weathering will cause the carbon stored in the rock to be released to the atmosphere as carbon dioxide.

In summary, carbon moves among all four of Earth's major spheres. It is essential to every living thing in the biosphere. In the atmosphere, carbon dioxide is an important greenhouse gas. In the hydrosphere, carbon dioxide is dissolved in lakes, rivers, and the ocean. In the geosphere, carbon is contained in carbonate sediments and sedimentary rocks and is stored as organic matter dispersed through sedimentary rocks and as deposits of coal and petroleum.

Figure 7.31

Carbon cycle This simplified diagram emphasizes the flow of carbon between the atmosphere and the hydrosphere, geosphere, and biosphere. The arrows show whether the flow of carbon is into or out of the atmosphere.

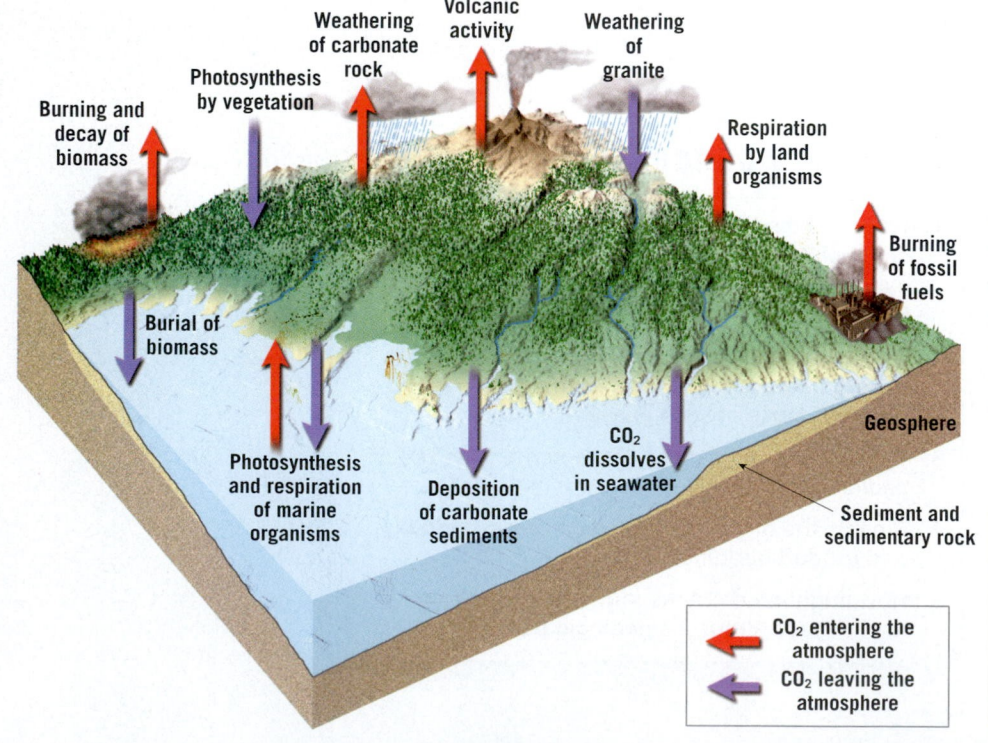

Burning and decay of biomass

Photosynthesis by vegetation

Weathering of carbonate rock

Volcanic activity

Weathering of granite

Respiration by land organisms

Burning of fossil fuels

Burial of biomass

Photosynthesis and respiration of marine organisms

Deposition of carbonate sediments

CO_2 dissolves in seawater

Geosphere

Sediment and sedimentary rock

CO_2 entering the atmosphere

CO_2 leaving the atmosphere

7.8 Concept Checks

1. Describe how chemical weathering and the formation of biochemical sediment remove carbon from the atmosphere and store it in the geosphere.

2. Provide an example by which carbon moves from the geosphere to the atmosphere.

7.1 An Introduction to Sedimentary Rocks

Explain the importance of sedimentary rocks and summarize the part of the rock cycle that pertains to sediments and sedimentary rocks. List the three categories of sedimentary rocks.

KEY TERMS sedimentary rock, detrital sedimentary rock, chemical sedimentary rock, organic sedimentary rock

- Though igneous and metamorphic rocks make up most of Earth's crust by volume, sediment and sedimentary rocks are concentrated near the surface. There, at the interface between Earth's four spheres, sediments and the rock layers they eventually form make a record of past conditions and events at the surface. Sedimentary rocks contain fossils that show the evolution of life over time.
- Numerous geologic resources are restricted to sedimentary rocks, such as coal, oil, uranium, and several major ores of metals.

- Sediments are the raw materials from which sedimentary rocks are made. Sediments are produced from the weathering of preexisting rocks. Both solid particles of many sizes and chemical residues, including ions in solution, qualify as sediments.
- Once produced, sediments are transported from their source area by water currents, wind, glacial ice, or simply downhill movement under the influence of gravity. Eventually, these will be deposited at a new site where they will be compacted and/or cemented so that the individual sediments become bonded together into a sedimentary rock.
- There are three main varieties of sedimentary rocks: detrital, chemical, and organic.

Q Of the two main sources of energy that drive the rock cycle—Earth's internal heat and solar energy—which one is primarily responsible for the formation of sediment and sedimentary rocks? Briefly explain.

7.2 Detrital Sedimentary Rocks

Discuss the primary basis for distinguishing among detrital rocks and describe how the origin and history of such rocks might be determined.

KEY TERMS shale, fissility, sandstone, sorting, conglomerate, breccia

- Detrital sedimentary rocks are made of solid particles, mostly quartz grains and microscopic clay minerals. Quartz and clay dominate because unlike most other minerals, they are stable at Earth's surface. Feldspars and micas are conspicuous additions to certain detrital sediments, indicating a relatively short time in the chemical weathering environment; these were mechanically weathered, transported a relatively short distance, and deposited with minimal decomposition.
- Detrital sedimentary rocks are classified mainly based on the size of the sedimentary grains of which they are composed. Grain size is a clue about how energetic the environment of deposition was. Bigger particles indicate more powerful transporting currents; finer grains can be deposited only where current energy is relatively low.
- Shale is made mostly of small grains of clay minerals that accumulate in low-energy depositional environments such as the deep sea, lake bottoms, and floodplains adjacent to rivers. Shale is fissile due to the alignment of microscopic clay flakes parallel to bedding. Shale containing a lot of organic material forms in low-oxygen environments and is characterized by a black color.
- Sandstone is dominated by sand-sized grains and may exhibit various degrees of sorting. Sorting is a reflection of how abruptly or

gradually the sand was deposited. Rounding of the individual sand grains is another important aspect of a sandstone's texture: Rounder grains signify further transport, while angular grains imply a shorter transport distance. Sandstones also vary in their composition: The presence of minerals that are relatively unstable at the surface (like feldspar) implies that the sand did not undergo much chemical weathering prior to deposition. The greater the proportion of quartz, the more the source sediment was chemically weathered before deposition. Three main varieties of sandstone worth knowing are quartz sandstone, arkose, and graywacke.

- Conglomerate and breccia are characterized by a high proportion of gravel-sized grains. If deposited by water, a conglomerate implies a very energetic current. Grains in a breccia are angular, indicating that the material was deposited closer to its source area. Conglomerate is made of rounded grains, implying a significant amount of transport before deposition.

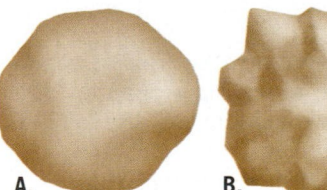 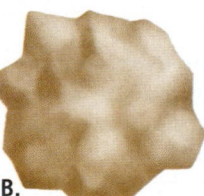 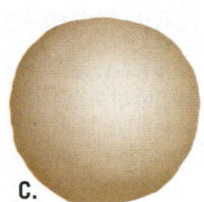

A.　　　B.　　　C.

Q Examine these sketches of sediment particles. Which particle—A, B, or C—has traveled farthest from its source? Explain your reasoning.

7.3 Chemical Sedimentary Rocks

Explain the processes involved in the formation of chemical sedimentary rocks and describe several examples.

KEY TERMS biochemical, limestone, dolostone, chert, evaporite, salt flat

- Chemical sedimentary rocks are formed when ions dissolved in solution link together to form mineral crystals. Sometimes this happens inorganically (without life being involved), and other times living

organisms biochemically extract the ions to precipitate mineral matter as bone or shell.

- Limestone is the most common chemical sedimentary rock. It forms mainly in shallow, warm ocean settings. Limestone is dominated by calcium carbonate. This is the material from which corals construct reefs. Coquina and chalk are also examples of biochemical limestone. Travertine and oolitic limestone are examples of inorganic limestone.
- Dolostone is a chemical sedimentary rock that is dominated by the mineral dolomite. Like calcite, dolomite is a carbonate mineral, but about half of its calcium ions have been replaced by magnesium ions.

- Chert is the general term for rocks made of microcrystalline silica. If the chert is red, it is called jasper. Black chert is flint. Agate is multicolored. When silica replaces plant matter to make petrified wood, often a variety of colors are present.
- Evaporite deposits form when minerals precipitate from an ever-more-concentrated solution of dissolved ions. This is how the vast salt flats of the American West formed. Digging into these deposits reveals rock salt, rock gypsum, and the potassium salt sylvite.

Dennis Tasa

Q Does this rock salt more likely have a biochemical origin or an inorganic origin?

7.4 Coal: An Organic Sedimentary Rock

Outline the successive stages in the formation of coal.

KEY TERM coal

- Coal forms from large amounts of plant matter buried in low-oxygen depositional environments such as swamps and bogs. Through compression, the peat formed becomes compressed into a low-grade form of coal called lignite. Lignite can be compressed further, driving out volatile components and concentrating carbon to make higher-grade bituminous coal. Metamorphism accompanying mountain building can take this concentration process even further, producing the highest-grade coal, anthracite.

Q Considering that it comes from plants, what is the ultimate source of the energy in coal?

7.5 Turning Sediment into Sedimentary Rock: Diagenesis and Lithification

Describe the processes that convert sediment into sedimentary rock and other changes associated with burial.

KEY TERMS diagenesis, lithification, compaction, cementation

- When sediments are buried to relatively shallow depths (the upper few kilometers of the crust), changes in temperature and pressure trigger a variety of processes called diagenesis, a collective term for all the chemical, physical, and biological changes that occur after sediments are deposited and during and after lithification.
- The transformation of sediment into sedimentary rock is called lithification. The two main processes contributing to lithification are compaction (a reduction in pore space by packing grains more tightly together) and cementation (a reduction in pore space by adding new mineral material that acts as a "glue" to bind the grains to each other).

7.6 Classification of Sedimentary Rocks

Summarize the criteria used to classify sedimentary rocks.

KEY TERMS clastic, nonclastic (crystalline)

- Sedimentary rocks are classified primarily based on whether they are detrital, chemical, or organic. Detrital rocks are subdivided by grain size, while among the chemical rocks, mineral composition is the key distinguishing characteristic. An additional characteristic is whether the rocks exhibit a clastic or nonclastic (crystalline) texture.
- Crystalline texture is common to nonclastic sedimentary rocks and igneous rocks and is particularly apparent under magnification. However, the minerals involved are totally different and allow the two types of rock to be distinguished.

Q Would a limestone made of shell fragments be detrital or chemical? Would its texture be clastic or crystalline (nonclastic)?

7.7 Sedimentary Rocks Represent Past Environments

Distinguish among three broad categories of sedimentary environments and provide an example of each. List several sedimentary structures and explain why these features are useful to geologists.

KEY TERMS environment of deposition (sedimentary environment), facies, strata (beds), bedding plane, cross-bedding, graded bed, ripple mark, mud crack, fossils

- Different combinations of tectonic, climatic, and biological conditions result in different types of sediment accumulating. The principle of uniformity suggests that the sedimentary record can be interpreted in light of modern depositional environments. Continental, marine, and transitional (shoreline) environments all have distinctive characteristics that allow geologists to identify sedimentary rocks formed in those environments.
- Sedimentary facies are lateral equivalents that represent different depositional conditions operating in adjacent areas at the same time. For instance, a beach today may be depositing sand, while a kilometer or two offshore only mud is being deposited, and further beyond that, carbonate minerals may be precipitating. All are the same age but represent neighboring areas governed by different conditions.
- Sedimentary structures are patterns that form in sedimentary rock at the time of deposition (or shortly thereafter), before the sediments

become lithified. They can provide powerful clues to the conditions under which the sediment accumulated.

- Beds (or strata) are sheets of sediment deposited in a more-or-less continuous layer. Sometimes cross-beds are preserved within bedding, allowing geologists to deduce the direction of the depositional current. Ripple marks, small waves created by wind or water on the surface of a sediment layer, also provide useful clues. Graded beds indicate depositional currents that quickly lost their energy. Larger grains settled out first, and the smallest grains settled out last. Mud cracks form when mud contracts upon drying out, and they indicate that the sediment was exposed to air. Fossils serve as useful tools that allow geologists to date and interpret past environmental conditions.

Q The accompanying photo shows Los Osos Creek entering California's Morro Bay. Sediment transported by the creek has been deposited as a delta. Which broad category of sedimentary environment is represented in this scene?

Michael Collier

7.8 The Carbon Cycle and Sedimentary Rocks

Relate weathering processes and sedimentary rocks to the carbon cycle.

KEY TERM carbon cycle

• Carbon is a vital component of the atmosphere, biosphere, geosphere, and hydrosphere. The same atom of carbon that is currently part of your nose may have entered your body in a piece of bread. Prior to that, it may once have been part of a plant, and before that it may have been a carbon dioxide molecule in the atmosphere. How did it get to the atmosphere? Perhaps it escaped into the air after originally being dissolved in the ocean, and perhaps it got to the ocean by being weathered from limestone and then washed down a stream. While individual details vary, the key point is that carbon is a reactive element that is equally at home in rocks, water, the air, and living tissue.

Q Follow the history of a single carbon atom as it moves from one part of the Earth system to another. Using any order you choose, include as many of the following sites as possible: a cave, a river, a dinosaur, a coal-fired power plant, a volcanic eruption, a swamp, a can of beer, and the coral reef pictured here.

Exactostock/SuperStock

Give It Some *Thought*

1. Develop a geologic "life history" of a sedimentary rock. Begin with a mass of igneous bedrock in a mountain area and end with your sedimentary rock being collected by a future geology student. Be as complete as possible.

2. How is the use of the term *clay* in Chapter 3 different from the use of the term in Figure 7.3? How are these two uses of the term related?

3. If you hiked to a mountain peak and found limestone at the top, what would that indicate about the likely geologic history of the rock atop the mountain?

4. Why is it *not* necessary to indicate the texture of detrital rocks on the identification chart for sedimentary rocks (see Figure 7.22)?

5. During a hike in Utah's Zion National Park, you pick up a sedimentary rock sample. When you examine the sample with your hand lens, you see that the rock consists mainly of rounded glassy particles that appear to be quartz. To be sure, you conduct two basic tests. When you check for hardness, the rock easily scratches glass, which is what quartz would do. However, when you place a drop of acid on the sample, it fizzes. Explain how a rock that appears to be rich in quartz could effervesce with acid.

6. Examine the accompanying sketch, which shows three sediment layers on the ocean floor. What term is applied to such layers? What process was responsible for creating these layers? Are these layers more likely part of an offshore lagoon or a deep-sea fan?

7. In which of the environments illustrated in Figure 7.25 would you expect to find:

a. An evaporite deposit?   **b.** A well-sorted sand deposit?

c. A deposit that includes a high percentage of partially decomposed plant material?

d. A jumbled mix of many sediment sizes?

8. This image shows the surface of a sand dune. What term is applied to the wave-like ridges on the surface? Be as specific as possible. Is the prevailing wind direction most likely from the left or from the right? Explain.

Michael Collier

9. While on a field trip with your geology class, you stop at an outcrop of sandstone. An examination with a hand lens shows that the sandstone is poorly sorted and rich in feldspar and quartz. Your instructor tells you that the sediment was derived from one of two sites in the area:

Site 1: A nearby exposure of weathered basaltic lava flows.

Site 2: An outcrop of granite at the previous field trip stop up the road.

Select the most likely site and explain your choice. What name is given to this type of sandstone?

10. This rock sample consists of intergrown crystals. How would you determine whether the rock is sedimentary or igneous? If it is sedimentary, what term describes its texture?

E. J. Tarbuck

8

Metamorphism and Metamorphic Rocks

Metamorphic rocks of Pemaquid Point, Bristol, Maine.
(Photo by George Oze Photography/SuperStock)

Each statement represents the primary **LEARNING OBJECTIVE** for the corresponding major heading within the chapter. After you complete the chapter, you should be able to:

8.1 Compare and contrast the environments that produce metamorphic, sedimentary, and igneous rocks.

8.2 List and distinguish among the four agents that drive metamorphism.

8.3 Explain how foliated and nonfoliated textures develop.

8.4 List and describe the most common metamorphic rocks.

8.5 Write a description for each of these metamorphic environments: contact metamorphism, hydrothermal metamorphism, subduction zone metamorphism, and regional metamorphism.

8.6 Explain how index minerals are used to establish the metamorphic grade of a rock body.

8.7 Describe the temperature and pressure conditions associated with the following metamorphic facies: blueschist facies, hornfels facies, and zeolite facies.

The folded and metamorphosed rocks shown in the chapter opening photo were once flat-lying sedimentary strata. Compressional forces of unimaginable magnitude, combined with temperatures hundreds of degrees above surface conditions, prevailed for perhaps thousands or millions of years to produce the deformation displayed by these rocks. Under such extreme conditions, solid rock responds by folding, fracturing, and often flowing. This chapter looks at the tectonic forces that forge metamorphic rocks and how these rocks change in appearance, mineralogy, and sometimes overall chemical composition.

8.1 | What Is Metamorphism?

Compare and contrast the environments that produce metamorphic, sedimentary, and igneous rocks.

Unlike some igneous and sedimentary processes that occur in surface or near-surface environments, metamorphism most often occurs deep within Earth, beyond our direct observation. Notwithstanding this significant obstacle, geologists have developed techniques that allow them to learn about the conditions under which metamorphic rocks form. In turn, the study of metamorphic rocks provides important insights into how tectonic processes operate to alter the structure and composition of Earth's crust.

Figure 8.1
Metamorphic grade A. Low-grade metamorphism illustrated by the transformation of the common sedimentary rock shale to the more compact metamorphic rock slate. **B.** High-grade metamorphic environments obliterate the existing texture and often change the mineralogy of the parent rock. High-grade metamorphism occurs at temperatures that approach those at which rocks melt. (Photos by Dennis Tasa)

A. Parent rock (Shale)
Loosely packed clay minerals

Low-grade metamorphism
Low temperatures and pressures

Metamorphic rock (Slate)
Tightly packed chlorite and mica minerals

B. Parent rock (Granodiorite)
Randomly oriented minerals

High-grade metamorphism
Strong compressional forces, high temperatures and pressures

Metamorphic rock (Folded gneiss)
Deformed layers of segregated minerals

Recall from the discussion of the rock cycle in Chapter 1 that metamorphism is the transformation of one rock type into another rock type. Metamorphic rocks are produced from preexisting sedimentary and igneous rocks, as well as from other metamorphic rocks. Thus, every metamorphic rock has a **parent rock**—the rock from which it was formed.

Metamorphism, which means to "change form," is a process that transforms the mineralogy, texture, and sometimes the chemical composition of the parent rock. The **mineralogy** (mineral constituents of a rock) changes because the rock is subjected to new conditions, usually elevated temperatures and pressures, which are significantly different from those in which it initially formed. For example, clay minerals, which are the most common minerals in sedimentary rocks, are stable only at Earth's surface. (Kaolinite is one example of a clay mineral; refer to Figure 3.37, page 97.) When clay minerals are buried to a depth where temperatures exceed 200°C (nearly 400°F), they are transformed into the minerals chlorite and/or muscovite mica. (Chlorite is a mica-like mineral formed by the metamorphism of dark iron- and magnesium-rich silicate minerals.) Under more extreme conditions, chlorite becomes biotite mica. Metamorphism also alters a rock's texture, producing larger crystals and sometimes a distinct layered or banded appearance.

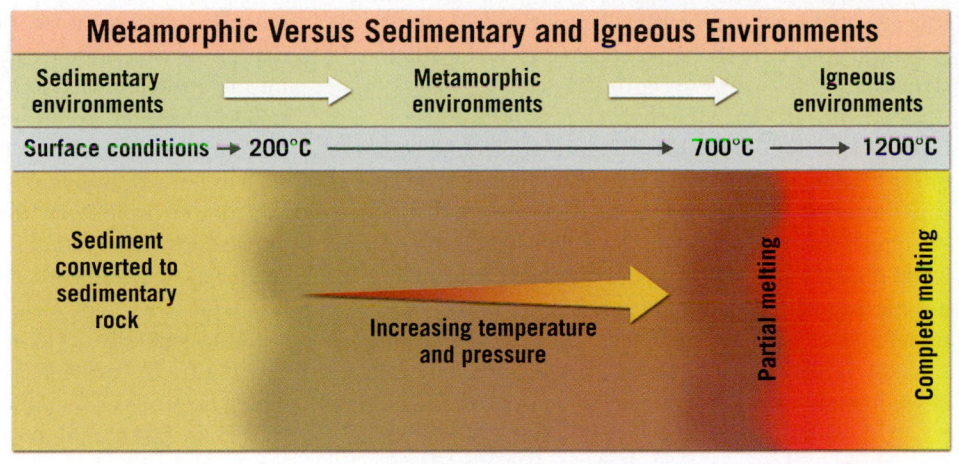

Metamorphic Versus Sedimentary and Igneous Environments

Sedimentary environments → Metamorphic environments → Igneous environments

Surface conditions → 200°C ——————————— 700°C —— 1200°C

Sediment converted to sedimentary rock

Increasing temperature and pressure

Partial melting

Complete melting

Figure 8.2
Metamorphic environments Metamorphism occurs over a range of temperatures that lie between those experienced in sedimentary environments and temperatures that approach those at which rocks melt. Pressure, which includes confining pressure and differential stress, also plays a major role in metamorphism.

The degree to which a parent rock changes during metamorphism is called its **metamorphic grade**, and it varies from low grade (low temperatures and pressures) to high grade (high temperatures and pressures). For example, in low-grade metamorphic environments, the common sedimentary rock shale becomes the more compact metamorphic rock slate. During this transformation, the clay minerals in shale are transformed into tiny chlorite and muscovite mica flakes. Hand samples of these rocks are sometimes difficult to distinguish from one another, illustrating that the transition from sedimentary to metamorphic rock is often gradual and the change subtle (**Figure 8.1A**).

In environments where temperatures and pressures are more extreme, metamorphism causes a transformation so complete that the identity of the parent rock cannot be easily determined. In high-grade metamorphism, such features as bedding planes, fossils, and vesicles that existed in the parent rock are obliterated. Further, when rocks deep in the crust are subjected to compressional stress (like being placed in a giant vise), the entire mass may be deformed, usually by folding (**Figure 8.1B**).

Figure 8.2 illustrates the relationships among metamorphic, sedimentary, and igneous environments.

Metamorphism occurs over a range of temperatures that lie between those experienced during the formation of sedimentary rocks (up to about 200°C [400°F]) and temperatures approaching those at which rocks begin to melt (about 700°C [1300°F]). However, *during metamorphism, the rock remains essentially solid.* If complete melting occurs, the rock has entered the realm of igneous activity, as discussed in Chapter 4.

Pressure also plays an important role in metamorphism. Although confining pressure acts to compact sediment to form sedimentary rocks, the pressures involved in metamorphism are even greater—sufficient to convert mineral matter into denser forms having more compact crystalline structures. Thus, metamorphism involves the formation of new minerals from preexisting ones.

8.1　Concept Checks

1. *Metamorphism* means to "change form." Describe how a rock may change during metamorphism.

2. What is meant by the statement "Every metamorphic rock has a *parent rock*"?

3. Define *metamorphic grade.*

8.2 | What Drives Metamorphism?

List and distinguish among the four agents that drive metamorphism.

The agents of metamorphism include *heat, pressure, directional stress,* and *chemically active fluids.* During metamorphism, rocks may be subjected to all four metamorphic agents simultaneously. However, the degree of metamorphism and the contribution of each agent vary greatly from one environment to another.

Heat as a Metamorphic Agent

The most important factor driving metamorphism is *heat* because it provides the energy needed to produce the chemical reactions that result in the recrystallization of existing minerals. Recall from the discussion of igneous rocks in Chapter 4 that an increase in temperature causes the atoms within a mineral to vibrate more rapidly. Even in a crystalline solid, where atoms are strongly bonded, this elevated level of activity allows individual atoms to migrate more freely between sites in the crystalline structure.

Changes Caused by Heat The formation of new mineral grains that tend to be larger than the original grains is called **recrystallization**. During this process, the mineralogy of the rock may or may not change. For example, when quartz sandstone is metamorphosed to form quartzite, the mineralogy does not change; the quartz grains remain quartz. By contrast, when shale is metamorphosed to slate, the clay minerals are recrystallized and become new minerals—usually chlorite and muscovite.

Although the mineralogy changes in the transition from shale to slate, the overall chemical composition remains essentially unchanged. Instead, the existing atoms are rearranged into new crystalline structures that are more stable in the new environment. (In some environments, ions may actually migrate into or out of a rock, thereby changing its overall chemical composition.)

What Is the Source of Heat? There are two primary sources of heat within Earth. One is the increasing temperature that occurs as we travel deeper into Earth's interior. The second is heat being released to the surrounding rocks as a magma body cools.

Earth's interior is extremely hot, mainly because of heat released from the repeated collision of asteroid-size bodies during the formation of our planet as well as from energy being continually released by the decay of radioactive elements. The rate of increase in temperature with depth is known as the **geothermal gradient**. In the upper crust, this increase in temperature averages about 25°C (77°F) per kilometer (**Figure 8.3**). Thus, rocks that formed at Earth's surface will experience a gradual

increase in temperature if they are transported to greater depths. As described earlier, clay minerals tend to become unstable when buried to a depth of about 8 kilometers (5 miles), where temperatures are about 200°C (400°F). The clay minerals begin to recrystallize into new minerals, such as chlorite and muscovite, both of which are stable in this new environment. However, many silicate minerals, particularly those found in crystalline igneous rocks—such as quartz and feldspar—remain stable at these temperatures. Thus, metamorphic changes in quartz and feldspar generally occur at higher temperatures.

Figure 8.3 provides several examples of conditions in which heat drives metamorphism. Environments where rocks may be carried to great depths and heated include convergent plate boundaries where slabs of sediment-laden oceanic crust are being subducted. Rocks may also become deeply buried in large basins where gradual subsidence results in thick accumulations of sediment. These basins, exemplified by the Gulf of Mexico, are known to develop low-grade metamorphic conditions near the base of the pile. In addition, continental collisions, which result in mountain building, cause some rocks to be uplifted while others are thrust downward, where elevated temperatures and pressures trigger metamorphism.

Heat may also be transported from the mantle into the shallowest layers of the crust. Rising mantle plumes, upwelling at mid-ocean ridges, and magma generated by partial melting of mantle rock at subduction zones are three such examples (see Figure 8.3). When magma intrudes rocks at shallow depths, the magma cools and releases heat, which "bakes" the surrounding host rock.

Confining Pressure

Pressure, like temperature, increases with depth because the thickness of the overlying rock increases. Buried rocks are subjected to **confining pressure**, which is analogous to water pressure, in which the forces are applied equally in all directions (**Figure 8.4A**). For example, the deeper scuba divers go, the greater the confining pressure.

Confining pressure causes the spaces between mineral grains to close, producing more compact rocks that have greater densities. If the pressure becomes extreme enough, it can cause the atoms in a mineral to pack more closely together to produce a new, denser mineral. Recall from

SmartFigure 8.3
Sources of heat for metamorphism The two main sources of heat for metamorphism are the increasing temperature that occurs as we travel deeper into Earth's interior and heat released to the surrounding rocks when a magma body cools. (https://goo.gl/JzqfLx)

Tutorial

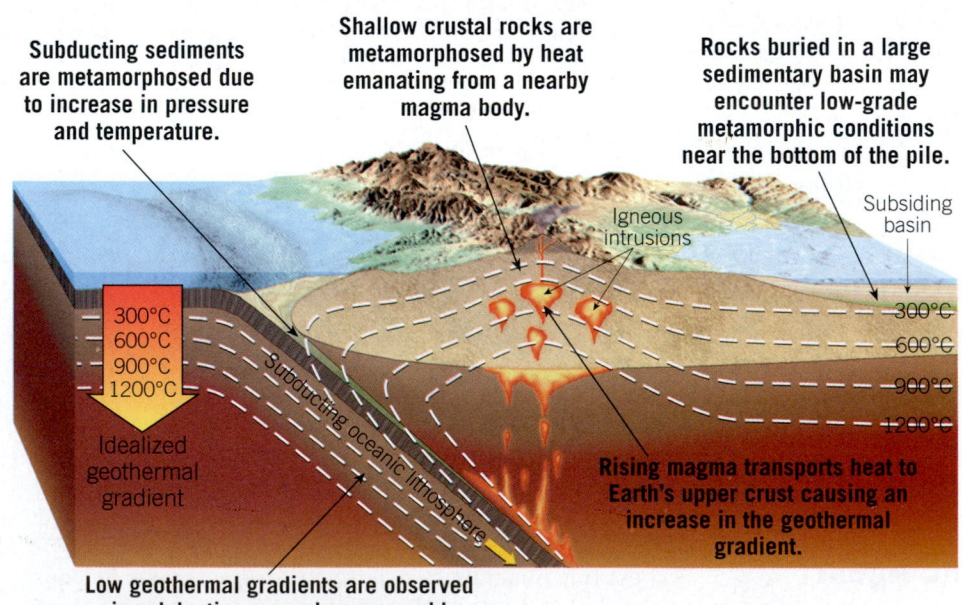

Subducting sediments are metamorphosed due to increase in pressure and temperature.

Shallow crustal rocks are metamorphosed by heat emanating from a nearby magma body.

Rocks buried in a large sedimentary basin may encounter low-grade metamorphic conditions near the bottom of the pile.

Igneous intrusions

Subsiding basin

300°C
600°C
900°C
1200°C
Idealized geothermal gradient

Subducting oceanic lithosphere

300°C
600°C
900°C
1200°C

Rising magma transports heat to Earth's upper crust causing an increase in the geothermal gradient.

Low geothermal gradients are observed in subduction zones because cold oceanic crust and overlying sediments are descending into the mantle.

Chapter 3 that the transformation of one mineral (polymorph) to another is called a *phase change*. Confining pressure does *not*, however, fold or fracture rocks.

Differential Stress

In addition to confining pressure, rocks may be subjected to directed pressure. This occurs, for example, at convergent plate boundaries where slabs of lithosphere collide. Here the forces that deform rock are unequal in different directions and are referred to as **differential stress**. (A discussion of various types of *differential stress* is provided in Chapter 10.) Unlike confining pressure, which "squeezes" rock equally in all directions, differential stresses are greater in one direction than in others.

Differential stress that squeezes a rock mass as if it were placed in a vise is termed **compressional stress**. As shown in **Figure 8.4B**, rocks subjected to compressional stress are shortened in the direction of greatest stress and elongated, or lengthened, in the direction perpendicular to that stress. Along convergent plate boundaries, the greatest differential stress is directed horizontally in the direction of plate motion. Consequently, in these settings, the crust is greatly shortened (horizontally) and thickened (vertically), resulting in mountainous topography.

In high-temperature, high-pressure environments, rocks are *ductile*, which allows their mineral grains to flatten (like what happens when you step on a tennis ball) when subjected to differential stress. The *metaconglomerate*, also called a stretch pebble conglomerate, shown in **Figure 8.5** illustrates this tendency. The parent rock, a conglomerate, consisted of nearly spherical pebbles that have been flattened into elongated structures by differential stress. On a larger scale, rocks that are ductile deform by flowing rather than breaking or fracturing. As a result, deeply buried rocks will develop intricate folds when deformed by differential stress (**Figure 8.6**).

By contrast, in near-surface environments where temperatures and pressure are comparatively low, rocks are *brittle* and tend to fracture when subjected to differential stress. Continued deformation grinds and pulverizes the mineral grains into smaller and smaller fragments.

Chemically Active Fluids

Water is abundant in Earth's crust. In the upper crust, water occurs as groundwater. At depth, hot water is released into the surrounding rocks when a magma body cools and solidifies. In addition, many minerals, including clays, micas, and amphiboles, are hydrated—which means they contain water in their crystalline structures. Elevated temperatures and pressures cause these minerals to dehydrate, expelling hot, mineral-laden water.

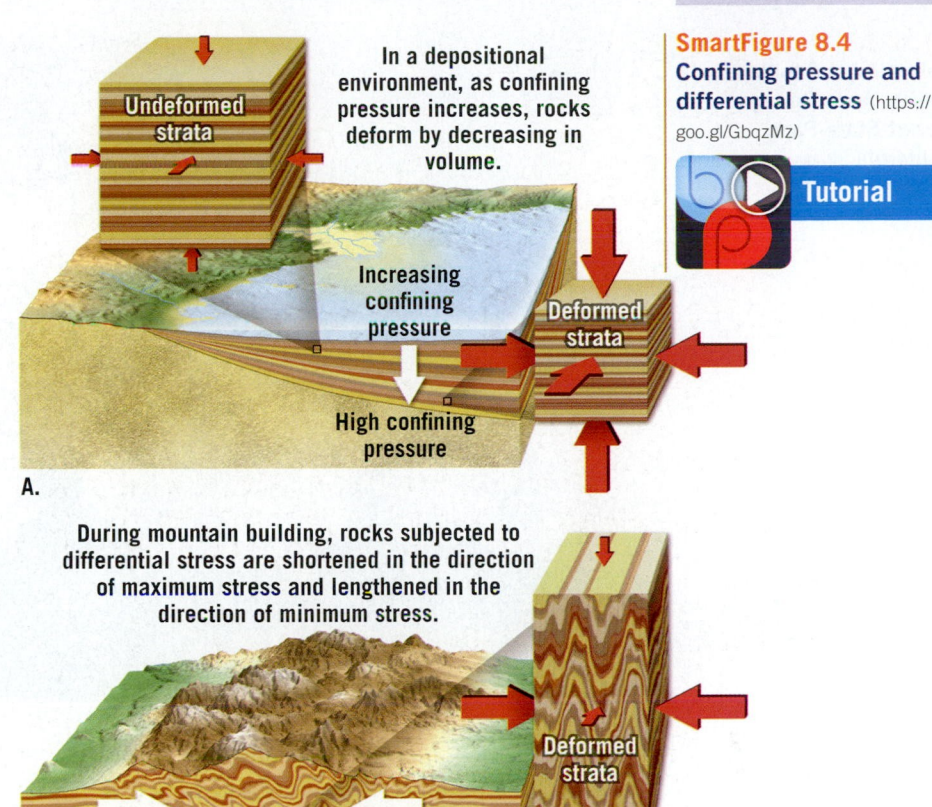

SmartFigure 8.4
Confining pressure and differential stress (https://goo.gl/GbqzMz)

Tutorial

A. In a depositional environment, as confining pressure increases, rocks deform by decreasing in volume.

Undeformed strata

Increasing confining pressure

High confining pressure

Deformed strata

B. During mountain building, rocks subjected to differential stress are shortened in the direction of maximum stress and lengthened in the direction of minimum stress.

Deformed strata

Hot, chemically active fluids enhance metamorphism by dissolving and transporting ions from one site in the crystal structure to another, thereby facilitating the process of recrystallization. In increasingly hotter environments, these fluids become correspondingly more reactive.

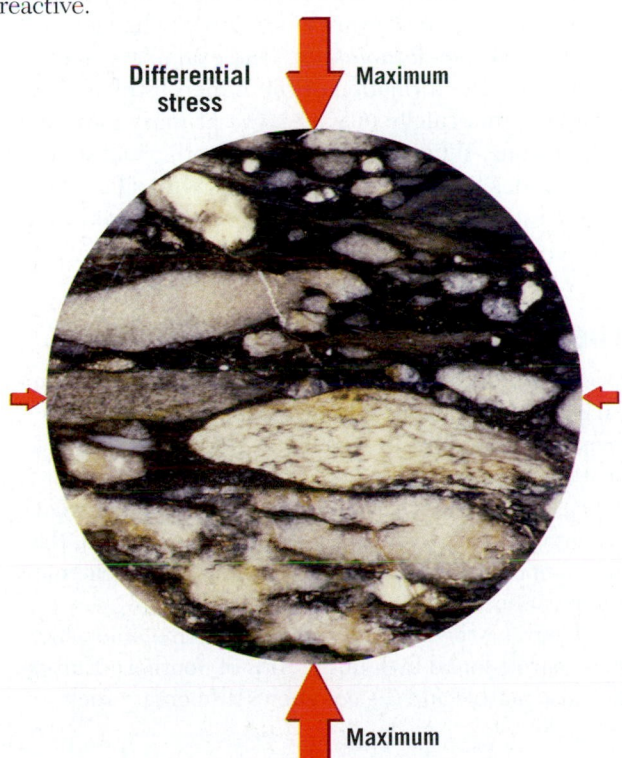

Differential stress

Maximum

Maximum

Figure 8.5
Metaconglomerate, also called stretched pebble conglomerate This metaconglomerate is made of once nearly spherical pebbles that have been heated and flattened into elongated structures by differential stress. (Photo by E. J. Tarbuck)

Figure 8.6
Deformed and folded gneiss, Anza Borrego Desert State Park, California (Photo by A. P. Trujillo/APT Photos)

In some metamorphic environments, hot fluids transport mineral matter over considerable distances. This occurs, for example, when hot, ion-rich fluids, called a *hydrothermal solution*, are expelled from a magma body as it cools and solidifies. If the rocks that surround the pluton differ markedly in chemical composition from the invading fluids, there may be an exchange of ions between these fluids and host rocks. In other words, these fluids bring in new atoms or take out atoms rather than simply reorganize what is already present. When this occurs, the overall chemical composition of the surrounding rock changes, in a process called *metasomatism*. One example of metasomatism is the formation of the mineral wollastonite $(CaSiO_3)$ from calcite $(CaCO_3)$, the primary ingredient in limestone. When a silica-rich hydrothermal solution invades limestone, calcite reacts with silica (SiO_2) to generate wollastonite, and the gas carbon dioxide (CO_2) is driven off.

The Importance of Parent Rock

Most metamorphic rocks have the same overall chemical composition as the parent rocks from which they formed, except for the possible loss or acquisition of volatiles such as water (H_2O) and carbon dioxide (CO_2). Therefore, when we try to establish from what parent material metamorphic rocks were derived, the most important clue comes from their overall chemical composition.

Consider the large exposures of the metamorphic rock marble found high in the Alps of Southern Europe. Because marble and the common sedimentary rock limestone have the same mineralogy (calcite), it seems reasonable to conclude that limestone is the parent rock of marble. Furthermore, because limestone usually forms in warm, shallow marine environments, we can surmise that considerable deformation must have occurred to convert limy deposits in a shallow sea into marble crags in the lofty Alps.

The mineral makeup of the parent rock also largely determines the degree to which each metamorphic agent will cause change. For example, when magma forces its way into an existing body of rock, high temperatures and hot fluids may alter the host rock. If the host rock is composed of minerals that are comparatively nonreactive, such as quartz, any alterations that may occur will be confined to a narrow zone next to the igneous intrusion. However, when the host rock is limestone, which is highly reactive, the zone of metamorphism may extend far from the intrusion.

8.2   Concept Checks

1. List four agents that drive metamorphism.

2. Why is heat considered the most important agent of metamorphism?

3. How is confining pressure different from differential stress?

4. What role do chemically active fluids play in metamorphism?

5. What characteristic of a metamorphic rock is determined primarily by its parent rock?

8.3 | Metamorphic Textures

Explain how foliated and nonfoliated textures develop.

The term **texture** is used to describe the size, shape, and arrangement of the mineral grains within a rock. Recall that texture is one way in which igneous and sedimentary rocks are classified. Most igneous and many sedimentary rocks consist of mineral grains or crystals that have a random orientation and thus appear uniform when viewed from any direction. By contrast, metamorphic rocks that contain platy minerals (such as micas) and/or elongated minerals (such as amphiboles) typically display some kind of *preferred orientation* in which the mineral grains exhibit a parallel to subparallel alignment. Like a fistful of pencils, rocks containing elongated mineral grains that are oriented parallel to each other appear different when viewed from the side than when viewed head-on. A rock that exhibits a preferred orientation of its mineral constituents is said to possess *foliation*.

Foliation

The term **foliation** refers to any planar (nearly flat) arrangement of mineral grains or crystals within a rock. Although foliation may occur in some sedimentary and even a few types of igneous rocks, it is a fundamental characteristic of metamorphosed rocks that have been strongly deformed, mainly by folding. In metamorphic environments, foliation is ultimately driven by compressional stress that shortens rock units, causing mineral grains in preexisting rocks to develop parallel, or nearly parallel, alignments. Examples of foliation include the parallel alignment of platy minerals through *rotation*, *recrystallization*, and *flattening of mineral grains or pebbles*. Foliated textures include *rock cleavage* in which rocks can be easily split into tabular slabs, and *compositional banding* in which the separation of dark and light minerals generates a layered appearance. These diverse types of foliation can form in many different ways.

Rotation of Platy Mineral Grains The rotation of existing mineral grains is the easiest of the foliation mechanisms to envision. **Figure 8.7** illustrates the mechanics by which platy or elongated mineral grains are rotated. Note that the new alignment is roughly perpendicular to the direction of maximum stress. Although physical rotation of platy minerals contributes to the development of foliation in low-grade metamorphism, other mechanisms dominate in more extreme environments.

Recrystallization That Produces New Minerals Recall that *recrystallization* is the creation of new mineral grains from preexisting ones. When recrystallization occurs as rock is being subjected to differential stress, any elongated minerals (such as amphiboles) and platy minerals (such as micas) that form tend to recrystallize perpendicular to the direction of maximum stress. Thus,

the newly formed mineral grains exhibit a distinct layering, and the metamorphic rocks containing them exhibit foliation.

Flattening Spherically Shaped Grains Mechanisms that flatten existing mineral grains are important in the metamorphism of rocks that contain minerals such as quartz, calcite, and olivine. These minerals normally develop roughly spherical crystals and have a rather simple chemical composition.

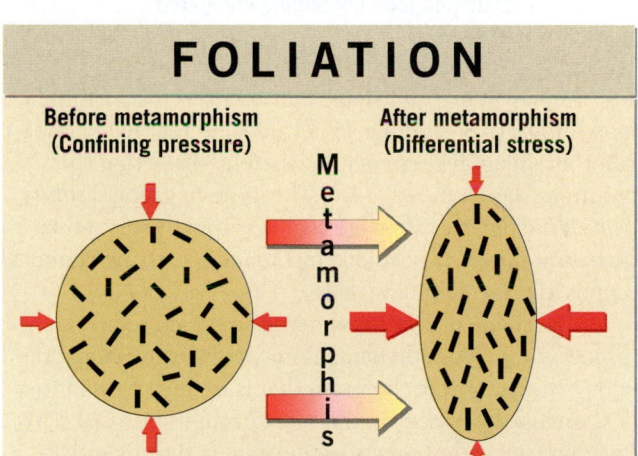

FOLIATION

Before metamorphism (Confining pressure)

After metamorphism (Differential stress)

Metamorphism

SmartFigure 8.7
Mechanical rotation of platy mineral grains to produce foliation
(https://goo.gl/vaWv6l)

Animation

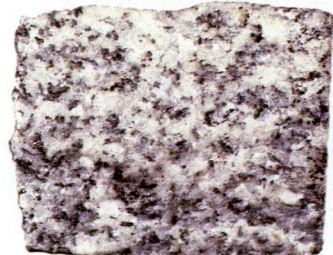

Platy and elongated mineral grains having random orientation.

Mineral grains that are aligned roughly perpendicular to the direction of maximum differential stress.

Figure 8.8

Solid-state flow of mineral grains Mineral grains can be flattened by solid-state flow when units of a mineral's crystalline structure slide relative to each other. This mechanism involves breaking existing chemical bonds and forming new ones.

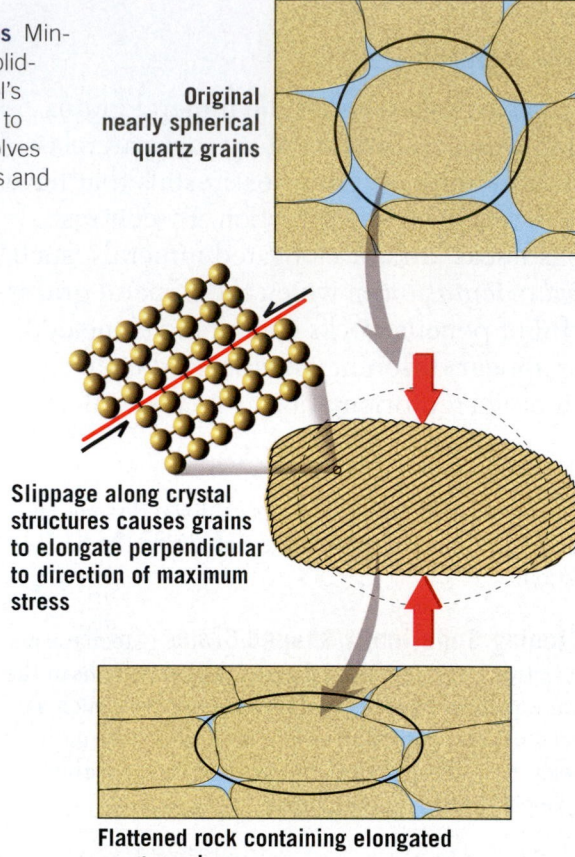

Original nearly spherical quartz grains

Slippage along crystal structures causes grains to elongate perpendicular to direction of maximum stress

Flattened rock containing elongated quartz grains

A change in grain shape can occur as distinct units of a mineral's crystalline structure slide relative to one another along discrete planes, thereby distorting the grain, as shown in **Figure 8.8**. This type of gradual *solid-state flow* involves slippage that disrupts the crystal lattice as atoms shift positions by breaking existing chemical bonds and forming new ones.

The shape of a mineral may also be altered by a process in which individual atoms move from a location along the margin of the grain that is highly stressed to a less-stressed position on the same grain (**Figure 8.9**). This mechanism, called *pressure solution*, is significantly aided by hot, ion-rich water. Mineral matter (ions) dissolves where grains are in contact with each other (areas of high stress) and is deposited in pore spaces (areas of low stress). As a result, the mineral grains tend to become shortened in the direction of maximum stress and elongated in the direction of minimum stress. While both of these mechanisms flatten mineral grains, the mineralogy remains the same.

Foliated Textures

Various types of foliation exist, depending largely upon the grade of metamorphism and the mineralogy of the parent rock. We will look at three: *rock, or slaty, cleavage*; *schistosity*; and *gneissic texture*, or *banding*.

Rock, or Slaty, Cleavage Rocks that split into thin slabs when hit with a hammer exhibit **rock cleavage**. Rock cleavage develops in various metamorphic rocks but is best displayed in slates that exhibit an excellent splitting property called **slaty cleavage** (**Figure 8.10**). Because it splits easily, slate is used for building materials such as roof and floor tiles as well as billiard table surfaces.

In low-grade metamorphic environments, slaty cleavage is known to develop where beds of shale (and related sedimentary rocks) are strongly folded and metamorphosed to form slate (**Figure 8.11**). The process begins when compressional stress begins to deform rock units, producing broad folds. With further deformation the clay minerals in shale, which initially aligned roughly parallel to the bedding surfaces, begin to recrystallize into tiny flakes of chlorite and mica. However, these new platy mineral grains grow so they are aligned roughly

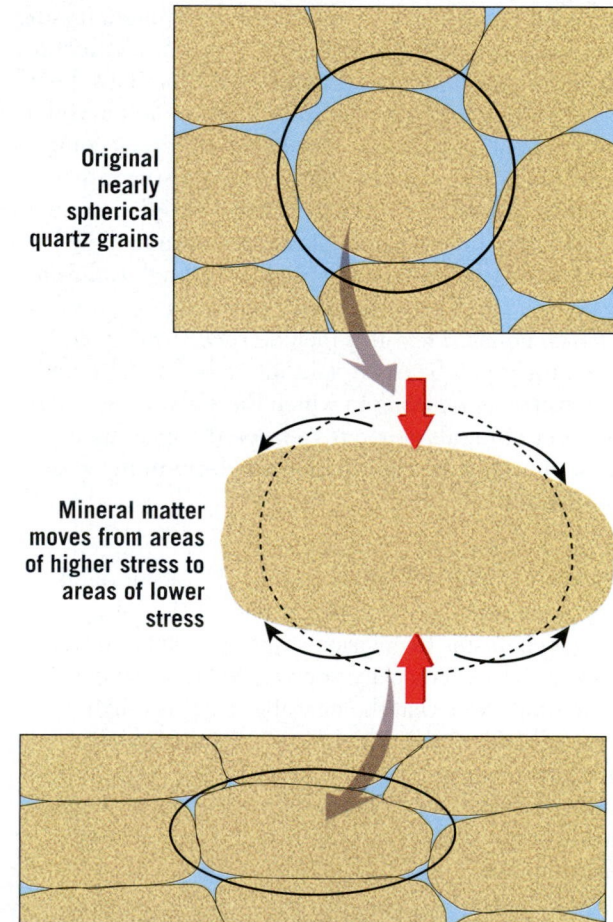

Original nearly spherical quartz grains

Mineral matter moves from areas of higher stress to areas of lower stress

Flattened rock containing elongated quartz grains

Figure 8.9

Pressure solution results in flattened mineral grains This mechanism flattens mineral grains by dissolving ions from areas of high stress and depositing ions at sites of low stress. This mechanism, as well as the one shown in Figure 8.8, changes the shape of mineral grains but not their volume and mineralogy.

perpendicular to the maximum directional stress, as shown in Figure 8.11B.

Because slate typically forms during the low-grade metamorphism of shale, evidence of the original sedimentary bedding surfaces is often preserved. However, as Figure 8.11C illustrates, the orientation of slate's cleavage usually develops at an angle to the sedimentary beds. Thus, unlike shale, which splits *along* bedding planes, slate often splits *across* bedding surfaces. Other metamorphic rocks, such as schists and gneisses, sometimes split along planar surfaces and exhibit rock cleavage.

Schistosity At higher temperatures and pressures, the minute mica and chlorite flakes in slate begin to recrystallize into larger muscovite and biotite crystals. When these platy minerals are large enough to be discernible with the unaided eye, they exhibit planar or layered structures called **schistosity**. Rocks that have this type of foliation are referred to as *schist*. In addition to containing platy minerals, schist often contains deformed quartz and feldspar crystals that appear flattened or lens-shaped and are embedded among the mica grains.

Gneissic Texture, or Banding During high-grade metamorphism, ion migration can result in the segregation of minerals, as shown in **Figure 8.12**. Notice that the dark biotite and amphibole crystals and light silicate minerals (quartz and feldspar) have separated, giving the rock a banded appearance called **gneissic texture**, or **gneissic banding**. Metamorphic rocks with this texture are called *gneiss* (pronounced "nice"). Although they are foliated, gneisses do not usually split as easily as slates and some schists.

Figure 8.10
Excellent slaty cleavage Slaty cleavage is exhibited by the rock in this slate quarry. Because slate breaks into flat slabs, it has many uses. (Photo by Fred Bruemmer/Photolibrary) The inset photo shows the use of slate for the roof of this house in Switzerland. (Photo by E. J. Tarbuck)

Other Metamorphic Textures

Metamorphic rocks that *do not* exhibit foliated textures are referred to as **nonfoliated**. Nonfoliated metamorphic rocks typically develop in environments where deformation is minimal and the parent rocks are composed of minerals

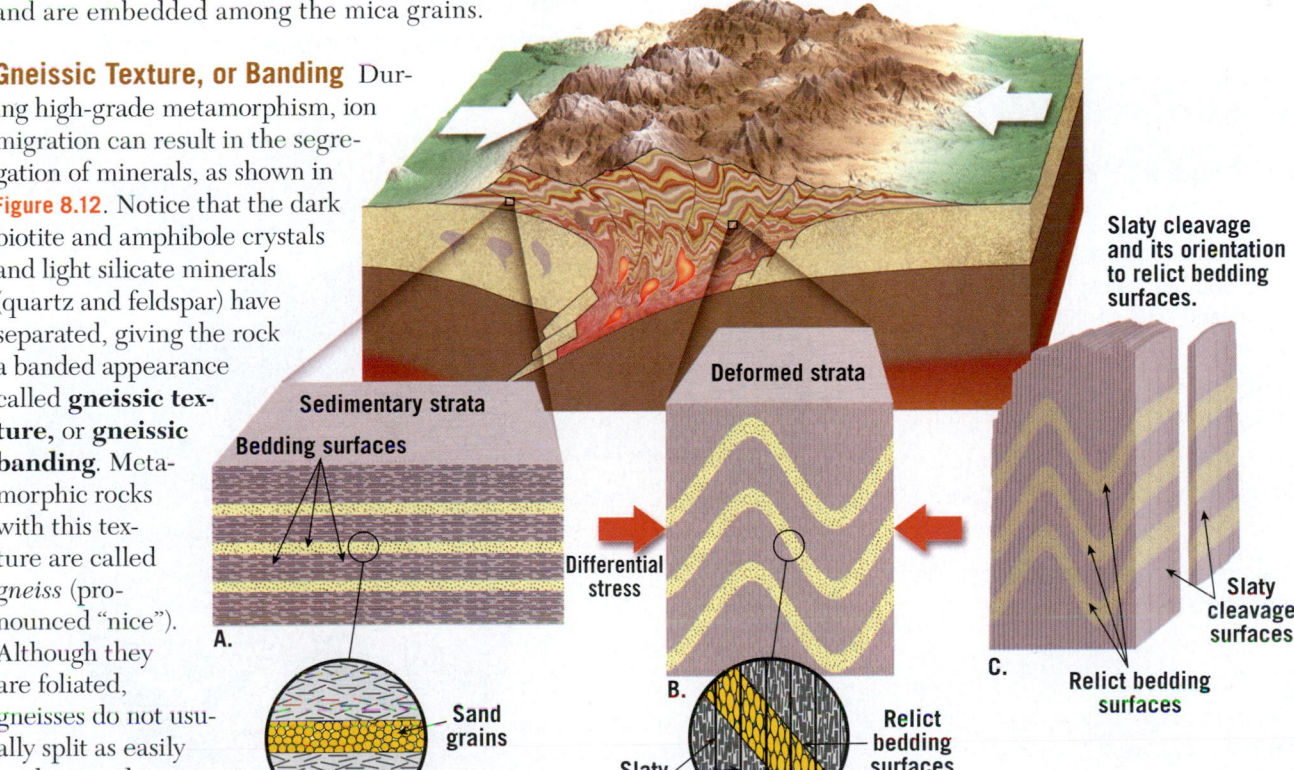

SmartFigure 8.11
Development of rock cleavage When shale that is interbedded with sandstone is strongly folded and metamorphosed, the clay minerals begin to recrystallize into tiny flakes of chlorite and mica. These new platy minerals grow so they are aligned roughly perpendicular to the directed stress, which gives slate its foliation. (https://goo.gl/V2tgtQ)

Figure 8.12
Development of gneissic banding Gneissic banding develops through the migration of ions that cause felsic and mafic minerals to grow in separate layers.

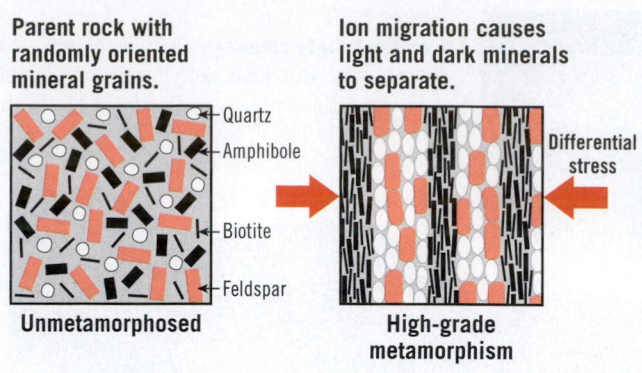

Parent rock with randomly oriented mineral grains.

- Quartz
- Amphibole
- Biotite
- Feldspar

Unmetamorphosed

Ion migration causes light and dark minerals to separate.

Differential stress

High-grade metamorphism

Dennis Tasa

Gneissic texture

that exhibit equidimensional crystals, such as quartz or calcite. For example, when a fine-grained limestone (made of calcite) is metamorphosed by the intrusion of a hot magma body, the small calcite grains recrystallize to form larger interlocking crystals. The resulting rock, *marble*, exhibits

large, intergrown calcite crystals, similar in appearance to those in coarse-grained igneous rocks.

Metamorphic rocks may also contain some unusually large grains, called *porphyroblasts*, that are surrounded by a fine-grained matrix of other minerals. **Porphyroblastic textures** develop in a wide range of rock types and metamorphic environments when minerals in the parent rock recrystallize to form new minerals. During recrystallization, certain metamorphic minerals, such as garnet, tend to develop *a small number of very large crystals*. By contrast, minerals such as muscovite and biotite typically form *a large number of smaller grains*. As a result, metamorphic rocks that contain large crystals (porphyroblasts) of, for example, garnet embedded in a finer-grained matrix of biotite and muscovite, are relatively common (**Figure 8.13**).

Figure 8.13
Garnet–mica schist The dark red garnet crystals (porphyroblasts) are embedded in a matrix of fine-grained micas. (Photo by E. J. Tarbuck)

Close up of porphyroblast

Porphyroblasts

8.3   Concept Checks

1. Define *foliation*.
2. Briefly describe three ways in which the mineral grains in a rock develop a preferred orientation (that is, foliation).
3. Distinguish among *slaty cleavage, schistosity,* and *gneissic* textures.
4. What is meant by *nonfoliated texture*? Name one rock that exhibits this texture.

EYE ON EARTH 8.1

This metamorphic rock outcrop (bedrock exposed at the surface) is found in the Southern Alps, located on the South Island of New Zealand. The continued growth of the Southern Alps is somewhat unique in that these mountains lie where the Pacific and Australian plates collide and simultaneously slide past one another along a large transform fault called the Alpine Fault.

QUESTION 1 Do the rocks in this outcrop display foliation? Explain.

QUESTION 2 Do these rocks appear to have experienced high-grade or low-grade metamorphism? Explain.

Shutterstock

E. J. Tarbuck

8.4 | Common Metamorphic Rocks

List and describe the most common metamorphic rocks.

Most metamorphic rocks that we observe at Earth's surface were derived from the three most common sedimentary rocks—shale, limestone, and quartz sandstone. Shale is the most likely parent of most slate, phyllite, schist, and gneiss. This sequence of metamorphic rocks reflects an increase in grain size, a change in rock texture, and a change in mineralogy.

Limestone, which is composed of the mineral calcite ($CaCO_3$), is the parent rock of marble, while quartz (SiO_2) sandstone is the parent of quartzite. Because calcite and quartz are simple chemical compounds compared to clay minerals, their mineralogy does not change during metamorphism; calcite usually remains calcite, and quartz remains quartz. Rather, these minerals recrystallize to produce larger fused grains that are the main constituents of marble and quartzite, respectively.

The major characteristics of the most common metamorphic rocks are summarized in **Figure 8.14**. Notice that metamorphic rocks can be broadly classified by the type of foliation exhibited and, to a lesser extent, the chemical composition of the parent rock. It is worth noting that certain rock *names* (slate, schist, and gneiss) are also used to describe rock *texture*.

Foliated Metamorphic Rocks

Slate A very fine-grained (less than 0.5-millimeter) foliated rock composed mainly of minute chlorite and mica flakes (too small to be visible to the human eye) is termed **slate**. Slate may also contain tiny quartz and feldspar crystals. Thus, slate generally appears dull and closely resembles shale. A noteworthy characteristic of slate is its excellent rock cleavage, or tendency to break into flat slabs (see Figure 8.10).

Figure 8.14
Classification of common metamorphic rocks (Photos by E. J. Tarbuck)

Metamorphic Rock		Texture	Comments	Parent Rock
Slate	Foliated		Composed of tiny chlorite and mica flakes, breaks in flat slabs called slaty cleavage, smooth dull surfaces	Shale, mudstone, or siltstone
Phyllite			Fine-grained, glossy sheen, breaks along wavy surfaces	Shale, mudstone, or siltstone
Schist			Medium- to coarse-grained, scaly foliation, micas dominate	Shale, mudstone, or siltstone
Gneiss			Coarse-grained, compositional banding due to segregation of light and dark colored minerals	Shale, granite, or volcanic rocks
Marble	Nonfoliated		Medium- to coarse-grained, relatively soft (3 on the Mohs scale), interlocking calcite or dolomite grains	Limestone, dolostone
Quartzite			Medium- to coarse-grained, very hard, massive, fused quartz grains	Quartz sandstone
Hornfels			Very fine-grained, often exceedingly tough and durable, usually dark colored	Often shale, but can have any composition

Slate is most often generated by the low-grade metamorphism of shale, mudstone, or siltstone. Less frequently it is produced when volcanic ash is metamorphosed. Slate's color depends on its mineral constituents:

Figure 8.15
Mica schist This sample of schist is composed mostly of muscovite and biotite and exhibits foliation. (Photo by E. J. Tarbuck)

Mica schist

Parallel alignment of mineral grains

Black (carbonaceous) slate contains organic material, red slate gets its color from iron oxide, and green slate usually contains a lot of the mineral chlorite.

Phyllite **Phyllite** represents a degree of metamorphism between slate and schist. Its constituent platy minerals are larger than those in slate but not large enough to be readily identifiable with the unaided eye. Although phyllite appears similar to slate, it can be easily distinguished from slate by its glossy sheen and wavy surface (see Figure 8.14). Phyllite exhibits rock cleavage and is composed mainly of very fine crystals of mainly muscovite, chlorite, or both.

Schist Medium- to coarse-grained metamorphic rocks in which platy minerals are dominant are called **schists**. These flat components commonly include muscovite and biotite that display parallel alignments that give the rock its foliated texture (**Figure 8.15**). In addition, schists contain smaller amounts of other minerals, often quartz and feldspar. Some schists are composed mostly of dark minerals (such as amphiboles). As with slate, the parent rock of most schists is shale that has undergone medium- to high-grade metamorphism during a major mountain-building episode.

Figure 8.16
Banded gneiss found in the Adirondacks, New York (Photo by Michael Collier)

As you learned in the previous section, the term *schist* describes the texture of rocks, and as such it is used to name rocks that have a wide variety of chemical compositions. To indicate composition, mineral names are added. For example, schists composed primarily of muscovite and biotite are called *mica schist* (see Figure 8.15). Mica schists often contain *accessory minerals*, some of which are unique to metamorphic rocks. Some common accessory minerals that occur as porphyroblasts include *garnet*, *staurolite*, and *andalusite*, in which case the rock is called *garnet-mica schist*, *staurolite-mica schist*, or *andalusite-mica schist* (see Figure 8.13).

In addition, schists may be composed largely of the minerals chlorite or talc, in which case they are called *chlorite schist* and *talc schist*, respectively. Both chlorite and talc schists can form when rocks having basaltic compositions undergo metamorphism.

Gneiss **Gneiss** is the term applied to medium- to coarse-grained banded metamorphic rocks in which granular and elongated (as opposed to platy) minerals predominate. The most common minerals in gneiss are quartz, potassium feldspar, and plagioclase feldspar. Most gneisses also contain lesser amounts of biotite, muscovite, and amphibole. Some gneisses will split along the layers of platy minerals, but most break in an irregular fashion.

Recall that during high-grade metamorphism, the light and dark components separate, giving gneisses their characteristic banded or layered appearance. Thus, most gneisses consist of alternating bands of white or reddish feldspar-rich zones and layers of dark ferromagnesian minerals (**Figure 8.16**). These banded gneisses often exhibit evidence of deformation, including folds and sometimes faults.

Gneisses having a felsic composition may be derived from granite or its fine-grained equivalent, rhyolite. However, most gneisses are generated through high-grade metamorphism of shale. Therefore, gneiss represents the highest-grade metamorphic rock in the sequence of shale, slate, phyllite, schist, and gneiss. Like schists, gneisses may also include large crystals of accessory minerals such as garnet. Gneisses made up primarily of dark minerals also occur. For example, an amphibole-rich rock that exhibits a gneissic texture is called *amphibolite*.

Nonfoliated Metamorphic Rocks

Marble The metamorphism of limestone or dolostone produces the crystalline metamorphic rock called **marble** (see Figure 8.14). Pure marble is white and composed essentially of the mineral calcite. Because of its relative softness (3 on the Mohs scale), marble is easy to cut and shape. White marble is particularly prized as a stone from which monuments and statues are carved, such as the Lincoln Memorial in Washington, DC, and the Taj Mahal in India (see GEOgraphics 8.1). Unfortunately, when marble is exposed to acid rain, its composition (calcium carbonate) makes it susceptible to chemical weathering.

The parent rocks of most marbles contain impurities that color the stone. Thus, marble can be pink, gray, green, or even black and may contain a variety of accessory minerals (such as chlorite, mica, garnet, and wollastonite). When marble forms from limestone interbedded with shales, it appears banded and exhibits visible foliation. When deformed, these banded marbles may develop highly contorted mica-rich folds that enhance the rocks' artistic appearance. These decorative marbles have been used as building stones since prehistoric times.

Quartzite Quartzite is a very hard metamorphic rock formed from quartz sandstone (**Figure 8.17**). Under moderate- to high-grade metamorphism, the quartz grains in sandstone fuse together (see the inset in Figure 8.17). Recrystallization is often so complete that, when broken, quartzite splits across the original quartz grains rather than along their boundaries. In some instances, sedimentary features such as cross-bedding are preserved and give the rock a banded appearance. Pure quartzite is white, but iron oxide may produce reddish or pinkish stains, while dark mineral grains may impart shades of green or gray.

Hornfels Hornfels is a fine-grained nonfoliated metamorphic rock, and unlike marble and quartzite, it has a variable mineral composition. The parent rock of most hornfels is shale or another clay-rich rock, which has been "baked" by a hot intruding magma body. Hornfels tends to be gray to black in color and quite hard, and it may display conchoidal fracture.

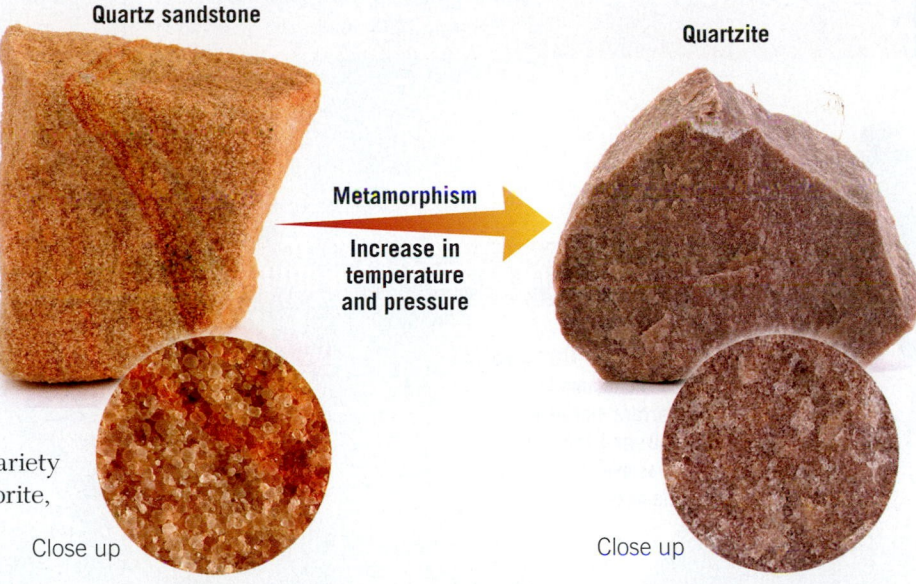

Quartz sandstone

Metamorphism

Increase in temperature and pressure

Quartzite

Close up

Close up

Figure 8.17
Quartzite Quartzite is a nonfoliated metamorphic rock formed from quartz sandstone. These close-up images show the interlocking quartz grains typical of quartzite as compared to the loosely bound grains in this quartz sandstone.
(Photos by Dennis Tasa)

8.4 Concept Checks

1. Although slate and phyllite resemble each other, how can they be differentiated?

2. In the rock mica schist, what does *mica* indicate, and what does *schist* indicate?

3. Briefly describe the appearance of the metamorphic rock gneiss and explain how it forms.

4. Describe slate, phyllite, schist, and gneiss in terms of texture and grain size.

5. Compare and contrast marble and quartzite.

8.5 Metamorphic Environments

Write a description for each of these metamorphic environments: contact metamorphism, hydrothermal metamorphism, subduction zone metamorphism, and regional metamorphism.

Metamorphism occurs in many different environments. Most of these environments occur in the vicinity of plate margins, and several are associated with igneous activity. We will consider the basic metamorphic environments—*contact* or *thermal metamorphism, hydrothermal metamorphism, burial* and *subduction zone metamorphism,* and *regional metamorphism*—as well as a few types of metamorphism that generate relatively small quantities of metamorphic rock.

Marble

Marble is crystalline metamorphic rock whose parent rock was limestone or dolostone. Pure marble is white; however, most marble contains impurities such as iron oxide, chlorite, and organic debris which renders the marble pink, green, gray or sometimes black.

Pure white marbles are prized for sculpture

Statue of David carved by Michelangelo from pure Carrara marble between 1501 and 1504. The statue rises over 17 feet above the base.

Venus de Milo is an ancient Greek statue carved out of marble sometime between 130 and 100 B.C. This statue is on permanent display at the Louvre in Paris, France.

Carrara Marble quarries located in the mountains near Carrara, Italy have produced some of the finest marbles for more than two thousand years.

These marble quarries located near Carrara, Italy are so large they can be seen from space.

Fotigrafiche/Shutterstock

NASA

Ray Roberts/Alamy Images

Circumnavigation/Fotolia

Marble, because of its workability, is a widely used building stone

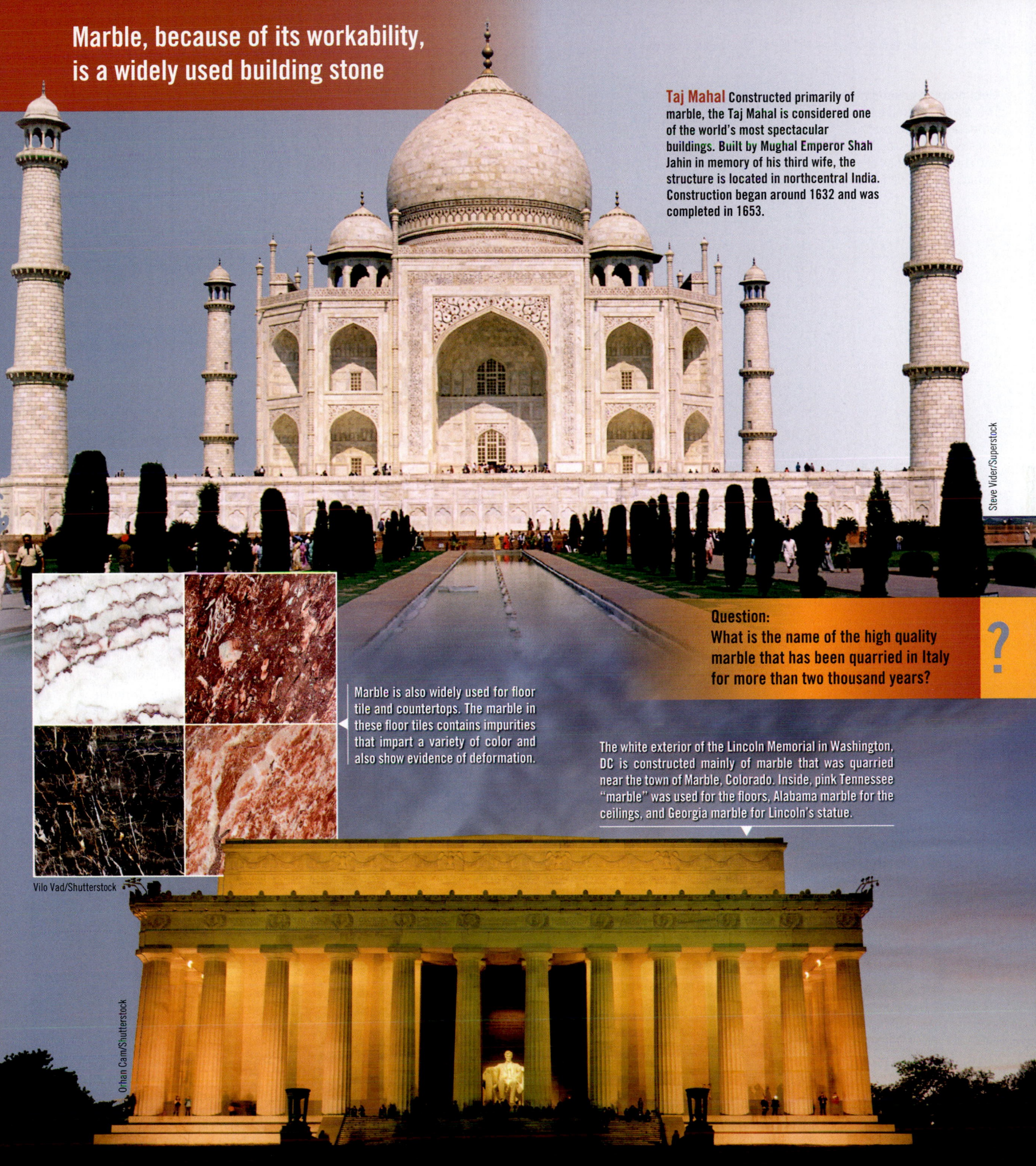

Taj Mahal Constructed primarily of marble, the Taj Mahal is considered one of the world's most spectacular buildings. Built by Mughal Emperor Shah Jahin in memory of his third wife, the structure is located in northcentral India. Construction began around 1632 and was completed in 1653.

Steve Vider/Superstock

Question:
What is the name of the high quality marble that has been quarried in Italy for more than two thousand years?

?

Marble is also widely used for floor tile and countertops. The marble in these floor tiles contains impurities that impart a variety of color and also show evidence of deformation.

The white exterior of the Lincoln Memorial in Washington, DC is constructed mainly of marble that was quarried near the town of Marble, Colorado. Inside, pink Tennessee "marble" was used for the floors, Alabama marble for the ceilings, and Georgia marble for Lincoln's statue.

Vilo Vad/Shutterstock

Orhan Cam/Shutterstock

Figure 8.18

Figure 8.18
Metamorphic environments
This graph illustrates the temperatures and pressures typically associated with the major types of metamorphic environments.

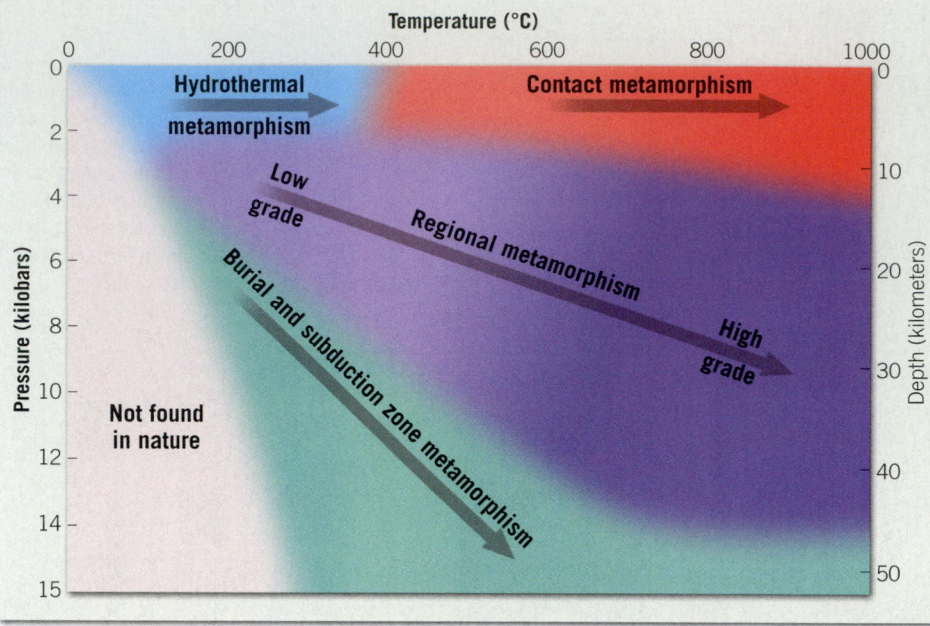

As **Figure 8.18** illustrates, each type of metamorphism occurs over a range of temperatures and pressures. On this graph, temperature, which increases from surface conditions to about 1000°C (1800°F), is displayed along the top, and pressure is shown along the vertical axis. Notice that hydrothermal metamorphism occurs in relatively low-temperature and low-pressure environments, whereas burial and subduction metamorphism occur under relatively low-temperature and high-pressure conditions. Although not shown on this graph, regional and subduction metamorphism involve directional stress, whereas hydrothermal and contact metamorphism do not.

Contact, or Thermal, Metamorphism

Contact metamorphism, or **thermal metamorphism**, occurs in Earth's upper crust (low pressure), when rocks immediately surrounding a molten igneous body are "baked" (high temperature). Because contact metamorphism does not involve directional stress, the resulting metamorphic rocks are not foliated.

Contact metamorphism alters rocks in a discrete zone adjacent to the heat source, called an **aureole** (**Figure 8.19**). Small intrusions such as dikes and sills (see Chapter 4) typically form aureoles only a few centimeters thick. By contrast, large molten bodies that eventually cool to form batholiths can produce aureoles that extend outward for several kilometers. These large aureoles often consist of distinct *zones of metamorphism*. Close

SmartFigure 8.19
Contact metamorphism Contact metamorphism produces a zone of alteration called an *aureole* around an intrusive igneous body. In the photo, the dark layer is a type of metamorphic aureole called a *roof pendant*, which consists of metamorphosed host rocks that are in contact with the upper part of the light-colored igneous pluton. The term *roof pendant* implies that the rock was once the roof of a magma chamber. The photo shows the Sierra Nevada, near Bishop, California. (Photo by Michael Collier) (https://goo.gl/MWBhsm)

Tutorial

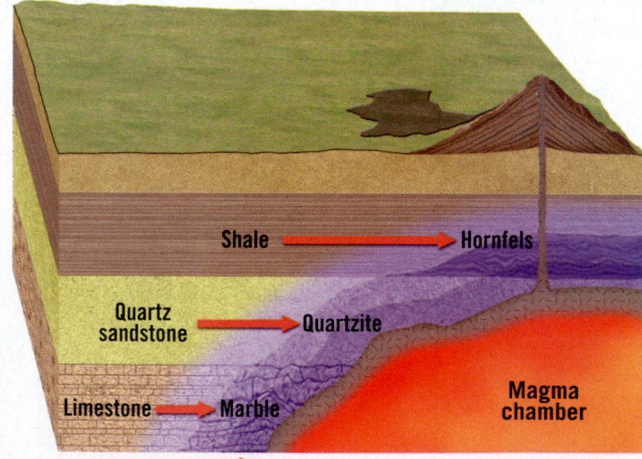

Figure 8.20
Rocks produced by contact metamorphism Contact metamorphism of shale yields hornfels, while contact metamorphism of quartz sandstone and limestone produces quartzite and marble, respectively.

to the magma body, high-temperature minerals such as garnet may form, whereas farther away, low-grade minerals such as chlorite are produced.

Depending mainly on the composition of the parent rock, a variety of metamorphic rocks can form in the same setting (**Figure 8.20**). For example, during contact metamorphism of mudstones and shales, the clay minerals are baked, much like clay is baked in a kiln to make pottery. The result is a very hard, fine-grained metamorphic rock called *hornfels* (see Figure 8.14). Hornfels can also form from a variety of other materials, including volcanic ash and basaltic rocks. Other metamorphic rocks that are produced by contact metamorphism are marble and quartzite (see Figure 8.20). Recall that limestone is the parent of marble and that the metamorphism of quartz sandstone produces quartzite.

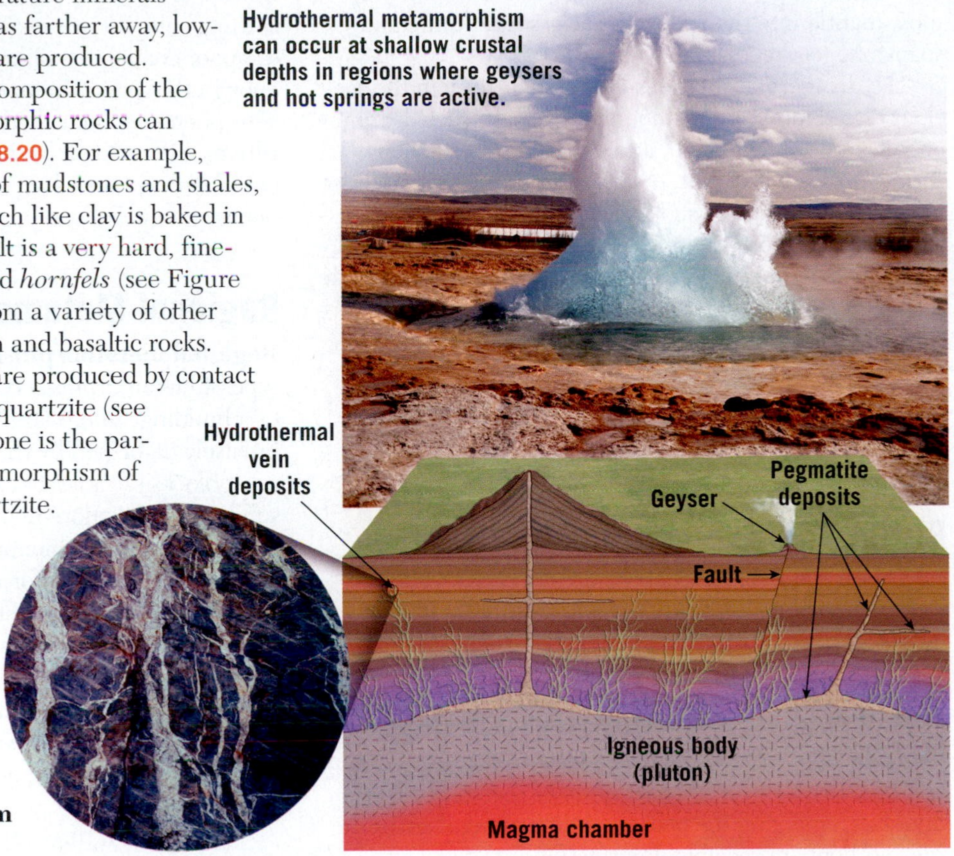

Hydrothermal Metamorphism

When hot, ion-rich water circulates through pore spaces or fractures in rock, a chemical alteration called **hydrothermal metamorphism** may occur (**Figure 8.21**). Recall that hot mineral-laden fluids called *hydrothermal solutions* contribute to metamorphism by enhancing the recrystallization of existing minerals. In addition, hot ion-rich fluids facilitate the movement of mineral matter into and out of rock bodies, thereby changing their overall chemical composition.

The water for hydrothermal metamorphism can be groundwater that has percolated down from the surface, where it is heated and circulates upward. This type of metamorphism tends to occur at low pressures (shallow depth) and relatively low to moderate temperatures.

Water that drives hydrothermal metamorphism may also arise from igneous activity. As large magma bodies cool and solidify, ion-rich water is driven into the surrounding host rocks. When the host rock is porous or highly fractured, mineral matter contained in these fluids may precipitate to form important deposits of copper, silver, and gold. These ion-rich fluids can also generate pegmatites—very coarse-grained granitic (felsic) igneous rocks.

As scientific understanding of plate tectonics developed, it became clear that the most widespread occurrence of hydrothermal metamorphism is along the axis of the mid-ocean ridge system (**Figure 8.22**). As plates move apart, upwelling magma from the mantle generates new seafloor. Seawater percolating through the young, hot oceanic crust is heated and chemically reacts with the newly formed basaltic rocks. The result is the

Hydrothermal metamorphism can occur at shallow crustal depths in regions where geysers and hot springs are active.

Hydrothermal vein deposits

Geyser

Pegmatite deposits

Fault →

Igneous body (pluton)

Magma chamber

Figure 8.21
Hydrothermal metamorphism associated with an intrusive igneous body Pegmatites and hydrothermal mineral deposits form adjacent to an igneous intrusion (pluton). (Photo by Pavel Svofoda/Fotolia)

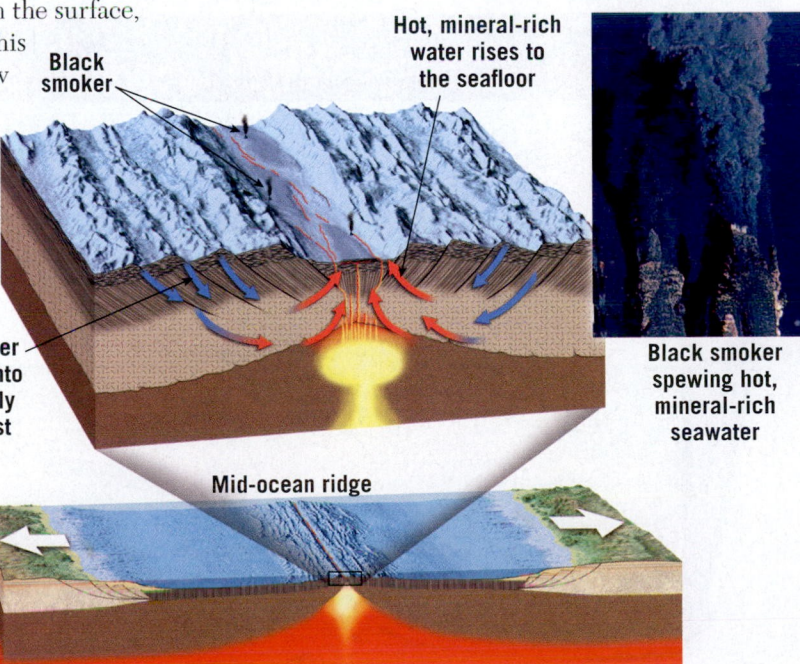

Hot, mineral-rich water rises to the seafloor

Black smoker

Cold seawater percolates into the hot newly formed crust

Black smoker spewing hot, mineral-rich seawater

Mid-ocean ridge

Figure 8.22
Hydrothermal metamorphism along a mid-ocean ridge (Photo by Fisheries and Oceans Canada/Uvic-Verena Tunnicliffe/Newscom)

Serpentinite

Soapstone

Figure 8.23
Serpentinite and soapstone These metamorphic rocks are produced by hydrothermal alteration of mafic rocks along the mid-ocean ridge system. (Photos by Dennis Tasa)

conversion of mafic rocks of the oceanic crust and uppermost mantle into the hydrated rocks serpentinite and soapstone (**Figure 8.23**).

Hydrothermal solutions circulating through the seafloor also remove large amounts of metals, such as iron, cobalt, nickel, silver, gold, and copper, from the newly formed crust. These hot, metal-rich fluids eventually rise along fractures and gush from the seafloor at temperatures of about 350°C (660°F), generating particle-filled clouds called *black smokers*. Upon mixing with the cold seawater, sulfides and carbonate minerals containing these heavy metals precipitate to form metallic deposits. Geologists credit this process with the formation of the copper ores mined today on the Mediterranean island of Cyprus.

Burial and Subduction Zone Metamorphism

Burial metamorphism tends to occur where massive amounts of sedimentary or volcanic material accumulate in a subsiding basin such as the Gulf of Mexico (see Figure 8.3). Here, low-grade metamorphic conditions may be reached within the deepest layers. Confining pressure and heat drive the recrystallization of the constituent minerals, changing the texture and/or mineralogy of the rock without appreciable deformation.

Rocks and sediments can also be carried to great depths along convergent boundaries where oceanic

lithosphere is being subducted (see Figure 8.3). In this setting, cold, dense oceanic crust and sediments, which are poor conductors of heat, are subducting rapidly enough that pressure increases faster than temperature. This phenomenon, called **subduction zone metamorphism**, differs from burial metamorphism in that differential stress plays a major role in deforming rock as it is being metamorphosed.

Regional Metamorphism

Regional metamorphism is a common, widespread type of metamorphism typically associated with mountain building, where large segments of Earth's crust are intensely deformed by the collision of two continental crust blocks (**Figure 8.24**). Whereas denser oceanic crust subducts under more buoyant continental crust, with the collision of continental crust blocks, two landmasses instead collide and deform. Sediments and crustal rocks that form the margins of the colliding continents are folded and faulted and, as a result, shorten and thicken like a rumpled carpet. Continental collisions may also cause crystalline basement rocks lying under sedimentary layers, as well as slices of oceanic crust that once floored the intervening ocean basin, to be uplifted and deformed.

The general thickening of the crust that occurs during mountain building results in buoyant lifting, in which deformed rocks are elevated high above sea level. Crustal thickening also results in the deep burial of large quantities of rock as one crustal block is thrust beneath another. Deep in the roots of mountains, elevated temperatures caused by deep burial are responsible for the most intense metamorphic activity within a mountain belt.

In some settings, deeply buried rocks become heated beyond their melting points, producing magma. When these magma bodies grow large enough to buoyantly rise, they intrude the overlying metamorphic and sedimentary rocks (see Figure 8.24). Consequently, the cores of many mountain belts consist of folded and faulted metamorphic rocks, often intertwined with igneous bodies. Over time, these deformed rock masses are uplifted, and erosion removes the overlying material to expose the igneous and metamorphic core of the mountain range.

Regional metamorphism produces some of the most common metamorphic rocks.

SmartFigure 8.24
Regional metamorphism Regional metamorphism is often associated with a continental collision where rocks are squeezed between two converging plates, resulting in mountain building. (https://goo.gl/kkykbZ)

 **Animation**

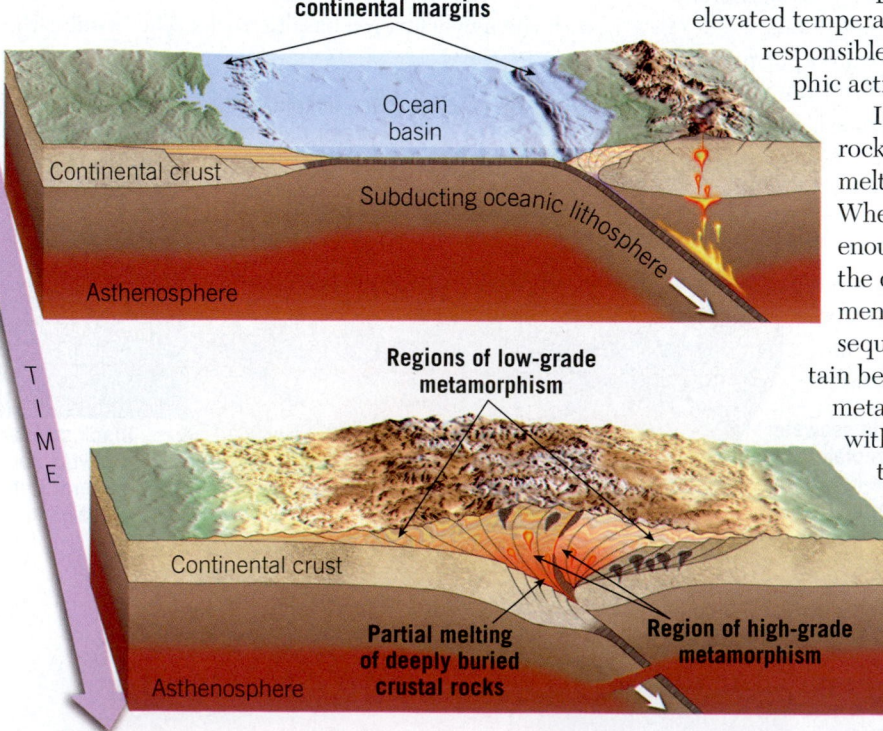

Sediments deposited on continental margins

Ocean basin

Continental crust

Subducting oceanic lithosphere

Asthenosphere

TIME

Regions of low-grade metamorphism

Continental crust

Partial melting of deeply buried crustal rocks

Region of high-grade metamorphism

Asthenosphere

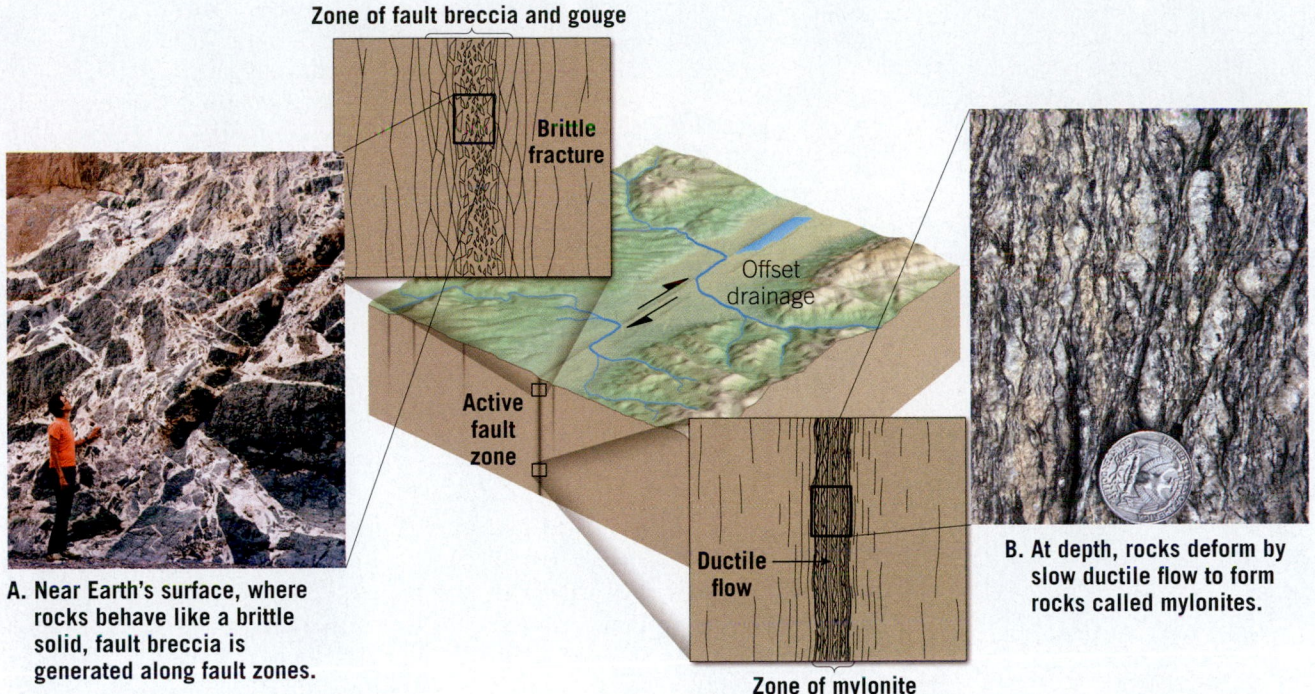

Zone of fault breccia and gouge

Brittle fracture

Offset drainage

Active fault zone

Ductile flow

Zone of mylonite

A. Near Earth's surface, where rocks behave like a brittle solid, fault breccia is generated along fault zones.

B. At depth, rocks deform by slow ductile flow to form rocks called mylonites.

SmartFigure 8.25
Metamorphism along a fault zone (Photo A by A. P. Trujillo; photo B by Ann Bykerk-Kauffman) (https://goo.gl/sV1uW3)

Tutorial

In this process, shale is metamorphosed to produce the sequences of slate, phyllite, schist, and gneiss. In addition, quartz sandstone and limestone are metamorphosed into quartzite and marble.

Other Metamorphic Environments

Other, less-common types of metamorphism generate relatively small amounts of metamorphic rock that tends to be geographically localized.

Metamorphism Along Fault Zones Near Earth's surface, rock behaves like a brittle solid. Consequently, movement along a fault zone fractures and pulverizes rock (**Figure 8.25A**). The result is a loosely coherent rock called *fault breccia* composed of broken and crushed rock fragments. Displacements along California's San Andreas Fault have created a zone of fault breccia and related rock types more than 1000 kilometers (600 miles) long and up to 3 kilometers (nearly 2 miles) wide.

Much of the deformation associated with fault zones occurs at great depth and thus at high temperatures. In this environment, preexisting minerals deform by ductile flow (**Figure 8.25B**). As large slabs of rock move in opposite directions, the minerals in the fault zone between them tend to form elongated grains that give the rock a foliated and/or lineated appearance. Rocks formed in these zones of intense ductile deformation are termed *mylonites* (*mylo* = a mill, *ite* = a stone).

Impact Metamorphism **Impact**, or **shock**, **metamorphism** occurs when high-speed projectiles called

meteoroid (fragments of comets or asteroids) strike Earth's surface (see GEOgraphics 8.2). Upon impact, the energy of the once rapidly moving meteorite is transformed into heat energy and shock waves that pass through the surrounding rocks. The result is pulverized, shattered, and sometimes melted rock.

The products of these impacts, called *impactiles*, include mixtures of fused fragmented rock plus glass-rich ejecta that resemble volcanic bombs. In some cases, a very dense form of quartz (*coesite*) and minute *diamonds* are found. The existence of these high-pressure minerals provides convincing evidence that pressures and temperatures involved in impact metamorphism can be as great as those found in the upper mantle.

8.5   Concept Checks

1. In which type of metamorphism does compressional stress play a major role?

2. Name three rocks that are produced by contact metamorphism.

3. What is an *aureole*?

4. What is the agent of hydrothermal metamorphism?

5. Which type of plate boundary is associated with regional metamorphism?

6. List the common metamorphic rocks generated by regional metamorphism.

Impact Metamorphism

Recently it became clear that comets and asteroids have collided with Earth far more frequently than was once assumed. The evidence: More than 100 giant impact structures have been identified to date.

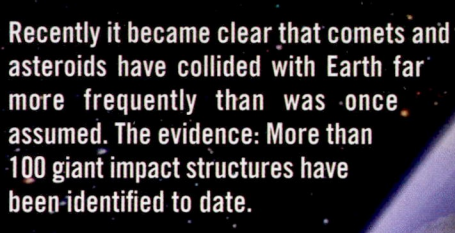

Impact breccia
Leigh Anne Del Ray-Crowell/Aerolite Meteorites LLC

One signature of impact craters is shock metamorphism

When high-velocity projectiles (comets, asteroids) impact Earth's surface, pressures are extreme and temperatures momentarily exceed 2000 °C. The result is pulverized and shattered rock that may form impact breccia.

Tektites: Products of impact metamorphism

Tektites are silica-rich glass beads no more than a few centimeters across that are jet black, dark green, or yellowish in color. Most researchers agree that tektites are the result of impacts of large projectiles that are capable of melting crustal rock. In Australia, millions of tektites are strewn over an area seven times the size of Texas. Several other tektite groupings (called strewnfields) have been identified worldwide.

Tektites recovered from Nullarbor Plain, Australia.

Brian Mason/Smithsonian

Meteor Crater, Arizona

Comparatively young impact craters, such as Meteor Crater located west of Winslow, Arizona, appear fresh with rock fragments (ejecta) that ring the impact site.

Most meteorite impacts occurred millions of years ago.

Michael Collier

? **Question:**
Which of these is not evidence that an impact event occurred; tektites, migmatites, impact breccia, or shock metamorphism?

8.6 | Metamorphic Zones

Explain how index minerals are used to establish the metamorphic grade of a rock body.

In areas affected by metamorphism, geologists can observe the usually systematic variations in the mineralogy and texture of the altered rocks. These differences are clearly related to variations in the degree of metamorphism that takes place in each metamorphic zone.

Textural Variations

Across areas where regional metamorphism has occurred, rock textures vary based on the intensity of metamorphism. If we begin with a clay-rich sedimentary rock such as shale or mudstone, a gradual increase in metamorphic intensity from low grade to high grade is accompanied by a general coarsening of the grain size. **Figure 8.26** illustrates that as metamorphic intensity increases, shale changes to a fine-grained slate, which then forms phyllite, which, through continued recrystallization, generates a medium-grained schist. Under more intense conditions, a gneissic texture that exhibits layers of dark and light minerals may develop. This systematic transition in metamorphic textures can be observed as we approach the Appalachian Mountains from the west. Beds of shale, which once extended over large areas of the eastern United States, still occur as nearly flat-lying

strata in Ohio. However, in the broadly folded Appalachians of central Pennsylvania, the rocks that once formed flat-lying beds are folded and display a preferred orientation of platy mineral grains, as exhibited by well-developed slaty cleavage. As we move further east toward the intensely deformed crystalline Appalachians, we find large exposures of schists. Some of the most intense zones of metamorphism are found in Vermont and New Hampshire, where gneissic rocks are exposed at the surface.

Index Minerals and Metamorphic Grade

In addition to textural changes, we encounter corresponding changes in mineralogy as we shift from areas of low-grade metamorphism to those of high-grade metamorphism. An idealized transition in mineralogy

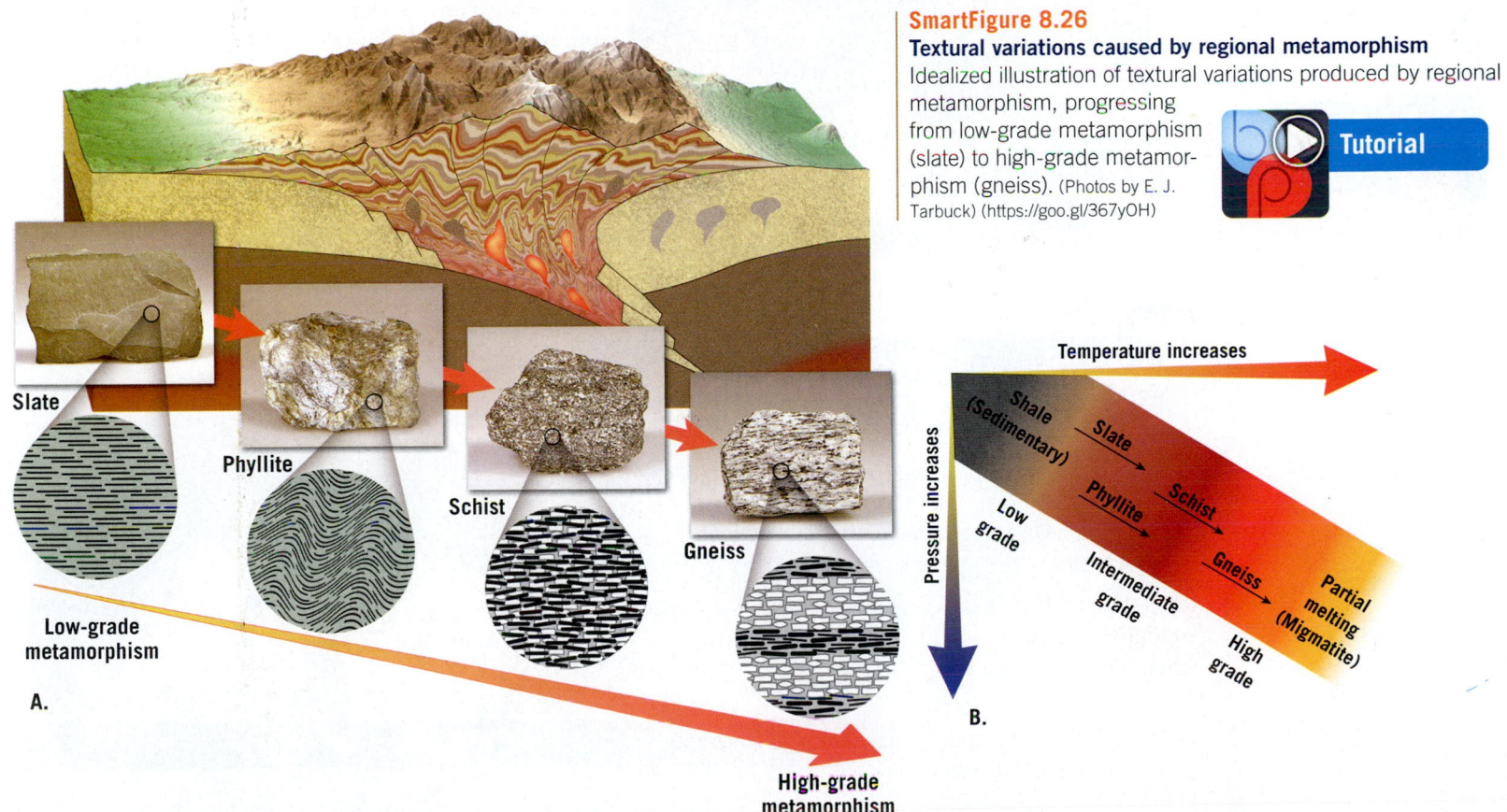

SmartFigure 8.26

Textural variations caused by regional metamorphism
Idealized illustration of textural variations produced by regional metamorphism, progressing from low-grade metamorphism (slate) to high-grade metamorphism (gneiss). (Photos by E. J. Tarbuck) (https://goo.gl/367yOH)

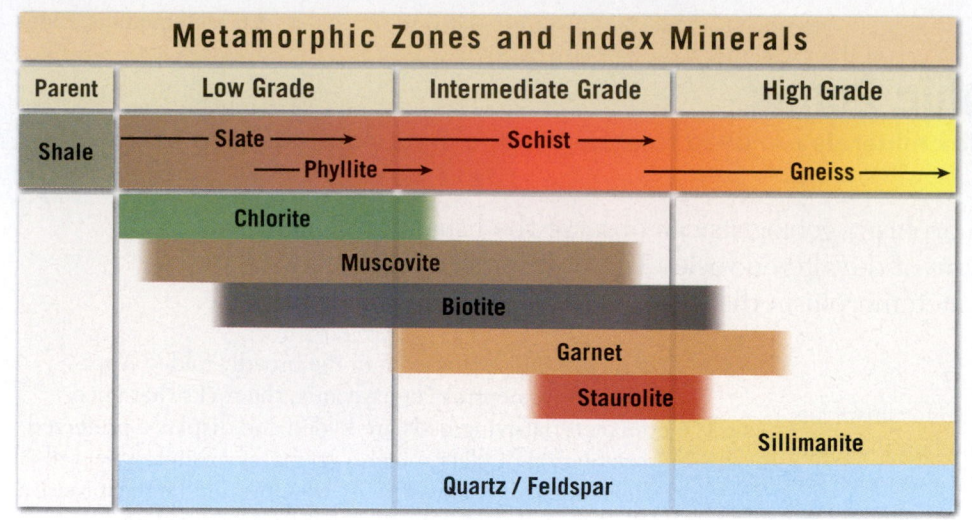

Metamorphic Zones and Index Minerals

Parent	Low Grade	Intermediate Grade	High Grade
Shale	Slate → Schist →		Gneiss →
	→ Phyllite →		

Chlorite

Muscovite

Biotite

Garnet

Staurolite

Sillimanite

Quartz / Feldspar

Figure 8.27
Metamorphic zones and index minerals This is a typical transition of various index minerals associated with the progression from low-grade to high-grade metamorphism of the rock shale.

that results from the regional metamorphism of shale is shown in **Figure 8.27**. The first new mineral to form as shale changes to slate is chlorite. At higher temperatures, flakes of muscovite and biotite begin

to dominate. Under more extreme conditions, metamorphic rocks may contain garnet and staurolite crystals (**Figure 8.28**). At temperatures approaching the melting point of rock, sillimanite forms. Sillimanite is a high-temperature metamorphic mineral used to make porcelains used in extreme environments, such as for spark plugs.

Through the study of metamorphic rocks in their natural settings (called *field studies*) and through experimental studies, researchers have learned that certain minerals, such as those listed in Figure 8.27, are good indicators of the metamorphic environment in which they formed. Using these **index minerals**, geologists distinguish among different zones of regional metamorphism. For example, the mineral chlorite begins to form when temperatures are relatively low—less than 200°C (400°F; see Figure 8.27). Thus, rocks containing chlorite (usually slates) are categorized as *low grade*. By contrast, the mineral sillimanite forms only in extreme environments where temperatures exceed 600°C (1100°F), and rocks containing it are considered *high grade*. By mapping the occurrences of index minerals, geologists can identify zones of varying metamorphic grades (**Figure 8.29**).

Migmatites In the most extreme environments, even the highest-grade metamorphic rocks undergo change. For example, gneissic rocks may be heated sufficiently to trigger melting. However, minerals melt at different temperatures. The light-colored silicates, usually quartz and potassium feldspar, have the lowest melting temperatures

SmartFigure 8.28
Garnet, an index mineral, provides evidence of intermediate- to high-grade metamorphism These garnet porphyroblasts are found in a gneiss in the Adirondacks, New York. (Photo by Michael Collier) (http://goo.gl/KrkufS)

Mobile Field Trip

Garnet

EYE ON EARTH 8.2

This rock outcrop, located in Joshua Tree National Park, California, consists of dark-colored metamorphic rocks that overlie light-colored igneous rocks.

QUESTION 1 Name the type of metamorphism—contact, hydrothermal, burial, subduction zone, or regional metamorphism—that likely produced these metamorphic rocks.

QUESTION 2 Write a brief statement that outlines the geologic history of this area, based on what you observe in this image.

E.J. Tarbuck

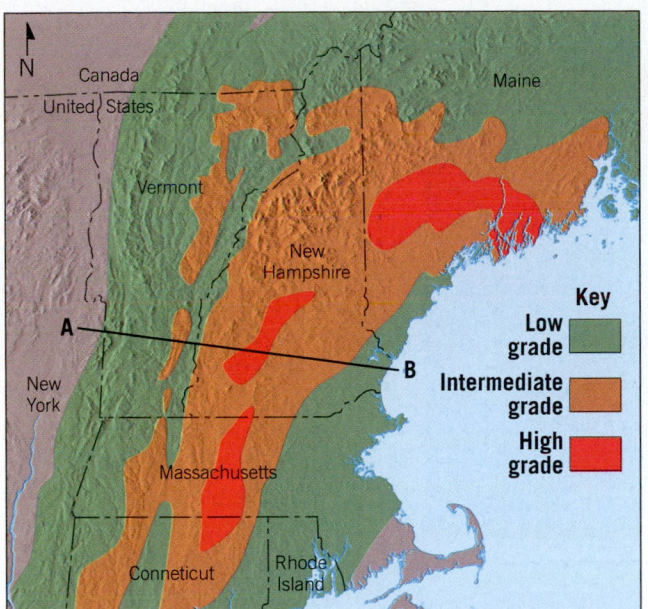

Figure 8.29
Zones of metamorphic intensities in New England
Highly generalized map that shows areas of low- to high-grade metamorphism in New England.

Figure 8.30
Migmatite Under high-grade metamorphism, light-colored (felsic) minerals in a gneiss may begin to melt, while the dark-colored (mafic) minerals remain solid. If this melt solidifies in place, the rock—called a migmatite—will contain light-colored igneous rock intermixed with metamorphic rock composed of dark-colored (mafic) minerals. (Photo by Harlan H. Roepke)

and begin to melt first, whereas the mafic silicates, such as amphibole and biotite, remain solid. When the partially melted rock cools, the light bands will be composed of igneous or igneous-appearing components, while the dark bands will consist of unmelted metamorphic material. Rocks of this type are called **migmatites** (*migma* = mixture, *ite* = a stone) (**Figure 8.30**). The bands in migmatites often form intricate folds and may contain tabular inclusions of the dark components. Migmatites serve to illustrate the fact that some rocks are considered transitional and do not fit neatly into any of the three basic rock groups.

8.6 | **Concept Checks**

1. Describe the different grades of metamorphism that might be encountered moving west to east from Ohio to the crystalline core of the Appalachians.

2. How do geologists use index minerals?

3. Explain why migmatites are difficult to place into any one of the three basic rock groups.

8.7 | Interpreting Metamorphic Environments

Describe the temperature and pressure conditions associated with the following metamorphic facies: blueschist facies, hornfels facies, and zeolite facies.

Nearly a century ago, geologists realized that groups of associated minerals could be used to determine the pressures and temperatures at which rocks undergo metamorphism. This discovery led Finnish geologist Pentti Eskola to propose the concept of metamorphic facies. Simply, metamorphic rocks containing the same assemblage of minerals belong to the same **metamorphic facies**—implying that they formed in very similar metamorphic environments. Using metamorphic facies to determine a metamorphic environment is analogous to using a group of plants to define a climatic zone, areas that experience similar precipitation and temperature conditions. For instance, sparsely vegetated regions dominated by cacti identify the desert climate zone, characterized by low precipitation and high temperatures.

Common Metamorphic Facies

The common metamorphic facies are shown in **Figure 8.31**. These include the *hornfels, zeolite, greenschist, amphibolite, granulite, blueschist,* and *eclogite facies.*

Facies names are based on the minerals that define them. For example, rocks of the amphibolite facies are characterized by hornblende (a common amphibole); the greenschist facies consists of schists in which the

Figure 8.31
Metamorphic facies and corresponding temperature and pressure conditions Note the metamorphic rocks produced from regional metamorphism of basalt versus shale under similar conditions of temperature and pressure.

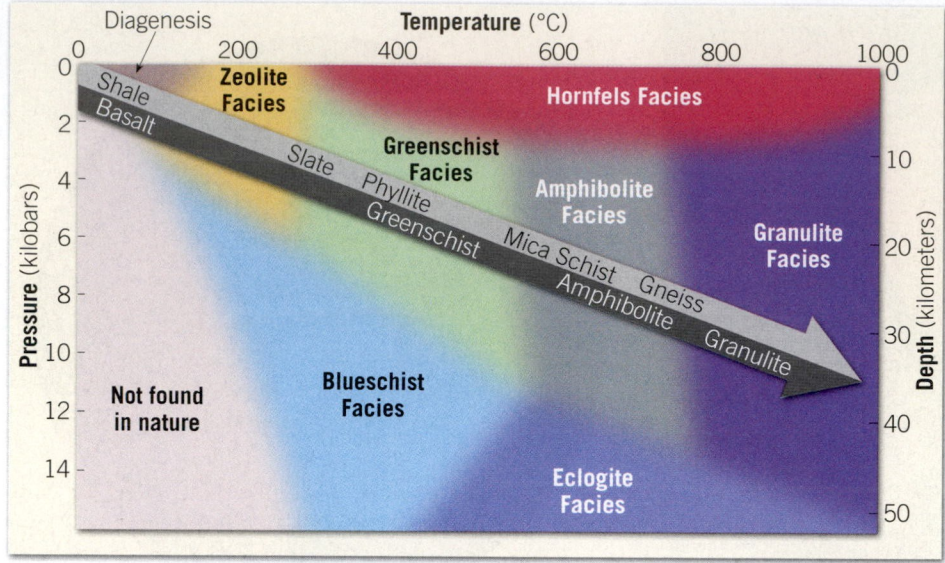

green minerals chlorite, epidote, and serpentine are prominent. Similar groups of minerals are found in rocks of all ages and in all parts of the world. Thus, the concept of metamorphic facies is useful in interpreting Earth's history. Rocks belonging to the same metamorphic facies all formed under the same conditions of temperature and pressure, and therefore in similar tectonic settings, regardless of their location or age.

It should be noted that the name for each metamorphic facies refers to a metamorphic rock derived specifically from a basaltic parent. This is because Pentti Eskola concentrated his work on the metamorphism of basalts, and his basic terminology, although now slightly modified, remains. The names of Eskola's facies serve as convenient labels for particular combinations of temperatures and pressures, no matter what the mineral composition. In other words, even if a nonbasaltic parent rock produces different indicator minerals under a given set of metamorphic conditions, the facies names shown in Figure 8.31 are used to denote the temperature and pressure ranges embodied by that metamorphic rock. For example, mica schist belongs to the amphibolite facies, despite the fact that mica schist was derived from shale—a nonbasaltic parent rock.

Metamorphic Facies and Plate Tectonics

Figure 8.32 shows how the concept of facies fits into the context of plate tectonics. Near deep-ocean trenches, slabs of relatively cool oceanic lithosphere and the overlying crust are subducted. As the lithosphere descends, sediments and crustal rocks are subjected to steadily increasing temperatures and pressures (Figure 8.32A). However, temperatures in the slab remain cooler than the surrounding mantle because rock is a poor conductor of heat and therefore warms slowly. The metamorphic facies associated with this type of high-pressure, low-temperature environment is called the *blueschist facies* because of the presence of the blue-colored variety of amphibole called *glaucophane* (**Figure 8.33A**). The rocks of

Figure 8.32
Metamorphic facies and plate tectonics These block diagrams show various metamorphic facies and the tectonic environments that generate them.

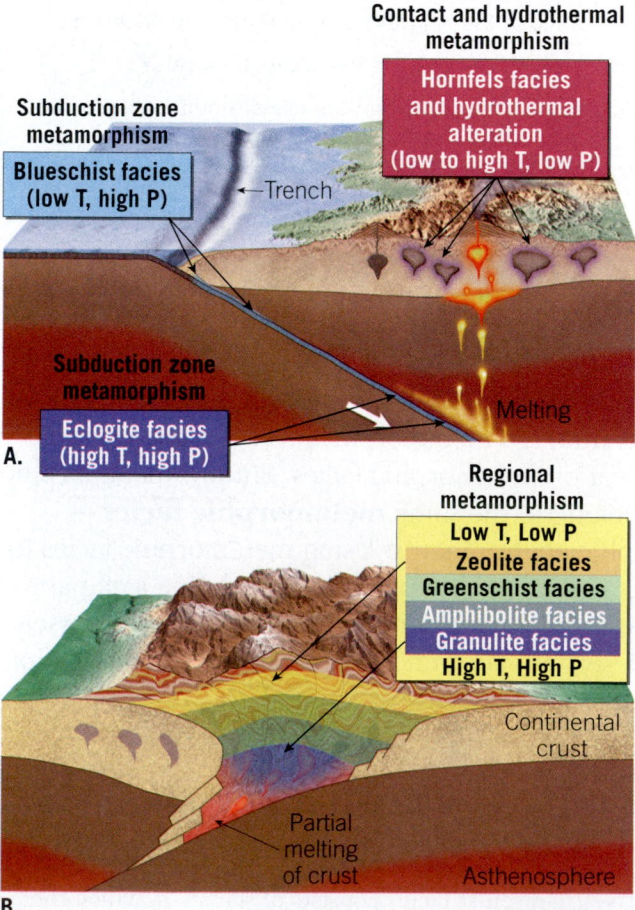

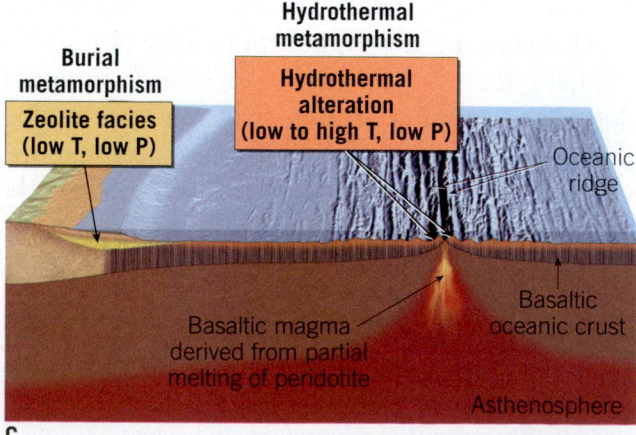

the Coast Range of California belong to the blueschist facies; these highly deformed rocks were once deeply buried but have been uplifted because of a change in the plate boundary. In some areas, subduction carries rocks to even greater depths, producing the *eclogite facies* that is diagnostic of very high temperatures and pressures (Figure 8.33B).

Along some convergent zones, continental plates collide to form extensive mountain belts (see Figure 8.32B). This activity results in large areas of regional metamorphism that often include zones of contact and hydrothermal metamorphism. The increasing temperatures and pressures associated with regional metamorphism are recorded by the *zeolite–greenschist–amphibolite–granulite facies* sequence shown in Figure 8.31.

A. Blueschist forms in low-temperature, high-pressure environments

B. Eclogite forms in high-temperature and extreme high-pressure environments

Figure 8.33
Rocks produced by subduction zone metamorphism A. Blueschist has a blue hue because of the blue-colored amphibole called glaucophane. **B.** This sample of eclogite contains reddish grains of garnet and green grains of pyroxene. (Photos by Dennis Tasa)

Mineral Stability and Metamorphic Environments

In most tectonic environments, such as along subduction zones, rocks experience an increase in *both* pressure and temperature simultaneously. An increase in pressure causes minerals to contract, which favors the formation of high-density minerals. However, increased temperature results in expansion, so mineral phases that occupy greater volume (are less dense) tend to be more stable at high temperatures. Thus, determining the conditions of temperature and pressure at which a mineral is stable (does not change) is not an easy task. To help in this endeavor, researchers have turned to the laboratory. Here materials of various compositions are heated and placed under pressures that approximate conditions at various depths within Earth. From such experiments, we can determine which minerals are likely to form in various metamorphic environments.

Some minerals, such as quartz, are stable over a wide range of metamorphic settings and therefore are not useful in determining metamorphic environments. Fortunately, other groups of related minerals do provide useful estimates of conditions during metamorphism. One of the most important of these groups includes the

EYE ON EARTH 8.3

The Matterhorn, located on the border between Switzerland and Italy, is one of the highest peaks in the Alps. Except for its base, the Matterhorn is composed mainly of gneiss.

QUESTION 1 To which metamorphic grade do the rocks of the upper Matterhorn belong?

QUESTION 2 What type of plate boundary—divergent, convergent, or transform fault boundary—is associated with the formation of the Matterhorn and related mountainous structures of the Alps?

QUESTION 3 Which type of metamorphism—contact, hydrothermal, burial, subduction zone, or regional metamorphism—likely produced the Matterhorn?

Vaclav Volrab/Shutterstock

Figure 8.34
Group of minerals useful in determining metamorphic environments This phase diagram illustrates the conditions of pressure and temperature at which the three polymorphs of Al_2SiO_5 (andalusite, kyanite, and sillimanite) are stable. (Photo A by Harry Taylor/Dorling Kindersley Media Library; photo B by Dennis Tasa; photo C by Biophoto Associates/Photo Researchers, Inc.)

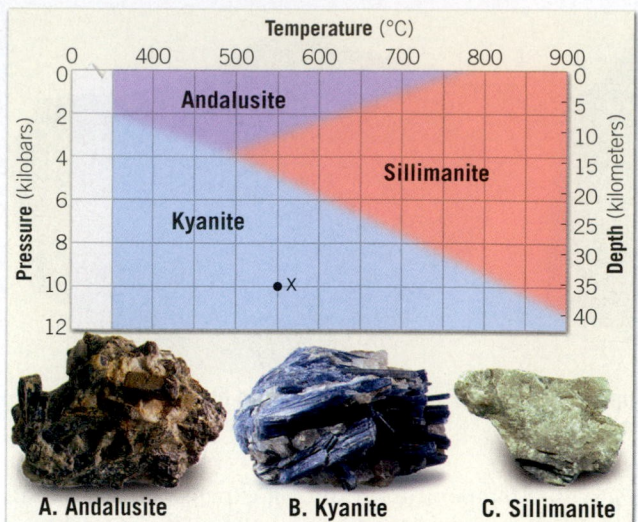

A. Andalusite   B. Kyanite   C. Sillimanite

minerals *kyanite, andalusite,* and *sillimanite*. These three minerals have identical chemical compositions (Al_2SiO_5) but different crystalline structures, which makes them *polymorphs* (see Chapter 3). **Figure 8.34** is a *phase diagram* that shows the specific range of pressures and temperatures at which each of these aluminum-rich silicates is stable.

Because shales and mudstones, which are very common, contain the elements found in these minerals, metamorphic products of shale (slate, phyllite, schist, and gneiss) often contain varying amounts of kyanite, andalusite, or sillimanite. For example, if shale were buried to a depth of about 35 kilometers (10 kilobars), at a

temperature of 550°C, the mineral kyanite would form (see *x* in Figure 8.34).

In general, andalusite is produced by contact metamorphism in near-surface environments where temperatures are high but pressures are relatively low. Kyanite is the high-pressure polymorph that forms during the subduction and deep burial associated with mountain building. Sillimanite, on the other hand, forms only at high temperatures, as a result of contact with a very hot magma body and/or very deep burial. Knowing the ranges of temperatures and pressures a rock experienced during metamorphism provides geologists with valuable data needed to interpret past tectonic environments.

8.7   Concept Checks

1. Define *metamorphic facies*.

2. What two physical conditions vary within Earth to produce different metamorphic environments?

3. The rocks of the Coast Range of California belong to the blueschist facies. What does this tell you about the environment in which these rocks formed?

4. Which process would most likely form minerals belonging to the hornfels facies—contact metamorphism or regional metamorphism? Explain.

8.1 What Is Metamorphism?

Compare and contrast the environments that produce metamorphic, sedimentary, and igneous rocks.

KEY TERMS parent rock, metamorphism, mineralogy, metamorphic grade

- Rocks subjected to elevated temperatures and pressures can react and "change form" to produce metamorphic rocks. Every metamorphic rock has a parent rock—the rock it was prior to metamorphism. New minerals form when minerals in parent rocks undergo metamorphosis. Metamorphic reactions tend to produce larger crystals and may generate layers or bands of minerals aligned in a common direction.
- Metamorphic grade describes the intensity of metamorphosis. Low-grade metamorphic rocks strongly resemble their parent rock. High-grade metamorphism destroys textures of parent rocks and features such as fossils.
- Metamorphism takes place in the solid state and in most cases does not involve any melting.

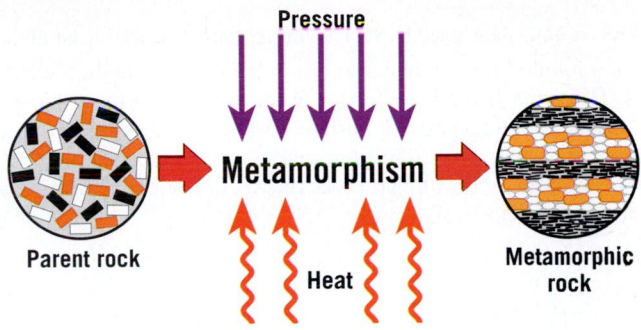

Q Compare the processes that produce metamorphic rocks to those that produce igneous or sedimentary rocks.

8.2 What Drives Metamorphism?

List and distinguish among the four agents that drive metamorphism.

KEY TERMS recrystallization, geothermal gradient, confining pressure, differential stress, compressional stress

- Heat, pressure, directional stress, and chemically active fluids are four agents that drive metamorphic reactions. Any one alone may trigger metamorphism, or all four may exert influence simultaneously.
- The burial of rock or the intrusion of a nearby magma body will raise the temperature of a rock. Heat provides energy that drives chemical reactions and results in the recrystallization of the existing minerals. Different minerals have different levels of susceptibility to recrystallization: Some crystals just grow larger, while others react to form new minerals. Quartz is stable across a wide range of temperatures, whereas clay minerals are stable only at low (near-surface) temperatures.

- Confining pressure resulting from burial is of the same magnitude in every direction. An increase in confining pressure causes rocks and minerals to compact into more dense configurations.
- Differential stress results from tectonic forces, where the pressure is greater in some directions than in others. Rocks subjected to differential stress deep in Earth's crust tend to shorten in the direction of greatest stress and elongate in the direction(s) of least stress, producing flattened or stretched grains. If the same differential stress is applied to a rock in the shallow crust, it may deform in a brittle fashion instead, breaking into smaller pieces.
- Water is an important chemically active fluid in Earth's crust. Hot water can facilitate numerous chemical reactions and may transport dissolved mineral matter great distances to new locations. This may alter the composition of the metamorphosing rock by introducing or removing certain elements.

Q Draw a sketch illustrating the difference between confining pressure and differential stress.

8.3 Metamorphic Textures

Explain how foliated and nonfoliated textures develop.

KEY TERMS texture, foliation, rock cleavage, slaty cleavage, schistosity, gneissic texture (gneissic banding), nonfoliated, porphyroblastic texture

- Texture can reveal the orientation of stresses that helped form a metamorphic rock. A common metamorphic texture is foliation, the planar arrangement of mineral grains. Foliation forms perpendicular to the direction of maximum differential stress, through a combination of processes: rotation of mineral grains, recrystallization and growth of new mineral grains, and flattening of grains by solid-state flow or pressure solution.
- Several arrangements of mineral grains are classified as varieties of foliation, including slaty (rock) cleavage, schistosity, and gneissic banding.
- Nonfoliated metamorphic rocks recrystallize under confining pressure, so the mineral grains exhibit a random orientation. Marble, for instance, forms when the calcite in a parent limestone grows into a crystalline mass.
- Porphyroblasts are large crystals of certain minerals, such as garnet, that form in some metamorphic rocks. They are larger than other grains in the same rock and are distributed through the rock with some space between them, like chocolate chips in a cookie.

Dennis Tasa

Q Examine the accompanying photograph. Is this rock foliated or nonfoliated? Did it form under confining pressure or differential stress? Which pair of arrows shows the direction of maximum stress?

8.4 Common Metamorphic Rocks

List and describe the most common metamorphic rocks.

KEY TERMS slate, phyllite, schist, gneiss, marble, quartzite, hornfels

- Common foliated metamorphic rocks include slate, phyllite, schist, and gneiss. The order listed is the order of increasing metamorphic grade: Slate is least metamorphosed, and gneiss is the most metamorphosed.
- Common nonfoliated metamorphic rocks include quartzite, marble, and hornfels, the recrystallized rocks that form from quartz sandstone, limestone, and shale, respectively.

Q If you were on a class field trip and found an outcrop of nonfoliated light-colored rock, how would you determine whether it is quartzite or marble? How would you distinguish quartzite from quartz sandstone?

Dennis Tasa

8.5 Metamorphic Environments

Write a description for each of these metamorphic environments: contact metamorphism, hydrothermal metamorphism, subduction zone metamorphism, and regional metamorphism.

KEY TERMS contact (thermal) metamorphism, aureole, hydrothermal metamorphism, burial metamorphism, subduction zone metamorphism, regional metamorphism, impact (shock) metamorphism

- Metamorphism can be triggered by diverse geologic situations. Convergent and divergent plate boundaries are both sites of metamorphism, as are deep stacks of sediments on passive margins.
- Contact metamorphism occurs when heat is the dominant metamorphic variable. Typically, this heat emanates from a nearby body of magma that "bakes" the surrounding rock a short distance away.
- Hydrothermal metamorphism occurs in relatively limited areas, with hot water rather than conduction as the agent of heat transfer. The hot water

readily dissolves various ions, which may react with the surrounding rock to form new minerals. Hydrothermal metamorphism produces economically important deposits of metal ore.
- Burial metamorphism occurs when rocks are buried beneath kilometers of overlying rock. Under confining pressure, elevated temperatures cause metamorphic reactions. Subduction zone metamorphism is similar, but with the added influence of differential stress.
- Plate collisions produce regional metamorphism. Rocks at these sites of crustal thickening are subjected to both high temperatures and differential pressure. The resulting belts of metamorphic rock (and associated igneous intrusions) mark zones where continental blocks collided to produce mountains. Long after the mountains have worn away, the collision site is marked by a deformed zone of regional metamorphic rocks that may include slate, schist, marble, and gneiss.
- Other distinctive varieties of metamorphism are associated with faults and meteorite impact structures.

8.6 Metamorphic Zones

Explain how index minerals are used to establish the metamorphic grade of a rock body.

KEY TERMS index mineral, migmatite

- The amount of change in metamorphic rocks increases as the rocks are exposed to higher temperatures and pressures. Metamorphic rocks reveal the degree of metamorphism through their textures and the minerals they contain.

Porphyroblast of sillimanite

Marli Miller

- Grain size increases with higher levels of metamorphism. Under progressively higher temperatures and pressures, shale may transform first to slate, then to phyllite, to schist, and to gneiss.
- Certain minerals can act as indicators of the environment (temperature/pressure) that a metamorphic rock experienced. Chlorite is a mineral associated with low-grade metamorphism, whereas garnet and staurolite are index minerals for intermediate metamorphic grades. Sillimanite indicates a relatively high metamorphic grade.
- In extreme metamorphic environments, light-colored silicate minerals, usually quartz and potassium feldspar, which have the lowest melting points, can turn to magma. The resulting rocks, called migmatites, exhibit evidence of partial melting and consist of bands of light-colored silicates mixed with dark bands of unmelted dark-colored metamorphic rock.

Q When exposed to metamorphic conditions over sufficient time, shale becomes schist. This sample of schist contains porphyroblasts of sillimanite. Do these porphyroblasts indicate low-, intermediate-, or high-grade metamorphism?

8.7 Interpreting Metamorphic Environments

Describe the temperature and pressure conditions associated with the following metamorphic facies: blueschist facies, hornfels facies, and zeolite facies.

KEY TERM metamorphic facies

- Similar metamorphic conditions encourage the growth of similar metamorphic minerals. The assemblage of minerals in a metamorphic rock is evidence of the specific temperature/pressure conditions the rock reached during metamorphism.

- Blueschist and eclogite facies metamorphism is typical of subduction zones, where pressure is the dominant driver of change. The zeolite–greenschist–amphibolite–granulite sequence of metamorphic facies is typical of mountain-building activity at convergent plate boundaries, where both temperature and pressure conditions are important. This regional metamorphism results in these facies grading into one another across the trend of a mountain belt. The hornfels facies is typical of metamorphic rocks that get "baked" at high temperatures but relatively low pressures.

Give It Some Thought

1. Each of the following statements describes one or more characteristics of a particular metamorphic rock. For each statement, identify the metamorphic rock that is being described:

   a. calcite rich and nonfoliated

   b. loosely coherent and composed of broken fragments that formed along a fault zone

   c. represents a grade of metamorphism between slate and schist

   d. composed of tiny chlorite and mica grains and displaying excellent rock cleavage

   e. foliated and composed predominantly of platy materials

   f. composed of alternating bands of light and dark silicate minerals

   g. hard and nonfoliated, often produced by contact metamorphism

2. One of the rock outcrops in the accompanying photos consists mainly of metamorphic rock. Which do you think it is? Explain why you ruled out the other rock bodies. (Photos by E. J. Tarbuck)

A.   B.

C.

3. Describe the textural changes that occur as shale goes from low- to high-metamorphic grade to form the rocks slate, schist, and gneiss.

4. Examine the accompanying photos that show the geology of the Grand Canyon. Notice that most of the canyon consists of layers of sedimentary rocks, but if you were to hike into the inner gorge, you would encounter the Vishnu schist, a metamorphic rock.

A. Inner Gorge of the Grand Canyon

B. Close up of Vishnu Schist (dark color)

   a. What process might have been responsible for the formation of the Vishnu schist?

   b. What does the Vishnu schist tell you about the history of the Grand Canyon prior to the formation of the canyon itself?

   c. Why is the Vishnu schist visible at Earth's surface?

   d. Is it likely that rocks similar to the Vishnu schist exist elsewhere but are not exposed at Earth's surface? Explain.

5. Examine the accompanying close-up images of a conglomerate and a metaconglomerate.

A. Conglomerate

B. Metaconglomerate

Photos by Dennis Tasa

a. Describe how the conglomerate is different from the metaconglomerate.

b. Does this metaconglomerate appear to have been exposed to significant differential stress? Explain.

6. Refer to the accompanying diagram and match each labeled area with the appropriate environment listed above right:

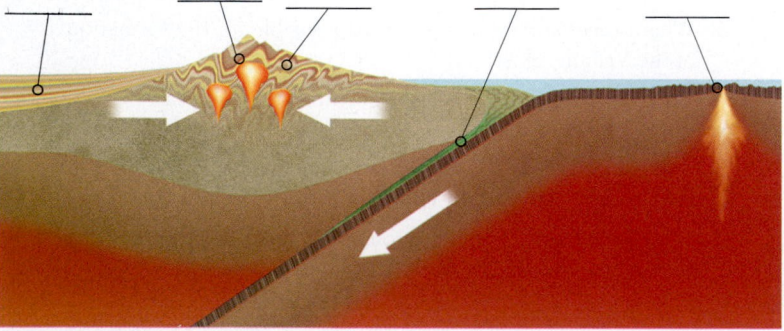

a. contact metamorphism

b. subduction metamorphism

c. regional metamorphism

d. burial metamorphism

e. hydrothermal metamorphism

7. Based on the information provided in Figure 8.29, complete the following:

a. Describe how the metamorphic grade changes from west to east across New England along line A–B.

b. How might these metamorphic rocks have formed?

c. Are these zones of metamorphism consistent with the current tectonic setting of New England? Explain.

8. Examine the accompanying close-up images of six different rocks labeled A–F. Classify them as igneous, sedimentary, or metamorphic, based on texture. (*Hint:* There are two of each rock type.)

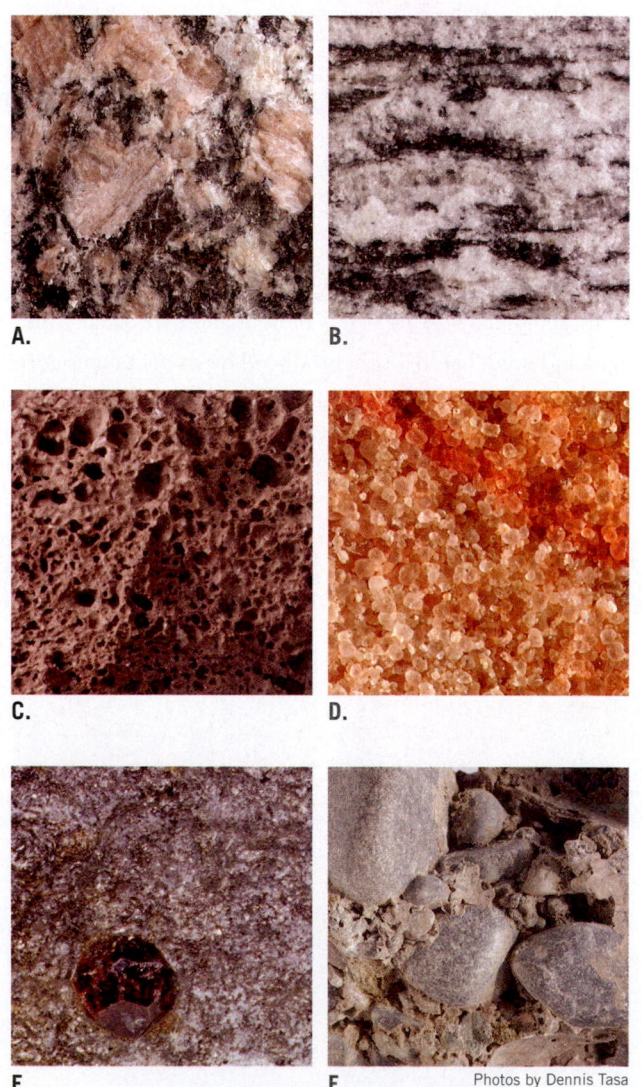

A. **B.**

C. **D.**

E. **F.** Photos by Dennis Tasa

9. The accompanying image shows a metamorphic rock located in Purgatory Chasm in Newport, Rhode Island, that is made of elongated cobbles composed mainly of quartz.

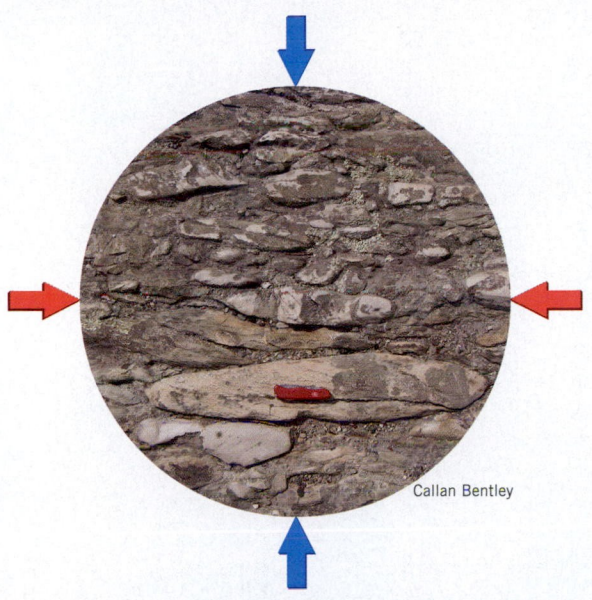

Callan Bentley

a. What name would you give to this metamorphic rock?

b. Which set of arrows (red or blue) best represents the direction of maximum differential stress?

c. Is this type of deformation best described as ductile or brittle?

10. Suppose that you are an exploration geologist hired to interpret the metamorphic rocks of two islands to determine the prospects for mining metals. You are looking for hydrothermal deposits, a common source of metal ores. Island 1 shows a circular ring of disturbed rock with coesite (a dense form of quartz), tiny diamonds, and glass, surrounded by ejected material. Island 2 is dominated by parallel bands of phyllite and garnet-mica schist. Are the rocks on either island typical of hydrothermal metamorphism? Explain.

MasteringGeology™

www.masteringgeology.com

Looking for additional review and test prep materials? With individualized coaching on the toughest topics of the course, MasteringGeology offers a wide variety of ways for you to move beyond memorization to begin thinking like a geologist. Visit the Study Area in www.masteringgeology.com to find practice quizzes, study tools, and multimedia that will improve your understanding of this chapter's content. Sign in today to enjoy the following features: **Self Study Quizzes, SmartFigures: Tutorials/Animations/Condor Videos/Mobile Field Trips, Geoscience Animation Library, GEODe, RSS Feeds, Digital Study Modules,** and an optional **Pearson eText.**

9

Geologic Time

Rafting the Colorado river in Arizona's Marble Canyon. The rocks exposed in the walls of the canyon contain clues to millions of years of Earth history. (Photo by Michael Collier)

Each statement represents the primary **LEARNING OBJECTIVE** for the corresponding major heading within the chapter. After you complete the chapter, you should be able to:

9.1 Distinguish between numerical and relative dating and apply relative dating principles to determine a time sequence of geologic events.

9.2 Define *fossil* and discuss the conditions that favor the preservation of organisms as fossils. List and describe various types of fossils.

9.3 Explain how rocks of similar age that are in different places can be matched up.

9.4 Discuss three types of radioactive decay and explain how radioactive isotopes are used to determine numerical dates.

9.5 Distinguish among the four basic time units that make up the geologic time scale and explain why the time scale is considered to be a dynamic tool.

9.6 Explain how reliable numerical dates are determined for layers of sedimentary rock.

In the late eighteenth century, James Hutton recognized the immensity of Earth history and the importance of time as a component in all geologic processes. In the nineteenth century, Sir Charles Lyell and others effectively demonstrated that Earth had experienced many episodes of mountain building and erosion, which must have required great spans of geologic time. Although these pioneering scientists understood that Earth was very old, they had no way of determining its age in years. Was it tens of millions, hundreds of millions, or even billions of years old? Long before geologists could establish a geologic time scale that included numerical dates in years, they gradually assembled a time scale using relative dating principles. What are these principles? What part do fossils play? With the discovery of radioactivity and the development of radiometric dating techniques, geologists now can assign quite accurate dates to many of the events in Earth history. What is radioactivity? Why is it a good "clock" for dating the geologic past?

9.1 | Creating a Time Scale: Relative Dating Principles

Distinguish between numerical and relative dating and apply relative dating principles to determine a time sequence of geologic events.

Figure 9.1 shows a hiker resting atop the Permian-age Kaibab Formation at Cape Royal, on the Grand Canyon's North Rim. Beneath him are thousands of meters of sedimentary strata that go as far back as Cambrian time, more than 540 million years ago. These strata rest atop even older sedimentary, metamorphic, and igneous rocks from a span known as the Precambrian. Some of these rocks are 2 billion years old. Although the Grand Canyon's rock record has numerous interruptions, the rocks beneath the hiker contain clues to great spans of Earth history.

The Importance of a Time Scale

Like the pages in a long and complicated history book, rocks record the geologic events and changing life-forms of the past. The book, however, is not complete. Many pages, especially in the early chapters, are missing. Others are tattered, torn, or smudged. Yet enough of the book remains to allow much of the story to be deciphered.

Interpreting Earth history is an important goal of the science of geology. Like a modern-day sleuth, a geologist must interpret the clues found preserved in the rocks. By studying rocks, especially sedimentary rocks, and the features they contain, geologists can unravel the complexities of the past.

Geologic events by themselves, however, have little meaning until they are put into a time perspective. Studying history, whether it is the Civil War or the age of dinosaurs, requires a calendar. Among geology's major contributions to human knowledge are the *geologic time scale* and the discovery that Earth history is exceedingly long.

perceive our planet. They learned that Earth is much older than anyone had previously imagined, and they learned that its surface and interior have been changed over and over again by the same geologic processes that operate today.

Numerical Dates During the late 1800s and early 1900s, attempts were made to determine Earth's age. Although some of the methods appeared promising at the time, none of those early efforts proved to be reliable. What those scientists were seeking was a **numerical date**. Such dates specify the actual number of years that have passed since an event occurred. Today, our understanding of radioactivity allows us to accurately determine numerical dates for rocks that represent important events in Earth's distant past. We will study radioactivity later in this chapter. Prior to the discovery of radioactivity, geologists had no reliable method of carrying out numerical dating and had to rely solely on relative dating.

Numerical and Relative Dates

The geologists who developed the geologic time scale revolutionized the way people think about time and

Relative Dates When we place rocks in their proper *sequence of formation*—indicating which formed first, second, third, and so on—we are establishing **relative dates**. Such dates cannot tell us how long ago something took place, only that it followed one event and preceded another. The relative dating techniques that were developed are valuable and still widely used. Numerical dating methods did not replace those techniques but supplemented them. To establish a relative time scale, a few basic principles had to be discovered and applied. They were major breakthroughs in thinking at the time, and their discovery was an important scientific achievement.

Principle of Superposition

Nicolas Steno (1638–1686), a Danish anatomist, geologist, and priest, was the first to recognize a sequence of historical events in an outcrop of sedimentary rock layers. Working in the mountains of western Italy, Steno applied a very simple rule that has become the most basic principle of relative dating—the **principle of superposition** (*super* = above; *positum* = to place). This principle simply states that in an undeformed sequence of sedimentary rocks, each bed is older than the one above and younger than the one below. Although it may seem obvious that a rock layer could not be deposited with nothing beneath it for support, it was not until 1669 that Steno clearly stated this principle.

This rule also applies to other surface-deposited materials, such as lava flows and beds of ash from volcanic eruptions. Applying the law of superposition to the beds exposed in the upper portion of the Grand Canyon, we can easily place the layers in their proper order. Among those pictured in **Figure 9.2**, the sedimentary rocks in the Supai Group are the oldest, followed in order by the Hermit Shale, Coconino Sandstone, Toroweap Formation, and Kaibab Limestone.

Figure 9.1
Contemplating geologic time This hiker is resting atop the Kaibab Formation, the uppermost layer in the Grand Canyon. (Photo by Michael Collier)

Figure 9.2
Superposition Applying the principle of superposition to these layers in the upper portion of the Grand Canyon, the Supai Group is oldest, and the Kaibab Limestone is youngest.

Youngest

Kaibab Limestone: shallow marine limestone that rims much of the canyon

Toroweap Formation: shallow marine, thin-to-medium bedded sandy limestone

Coconino Sandstone: cliff-forming cross-bedded sandstone

Hermit Shale: red, slope-forming thinly-bedded shales and siltstones

Supai Group: alternating layers of sandstone, siltstone and shale

Oldest

Geologist's Sketch

Dennis Tasa

Figure 9.3
Original horizontality Most layers of sediment are deposited in a nearly horizontal position. When we see strata that are folded or tilted, we can assume that they were moved into that position by crustal disturbances *after* their deposition. (Photo by Marco Simoni/Robert Harding World Imagery)

Principle of Original Horizontality

Steno is also credited with recognizing the importance of another basic rule, the **principle of original horizontality**, which says that layers of sediment are generally deposited in a horizontal position. Thus, if we observe rock layers that are flat, it means they have not been disturbed and still have their *original* horizontality. The layers in the Grand Canyon illustrate this in Figures 9.1 and 9.2. But if they are folded or inclined at a steep angle (discussed in detail in Chapter 10), they must have been moved into that position by crustal disturbances sometime *after* their deposition (**Figure 9.3**).

Principle of Lateral Continuity

The **principle of lateral continuity** refers to the fact that sedimentary beds originate as continuous layers that extend in all directions until they eventually grade into a different type of sediment or until they thin out at the edge of the basin of deposition (**Figure 9.4A**). For example, when a river creates a canyon, we can assume that identical or similar strata on opposite sides once spanned the canyon (**Figure 9.4B**). Although rock outcrops may be separated by a considerable distance, the principle of lateral continuity tells us those outcrops once formed a continuous layer (**Figure 9.4C**). This principle allows geologists to relate rocks in isolated outcrops to one another. Combining the principles of lateral continuity and superposition lets us extend relative age relationships over broad areas. This process, called *correlation*, is examined in Section 9.3.

Principle of Cross-Cutting Relationships

Figure 9.5 shows layers of rock that have been offset by a *fault*, a fracture in rock along which displacement occurs. It is clear that the strata must be older than the fault that broke them. The **principle of cross-cutting relationships** states that geologic features that cut across rocks must form *after* the rocks they cut through. Igneous intrusions (see Chapter 4) provide another example. The dikes shown in **Figure 9.6** are tabular masses of igneous rock that cut through the surrounding rock. The magmatic heat from igneous intrusions often creates a narrow "baked" zone of contact metamorphism on the adjacent rock, also indicating that the intrusion occurred *after* the surrounding rocks were in place.

EYE ON EARTH 9.1

This image shows West Cedar Mountain in southern Utah. The gray rocks are shale that originated as muddy river delta deposits. The sediments composing the orange sandstone were deposited by a river.

QUESTION 1 Place the events related to the geologic history of this area in proper sequence. Explain your logic. Include the following: uplift, sandstone, erosion, and shale.

QUESTION 2 What term is applied to the type of dates you established for this site?

Shale

Sandstone

Michael Collier

Principle of Inclusions

Sometimes inclusions can aid in the relative dating process. *Inclusions* are fragments of one rock unit that have been enclosed within another. The basic **principle of inclusions** is logical and straightforward. The rock mass adjacent to the one containing the inclusions must have been there first in order to provide the rock fragments. Therefore, the rock mass that contains inclusions is the younger of the two. For example, when magma intrudes into surrounding rock, blocks of the surrounding rock may become dislodged and incorporated into the magma. If these pieces do not melt, they remain as inclusions known as *xenoliths* (see Chapter 4). In another example, when sediment is deposited atop a weathered mass of bedrock, pieces of the weathered rock become incorporated into the younger sedimentary layer (**Figure 9.7**).

Unconformities

When we observe layers of rock that have been deposited essentially without interruption, we call them **conformable**. Particular sites exhibit conformable beds representing certain spans of geologic time. However, no place on Earth has a complete set of conformable strata.

Throughout Earth history, the deposition of sediment has been interrupted over and over again. All such breaks in the rock record are termed *unconformities*. An **unconformity** represents a long period during which deposition ceased, erosion removed previously formed rocks, and then deposition resumed. In each case, uplift and erosion were followed by subsidence and renewed sedimentation.

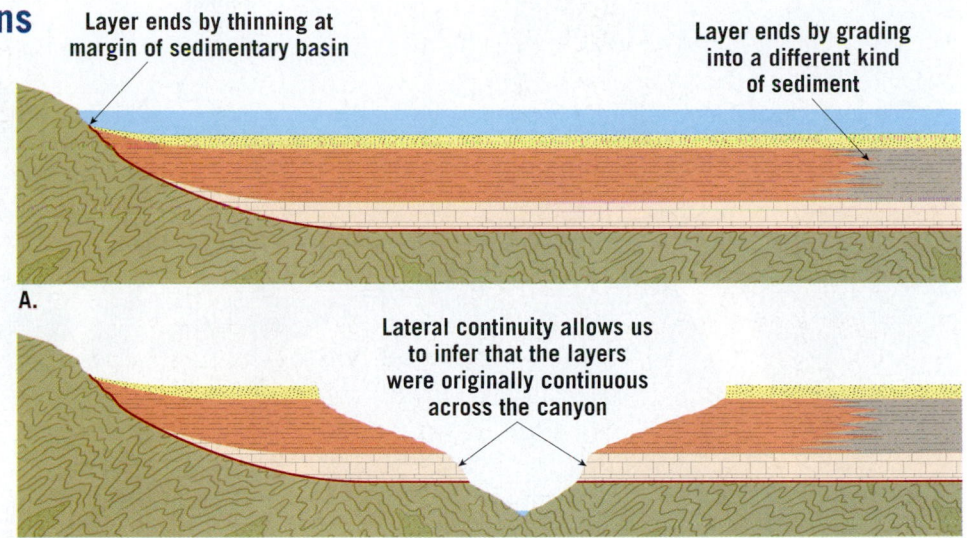

A. Layer ends by thinning at margin of sedimentary basin

Layer ends by grading into a different kind of sediment

B. Lateral continuity allows us to infer that the layers were originally continuous across the canyon

Coconino Sandstone

Redwall Limestone

C.

Figure 9.4
Lateral continuity
A. Sediments are deposited over a large area in a continuous sheet. Sedimentary strata extend continuously in all directions until they thin out at the edge of a depositional basin or grade into a different type of sediment.
B. Although rock exposures are separated by many miles, we can infer that they were once continuous.
C. The idea depicted in B is illustrated in this image of the Grand Canyon. (Photo by bcampbell65/Shutterstock)

Fault

SmartFigure 9.5
Cross-cutting fault The rocks are older than the fault that displaced them. (Morley Read/Alamy) (https://goo.gl/BiFVHa)

Condor Video

Figure 9.6
Cross-cutting dikes This igneous intrusion is younger than the rocks that are intruded. (Photo by Jonathan.S kt)

Dikes

Unconformities are important features because they represent significant geologic events in Earth history. There are three basic types of unconformities, and geologists can use them to identify what intervals of time are not represented by strata and thus are missing from the geologic record.

Angular Unconformity   Perhaps the most easily recognized unconformity is an **angular unconformity**. It consists of tilted or folded sedimentary rocks that are overlain by younger, more flat-lying strata. An angular unconformity indicates that during the pause in deposition, a period of deformation (folding or tilting) and erosion occurred (**Figure 9.8**).

SmartFigure 9.7
Inclusions The rock containing inclusions is younger than the inclusions. (https://goo.gl/Okfrm6)

Tutorial

These inclusions of igneous rock contained in the adjacent sedimentary layer indicate the sediments were deposited atop the weathered igneous mass and thus are younger.

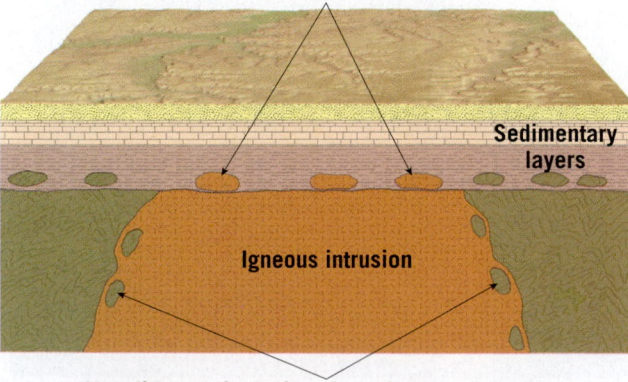

Sedimentary layers

Igneous intrusion

Xenoliths are inclusions in an igneous intrusion that form when pieces of surrounding rock are incorporated into magma.

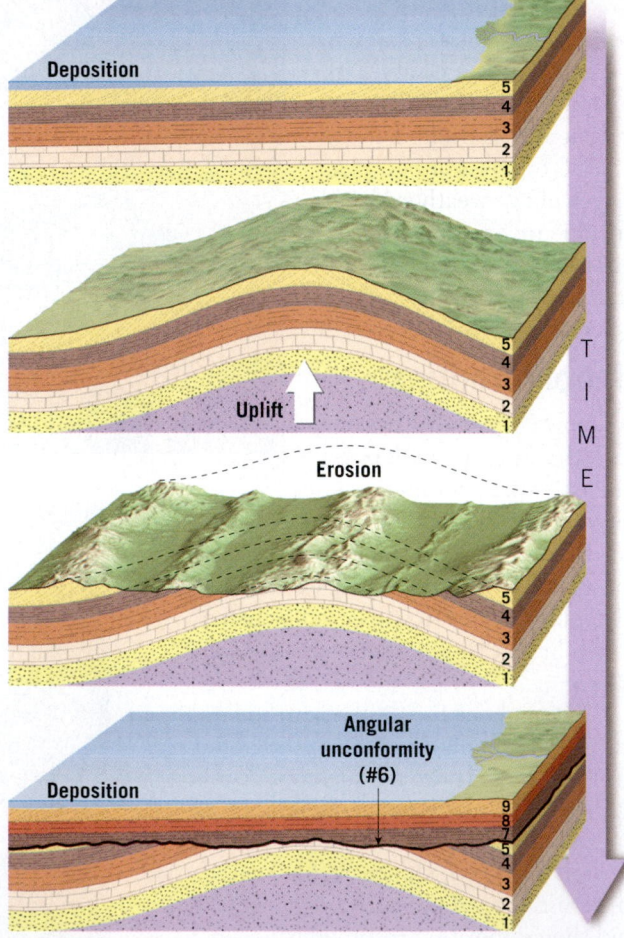

Deposition

Uplift

T I M E

Erosion

Angular unconformity (#6)

Deposition

SmartFigure 9.8
Formation of an angular unconformity An angular unconformity represents an extended period during which deformation and erosion occurred. (https://goo.gl/arrwhC)

Tutorial

Figure 9.9
Siccar Point, Scotland James Hutton studied this famous unconformity in the late 1700s. (Photo by Marli Miller)

Above the unconformity lie gently dipping beds of reddish sandstone and conglomerate

Angular unconformity

Rock hammer

Below the unconformity lie nearly vertical beds of sandstone and shale

When James Hutton studied an angular unconformity in Scotland more than 200 years ago,* he understood that it represented a major episode of geologic activity (**Figure 9.9**). He and his colleagues also appreciated the immense time span implied by such relationships; a companion later wrote of their visit to this site, "the mind seemed to grow giddy by looking so far into the abyss of time."

Disconformity A **disconformity** is a gap in the rock record that represents a period of erosion rather than deposition. Imagine that a series of sedimentary layers is deposited in a shallow marine setting. Following this period of deposition, sea level falls or the land rises, exposing some the sedimentary layers. During this span, when the sedimentary beds are above sea level, no new sediment accumulates, and some of the existing layers are eroded away. Later, sea level rises or the land subsides, submerging the landscape. Now the surface is again below sea level, and a new series of sedimentary beds is deposited. The boundary separating the two sets of beds is a disconformity—a span for which there is no rock record (**Figure 9.10**). Because the layers above and below a disconformity are parallel, these features are sometimes difficult to identify unless you notice evidence of erosion such as a buried stream channel.

*This pioneering geologist is discussed in the section on the birth of modern geology in Chapter 1.

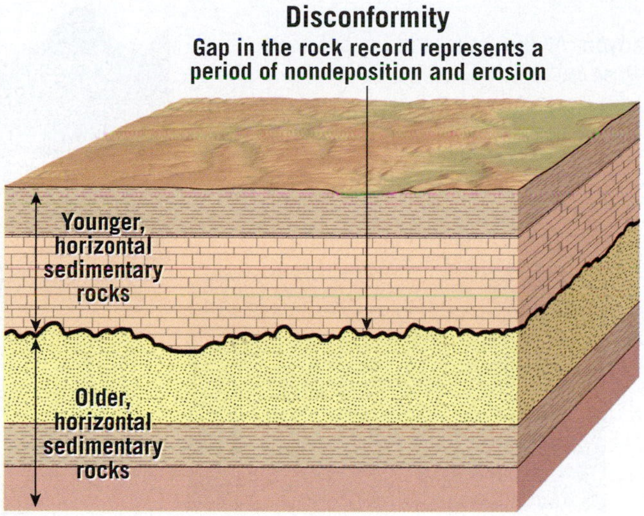

Disconformity
Gap in the rock record represents a period of nondeposition and erosion

Younger, horizontal sedimentary rocks

Older, horizontal sedimentary rocks

Figure 9.10
Disconformity The layers on both sides of this gap in the rock record are essentially parallel.

Nonconformity The third basic type of unconformity is a **nonconformity**, in which younger sedimentary strata overlie older metamorphic or intrusive igneous rocks (**Figure 9.11**). Just as angular unconformities and some disconformities imply crustal movements, so too do nonconformities. Intrusive igneous masses and metamorphic rocks originate far below the surface. Thus, for a nonconformity to develop, there must be a period of uplift and erosion of overlying rocks. Once exposed at the surface, the igneous or metamorphic rocks are subjected to weathering and erosion, then undergo subsidence and renewed sedimentation.

Unconformities in the Grand Canyon The rocks exposed in the Grand Canyon of the Colorado River represent a tremendous span of geologic history. It is a wonderful place to take a trip through time. The canyon's colorful strata record a long history of sedimentation in a variety of environments—advancing seas, rivers and deltas, tidal flats and sand dunes. But the record is not

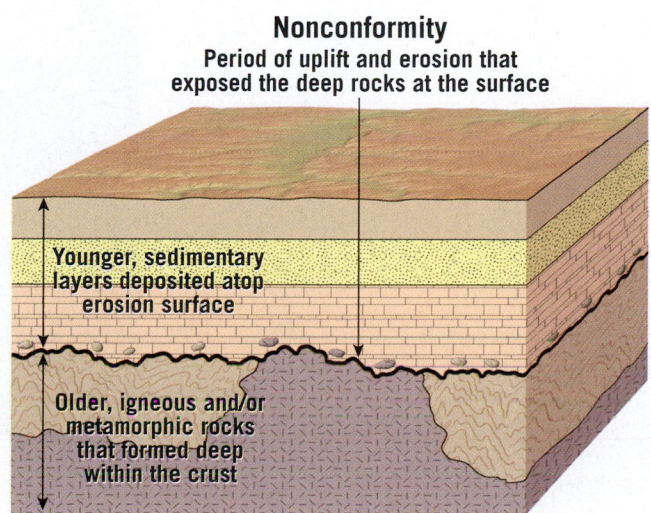

Nonconformity
Period of uplift and erosion that exposed the deep rocks at the surface

Younger, sedimentary layers deposited atop erosion surface

Older, igneous and/or metamorphic rocks that formed deep within the crust

Figure 9.11
Nonconformity Younger sedimentary rocks rest atop older metamorphic or igneous rocks.

Figure 9.12

Cross section of the Grand Canyon All three types of unconformities are present. (Center photo by Marli Miller; other photos by E. J. Tarbuck)

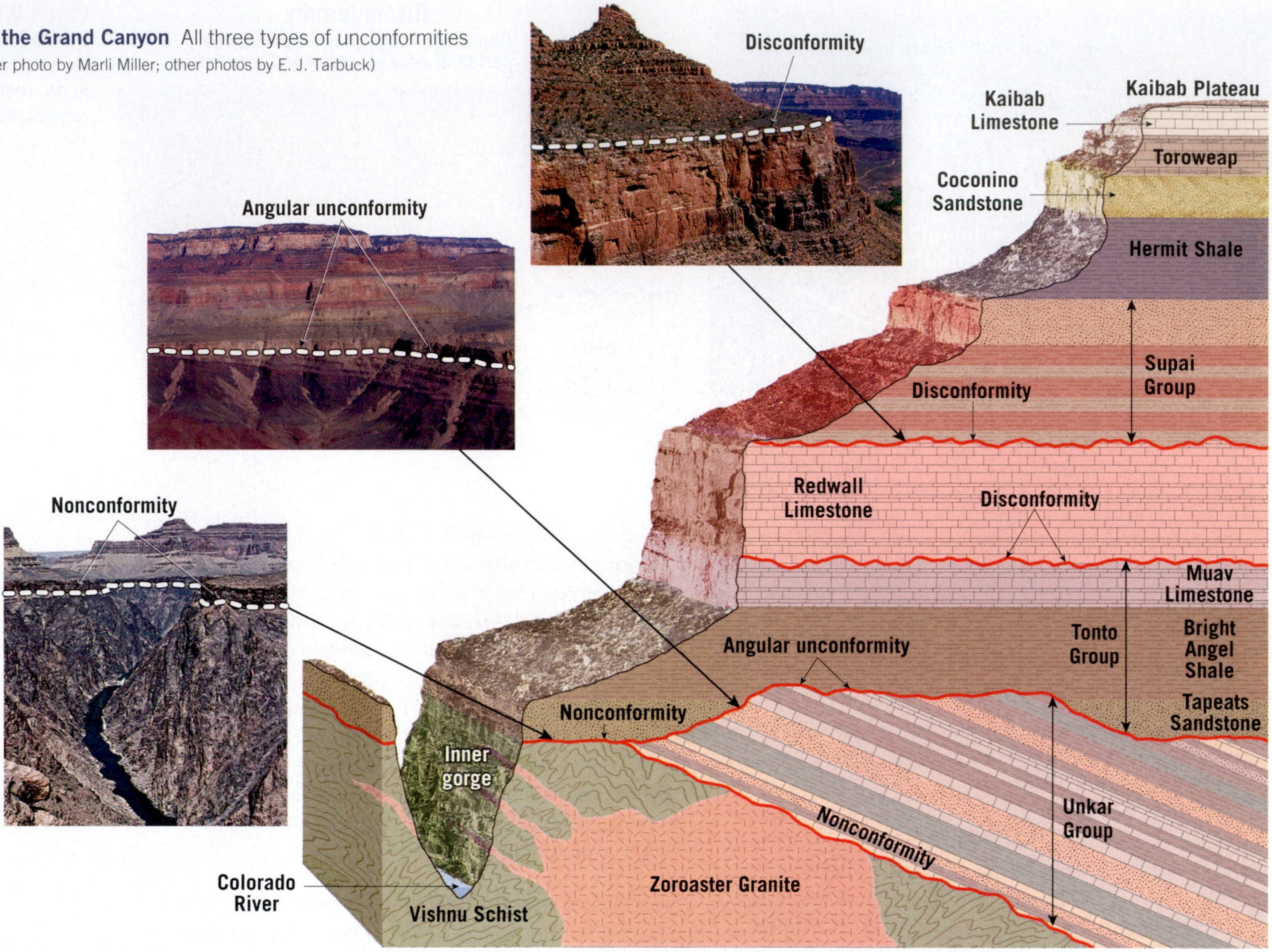

Disconformity

Kaibab Plateau

Kaibab Limestone

Toroweap

Coconino Sandstone

Hermit Shale

Angular unconformity

Supai Group

Disconformity

Redwall Limestone

Disconformity

Muav Limestone

Nonconformity

Tonto Group

Bright Angel Shale

Tapeats Sandstone

Angular unconformity

Nonconformity

Inner gorge

Unkar Group

Colorado River

Vishnu Schist

Zoroaster Granite

Nonconformity

EYE ON EARTH 9.2

This close-up shows pieces of diorite (darkest rock) in granite. The thin white line is a vein of quartz. Think of a vein as a tiny dike.

QUESTION 1 What term is applied to the pieces of diorite?

QUESTION 2 Place the quartz vein, diorite, and granite in order from oldest to youngest.

Marli Miller

SmartFigure 9.13
**Applying principles
of relative dating** (https://
goo.gl/w4HtAw)

Tutorial

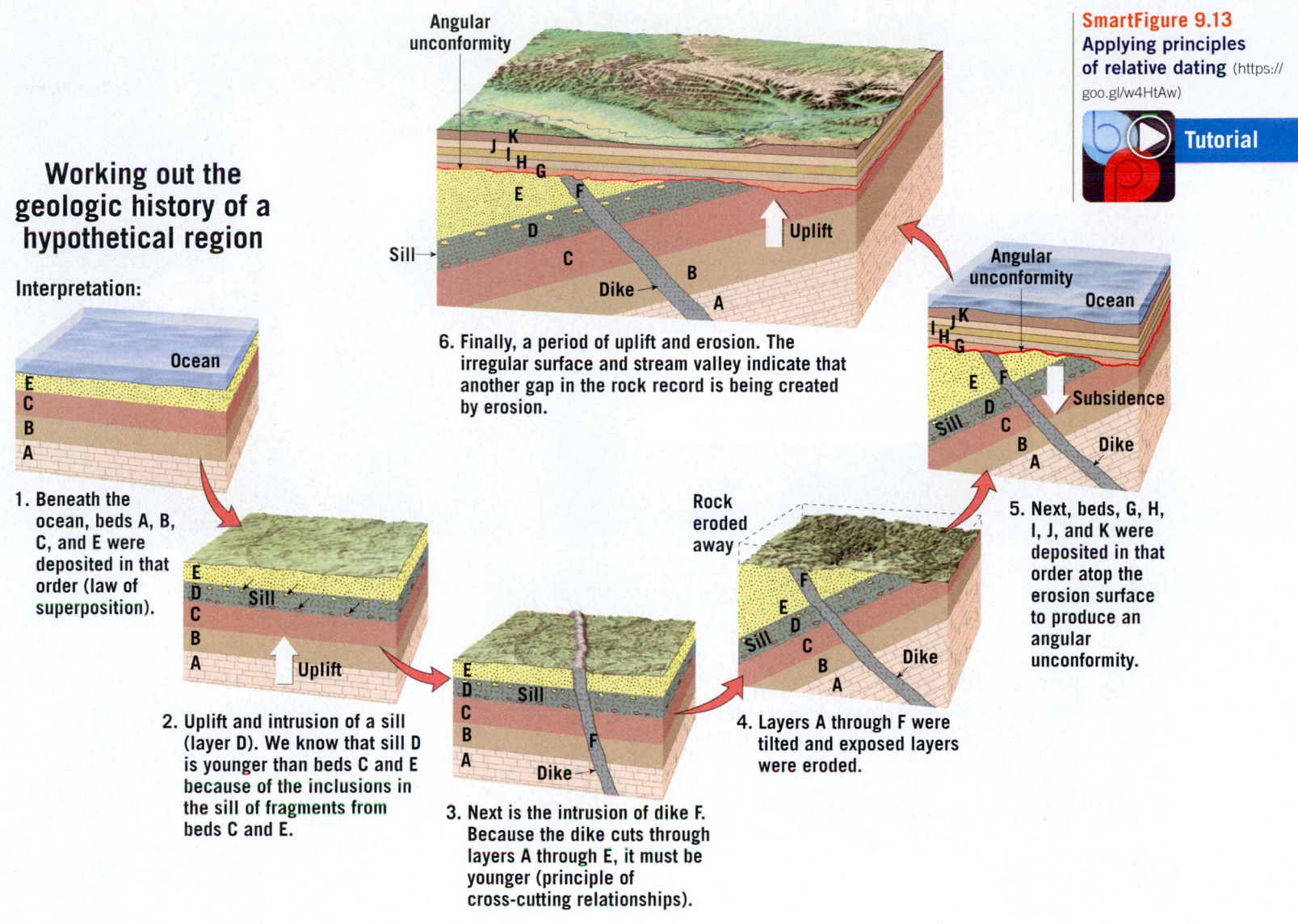

Working out the geologic history of a hypothetical region

Interpretation:

1. Beneath the ocean, beds A, B, C, and E were deposited in that order (law of superposition).

2. Uplift and intrusion of a sill (layer D). We know that sill D is younger than beds C and E because of the inclusions in the sill of fragments from beds C and E.

3. Next is the intrusion of dike F. Because the dike cuts through layers A through E, it must be younger (principle of cross-cutting relationships).

4. Layers A through F were tilted and exposed layers were eroded.

5. Next, beds, G, H, I, J, and K were deposited in that order atop the erosion surface to produce an angular unconformity.

6. Finally, a period of uplift and erosion. The irregular surface and stream valley indicate that another gap in the rock record is being created by erosion.

continuous. Unconformities represent vast amounts of time that have not been recorded in the canyon's layers. **Figure 9.12** is a geologic cross section of the Grand Canyon. All three types of unconformities can be seen in the canyon walls.

Applying Relative Dating Principles

If you apply the principles of relative dating to the hypothetical geologic cross section in **Figure 9.13**, you can place in proper sequence the rocks and the events they represent. The statements within the figure summarize the logic used to interpret the cross section.

In this example, we establish a relative time scale for the rocks and events in the area of the cross section. Remember that this method gives us no idea how many years of Earth history are represented, for we have no

numerical dates. Nor do we know how this area compares to any other. See GEOgraphics 9.1 for another example of applying relative dating principles.

9.1 | Concept Checks

1. Distinguish between numerical dates and relative dates.

2. Sketch and label four simple diagrams that illustrate each of the following: superposition, original horizontality, lateral continuity, and cross-cutting relationships.

3. What is the significance of an unconformity?

4. Distinguish among angular unconformity, disconformity, and nonconformity.

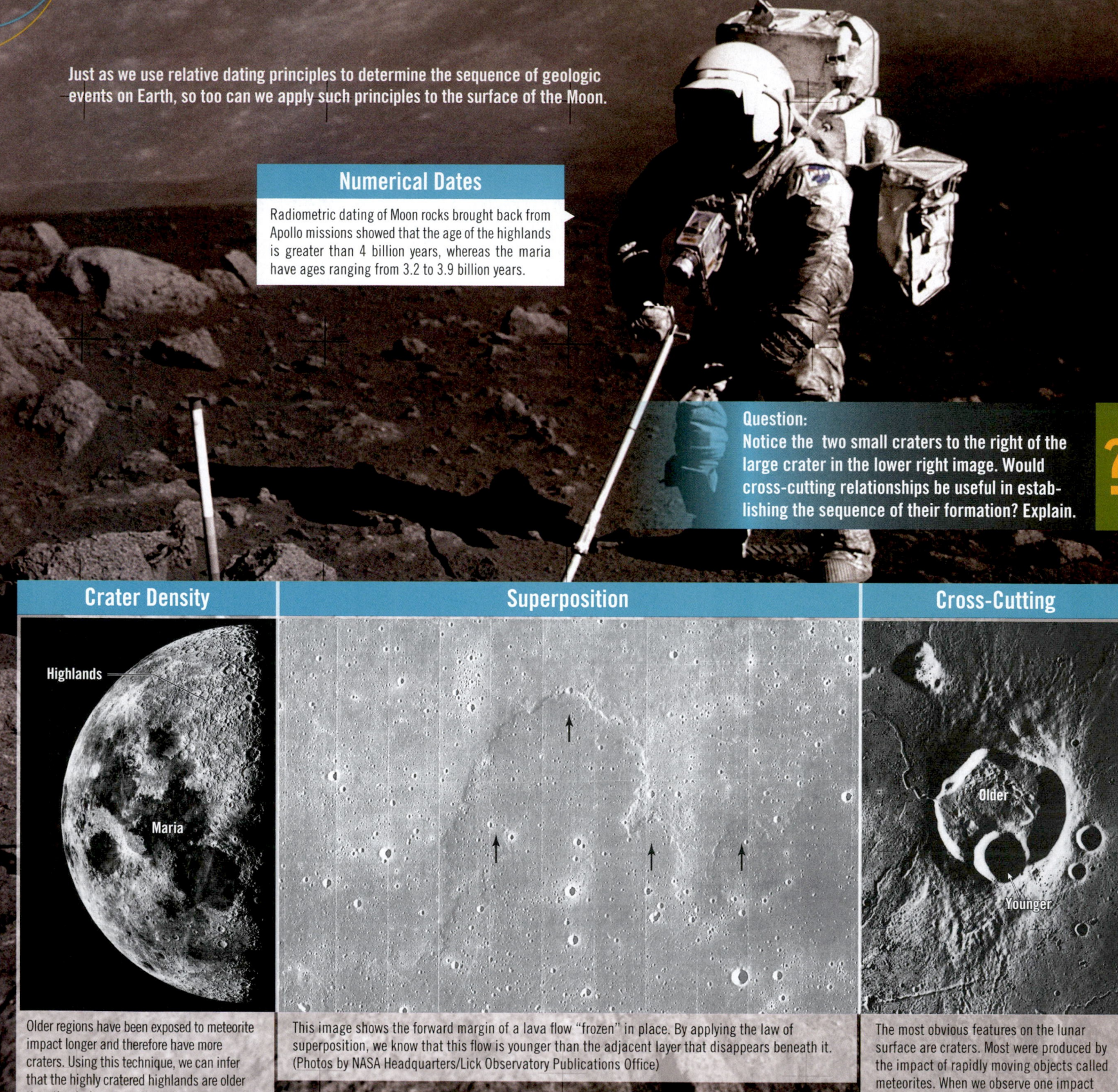

Dating the Lunar Surface

Just as we use relative dating principles to determine the sequence of geologic events on Earth, so too can we apply such principles to the surface of the Moon.

Numerical Dates

Radiometric dating of Moon rocks brought back from Apollo missions showed that the age of the highlands is greater than 4 billion years, whereas the maria have ages ranging from 3.2 to 3.9 billion years.

Question:
Notice the two small craters to the right of the large crater in the lower right image. Would cross-cutting relationships be useful in establishing the sequence of their formation? Explain.

?

Crater Density

Highlands

Maria

Older regions have been exposed to meteorite impact longer and therefore have more craters. Using this technique, we can infer that the highly cratered highlands are older than the dark areas called maria. The number of craters per unit area (called crater density) is obviously much greater in the highlands.

Superposition

This image shows the forward margin of a lava flow "frozen" in place. By applying the law of superposition, we know that this flow is younger than the adjacent layer that disappears beneath it. (Photos by NASA Headquarters/Lick Observatory Publications Office)

Cross-Cutting

Older

Younger

The most obvious features on the lunar surface are craters. Most were produced by the impact of rapidly moving objects called meteorites. When we observe one impact crater that overlaps another, we know the continuous unbroken crater came after the one it cuts across.

9.2 | Fossils: Evidence of Past Life

Define *fossil* and discuss the conditions that favor the preservation of organisms as fossils. List and describe various types of fossils.

Fossils, the remains or traces of prehistoric life, are inclusions in sediment and sedimentary rocks. They are basic and important tools for interpreting the geologic past. The scientific study of fossils is called **paleontology**. It is an interdisciplinary science that blends geology and biology in an attempt to understand all aspects of the succession of life over the vast expanse of geologic time (see GEOgraphics 9.2). Knowing the nature of the life-forms that existed at a particular time helps researchers understand past environmental conditions. Further, fossils are important time indicators and play a key role in correlating rocks of similar ages that are from different places.

Types of Fossils

Fossils are of many types. The remains of relatively recent organisms may not have been altered at all. Such objects as teeth, bones, and shells are common examples (**Figure 9.14**). Far less common are entire animals, flesh included, that have been preserved because of rather unusual circumstances. Remains of prehistoric elephants called mammoths that were frozen in the Arctic tundra of Siberia and Alaska are examples, as are the mummified remains of sloths preserved in a dry cave in Nevada.

Permineralization When mineral-rich groundwater permeates porous tissue such as bone or wood, minerals precipitate out of solution and fill pores and empty spaces, a process called *permineralization*. The formation of *petrified wood* involves permineralization with silica, often from a volcanic source such as a surrounding layer of volcanic ash. The wood is gradually transformed into chert, sometimes with colorful bands from impurities such as iron or carbon (**Figure 9.15A**). The word *petrified* literally means "turned into stone." Sometimes the microscopic details of the petrified structure are faithfully retained.

Molds and Casts Another common class of fossils is *molds* and *casts*. When a shell or other structure is buried in sediment and then dissolved by underground water, a *mold* is created. The mold faithfully reflects only the shape and surface marking of the organism; it does not reveal any information concerning its internal structure. If these hollow spaces are subsequently filled with mineral matter, *casts* are created (**Figure 9.15B**).

Skeleton of a mammoth, a prehistoric relative of modern elephant, from the La Brea tar pits.

Excavating bones from pit 91. It is a site rich in unaltered Ice Age organisms. Scientists have been excavating here since 1915.

Figure 9.14
La Brea tar pits The fossils here are actual (unaltered) remains.
(Excavation photo by Reed Saxon/ AP Wide World Photo; skeleton photo by Martin Shields/Alamy)

Figure 9.15
Types of fossils (Photo A by Bernhard Edmaier/Science Source; photos B, D, and F by E. J. Tarbuck; photo C by Florissant Fossil Beds National Monument; photo E by Colin Keates/Dorling Kindersley Media Library)

A. Petrified wood in Arizona's Petrified Forest National Park

B. This trilobite photo illustrates mold and cast

C. A fossil bee preserved as a thin carbon film

D. Impressions are common fossils and often show considerable detail

E. Spider in amber

F. Coprolite is fossil dung

and protected the remains from damage by water and air. As the resin hardened, a protective pressure-resistant case was formed.

Trace Fossils   In addition to the fossils already mentioned, there are numerous other types, many of them only traces of prehistoric life. Examples of such indirect evidence include:

* Tracks—animal footprints made in soft sediment that later turned into sedimentary rock.
* Burrows—tubes in sediment, wood, or rock made by an animal. These holes may later become filled with mineral matter and preserved. Some of the oldest-known fossils are believed to be worm burrows.
* Coprolites—fossil dung and stomach contents that can provide useful information pertaining to the size and food habits of organisms (**Figure 9.15F**).
* Gastroliths—highly polished stomach stones that were used in the grinding of food by some extinct reptiles.

Carbonization and Impressions   A type of fossilization called *carbonization* is particularly effective at preserving leaves and delicate animal forms. It occurs when fine sediment encases the remains of an organism. As time passes, pressure squeezes out the liquid and gaseous components and leaves behind a thin residue of carbon (**Figure 9.15C**). Black shale deposited as organic-rich mud in oxygen-poor environments often contains abundant carbonized remains. If the film of carbon is lost from a fossil preserved in fine-grained sediment, a replica of the surface, called an *impression*, may still show considerable detail (**Figure 9.15D**).

Amber   Delicate organisms, such as insects, are difficult to preserve, and consequently they are relatively rare in the fossil record. Not only must they be protected from decay, but they must not be subjected to any pressure that would crush them. One way in which some insects have been preserved is in *amber*, the hardened resin of ancient trees. The spider in **Figure 9.15E** was preserved after being trapped in a drop of sticky resin. Resin sealed off the insect from the atmosphere

Conditions Favoring Preservation

Only a tiny fraction of the organisms that have lived during the geologic past have been preserved as fossils. Normally, the remains of an animal or a plant are destroyed. Under what circumstances are they preserved? Two special conditions appear to be necessary: rapid burial and the possession of hard parts.

When an organism perishes, its soft parts usually are quickly eaten by scavengers or decomposed by bacteria. Occasionally, however, the remains are buried by sediment. When this occurs, the remains are protected from the surface environment, where destructive processes operate. Rapid burial, therefore, is an important condition favoring preservation.

In addition, animals and plants have a much better chance of being preserved as part of the fossil record if they have hard parts. Although traces and imprints of soft-bodied animals such as jellyfish, worms, and insects

How is paleontology different from archaeology?

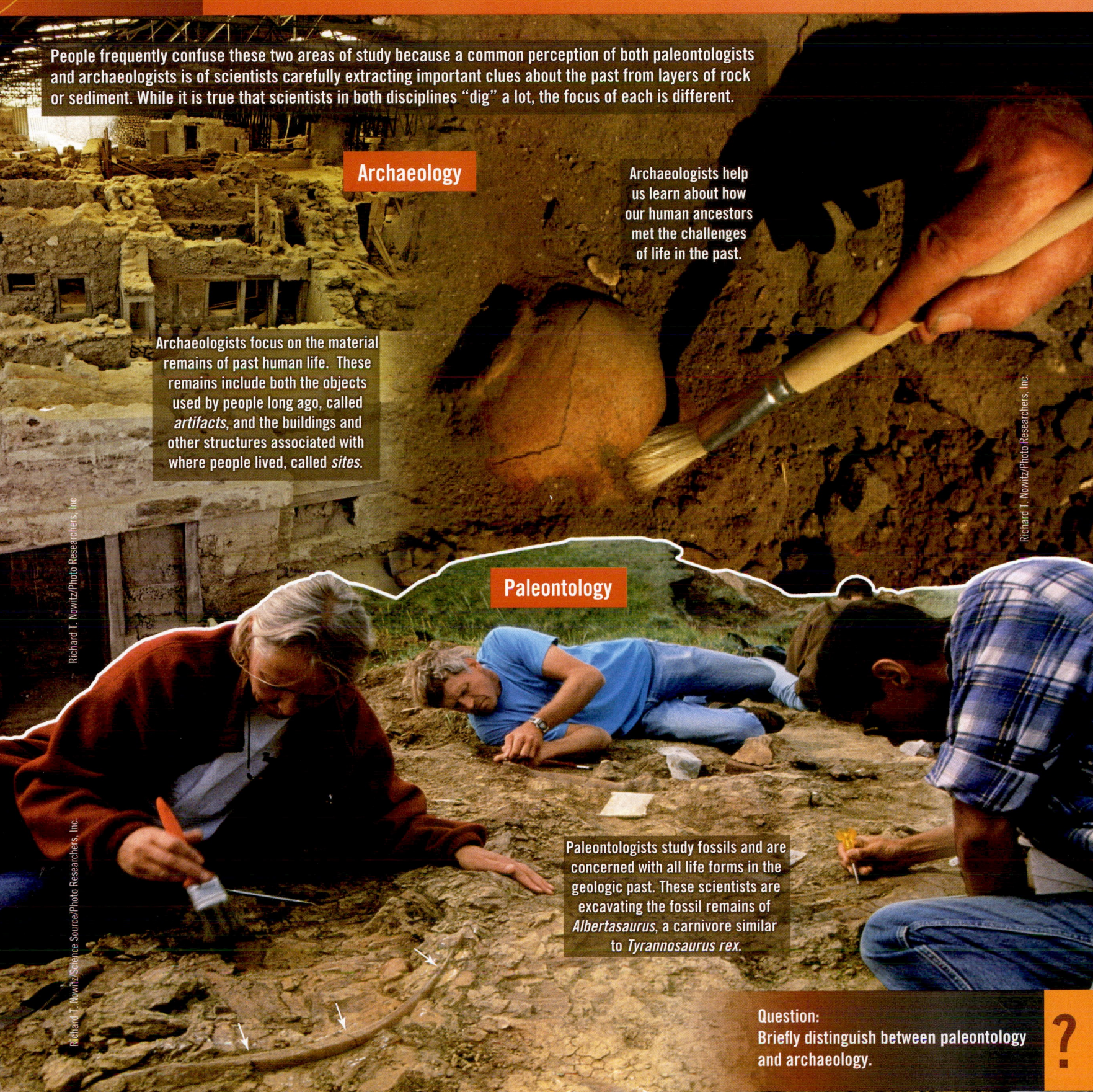

People frequently confuse these two areas of study because a common perception of both paleontologists and archaeologists is of scientists carefully extracting important clues about the past from layers of rock or sediment. While it is true that scientists in both disciplines "dig" a lot, the focus of each is different.

Archaeology

Archaeologists help us learn about how our human ancestors met the challenges of life in the past.

Archaeologists focus on the material remains of past human life. These remains include both the objects used by people long ago, called *artifacts*, and the buildings and other structures associated with where people lived, called *sites*.

Paleontology

Paleontologists study fossils and are concerned with all life forms in the geologic past. These scientists are excavating the fossil remains of *Albertasaurus*, a carnivore similar to *Tyrannosaurus rex*.

Richard T. Nowitz/Photo Researchers, Inc.

Richard T. Nowitz/Photo Researchers, Inc.

Richard T. Nowitz/Science Source/Photo Researchers, Inc.

Question:
Briefly distinguish between paleontology and archaeology.

?

exist, they are not common. Flesh usually decays so rapidly that preservation is exceedingly unlikely. Hard parts such as shells, bones, and teeth predominate in the record of past life.

Because preservation is contingent on special conditions, the record of life in the geologic past is biased. The fossil record of those organisms with hard parts that lived in areas of sedimentation is quite abundant. However, we get only an occasional glimpse of the vast array of other life-forms that did not meet the special conditions favoring preservation.

9.3   Correlation of Rock Layers

Explain how rocks of similar age that are in different places can be matched up.

To develop a geologic time scale that is applicable to the entire Earth, rocks of similar age in different regions must be matched up. Such a task is called **correlation**. By correlating the rocks from one place to another, a more comprehensive view of the geologic history of a region is possible. **Figure 9.16**, for example, shows the correlation of strata at three sites on the Colorado Plateau in southern Utah and northern Arizona. No single locale exhibits the entire sequence, but correlation reveals a more complete picture of the sedimentary rock record.

Correlation Within Limited Areas

Within a limited area, geologists can correlate rocks of one locality with those of another simply by walking along the outcropping edges, but this may not be possible when the rocks are mostly concealed by soil and vegetation. Correlation over short distances is often achieved by noting the position of a bed in a sequence of strata. Or a layer may be identified in another location if it is composed of distinctive or uncommon minerals.

Many geologic studies involve relatively small areas. Although they are important in their own right, their full value is realized only when they are correlated with other regions. Although the methods just described are sufficient to trace a rock formation over relatively short distances, they are not adequate for matching up rocks separated by great distances. When correlation between widely separated areas or between continents is the objective, geologists must rely on fossils.

Fossils and Correlation

The existence of fossils had been known for centuries, yet it was not until the late 1700s and early 1800s that their significance as geologic tools was made evident. During this period, English engineer and canal builder William Smith discovered that each rock formation in the canals he worked on contained fossils unlike those in the beds either above or below. Further, he noted that sedimentary strata in widely separated areas could be identified—and correlated—based on their distinctive fossil content.

Principle of Faunal Succession   Based on Smith's classic observations and the findings of many later geologists, one of the most important basic principles in historical geology was formulated: *Fossil organisms succeed one another in a definite and determinable order, and therefore any time period can be recognized by its fossil content.* This is known as the **principle of fossil succession**. In other words, when fossils are arranged according to their age, they do not present a random or haphazard picture. To the contrary, fossils document the evolution of life through time.

For example, an Age of Trilobites is recognized quite early in the fossil record. Then, in succession, paleontologists recognize an Age of Fishes, an Age of Coal Swamps, an Age of Reptiles, and an Age of Mammals. These "ages" pertain to groups that were especially plentiful and characteristic during particular time periods. Within each of the "ages" are many subdivisions, based, for example, on certain species of trilobites and certain types of fish, reptiles, and so on. This same succession of dominant organisms, never out of order, is found on every continent.

Index Fossils and Fossil Assemblages   When fossils were found to be time indicators, they became the most useful means of correlating rocks of similar age in

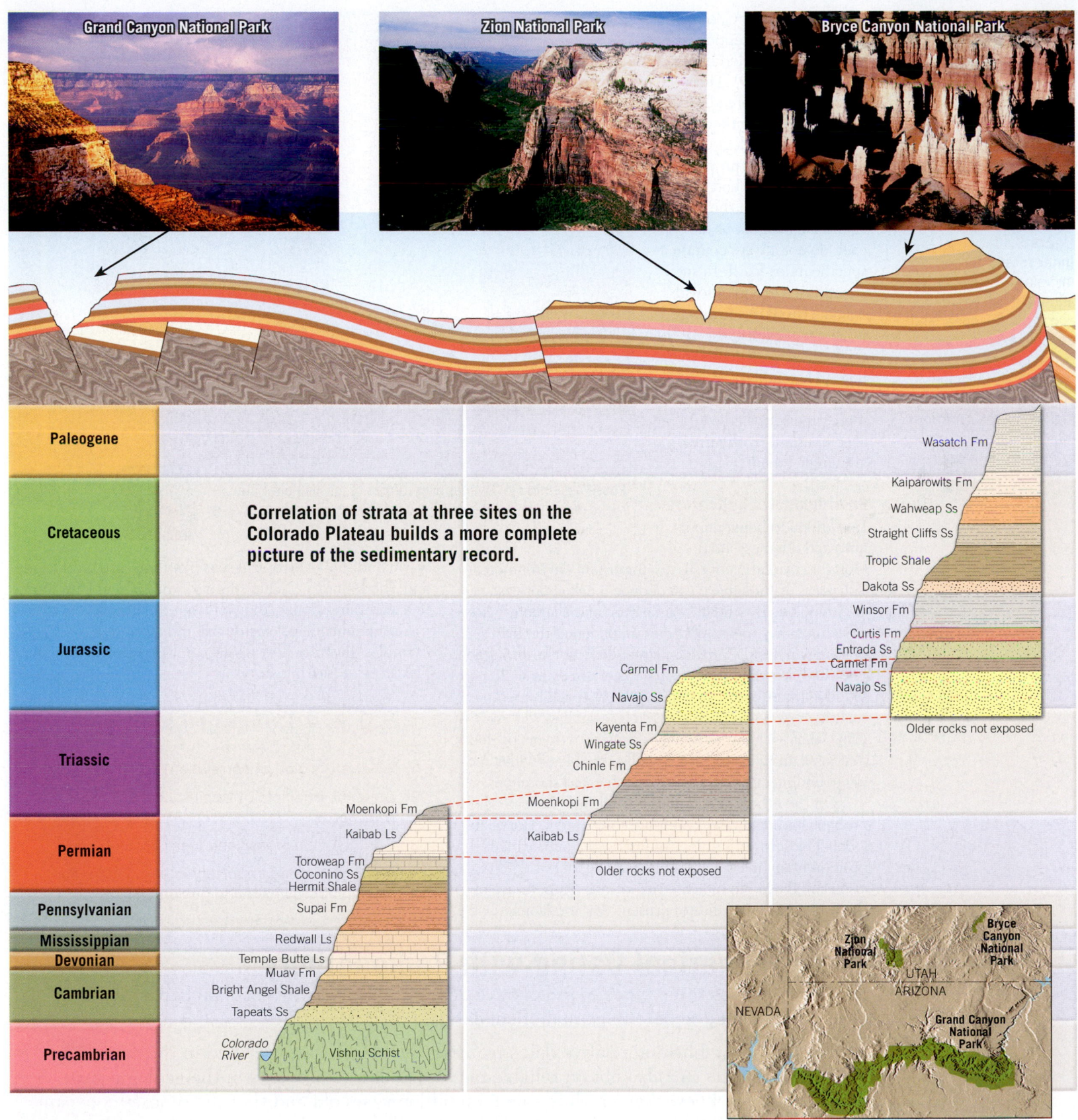

Grand Canyon National Park

Zion National Park

Bryce Canyon National Park

Correlation of strata at three sites on the Colorado Plateau builds a more complete picture of the sedimentary record.

Paleogene	Wasatch Fm
Cretaceous	Kaiparowits Fm Wahweap Ss Straight Cliffs Ss Tropic Shale Dakota Ss
Jurassic	Winsor Fm Curtis Fm Entrada Ss Carmel Fm Navajo Ss

Carmel Fm
Navajo Ss

Older rocks not exposed

Kayenta Fm
Wingate Ss
Chinle Fm

Moenkopi Fm

Kaibab Ls

Older rocks not exposed

Moenkopi Fm
Kaibab Ls
Toroweap Fm
Coconino Ss
Hermit Shale
Supai Fm
Redwall Ls
Temple Butte Ls
Muav Fm
Bright Angel Shale
Tapeats Ss
Colorado River
Vishnu Schist

Triassic
Permian
Pennsylvanian
Mississippian
Devonian
Cambrian
Precambrian

Bryce Canyon National Park
Zion National Park
UTAH
ARIZONA
NEVADA
Grand Canyon National Park

Figure 9.16
Correlation Matching strata at three locations on the Colorado Plateau. (Photos by E. J. Tarbuck)

different regions. Geologists pay particular attention to certain fossils called **index fossils** (**Figure 9.17**). These fossils are widespread geographically but limited to a short span of geologic time, so their presence provides an important method of matching rocks of the same age. Rock formations, however, do not always contain a specific index fossil. In such situations, a group of fossils, called a **fossil assemblage**, is used to establish the age of the bed. **Figure 9.18** illustrates how an assemblage of fossils may be used to date rocks more precisely than could be accomplished by the use of any single fossil.

Figure 9.17
Index fossils Since microfossils are often very abundant, widespread, and quick to appear and become extinct, they constitute ideal index fossils. This scanning electron micrograph shows marine microfossils from the Miocene epoch. (Photo by Biophoto Associates/Science Source)

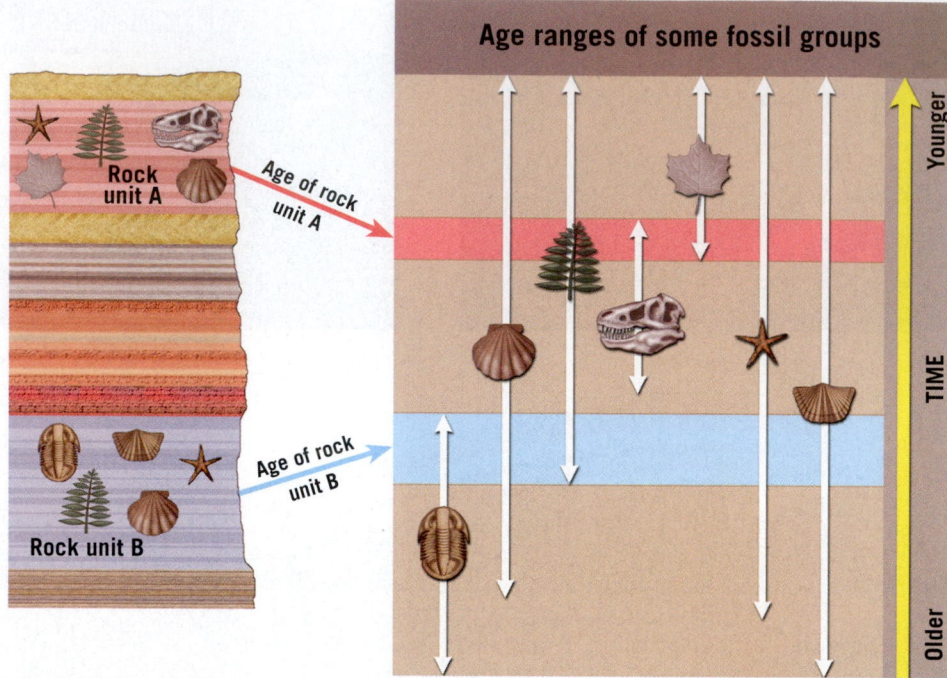

SmartFigure 9.18
Fossil assemblage Overlapping ranges of fossils help date rocks more exactly than using a single fossil. (https://goo.gl/dUqgP3)

Environmental Indicators

In addition to being important, and often essential, tools for correlation, fossils are important environmental indicators. Although we can deduce much about past environments by studying the nature and characteristics of sedimentary rocks, a close examination of the fossils present can usually provide a great deal more information. For example, when the remains of certain clam shells are found in limestone, a geologist quite reasonably assumes that the region was once covered by a shallow sea. Also, by using what we know of living organisms, we can conclude that fossil animals with thick shells, capable of withstanding pounding and surging waves, inhabited shorelines.

On the other hand, animals with thin, delicate shells probably indicate deep, calm offshore waters. Hence, by looking closely at the types of fossils, the approximate position of an ancient shoreline may be identified. Fossils also can be used to indicate the former temperature of the water. Certain kinds of present-day corals must live in warm and shallow tropical seas like those around Florida and The Bahamas. When similar types of coral are found in ancient limestones, they indicate the marine environment that must have existed when they were alive. These examples illustrate how fossils can help unravel the complex story of Earth history.

9.3 Concept Checks

1. What is the goal of correlation?
2. State the principle of fossil succession in your own words.
3. Contrast index fossil and fossil assemblage.
4. Along with being important in correlation, how else are fossils useful to geologists?

9.4 Numerical Dating with Radioactivity

Discuss three types of radioactive decay and explain how radioactive isotopes are used to determine numerical dates.

In addition to establishing relative dates by using the principles described in the preceding sections, scientists can also obtain reliable numerical dates for events in the geologic past. For example, we know that Earth is about 4.6 billion years old and that the dinosaurs became extinct about 65 million years ago. As discussed in Chapter 1, dates expressed in millions and billions of years truly stretch our imagination because our personal calendars involve time measured in hours, weeks, and years. Nevertheless, the vast expanse of geologic time is a reality, and radiometric dating allows us to measure it. In this section you will learn about radioactivity and its application in radiometric dating.

Reviewing Basic Atomic Structure

Recall from Chapter 3 that each atom has a *nucleus* that contains protons and neutrons and is orbited by electrons. *Electrons* have a negative electrical charge, and *protons* have a positive charge. A *neutron* is actually a proton and an electron combined, so it has no charge (it is neutral).

The *atomic number* is the number of protons in the nucleus. Every element has a different number of protons and thus a different identifying atomic number (hydrogen = 1, carbon = 6, oxygen = 8, uranium = 92, etc.). Atoms of the same element always have the same number of protons, so the atomic number stays constant.

Practically all of an atom's mass (99.9 percent) is in the nucleus, indicating that electrons have virtually no mass at all. So, by adding the protons and neutrons in an atom's nucleus, we derive the atom's *mass number*. The number of neutrons can vary, and these variants, or *isotopes*, have different mass numbers.

To summarize with an example, uranium's nucleus always has 92 protons, so its atomic number is always 92. But its neutron population varies, and uranium has three isotopes: uranium-234 (protons + neutrons = 234), uranium-235, and uranium-238. All three isotopes are mixed in nature. They look the same and behave the same in chemical reactions.

Radioactivity

The forces that bind protons and neutrons together in the nucleus are usually strong. However, in some isotopes, the nuclei are unstable because the forces that bind protons and neutrons together are not strong enough. As a result, the nuclei spontaneously break apart, or decay, in a process called **radioactivity**.

What happens when unstable nuclei break apart? Three common types of radioactive decay are illustrated in **Figure 9.19** and can be summarized as follows:

- *Alpha particles* (α particles) may be emitted from the nucleus. An alpha particle is composed of 2 protons and 2 neutrons. Thus, the emission of an alpha particle means that the mass number of the isotope is reduced by 4, and the atomic number is lowered by 2.
- When a *beta particle* (β particle), or an electron, is given off from a nucleus, the mass number remains unchanged because electrons have practically no mass. However, because the electron has come from a neutron (remember that a neutron is a combination of a proton and an electron), the nucleus contains 1 more proton than before. Therefore, the atomic number increases by 1.
- Sometimes an electron is captured by the nucleus. The electron combines with a proton and forms a neutron. As in the last example, the mass number remains unchanged. However, because the nucleus now contains 1 fewer proton, the atomic number decreases by 1.

An unstable radioactive isotope is called the *parent*, and isotopes resulting from the decay of the parent

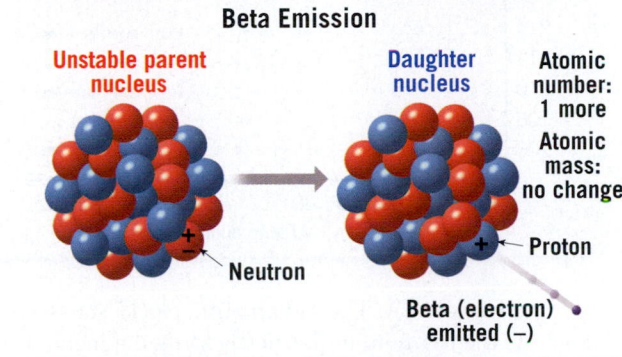

Alpha Emission

Unstable parent nucleus → Daughter nucleus
Atomic number: 2 fewer
Atomic mass: 4 fewer
Proton · Neutron
Alpha particle emitted

Beta Emission

Unstable parent nucleus → Daughter nucleus
Atomic number: 1 more
Atomic mass: no change
Neutron · Proton
Beta (electron) emitted (−)

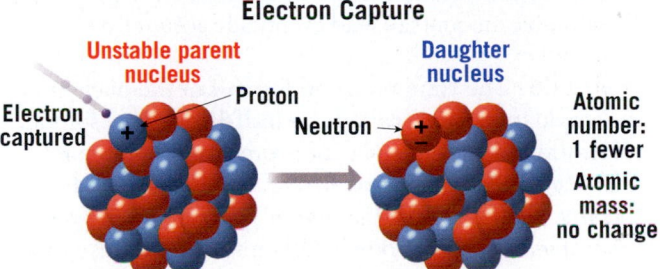

Electron Capture

Unstable parent nucleus → Daughter nucleus
Electron captured · Proton · Neutron
Atomic number: 1 fewer
Atomic mass: no change

Figure 9.19
Common types of radioactive decay Notice that in each example, the number of protons (atomic number) in the nucleus changes, thus producing a different element.

are termed the *daughter products*. **Figure 9.20** provides an example of radioactive decay. When the radioactive parent, uranium-238 (atomic number 92, mass number 238) decays, it follows a number of steps, emitting 8 alpha particles and 6 beta particles before finally becoming the stable daughter product lead-206 (atomic number 82, mass number 206). One of the unstable daughter products produced during this decay series is radon.

Radiometric Dating

Certainly among the most important results of the discovery of radioactivity is that it provides a reliable means of calculating the ages of rocks and minerals that contain particular radioactive isotopes. The procedure is called **radiometric dating**. Radiometric dating is reliable because the rates of decay for many isotopes have been precisely measured and do not vary under the physical conditions that exist in Earth's outer layers. Therefore, each radioactive isotope used for dating has been decaying at a fixed rate since the formation of the minerals in which it occurs, and the products of decay have been accumulating at a corresponding rate. For example, when uranium is incorporated into a mineral that crystallizes from magma, there is no lead (the stable daughter product)

Figure 9.20
Decay of U-238
Uranium-238 is an example of a radioactive decay series. Before the stable end product (Pb-206) is reached, many different isotopes are produced as intermediate steps.

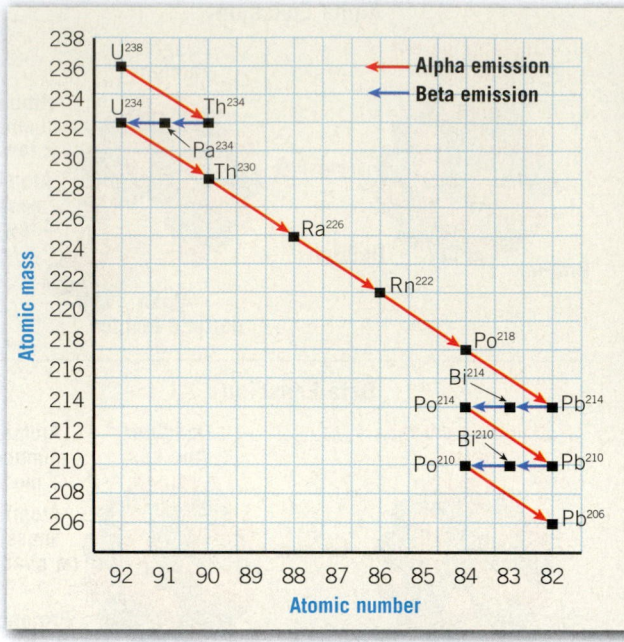

to daughter atoms is 1:7 (one parent atom for every seven daughter atoms).

If the half-life of a radioactive isotope is known and the parent/daughter ratio can be determined, the age of the sample can be calculated. For example, assume that the half-life of a hypothetical unstable isotope is 1 million years, and the parent/daughter ratio in a sample is 1:15. This ratio indicates that four half-lives have passed and that the sample must be 4 million years old.

Notice that the *percentage* of radioactive atoms that decay during one half-life is always the same: 50 percent. However, the *actual number* of atoms that decay with the passing of each half-life continually decreases. Thus, as the percentage of radioactive parent atoms declines, the proportion of stable daughter atoms rises, with the increase in daughter atoms just matching the drop in parent atoms. This fact is the key to radiometric dating.

from previous decay. The radiometric "clock" starts at this point. As the uranium in this newly formed mineral disintegrates, atoms of the daughter product are trapped, and measurable amounts of lead eventually accumulate.

Half-Life The time required for half of the nuclei in a sample to decay is called the **half-life** of the isotope. Half-life is a common way of expressing the rate of radioactive disintegration. **Figure 9.21** illustrates what occurs when a radioactive parent decays directly into its stable daughter product. When the quantities of parent and daughter are equal (ratio 1:1), we know that one half-life has transpired. When one-quarter of the original parent atoms remain and three-quarters have decayed to the daughter product, the parent/daughter ratio is 1:3, and we know that two half-lives have passed. After three half-lives, the ratio of parent atoms

Using Radioactive Isotopes

Of the many radioactive isotopes that exist in nature, five have proved particularly useful in providing radiometric ages for ancient rocks (**Table 9.1**). Rubidium-87, thorium-232, and the two isotopes of uranium are used only for dating rocks that are millions of years old, but potassium-40 is more versatile.

Potassium-Argon Dating Although the half-life of potassium-40 is 1.3 billion years, analytical techniques make it possible to detect tiny amounts of its stable daughter product, argon-40, in some rocks that are younger than 100,000 years. Another important reason for its frequent use is that potassium is an abundant constituent of many common minerals, particularly micas and feldspars.

Although potassium (K) has three natural isotopes, ^{39}K, ^{40}K, and ^{41}K, only ^{40}K is radioactive. When ^{40}K decays, it does so in two ways. About 11 percent changes to argon-40 (^{40}Ar) by means of electron capture (see Figure 9.19, bottom). The remaining 89 percent of ^{40}K decays to calcium-40 (^{40}Ca) by beta emission (see Figure 9.19, middle). The decay of ^{40}K to ^{40}Ca, however, is not useful for radiometric dating because the ^{40}Ca produced by radioactive disintegration cannot be distinguished from calcium that may have been present when the rock formed.

The potassium-argon clock begins when potassium-bearing minerals crystallize from a magma or form

SmartFigure 9.21
Radioactive decay curve
Change is exponential. Half of the radioactive parent remains after one half-life. After a second half-life, one-quarter of the parent remains, and so forth.
(https://goo.gl/DFHxZg)

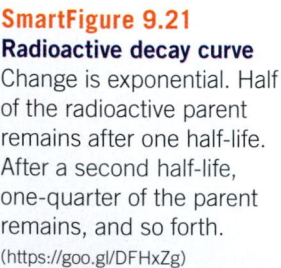

Tutorial

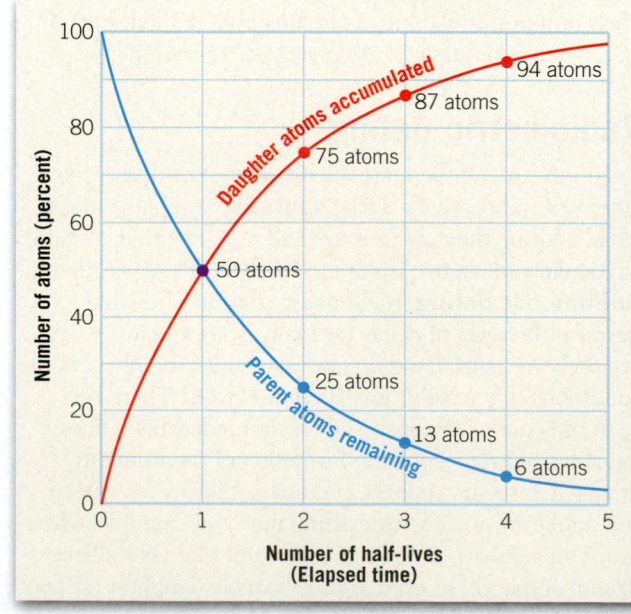

TABLE 9.1	Isotopes Frequently Used in Radiometric Dating	
Radioactive Parent	**Stable Daughter Product**	**Currently Accepted Half-Life Values**
Uranium-238	Lead-206	4.5 billion years
Uranium-235	Lead-207	704 million years
Thorium-232	Lead-208	14.1 billion years
Rubidium-87	Strontium-87	47.0 billion years
Potassium-40	Argon-40	1.3 billion years

within a metamorphic rock. At this point, the new minerals will contain ^{40}K but will be free of ^{40}Ar because this element is an inert gas that does not chemically combine with other elements. As time passes, the ^{40}K steadily decays by electron capture. The ^{40}Ar produced by this process remains trapped within the mineral's crystal lattice. Because no ^{40}Ar was present when the mineral formed, all of the daughter atoms trapped in the mineral must have come from the decay of ^{40}K. To determine a sample's age, the $^{40}K/^{40}Ar$ ratio is measured precisely and the known half-life for ^{40}K applied.

A Complex Process

Bear in mind that although the basic principle of radiometric dating is simple, the actual procedure is quite complex. The analysis that determines the quantities of parent and daughter must be painstakingly precise. In addition, some radioactive materials do not decay directly into the stable daughter product, as was the case with our hypothetical example, and this fact may further complicate the analysis. In the case of uranium-238, there are 13 intermediate unstable daughter products formed before the 14th and last daughter product, the stable isotope lead-206, is produced (see Figure 9.20).

Sources of Error

It is important to understand that an accurate radiometric date can be obtained only for a mineral that remained a closed system from its formation to the present. Obtaining a correct date is not possible unless there was neither addition nor loss of parent or daughter isotopes. This is not always the case, and an important limitation of the potassium-argon method arises from the fact that argon is a gas, and it may leak from minerals, throwing off measurements. Indeed, losses can be significant if the rock is subjected to relatively high temperatures.

Of course, a reduction in the amount of daughter product, ^{40}Ar, would lead to underestimating the rock's actual age. Sometimes temperatures remain high enough over a sufficiently long period so that all argon escapes. When this happens, the potassium-argon clock is reset, and dating the sample gives only the time of thermal resetting, not the true age of the rock. For other radiometric clocks, a loss of daughter atoms can occur if the rock has been subjected to weathering or leaching. To avoid such a problem, one simple safeguard is to use only fresh, unweathered material and not samples that exhibit signs of chemical alteration.

To guard against error in radiometric dating, scientists often use cross-checks. This involves subjecting a sample to two different methods. If the results agree, the likelihood is high that the date is reliable. If the results are appreciably different, other cross-checks must be employed to determine which, if either, is correct.

Earth's Oldest Rocks

Radiometric dating methods have produced literally thousands of dates for events in Earth history. Rocks exceeding 3.5 billion years in age are found on all of the continents. Earth's oldest rocks (so far) may be as old as 4.28 billion years (b.y.). Discovered in northern Quebec, Canada, on the shores of Hudson Bay, these rocks may be remnants of Earth's earliest crust. Rocks from western Greenland have been dated at 3.7 to 3.8 b.y., and rocks nearly as old are found in the Minnesota River Valley and northern Michigan (3.5 to 3.7 b.y.), in southern Africa (3.4 to 3.5 b.y.), and in western Australia (3.4 to 3.6 b.y.). Tiny crystals of the mineral zircon having radiometric ages as old as 4.3 b.y. have been found in younger sedimentary rocks in western Australia. The source rocks for these tiny durable grains either no longer exist or have not yet been found.

Radiometric dating has vindicated the ideas of Hutton, Darwin, and others, who more than 150 years ago inferred that geologic time must be immense. Indeed, modern dating methods prove that there has been enough time for the processes we observe to have accomplished tremendous tasks.

Dating with Carbon-14

To date relatively recent events, carbon-14 is used (**Figure 9.22**). Carbon-14 is the radioactive isotope of carbon. The process is often called **radiocarbon dating**. Because the half-life of carbon-14 is only 5730 years, radiocarbon dating can be used to date events from the historic past as well as those from very recent geologic history. In some cases carbon-14 can be used to date events as far back as 70,000 years.

Carbon-14 is continuously produced in the upper atmosphere as a result of cosmic-ray bombardment. Cosmic rays (high-energy nuclear particles) shatter the nuclei of gas atoms, releasing neutrons. Some of the neutrons

Production of carbon-14

Nitrogen-14
atomic number 7
atomic mass 14

Neutron capture

Carbon-14
atomic number 6
atomic mass 14

Proton emitted

A.

Decay of carbon-14

Neutron

Proton

(–)

Beta (electron) emitted

Carbon-14
atomic number 6
atomic mass 14

Nitrogen-14
atomic number 7
atomic mass 14

B.

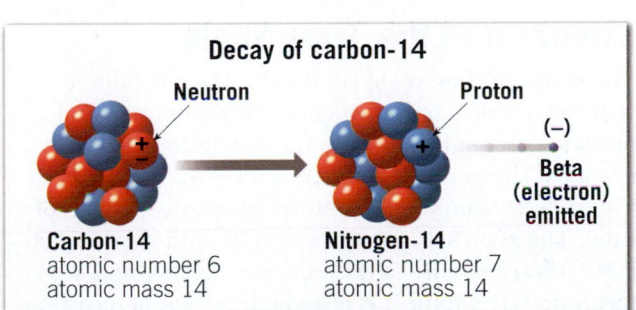

Figure 9.22
Carbon-14 Production and decay of radiocarbon. These sketches represent the nuclei of the respective atoms.

Figure 9.23
Cave art Chauvet Cave in southern France, discovered in 1994, contains some of the earliest-known cave paintings. Radiocarbon dating indicates that most of the images were drawn between 30,000 and 32,000 years ago.
(Photo by Javier Trueba/MSF/Science Source)

carbon-14 gradually decreases as it decays to nitrogen-14 by beta emission (Figure 9.22B). By comparing the proportions of carbon-14 and carbon-12 in a sample, radiocarbon dates can be determined. It is important to emphasize that carbon-14 can only be used to date organic materials, such as wood, charcoal, bones, flesh, and cloth.

Although carbon-14 is only useful in dating the last small fraction of geologic time, it is a valuable tool for anthropologists, archaeologists, and historians, as well as for geologists who study very recent Earth history (**Figure 9.23**). In fact, the development of radiocarbon dating was considered so important that the chemist who discovered this application, Willard F. Libby, received a Nobel Prize in 1960.

are absorbed by nitrogen atoms (atomic number 7, mass number 14), causing each nucleus to emit a proton. As a result, the atomic number decreases by 1 (to 6), and a different element, carbon-14, is created (Figure 9.22A). This isotope of carbon quickly becomes incorporated into carbon dioxide, which circulates in the atmosphere and is absorbed by living matter. As a result, all organisms—including you—contain a small amount of carbon-14.

As long as an organism is alive, the decaying radiocarbon is continually replaced, and the proportions of carbon-14 and carbon-12 remain constant. Carbon-12 is the stable and most common isotope of carbon. However, when any plant or animal dies, the amount of

9.4 Concept Checks

1. List three types of radioactive decay. For each type, describe how the atomic number and atomic mass change.

2. Sketch a simple diagram to explain the idea of half-life.

3. Why is radiometric dating a reliable method for determining numerical dates?

4. For what time span does radiocarbon dating apply?

9.5 | The Geologic Time Scale

Distinguish among the four basic time units that make up the geologic time scale and explain why the time scale is considered to be a dynamic tool.

Geologists have divided the whole of geologic history into units of varying length. Together, they compose the **geologic time scale** of Earth history (**Figure 9.24**). The major units of the time scale were delineated during the nineteenth century, principally by scientists in Western Europe and Great Britain. Because radiometric dating was unavailable at that time, the entire time scale was created using methods of relative dating. It was only in the twentieth century that radiometric methods permitted numerical dates to be added.

Structure of the Time Scale

The geologic time scale subdivides the 4.6-billion-year history of Earth into many different units and provides a meaningful time frame within which the events of the geologic past are arranged. As shown in Figure 9.24, **eons** represent the greatest expanses of time. The eon that began about 542 million years ago is the **Phanerozoic**, a term derived from Greek words meaning "visible life." It is an appropriate description because the rocks and deposits of the Phanerozoic eon contain abundant fossils that document major evolutionary trends.

Another glance at the time scale reveals that eons are divided into **eras**. The Phanerozoic eon consists of the **Paleozoic era** (*paleo* = ancient, *zoe* = life), the **Mesozoic era** (*meso* = middle, *zoe* = life), and the **Cenozoic era** (*ceno* = recent, *zoe* = life). As the names imply, these eras are bounded by profound worldwide changes in life-forms.*

*Major changes in life-forms are discussed in Chapter 22.

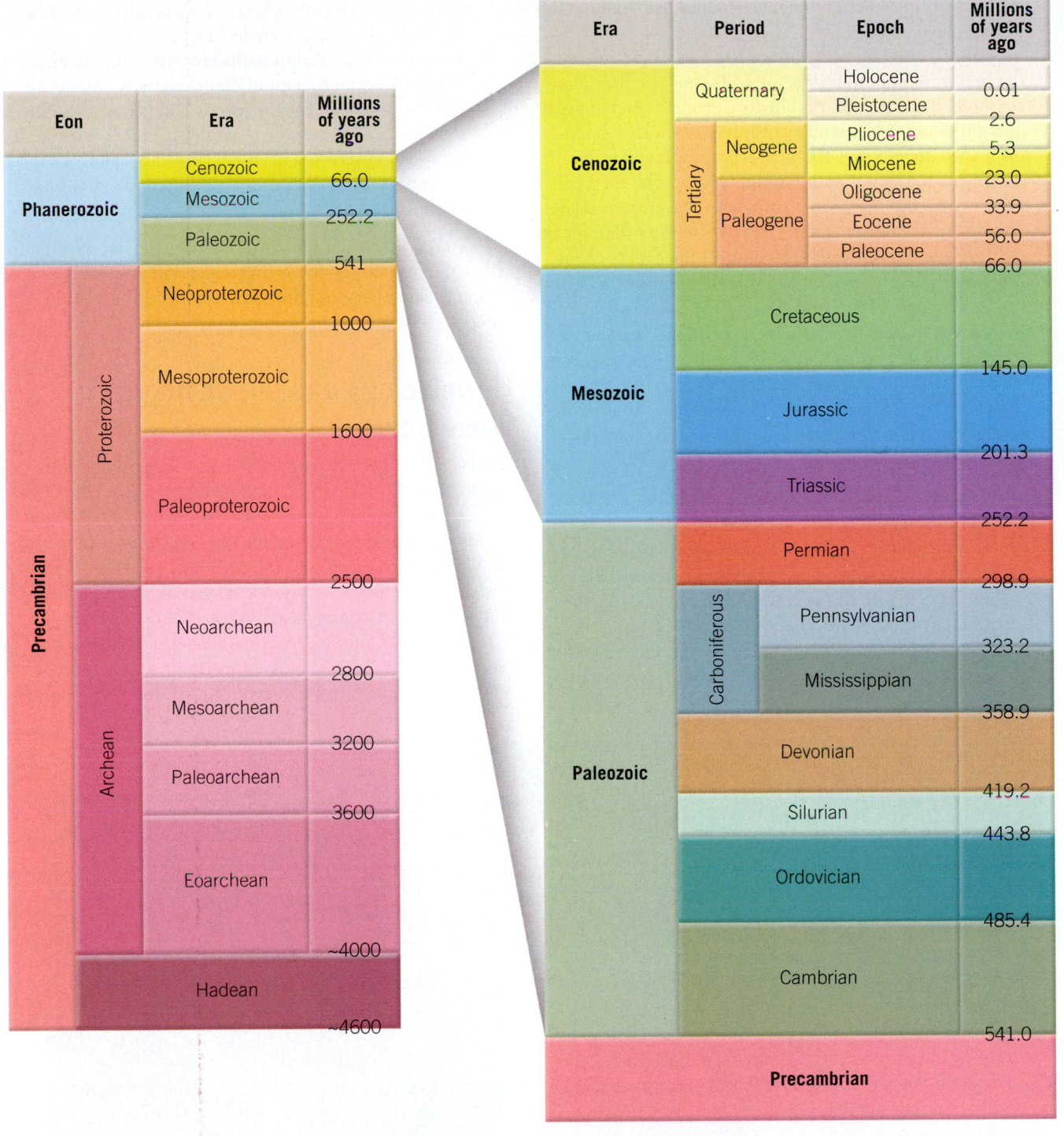

Figure 9.24
Geologic time scale
Numbers on the time scale represent time in millions of years before present. Numerical dates were added long after the time scale was established using relative dating techniques. The dates on this time scale are those currently accepted by the International Commission on Stratigraphy (ICS) in 2014. The color scheme used on this chart was selected because it is similar to that used by the ICS.

Each era of the Phanerozoic eon is further divided into time units known as **periods**. The Paleozoic has seven, and the Mesozoic and Cenozoic each have three. Each of these periods is characterized by a somewhat less profound change in life-forms as compared with the eras.

Each of the periods is divided into still smaller units called **epochs**. As you can see in Figure 9.24, seven epochs have been named for the periods of the Cenozoic. The epochs of other periods usually are simply termed *early*, *middle*, and *late*.

Precambrian Time

Notice that the detailed portion of the geologic time scale does not begin until about 542 million years ago, the date for the beginning of the Cambrian period. The nearly 4 billion years prior to the Cambrian are divided into two eons, the **Archean** (*archaios* = ancient) and the **Proterozoic** (*proteros* = before, *zoe* = life). It is also common for this vast expanse of time to simply be referred to as the **Precambrian**. Although it represents about 88 percent of Earth history, the Precambrian is not divided into nearly as many smaller time units as the Phanerozoic eon.

Why is the huge expanse of Precambrian time not divided into numerous eras, periods, and epochs? The reason is that Precambrian history is not known in great enough detail. The quantity of information that geologists have deciphered about Earth's past is somewhat analogous to the detail of human history. The further back we go, the less that is known. Certainly more data and information exist about the past 10 years than for the first decade of the twentieth century; the events of the nineteenth century have been documented much better than the events of the first century C.E.; and so on. So it is with Earth history. The more recent past has the freshest, least disturbed, and most observable record. The further back in time a geologist goes, the more fragmented the record and clues become. Other reasons to explain our lack of a detailed time scale for this vast segment of Earth history include:

- The first abundant fossil evidence does not appear in the geologic record until the beginning of the Cambrian period. Prior to the Cambrian, simple life-forms such as algae, bacteria, fungi, and worms predominated. All of these organisms lack hard parts, an important condition favoring preservation. For this reason, there is only a meager Precambrian fossil record. Many exposures of Precambrian rocks have

been studied in some detail, but correlation can be difficult when fossils are lacking.

- Because Precambrian rocks are very old, most have been subjected to a great many changes. Much of the Precambrian rock record is composed of highly distorted metamorphic rocks. This makes it difficult to interpret past environments because many of the clues present in the original sedimentary rocks have been destroyed.

Radiometric dating has provided a partial solution to the troublesome task of dating and correlating Precambrian rocks. But untangling the complex Precambrian record still remains a daunting task.

Terminology and the Geologic Time Scale

Some terms that are associated with the geologic time scale are not "officially" recognized as being a part of it. The best known, and most common, example is *Precambrian*—the informal name for the eons that came before the current Phanerozoic eon. Although the term *Precambrian* has no formal status on the geologic time scale, it has been traditionally used as though it does.

Hadean is another informal term that is found on some versions of the geologic time scale and is used by many geologists. It refers to the earliest interval (eon) of

EYE ON EARTH 9.3

This image is from the bottom of the Grand Canyon, a zone called the Inner Gorge. The dark rock is the Vishnu Schist. The light-colored rock, called the Zoroaster Granite, is a series of dikes. Both rocks date from Precambrian time. Suppose you are on a raft trip through the Grand Canyon. Your companions are bright and curious but are not trained in geology, as you are. Sitting around the campfire the night before you reached the site pictured here, you and your fellow rafters discussed the geologic time scale and the magnitude of geologic time.

QUESTION 1 When you arrive at this site, someone asks why Precambrian history seems so sketchy and why this time span is not divided into nearly as many subdivisions as Phanerozoic time. Use this setting as you answer these questions.

QUESTION 2 Which is older, the Vishnu Schist or the Zoroaster Granite? Explain.

Rafting the Colorado River

Michael Collier

Earth history—before the oldest-known rocks. When the term was coined in 1972, the age of Earth's oldest rocks was about 3.8 billion years. Today that number stands at slightly greater than 4 billion, and, of course, is subject to revision. The name *Hadean* derives from *Hades*, Greek for *underworld*—a reference to the "hellish" conditions that prevailed on Earth early in its history.

Effective communication in the geosciences requires that the geologic time scale consist of standardized divisions and dates. So, who determines which names and dates on the geologic time scale are "official"? The organization that is largely responsible for maintaining and updating this important document is the International Commission on Stratigraphy (ICS), a committee of the International Union of Geological Sciences. Advances in the geosciences require that the scale be periodically updated to include changes in unit names and boundary age estimates.

An ICS geologic time scale from just a few years ago would have shown the Cenozoic era divided into the Tertiary and Quaternary periods. However, more recent versions divide the Tertiary period into the Paleogene and Neogene periods. As our understanding of this time span has changed, so too has its designation on the geologic time scale. Today, the Tertiary period is considered a "historic" name and has no official status on the ICS version of the time scale. Many time scales still contain references to the Tertiary period, though, including Figure 9.24. One reason for this is that a great deal of past (and some current) geologic literature uses this name.

Some scientists have suggested that the Holocene epoch has ended and that we have entered a new epoch called the *Anthropocene*. It is considered to be the span (beginning in the early 1800s) in which the global environmental effects of increased human population and economic development have dramatically transformed Earth's surface. Although this designation is currently used as an informal metaphor for human-caused global environmental change, a number of scientists feel that there is merit in recognizing the Anthropocene as a new "official" geologic epoch.

For those studying historical geology, it is important to realize that the geologic time scale is a dynamic tool that continues to be refined as our knowledge and understanding of Earth history evolves.

9.5 | Concept Checks

1. List the four basic units that make up the geologic time scale.

2. Why is *zoic* part of so many names on the geologic time scale?

3. What term applies to *all* of geologic time prior to the Phanerozoic eon? Why is this span *not* divided into as many smaller time units as the Phanerozoic eon?

4. To what does the term *Hadean* apply? Is it an "official" part of the geologic time scale?

9.6 | Determining Numerical Dates for Sedimentary Strata

Explain how reliable numerical dates are determined for layers of sedimentary rock.

Although reasonably accurate numerical dates have been worked out for the periods of the geologic time scale (see Figure 9.24), this is not an easy task. The primary difficulty in assigning numerical dates to units of time is the fact that not all rocks can be dated by using radiometric methods. For a radiometric date to be useful, all the minerals in the rock must have formed at about the same time. For this reason, radioactive isotopes can be used to determine when minerals in an igneous rock crystallized and when pressure and heat created new minerals in a metamorphic rock.

However, samples of sedimentary rock can only rarely be dated directly by radiometric means. Although a detrital sedimentary rock may include particles that contain radioactive isotopes, the rock's age cannot be accurately determined because the grains composing the rock are not the same age as the rock in which they occur. Rather, the sediments have been weathered from rocks of diverse ages.

Radiometric dates obtained from metamorphic rocks may also be difficult to interpret because the age of a particular mineral in a metamorphic rock does not necessarily represent the time when the rock initially formed. Instead, the date might indicate any one of a number of subsequent metamorphic phases.

If samples of sedimentary rocks rarely yield reliable radiometric ages, how can numerical dates be assigned

Figure 9.25
Dating sedimentary strata Numerical dates for sedimentary layers are usually determined by examining their relationship to igneous rocks.

Rocks of Paleogene age
Rocks of Cretaceous age
Rocks of Jurassic age

Wasatch Formation
Mesaverde Formation
Mancos Shale
Dakota Sandstone
Volcanic ash bed dated at 160 million years
Morrison Formation
Summerville Formation

Igneous dike dated at 66 million years

to sedimentary layers? Usually a geologist must relate the strata to datable igneous masses, as in **Figure 9.25**. In this example, radiometric dating has determined the ages of the volcanic ash bed in the Morrison Formation and the dike cutting the Mancos Shale and Mesaverde Formation. The sedimentary beds below the ash are obviously older than the ash, and all the layers above the ash are younger. The dike is younger than the Mancos Shale and the Mesaverde Formation but older than the

Wasatch Formation because the dike does not intrude these Paleogene-age rocks. From this kind of evidence, geologists estimate that the last part of the Morrison Formation was deposited about 160 million years ago, as indicated by the ash bed. Further, they conclude that the Paleogene period began after the intrusion of the dike, 66 million years ago. This is just one example of literally thousands that illustrate how datable materials are used to bracket the various episodes in Earth history within specific time periods. It shows the necessity of combining laboratory dating methods with relative dating principles applied to field observations of rocks.

9.6 Concept Checks

1. Briefly explain why it is often difficult to assign a reliable numerical date to a sample of sedimentary rock.

2. How might a numerical date for a layer of sedimentary rock be determined?

EYE ON EARTH 9.4

This is a close-up of a sample of the detrital sedimentary rock conglomerate. The rock contains radioactive isotopes that will yield numerical dates.

QUESTION 1 Although radioactive isotopes are present, a reliable numerical date for this conglomerate cannot be *accurately* determined. Explain.

QUESTION 2 How might a numerical age range be established for the conglomerate layer?

E.J. Tarbuck

9.1 Creating a Time Scale: Relative Dating Principles

Distinguish between numerical and relative dating and apply relative dating principles to determine a time sequence of geologic events.

KEY TERMS numerical date, relative date, principle of superposition, principle of original horizontality, principle of lateral continuity, principle of cross-cutting relationships, principle of inclusions, conformable, unconformity, angular unconformity, disconformity, nonconformity

- The two types of dates that geologists use to interpret Earth history are (1) relative dates, which put events in their proper sequence of formation, and (2) numerical dates, which pinpoint the time in years when an event took place.
- Relative dates can be established using the principles of superposition, original horizontality, cross-cutting relationships, and inclusions.

Unconformities, gaps in the geologic record, may be identified during the relative dating process.

Q The accompanying photo shows four features. Place the features in the proper sequence, from oldest to youngest. Explain your reasoning.

Basalt xenolith — Joint — Granite dike — Granite

Mike Beauregard

9.2 Fossils: Evidence of Past Life

Define *fossil* and discuss the conditions that favor the preservation of organisms as fossils. List and describe various types of fossils.

KEY TERMS fossil, paleontology

- Fossils are remains or traces of ancient life. Paleontology is the branch of science that studies fossils.
- Fossils can form through many processes. For an organism to be preserved as a fossil, it usually needs to be buried rapidly. Also, an organism's hard parts are most likely to be preserved because soft tissue decomposes rapidly in most circumstances.

Q What term is used to describe the type of fossil that is shown here? Briefly, how did it form?

E.J. Tarbuck

9.3 Correlation of Rock Layers

Explain how rocks of similar age that are in different places can be matched up.

KEY TERMS correlation, principle of fossil succession, index fossil, fossil assemblage

- Matching up exposures of rock that are the same age but are in different places is called correlation. By correlating rocks from around the world, geologists developed the geologic time scale and obtained a fuller perspective on Earth history.
- Fossils can be used to correlate sedimentary rocks in widely separated places by using the rocks' distinctive fossil content and applying the principle of fossil succession. The principle states that fossil organisms succeed one another in a definite and determinable order, and, therefore, a time period can be recognized by examining its fossil content.
- Index fossils are particularly useful in correlation because they are widespread and associated with a relatively narrow time span. The overlapping ranges of fossils in an assemblage may be used to establish an age for a rock layer that contains multiple fossils.
- Fossils may be used to establish ancient environmental conditions that existed when sediment was deposited.

Q Which would make a better index fossil for modern times: a penguin or a pigeon? Why?

9.4 Numerical Dating with Radioactivity

Discuss three types of radioactive decay and explain how radioactive isotopes are used to determine numerical dates.

KEY TERMS radioactivity, radiometric dating, half-life, radiocarbon dating

- Radioactivity is the spontaneous breaking apart (decay) of certain unstable atomic nuclei. Three common forms of radioactive decay are (1) emission of an alpha particle from the nucleus, (2) emission of a beta particle (electron) from the nucleus, and (3) capture of an electron by the nucleus.

- An unstable radioactive isotope, called a parent, will decay and form daughter products. The length of time for one-half of the nuclei of a radioactive isotope to decay is called the half-life of the isotope. If the half-life of an isotope is known, and the parent/daughter ratio can be measured, the age of the sample can be calculated.

Q Measurements of zircon crystals containing trace amounts of uranium from a specimen of granite yield parent/daughter ratios of 25 percent parent (uranium-235) and 75 percent daughter (lead-206). The half-life of uranium-235 is 704 million years. How old is the granite?

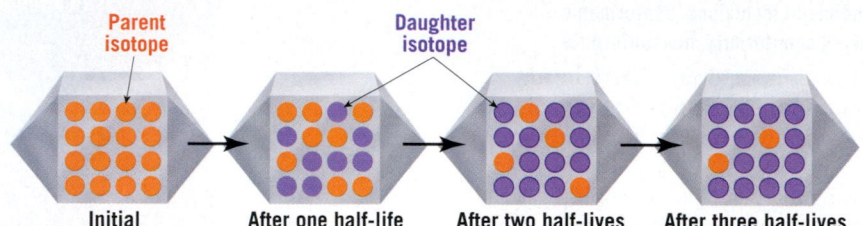

Parent isotope / Daughter isotope

Initial | After one half-life | After two half-lives | After three half-lives

9.5 The Geologic Time Scale

Distinguish among the four basic time units that make up the geologic time scale and explain why the time scale is considered to be a dynamic tool.

KEY TERMS geologic time scale, eon, Phanerozoic, era, Paleozoic era, Mesozoic era, Cenozoic era, period, epoch, Archean, Proterozoic, Precambrian

- Earth history is divided into units of time on the geologic time scale. Eons are divided into eras, which each contain multiple periods. Periods are divided into epochs.

- Precambrian time includes the Archean and Proterozoic eons. It is followed by the Phanerozoic eon, which is well documented by abundant fossil evidence, resulting in many subdivisions.
- The geologic time scale is a work in progress, continually being refined as new information becomes available.

Q Is the Mesozoic an example of an eon, an era, a period, or an epoch? What about the Jurassic?

9.6 Determining Numerical Dates for Sedimentary Strata

Explain how reliable numerical dates are determined for layers of sedimentary rock.

- Sedimentary strata are usually not directly datable using radiometric techniques because they consist of the material produced by the weathering of other rocks. A particle in a sedimentary rock comes from some older source rock. If you were to date the particle using isotopes, you would get the age of the source rock, not the age of the sedimentary deposit.
- One way geologists assign numerical dates to sedimentary rocks is to use relative dating principles to relate them to datable igneous masses, such as dikes and volcanic ash beds. A layer may be older than one igneous feature and younger than another.

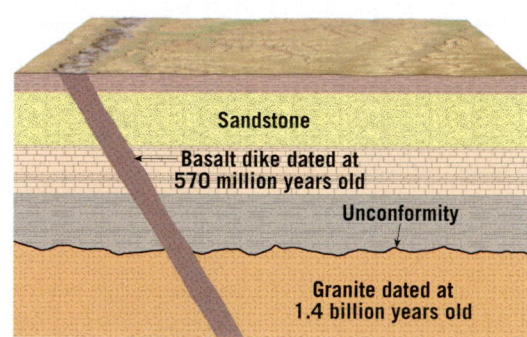

Sandstone

Basalt dike dated at 570 million years old

Unconformity

Granite dated at 1.4 billion years old

Q Express the numerical age of the sandstone layer in the diagram as accurately as possible.

Give It Some Thought

1. The accompanying image shows the metamorphic rock gneiss, a basaltic dike, and a fault. Place these three features in their proper sequence (which came first, second, and third) and explain your logic.

Marli Miller

2. A mass of granite is in contact with a layer of sandstone. Using a principle described in this chapter, explain how you might determine whether the sandstone was deposited on top of the granite or whether the magma that formed the granite was intruded after the sandstone was deposited.

3. This scenic image is from Monument Valley in the northeastern corner of Arizona. The bedrock in this region consists of layers of sedimentary rocks. Although the prominent rock exposures ("monuments") in this photo are widely separated, we can infer that they represent a once-continuous layer. Discuss the principle that allows us to make this inference.

Michael Collier

4. The accompanying photo shows two layers of sedimentary rock. The lower layer is shale from the late Mesozoic era. Note the old river channel that was carved into the shale after it was deposited. Above is a younger layer of boulder-rich breccia. Are these layers conformable? Explain why or why not. What term from relative dating applies to the line separating the two layers?

Breccia

Former streambed

Shale

Callan Bentley

5. These polished stones are called *gastroliths*. Explain how such objects can be considered fossils. What category of fossil are they? Name another example of a fossil in this category.

0 1 2
Centimeters

Francois Gohier/Photo Researchers, Inc.

6. If a radioactive isotope of thorium (atomic number 90, mass number 232) emits 6 alpha particles and 4 beta particles during the course of radioactive decay, what are the atomic number and mass number of the stable daughter product?

7. A hypothetical radioactive isotope has a half-life of 10,000 years. If the ratio of radioactive parent to stable daughter product is 1:3, how old is the rock that contains the radioactive material?

8. This scene in Montana's Glacier National Park shows layers of Precambrian sedimentary rocks. The darker layer contained within the sedimentary layers is igneous. The narrow, light-colored areas adjacent to the igneous rock were created when molten material that formed the igneous rock baked the adjacent rock.

Marli Miller

a. Is the igneous layer more likely a lava flow that was laid down at the surface prior to the deposition of the layers above it or a sill that was intruded after all the sedimentary layers were deposited? Explain.

b. Is it likely that the igneous layer will exhibit a vesicular texture? Explain.

c. To which group (igneous, sedimentary, or metamorphic) does the light-colored rock belong? Relate your explanation to the rock cycle.

9. Solve the problems related to the magnitude of Earth history below. To make calculations easier, round Earth's age to 5 billion years.

a. What percentage of geologic time is represented by recorded history? (Assume 5000 years for the length of recorded history.)

b. Humanlike ancestors (hominids) have been around for roughly 5 million years. What percentage of geologic time is represented by these ancestors?

c. The first abundant fossil evidence does not appear until the beginning of the Cambrian period, about 540 million years ago. What percentage of geologic time is represented by abundant fossil evidence?

10. A portion of a popular college text in historical geology includes 10 chapters (281 pages) in a unit titled "The Story of Earth." Two chapters (49 pages) are devoted to Precambrian time. By contrast, the last two chapters (67 pages) focus on the most recent 23 million years, with 25 of those pages devoted to the Holocene Epoch, which began 10,000 years ago.

a. Compare the percentage of pages devoted to the Precambrian to the percentage of geologic time that this span represents.

b. How does the number of pages about the Holocene compare to its percentage of geologic time?

c. Suggest some reasons the text seems to have such an unequal treatment of Earth history.

11. The accompanying diagram is a cross section of a hypothetical area. Place the lettered features in the proper sequence, from oldest to youngest. Where in the sequence can you identify an unconformity?

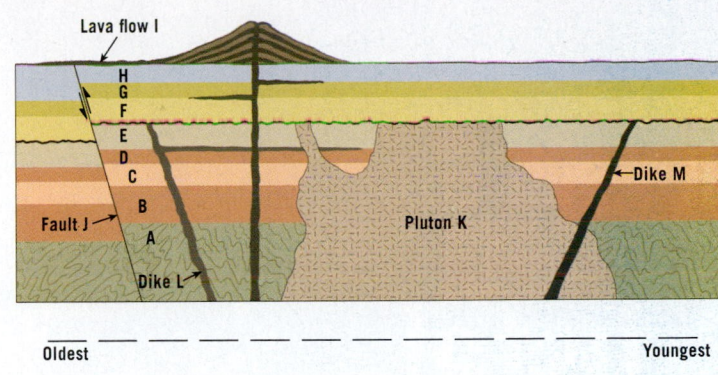

MasteringGeology™

10

Crustal Deformation

Wyoming's Grand Tetons are the result of crustal deformation created by tectonic forces that operate in Earth's interior. (Photo by Michael Collier)

FOCUS ON CONCEPTS

Each statement represents the primary **LEARNING OBJECTIVE** for the corresponding major heading within the chapter. After you complete the chapter you should be able to:

10.1 Describe the three types of differential stress and name the type of plate boundary most commonly associated with each.

10.2 Compare and contrast brittle and ductile deformation.

10.3 List and describe five common folded structures.

10.4 Sketch and briefly describe the relative motion of rock bodies located on opposite sides of normal, reverse, and thrust faults as well as both types of strike-slip faults.

10.5 Explain how strike and dip are measured and how these measurements tell geologists about the orientations of rock structures located mainly below Earth's surface.

Earth is a dynamic planet. Shifting lithospheric plates gradually change the face of our planet by moving continents across the globe. The results of this tectonic activity are perhaps most strikingly apparent in Earth's major mountain belts. Rocks containing fossils of marine organisms are found thousands of meters above sea level, and massive rock units are bent, contorted, overturned, and sometimes riddled with fractures.

This chapter explores the forces that deform rock and the rock structures that result. Foliation and rock cleavage were examined in Chapter 8; this chapter is devoted to the other major structural features of Earth's crust and the tectonic forces that produce them.

10.1 | What Causes Rock to Deform?

Describe the three types of differential stress and name the type of plate boundary most commonly associated with each.

Tutorial

Every body of rock, no matter how strong, has a point at which it will deform by bending or breaking. **Deformation** (*de* = out, *forma* = form) is a general term for the changes in the shape or position of a rock body in response to differential stress. Most crustal deformation occurs along plate boundaries. Plate motions and plate interactions along their margins generate the tectonic forces that cause rock to deform.

The basic geologic features that form as a result of the forces generated by the interactions of tectonic plates are called **rock structures**, or **geologic structures** (**Figure 10.1**). Rock structures include *folds* (wave-like undulations), *faults* (fractures along which one rock body slides past another), *joints* (cracks), and small-scale structures associated with metamorphic rocks such as *foliation* and *rock cleavage* that were introduced in Chapter 8.

Stress: The Force That Deforms Rocks

From everyday experience, you know that if a door is stuck, you must expend energy, called *force*, to open it. Geologists use the term **stress** to describe the forces that deform rocks. Whenever the stresses acting on a rock body exceed its strength, the rock will deform—usually by one or more of the following processes: folding, flowing, fracturing, or faulting.

The magnitude of stress is not simply a function of the amount of force applied but also relates to the area on which the force acts. For example, if you are walking barefoot on a hard surface, the force (weight) of your

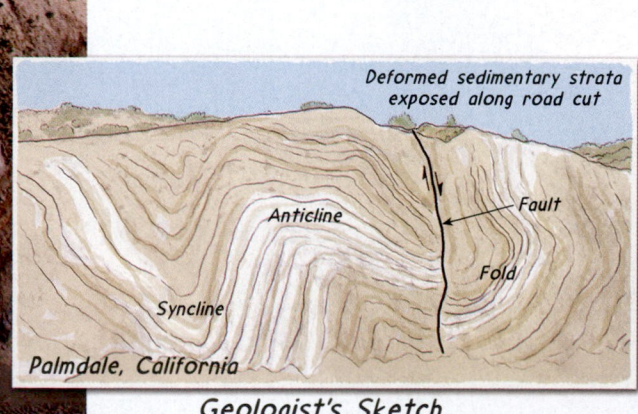

Deformed sedimentary strata exposed along road cut

Anticline

Fault

Fold

Syncline

Palmdale, California

Geologist's Sketch

body is distributed across your entire foot, so the stress acting on any one point of your foot is low. However, if you step on a small pointed rock (ouch!), the stress concentration at that point on your foot will be high. Thus, you can think of stress as a measure of how much force is applied over a particular area.

As you saw in Chapter 8, when stress is applied uniformly in all directions, it is called **confining pressure** (**Figure 10.2A**). Confining pressure causes the spaces between mineral grains to close, producing more compact rocks that have greater densities. Confining pressure does not, however, change the shape or orientation of a rock body. By contrast, when stress is applied unequally in different directions, it is termed **differential stress**. We will consider three types of differential stress:

1. **Compressional stress.** Differential stress that squeezes a rock mass as if placed in a vise is known as **compressional stress** (*com* = together, *premere* = to press) (**Figure 10.2B**). Compressional stresses are most often associated with convergent plate boundaries. When plates collide, Earth's crust is generally shortened horizontally and thickened vertically. Over millions of years, this deformation produces mountainous terrains.

2. **Tensional stress.** Differential stress that pulls apart or elongates rock bodies is known as **tensional stress** (*tendere* = to stretch) (**Figure 10.2C**). Along divergent plate boundaries where plates are moving apart, tensional stresses stretch and lengthen rock bodies. For example, in the Basin and Range Province in the western United States, tensional forces have stretched and fractured the crust by as much as twice its original width.

3. **Shear stress.** Differential stress can also cause rock to *shear*, which involves the movement of one part of a rock body past another (**Figure 10.2D**). Shear is similar to the slippage that occurs between individual playing cards when the top of the deck is moved relative to the bottom (**Figure 10.3**). Small-scale deformation of rocks by shear stresses occurs along closely spaced parallel surfaces of weakness, such as foliation surfaces and microscopic fractures, where slippage changes the shape of rocks. By contrast, at transform fault boundaries, such as the San Andreas Fault, **shear stress** causes large segments of Earth's crust to slip horizontally past one another.

Strain: A Change in Shape Caused by Stress

When flat-lying sedimentary layers are uplifted and tilted, their orientations change, but their shapes are often retained. This type of deformation results in a change in orientation (tilting) of the once flat-lying sedimentary layers and is called *rotation*. Deformation may also cause a rock body to change location—a process called *displacement*. For example, displacement occurs during faulting when the rocks on one side of the fault move relative to the rocks on the opposite side. This type of deformation changes the location of a rock body but does not substantially change its shape or orientation.

Differential stress can also change the shape of a rock body, a type of deformation that geologists call **strain**. Like the circle shown in Figure 10.3, *strained bodies lose their original configuration during deformation*. Strain is illustrated by the deformed trilobite fossil show in **Figure 10.4**. When we compare the deformed specimen to a trilobite fossil of the same species that has

A. Confining pressure

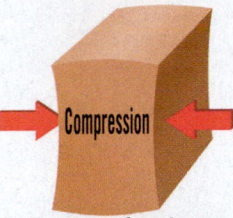

B. Compressional stress (shortening)

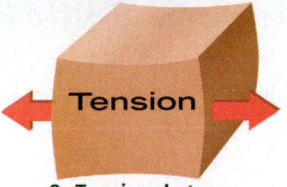

C. Tensional stress (stretching)

D. Shear stress (sliding and tearing)

Figure 10.2
Confining pressure versus three types of stresses
Confining pressure **A.** changes the volume of a block of rock but not its general shape. In contrast, compressional **B.**, tensional **C.**, and shear **D.** stresses change the shape of a rock body.

By sliding the top of the deck relative to the bottom, we can illustrate the type of shearing that commonly occurs along closely spaced planes of weakness in rocks.

Shear stress causes the circle in this deck of cards to become an ellipse, which can be used to measure the amount and type of strain.

Figure 10.3
Shearing and the resulting deformation (strain)
An ordinary deck of playing cards with a circle embossed on its side illustrates shearing and the resulting strain.

Figure 10.4
Deformed trilobite Compare the fossil of this common Paleozoic life-form with the fossil in Figure 9.15B, page 284, to see the extent of change. (Photo by Marli Miller)

not been deformed (see Figure 9.15B), the type of strain that occurred to the surrounding rock body can be determined. *Stress* is the force that acts to deform rock bodies, while *strain* is the resulting distortion, or change in the shape of the rock body.

10.1 | Concept Checks

1. What is rock deformation? How might a rock body change during deformation?

2. List the three types of differential stress and briefly describe the changes they impart to rock bodies.

3. Identify the plate boundary that is most commonly associated with each of the three types of differential stresses.

4. How is strain different from stress?

10.2 | How Do Rocks Deform?

Compare and contrast brittle and ductile deformation.

When rocks are subjected to stresses that exceed their strength, they deform, usually by bending or breaking. It is easy to visualize how rocks break because we normally think of rocks as being brittle. But how can large masses of rock be *bent* into intricate folds without fracturing in the process? To answer this question, geologists performed experiments in which they subjected rocks to differential stress under conditions experienced at various depths within Earth's crust.

Types of Deformation

Through these experiments to determine how large masses of rock can be *bent* into intricate folds without fracturing, geologists determined that although different rocks deform under somewhat different conditions and rates, rocks experience three types of deformation: *elastic*, *brittle*, and *ductile*.

Elastic Deformation When stress is applied *gradually*, rocks initially respond by deforming elastically. Changes that result from **elastic deformation** are recoverable; that is, like a rubber band, the rock will snap back to nearly its original size and shape when the stress is removed. During elastic deformation, the chemical bonds of the minerals within a rock are stretched but do not break. When the stress is removed, the bonds snap back to their original length. You will see in Chapter 11 that the energy for most earthquakes comes from the release of stored elastic energy as the rock snaps back to its original shape.

Brittle Deformation When stress exceeds the elastic limit (strength) of a rock, the rock either bends or breaks. Rocks that break into smaller pieces exhibit **brittle deformation** (*bryttian* = to shatter). From our everyday experience, we know that glass objects, wooden pencils, china plates, and even our bones exhibit brittle failure once their strength is surpassed. Brittle deformation occurs when stress breaks the chemical bonds that hold a material together.

Ductile Deformation **Ductile deformation** is a process in which a rock body is forced to flow in a solid state so that it changes shape without fracturing. Ordinary objects that display ductile behavior include modeling clay, beeswax, taffy, and some metals. For example, an automobile's fender may be dented but not broken when hit by another vehicle.

Ductile deformation may occur by slippage along planes of weakness such as bedding surfaces and foliation in rock bodies. Recall from Chapter 8 that individual mineral grains may also change shape through the process of solid-state flow (see Figure 8.8).

Factors That Affect Rock Strength

The major factors that influence the strength of a rock and how it will deform include temperature, confining pressure, rock type, and time.

The Role of Temperature The effect of temperature on the strength of a material can be easily demonstrated with a piece of glass tubing commonly found in a chemistry lab. If the tubing is dropped on a hard surface, it will shatter. However, if the tubing is heated over a Bunsen burner, it can be easily bent into a variety of shapes. Rocks respond to heat similarly. Where temperatures are high (deep in Earth's crust), rocks tend to soften and become more malleable, so they deform by folding or flowing (ductile deformation). Likewise, where temperatures are low (at or near the surface), rocks tend to behave like brittle solids and fracture.

The Role of Confining Pressure Recall from Chapter 8 that pressure, like temperature, increases with depth as the thickness of the overlying rock increases. Buried rocks are subjected to confining pressure, which is much like water pressure, where the forces are applied equally in all directions. Confining pressure "squeezes" the materials in Earth's crust, making the materials stronger and thus harder to break. Therefore, rocks that are deeply buried are "held together" by the immense pressure and tend to bend rather than fracture.

The Influence of Rock Type In addition to the impact of the physical environment, deformation of rock is greatly influenced by its mineral composition and texture. For example, igneous and some metamorphic rocks (quartzite, for example) are composed of minerals that have strong internal chemical bonds. These strong, brittle rocks tend to fail by fracturing when subjected to stresses that exceed their strength. By contrast, sedimentary rocks that are weakly cemented or metamorphic rocks that contain zones of weakness, such as foliation, are more susceptible to ductile deformation and tend to fold or flow (**Figure 10.5**). Weak rocks that are most likely to behave in a ductile manner (bend or flow) when subjected to differential stress include rock salt, shale, limestone, and schist. In fact, rock salt is so weak that large masses of it often rise through overlying beds of sedimentary rocks, much as hot magma rises toward Earth's surface. Perhaps the weakest naturally occurring solid to exhibit ductile flow is glacial ice. At increasing

Figure 10.5
Rocks exhibiting ductile deformation
These sedimentary rock layers were deformed at great depth, under high temperatures and high pressures. Consequently, they bend without breaking—a type of ductile deformation. (Photo by Michael Collier)

depths, however, the strength of all rock types decreases significantly.

Some **outcrops**—sites where bedrock is exposed at the surface—consist of a sequence of interbedded weak and strong rock layers, such as shale and well-cemented sandstone beds, that have been moderately deformed by folding. In these settings, the strong sandstone layers are often highly fractured, while the weak shale beds form broad undulating folds. The formation of diverse structures (one brittle and one ductile) within the same rock mass can be illustrated by placing a Milky Way or similar chocolate-over-caramel candy bar into a refrigerator. When the cool candy bar is slowly bent, it will exhibit brittle deformation in the chocolate and ductile deformation in the caramel.

Time as a Factor One key factor that researchers are unable to duplicate in the laboratory is how rocks respond to small stresses applied gradually over long spans of *geologic time*. However, we can observe the effects of time on deformation in everyday settings. For example, marble benches have been known to sag under their own weight over a time span of 100 years or so, and wooden bookshelves may bend within a few months after being loaded with books.

In general, when tectonic forces are applied gradually over long spans of time, rocks tend to display ductile behavior and deform by bending or flowing. However, the same rocks may shatter if force is applied suddenly. An analogous situation occurs when you slowly move the two ends of a taffy bar together. The taffy will deform by folding. However, if you hit the taffy swiftly against the edge of a table, it will break into two or more pieces, exhibiting brittle failure.

Figure 10.6
Deformation caused by three types of stress Brittle deformation (fracturing and faulting) dominates in the upper crust, where the temperatures are comparatively cool. By contrast, at depths greater than about 10 kilometers (6 miles), where temperatures are high, rock deforms by ductile flow and folding.

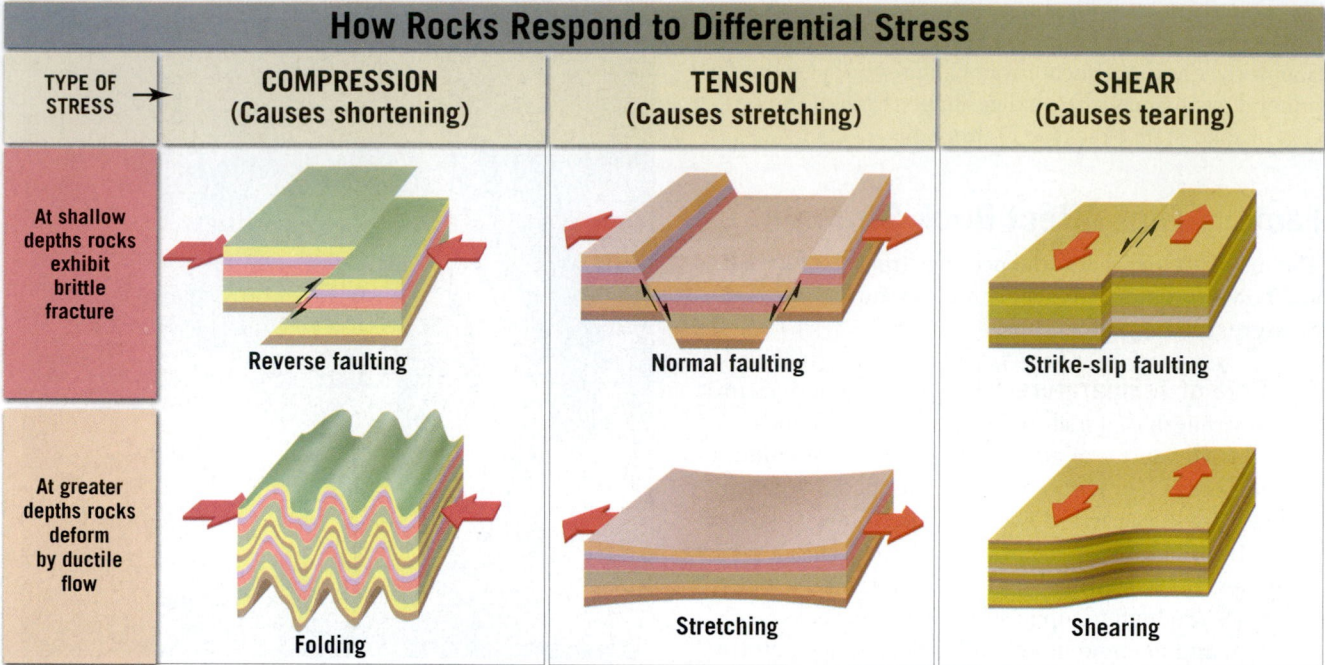

How Rocks Respond to Differential Stress

TYPE OF STRESS →	COMPRESSION (Causes shortening)	TENSION (Causes stretching)	SHEAR (Causes tearing)
At shallow depths rocks exhibit brittle fracture	Reverse faulting	Normal faulting	Strike-slip faulting
At greater depths rocks deform by ductile flow	Folding	Stretching	Shearing

Ductile Versus Brittle Deformation and the Resulting Rock Structures

You have probably seen a drinking glass drop on a hard surface and shatter into pieces. What you witnessed is analogous to brittle deformation that occurs in Earth's crust. In nature, brittle deformation occurs when stresses exceed the strength of a rock, causing it to break or fracture.

In the upper crust, to depths of about 10 kilometers (6 miles), most rocks tend to exhibit brittle behavior by fracturing or faulting. When tectonic forces cause upwarping of the crust, rocks near the surface are stretched and pulled apart, forming fractures called *joints* (see Figure 10.26, page 321). *Faults*, by contrast, are fractures in Earth's crust where the rocks on one side of the fault surface are displaced relative to the rocks on the other side.

Folds, common features of deformed sedimentary rock layers, provide evidence that rocks can bend without breaking (see Figure 10.5). Rocks that display evidence of ductile deformation usually were generated at depth in high-temperature and high-pressure environments. The resulting rocks often exhibit contorted folds that give the impression that the strength of the rock was akin to that of soft putty. Weak rocks, however, can bend or flow without breaking in less extreme environments.

Figure 10.6 illustrates deformation of Earth's crust as a result of the three types of stress: compressional, tensional, and shear stress. Compressional stresses associated with convergent plate boundaries tend to *shorten* and *thicken* Earth's crust by folding, flowing, and faulting, whereas tensional stresses along divergent plate boundaries tend to *lengthen* rock bodies by displacement along faults in the upper crust and by ductile flow at depth. Shearing, often associated with transform plate boundaries, tends to produce fault zones in the upper crust and ductile flow at depth. Brittle deformation—fracturing and faulting—dominates in the upper crust, where temperatures are relatively cool. By contrast, at greater depths where temperatures and pressures are high, rocks exhibit ductile behavior and deform by folding or flowing.

10.2 Concept Checks

1. Describe elastic deformation.

2. Compare and contrast brittle and ductile deformation and describe the conditions that favor one over the other.

3. List and describe the four factors that affect rock strength.

10.3 | Folds: Rock Structures Formed by Ductile Deformation

List and describe five common folded structures.

Along convergent plate boundaries, flat-lying sedimentary strata, tabular intrusions, and volcanic rocks are often bent into a series of wavelike undulations called **folds**. Folds in sedimentary strata are much like those that would form if you were to hold the ends of a sheet of paper on a flat surface and then push them together. In nature, folds come in a wide variety of sizes and configurations. Some folds are broad flexures in which strata hundreds of meters thick have been slightly warped. Others are very tight microscopic structures found in metamorphic rocks. Size differences notwithstanding, most folds result from compressional stresses that result in a shortening and thickening of the crust.

To aid our understanding of folds and folding, it is important to become familiar with the terminology used to name the parts of a fold. Folds are geologic structures consisting of stacks of originally horizontal surfaces, such as sedimentary strata, that have been bent as a result of permanent deformation. Each layer is bent around an imaginary axis called a *hinge line*, or simply a *hinge* (Figure 10.7). In some folds, the hinge lines are horizontal. However, hinge lines can be inclined at an angle, in which case the fold is known as a *plunging fold* (see Figure 10.9).

Folds are also described by their *axial plane*, which is a surface that connects all the hinge lines of the folded strata. In simple folds, the axial plane is vertical and divides the fold into two roughly symmetrical *limbs*. However, the axial plane often leans to one side so that one limb is steeper and shorter than the other. In extreme cases, the axial plane can be tilted such that it has a horizontal orientation, making the fold appear to be "lying on its side."

Anticlines and Synclines

The two most common types of folds are anticlines and synclines (Figure 10.8). **Anticlines** occur when compressional stresses squeeze sedimentary layers into arch-like folds. These upfolded structures are sometimes spectacularly displayed along highway roadcuts that pass through deformed strata. Typically found in association with anticlines are trough-like structures called **synclines**. Notice in Figure 10.8 that the right limb of the syncline is also the left limb of the adjacent anticline.

Depending on their orientation, these basic folds are described as *symmetrical* when the limbs are mirror images of each other and *asymmetrical* when they are not. An asymmetrical fold is said to be *overturned* if one or both limbs are tilted beyond the vertical. An overturned fold can also "lie on its side" so the axial plane is horizontal. These *recumbent* folds are common

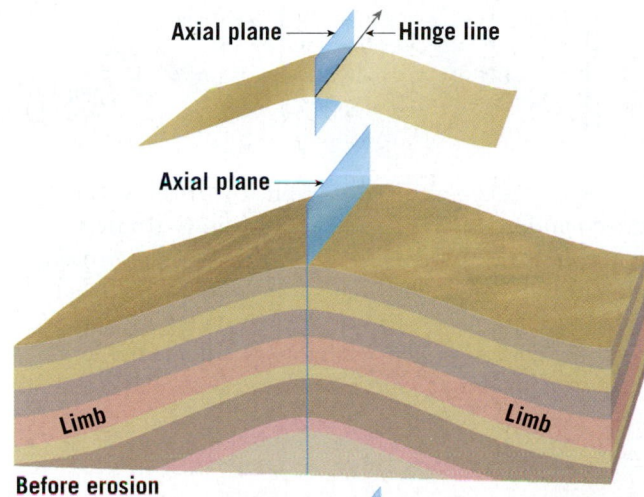

Before erosion

After erosion

SmartFigure 10.7
Features associated with symmetrical folds The axial plane divides a fold as symmetrically as possible, while the hinge line traces the points of maximum bending of any layer.
(https://goo.gl/S9h2Ic)

Condor Video

in highly deformed mountainous regions such as the Alps.

Folds also can be tilted by tectonic forces that cause their hinge lines to slope rather than have a horizontal orientation (Figure 10.9A). Folds of this type are called *plunging folds* because the hinge lines of the fold dip downward (plunge) and penetrate Earth's surface. Figure 10.9B shows the outcrop pattern produced when erosion removes the upper layers of a plunging anticline and exposes its interiors. Note that the outcrop pattern of an eroded anticline "points" in the direction it is plunging. The opposite is true for a syncline, which exhibits an outcrop pattern that points in the opposite direction of its plunge.

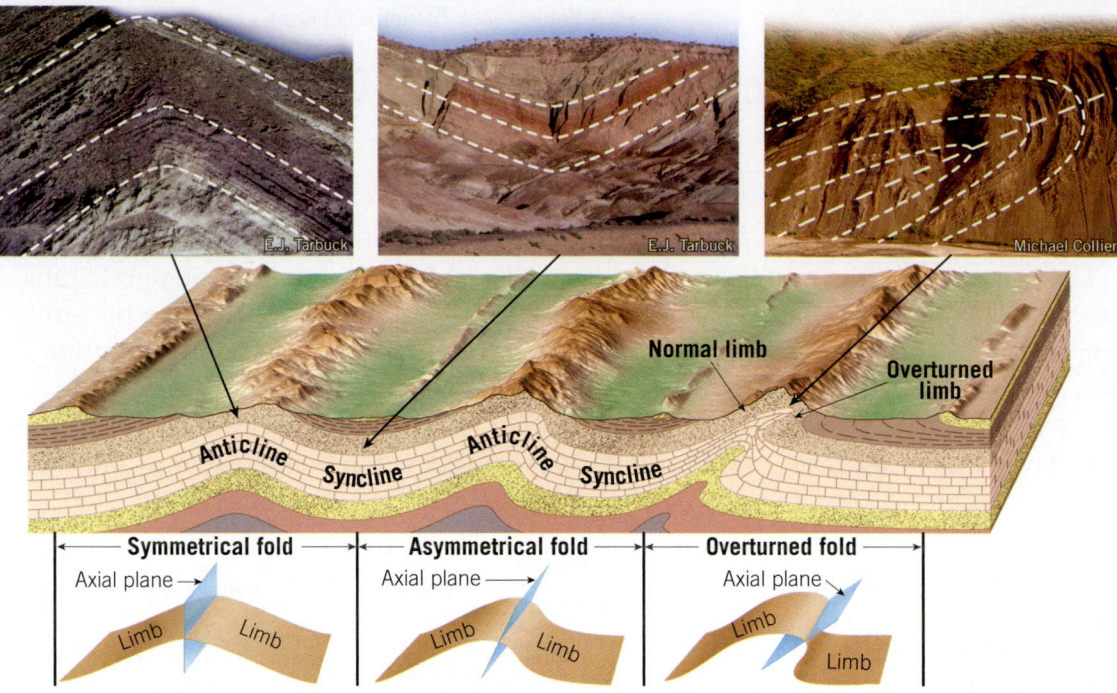

SmartFigure 10.8
Common types of folds When sedimentary rock layers are folded in an arch-like manner, the structure is called an *anticline*. By contrast, trough-like structures are called *synclines*. Notice that the limb of an anticline is also the limb of the adjacent syncline.
(https://goo.gl/Xz0acg)

Tutorial

A good example of topography that results when erosional forces attack folded sedimentary strata is found in the Valley and Ridge Province of the Appalachians (Figure 10.10). It is important to recognize that ridges are not necessarily associated with anticlines, nor are valleys related to synclines. Rather, ridges and valleys result because of differential weathering and erosion. In the Valley and Ridge Province, for instance, resistant sandstone beds remain as imposing ridges separated by valleys cut into more easily eroded shale or limestone beds.

SmartFigure 10.9
A plunging anticline A. A fold is said to plunge if the axis (hinge line) of the fold is tilted rather than horizontal. **B.** The same fold shown in **A**, as it would appear after being eroded.
(http://goo.gl/kgB5Qg)

Mobile Field Trip

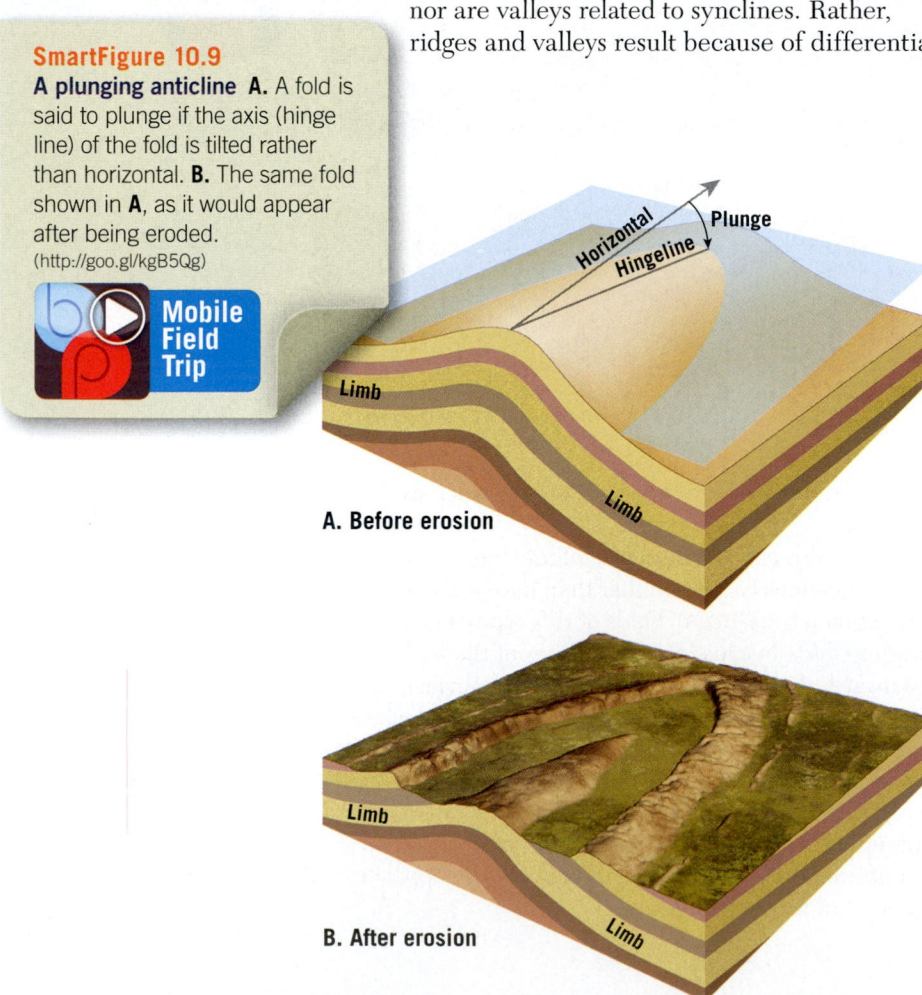

SmartFigure 10.10
Topography that results when erosional forces attack folded sedimentary strata This satellite image, taken over south-central Pennsylvania, shows a portion of the Valley and Ridge Province of the Appalachians. Can you identify a plunging fold in this image? (Photo courtesy of NASA) (https://goo.gl/l1c66S)

Tutori

Domes and Basins

Broad uplifting of crystalline basement rock may deform the overlying cover of sedimentary strata and generate large folds. When this upwarping produces a circular or slightly elongated structure, the feature is called a **dome** (**Figure 10.11A**). The Black Hills of western South Dakota is a very large structural dome generated by upwarping. Here erosion has stripped away the highest portions of the overlying sedimentary beds, exposing older igneous and metamorphic rocks in the center (**Figure 10.12**).

Structural domes can also be formed by the intrusion of magma (laccoliths), as shown in Figure 4.28, page 130. In addition, the upward migration of buried salt

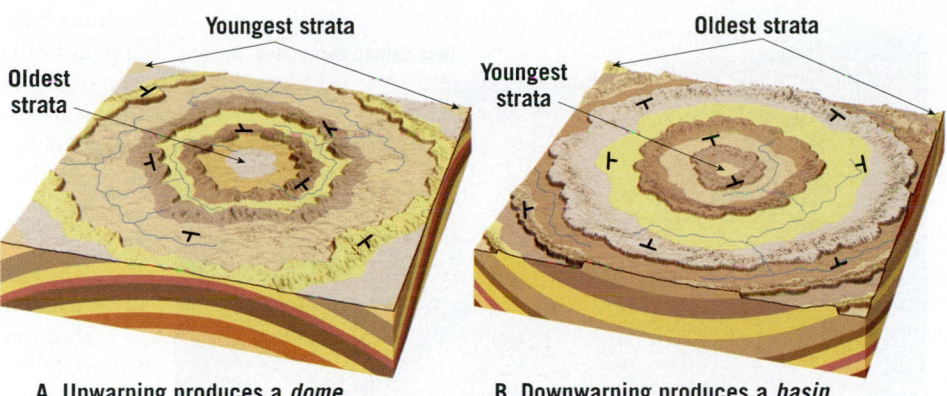

A. Upwarping produces a *dome*.

B. Downwarping produces a *basin*.

SmartFigure 10.11
Domes versus basins
Gentle upwarping and downwarping of crustal rocks produce **A.** domes and **B.** basins. Erosion of these structures results in an outcrop pattern that is roughly circular or slightly elongate. (https://goo.gl/9PJtjY)

Tutorial

deposits can produce salt domes like those surrounding the coastal areas of the Gulf of Mexico. Salt domes are economically important rock structures because when salt migrates upward, the surrounding oil-bearing sedimentary layers are often bent upward to form oil reservoirs (see Figure 23.5C, page 709).

Downwarped structures having a bowl-like shape are termed **basins** (**Figure 10.11B**). Several large basins exist in the United States (**Figure 10.13**). The basins of Michigan and Illinois have gently sloping beds similar to saucers. These basins resulted from large accumulations of sediment, the weight of which caused the underlying crust to subside (see the section on isostasy in Chapter 14). In addition, a few structural basins resulted from giant meteorite impacts, including Sudbury Basin, Canada.

Because large basins contain sedimentary beds sloping at low angles, they are usually identified by the age of the rocks composing them. The youngest rocks of a

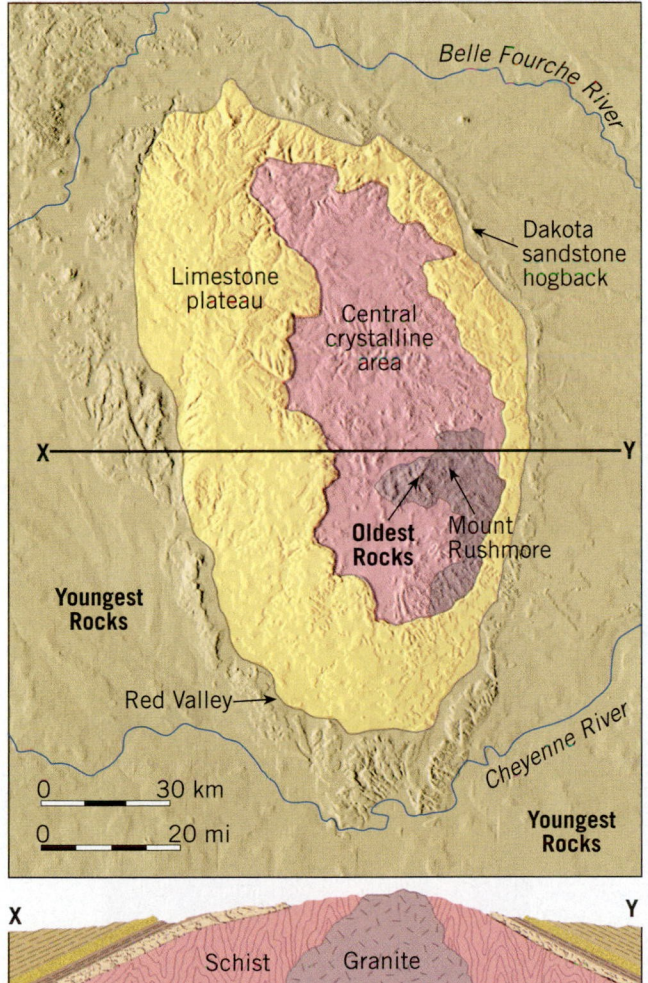

Figure 10.12
The Black Hills of South Dakota, a large structural dome The central core of the Black Hills is composed of resistant Precambrian-age igneous and metamorphic rocks. The surrounding rocks are mainly younger limestones and sandstones.

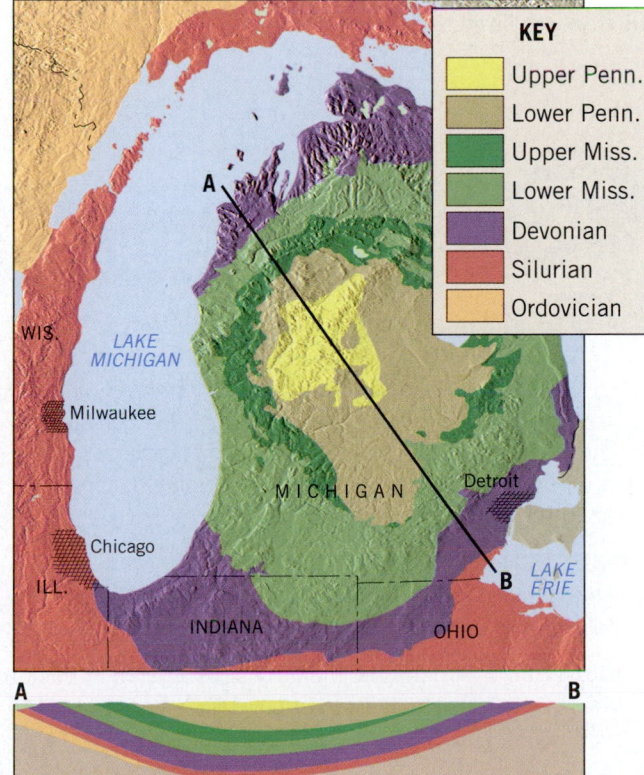

KEY

- Upper Penn.
- Lower Penn.
- Upper Miss.
- Lower Miss.
- Devonian
- Silurian
- Ordovician

Figure 10.13
The bedrock geology of the Michigan Basin The youngest rocks are centrally located, whereas the oldest beds flank this structure.

Monocline

East Kaibab Monocline, Arizona

Monocline

Sedimentary strata removed by erosion

Fault in basement rock

sedimentary strata (**Figure 10.14**). These folds appear to have resulted from the reactivation of ancient, steep-dipping faults located in basement rocks beneath the plateau. As large blocks of basement rock were displaced upward, the comparatively ductile sedimentary strata above responded by draping over the fault like clothes hanging over a bench. Displacement along these reactivated faults often exceeds 1 kilometer (0.6 miles).

Examples of monoclines found on the Colorado Plateau include the East Kaibab Monocline, the Raplee Anticline, the Waterpocket Fold, and the San Rafael Swell. The inclined strata shown in Figure 10.14 once extended over the sedimentary layers now exposed at the surface—evidence that tremendous volumes of rock have been eroded from this area.

basin are found near the center, and the oldest rocks are at the flanks. This is just the opposite of a dome, such as the Black Hills, where the oldest rocks are found in the interior.

Monoclines

Although we discuss folds and faults in separate sections, folds can be uniquely coupled with faults. Examples of this close association are broad, regional features called *monoclines*. Particularly prominent features of the Colorado Plateau, **monoclines** (*mono* = one, *kleinen* = incline) are large, step-like folds in otherwise horizontal

10.3   Concept Checks

1. Distinguish between anticlines and synclines, between domes and basins, and between anticlines and domes.

2. Draw a cross-sectional view of a symmetrical anticline. Include a line to represent the axial plane and label both limbs.

3. The Black Hills of South Dakota are a good example of what type of geologic structure?

4. Where are the youngest rocks in an eroded basin outcrop: near the center or near the flanks?

5. How do monoclines form?

EYE ON EARTH 10.1

This image features a large geologic structure, called Sheep Mountain, that outcrops in north-central Wyoming. It consists of sedimentary layers that were folded about 60 million years ago during the Laramide mountain-building episode. This episode generated a portion of the Rocky Mountains that stretches from southern Montana to New Mexico.

QUESTION 1 Based on the direction that these folded sedimentary layers are plunging, would you describe this fold as a *plunging anticline* or a *plunging syncline*? Explain.

QUESTION 2 Are the sedimentary beds located at point **A** younger or older than those at point **B**?

A   B

Direction of plunge

Michael Collier

10.4 | Faults and Joints: Rock Structures Formed by Brittle Deformation

Sketch and briefly describe the relative motion of rock bodies located on opposite sides of normal, reverse, and thrust faults as well as both types of strike-slip faults.

Faults form where brittle deformation leads to fracturing and displacement of Earth's crust. Occasionally, small faults can be recognized in roadcuts where sedimentary beds have been offset less than a meter, as shown in **Figure 10.15**. Faults of this scale usually occur as single discrete breaks. By contrast, large faults, like the San Andreas Fault in California, have displacements of hundreds of kilometers and consist of many interconnecting fault surfaces. These structures, described as *fault zones*, can be several kilometers wide and are often easier to identify from aerial photographs than at ground level.

Because faults have nearly planar surfaces, geologists describe their orientations by their strike and dip. *Strike* can be defined as the direction of a horizontal line on an inclined surface. For example, the strike (or *trend*) of the San Andreas Fault is in a northwesterly direction. *Dip* is the angle of inclination of a surface measured from a horizontal plane. The San Andreas is nearly vertical, so it has a dip, or *inclination*, of about 90 degrees. By contrast, some faults are nearly horizontal, with dips of 10 degrees or less.

Strike and dip can be used to describe the orientation of sedimentary layers, dikes, joints, and other planar structures, in addition to faults. For example, the hinge and limbs of an anticline can be described by their strike and dip. How strike and dip are used to map geologic structures is considered later in this chapter.

Dip-Slip Faults

Faults in which movement is primarily parallel to the dip (inclination) of the fault surface are called **dip-slip faults**. Geologists identify the rock surface immediately above the fault surface as the **hanging wall block** and the rock below the fault as the **footwall block** (**Figure 10.16**). These names were first used by prospectors and miners who excavated metallic ore deposits such as gold that had precipitated from hydrothermal solutions along inactive fault zones. The miners would walk on the rocks below the mineralized fault zone (*the footwall block*) and hang their lanterns on the rocks above (*the hanging wall block*).

Normal Faults Dip-slip faults are classified as **normal faults** when the hanging wall block moves *down* relative to the footwall block (**Figure 10.17**). Normal faults are found in a variety of sizes. Some are small, having displacements of only a meter or so, like the one shown in the roadcut in Figure 10.15. Others, however, extend for tens of kilometers and may sinuously trace the boundary of a mountain front. Most large normal faults have relatively steep dips that tend to flatten out with depth. Normal faults tend to be associated with tensional stresses that pull rock units apart, thereby lengthening the

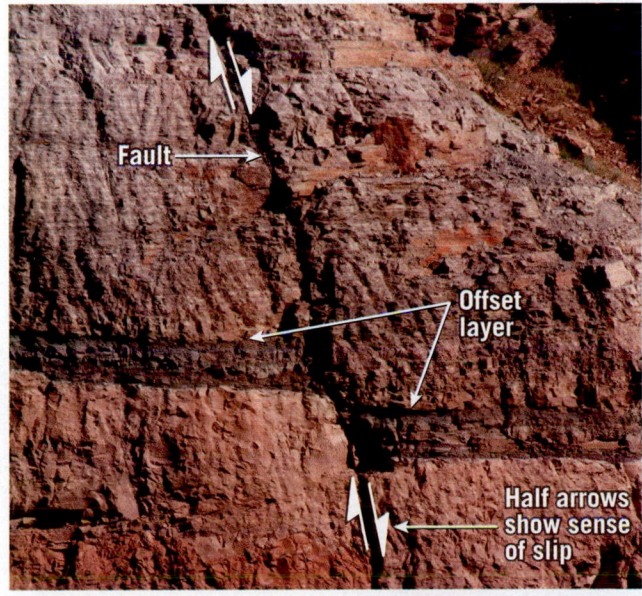

SmartFigure 10.15
Faults are fractures along which slip has occurred (Photo by E. J. Tarbuck) (https://goo.gl/vFFfSV)

Condor Video

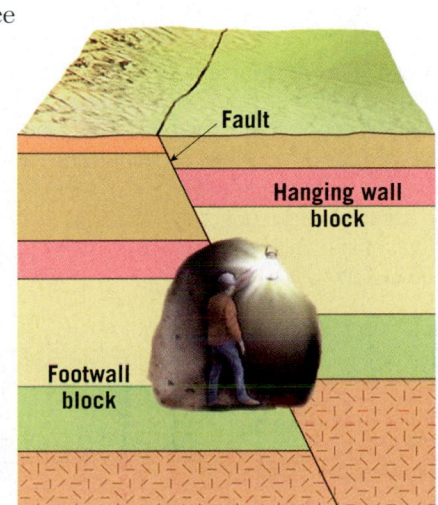

SmartFigure 10.16
Hanging wall block and footwall block The rock immediately above a fault surface is the *hanging wall block,* and the one below is the *footwall block.* These names came from miners who excavated ore deposits that formed along fault zones. The miners hung their lanterns on the rocks above the fault trace (hanging wall block) and walked on the rocks below the fault trace (footwall block). (http://goo.gl/RbIGZH)

Animation

SmartFigure 10.17
Normal dip-slip faults The upper diagram illustrates the relative displacement that occurs between the blocks on either side of a normal dip-slip fault; the hanging wall block drops relative to the footwall block. The lower diagram shows how erosion may alter the up-faulted block. (https://goo.gl/FzUys7)

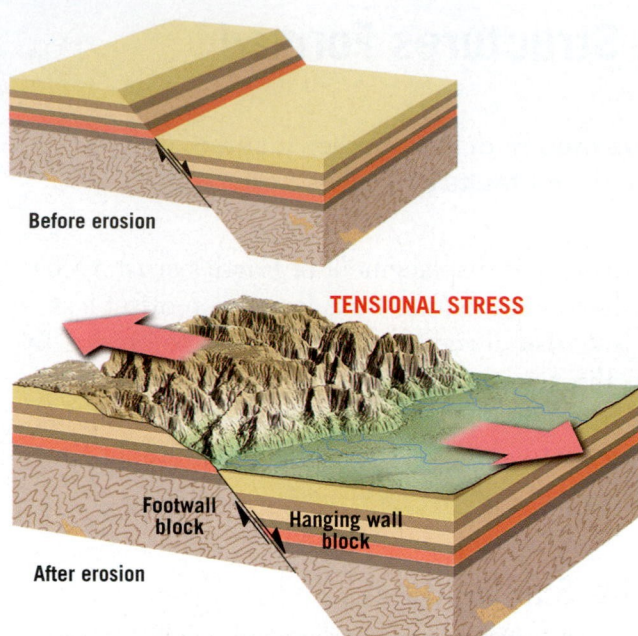

Before erosion

TENSIONAL STRESS

Footwall block

Hanging wall block

After erosion

Normal faults are dip-slip faults in which the hanging wall block moves down relative to the footwall block.

Tutorial

crust. This "pulling apart" can be accomplished either by uplifting that causes the surface to stretch and break or by horizontal forces that have opposing orientations.

In the western United States, large normal faults are associated with structures called **fault-block mountains**. Excellent examples of fault-block mountains are found in the Basin and Range Province, a region that encompasses Nevada and portions of the surrounding states (**Figure 10.18**). Here the crust has been elongated and broken to create more than 200 relatively small mountain ranges. Averaging about 80 kilometers (50 miles) in length, the ranges rise 900 to 1500 meters (3000

to 5000 feet) above the adjacent down-faulted basins. One exception is the Panamint Range, which rises 3368 meters above sea level—more than 11,000 feet above the floor of Death Valley, which lies directly to the east.

Basin and Range Province topography evolved in association with a system of normal faults trending roughly north–south. Movements along these faults produced alternating uplifted fault blocks called **horsts** (*horst* = hill) and down-dropped blocks called **grabens** (*graben* = ditch). Horsts form the ranges and are the source of sediments that have accumulated in the basins created by the grabens. As Figure 10.18 illustrates, structures called **half-grabens**, which are tilted fault blocks, also contribute to the alternating topographic highs and lows in the Basin and Range Province.

Also notice in Figure 10.18 that the slopes of the large normal faults associated with the Basin and Range Province decrease with depth and eventually join to form a nearly horizontal fault called a **detachment fault**. These faults represent a major boundary between the rocks below, which exhibit ductile deformation, and the rocks above, which exhibit mainly brittle deformation.

Reverse and Thrust Faults Dip-slip faults in which the hanging wall block moves *up* relative to the footwall block are called **reverse faults** (**Figure 10.19**). A type of reverse fault having a dip less than 45 degrees is called a **thrust fault**. Reverse and thrust faults result from

SmartFigure 10.18
Normal faulting in the Basin and Range Province Here, tensional stresses have elongated and fractured the crust into numerous blocks. Movement along these faults has tilted the blocks, producing parallel mountain ranges called *fault-block mountains*. The down-faulted blocks (*grabens*) form basins, whereas the up-faulted blocks (*horsts*) erode to form rugged mountainous topography. In addition, numerous tilted blocks (*half-grabens*) form both basins and mountains. (Photo by Michael Collier) (http://goo.gl/FbBhca)

Mobile Field Trip

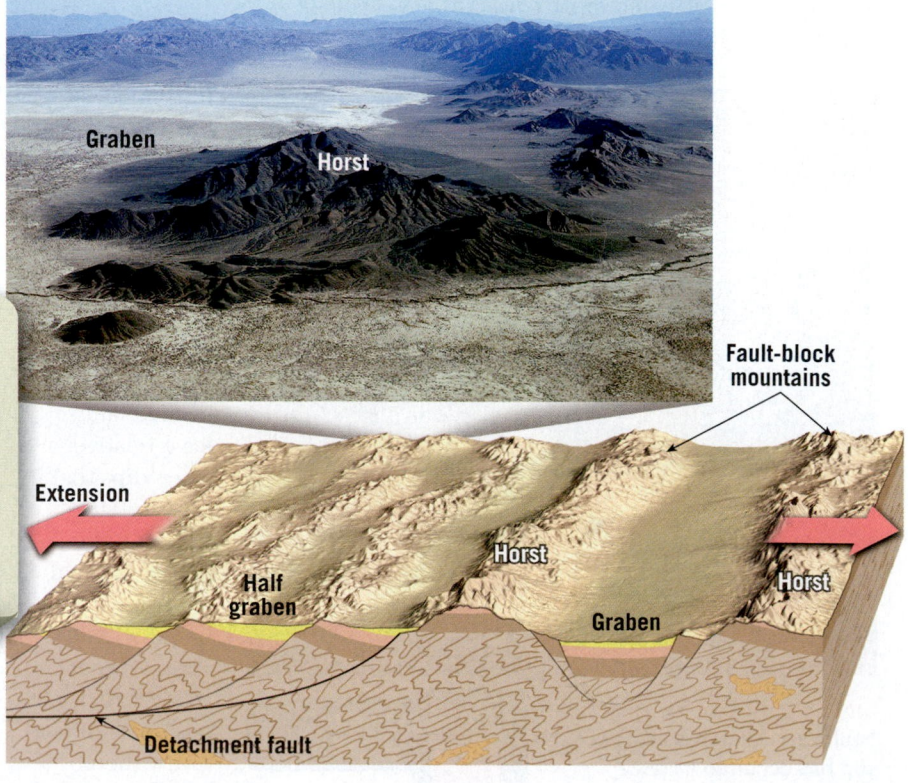

Graben

Horst

Fault-block mountains

Extension

Half graben

Horst

Graben

Horst

Detachment fault

compressional stresses that produce horizontal shortening of the crust.

Most high-angle reverse faults are small and accommodate local displacements in regions dominated by other types of faulting. Thrust faults, on the other hand, exist at all scales, with some large thrust faults having displacements ranging from tens to hundreds of kilometers. When thrust faults are oriented at a very low angle, the hanging wall block can be thrust nearly horizontally over the footwall block.

Thrust faulting is most pronounced along convergent plate boundaries. Compressional forces associated with colliding plates generally create folds as well as thrust faults that thicken and shorten the crust to produce mountainous topography. Examples of mountainous belts produced by this type of compressional tectonics include the Alps, Northern Rockies, Himalayas, and Appalachians.

Montana's Glacier National Park, located in the Northern Rockies, is a classic site of thrust faulting (**Figure 10.20**). Mountain peaks that provide the park's majestic scenery have been carved mainly from a thick sheet of Precambrian limestones that were displaced, as essentially one unit, over much younger Cretaceous shale deposits. This block of sedimentary rock was 6 kilometers (4 miles) thick and slid a distance of about 100 kilometers (60 miles) along the Lewis Thrust Fault. At the eastern edge of Glacier National Park is a peak called Chief Mountain. This well-known landmark is an isolated remnant of a thrust sheet that was severed by the erosional forces of glacial ice and

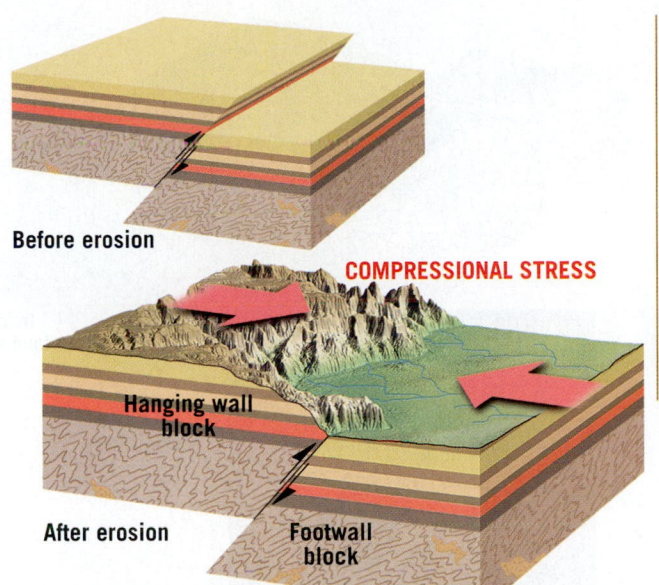

Before erosion

COMPRESSIONAL STRESS

Hanging wall block

After erosion

Footwall block

Reverse faults are dip-slip faults in which the hanging wall block moves up relative to the footwall block.

running water. An isolated block, like Chief Mountain, is called a **klippe** (*klippe* = cliff) (Figure 10.20D).

Strike-Slip Faults

A fault in which the dominant displacement is horizontal and parallel to the strike (trend) of the fault surface is called

SmartFigure 10.19
Reverse faults Reverse faults are a type of dip-slip fault generated by compressional stresses that displace the hanging wall block upward relative to the footwall block. (https:// goo.gl/6ngGC7)

 **Animation**

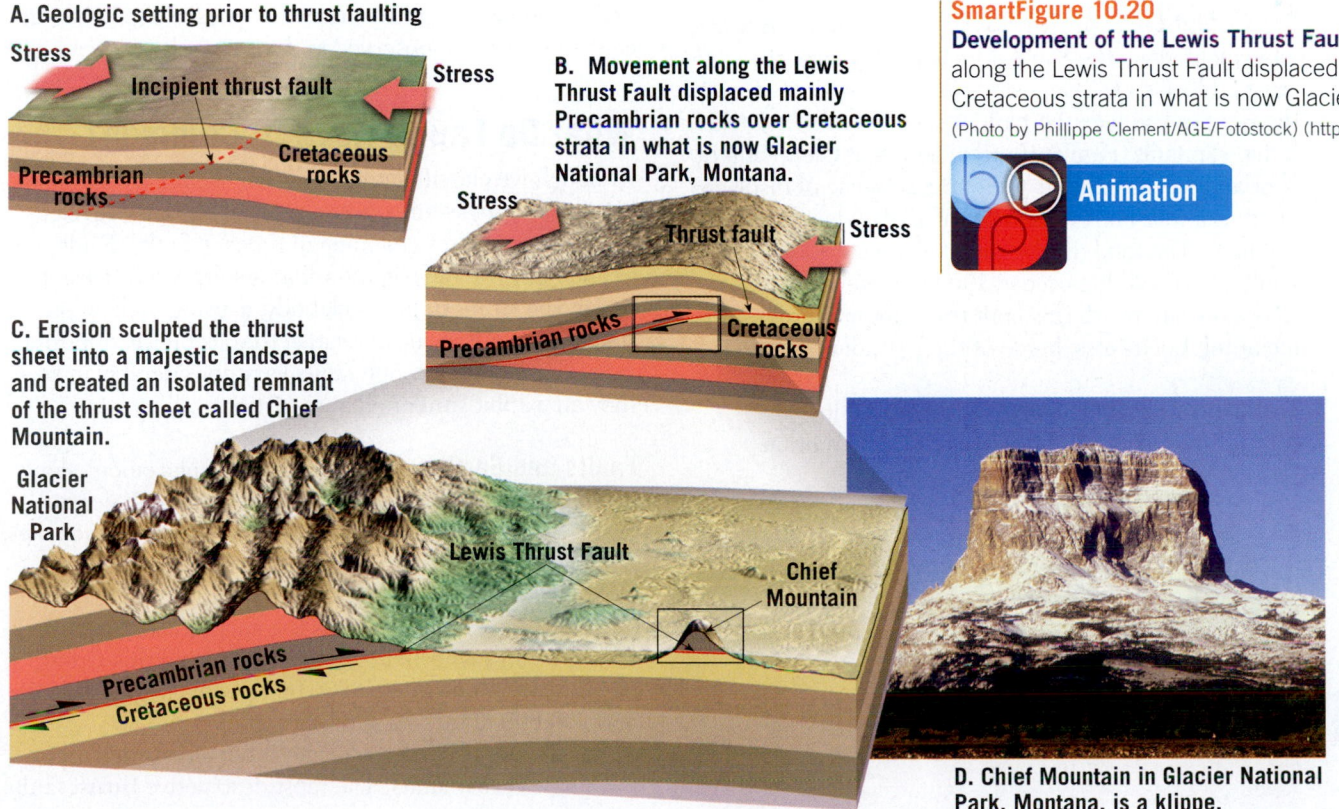

A. Geologic setting prior to thrust faulting

Stress

Stress

Incipient thrust fault

Cretaceous rocks

Precambrian rocks

B. Movement along the Lewis Thrust Fault displaced mainly Precambrian rocks over Cretaceous strata in what is now Glacier National Park, Montana.

Stress

Thrust fault

Stress

Precambrian rocks   **Cretaceous rocks**

C. Erosion sculpted the thrust sheet into a majestic landscape and created an isolated remnant of the thrust sheet called Chief Mountain.

Glacier National Park

Lewis Thrust Fault

Chief Mountain

Precambrian rocks
Cretaceous rocks

D. Chief Mountain in Glacier National Park, Montana, is a klippe.

SmartFigure 10.20
Development of the Lewis Thrust Fault Large-scale movement along the Lewis Thrust Fault displaced Precambrian rock over Cretaceous strata in what is now Glacier National Park, Montana. (Photo by Phillippe Clement/AGE/Fotostock) (https://goo.gl/0gl6HB)

Animation

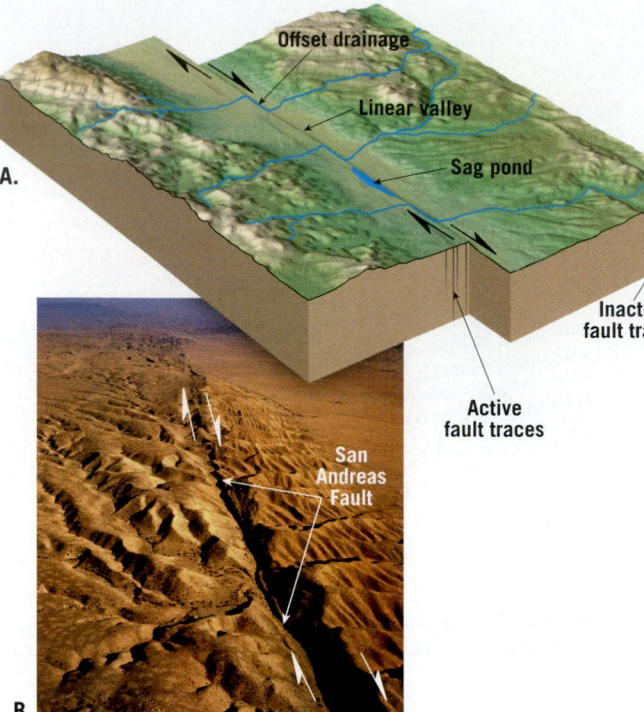

a **strike-slip fault** (Figure 10.21). The earliest scientific records of strike-slip faulting were made following surface ruptures that produced large earthquakes. One of the most noteworthy of these was the great San Francisco earthquake of 1906. During this strong earthquake, structures such as fences and roads that were built across the San Andreas Fault were displaced as much as 4.7 meters (15 feet).

Right- and Left-Lateral Strike-Slip Faults Because movement along the San Andreas Fault causes the crustal block on the opposite side of the fault to move to the right as you face the fault, it is called a *right-lateral* strike-slip fault (Figure 10.21). The Great Glen Fault in Scotland, which exhibits the opposite sense of displacement, is a well-known example of a *left-lateral* strike-slip fault. The total displacement along the Great Glen Fault is estimated to exceed 100 kilometers (60 miles). Also associated with this fault trace are numerous lakes, including Loch Ness, home of the legendary monster.

Transform Faults Some strike-slip faults slice through Earth's crust and accommodate motion

between two tectonic plates. These large strike-slip faults are called **transform faults** (*trans* = across, *forma* = form). Numerous transform faults cut the oceanic lithosphere and link spreading oceanic ridges. Others accommodate displacement between continental blocks that slip horizontally past each other. Some of the best-known transform faults include California's San Andreas Fault, New Zealand's Alpine Fault, the Middle East's Dead Sea Fault, and Turkey's North Anatolian Fault. Large transform faults, like the San Andreas, accommodate relative displacements of up to several hundred kilometers (see GEOgraphics 10.1).

Rather than being a single fracture along which movement takes place, most continental transform faults consist of a zone of roughly parallel fractures. The zone may be up to several kilometers wide. The most recent movement, however, is often along a strand only a few meters wide, which may offset features such as stream channels (Figure 10.21). Crushed and broken rocks produced during faulting are more easily eroded, so depressions such as linear valleys or sag ponds often mark the locations of large strike-slip faults.

Oblique-Slip Faults

Faults that exhibit both dip-slip and strike-slip movement, called **oblique-slip faults**, are caused by a combination of shearing and tensional or compressional stress (Figure 10.22). Nearly all faults have minor components of both dip-slip and strike-slip movement, so defining a fault as oblique requires that both types of slip be significant enough to be observed and measured.

What Do Faults Have in Common?

Although we classify faults based on the relative movement of the rocks on opposite sides of the fault surface, the same processes generate all types of faults. Faults are a type of brittle deformation that results when stress surpasses the strength of a rock body, causing rock on one side of the fault to move relative to rock on the opposite side. Therefore, although faults vary considerably in size, they all exhibit similar characteristics.

Faults and Earthquakes Sudden displacement along any fault can produce an earthquake, although normal faults are not generally associated with large earthquakes. Some of the most destructive earthquakes occur along large strike-slip faults, like the San Andreas. One of the most noteworthy of these was the earthquake that struck near Port-au-Prince, Haiti, in 2010. During this strong earthquake, an estimated 230,000 people lost their lives, and 300,000 homes were destroyed (Figure 10.23).

Even larger earthquakes have been recorded along low-angle thrust faults. The most destructive thrust faults

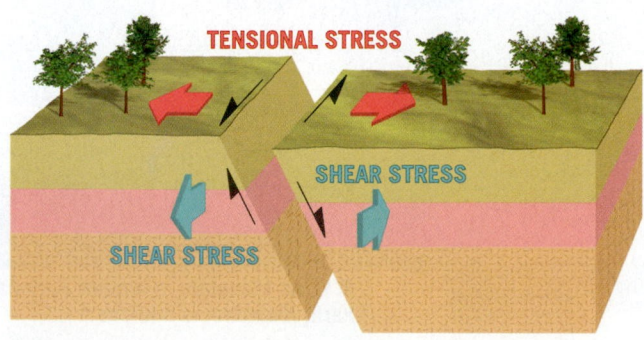

Figure 10.23
Destruction in Port-au-Prince, Haiti, following a devastating earthquake in January 2010. (Photo by Orlando Barria/EFE/Newscom)

form a convergent plate boundary between a subducting slab of oceanic lithosphere and the overlying plate. This type of thrust fault is so large and destructive it is called a **megathrust fault**. Megathrust faults lie beneath the ocean floor and can displace the overlying water, generating destructive *tsunamis*. Megathrust faults have produced the majority of Earth's most powerful earthquakes, including the 2011 Japan quake and the 2004 Indian Ocean quake near Sumatra.

Fault Scarps Vertical displacements along faults may produce long, low cliffs called **fault scarps** (*scarpe* = slope). Fault scarps, such as the one shown in **Figure 10.24**, are usually generated by rapid vertical displacements associated with

earthquakes. Occasionally, horizontal movement along a strike-slip fault produces a fault scarp when an area of higher ground is displaced next to lower terrain.

Fault scarp

Up-faulted

Down-faulted

USGS

Figure 10.24
Fault scarp This fault scarp was created during the Alaska earthquake of 1964. (Photo courtesy of the USGS)

The San Andreas Fault System

The San Andreas Fault system is a 1,300 kilometer (780 miles) long strike-slip fault that slices through two thirds of the length of California. Geologists have determined that this fault zone measures tens of kilometers wide in some locations, and extends to a depth of at least 18 kilometers.

Photo: Sacramento Bee/John Trotter/Newscom

Although the 1989 Loma Prieta earthquake was centered in a remote section of the Santa Cruz Mountains, major damage occurred in the Marina District of San Francisco.

Map: USGS

Because of its great length and complexity, the San Andreas Fault is more appropriately referred to as a "fault system" which consists of the San Andreas Fault and several other major branches.

In the Los Angeles area several low-angle thrust faults have uplifted large crustal blocks including the San Gabriel Mountains. Because some thrust faults do not penetrate Earth's surface, they are called blind thrusts. Movement on such a blind thrust produced the 1994 earthquake in the Northridge area of Los Angeles.

Photo: Chad Ehlers

CALIFORNIA

Hayward Fault

San Francisco

Calaveras Fault

San Andreas Fault

Garlock Fault

Eastern California Shear Zone

Los Angeles

San Jacinto Fault

Imperial Fault

San Francisco

Oakland

Hayward Fault

Greenville Fault

San Jose

Calaveras Fault

Range Front Fault

San Andreas Fault

San Gregorio Fault

Sargent Fault

Simi Fault

Northridge Hills Fault

San Fernando Valley

Thousand Oaks

Santa Monica Fault

Santa Monica

Los Angeles

San Gabriel Fault zone

SAN ANDREAS FAULT

Sierra Madre Fault zone

Newport Inglewood Fault zone

Whittier Fault

Torrance

Palos Verdes Fault

Irvine

Mojave Desert

San Andreas Fault Zone

San Gabriel Mountains

Sierra Madre Thrust Fault

Los Angeles

Stress

Stress

BRITTLE

DUCTILE

CRUS

◄ Tomales Bay is a narrow inlet located north of San Francisco that nearly separates Point Reyes Peninsula from mainland North America. This linear depression, now filled with seawater, formed as a result of displacement along roughly parallel fault traces of the San Andreas Fault. Tomales Bay is more than 1 mile wide, suggesting that the active zone of faulting along this section of the San Andreas is equally wide.

Tomales Bay

San Andreas Fault

Drakes Bay

Point Reyes

N

Photo: Michael Collier

Question:
1. What type of fault is the San Andreas?
2. List three major branches of the San Andreas Fault located in the San Francisco Bay area.

?

Photo: NASA

Along the San Andreas Fault, the Pacific plate grinds horizontally past the North American plate at a rate of about 5 centimeters per year. At its southern end, the San Andreas connects with a spreading center located in the Gulf of California.

Basin and Range Province

Extension

North American plate

Pacific plate

Great Valley

San Andreas Fault

Photo: Michael Collier

Calaveras Fault

Rather than generating earthquakes, some segments of the San Andreas Fault experience creep, a somewhat continuous movement along the fault without measurable earthquake activity. This once straight sidewalk and concrete wall were built decades ago in Hollister, California, and show effects of creep along the Calaveras Fault.

The Carrizo Plain is an expansive grassland that is bisected by the San Andreas Fault. For nearly 50 miles, a linear trough, offset stream channels, narrow ridges, and sag ponds mark the trace of this iconic fault.

Photo: Michael Collier

Pinnacles National Monument lies near the San Andreas Fault about 80 miles south of the San Francisco Bay area. It is composed of highly eroded volcanic rocks that spewed from Earth's interior about 23 million years ago. A strikingly similar group of rocks called the Neenach Volcanic Formation is located to the southeast, near Tejon Pass. Pinnacles were sliced from the Neenach Volcanic Formation and transported by the Pacific plate 195 miles toward the northeast. By matching these and other rock units across the fault, geologists have determined that the total displacement along the San Andreas Fault system from earthquakes and creep exceeds 340 miles.

Photo: Kevin Schafer/AGE Fotostock

Figure 10.25
Slickensides As two rock bodies slide past one another, their fault surfaces often become polished and grooved into *slickensides*. (Photo by E. J. Tarbuck)

Slickensides

The majority of fault surfaces that have been uplifted and exposed by erosion are remnants of past deformation. On some of these fault surfaces, the rocks became highly polished and striated, or grooved, as the crustal blocks slide past one another. These polished and striated surfaces, called **slickensides** (*sliken* = smooth), provide geologists with evidence for the direction of the most recent displacement along the fault (**Figure 10.25**). By swiping your hand back and forth across these surfaces, the direction of relative movement should be obvious.

Joints

Among the most common rock structures are fractures called joints. Unlike faults, **joints** are fractures along which no appreciable displacement has occurred. Although some joints have a random orientation, most occur in roughly parallel groups.

We have already considered two types of joints. Recall from Chapter 4 that *columnar joints* form when igneous rocks cool and develop shrinkage fractures that produce elongated, pillarlike columns (see Figure 4.31, page 132). In addition, recall from Chapter 6 that sheeting produces a pattern of gently curved joints that develop more or less parallel to the surface of large exposed igneous bodies such as batholiths. Here the jointing results from the gradual expansion that occurs when erosion removes the overlying load and reduces pressure on the rock.

Most joints are produced when rocks in the outermost crust are deformed, causing the rock to fail by brittle fracture. Extensive joint patterns often develop in response to relatively subtle and often barely perceptible regional upwarping and downwarping of the crust. In many cases, the cause of jointing at a particular locale is not particularly obvious.

Many rocks are broken by two or even three sets of intersecting joints that slice the rock into numerous regularly shaped blocks. These *joint sets* often exert a strong influence on other geologic processes. For example, chemical weathering tends to be concentrated along joints, and joint patterns influence how groundwater moves through the crust (**Figure 10.26**).

Joints may also be significant from an economic standpoint. Some of the world's largest and most important mineral deposits are located along joint systems. Hydrothermal solutions (mineralized fluids) can migrate into fractured host rocks and precipitate economically significant amounts of copper, silver, gold, zinc, lead, and uranium.

Highly jointed rocks present a risk to construction projects, including bridges, highways, and dams. On June 5, 1976, the Teton Dam in Idaho failed, taking 14 lives and causing nearly $1 billion in property damage. This earthen dam, constructed of very erodible clays and silts, was situated on highly fractured volcanic rocks. Although attempts were made to fill the voids in the jointed rock, water gradually penetrated the subsurface fractures and undermined the dam's foundation. Eventually the moving water cut a tunnel into the easily erodible clays and silts. Within minutes the dam failed, sending a 20-meter- (65-foot-) high wall of water down the Teton and Snake Rivers.

EYE ON EARTH 10.2

This satellite image shows an area just south of the Tien Shan Mountains, China. The distinctive red, green, and cream-colored bands are sedimentary rock layers that were tilted by compressional forces as various landmasses collided with southern Asia. Also visible in this image is the Piqiang Fault, which runs roughly perpendicular to the colored sedimentary layer for about 70 kilometers (40 miles).

QUESTION 1 What type of fault is the Piqiang Fault?

QUESTION 2 Is it a right-lateral or left-lateral fault?

Figure 10.26 Nearly parallel joints in the Navajo Sandstone, Arches National Park, Utah Weathering and erosion along the joints have produced a topography known in the park as "fins." (Photo by Michael Collier)

10.5 | Mapping Geologic Structures

Explain how strike and dip are measured and how these measurements tell geologists about the orientations of rock structures located mainly below Earth's surface.

By studying the orientations of faults, folds, and tilted sedimentary formations, geologists can often reconstruct the original geologic setting and determine the nature of the forces that generated these structures. In this way, the complex events of Earth's geologic history are unraveled.

Frequently, structures are so large that only a small portion is visible from any particular vantage point. In many situations, most of the bedrock is concealed by vegetation or buried by recent sedimentation. Consequently, reconstruction must be done using data gathered from a limited number of outcrops of exposed bedrock. Despite these challenges, a number of mapping techniques enable geologists to infer the shape and orientation of rock structures below the surface. In recent years this work has been aided by advances in aerial photography, satellite imagery, and the development of the Global Positioning System (GPS). In addition, seismic reflection profiling (see Chapter 12) and drill holes provide data on the composition and orientation of rock structures that lie at depth.

In the previous section you learned that geologists use *strike* (trend) and *dip* (inclination) to describe the orientation of planar rock features, such as sedimentary bedding and fault surfaces (**Figure 10.27**). In addition, by knowing the strike and dip of rocks at the surface, geologists can predict the nature and structure of rock units hidden beneath the surface.

Strike is the compass bearing (direction) of the line produced by the intersection of an inclined rock layer (or fault) with a horizontal surface (Figure 10.27). The strike is generally expressed as an angle relative to

Strike and Dip

Sedimentary rock units, because they are usually deposited in horizontal layers, are most useful when studying rock deformation. If the sedimentary rock layers are horizontal, geologists infer that the area is undisturbed structurally. Inclined, bent, or broken strata indicate that a period of deformation occurred following deposition.

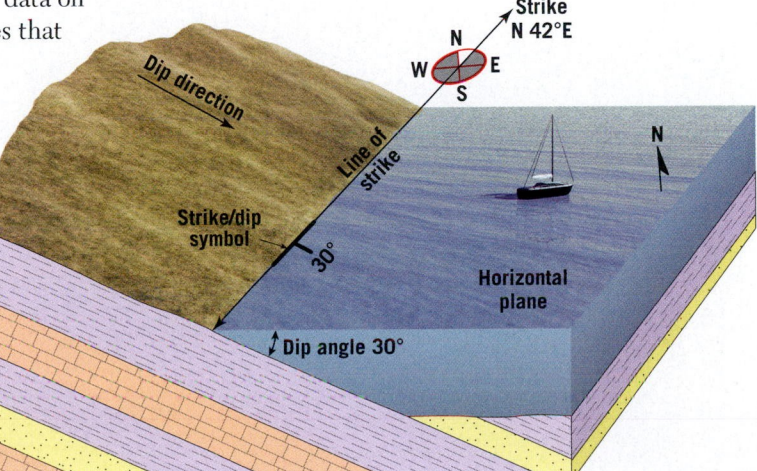

Figure 10.27 Strike and dip of rock layers

Strike N 42°E
Dip direction
Line of strike
Strike/dip symbol
30°
Dip angle 30°
Horizontal plane

Figure 10.28
Mapping geologic structures By establishing the strike and dip of outcropping sedimentary beds and placing symbols on a map, geologists can infer the orientation of the rock structures below the surface.

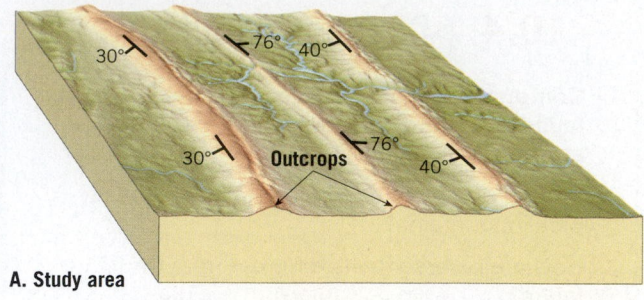

A. Study area

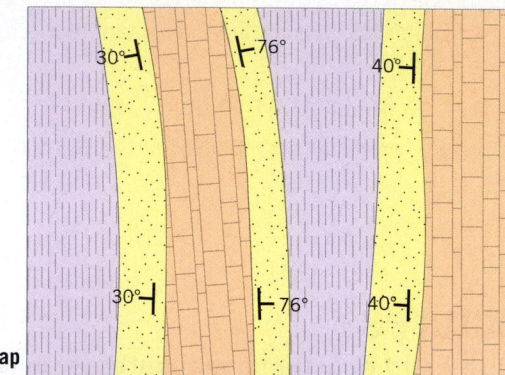

B. Geologic map

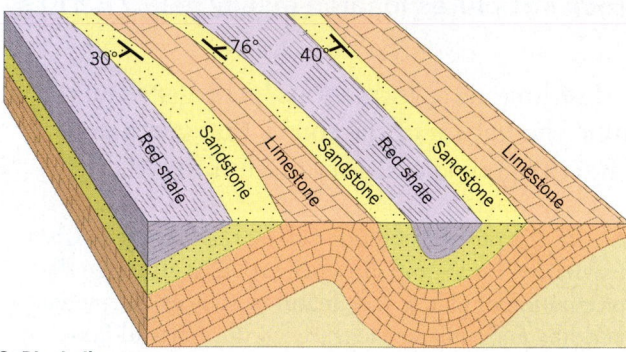

C. Block diagram

north. For example, N10°E means the line of strike is 10 degrees to the east of north.

Dip is the angle of inclination of the surface of a rock unit or fault, measured from a horizontal surface. Dip includes both an angle of inclination and a direction toward which the rock is inclined. In Figure 10.27 the dip angle of the rock layer is 30 degrees. The direction of dip is always at a 90-degree angle to the strike.

When doing field research, geologists measure the strike and dip of sedimentary strata at as many outcrops as practical (**Figure 10.28A**). These data are then plotted on a topographic map or an aerial photograph, using T-shaped symbols. In laboratories, this information, as well as notes about the rock units, is used to prepare graphical depictions of the study area, called **geologic maps**. **Figure 10.28B** shows a simplified geologic map of the study area. Geologic maps are valuable tools used to infer the orientation and shape of buried rock structures (**Figure 10.28C**). Using this information, geologists can reconstruct the pre-erosional structures and begin to interpret the region's geologic history. Geologic maps are also used to aid in extracting mineral resources and assessing hazards, such as the potential for groundwater contamination from a proposed disposal site.

10.5   Concept Checks

1. Distinguish between the two measurements used to establish the orientation of deformed strata.

2. Briefly describe the method geologists use to infer the orientation of rock structures that lie mainly below Earth's surface.

10.1 What Causes Rock to Deform

Describe the three types of differential stress and name the type of plate boundary most commonly associated with each.

KEY TERMS deformation, rock structure (geologic structure), stress, confining pressure, differential stress, compressional stress, tensional stress, shear stress, strain

- Rock structures are generated when rocks are deformed by bending or breaking due to differential stress. Crustal deformation produces geologic structures that include folds, faults, joints, foliation, and rock cleavage.

- Stress is the force that drives rock deformation. When stress has the same magnitude in every direction, it is called confining pressure. Alternatively, when the amount of stress coming from one direction is greater in magnitude than the stress coming from another direction, we call it differential stress. There are three main types of differential stress: compressional, tensional, and shear stress.
- Strain is the change in the shape of a rock body caused by stress.

Q Classify the following everyday situations as illustrating confining pressure, compressional stress, tensional stress, or shear stress: (a) a watermelon being run over by a steamroller, (b) a person diving to the bottom of the deep end of a swimming pool, (c) playing a game of tug-of-war, (d) kneading bread dough, and (e) slipping on a banana peel.

10.2 How Do Rocks Deform?

Compare and contrast brittle and ductile deformation.

KEY TERMS elastic deformation, brittle deformation, ductile deformation, outcrop

- There are several types of deformation. Elastic deformation is a temporary stretching of the chemical bonds in a rock. When the stress is released, the bonds snap back to their original lengths. When stress is greater than the strength of the bonds, the rock deforms in either a brittle or ductile fashion. Brittle deformation occurs when rocks break into smaller pieces, whereas ductile deformation is a solid-state flow that allows a rock to bend without fracturing.
- The type of deformation (ductile or brittle) that occurs depends mainly on temperature, confining pressure, rock type, and time. In Earth's upper crust where temperatures and pressure are low, rocks tend to exhibit brittle behavior and break or fracture. At depth rocks tend to deform by flowing or bending.
- Igneous rocks tend to be strong and are more likely to deform in a brittle fashion, whereas sedimentary rocks are weaker and usually deform in a ductile fashion.

- The rate at which differential stress is applied also affects how rocks deform. Silly Putty provides a good analogy: If pulled apart quickly, Silly Putty tends to break, whereas if pulled apart slowly, it tends to stretch (ductile flow) without breaking.

Q Examine the accompanying illustration of a collision between two tectonic plates. At which location (A or B) would brittle deformation be more prevalent than ductile deformation?

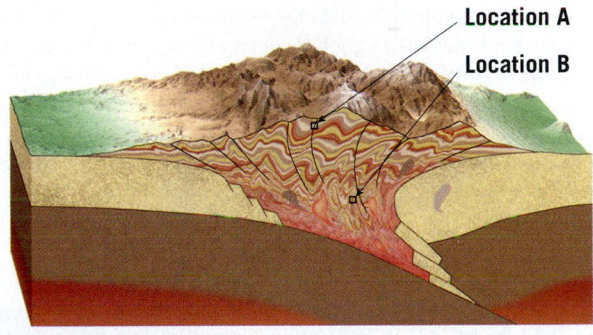

Location A

Location B

10.3 Folds: Rock Structures Formed by Ductile Deformation

List and describe five common folded structures.

KEY TERMS fold, anticline, syncline, dome, basin, monocline

- Folds are wavelike undulations in layered rocks that develop through ductile deformation in rocks undergoing compressional stress.
- Folds may be described in terms of their geometric configuration: If the limbs of a fold dip down from the hinge, the fold has an arch-like structure and is called an anticline. If the limbs of a fold dip upward, the fold has a trough-like structure and is called a syncline. Anticlines and synclines may be symmetrical, asymmetrical, overturned, or recumbent.

- The shape of a fold does not necessarily correlate to the shape of the landscape above it. Rather, surface topography usually reflects patterns of differential weathering.
- A fold is said to plunge when its axis penetrates the ground at an angle. This results in a V-shaped outcrop pattern of the folded layers.
- Domes and basins are large folds that produce nearly circular-shaped outcrop patterns. The overall shape of a dome or basin is like a saucer or a bowl, either right-side-up (basin) or inverted (dome).
- Monoclines are large steplike folds in otherwise horizontal strata that result from subsurface faulting. Imagine a carpet draped over a short staircase to envision how the strata can go from horizontal to tilted and back to horizontal again.

10.4 Faults and Joints: Structures Formed by Brittle Deformation

Sketch and briefly describe the relative motion of rock bodies located on opposite sides of normal, reverse, and thrust faults as well as both types of strike-slip faults.

KEY TERMS fault, dip-slip fault, hanging wall block, footwall block, normal fault, fault-block mountain, horst, graben, half-graben, detachment fault, reverse fault, thrust fault, klippe, strike-slip fault, transform fault, oblique-slip fault, megathrust fault, fault scarp, slickenside, joint

- Faults are fractures along which one rock body slides past another.
- The direction of offset on a fault may be determined by comparing the blocks of rock on either side of the fault surface. Faults in which movement is primarily parallel to the dip of the fault surface are called dip-slip faults. Dip-slip faults are classified as normal faults if the hanging wall moves *down* relative to the footwall and as reverse faults if the hanging wall moves *up* relative to the footwall. Large

normal faults with low dip angles are detachment faults, whereas reverse faults with low dip angles are called thrust faults.

- Areas of tectonic extension, such as the Basin and Range Province, produce fault-block mountains—horsts separated by neighboring grabens or half-grabens.
- Areas of tectonic compression, such as mountain belts, are dominated by reverse faults that shorten the crust horizontally while thickening it vertically.
- Most of the movement along a strike-slip fault is along the strike or trend of the fault trace. Transform faults are large strike-slip faults that serve as tectonic boundaries between lithospheric plates.
- Oblique-slip faults display characteristics of both dip-slip and strike-slip faults.
- Joints are fractures in rocks along which no appreciable movement has occurred.

Q What type of rock structure is shown in each of the accompanying photos: faults or joints? Explain how you arrived at your answer.

A.   **B.**   E.J. Tarbuck

10.5 Mapping Geologic Structures

Explain how strike and dip are measured and how these measurements tell geologists about the orientations of rock structures located mainly below Earth's surface.

KEY TERMS strike, dip, geologic map

- Geologic maps are valuable tools for establishing the shape and orientation of subterranean rock structures, mainly folds and faults—an important step for resource extraction and hazard assessment. Geologic maps are also used to reconstruct the geologic history of a region.
- Strike and dip are measurements of the orientation of planar rock features such as sedimentary bedding and fault surfaces. Strike is the compass direction of the line at the intersection between a horizontal plane and the geologic surface. Dip is the angle of inclination between the horizontal plane and a line perpendicular to the strike.

Q The accompanying simplified geologic map shows the strike and dip of outcropping sedimentary beds. Based on these measurements, what rock structure is most likely beneath Earth's surface: anticline, syncline, dome, or basin?

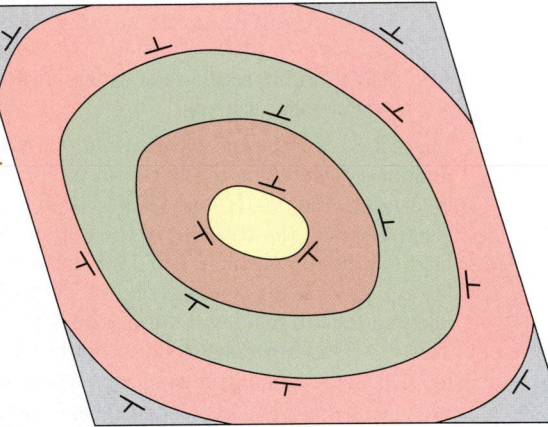

Give It Some Thought

1. Is granite or mica schist more likely to fold or flow rather than fracture when subjected to differential stress? Explain.

2. What type of deformation is illustrated by a coin that was run over by a passing train?

Anthony Pleva/Alamy

3. Refer to the accompanying photo to answer the following questions:
   a. Name the type of fold shown.
   b. Would you describe this fold as symmetrical or asymmetrical?

E.J. Tarbuck

c. What name is given to the part of the fold labeled A?

d. Is the white dot labeled B located along the fold line, hinge line, or dip line of this particular fold?

4. Refer to the accompanying diagrams to answer the following:

   a. What type of dip-slip fault is shown in Diagram 1? Were the dominant forces during faulting tensional, compressional, or shear?

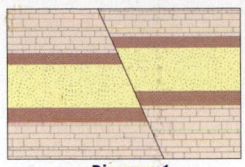
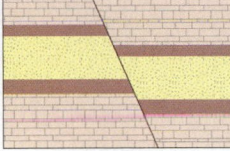

Diagram 1
(cross section)

Diagram 2
(cross section)

Diagram 3

b. What type of dip-slip fault is shown in Diagram 2? Were the dominant forces during faulting tensional, compressional, or shear?

c. Match the correct pair of arrows in Diagram 3 to the faults in Diagrams 1 and 2.

5. Refer to the accompanying photo to answer the following:

a. The white line shows the approximate location of a fault that displaced these furrows created by a plow. What type of fault caused the offset shown?

b. Is this a right-lateral or left-lateral fault? Explain.

USGS

6. With which of the three types of plate boundaries does normal faulting predominate? Thrust faulting? Strike-slip faulting?

7. Write a brief statement describing each of the accompanying photos, using terms from the following list: strike-slip, dip-slip, normal, reverse, right-lateral, left-lateral.

Fault

Fault

Marli Miller

Marli Miller

A.

B.

8. The accompanying photo, taken near the bottom of the Grand Canyon, shows a quartz vein that has been deformed.

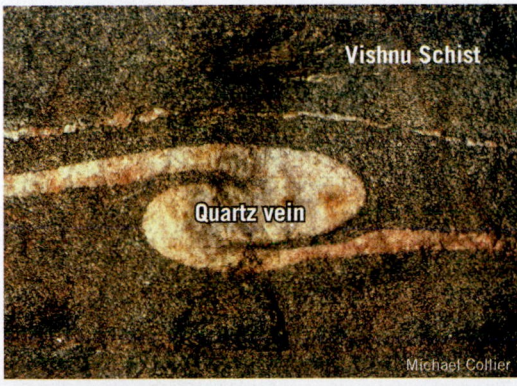

Vishnu Schist

Quartz vein

Michael Collier

a. What type of deformation is exhibited—ductile or brittle?

b. Did this deformation most likely occur near Earth's surface or at great depth?

9. The accompanying simplified geologic map shows the strike and dip of outcropping sedimentary beds. From these measurements, describe the rock structure that is most likely beneath Earth's surface, using terms from the following list: anticline, syncline, dome, basin, symmetrical, asymmetrical.

Map view

41° 62° 40° 42° 60° 61°

10. Examine the diagrams depicting the structure of East Africa and the Canadian Rockies.

a. Characterize the type of faulting found at each location and identify the type of differential stress that resulted in these landforms.

b. Along what type of plate boundary did each of these structures form?

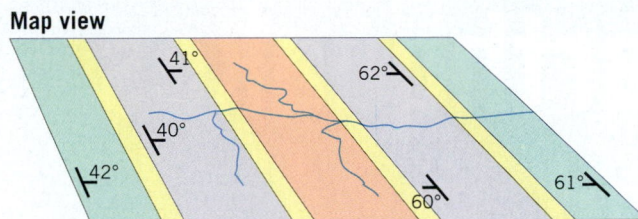

East Africa

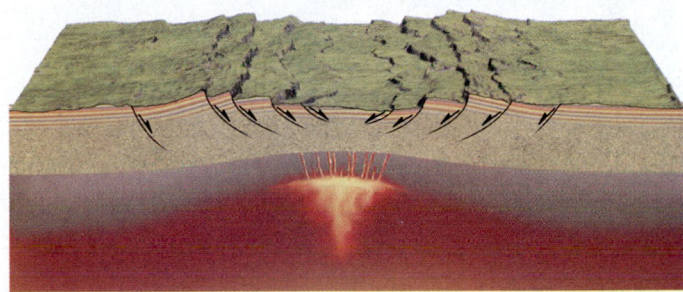

Canadian Rockies

11

Earthquakes and Earthquake Hazards

Tsunami striking the coast of Japan on March 11, 2011.
(Photo by Sadatsugu Tomizawa/AFP/Getty Images)

On January 12, 2010, an estimated 316,000 people lost their lives when a magnitude 7.0 earthquake struck the small Caribbean nation of Haiti, the poorest country in the Western Hemisphere. In addition to the staggering death toll, more than 300,000 people were injured, and more than 280,000 houses were destroyed or damaged. The quake originated only 25 kilometers (15 miles) from the country's densely populated capital city of Port-au-Prince (Figure 11.1). It occurred along a San Andreas–like fault system at a depth of just 10 kilometers (6 miles). Because of the quake's shallow depth, ground shaking was exceptional for an event of this magnitude.

Other factors that contributed to the Port-au-Prince disaster included the city's geologic setting and the nature of its buildings. The city is built on sediment, which is quite susceptible to ground shaking during an earthquake. More importantly, inadequate or nonexistent building codes meant that buildings collapsed far more readily than they should have. At least 52 aftershocks, measuring magnitude 4.5 or greater, jolted the area and added to the trauma that survivors experienced for days after the original quake. An earthquake's *magnitude* (abbreviation: *M*) is a measure of earthquake strength that will be discussed later in this chapter.

Figure 11.1
Presidential palace damaged during the 2010 Haiti earthquake (Photo by Luis Acosta/AFP/Getty Images)

11.1 | What Is an Earthquake?

Sketch and describe the mechanism that generates most earthquakes.

An **earthquake** is ground shaking caused by the sudden and rapid movement of one block of rock slipping past another along fractures in Earth's crust called **faults**. Most faults are *locked*, except for brief, abrupt movements when sudden slippage produces an earthquake. Faults are locked because the confining pressure exerted by the overlying crust is enormous, causing these fractures in the crust to be "squeezed shut."

Earthquakes tend to occur along preexisting faults where internal stresses cause the crustal rocks to rupture or break into two or more units. The location where slippage begins is called the **hypocenter, or focus**. Earthquake waves radiate from this spot outward into the surrounding rock. The point on Earth's surface directly above the hypocenter is called the **epicenter** (**Figure 11.2**).

Large earthquakes release huge amounts of stored-up energy as **seismic waves**—a form of energy that travels through the lithosphere and Earth's interior. The energy carried by these waves causes the material that transmits them to shake. Seismic waves are analogous to waves produced when a stone is dropped into a calm pond. Just as the impact of the stone creates a pattern of circular waves, an earthquake generates waves that radiate outward in all directions from the hypocenter. Although seismic energy dissipates rapidly as it moves away from the quake's hypocenter, sensitive instruments can detect earthquakes even when they occur on the opposite side of Earth.

Thousands of earthquakes occur around the world every day. Fortunately, most are small and cannot be detected by people. Of these, only about 15 strong earthquakes (magnitude 7 or greater) are recorded each year, many of them occurring in remote regions. Occasionally, a large earthquake is triggered near a major population center. Such events are among the most destructive

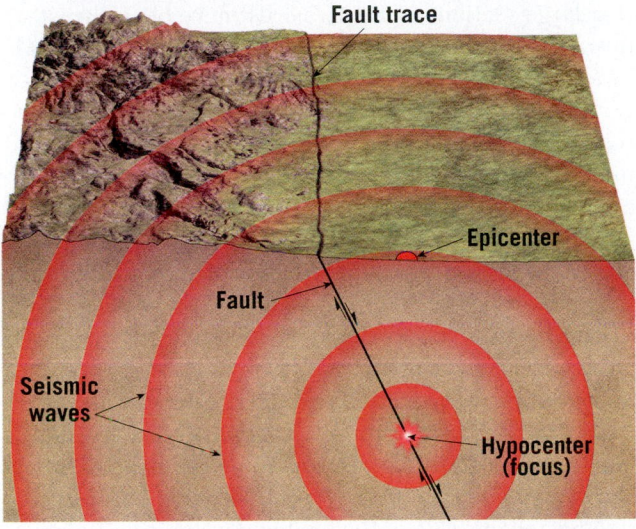

Figure 11.2
Earthquakes hypocenter and epicenter The *hypocenter* is the zone at depth where the initial displacement occurs. The *epicenter* is the surface location directly above the hypocenter.

natural forces on Earth. The shaking of the ground, coupled with the liquefaction of soils, wreaks havoc on buildings, roadways, and other structures. In addition, when a quake occurs in a populated area, power and gas lines are often ruptured, causing numerous fires. In the famous 1906 San Francisco earthquake, much of the damage was caused by fires that became uncontrollable when broken water mains left firefighters with only trickles of water (**Figure 11.3**).

San Francisco in flames following the 1906 quake. Broken water mains left firefighters without water.

Figure 11.3
Earthquakes can trigger fires
(Reproduced from the collection of the Library of Congress; inset photo by Hal Garb/AFP/Getty Images)

Fire triggered when a gas line ruptured during the Northridge earthquake in southern California in 1994.

Discovering the Causes of Earthquakes

The energy released by volcanic eruptions, massive landslides, and meteorite impacts can generate earthquake-like waves, but these events are usually weak. What mechanism produces a destructive earthquake? As you have learned, Earth is not a static planet. Because fossils of marine organisms have been discovered thousands of meters above sea level, we know that large sections of Earth's crust have been thrust upward. Other regions, such as California's Death Valley, exhibit evidence of extensive subsidence. In addition to these vertical displacements, offsets in fences, roads, and other structures indicate that horizontal movements between blocks of Earth's crust are also common (**Figure 11.4**).

The actual mechanism of earthquake generation eluded geologists until H. F. Reid conducted a landmark study following the 1906 San Francisco earthquake. This earthquake was accompanied by horizontal surface displacements of several meters along the northern portion of the San Andreas Fault. Field studies determined that during this single earthquake, the Pacific plate lurched as much as 9.7 meters (32 feet) northward past the adjacent North American plate. To better visualize this, imagine standing on one side of the fault and watching a person on the other side suddenly slide horizontally 32 feet to your right.

What Reid concluded from his investigations is illustrated in **Figure 11.5**. Over tens to hundreds of years, differential stress slowly bends the crustal rocks on both sides of a fault. This is much like a person bending a limber wooden stick, as shown in **Figure 11.5A,B**. Frictional resistance keeps the fault from rupturing and slipping. (Friction inhibits slippage and is enhanced by irregularities that occur along the fault surface.) At some point, the stress along the fault overcomes the frictional resistance, and slip initiates. Slippage allows the deformed (bent) rock to "snap back" to its original, stress-free, shape; a series of earthquake waves radiate as it slides (**Figure 11.5C,D**). Reid termed this "springing back" **elastic rebound** because the rock behaves elastically, much as a stretched rubber band does when it is released.

Aftershocks and Foreshocks

Strong earthquakes are followed by numerous earthquakes of lesser magnitude, called **aftershocks,** which result from crust along the fault surface adjusting to the displacement caused by the main shock. Aftershocks gradually diminish in frequency and intensity over a period of several months following an earthquake. In a little more than a month following the 2010 Haiti earthquake, the U.S. Geological Survey detected nearly 60 aftershocks with magnitudes of 4.5 or greater. The two largest aftershocks had magnitudes of 6.0 and 5.9, both large enough to inflict damage. Hundreds of minor tremors were also felt.

Although aftershocks are weaker than the main earthquake, they often trigger the

Figure 11.4
Displacement of structures along a fault (Color photo by John S. Shelton/University of Washington Libraries; inset photo by G. K. Gilbert/USGS)

Slippage along a fault produced this offset in an orange grove east of Calexico, California.

This fence was offset 2.5 meters (8.5 feet) during the 1906 San Francisco earthquake.

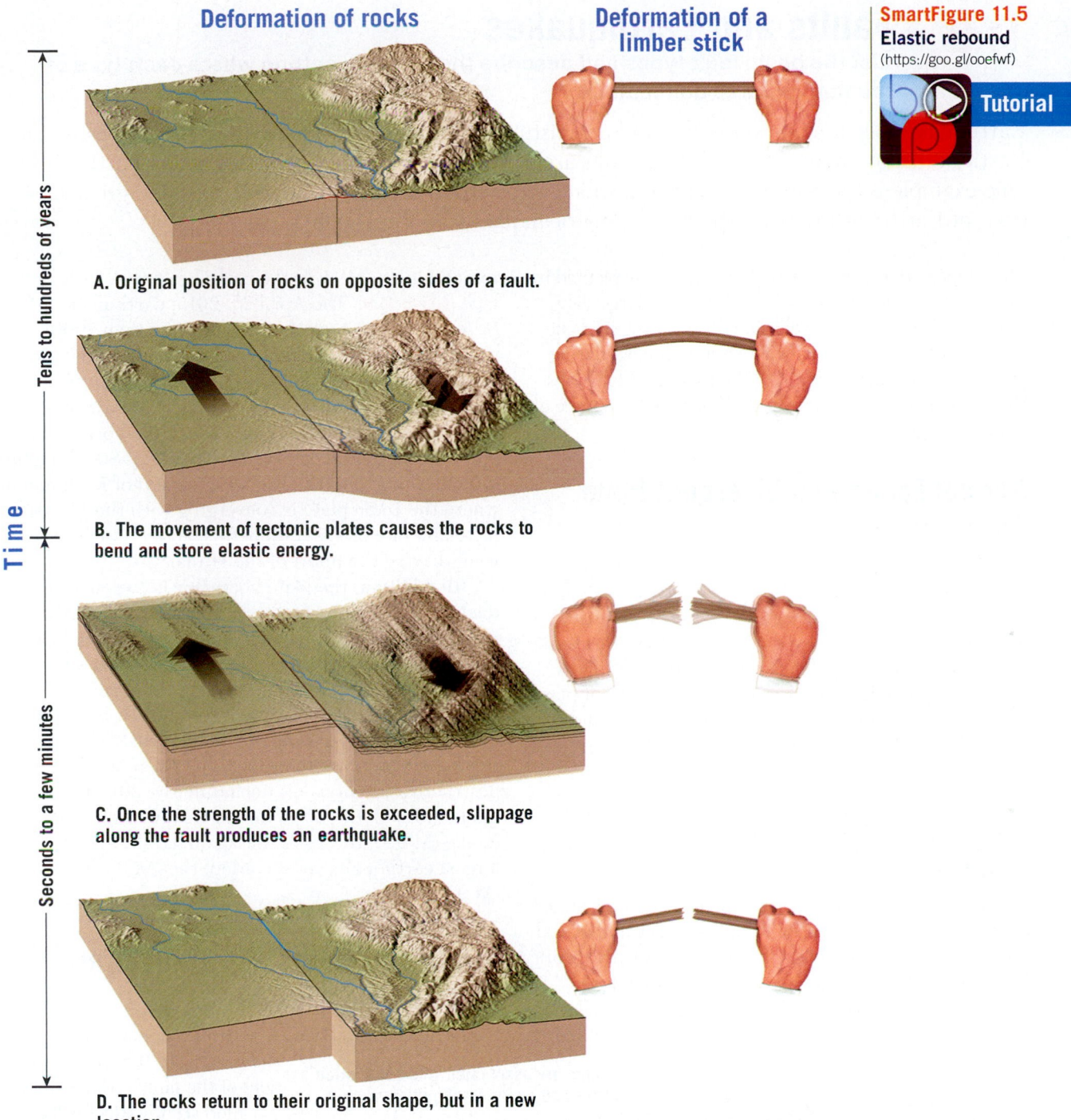

Deformation of rocks

Deformation of a limber stick

Time

Tens to hundreds of years

A. Original position of rocks on opposite sides of a fault.

B. The movement of tectonic plates causes the rocks to bend and store elastic energy.

Seconds to a few minutes

C. Once the strength of the rocks is exceeded, slippage along the fault produces an earthquake.

D. The rocks return to their original shape, but in a new location.

SmartFigure 11.5
Elastic rebound
(https://goo.gl/ooefwf)

Tutorial

destruction of already weakened structures. For example, in northwestern Armenia in 1988, where many people lived in large apartment buildings constructed of brick and concrete slabs, a moderate earthquake of magnitude 6.9 weakened many structures, and a strong aftershock of magnitude 5.8 completed the demolition.

In contrast to aftershocks, small earthquakes called **foreshocks** often, but not always, precede major earthquakes by days or, in some cases, several years. Monitoring of foreshocks to predict forthcoming earthquakes has been attempted with only limited success.

11.1   Concept Checks

1. What is an earthquake? Under what circumstances do most large earthquakes occur?

2. How are faults, hypocenters, and epicenters related?

3. Who was first to explain the mechanism by which most earthquakes are generated?

4. Explain what is meant by *elastic rebound*.

11.2 | Faults and Earthquakes

List the basic fault types and describe the tectonic setting where each type of faulting tends to dominate.

Earthquakes occur along both new and preexisting faults, in places where differential stresses cause Earth's crust to break. Some of these faults are large and capable of generating major earthquakes. One example is the San Andreas Fault, which is the strike-slip fault that separates two great sections of Earth's lithosphere: the North American plate and the Pacific plate.

The slippage that occurs along faults can be explained by the plate tectonics theory, which states that large slabs of Earth's lithosphere are continually grinding past one another. These mobile plates interact with neighboring plates, straining and deforming the rocks at their edges. Faults associated with plate boundaries are the source of most earthquakes.

Normal Faults and Divergent Plate Boundaries

Recall from Chapter 10 that there are three basic types of faults: *normal faults*, *reverse faults* (including *thrust faults*), and *strike-slip faults*. **Normal faults** occur where tensional stress causes Earth's crust to be stretched and elongated. Therefore, most earthquakes along normal faults are associated with divergent plate boundaries, mainly seafloor spreading centers and continental rifts. Although common, normal faults are not generally associated with large earthquakes.

Thrust Faults and Convergent Plate Boundaries

Strong earthquakes also occur along large faults associated with convergent plate boundaries. Compressional forces associated with continental collisions that result in mountain building generate numerous **reverse faults** and **thrust faults**. Displacement along reverse and thrust faults result in the rocks above a fault being

forced (*or thrust*) over the rocks below the fault surface (**Figure 11.6**). The April 25, 2015, earthquake (M7.8) in Nepal, which took the lives of more than 7000 people, is an example of an earthquake that occurred along a thrust fault. This thrust fault is on or near the plate boundary between the subducting India plate and the overriding Eurasian plate to the north. The epicenter of the quake was located about 80 kilometers (50 miles) north of the Nepalese capital of Kathmandu, where the India plate is converging with the Eurasian plate at a rate of 4.5 centimeters (about 2 inches) per year, driving the uplift of the Himalayas.

In addition, the plate boundary between a subducting slab of oceanic lithosphere and the overlying plate form a fault referred to as a **megathrust fault** (see Figure 11.6). Because these large thrust faults lie partially beneath the ocean floor, movement along these faults may displace the overlying seawater, generating destructive *tsunamis*. Megathrust faults have produced the majority of Earth's most powerful and destructive earthquakes, including the 2011 Japan quake (M 9.0), the 2004 Indian Ocean (Sumatra) quake (M 9.1), the 1964 Alaska quake (M 9.2), and the largest earthquake yet recorded, the 1960 Chile quake (M 9.5).

Strike-slip Faults and Transform Plate Boundaries

Faults in which the dominant displacement is horizontal and parallel to the *strike* (direction) of the fault surface are called **strike-slip faults**. Large strike-slip faults that slice through Earth's lithosphere and accommodate motion between two tectonic plates are called **transform faults**. For example, the San Andreas Fault is a large strike-slip fault that separates the North American plate and the Pacific plate.

Most transform faults, including the San Andreas Fault, are not perfectly straight

Figure 11.6
Megathrust faults are the sites of Earth's largest earthquakes Convergent plate boundaries are the sites where one plate is subducting beneath another, producing some of Earth's largest earthquakes.

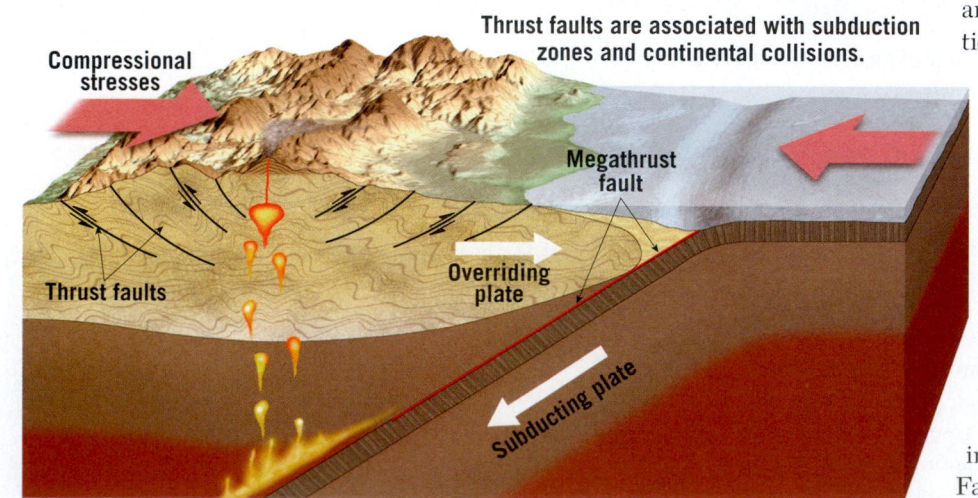

Compressional stresses

Thrust faults are associated with subduction zones and continental collisions.

Megathrust fault

Thrust faults

Overriding plate

Subducting plate

or continuous; instead, they consist of numerous branches and smaller fractures that display kinks and offsets (**Figure 11.7**). In addition, geologists have learned that displacement along transform faults occurs in discrete segments that often behave differently from one another (see GEOgraphics 10.1, page 318). Some sections of the San Andreas Fault exhibit slow, gradual displacement known as **fault creep** and produce little seismic shaking. Other segments slip at relatively closely spaced intervals, producing numerous small to moderate earthquakes. Still other segments remain locked and store elastic energy for a few hundred years before they break loose. Ruptures on these segments usually result in major earthquakes.

Earthquakes that occur along locked segments of the San Andreas Fault tend to be repetitive. Soon after an earthquake occurs, strain begins accumulating due to the continuous motion of the plates on both sides of the fault. Decades or centuries later, the fault fails again.

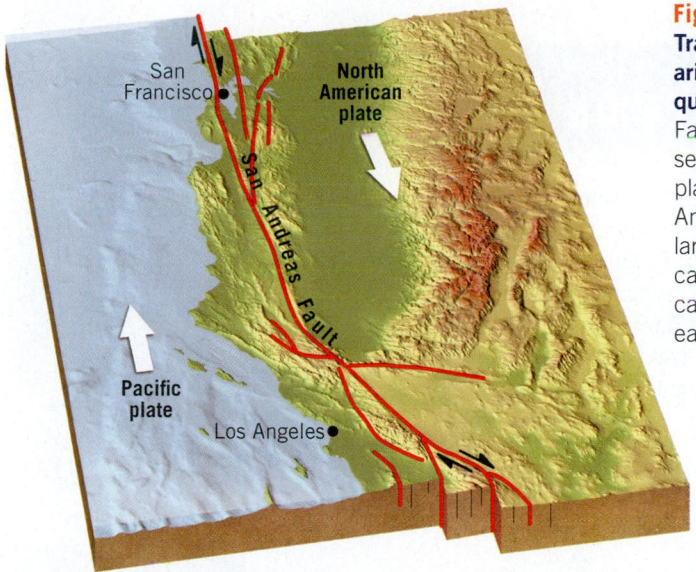

Figure 11.7
Transform plate boundaries and large earthquakes The San Andreas Fault is a large fault system separating the Pacific plate from the North American plate. These large strike-slip faults are called transform faults and can generate destructive earthquakes.

Fault Rupture and Propagation

Slippage along large faults does not occur instantaneously. The initial slip begins at the hypocenter and propagates (travels) along the fault surface, at 2 to 4 kilometers per second—faster than a rifle shot (**Figure 11.8**). Slippage on one section of the fault adds strain to the adjacent segment, which may also slip. As this zone of slippage advances, it can slow down, speed up, or even jump to a nearby fault segment. The propagation of the rupture zone along a fault that is 300 kilometers (200 miles) long, for example, takes about 1.5 minutes, and it takes about 30 seconds for a fault that is 100 kilometers (60 miles) long. Earthquake waves are generated at every point along the fault as that portion of the fault begins to slip.

Although a rupture can propagate in both horizontal directions from the hypocenter, it often travels mainly in one direction. For example, during the 1906 San Francisco earthquake, the rupture began near the southern end of the 477-kilometer (296-mile) rupture zone of the San Andreas Fault and traveled mainly northward. In contrast, the 1857 Fort Tejon earthquake, one of the largest U.S. earthquakes, began at the northern end of the rupture zone, and the zone of slippage propagated southward.

Moderate earthquakes result from slippage along relatively small faults or small segments of large faults. The amount of displacement on the fault surface, called fault slip, is typically no more than a few meters. By contrast, during the largest earthquakes, fault slips of 50 meters (160 feet) have been observed. The maximum displacement during the 2011 earthquake in Japan is estimated to have exceeded 40 meters (130 feet), one of the largest yet measured. In addition, the fault surface that ruptures during a large quake can exceed 1000 kilometers (620 miles) in length. Therefore, the ground shaking produced by large earthquakes is not only

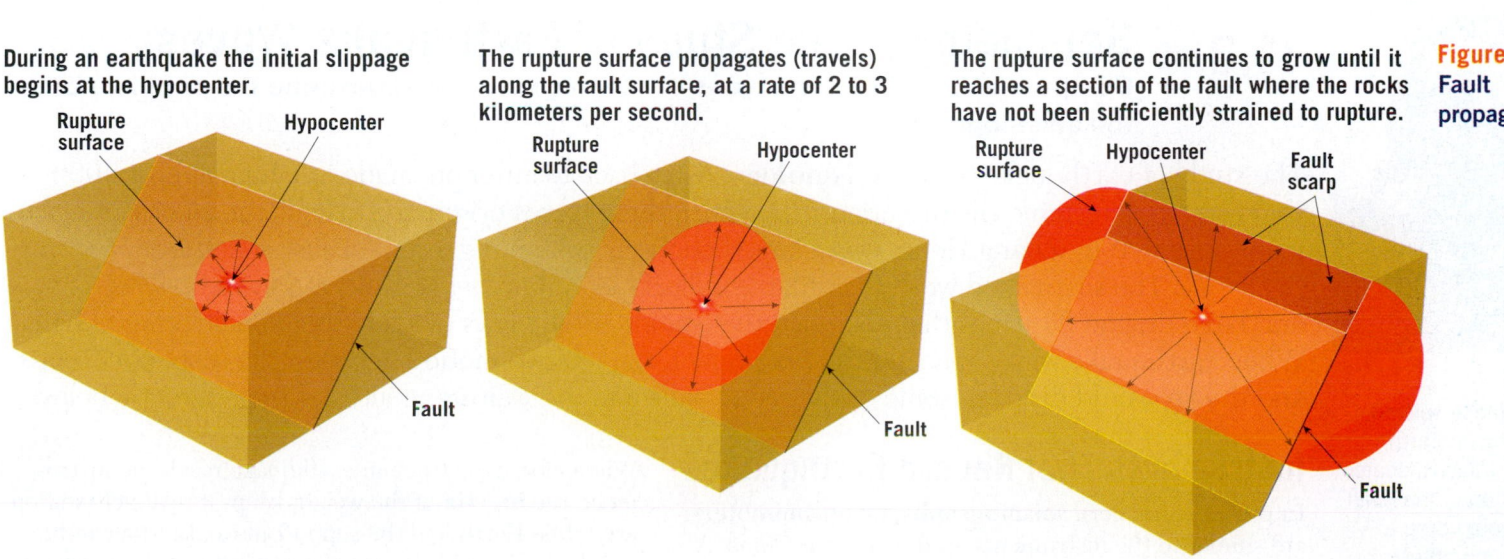

During an earthquake the initial slippage begins at the hypocenter.

The rupture surface propagates (travels) along the fault surface, at a rate of 2 to 3 kilometers per second.

The rupture surface continues to grow until it reaches a section of the fault where the rocks have not been sufficiently strained to rupture.

Figure 11.8
Fault propagation

The Calaveras Fault, a branch of the San Andreas Fault system, cuts directly through the town of Hollister, California. Rather than being "locked," this section of the fault is slowly slipping, producing noticeable offsets and damage to curbs, sidewalks, roads, and buildings. The concrete wall and sidewalk shown here were straight when they were originally constructed.

QUESTION 1 What term is used to describe the phenomenon observed in Hollister along the Calaveras Fault?

QUESTION 2 Are faults that exhibit this type of slippage considered likely to generate a major earthquake? Explain.

QUESTION 3 Based on this image, does this strike-slip fault exhibit right-lateral or left-lateral motion? (*Hint:* See Chapter 10.)

Calaveras Fault

Michael Collier

stronger but also lasts longer than the vibrations produced by slippage along small fault segments.

Why do earthquake ruptures stop rather than continue along the entire fault? Evidence suggests that slippage usually stops when the rupture reaches a section of the fault where the rocks have not been sufficiently strained to overcome frictional resistance, such as in a section of the fault that has recently experienced an earthquake. The rupture may also stop if it encounters a large kink, or an offset along the fault surface.

11.2 | Concept Checks

1. What type of fault tends to produce the most destructive earthquakes?

2. During an earthquake the entire length of a fault slips in a single event that lasts less than 1 second. True or false?

3. Defend or rebut this statement: Faults that do not experience fault creep may be considered safe.

11.3 | Seismology: The Study of Earthquake Waves

Compare and contrast the types of seismic waves and describe the principle of the seismograph.

The study of earthquake waves, **seismology**, dates back to attempts made in China almost 2000 years ago to determine the direction from which these waves originated. The earliest-known instrument, invented by Zhang Heng, was a large hollow jar containing a weight suspended from the top (**Figure 11.9**). The suspended weight (similar to a clock pendulum) was connected to the jaws of several large dragon figurines that encircled the container. The jaws of each dragon held a metal ball. When earthquake waves reached the instrument, the relative motion between the suspended mass and the jar would dislodge some of the metal balls into the waiting mouths of frogs directly below.

Ball dropping

Figure 11.9
Ancient Chinese seismograph During an Earth tremor, the dragons located in the direction of the main vibrations would drop a ball into the mouth of a frog below. (Photo by James E. Patterson Collection)

Instruments That Record Earthquakes

In principle, modern **seismographs,** or **seismometers**, are similar to the instruments used in ancient China. A seismograph has a weight freely suspended from a support that is securely attached to bedrock (**Figure 11.10**).

When vibrations from an earthquake reach the instrument, the **inertia** of the weight keeps it relatively stationary, while Earth and the support move. Inertia can be simply described by this statement: *Objects at rest tend to stay at rest, and objects in motion tend to remain in*

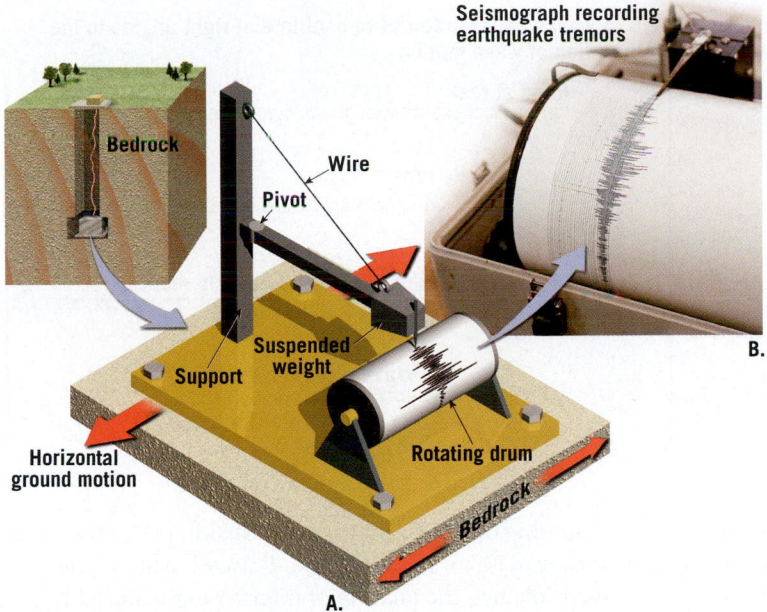

Seismograph recording earthquake tremors

Bedrock
Wire
Pivot
Support
Suspended weight
Rotating drum
Horizontal ground motion
Bedrock
A.
B.

Spring
Support moves
Rotating drum
Hinge
Suspended weight
Bedrock   **Vertical ground motion**

Figure 11.11
Seismograph designed to record vertical ground motion

SmartFigure 11.10
Principle of the seismograph The inertia of the suspended weight tends to keep it motionless while the recording drum, which is anchored to bedrock, vibrates in response to seismic waves. The stationary weight provides a reference point from which to measure the amount of displacement occurring as a seismic wave passes through the ground. (Photo courtesy of Zephyr/Science Source) (https://goo.gl/IucIEa)

Animation

motion, unless acted upon by an outside force. You have experienced inertia when you have tried to stop your automobile quickly and your body continued to move forward.

Earthquakes cause both vertical and horizontal ground motion; therefore, more than one type of seismograph is needed to accurately describe the intensity of shaking. The instrument shown in Figure 11.10 detects horizontal ground motion, while vertical ground motion is detected if the weight is suspended from a spring, as shown in **Figure 11.11**.

To detect very weak earthquakes or a great earthquake that has occurred in another part of the world, most seismographs are designed to amplify ground motion. In earthquake-prone areas, the instruments used are designed to withstand the violent shaking that can occur near a quake's epicenter.

Seismic Waves

The records obtained from seismographs, called **seismograms**, provide useful information about the nature of seismic waves. Seismograms reveal that two main types of seismic waves are generated by the slippage of a rock mass. One of these wave types, called **body waves**, travel through Earth's interior. The other type, called **surface waves**, travel in the rock layers just below Earth's surface (**Figure 11.12**).

Body Waves Body waves are further divided into two types—called **primary waves,** or **P waves**, and **secondary waves,** or **S waves**—and are identified by their mode of travel through intervening materials. P waves are "push/pull" waves; they momentarily push (compress) and

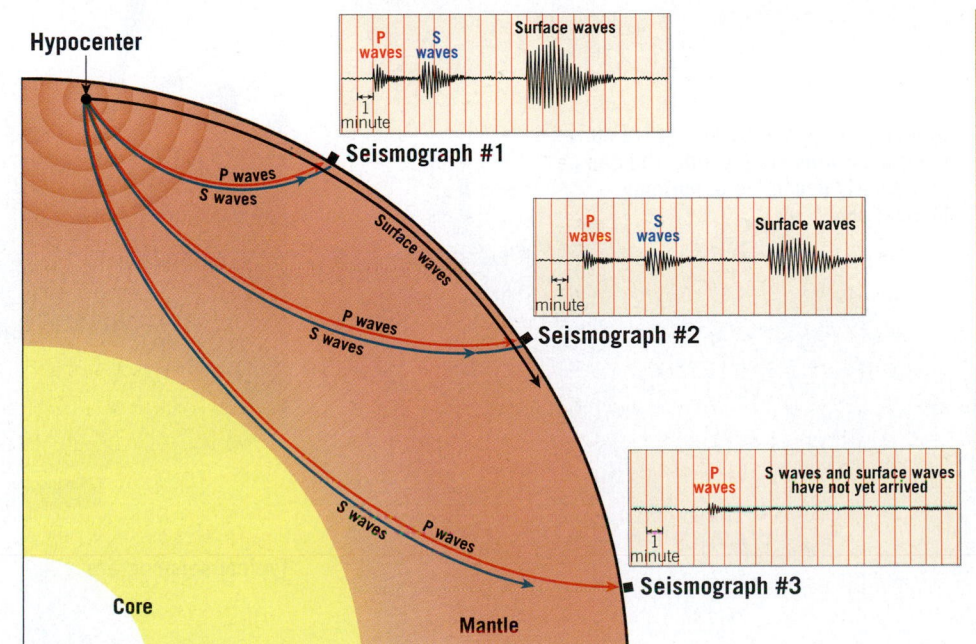

Hypocenter
P waves / S waves / Surface waves / minute
Seismograph #1
P waves / **S waves** / Surface waves
P waves / S waves
Surface waves
P waves / S waves / minute
Seismograph #2
S waves / P waves
P waves / S waves and surface waves have not yet arrived / 1 minute
Seismograph #3
Core   **Mantle**

SmartFigure 11.12
Body waves (P and S waves) versus surface waves P and S waves travel through Earth's interior, while surface waves travel in the layer directly below the surface. P waves are the first to arrive at a seismic station, followed by S waves, and then surface waves. (https://goo.gl/y6owpc)

Tutorial

A. As illustrated by a toy Slinky, P waves alternately compress and expand the material through which they pass.

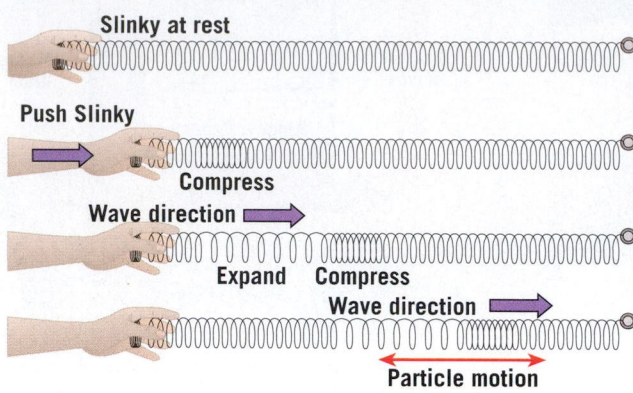

B. S waves cause material to oscillate at right angles to the direction of wave motion.

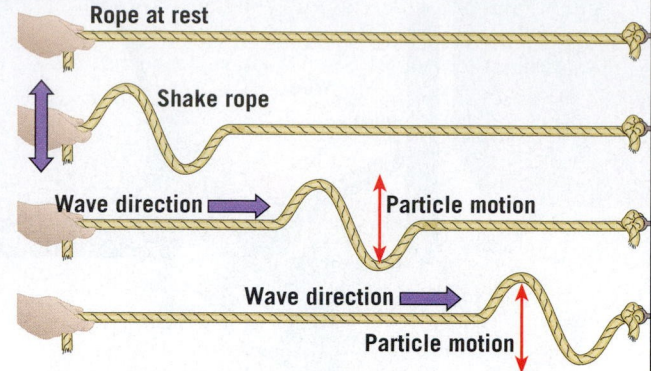

pull (stretch) rocks in the direction the wave is traveling (**Figure 11.13A**). This wave motion is similar to that generated by human vocal cords as they move air back and forth to create sound. Solids, liquids, and gases resist stresses that change their volume when compressed and, therefore, elastically spring back once the stress is removed. Therefore, P waves can travel through all these materials.

By contrast, S waves "shake" the particles at right angles to their direction of travel. This can be illustrated

by fastening one end of a rope and shaking the other end, as shown in **Figure 11.13B**. Unlike P waves, which temporarily change the *volume* of intervening material by alternately squeezing and stretching it, S waves change the *shape* of the material that transmits them. Because fluids (gases and liquids) do not resist stresses that cause changes in shape—meaning fluids do not return to their original shape once the stress is removed—liquids and gases do not transmit S waves.

Surface Waves There are two types of surface waves. One type causes Earth's surface and anything resting on it to move, much as ocean swells toss a ship (**Figure 11.14A**). The second type of surface wave causes Earth's materials to move side to side. This motion is particularly damaging to the foundations of structures (**Figure 11.14B**).

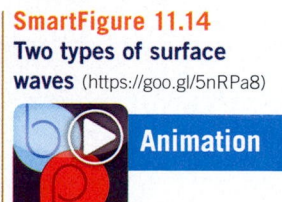
A. One type of surface wave travels along Earth's surface similar to rolling ocean waves. The red arrows show the movement of rock as the wave passes.

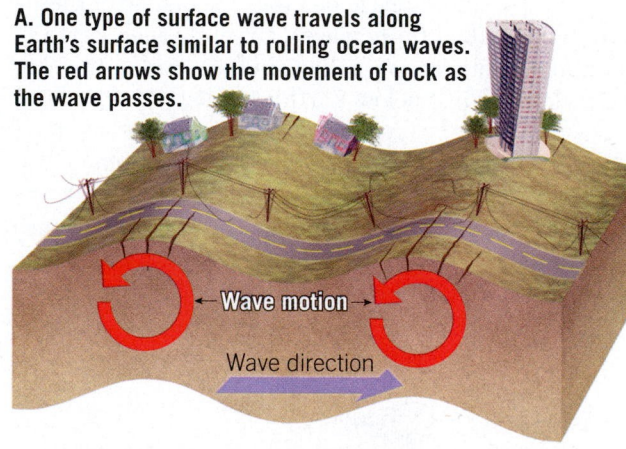

B. A second type of surface wave moves the ground from side to side and can be particularly damaging to building foundations.

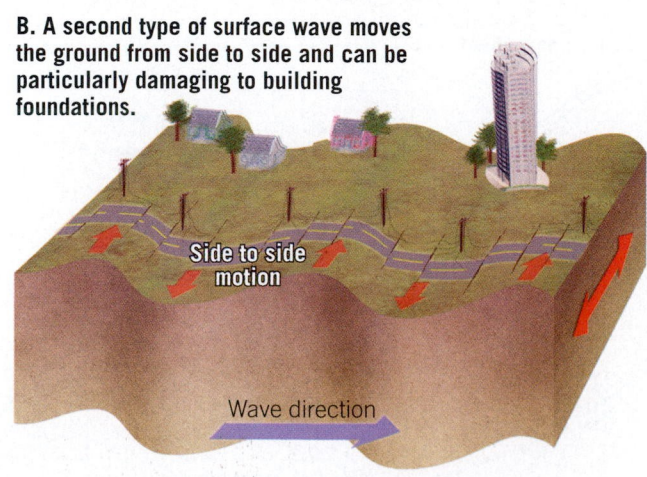

Note the time interval (about 5 minutes) between the arrival of the first P wave and the arrival of the first S wave.

Figure 11.15
Typical seismogram

Body Waves Versus Surface Waves By examining the seismogram shown in **Figure 11.15**, you can see that the major difference among seismic waves is their speed of travel. P waves are the first to arrive at a recording station, then S waves, and finally surface waves. Generally, in any solid Earth material, P waves travel about 1.7 times faster than S waves, and S waves are roughly 10 percent faster than surface waves.

In addition to the velocity differences in the waves, notice in Figure 11.15 that the height, or *amplitude*, of these wave types also varies. S waves have slightly greater amplitudes than P waves, and surface waves exhibit even greater amplitudes. Surface waves also retain their maximum amplitude longer than P and S waves. As a result, surface waves tend to cause greater ground shaking and, hence, greater property damage, than either P or S waves.

11.3 | **Concept Checks**

1. Describe how a seismograph works.
2. List the major differences between P, S, and surface waves.
3. Which type of seismic waves tends to cause the greatest destruction to buildings?

11.4 | # Locating the Source of an Earthquake

Explain how seismographs are used to locate the epicenter of an earthquake.

When seismologists analyze an earthquake, they first determine its *epicenter*, the point on Earth's surface directly above the focus (see Figure 11.2). The method used for locating an earthquake's epicenter relies on the fact that P waves travel faster than S waves.

The traveling waves are analogous to two racing automobiles, one faster than the other. The first P wave, like the faster automobile, always wins the race, arriving ahead of the first S wave. The greater the length of the race, the greater the difference in their arrival times at the finish line (the seismic station). Therefore, the greater the interval between the arrival of the first P wave and the arrival of the first S wave, the greater the distance to the epicenter. **Figure 11.16** shows three simplified seismograms for the same earthquake. Based on the P–S interval, which city—New York, Nome, or Mexico City—is farthest from the epicenter?

The system for locating earthquake epicenters was developed by using seismograms from earthquakes whose epicenters could be easily pinpointed from physical evidence. From these seismograms, travel–time graphs were constructed (**Figure 11.17**). Using the sample seismogram for New York in Figure 11.16 and the travel–time curve in Figure 11.17, we can determine the distance separating the recording station from the earthquake in three steps:

1. Using the seismogram for New York, we determine that the time interval between the arrival of the first P wave and the arrival of the first S wave is 5 minutes.
2. Using the travel–time graph, we find the location where the vertical separation between P and S curves is equal to the P–S time interval; in this example, it is 5 minutes.
3. From the position in step 2, we draw a vertical line that extends to the bottom of the graph and read the distance to the epicenter.

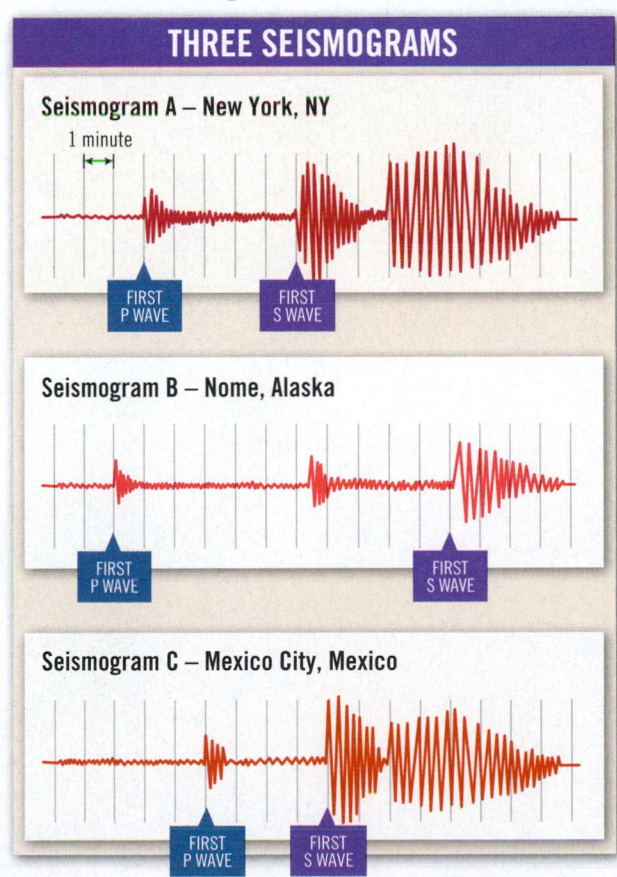

Figure 11.16
Seismograms of the same earthquake recorded at three different locations

THREE SEISMOGRAMS

Seismogram A – New York, NY
1 minute
FIRST P WAVE
FIRST S WAVE

Seismogram B – Nome, Alaska
FIRST P WAVE
FIRST S WAVE

Seismogram C – Mexico City, Mexico
FIRST P WAVE
FIRST S WAVE

Figure 11.17
Travel–time graph A travel–time graph is used to determine the distance to an earthquake's epicenter. The difference in arrival times of the first P and S waves in the example shown is 5 minutes.

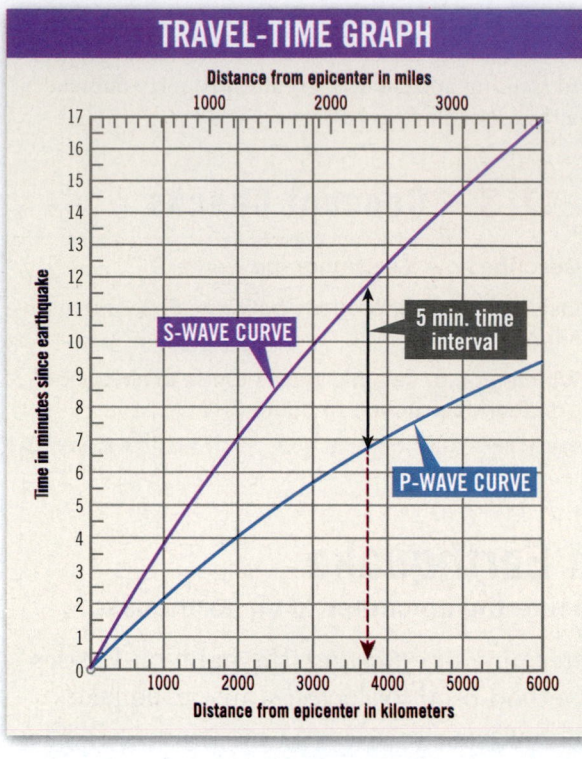

Using these steps, we determine that the earthquake occurred 3700 kilometers (2300 miles) from the recording instrument in New York City.

Now we know the *distance*, but what about *direction*? The epicenter could be in any direction from the seismic station. Using a method called *triangulation*, we can determine the location when we know the distance from two or more additional seismic stations (**Figure 11.18**). On a map or globe, a circle is drawn around each seismic station. The radius of these circles is equal to the distance from the seismic station to the epicenter. The point where the three circles intersect is the approximate epicenter of the quake.

11.4 Concept Checks

1. What information does a travel–time graph provide?

2. Briefly describe the *triangulation* method used to determine the epicenter of an earthquake.

Figure 11.18
Triangulation to locate an earthquake This method involves using the distance obtained from three or more seismic stations to establish the location of an earthquake.

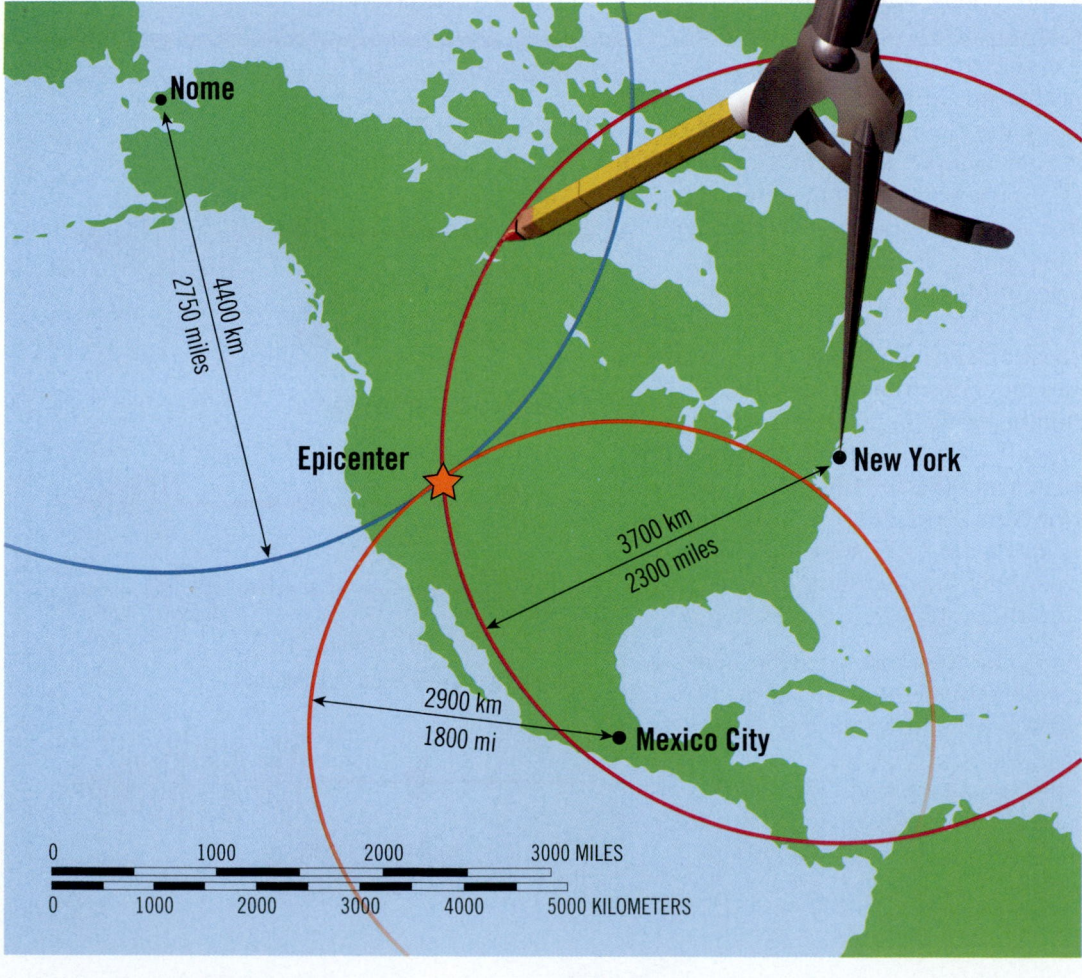

11.5 | Determining the Size of an Earthquake

Distinguish between intensity scales and magnitude scales.

Seismologists use a variety of methods to determine two fundamentally different measures that describe the size of an earthquake: *intensity* and *magnitude*. The first of these to be used was **intensity**—a measure of the amount of ground shaking at a particular location, based on observed property damage. Later, the development of seismographs allowed scientists to measure ground motion using instruments. This quantitative measurement, called **magnitude**, relies on data gleaned from seismic records to estimate the amount of energy released at an earthquake's source.

Intensity Scales

Until the mid-1800s, historical records provided the only accounts of the severity of earthquake shaking and destruction. Perhaps the first attempt to scientifically describe the aftermath of an earthquake came following the great Italian earthquake of 1857. By systematically mapping effects of the earthquake, a measure of the intensity of ground shaking was established. The map generated by this study used lines to connect places of equal damage and hence equal ground shaking. Using this technique, zones of intensity were identified, with the zone of highest intensity located near the center of maximum ground shaking and often (but not always) the earthquake epicenter.

In 1902, Giuseppe Mercalli developed a more reliable intensity scale, which is still used today in a modified form. The **Modified Mercalli Intensity scale**, shown in **Table 11.1**, was developed using California buildings as its standard. For example, on the 12-point Mercalli Intensity scale, when some well-built wood structures and most masonry buildings are destroyed by an earthquake, the affected area is assigned a Roman numeral X (10).

More recently, the U.S. Geological Survey has developed a webpage called "Did You Feel It," where Internet users enter their zip code and answer questions such as "Did objects fall off shelves?" Within a few hours, a Community Internet Intensity Map, like the one in **Figure 11.19** for the 2011 central Virginia earthquake (M5.8), is generated. Shaking was reported from Maine to Florida, an area occupied by one-third of the U.S. population. Several national landmarks were damaged, including the Washington Monument and the National Cathedral located about 130 kilometers (80 miles) away from the epicenter. Because the crustal rocks east of the Rocky Mountains are cool and strong, earthquakes are felt over a much larger area than those of similar magnitudes in the west (see Figure 11.19).

Magnitude Scales

To more accurately compare earthquakes around the globe, scientists searched for a way to describe the energy released by earthquakes that did not rely on factors such as building practices, which vary considerably from one part of the world to another. As a result, several magnitude scales were developed.

Richter Magnitude In 1935 Charles Richter of the California Institute of Technology developed the first magnitude scale to use seismic records. As shown in **Figure 11.20** (top), the **Richter scale** is calculated by measuring the amplitude of the largest seismic wave (usually an S wave or a surface wave) recorded on a seismogram. Because seismic waves weaken as the distance between the hypocenter and the seismograph increases, Richter developed a method that accounts for the decrease in wave

TABLE 11.1	Modified Mercalli Intensity Scale
I	Not felt except by a very few under especially favorable circumstances.
II	Felt only by a few persons at rest, especially on upper floors of buildings.
III	Felt quite noticeably indoors, especially on upper floors of buildings, but many people do not recognize it as an earthquake.
IV	During the day felt indoors by many, outdoors by few. Sensation like heavy truck striking building.
V	Felt by nearly everyone, many awakened. Disturbances of trees, poles, and other tall objects sometimes noticed.
VI	Felt by all; many frightened and run outdoors. Some heavy furniture moved; few instances of fallen plaster or damaged chimneys. Damage slight.
VII	Everybody runs outdoors. Damage negligible in buildings of good design and construction; slight to moderate in well-built ordinary structures; considerable in poorly built or badly designed structures.
VIII	Damage slight in specially designed structures; considerable in ordinary substantial buildings with partial collapse; great in poorly built structures. (Fall of chimneys, factory stacks, columns, monuments, walls.)
IX	Damage considerable in specially designed structures. Buildings shifted off foundations. Ground cracked conspicuously.
X	Some well-built wooden structures destroyed. Most masonry and frame structures destroyed. Ground badly cracked.
XI	Few, if any, (masonry) structures remain standing. Bridges destroyed. Broad fissures in ground.
XII	Damage total. Waves seen on ground surfaces. Objects thrown upward into air.

SmartFigure 11.19

USGS Community Internet Intensity Map Maps like this one are prepared using data collected over the Internet from people responding to questions such as "Did objects fall off shelves?" (https://goo.gl/Pdseso)

Tutorial

The green dots on the national map show locations of people who reported feeling earthquakes of similar magnitude, one that occurred in California versus one that occurred in Virginia. The difference is attributable to the rigidity of the bedrock.

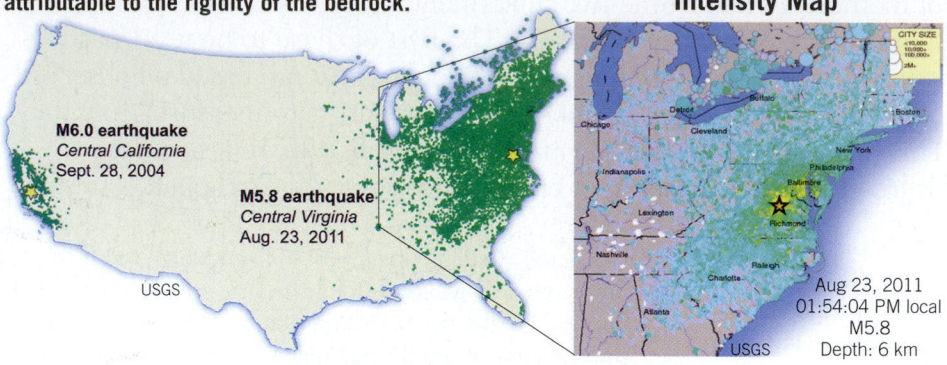

Key for USGS Community Internet Intensity Map

INTENSITY	I	II–III	IV	V	VI	VII	VIII	IX	X
SHAKING	Not felt	Weak	Light	Moderate	Strong	Very strong	Severe	Violent	Extreme
DAMAGE	none	none	none	Very light	Light	Moderate	Moderate/Heavy	Heavy	Very Heavy

Figure 11.20
Determining the Richter magnitude of an earthquake

1. Measure the height (amplitude) of the largest wave on the seismogram (23 mm) and plot it on the amplitude scale (right).

2. Determine the distance to the earthquake using the time interval separating the arrival of the first P wave and the arrival of the first S wave (24 seconds) and plot it on the distance scale (left).

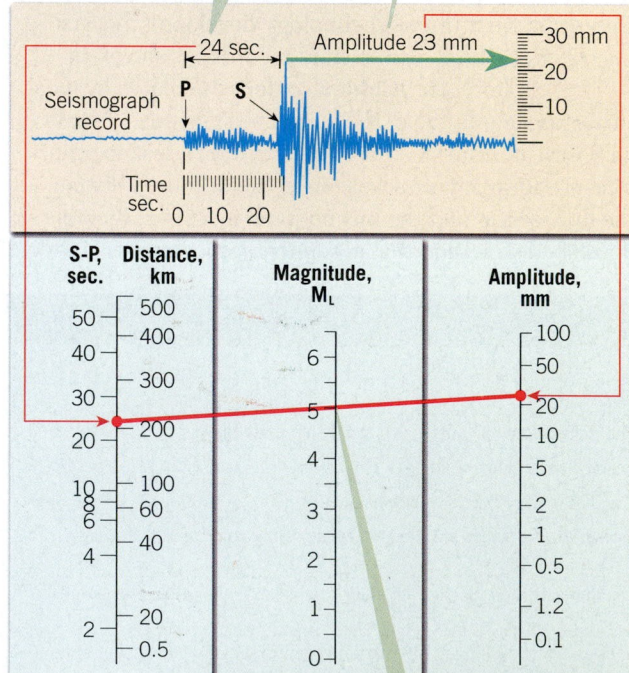

3. Draw a line connecting the two plots and read the Richter magnitude (M_L 5) from the magnitude scale (center).

amplitude with increasing distance. Theoretically, as long as equivalent instruments are used, monitoring stations at different locations will obtain the same Richter magnitude for each recorded earthquake. In practice, however, different recording stations often obtain slightly different magnitudes for the same earthquake—a result of the variations in rock types through which the waves travel.

Earthquakes vary enormously in strength, and great earthquakes produce wave amplitudes thousands of times larger than those generated by weak tremors. To accommodate this wide variation, Richter used a *logarithmic scale* to express magnitude, in which a *10-fold* increase in wave amplitude corresponds to an increase of 1 on the magnitude scale. Thus, the intensity of ground shaking for a magnitude 5 earthquake is 10 times greater than that produced by an earthquake having a Richter magnitude (M_L) of 4 (**Figure 11.21**).

In addition, each unit of increase in Richter magnitude equates to roughly a *32-fold increase in the energy released*. Thus, an earthquake with a magnitude of 6.5 releases 32 times more energy than one with a magnitude of 5.5 and roughly 1000 times (32 × 32) more energy than a magnitude 4.5 quake. A major earthquake with a magnitude of 8.5 releases millions of times more energy than the smallest earthquakes felt by humans (**Figure 11.22**).

The convenience of describing the size of an earthquake by a single number that can be calculated quickly from seismograms makes the Richter scale a powerful

Magnitude vs. Ground Motion and Energy

Magnitude Change	Ground Motion Change (amplitude)	Energy Change (approximate)
4.0	10,000 times	1,000,000 times
3.0	1000 times	32,000 times
2.0	100 times	1000 times
1.0	10 times	32 times
0.5	3.2 times	5.5 times
0.1	1.3 times	1.4 times

Figure 11.21
Magnitude versus ground motion and energy released An earthquake that is one magnitude stronger than another (M 6 versus M 5) produces seismic waves that have a maximum amplitude 10 times greater, and they release about 32 times more energy than the waves of the weaker quake.

Frequency and Energy Released by Earthquakes of Different Magnitudes

Magnitude (Mw)	Average Per Year	Description	Examples	Energy Release (equivalent kilograms of explosive)
<1		**Largest recorded earthquakes–** destruction over vast area massive loss of life possible	Chile, 1960 (M 9.5); Alaska, 1964 (M 9.0); Japan, 2011 (M 9.0)	56,000,000,000,000
	1	**Great earthquakes–** severe economic impact large loss of life	Sumatra, 2004 (M 9.1); Mexico City, 1980 (M 8.1)	1,800,000,000,000
	15	**Major earthquakes–** damage ($ billions) loss of life	New Madrid, Missouri 1812 (M 7.7); Haiti, 2012 (M 7.0); Charleston, South Carolina, 1886 (M 7.3)	56,000,000,000
	134	**Strong earthquakes–** can be destructive in populated areas	Kobe, Japan, 1995 (M 6.9); Loma Prieta, California, 1989 (M 6.9); Northridge, California, 1994 (M 6.7)	1,800,000,000
	1319	**Moderate earthquakes–** property damage to poorly constructed buildings	Mineral, Virginia, 2011 (M 5.8); Northern New York, 1994 (M 5.8); East of Oklahoma City, Oklahoma, 2011 (M5.6)	56,000,000
	13,000	**Light earthquakes–** noticable shaking of items indoors, some property damage	Western Minnesota, 1975 (M 4.6); Arkansas, 2011 (M 4.7)	1,800,000
	130,000	**Minor earthquakes–** felt by humans, very light property damage, if any	New Jersey, 2009 (M3.0); Maine, 2006 (M3.8); Texas, 2015 (3.6)	56,000
	1,300,000	**Very minor earthquakes–** felt by humans, no property damage		1,800
	Unknown	**Very minor earthquakes–** generally not felt by humans, but may be recorded		56

Data from USGS

Figure 11.22
Annual occurrence of earthquakes with various magnitudes

tool. Seismologists have since modified Richter's work and developed other Richter-like magnitude scales.

Despite its usefulness, the Richter scale is not adequate for describing very large earthquakes. For example, the 1906 San Francisco earthquake and the 1964 Alaska earthquake have roughly the same Richter magnitudes. However, based on the relative size of the affected areas and the associated tectonic changes, the Alaska earthquake released considerably more energy than the San Francisco quake. Thus, the Richter scale is considered *saturated* for major earthquakes because it cannot distinguish among them. Despite this shortcoming, Richter-like scales are still used because they can be calculated quickly.

Moment Magnitude   For measuring medium and large earthquakes, seismologists now favor a newer scale, called **moment magnitude** (M$_W$), which measures the total energy released during an earthquake. Moment magnitude is calculated by determining the average amount of slip on the fault, the area of the fault surface that slipped, and the strength of the faulted rock.

Moment magnitude can also be calculated by modeling data obtained from seismograms. The results are converted to a magnitude number, as in other magnitude scales. As with the Richter scale, each unit increase in moment magnitude equates to roughly a 32-fold increase in the energy released.

Because moment magnitude estimates the total energy released, it is better than the Richter scale for measuring very large earthquakes. Seismologists have used the moment magnitude scale to recalculate the magnitudes of older strong earthquakes. For example, the 1964 Alaska earthquake, originally given a Richter magnitude of 8.3, has since been recalculated using the moment magnitude scale, resulting in an upgrade to M$_W$ 9.2. Conversely, the 1906 San Francisco earthquake that was given a Richter magnitude of 8.3 was downgraded to M$_W$ 7.9. The strongest earthquake on record is the 1960 Chilean subduction zone earthquake, with a moment magnitude of 9.5.

11.5 | Concept Checks

1. What does the Modified Mercalli Intensity scale tell us about an earthquake?

2. What information is used to establish the lower numbers on the Mercalli scale?

3. How much more energy does a magnitude 7.0 earthquake release than a magnitude 6.0 earthquake?

4. Why is the moment magnitude scale favored over the Richter scale for large earthquakes?

11.6 | Earthquake Destruction

List and describe the major destructive forces that earthquake vibrations can trigger.

The most violent earthquake ever recorded in North America—the Alaska earthquake—occurred at 5:36 P.M. on March 27, 1964. Felt over most of the state, the earthquake had a moment magnitude (M_W) of 9.2 and lasted 3 to 4 minutes. This event left 128 people dead and thousands homeless, and it badly disrupted the state's economy. Within 24 hours of the initial shock, 28 aftershocks were recorded, 10 of them exceeding magnitude 6. The epicenter and the towns hardest hit by the quake are shown in **Figure 11.23**.

Figure 11.23
Region most affected by the Alaska earthquake, 1964

Many factors determine the degree of destruction that accompanies an earthquake. The most obvious is the *magnitude of the earthquake and its proximity to a populated area*. During an earthquake, the region within 20 to 50 kilometers (12 to 30 miles) of the epicenter tends to experience roughly the same degree of ground shaking, and beyond that limit, vibrations usually diminish rapidly. Earthquakes that occur in the stable continental interior, such as the New Madrid, Missouri,

earthquakes of 1811–1812, are generally felt more over a much larger area than those in earthquake-prone areas such as California.

Destruction from Seismic Vibrations

The 1964 Alaska earthquake provided geologists with insights into the role of ground shaking as a destructive force. As the energy released by an earthquake travels along Earth's surface, it causes the ground to vibrate in a complex manner involving up-and-down as well as side-to-side motion. The amount of damage to human-made structures attributable to the vibrations depends on several factors, including (1) *intensity* and (2) *duration of the vibrations*, (3) the *nature of the material on which structures rest*, and (4) the *nature of building materials and construction practices of the region*.

All the multistory structures in Anchorage were damaged by the vibrations. The more flexible wood-frame residential buildings fared best. A striking example of how construction variations affect earthquake damage is shown in **Figure 11.24**. You can see that the steel-frame building on the left withstood the vibrations, whereas the poorly designed JCPenney building was badly damaged. Engineers have learned that buildings constructed of blocks and bricks that are not reinforced with steel rods are the most serious safety threats in earthquakes. Unfortunately, most of the structures in the developing world are constructed of unreinforced concrete slabs and bricks made of dried mud—a primary reason the death toll in poor countries such as Haiti and Nepal is usually higher than for earthquakes of similar size in the United States.

The 1964 Alaska earthquake damaged most large structures in Anchorage, even though they were built according to the earthquake provisions of the Uniform Building Code. Perhaps some of that destruction can be attributed to the unusually long duration of the earthquake. Most quakes involve tremors that

Figure 11.24
Comparing damage to structures The poorly designed five-story JCPenney building in Anchorage, Alaska, sustained extensive damage. The steel-frame adjacent building incurred very little structural damage. (Courtesy of NOAA/Seattle)

last less than a minute. For example, the 1994 Northridge earthquake was felt for about 40 seconds, and the strong vibrations of the 1989 Loma Prieta earthquake lasted less than 15 seconds. But the Alaska quake reverberated for 3 to 4 minutes.

Amplification of Seismic Waves Although the region near the epicenter experiences about the same intensity of ground shaking, destruction may vary considerably in this area. These differences are usually attributable to the nature of the ground on which the structures are built. Soft sediments, for example, generally amplify the vibrations more than solid bedrock. Thus, the buildings in Anchorage that were situated on unconsolidated sediments experienced heavy structural damage (**Figure 11.25**). In contrast, most of the town of Whittier, though much nearer the epicenter, rested on a firm foundation of solid bedrock and, therefore, suffered much less damage from seismic vibrations.

Liquefaction The intense shaking of an earthquake can cause loosely packed water-logged materials, such as sandy stream deposits or fill, to be transformed into a substance that acts like fluids. The phenomenon of transforming a somewhat stable soil into mobile material capable of rising toward Earth's surface is known as **liquefaction**. When liquefaction occurs, the ground may not be capable of supporting buildings, and underground storage tanks and sewer lines may literally float toward the surface (**Figure 11.26**).

During the 1989 Loma Prieta earthquake, in San Francisco's Marina District, foundations failed, and geysers of sand and water shot from the ground, evidence that liquefaction had occurred (**Figure 11.27**). Liquefaction also contributed to the damage inflicted on San Francisco's water system during the 1906 earthquake. During the 2011 Japan earthquake, liquefaction caused entire buildings to sink several feet.

Seiches The effects of great earthquakes may be felt thousands of kilometers from their source. Ground motion may generate **seiches**, the rhythmic sloshing of water in lakes, reservoirs, and enclosed basins such as the Gulf of Mexico. The 1964 Alaska earthquake, for example, generated 2-meter (7-foot) waves off the coast of Texas and damaged small craft, while much smaller waves were noticed in swimming pools in both Texas and Louisiana.

Seiches can be particularly dangerous when they occur in reservoirs retained by earthen dams. These waves have been known to slosh over reservoir walls and weaken the structures, thereby endangering property as well as the lives of those downstream.

Figure 11.25
Ground failure caused this street in Anchorage, Alaska, to collapse (Photo by USGS)

Downtown Anchorage following the 1964 Alaskan earthquake.

Landslides and Ground Subsidence

The greatest earthquake-related damage to structures is often caused by landslides and ground subsidence triggered by earthquake vibrations. This was the case during the 1964 Alaska earthquake in Valdez and Seward, where the violent shaking caused coastal sediments to slump, carrying away both waterfronts. In Valdez, 31 people died when a dock slid into the sea. Because of the threat

Figure 11.26
Effects of liquefaction on buildings (Photo courtesy of USGS)

These tilted buildings rested on unconsolidated sediment that behaved like quicksand during the 1964 Niigata, Japan, earthquake.

Sand volcanoes

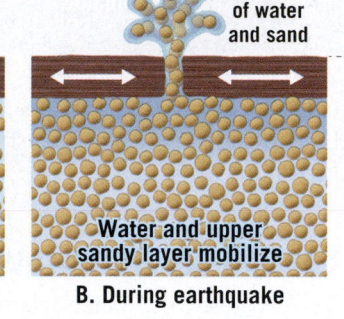

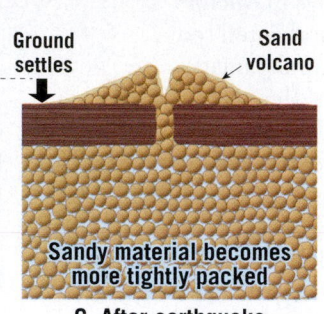

Eruption of water and sand

Ground settles

Sand volcano

Well packed soil

A. Before earthquake — Loose, saturated, sandy material

B. During earthquake — Water and upper sandy layer mobilize

C. After earthquake — Sandy material becomes more tightly packed

of recurrence, the entire town of Valdez was relocated to more stable ground about 7 kilometers (4.5 miles) away.

Much of the damage in Anchorage was attributed to landslides. Homes were destroyed in Turnagain Heights when a layer of clay lost its strength and over 200 acres of land slid toward the ocean (**Figure 11.28**). A portion of this spectacular landslide was left in its natural condition, as a reminder of this destructive event. The site was appropriately named "Earthquake Park." Downtown Anchorage was also disrupted as sections of the main business district dropped by as much as 3 meters (10 feet).

Fire

More than a century ago, San Francisco was the economic center of the western United States, largely because of gold and silver mining. Then, at dawn on April 18, 1906, a violent earthquake struck, triggering

an enormous firestorm (see Figure 11.3). Much of the city was reduced to ashes and ruins. It is estimated that 3000 people died and more than half of the city's 400,000 residents were left homeless.

The historic San Francisco earthquake reminds us of the formidable threat of fire, which started during that quake when gas and electrical lines were severed. The initial ground shaking broke the city's water lines into hundreds of disconnected pieces, which made controlling the fires virtually impossible. The fires, which raged out of control for 3 days, were finally contained when expensive houses along Van Ness Avenue were dynamited to provide a fire break, similar to the strategy used in fighting forest fires.

While few deaths were attributed to the San Francisco fires, other earthquake-initiated fires have been more destructive, claiming many more lives. For example, the 1923 earthquake in Japan triggered an estimated 250 fires, devastating the city of Yokohama and destroying more than half the homes in Tokyo. More than 100,000 deaths were attributed to the fires, which were driven by unusually high winds.

B. Blocks of land began to slide toward the sea on a weak layer called the Bootlegger Cove clay and in less than 5 minutes, as much as 200 meters of the Turnagain Heights bluff area had been destroyed.

Original profile — 200 meters

Tutorial

A. Vibrations from the Alaskan earthquake caused cracks to appear near the edge of the Turnagain Heights bluff.

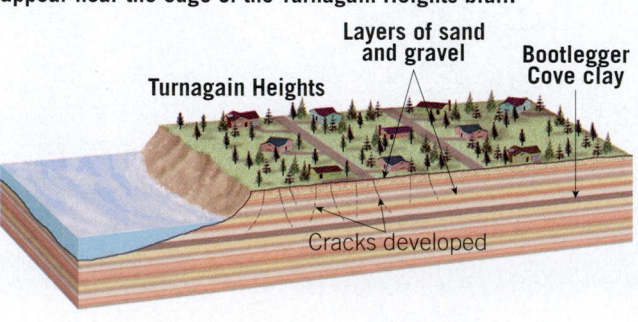

Layers of sand and gravel   Bootlegger Cove clay

Turnagain Heights

Cracks developed

C. Photo of a small area of destruction caused by the Turnagain Heights slide.

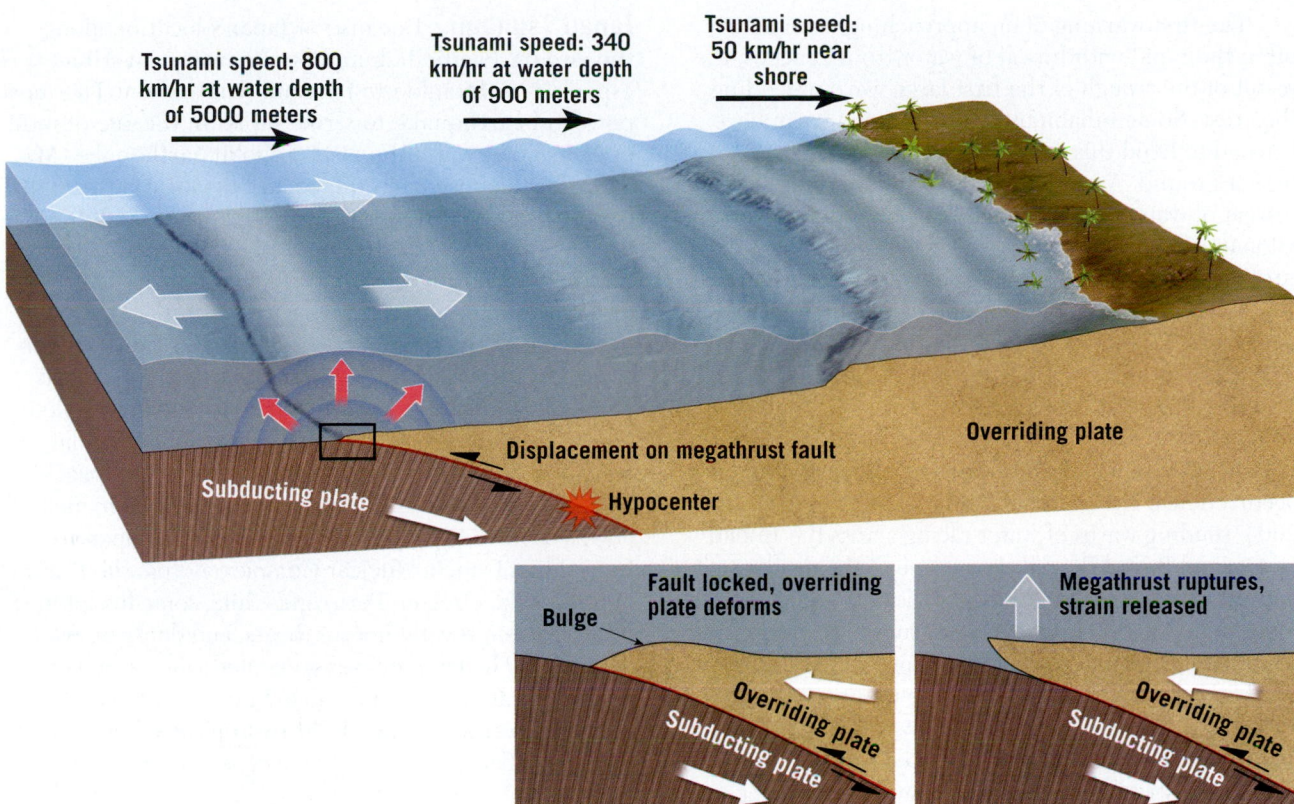

Tsunami speed: 800 km/hr at water depth of 5000 meters

Tsunami speed: 340 km/hr at water depth of 900 meters

Tsunami speed: 50 km/hr near shore

Overriding plate

Displacement on megathrust fault

Subducting plate

Hypocenter

Bulge

Fault locked, overriding plate deforms

Overriding plate

Subducting plate

Megathrust ruptures, strain released

Overriding plate

Subducting plate

SmartFigure 11.29
Tsunami generated by displacement of the ocean floor The speed of a wave correlates with ocean depth. Waves moving in deep water advance at speeds exceeding 800 kilometers (500 miles) per hour. Speed gradually slows to 50 kilometers (30 miles) per hour at depths of 20 meters (65 feet). As waves slow in shallow water, they grow in height until they rush onto shore with tremendous force. The size and spacing of the swells shown here are not to scale. (https://goo.gl/3Iphb3)

Tutorial

What Is a Tsunami?

Major undersea earthquakes occasionally set in motion a series of large ocean waves that are known by the Japanese name **tsunami** ("harbor wave"). Most tsunamis are generated by displacement along a megathrust fault that suddenly lifts a large slab of seafloor (**Figure 11.29**). Once generated, a tsunami resembles a series of ripples formed when a pebble is dropped into a pond. In contrast to ripples, tsunamis advance across the ocean at amazing speeds, about 800 kilometers (500 miles) per hour—equivalent to the speed of a commercial airliner. Despite this striking characteristic, a tsunami in the open ocean can pass undetected because its height (amplitude) is usually less than 1 meter (3 feet), and the distance separating wave crests ranges from 100 to 700 kilometers (60 to 450 miles). However, upon entering shallow coastal waters, these destructive waves "feel bottom" and slow, causing the water to pile up (see Figure 11.29). A few exceptional tsunamis have exceeded 30 meters (100 feet) in height. As the crest of a tsunami approaches the shore, it appears as a rapid rise in sea level with a turbulent and chaotic surface; it does not resemble a breaking wave (**Figure 11.30**).

SmartFigure 11.30
Tsunami generated off the coast of Sumatra, 2004
(AFP/Getty Images, Inc.) (https://goo.gl/b8OPkN)

Animation

The first warning of an approaching tsunami is often the rapid withdrawal of water from beaches, the result of the trough of the first large wave preceding the crest. Some inhabitants of the Pacific basin have learned to heed this warning and quickly move to higher ground. Approximately 5 to 30 minutes after the retreat of water, a surge capable of extending several kilometers inland occurs. In a successive fashion, each surge is followed by a rapid oceanward retreat of the sea. Therefore, people experiencing a tsunami should not return to the shore when the first surge of water retreats.

Tsunami Damage from the 2004 Indonesia Earthquake

A massive undersea earthquake of M_W 9.1 occurred near the island of Sumatra on December 26, 2004, sending waves of water racing across the Indian Ocean and Bay of Bengal. It was one of the deadliest natural disasters of any kind in modern times, claiming more than 230,000 lives. As water surged several kilometers inland, cars and trucks were flung around like toys in a bathtub, and fishing boats were rammed into homes. In some locations, the backwash of water dragged bodies and huge amounts of debris out to sea.

The destruction was indiscriminate, destroying luxury resorts as well as poor fishing hamlets along the Indian Ocean. Damages were reported as far away as the coast of Somalia in Africa, 4100 kilometers (2500 miles) west of the earthquake epicenter.

Japan Tsunami

Because of Japan's location along the circum-Pacific belt and its expansive coastline, it is especially vulnerable to tsunami destruction. The most powerful earthquake to strike Japan in the age of modern seismology was the 2011 Tohoku earthquake (M_W 9.0). This historic earthquake and devastating tsunami resulted in at least 15,890 deaths, more than 3000 people missing, and 6107 injured. Nearly 400,000 buildings, 56 bridges, and 26 railways were destroyed or damaged.

The majority of human casualties and damage were caused by a Pacific-wide tsunami that reached a maximum height of about 40 meters (130 feet) and travelled inland 10 kilometers (6 miles) in the region of Sendai, Japan (**Figure 11.31**). The chapter-opening photo (pages 326–327) shows this dramatic event. In addition, meltdowns occurred at three nuclear reactors in Japan's Fukushima Daiichi Nuclear Complex. Across the Pacific in California, Oregon, Peru, and Chile, some loss of life occurred, and several houses, boats, and docks were destroyed. The tsunami was generated when a slab of seafloor located 60 kilometers (37 miles) off the east coast of Japan was suddenly "thrust up" an estimated 5 to 8 meters (16 to 26 feet).

Tsunami Warning System

In 1946, a large tsunami struck the Hawaiian Islands without warning. A wave more than 15 meters (50 feet) high left several coastal villages in shambles. This destruction motivated the

Figure 11.31
Aftermath of the Japan tsunami, March 2011 This tsunami devastated parts of Natori City. (Photo by Mike Clark/AFP/Getty Images)

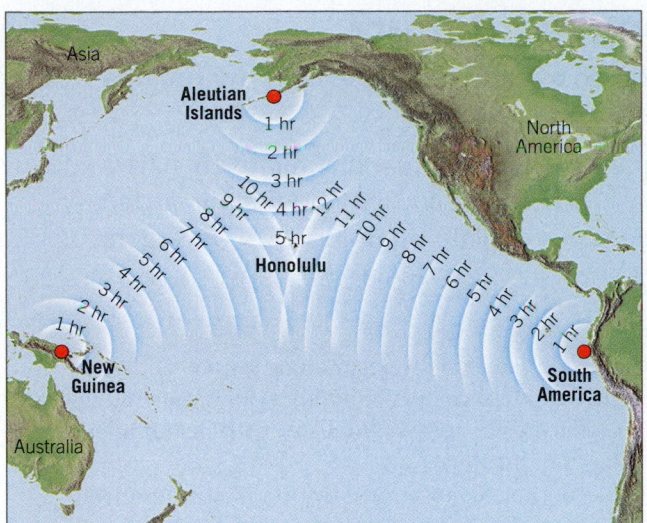

Figure 11.32
Tsunami travel times Travel times to Honolulu, Hawaii, from selected locations throughout the Pacific. (Data from NOAA)

U.S. Coast and Geodetic Survey to establish a tsunami warning system for coastal areas of the Pacific that today includes 26 countries. Seismic observatories throughout the region report large earthquakes to the Tsunami Warning Center in Honolulu. Scientists at

the center use deep-sea buoys equipped with pressure sensors to detect energy released by an earthquake. In addition, tidal gauges measure the rise and fall in sea level that accompany tsunamis, and warnings are issued within an hour. Although tsunamis travel very rapidly, there is sufficient time to warn all except those in the areas nearest the epicenter. For example, a tsunami generated near the Aleutian Islands would take 5 hours to reach Hawaii, and one generated near the coast of Chile would travel 15 hours before reaching the shores of Hawaii (**Figure 11.32**).

11.6 Concept Checks

1. List four factors that influence the amount of destruction that seismic vibrations cause to human-made structures.

2. In addition to the destruction created directly by seismic vibrations, list three other types of destruction associated with earthquakes.

3. What is a tsunami? How are tsunamis generated?

4. List at least three reasons an earthquake with a magnitude of 7.0 might result in more death and destruction than a quake with a magnitude of 8.0.

EYE ON EARTH 11.2

Water-saturated sandy soil provides students an opportunity to experience the phenomenon of liquefaction. Liquefaction may occur when ground shaking causes a layer of saturated material to lose strength and act like a fluid.

QUESTION 1 Describe what you think would happen to a structure built on sandy soil that suddenly experienced liquefaction during an earthquake.

QUESTION 2 How might a nearly empty underground storage tank be affected by liquefaction of the surrounding soil?

11.7 | Where Do Most Earthquakes Occur?

Locate Earth's major earthquake belts on a world map and label the regions associated with the largest earthquakes.

About 95 percent of the energy released by earthquakes originates in a few relatively narrow zones, shown in **Figure 11.33**. These zones of earthquake activity are located along fault surfaces where tectonic plates interact along one of the three types of plate boundaries—convergent, divergent, and transform plate boundaries.

Strong earthquakes can occur away from plate boundaries, although this is less common. Examples include the devastating Gujarat, India, earthquake of 2001, the 1811 and 1812 New Madrid, Missouri, earthquakes, and the Charleston, South Carolina, earthquake of 1886. Intraplate earthquakes occur when internal stresses build due to interaction of neighboring plates or because of loading or unloading as, for example, the melting of the glacial ice sheets following the last ice age. These stresses may cause failure along once-inactive fault zones.

Earthquakes Associated with Plate Boundaries

The zone of greatest seismic activity, called the **circum-Pacific belt**, encompasses the coastal regions of Chile, Central America, Indonesia, Japan, and Alaska, including the Aleutian Islands (see Figure 11.33). Most earthquakes in the circum-Pacific belt occur along convergent plate boundaries, where one plate slides at a low angle beneath another. The contacts between the subducting and overlying plates are *megathrust faults*, along which Earth's largest earthquakes are generated (see Figure 11.6).

There are more than 40,000 kilometers (25,000 miles) of subduction boundaries in the circum-Pacific belt where displacement is dominated by thrust faulting.

Ruptures occasionally occur along segments that are nearly 1000 kilometers (600 miles) long, generating catastrophic *megathrust earthquakes* with magnitudes of (M_W) 8 or greater.

Another major concentration of strong seismic activity, referred to as the *Alpine–Himalayan belt*, runs through the mountainous regions that flank the Mediterranean Sea and extends past the Himalayas (see Figure 11.33). Tectonic activity in this region is mainly attributed to collisions of the African plate with the Eurasian plate and of the Indian plate with Southeast Asia. These plate interactions created many thrust and strike-slip faults that remain active. In addition, numerous faults located a considerable distance from these plate boundaries have been reactivated as India continues its northward advance into Asia. For example, slippage on a complex fault system in 2008 in the Sichuan Province of China killed an estimated 87,000 people and left 1.5 million others homeless. The cause was the Indian subcontinent continually shoving the Tibetan Plateau eastward against the rocks of the Sichuan basin.

Figure 11.33 shows another continuous earthquake belt that extends thousands of kilometers through the world's oceans. This zone coincides with the oceanic ridge system, which is an area of frequent but weak seismic activity. As tensional forces pull the plates apart during seafloor spreading, displacement along normal faults generates most of the earthquakes in this zone. The remaining seismic activity is associated with slippage along transform faults located between ridge segments.

Transform faults and smaller strike-slip faults also run through continental crust, where they may generate large earthquakes that tend to occur on a cyclical basis. Examples include

Figure 11.33
Global earthquake belts Distribution of nearly 15,000 earthquakes with magnitudes equal to or greater than 5 for a 10-year period. (Data from USGS)

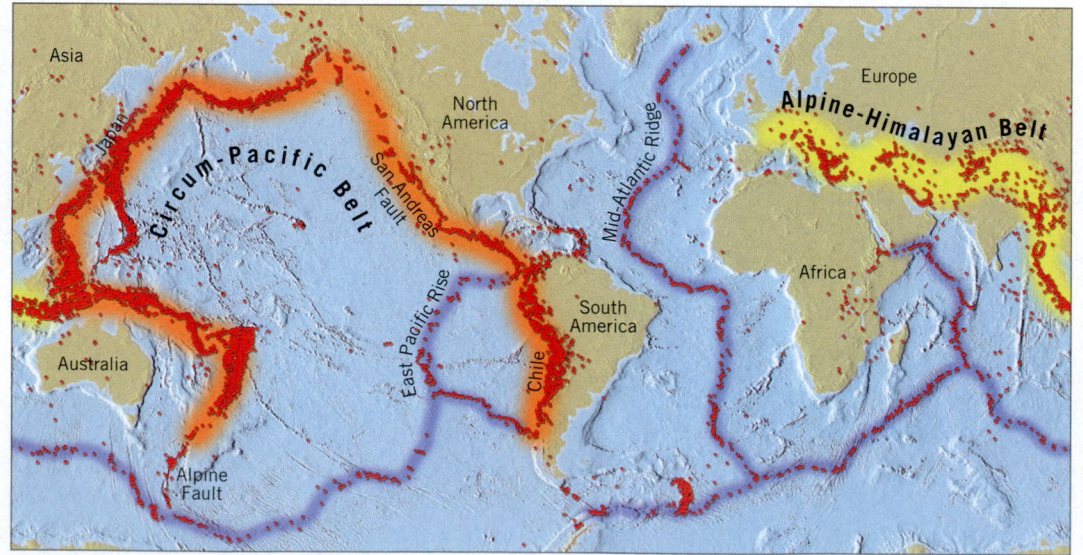

Large earthquakes are relatively uncommon in the middle of continents, far from the places where plates collide or grind past one another, or where one plate slides beneath another. Nevertheless, several damaging earthquakes have occurred in the central and eastern United States since colonial times.

Figure 11.34

Historical earthquakes east of the Rockies Large earthquakes are uncommon in the middle of continents, far from the places where plates collide or grind past one another, or where one plate slides beneath another. Nevertheless, several damaging earthquakes have occurred in the central and eastern United States since colonial times.

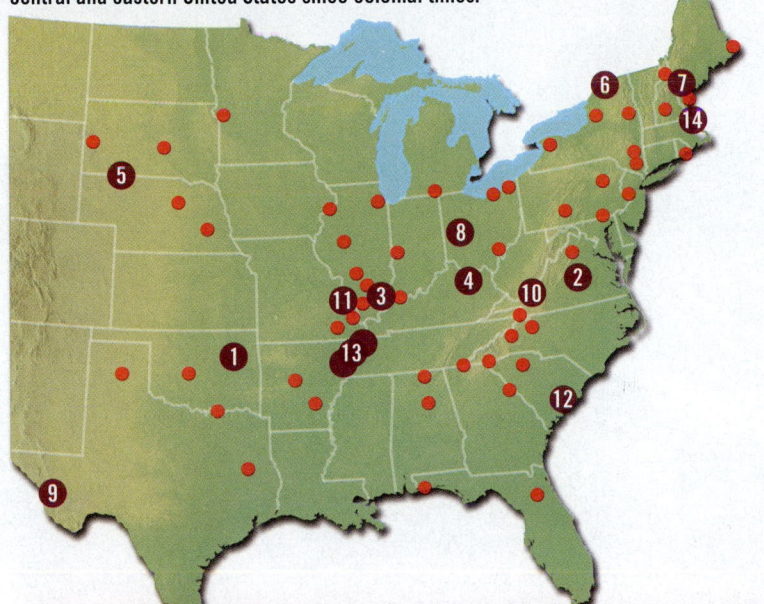

Historic Earthquakes East of the Rockies 1755–2011

	LOCATION	DATE	INTENSITY	MAGNITUDE*	COMMENTS
1	East of Oklahoma City	2011	VII	5.6	Fourteen homes destroyed
2	Mineral, Virginia	2011	VII	5.8	Felt by many due to its proximity to large population centers
3	Southeastern Illinois	2008	VII	5.4	Occurred along the Wabash Valley Seismic Zone
4	Northeast Kentucky	1980	VII	5.2	Largest earthquake ever recorded in Kentucky
5	Merriman, Nebraska	1964	VII	5.1	Largest earthquake ever recorded in Nebraska
6	Northern New York	1944	VIII	5.8	Left several structures unsafe for occupancy
7	Ossipee Lake, New Hampshire	1947	VII	5.5	Two earthquakes occurred four days apart
8	Western Ohio	1937	VIII	5.4	Extensive damage to chimneys and plaster walls
9	Valentine, Texas	1931	VIII	5.8	Brick buildings were severely damaged
10	Giles County, Virginia	1897	VIII	5.9	Changed the flow of natural springs
11	Charleston, Missouri	1895	VIII	6.6	Structural damage and liquefaction reported
12	Charleston, South Carolina	1886	X	7.3	Caused 60 deaths, destroyed many buildings
13	New Madrid, Missouri	1811–1812	X	7.7	Three strong earthquakes occurred in remote areas
14	Cape Ann, Massachusetts	1755	VIII	?	Chimneys leveled and brick buildings damaged in Boston

Source: U.S. Geological Survey
*Intensity and magnitudes have been estimated for many of these events.

California's San Andreas Fault, New Zealand's Alpine Fault, and Turkey's North Anatolian Fault, which produced a deadly earthquake in 1999.

Damaging Earthquakes East of the Rockies

When you think "earthquake," you probably think of California and Japan. However, six major earthquakes and several others that inflicted considerable damage have occurred in the central and eastern United States since colonial times (**Figure 11.34**).

Three of these quakes had estimated Richter magnitudes of 7.5, 7.3, and 7.8 and were centered near the Mississippi River valley in southeastern Missouri. Occurring on December 16, 1811; January 23, 1812; and February 7, 1812, these earthquakes, plus numerous smaller tremors, destroyed the town of New Madrid, Missouri, triggered massive landslides, and caused damage over a six-state area. The course of the Mississippi River was altered, and Tennessee's Reelfoot Lake was enlarged. The distances over which these earthquakes were felt are truly remarkable. Chimneys were reported downed in Cincinnati, Ohio, and Richmond, Virginia, and even Boston residents, located 1770 kilometers (1100 miles) to the northeast, felt the tremor.

Despite the history of the New Madrid earthquake, Memphis, Tennessee, the largest population center in the area, does not have adequate earthquake provisions in its building code. Further, because Memphis is located on

unconsolidated floodplain deposits, buildings are more susceptible to damage than are similar structures built on bedrock. It has been estimated that if an earthquake the size of the 1811–1812 New Madrid event were to strike in the next decade, it would result in casualties in the thousands and damages in tens of billions of dollars. Damaging earthquakes that occurred in Aurora, Illinois (1909), and Valentine, Texas (1931), remind us that other areas in the central United States are vulnerable.

The greatest historical earthquake in the eastern states occurred on August 31, 1886, in Charleston, South Carolina. The event, which spanned 1 minute, caused 60 deaths, numerous injuries, and great economic loss in a radius of 200 kilometers (120 miles) of Charleston. Within 8 minutes, effects were felt as far away as Chicago and St. Louis, where strong vibrations shook the upper floors of buildings, causing people to rush outdoors. In Charleston alone, more than 100 buildings were destroyed and 90 percent of the remaining structures were damaged (**Figure 11.35**).

Numerous other strong earthquakes have been recorded in the eastern United States. New England and adjacent areas have experienced sizable shocks since colonial times. The first reported earthquake in the Northeast took place in Plymouth, Massachusetts, in 1683 and was followed in 1755 by the destructive Cambridge, Massachusetts, quake. New York State has experienced more than 300 earthquakes large enough to be felt since record-keeping began.

Earthquakes in the central and eastern United States occur far less frequently than in California, yet history indicates that the East is vulnerable. Further,

these shocks east of the Rockies have generally produced structural damage over a larger area than counterparts of similar magnitude in California. This is because the underlying bedrock in the central and eastern United States is older and more rigid. As a result, seismic waves can travel greater distances with less attenuation than in the western United States. It is estimated that for earthquakes of similar magnitude, the region of maximum ground motion in the East may be up to 10 times larger than in the West (see Figure 11.19). Consequently, the higher rate of earthquake occurrence in the western United States is balanced somewhat by the fact that central and eastern U.S. quakes can damage larger areas.

11.7 Concept Checks

1. What is the zone of the greatest amount of seismic activity?

2. What type of plate boundary is associated with Earth's largest earthquakes?

3. Describe another geologic setting where strong earthquake activity occurs.

4. List two reasons why a repeat of the 1811–1812 New Madrid, Missouri, earthquakes could be destructive to the Memphis, Tennessee, metropolitan area.

5. Explain why an earthquake east of the Rockies may produce damage over a larger area than one of similar magnitude in California.

11.8 | Can Earthquakes Be Predicted?

Compare and contrast the goals of short-range earthquake predictions and long-range forecasts.

The vibrations that shook the San Francisco area in 1989 caused 63 deaths, heavily damaged the Marina District, and caused the collapse of a double-decked section of I-880 in Oakland, California (**Figure 11.36**). This level of destruction was the result of an earthquake of moderate magnitude (M_W 6.9). Seismologists warn that other earthquakes of comparable or greater strength can be expected along the San Andreas system, which cuts a nearly 1300-kilometer (800-mile) path through the western one-third of the state (see GEOgraphics 11.1). An obvious question is: Can these earthquakes be predicted?

Short-Range Predictions

The goal of short-range earthquake prediction is to provide a warning of the location and magnitude of a large earthquake within a narrow time frame (**Table 11.2**). Substantial efforts to achieve this objective have been attempted in Japan, the United States, China, and Russia—countries where earthquake risks are high. This research has concentrated on monitoring possible **precursors**—events or changes that precede a forthcoming earthquake and thus may provide warning. In California, for example, seismologists monitor changes in ground elevation and variations in strain levels near active faults. Other researchers measure changes in groundwater levels, while still others try to predict earthquakes based on an increase in the frequency of foreshocks that precede some, but not all, earthquakes.

Japanese and Chinese scientists have tried to monitor anomalous animal behavior. A few days before the May 12, 2008, earthquake in China's Sichuan Province, the streets of a village near the fault were filled with toads migrating from the mountains. Was this a warning? Perhaps. Walter Mooney, a USGS seismologist, put it best: "Everyone hopes that animals can tell us something we don't know . . . but animal behavior is way too unreliable." Although precursors may exist, we have yet to determine effective ways to interpret and utilize the information.

One claim of a successful short-range prediction, based on an increase in foreshocks, was made by the Chinese government after the February 4, 1975, earthquake in Liaoning Province. According to reports, very few people were killed—even though more than

Figure 11.36
Collapse of the double-decked section of I-880 This section of a double-decked highway, known as the Cypress Viaduct, collapsed during the 1989 Loma Prieta earthquake. (Photo by Paul Sakuma/AP Photo)

Seismic Risks on the San Andreas Fault System

California's San Andreas Fault runs diagonally from southeast to northwest for nearly 1300 kilometers (800 miles) through much of the western part of the state. For years researchers have been trying to predict the location of the next "Big One"—an earthquake with a magnitude of 8 or greater—along this fault system.

CALIFORNIA

The 1906 San Francisco earthquake caused displacement on the northernmost section of the fault. This event, likely relieved much of the strain that had been building during the previous 200 years or so.

1906 epicenter
San Francisco

Located just south of the 1906 rupture is a section of the San Andreas Fault that exhibits fault creep. When plates gradually slide past each other, less strain accumulates than when the fault is locked.

1857 epicenter

This 300-kilometer-long section of the San Andreas Fault System produced the Fort Tejon earthquake of 1857. Because a portion of the fault has likely accumulated considerable strain since the Fort Tejon quake, the U.S. Geological Survey gives it a 60 percent probability of producing a major earthquake in the next 30 years.

● Los Angeles

The next major quake on the San Andreas may well be on its southernmost 200 kilometers—an area that has not produced a large event in about 300 years.

The 1906 San Francisco earthquake was the most devastating in California's history. The quake and resulting fires caused an estimated 3000 deaths and extensively damaged buildings throughout the city.

Adam Teitelbaum/AFP/Getty Images

On October 17, 1989, millions of television viewers around the world were settling in to watch the third game of the World Series. Instead, they saw their TVs go to black as tremors hit San Francisco's Candlestick Park. Although the Loma Prieta earthquake was centered in a remote section of the Santa Cruz Mountains (100 miles to the south) major damage occurred in the Marina District of San Francisco.

Question:
The section of the San Andreas Fault located south of the San Francisco Bay area exhibits fault creep. Does that section of the fault have a high or low probability of generating a large earthquake? Explain.

?

The U. S. Geological Survey concluded that between 2003 and 2032 there is a 62 percent probability of at least one magnitude 6.7 or greater earthquake striking somewhere in the San Francisco Bay area.

% Probability of magnitude 6.7 or greater quake before 2032 on the indicated faults

Increasing probability along fault segments

In mid-January 1994, less than five years after the Loma Prieta event, the Northridge earthquake struck an area slightly north of Los Angeles. This moderate 6.7-magnitude earthquake claimed the lives of 57 people. Nearly 300 schools were severely damaged, and one dozen major roadways buckled. Among these were two of California's major arteries—sections of the Santa Monica Freeway and the Golden State Freeway (I-5) where an overpass collapsed completely and blocked the highway.

Rogers Creek fault

San Andreas fault

Concord-Green Valley fault

Greenville fault

Hayward fault

Mt. Diablo thrust fault

Calaveras fault

San Gregorio fault

Napa

Walnut Creek

San Francisco

Oakland

San Francisco Bay

Palo Alto

San Jose

Monterey Bay

Monterey

PACIFIC OCEAN

27%

4%

21%

3%

3%

10%

11%

EXTENT OF RUPTURE IN THE LOMA PRIETA QUAKE

I-5 NORTH 14 Autos

Tom McHugh/Photo Researchers, Inc.

TABLE 11.2 Some Notable Earthquakes

Year	Location	Deaths (est.)	Magnitude*	Comments
856	Iran	200,000		
893	Iran	150,000		
1138	Syria	230,000		
1268	Asia Minor	60,000		
1290	China	100,000		
1556	Shensi, China	830,000		Possibly the greatest natural disaster
1667	Caucasia	80,000		
1727	Iran	77,000		
1755	Lisbon, Portugal	70,000		Tsunami damage extensive
1783	Italy	50,000		
1908	Messina, Italy	120,000		
1920	China	200,000	7.5	Landslide buried a village
1923	Tokyo, Japan	143,000	7.9	Fire caused extensive destruction
1948	Turkmenistan	110,000	7.3	Almost all brick buildings near epicenter collapsed
1960	Southern Chile	5700	9.5	The largest-magnitude earthquake ever recorded
1964	Alaska	131	9.2	Greatest-magnitude North American earthquake
1970	Peru	70,000	7.9	Great rockslide
1976	Tangshan, China	242,000	7.5	Estimates for the death toll are as high as 655,000
1985	Mexico City	9500	8.1	Major damage occurred 400 km from epicenter
1988	Armenia	25,000	6.9	Poor construction practices
1990	Iran	50,000	7.4	Landslides and poor construction practices led to great damage
1993	Latur, India	10,000	6.4	Located in stable continental interior
1995	Kobe, Japan	5472	6.9	Damages estimated to exceed $100 billion
1999	Izmit, Turkey	17,127	7.4	Nearly 44,000 injured and more than 250,000 displaced
2001	Gujarat, India	20,000	7.9	Millions homeless
2003	Bam, Iran	31,000	6.6	Ancient city with poor construction
2004	Indian Ocean (Sumatra)	230,000	9.1	Devastating tsunami damage
2005	Pakistan/Kashmir	86,000	7.6	Many landslides; 4 million homeless
2008	Sichuan, China	87,000	7.9	Millions homeless, some towns will not be rebuilt
2010	Port-au-Prince, Haiti	316,000	7.0	More than 300,000 injured and 1.3 million homeless
2011	Japan	16,000	9.0	Majority of the casualties due to a tsunami

*Widely differing magnitudes have been estimated for some of these earthquakes. When available, moment magnitudes are used.
Source: U.S. Geological Survey.

1 million lived near the epicenter—because the earthquake was "predicted," and the residents were evacuated. Some Western seismologists have questioned this claim and suggest instead that an intense swarm of foreshocks, which began 24 hours before the main earthquake, may have caused many people to evacuate of their own accord.

One year after the Liaoning earthquake, an estimated 240,000 people perished in the Tangshan, China, earthquake, which was *not* predicted. There were no foreshocks. Predictions can also lead to false alarms. In a province near Hong Kong, people reportedly evacuated their dwellings for over a month, but no earthquake followed.

In order for a short-range prediction scheme to be generally accepted, it must be both accurate and reliable. Thus, *it must have a small range of uncertainty in regard to location and timing, and it must produce few failures or false alarms.* Can you imagine the debate that would precede an order to evacuate a large U.S. city, such as Los Angeles or San Francisco? The cost of evacuating millions of people, arranging for living accommodations, and providing for their lost work time and wages would be staggering.

Currently, no reliable method exists for making short-range earthquake predictions. In fact, leading seismologists in the past 100 years have generally concluded that short-range earthquake prediction is not feasible.

Figure 11.37

Seismic gaps: Tools for forecasting earthquakes Seismic gaps are "quiet zones," thought to be inactive, that are storing elastic strain that will eventually produce major earthquakes. This seismic gap occurs along a patch of the megathrust fault where oceanic lithosphere is being subducted beneath Sumatra, near Padang, a low-lying coastal city with a population of 800,000 people.

Long-Range Forecasts

In contrast to short-range predictions, which aim to predict earthquakes within a time frame of hours or days, long-range forecasts are estimates of how likely it is for an earthquake of a certain magnitude to occur on a time scale of 30 to 100 years or more. These forecasts give statistical estimates of the expected intensity of ground motion for a given area over a specified time frame. Although long-range forecasts are not as informative as we might like, these data are useful for providing important guides for building codes so that buildings, dams, and roadways are constructed to withstand expected levels of ground shaking.

Most long-range forecasting strategies are based on evidence that many large faults break in a cyclical manner, producing similar quakes at roughly similar intervals. In other words, as soon as a section of a fault ruptures, the continuing motions of Earth's plates begin to build strain in the rocks again until they fail once more. Seismologists have therefore studied historical

records of earthquakes to see if there are any discernible patterns so that they can establish the probability of recurrence.

Seismic Gaps Seismologists began to plot the distribution of rupture zones associated with great earthquakes around the globe. The maps revealed that individual rupture zones tend to occur adjacent to one another, without appreciable overlap, thereby tracing out a plate boundary. Because plates are moving at known velocities, the rate at which strain builds can also be estimated.

When these researchers studied historical records, they discovered that some seismic zones had not produced a large earthquake in more than a century or, in some locations, for several centuries. These quiet zones, called **seismic gaps**, are believed to be zones that are storing strain that will be released during a future earthquake. **Figure 11.37** shows a patch (seismic gap) of the megathrust fault that lies offshore of Padang, a low-lying city of 800,000 people off the coast of Sumatra that has not ruptured since 1797. Scientists are particularly concerned about this seismic gap because, in 2004, a rupture of an adjacent segment of this megathrust fault that lies to the north generated a tsunami that claimed 230,000 lives.

Paleoseismology Another method of long-term forecasting involves **paleoseismology** (*paleo* = ancient, *seismos* = shake), the study of the timing, location, and size of prehistoric earthquakes. Paleoseismology studies are often conducted by digging a trench across a suspected fault zone and then looking for evidence of ancient faulting, such as offset sedimentary strata or mud volcanoes (**Figure 11.38**). A large vertical offset of the layers of sediments indicates a large earthquake. Sometimes buried plant debris can be carbon dated, allowing for the timing of recurrence to be established.

One investigation that used this method focused on a segment of the San Andreas Fault that lies north and east of Los Angeles. At this site, the drainage of Pallet Creek has been repeatedly disturbed by successive ruptures along the fault zone. Trenches excavated across the creek bed have exposed sediments that have been displaced by several large earthquakes over a span of 1500 years. From these data, it was determined that strong earthquakes occur an average of once every 135 years. The last major event, the Fort Tejon earthquake, occurred on this segment of the San Andreas Fault in 1857, roughly 150 years ago. Because earthquakes occur on a cyclical basis, a major event in southern California may be imminent.

Using other paleoseismology techniques, researchers determined that several powerful earthquakes (magnitude 8 or larger) have repeatedly struck the

Figure 11.38
Paleoseismology: The study of prehistoric earthquakes These studies are often conducted by digging a trench across a fault zone and then looking for evidence of ancient displacements, such as offset sedimentary strata. This simplified diagram shows that vertical displacement occurred on this fault three different times, with each event producing an earthquake. Based on the size of the vertical displacement, these ancient earthquakes had estimated magnitudes of between 6.8 and 7.4. (Photo courtesy of USGS)

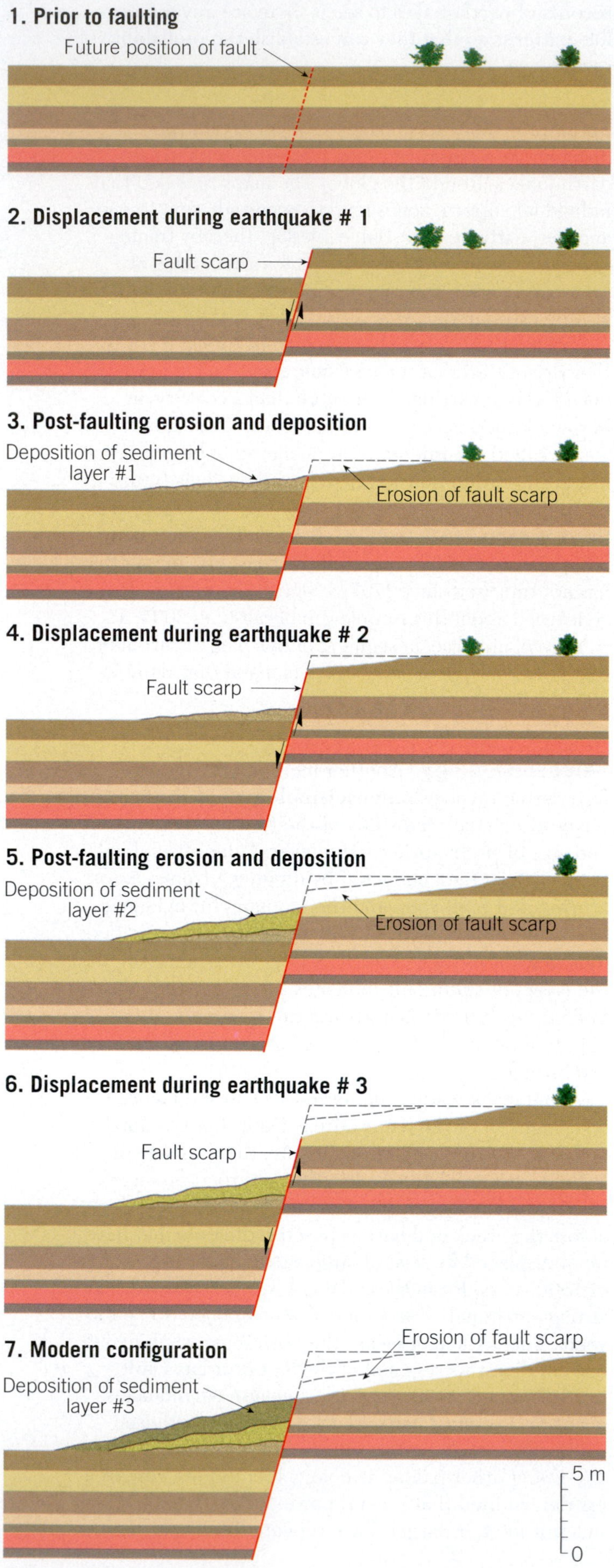

1. Prior to faulting
Future position of fault

2. Displacement during earthquake # 1
Fault scarp

3. Post-faulting erosion and deposition
Deposition of sediment layer #1
Erosion of fault scarp

4. Displacement during earthquake # 2
Fault scarp

5. Post-faulting erosion and deposition
Deposition of sediment layer #2
Erosion of fault scarp

6. Displacement during earthquake # 3
Fault scarp

7. Modern configuration
Erosion of fault scarp
Deposition of sediment layer #3

5 m

0

A.

B. The events depicted in the accompanying diagrams were deciphered by digging a trench (shown here) across the fault zone and studying the displaced sedimentary beds.

coastal Pacific Northwest over the past several thousand years. The most recent event, which occurred about 300 years ago, generated a destructive tsunami. As a result of these findings, public officials have taken steps to strengthen some of the region's existing buildings, dams, bridges, and water systems. Even the private sector responded. The U.S. Bancorp building in Portland, Oregon, was strengthened at a cost of $8 million.

11.8   Concept Checks

1. Are accurate, short-range earthquake predictions currently possible using modern seismic instruments? Explain.

2. What is the value of long-range earthquake forecasts?

11.1 What Is an Earthquake?

Sketch and describe the mechanism that generates most earthquakes.

KEY TERMS earthquake, fault, hypocenter (focus), epicenter, seismic wave, elastic rebound, aftershock, foreshock

- The sudden movements of large blocks of rock on opposite sides of faults cause most earthquakes. The location where the rock begins to slip is called the hypocenter, or focus. During an earthquake, seismic waves radiate from the hypocenter outward into the surrounding rock. The point on Earth's surface directly above the hypocenter is the epicenter.
- Over tens to hundreds of years, differential stresses gradually bend Earth's crust. Frictional resistance keeps the rock from rupturing and slipping. At some point, the stress overcomes the frictional resistance, and slippage allows the deformed (bent) rock to "spring back" to its original shape, generating an earthquake. The springing back is called elastic rebound.
- Foreshocks are smaller earthquakes that precede larger earthquakes. Aftershocks are smaller earthquakes that happen after large

earthquakes, as the crust readjusts to the new, post-earthquake conditions.

Q Label the blanks on the diagram to show the relationship between earthquakes and faults.

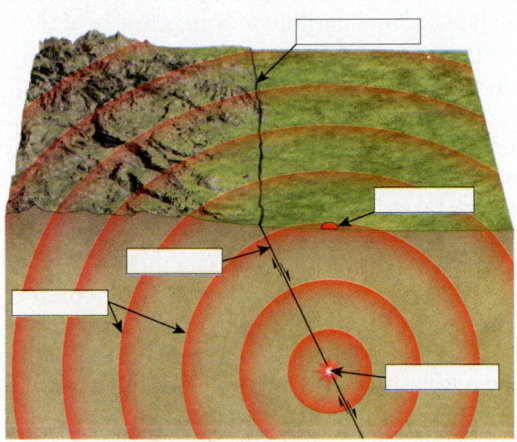

11.2 Faults and Earthquakes

List the basic fault types and describe the tectonic setting where each type of faulting tends to dominate.

KEY TERMS normal fault, reverse fault, thrust fault, megathrust fault, strike-slip fault, transform fault, fault creep

- The San Andreas Fault in California is an example of a large strike-slip fault that forms a transform plate boundary capable of generating destructive earthquakes.
- Convergent plate boundaries and associated subduction zones are marked by megathrust faults. These large faults are responsible for most of the largest earthquakes in recorded history. Megathrust faults are also capable of generating tsunamis.

11.3 Seismology: The Study of Earthquake Waves

Compare and contrast the types of seismic waves and describe the principle of the seismograph.

KEY TERMS seismology, seismograph (seismometer), inertia, seismogram, body waves, surface waves, P waves (primary waves), S waves (secondary waves)

- Seismology is the study of seismic waves. A seismograph measures these waves, using the principle of inertia. While the body of the instrument moves with the waves, the inertia of a suspended weight keeps a sensor stationary to record the relative difference between the two. The resulting record of the waves is called a seismogram.
- Seismograms reveal two main categories of earthquake waves: body waves (P waves and S waves) capable of moving through Earth's interior,

and surface waves, which travel only along the upper layers of the crust. P waves are the fastest, S waves are intermediate in speed, and surface waves are the slowest. However, surface waves tend to have the greatest amplitude, S waves are intermediate, and P waves have the lowest amplitude. Large-amplitude waves produce the most shaking, so surface waves usually account for most damage during earthquakes.

- P waves and S waves exhibit different kinds of motion. P waves momentarily push (compress) and pull (stretch) rocks as they travel through a rock body, thereby changing the volume of the rock. S waves impart a shaking motion as they pass through rock, changing the rock's shape but not its volume. Because fluids do not resist forces that change their shape, S waves cannot travel through fluids, whereas P waves can.

Q How could you physically demonstrate the difference between P waves and S waves to a friend who hasn't taken a geology course? (*Caution:* Don't hurt your friend!)

11.4 Locating the Source of an Earthquake

Explain how seismographs are used to locate the epicenter of an earthquake.

- Using the difference in arrival times between P and S waves, the distance separating a recording station from an earthquake's epicenter can be determined. When the distances are known from three or more seismic stations, the epicenter can be located using a method called triangulation.

11.5 Determining the Size of an Earthquake

Distinguish between intensity scales and magnitude scales.

KEY TERMS intensity, magnitude, Modified Mercalli Intensity scale, Richter scale, moment magnitude

- Intensity and magnitude are different measures of earthquake strength. Intensity measures the amount of ground shaking a place experiences due to an earthquake, and magnitude is an estimate of the actual amount of energy released during an earthquake.
- The Modified Mercalli Intensity scale is a tool for measuring an earthquake's intensity at different locations. The scale is based on verifiable physical evidence that is used to quantify intensity on a 12-point scale.
- The Richter scale takes into account both the maximum amplitude of the seismic waves measured at a given seismograph and the distance of that seismograph from the earthquake. The Richter scale is logarithmic, meaning that the next higher number on the scale represents seismic amplitudes that are 10 times greater than those represented by the number below. Furthermore, each larger number on the Richter scale represents the release of about 32 times more energy than the number below it.
- Because the Richter scale does not effectively differentiate between very large earthquakes, the moment magnitude scale was devised. This scale measures the total energy released from an earthquake by considering the strength of the faulted rock, the amount of slippage, and the area of the fault that slipped. Moment magnitude is the modern standard for measuring the size of earthquakes.

Q On the Richter scale diagram, first determine the Richter magnitude (M_L) for an earthquake at 400 kilometers distance, with a maximum amplitude of 0.5 millimeter. Second, for this same earthquake (same M_L), determine the amplitude of the biggest waves for a seismograph 40 kilometers from the hypocenter.

11.6 Earthquake Destruction

List and describe the major destructive forces that earthquake vibrations can trigger.

KEY TERMS liquefaction, seiche, tsunami

- Factors influencing how much destruction an earthquake might inflict on a human-made structure include (1) intensity of the shaking, (2) how long shaking persists, (3) the nature of the ground that underlies the structure, and (4) building construction standards. Buildings constructed of unreinforced bricks and blocks are more likely than other types of structures to be severely damaged in a quake.
- In general, bedrock-supported buildings fare best in an earthquake, as loose sediments amplify seismic shaking.
- Liquefaction may occur when water-logged sediment or soil is severely shaken during an earthquake. Liquefaction can reduce the strength of the ground to the point that it may not support buildings.

- A seiche is a "sloshing" motion of water caused by earthquake waves. Seiches are potentially dangerous for structures located along shorelines, and they can also cause earthen dams to fail.
- Earthquakes may also trigger landslides or ground subsidence, and they may break gas lines, which can initiate devastating fires.
- Tsunamis are large ocean waves that form when water is displaced, usually by a megathrust fault rupturing on the seafloor. Traveling at the speed of a jet, a tsunami is hardly noticeable in deep water. However, upon arrival in shallower coastal waters, the tsunami slows down and piles up, producing a wall of water sometimes more than 30 meters (100 feet) in height. Tsunamis cause major destruction in coastal areas if they strike the shoreline. Tsunami warning systems have been established in most of the large ocean basins.

Q Of the secondary earthquake hazards discussed in this section, which is (are) the greatest concern in the region where you live? Why?

11.7 Where Do Most Earthquakes Occur?

Locate Earth's major earthquake belts on a world map and label the regions associated with the largest earthquakes.

KEY TERMS circum-Pacific belt

- Most earthquake energy is released in the circum-Pacific belt, the ring of megathrust faults rimming the Pacific Ocean. Another earthquake belt is the Alpine–Himalayan belt, which runs along the zone where the Eurasian plate and the Indian–Australian and African plates collide.
- Earth's oceanic ridge system produces another belt of earthquake activity. Here seafloor spreading generates many frequent, small-magnitude quakes. Transform faults in the continental crust, including the San Andreas Fault, can produce large earthquakes.
- Although most destructive earthquakes are produced along plate boundaries, some occur at considerable distances from plate boundaries.

Examples include the 1811–1812 New Madrid, Missouri, earthquakes and the 1886 Charleston, South Carolina, earthquake.

Q Outline Earth's major earthquake belts on the accompanying map that has the plate boundaries drawn in red. Describe the type of plate boundary responsible for producing each belt.

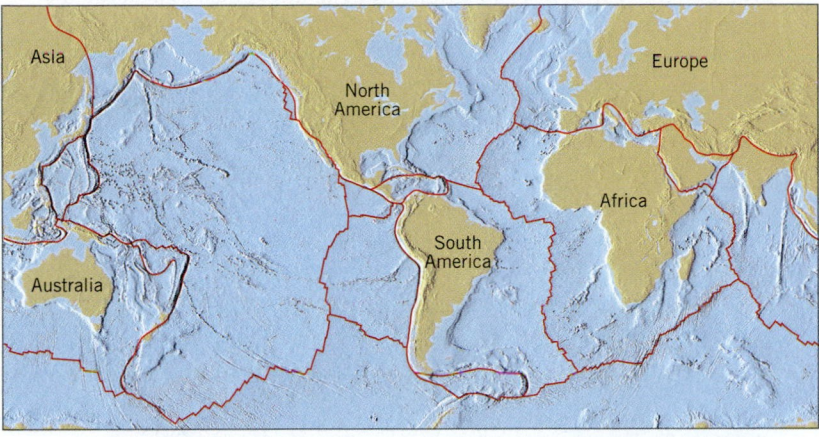

11.8 Can Earthquakes Be Predicted?

Compare and contrast the goals of short-range earthquake predictions and long-range forecasts.

KEY TERMS precursor, seismic gap, paleoseismology

- Successful earthquake prediction has been an elusive goal of seismology for many years. Shorter-range predictions (for hours or days) are based on precursor events such as changes in ground elevation or in strain levels near a fault. Unfortunately, this type of monitoring is not reliable.
- Long-range forecasts (for time scales of 30 to 100 years) are statistical estimates of the likelihood that an earthquake of a given magnitude will occur. Long-range forecasts are useful because they can guide development of building codes and infrastructure.
- Scientists have identified seismic gaps, portions of faults that have been storing strain for a long time, meaning these sites have great potential for experiencing an earthquake in the "not-too-distant future." Paleoseismology is another tool used to make long-range forecasts. Because earthquakes occur on a cyclical basis, determining how frequently they have occurred in the past can give some insight into when they are most likely to occur again.

Q If you were considering moving to a city located in a seismic gap, how would you determine whether it was safe or foolhardy? Which factors would be most influential in your decision?

Give It Some *Thought*

1. Describe the concept of elastic rebound. Develop an analogy other than a rubber band to illustrate this concept.

2. The accompanying map shows the locations of many of the largest earthquakes in the world since 1900. Refer to the map of Earth's plate boundaries in Figure 2.11 (page 45) and determine which type of plate boundary is most often associated with these destructive events.

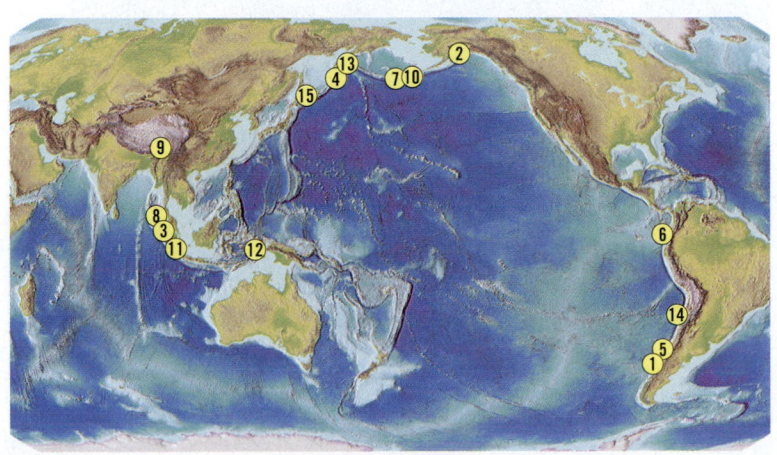

3. Use the accompanying seismogram to answer the following questions:

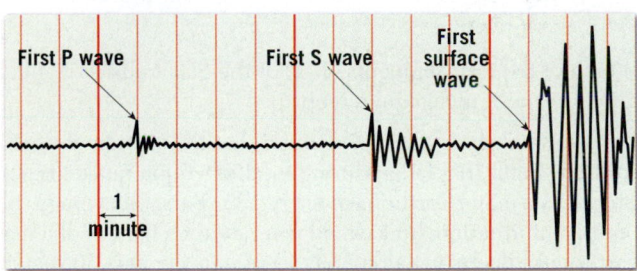

a. Which of the three types of seismic waves reached the seismograph first?

b. What is the time interval between the arrival of the first P wave and the arrival of the first S wave?

c. Use your answer from Question b and the travel–time graph in Figure 11.17 to determine the distance from the seismic station to the earthquake.

d. Which of the three types of seismic waves had the highest amplitude when they reached the seismic station?

4. You go for a jog on a beach and choose to run near the water, where the sand is well packed and solid under your feet. With each step, you notice that your footprint quickly fills with water but not water coming in from the ocean. What is this water's source? For what earthquake-related hazard is this phenomenon a good analogy?

5. Why is it possible to issue a tsunami warning but not a warning for an impending earthquake? Describe a scenario in which a tsunami warning would be of little value.

6. Using the accompanying map of the San Andreas Fault, answer the following questions:

a. Which of the four segments (1–4) of the San Andreas Fault do you think is experiencing fault creep?

b. Paleoseismology studies have found that the section of the San Andreas Fault that failed during the Fort Tejon quake (segment 3) produces a major earthquake every 135 years, on average. Based on this information, how would you rate the chances of a major earthquake occurring along this section in the next 30 years? Explain.

c. Do you think San Francisco or Los Angeles has the greater risk of experiencing a major earthquake in the near future? Defend your selection.

7. The accompanying image shows a double-decked section of Interstate 880 (the Nimitz Freeway) that collapsed during the 1989 Loma Prieta earthquake and caused 42 deaths. About 1.4 kilometers of this freeway section, called the Cypress Viaduct, collapsed, while a similar section survived the vibration. Both sections were subsequently demolished and rebuilt as a single-level structure, at a cost of $1.2 billion. Examine the map and seismograms from an aftershock that shows the intensity of shaking observed at three nearby locations to answer the following questions:

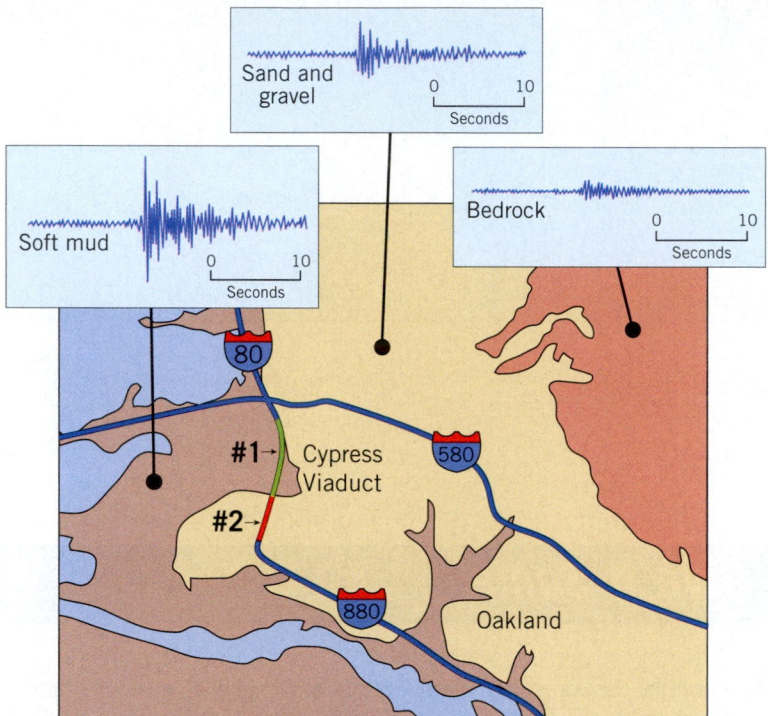

a. What type of ground material experienced the least amount of shaking during the aftershock?

b. What type of ground materials experienced the greatest amount of ground shaking during the same event?

c. Which of the two sections of the Cypress Viaduct shown on the map do you think collapsed? Explain.

8. Strike-slip faults, like the San Andreas Fault, are not perfectly straight but bend gradually back and forth. In some locations, the bends are oriented such that blocks on opposite sides of the fault pull away from each other, as shown in the accompanying sketch. As a result, the ground between the bends sags, forming a depression or basin. These depressions often fill with water.

a. What name is given to the depression in the accompanying photo? (*Hint:* Refer to Chapter 10.)

b. Describe what would result if these two blocks where moving in opposite directions.

Before offset   After offset

MasteringGeology™   www.masteringgeology.com

Looking for additional review and test prep materials? With individualized coaching on the toughest topics of the course, MasteringGeology offers a wide variety of ways for you to move beyond memorization to begin thinking like a geologist. Visit the Study Area in www.masteringgeology.com to find practice quizzes, study tools, and multimedia that will improve your understanding of this chapter's content. Sign in today to enjoy the following features: **Self Study Quizzes, SmartFigures: Tutorials/Animations/Condor Videos/Mobile Field Trips, Geoscience Animation Library, GEODe, RSS Feeds, Digital Study Modules,** and an optional **Pearson eText.**

12

Earth's Interior*

The Japanese high-tech drilling ship *Chikyu* is capable
of drilling 7000 meters (nearly 4.5 miles) into the ocean floor.
(Photo by Kyodo/Landov)

*This chapter was originally prepared by Professor Michael Wysession,
Washington University.

Each statement represents the primary **LEARNING OBJECTIVE** for the corresponding major heading within the chapter. After you complete the chapter, you should be able to:

12.1 Explain how Earth acquired its layered structure.

12.2 Describe how seismic waves are used to probe Earth's interior.

12.3 List and describe each of the layers of Earth's interior.

12.4 Describe the processes of heat transfer that operate within Earth's interior and where each of these processes dominates.

12.5 Discuss what studies of Earth's gravity, seismic tomography, and Earth's magnetic field have revealed about variations in the layers of our planet.

f you could slice Earth in half, the first thing you would notice is that it has distinct layers. The heaviest materials (metals) appear in the center. Lighter solids (rocks) make up the middle layers, and less dense liquids and gases make up the outer layer. Within Earth, we know these layers as the iron core, the rocky mantle and crust, the liquid ocean, and the gaseous atmosphere. More than 95 percent of the variations in composition and temperatures in Earth are due to this seemingly simple layered structure. However, this is not the end of the story. If it were, Earth would be a dead, lifeless cinder floating in space.

12.1 | Earth's Internal Structure

Explain how Earth acquired its layered structure.

In Chapter 1 you learned that Earth's interior consists of three major layers defined by their chemical composition—the crust, mantle, and core. In addition, Earth's three compositionally distinct shells can be further subdivided into layers, based on physical properties that include whether the layer is solid or liquid and how weak or strong it is. Knowledge of both types of layers is essential to our understanding of basic geologic processes, such as volcanism, earthquakes, and mountain building (**Figure 12.1**).

Gravity and Earth's Layers

If a bottle filled with clay, iron filings, water, and air were shaken, it would appear to have a single, muddy composition. If that bottle were then allowed to sit undisturbed, the different materials would separate and settle into layers. The iron filings, which are the densest, would be the first to sink to the bottom. Above the iron would be a layer of clay, then water, and, finally, air.

The force of gravity is responsible for the layering in the bottle of muddy water, as well as the layering we detect in Earth's interior.

Formation of Earth's Layered Structure

As material accumulated to form Earth (and for a short period afterward), the high-velocity impact of nebular debris and the decay of radioactive elements caused the temperature of our planet to steadily increase. During this time of intense heating, Earth became hot enough that elements including iron and nickel began to melt. Melting produced liquid blobs of heavy metal that gravitationally sank toward the center of the planet. This process occurred rapidly on the scale of geologic time and produced Earth's dense iron-rich core.

The early period of heating resulted in another process of chemical differentiation, whereby melting formed masses of less-dense, molten rock that buoyantly rose toward the surface and solidified to produce a primitive crust. These rocky materials were rich in oxygen and "oxygen-seeking" elements, particularly silicon and aluminum, along with lesser amounts of calcium, sodium, potassium, iron, and magnesium.

In addition, some heavy metals such as gold, lead, and uranium, which have low melting points or were highly soluble in the ascending molten masses, were scavenged from Earth's interior and concentrated in the developing crust. This early period of chemical segregation established the three basic divisions of Earth's interior: (1) the iron-rich *core*, (2) the thin *primitive crust*, and (3) Earth's largest layer, called the *mantle*, which is located between the core and crust. It was from the primitive mantle that less-dense materials rose to form the crust and more dense metals sank to the core.

Horizontal Variations in Composition and Temperature

In addition to Earth's layers, there are small horizontal variations in mineral composition and temperature with depth. These differences, although small, indicate that the interior of our planet remains very dynamic. The rocks of the mantle and crust are in constant motion, not only moving about through plate tectonics but also continuously recycling between the surface and the interior.

One surface manifestation of this ongoing process is volcanism—which carries water and gases from Earth's interior to replenish our oceans and atmosphere, allowing life to exist at the surface.

How Does Gravity Affect Density?

Gravity affects the density of rocks because rocks are compressed by the weight of material above them. As a result, the density of rocks increases with depth. Rocks in

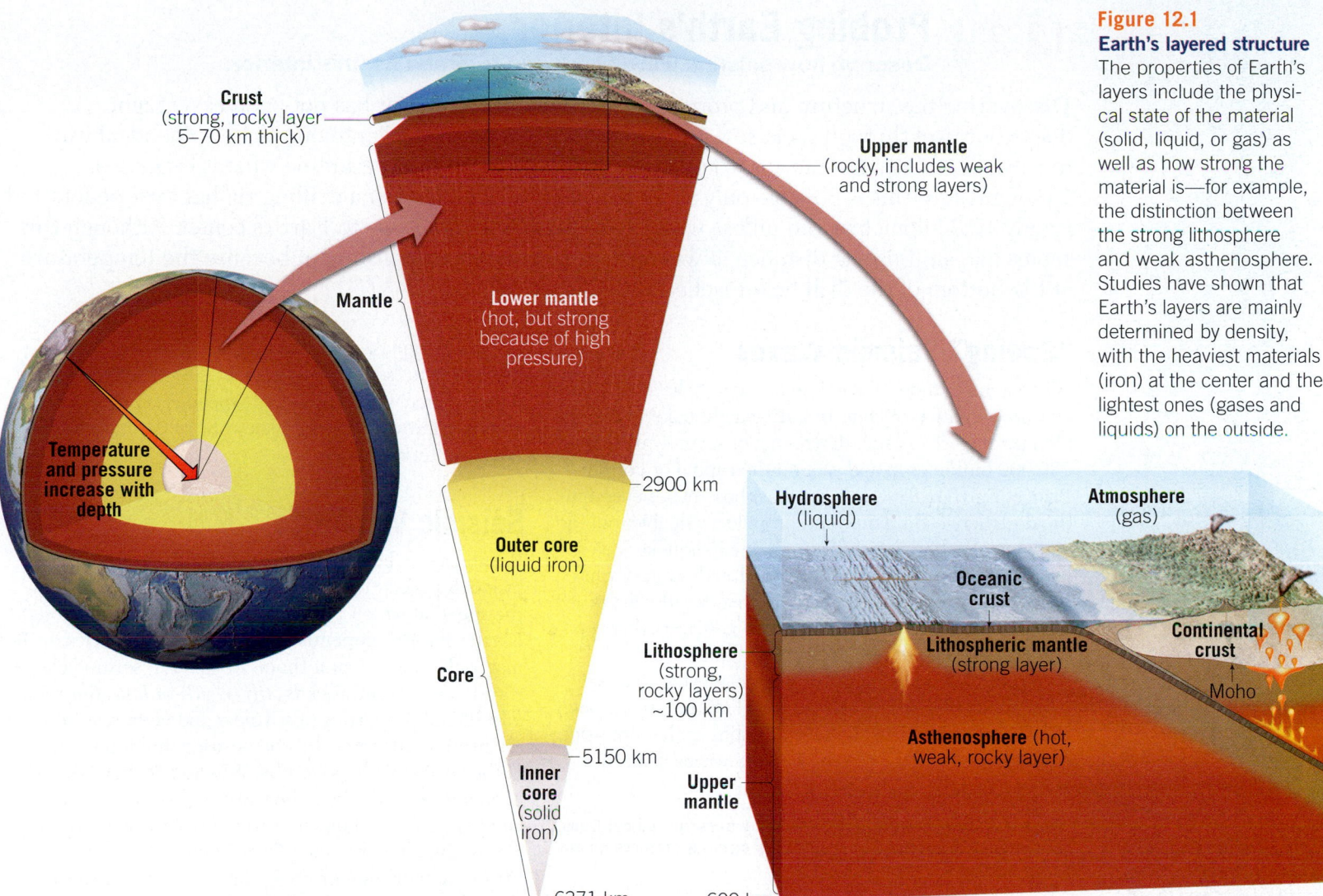

Figure 12.1
Earth's layered structure
The properties of Earth's layers include the physical state of the material (solid, liquid, or gas) as well as how strong the material is—for example, the distinction between the strong lithosphere and weak asthenosphere. Studies have shown that Earth's layers are mainly determined by density, with the heaviest materials (iron) at the center and the lightest ones (gases and liquids) on the outside.

the upper mantle have a density of about 3.3 grams per cubic centimeter (g/cm³), whereas the density of rocks with the same chemical composition found at the base of the mantle is about 5.6 g/cm³—nearly twice as great. This increase in density occurs partly because atoms (ions) shrink and occupy less space when subjected to immense pressure.

However, not all atoms compress at the same rate. For example, it is easier to compress negative ions than positive ions. Negative ions have more electrons than protons and tend to be "fluffier" than positive ions. As a result, when rocks are squeezed, the negative ions (such as O^{-2}) compress more easily than the positive ions (such as Si^{+4} and Mg^{+2}), so the ratios of ionic sizes change. As these ratios change, the atomic structure of a mineral may become unstable, causing the atoms to rearrange into a more stable and denser atomic structure. This process was discussed in Chapter 3 and is called a **phase change**.

For example, at depths between 300 and 400 kilometers (190 and 250 miles), the intense pressure of the overlying rocks causes the mineral *olivine*, a common constituent in mantle rocks, to become unstable. As a result, the atoms in olivine rearrange into a denser and more stable crystalline structure. The increase in density of mantle rocks with depth is due both to the compression of existing minerals and to the formation of new "high-pressure" minerals that have greater density.

12.1   Concept Checks

1. List the three compositionally distinct layers of Earth's interior.

2. Describe the process that produced layering during Earth's formation.

3. What causes a mineral phase change?

12.2 | Probing Earth's Interior

Describe how seismic waves are used to probe Earth's interior.

Discovering the structure and properties of Earth's deep interior has not been easy. Light does not travel through rock, so we must find other ways to "see" into our planet. The ideal way to learn about Earth's interior would be to dig or drill a hole and examine what is extracted. Unfortunately, this is possible only at shallow depths. The deepest a drilling rig has ever penetrated is only 12.3 kilometers (7.6 miles), which is about 1/500 of the way to Earth's center. Although this seems like a miniscule distance, it was an extraordinary accomplishment because the temperature at the bottom of the drill hole reached 180°C (356°F).

"Seeing" Seismic Waves

When a magnitude 7.0 earthquake struck the small Caribbean nation of Haiti in 2010, earthquake waves rippled through the island, destroying or severely damaging 280,000 buildings in and around the capital of Port-au-Prince. Earthquake waves, such as those experienced in the destructive Haitian event, provide a valuable tool for studying Earth's interior. The use of earthquake waves, also called **seismic waves**, to study Earth's interior has been greatly enhanced during the past decade due to the growing number of *seismograph networks* that can detect these waves from distant earthquakes.

About 3000 earthquakes occur each year that are large enough (about M$_w$ 5.5) to travel all the way through Earth and be recorded by seismographs at the other side of the globe (**Figure 12.2**). The P and S waves from these strong earthquakes act like medical X-rays and provide the means to "see" into our planet. Detailed studies of these seismic waves using high-speed computers have greatly improved our understanding of the nature of Earth's interior.

Seismic Velocities

Recall from Chapter 11 that seismic waves generated by earthquakes travel at different speeds. In addition, the speed at which P waves and S waves travel through Earth's interior depends largely on the properties of the materials that transmit them. In general, seismic waves travel fastest when rock is *stiff* (rigid) or *less compressible*. These properties of stiffness and compressibility are used to interpret the composition and temperature of the rock at various depths. When rock is heated, it becomes less stiff (imagine warming a frozen chocolate bar), and earthquake waves travel through it more slowly. Specifically, when P waves travel through molten or partially molten rock, they travel much slower than if the rock was solid. Furthermore, when S waves enter the outer core, which is liquid, the waves are not

SmartFigure 12.2
Seismic waves provide a way to "see" into our planet Illustration of seismic waves traveling through Earth's interior, assuming uniform materials along the path. (https://goo.gl/crzMb0)

Tutorial

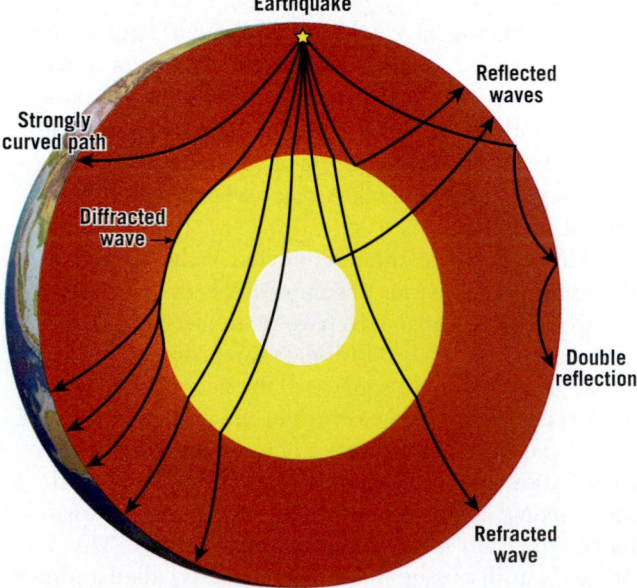

When traveling through Earth, seismic waves spread out from an earthquake source (hypocenter) as circular features called *wave fronts*.

Earthquake

Wave fronts

Rays

The paths taken by these waves can also be considered *seismic rays*, lines drawn perpendicular to the wave front as shown here.

Earthquake

Reflected waves

Strongly curved path

Diffracted wave

Double reflection

Refracted wave

Figure 12.3
Possible paths that seismic rays follow through Earth

When seismic waves (rays) encounter a boundary between materials with different properties, such as air and water, the energy splits into reflected and refracted (bent) waves.

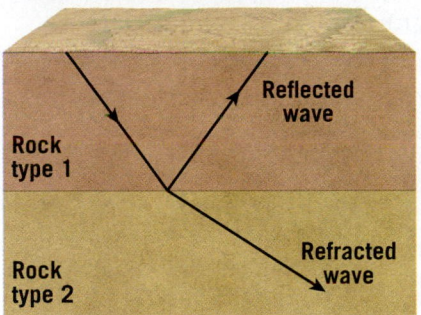

When the velocity of seismic waves increases as they pass from one layer into another, the waves refract (bend) toward the boundary separating the layers.

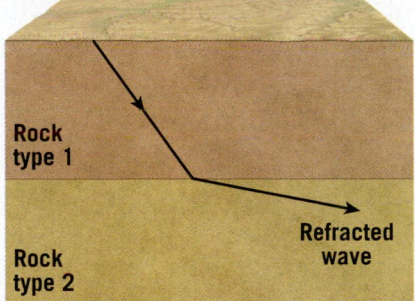

When the velocity of seismic waves decreases as they pass from one layer into another, the waves refract (bend) away from the boundary separating them.

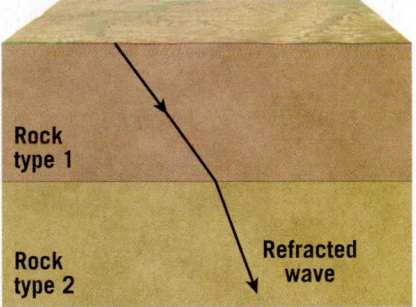

Figure 12.4
Reflection and refraction

transmitted at all because S waves do not travel through liquids (see Figure 12.6).

Likewise, waves travel at different speeds through Earth materials that have different compositions. For example, seismic waves travel faster through oceanic crust, which is composed of basalt, than through the continental crust, which has an overall composition akin to that of granite. Thus, the speed at which seismic waves travel has helped researchers determine both the types of rocks found within Earth and how hot they are.

Interactions Between Seismic Waves and Earth's Layers

Interpreting the waves recorded on seismograms in order to identify Earth's layered structure is challenging. Seismic waves do not travel along straight paths; instead, they are *reflected*, *refracted*, and *diffracted* as they pass through our planet (**Figure 12.3**). You are familiar with reflected sound waves we call echoes. When a seismic wave hits a boundary between different Earth materials, such as the boundary between the crust and the mantle, some of the waves are **reflected** back toward the surface (**Figure 12.4**). The remaining energy passes though the boundary and is **refracted** (bent). This is similar to how light is refracted (bent) as it passes from air to water. Seismic waves can also bend when they pass by a curved surface such as a boundary between two compositionally different layers—a process called **diffraction**. These different wave behaviors have been used to identify the boundaries that exist within Earth.

One of the most noticeable behaviors of seismic waves is that they follow strongly curved (refracted) paths (**Figure 12.5**). This occurs because the velocity of seismic waves generally increases with depth—the result of increasing pressure that squeezes the rock into a more compact, rigid material.

Within Earth's interior, where there are both distinct boundaries and gradual seismic velocity changes caused by changes in mineral properties, the pattern of seismic waves generated by an earthquake is complex. Nevertheless, researchers who study seismograms from large earthquakes obtained from around the globe can identify Earth's major internal structures.

12.2 Concept Checks

1. What characteristic of seismic waves makes them useful for probing Earth's interior?

2. What do reflected waves tell us about the composition of Earth's interior?

3. Why do seismic waves travel along curved paths through the mantle?

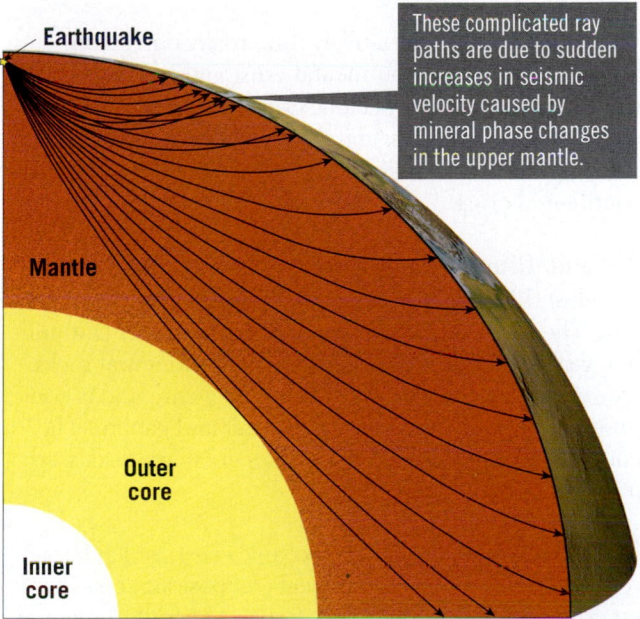

These complicated ray paths are due to sudden increases in seismic velocity caused by mineral phase changes in the upper mantle.

Figure 12.5
Paths of seismic waves though the mantle The rays follow curved (refracting) paths rather than straight paths because the seismic velocity of rocks increases with depth as a result of a steady increase in pressure.

12.3 | Earth's Layers

List and describe each of the layers of Earth's interior.

Combining the data obtained from seismological studies and mineral physics experiments has given us a layer-by-layer understanding of Earth's composition (see GEOgraphics 12.1). The variations in seismic velocities with depth are shown in **Figure 12.6**. By examining the behavior of a variety of rocks at the pressures corresponding to these depths, geologists have made important discoveries about the compositions of Earth's crust, mantle, and core.

SmartFigure 12.6
Average velocities of P and S waves at each depth
S waves are an indication of how rigid (stiff) the material is; faster velocities indicate greater rigidity. The inner core is less rigid than the mantle, and the liquid outer core has no rigidity.
(https://goo.gl/8dP6DM)

 **Tutorial**

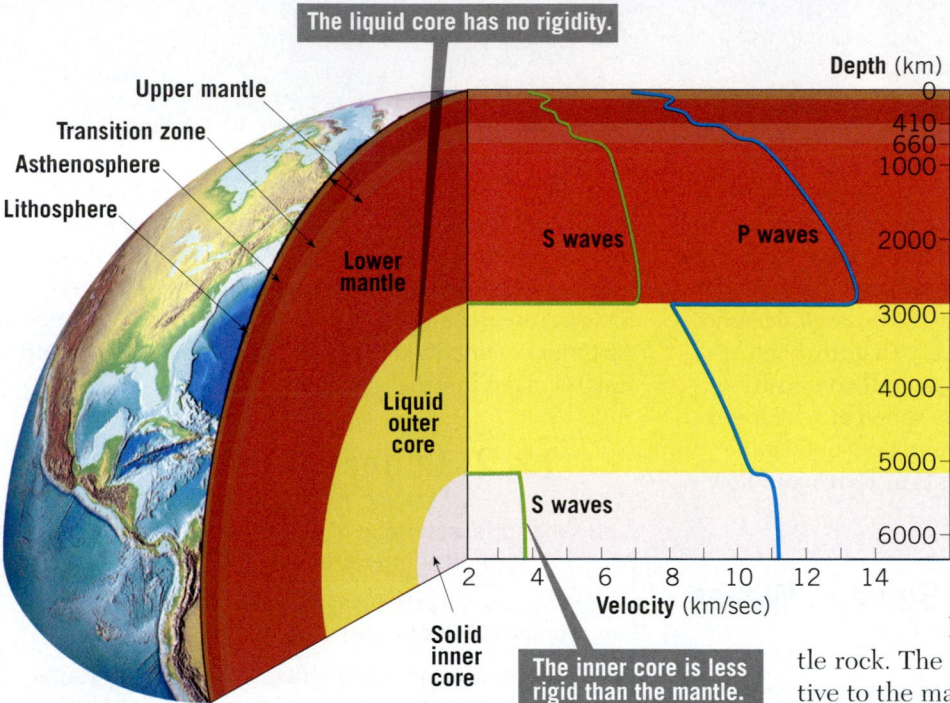

Continental crust averages about 40 kilometers (25 miles) thick but can be more than 70 kilometers (40 miles) thick in mountainous regions such as the Himalayas and the Andes. The thickest North American crust, beneath the Rockies, is more than 50 kilometers (30 miles) thick. By contrast, the thinnest crust in North America is beneath the Basin and Range region in the western United States, where the crust is as thin as 20 kilometers (12 miles).

Continental crust has an average density of about 2.7 g/cm^3, which is much lower than the density of mantle rock. The low density of continents relative to the mantle explains why continents are buoyant—acting like giant rafts, floating atop tectonic plates—and why they cannot be readily subducted into the mantle. Because continental rocks cannot be easily recycled into the mantle, continental rocks that exceed 4 billion years in age have been found.

Crust

The **crust** is Earth's relatively thin, rocky outer skin, and two types exist: continental crust and oceanic crust. Continental crust and oceanic crust have very different compositions, histories, and ages. In fact, oceanic crust is compositionally more similar to Earth's mantle than to its continental crust.

Oceanic Crust   The ocean crust is about 7 kilometers (4 miles) thick and forms along the mid-ocean ridge system. The rocks of the oceanic crust are younger (180 million years old or less) and denser than continental rocks. Ocean crust has a density of about 3.0 g/cm^3, and is composed of the dark igneous rocks basalt and gabbro. The composition and formation of ocean crust are discussed in greater detail in Chapter 13.

Continental Crust   Unlike oceanic crust, which has a relatively homogeneous chemical composition, continental crust consists of many rock types. Although the upper crust has an average composition of a *granitic rock* called *granodiorite*, its composition and structure vary considerably from place to place.

Discovering Boundaries: The Moho   The boundary between the crust and mantle, called the **Moho**, was one of the first features of Earth's interior discovered using seismic waves. In 1909, Croatian seismologist Andrija Mohorovičić discovered this boundary that now bears his name. At the base of the continents, P waves travel about 6 kilometers per second (km/s), but they abruptly increase in speed to 8 km/s at a slightly greater depth.

Mohorovičić cleverly used this large jump in seismic velocity to make his discovery. He noticed that two different sets of seismic waves were recorded at seismographs located within a few hundred kilometers of an earthquake. One set of waves moved through the ground at about 6 km/s, while the other set of waves traveled about 8 km/s—allowing Mohorovičić to correctly determine that the different waves were traveling through two different layers.

Recreating the Deep Earth

Seismology alone cannot determine the nature of the materials deep in Earth's interior. Additional information must be obtained by other techniques. Mineral physics experiments can measure physical properties of rocks and minerals such as stiffness, compressibility, and density while simulating the extreme conditions of the mantle and core.

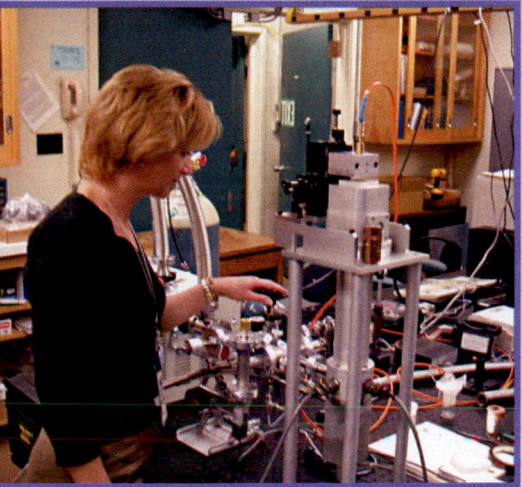

C.Arache, D. Jackson and S.T. Weir/Lawrence Livermore National Laboratory

Most mineral physics experiments are conducted using diamond-anvil presses like the one shown here. These take advantage of two important properties of diamonds—hardness and transparency. The tips of two diamonds are cut off, and a small mineral sample is placed between them. By squeezing two diamonds together, pressures as high as our planet's interior have been simulated. High temperatures are achieved by firing a laser beam through the diamond and into the mineral sample.

Douglas L. Peck Photography

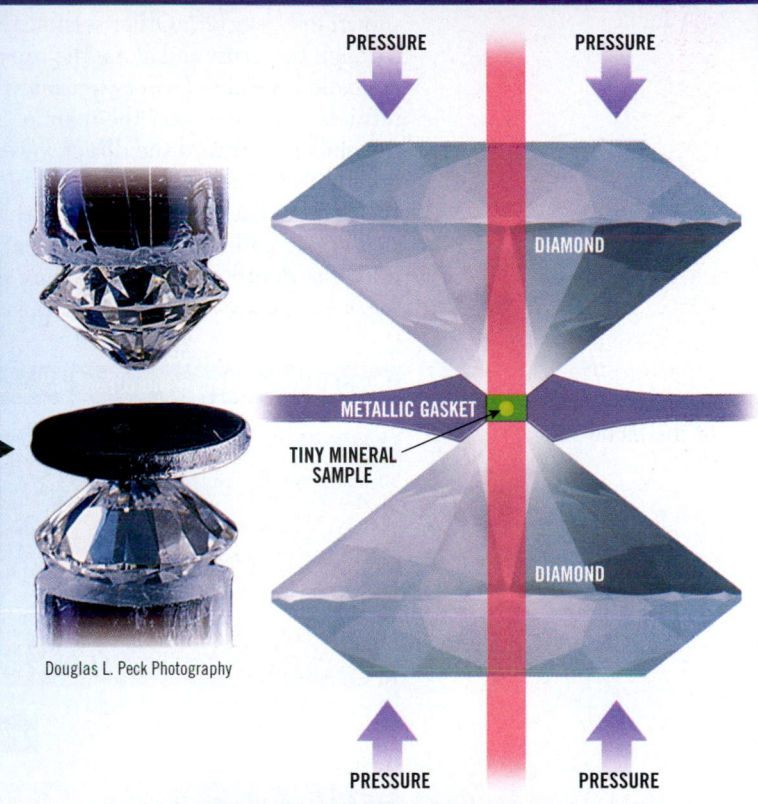

PRESSURE PRESSURE

DIAMOND

METALLIC GASKET

TINY MINERAL SAMPLE

DIAMOND

PRESSURE PRESSURE

One experiment examines the temperatures and pressures at which one mineral phase will become unstable and convert into a new "high-pressure" phase. These experiments are useful because they help identify where phase changes take place within Earth.

Question:
What two properties of diamonds make them ideal for use in a diamond-anvil press?

?

These experiments have also helped identify where changes in temperature, pressure, and density occur in Earth's interior, as shown in the graphs below.

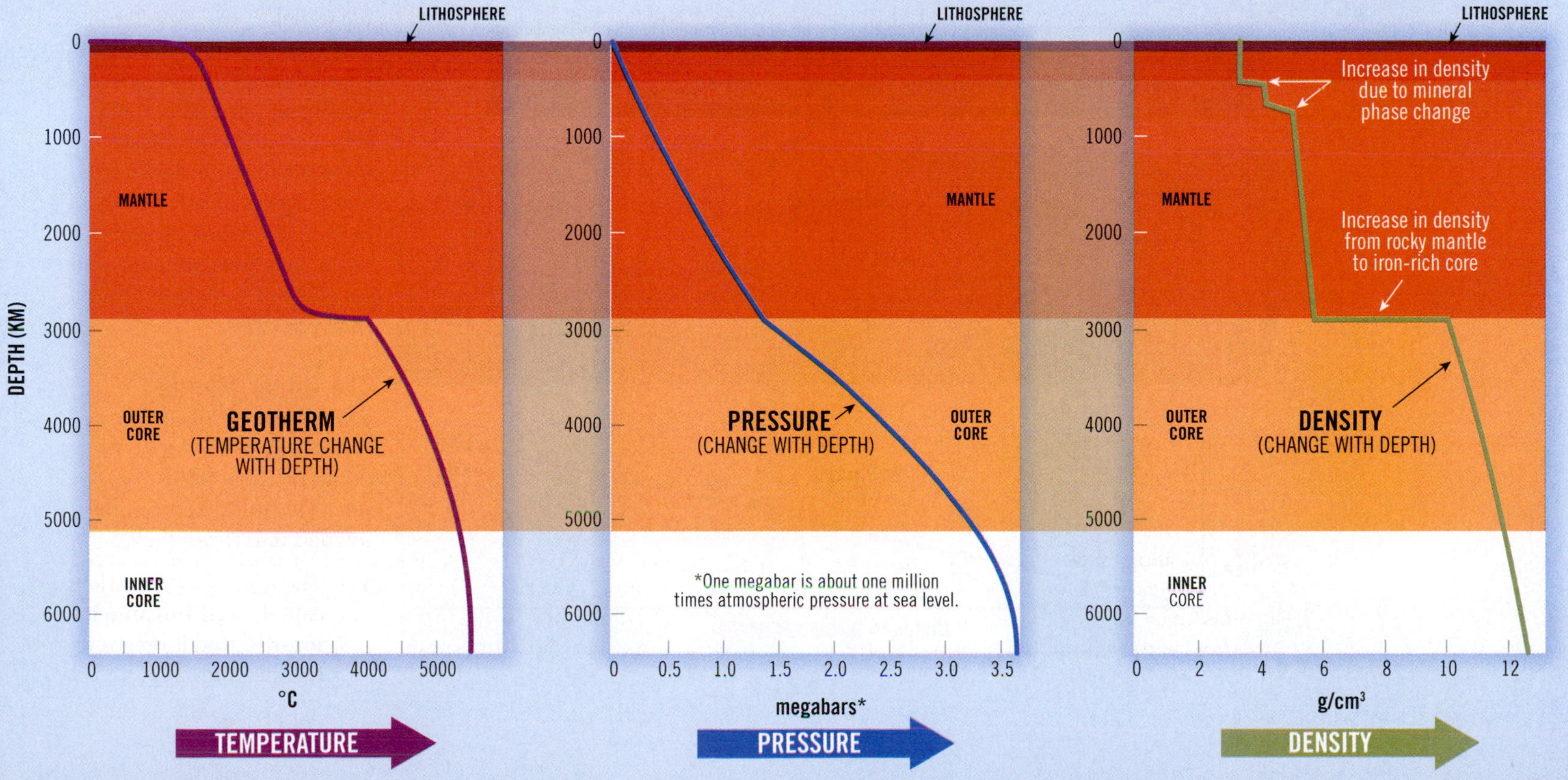

LITHOSPHERE

DEPTH (KM)

MANTLE

OUTER CORE

GEOTHERM
(TEMPERATURE CHANGE WITH DEPTH)

INNER CORE

0   1000   2000   3000   4000   5000
°C

TEMPERATURE

LITHOSPHERE

MANTLE

PRESSURE
(CHANGE WITH DEPTH)

OUTER CORE

*One megabar is about one million times atmospheric pressure at sea level.

0   0.5   1.0   1.5   2.0   2.5   3.0   3.5
megabars*

PRESSURE

LITHOSPHERE

Increase in density due to mineral phase change

MANTLE

Increase in density from rocky mantle to iron-rich core

OUTER CORE

DENSITY
(CHANGE WITH DEPTH)

INNER CORE

0   2   4   6   8   10   12
g/cm³

DENSITY

During a shallow earthquake, *direct waves* travel along a nearly straight path through the crust, as shown in **Figure 12.7**. Other seismic waves follow a path through the crust and along the top of the mantle. These are called *refracted waves* because they are bent, or refracted, as they enter the mantle. Seismographs near the epicenter record the direct waves first. However, seismographs farther from the epicenter record the refracted waves first. The point at which both waves arrive at the same time, called the *cross-over*, can be used to determine the depth of the Moho. Thus, using data from these two sets of waves and seismographs at various distances from an earthquake's epicenter, the thickness of the crust for any location can be calculated.

The difference between travel times for direct and refracted waves is comparable to the difference between driving to a destination on local roads versus on interstate highways. For short distances, you typically arrive sooner if you drive the most direct route, using local roads. For long distances, the trip may take less time if you take a less direct route that involves mostly interstate highways. The cross-over point, where both routes take an equal amount of time, is directly related to how far you must drive before reaching the interstate highway. Applied to determining the depth of the Moho, the cross-over is related to how far seismic waves travel through the crust (slow layer) before they reach the mantle (fast layer): The greater the cross-over distance, the deeper the Moho. The Moho lies about 25 to 70 kilometers (15 to 45 miles) beneath the continents and about 7 kilometers (4 miles) below the ocean floor.

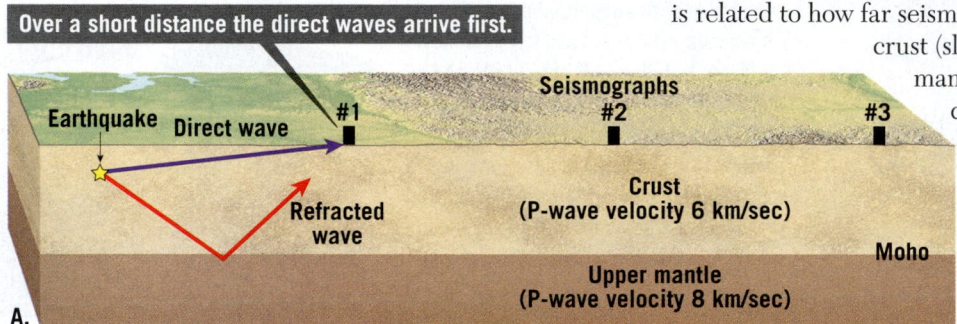

A.

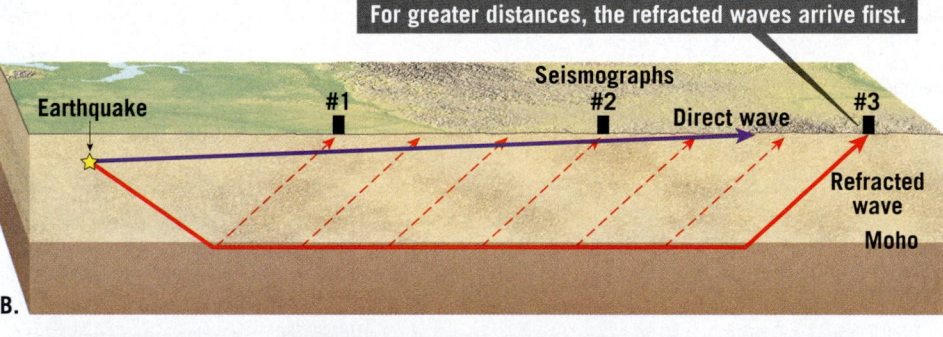

B.

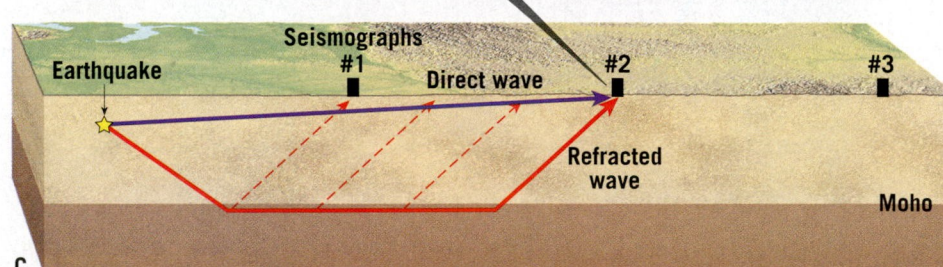

C.

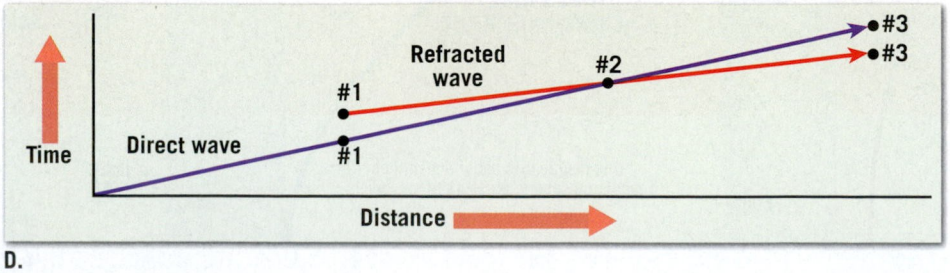

D.

Mantle

Beneath Earth's crust lies the mantle. More than 82 percent of Earth's volume is contained within the **mantle**, a nearly 2900-kilometer- (1800-mile-) thick shell extending from the base of the crust (Moho) to the liquid outer core (see Figure 12.1). Because S waves readily travel through the mantle, we know that it is a solid rocky layer composed of silicate minerals that are rich in iron and magnesium. However, despite its solid nature, rock in the mantle is quite hot and capable of flow, albeit at very slow velocities.

The Upper Mantle

Earth's **upper mantle** extends from the Moho to a depth of about 660 kilometers (410 miles) and can be divided into three shells:

1. The uppermost mantle is called the **lithospheric mantle**, and it ranges in thickness from only a few

kilometers under the mid-oceanic ridges to perhaps as much as 200 kilometers (125 miles) under the stable continental interiors. The uppermost mantle and the crust make up Earth's rigid outer shell, called the **lithosphere**.

2. Beneath the lithospheric mantle is a weak layer called the **asthenosphere**. The lithospheric mantle and asthenosphere are compositionally similar; however, the uppermost mantle is strong, whereas the asthenosphere is weak, as a result of Earth's temperature structure—a topic we will consider later.

3. The lower portion of the upper mantle, at depths between 410 and 660 kilometers, is called the **transition zone**.

Figure 12.8
Peridotite: A rock from Earth's mantle This sample of peridotite (green rock that is olivine rich) was carried up from the mantle and provides clues to the composition of Earth's interior. The mantle fragment (xenolith) was contained within a volcanic bomb from Vulkaneifel, Germany. €1 coin for scale. (Photo by Woudloper)

Rocks brought to the surface by volcanism and other geologic processes have provided geologists with valuable information about the composition of the upper mantle, which is composed mainly of the rock *peridotite* (**Figure 12.8**). Peridotite is an ultramafic rock that consists of the minerals *olivine* and *pyroxene*, minerals that are rich in iron and magnesium. As a result, the mantle is denser than either the continental crust or the oceanic crust that lie above it.

At the depth (and pressures) of the transition zone, olivine, which is stable in the uppermost mantle, is subjected to greater pressure and collapses into denser structures. In the top half of the transition zone, olivine converts to a more compact structure similar to the mineral spinel, and pyroxene converts to a garnet-like structure.

The Lower Mantle   The **lower mantle** lies between the transition zone (660 kilometers) and the liquid core (2900 kilometers). Within the lower mantle, both olivine and pyroxene take the structure of a dense mineral called *perovskite* (recently re-named *bridgemanite*) and other related minerals. Because the lower mantle is undoubtedly Earth's largest layer, occupying 56 percent of the volume of the planet, perovskite-structured silicate minerals are the single most abundant material within Earth.

The D″ Layer   In the lowest few hundred kilometers of the mantle is a highly variable and unusual region called D″ (pronounced "dee double-prime"). The **D″ layer**, the boundary layer between the rocky mantle and the liquid iron outer core, is thought to have large variations in composition as well as temperature (**Figure 12.9**). Cool areas in the D″ layer are thought to be the graveyard of subducted oceanic lithosphere, whereas the hot areas are the birthplace of deep mantle plumes.

The very base of D″, the part of the mantle in direct contact with the hot liquid iron core, is like

Earth's surface in that there are "upside-down mountains" of rock that protrude into the core. Furthermore, in some regions of the core–mantle boundary, the base of D″ may be hot enough to be partially molten. Evidence for partial melting comes from zones at the very base of the mantle where S-wave velocities decrease by 30 percent, an indication that the material there is quite weak.

Discovering Boundaries: The Core–Mantle Boundary   Evidence that Earth has a distinct central core was uncovered in 1906 by British geologist Richard Dixon Oldham. At locations beyond approximately 100 degrees from the epicenter of a large earthquake, Oldham observed that P and S waves were absent or very weak. In other words, Oldham found evidence for a central core that produced a **shadow zone** for seismic waves (**Figure 12.10**). In 1914, seismologist Beno

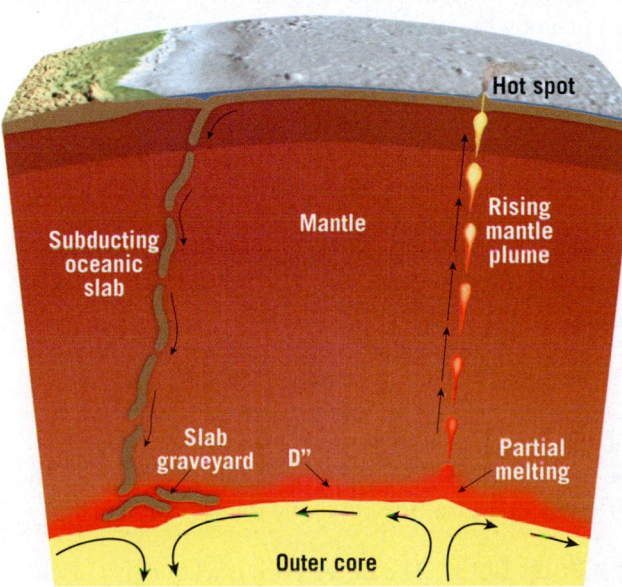

Figure 12.9
The variable and unusual D″ layer lies at the base of the mantle The D″ layer contains large, horizontal variations in both temperature and composition. Many geologists think that the D″ layer is the graveyard of subducted oceanic lithosphere and the birthplace of some mantle plumes.

Hot spot

Subducting oceanic slab

Mantle

Rising mantle plume

Slab graveyard

D″

Partial melting

Outer core

A. The P-wave shadow zone exists because P waves interact with the low-velocity liquid iron of the outer core, which causes their rays to be refracted downward. This creates a shadow zone where no direct P waves are recorded (although reflected P waves travel there).

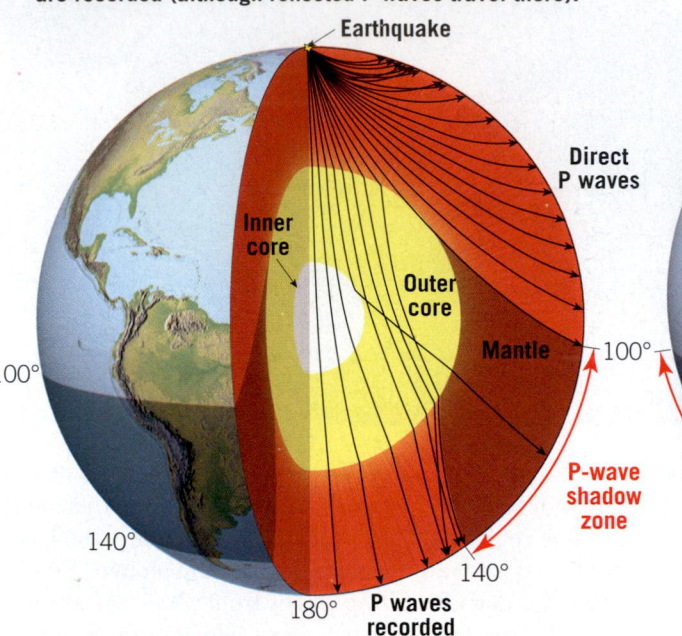

B. The core is an obstacle to S waves, because they cannot pass through liquids. Therefore, a large shadow zone exists for direct S waves.

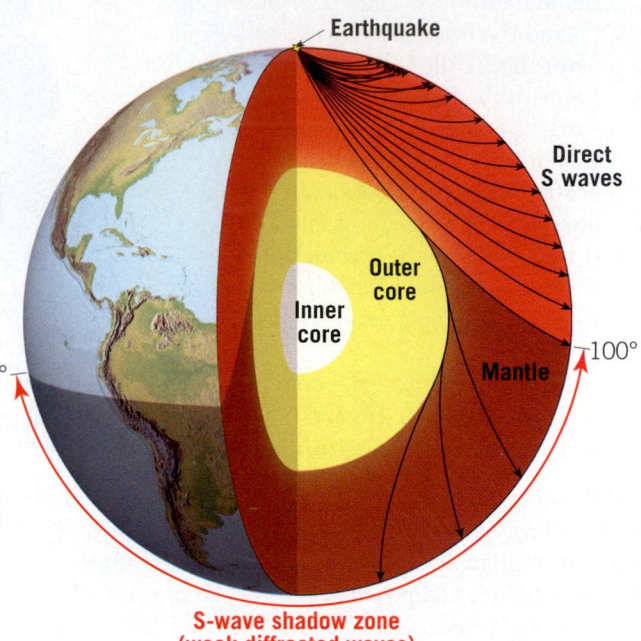

Gutenberg calculated 2900 kilometers (1800 miles) to the core boundary depth, which remains the accepted value.

As Oldham predicted, Earth's core exhibits markedly different properties from the mantle above, which causes considerable refraction of P waves—similarly to how light is refracted as it passes from air to water. In addition, because the outer core is liquid iron, it blocks the transmission of S waves, which do not travel through liquids.

The locations of the P- and S-wave shadow zones and how their paths are affected by the core are shown in Figure 12.10. Whereas some P and S waves still arrive in the shadow zone, they differ greatly from those expected in a planet without a core.

Core

The composition of the **core** is thought to be an iron–nickel alloy, with minor amounts of oxygen, silicon, and sulfur. Because of the extreme pressure found in the core, this iron-rich material has an average density of more than 10 g/cm³ and is about 13 g/cm³ (13 times the density of water) at Earth's center.

The core accounts for only about one-sixth of Earth's volume but one-third of its mass. This difference is because the core is composed mostly of iron, which has the greatest density of the common elements. Furthermore, the core consists of two zones that have markedly different strengths, the *outer* and *inner core*.

The Outer Core   The **outer core** is a liquid iron-rich layer 2270 kilometers (1410 miles) thick. The nature of

the outer core was discovered when researchers noticed that P wave velocities drop dramatically as they cross the core–mantle boundary, and S waves do not penetrate the outer core. Because S waves do not pass through liquids, their absence in the outer core indicates its liquid state. The movement of metallic iron within this zone generates Earth's magnetic field.

The Inner Core   At Earth's center lies the **inner core**, a solid dense sphere with a radius of 1216 kilometers (754 miles). Because the inner core is a sphere, whereas Earth's other layers are shells, drawings make the inner core appear much larger than it really is (see Figure 12.1). The inner core is actually relatively small, only 1/142 of the volume of Earth (less than 1 percent). Despite its higher temperature, the iron in the inner core is solid due to the immense pressures that exist in the center of the planet.

The inner core did not exist early in Earth's history, when our planet was hotter. However, as Earth cooled, iron began to crystallize at the center to form the solid inner core. Even today, the inner core continues to grow larger—at the expense of the outer core.

Separated from the mantle by the liquid outer core, the solid inner core is free to rotate independently from Earth's outer layers. It is thought that the inner core rotates faster than the crust and mantle, lapping them every few hundred years (**Figure 12.11**).

A recent study found a distinct sphere within the inner core that is about one-half the diameter of the inner core. While the iron crystals in the outer-inner core align in roughly a north–south direction, those in

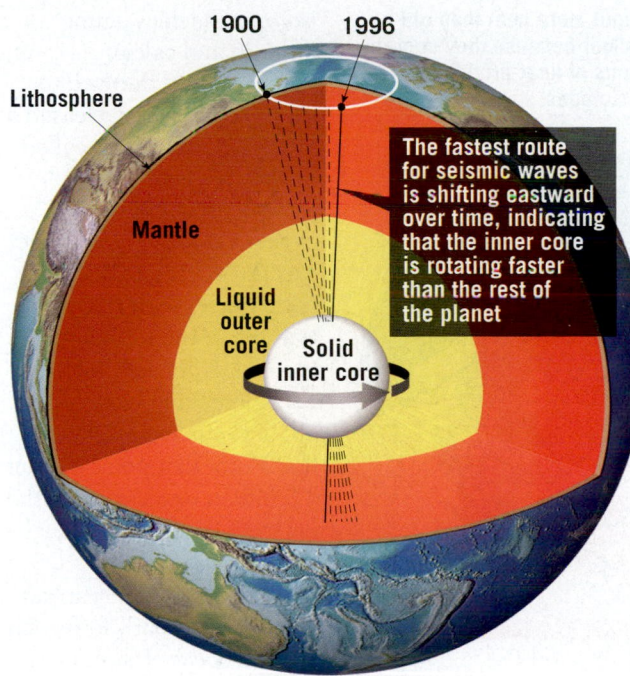

1900   1996

Lithosphere

Mantle

Liquid
outer
core

Solid
inner core

The fastest route
for seismic waves
is shifting eastward
over time, indicating
that the inner core
is rotating faster
than the rest of
the planet

Figure 12.11
**The solid inner core moves independently of Earth's other
layers** Slight variations in the travel times of seismic waves
through the core, measured over many decades, suggest that
the inner core actually rotates faster than the mantle. The
reason for this is not yet understood.

the inner-inner core align in an east–west direction. This
finding implies that some dramatic change may have
occurred during the formation of the inner core to flip its
orientation.

**Discovering Boundaries: The Inner Core–Outer Core
Boundary**   The boundary between the solid inner core
and liquid outer core was discovered in 1936 by Danish
seismologist Inge Lehman. By examining seismograph
records, Lehman discovered that some P waves were
strongly refracted (bent) by a sudden increase in seismic
velocities at a boundary within Earth's core. The waves
she observed were bent enough to arrive within the
P-wave shadow zone, as shown in Figure 12.10A.

12.3 | Concept Checks

1. How do continental crust and oceanic crust
   differ?

2. What is the Moho? How were seismic waves used
   to discover the Moho and determine its depth?

3. List and briefly describe the composition and
   physical properties of the layers of the mantle.

4. How are Earth's inner and outer cores different?
   How are they similar?

12.4 | Earth's Temperature

**Describe the processes of heat transfer that operate within Earth's interior and
where each of these processes dominates.**

Deciphering Earth's temperature structure and how temperature changes with depth are important for determining the movements of rock within our planet. As you are probably aware, heat
flows from hotter regions toward colder regions. Earth is about 5500°C at its center and 15°C at its
surface. As a result, heat flows toward Earth's surface.

The rate at which Earth is cooling can be estimated
by determining the rate at which heat escapes Earth's
surface—a mere 87 milliwatts per square meter. At
this rate, it would require all the energy emitted from
about 690 square meters, roughly the size of a baseball
diamond, to power one 60-watt light bulb. However,
because Earth's surface is so large, heat leaves at a rate
about three times greater than the world rate of energy
consumption.

Figure 12.12 illustrates that heat does not escape
Earth's surface at the same rate in all locations. Heat
flow is highest near mid-ocean ridges, where hot magma
is consistently rising toward the surface. The rate of heat
flow is also high in continental regions, where the rocks
are enriched in radioactive isotopes. Heat flow is lowest in
the deep abyssal plains, which are areas of old, cold oceanic seafloor.

How Did Earth Get So Hot?   Like all other planets in our solar system, Earth has experienced two
thermal stages (Figure 12.13). The first stage occurred
during Earth's formation and lasted about 50 million
years—a relatively short time span in geologic terms—
and involved a rapid increase in internal temperature.
The second stage has been the very slow process of
cooling over the remaining 4.5 billion years of Earth
history.

Several factors contributed to the early increase in
temperature. Earth formed through a very violent process involving the collisions of countless planetesimals
("baby planets") during the birth of our solar system.
With each collision, the kinetic energy of motion was
converted into thermal energy. As the early Earth grew
in size, its temperature rapidly increased. Our young
planet also contained many short-lived radioactive

Figure 12.12
Rate of heat flow at Earth's surface A map of the rate of heat flow out of Earth as it gradually cools over time, measured in milliwatts per square meter.

Earth loses most of its heat near mid-ocean ridges, where magma rises to fill the cracks formed when tectonic plates pull apart.

Continents emit more heat than old oceanic seafloor because they contain higher amounts of heat-producing radioactive isotopes.

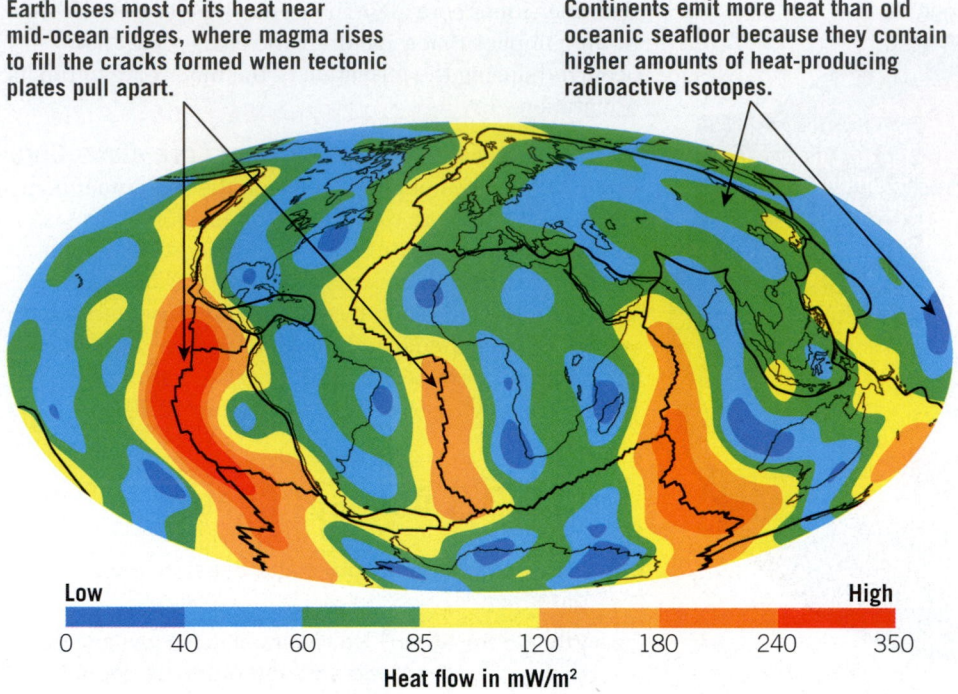

Low   High

| 0 | 40 | 60 | 85 | 120 | 180 | 240 | 350 |

Heat flow in mW/m²

isotopes, such as aluminum-26 and calcium-41. As these isotopes decayed to stable forms, they released a great deal of energy, called *radiogenic heat*.

Another significant event that heated our planet was the collision of a Mars-sized object with Earth that led to the formation of the Moon. At that time the entire core and most, if not all, of the mantle was molten. From that point—about 4.5 billion years ago—to the present, Earth has gradually cooled.

If Earth's only heat had come from its early formation and the decay of short-lived radioactive isotopes, our planet would have cooled to a frozen cinder long ago. However, the mantle and crust also contain long-lived radioactive isotopes that keep our planet cooking as if on a slow burner. As shown in Table 9.1 (page 290), the half-lives of the four main isotopes—uranium-235, uranium-238, thorium-232, and potassium-40—are billions of years long.

Radioactivity plays two vital roles in geology. It is the means for determining the ages of rocks, as discussed in Chapter 9. More importantly, it is the source of radiogenic heat that drives mantle convection and, ultimately, the movement of Earth's tectonic plates.

SmartFigure 12.13
Earth's thermal history through time (https://goo.gl/roXiv2)

Tutorial

The young Earth was hot because of the heat generated by countless collisions with planetesimals, and because of heat released by the decay of short-lived radioactive isotopes.

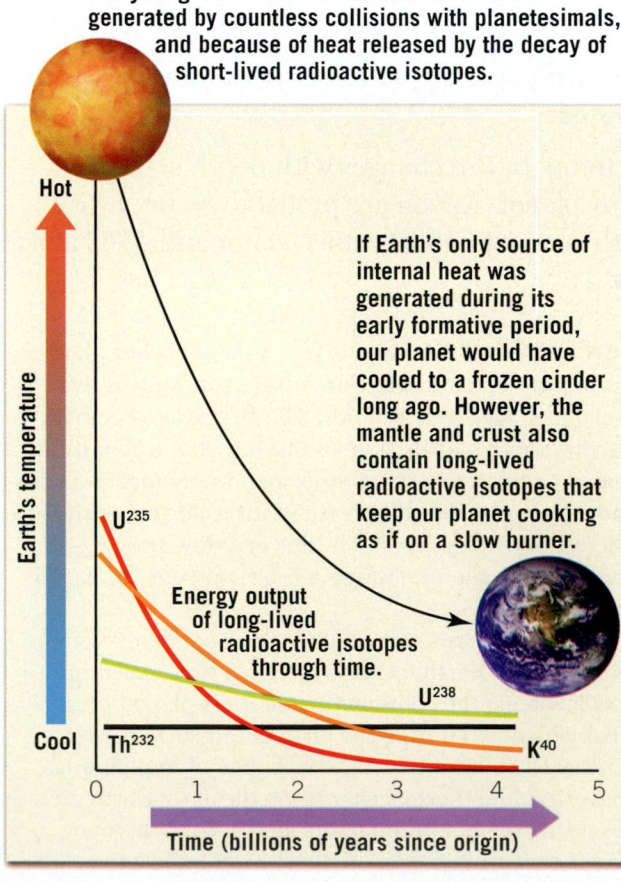

If Earth's only source of internal heat was generated during its early formative period, our planet would have cooled to a frozen cinder long ago. However, the mantle and crust also contain long-lived radioactive isotopes that keep our planet cooking as if on a slow burner.

Energy output of long-lived radioactive isotopes through time.

Hot

Earth's temperature

U^{235}

U^{238}

Cool Th^{232} K^{40}

| 0 | 1 | 2 | 3 | 4 | 5 |

Time (billions of years since origin)

Heat Flow

Heat travels from Earth's interior to space via three different mechanisms: *convection*, *conduction*, and *radiation*. The layers where these mechanisms contribute significantly to the outward flow of thermal energy are shown in **Figure 12.14**. Only two of these processes, convection and conduction, operate within Earth's interior. The third, radiation, transports heat away from Earth's surface and eventually back to space.

Convection The transfer of heat in a fluid-like manner, where hot materials displace cooler materials (or vice versa), is called **convection**. It is the primary means of heat transfer within Earth. You are familiar with convection if you have ever watched boiling water in a pot. The water appears to be rolling—rising in the middle and sinking down the sides of the pot. This pattern, called a *convection cycle*, occurs within Earth's mantle and outer core, and possibly within the inner core as well.

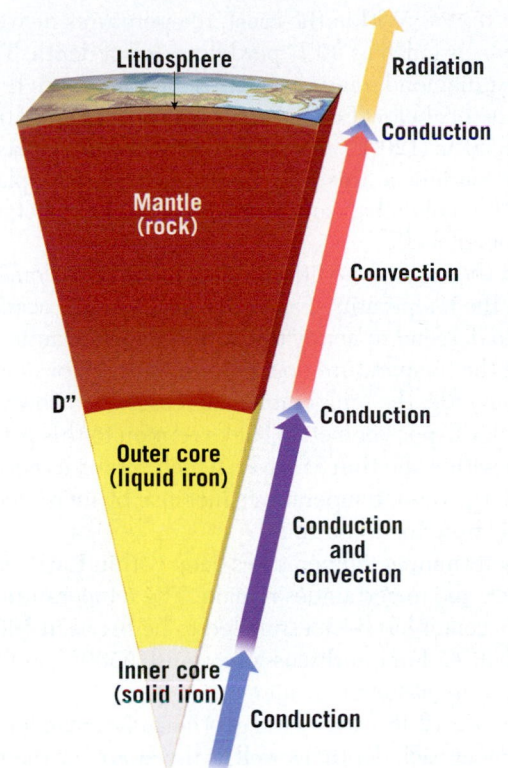

Figure 12.14

Dominant types of heat transfer at various depths Heat travels from Earth's interior to the surface through the processes of convection and conduction. However, Earth ultimately loses its heat to space through radiation.

Convection occurs because of several factors: thermal expansion, gravity-induced buoyancy, and fluidity. When water at the bottom of a pot is heated, it expands and rises, replacing the cooler, denser water at the top. Gravity is the driving force for convection. Hot water is less dense (more buoyant) than cold water, so it rises, and the cold water sinks to take its place. If you tried to boil water in outer space, where there is no strong gravity, you would find that the pot of water would not convect.

Materials must also be weak enough to flow. Scientists usually measure a material's fluidity in terms of its *resistance* to flow, called its **viscosity**. Water flows easily and has a low viscosity. The liquid iron of Earth's outer core likely has a viscosity similar to that of water, and it convects easily as well. Materials that are viscous do not

flow as easily as water but can still flow. For example, ketchup flows, although it is thousands of times more viscous than water. Rock in the lower mantle is 10 trillion trillion (10^{25}) times more viscous than water, but yet it flows.

Temperature differences between the top and bottom of a convection cycle determine how vigorous the convection will be. Earth's surface is extremely cold compared to the interior, so newly formed oceanic lithosphere cools rapidly (**Figure 12.15**). This causes oceanic lithosphere to contract, become denser and heavier, and, in time, sink back into the mantle at subduction zones. As these cold sinking slabs descend, they absorb heat. Some of this crustal material eventually becomes warm and buoyant enough to return to the surface. Cool oceanic lithosphere can therefore be thought of as the top of the mantle convection cycle, whereas rising mantle rock below spreading centers and the warm buoyant rocks called **mantle plumes** are the upward-flowing arms of this convection cycle (see Figure 12.15).

Convection sometimes occurs when differences in density arise through chemical rather than thermal means. *Chemical convection* is an important mechanism operating in the outer core. As iron crystallizes and sinks to form the solid inner core, it leaves behind a molten material that contains a higher percentage of lighter elements. Because this liquid is more buoyant than the surrounding iron-rich material, it rises and contributes to convective flow in the outer core.

Conduction   The flow of heat *through a material* is called **conduction**. Heat conducts in two ways: (1) through the collisions of atoms and (2) through the flow of electrons. In rocks, atoms are locked in place but are constantly oscillating. If one side of a rock is heated, the atoms on that side will oscillate more energetically. This increases the intensity of the collisions with neighboring atoms, and in a domino effect, the energy

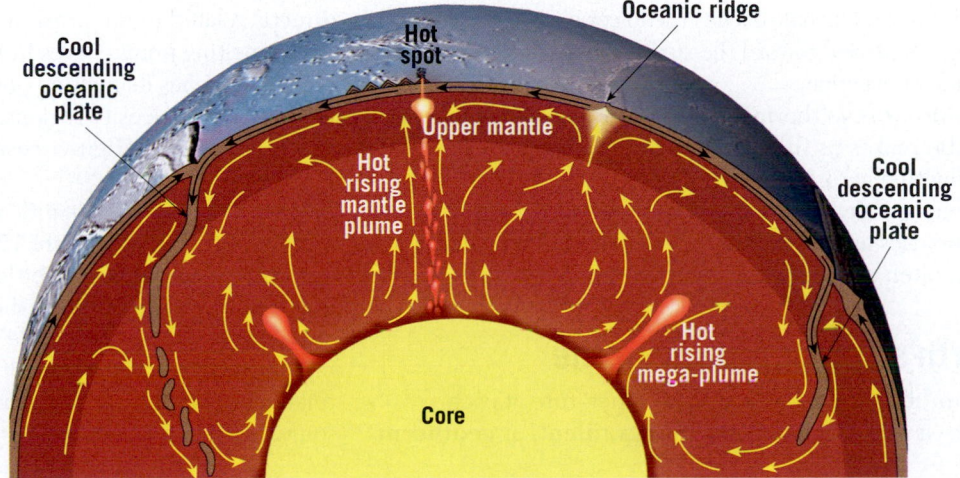

Figure 12.15

Whole-mantle convection According to this model, the entire mantle is in motion, driven by the sinking of cold oceanic lithosphere back into the deep mantle. This is like stirring a pot of stew with downward strokes of a spoon. The upward flow of rock likely occurs through a combination of mantle plumes and a broad return flow of rock to replace the oceanic lithosphere that leaves the surface at subduction zones. Not all scientists agree with this model.

slowly propagates through the rock. Conduction occurs much more quickly in metals than in rocky substances. Although the atoms in metals are also locked in place, some of their electrons are free to move through the material, and these electrons can carry heat quickly from one side of a metallic object to another.

Materials conduct at vastly different rates. For example, heat conducts about 40,000 times more easily through a diamond than through air. Most rocks are poor conductors of heat, so conduction is not an efficient way to move heat through most of Earth. However, it is an important mechanism in places such as the lithosphere, the D″ layer, and the core.

Heat Flow in Earth's Interior   The dominant types of heat transfer by layer in Earth's interior are illustrated in Figure 12.14. Conduction is thought to be the most important process in the solid inner core. When heat conducts from the inner core to the outer core, convection begins to play a more significant role in carrying heat to the top of the core.

Convective flow in the outer core is thought to be driven by three main mechanisms:

1. As heat conducts out of the core into the overlying mantle, the material in the outermost core cools, becomes denser, and sinks. This is a form of top-down, thermally driven convection.
2. Crystallization of solid iron near the bottom of the outer core sinks to form the inner core and leaves behind fluids that are less dense because they are depleted of iron. As this buoyant fluid rises, away from the inner core boundary, it drives a type of chemical convection.
3. There are thought to be radioactive isotopes, such as potassium-40, within the core that could provide additional heat to drive thermal convection.

The transfer of energy from the core to the mantle occurs by conduction because the iron-rich material of the core is too dense to intrude (convect) into the less-dense mantle rocks. For thermal energy to leave the core, it must conduct across the core–mantle boundary and up through the D″ layer. Once it reaches the lower mantle, thermal energy is carried toward the surface through slow, sluggish mantle convection.

Most of the thermal energy that reaches the upper mantle makes its final journey to the surface by slowly conducting across the stiff, rigid lithosphere. The remaining energy is carried to the surface along divergent plate boundaries and other sites of volcanic activity where rising molten rock erupts.

Earth's Temperature Profile

The profile of Earth's average temperature at each depth is called the **geothermal gradient**, or **geotherm**

(**Figure 12.16A**). Within the crust, temperatures increase rapidly—as much as 30°C per kilometer of depth. The deepest diamond mines in South Africa are more than 3 kilometers below Earth's surface, where temperatures exceed 50°C (120°F). However, temperature increases do not continue at this rapid rate; if they did, our planet would be molten ball surrounded by a 100-kilometer-thick outer shell.

At the base of the lithosphere, about 100 kilometers down, the temperature is roughly 1400°C. You would need to descend to nearly the bottom of the mantle before the temperature doubled to 2800°C. For most of the mantle, the temperature increases very slowly— about 0.3°C per kilometer. The exception to this pattern is within the thin D″ layer, which acts as a thermal boundary, where temperatures increase by more than 1000°C from top to bottom.

Determining temperatures deep within Earth is difficult, and uncertainties remain. The temperature at Earth's center has been estimated to be between 5000° and 8000°C. For our discussion, we use 5500°C as the temperature at Earth's center.

Figure 12.16A shows the geotherm (average temperature at each depth) as well as the curve for the melting point of material at each depth. Note that the melting point curves generally increase gradually with depth as a result of the continual increase in pressure exerted by the overlying material. Squeezing a material makes it more difficult to melt because in their liquid form, materials usually take up more volume than in their solid form. As a result, higher pressures result in higher melting temperatures.

Considered together, the geotherm and the melting point curve are valuable tools for investigating the behavior of Earth's materials. In layers where the geotherm (temperature at depth) is greater than the melting temperature, the material is molten. As shown in Figure 12.16A, this situation occurs in the outer core.

The relationship between the geotherm and the melting temperature not only determines whether a material is molten but also indicates its stiffness, or viscosity. Notice how viscosity, shown in **Figure 12.16B**, is directly related to the proximity of the geotherm curves to the melting point curves in Figure 12.16A. When rock approaches its melting point, it begins to soften and weaken. Low-viscosity regions, such as the asthenosphere and D″, are weak. High-viscosity regions, such as the lithosphere, are rigid.

Most of the lower mantle is very stiff, so rock moves sluggishly there (see Figure 12.16B). Researchers have determined that convective flow is several times slower in the lower mantle than in the upper mantle. However, at the very base of the mantle, temperature increases rapidly with depth. As a result, rock in the D″ layer is relatively weak, flows more easily, and may experience some melting.

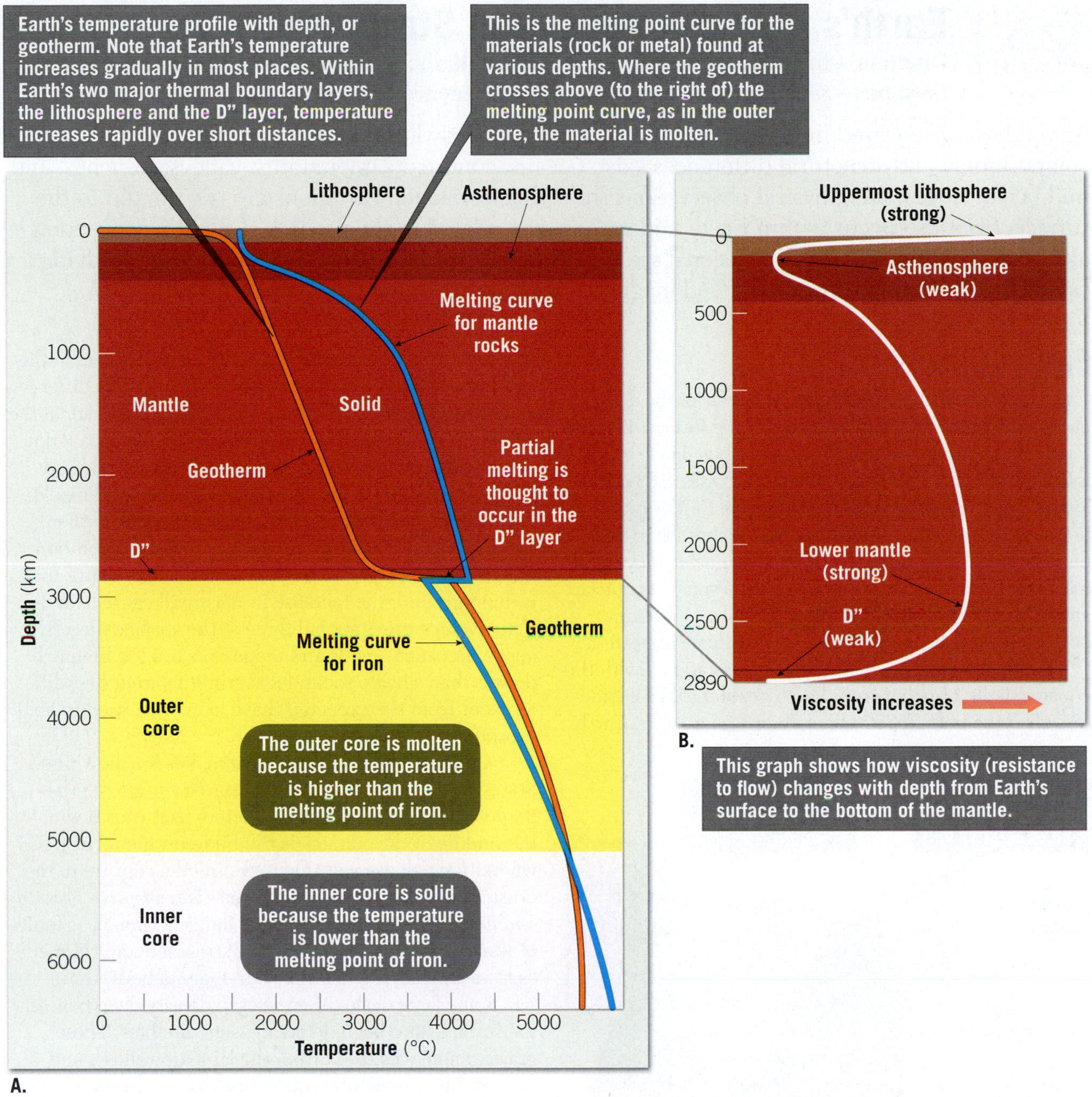

A.

Earth's temperature profile with depth, or geotherm. Note that Earth's temperature increases gradually in most places. Within Earth's two major thermal boundary layers, the lithosphere and the D" layer, temperature increases rapidly over short distances.

This is the melting point curve for the materials (rock or metal) found at various depths. Where the geotherm crosses above (to the right of) the melting point curve, as in the outer core, the material is molten.

Lithosphere   Asthenosphere

Melting curve for mantle rocks

Mantle   Solid

Geotherm

Partial melting is thought to occur in the D" layer

D"

Melting curve for iron   Geotherm

Outer core

The outer core is molten because the temperature is higher than the melting point of iron.

Inner core

The inner core is solid because the temperature is lower than the melting point of iron.

Depth (km): 0, 1000, 2000, 3000, 4000, 5000, 6000

Temperature (°C): 0, 1000, 2000, 3000, 4000, 5000

B.

Uppermost lithosphere (strong)

Asthenosphere (weak)

Lower mantle (strong)

D" (weak)

Viscosity increases

Depth: 0, 500, 1000, 1500, 2000, 2500, 2890

This graph shows how viscosity (resistance to flow) changes with depth from Earth's surface to the bottom of the mantle.

SmartFigure 12.16
The geotherm and melting point curve are valuable tools for investigating Earth materials at depth
If you compare these two graphs, you can see that rocks are weakest and flow more easily at depths where the temperatures of rocks are close to melting (the asthenosphere and D" layer). High viscosities, like those of the crust and lithosphere, show rock that is stiffer and flows less easily. (https://goo.gl/w6wQl3)

Tutorial

In the core, temperature increases much more slowly than pressure. From the core–mantle boundary to Earth's center, the temperature increases by only about 40 percent, or from about 4000° to 5500°C. Pressure, however, nearly triples, increasing from 1.36 to 3.64 megabars (1 megabar is 1 million times greater than standard atmospheric pressure). Although the temperature is cooler in the outer core than it is in the inner core, the outer core remains a liquid because it is under less pressure. Conversely, iron in the inner core remains solid at these high temperatures because its melting temperature increases dramatically at these extreme pressures.

12.4   Concept Checks

1. In general, how does temperature change between Earth's surface and its core?

2. What were the two main sources of Earth's original internal heat?

3. List the mechanisms of heat transfer that operate inside Earth.

4. Why is the asthenosphere weaker than the lithosphere?

5. Explain why the inner core is solid even though it is hotter than the liquid outer core.

12.5 | Earth's Three-Dimensional Structure

Discuss what studies of Earth's gravity, seismic tomography, and Earth's magnetic field have revealed about variations in the layers of our planet.

As you have discovered in this chapter, Earth is not perfectly layered. At the surface, a variety of compositional and structural differences exist: oceans, continents, mountains, valleys, trenches, and mid-ocean ridges. Geophysical observations show that horizontal variations are not limited to the surface; they also occur within Earth and are directly related to the process of mantle convection and plate tectonics. Three-dimensional structures within Earth have been identified by studying variations in Earth's gravitational and magnetic fields and imaging called *seismic tomography*.

Earth's Gravity

Earth's rotation is the most significant cause for the differences in the force of gravity observed at the surface. Because Earth rotates around its axis, the acceleration due to gravity* is less at the equator (9.78 m/s^2) than at the poles (9.83 m/s^2). Two reasons account for this phenomenon. Earth's rotation causes a centrifugal force that is in proportion to the distance from the axis of rotation. In a manner similar to the force that throws you sideways in a vehicle going too quickly around a curve or corner, centrifugal force acts to throw objects outward at the equator, where the force is greatest.

Earth's shape is also affected by its rotation—with the equator slightly further from Earth's center (6378 kilometers) than the poles (6357 kilometers) (**Figure 12.17**). Earth, therefore, is not a perfect sphere but instead bulges at the equator—a shape called an *oblate ellipsoid*. This difference causes the force of gravity to be slightly weaker at the equator than at the poles because gravitational attraction is less when objects are further apart. In fact, your body weight is 0.5 percent less at the equator than at the poles.

Gravity measurements show that there are other variations that cannot be explained by Earth's rotation. For instance, when a large body of unusually dense rock is underground, the increase in mass will cause a larger-than-average gravitational force at the surface directly above. Because metals and metal ores tend to be much denser than silicate rocks, local *gravity anomalies* (differences from the expected) have long been used to help prospect for ore deposits.

A map of regional gravity anomalies for the United States is shown in **Figure 12.18**. A narrow *positive gravity anomaly* (stronger than expected) that runs down the middle of the country is the mid-continent rift (red), where thick, dense volcanic rocks filled a rupture in the crust more than 1 billion years ago. The *negative gravity anomaly* (blue) in the Basin and Range region is a result of warm, low-density crust being stretched and thinned as it was intruded by hot buoyant magma bodies.

Some large-scale differences in density deep beneath the surface have been detected using satellites. These gravity anomalies result from the large upwellings and downwellings of mantle convection. Areas of upwelling are associated with hot mantle plumes, whereas downwelling occurs where cold oceanic slabs descend into the mantle.

Seismic Tomography

The three-dimensional changes in composition and density that are detected with gravity measurements can also be viewed using seismology. The technique, called **seismic tomography**, involves collecting signals from many different earthquakes recorded at many seismograph stations, in order to "see" all parts of Earth's interior. Seismic tomography is similar to medical tomography, in which doctors use technology such as CT scans to make three-dimensional images of humans' internal organs.

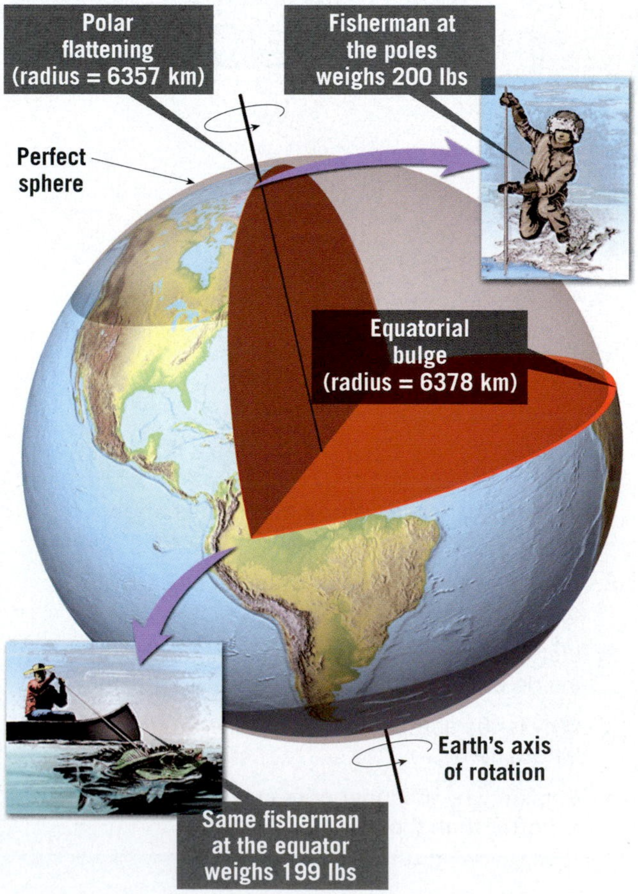

Figure 12.17
Earth is not a sphere but an oblate spheroid
Because Earth rotates, it bulges at the equator and flattens at the poles.

Polar flattening (radius = 6357 km)

Fisherman at the poles weighs 200 lbs

Perfect sphere

Equatorial bulge (radius = 6378 km)

Earth's axis of rotation

Same fisherman at the equator weighs 199 lbs

*The force of gravity causes objects, such as apples, to accelerate as they fall to the ground, hence the expression "acceleration due to gravity."

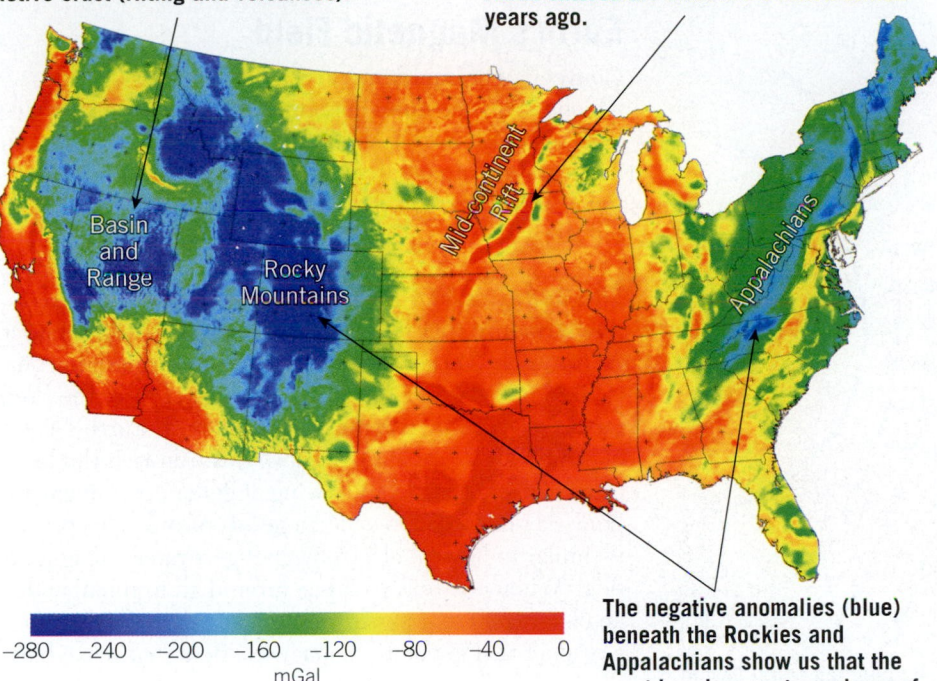

The negative anomaly (blue) in the Basin and Range Province is the result of hotter, less dense, and tectonically active crust (rifting and volcanoes).

The narrow positive anomaly (red) that runs in a line down the middle of the country is the mid-continent rift, where dense volcanic rocks entered the crust more than a billion years ago.

Figure 12.18
Gravity anomalies beneath the continental United States

The negative anomalies (blue) beneath the Rockies and Appalachians show us that the crust has deep roots made up of less dense rock beneath the mountains.

−280 −240 −200 −160 −120 −80 −40 0
mGal

Figure 12.19
A seismic tomographic slice, showing the structure of the mantle
Colors show variations in the speed of S waves from their average value. Cool, strong rock produces fast S waves and is shown in blue, while warm, weak rock produces slow S waves and is shown in red.

Seismic tomography identifies regions where P or S waves travel faster or slower than average for a particular depth. These *seismic velocity anomalies* are then interpreted as variations in material properties such as temperature, composition, mineral phase, or water content. For instance, increasing the temperature of rock about 100°C can decrease S-wave velocities about 1 percent, so images from seismic tomography are often interpreted in terms of temperature variations.

A seismic tomography cross section for the mantle centered beneath North America is shown in **Figure 12.19**. Regions where waves travel slower than average (negative anomalies) are red, and regions where waves travel faster than average (positive anomalies) are blue. Significant patterns can be observed in this diagram. For example, the lithosphere beneath the interiors of North America and Africa exhibits faster seismic velocities than oceanic lithosphere because continental lithosphere is older and has been cooling for billions of years. Seismic imaging also shows that continental lithosphere (deep blue areas) can be quite thick, extending more than 300 kilometers (180 miles) into the mantle. Conversely, oceanic ridges such as the Mid-Atlantic Ridge exhibit slow seismic velocities because they are extremely hot (see Figure 12.19).

In the mantle beneath North America, you can see a sloping zone of fast seismic velocities (blue/green) representing a sheet of descending oceanic lithosphere known as the Farallon plate. The segment of this former slab of seafloor seen in Figure 12.19 is currently sinking and warming as it moves toward the core–mantle boundary. Given enough time, this slab may become hot and buoyant enough to begin to rise back to the surface.

The western United States is tectonically active, making that portion of the continent warmer and weaker (shown by the color red), which slows S waves.

The large blue structure extending far beneath North America is likely a sheet of cold, dense, ancient Pacific seafloor that is sinking toward the base of the mantle.

Older portions of continents, North America and Africa, are cold and strong—the blue color indicates fast S wave speeds.

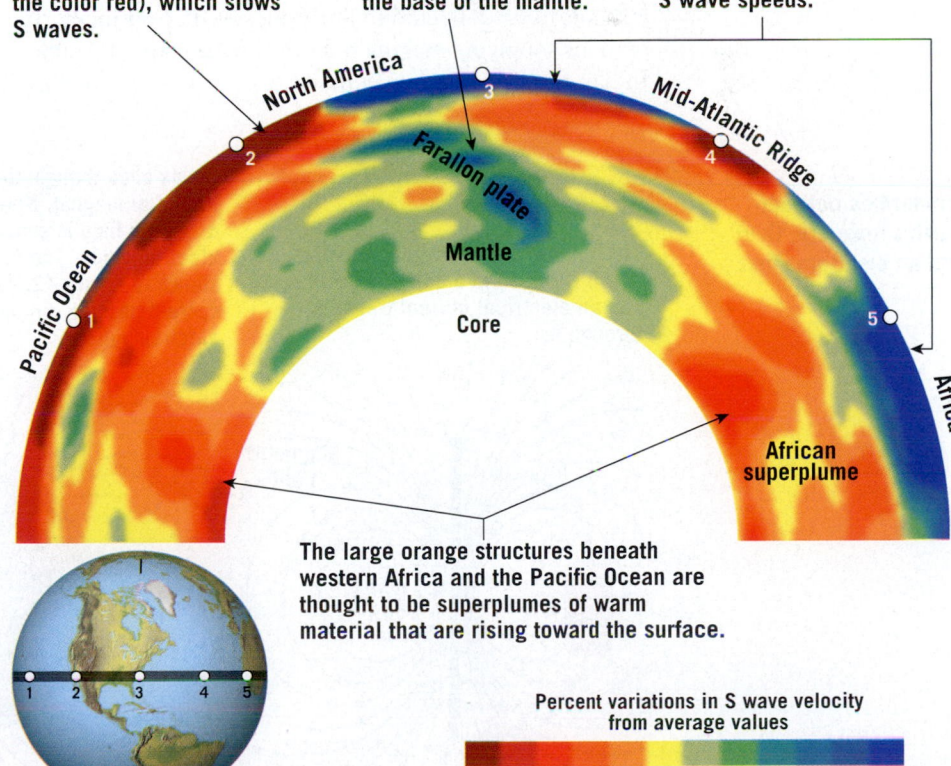

The large orange structures beneath western Africa and the Pacific Ocean are thought to be superplumes of warm material that are rising toward the surface.

Percent variations in S wave velocity from average values

−1.5% −1.0% −0.5% 0 0.5% 1.0% 1.5%
Slow   Fast

Figure 12.20
How Earth's magnetic field is generated in the liquid, iron-rich outer core

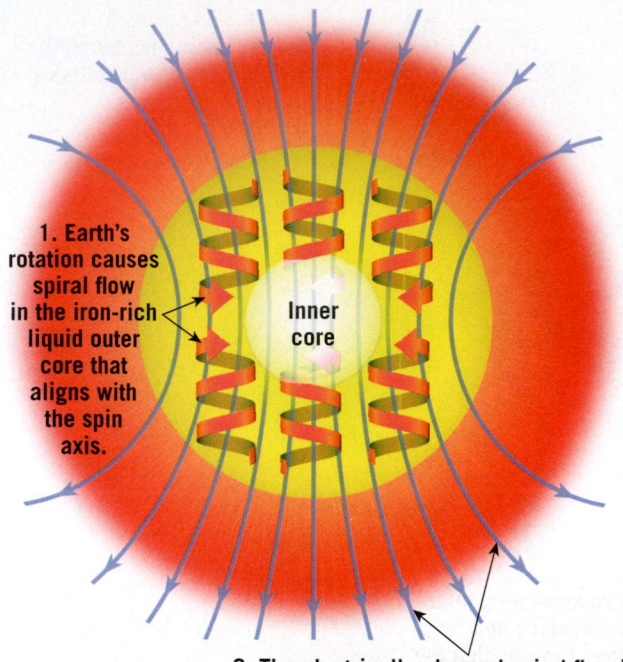

1. Earth's rotation causes spiral flow in the iron-rich liquid outer core that aligns with the spin axis.

Inner core

2. The electrically-charged spiral flow in the outer core generates Earth's magnetic field. This is similar to how an electromagnet works.

The large region of slow seismic velocities beneath Africa (the large reddish orange region at the lower right of Figure 12.19) is called the *African superplume*—a region of upward flow in the mantle. These slow velocities are likely due to both unusually high temperatures and rock that is highly enriched in iron. The rising rock cannot easily break through the thick African crust, so it seems to be deflected to both sides of the continent, perhaps supplying magma to both the Mid-Atlantic and Indian Ocean spreading centers.

Earth's Magnetic Field

Convection of liquid iron in the outer core is vigorous, which makes the outer core *appear* uniform when viewed with seismic waves. In reality, however, patterns of flow in the outer core create variations in Earth's magnetic field that are measurable at the surface and tell a different story.

The Geodynamo   Earth's **magnetic field** has a complex structure that reflects the processes that generate it. As iron-rich fluid in the outer core rises, its path becomes twisted because of Earth's rotation. As a result, the fluid moves in spiraling columns that align with Earth's axis of rotation (**Figure 12.20**). Because this iron-rich fluid is electrically charged and flowing, it generates a magnetic field—a phenomenon called a **geodynamo**. This process is similar to how an electromagnet generates a magnetic field. When a wire is wrapped around an iron nail and an electric current passes through it, the nail becomes a magnet that generates a magnetic field that resembles the one surrounding a bar magnet (**Figure 12.21A**). This type of magnetic field is called a *dipolar field* because it has two poles (north and south magnetic poles). The magnetic field that emanates from Earth's outer core has the same dipolar form (**Figure 12.21B**).

However, the convection in the outer core is considerably more complex. More than 90 percent of Earth's magnetic field is dipolar, but the remainder is a result of more complicated patterns of convection. In addition, some of the features of Earth's magnetic field change over time. For centuries, sailors have used compasses to navigate, primarily keeping track of the direction toward which compass needles point. From compass observations, it was determined that the positions of the magnetic poles gradually change. Understanding this phenomenon requires an examination of how the magnetic field is measured.

Figure 12.21
Similarities between Earth's magnetic field and an electromagnet

This simple electromagnet consists of a nail wrapped in a coil of wire that has an electrical current traveling through it.

It was once thought that Earth's core acted like a large bar magnet. Now scientists think that Earth's magnetic field is similar to that produced by an electromagnet. The cylinders of spiraling liquid iron shown in Figure 12.20 behave like the electric current passing through the coil wire of an electromagnet.

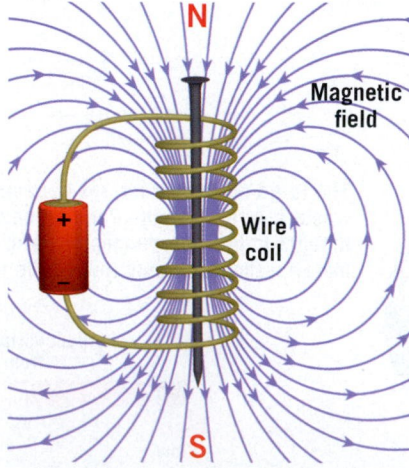

A. Electromagnet (Dipolar field)

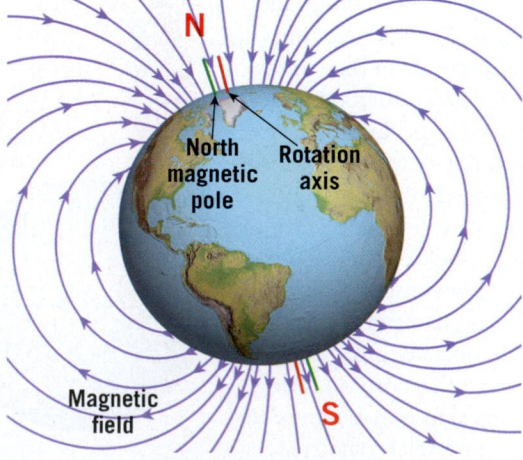

B. Earth's magnetic field (Dipolar field)

Measuring Earth's Magnetic Field and Its Changes   Anywhere on Earth's surface, the direction of the magnetic field is measured with two angles, called *declination* and *inclination*. The declination measures the direction to the magnetic north pole with respect to the direction to the geographic North Pole (Earth's axis of rotation). The inclination measures the downward tilt of the magnetic lines of force at any location—what a

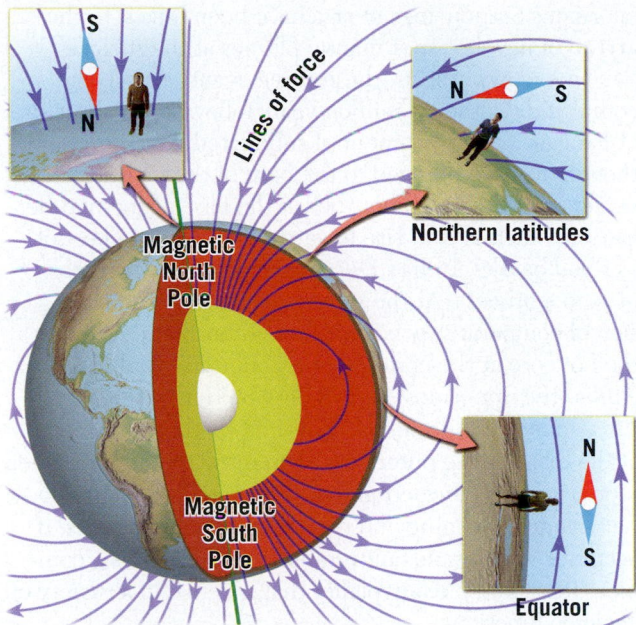

Figure 12.22
Inclination or dip of the magnetic field at different locations
Although a compass measures only the horizontal direction of the magnetic field (the *declination*), at most locations, the field also dips in or out of the surface at various angles (*inclination*).

Figure 12.23
Locations of the north magnetic pole over time The patterns of convection within the outer core change fast enough that we can see the strength and position of the magnetic field change significantly over our lifetimes.

compass would show if tilted on its side. At the magnetic north pole, the field points directly downward, while at the equator, it is horizontal (**Figure 12.22**). In the central United States it tilts downward at an intermediate angle.

Recent studies have shown that the locations of the magnetic poles change significantly over time. Earth's magnetic north pole was previously located in Canada but moved northward into the Arctic Ocean during the past decade. Currently, it is moving northwestward at over 70 kilometers (40 miles) per year (**Figure 12.23**). While the magnetic north pole has been moving toward the geographic North Pole, the magnetic south pole has been moving *away* from the geographic South Pole, passing from Antarctica to the Pacific Ocean.

Magnetic Reversals   Although the pattern of convection in the core changes over time, causing the magnetic poles to move, the locations of the magnetic poles averaged over thousands of years align with Earth's axis of rotation (geographic poles). One major exception occurs during periods of **magnetic reversals**. At apparently random times, Earth's magnetic field reverses polarity so that the *north* needle on a compass points *south*. (The importance of these reversals in the study of paleomagnetism is described in Chapter 2.) During a reversal, the strength of the magnetic field decreases to about 10 percent of normal, and the locations of the poles begin to wander, going so far as to cross the equator (**Figure 12.24**). When the strength of the magnetic field returns to normal levels, the field is regenerated with reverse polarity. The entire process takes only a few thousand years.

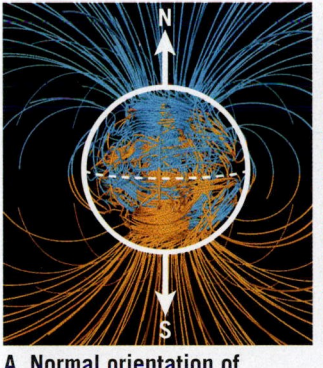
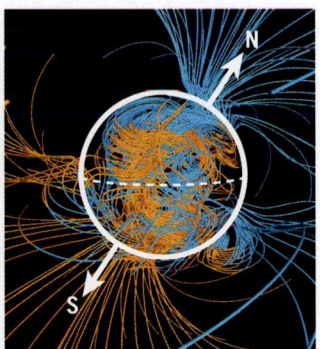

A. Normal orientation of magnetic field

B. Magnetic field weakens and poles begin to wander

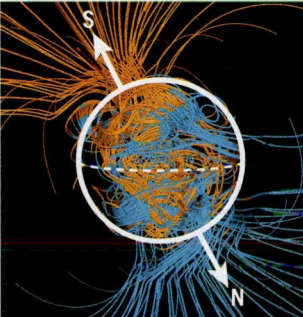
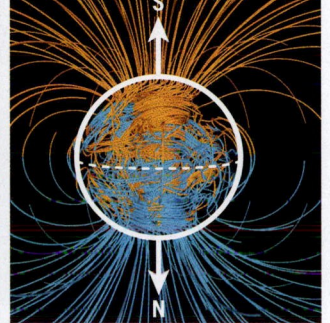

C. Poles wander across the equator

D. Reversal complete with north pole pointing south

Figure 12.24
Computer simulations showing how Earth's magnetic field might reverse direction The white circle represents the core–mantle boundary, and the dashed white line represents the projected position of the equator. The arrows point to the north (N) and south (S) magnetic poles. During a reversal, the strength of the magnetic field weakens, and the poles begin to wander greatly, going so far as to cross the equator. When the strength of the field returns to normal levels, the field is regenerated with reverse polarity.

The rate at which the magnetic field reverses provides evidence that convection patterns in the outer core change over relatively short time spans. This complex process is now being modeled using high-speed computers. In addition, Figure 12.24 illustrates how the magnetic field lines twist in a complex manner before returning to a more uniform, simpler dipolar pattern.

The existence of magnetic reversals has been extremely important to geoscientists in providing the foundation for the theory of plate tectonics. However, magnetic reversals have potentially harmful consequences for Earth's land dwellers. An atmospheric magnetic layer, known as the *magnetosphere*, surrounds our planet and protects Earth's surface from bombardment by ionized particles, called *solar wind*, emitted by the Sun. If the strength of the magnetic field decreases significantly during a reversal, the increased amounts of ionized particles reaching Earth's surface could cause health hazards for humans and other life-forms.

Global Dynamic Connections   The layers of planet Earth are not isolated from one another; rather, they are connected by their thermally driven motions. These connected motions do not necessarily interact in a predictable, steady manner; instead, they tend to occur episodically or in pulses. One example is the possible connection between magnetic reversals, hot-spot volcanism, and the breakup of the supercontinent Pangaea.

Pangaea began to break up about 200 million years ago, and the accompanying increase in plate motion resulted in the subduction of large amounts of seafloor. About 80 million years later, the magnetic reversal process shut down, preventing Earth's magnetic field from reversing for 35 million years. During this period, several enormous outpourings of lava have been linked to the arrival of new hot-spot mantle plumes at the surface.

In one hypothesis, these three events are closely connected. In the 80 million years following the breakup of Pangaea, a large amount of subducted lithosphere is thought to have plunged to the base of the mantle. This would have displaced hot rock at the base of the mantle, causing much of it to rise toward the surface and erupt as flood basalts. India's Deccan Traps are an example of such a process. At the same time, the sudden infusion of comparatively cold oceanic lithosphere next to the hot core at the core–mantle boundary would have chilled the uppermost core. This would have increased the temperature gradient—hotter at the bottom, cooler at the top—in the outer core. The result would have been more vigorous convection that prevented the magnetic field from weakening and reversing. This hypothesis, if accurate, is an important reminder that Earth is a complex, churning, pulsing planet that is active in a variety of geologic functions.

12.5   Concept Checks

1. Is the force of gravity the same over Earth's entire surface? Explain why or why not.

2. Give examples of tectonic features of Earth's three-dimensional structure that are revealed by seismic tomography.

3. What is thought to produce Earth's magnetic field?

4. Describe how the positions of the magnetic poles change through time.

12.1 Earth's Internal Structure

Explain how Earth acquired its layered structure.

KEY TERMS phase change

- The layered internal structure of Earth developed due to gravitational sorting of Earth materials early in the planet's history. The densest material sunk to the center, and the least dense material rose to the exterior.

- Within layers such as the mantle, atoms in a mineral may rearrange into a more compact crystal structure. Given sufficient pressure, such a mineral phase change can pack the same amount of matter into a smaller volume. This results in increased density. For example, olivine undergoes a phase change to become a more compact mineral at a depth of about 400 kilometers (250 miles).

Q Arrange the following Earth layers in order from most dense to least dense: crust, atmosphere, core, mantle, oceans.

12.2 Probing Earth's Interior

Describe how seismic waves are used to probe Earth's interior.

KEY TERMS seismic wave, reflection, refraction, diffraction

- Seismic waves allow geoscientists to "look" into Earth's interior, which would otherwise be invisible to scientific investigation. Like the X-rays used to image human bodies, seismic waves generated by large earthquakes reveal details about Earth's layered structure.
- Seismic waves travel most rapidly through stiff, elastic materials, so when P or S waves cross a mineral phase change at depth, seismographs detect their increased velocity. In general, seismic waves follow curved paths because their velocities generally increase with depth. When seismic waves pass through partially molten material, the seismic waves slow down because magma is less rigid and less elastic than solid rock. S waves cannot be transmitted through liquids, but P waves can.
- When seismic waves encounter a different layer, they may either reflect or refract. Reflection is like "bouncing"—a change in the direction of the wave away from the new layer. With refraction, the wave continues into the new layer, but it bends, changing its path to a different trajectory.

Q Explain how S waves "tell" us Earth's outer core is liquid.

12.3 Earth's Layers

List and describe each of the layers of Earth's interior.

KEY TERMS crust, Moho, mantle, upper mantle, lithospheric mantle, lithosphere, asthenosphere, transition zone, lower mantle, D″ layer, shadow zone, core, outer core, inner core

- Earth has two distinct kinds of crust: oceanic and continental. Oceanic crust is thinner, denser, and younger than continental crust. Oceanic crust also readily subducts, whereas the less-dense continental crust does not.
- The base of the crust is the Mohorovičić discontinuity, or Moho. It was discovered because seismic waves travel faster below the Moho (in the mantle) than they do above it (in the crust). Beneath the continents, the Moho lies 25 to 70 kilometers below the surface. In oceanic lithosphere, the Moho occurs at a depth of about 7 kilometers.

- Earth's mantle may be divided by density into upper (3.3 g/cm^3) and lower (5.6 g/cm^3) portions. The uppermost mantle makes up the bulk of rigid lithospheric plates, while a relatively weak layer, the asthenosphere, lies beneath it. There is a transition zone at the bottom of the upper mantle, between 410 and 660 kilometers in depth. The lower mantle lies below the transition zone. At its base, just above the core, is the unusual layer called D″. Seismic wave velocities slow dramatically in the D″ layer, indicating that it may be partially molten. This may be where subducted mantle slabs go before being recycled as mantle plumes.
- The composition of Earth's core is likely a mix of iron, nickel, and lighter elements. The outer core is dense (around 10 times the density of water) and liquid, as S waves cannot pass through it. The inner core is very dense (more than 13 times the density of water) and solid.

12.4 Earth's Temperature

Describe the processes of heat transfer that operate within Earth's interior and where each of these processes dominates.

KEY TERMS convection, viscosity, mantle plume, conduction, geothermal gradient (geotherm)

- Heat flows from Earth's interior outward toward the surface. Heat flow through the crust is higher in places where the crust is thin and where magma is close to the surface. There are two main sources of heat in Earth's interior. First is residual kinetic energy left over from the amalgamation of the planet during the formation of the solar

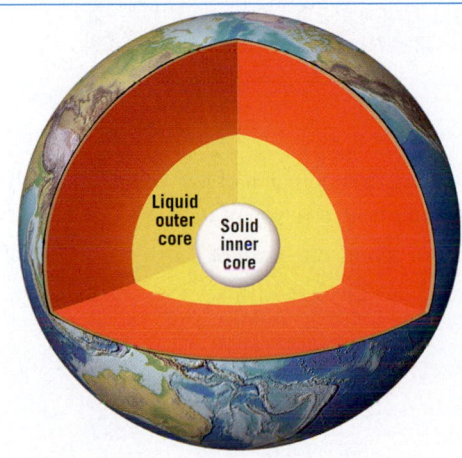

system. The second is radioactivity. Early in Earth's history, short-lived isotopes (now extinct) generated additional heat through radioactive decay. Radioactivity of long-lived isotopes (such as uranium-235 and potassium-40) continues to add heat at a lower rate today.

- Heat flows through Earth's interior via several mechanisms. Convection is the most important mechanism of heat transfer in Earth's interior, occurring in the liquid outer core and in the mantle. Conduction, on the other hand, is the mechanism by which the core heats the mantle. Heat eventually reaches Earth's surface, and some of it radiates into space.

- The geothermal gradient is the profile of Earth's temperature with depth. Temperature increases rapidly relative to depth in both the lithosphere and the D″ layer. Other layers experience a relatively gentle geotherm of 0.3°C per kilometer. The position of the melting point curve relative to the geotherm is a key to the behavior of material in Earth's interior layers. In layers where the geotherm is greater than the melting point curve, the material will be partially or totally molten.

Q How can Earth's outer core be molten if the solid inner core is even hotter? Explain this "paradox."

12.5 Earth's Three-Dimensional Structure

Discuss what studies of Earth's gravity, seismic tomography, and Earth's magnetic field have revealed about variations in the layers of our planet.

KEY TERMS seismic tomography, magnetic field, geodynamo, magnetic reversal

- Earth exhibits not only vertical variations (layering) but horizontal variations as well (differences within a given layer). The composition and thickness of the mantle and core vary from place to place, just like the crust.

- Gravity anomalies occur where rock is more dense (often because it is cold) and less dense (often because it is warm).
- Seismic tomography is a technique for imaging Earth's interior based on whether seismic waves speed up or slow down when they pass through it. Slower waves indicate relatively warm rock, which is likely rising due to convection.
- Convection of the liquid outer core generates a magnetic field that is larger than the entire planet. Over time, Earth's magnetic field changes in both the polarity and orientations of its poles.

Give It Some Thought

1. What are the two major reasons for the increase in density with depth in Earth's interior?

2. The accompanying diagram shows one incoming ray and three possible outgoing ray paths, labeled 1, 2, and 3. Match each of the outgoing ray paths with one of these descriptions:

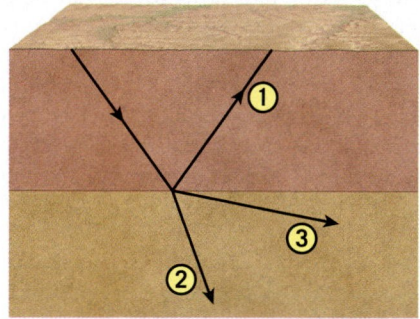

a. Reflected ray
b. Refracted ray going from a region of low velocity to high velocity
c. Refracted ray going from a region of high velocity to low velocity

3. Explain how the core is only one-sixth of Earth's volume, yet it is one-third of Earth's mass.

4. The accompanying diagram shows the internal structures of Earth, Mars, and Earth's Moon. Based on what you know about the composition of the layers shown and how density increases with depth, list these three bodies in order from the most dense to the least dense.

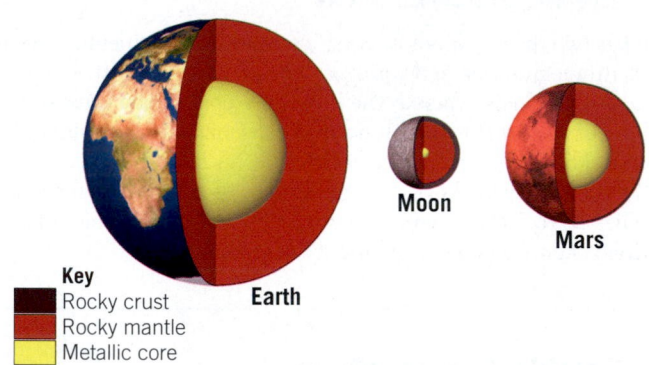

Key
Rocky crust
Rocky mantle
Metallic core

5. Describe how Earth's inner core grows in size.

6. Explain why convection is an inefficient means of heat transfer in materials with high viscosity.

7. Imagine that you are exploring another planet and using seismic data to determine the planet's internal structure. Based on the ray paths for the S waves shown in the accompanying diagram, answer the following questions:

   a. How many distinct layers does this planet appear to have (not including its outer surface)?

   b. Which of these layers is the thickest? Which is the thinnest?

   c. Does this planet have a liquid layer? If so, which one?

8. Use the depth data from this figure to draw a model of Earth's layers. Use a metric measuring tape and a scale of 1 centimeter in the model to represent each kilometer of depth in the actual Earth.

   a. Mark each major boundary.

   b. Take an imaginary "walk" from the surface to the core and describe the changes you would encounter along the way.

9. Earth once rotated much faster than it currently does. How would Earth's shape have been different in the past than it is today?

10. Earthquakes below the Yellowstone caldera originate at very shallow depths, about 4 kilometers on average. Below this depth, the rocks are at about 400°C, too hot and weak to store elastic energy. Based on this data, answer the following questions:

(Photo by Quasarphoto/Fotolia.com)

   a. What is the average geothermal gradient in the first 4 kilometers beneath the Yellowstone caldera, assuming an average surface temperature of 0°C (32°F) and assuming that the temperature at 4 kilometers is 400°C (752°F)?

   b. At about what depth is the groundwater below the Yellowstone caldera hot enough to "boil" and therefore capable of generating a geyser?

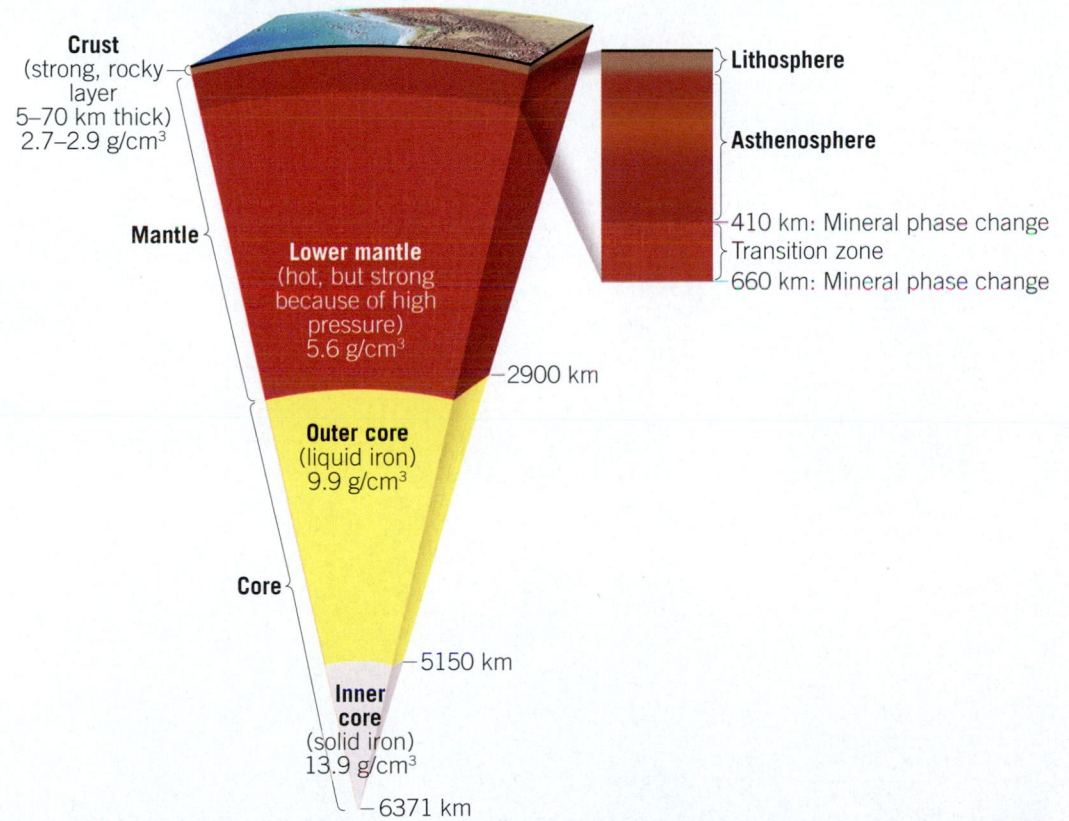

Crust (strong, rocky layer 5–70 km thick) 2.7–2.9 g/cm³

Mantle

Lower mantle (hot, but strong because of high pressure) 5.6 g/cm³

Lithosphere

Asthenosphere

410 km: Mineral phase change
Transition zone
660 km: Mineral phase change

2900 km

Outer core (liquid iron) 9.9 g/cm³

Core

5150 km

Inner core (solid iron) 13.9 g/cm³

6371 km

13

Origin and Evolution of the Ocean Floor

Aerial view of Bora Bora, an island in French Polynesia that is a remanent of an extinct volcano surrounded by a barrier reef. (Photo by Michel Renaudeau/AGE Fotostock)

Each statement represents the primary **LEARNING OBJECTIVE** for the corresponding major heading within the chapter. After you complete the chapter, you should be able to:

13.1 Define *bathymetry* and describe the various bathymetric techniques used to map the ocean floor.

13.2 Compare and contrast a passive continental margin with an active continental margin and list the major features of each.

13.3 List and describe the major features of the deep-ocean basins.

13.4 Sketch and label a cross-sectional view of the Mid-Atlantic Ridge. Explain how a cross section of the East Pacific Rise would look different.

13.5 Write a statement describing how spreading rates affect ridge topography.

13.6 List the four layers of oceanic crust and explain how oceanic crust forms and how it differs from continental crust.

13.7 Outline the steps by which continental rifting results in the formation of new ocean basins.

13.8 Compare and contrast spontaneous subduction and forced subduction.

The ocean is Earth's most prominent feature, covering more than 70 percent of its surface. Yet, prior to the 1950s, information about the ocean floor was extremely limited. With the development of modern instruments, our understanding of the diverse topography of the ocean floor improved dramatically. Particularly significant was the discovery of the global oceanic ridge system, a broad elevated landform that stands 2 to 3 kilometers higher than the adjacent deep-ocean basins and is the longest topographic feature on Earth.

In this chapter we will examine the topography of the ocean floor and look at the processes that generate its varied features.

13.1 | An Emerging Picture of the Ocean Floor

Define *bathymetry* and describe the various bathymetric techniques used to map the ocean floor.

If all water could be drained from the ocean basins, a great variety of features would be observed, including volcanic peaks, deep trenches, extensive plains, linear ridges, and large plateaus. In fact, the topography would be nearly as diverse as that on the continents.

Mapping the Seafloor

The complex nature of ocean-floor topography did not unfold until the historic 3½-year voyage of the HMS *Challenger* (**Figure 13.1**). From December 1872 to May 1876, the *Challenger* expedition made the first comprehensive study of the global ocean ever attempted. During the 127,500-kilometer (79,200-mile) voyage, the ship and its crew of scientists traveled to every ocean

SmartFigure 13.1
HMS *Challenger*
The first systematic bathymetric survey of the ocean was made aboard the HMS *Challenger* during its historic 3½-year voyage. (Photo courtesy of the Library of Congress) (https://goo.gl/F3dYJ2)

Tutorial

except the Arctic. Throughout the voyage, they sampled a multitude of ocean properties, including water depth, which was accomplished by laboriously lowering long weighted lines overboard. The HMS *Challenger* measured the depth of the deepest spot on the seafloor in 1875. Today, this spot is called the Challenger Deep and is about 10,994 meters (36,070 feet) deep.

Modern Bathymetric Techniques The measurement of ocean depths and the charting of the shape or topography of the ocean floor is known as **bathymetry** (*bathos* = depth, *metry* = measurement). Today, sound energy is often used to measure water depths. The basic approach employs **sonar**, an acronym for *sound navigation and ranging*. The first devices that used sound to measure water depth, called **echo sounders**, were developed early in the twentieth century. Echo sounders work by transmitting a sound wave (called a *ping*) into the water in order to produce an echo when it bounces off any object, such as a large marine organism or the ocean floor (**Figure 13.2**). A sensitive receiver intercepts the reflected echo, and a clock precisely records the travel time to fractions of a second. Using the speed of sound waves in water—about 1500 meters (4900 feet) per second—and measuring the time required for the energy pulse to reach the ocean floor and return, the depth can be calculated: Depth = ½ (1500 m/sec × echo travel time). Bathymetry determined from continuous monitoring of these echoes is plotted to obtain a profile of the ocean floor. By laboriously combining numerous profiles, a detailed chart of the seafloor has been produced.

Following World War II, the U.S. Navy developed *sidescan sonar* to look for explosive devices that had been deployed in shipping lanes (**Figure 13.3A**). Torpedo-shaped instruments towed behind ships send out a fan

of sound extending on either side of the ship's path. By combining swaths of sidescan sonar data, oceanographers produced the first photograph-like images of the seafloor (**Figure 13.4**).

Although sidescan sonar provides valuable views of the seafloor, it does not provide bathymetric (water depth) data. This drawback was resolved with the development of *high-resolution multibeam* instruments (see Figure 13.3A). These systems use hull-mounted sound sources that send out a fan of sound and then record reflections from the seafloor through a set of narrowly focused receivers aimed at different angles. Rather than obtain the depth of a single point every few seconds, this technique allows a survey ship to map a swath of

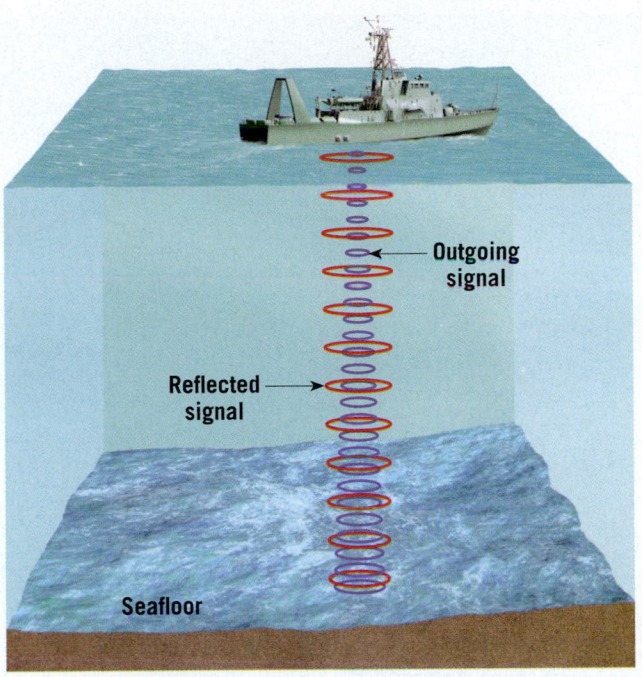

Figure 13.2
Echo sounder An echo sounder determines the water depth by measuring the time interval required for an acoustic wave to travel from a ship to the seafloor and back.

Outgoing signal

Reflected signal

Seafloor

A. Sidescan sonar and multibeam sonar operating from the same research vessel.

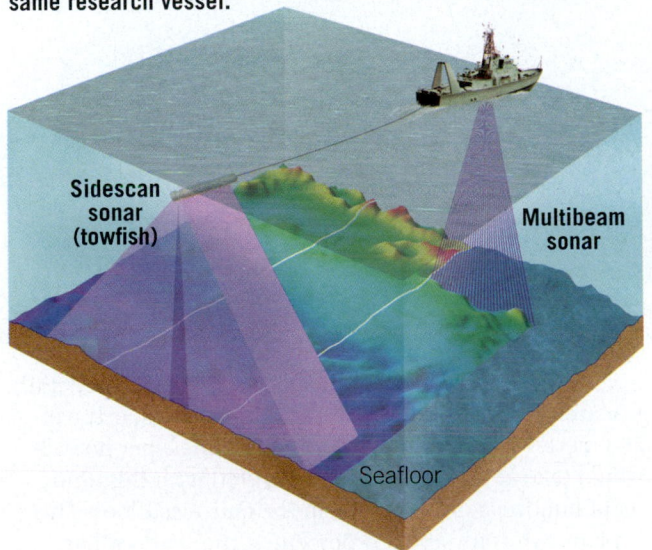

Sidescan sonar (towfish)

Multibeam sonar

Seafloor

B. Color-enhanced perspective map of the seafloor and coastal landforms in the Los Angeles area of California.

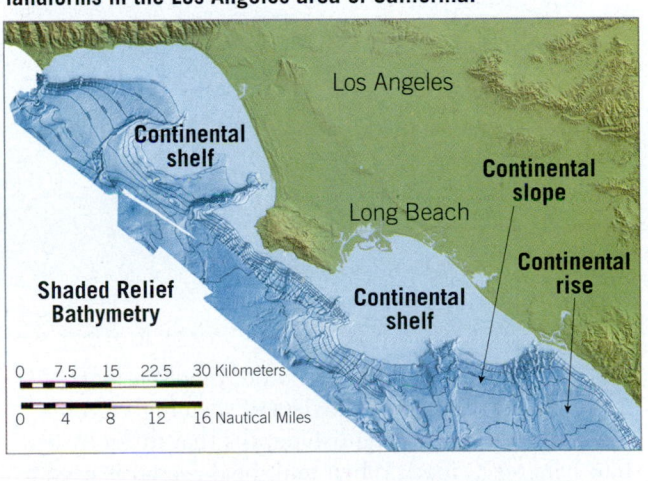

Los Angeles

Continental shelf

Continental slope

Long Beach

Continental rise

Shaded Relief Bathymetry

Continental shelf

0   7.5   15   22.5   30 Kilometers

0   4   8   12   16 Nautical Miles

Figure 13.3
Sidescan and multibeam sonar

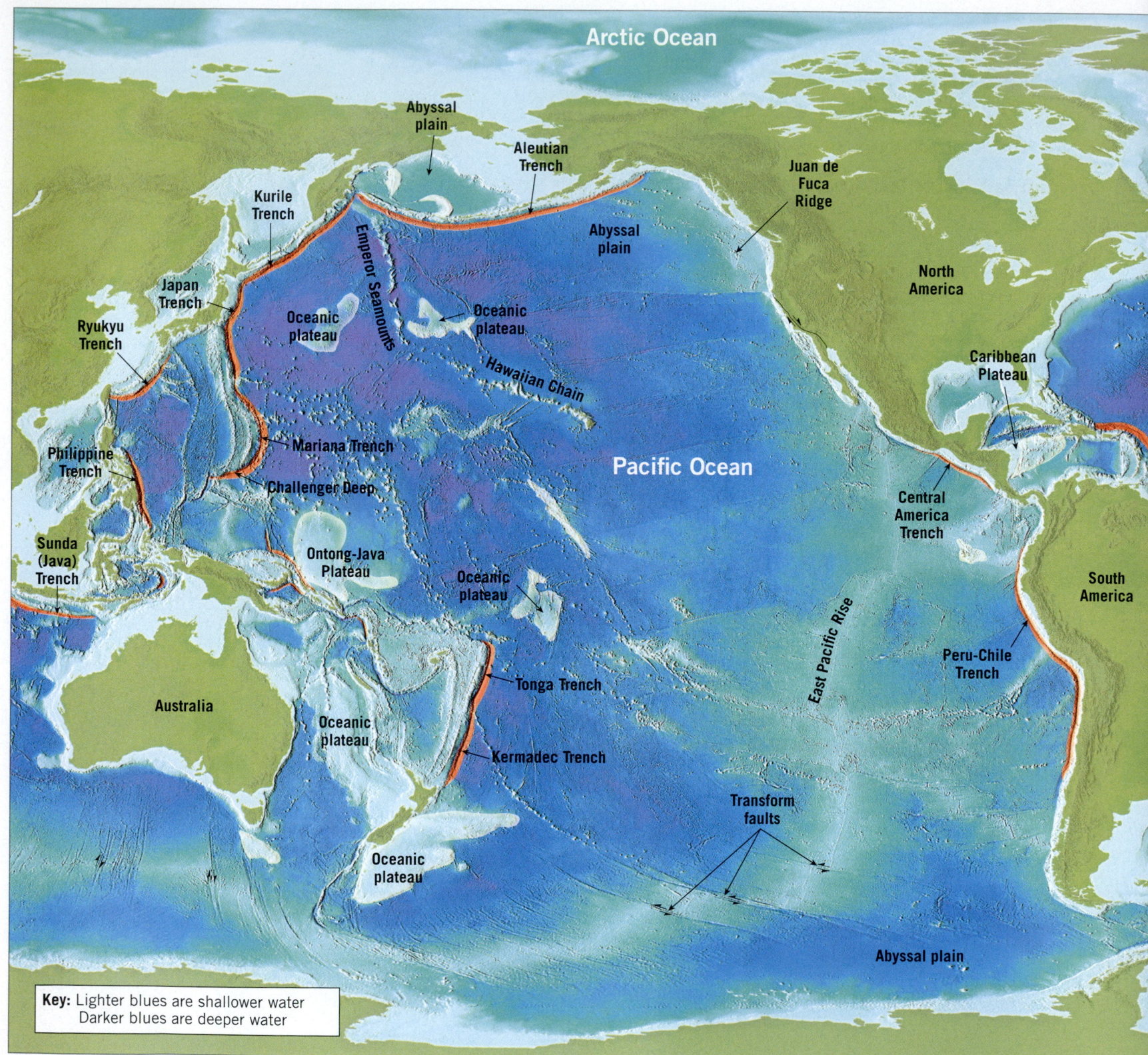

Arctic Ocean

Abyssal plain

Aleutian Trench

Kurile Trench

Juan de Fuca Ridge

Emperor Seamounts

Abyssal plain

North America

Japan Trench

Oceanic plateau

Oceanic plateau

Caribbean Plateau

Ryukyu Trench

Hawaiian Chain

Philippine Trench

Mariana Trench

Pacific Ocean

Challenger Deep

Central America Trench

Sunda (Java) Trench

Ontong-Java Plateau

Oceanic plateau

South America

Australia

Oceanic plateau

East Pacific Rise

Peru-Chile Trench

Tonga Trench

Kermadec Trench

Oceanic plateau

Transform faults

Oceanic plateau

Abyssal plain

Key: Lighter blues are shallower water
Darker blues are deeper water

Figure 13.4
Major features of the seafloor

ocean floor tens of kilometers wide. In addition, these systems collect bathymetric data of such high resolution that they can distinguish depths that differ by less than 1 meter (3 feet). When multibeam sonar is used to map sections of seafloor, the ship travels in a regularly spaced back-and-forth pattern known as "mowing the lawn."

Despite their greater efficiency and enhanced detail, research vessels equipped with multibeam sonar travel at a mere 10 to 20 kilometers (6 to 12 miles) per hour. It would take at least 100 vessels outfitted with this equipment hundreds of years to map the entire seafloor. This explains why only about 5 percent of the seafloor has been mapped in detail (**Figure 13.3B**).

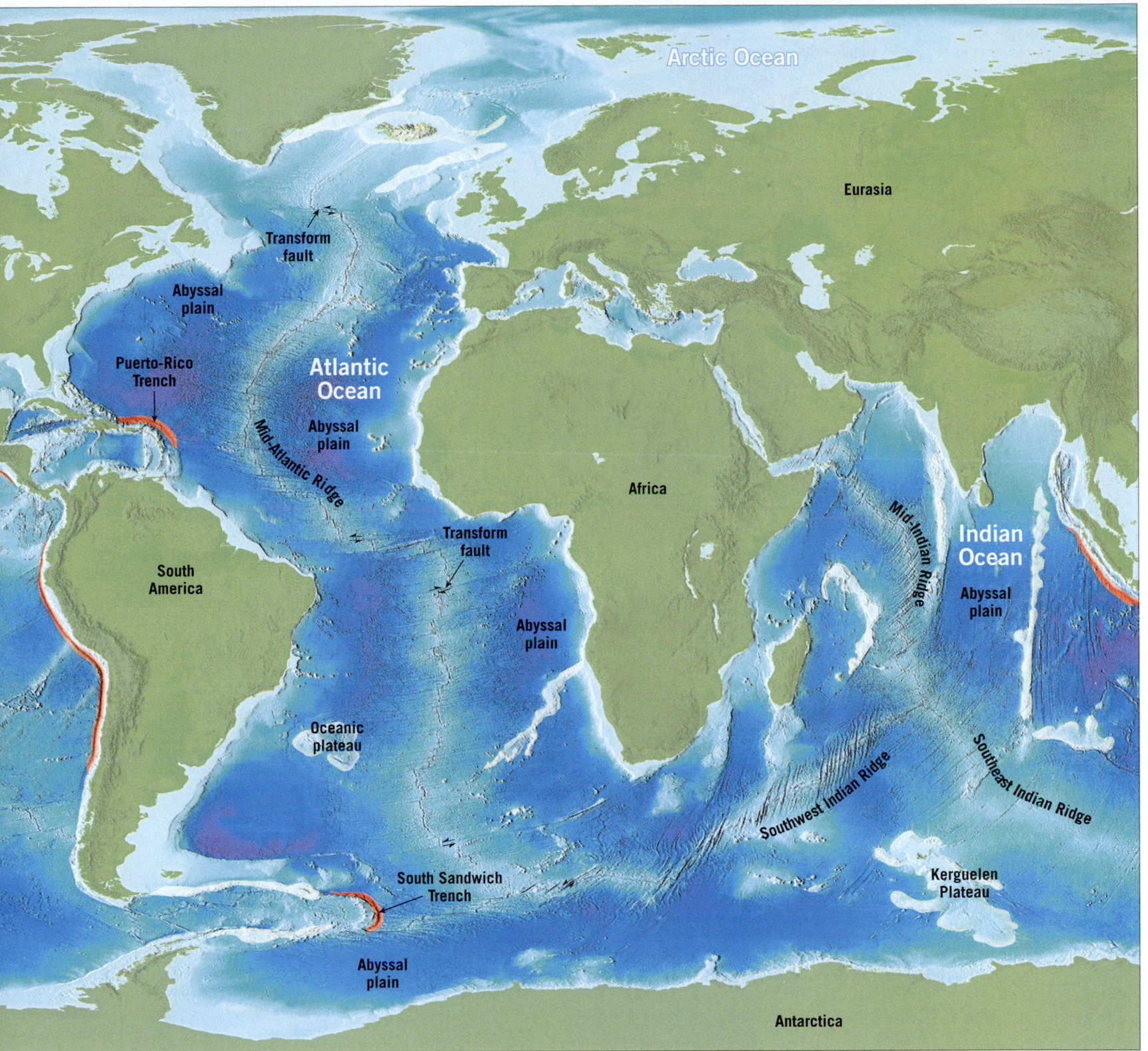

Mapping the Ocean Floor from Space Another technological breakthrough that led to an enhanced understanding of the seafloor involves measuring the shape of the ocean surface from space. After compensating for waves, tides, currents, and atmospheric effects, scientists discovered that the water's surface is not perfectly "flat." Because massive structures such as seamounts and ridges exert stronger-than-average gravitational attraction, they produce elevated areas on the ocean surface. Conversely, canyons and trenches create slight depressions.

Satellites equipped with *radar altimeters* are able to measure subtle differences in sea level by bouncing microwaves off the sea surface (**Figure 13.5**). These devices can measure variations as small as a few centimeters.

Figure 13.5
Satellite altimeter A satellite altimeter measures the variation in sea-surface elevation, which is caused by gravitational attraction and mimics the shape of the seafloor. The sea-surface anomaly is the difference between the measured ocean surface and the theoretical ocean surface.

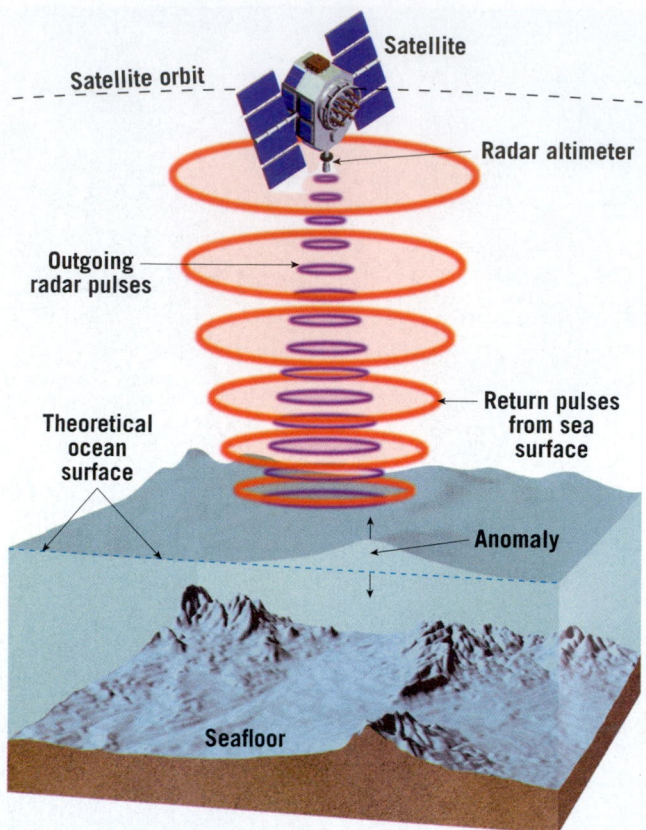

Such data have added greatly to our knowledge of ocean-floor topography. Combined with traditional sonar depth measurements, the data are used to produce detailed ocean-floor maps, such as the one in Figure 13.4.

Provinces of the Ocean Floor

Oceanographers studying the topography of the ocean floor identify three major areas: *continental margins, deep-ocean basins,* and *oceanic (mid-ocean) ridges.* The map in **Figure 13.6** outlines these provinces for the North Atlantic Ocean, and the profile at the bottom shows the varied topography. Profiles of this type usually have their vertical dimension exaggerated many times (40 times, in this case) to make topographic features more conspicuous. Vertical exaggeration, however, makes slopes appear *much* steeper than they actually are.

13.1   Concept Checks

1. Define *bathymetry.*

2. Assuming that the average speed of sound waves in water is 1500 meters (4900 feet) per second, determine the water depth if the signal sent out by an echo sounder requires 6 seconds to strike bottom and return to the recorder.

3. Describe how satellites orbiting Earth can determine features on the seafloor without being able to directly observe them beneath several kilometers of seawater.

4. What are the three major topographic provinces of the ocean floor?

Figure 13.6
Major topographic divisions of the North Atlantic The accompanying profile extends from New England to the coast of North Africa and provides a glimpse of the various structures found across the Atlantic basin.

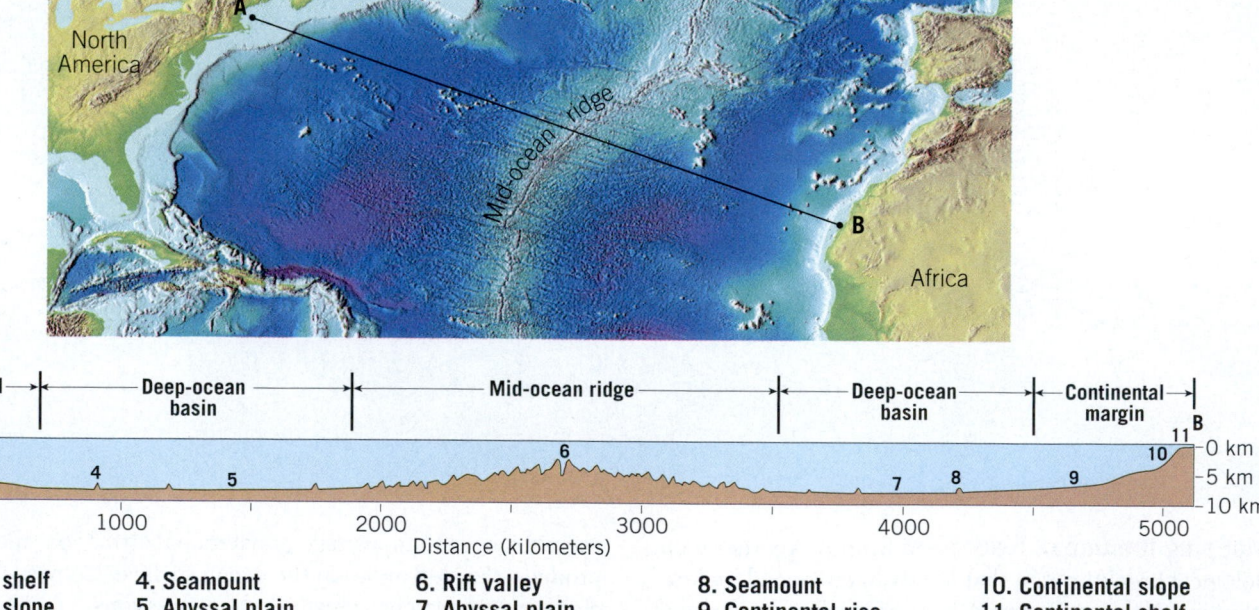

1. Continental shelf
2. Continental slope
3. Continental rise
4. Seamount
5. Abyssal plain
6. Rift valley
7. Abyssal plain
8. Seamount
9. Continental rise
10. Continental slope
11. Continental shelf

13.2 | Continental Margins

Compare and contrast a passive continental margin with an active continental margin and list the major features of each.

As the name implies, the **continental margins** are the outer margins of the continents, where continental crust transitions to oceanic crust. Two types of continental margin have been identified—*passive* and *active*. Nearly the entire Atlantic Ocean and a large portion of the Indian Ocean are surrounded by passive continental margins (see Figure 13.4). By contrast, most of the Pacific Ocean is bordered by active continental margins (subduction zones), as shown in **Figure 13.7**. Notice that some of those active subduction zones lie far beyond the margins of the continents.

Passive Continental Margins

Passive continental margins are geologically inactive regions located some distance from plate boundaries. As a result, they are not associated with strong earthquakes or volcanic activity. Passive continental margins develop when continental blocks rift apart and are separated by continued seafloor spreading. As a result, the continental blocks are firmly attached to the adjacent oceanic crust.

Most passive margins are relatively wide and are sites where large quantities of sediments are deposited. The features comprising passive continental margins include the continental shelf, the continental slope, and the continental rise (**Figure 13.8**).

Continental Shelf The gently sloping, submerged surface that extends from the shoreline toward the deep-ocean basin is called the **continental shelf**. It consists mainly of continental crust, capped with sedimentary rocks and sediments eroded from adjacent landmasses.

The continental shelf varies greatly in width. The shelf is almost nonexistent along some continents, and it extends seaward more than 1500 kilometers (930 miles) along others. The average inclination of the continental

shelf is only about one-tenth of 1 degree, a slope so slight that it would appear to an observer to be a horizontal surface.

The continental shelf tends to be relatively featureless; however, some areas are mantled by extensive glacial deposits and thus are quite rugged. In addition, some continental shelves are dissected by large valleys running from the coastline into deeper waters. Many of these

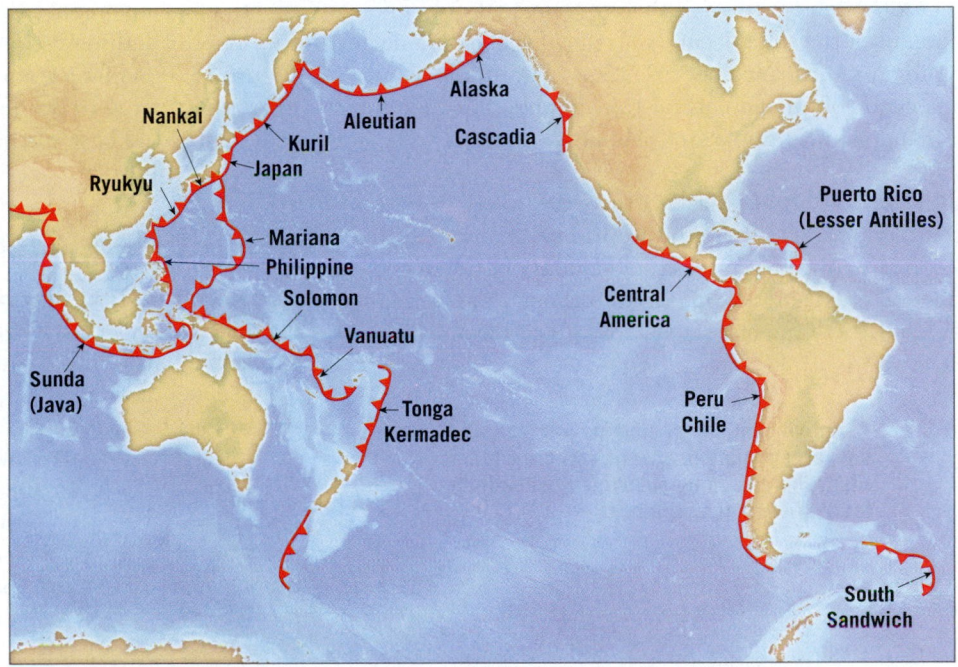

Figure 13.7
Distribution of Earth's subduction zones Notice that most of Earth's active subduction zones surround the Pacific basin.

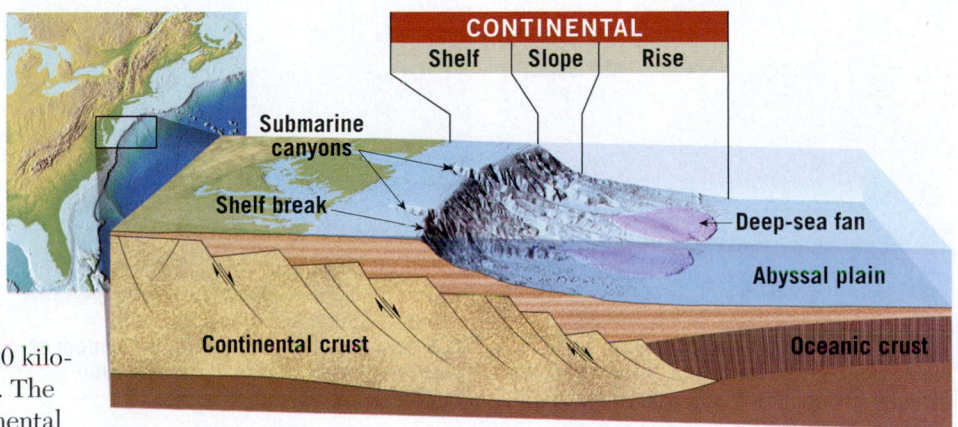

SmartFigure 13.8
Passive continental margins The slopes shown for the continental shelf and continental slope are greatly exaggerated. The continental shelf has an average slope of one-tenth of 1 degree, while the continental slope has an average slope of about 5 degrees. (https://goo.gl/FkqFPL)

shelf valleys are the seaward extensions of river valleys on the adjacent landmass. They were eroded during the last Ice Age (Quaternary period), when enormous quantities of water were stored in vast ice sheets on the continents, which lowered sea level by at least 100 meters (330 feet). Because of this sea-level drop, rivers extended their courses, and land-dwelling plants and animals migrated to the newly exposed portions of the continents. Dredging off the coast of North America has retrieved the ancient remains of numerous land dwellers, including mammoths, mastodons, and horses—providing further evidence that portions of the continental shelves were once above sea level.

Although continental shelves represent only 7.5 percent of the total ocean area, they have economic and political significance because they contain important reservoirs of oil and natural gas (discussed in Chapter 23) and support important fishing grounds.

Continental Slope Marking the seaward edge of the continental shelf is the **continental slope**, a relatively steep structure that marks the boundary between continental crust and oceanic crust. Although the inclination of the continental slope varies greatly from place to place, it averages about 5 degrees and in places exceeds 25 degrees.

Continental Rise The continental slope merges into a more gradual incline known as the **continental rise**, which may extend seaward for hundreds of kilometers. The continental rise consists of a thick accumulation of sediment that has moved down the continental slope and onto deep-ocean floor. Most of the sediments are delivered to the seafloor by **turbidity currents**, mixtures of sediment and water that periodically flow down **submarine canyons** (see Figure 7.29, page 234). When these muddy slurries emerge from the mouth of a canyon onto the relatively flat ocean floor, they deposit sediment that forms a **deep-sea fan** (see Figure 13.8). As fans from adjacent submarine canyons grow, they merge to produce a continuous wedge of sediment at the base of the continental slope, forming the continental rise.

Active Continental Margins

Active continental margins are located along convergent plate boundaries, where oceanic lithosphere is being subducted beneath the leading edge of a continent (**Figure 13.9A**). Deep-ocean trenches are the major topographic expression at convergent plate boundaries. Most of these deep, narrow furrows surround the Pacific basin. One exception is the Puerto Rico trench, which forms the boundary between the Caribbean Sea and the Atlantic Ocean (**Figure 13.10**).

Along some subduction zones, sediments from the ocean floor and pieces of oceanic crust are scraped from the descending oceanic plate and plastered against the edge of the overriding plate (**Figure 13.9B**). This chaotic accumulation of deformed sediment and scraps of oceanic crust is called an **accretionary wedge** (*ad* = toward, *crescere* = to grow). Prolonged plate subduction can produce massive accumulations of sediment along active continental margins.

The opposite process, known as **subduction erosion**, characterizes many other subduction zones (see Figure 13.9B). Rather than sediment accumulating along the front of the overriding plate,

Figure 13.9
Active continental margins

A. Active continental margins are located along convergent plate boundaries where oceanic lithosphere is being subducted beneath the leading edge of a continent.

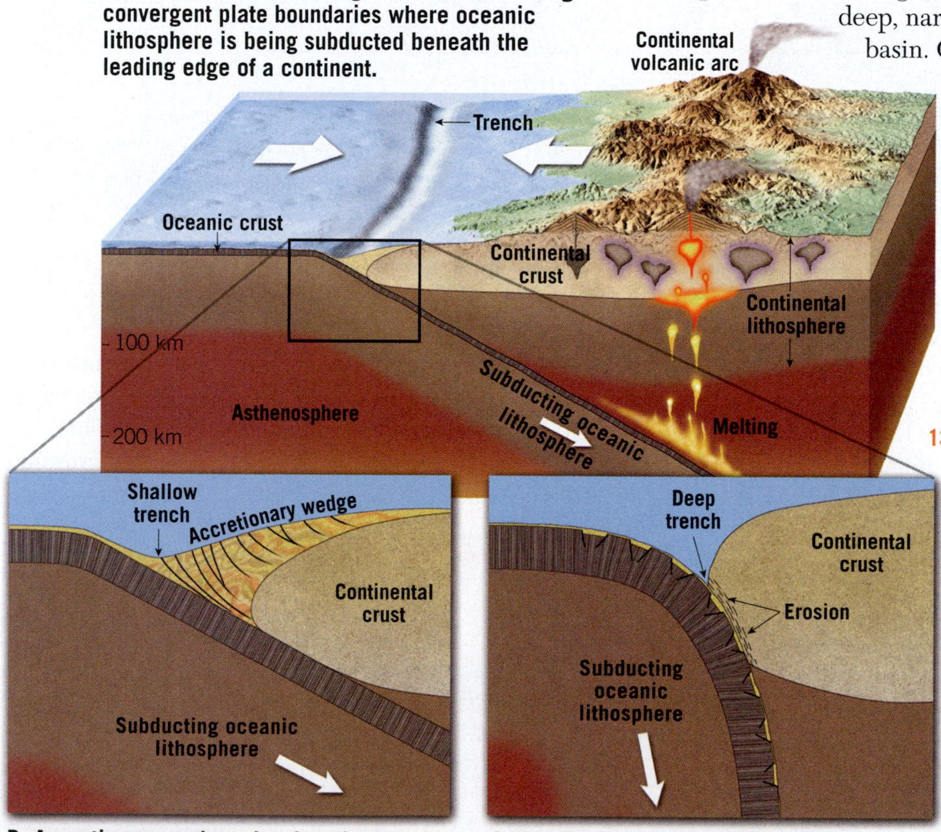

B. Accretionary wedges develop along subduction zones where sediments from the ocean floor are scraped from the descending oceanic plate and pressed against the edge of the overriding plate.

C. Subduction erosion occurs where sediment and rock scraped off the bottom of the overriding plate are carried into the mantle by the subducting plate.

sediment and rock are scraped off the bottom of the overriding plate and transported down into the mantle by the subducting plate. Subduction erosion is particularly effective where cold, dense oceanic lithosphere subducts at a steep angle, as exemplified by the Mariana trench. Sharp bending of the subducting plate causes faulting in the ocean crust and a rough surface, as shown in **Figure 13.9C**.

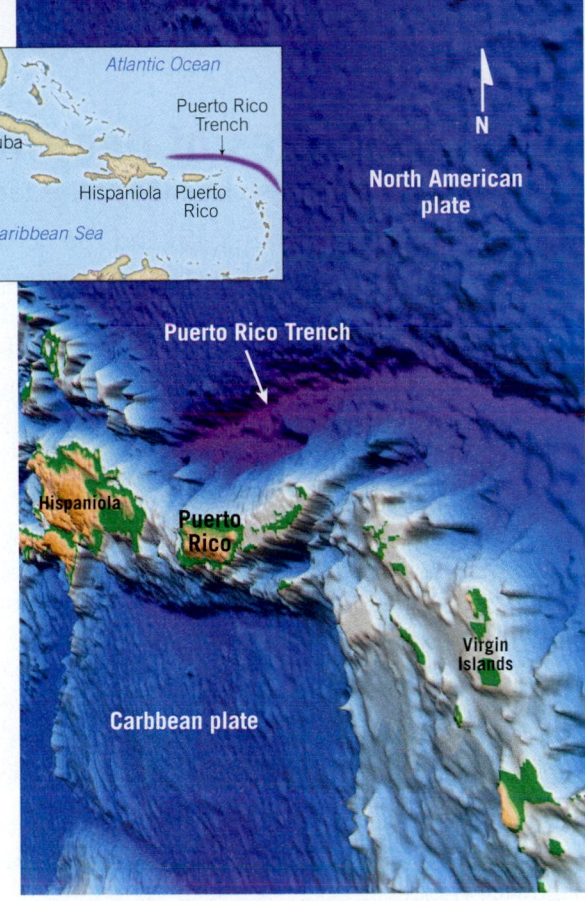

Figure 13.10
The Puerto Rico trench
This trench separates the Caribbean Sea from the Atlantic Ocean. (Photo by USGS)

13.2 Concept Checks

1. List the three major features of a passive continental margin. Which of these features is considered a flooded extension of the continent? Which one has the steepest slope?

2. Describe the differences between active and passive continental margins and give a geographic example of each.

3. How are active continental margins related to plate tectonics?

4. Briefly explain how an accretionary wedge forms.

5. What is meant by *subduction erosion*?

EYE ON EARTH 13.1

This perspective view of the Pacific margin is looking southwest toward the Palos Verdes Peninsula, with Los Angeles in the background.

QUESTION 1 Match the features labeled 1 to 4 with the following terms: continental shelf, continental slope, continental rise, and submarine canyon.

QUESTION 2 Based on the features shown, describe the type of continental margin in the image.

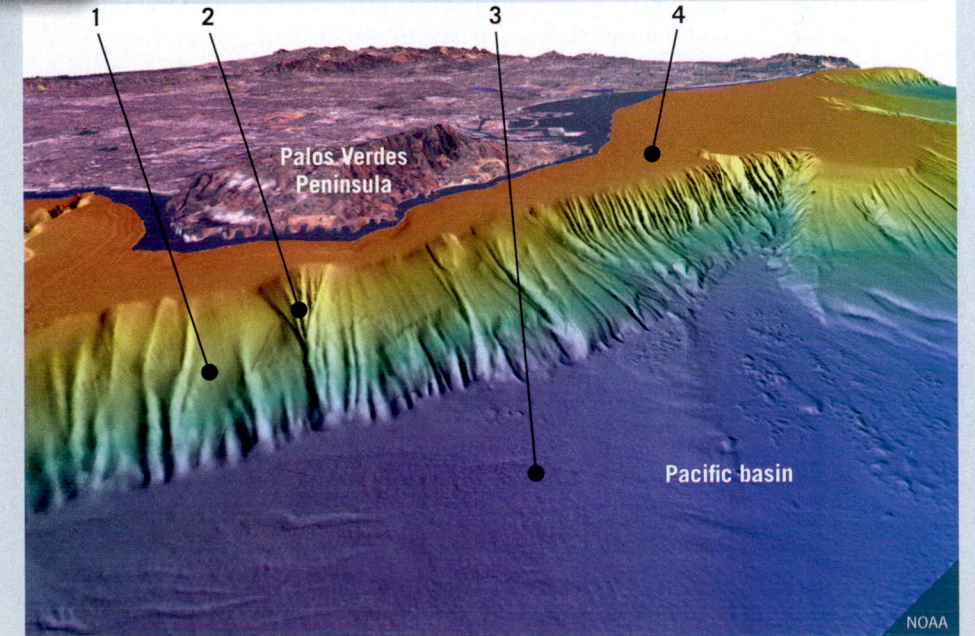

13.3 | Features of Deep-Ocean Basins

List and describe the major features of the deep-ocean basins.

Between the continental margin and the oceanic ridge lies the **deep-ocean basin** (see Figure 13.6). The size of this region—almost 30 percent of Earth's surface—is roughly comparable to the percentage of land above sea level. This region contains *deep-ocean trenches*, which are extremely deep linear depressions in the ocean floor; remarkably flat areas known as *abyssal plains*; tall volcanic peaks called *seamounts* and *guyots*; and large, elevated flood basalt provinces called *oceanic plateaus*. Another ocean feature with a biological origin is the *coral atoll*, featured in GEOgraphics 13.1.

Deep-Ocean Trenches

Deep-ocean trenches are long, narrow creases in the seafloor that are the deepest parts of the ocean floor. Most trenches are located along the margins of the Pacific Ocean, where many exceed 10 kilometers (6 miles) in depth (see Figure 13.4). The Challenger Deep, located in the Mariana trench, has been measured at 10,994 meters (36,069 feet) below sea level, making it the deepest known part of the world ocean (**Figure 13.11**). Only two trenches are located in the Atlantic—the Puerto Rico trench adjacent to the Lesser Antilles arc (see Figure 13.10) and the South Sandwich trench.

Although deep-ocean trenches represent a very small portion of the area of the ocean floor, they are nevertheless significant geologic features. Trenches are sites of plate convergence where slabs of oceanic lithosphere subduct and plunge back into the mantle. In addition to generating earthquakes as one plate "scrapes" against another, plate subduction also triggers volcanic activity. As a result, most trenches run parallel to an arc-shaped row of active volcanoes called a **volcanic island arc** (see Figure 2.18, page 52). Furthermore, **continental volcanic arcs**, such as those making up portions of the Andes and the Cascades, are located parallel to subduction zones that lie adjacent to continental margins. The volcanic activity associated with the trenches that surround the Pacific Ocean explains why this region is called the *Ring of Fire*.

Abyssal Plains

Abyssal plains (*a* = without, *byssus* = bottom) are flat features of the deep-ocean floor; in fact, they are likely the most level places on Earth (see Figure 13.4). The abyssal plain found off the coast of Argentina, for example, has less than 3 meters (10 feet) of relief over a distance exceeding 1300 kilometers (800 miles). The monotonous topography of abyssal plains is occasionally interrupted by the protruding summit of a partially buried volcanic peak (seamount).

Using **seismic reflection profilers**, researchers have determined that the relatively featureless topography of abyssal plains is due to thick accumulations of sediment that have buried an otherwise rugged ocean floor (**Figure 13.12**). The nature of the sediment indicates that these plains consist primarily of three materials:

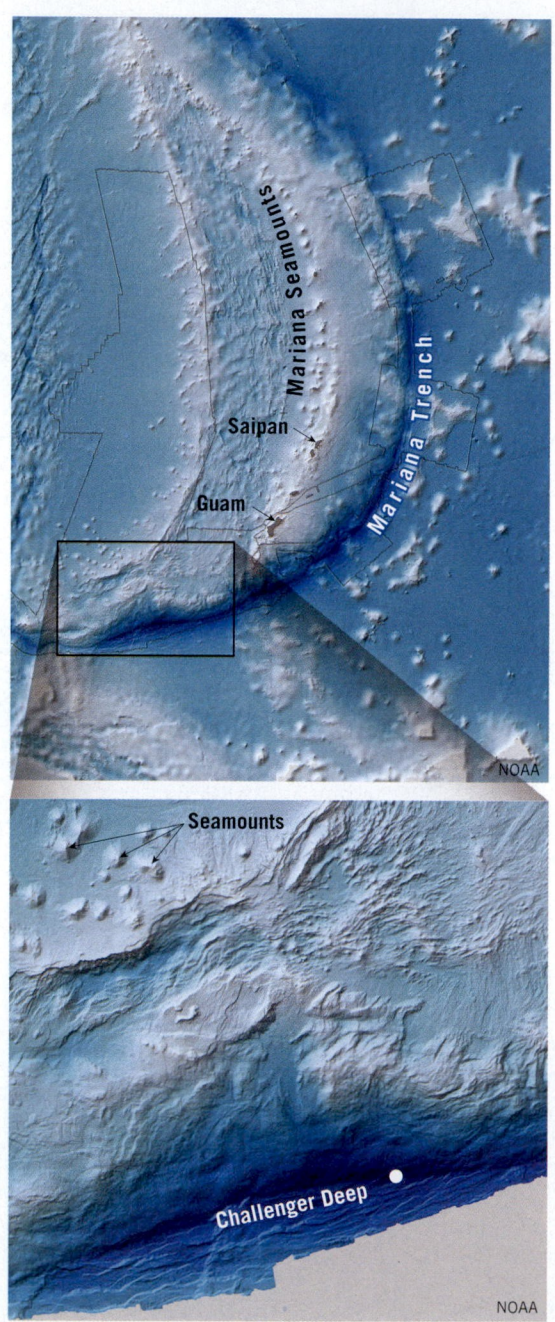

Figure 13.11
The Challenger Deep Located near the southern end of the Mariana trench, the Challenger Deep is the deepest place in the global ocean, about 10,994 meters deep. Film director James Cameron (*Titanic* and *Avatar*) made news in March 2012 as the first person to dive to the bottom of the Challenger Deep in more than 50 years.

(1) fine sediments transported far out to sea by turbidity currents, (2) mineral matter that has precipitated out of seawater, and (3) shells and skeletons of microscopic marine organisms.

Abyssal plains are found in all oceans. However, the Atlantic Ocean has the most extensive abyssal plains because it has few trenches to act as traps for sediment carried down the continental slope.

Volcanic Structures on the Ocean Floor

Dotting the seafloor are numerous volcanic structures of various sizes. Many occur as isolated features that resemble volcanic cones on land. Others occur in nearly linear chains that stretch for thousands of kilometers, while still others are massive structures that cover areas the size of Texas.

Seamounts and Volcanic Islands Submarine volcanoes, called **seamounts**, may rise hundreds of meters above the surrounding topography. It is estimated that more than a million seamounts exist. Some grow large enough to become oceanic islands, but most do not have a sufficiently long eruptive history to build a structure above sea level. Although seamounts are found on the floors of all the oceans, they are most common in the Pacific.

Some, like the Hawaiian Island–Emperor Seamount chain, which stretches from the Hawaiian Islands to the Aleutian trench, form over volcanic hot spots (see Figure 2.27, page 59). Others are born near oceanic ridges.

If a volcano grows large enough before plate motion carries it away from its magma source, the structure may emerge as a **volcanic island**. Examples of volcanic islands include Easter Island, Tahiti, Bora Bora, the Galapagos Islands, and the Canary Islands.

Guyots Inactive volcanic islands are gradually but inevitably lowered to near sea level by the forces of weathering and erosion. As a moving plate slowly carries volcanic islands away from the elevated oceanic ridge or hot spot over which they formed, they gradually sink and disappear below the water surface. Submerged, flat-topped seamounts that formed in this manner are called **guyots.***

*The term *guyot* is named after Arnold Guyot, Princeton University's first geology professor. It is pronounced "GEE-oh" with a hard *g*, as in "give."

Seismic reflection profile

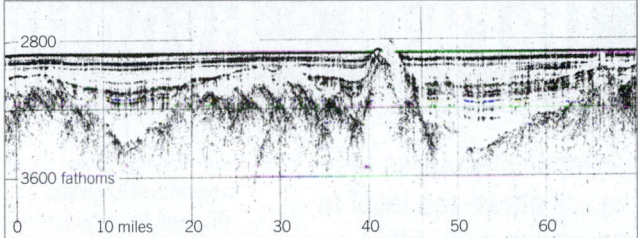

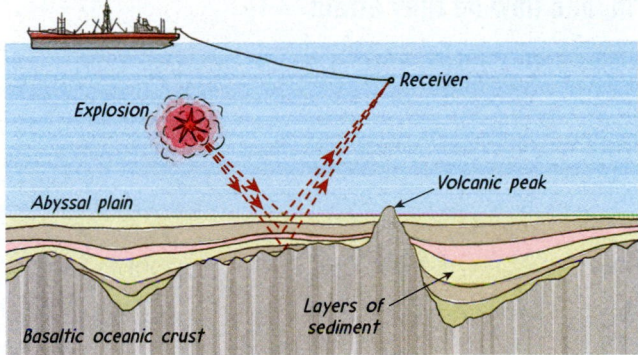

Geologist's Sketch

Figure 13.12
Seismic profile of the ocean floor This seismic cross section and matching sketch across a portion of the Madeira abyssal plain in the eastern Atlantic Ocean show the irregular oceanic crust buried by sediments. (Image courtesy of Charles Hollister, Woods Hole Oceanographic Institution)

Oceanic Plateaus The ocean floor also contains several massive **oceanic plateaus**. Resembling the flood basalt provinces found on continents, oceanic plateaus are thought to form when the bulbous head of a rising mantle plume melts, producing a vast outpouring of basaltic lava (see Figure 5.33, page 173).

Some oceanic plateaus appear to have formed quickly in geologic terms. Examples include the Ontong Java Plateau, which formed in less than 3 million years, and the Kerguelen Plateau, which formed in 4.5 million years (see Figure 13.4).

13.3 Concept Checks

1. Explain how deep-ocean trenches are related to plate boundaries.

2. Why are abyssal plains more extensive on the floor of the Atlantic than on the floor of the Pacific?

3. How does a flat-topped seamount, called a *guyot*, form?

4. What features on the ocean floor most resemble flood basalt provinces on the continents?

Explaining Coral Atolls: Darwin's Hypothesis

Coral atolls are ring-shaped structures that often extend from slightly above sea level to depths of several thousand meters. What causes atolls to form, and how do they attain such great thicknesses?

Reef-building corals only grow in warm, clear sunlit water. The depth of most active reef growth is limited to warm tropical water no more than about 45 meters (150 feet) deep. The strict environmental conditions required for coral growth create an interesting paradox: How can corals—which require warm, shallow, sunlit water no deeper than a few dozen meters—create massive structures such as coral atolls that extend to great depths?

Most corals secrete a hard external skeleton made of calcium carbonate. Colonies of these corals build large calcium carbonate structures, called reefs, where new colonies grow atop the strong skeletons of previous colonies. Sponges and algae may attach to the reef, further enlarging the structure.

Corals: Tiny colonial animals Corals are marine animals that typically live in compact colonies of many identical organisms called polyps.

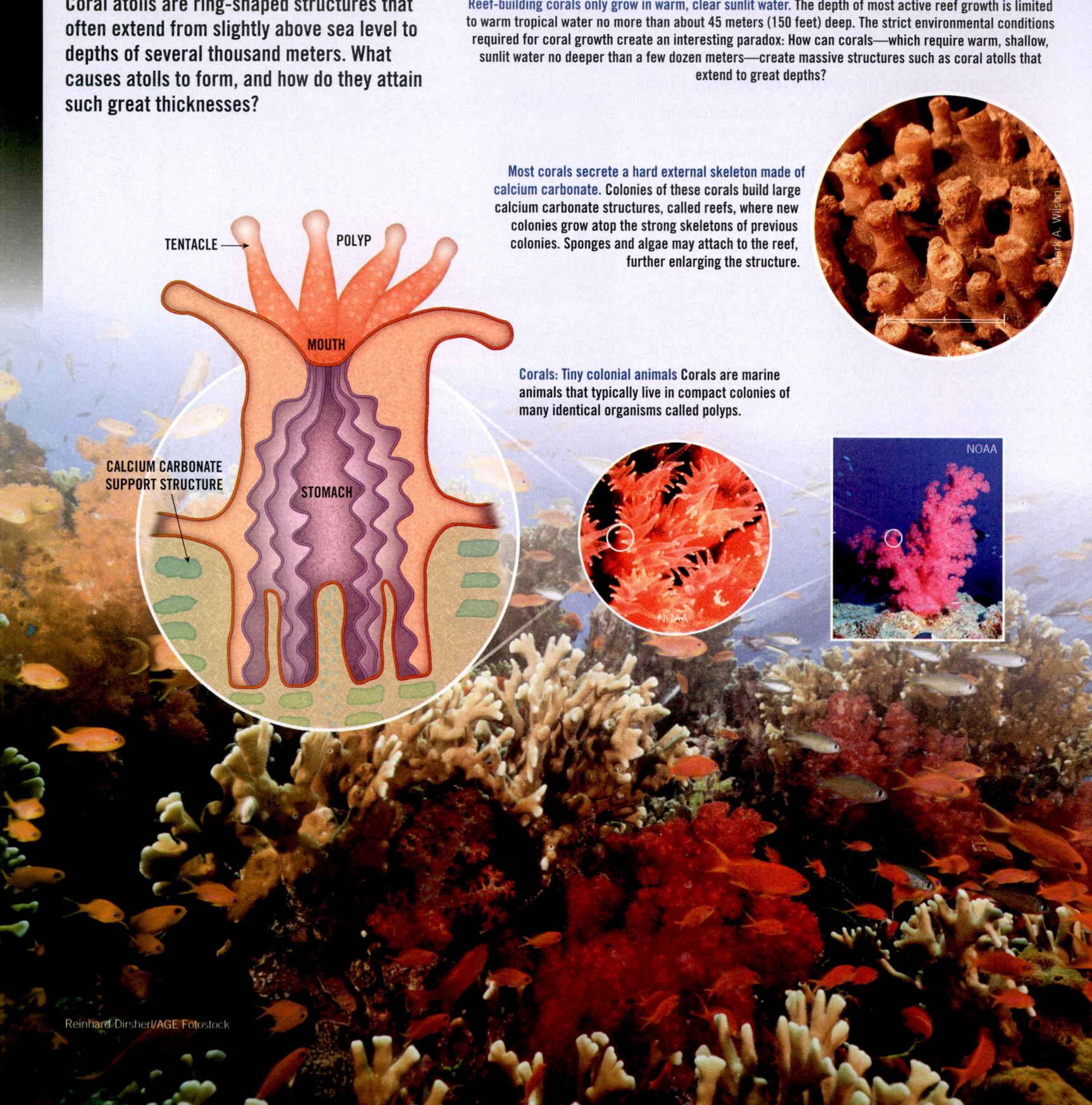

TENTACLE

POLYP

MOUTH

CALCIUM CARBONATE SUPPORT STRUCTURE

STOMACH

Mark A. Wilson

NOAA

NOAA

Reinhard Dirsherl/AGE Fotostock

Fringing reef   Barrier reef   Atoll

David Hiser/Getty Images   Frans Lanting/Mint Images/Getty Images   imagebroker/Alamy Images

What Darwin observed Naturalist Charles Darwin was one of the first to formulate a hypothesis on the origin of these ringed-shaped atolls. Aboard the British ship HMS *Beagle* during its famous global circumnavigation, Darwin noticed a progression of stages in coral reef development from (1) a fringing reef along the margins of a volcano to (2) a barrier reef with a volcano in the middle to (3) an atoll, consisting of a continuous or broken ring of coral reef surrounding a central lagoon.

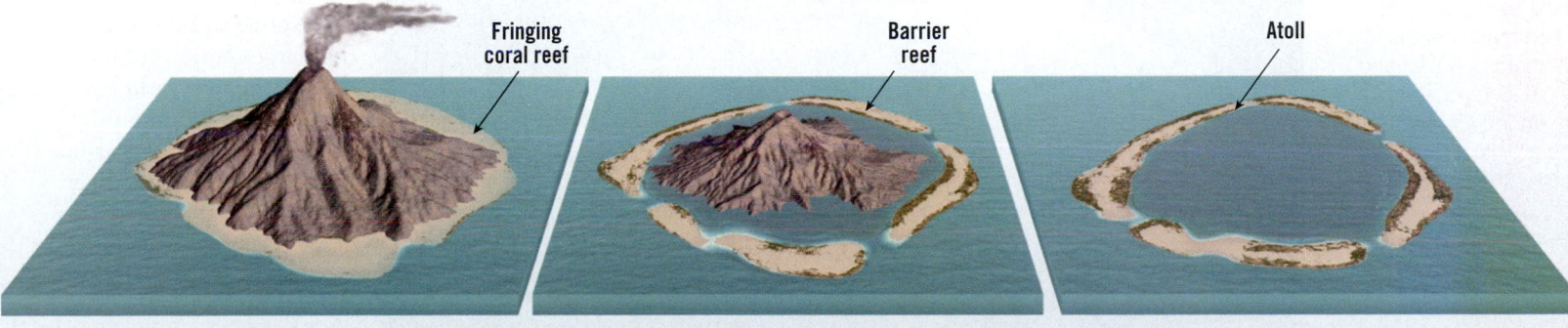

Fringing coral reef   Barrier reef   Atoll

Darwin's hypothesis Darwin's hypothesis asserts that, in addition to being lowered by erosional forces, many volcanic islands gradually sink. Darwin also suggested that corals respond to the gradual change in water depth caused by the subsiding volcano by building the reef complex upward. During Darwin's time, however, there was no plausible mechanism to account for how or why so many volcanic islands sink.

Plate tectonics and coral atolls The plate tectonics theory provides the most current scientific explanation regarding how volcanic islands become extinct and sink to great depths over long periods of time. Some volcanic islands form over relatively stationary mantle plumes, causing the lithosphere to be buoyantly uplifted. Over spans of millions of years, these volcanic islands gradually sink as moving plates carry them away from the region of hot-spot volcanism.

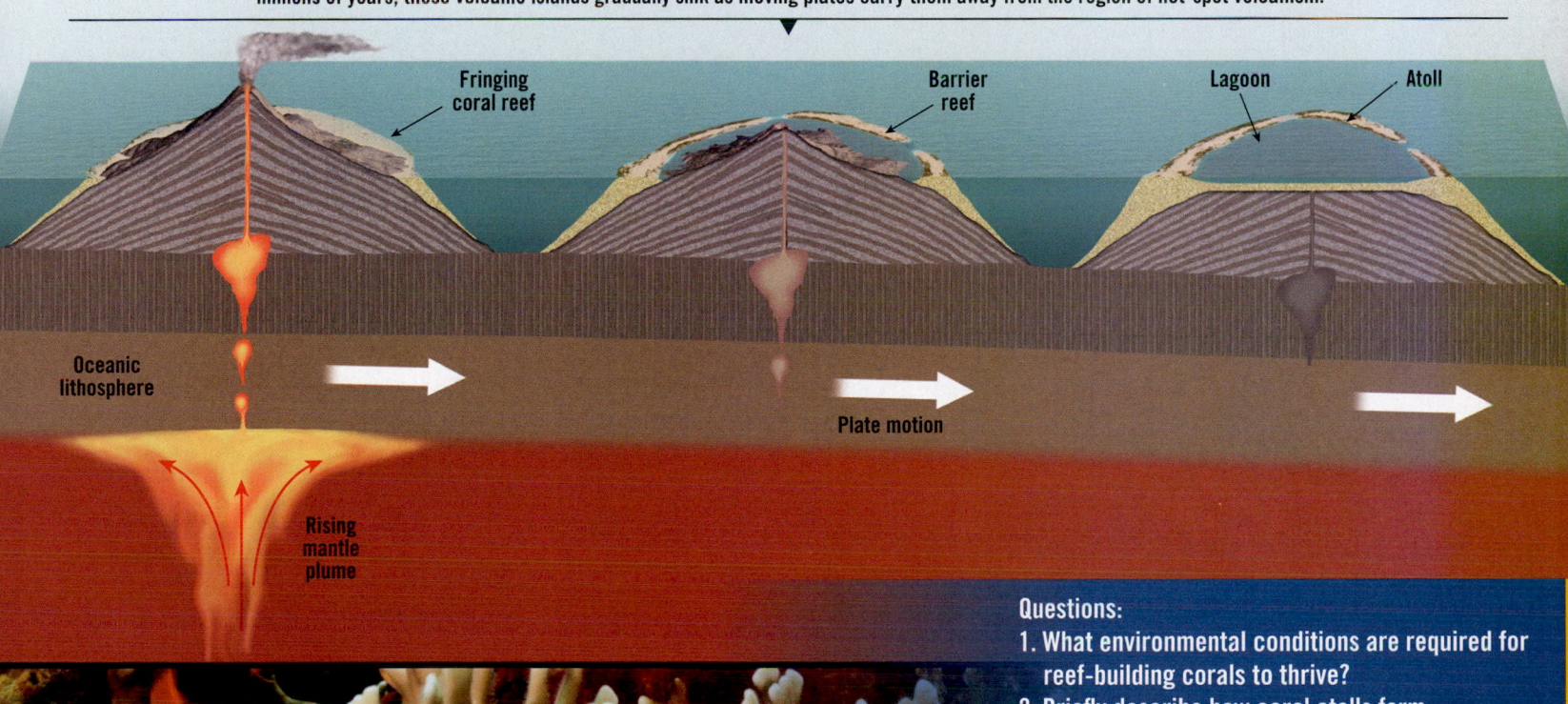

Fringing coral reef   Barrier reef   Lagoon   Atoll

Oceanic lithosphere

Plate motion

Rising mantle plume

Questions:
1. What environmental conditions are required for reef-building corals to thrive?
2. Briefly describe how coral atolls form.

?

13.4 | **Anatomy of the Oceanic Ridge**

Sketch and label a cross-sectional view of the Mid-Atlantic Ridge. Explain how a cross section of the East Pacific Rise would look different.

Along well-developed divergent plate boundaries, the seafloor is elevated, forming a broad linear swell called the **oceanic ridge,** or **mid-ocean ridge** or **rise.** Our knowledge of the oceanic ridge system comes from soundings of the ocean floor, core samples from deep-sea drilling, visual inspection using deep-diving submersibles (**Figure 13.13**), and firsthand inspection of slices of ocean floor that have been thrust onto dry land during continental collisions.

Figure 13.13

The deep-diving submersible *Alvin* This submersible is 7.6 meters long, weighs 16 tons, has a cruising speed of 1 knot, and can reach depths of 4000 meters. A pilot and two scientific observers are along during a normal 6- to 10-hour dive. (Photo by Rod Catanach KRT/Newscom)

3 kilometers above the adjacent deep-ocean basins and marks the plate boundary, where new oceanic crust is created.

Notice in Figure 13.14 that large sections of the oceanic ridge system have been named based on their locations within the various ocean basins. A ridge may run through the middle of an ocean basin, where it is appropriately called a *mid-ocean ridge.* The Mid-Atlantic Ridge and the Mid-Indian Ridge are examples. By contrast, the East Pacific Rise is *not* a "mid-ocean" feature. Rather, as its name implies, it is located in the eastern Pacific, far from the center of the ocean.

The oceanic ridge system winds through all major oceans in a manner similar to the seam on a baseball, and it is the longest topographic feature on Earth, exceeding 70,000 kilometers (43,000 miles) in length (**Figure 13.14**). The crest of the ridge typically stands 2 to

The term *ridge* is somewhat misleading because these features are not narrow and steep, as the term

Figure 13.14

Distribution of the oceanic ridge system The map shows ridge segments that exhibit slow, intermediate, and fast spreading rates.

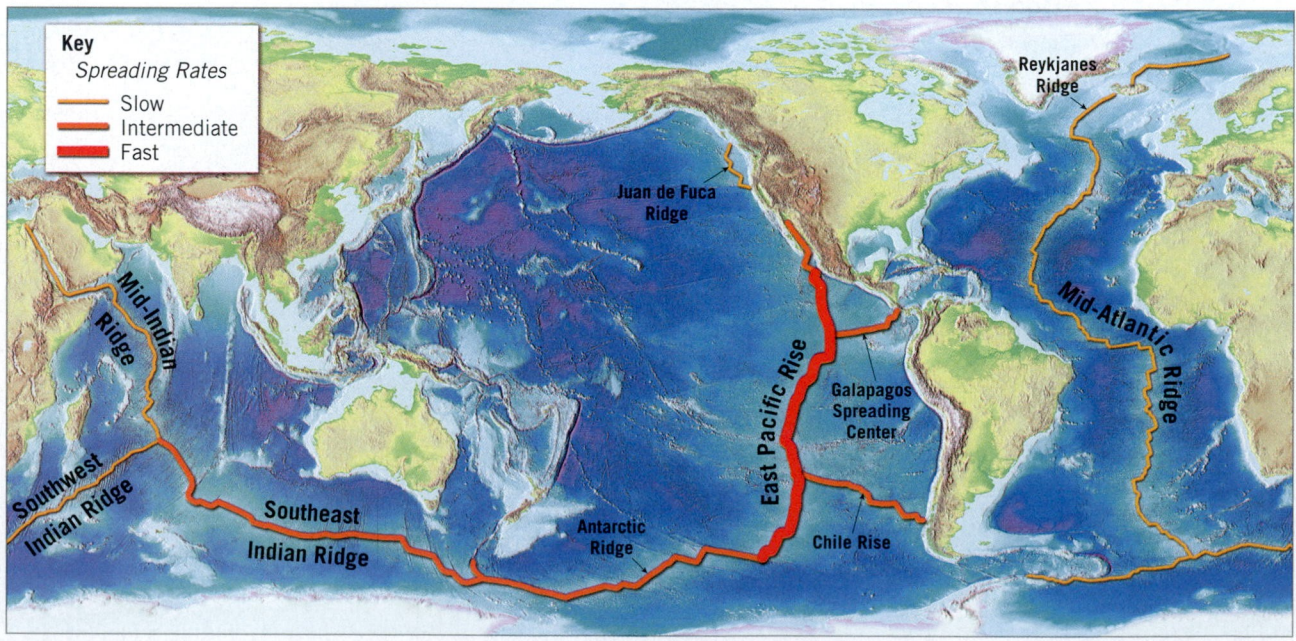

implies, but have widths from 1000 to 4000 kilometers (600 to 2500 miles) and the appearance of broad, elongated swells that exhibit varying degrees of ruggedness. Furthermore, the ridge system is broken into segments that range from a few tens to hundreds of kilometers in length. Each ridge segment is offset from the adjacent segment by a transform fault.

Oceanic ridges are as high as some mountains on the continents, but the similarities end there. Whereas most mountain ranges on land form when the compressional forces associated with continental collisions fold and metamorphose thick sequences of sedimentary rocks, oceanic ridges form where upwelling from the mantle generates new oceanic crust. Oceanic ridges consist of layers and piles of newly formed basaltic rocks that are buoyantly uplifted by the hot mantle rocks from which they formed.

Along the axis of some segments of the oceanic ridge system are deep, down-faulted structures called **rift valleys** because of their striking similarity to the continental rift valleys found in East Africa (**Figure 13.15**). Some rift valleys, including those along the rugged Mid-Atlantic Ridge, are typically 30 to 50 kilometers (20 to 30 miles) wide and have walls that tower 500 to 2500 meters (1600 to 8200 feet) above the valley floor. This makes them comparable to the deepest and widest part of Arizona's Grand Canyon.

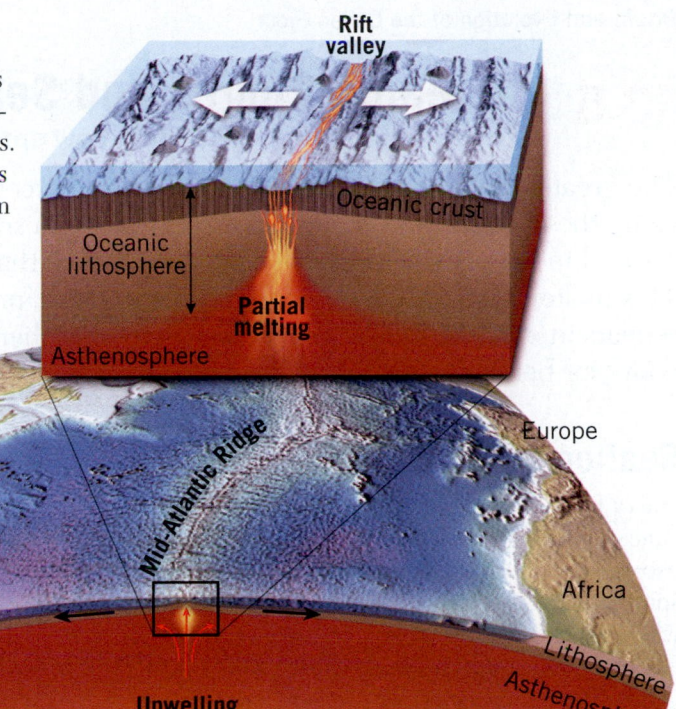

SmartFigure 13.15
Rift valleys The axes of some segments of the oceanic ridge system contain deep down-faulted structures called *rift valleys* that may exceed 30 to 50 kilometers in width and 500 to 2500 meters in depth. (https://goo.gl/fFLJ9V)

Tutorial

13.4 Concept Checks

1. Briefly describe the oceanic ridge system.

2. Although oceanic ridges can be as tall as some mountains found on the continents, list some ways in which the features are different.

3. Where do rift valleys form along the oceanic ridge system?

EYE ON EARTH 13.2

Offshore drilling rigs are used to tap oil and natural gas reserves on the continental shelf and beyond. Notable offshore fields are found in the Gulf of Mexico, the North Sea, and the Persian Gulf, as well as off the coasts of West Africa, Brazil, and Newfoundland. Drilling rigs can be fixed to the seafloor or may float. The latter group includes semi-submersibles, anchored platforms that are partially filled with seawater, and drill ships that typically use dynamic positioning to remain over the drilling location.

QUESTION 1 What was the name of the drilling rig that caused a major oil spill in the Gulf of Mexico and sank in 2010?

QUESTION 2 One of the world's deepest platforms, the Perdito, located in the Gulf of Mexico, is floating in 2,438 meters (7997 feet) of water. Explain at least one logistic issue or difficulty posed by drilling in water this deep.

Peter Bowater/Photo Researchers, Inc.

13.5 | Oceanic Ridges and Seafloor Spreading

Write a statement describing how spreading rates affect ridge topography.

The greatest volume of magma (more than 60 percent of Earth's total yearly output) is produced along the oceanic ridge system in association with seafloor spreading. As plates diverge, fractures created in the oceanic crust fill with molten rock that gradually wells up from the hot mantle below. This molten material slowly cools and crystallizes, producing new slivers of seafloor. This process repeats in episodic bursts, generating new lithosphere that moves away from the ridge crest in a conveyor belt fashion.

Seafloor Spreading

Harry Hess of Princeton University formulated the concept of seafloor spreading in the early 1960s. Later, geologists were able to verify Hess's view that seafloor spreading occurs along the crests of oceanic ridges, where hot mantle rock rises to replace the material that has shifted horizontally (see Figure 13.15). Recall from Chapter 4 that as rock rises, it experiences a decrease in confining pressure that may lead to *decompression melting*.

Partial melting of ultramafic mantle rock produces basaltic magma that has a surprisingly consistent chemical composition. The newly formed melt separates from the mantle rock and rises toward the surface. Along some ridge segments, the melt collects in small, elongated reservoirs located just beneath the ridge crest. Eventually, about 10 to 20 percent of the melt migrates upward along fissures and erupts as lava flows on the ocean floor, while the remainder crystallizes at depth to form the lower crust. This activity continuously adds new basaltic rock to diverging plate margins, temporarily welding them together; they are then broken as spreading continues. Along some ridges, outpourings of pillow lavas build submerged shield volcanoes (seamounts) as well as elongated lava ridges. At other locations, more voluminous lava flows pave the surface to create a relatively subdued topography.

Why Are Oceanic Ridges Elevated?

The primary reason for the elevated position of the ridge system is that newly created oceanic lithosphere is hot and therefore less dense than cooler rocks of the deep-ocean basin. As the newly formed basaltic crust travels away from the ridge crest due to seafloor spreading, it is cooled from above as seawater circulates through the pore spaces and fractures in the rock. Further cooling occurs as it gets farther and farther from the zone of hot mantle upwelling. As a result, the lithosphere gradually cools, contracts, and becomes denser. This thermal contraction accounts for the greater ocean depths that occur away from the ridge. After about 80 million years of cooling and contraction, rock that was once part of an elevated ocean-ridge system becomes part of the deep-ocean basin.

As lithosphere is displaced away from the ridge crest, cooling also causes a gradual increase in lithospheric thickness. This happens because the boundary between the lithosphere and asthenosphere is a thermal (temperature) boundary. Recall from Chapter 12 that the lithosphere is Earth's cool, stiff outer layer, whereas the asthenosphere is a comparatively hot and weak layer. As material in the uppermost asthenosphere ages (cools), it becomes stiff and rigid. Thus, the upper portion of the asthenosphere is gradually converted to lithosphere simply by cooling. Oceanic

Figure 13.16
Topography of slow and fast spreading centers

A. At slow spreading rates a prominent central rift valley develops along the ridge crest, and the topography of the ridge is typically rugged.

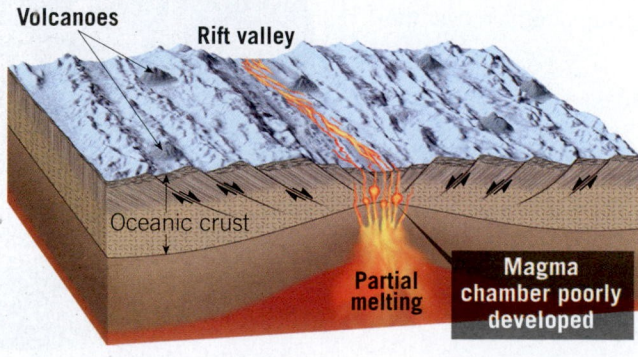

Volcanoes · Rift valley · Oceanic crust · Partial melting · Magma chamber poorly developed

B. Along fast spreading centers, medial rift valleys do not develop and the topography is comparatively smooth.

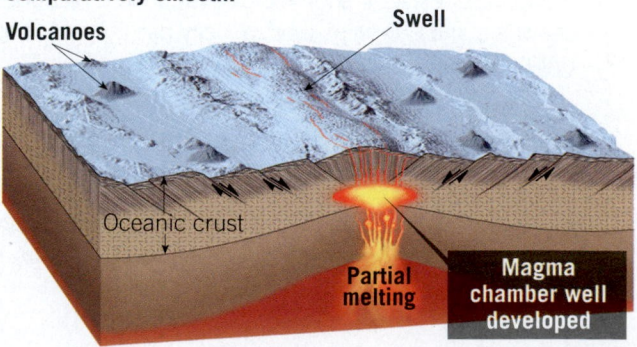

Volcanoes · Swell · Oceanic crust · Partial melting · Magma chamber well developed

lithosphere continues to thicken until it is about 80 to 100 kilometers (50 to 60 miles) thick. Thereafter, its thickness remains relatively unchanged until it is subducted.

Ridges that have fast spreading rates are characterized by smooth topography, gently sloping flanks, and a central swell.

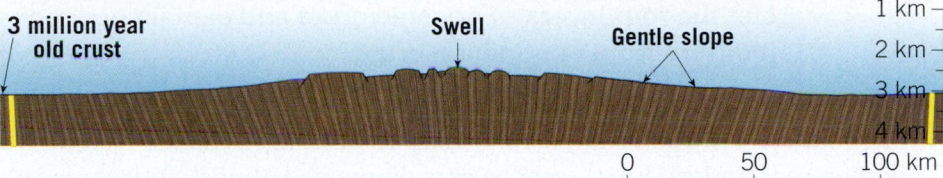

Ridges that have intermediate spreading rates are characterized by small rift valleys (less than 500 meters deep) and moderately sloping flanks.

Ridges that have slow spreading rates are characterized by rugged topography, well-developed rift valleys (500-2500 meters deep) and steeply sloping flanks.

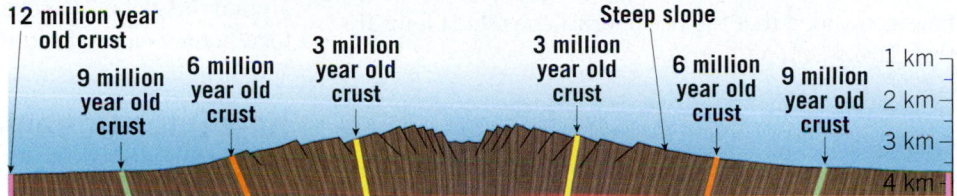

Spreading Rates and Ridge Topography

When researchers studied various segments of the oceanic ridge system, it became clear that there were topographic differences. These differences appear to result from differences in spreading rates—which largely determine the amount of melt generated at a rift zone. More magma wells up from the mantle at fast spreading centers than at slow spreading centers. This difference in output causes differences in the structure and topography of various ridge segments.

Oceanic ridges that exhibit slow spreading rates from 1 to 5 centimeters per year have prominent rift valleys and rugged topography (**Figure 13.16A**). The Mid-Atlantic and Mid-Indian Ridges are examples. The vertical displacement of large slabs of oceanic crust along normal faults is responsible for the steep walls of the rift valleys. Furthermore, volcanism produces numerous cones in the rift valley, which enhance the rugged topography of the ridge crest.

By contrast, along the Galapagos Ridge, an intermediate spreading rate of 5 to 9 centimeters per year is the norm. As a result, the rift valleys that develop are relatively shallow—often less than 200 meters (660 feet) deep. In addition, their topography is more subdued compared to ridges that have slower spreading rates.

At fast spreading centers (greater than 9 centimeters per year), such as along much of the East Pacific Rise, rift valleys are generally absent (**Figure 13.16B**). Instead, the ridge axis is elevated. These elevated structures, called *swells*, are built from lava flows up to 10 meters (30 feet) thick that have incrementally paved

the ridge crest with volcanic rocks (see Figure 13.16B). In addition, because the depth of the ocean depends largely on the age of the seafloor, ridge segments that exhibit faster spreading rates tend to have more gradual profiles than ridges that have slower spreading rates (**Figure 13.17**). Because of these differences in topography, the gently sloping, less rugged portions of fast-spreading ridges are called *rises*.

13.5 Concept Checks

1. What is the source of magma for seafloor spreading?

2. What is the primary reason for the elevated position of the oceanic ridge system?

3. Why does the lithosphere thicken as it moves away from the ridge as a result of seafloor spreading?

4. Compare and contrast a slow spreading center such as the Mid-Atlantic Ridge with one that exhibits a faster spreading rate, such as the East Pacific Rise.

13.6 | The Nature of Oceanic Crust

List the four layers of oceanic crust and explain how oceanic crust forms and how it differs from continental crust.

An interesting aspect of oceanic crust is that its thickness and structure are remarkably consistent throughout the entire ocean basin. Seismic soundings indicate that its thickness averages only about 7 kilometers (5 miles). Furthermore, it is composed almost entirely of mafic rocks that are underlain by a layer of the ultramafic rock peridotite, which forms the lithospheric mantle.

Although most oceanic crust forms out of view, far below sea level, geologists have been able to examine the structure of the ocean floor firsthand. In locations such as Newfoundland, Cyprus, Oman, and California, slivers of oceanic crust and underlying mantle have been thrust high above sea level. From these exposures and core samples collected by deep-sea drilling ships, researchers have concluded that the ocean crust consists of four distinct layers (Figure 13.18):

- **Layer 1.** The upper layer is a sequence of deep-sea sediments or sedimentary rocks. Sediments are very thin near the axes of oceanic ridges but may be several kilometers thick next to continents.
- **Layer 2.** Below the layer of sediments is a rock unit composed mainly of basaltic lavas that contain abundant pillowlike structures called *pillow lavas.*
- **Layer 3.** The middle, rocky layer is made up of numerous interconnected dikes that have a nearly vertical orientation, called the *sheeted dike complex.* These dikes are former pathways where magma rose to feed pillow basalts on the ocean floor.

- **Layer 4.** The lowest unit is mainly gabbro, the coarse-grained equivalent of basalt, which crystallized deeper in the crust without erupting.

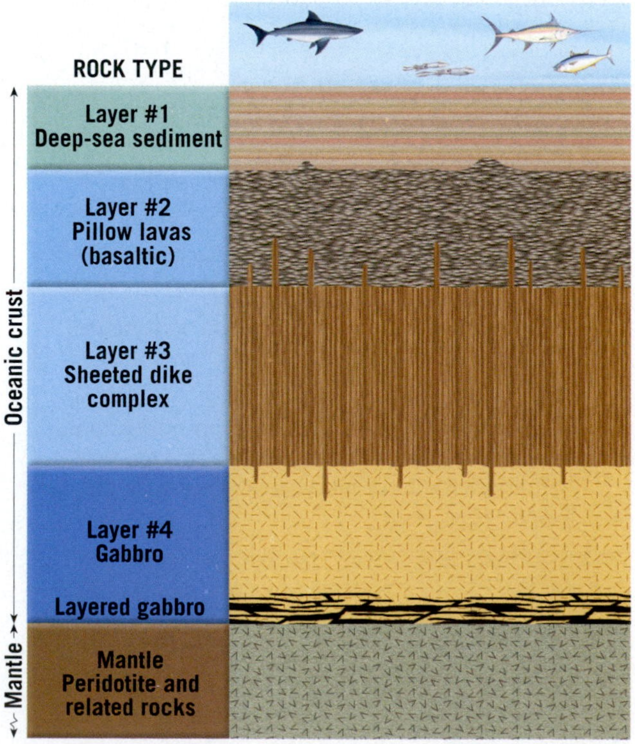

When fragments of oceanic crust and the underling mantle are discovered on land, they are called an **ophiolite complex**. From studies of various ophiolite complexes around the globe and related data, geologists have pieced together a scenario for the formation of the ocean floor.

How Does Oceanic Crust Form?

The molten rock that forms new oceanic crust originates from partial melting of the ultramafic mantle rock. This process generates basaltic melt that is less dense than the surrounding solid rock. The newly formed melt rises through the upper mantle along thousands of tiny conduits that feed into a few dozen larger, elongated channels, perhaps 100 meters (300 feet) or more wide. These structures, in turn, feed lens-shaped magma chambers located directly beneath the ridge crest. The addition of melt from below steadily increases the pressure inside the magma chambers. As a result, the rocks above these reservoirs periodically fracture, allowing the melt to ascend along numerous vertical fractures that develop in the ocean crust. Some of the melt cools and solidifies to form dikes. New dikes intrude older dikes, which are still warm and weak, to form a **sheeted dike complex**. This portion of the oceanic crust is usually 1 to 2 kilometers thick.

Roughly 10 to 20 percent of the melt eventually erupts on the ocean floor. Because the surface of these submarine lava flows is chilled quickly by seawater, these flows generally travel no more than a few kilometers before completely solidifying. The forward motion occurs as lava accumulates behind the congealed margin and then breaks through. This process occurs repeatedly, as molten basalt is extruded—like toothpaste from a tightly squeezed tube (see Figure 5.6, page 148). The result is protuberances resembling large bed pillows stacked one atop the other, hence the name **pillow lavas** (Figure 13.19).

In some settings, pillow lavas may build volcano-size mounds that resemble shield volcanoes, whereas in other situations they form elongated ridges tens of kilometers long. These structures are eventually separated from their supply of magma as they are carried away from the ridge crest by seafloor spreading.

Figure 13.18
Ophiolite complex: Structure of the oceanic crust This view of the layered structure of oceanic crust is based on data obtained from ophiolite complexes, seismic profiling, and core samples obtained from deep-sea drilling expeditions.

ROCK TYPE

Oceanic crust
Layer #1 Deep-sea sediment
Layer #2 Pillow lavas (basaltic)
Layer #3 Sheeted dike complex
Layer #4 Gabbro
Layered gabbro

Mantle
Mantle Peridotite and related rocks

The lowest unit of the ocean crust develops from crystallization within the central magma chamber itself. The first minerals to crystallize are olivine, pyroxene, and occasionally chromite (chromium oxide), which settle through the magma to form a layered zone near the floor of the reservoir. The remaining melt tends to cool along the walls of magma chambers to form massive amounts of coarse-grained gabbro. This portion accounts for up to 5 of the 7 kilometers (3 to 4.5 miles) of ocean crust thickness.

Although molten rock rises continuously from the mantle toward the surface, seafloor spreading occurs in pulse-like bursts. As the melt begins to accumulate in the lens-shaped reservoirs, it is blocked from continuing upward by the stiff overlying rocks. As the amount of melt entering the magma reservoirs increases, pressure rises. Periodically, the pressure exceeds the strength of the overlying rocks, which fracture and initiate a short episode of seafloor spreading.

Interactions Between Seawater and Oceanic Crust

In addition to serving as a mechanism for dissipating Earth's internal heat, the interaction between seawater and the newly formed basaltic crust alters both the seawater and the crust. The permeable and highly fractured lava of the upper oceanic crust allows seawater to penetrate to depths of 2 to 3 kilometers (1 to 2 miles). Seawater circulating through the hot crust is heated and chemically reacts with the basaltic rock in a process called *hydrothermal* (hot water) *metamorphism* (see Figure 8.22, page 257). This alteration causes the dark silicates (olivine and pyroxene) in basalt to form new metamorphic minerals such as chlorite and serpentine. Simultaneously, the hot seawater dissolves ions of silica, iron, copper, and occasionally silver and gold from the hot basalts. When the water temperature reaches a few hundred degrees Celsius,

Figure 13.19
Cross-sectional view of pillow lava This pillow lava is exposed along a sea cliff at Cape Wanbrow, New Zealand. Notice that each "pillow" shows an outer dark glassy layer created by rapid cooling enclosing a dark gray basalt interior.
(Photo by GR Roberts/Photo Researchers, Inc.)

these mineral-rich fluids buoyantly rise along fractures and eventually spew out on the ocean floor.

Studies conducted by submersibles along several segments of the oceanic ridge have photographed these metallic-rich solutions gushing from the seafloor to form particle-filled clouds called **black smokers** (see GEOgraphics 13.2). As the hot liquid (up to 400°C [750°F]) mixes with the cold, mineral-laden seawater, the dissolved minerals precipitate to form massive metallic sulfide deposits, some of which are economically important. Occasionally these deposits grow upward to form underwater chimney-like structures equivalent in height to skyscrapers.

13.6 | Concept Checks

1. Briefly describe the four layers of the ocean crust.

2. How does a sheeted dike complex form?

3. How does hydrothermal metamorphism alter the basaltic rocks that make up the seafloor? How is seawater changed during this process?

4. What is a black smoker?

13.7 | Continental Rifting: The Birth of a New Ocean Basin

Outline the steps by which continental rifting results in the formation of new ocean basins.

The breakup of Pangaea nearly 200 million years ago opened a new ocean basin—the Atlantic. Although geoscientists still debate what initiated this event, the breakup of Pangaea illustrates that ocean basins originate when large landmasses break apart.

Evolution of an Ocean Basin

The opening of a new ocean basin begins with the formation of a **continental rift**, an elongated depression along which the entire lithosphere is stretched and thinned. Where the lithosphere is thick, cool, and strong, rifts tend to be narrow—often less than a few hundred kilometers wide. Modern examples of narrow continental rifts include the East African Rift, the Rio Grande Rift (southwestern United States), the Baikal Rift (south-central Siberia), and the Rhine Valley (northwestern

Deep-Sea Hydrothermal Vents

Deep-sea hydrothermal vents are openings in the oceanic crust from which geothermally heated water rises. They are found mainly along the oceanic ridge system where tectonic plates rift apart, resulting in the production of new seafloor by upwelling magma.

When these hot, mineral-rich fluids reach the seafloor their temperatures can exceed 350 °C, but because of the extremely high pressures exerted by the water column above, they do not boil. When this hydrothermal fluid comes into contact with the much colder chemical-rich seawater, mineral matter rapidly precipitates to form shimmering smoke-like clouds called "*black smokers*." The particles that compose the black smokers eventually settle out of the seawater. These deposits may contain economically significant amounts of iron, copper, zinc, lead, and occasionally silver and gold.

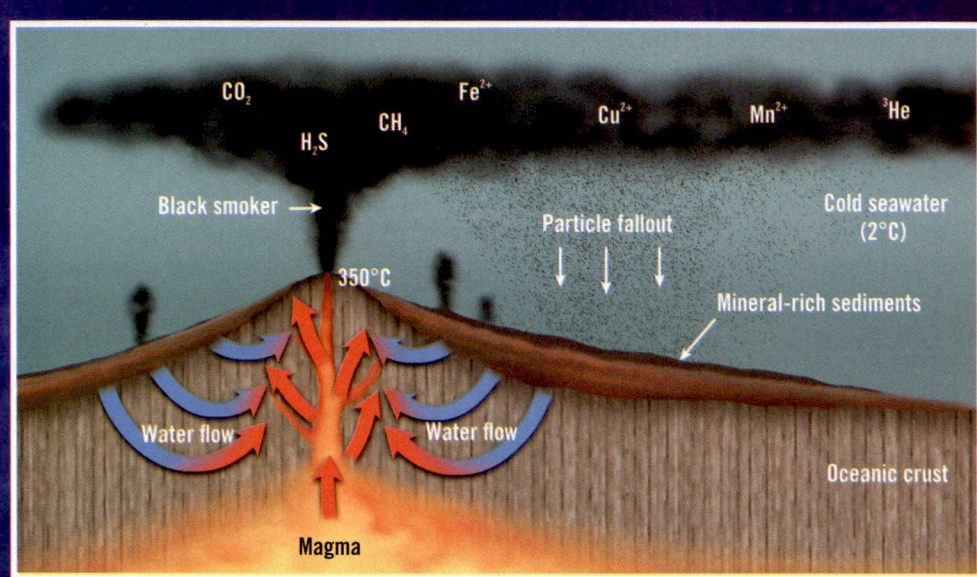

CO_2 Fe^{2+} Cu^{2+} Mn^{2+} ^3He

H_2S CH_4

Black smoker →

Particle fallout

Cold seawater (2°C)

350°C

Mineral-rich sediments

Water flow   Water flow

Oceanic crust

Magma

At oceanic ridges, cold seawater circulates hundreds of meters down into the highly fractured basaltic crust, where it is heated by magmatic sources. Along the way, the hot water strips metals and other elements such as sulfur from surrounding rock. This heated fluid eventually becomes hot and buoyant enough to rise along conduits and fractures toward the surface.

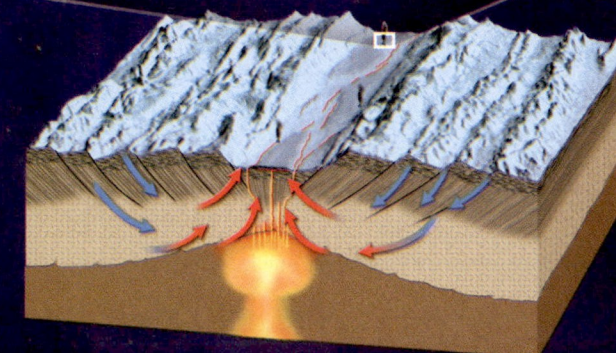

Some minerals immediately solidify and contribute to the formation of spectacular chimney-like structures, which can be as tall as a 15-story building, and are appropriately given names like *Godzilla* and *Inferno*.

NOAA

Tubeworms

Fisheries and Oceans Canada/UVic-Verena Tunnicliffe/Newscom

Murton/Southampton Oceanography Centre/Photo Researchers Inc.

Hydrothermal vents are also remarkable for the marine biology they support. In these environments, completely devoid of sunlight, microorganisms utilize mineral-rich hydrothermal fluid to perform chemosynthesis—the conversion of carbon atoms into organic compounds without sunlight for energy. The microbial communities, in turn, support larger, more complex animals such as fish, crabs, worms, mussels, clams, and perhaps the most unique, the tubeworm. With their white chitinous tubes and bright red plumes, these conspicuous creatures rely entirely on bacteria growing in their trophosome, an internal organ designed for harvesting bacteria. These symbiotic bacteria rely on the tubeworm to provide them with a suitable habitat and, in return, they use chemosynthesis to provide carbon-based building blocks (nutrients) to the tubeworms.

Question:
Where are hydrothermal vents found?

?

Most hydrothermal vents are found around the oceanic ridge system including some small spreading centers such as the Juan de Fuca Ridge and Galapagos Rift, as well as in the back arc basins that lie behind subduction zones.

KEY
- 🔴 Active
- 🔵 Unconfirmed

GLOBAL DISTRIBUTION OF HYDROTHERMAL VENT FIELDS

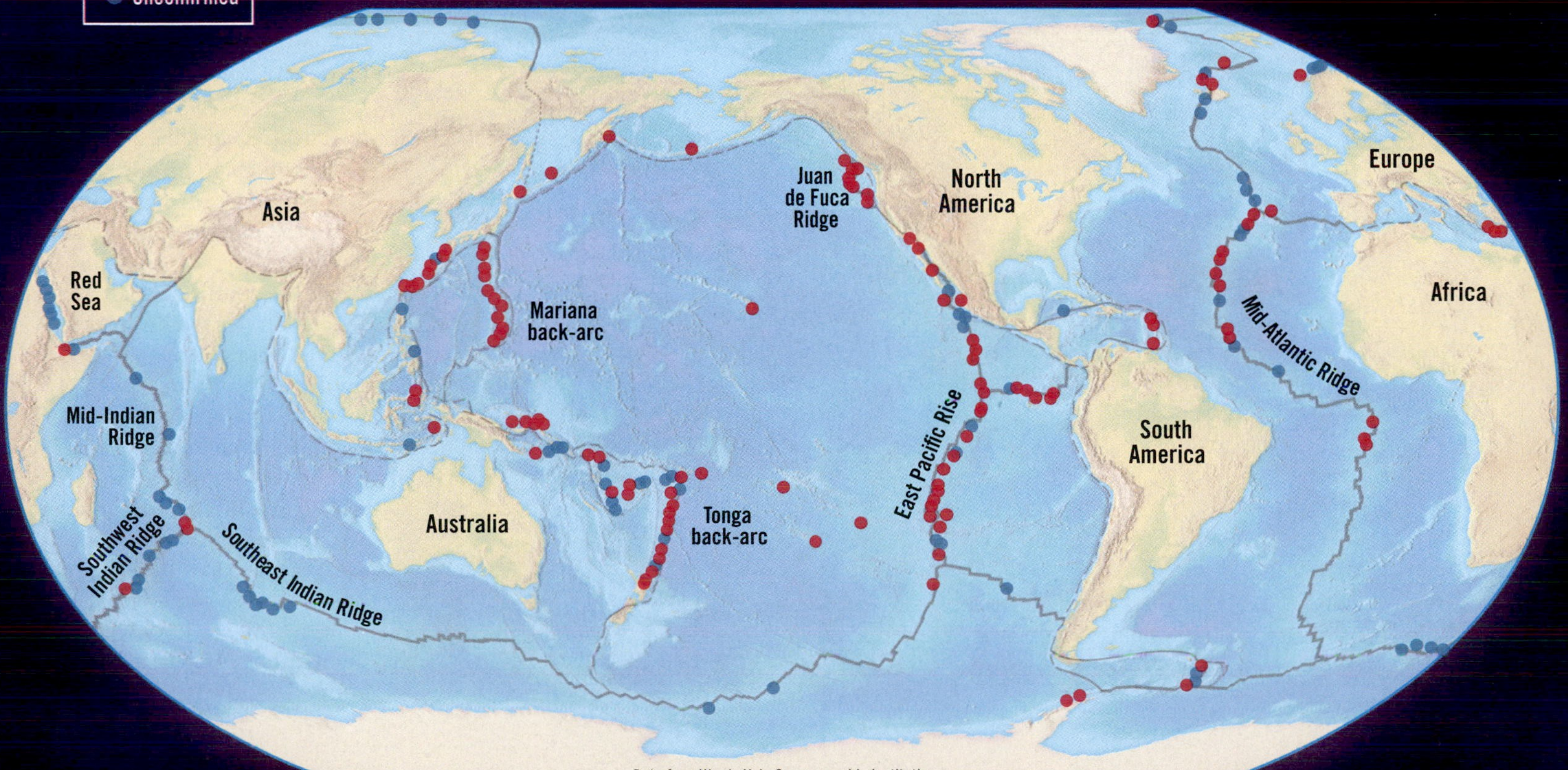

Asia

Red Sea

Mid-Indian Ridge

Southwest Indian Ridge

Southeast Indian Ridge

Australia

Mariana back-arc

Tonga back-arc

Juan de Fuca Ridge

North America

East Pacific Rise

South America

Europe

Africa

Mid-Atlantic Ridge

Data from Woods Hole Oceanographic Institution

Europe). By contrast, where the crust is thin, hot, and weak, rifts can be more than 1000 kilometers (600 miles) wide, as exemplified by the Basin and Range region in the western United States.

In settings where rifting continues, the rift system evolves into a young, narrow ocean basin, such as the present-day Red Sea. Continued seafloor spreading eventually results in the formation of a mature ocean basin bordered by rifted continental margins. The Atlantic Ocean is such a feature.

What follows is an overview of ocean basin evolution, using modern examples to represent the various stages of rifting.

East African Rift The East African Rift is a continental rift that extends through eastern Africa for approximately 3000 kilometers (2000 miles). It consists of several interconnected rift valleys that split into eastern and western sections around Lake Victoria (**Figure 13.20**). Whether this rift will eventually develop into a spreading center, with the Somali subplate separating from the rest of the continent of Africa, is uncertain.

The most recent period of rifting began about 20 million years ago, as upwelling in the mantle intruded the base of the lithosphere (**Figure 13.21A**). Buoyant uplifting of the heated lithosphere led to doming and stretching of the crust. Consequently, the upper crust was broken along high-angle normal faults, producing down-faulted blocks, or *grabens*, while the lower crust deformed by ductile stretching (**Figure 13.21B**).

In the early stages of rifting, magma generated by decompression melting of the rising mantle rocks intruded the crust. Occasionally, some of the magma migrated upward along fractures and erupted at the surface. This activity produced extensive basaltic flows within the rift as well as volcanic cones—some forming more than 100 kilometers (60 miles) from the rift axis. Examples include Mount Kenya and Mount Kilimanjaro, the highest point in Africa, which rises almost 6000 meters (20,000 feet) above the Serengeti Plain.

Red Sea Gradually, a rift valley will lengthen and deepen, eventually extending to the margin of the continent (**Figure 13.21C**). At this point, the continental rift becomes a narrow linear sea with an outlet to the ocean, similar to the Red Sea.

The Red Sea formed when the Arabian Peninsula rifted from Africa beginning about 30 million years ago (see Figure 13.20A). Steep fault scarps that rise as much as 3 kilometers (2 miles) above sea level flank the margins of this water body. Thus, the escarpments

SmartFigure 13.20
East African Rift Valley
A. Map showing the extent of the East African Rift Valley, as well as the rifted valley that is now occupied by the Red Sea. **B.** Satellite-generated image of a small section of the eastern branch of the rift valley located west of Mount Kilimanjaro. Higher uplifted areas appear reddish brown and include Serengeti National Park, which supports the world's greatest concentration of large mammals. The dark green areas and the large lakes are situated on the floor of the rift valley. Volcanoes formed during the rifting can also be seen, including the caldera of the volcano Ngorongoro that is thought to be extinct. (https://goo.gl/7fRZzy)

Tutorial

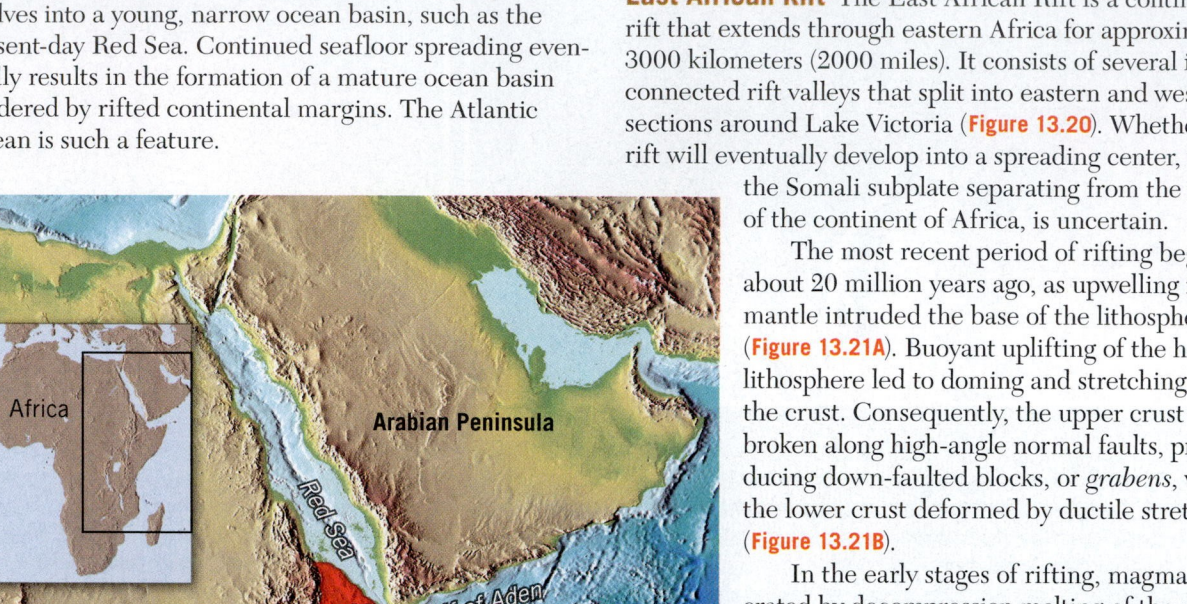

A.

B.

A. Tensional forces and buoyant uplifting of the heated lithosphere cause the upper crust to be broken along normal faults, while the lower crust deforms by ductile stretching.

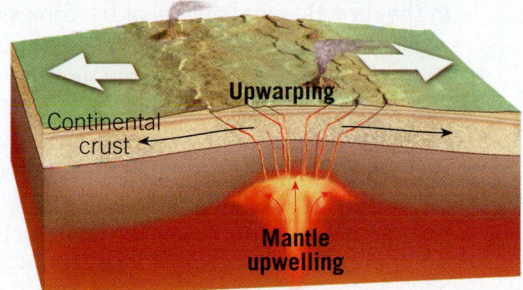

Illustration of the separation of South America and Africa to form the South Atlantic.

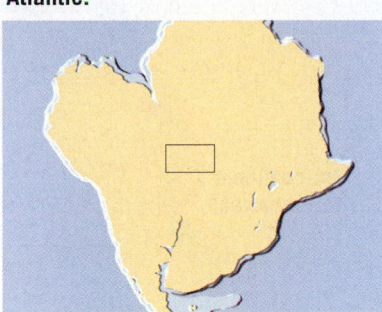

SmartFigure 13.21
Formation of an ocean basin (https://goo.gl/rkXCfv)

B. As the crust is pulled apart, large slabs of rock sink, generating a rift valley.

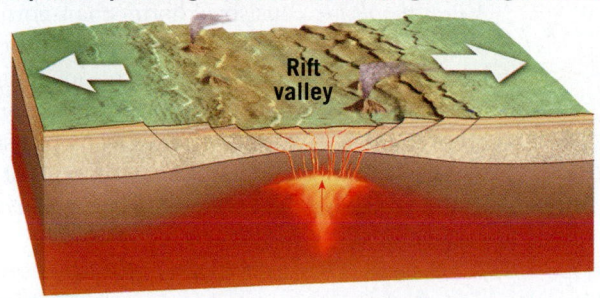

C. Further spreading generates a narrow sea.

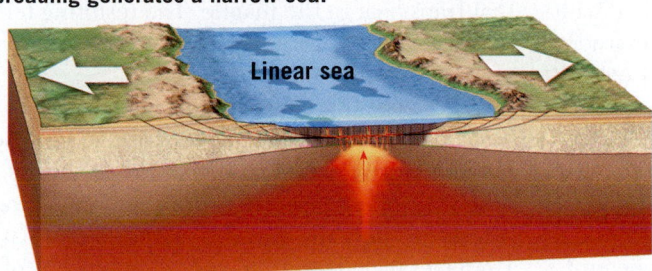

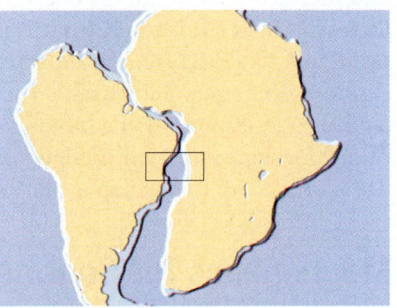

D. Eventually, an expansive ocean basin and ridge system are created.

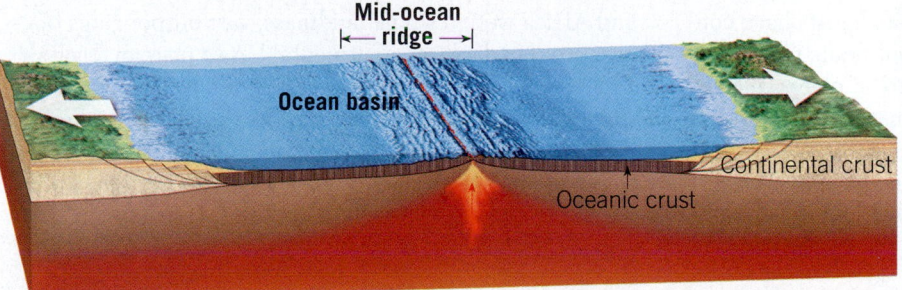

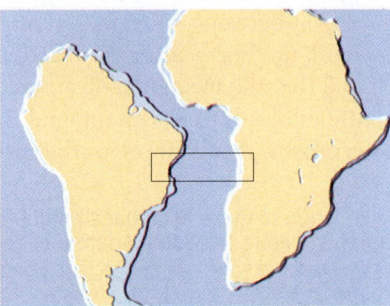

surrounding the Red Sea are similar to the steep cliffs that border the East African Rift. Although the Red Sea reaches oceanic depths (up to 5 kilometers [3 miles]) in only a few locations, symmetrical magnetic stripes indicate that typical seafloor spreading has been occurring for at least the past 5 million years.

Atlantic Ocean If spreading continues, the Red Sea will grow wider and develop an elevated oceanic ridge similar to the Mid-Atlantic Ridge (**Figure 13.21D**). The Atlantic

Ocean shows what the Red Sea could eventually become over tens of millions of years. As the Atlantic formed, new oceanic crust was added to the diverging plates, and the rifted continental margins gradually receded from the region of upwelling. As a result, they cooled, contracted, and sank.

Over time, continental margins subsided below sea level, and material that had eroded from the adjacent highlands blanketed this once-rugged topography. The result was a *passive continental margin* on both sides of

Figure 13.22
Midcontinent rift
Map showing the location of a failed rift extending from the Great Lakes region to Kansas.

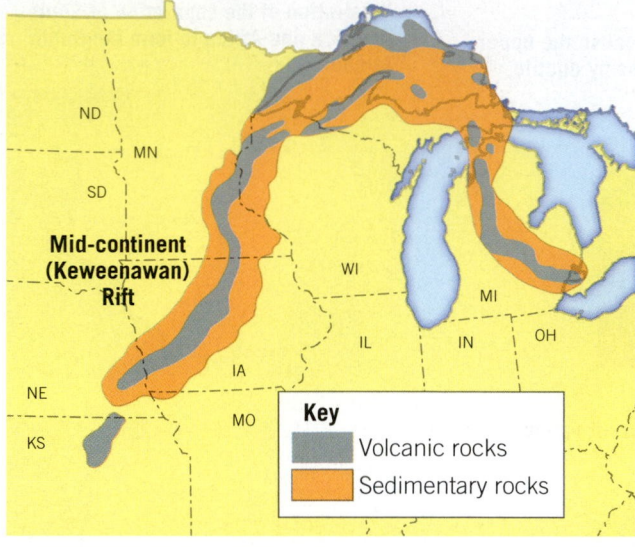

the Atlantic, consisting of rifted continental crust that has been covered by a thick wedge of relatively undisturbed sediment and sedimentary rock.

Failed Rifting Not all continental rift valleys develop into full-fledged spreading centers. In the central United States, a failed rift extends from Lake Superior into Kansas (**Figure 13.22**). This once-active rift valley is filled with clastic sedimentary and basaltic rocks that were extruded onto the crust more than a billion years ago. Why one rift valley develops into a full-fledged active spreading center while others fail to develop is not fully understood.

Mechanisms for Continental Rifting

At least two supercontinents existed in the geologic past. Pangaea, the most recent, was assembled into a supercontinent between 450 and 230 million years ago, only to break up shortly after it formed. Geologists have concluded that the formation of supercontinents followed by continental splitting is an integral part of plate tectonics. This process, which involves the formation and dispersal of supercontinents, is called the **supercontinent cycle** and is described in detail in Chapter 22.

The supercontinent cycle must involve major changes in the direction and nature of the forces that drive plate motion. In other words, over long periods of geologic time, the forces that drive plate motions tend to organize crustal fragments into a single supercontinent, only to change directions and disperse them again. Mechanisms that are thought to contribute to continental rifting include plumes of hot mobile rock rising from deep in the mantle, upwelling from shallow levels in the asthenosphere, and forces that arise from plate motions.

Mantle Plumes and Hot-Spot Volcanism

A *mantle plume* consists of hotter-than-normal mantle rock that has a large mushroom-shaped head hundreds of kilometers in diameter attached to a long, narrow, trailing tail. As the plume head nears the base of the rigid lithosphere, it spreads laterally. Decompression melting within the plume generates huge volumes of basaltic magma that rises and triggers *hot-spot volcanism* at the surface.

Research suggests that mantle plumes tend to concentrate beneath a supercontinent because once assembled, a large landmass forms an insulating "blanket" that traps heat in the mantle. The resulting temperature increase leads to the formation of mantle plumes that serve to dissipate heat.

Evidence that mantle plumes play a role in the breakup of at least some landmasses can be observed in modern passive continental margins. In several regions on both sides of the Atlantic, continental rifting was preceded by crustal uplift and massive outpourings of basaltic lava. Examples include the Etendeka flood basalts of southwest Africa and the Paraná basalt province of South America.

About 130 million years ago, when South America and Africa were a single landmass, vast outpourings of lava produced a large continental basalt plateau (**Figure 13.23A**). Next, the South Atlantic began to open, splitting the basalt province into two parts—the Etendeka and

Figure 13.23
The role that mantle plumes might play in continental rifting

A. Location of these flood-basalt plateaus 130 million years ago, just before the South Atlantic began to open.

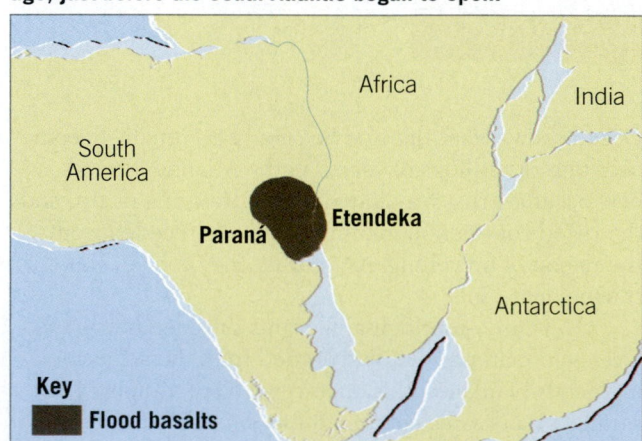

B. Relationship of the Paraná and Etendeka basalt plateaus to the Tristan da Cunha hotspot.

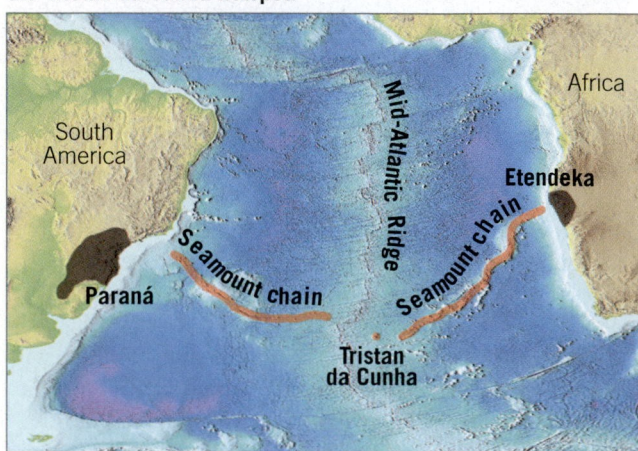

Approximate surface location of mantle plumes prior to the breakup of Pangaea. The location of the plume that produced the Central Atlantic Province is unknown and may have involved a superplume that was deflected by the unusually thick lithosphere beneath western Africa. The Central Atlantic Province includes lava flows, sills, and dikes in northeastern South America, northwestern Africa, southwestern Europe, and eastern North America.

Timing of the breakup of Pangaea along various rift zones and the plume volcanism that was associated with each period of continental fragmentation. In most cases, volcanism appears to precede breakup by a few million years, or more.

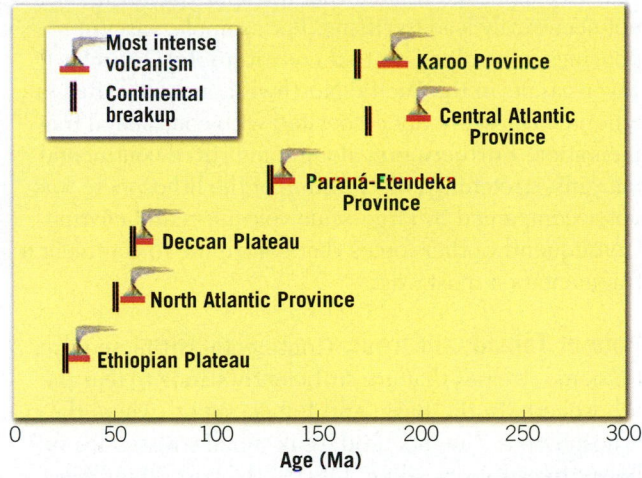

Figure 13.24
The possible role of mantle plumes in the breakup of Pangaea (Data after Courtillot et al.)

Ethiopia called the Afar Lowlands, an area of extensive volcanism (see Figure 13.20). This plume generated a typical rift system consisting of three arms that meet at a triple junction. Two of these rifts, the Red Sea and the Gulf of Aden, are active spreading centers. The third arm is the East African Rift, which may represent the initial stage in the breakup of a continent, as described earlier, or it may be destined to become a failed rift.

Figure 13.24 illustrates the location of a few mantle plumes that have generated large flood basalt plateaus and were presumably involved in the breakup of Pangaea. One mantle plume is currently located beneath Iceland, near the crest of the Mid-Atlantic Ridge (**Figure 13.25**). Vast outpourings of basaltic lava began about 55 million years ago; evidence of these outpourings occurs in eastern Greenland, as well as across the Atlantic, in the Hebrides Islands of northern Scotland. The oldest magnetic stripes

Paraná basalt plateaus. As the ocean basin grew, the tail of the plume produced a string of seamounts on each side of the newly formed ridge (**Figure 13.23B**). The modern area of hot-spot activity is centered around the volcanic island of Tristan da Cunha, on the Mid-Atlantic Ridge.

When hot, buoyant mantle plumes reach the base of the lithosphere, they cause the overlying crust to dome and weaken. Doming—of perhaps as much as 1000 meters (3300 feet)—tends to produce three *rifts*, or *arms*, that join in the area above the rising plume, called a **triple junction**. Frequently, continental breakup and the formation of an ocean basin occur along two of the rift arms, whereas the third arm may be less developed and constitute a failed rift that becomes filled with sediments. For example, the Afar plume, associated with the split of the Arabian Peninsula from Africa, is located beneath a region of northeastern

Figure 13.25
Eruption of fluid basaltic lava, Iceland, 2010 Iceland is located over a mantle plume that began to build this large volcanic island more than 20 million years ago. (Photo by Pakos Photography/Alamy)

between Greenland and Europe are the same age, supporting the connection between the emergence of the Icelandic plume and seafloor spreading in the North Atlantic.

It is important to note that hot-spot volcanism does not necessarily lead to rifting. For example, vast outpourings of basaltic lava that constitute the Columbia River basalts in the Pacific Northwest, as well as Russia's Siberian Traps, are not associated with continental fragmentation. Furthermore, along some rifted continental margins, stretching and thinning of the lithosphere was not accompanied by large-scale volcanism and melting. Consequently, other forces that contribute to continental fragmentation must exist.

Role of Tensional Stress Continental rifting requires tensional stresses that are sufficiently strong to tear the lithosphere. In the Basin and Range region, where the lithosphere is thin, hot, and weak, small stresses are sufficient to cause spreading. During the past 20 million years, a broad zone of upwelling within the asthenosphere is thought to have caused considerable stretching and thinning of the crust in this region (see Figure 14.17,

page 434). In such settings, rifting is accompanied by large-scale melting and volcanism.

Tensional stresses resulting from plate motions are also thought to be particularly significant in continental rifting. In settings where a continent is attached to a subducting slab of oceanic lithosphere, the continental crust is pulled along by the descending slab. However, this continental lithosphere is thick and tends to resist being towed, which creates tensional stresses that may be sufficient to tear the landmass. The zones of rifting in the fragmentation of a supercontinent may be influenced by a preexisting weakness, such as sutures—the sites where continents once collided to form the supercontinent.

13.7 | Concept Checks

1. Name a modern example of a continental rift.
2. Briefly describe each of the four stages in the evolution of an ocean basin.
3. What role do hot spots and mantle plumes play in the breakup of a supercontinent?

13.8 | Destruction of Oceanic Lithosphere
Compare and contrast spontaneous subduction and forced subduction.

Although new lithosphere is continually being produced at divergent plate boundaries, Earth's surface area is not growing larger. In order to balance the amount of newly created lithosphere, there must be a process whereby oceanic lithosphere is destroyed.

Why Oceanic Lithosphere Subducts

The process of plate subduction is complex, and the ultimate fate of oceanic lithosphere is still being debated. Does subducted oceanic crust pile up at the boundary between the upper mantle and the lower mantle, where it is assimilated into the mantle? Or do most plates descend to the core–mantle boundary, where they are heated and eventually rise back toward Earth's surface as mantle plumes?

What is known with some certainty is that oceanic lithosphere will resist subduction unless its overall density is greater than that of the underlying asthenosphere. It takes at least 15 million years for a young slab of oceanic lithosphere to cool sufficiently to become denser than the supporting asthenosphere.

Spontaneous Subduction Subduction zones can be divided into two basic types, based on the nature of the subducting plate. The first type, referred to as a *Mariana-type subduction zone*, is characterized by old, dense lithosphere sinking into the mantle by its own weight. The lithosphere entering the Mariana trench is about 185 million years old, some of the oldest and densest lithosphere in today's oceans. Along this trench, the subducting slab descends into the

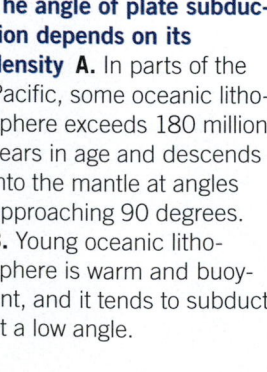

Figure 13.26
The angle of plate subduction depends on its density A. In parts of the Pacific, some oceanic lithosphere exceeds 180 million years in age and descends into the mantle at angles approaching 90 degrees. **B.** Young oceanic lithosphere is warm and buoyant, and it tends to subduct at a low angle.

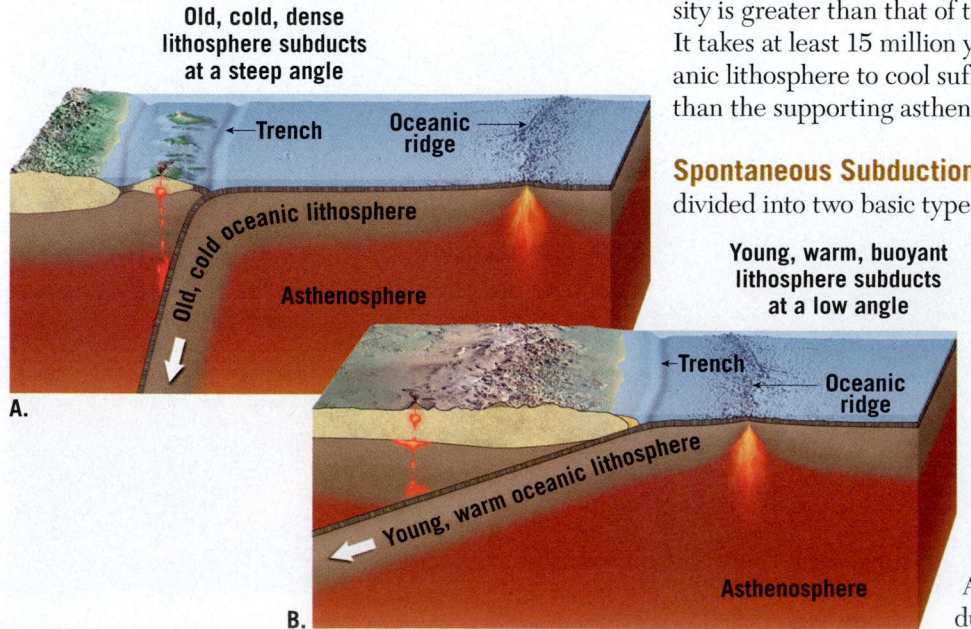

mantle at a steep angle that approaches 90 degrees (**Figure 13.26A**). Steep subduction angles produce deep trenches, which account in part for the depth of the Challenger Deep, located at the southern limb of the Mariana trench. The Mariana and most of the other subduction zones in the western Pacific involve cold, dense lithosphere and therefore exhibit **spontaneous subduction**.

It is important to note that the *lithospheric mantle*, which makes up about 80 percent of the descending oceanic slab, drives subduction. Even when the *oceanic crust* is quite old, its density is still less than that of the underlying asthenosphere. Subduction, therefore, depends on lithospheric mantle that is colder and therefore denser than the underlying asthenosphere.

When an oceanic slab descends to about 400 kilometers (250 miles), mineral phase changes (transitions from low-density mineral to high-density mineral; see Chapter 12) enhance subduction. At this depth, the transition of olivine to its compact, much denser spinel structure increases the density of the slab, which helps pull the plate into the subduction zone.

Forced Subduction The second type of subduction zone, called the *Peru–Chile–type subduction*, is characterized by younger, hotter, and less-dense lithosphere that dips at shallower angles (**Figure 13.26B**). Along Peru–Chile–type boundaries, the lithosphere is too buoyant to subduct spontaneously; rather, it is *forced* beneath the overlying plate by compressional forces.

In areas where **forced subduction** occurs, a strong coupling develops between the overlying plate and the subducting plate, which can result in particularly strong and frequent earthquakes. Stated another way, plate motion generates horizontal compressional forces that cause the upper plate and underlying plate to grind against each other. The result can be folding and thickening of the upper plate and sometimes the formation of mountainous terrains like we see today in the Andes.

Shallow subduction and strong coupling have also been observed in the past decade along the Sunda subduction zone, off the coast of Sumatra, another region that has experienced several major earthquakes.

Researchers have also determined that unusually thick units of oceanic crust, those that approach 30 kilometers (20 miles) in thickness, are likely to resist subduction. The Ontong Java Plateau, for example, is a thick oceanic plateau, about the size of Alaska, located in the western Pacific (see Figure 13.4). About 20 million years ago, this plateau reached the trench that forms the boundary between the subducting Pacific plate and the overriding Australian–Indian plate. Apparently too buoyant to subduct, the Ontong Java Plateau clogged the trench. We will consider the fate of crustal fragments that are too buoyant to subduct in the next chapter.

Subducting Plates: The Demise of Ocean Basins

In the 1970s, geologists began using magnetic stripes and fracture zones on the ocean floor to reconstruct the past 200 million years of plate movement. This research showed that parts of, or even entire, ocean basins have been destroyed along subduction zones. For example, during the breakup of Pangaea shown in Figure 2.23 on page 56, notice that the African plate moved northward, eventually colliding with Eurasia. During this event, the floor of the intervening Tethys Ocean was almost entirely consumed into the mantle, leaving behind a few small remnants—the Eastern Mediterranean Sea and the Black Sea.

Reconstructions of the breakup of Pangaea also helped investigators understand the demise of the Farallon plate—a large oceanic plate that once occupied much of the eastern Pacific basin. At the time of the breakup of Pangaea, the Farallon plate was situated on the eastern side of a spreading center, as shown in **Figure 13.27A**.

SmartFigure 13.27
The demise of the Farallon plate Because the Farallon plate was subducting faster than it was being generated, it continually got smaller and smaller. The remaining fragments of the once-mighty Farallon plate are the Juan de Fuca, Cocos, and Nazca plates.
(https://goo.gl/dnxck4)

Tutorial

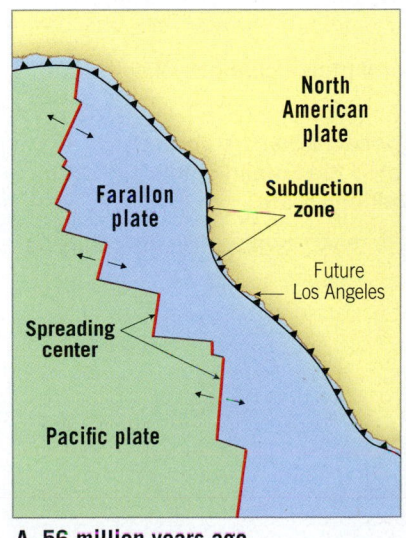

A. 56 million years ago

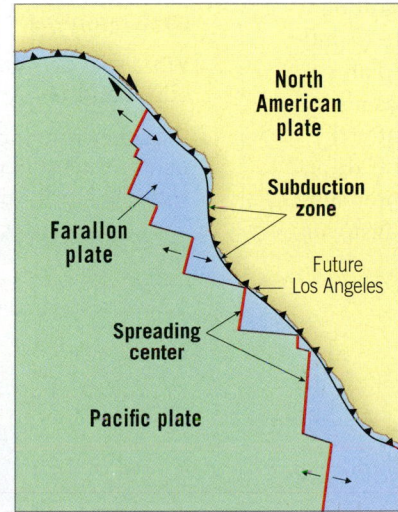

B. 30 million years ago

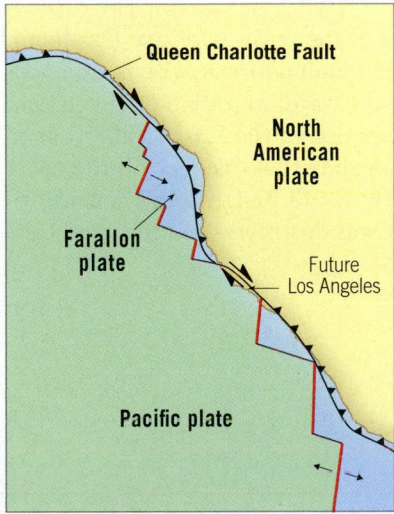

C. 20 million years ago

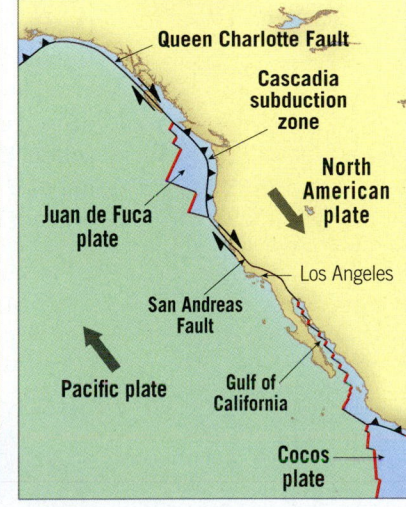

D. Today

Figure 13.28
The separation between the Baja Peninsula and North America (SeaWiFS Project/ORBIGMAGE/NASA/Goddard Space Flight Center)

The spreading center, which generated both the Farallon and Pacific plates, is the East Pacific Rise.

Beginning about 180 million years ago, the Americas were propelled westward as Pangaea broke up and the Atlantic Ocean started to open. As a result, the Farallon plate began subducting beneath the Americas faster than it was being generated, causing it to decrease in size (**Figure 13.27B**). The three remaining fragments of the once-extensive Farallon plate include the modern Juan de Fuca, Cocos, and Nazca plates.

The westward migration of North America also caused a section of the East Pacific Rise to enter the subduction zone that once lay off the coast of California (see Figure 13.27B). As this spreading center subducted, it was destroyed and replaced by a transform fault system that currently accommodates the differential motion between the North American and Pacific plates. Because of this change in plate geometry, the Pacific plate has captured a sliver of North America (the Baja Peninsula and a portion of southern California) and is carrying it northwestward toward Alaska at a rate of about 6 centimeters (2.5 inches) per year.

As more of the ridge subducted, the transform fault system, which we now call the San Andreas Fault, increased in length (**Figure 13.27C**). Today, the southern end of the San Andreas Fault connects to a young spreading center that is generating the Gulf of California (**Figure 13.28**). A similar event generated the Queen Charlotte transform fault, located off the west coast of Canada and southeastern Alaska.

13.8   Concept Checks

1. Explain why oceanic lithosphere subducts even though the oceanic crust is less dense than the underlying asthenosphere.

2. Compare spontaneous subduction and forced subduction. Provide examples of places where each operates.

3. What role do mineral phase changes play in plate subduction?

4. Explain what happened when the spreading center that generated the Farallon plate collided with the North American plate.

13.1 An Emerging Picture of the Ocean Floor

Define *bathymetry* and describe the various bathymetric techniques used to map the ocean floor.

KEY TERMS bathymetry, sonar, echo sounder

- The seafloor is mapped with sonar—shipboard instruments that emit pulses of sound that "echo" off the bottom. Satellites are also used to map the ocean floor. Their instruments measure slight variations in sea level that result from differences in the gravitational pull of features on the seafloor.
- Mapping efforts have revealed three major areas of the ocean floor: continental margins, deep-ocean basins, and oceanic ridges.

13.2 Continental Margins

Compare and contrast a passive continental margin with an active continental margin and list the major features of each.

KEY TERMS continental margin, passive continental margin, continental shelf, continental slope, continental rise, turbidity current, submarine canyon, deep-sea fan, active continental margin, accretionary wedge, subduction erosion

- Continental margins are transitional zones between continental and oceanic crust. Active margins occur along continental plate boundaries, where an oceanic slab is subducting beneath a continent. Passive continental margins are on the trailing edge of continents, far from plate boundaries.
- Heading offshore from the shoreline of a passive margin, a submarine traveler first encounters the gently sloping continental shelf and then the steeper continental slope marking the end of the continental crust and the beginning of the oceanic crust. Beyond the continental slope is the continental rise, which is made of sediment transported by turbidity currents.
- At an active continental margin, material may be added to the leading edge of a continent in the form of an accretionary wedge (common at shallow-angle subduction zones), or material may be scraped off the edge of a continent by subduction erosion (common at steeply dipping subduction zones).

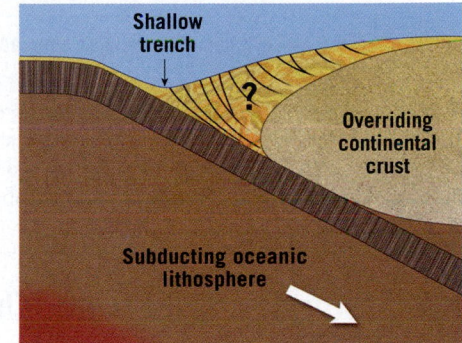

Q What type of continental margin is depicted in this diagram? Name the feature indicated by the question mark.

13.3 Features of Deep-Ocean Basins

List and describe the major features of the deep-ocean basins.

KEY TERMS deep-ocean basin, deep-ocean trench, volcanic island arc, continental volcanic arc, abyssal plain, seismic reflection profiler, seamount, volcanic island, guyot, oceanic plateau

- The deep-ocean basin makes up about half of the ocean floor's area. Much of it is abyssal plain (deep, featureless sediment-draped crust). Subduction zones and deep-ocean trenches also occur in deep-ocean basins. Paralleling trenches are volcanic island arcs (if an oceanic slab is subducted under oceanic lithosphere) or continental volcanic arcs (if an oceanic slab is subducted under an overriding continental plate).
- A variety of volcanic structures poke up from the abyssal plain. Seamounts are submarine volcanoes; if they emerge at the ocean's surface, we call them volcanic islands. Guyots are old volcanic islands that have had their tops eroded off before sinking below the sea. Oceanic plateaus are submarine flood basalt provinces—unusually thick sections of oceanic crust formed by massive underwater eruptions of lava.

Q On this cross-sectional view of a deep-ocean basin, label the following features: a seamount, a guyot, a volcanic island, an oceanic plateau, and an abyssal plain.

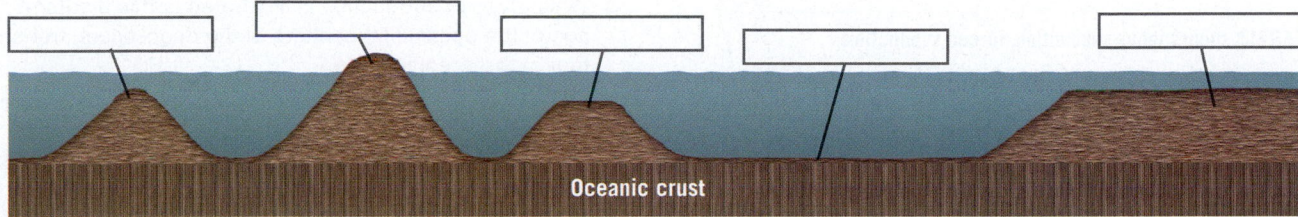

13.4 Anatomy of the Oceanic Ridge

Sketch and label a cross-sectional view of the Mid-Atlantic Ridge. Explain how a cross section of the East Pacific Rise would look different.

KEY TERMS oceanic ridge (mid-ocean ridge or rise), rift valley

- The oceanic ridge system is the longest topographic feature on Earth, wrapping around the world through all major ocean basins. It is a few kilometers tall, a few thousand kilometers wide, and a few tens of thousands of kilometers long. The ridge summit is where new oceanic crust is generated, often marked by a rift valley.

13.5 Oceanic Ridges and Seafloor Spreading

Write a statement describing how spreading rates affect ridge topography.

- Oceanic ridges form due to seafloor spreading: As plates of oceanic lithosphere move apart, the warm mantle beneath rises and undergoes decompression melting. The resulting magma is basaltic in composition. Some of it erupts along the ridge axis, but much of it crystallizes at depth. The freshly cooled basalt or gabbro makes up new oceanic crust.

- Oceanic ridges are elevated features because they are warm and therefore less dense than older, colder oceanic lithosphere that makes up the deep-ocean basins. Heat loss causes the oceanic crust to subside and be buried. After 80 million years, crust that was once an oceanic ridge can become an abyssal plain.
- The rate at which seafloor spreading occurs determines the shape of the oceanic ridge. Ridges with slow spreading rates (1 to 5 centimeters per year) have prominent rift valleys and rugged topography. Those with fast spreading rates (greater than 9 centimeters per year) lack rift valleys and show a smoother, more subdued topography.

13.6 The Nature of Oceanic Crust

List the four layers of oceanic crust and explain how oceanic crust forms and how it differs from continental crust.

KEY TERMS ophiolite complex, sheeted dike complex, pillow lava, black smoker

- Ophiolite complexes contain slices of oceanic crust that have been thrust above sea level. They have four distinct layers (from the top down): (1) deep-sea sediment, (2) pillow lava flows, (3) the sheeted dike complex, and (4) the lowermost gabbro layer.

- As two divergent plates move apart, fractures open perpendicular to the stretching direction, and lava moves up through these cracks, toward the seafloor. Once the lava has cooled and sealed these fractures shut, they become preserved as dikes of the sheeted dike complex. Magma that cools at depth crystallizes to become gabbro. Any lava that makes it to the seafloor is erupted as pillow lavas, which are gradually buried by deep-sea sedimentation.
- Along mid-ocean ridges, seawater flows through fissures in the oceanic crust and is heated by nearby pockets of magma. The warmer the seawater, the more chemically active it becomes. The hot water causes hydrothermal metamorphism and dissolves metal ions. These hot, dark solutions may spew out of the crust as black smokers.

13.7 Continental Rifting: The Birth of a New Ocean Basin

Outline the steps by which continental rifting results in the formation of new ocean basins.

KEY TERMS continental rift, supercontinent cycle, triple junction

- When continents rift apart, new ocean basins may form. The East African Rift is an example of the initial stages of continental breakup, with rift valleys that are sites of basaltic volcanism and deposition of detrital sediment. The Red Sea is an ongoing example of more advanced rifting in which seafloor spreading is occurring and the rift is submerged below sea level. Over time, the rift may widen through seafloor spreading, forming an ocean basin flanked by passive continental margins. A modern example of this stage is the Atlantic Ocean.
- Continental rifting may be initiated by mantle plumes, or perhaps continents are more likely to break when sites of preexisting mechanical weakness are subjected to tensional stresses. The assembly and breakup of supercontinents is called the supercontinent cycle. At least two supercontinents have formed in Earth's past.

Q Consider the Baja Peninsula of Mexico (see Figure 13.28). Which stage of continental rifting does the Gulf of California most closely match?

13.8 Destruction of Oceanic Lithosphere

Compare and contrast spontaneous subduction and forced subduction.

KEY TERMS spontaneous subduction, forced subduction

- Most of the oceanic lithosphere produced by seafloor spreading is matched by an equivalent amount that is destroyed though subduction. Subduction carries oceanic lithosphere into the mantle, where its ultimate fate is still uncertain.
- When oceanic lithosphere is sufficiently old (and therefore cold), it may begin to sink because of its increased density. The resulting spontaneous subduction zone will plunge at an angle near 90 degrees and be marked by a deep trench. The Mariana trench is an example.
- Forced subduction shoves relatively buoyant oceanic lithosphere underneath another plate. This results in shallow subduction angles and large earthquakes along megathrust faults. The Peru–Chile trench is an example.

- Regardless of its initial cause, subduction serves to close ocean basins. This may eventually bring once widely separated landmasses into contact with one another.

Q Which type of subduction is illustrated in this diagram? Describe the age of the oceanic lithosphere at the deep-ocean trench compared to that at the oceanic ridge.

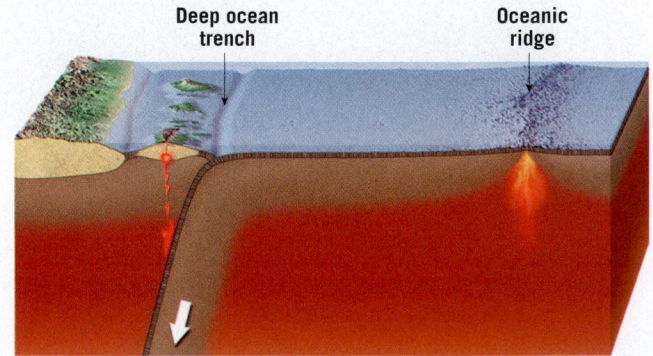

Give It Some Thought

1. How many seconds would it take an echo sounder's ping to make the trip from a ship to the Challenger Deep (10,994 meters) and back? Recall that depth = ½ (1500 m/sec × echo travel time).

2. Refer to the accompanying map, which shows the eastern seaboard of the United States to complete the following:

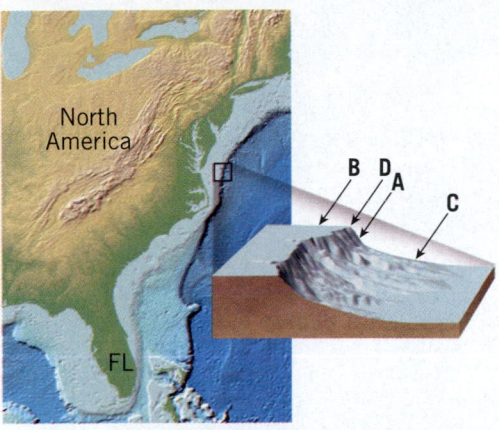

a. Which letter is associated with each of the following: continental shelf, continental rise, and shelf-break?

b. How does the size of the continental shelf that surrounds the state of Florida compare with the size of the Florida peninsula?

c. Why are there no deep-ocean trenches on this map?

3. Mataiva is a small atoll in Tuamotus Archipelago, French Polynesia. Using the accompanying map, determine the approximate number of atolls in this small region of the western Pacific. What conclusion can you draw about the abundance of atolls in the western Pacific?

Tuamotus Archipelago

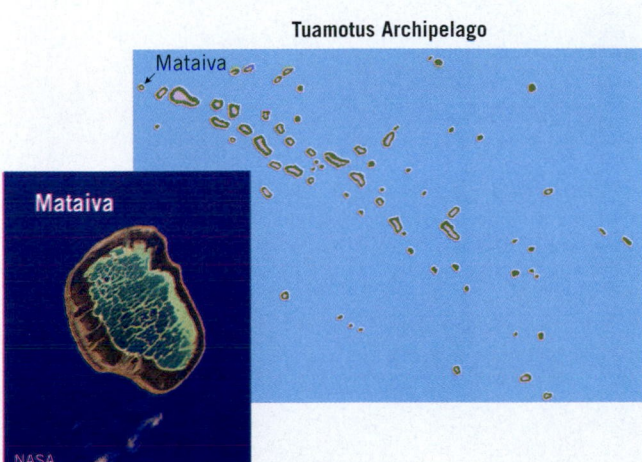

4. Referring to Figure 13.17, compare and contrast the topography of the crest of an oceanic ridge that exhibits a slow spreading rate with one that exhibits a fast spreading rate. Give examples of each.

5. Briefly explain why the ocean floor generally gets deeper the farther one travels from the ridge crest.

6. The accompanying photo is a false-color sonar image that shows the ridge crest (linear pink area) of a section of a spreading center.

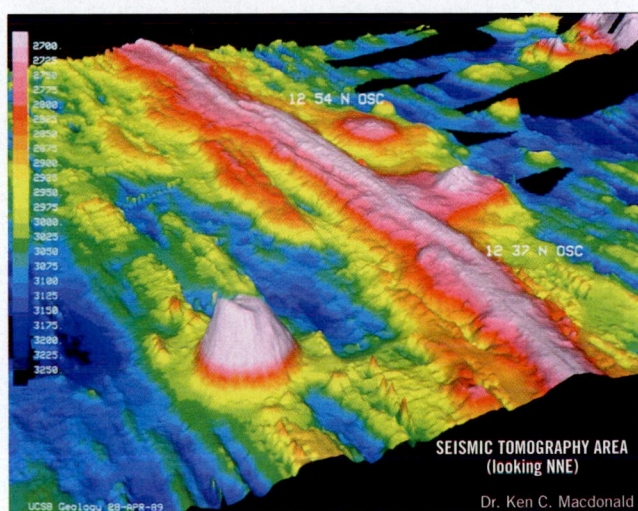

a. Is the structure along the ridge crest characteristic of a fast or slow spreading center? Explain.

b. What name is given to the submerged conical-shaped structure in the lower-left portion of this image?

7. Name a location where a new ocean basin may develop in the future. Explain why you chose that site.

8. Refer to Figure 13.4. At which of the following locations would you expect to have the highest angle of subduction: the Puerto Rico subduction zone or the Cascadia subduction zone? Explain.

9. Refer to Figure 13.27. Predict the fate of the Juan de Fuca plate. What type of boundary might the Cascadia subduction zone become in the future? Explain.

10. Explain this statement: Oceans have existed on Earth for over 4 billion years, but the oldest ocean basin is only about 200 million years old.

11. This image shows lava at a temperature of about 1200°C erupting on the seafloor west of the Tonga trench. What name is given to lava flows like these that erupt underwater?

14

Mountain Building

Mount Rundle consists mainly of limestones and shales that were trust over younger rocks during the formation of the Canadian Rockies. Mount Rundle is located in Banff National Park overlooking the town of Banff, Canada. (Photo by Mike Grandmaison/AGE Fotostock)

Each statement represents the primary **LEARNING OBJECTIVE** for the corresponding major heading within the chapter. After you complete the chapter, you should be able to:

14.1 Locate and name Earth's major mountain belts on a world map.

14.2 List and describe the four major features associated with subduction zones.

14.3 Sketch a cross-section of an Andean-type mountain belt and describe how its major features are generated.

14.4 Compare and contrast the formation of a Cordilleran-type mountain belt with that of an Alpine-type mountain belt.

14.5 Summarize the stages in the formation of a fault-block mountain range.

14.6 Suggest several processes other than tectonic activity that can affect the elevation of a region.

Mountains provide some of the most spectacular scenery on our planet. Poets, painters, and songwriters have captured their splendor. Geologists understand that at some time, all continental regions were mountainous masses and that continents grow through the addition of mountains to their flanks. As geologists unravel the secrets of mountain formation, they gain a deeper understanding of the evolution of Earth's continents. If continents do indeed grow by adding mountains to their flanks, then why do mountains exist in the interior of landmasses? To answer this and related questions, this chapter pieces together the sequence of events that generate these lofty structures.

14.1 | Mountain Building

Locate and name Earth's major mountain belts on a world map.

Mountain building has occurred in the recent geologic past at several locations around the world. Young mountain belts include the American Cordillera (*cordillera* means "spine" or "backbone"), which runs along the western margin of the Americas from Cape Horn at the tip of South America to Alaska and includes the Andes and Canadian Rockies; the Alpine–Himalaya chain, which extends along the margin of the Mediterranean, through Iran to northern India and into Indochina; and the mountainous terrains of the western Pacific, which include volcanic island arcs that comprise Japan, the Philippines, and Sumatra. Most of these young mountain belts have come into existence within the past 100 million years (**Figure 14.1**). Some, including the Himalayas, began their growth as recently as 50 million years ago. The connection between convergent plate boundaries and young mountain belts can be observed by comparing Figure 14.1 with the map of plate boundaries shown in Figure 2.11 (page 45).

In addition to these young mountain belts, there are several chains of Paleozoic-age mountains on Earth. Although these older mountain belts are deeply eroded and topographically less prominent, they exhibit the same structural features found in younger mountains. The Appalachians in the eastern United States and the Urals in Russia are classic examples of this group of older, well-worn mountain belts.

The term for the processes that collectively produce a mountain belt is **orogenesis** (*oros* = mountain, *genesis* = to come into being). Most major mountain belts display

Figure 14.1

Earth's major mountain belts Notice the east–west trend of major mountain belts in Eurasia in contrast to the north–south trend of the North and South American Cordillera. The shields and stable platforms shown are composed of old crustal rocks that have been highly deformed during ancient mountain-building events. These are discussed in Chapter 21.

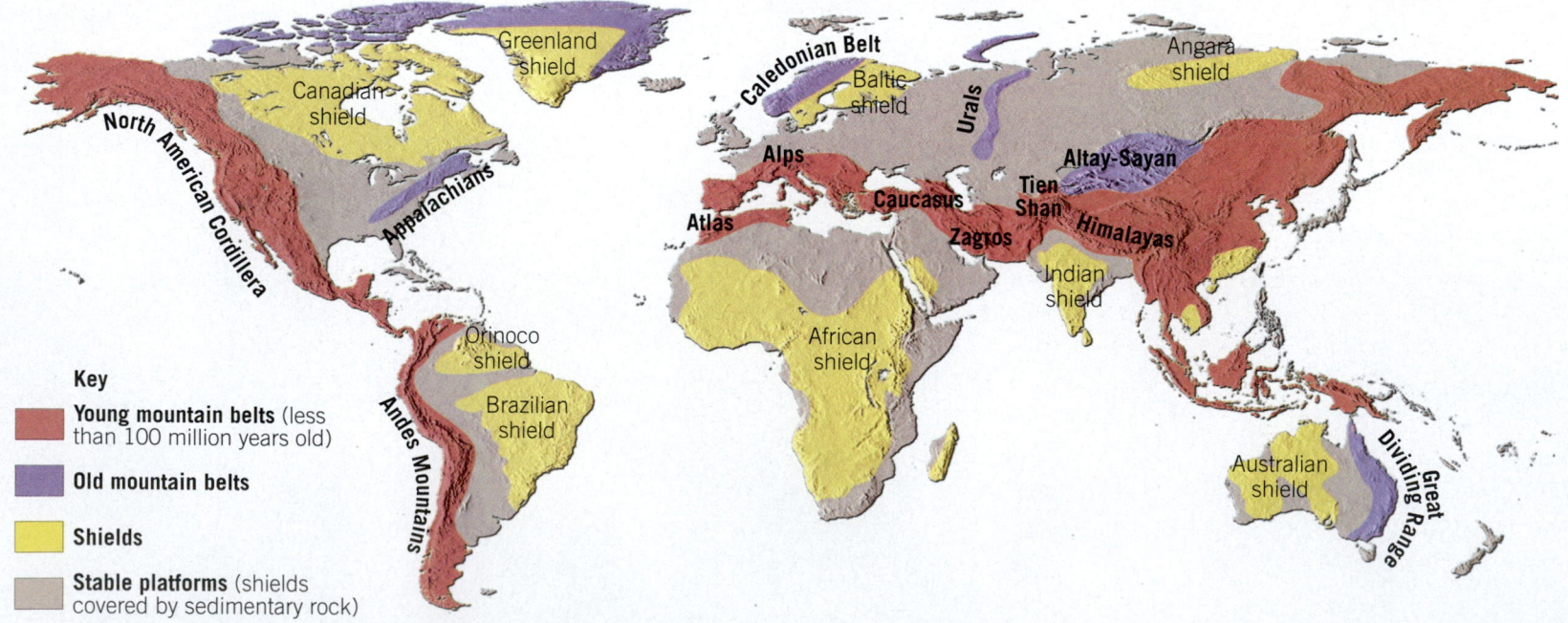

Key

Young mountain belts (less than 100 million years old)

Old mountain belts

Shields

Stable platforms (shields covered by sedimentary rock)

striking visual evidence of great compressional forces that have shortened the crust horizontally while thickening it vertically. These **compressional mountains** contain large quantities of preexisting sedimentary and crystalline rocks that have been faulted and contorted into a series of folds (**Figure 14.2**). Although folding and thrust faulting are often the most conspicuous signs of orogenesis, varying degrees of metamorphism and igneous activity are always present.

How do mountain belts form? Since the time of the ancient Greeks, this question has intrigued philosophers and scientists. One early proposal suggested that mountains are simply wrinkles in Earth's crust, produced as the planet cooled from its original semimolten state. According to this idea, Earth contracted and shrank as it lost heat, which caused the crust to deform in a manner similar to how an orange peel wrinkles as the fruit dries out. However, neither this nor any other early hypothesis withstood scientific scrutiny.

The theory of plate tectonics provides a model for orogenesis with excellent explanatory power and accounts for the origin of virtually all the present mountain belts and most of the ancient ones. According to this model, the tectonic processes that generate Earth's major mountainous terrains occur along convergent plate boundaries, where oceanic lithosphere subducts into the mantle. We will first look at the nature of convergent plate boundaries and then examine how the

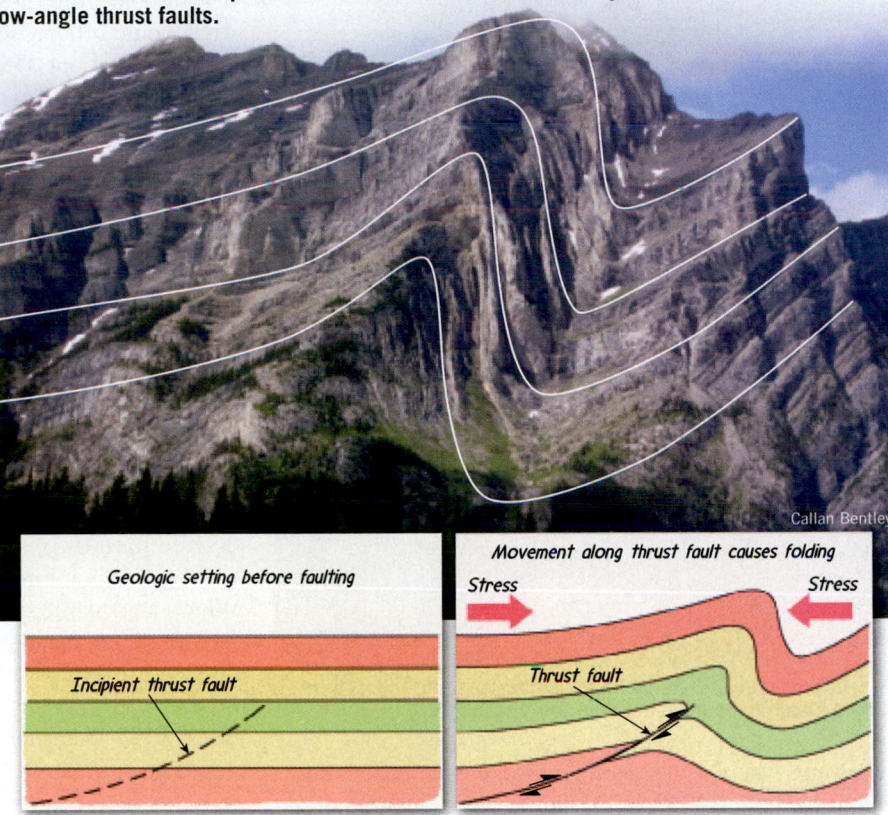

Highly deformed sedimentary strata exposed on the face of Alberta's Mount Kidd. These sedimentary rocks are continental shelf deposits that were folded and displaced toward the interior of Canada by low-angle thrust faults.

Callan Bentley

Geologic setting before faulting

Incipient thrust fault

Movement along thrust fault causes folding

Stress   Stress

Thrust fault

Geologist's Sketch

SmartFigure 14.2
Mount Kidd, Alberta, Canada (https://goo.gl/cPwoXR)

Tutorial

process of subduction has driven mountain building around the globe.

14.1   Concept Checks

1. Define *orogenesis*.

2. In the plate tectonics model, which type of plate boundary is most directly associated with Earth's major mountain belts?

14.2 | Subduction Zones

List and describe the four major features associated with subduction zones.

In their ongoing quest to unravel the events that produce mountains, researchers examine ancient mountain belts as well as sites where orogenesis is currently active. Of particular interest are convergent plate boundaries, where lithospheric plates subduct (see Figure 13.7, page 393). The subduction of oceanic lithosphere generates Earth's strongest earthquakes and most explosive volcanic eruptions, and it plays a pivotal role in generating most of Earth's mountain belts.

Features of Subduction Zones

Subduction zones can be roughly divided into four regions: (1) a *volcanic arc*, which is built on the overlying plate; (2) a *deep-ocean trench*, which forms where subducting slabs of oceanic lithosphere bend and descend into the asthenosphere; (3) a *forearc region*, which is located between a trench and a volcanic arc; and (4) a *back-arc region*, which is located on the side of the volcanic arc opposite the trench.

Volcanic Arcs   Perhaps the most obvious structure generated by subduction is a *volcanic arc* (**Figure 14.3A**).

Figure 14.3
**Development of two
types of volcanic arcs**
A. Volcanic island arcs form
where one slab of oceanic
lithosphere is subducted
beneath another slab of
oceanic lithosphere.
B. Continental volcanic
arcs are generated when a
slab of oceanic lithosphere
subducts beneath a block
of continental crust.

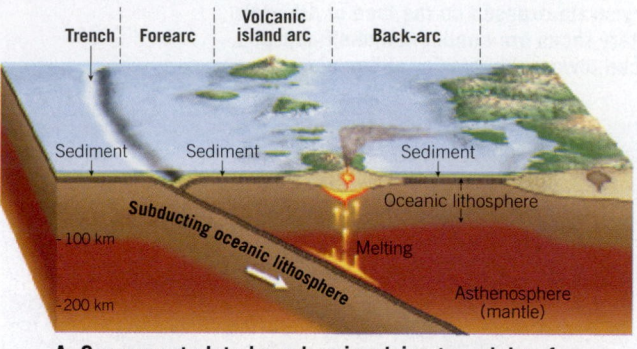

**A. Convergent plate boundary involving two slabs of
oceanic lithosphere.**

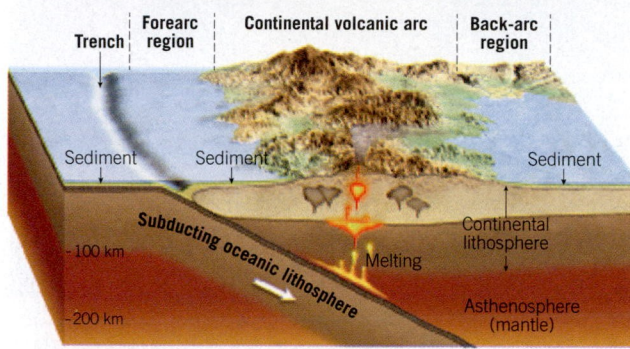

**B. Convergent plate boundary where oceanic lithosphere is
subducting beneath continental lithosphere.**

In settings where two oceanic slabs converge, one is subducted beneath the other, initiating partial melting of the mantle wedge located above the subducting plate. This molten rock rises and eventually leads to the growth of a **volcanic island arc,** or simply an **island arc,** on the ocean floor. Examples of active island arcs include the Mariana, Tonga, and Aleutian arcs in the Pacific (see Figure 2.18, page 52).

By contrast, when oceanic lithosphere is subducted beneath a continental block, a **continental volcanic arc** results (**Figure 14.3B**). Continental volcanic arcs build on the topography of older, thicker continental blocks, resulting in volcanic peaks that may reach 6000 meters (nearly 20,000 feet) above sea level. The Cascade Range of the Pacific Northwest is a classic example.

Deep-Ocean Trenches

Deep-ocean trenches are created where oceanic lithosphere bends as it descends into the mantle. Trench depth is strongly related to the age, and therefore the temperature and density, of the subducting oceanic slab. In the western Pacific, where oceanic lithosphere is cold and dense, oceanic slabs descend into the mantle at steep angles, producing trenches with average depths of about 8 kilometers (5 miles) below sea level. A well-known example is the Mariana trench, where the deepest area is an amazing 10,994 meters (36,069 feet) below sea level.

By contrast, the Cascadia subduction zone off the coasts of Washington and Oregon lacks a well-defined trench, partly because the warm, buoyant Juan de Fuca plate subducts at a very low angle. Trench depth is also related to the availability of sediments. A massive amount of sediment from the Columbia River basin fills most of what would otherwise be a shallow trench in this subduction zone—about 3 kilometers (2 miles) deep.

Forearc

The **forearc** region of a subduction zone is located between a deep-ocean trench and the associated volcanic arc (see Figure 14.3). Here pyroclastic material from the volcanic arc, as well as sediments eroded from the adjacent landmass, accumulate. Ocean-floor sediments are also carried to forearc regions by subducting plates.

The amount of sediment carried to a forearc region varies. The forearc region adjacent to the Mariana trench, for example, contains minimal sediment, partially because of its distance from a significant source of sediment. By contrast, the forearc region adjacent to the Cascadia subduction zone is choked with sediment derived from the nearby outlet of the Columbia River.

In addition, forearc width can vary significantly. Where an oceanic slab subducts at a steep angle, the forearc region is quite narrow, but when the angle of subduction is low, the forearc tends to be broad.

Back-Arc

Another site where sediments and volcanic debris accumulate is the **back-arc,** which is located on the backside of the volcanic arc when viewed from the trench (see Figure 14.3). In these regions tensional forces tend to prevail, causing Earth's crust to stretch and thin and resulting in the formation of a down-faulted basin. The reason for this development is considered in the next section.

Back-arc regions associated with volcanic island arcs tend to be long linear seas, such as the Sea of Japan and the Java Sea. In continental settings, the back-arc regions are located landward of the continental volcanic arc. Here stretching of the crust usually results in subsidence, forming basins that quickly fill with volcanic ash and sediments derived from the growing volcanic structures.

Extension and Back-Arc Spreading

Because subduction zones form where two plates converge, it is logical to assume that large compressional forces deform the plate margins. However, convergent margins are *not necessarily* regions dominated by compressional forces. As mentioned above, tensional stresses act on the overlying plates along some convergent plate margins and cause extension—stretching and thinning—of the crust. But how do extensional processes operate where two plates are moving together?

The age of the subducting oceanic slab is thought to play a significant role in determining the dominant forces acting on the overriding plate. When a relatively cold, dense slab subducts, it does *not* follow a fixed path into

Figure 14.4
Formation of a back-arc basin

A. Subduction and "roll back" of an oceanic slab creates flow in the mantle, called slab suction, that "pulls" the upper (non-subducting) plate toward the retreating trench.

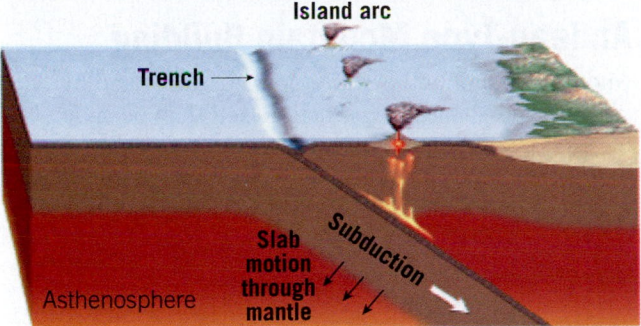

B. Slab suction causes the overlying plate to become elongated and thinned, often resulting in the formation of a basin and spreading center behind the volcanic arc.

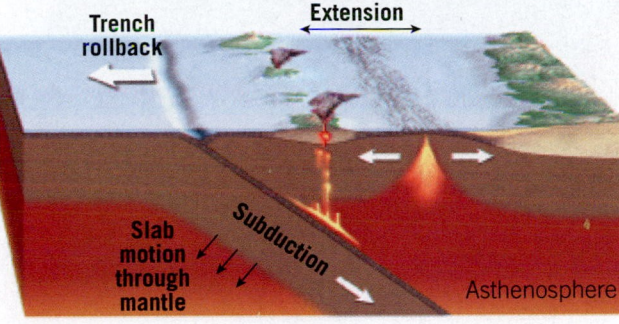

C. Some back-arc basins develop into well-developed spreading centers that generate a deep ocean basin behind the volcanic arc.

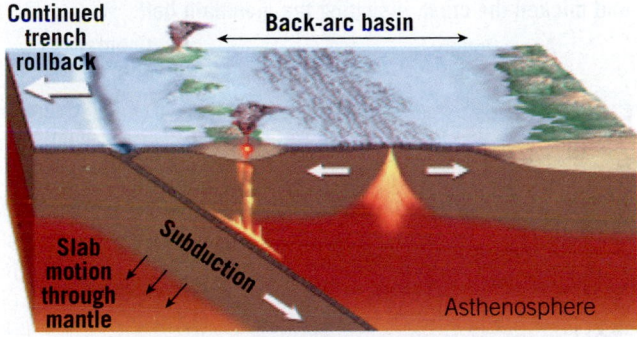

the asthenosphere (**Figure 14.4A**). Rather, it sinks vertically as it descends along an angled path. This causes the trench to retreat, or "roll back," as shown in **Figure 14.4B**. As the subducting plate sinks, it creates a flow in the asthenosphere called *slab suction* that "pulls" the upper plate toward the retreating trench. (Visualize what would have happened if you were in a lifeboat, unable to move away from the *Titanic* as it sank!)

Slab suction, in turn, produces tensional stress that elongates and thins the overriding plate, most often creating a basin in the region behind the volcanic arc (**Figure 14.4C**). Thinning of the crust results in upwelling of hot mantle rock and accompanying decompression melting. Continued extension may initiate seafloor spreading, which increases the size of the newly formed basin. Basins of this type within a back-arc region are termed **back-arc basins** (see Figure 14.4C). Seafloor spreading is currently enlarging the back-arc basins found landward of the Mariana and Tonga volcanic island arcs.

| 14.2 | Concept Checks |

1. List the four major features of subduction zones.
2. Briefly describe how back-arc basins form.

14.3 | Subduction and Mountain Building

Sketch a cross-section of an Andean-type mountain belt and describe how its major features are generated.

The subduction of oceanic lithosphere is the driving force in orogenesis. Oceanic lithosphere subducting beneath an oceanic plate results in a volcanic island arc and related tectonic features. Subduction beneath a continental block, on the other hand, forms a continental volcanic arc and mountainous topography along the margin of a continent. In addition, volcanic island arcs and other crustal fragments "drift" across the ocean basin until they reach a subduction zone, where they collide and become welded to another crustal fragment or a larger continental block. If subduction continues long enough, it can ultimately lead to the collision of two continental blocks, triggering a major mountain-building event.

Island Arc–Type Mountain Building

Island arcs result from the steady subduction of oceanic lithosphere, which may last for 200 million years or more.

Periodic volcanic activity, the emplacement of igneous plutons at depth, and the accumulation of sediment that is scraped from the subducting plate gradually increase the volume of crustal material capping the upper plate

**Figure 14.5
Andean-type mountain
building**

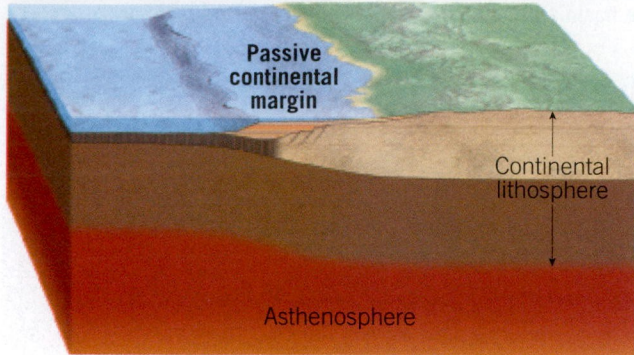

A. Passive continental margin with an extensive platform of
sediments and sedimentary rocks.

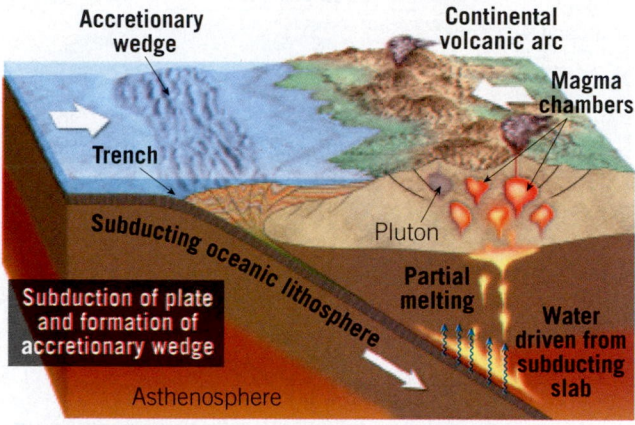

B. Plate convergence generates a subduction zone, and
partial melting produces a continental volcanic arc.
Compressional forces and igneous activity further deform
and thicken the crust, elevating the mountain belt.

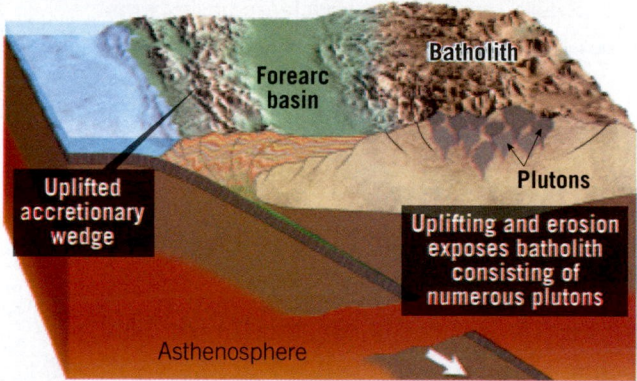

C. Subduction ends and is followed by a period of uplift.

(see Figure 14.3A). Some large volcanic island arcs, such
as Japan, owe their size to having been built on fragments
of continental crust that have rifted from a large landmass or the joining of multiple island arcs over time.

The continued growth of a volcanic island arc can
generate mountainous topography consisting of nearly
parallel belts of igneous and metamorphic rocks. This
activity, however, is viewed as just one phase in the
development of Earth's major mountain belts. As you will
see later, some volcanic arcs are carried by subducting
plates to the margin of large continental blocks, where

they become involved in large-scale mountain-building
episodes.

Andean-Type Mountain Building

Andean-type mountain building is characterized by
subduction beneath a continent rather than oceanic
lithosphere. Subduction along these active continental
margins is associated with long-lasting magmatic activity
that builds continental volcanic arcs. The result is crustal
thickening, with the crust reaching thicknesses of more
than 70 kilometers (45 miles).

The first stage in the development of Andean-type
mountain belts, named after the Andes Mountains of
South America, occurs along *passive continental margins*
(**Figure 14.5A**). Recall that passive continental margins are
geologically inactive regions located some distance from
plate boundaries. Passive continental margins develop
when continental blocks rift apart and are separated
by seafloor spreading that occurs along a mid-ocean
ridge. The margins on both sides of the Atlantic provide
modern examples of passive continental margins, where
continued deposition of sediments has produced a thick
platform of shallow-water sandstones, limestones, and
shales (see Figure 13.6, page 392).

The formation of an Andean-type mountain begins
when the forces that drive plate motions change and a
subduction zone develops along the margin of a continent. In order for a new subduction zone to form, oceanic
lithosphere must be old and dense enough to create a
downward force capable of shearing the lithosphere.
Alternatively, strong compressional forces tear off the
oceanic lithosphere along a continental margin, thereby
initiating subduction.

Building Volcanic Arcs   Recall that as oceanic lithosphere descends into the mantle, increasing temperatures
and pressures drive volatiles (mostly water and carbon
dioxide) from the crustal rocks. These mobile fluids
migrate upward into the wedge-shaped region of mantle
between the subducting slab and upper plate. At a depth
of about 100 kilometers (60 miles), these water-rich fluids
sufficiently reduce the melting point of hot mantle rock
to trigger some melting (**Figure 14.5B**). Partial melting of
the ultramafic rock peridotite generates *primary magmas*,
with mafic (basaltic) compositions. Because these newly
formed basaltic magmas are less dense than the rocks
from which they originated, they will buoyantly rise. Upon
reaching the base of the low-density materials of the continental crust, they typically collect, or pond.

Continued ascent through the thick continental crust
is generally achieved through magmatic differentiation, in
which heavy ferromagnesian minerals crystallize and settle
out of the magma, leaving the remaining melt enriched in
silica and other "light" components (see Chapter 4). Hence,
through magmatic differentiation, a comparatively dense

basaltic magma can generate low-density, buoyant *secondary magma* of intermediate and/or felsic composition.

Emplacement of Batholiths

Because of its low density and great thickness, continental crust significantly impedes the ascent of molten rock. Consequently, a high percentage of the magma that intrudes the crust never reaches the surface; instead, it crystallizes at depth to form massive igneous plutons called *batholiths* (see Figure 4.28C, page 130). The result of this activity is thickening of Earth's crust.

Eventually, uplifting and erosion exhume the batholiths consisting of numerous interconnected plutons (**Figure 14.5C**). The American Cordillera contains several large batholiths, including the Sierra Nevada of California, the Coast Range Batholith of western Canada, and several large igneous bodies in the Andes (**Figure 14.6**). Most batholiths consist of intrusive igneous rocks that range in composition from granite to diorite.

Development of an Accretionary Wedge

During the development of volcanic arcs, unconsolidated sediments that are carried on the subducting plate, as well as fragments of oceanic crust, may be scraped off and plastered against the edge of the overriding plate. The resulting chaotic accumulation of deformed and thrust-faulted sediments and scraps of ocean crust is called an **accretionary wedge** (see Figure 14.5B). The processes that deform these sediments are comparable to a wedge of soil being scraped and pushed in front of an advancing bulldozer.

Some of the sediments that comprise an accretionary wedge are muds that accumulated on the ocean floor and were subsequently carried to the subduction zone by plate motion. Additional materials are derived from an adjacent continent or volcanic arc and consist of volcanic debris and products of weathering and erosion.

In regions where sediment is plentiful, prolonged subduction may thicken a developing accretionary wedge sufficiently that it protrudes above sea level. This has occurred along the southern end of the Puerto Rico trench, where the Orinoco River basin of Venezuela is a major source of sediments. The resulting wedge emerges to form the island of Barbados. By contrast, some subduction zones have small accretionary wedges or none at all.

Forearc Basins

As an accretionary wedge grows upward, it tends to act as a barrier to the movement of sediment from the volcanic arc to the trench. As a result, sediments begin to collect between the accretionary wedge and the volcanic arc. This region, which is composed of relatively undeformed layers of sediment and sedimentary rocks, is called a **forearc basin** (see Figure 14.5C). Subsidence and continued sedimentation in forearc basins can generate a sequence of nearly horizontal sedimentary strata that can attain thicknesses of several kilometers.

Light rock is a batholith composed of granite

Dark rock is roof pendant–metamorphosed host rock

Torres de Plaine, southern Chile

Geologist's Sketch

Figure 14.6
Torres del Paine National Park, Chile This mountainous area, east of the high southern Andes, consists mainly of large granite plutons (light color) between layers of sedimentary and metamorphic rocks (dark color). Magma intruded and metamorphosed the adjacent sedimentary host rock to form these rock bodies that were uplifted and exhumed by erosional forces. (Photo by Michael Collier)

Sierra Nevada, Coast Ranges, and Great Valley

California's Sierra Nevada, Coast Ranges, and Great Valley are excellent examples of the tectonic structures typically generated along an Andean-type subduction zone (**Figure 14.7**). These structures were produced by the

Figure 14.7
Tectonic structures of California generated by an Andean-type subduction zone

EYE ON EARTH 14.1

These interbedded layers of chert and shale have been strongly folded during the growth of an accretionary wedge. A recent period of uplift has exposed these deformed strata near Marin Headlands, north of San Francisco, California.

QUESTION 1 What is the nature of the stress that most likely generated these highly folded strata: compressional or tensional?

QUESTION 2 Along what type of plate boundaries do accretionary wedges form?

QUESTION 3 What type of plate boundary is found today in the San Francisco Bay area?

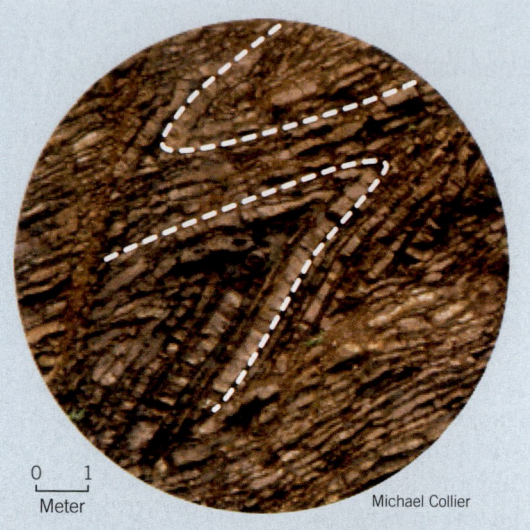

0   1
Meter

Michael Collier

---

subduction of a portion of the Pacific basin (the Farallon plate) under the western margin of California (see Figure 13.28, page 414). The Sierra Nevada Batholith is a remnant of the continental volcanic arc that was produced by many surges of magma during a time span that exceeded 100 million years. The Coast Ranges were built from the vast accumulation of sediments (accretionary wedge) that collected along the continental margin, or perhaps an island arc that lay offshore.

Beginning about 30 million years ago, subduction gradually ceased along much of the margin of North America, as the spreading center that produced the Farallon plate entered the California trench. The uplifting and erosion that followed removed most of the evidence of past volcanic activity and exposed a core of crystalline igneous and associated metamorphic rocks that make up the Sierra Nevada (see Figure 14.5C). The Coast Ranges were uplifted only recently, as evidenced by the young, unconsolidated sediments that currently blanket portions of these highlands.

California's Great Valley is a remnant of the forearc basin that formed between the Sierra Nevada and the accretionary prism and trench that lay offshore. Throughout much of its history, portions of the Great Valley lay below sea level. This sediment-laden basin contains thick marine deposits and debris eroded from the adjacent continental volcanic arc.

14.3 | Concept Checks

1. The formation of mountainous topography at a volcanic island arc is considered just one phase in the development of a major mountain belt. Explain.

2. How do mantle-derived magmas that have a basaltic composition generate magmas that exhibit an intermediate to felsic composition?

3. What is a batholith? In what modern tectonic setting are batholiths being generated?

4. What is an accretionary wedge? Briefly describe its formation.

5. In what ways are the Sierra Nevada and the Andes ranges similar?

14.4 | Collisional Mountain Belts

Compare and contrast the formation of a Cordilleran-type mountain belt with that of an Alpine-type mountain belt.

Most major mountain belts are generated when one or more buoyant crustal fragments collide with a continental margin as a result of subduction. Whereas oceanic lithosphere, which is relatively dense, readily subducts, continental lithosphere contains significant amounts of low-density crustal rocks and is therefore too buoyant to undergo subduction. Consequently, the arrival of a crustal fragment at a trench results in a collision and usually ends further subduction.

Cordilleran-Type Mountain Building

A Cordilleran-type orogeny, named after the North American Cordillera, is associated with a Pacific-like ocean—in that unlike the Atlantic, the Pacific may never close. The rapid rate of seafloor spreading in the Pacific basin is balanced by a high rate of subduction. In this setting, it is highly likely that island arcs or small crustal fragments will be carried along until they collide with an active continental margin. The process of collision and accretion (joining together) of comparatively small crustal fragments to a continental margin has generated many of the mountainous regions that rim the Pacific. Geologists call these accreted crustal blocks *terranes*. The term **terrane** is used to describe a crustal fragment consisting of a distinct and recognizable series of rock formations that has been transported by plate tectonic processes. By contrast, the term *terrain* is used when describing the shape of the surface topography, or "lay of the land."

The Nature of Terranes   What is the nature of crustal fragments, and where did they originate? Research suggests that prior to their accretion to a continental block, some of these fragments may have been **microcontinents** similar to the modern-day island of Madagascar, located east of Africa in the Indian Ocean. Many others were island arcs similar to Japan, the Philippines, and the Aleutian Islands. Still others may have been submerged oceanic plateaus created by massive outpourings of basaltic lavas associated with mantle plumes (see Figure 13.4, pages 390–391). More than 100 of these relatively small crustal fragments, most of which are in the Pacific, are known to exist.

Accretion and Orogenesis   As oceanic plates move, they carry embedded oceanic plateaus, volcanic island arcs, and microcontinents to an Andean-type subduction zone. When an oceanic plate contains small seamounts, these structures are generally subducted along with the descending oceanic slab. However, large, thick units of oceanic crust, such as the Ontong Java Plateau, which is the size of Alaska, or an island arc composed of abundant "light" igneous rocks, render the oceanic lithosphere too buoyant to subduct. In these situations, a collision between the crustal fragment and the continental margin occurs.

The sequence of events that happen when small crustal fragments reach a Cordilleran-type margin is shown in **Figure 14.8**. Rather than subduct, the upper crustal layers of these thickened zones are "peeled" from the descending plate and thrust in relatively thin sheets on the adjacent continental block. Convergence does not generally end with the accretion of a crustal fragment. Rather, new subduction zones typically form, and they can carry other island arcs or microcontinents toward a collision with the continental margin. Each collision displaces earlier accreted terranes further inland, adding to the zone of deformation as well as to the thickness and lateral extent of the continental margin.

A. A microcontinent and a volcanic island arc are being carried toward a subduction zone.

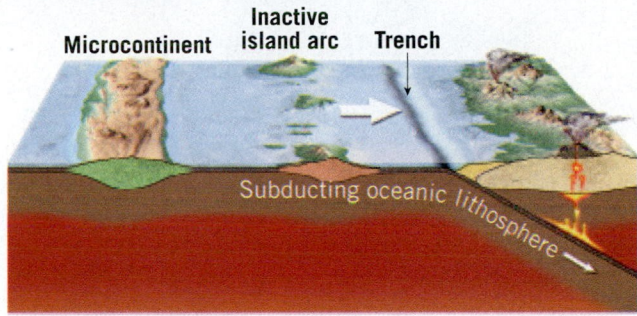

B. The volcanic island arc is sliced off the subducting plate and thrust onto the continent.

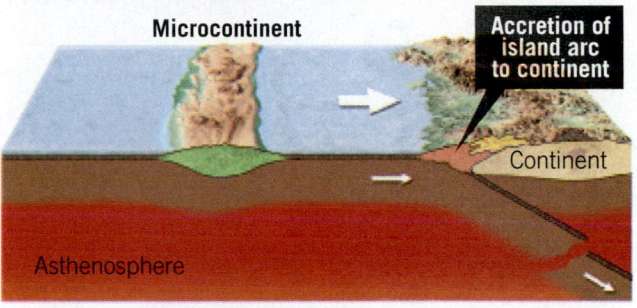

C. A new subduction zone forms seaward of the old subduction zone.

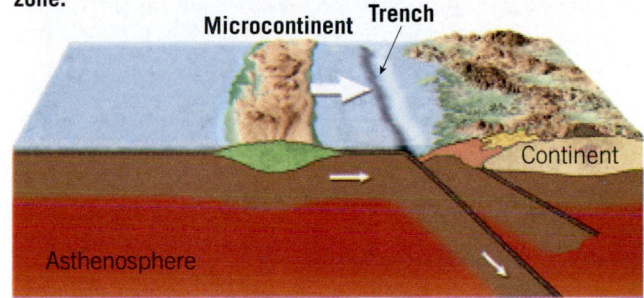

D. The accretion of the microcontinent to the continental margin shoves the remnant island arc further inland and grows the continental margin seaward.

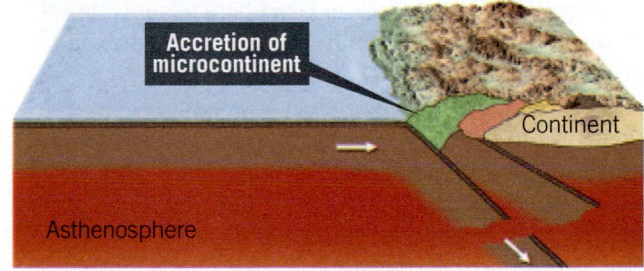

**SmartFigure 14.8
Collision and accretion of small crustal fragments to a continental margin** (https://goo.gl/RtF14w)

Tutorial

The North American Cordillera   The correlation between mountain building and the accretion of crustal fragments arose primarily from studies conducted in the North American Cordillera (**Figure 14.9**). Researchers determined that some of the rocks in the orogenic belts of Alaska and British Columbia contain fossil and paleomagnetic evidence indicating that these strata previously lay much closer to the equator.

▶ Animation

where two continental masses collide. This type of orogeny may also involve the accretion of continental fragments or island arcs that occupied the ocean basin that once separated the two continental blocks. Mountain belts formed by the closure of major ocean basins include the Himalayas, Appalachians, Urals, and Alps (see Figure 14.1). Continental collisions result in the development of mountains characterized by shortened and thickened crust, achieved through folding and large-scale thrust faulting.

The zone where two continents collide and are "welded" together is called the **suture**. This portion of the mountain belt often preserves slivers of oceanic lithosphere, or *ophiolites*, that were trapped between the colliding plates. The unique structure of these pieces of oceanic lithosphere (see Figure 13.18, page 404) helps identify the collision boundary.

Noteworthy features of most collisional mountain ranges are **fold-and-thrust belts**. These mountainous zones result from the deformation of thick sequences of shallow marine sedimentary rocks, like those currently found along passive continental margins of the Atlantic. During continental collisions, sedimentary rocks are pushed inland, away from the core of the developing mountain belt and over the stable continental interior. In essence, crustal shortening is achieved by displacement along thrust faults where once relatively flat-lying strata are "sliced" into thick layers that are eventually stacked one upon another. During this displacement, material caught between the thrust faults is often folded, thereby forming the other major structure of a fold-and-thrust belt. Excellent examples of fold-and-thrust belts are found in the Appalachian Valley and Ridge province, the Canadian Rockies, the Lesser (southern) Himalayas, and the northern Alps.

In the next section we take a closer look at two examples of collisional mountains: the Himalayas and the Appalachians. The Himalayas, Earth's youngest collisional mountains, are still rising. By contrast, the Appalachians are a much older mountain belt, in which active mountain building ceased about 250 million years ago.

It is now known that many of the terranes that make up the North American Cordillera were scattered throughout the eastern Pacific, like the island arcs and oceanic plateaus currently distributed in the western Pacific. During the breakup of Pangaea, the eastern portion of the Pacific basin (Farallon plate) began to subduct under the western margin of North America. This activity resulted in the piecemeal addition of crustal fragments to the entire Pacific margin of the continent—from Mexico's Baja Peninsula to northern Alaska (see Figure 14.9). Geologists expect that many modern microcontinents will likewise be accreted to active continental margins surrounding the Pacific, producing new orogenic belts.

Alpine-Type Mountain Building: Continental Collisions

Alpine-type orogenies, named after the Alps that have been intensively studied for more than 200 years, occur

The Himalayas

The mountain-building episode that created the Himalayas began roughly 50 million years ago, when India began to collide with Asia. Prior to the breakup of Pangaea, India was located between Africa and Antarctica in the Southern Hemisphere. As Pangaea fragmented, India moved rapidly, geologically speaking, a few thousand kilometers in a northward direction.

The subduction zone that facilitated India's northward migration was near the southern margin of Asia (**Figure 14.10A**). Continued subduction along Asia's margin created an Andean-type plate margin containing a well-developed continental volcanic arc and an accretionary wedge. India's northern margin, on the other hand, was a passive continental margin consisting of a thick platform of shallow-water sediments and sedimentary rocks.

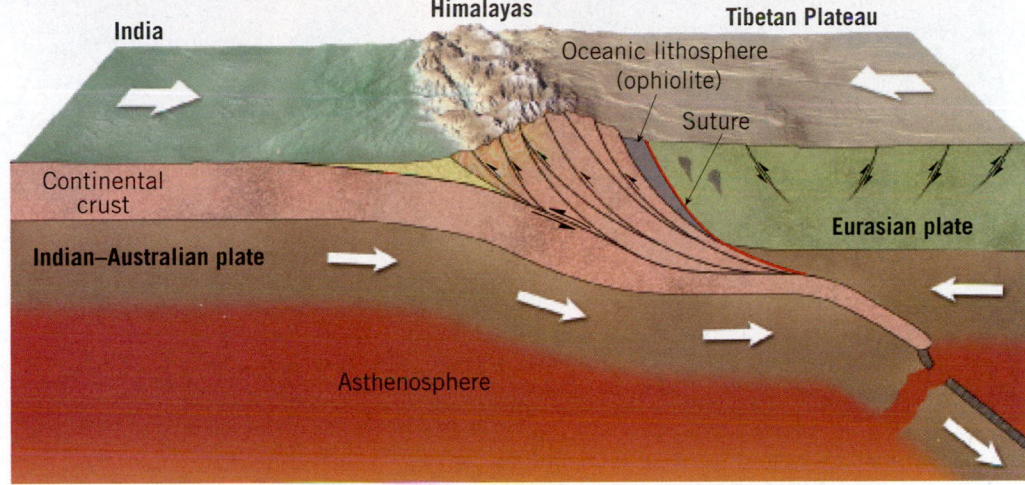

Thick sequence of shallow marine sedimentary rocks

Developing accretionary wedge

Continental volcanic arc

India

Asia

Oceanic crust

Continental crust

Continental crust

Subducting oceanic lithosphere

Asthenosphere

A. Prior to the collision of India and Asia, India's northern margin consisted of a thick platform of continental shelf sediments, whereas Asia's was an active continental margin with a well developed accretionary wedge and volcanic arc.

Himalayas

India

Tibetan Plateau

Oceanic lithosphere (ophiolite)

Suture

Continental crust

Indian–Australian plate

Eurasian plate

Asthenosphere

B. The continental collision folded and faulted the crustal rocks that lay along the margins of these continents to form the Himalayas. This event was followed by the gradual uplift of the Tibetan Plateau as the subcontinent of India was shoved under Asia.

SmartFigure 14.10
Continental collision, the formation of the Himalayas These diagrams illustrate the collision of India with the Eurasian plate that produced the spectacular Himalayas. (https://goo.gl/WQCj6g)

Tutorial

Geologists have determined that one or perhaps more small crustal fragments were positioned on the subducting plate somewhere between India and Asia. During the closing of the intervening ocean basin, a small crustal fragment, which now forms southern Tibet, reached the trench. This event was followed by the docking of India to Eurasia. The tectonic forces involved in the collision of India with Asia were immense, causing the more deformable materials located on the seaward edges of these landmasses to become highly folded and faulted (**Figure 14.10B**). The shortening and thickening of the crust elevated great quantities of crustal material, thereby generating the spectacular Himalaya Mountains (**Figure 14.11**). As a result, tropical marine limestones that formed along the continental shelf now lie at the summit of Mount Everest.

In addition to uplift, crustal shortening caused rocks at the "bottom of the pile" to become deeply buried—an environment where these rocks experienced elevated temperatures and pressures (see Figure 14.10B). Partial melting within the deepest and

most-deformed region of the developing mountain belt produced magmas that intruded the overlying rocks. These environments generate the metamorphic and igneous cores of collisional mountains.

The formation of the Himalayas was followed by a period of uplift that raised the Tibetan Plateau. Seismic evidence suggests that a portion of the Indian subcontinent was thrust beneath Tibet—a distance of perhaps 400 kilometers (250 miles). If this occurred, the added crustal thickness would account for the lofty landscape of southern Tibet, which has an average elevation higher than Mount Whitney, the highest point in the contiguous United States.

The collision with Asia slowed but did not stop the northward migration of India, which has since penetrated at least 2000 kilometers (1200 miles) into the mainland of Asia. Crustal shortening accommodated some of this motion. Much of the remaining penetration into Asia caused lateral displacement of large blocks of the Asian crust by a mechanism described as *continental escape*. As shown in **Figure 14.12**, when

Figure 14.11
Bold peaks of the Karakoram Range, part of the "Greater Himalayas" The second highest peak in the world, K2, is located in the Karakoram Range. (Photo by Jimmy Chin/National GeographicSociety/Getty Images)

India continued its northward trek, parts of Asia were "squeezed" eastward, out of the collision zone. These displaced crustal blocks include much of Southeast Asia (the region between India and China) and sections of China.

Why was the interior of Asia deformed to such a large extent, while India has remained essentially intact? The answer lies in the nature of these diverse crustal blocks. Much of India is a continental shield composed mainly of Precambrian rocks (see Figure 14.1). This thick, cold slab of crustal material has been intact for more than 2 billion years and is mechanically strong as a result. By contrast, Southeast Asia was assembled more recently, from the collision of several smaller crustal fragments. Consequently, it is still relatively "warm and weak" from recent periods of mountain building.

The Appalachians

The Appalachian Mountains provide great scenic beauty near the eastern margin of North America, from Alabama to Newfoundland. Mountain belts of similar origin that formed during the same period and were once contiguous are found in the British Isles, Scandinavia, northwestern Africa, and Greenland (see Figure 2.6, page 41). The orogeny that generated this extensive mountain system lasted a few hundred million years and was one of the stages in assembling the supercontinent of Pangaea. Detailed studies of the Appalachians indicate that the formation of this mountain belt was complex and resulted from three distinct episodes of mountain building.

Our simplified overview begins roughly 750 million years ago, with the breakup of a supercontinent called Rodinia that predates Pangaea. Similar to the breakup of Pangaea, this episode of continental rifting and seafloor spreading generated a new ocean between the rifted continental blocks. Located within this developing ocean basin was an active volcanic arc that lay off the coast of what would later become North America and a microcontinent situated closer to Africa (**Figure 14.13A**).

Figure 14.12
India's continued northward migration severely deformed much of China and Southeast Asia

Map view showing the southeastward displacement of China and the mainland of Southeast Asia as India plowed into Asia.

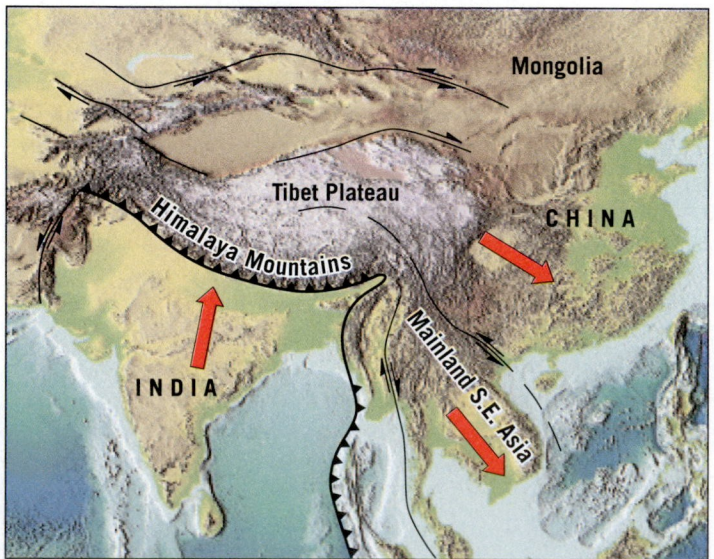

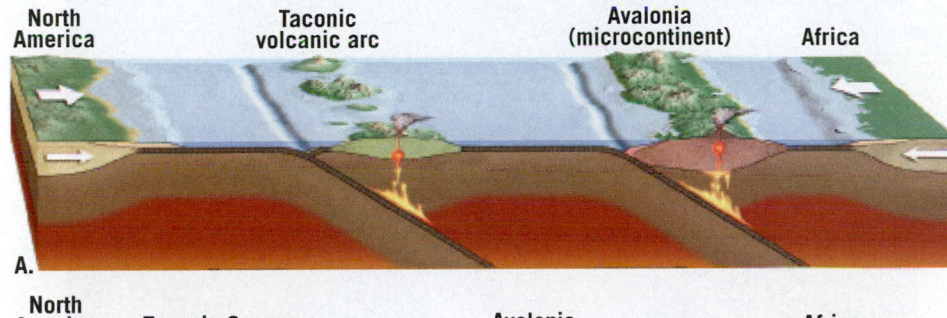

SmartFigure 14.13
Formation of the Appalachian Mountains The Appalachians formed during the closing of a precursor to the Atlantic Ocean. This event involved three separate stages of mountain building that spanned more than 300 million years. (Based on Zve Ben Avraham, Jack Oliver, Larry Brown, and Frederick Cook) (https://goo.gl/b2Deug)

Closing of an Ocean Basin
About 600 million years ago, the precursor to the North Atlantic began to close. Located within this ocean basin was an active volcanic arc that lay off the coast of North America and a microcontinent situated closer to Africa.

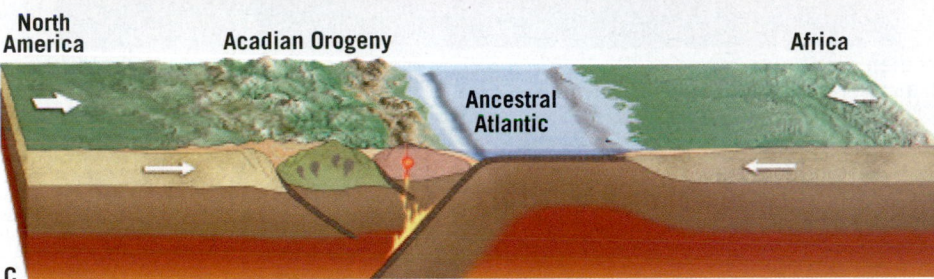

Taconic Orogeny Around 450 million years ago, the marginal sea between the volcanic island arc and North America closed. The collision, called the Taconic Orogeny, thrust the island arc over the eastern margin of North America.

Acadian Orogeny A second episode of mountain building, called the Acadian Orogeny, occurred about 350 million years ago and involved the collision of a microcontinent with North America.

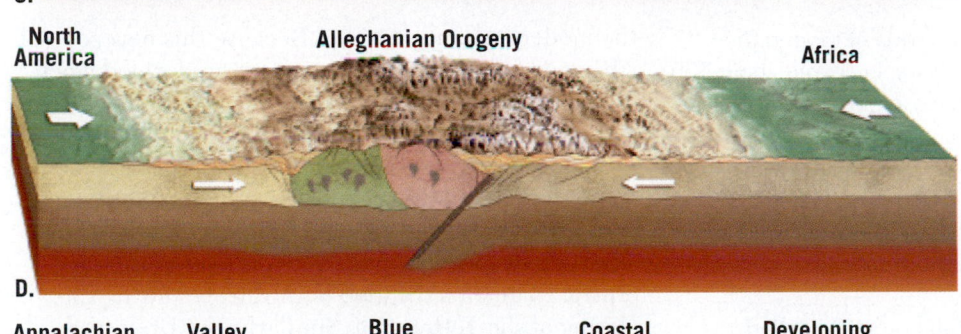

Alleghanian Orogeny The final event, the Alleghanian Orogeny, occurred between 250 and 300 million years ago, when Africa collided with North America. The result was the formation of the Appalachian Mountains, perhaps once as majestic as the Himalayas. The Appalachians lay in the interior of the newly assembled supercontinent of Pangaea.

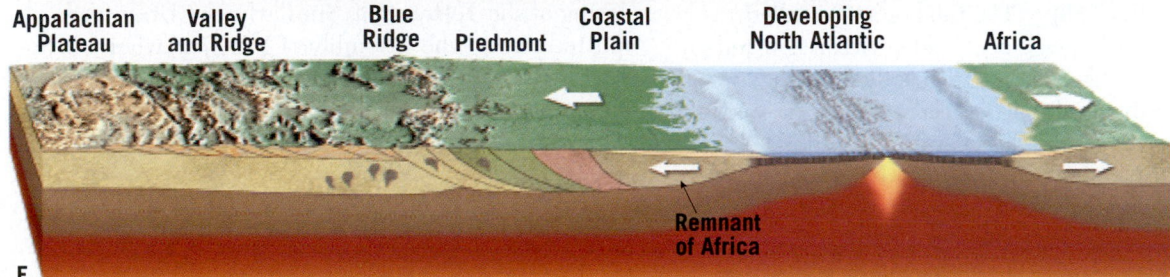

Rifting of Pangaea About 180 million years ago, Pangaea began to break into smaller fragments, a process that ultimately created the modern Atlantic Ocean. Because this new zone of rifting occurred east of the suture that formed when Africa and North America collided, remnants of African crust remain "welded" to the North American plate.

About 600 million years ago, for reasons geologists do not completely understand, plate motion changed dramatically, and the ancient ocean basin began to close. This led to three main orogenic events that culminated in the collision of North America and Africa.

Taconic Orogeny Around 450 million years ago, the marginal sea between the volcanic island arc and ancestral North America began to close. The collision that ensued, called the *Taconic Orogeny*, caused the volcanic arc along with ocean sediments located on the upper plate to be thrust over the larger continental block. The remnants of this volcanic arc and oceanic sediments are recognized today as the metamorphic rocks found across much of the western Appalachians, especially in New York (**Figure 14.13B**). In addition to the pervasive regional metamorphism, numerous magma bodies intruded the crustal rocks along the entire continental margin.

SmartFigure 14.14
The Valley and Ridge province This region of the Appalachian Mountains consists of folded and faulted sedimentary strata that were displaced landward along thrust faults as a result of the collision of Africa with North America. (NASA/GSFC/JPL, MISR Science Team) (http://goo.gl/sMnjA6)

Mobile Field Trip

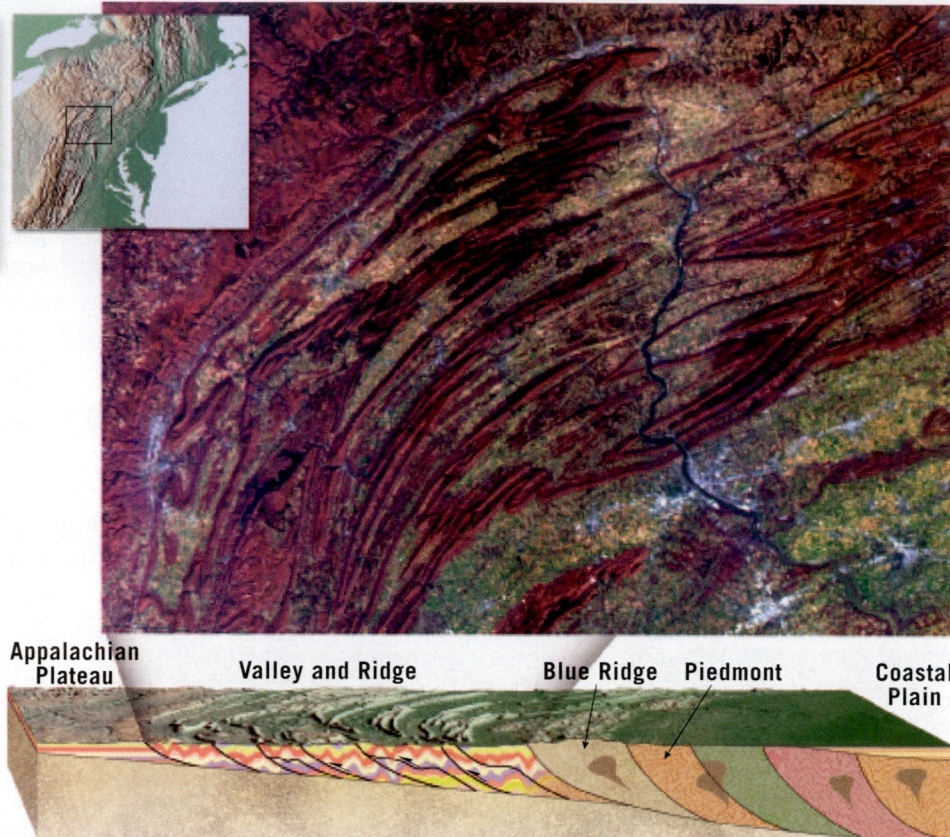

Appalachian Plateau   Valley and Ridge   Blue Ridge   Piedmont   Coastal Plain

Acadian Orogeny   A second episode of mountain building, called the *Acadian Orogeny*, occurred about 350 million years ago. The continued closing of this ancient ocean basin led to the collision of a microcontinent with North America (Figure 14.13C). This orogeny involved thrust faulting, metamorphism, and the intrusion of several large granite bodies. In addition, this event added substantially to the width of North America.

Alleghanian Orogeny   The final orogeny, called the *Alleghanian Orogeny*, occurred between 250 and 300 million years ago, when Africa collided with North America. The material that was accreted earlier was displaced by as much as 250 kilometers (155 miles) toward the interior of North America. This event also displaced and further deformed the shelf sediments and sedimentary rocks that had once flanked the eastern margin of North America (Figure 14.13D). Today these folded and thrust-faulted sandstones, limestones, and shales make up the largely unmetamorphosed rocks of the Valley and Ridge province (Figure 14.14). Outcrops of the folded and thrust-faulted structures that characterize collisional mountains are found as far inland as central Pennsylvania and West Virginia.

With the collision of Africa and North America, the majestic Appalachians were located in the interior of Pangaea. Then, about 180 million years ago, this newly formed supercontinent began to break into smaller fragments, a process that ultimately created the modern Atlantic Ocean. Because this new zone of rifting occurred east of the suture that formed when Africa and North America collided, remnants of Africa remain "welded" to the North American plate (Figure 14.13E). The crust underlying Florida is an example.

Other mountain ranges that exhibit evidence of continental collisions include the Alps and the Urals. The Alps formed as Africa and at least two smaller crustal fragments collided with Europe during the closing of the Tethys Sea. Similarly, the Urals were uplifted during the assembly of Pangaea, when northern Europe and northern Asia collided, forming a major portion of Eurasia.

14.4   Concept Checks

1. Differentiate between *terrane* and *terrain*.

2. During the formation of the Himalayas, the continental crust of Asia was deformed more than India proper. Why was this the case?

3. How does the plate tectonics theory help explain the existence of fossil marine life in rocks atop compressional mountains?

4. How can the Appalachian Mountains be considered a collision-type mountain range when the nearest continent is 5000 kilometers (3000 miles) away?

14.5 | Fault-Block Mountains

Summarize the stages in the formation of a fault-block mountain range.

Most mountain belts form in compressional environments, as evidenced by the predominance of large thrust faults and folded strata. However, other tectonic processes, such as continental rifting, can also produce mountainous terrain. Recall that continental rifting occurs when tensional forces stretch and thin the lithosphere, resulting in upwelling of hot mantle rock. Upwelling heats the thinned lithosphere, which becomes less dense (more buoyant) and rises. This accounts, in part, for the elevated topography associated with continental rifts. Simultaneously, stretching elongates the rigid upper crust, which breaks into large crustal blocks that are bounded by high-angle normal faults. Continued rifting causes the blocks to tilt, with one edge rising as the other drops (see Figure 10.18, page 314). Mountains that form in these tectonic settings are termed **fault-block mountains**.

The Teton Range in western Wyoming is an excellent example of fault-block mountains. This lofty structure was faulted and uplifted along its eastern flank as the block tilted downward to the west. Looking west from Jackson Hole, Wyoming, the eastern front of this mountain rises more than 2 kilometers (1.2 miles) above the valley, making it one of the most imposing mountain fronts in the United States (**Figure 14.15**).

The Basin and Range Province

Located directly east of the Sierra Nevada is the Basin and Range province—one of Earth's largest regions of fault-block mountains. This region extends in a roughly north–south direction for nearly 3000 kilometers (2000 miles) and encompasses all of Nevada and portions of the surrounding states, as well as a large area of western Mexico (**Figure 14.16**). In the Basin and Range province, Earth's brittle upper crust has been broken into hundreds of fault blocks. Uplifting and tilting of these faulted

Figure 14.16
Map of the Basin and Range province in the United States

A.

Grand Teton Range

Jackson Hole Valley

B.

Figure 14.15
Wyoming's Teton Range, an example of fault-block mountains (Photo by Rob Marmion/Shutterstock)

structures produced nearly parallel mountain ranges, averaging about 80 kilometers (50 miles) in length, which rise above adjacent sediment-filled basins (see Figure 10.18, page 314).

Geologists have proposed several hypotheses to explain the events that generated the Basin and Range region. The most widely accepted view is that a change in the nature of the plate boundary along California's western margin led to the formation of this region (**Figure 14.17**). About 30 million years ago, the dominant forces acting on the western margin of North America were compressional, caused by the buoyant subduction of a segment of the Farallon plate (see **Figure 14.17A**). Starting about 25 million years ago, subduction gradually ceased along the California coast as the spreading center that separated the Farallon plate from the Pacific began to subduct beneath North America.

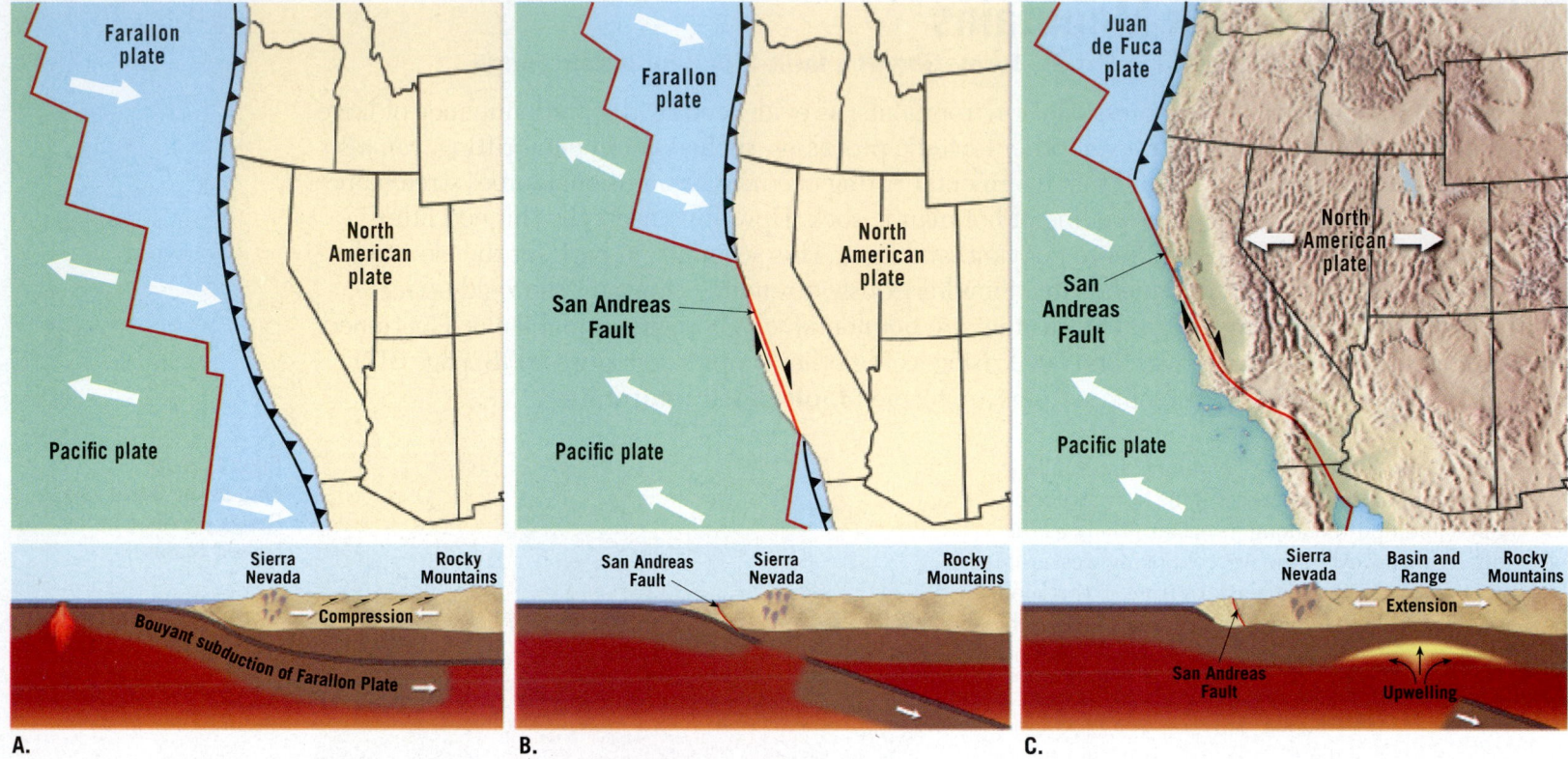

Figure 14.17
Proposed model for the formation of the Basin and Range province The Basin and Range region consists of more than 100 fault-block mountains generated during the past 20 million years. Upwelling of hot mantle rock and gravitational collapse (crustal sliding) may have contributed to considerable stretching and thinning of the crust.

This event spawned the San Andreas Fault, with its strike-slip motion, that currently separates the Pacific plate from the North American plate (see **Figure 14.17B**). According to this hypothesis, the northwestward motion of the Pacific plate produced tensional forces that stretched and fractured the crust of the North American plate to produce the fault-block mountains of the Basin and Range province (see **Figure 14.17C**). Stretching and thinning of the lithosphere also led to mantle upwelling, which accounts for the higher-than-average elevation of this region.

Another model contends that about 20 million years ago, the cold, dense lithospheric mantle located beneath the Basin and Range region decoupled (separated) from the overlying crustal layer and slowly sank into the mantle. This process, called *delamination*, resulted in upwelling and lateral spreading of hot mantle rock that produced tensional forces that stretched and thinned the

overlying crust. According to this view, these elevated crustal blocks began to gravitationally slide from their lofty perches to generate the fault-block topography of the Basin and Range province.

14.5 | Concept Checks

1. Name one process that accounts for the elevated topography of at least some fault-block mountains.

2. In what way does the formation of fault-block mountains differ from the processes that generate most other major mountain belts?

3. Briefly describe the basic structure of the Basin and Range province and identify its geographic extent.

14.6 | What Causes Earth's Varied Topography?

Suggest several processes other than tectonic activity that can affect the elevation of a region.

The causes of Earth's varied topography are complex and cumulative; see, for example, GEOgraphics 14.1. Geologists know that colliding plates provide the tectonic forces that thicken and elevate crustal rocks during mountain building. Adjustments in Earth's gravitational balance can affect elevation, and up-and-down motions within the mantle can also change the elevation of a region. Furthermore, whenever land is uplifted, weathering and erosion sculpt Earth's surface into a vast array of landforms. In this section large-scale regional uplift and subsidence are discussed; the specific ways in which weathering and erosion alter Earth's surface are covered in subsequent chapters.

The Principle of Isostasy

During the 1840s, researchers discovered that Earth's low-density crust "floats" on top of the high-density, deformable rocks of the mantle. The concept of a floating crust in gravitational balance is called **isostasy**. The principle of isostasy helps us understand many large-scale variations on Earth' surface—from towering mountains to deep-ocean basins.

One way to explore the concept of isostasy is to envision a series of wooden blocks of different heights floating in water, as shown in **Figure 14.18**. Note that the thicker wooden blocks float higher than the thinner blocks. Similarly, compressional mountains stand high above the surrounding terrain because crustal thickening creates buoyant crustal "roots" that extend deep into the supporting material below. Thus, lofty mountains such as the Himalayas are much like the thicker wooden blocks shown in Figure 14.18.

How Is Isostasy Related to Changes in Elevation?

Visualize what would happen if another small block of wood were placed atop one of the blocks in Figure 14.18. The combined block would sink until it reached a new isostatic (gravitational) balance. At this point, the top of the combined block would be higher than before, and the bottom would be lower. This process of establishing a new level of gravitational balance by loading or unloading is called **isostatic adjustment**.

Applying the concept of isostatic adjustment, we should expect that when weight is added to the crust, it will respond by subsiding and will rebound when weight is removed. (Visualize what happens when a ship's cargo is loaded or unloaded.) Evidence for crustal subsidence followed by crustal rebound is provided by Ice Age glaciers. When continental ice sheets occupied portions of North America during the Pleistocene epoch, the added weight of 3-kilometer- (2-mile-) thick masses of ice caused downwarping of Earth's crust by hundreds of meters. In the 8000 years since this last ice sheet melted, gradual uplift of more than 300 meters (1000 feet) has occurred in Canada's Hudson Bay region, where the thickest ice had accumulated.

One of the consequences of isostatic adjustment is that, as erosion lowers a mountain range, the crust

rises in response to the reduced load (**Figure 14.19**). The processes of uplift and erosion continue until the mountain block reaches "normal" crustal thickness. When this occurs, these once-elevated structures will be near sea level, and the once–deeply buried interior of the mountain will be exposed at the surface. In addition, as mountains are worn down, the eroded sediment is deposited on adjacent landscapes, causing these areas to subside (see Figure 14.19).

When compressional mountains are young they are composed of thick, low density crustal rocks that float on the denser mantle below.

As erosion lowers the mountains, the crust rises in response to the reduced load in order to maintain isostatic balance.

Erosion and uplift continue until the mountains reach "normal" crustal thickness.

 Animation

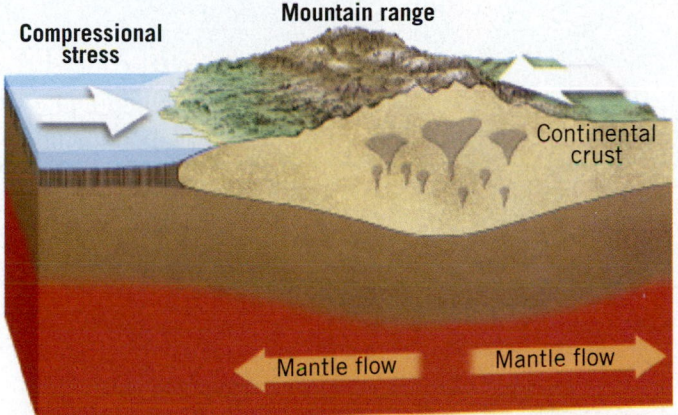
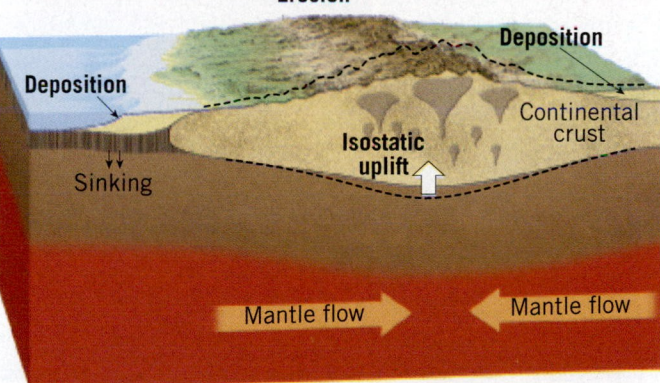
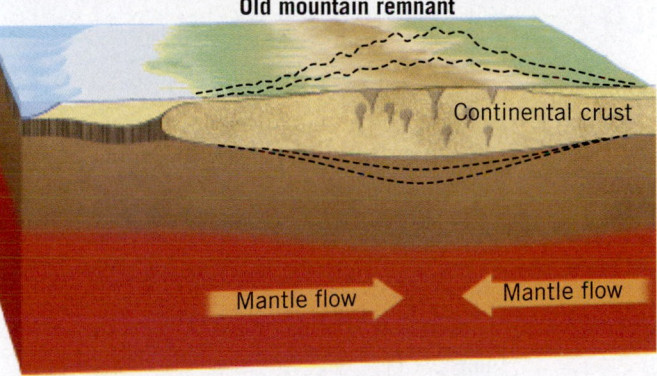

 Tutorial

The Laramide Rockies

The portion of the Rocky Mountains that extends from southwestern Montana to New Mexico was produced during a period of deformation known as the Laramide Orogeny. This event, which created some of the most picturesque scenery in the United States, peaked about 60 million years ago.

Colorado Rockies near Steamboat Springs
(Photo by Michael Collier)

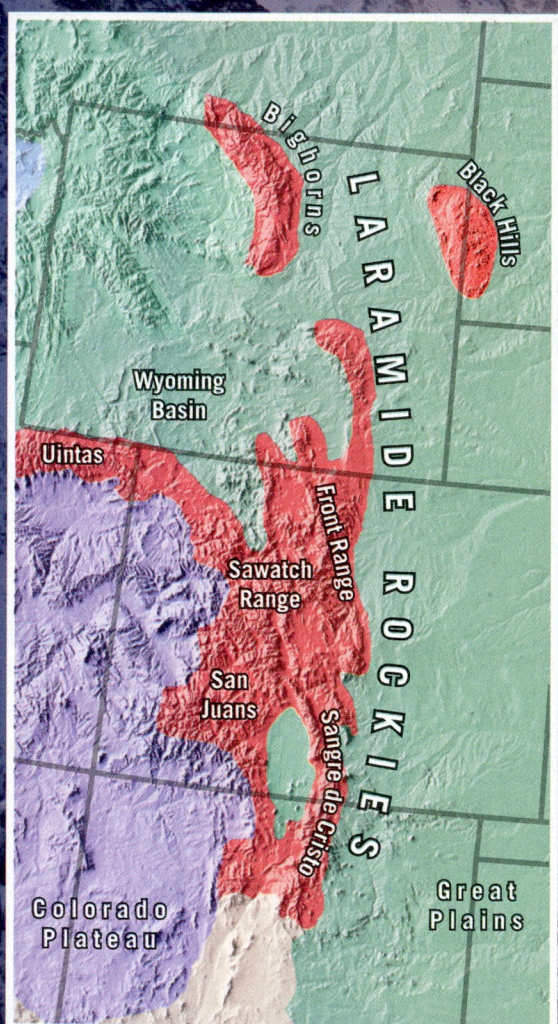

Where are the Laramide Rockies?

Sometimes called the Central and Southern Rockies, this mountain belt lies to the east of the Colorado Plateau and includes the Bighorns of Wyoming, the Front Range of Colorado, the Uintas of Utah, and the Sangre de Cristo of southern Colorado and New Mexico.

What is the geologic history of the Laramide Rockies?

These mountain ranges formed when Precambrian age basement rocks were uplifted nearly vertically along reverse and thrust faults, upwarping the overlying layers of younger sedimentary rocks. Uplifting accelerated the processes of weathering and erosion, which removed much of the younger sedimentary cover from the highest portions of the uplifted blocks. Intrusion of igneous plutons and volcanism occurred simultaneously with this period of mountain building.

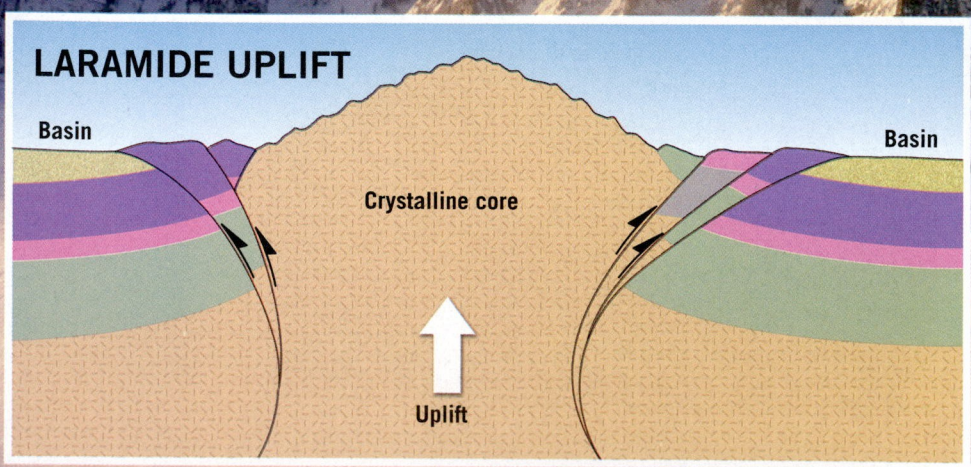

LARAMIDE UPLIFT

Basin

Crystalline core

Basin

Uplift

KEY
- Paleogene sedimentary rocks
- Mesozoic sedimentary rocks
- Paleozoic sedimentary rocks
- Precambrian sedimentary rocks
- Precambrian crystalline rocks

What is responsible for the steep, rugged form of these mountains?

The resulting mountainous topography consists mainly of individual, elongated blocks of igneous and metamorphic rocks flanked by upturned sedimentary strata and separated by sediment filled basins. The cores of these large uplifted blocks include some of the highest and most scenic topography in the west, including Long's Peak, Maroon Bells, and Mount Sneffels, all located in Colorado. Much of the scenic beauty of these structures can be credited to the work of glaciers.

Long's Peak
photo by Daniel H. Bailey/Alamy

Maroon Bells
photo by Raymond Forbes/Superstock

How did this mountain belt form?

Because the Laramide Rockies lie more than 1500 kilomers from the plate boundary that produced the Cordilleran Orogeny, it has been difficult to identify the mechanism which produced them. The most widely accepted model proposes that a thickened slab of the Farallon Plate began to subduct under the west coast of California about 85 million years ago. Because of its thickness, this buoyant slab resisted subduction as it was shoved beneath the continent. As the thick slab continued eastward under the Colorado Plateau it triggered uplift of a least 2 kilometers and built several monoclines. Upon reaching the Rockies compressional forces squeezed the crust, which responded by developing high angle faults along which igneous and metamorphic basement rock was uplifted. These crystalline blocks form the cores of many of the high peaks that comprise this mountain belt.

Sierra Nevada
Colorado Plateau
Laramide Rockies
Great Plains
Bouyant subduction of Farallon plate
Compressional stress
Partial melting
Asthenosphere
Over-thickened slab of oceanic crust

Questions:
1. List the major mountain ranges that comprise the Laramide Rockies.
2. Why isn't the formation of the Laramide Rockies easily explained using the plate tectonic model?

Figure 14.20
Gravitational collapse
Without compressional forces to support them, mountains gradually collapse under their own weight. Gravitational collapse involves normal faulting in the upper, brittle portion of the crust and ductile spreading in the warm, weak rocks at depth.

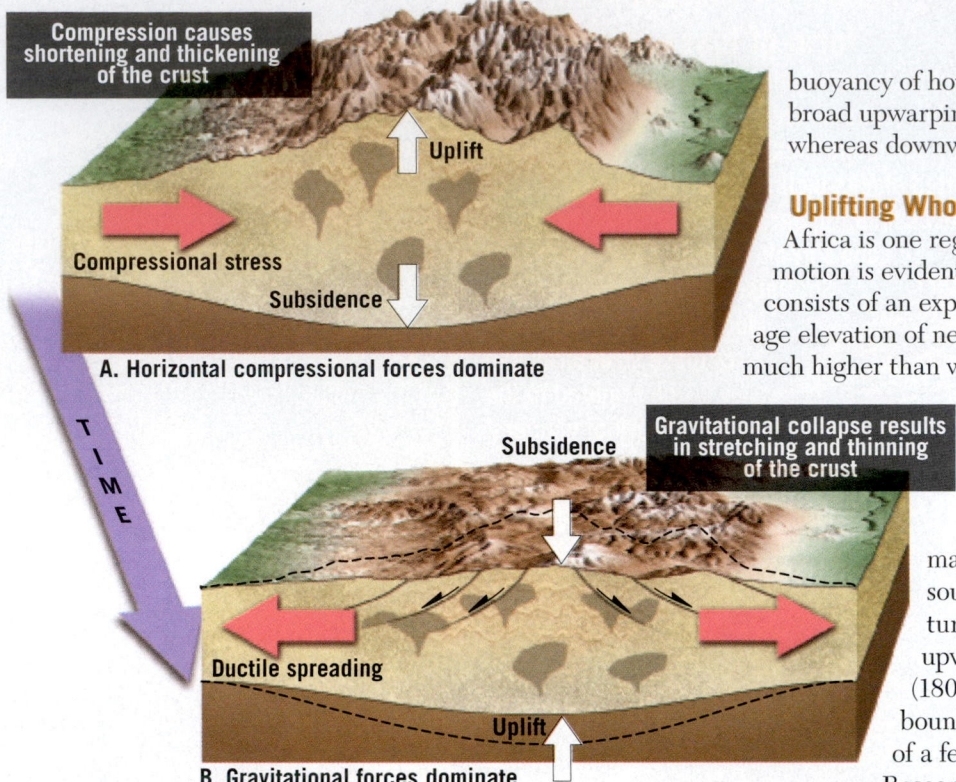

A. Horizontal compressional forces dominate

Compression causes shortening and thickening of the crust

Uplift

Compressional stress

Subsidence

TIME

Gravitational collapse results in stretching and thinning of the crust

Subsidence

Ductile spreading

Uplift

B. Gravitational forces dominate

How High Is Too High?

Where compressional forces are great, such as those driving India into Asia, lofty mountains such as the Himalayas result. Is there a limit on how high a mountain can rise? As mountaintops are elevated, gravity-driven processes such as erosion and mass wasting are accelerated, carving the deformed strata into rugged landscapes. Equally important, however, is the fact that gravity also acts on the rocks within these massive structures. The higher the mountain, the greater the downward force on rocks near the base. Eventually, the rocks deep within the developing mountain, which are relatively warm and weak, begin to flow laterally, as shown in **Figure 14.20**. This is analogous to what happens when a ladle of very thick pancake batter is poured onto a hot griddle. Similarly, mountains are altered by a process called **gravitational collapse**, which involves ductile spreading at depth and normal faulting and subsidence in the upper, brittle portion of Earth's crust.

Considering these factors, a seemingly logical question follows: "What keeps the Himalayas standing?" Simply, the horizontal compressional forces that are driving India into Asia are greater than the vertical force of gravity. However, when India's northward trek ends, the downward pull of gravity, as well as weathering and erosion, will become the dominant forces acting on this mountainous region.

Mantle Convection: A Cause of Vertical Crustal Movement

Based on studies of Earth's gravitational field, it became apparent that up-and-down convective flow in the mantle also affects the elevation of Earth's major landforms. The buoyancy of hot rising material accounts for broad upwarping in the overlying lithosphere, whereas downward flow causes downwarping.

Uplifting Whole Continents Southern Africa is one region where large-scale vertical motion is evident. The topography of the region consists of an expansive plateau that has an average elevation of nearly 1500 meters (5000 feet)—much higher than what would be predicted for a stable continental platform.

Evidence from seismic tomography (see Figure 12.19, page 379) indicates that a large mass of hot mantle rock is centered below the southern tip of Africa. This structure, called a *superplume*, extends upward about 2900 kilometers (1800 miles) from the core–mantle boundary and has a lateral expanse of a few thousand kilometers. Researchers have determined that the upward flow associated with this huge mantle plume is sufficient to elevate southern Africa.

Crustal Subsidence Extensive areas of downwarping also occur at Earth's surface. In the United States, large, nearly circular basins are found in Michigan and Illinois. Similar structures are known on other continents as well.

The cause of the downwarping that created these basins may be linked to the subduction of slabs of oceanic lithosphere. One proposal suggests that when subduction ceases, the descending slab detaches from the trailing lithosphere and continues its descent into the mantle. As this detached lithospheric slab sinks, it creates a downward flow in its wake that tugs at the base of the overriding continent. In some settings, the crust is apparently pulled down sufficiently to produce a large basin that eventually fills with sediments. As the oceanic slab sinks deeper into the mantle, the pull of the trailing wake weakens, and the continent "floats" back into isostatic balance.

14.6 Concept Checks

1. Define *isostasy*. What happens to a floating object when weight is added? Removed?

2. Briefly describe how the principle of isostatic adjustment accounts for changes in the elevation of mountains.

3. Give one example of evidence that supports the concept of crustal uplift.

4. Explain the process whereby mountainous regions experience gravitational collapse.

14.1 Mountain Building

Locate and name Earth's major mountain belts on a world map.

KEY TERMS orogenesis, compressional mountain

- Orogenesis is the making of mountains. Most orogenesis occurs along convergent plate boundaries, where compressional forces cause folding and faulting of the rock, thickening the crust vertically and shortening it horizontally. Some mountain belts are very old, while others are actively forming today.

Q Look at the South American plate on the accompanying map. Explain why the Andes Mountains are located on the western margin of South America rather than the eastern margin.

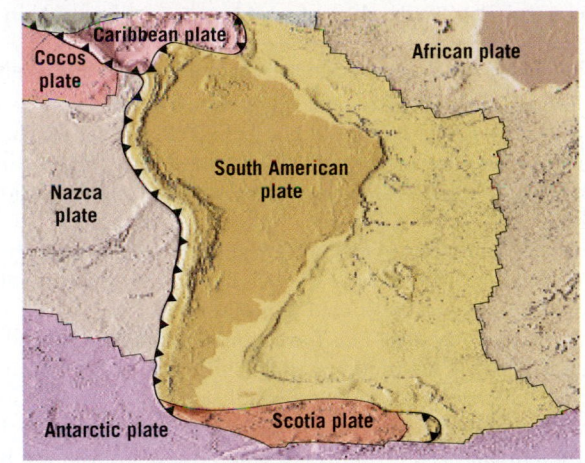

14.2 Subduction Zones

List and describe the four major features associated with subduction zones.

KEY TERMS volcanic island arc (island arc), continental volcanic arc, deep-ocean trench, forearc, back-arc, back-arc basin

- Sites of subduction are marked by volcanic arcs, deep-ocean trenches, and forearc basins located between them. A subduction zone may also contain a back-arc basin, a site of tectonic extension that forms due to trench rollback and the sinking of old, cold, dense oceanic lithosphere.

Q Match the areas represented by the letters A–D with the following labels: volcanic island arc, trench, forearc, and back-arc.

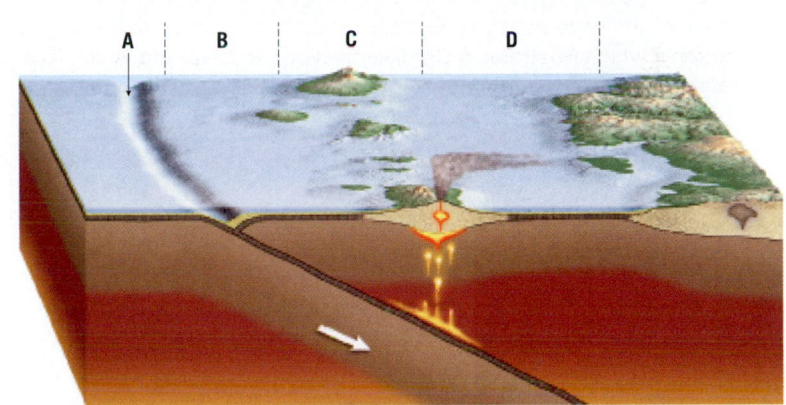

14.3 Subduction and Mountain Building

Sketch a cross-section of an Andean-type mountain belt and describe how its major features are generated.

KEY TERMS accretionary wedge, forearc basin

- Subduction leads to orogenesis. If a plate subducts beneath oceanic lithosphere, island arc–type mountain building results, with a thick accumulation of erupted volcanic rocks mixed with sediment scraped off the subducted plate. Andean-type mountain building occurs where subduction takes place beneath continental lithosphere.
- Release of water from a subducted slab triggers melting in the wedge of mantle above the slab. These primary magmas have basaltic (mafic) compositions. They rise to the base of the continental crust and pond. Fractional crystallization then separates ferromagnesian minerals from a melt that grows increasingly felsic in composition.
- The evolved magma rises further and may either cool and crystallize below the surface to form a batholith or erupt at the surface to form a continental volcanic arc.
- Sediment scraped off the subducting plate builds up in an accretionary wedge. The thickness of the wedge varies depending on the amount of sediment available. Between the accretionary wedge and the volcanic arc is a relatively calm site of sedimentary deposition, the forearc basin.
- The geography of central California preserves an accretionary wedge (Coast Ranges), a forearc basin (Great Valley), and the roots of a continental volcanic arc (Sierra Nevada).

14.4 Collisional Mountain Belts

Compare and contrast the formation of a Cordilleran-type mountain belt with that of an Alpine-type mountain belt.

KEY TERMS terrane, microcontinent, suture, fold-and-thrust belt

- Terranes are relatively small crustal fragments. Terranes may be accreted to continents when subduction brings them to a trench, but they cannot subduct due to their relatively low density. The terranes are "peeled off" the subducted slab and thrust onto the leading edge of the continent.
- The presence of ophiolites may mark the suture zone between two landmasses. Fold-and-thrust belts of folded and faulted sedimentary strata flank the sides of a collisional mountain belt.
- The Himalayas and Appalachians have similar origins in the collision of continents formerly separated by (now-subducted) ocean basins. The older Appalachians were caused by the collision of ancestral North America with ancestral Africa more than 250 million years ago. In contrast, the Himalayas are younger, formed by the collision of India and Eurasia starting around 50 million years ago. They are still rising today.

Q Which of the accompanying sketches best illustrates an Andean-type orogeny, which illustrates a Cordilleran-type orogeny, and which illustrates an Alpine-type orogeny?

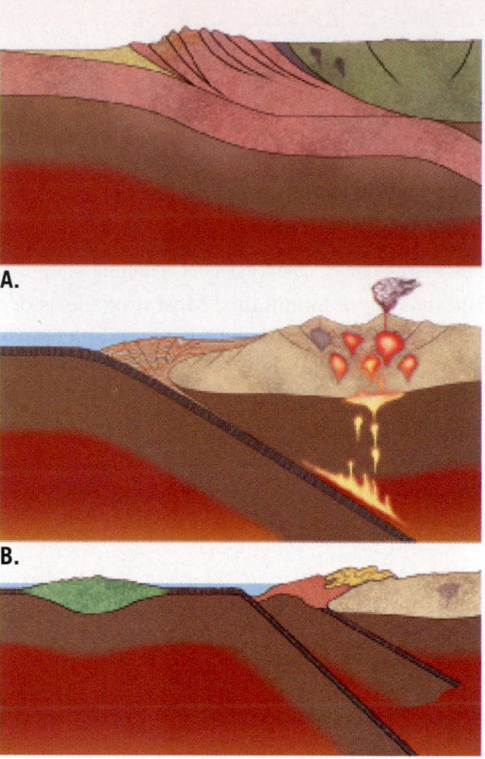

A.

B.

C.

14.5 Fault-Block Mountains

Summarize the stages in the formation of a fault-block mountain range.

KEY TERM: fault-block mountain

- Mountains can form in extensional tectonic settings. When the crust is thinned and stretched, normal faulting breaks the landscape into chunks, some of which slide down relative to their neighbors. Fault-block mountains are common in the Basin and Range province in the western United States. The Teton Range is another example.

14.6 What Causes Earth's Varied Topography?

Suggest several processes other than tectonic activity that can affect the elevation of a region.

KEY TERMS isostasy, isostatic adjustment, gravitational collapse

- Isostasy is the principle that the elevation of Earth's crust can change in accordance with gravitational balance. If additional weight is placed on the crust (a flood basalt, a thick layer of sediments, or an ice sheet), the crust will sink into the mantle until it comes to equilibrium with the extra load. If the new mass is removed, the crust will rebound upward.
- When mountains form and rock mass is pushed up to the height of a young compressional mountain belt such as the Himalayas, the rocks at the base of the mountains become warmer and weaker. As the overlying rock mass pushes down on them through gravity, these deeper rocks flow out of the way. This causes the mountain belt to collapse outward into a wider, lower-elevation region.
- Convection in the mantle can influence the vertical position of the crust. Areas of crust above sites of mantle upwelling may bulge upward, while the crust above sites of mantle downwelling may sink into broad basin-like structures.

Q Explain how a mountainous topography responds to the removal of material by erosion.

Give It Some Thought

1. Suppose that a sliver of oceanic crust were discovered in the interior of a continent. Would this refute the theory of plate tectonics? Explain.

2. Refer to the accompanying map, which shows the location of the Galapagos Rise and the Rio Grande Rise, to answer the following questions:

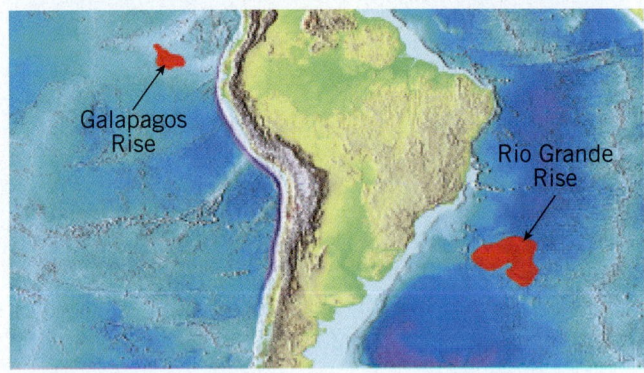

   a. Compare the continental margin of the west coast of South America with the continental margin along the east coast.

   b. Based on your answer to the question above, is the Galapagos Rise or the Rio Grande Rise more likely to end up accreted to a continent? Explain your choice.

   c. In the distant future, how might a geologist determine that this accreted landmass is distinct from the continental crust to which it accreted?

3. The Ural Mountains exhibit a north–south orientation through Eurasia. How does the theory of plate tectonics explain the existence of this mountain belt in the interior of an expansive landmass?

4. Briefly describe the major differences between the evolution of the Appalachian Mountains and the North American Cordillera.

5. Mount Moran (3842 meters elevation) in the Teton Range rises 1.8 kilometers above Jackson Hole, Wyoming, and is capped by a layer of Flathead Sandstone. On the other side of the Teton Fault, the same Flathead Sandstone lies 7 kilometers below Jackson Hole. The Teton Fault dips to the east at 45 to 75 degrees, but for the sake of simplicity, use a value of 60 degrees to estimate displacement. Calculate the total offset on the Teton Fault. (Use the diagram in the upper right as a guide.)

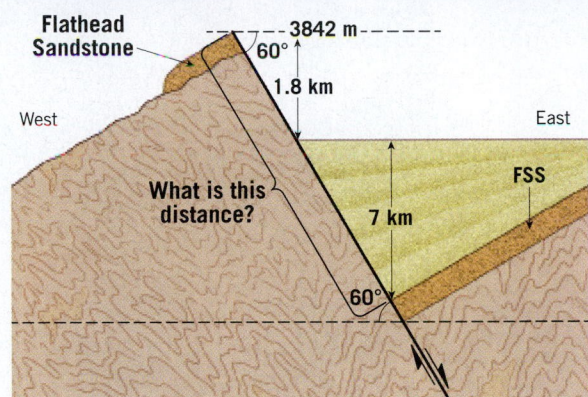

6. What processes (besides formation and melting of large ice sheets) could cause isostatic adjustments?

7. Locate Earth's young mountain belts on the plate map in Figure 2.11, page 45 and note their relationships to plate boundaries. (*Hint*: Use Figure 14.1 as a guide.)

8. Briefly explain the formation of the following geologic landforms: the Coast Ranges, Great Valley, Sierra Nevada, and Basin and Range province.

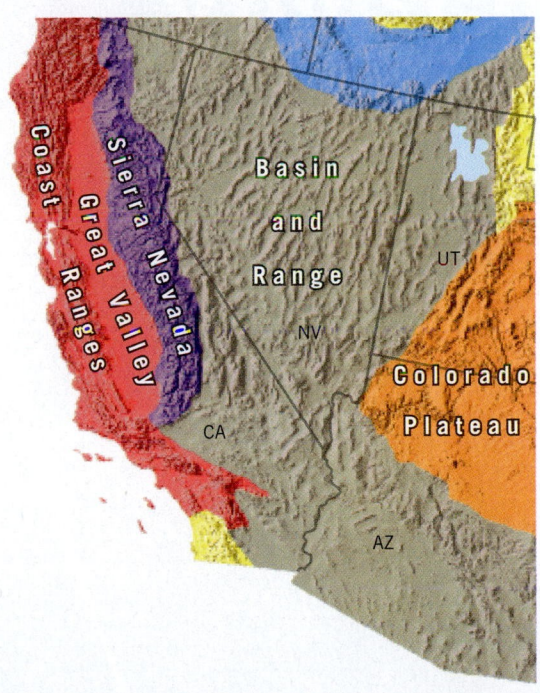

15

Mass Wasting: The Work of Gravity

On March 22 2014, a major landslide occurred near Oso, Washington. Triggered by heavy rains, a large mass of mud and debris buried a 2.6-square-kilometer (1-square-mile) area, damming a river, engulfing 49 buildings, and blocking a highway. The event claimed 43 lives.
(Photo by Michael Collier)

Each statement represents the primary **LEARNING OBJECTIVE** for the corresponding major heading within the chapter. After you complete the chapter, you should be able to:

15.1 Describe how mass-wasting processes can cause natural disasters and discuss the role that mass wasting plays in the development of landforms.

15.2 Summarize the factors that control and trigger mass-wasting processes.

15.3 List and explain the criteria that are commonly used to classify mass-wasting processes.

15.4 Distinguish among slump, rockslide, debris flow, and earthflow.

15.5 Review the general characteristics of slow mass-wasting processes and describe the unique issues associated with a permafrost environment.

Many of the preceding chapters focused on *internal processes* that are powered by energy from Earth's interior and produce volcanoes, earthquakes, ocean basins, and mountains. Beginning with this chapter and extending through Chapter 20, we turn our attention to *external processes* that sculpt and erode Earth's surface. Driven by gravity and energy from the Sun, external processes are responsible for shaping a wide variety of landforms and creating many distinctive landscapes.

15.1 | The Importance of Mass Wasting

Describe how mass-wasting processes can cause natural disasters and discuss the role that mass wasting plays in the development of landforms.

Earth's surface is not perfectly flat but instead consists of slopes of many different varieties. Some are steep and precipitous; others are moderate or gentle. Some are long and gradual; others are short and abrupt. Slopes can be mantled with soil and covered by vegetation or can consist of barren rock and rubble. Taken together, slopes are the most common elements in our physical landscape. Although most slopes may appear to be stable and unchanging, the force of gravity causes rock material to move downslope by way of processes we call *mass wasting*. At one extreme, the movement may be gradual and practically imperceptible. At the other extreme, it may consist of a roaring debris flow or a thundering rock avalanche.

Some mass-wasting processes are dangerous events that represent significant geologic hazards. Perhaps less well known is the fact that mass-wasting processes play an important role in the development and evolution of many of Earth's varied landforms.

Landslides as Geologic Hazards

The chapter-opening photo is a striking example of a phenomenon that many people would call a landslide. For most of us, the word *landslide* implies a sudden event in which large quantities of rock and soil plunge down steep slopes. When people and communities are in the way, a natural disaster may result. Landslides constitute major geologic hazards that each year in the United States cause billions of dollars in damages and the loss of dozens of lives. As you will see, many landslides occur in connection with other major natural disasters, including earthquakes, volcanic eruptions, wildfires, and severe storms.

If you were to view several images of events called landslides, you would likely notice that the term seems to refer to several different things—from mudflows to rock avalanches. This variety reflects the fact that although many people, including geologists, frequently use the word *landslide*, the term has no specific definition in geology. Rather, it is a popular, nontechnical word used to describe any or all relatively rapid forms of mass wasting.

Landslides are spectacular examples of a basic geologic process called mass wasting. **Mass wasting** refers to the downslope movement of rock, regolith, and soil under the direct influence of gravity. It is distinct from the erosional processes that are examined in subsequent chapters in that mass wasting does not require a transporting medium, such as water, wind, or glacial ice.

Landslides are not just events that occur in remote mountains and canyons. People frequently live where rapid mass-wasting events occur. Nevertheless, even in populated areas with steep slopes, catastrophic landslides are relatively rare. Consequently, people living in susceptible areas often do not appreciate the risks of living where they do. However, such events occur with some regularity around the world. GEOgraphics 15.1 provides some examples.

The Role of Mass Wasting in Landform Development

In the evolution of most landforms, mass wasting is the step that follows weathering, discussed in Chapter 6. By itself, weathering does not produce significant landforms. Rather, landforms develop as products of weathering are removed from the places where they originate. Once weathering weakens and breaks rock apart, mass wasting transfers the debris downslope, where a stream or glacier, acting as a conveyor belt, usually carries it away. Although there may be many intermediate stops along the way, the sediment is eventually transported to its ultimate destination: the sea.

The combined effects of mass wasting and running water produce stream valleys, which are among the most common and conspicuous of Earth's landforms. If streams alone were responsible for creating the valleys in which they flow, the valleys would be very narrow features resembling the canyon in **Figure 15.1**. However, the fact that most river valleys are much wider than they are deep is a strong indication of the significance of mass-wasting processes in supplying material to streams. This is illustrated by the Grand Canyon (**Figure 15.2**). The walls of the canyon extend far from the Colorado River due to the transfer of weathered debris downslope to the river and its tributaries by mass-wasting processes. In this manner, streams and mass wasting combine to modify and sculpt the surface. Of course, glaciers, groundwater, waves, and wind are also important agents in shaping landforms and developing landscapes.

Slopes Change Through Time

It is clear that if mass wasting is to occur, there must be slopes that rock, soil, and regolith can move down. Earth's mountain-building and volcanic processes, driven by plate tectonics, produce these slopes through sporadic changes in the elevations of landmasses and the ocean floor. If dynamic internal processes did not continually produce regions having higher elevations, the system that moves debris to lower elevations would gradually slow and eventually cease.

Most rapid and spectacular mass-wasting events occur in areas of rugged, geologically young mountains. Newly formed mountains are rapidly eroded by rivers and glaciers into regions characterized

Figure 15.1
Zion Narrows The Virgin River in southern Utah is responsible for cutting this narrow canyon through the Navajo Sandstone. (Photo by James Kay/SC Photos/Alamy Images)

Colorado River

SmartFigure 15.2
Excavating the Grand Canyon The walls of the canyon extend far from the channel of the Colorado River. This results primarily from the transfer of weathered debris downslope to the river and its tributaries by mass-wasting processes. (Photo by Bryan Brazil/Shutterstock) (https://goo.gl/9geHci)

Tutorial

Sedimentary layers

Material eroded by running water   ls
ss
Shale
ss
Shale
ls
Shale
ss
Shale

Metamorphic and igneous rocks

Weathered debris moved downslope by mass wasting

Colorado River

Geologist's Sketch

Landslides as Natural Disasters

Landslides don't just occur in remote mountains and canyons. People frequently live where rapid mass wasting events occur. Here are a few examples.

Even in areas with steep slopes, catastrophic landslides are relatively rare, thus, people living in susceptible areas often do not appreciate the risks of living where they do. This deadly event, **triggered by an earthquake in January 2001, buried 300 homes in Santa Tecla, El Salvador**. Many unsuspecting residents perished.

Ed Harp/USGS

On **March 1, 2012, this boulder broke loose** from a steep mountain slope in the **French Alps**, crushing a car and then crashing into this house. There was no warning and no obvious trigger for this event.

Thevenot Laurent/PHOTOPQR/LE PROGRES/Newscom

Question:
List two mass-wasting triggers represented in these images.

?

Weathered rock, precipitous slopes, and the shock of an earthquake combined to produce many **rockfalls and rock avalanches in and around Christchurch, New Zealand on February 22, 2011.**

Neil Sands/AFP/Getty Images

Marty Melville/AFP/Getty Images

In January 2011, following a period of torrential rains, this thick slurry of mud, appropriately called a mudflow, buried these cars in **Nova Friburgo, Brazil**. Heavy rains are an important trigger of mass wasting processes.

Ed Harp/USGS

by steep and unstable slopes. It is in such settings that massive destructive landslides, such as those described at the beginning of the chapter, occur. As mountain building subsides, mass-wasting and erosional processes lower the land. Through time, steep and rugged mountain slopes give way to gentler, more subdued terrain. Thus, as a landscape ages, massive and rapid mass-wasting processes give way to smaller, less dramatic downslope movements that are often imperceptibly slow.

15.2 | Controls and Triggers of Mass Wasting

Summarize the factors that control and trigger mass-wasting processes.

Gravity is the controlling force of mass wasting, but several factors play important roles in overcoming inertia and creating downslope movements. Long before a landslide occurs, various processes work to weaken slope material, gradually making it more and more susceptible to the pull of gravity. During this span, the slope remains stable but gets closer and closer to being unstable. Eventually, the strength of the slope is weakened to the point that something causes the slope to cross the threshold from stability to instability. Such an event that initiates downslope movement is called a **trigger**. Remember that a trigger is not the sole cause of a mass-wasting event but just the last of many causes. Among the common factors that trigger mass-wasting processes are saturation of material with water, oversteepening of slopes, removal of anchoring vegetation, and ground vibrations from earthquakes.

The Role of Water

Mass wasting is sometimes triggered when heavy rains or periods of snowmelt saturate surface materials. The water does not transport the material. Rather, it allows gravity to more easily set the material in motion. This was the case in March 2014, when a massive debris flow (popularly called a mudslide in the media) occurred on steep mountain slopes east of Oso, Washington (see chapter-opening photo). In September 2013, heavy rains on steep slopes over a 5-day span triggered more than 1135 debris flows over an area of 3430 square kilometers (1338 square miles) in and near Boulder, Colorado (**Figure 15.3**).

When the pores in sediment become filled with water, the cohesion among particles is destroyed, allowing them to move past one another with relative ease. For example, when sand is slightly moist, it sticks together quite well. However, if enough water is added to fill the openings between the grains, the sand will ooze out in all directions (**Figure 15.4**). Thus, saturation reduces the internal resistance of materials, which are then easily set in motion by the force of gravity. When clay is wetted, it

SmartFigure 15.3
Debris flow in the Colorado Front Range During the week of September 9–13, 2013, residents in and near Boulder, Colorado, received a harsh reminder of the dangers posed by debris flows. During that 5-day span, nearly continuous rainfall triggered numerous flash floods and more than 1100 debris flows in an area covering more than 3400 square kilometers (1300 square miles). Most occurred on steep slopes greater than 25 degrees. (Photo by Rick Wilking/Reuters) (http://goo.gl/M7j5H6)

Mobile Field Trip

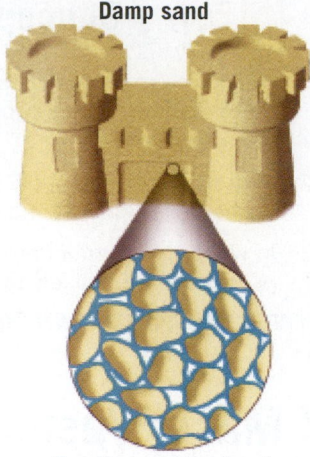
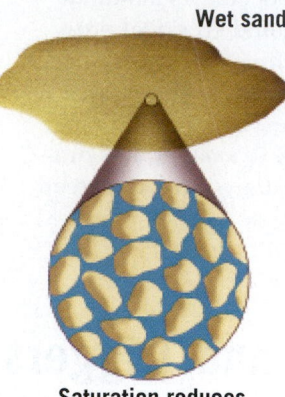

Dry sand

Damp sand

Wet sand

Dry sand grains are bound mainly by friction with one another

Small amounts of water increase the cohesion among sand grains

Saturation reduces friction and causes the sand to flow

becomes very slick—another example of the "lubricating" effect of water. Water also adds considerable weight to a mass of material. The added weight in itself may be enough to cause the material to slide or flow downslope.

Oversteepened Slopes

Oversteepening of slopes is another trigger of many mass movements. There are many situations in nature where oversteepening takes place. For example, as a stream cuts into a valley wall, it removes material from the base of the wall. This causes the slope to become too steep and material to fall or slide into the stream. Furthermore,

through their activities, people often create oversteepened and unstable slopes that become prime sites for mass wasting (**Figure 15.5**).

Loose, granular particles (sand-size or coarser) assume a stable slope called the **angle of repose** (*reposen* = to be at rest). This is the steepest angle at which material remains in place (**Figure 15.6**). Depending on the size and shape of the particles, the angle of repose varies from 25 to 40 degrees. The larger, more angular particles maintain the steepest slopes. If the angle is increased, the rock debris will adjust by moving downslope.

Oversteepening is not just important because it triggers movements of unconsolidated granular materials. Oversteepening also produces unstable slopes and mass movements in cohesive soils, regolith, and bedrock. The response will not be immediate, as with loose, granular material, but sooner or later, one or more mass-wasting processes will eliminate the oversteepening and restore stability to the slope.

Removal of Vegetation

Plants protect against erosion and contribute to the stability of slopes because their root systems bind soil and regolith together. In addition, plants shield the soil surface from the erosional effects of raindrop impact. Where plants are lacking, mass wasting is enhanced, especially

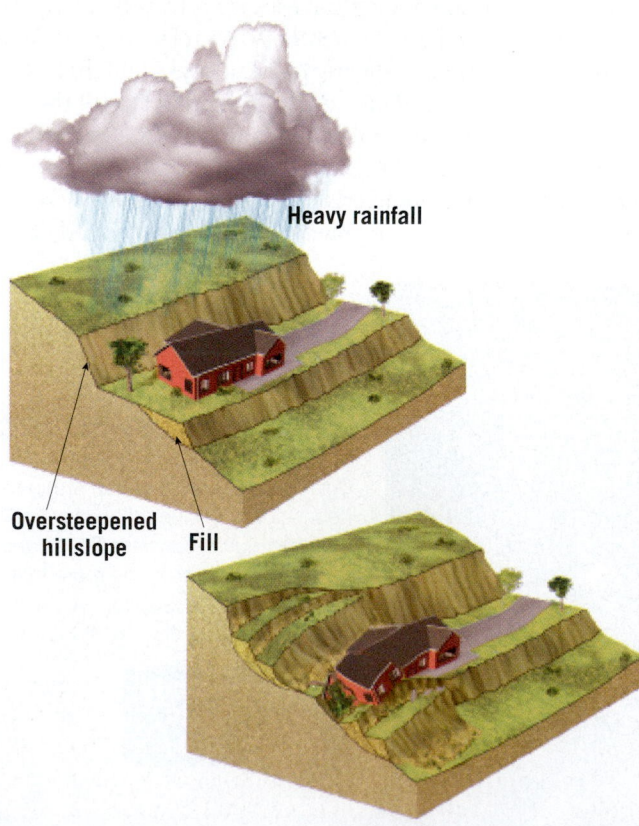

Heavy rainfall

Oversteepened hillslope　**Fill**

Angle of repose

Figure 15.6
Angle of repose The angle of repose is the steepest angle at which an accumulation of granular particles remains stable. Larger, more angular particles maintain the steepest slopes.
(Photo by G. Leavens/Science Source)

if slopes are steep and water is plentiful. When anchoring vegetation is removed by forest fires or by people (for timber, farming, or development), surface materials frequently move downslope.

An unusual example illustrating the anchoring effect of plants occurred several decades ago, on steep slopes near Menton, France. Farmers replaced olive trees, which have deep roots, with a more profitable but shallow-rooted crop: carnations. When the less stable slope failed, the landslide took 11 lives.

In July 1994 a severe wildfire swept Storm King Mountain, west of Glenwood Springs, Colorado, denuding the slopes of vegetation. Two months later, heavy rains resulted in numerous debris flows—rapid mass-wasting events involving water-saturated rock and soil. One debris flow blocked Interstate 70 and threatened to dam the Colorado River. A 5-kilometer (3-mile) length of the highway was inundated with tons of rock, mud, and burned trees. The closure of Interstate 70 imposed costly delays for travelers on this major highway.

Wildfires are inevitable in the western United States, and fast-moving, highly destructive debris flows triggered by intense rainfall are some of the most dangerous postfire hazards (**Figure 15.7**). Such events are particularly dangerous because they tend to occur with little warning. Their mass and speed make them particularly destructive. Postfire debris flows are most common in the 2 years after a fire. Some of the largest debris-flow events have been triggered by the very first intense rainfall following a wildfire. It takes much less rain to trigger debris flows in burned areas than in unburned areas. In southern California, as little as 7 millimeters (0.3 inch) of rain in 30 minutes has triggered debris flows.

How large can these flows be? According to the U.S. Geological Survey, documented debris flows from burned areas in southern California and other western states have ranged in volume from as small as 600 cubic meters (about 20,000 cubic feet) to as much as 300,000 cubic meters (more than 10 million cubic feet). This larger volume is enough material to cover a football field with mud and rocks to a depth of about 65 meters (almost 215 feet)!

In addition to eliminating plants that anchor the soil, fire can promote mass wasting in other ways. Following a wildfire, the upper part of the soil may become dry and loose. As a result, even in dry weather, the soil tends to move down steep slopes. Fire can also "bake" the ground, creating a water-repellent layer at a shallow depth. This nearly impermeable barrier prevents or slows the infiltration of water, resulting in increased surface runoff during rains. The consequence can be dangerous torrents of viscous mud and rock debris.

Figure 15.7
Wildfires contribute to mass wasting
In September 2014, this wildfire consumed 9000 acres of dense forest on steep terrain near Fish Pond, California. During the summer, wildfires are common occurrences in many parts of the western United States. Millions of acres are burned each year. The loss of anchoring vegetation sets the stage for accelerated mass wasting.
(Photo by Noah Berger/Reuters)

Earthquakes as Triggers

Conditions that favor mass wasting may exist in an area for a long time without movement occurring. An additional factor is sometimes necessary to trigger the movement. Among the most important and dramatic triggers are earthquakes. An earthquake and its aftershocks can dislodge enormous volumes of rock and unconsolidated material (**Figure 15.8**).

Examples from California and China A memorable U.S. example of an earthquake triggering mass wasting occurred in January 1994, when a quake struck the Los Angeles region of southern California. Named for its epicenter in the town of Northridge, the magnitude 6.7 event produced estimated losses of $20 billion. Some of the losses resulted from more than 11,000 landslides in an area of about 10,000 square kilometers (3900 square miles)

Figure 15.8
Earthquakes as triggers
A major earthquake in China's mountainous Sichuan Province in May 2008 triggered hundreds of landslides. The landslide shown here took 51 lives. (Photo by Lynn Highland/U.S. Geological Survey)

that were set in motion by the quake. Most were shallow rock falls and slides, but some were much larger and filled canyon bottoms with jumbles of soil, rock, and plant debris. The debris in canyon bottoms created a secondary threat because it can mobilize during rainstorms, producing debris flows. Such flows are common and often disastrous in southern California.

On May 12, 2008, a magnitude 7.9 earthquake struck near the city of Chengdu, in China's Sichuan Province. Compounding the tragic effects caused directly by the severe ground shaking were hundreds of landslides triggered by the quake and its many aftershocks. Rock avalanches and debris slides thundered down steep mountain slopes, burying buildings and blocking roads and rail lines. The landslides also dammed rivers, creating more than two dozen lakes. Earthquake-created lakes present a dual danger. Apart from the upstream floods that occur as the lake builds behind the natural dam, the piles of rubble that form the dam may be unstable. Another quake, or simply the pressure of water behind it, could burst the dam, sending a wall of water downstream. Such floods may also occur when water begins to cascade over the top of the dam. The largest of the lakes created by the May 12 earthquake, Tangjiashan Lake, threatened roughly 1.3 million people. In this instance, a disaster was avoided when Chinese engineers successfully breached the landslide dam and safely drained the lake.

Liquefaction Intense ground shaking during earthquakes can cause water-saturated surface materials to lose their strength and behave as fluidlike masses that flow (see Figure 11.26 and 11.27, pages 343-344). This process, called **liquefaction**, was a major cause of property damage in Anchorage, Alaska, during the massive 1964 Good Friday earthquake—the largest quake to strike North America in the twentieth century.

Landslides Without Triggers?

Do rapid mass-wasting events always require some sort of trigger, such as heavy rains or an earthquake? The answer is no. For example, on the afternoon of May 9, 1999, a landslide killed 10 hikers and injured many others at Sacred Falls State Park near Hauula, on the north shore of Oahu, Hawaii. The tragic event occurred when a mass of rock from a canyon wall plunged 150 meters (500 feet) down a nearly vertical slope to the valley floor. Because of safety concerns, the park was closed so that landslide specialists from the U.S. Geological Survey could investigate the site. Their study concluded that the landslide occurred *without triggering* by external conditions.

Many rapid mass-wasting events occur without a discernible trigger. Slope materials gradually weaken over time, under the influence of long-term weathering, infiltration of water, and other physical processes. Eventually, if the strength falls below what is necessary to maintain slope stability, a landslide will occur. The timing of such events is random, and thus accurate prediction is impossible.

GEOgraphics 15.2 includes two maps that show the potential for landslides in the contiguous United States and worldwide. All regions experience some damage from rapid mass-wasting processes, but it is obvious that not all regions have the same landslide potential.

15.2   Concept Checks

1. How does water affect mass-wasting processes?
2. Describe the significance of the angle of repose.
3. How might a wildfire influence mass wasting?
4. Describe the relationship between earthquakes and landslides.

EYE ON EARTH 15.1

During summer, wildfires are common occurrences in many parts of the western United States. Millions of acres are burned each year. This wildfire occurred near Santa Clarita, California, in July 2004. We have learned that Earth is a system in which various parts of its four major spheres interact in uncountable ways. Let's relate this idea to the situation pictured here.

QUESTION 1 What atmospheric conditions might have preceded and thus set the stage for this wildfire?

QUESTION 2 What might have ignited the blaze? Suggest a natural possibility and a human possibility.

QUESTION 3 Describe at least two ways that wildfires, like the one shown here, might influence future mass-wasting events in the area.

Joshua Gates Weisberg/EPA Newscom

15.3 | Classification of Mass-Wasting Processes

List and explain the criteria that are commonly used to classify mass-wasting processes.

Geologists use the term *mass wasting* to describe a broad array of different processes. Generally, the different types are classified based on the type of material involved, the kind of motion displayed, and the velocity of the movement.

Type of Material

Classifying mass-wasting processes on the basis of the material involved in the movement depends on whether the descending mass began as unconsolidated material or as bedrock. If soil and regolith dominate, terms such as *debris*, *mud*, or *earth* are used in the description. In contrast, when a mass of bedrock breaks loose and moves downslope, the term *rock* may be part of the description.

Type of Motion

In addition to characterizing the type of material involved in a mass-wasting event, the way in which the material moves may also be important. Generally, the kind of motion is described as either a fall, a slide, or a flow.

Fall When the movement in a mass-wasting event involves the free fall of detached individual pieces of any size, it is termed a **fall**. Fall is a common form of movement on slopes that are so steep that loose material cannot remain on the surface. The rock may fall directly to the base of the slope or may move in a series of leaps and bounds over other rocks along the way. Rock falls are the primary way in which **talus slopes** are built and maintained (**Figure 15.9**). Many falls result when freeze and thaw cycles and/or the action of plant roots loosen rock to the point that gravity takes over. Although signs along bedrock cuts on highways warn of falling rock, few of us have actually witnessed such an event in progress. However, as **Figure 15.10** illustrates, they do indeed occur.

When large masses of rock plunge from great heights, they hit the ground with enormous force and often trigger additional mass-wasting events. One especially deadly example occurred in Peru. In May 1970, an earthquake caused a huge mass of rock and ice to break free from the precipitous north face of Nevados Huascarán, the loftiest peak in the Peruvian Andes. The material plunged nearly a kilometer and pulverized on impact. The rock avalanche that followed rushed down the mountainside, made fluid by

Landslide Risks: United States and Worldwide

According to the U. S. Geological Survey, each year in the United States, landslides cost nearly $4 billion (2010 dollars) in damage repair and cause between 25 and 50 deaths. All states experience rapid mass-wasting processes, but not all areas have the same landslide potential. What's the risk where you live?

U.S. LANDSLIDE POTENTIAL

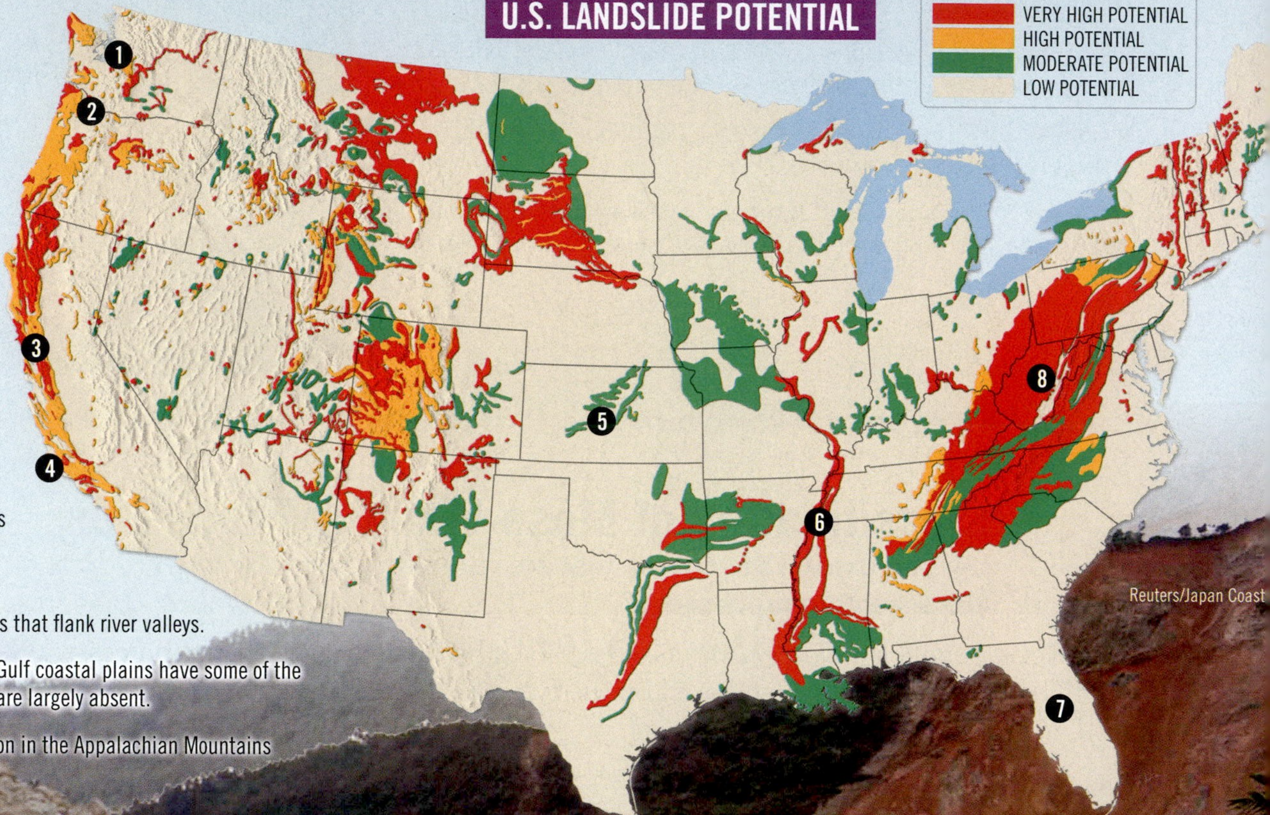

KEY
- VERY HIGH POTENTIAL
- HIGH POTENTIAL
- MODERATE POTENTIAL
- LOW POTENTIAL

1 In parts of the **Seattle area,** volcanic mudflows called lahars are a potential threat.

2 In the mountainous parts of the **Pacific Northwest**, heavy rains and melting snow often trigger rapid forms of mass wasting.

3 **Coastal California's** steep slopes have a high landslide potential often triggered by winter storms or ground shaking associated with earthquakes.

4 Strong wave activity undercuts and oversteepens coastal cliffs.

5 In the **center of the country**, the plains states are relatively flat, so landslide potential is mostly low-to-moderate.

6 High potential occurs along steep bluffs that flank river valleys.

7 **Florida** and the adjacent Atlantic and Gulf coastal plains have some of the lowest potential because steep slopes are largely absent.

8 In the **East**, landslides are most common in the Appalachian Mountains

Reuters/Japan Coast

GLOBAL LANDSLIDE RISKS

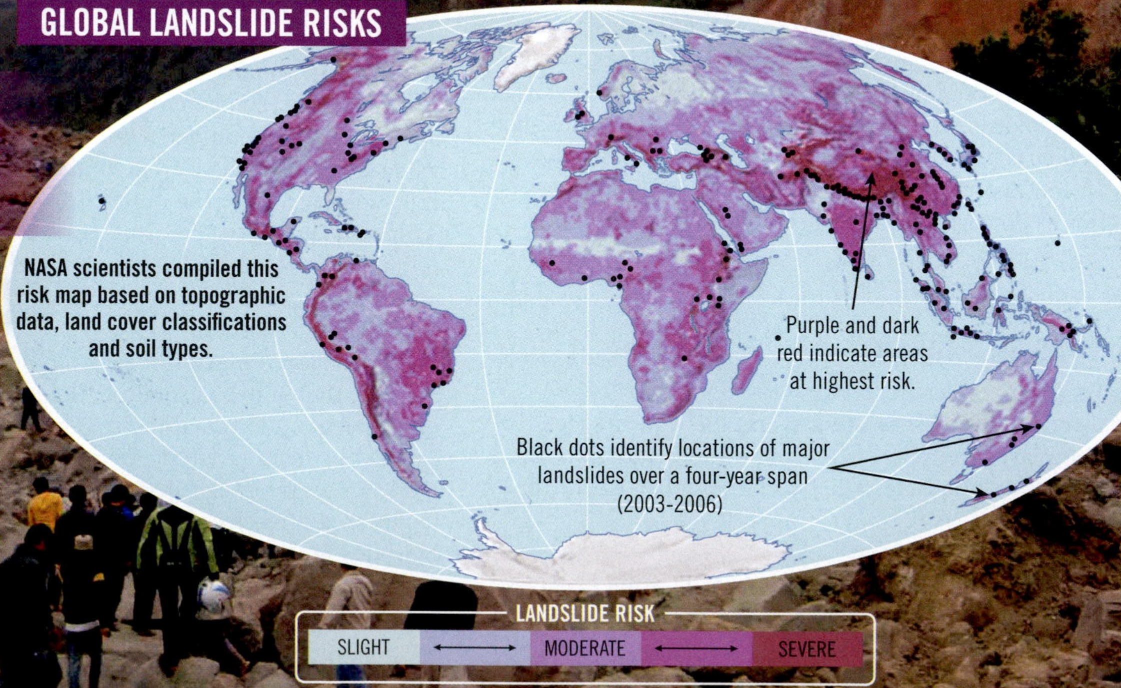

? Question:
What do areas with the highest landslide potential have in common?

NASA scientists compiled this risk map based on topographic data, land cover classifications and soil types.

Purple and dark red indicate areas at highest risk.

Black dots identify locations of major landslides over a four-year span (2003-2006)

LANDSLIDE RISK
SLIGHT ← → MODERATE ← → SEVERE

Xinhua/Photoshot/Newscom

In the mountains, mechanical weathering produces angular rock fragments that fall to the base of a cliff. Over time, a talus slope forms.

Talus Slopes

A.

The large angular rock fragments and high angle of repose make talus slopes a challenging climb.

B.

Figure 15.9
Talus slopes A. These large talus slopes (sometimes called *talus cones*) are in Canada's Banff National Park. (Photo by Marli Miller) **B.** This image of a hiker shows that talus slopes are steep features. (Photo by Whit Richardson/Alamy Images)

trapped air and ice. Along the way, it ripped loose millions of tons of additional debris that ultimately and tragically buried more than 20,000 people in the towns of Yungay and Ranrahirca.

A different effect triggered by a rock fall occurred in Yosemite National Park on July 10, 1996. When two large rock masses broke loose from steep cliffs and fell about 500 meters (1640 feet) to the floor of Yosemite Valley, the impacts were great enough to be recorded at seismic stations 200 kilometers (125 miles) from the site. As the dislodged rock masses struck the ground, they generated atmospheric pressure waves that were comparable in velocity to a tornado or hurricane. The force of the air blasts uprooted and snapped more than 1000 trees, including some that were 40 meters (130 feet) tall.

Slide Many mass-wasting processes are described as **slides**. The term refers to mass movements in which there is a distinct zone of weakness separating the slide material from the more stable underlying material. Two basic types of slides are recognized. *Rotational slides* are those in which the surface of rupture is a concave-upward curve resembling the shape of a spoon, and the descending material exhibits a downward and outward rotation. By contrast, a *translational slide* is one in which a mass of material moves along relatively flat surfaces such as joints, faults, or bedding planes. Such slides exhibit little rotation or backward tilting.

Flow The third type of movement common to mass-wasting processes is termed **flow**. Flow occurs when material moves downslope as a viscous fluid. Most flows are saturated with water and typically move as lobes or tongues. Frequently, when flows of saturated mud and debris occur, they are incorrectly described in the media as "mudslides."

SmartFigure 15.10
Watch out for falling rock! This rockfall blocked a Montana highway. (Main photo by Harris Shiffman/ Shutterstock; inset photo by AP Images) (http://goo.gl/QobZlc)

Animation

Figure 15.11
Blackhawk rock avalanche This prehistoric event is considered one of the largest-known landslides in North America. (Photo by Michael Collier)

San Bernardino Mountains

Between 9 and 30 meters thick

8 kilometers

The mass of rocky debris raced downslope on a cushion of compressed air.

Rate of Movement

Some of the events that have been described so far in this chapter involved very rapid rates of movement. For example, it is estimated that the debris that rushed down the slopes of Peru's Nevados Huascarán moved at speeds in excess of 200 kilometers (125 miles) per hour. This most rapid type of mass movement is termed a **rock avalanche**. Researchers now understand that rock avalanches, such as the one that produced the scene in **Figure 15.11**, must literally "float on air" as they move downslope. That is, high velocities result when air becomes trapped and compressed beneath the falling mass of debris, allowing it to move as a buoyant, flexible sheet across the surface.

Most mass movements, however, do not occur at the speeds of rock avalanches. In fact, a great deal of mass wasting is imperceptibly slow. One process we will examine later, termed *creep*, results in particle movements that are usually measured in millimeters or centimeters per year. Thus, as you can see, rates of movement can be spectacularly sudden or exceptionally gradual. Although various types of mass wasting are often classified as either *rapid* or *slow*, such a distinction is highly subjective because a wide range of rates exists between the two extremes. Even the velocity of a single process at a particular site can vary considerably.

15.3 Concept Checks

1. What terms are used to describe the way material moves during mass wasting?

2. Why can rock avalanches move at such great speeds?

15.4 | Rapid Forms of Mass Wasting

Distinguish among slump, rockslide, debris flow, and earthflow.

The common rapid mass-wasting processes discussed in this section are slump, rockslide, debris flow, and earthflow. They are most common where slopes are steep. The speed of movement varies from barely perceptible to very rapid.

Slump

Slump is an example of a rotational slide and refers to the downward sliding of a mass of rock or unconsolidated material moving as a unit along a curved surface (**Figure 15.12**). Usually the slumped material does not travel spectacularly fast, and it does not travel very far. Slump is a common form of mass wasting, especially in thick accumulations of cohesive materials such as clay. The ruptured surface is characteristically spoon shaped and concave upward or outward. As the movement occurs, a crescent-shaped scarp is created at the head, and the block's upper surface is sometimes tilted backward. Although slump may involve a single mass, it often consists of multiple blocks. Sometimes water collects between the base of the scarp and the top of the tilted block. As this water percolates downward along the surface of rupture, it may promote further instability and additional movement.

Slump commonly occurs because a slope has been oversteepened. The material on the upper portion of a slope is held in place by the material at the bottom of the slope. As this anchoring material at the base is removed, the material above is made unstable and reacts to the pull of gravity. One relatively common example is a valley wall that becomes oversteepened by a meandering river. The photo in **Figure 15.13** provides an example in which a coastal cliff has been undercut by wave action at its base. Slumping may also occur when a slope is overloaded, causing internal stress on the material below. This type of slump often occurs where weak, clay-rich material underlies layers of stronger, more resistant rock such as sandstone. The seepage of

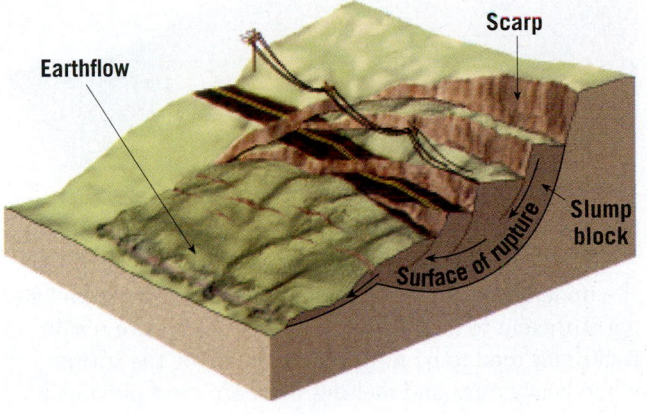

Figure 15.12
Slump Slump occurs when material slips downslope en masse along a curved surface of rupture. It is an example of a rotational slide. Earthflows frequently form at the base of the slump.

water through the upper layers reduces the strength of the clay below, resulting in slope failure.

Rockslide

Rockslides are translational slides that occur when blocks of bedrock break loose and slide down a slope (**Figure 15.14**). If the material involved is largely unconsolidated, the term **debris slide** is used instead. Such events are among the fastest and most destructive mass movements. Usually rockslides take place in a geologic setting where the rock strata are inclined, or where joints and fractures exist parallel to the slope. When such a rock unit is undercut at the base of the slope, it loses support, and the rock eventually gives way. Sometimes the rockslide is triggered when rain or melting snow lubricates

Figure 15.13
Slump at Point Fermin, California Waves undercut the base of the steep slope, making it unstable. (Photo by John S. Shelton/University of Washington Libraries)

Geologist's Sketch

Figure 15.14
Rockslide These rapid movements are classified as translational slides in which the material moves along a relatively flat surface with little or no rotation or backward tilting.

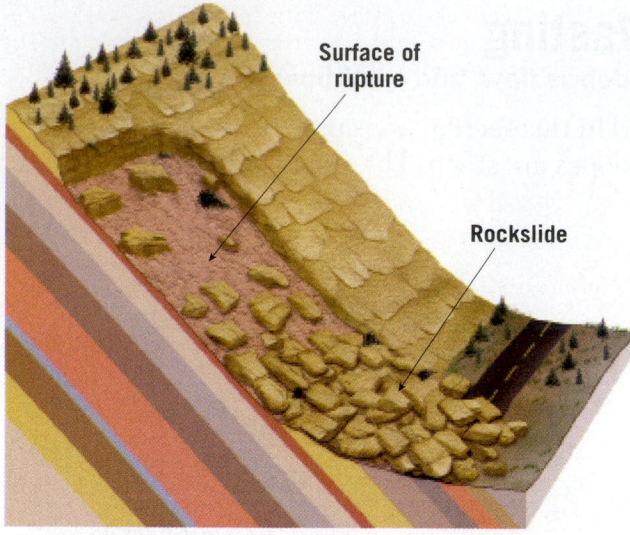

a severe earthquake west of Yellowstone National Park triggered a massive slide in the canyon of the Madison River in southwestern Montana. In a matter of moments, an estimated 27 million cubic meters of rock, soil, and trees slid into the canyon. The debris dammed the river and buried a campground and highway. More than 20 unsuspecting campers perished.

Heavy rains and melting snow, rather than an earthquake, triggered another major rockslide in the Yellowstone region. Not far from the site of the Madison Canyon slide, the legendary Gros Ventre rockslide occurred 34 years earlier. The Gros Ventre River flows west from the northernmost part of the Wind River Range in northwestern Wyoming, through Grand Teton National Park, and eventually empties into the Snake River. On June 23, 1925, a massive rockslide took place in its valley, just east of the small town of Kelly. In the span of only minutes, a great mass of sandstone, shale, and soil crashed down the south side of the valley, carrying with it a dense pine forest. The volume of debris, estimated at 38 million cubic meters (50 million cubic yards), created a dam on the Gros Ventre River 70 meters (230 feet) high. Because the river was completely blocked, a lake was formed. It filled so quickly that a house that had been 18 meters (60 feet) above the river was floated off its foundation 18 hours after the slide. In 1927, the lake overflowed the dam, partially draining the lake and resulting in a devastating flood downstream.

Why did the Gros Ventre rockslide take place? **Figure 15.15** shows a diagrammatic cross-sectional view of the geology of the valley. Notice that (1) the sedimentary strata in this area dip (tilt) 15 to 21 degrees; (2) underlying the bed of sandstone is a relatively thin layer of clay; and (3) at the bottom of the valley, the river had cut through much of the sandstone layer. During the spring of 1925, water from heavy rains and melting snow seeped through the sandstone, saturating the clay below. Because much of the sandstone layer had been cut through by the Gros Ventre River, the layer had virtually no support at the bottom of the slope. Eventually the sandstone could no longer hold its position on the wetted clay, and gravity pulled the mass down the side of the valley. The circumstances at this location were such that the event was inevitable.

the underlying surface to the point that friction is no longer sufficient to hold the rock unit in place. As a result, rockslides tend to be most common during the spring, when heavy rains and melting snow are most prevalent.

As mentioned earlier, earthquakes can trigger rockslides and other mass movements. There are many well-known examples. The 1811 earthquake at New Madrid, Missouri (see Chapter 11), caused slides in an area of more than 13,000 square kilometers (5000 square miles) along the Mississippi River valley. On August 17, 1959,

SmartFigure 15.15
Gros Ventre rockslide
This massive slide occurred on June 23, 1925, just east of the small town of Kelly, Wyoming. (Photo by Michael Collier) (https://goo.gl/el2Ecj)

Tutorial

The side of the mountain gave way when the tilted sandstone bed, that had been cut through by the river, could no longer maintain its position atop the saturated bed of clay.

Gros Ventre landslide debris

Former land surface

Scar

Lake

Clay bed

Sandstone

Limestone

Rupture surface

0   0.5   1
Kilometer

Even though the Gros Ventre rockslide occurred in 1925, the scar left on the side of Sheep Mountain is still a prominent feature.

Debris Flow

Debris flow is a relatively rapid type of mass wasting that involves a flow of soil and regolith containing a large amount of water (**Figure 15.16**). The events depicted in the chapter-opening photo and Figure 15.3 are examples. Debris flows are sometimes called **mudflows** when the material is primarily fine-grained. Although they can occur in many different climate settings, they tend to occur more frequently in semiarid mountainous regions. Because of their fluid properties, debris flows frequently follow canyons and stream channels. In populated areas, debris flows can pose a significant hazard to life and property.

The consistency of a debris flow ranges from that of wet concrete to a soupy mixture not much thicker than muddy water.

When the material is fine-grained, the event may be called a *mudflow*.

Figure 15.16
Debris flow A debris flow is a moving tongue of well-mixed mud, soil, rock, and water. (Photo by Michael Collier)

Debris Flows in Semiarid

Regions When a cloudburst or rapidly melting mountain snows create a sudden flood in a semiarid region, large quantities of soil and regolith are washed into nearby stream channels because there is usually little vegetation to anchor the surface material. The end product is a flowing tongue of well-mixed mud, soil, rock, and water. Its consistency may range from that of wet concrete to a soupy mixture not much thicker than muddy water. The rate of flow, therefore, depends not only on the steepness of the slope but also on the water content. Dense debris flows are capable of carrying or pushing large boulders, trees, and even houses with relative ease.

Debris flows pose a serious hazard to development in relatively dry mountainous areas such as southern California. The construction of homes on canyon hillsides and the removal of native vegetation by brush fires and other means have increased the frequency of these destructive events. Moreover, when a debris flow reaches the end of a steep, narrow canyon, it spreads out, covering the area beyond the mouth of the canyon with a mixture of wet debris. This material contributes to the buildup of fanlike deposits, called *alluvial fans*, at canyon mouths. The fans are relatively easy to build on, often have nice views, and are close to the mountains; in fact, like the nearby canyons, many have become preferred sites for development. Because debris flows occur only sporadically, the public is often unaware of the potential hazard of such sites.

Lahars Debris flows composed mostly of volcanic materials on the flanks of volcanoes are called **lahars**. The word originated in Indonesia, a volcanic region that has experienced many of these often-destructive events. Historically, lahars have been some of the deadliest volcano-related hazards. They can occur either during an eruption or when a volcano is quiet. They take place when highly unstable layers of ash and debris become saturated with water and flow down steep volcanic slopes, generally following existing stream channels (**Figure 15.17**). Heavy rainfalls often trigger these

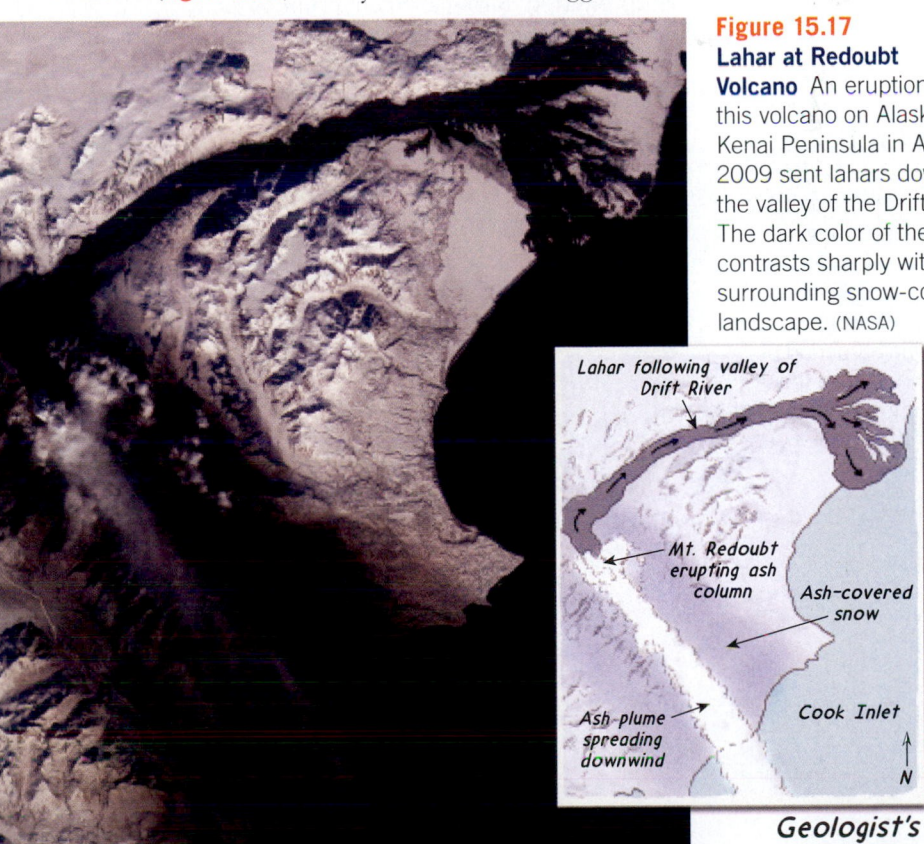

Figure 15.17
Lahar at Redoubt Volcano An eruption of this volcano on Alaska's Kenai Peninsula in April 2009 sent lahars down the valley of the Drift River. The dark color of the lahar contrasts sharply with the surrounding snow-covered landscape. (NASA)

Lahar following valley of Drift River

Mt. Redoubt erupting ash column

Ash-covered snow

Ash-plume spreading downwind

Cook Inlet

N

Geologist's Sketch

Figure 15.18

Lahar at Mount St. Helens The force of this viscous debris flow that followed the Toutle River tore the end section from a house and lodged it in the trees. Lahars generated by the Mount St. Helens eruption surged up valley walls as much as 110 meters (360 feet) and over hills as high as 76 meters (250 feet). Based on "bathtub-ring" mudlines, some lahars at their peak averaged 10 to 20 meters (33 to 66 feet) deep. (Photo by D. R. Crandell, U.S. Geological Survey, Denver)

the Toutle River at speeds that were often in excess of 30 kilometers (20 miles) per hour. Fortunately, the affected area was not densely settled. Nevertheless, more than 200 homes were destroyed or severely damaged (**Figure 15.18**). Most bridges met a similar fate.

In November 1985, lahars were produced during the eruption of Nevado del Ruiz, a 5300-meter (17,400-foot) volcano in the Andes Mountains of Colombia. The eruption melted much of the snow and ice that capped the uppermost 600 meters (2000 feet) of the peak, producing torrents of hot, viscous mud, ash, and debris. The lahars moved outward from the volcano, following the valleys of three rain-swollen rivers that radiate from the peak. The flow that moved down the valley of the Lagunilla River was the most destructive. It devastated the town of Armero, 48 kilometers (30 miles) from the mountain. Most of the more than 25,000 deaths caused by the event occurred in this once-thriving agricultural community.

Death and property damage due to the lahars also occurred in 13 other villages within the

flows. Others are initiated when large volumes of ice and snow are melted by heat flowing to the surface from within the volcano or by the hot gases and near-molten debris emitted during a violent eruption.

The eruption of Mount St. Helens in May 1980 created several lahars. The flows and accompanying floods raced down the valleys of the north and south forks of

Figure 15.19

Earthflow This small tongue-shaped earthflow occurred on a newly formed slope along a recently constructed highway in central Illinois. It formed in clay-rich material following a period of heavy rain. Notice the small slump at the head of the earthflow. (Photo by E. J. Tarbuck)

Geologist's Sketch

180-square-kilometer (70-square-mile) disaster area. Although a great deal of pyroclastic material was explosively ejected from Nevado del Ruiz, it was the lahars triggered by this eruption that made this such a devastating natural disaster. In fact, it was the worst volcanic disaster to occur since 28,000 people died following the 1902 eruption of Mount Pelée on the Caribbean island of Martinique.°

Earthflow

We have seen that debris flows are frequently confined to channels in semiarid regions. In contrast, **earthflows** most often form on hillsides in humid areas during times of heavy precipitation or snowmelt. When water saturates the soil and regolith on a hillside, the material may break away, leaving a scar on the slope and forming a tongue- or teardrop-shaped mass that flows downslope (**Figure 15.19**).

The materials most commonly involved are rich in clay and silt and contain only small proportions of sand and coarser particles. Earthflows range in size from bodies a few meters long, a few meters wide, and less than a meter deep to masses more than 1 kilometer long, several hundred meters wide, and more than 10 meters deep. Because earthflows are quite viscous, they generally

*A discussion of the Mount Pelée eruption, as well as additional material on lahars, can be found in Chapter 5.

move at slower rates than the more fluid debris flows described in the preceding section. They are characterized by a gradual movement that may go on for periods ranging from days to years. Depending on the steepness of the slope and the material's consistency, measured velocities range from less than a millimeter a day up to several meters a day. Over the time span during which an earthflow is active, movement is typically faster during wet periods than during drier times. In addition to occurring as isolated hillside phenomena, earthflows commonly take place in association with large slumps. In this situation, they may be seen as tongue-like flows at the base of the slump block.

15.4 Concept Checks

1. Without looking at Figures 15.12 and 15.13, sketch and label a simple cross-section (side view) of a slump.

2. Both slumps and rockslides move by sliding. How do these processes differ?

3. What factors led to the massive rockslide at Gros Ventre, Wyoming?

4. How is a lahar different from a debris flow that might occur in southern California?

5. Contrast earthflows and debris flows.

EYE ON EARTH 15.2

Heavy rains in late July 2010 triggered the mass wasting that occurred in this mountain valley near Durango, Colorado. Heavy equipment is clearing away material that blocked railroad tracks and significantly narrowed the adjacent stream channel.

QUESTION 1 What is the likely type of mass wasting that occurred here? Explain your choice.

QUESTION 2 Most of us are familiar with the phrase "One thing leads to another." It certainly applies to the Earth system. Suppose the material from the mass-wasting event had completely filled the stream. What other natural hazard might have developed?

Soaring Tree Adventures

15.5 | Slow Movements

Review the general characteristics of slow mass-wasting processes and describe the unique issues associated with a permafrost environment.

Movements such as rockslides, rock avalanches, and lahars are certainly the most spectacular and catastrophic forms of mass wasting. These dangerous events deserve intensive study to enable more effective prediction, timely warnings, and better controls to save lives. However, because of their large size and spectacular nature, they give us a false impression of their importance as mass-wasting processes. Indeed, sudden movements are responsible for moving less material than the slower and far more subtle action of creep. Whereas rapid types of mass wasting are characteristic of mountains and steep hillsides, creep takes place on both steep and gentle slopes and is thus much more widespread.

SmartFigure 15.20
Creep The repeated expansion and contraction of the surface material causes a net downslope migration of soil and rock particles. (https://goo.gl/nJO3Eh)

Tutorial

Figure 15.21
Effects of creep Although creep is an imperceptibly slow movement, its effects are often visible. (Photos by D. Bradley/NOAA, Marli Miller, and Tim McGuire Images, Inc./Alamy)

Creep

Creep is a type of mass wasting that involves the *gradual* downhill movement of soil and regolith. One factor that contributes to creep is the alternating expansion and contraction of surface material caused by freezing and thawing or wetting and drying. As shown in **Figure 15.20**, freezing or wetting lifts particles at right angles to the slope, and thawing or drying allows the particles to fall back to a slightly lower level. Each cycle therefore moves the material a tiny distance downslope. Creep is aided by anything that disturbs the soil, such as raindrop impact and disturbance by plant roots and burrowing animals. Creep is also promoted when the ground becomes saturated with water. Following a heavy rain or snowmelt, a water-logged soil may lose its internal cohesion, allowing gravity to pull the material downslope. Because creep is imperceptibly slow, the process cannot be observed in action. However, the effects of creep can be observed. Creep causes fences and utility poles to tilt and retaining walls to be displaced (**Figure 15.21**).

Solifluction

When soil is saturated with water, the soggy mass may flow downslope at a rate of a few millimeters or a few centimeters per day or per year. Such a process is called **solifluction** (literally "soil flow"). It is a type of mass wasting that is common wherever water cannot escape from the saturated surface layer by infiltrating to deeper levels. A dense clay hardpan in soil or an impermeable bedrock layer can promote solifluction.

Solifluction is also common in regions underlain by permafrost. *Permafrost* refers to the permanently frozen ground that occurs in association with Earth's harsh tundra

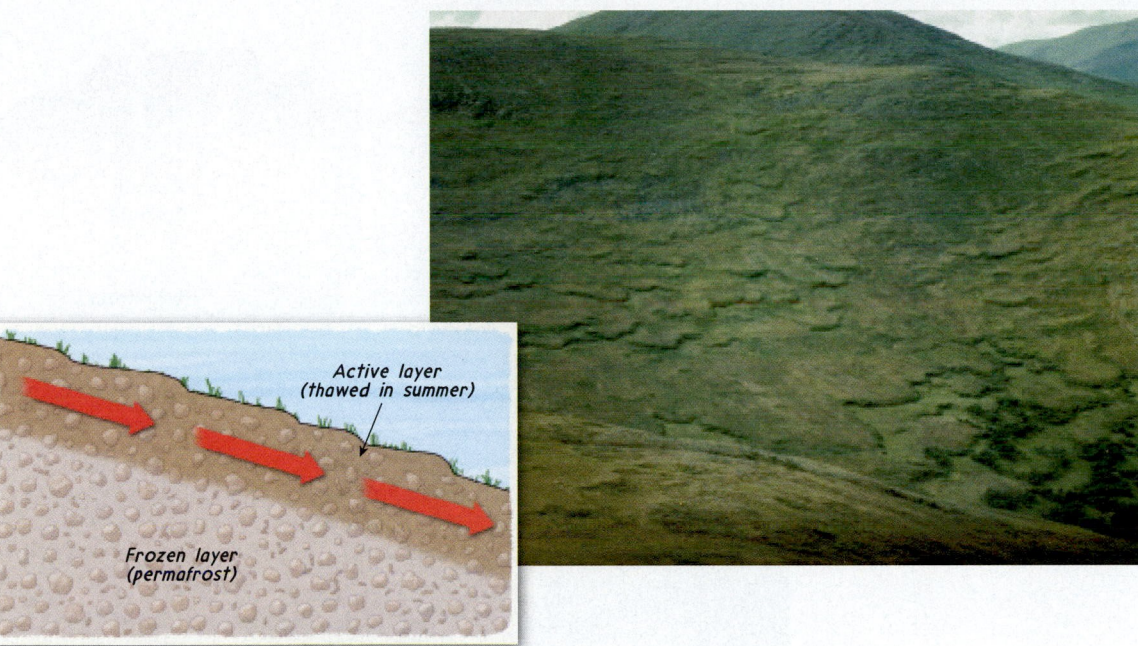

Geologist's Sketch

Figure 15.22
Solifluction lobes near the Arctic Circle in Alaska
Solifluction occurs in permafrost regions when the active layer thaws in summer. Because summers are cool and very short, frozen soils generally thaw to depths of less than 1 meter (3 feet). (Photo from the James E. Patterson Collection courtesy of F. K. Lutgens)

and subarctic climates. (There is more about permafrost in the next section.) Solifluction occurs in a zone above the permafrost called the *active layer*, which thaws to a depth of about a meter during the brief high-latitude summer and then refreezes in winter. During the summer season, water is unable to percolate into the impervious permafrost layer below. As a result, the active layer becomes saturated and slowly flows. The process can occur on slopes as gentle as 2 to 3 degrees. Where there is a well-developed mat of vegetation, a solifluction sheet may move in a series of well-defined lobes or as a series of partially overriding folds (**Figure 15.22**).

The Sensitive Permafrost Landscape

Many of the mass-wasting disasters described in this chapter had sudden and disastrous impacts on people. When the activities of people cause ice contained in permanently frozen ground to melt, the impact is more gradual and less deadly. Nevertheless, because permafrost regions are sensitive and fragile landscapes, the scars resulting from poorly planned actions can remain for generations.

Permanently frozen ground, known as **permafrost**, occurs where summers are too short and cool to melt more than a shallow surface layer. Deeper ground remains frozen year-round. Permafrost is extensive in the lands surrounding the Arctic Ocean (**Figure 15.23**). Strictly speaking, permafrost is defined only on the basis of temperature; that is, it is ground with temperatures that have remained below 0°C (32°F) continuously for 2 years or more. The degree to which ice is present in the ground strongly affects the behavior of

the surface material. Knowing how much ice is present and where it is located is very important when it comes to constructing roads, buildings, and other projects in areas underlain by permafrost.

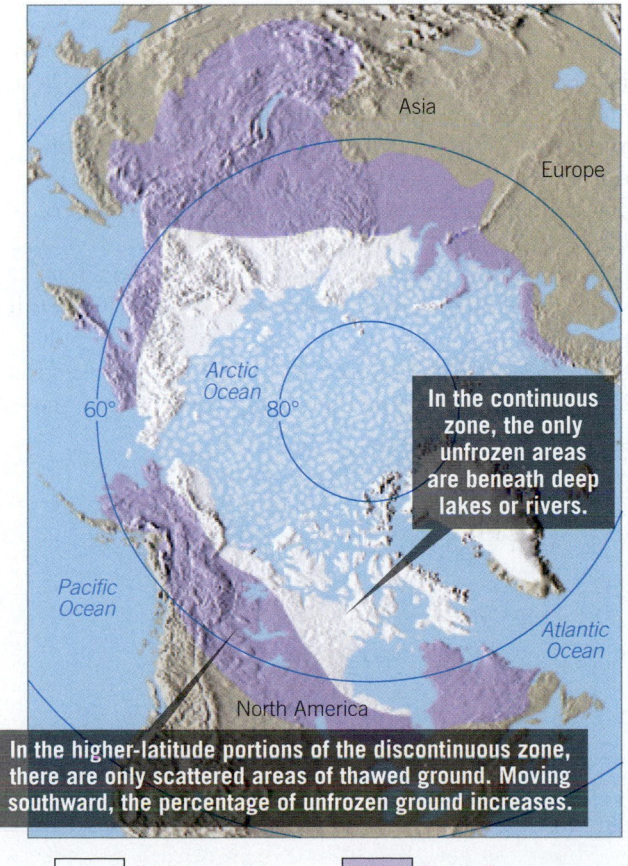

In the continuous zone, the only unfrozen areas are beneath deep lakes or rivers.

In the higher-latitude portions of the discontinuous zone, there are only scattered areas of thawed ground. Moving southward, the percentage of unfrozen ground increases.

☐ Continuous zone   ☐ Discontinuous zone

Figure 15.23
Distribution of permafrost in the Northern Hemisphere More than 80 percent of Alaska and about 50 percent of Canada are underlain by permafrost.

SmartFigure 15.24

When permafrost thaws This building, located south of Fairbanks, Alaska, subsided because of thawing permafrost. Notice that the right side, which was heated, settled much more than the unheated porch on the left. (Photo by Steve McCutcheon, U.S. Geological Survey) (https://goo.gl/PMHMpp)

When people disturb the surface, such as by removing the insulating vegetation mat or by constructing roads and buildings, the delicate thermal balance is disturbed, and ice within the permafrost can thaw. Thawing produces unstable ground that may slide, slump, subside, and undergo severe frost heaving. When a heated structure is built directly on permafrost that contains a high proportion of ice, thawing creates soggy material into which a building can sink (**Figure 15.24**). One solution is to place buildings and other structures on piles, like stilts. Such piles allow subfreezing air to circulate between the floor of the building and the soil and thereby keep the ground frozen.

15.5 Concept Checks

1. Describe the basic mechanisms that contribute to creep. How might you recognize that creep is occurring?

2. During what season does solifluction in the Arctic occur? Explain why it occurs only during that season.

3. What is permafrost? How might disturbing permafrost lead to unstable ground that may slide, flow, or subside?

15.1 The Importance of Mass Wasting

Describe how mass-wasting processes can cause natural disasters and discuss the role that mass wasting plays in the development of landforms.

KEY TERM mass wasting

DunnRight Photography

- After weathering breaks apart rock, gravity moves the debris downslope, in a process called mass wasting. Sometimes this occurs rapidly as a landslide, and at other times the movement is slower. Landslides are a significant geologic hazard, taking many lives and destroying property.
- Mass wasting serves an important role in landscape development. It widens stream valleys and helps tear down the mountains thrust up by plate tectonics.

Q How is mass wasting, such as the example shown here, different from erosional processes?

15.2 Controls and Triggers of Mass Wasting

Summarize the factors that control and trigger mass-wasting processes.

KEY TERMS trigger, angle of repose, liquefaction

- An event that initiates a mass-wasting process is referred to as a trigger. The addition of water, oversteepening of the slope, removal of vegetation, and shaking due to an earthquake are four important examples. Not all landslides are triggered by one of these four processes, but many are.
- Water added to a slope can expand the pores, separating grains and causing them to lose their cohesion. Water also lubricates the contacts between the grains and adds a significant amount of mass to a wetted slope.

- Granular materials can pile up to a certain angle of slope, but granular piles steeper than that critical angle will spontaneously collapse outward to form a gentler slope. For most geologic materials, this angle of repose varies between 25 and 40 degrees from horizontal. Oversteepened slopes are likely to fail with a landslide.
- The roots of plants (especially deep ones) act as a three-dimensional "net" that holds soil and regolith particles in place. When the plants die, the soil loses an important support structure. Plants may be removed naturally (through wildfires, for instance) or by humans harvesting timber, planting crops, or building structures.
- Earthquakes are significant triggers that deliver an energetic jolt to slopes poised on the brink of failure.

Q Do all mass wasting events have triggers? Explain.

15.3 Classification of Mass-Wasting Processes

List and explain the criteria that are commonly used to classify mass-wasting processes.

KEY TERMS fall, talus slope, slide, flow, rock avalanche

- There are a variety of Earth materials (rock, soil, and regolith) and a variety of rates of mass-wasting movement. The type of material and the nature of the motion are combined into classifying terms for the different types of mass wasting.
- Rock falls occur when pieces of bedrock detach and fall freely through the air, slamming into the ground below with tremendous force. Repeated rock falls generate a talus slope, the characteristic "apron" of angular rock debris that accumulates below mountain cliffs. Slides occur when discrete blocks of rock or unconsolidated material slip downslope on a planar or curved surface. Unlike in a fall, the material in a slide does not drop through the air. Flows occur when individual grains or particles move randomly in a slurry, a viscous mixture of water-saturated materials.

Taylor S. Kennedy/National Geographic RF/Glow Images

- Rock avalanches move incredibly rapidly over surprising distances (as much as tens of kilometers in a few minutes) because they ride along a layer of compressed air, in the same way that a hovercraft slides over land with minimal friction. Other forms of mass wasting are much slower, moving perhaps only millimeters per year.

Q Refer to the photo at the bottom of page 463. What term is applied to the pile of angular rock fragments at the base of this rugged peak? Describe the process that produced this feature. Did the feature form rapidly or gradually? What other chapter in this book discusses this feature?

15.4 Rapid Forms of Mass Wasting

Distinguish among slump, rockslide, debris flow, and earthflow.

KEY TERMS slump, rockslide, debris slide, debris flow, mudflow, lahar, earthflow

- A slump is a distinctive and common form of mass wasting in which coherent blocks of material move downhill on a spoon-shaped slip surface. Slumps are often marked by curved scarps that open up at their tops. They are frequently triggered by oversteepening, such as that caused by stream erosion of a valley wall.
- Rockslides are rapid mass-wasting events in which a coherent block of rock slides downhill along a planar surface. Often this is a preexisting structure such as a joint or a bedding plane. Situations where these surfaces dip into a valley at an angle are especially dangerous.
- Debris flows occur when unconsolidated soil or regolith becomes saturated with water and moves downhill in a slurry, picking up other objects (trees, houses, livestock) along the way. Varieties of debris flow include mudflows, which are dominated by small particle sizes, and lahars, which involve volcanic materials. Debris flows can move quickly—up to 30 kilometers (nearly 20 miles) per hour.
- Earthflow is characterized by a similar loss of coherence between grains in unconsolidated material, but it is much slower than a debris flow. Typically, sites of earthflow show an uphill scarp and a lobe of viscous soil on the downhill side.

Q This rockslide occurred in the rugged Himalayas of northern India. Identify a feature in the photo that may have been a factor that contributed to the slide. Speculate about what might have triggered the event.

Aji Jayachandran/Demotix/Corbis

15.5 Slow Movements

Review the general characteristics of slow mass-wasting processes and describe the unique issues associated with a permafrost environment.

KEY TERMS creep, solifluction, permafrost

- Creep is a widespread and important form of mass wasting that is very slow. It occurs when freezing (or wetting) causes soil particles to be pushed out away from the slope, only to drop down to a lower position following thawing (or drying). In contrast, solifluction is the gradual flow of a saturated surface layer that is underlain by an impermeable zone. In arctic regions, the impermeable zone is permafrost.
- Permafrost, permanently frozen ground, covers large portions of North America and Siberia. Constructing buildings and other infrastructure in such regions requires special planning. Leaking heat can melt permafrost, causing a loss of volume and triggering flow in the formerly frozen soil. The resulting subsidence can be devastating.

Q This pipeline carries heated oil from Alaska's North Slope to a port along the south coast. Notice that the pipeline is not buried but rather suspended above ground. Suggest a reason why the pipeline in this image is not buried.

Pipeline

Both photos by Michael Collier.

Give It Some Thought

1. Describe a type of mass wasting that might occur in your home area. Remember to consider characteristics such as climate, surface materials, and steepness of slopes. Does your example have a trigger?

2. Rivers, groundwater, glaciers, wind, and waves can all move and deposit sediment. Geologists refer to these phenomena as *agents of erosion*. Mass wasting also involves the movement and deposition of sediment, yet it is *not* classified as an agent of erosion. How is mass wasting different?

3. This image shows landslide debris atop Buckskin Glacier in Denali National Park in the rugged Alaska Range. The glacier feeds a river that flows into Cook Inlet, just west of the city of Anchorage. Cook Inlet is an arm of the North Pacific.

Michael Collier

a. Based on what you learned about sorting of sediment in Chapter 7 (Figure 7.7, page 217), would you expect the material deposited by the landslide to be well sorted? Why or why not?

b. Also referring to Chapter 7 and Figure 7.7, would you expect the particles in the land slide debris to be rounded or angular? Explain.

c. These mountains are clearly being sculpted by glaciers. However, other processes also have important roles. Briefly describe some of these processes and the role that each plays in the evolution of this mountainous landscape.

4. Describe at least one situation in which an internal process might cause or contribute to a mass-wasting event.

5. Do you think it is likely that landslides frequently occur on the Moon? Explain why or why not.

6. GEOgraphics 15.2 on page 452 includes a world map showing global landslide risks. What criteria were used to construct the map? What additional data could be considered that would make this risk map even more useful?

7. In August 1959, a strong earthquake struck an area along Montana's Madison River, just west of Yellowstone National Park. Based on what you see in the accompanying photo, prepare a brief summary that explains how the area changed as a result of the earthquake. In addition, suggest one or more *new* natural hazards that exist now but did not exist before the quake occurred.

Marli Miller

8. Mass wasting is influenced by many processes associated with all four spheres of the Earth system. Select two items from the list below. For each, outline a series of events that relate the item to various spheres and to a mass-wasting process. Here is an example which assumes that "frost wedging" is an item on the list: *Frost wedging involves rock (geosphere) being broken when water (hydrosphere) freezes. Freeze–thaw cycles (atmosphere) promote frost wedging. When frost wedging loosens a rock on a cliff, the fragment tumbles to the base of the cliff. This event, rock fall, is an example of mass wasting.* Now you give it a try. Use your imagination.
   - Deforestation
   - Spring thaw/melting snow
   - Highway roadcut
   - Crashing waves
   - Cavern formation (see Figure 17.36, page 525)

9. When the rail line in the accompanying photo was built in rural Alaska in the 1930s, the terrain was relatively level. Not long after the railroad was completed, a great deal of subsidence and shifting of the ground occurred, turning the tracks into the "roller coaster" shown here. As a result, the rail line had to be abandoned. Suggest a reason why the ground became unstable and shifted.

L. A. Yehle/O.J. Ferrians, Jr./USGS

Mastering Geology™

www.masteringgeology.com

16
Running Water

Extraordinary rains in September 2013, caused extensive flash flooding in the Colorado Rockies in and around the city of Boulder. During the height of the flood, the Big Thompson River washed out a portion of U.S. 34 and nearly did the same to the houses in the bottom of the image. (Photo by Andy Cross/*The Denver Post*/Getty Images)

Consider the bittersweet relationship we have with rivers. They are vital economic tools—used as highways to move goods and as sources of water for irrigation and energy—as well as prime locations for recreation. When considered as part of the Earth system, rivers and streams represent a fundamental link in the constant cycling of our planet's water. Yet running water is the dominant agent of landscape alteration, eroding more terrain and transporting more sediment than any other process. Soils and other materials can be washed away, only to be deposited elsewhere. Because human populations concentrate along rivers, flooding represents enormous potential for destruction and loss of life.

This chapter provides an overview of the hydrologic cycle, the nature of river systems, the types of river channels and the factors that produce them, the influences of running water on our planet's landscapes, and the nature of floods and their impact on people.

16.1 | Earth as a System: The Hydrologic Cycle

List the hydrosphere's major reservoirs and describe the different paths that water takes through the hydrologic cycle.

We live on a planet that is unique in the solar system—in just the right location and just the right size (see Chapter 22). If Earth were appreciably closer to the Sun, water would exist only as a vapor. Conversely, water would be forever frozen if our planet were much farther away. Moreover, Earth is large enough to have a hot mantle that supports convective flow, which carries water to the surface through volcanism. Water that rose from Earth's interior through mantle convection generated our planet's oceans and atmosphere. Thus, by coincidence of favorable size and location, Earth is the only planet in the solar system that has a global ocean and a hydrologic cycle.

Water is found almost everywhere on Earth—in the oceans, glaciers, rivers, lakes, air, soil, and living tissue. All these "reservoirs" constitute Earth's hydrosphere, which contains about 1.39 billion cubic kilometers (333 million cubic miles) of water. The vast majority of it, an estimated 96.5 percent, is stored in the global ocean. Ice sheets and glaciers account for another 1.74 percent, leaving less than 2 percent divided among lakes, streams, subsurface water, and the atmosphere (**Figure 16.1**).

Water is constantly moving among Earth's different spheres—the *hydrosphere*, the *atmosphere*, the *geosphere*, and the *biosphere*. This unending circulation of water is called the **hydrologic cycle** (**Figure 16.2**). Recall from Chapter 1 that the hydrologic cycle is one of many important subsystems that constitute the Earth system.

When precipitation falls on land, it either soaks into the ground, a process called **infiltration**, flows over the surface as **runoff**, or immediately

Figure 16.1
Distribution of Earth's water

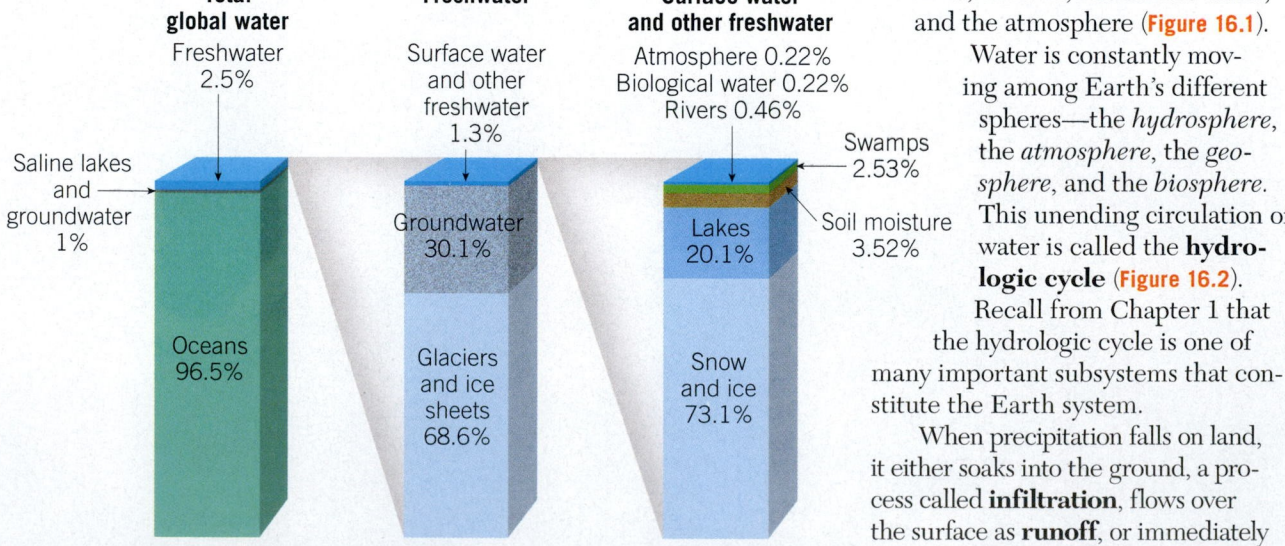

Total global water
- Freshwater 2.5%
- Saline lakes and groundwater 1%
- Oceans 96.5%

Freshwater
- Surface water and other freshwater 1.3%
- Groundwater 30.1%
- Glaciers and ice sheets 68.6%

Surface water and other freshwater
- Atmosphere 0.22%
- Biological water 0.22%
- Rivers 0.46%
- Swamps 2.53%
- Lakes 20.1%
- Soil moisture 3.52%
- Snow and ice 73.1%

evaporates. Much of the water that infiltrates or runs off eventually finds its way back to the atmosphere via evaporation from soil, lakes, and streams. In addition, some of the water that soaks into the ground is absorbed by plants, which later release it into the atmosphere. This process is called **transpiration** (*trans* = across, *spiro* = to breathe). Each year a field of crops may transpire the equivalent of a water layer 0.6 meter (2 feet) deep over the entire field. The same area of trees may pump twice this amount into the atmosphere. Because both evaporation and transpiration involve the transfer of water from the surface directly to the atmosphere, they are often considered together as the combined process of **evapotranspiration**.

Hydrologic Cycle

Evaporation

Precipitation
284,000 km³

Precipitation
96,000 km³

Precipitation

Evaporation/Transpiration
60,000 km³

Evaporation
320,000 km³

36,000 km³

Runoff

Infiltration

Oceans

SmartFigure 16.2
The hydrologic cycle
The primary movement of water through the cycle is shown by the large arrows. A number refers to the annual amount of water taking a particular path.
(https://goo.gl/8IRwFJ)

Tutorial

More water falls on land as precipitation than is lost by evapotranspiration. The excess is carried back to the ocean mainly by streams; less than 1 percent returns as groundwater. However, much of the water that flows in rivers is not transmitted directly into river channels after falling as precipitation. Instead, a large percentage first soaks into surface materials and then gradually flows as groundwater to river channels. In this manner, groundwater provides a form of storage that sustains the flow of streams between storms and during periods of drought. Groundwater is the focus of Chapter 17.

When precipitation falls in very cold areas—at high elevations or high latitudes—the water may not immediately soak in, run off, or evaporate. Instead, it becomes part of a snowfield or a glacier. In this way, glaciers store large quantities of water. If present-day glaciers were to melt and release their stored water, sea level would rise by several tens of meters worldwide and submerge many heavily populated coastal areas. As you will see in Chapter 18, over the past 2 million years, huge ice sheets have formed and melted on several occasions, each time changing the balance of the hydrologic cycle.

Figure 16.2 also shows that Earth's hydrologic cycle is balanced. This basically means that although water is constantly moving from one reservoir to another, the overall amount of water in the oceans and on land remains about the same. Each year, solar energy evaporates about 320,000 cubic kilometers of water from the oceans, but only 284,000 cubic kilometers return to the oceans as precipitation. A balance is achieved by the 36,000 cubic kilometers that are carried to the ocean as runoff. Although runoff makes up a small percentage of the total, running water is, nevertheless, *the single most important erosional agent sculpting Earth's land surface.*

16.1 | Concept Checks

1. Describe or sketch the movement of water through the hydrologic cycle. Once precipitation has fallen on land, what paths might the water take?

2. What is meant by the term *evapotranspiration*?

3. Over the oceans, evaporation exceeds precipitation, yet sea level does not drop. Explain why.

16.2 | Running Water

Describe the nature of drainage basins and river systems. Sketch and briefly explain four basic drainage patterns.

Recall that most of the precipitation that falls on land either enters the soil (infiltration) or remains at the surface, moving downslope as runoff. The amount of water that runs off rather than soaks into the ground depends on several factors: (1) the intensity and duration of rainfall, (2) the amount of water already in the soil, (3) the nature of the surface material, (4) the slope of the land, and (5) the extent and type of vegetation. When the surface material is highly impermeable or when it becomes saturated, runoff is the dominant process. Runoff is also high in urban areas because large areas are covered by impermeable buildings, roads, and parking lots.

Figure 16.3
Drainage basin and divide
A drainage basin is the area drained by a stream and its tributaries. Boundaries between basins are called divides.

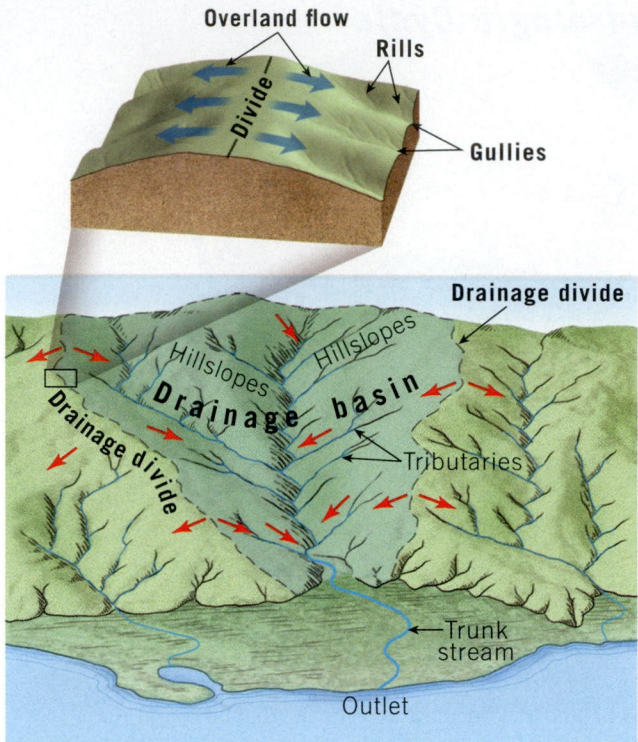

Runoff initially flows in broad, thin sheets across slopes in a process called *sheet flow*. This thin, unconfined flow eventually develops threads of current that form tiny channels called *rills*. Rills meet to form *gullies*, which join to form brooks, creeks, or streams. Then, when they reach an undefined size, they are called rivers. Although the terms *river* and *stream* are often used interchangeably, geologists define **stream** as water that flows in a channel, regardless of size. **River**, on the other hand, is a general term for streams that carry substantial amounts of water and have numerous tributaries.

In humid regions, the water to support streamflow comes from two sources: overland flow that sporadically enters the stream and groundwater that enters the channel. In areas where the bedrock is composed of soluble rocks such as limestone, large openings may exist that facilitate the transport of groundwater to streams. In arid regions, however, the *water table* may be below the level of the stream channel, in which case the stream loses water to the groundwater system by outflow percolating through the streambed.

Drainage Basins

Every stream drains an area of land called a **drainage basin,** or **watershed** (**Figure 16.3**). Each drainage basin is bounded by an imaginary line called a **divide**, which is clearly visible as a sharp ridge in some mountainous areas but can be more difficult to determine when the topography is subdued. The outlet, where the stream exits the drainage basin, is at a lower elevation than the rest of the basin.

Drainage divides range in scale from a small ridge separating two gullies on a hillside to a *continental divide* that splits an entire continent into enormous drainage basins. The Mississippi River has the largest drainage basin in North America, collecting and carrying 40 percent of the flow in the United States (**Figure 16.4**).

By looking at the drainage basin in Figure 16.3, it should be obvious that slopes cover most of the area. Water erosion on the hillsides is aided by the impact of falling rain and surface flow, moving downslope as sheets or in rills toward a stream channel. Hillslope erosion is the main source of fine particles (clays and fine sand) carried in stream channels.

If you could observe the streams in an area similar to that depicted in Figure 16.3 over several years, you would see many of them lengthen their courses by **headward erosion**—that is, by extending the heads of their valleys upslope. In their headwater sections,

SmartFigure 16.4
Mississippi River drainage basin The drainage basin of the Mississippi River forms a funnel that stretches from Montana and southern Canada in the west to New York State in the east that runs down to a spout in Louisiana. It consists of many smaller drainage basins. The drainage basin of the Yellowstone River is one of many that contribute water to the Missouri River, which, in turn, is one of many that make up the drainage basin of the Mississippi River.
(https://goo.gl/z6epSn)

Tutorial

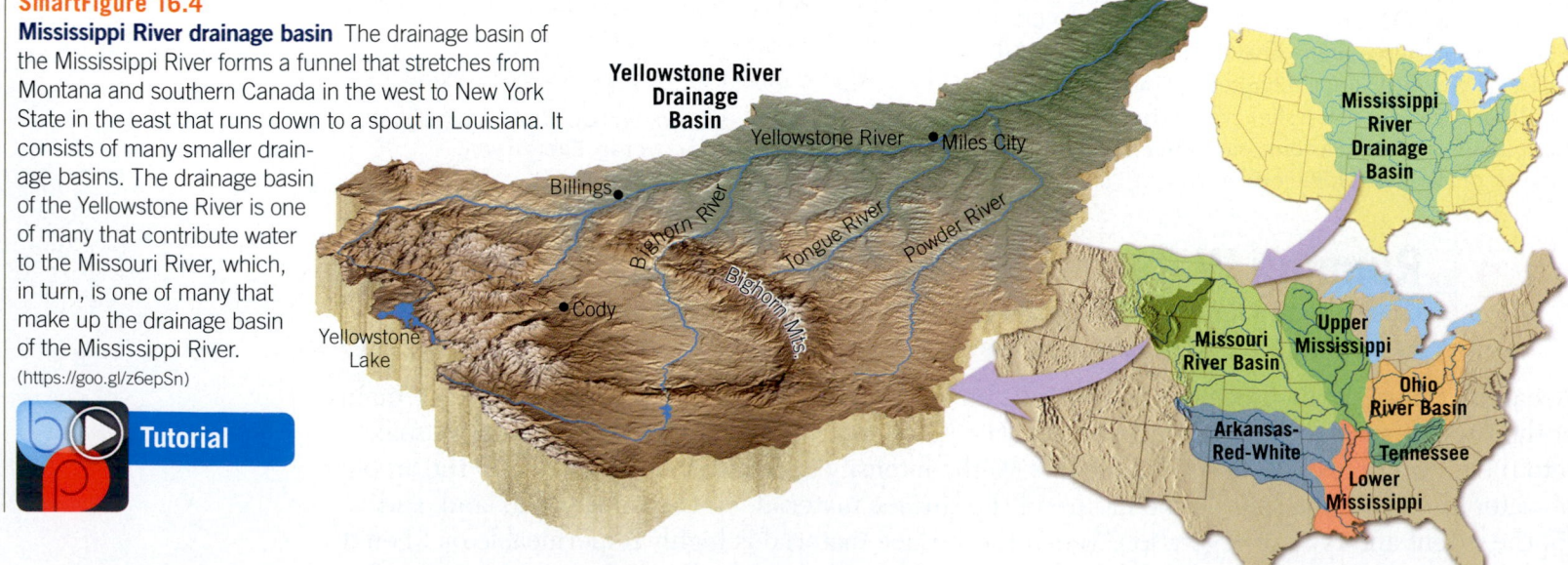

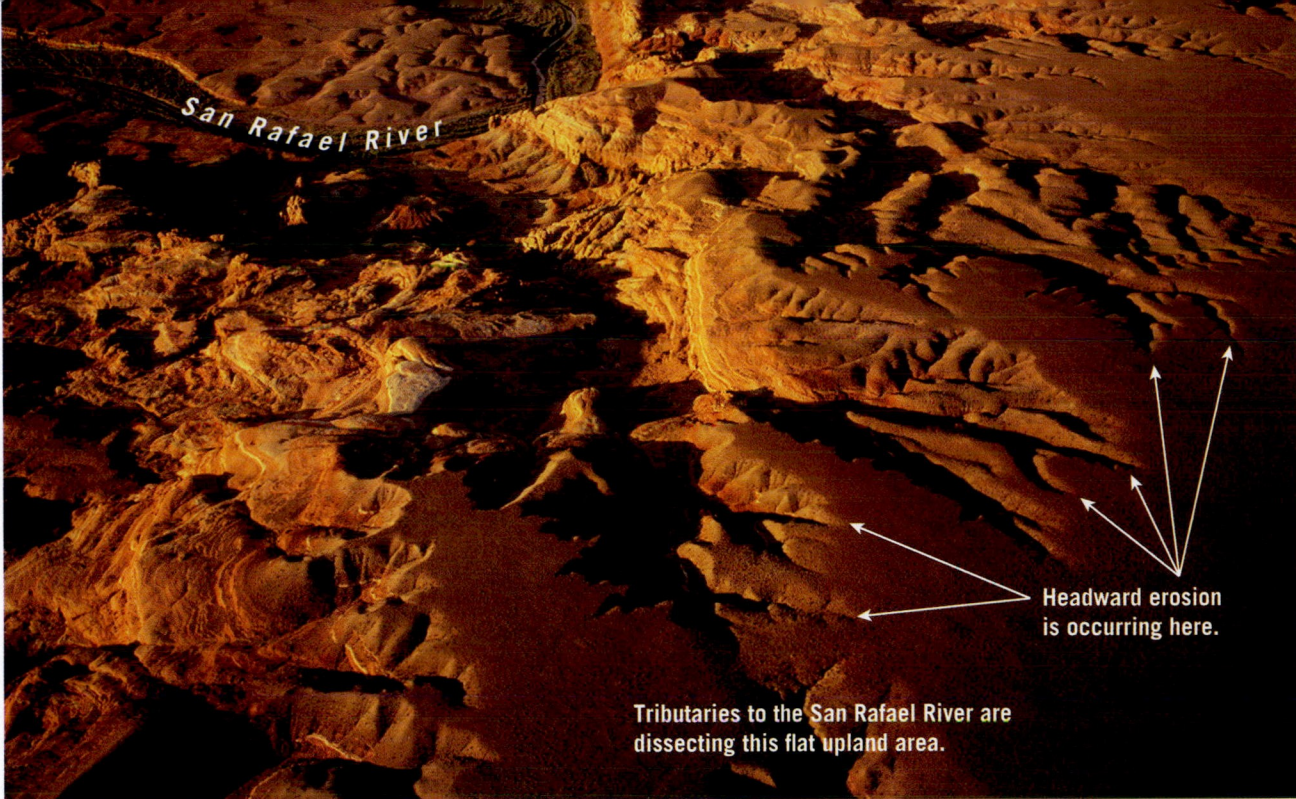

San Rafael River

Headward erosion
is occurring here.

Tributaries to the San Rafael River are
dissecting this flat upland area.

SmartFigure 16.5
Headward erosion
A stream lengthens its
course by extending the
head of its valley upslope
into previously undissected
terrain. (Photo by Michael
Collier) (https://goo.gl/hDFQip)

Tutorial

streams tend to erode their channels downward. This lowering of the stream leads to increased rates of erosion on the steeper slopes, which lie in the opposite direction of streamflow. Thus, through headward erosion, a valley is extended into previously undissected terrain. In **Figure 16.5**, the tributaries to Utah's San Rafael River illustrate this process.

River Systems

Rivers drain much of the land area, with the exception of extremely arid regions or polar areas that are permanently frozen. To a large extent, the variety of rivers that exist is a reflection of the different environments in which they are found. For example, although the Paraná–La Plata river system in South America drains an area roughly the same size as the Nile in Egypt, it carries nearly 10 times more water to the ocean. Because its drainage basin is entirely in a rainy tropical climate, the Paraná–La Plata system has a huge discharge. By contrast, the Nile, which also originates in a humid region, flows through an expansive arid landscape, where significant amounts of water evaporate or are withdrawn to sustain agriculture. Thus, climatic differences and human intervention can significantly influence the character of a river. Later, we will examine other factors that contribute to stream variability.

River systems not only involve a network of stream channels but the entire drainage basin. Based on the dominant processes operating within them, river systems can be divided into three zones: *sediment production*—where erosion dominates, *sediment transport,* and *sediment deposition* (**Figure 16.6**). It is important to recognize that sediment is being eroded, transported, and

deposited along the entire length of a stream, regardless of which process is dominant within each zone.

Sediment Production   The zone of *sediment production*, where most of the sediment is derived, is located in the headwaters region of the river system. Much of the sediment carried by streams begins as bedrock that is subsequently broken down by weathering and then moved downslope by mass wasting and by way of sheet flow and rills. Bank erosion can also contribute significant amounts of sediment. In addition, scouring of the channel bed deepens the channel and adds to the stream's sediment load.

Sediment Transport   Sediment acquired by a stream is transported through the channel network along sections called *trunk streams*. When trunk streams are in balance, the amount of sediment eroded from their banks equals the amount deposited elsewhere in the channel.

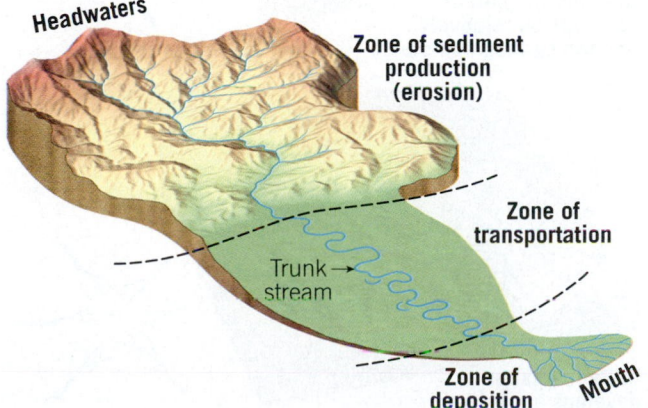

Headwaters

Zone of sediment
production
(erosion)

Zone of
transportation

Trunk
stream

Zone of
deposition   Mouth

Figure 16.6
Zones of a river
Each of the three zones
is based on the dominant
process that is operating
in that part of the river
system.

Although trunk streams rework their channels over time, they are not a source of sediment, nor do they accumulate or store it.

Sediment Deposition

When a river reaches the ocean or another large body of water, it slows, and the energy to transport sediment is greatly reduced. Most of the sediments accumulate at the mouth of the river to form a delta, are reconfigured by wave action to form a variety of coastal features, or are moved far offshore by ocean currents. Because coarse sediments tend to be deposited upstream, it is primarily the fine sediments (clay, silt, and fine sand) that eventually reach the ocean. Taken together, erosion, transportation, and deposition are the processes by which rivers move Earth's surface materials and sculpt landscapes.

Drainage Patterns

Drainage systems are interconnected networks of streams that form a variety of patterns. The nature of a drainage pattern can vary greatly from one type of terrain to another, primarily in response to the kinds of rock on which the streams developed and/or the structural pattern of joints, faults, and folds (see Chapter 10). **Figure 16.7** illustrates four drainage patterns.

Dendritic Pattern

The most commonly encountered drainage pattern is the **dendritic pattern** (Figure 16.7A). This pattern of irregularly branching tributary streams resembles the branching pattern of a deciduous tree. In fact, the word *dendritic* means "treelike." The dendritic pattern forms where the underlying material is relatively uniform. Because the surface material is essentially uniform in its resistance to erosion, it does not control the pattern of streamflow. Rather, the pattern is determined chiefly by the direction of slope of the land.

Radial Pattern

When streams diverge from a central area like spokes from the hub of a wheel, the pattern is said to be **radial** (Figure 16.7B). This pattern typically develops on isolated volcanic cones and domal uplifts.

Rectangular Pattern

A **rectangular pattern** exhibits many right-angle bends (Figure 16.7C). This pattern develops when the bedrock is crisscrossed by a series of joints and/or faults. Because fractured rock tends to weather and erode more easily than unbroken rock, the geometric pattern of joints guides the paths of streams as they carve their valleys.

Trellis Pattern

A **trellis pattern** is a rectangular drainage pattern in which tributary streams are nearly parallel to one another and have the appearance of a garden trellis (Figure 16.7D). This pattern forms in areas underlain by alternating bands of resistant and less-resistant rock and is particularly well displayed in the folded Appalachian Mountains, where both weak and strong strata outcrop in nearly parallel belts.

Figure 16.7
Drainage patterns
Networks of streams form a variety of patterns.

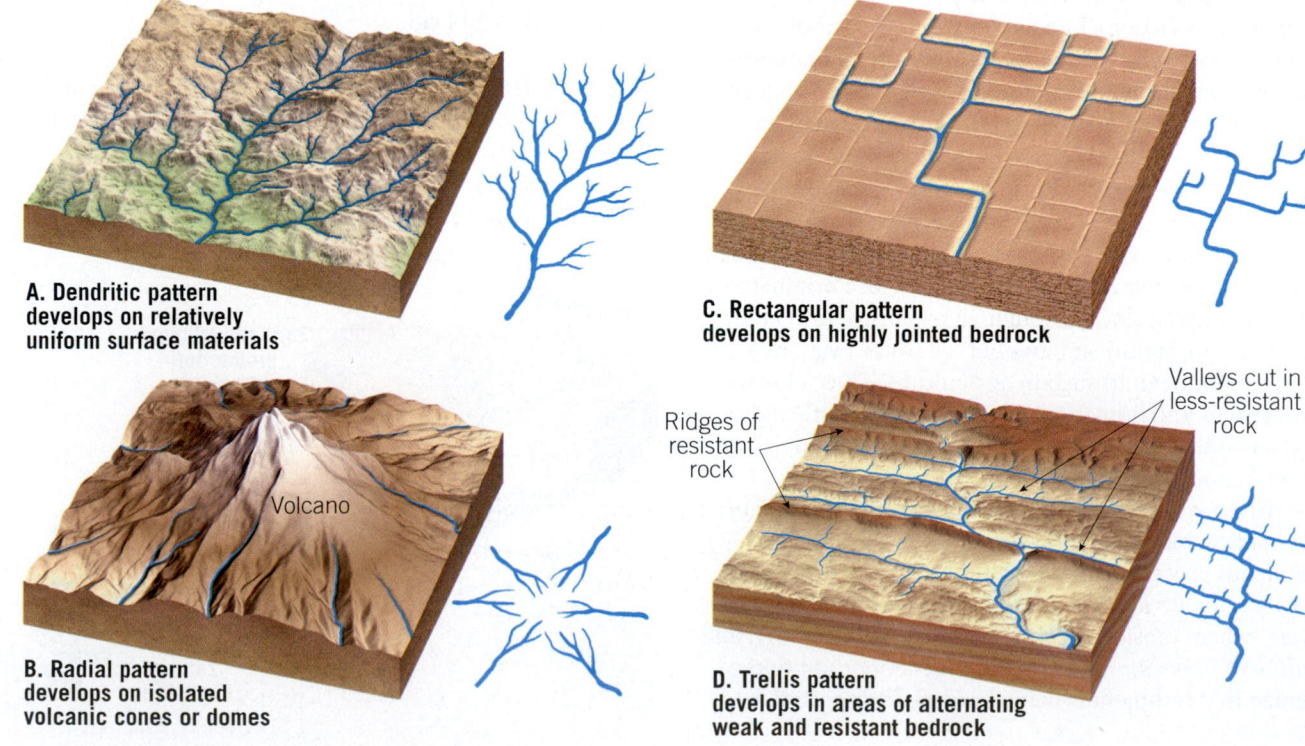

A. Dendritic pattern
develops on relatively
uniform surface materials

B. Radial pattern
develops on isolated
volcanic cones or domes

Volcano

C. Rectangular pattern
develops on highly jointed bedrock

Ridges of resistant rock

Valleys cut in less-resistant rock

D. Trellis pattern
develops in areas of alternating
weak and resistant bedrock

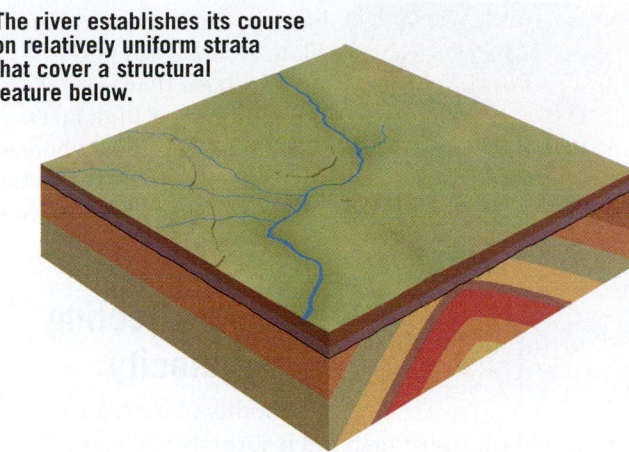

The river establishes its course on relatively uniform strata that cover a structural feature below.

As the stream erodes downward, it encounters and cuts through the resistant rock to create a water gap. By this process the river is superimposed on the ridge.

Water gap →

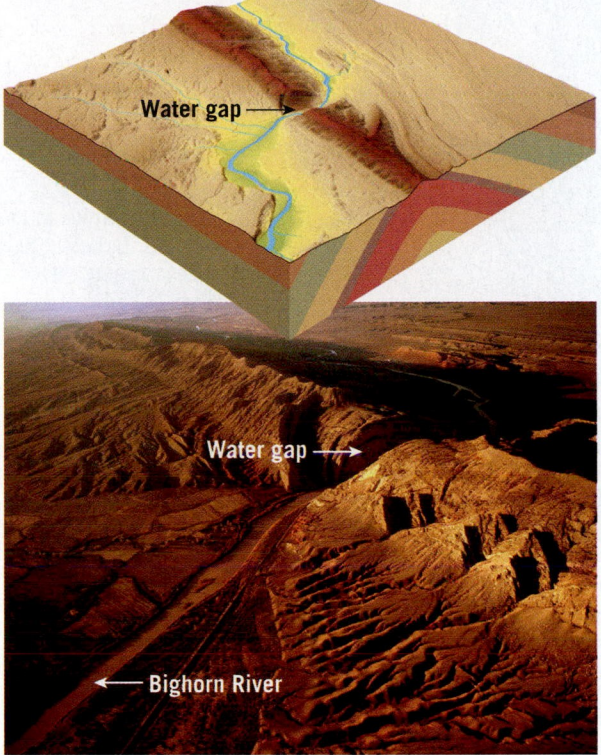

Water gap →→

← Bighorn River

The Bighorn River is a superposed stream that created a water gap through Wyoming's Sheep Mountain.

Figure 16.8
Development of a superposed stream This is one way in which a water gap can form. The Bighorn River is a good example. See SmartFigure Mobile Field Trip 10.9 for some excellent views and additional explanation of the Bighorn River and how it created the water gap shown here. (Photo by Michael Collier)

Formation of a Water Gap   To fully understand the drainage pattern displayed by a stream, it is often useful to consider a stream's entire history. For example, river valleys occasionally cut through a ridge or mountainous topography that lies across their path. This situation is illustrated by the trellis pattern in Figure 16.7D. The steep-walled notch followed by the river through a tectonic structure is called a **water gap** (Figure 16.8).

Why do streams cut *across* such structures and not flow *around* them? One possibility is that a stream existed before the ridge or mountain was uplifted. In this situation, the stream, called an **antecedent stream**, eroded its bed downward at a pace equal to the rate of uplift. That is, the stream would maintain its course as folding or faulting gradually raised the structure across its path.

A second possibility is that a **superposed stream** eroded its channel into an existing structure (see Figure 16.8). This can occur when folded beds or resistant rocks are buried beneath layers of relatively flat-lying sediments or sedimentary strata. Streams originating on the overlying strata establish their courses without regard to the structures below. Then, as the valley deepens, the river continues to cut its valley into the underlying structure. The folded Appalachians feature several superposed rivers, including the Potomac and the Susquehanna, which cut their channels through folded strata on their way to the Atlantic.

16.2   Concept Checks

1. List several factors that cause infiltration and runoff to vary from place to place and time to time.

2. Draw a simple sketch of a drainage basin and a divide and label each.

3. What are the three main parts (zones) of a river system?

4. Prepare a simple sketch of the four drainage patterns discussed in this section. Briefly describe why streams exhibit each pattern.

5. Contrast antecedent and superposed streams.

16.3 | Streamflow Characteristics
Discuss streamflow and the factors that cause it to change.

The water in river channels moves under the influence of gravity. In very slowly flowing streams, water moves in nearly straight-line paths parallel to the stream channel; this is called **laminar flow**. However, streams typically exhibit **turbulent flow**, in which the water moves in an erratic fashion characterized by a series of horizontal and vertical swirling motions (Figure 16.9). Strong turbulent behavior occurs in whirlpools and eddies, as well as in roiling whitewater rapids. Even streams that appear smooth on the surface often exhibit turbulent flow near the bottom and sides of the channel, where flow resistance is greatest. *Turbulence* contributes to a stream's ability to erode its channel because it acts to lift sediment from the streambed.

Figure 16.9
Laminar and turbulent flow Most often streamflow is turbulent. (Photos by Michael Collier)

A.

B.

Running the rapids in the Grand Canyon— an extreme example of turbulent flow.

This water is not standing still. It is moving slowly toward the bottom of the image. The flow in the foreground is primarily laminar.

An important factor influencing stream turbulence is the water's *flow velocity*. As the velocity of a stream increases, the flow becomes more turbulent. Flow velocities can vary significantly from place to place along a stream channel, as well as over time, in response to variations in the amount and intensity of precipitation. If you have ever waded into a stream, you may have noticed that the strength of the current increased as you moved into deeper parts of the channel. This is related to the fact that frictional resistance is greatest near the banks and bed of a stream channel.

Flow velocities are determined at stream *gaging stations* by averaging measurements taken at various

locations across the stream's channel. Some sluggish streams have flow velocities of less than 1 kilometer (0.62 mile) per hour, whereas stretches of some fast-flowing rivers may exceed 30 kilometers (19 miles) per hour.

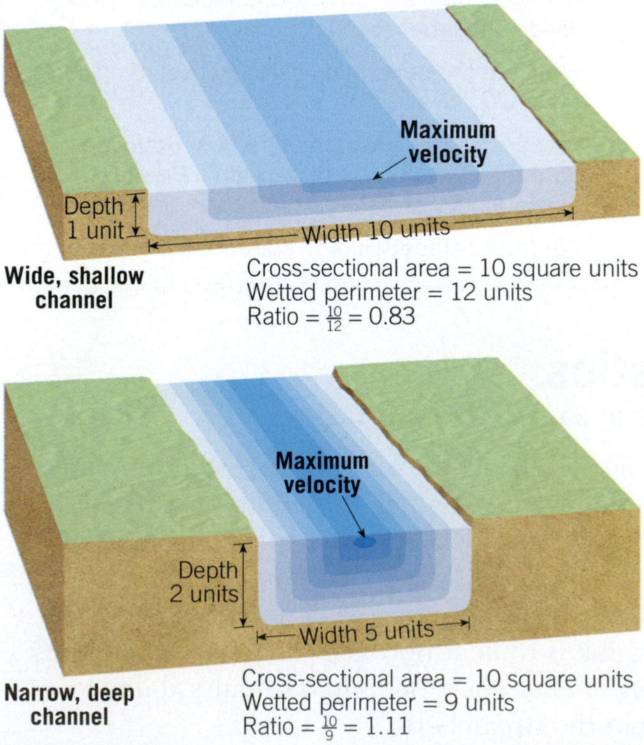

Factors Affecting Flow Velocity

The ability of a stream to erode and transport material is directly related to its flow velocity. Even slight variations in flow rate can lead to significant changes in the sediment load transported by a stream. Several factors influence flow velocities and, therefore, control a stream's potential to do its "work" of transporting sediment. These factors include (1) channel slope, or gradient, (2) channel cross-sectional shape, (3) channel size and roughness, and (4) discharge, or the amount of water flowing in the channel.

Gradient   The slope of a stream channel, expressed as the vertical drop of a stream over a specified distance, is called **gradient**. Portions of the lower Mississippi River have very low, or gentle, gradients, about 10 centimeters or less per kilometer. By contrast, some mountain streams have channels that drop at a rate of more than 40 meters per kilometer—a gradient 400 times steeper than that of the lower Mississippi. Gradient also varies along the length of a particular channel. When the gradient is steeper, more gravitational energy is available to drive channel flow.

Channel Shape   As water in a stream channel moves downslope, it encounters a significant amount of frictional resistance. The *cross-sectional shape* (a slice taken across the channel) determines, to a large extent, the amount of flow that comes in contact with the banks and bed of the channel. This measure is referred to as the **wetted perimeter**. The most efficient channel is one with the least wetted perimeter for its cross-sectional area. **Figure 16.10** compares two channels that differ only in shape: One is wide and shallow, the other is narrow and deep. Although the cross-sectional area of both is identical, the narrow and deep one has less water in contact with the channel (a smaller wetted perimeter) and therefore less frictional drag. As a result, if all other factors are equal, the water will flow more efficiently and at a higher velocity in a deep and narrow channel than in a wide and shallow channel.

Channel Size and Roughness   Water depth also affects the frictional resistance that the channel exerts on flow. Maximum flow velocity occurs when a stream is *bankfull*, before water starts to overflow its banks

Figure 16.10
Influence of channel shape on velocity The stream with the smaller wetted perimeter has less frictional drag and will flow more rapidly, all else being equal.

Maximum velocity

Depth 1 unit

Width 10 units

Wide, shallow channel

Cross-sectional area = 10 square units
Wetted perimeter = 12 units
Ratio = $\frac{10}{12}$ = 0.83

Maximum velocity

Depth 2 units

Width 5 units

Narrow, deep channel

Cross-sectional area = 10 square units
Wetted perimeter = 9 units
Ratio = $\frac{10}{9}$ = 1.11

and inundate the surrounding floodplain. At this stage, the channel's ratio of the cross-sectional area to wetted perimeter is highest and streamflow is most efficient. Therefore, greater water depth increases this ratio, which in turn increases channel efficiency. All other factors being equal, flow velocities are higher in large channels than in small channels.

Most streams have channels that can be described as *rough*. Elements such as boulders, irregularities in the channel bed, and woody debris create turbulence that significantly reduces flow velocity.

SmartFigure 16.11
The Mighty Mississippi near Memphis, Tennessee The Mississippi is North America's largest river. From head to mouth, it is nearly 3900 kilometers (2400 miles) long. Its watershed encompasses about 40 percent of the lower 48 states and includes all or parts of 31 states and 2 Canadian provinces. Average discharge at its mouth is about 16,800 cubic meters (593,000 cubic feet) per second. (Photo by Michael Collier) (http://goo.gl/LkgZXX)

Mobile Field Trip

Discharge   Streams vary in size from small headwater creeks less than a meter wide to large rivers with widths of several kilometers. The size of a stream channel is largely determined by the amount of water supplied from the drainage basin. The measure most often used to compare the size of streams is **discharge**—the volume of water flowing past a certain point in a given unit of time. Discharge, usually measured in cubic meters per second or cubic feet per second, is determined by multiplying a stream's cross-sectional area by its flow velocity. The largest river in North America, the Mississippi, has an average discharge at its mouth of about 16,800 cubic meters (593,000 cubic feet) per second (**Figure 16.11**). That figure is dwarfed by South America's mighty Amazon River, which discharges nearly 13 times more water than the Mississippi. GEOgraphics 16.1 presents more information on Earth's major rivers.

The discharge of a river system changes over time because of variations in the amount of precipitation received by the drainage basin. Studies show that when discharge increases, the width, depth, and flow velocity of the channel all increase predictably. As we saw earlier, when the size of the channel increases, proportionally less water is in contact with the bed and banks of the channel. This reduces friction, which acts to retard the flow, resulting in an increase in the rate of streamflow.

Monitoring Streamflow   The U.S. Geological Survey maintains a network of about 7500 stream gaging stations that collect basic data about the country's surface-water resources (**Figure 16.12**). Data collected include flow velocity, discharge, and river stage. *Stage* is the height of the water surface relative to a fixed reference point. This

The U.S. Geological Survey operates and maintains nearly 7500 stream gaging stations.

Sunday, October 14, 2012 13:30 ET

Streamflow

| Low | Much below normal | Below normal | Normal | Above normal | Much above normal | High |

Colors indicate whether the current river stage is above or below normal.

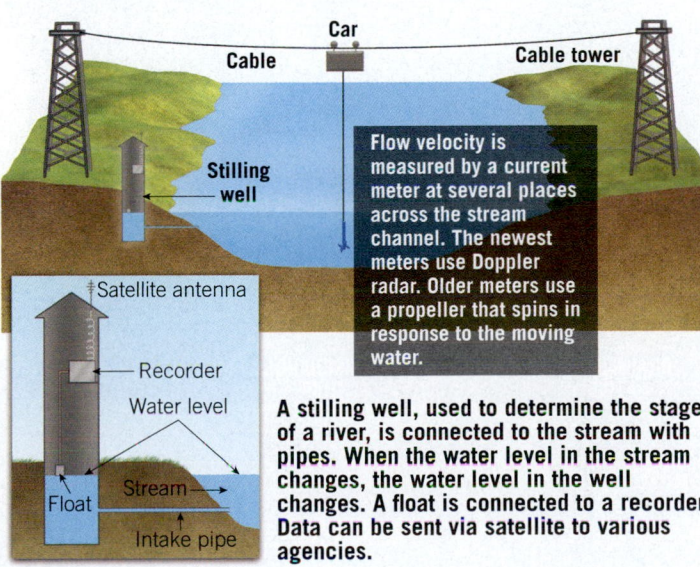

Cable   Car   Cable tower

Stilling well

Flow velocity is measured by a current meter at several places across the stream channel. The newest meters use Doppler radar. Older meters use a propeller that spins in response to the moving water.

Satellite antenna

Recorder

Water level

Float   Stream

Intake pipe

A stilling well, used to determine the stage of a river, is connected to the stream with pipes. When the water level in the stream changes, the water level in the well changes. A float is connected to a recorder. Data can be sent via satellite to various agencies.

Figure 16.12
Stream gaging stations The United States has a dense network of stream gaging stations. To determine a channel's discharge, it is necessary to survey the channel to determine its shape and calculate its area. A stilling well measures the stage, and a current meter determines flow velocity.

What Are the Largest Rivers?

When rivers are ranked, the criterion most often used is the amount of water the river delivers to the ocean expressed in cubic feet (ft³) or cubic meters (m³) per second.

The flow of the Amazon accounts for about 15 percent of all the fresh water that flows into the oceans. Much of its huge drainage basin is tropical rainforest that receives 80 inches or more of rainfall each year.

WORLD'S 10 LARGEST RIVERS

#6 YENISEI RIVER
Drainage basin: 1,000,000 square miles
Average discharge: 614,000 cubic feet per second

#9 LENA RIVER
Drainage basin: 936,000 square miles
Average discharge: 547,000 cubic feet per second

#8 MISSISSIPPI RIVER
Drainage basin: 1,150,000 square miles
Average discharge: 593,000 cubic feet per second

#7 ORINOCO RIVER
Drainage basin: 340,000 square miles
Average discharge: 600,000 cubic feet per second

#5 GANGES RIVER
Drainage basin: 409,000 square miles
Average discharge: 660,000 cubic feet per second

#3 YANGTZE RIVER
Drainage basin: 750,000 square miles
Average discharge: 770,000 cubic feet per second

#1 AMAZON RIVER
Drainage basin: 2,231,000 square miles
Average discharge: 7,500,000 cubic feet per second

#2 CONGO RIVER
Drainage basin: 1,550,000 square miles
Average discharge: 1,400,000 cubic feet per second

#4 BRAHMAPUTRA RIVER
Drainage basin: 361,000 square miles
Average discharge: 700,000 cubic feet per second

#10 PARANA RIVER
Drainage basin: 890,000 square miles
Average discharge: 526,000 cubic feet per second

The size of the drainage basin and the amount of rainfall it receives are the primary factors influencing discharge.

The drainage basin of the "Mighty Mississippi" covers about 40 percent of the lower 48 states and includes all or parts of 31 states and 2 Canadian provinces. The discharge of North America's largest river is just one-twelfth that of the Amazon.

LARGEST U.S. RIVERS

COLUMBIA RIVER
MISSOURI RIVER
ST. LAWRENCE RIVER
OHIO RIVER
MISSISSIPPI RIVER
TENNESSEE RIVER
YUKON RIVER

RANK	RIVER	DRAINAGE BASIN (1000 mi²)	AVERAGE DISCHARGE AT MOUTH (1000 ft³/sec)
1	Mississippi	1,150	593
2	St. Lawrence	396	348
3	Ohio	203	281
4	Columbia	258	265
5	Yukon	328	225
6	Missouri	529	76
7	Tennessee	41	68

Questions:
1. Why is the Amazon's discharge so large?
2. About how many times larger is the discharge of the Amazon than the second ranked river, the Congo?

Guido Alberto Rossi/AGE Fotostock

measurement is frequently reported by the media, especially when a river approaches or surpasses *flood stage*. Streamflow measurements are essential components of river models used to make flood forecasts and issue warnings. There are other applications as well. The data are used to make decisions related to water supply allocations, operation of wastewater treatment plants, design of highway bridges, and recreation activities.

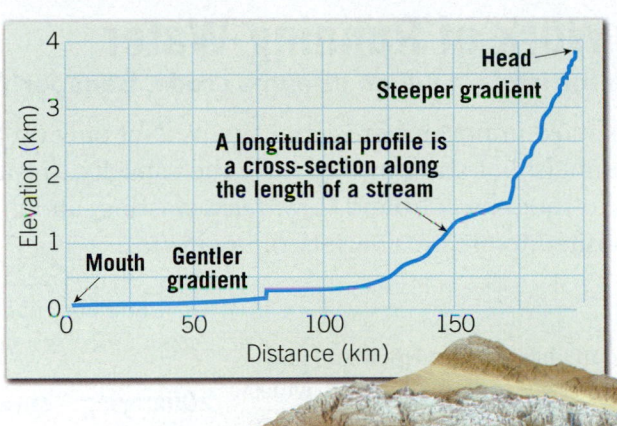

Figure 16.13
Longitudinal profile California's Kings River originates high in the Sierra Nevada and flows into the San Joaquin Valley.

Changes Downstream

One useful way of studying a stream is to examine its **longitudinal profile**. Such a profile is simply a cross-sectional view of a stream from its source area (called the **head,** or **headwaters**) to its **mouth**, the point downstream where it empties into another water body—a river, a lake, or the ocean. **Figure 16.13** shows that the most obvious feature of a typical longitudinal profile is its concave shape—a result of the decrease in slope that occurs from the headwaters to the mouth. In addition, local irregularities exist in the profiles of most streams; the flatter sections may be associated with lakes or reservoirs, and the steeper sections are sites of rapids or waterfalls.

The change in slope observed on most stream profiles is usually accompanied by an increase in discharge and channel size, as well as a reduction in sediment particle size (**Figure 16.14**). For example, data from successive gaging stations along most rivers show that, in humid regions, discharge increases toward the mouth. This should come as no surprise because, as we move downstream, more and more tributaries contribute water to the main channel. In the case of the Amazon, for example, about 1000 tributaries join the main river along its 6500-kilometer (4000-mile) course across South America.

In order to accommodate the growing volume of water, channel size typically increases downstream as well. Recall that flow velocities are higher in large channels than in small channels. Furthermore, observations show a general decline in sediment size downstream, making the channel smoother and more efficient.

Although the channel slope decreases—the gradient becomes less steep—toward a stream's mouth, the flow velocity generally increases. This fact contradicts our intuitive assumptions of swift, narrow headwater streams and wide, placid rivers flowing across more subtle topography. Increases in channel size and discharge and decreases in channel roughness that occur downstream compensate for the decrease in slope—thereby making the stream more efficient (see Figure 16.14). Thus, the average flow velocity is typically lower in headwater streams than in wide, placid-appearing rivers.

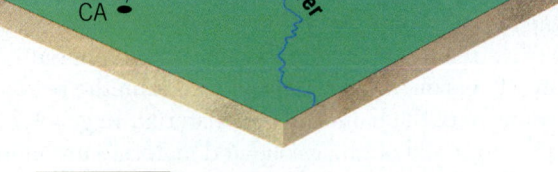

16.3 Concept Checks

1. Contrast laminar flow and turbulent flow.

2. Summarize the factors that influence flow velocity.

3. What is a longitudinal profile?

4. What typically happens to channel width, channel depth, flow velocity, and discharge between the headwaters and the mouth of a stream? Briefly explain why these changes occur.

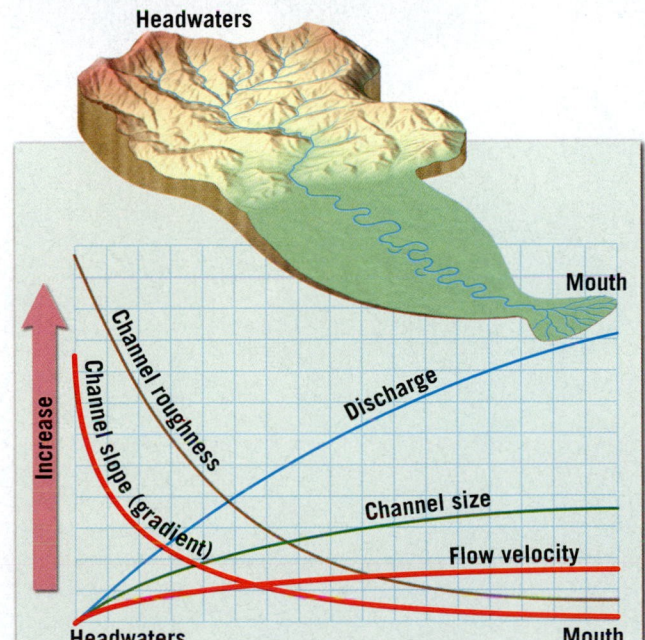

SmartFigure 16.14
Channel changes from head to mouth
Although the gradient decreases toward the mouth of a stream, increases in discharge and channel size and decreases in roughness more than offset the decrease in slope. Consequently, flow velocity usually increases toward the mouth. (https://goo.gl/6srX2s)

16.4 | The Work of Running Water

Outline the ways in which streams erode, transport, and deposit sediment.

Streams are Earth's most important erosional agents. Not only do they have the ability to deepen and widen their channels, but streams also have the capacity to transport enormous quantities of sediment delivered to them by overland flow, mass wasting, and groundwater. Eventually, much of this material is deposited to create a variety of landforms.

Stream Erosion

A stream's ability to accumulate and transport soil and weathered rock is aided by the work of raindrops, which knock sediment particles loose (see Figure 6.24 on page 202). When the ground is saturated, rainwater cannot infiltrate, so it flows downslope, transporting some of the material it dislodges. On barren slopes the flow of muddy water (sheet flow) often produces small channels (rills), which in time can evolve into larger gullies (see Figure 6.25 on page 203).

Once flow is confined within a channel, the erosional power of a stream is related to its slope and discharge. The rate of erosion, however, also depends on the relative resistance of the bank and bed material. In general, channels composed of unconsolidated materials are more easily eroded than channels cut into bedrock.

When a channel is sandy, the particles are easily dislodged from the bed and banks, and they are lifted into the moving water. Moreover, banks consisting of sandy material are often *undercut*, or eroded at their base, dumping even more loose debris into the water to be carried downstream. Banks that consist of coarse gravels or cohesive clay and silt particles tend to be relatively resistant to erosion. Thus, channels with cohesive silty banks are generally narrower than comparable ones with sandy banks.

Figure 16.15
Potholes The rotational motion of swirling pebbles acts like a drill to create potholes. (Photo by Storm Studio/Alamy Images)

Abrasion by long-lived whirlpools armed with sand and pebbles creates bowl-shaped bedrock depressions called potholes.

Streams cut channels into bedrock through three main processes: *quarrying*, *abrasion*, and *corrosion*.

Quarrying **Quarrying** involves the removal of blocks from the bed of a stream channel. This process is aided by fracturing and weathering that loosen the blocks sufficiently so they are movable during times of high flow rates. Quarrying is mainly a result of the impact forces exerted by flowing water.

Abrasion The process by which the bed and banks of a bedrock channel are ceaselessly bombarded by particles carried into the flow is termed **abrasion**. The individual sediment grains are also abraded by their many impacts with the channel and with one another. Thus, by scraping, rubbing, and bumping, abrasion erodes a bedrock channel and simultaneously smooths and rounds the abrading particles. That is why smooth, rounded cobbles and pebbles are found in streams. Abrasion also results in a reduction in the size of the sediments transported by streams.

Features common to some bedrock channels are circular depressions known as **potholes**, which are created by the abrasive action of particles swirling in fast-moving eddies (**Figure 16.15**). The rotational motion of sand and pebbles acts like a drill that bores the holes. As the particles wear down to nothing, they are replaced by new ones that continue to drill the streambed. Eventually, smooth depressions several meters across and deep may result.

Corrosion Bedrock channels formed in soluble rock such as limestone are susceptible to **corrosion**—a process in which rock is gradually dissolved by the flowing water. Corrosion is a type of chemical weathering between the solutions in the water and the mineral matter that makes up the bedrock.

Transport of Sediment by Streams

All streams, regardless of size, transport some weathered rock material (**Figure 16.16**). Streams also sort the solid sediment they transport because finer, lighter material is carried more readily than larger, heavier particles. Streams transport their load of sediment in three ways: (1) in solution (**dissolved load**), (2) in suspension (**suspended load**), and (3) by sliding, skipping, or rolling along the bottom (**bed load**).

Dissolved Load Most of the dissolved load is brought to a stream by groundwater and is dispersed throughout

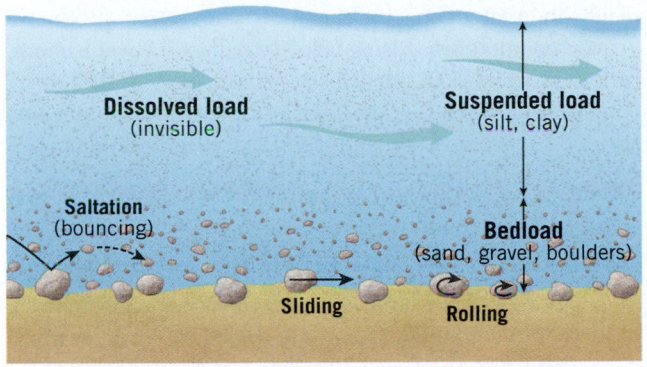

SmartFigure 16.16
Transport of sediment Streams transport their load of sediment in three ways. The dissolved and suspended loads are carried in the general flow. The bed load includes coarse sand, gravel, and boulders that move by rolling, sliding, and saltation. (https://goo.gl/F9TOv1)

Animation

Figure 16.17
Suspended load
An aerial view of the Colorado River in the Grand Canyon. Heavy rains washed sediment into the river. (Photo by Michael Collier)

This river's muddy appearance is a result of suspended sediment.

the flow. When water percolates through the ground, it acquires soluble soil compounds. Then it seeps through cracks and pores in bedrock, dissolving additional mineral matter. Eventually much of this mineral-rich water finds its way into streams.

The velocity of streamflow has essentially no effect on a stream's ability to carry its dissolved load; material in the solution goes wherever the stream goes. Precipitation of the dissolved mineral matter occurs when the chemistry of the water changes, when organisms create hard parts, or when the water enters an inland "sea," located in an arid climate where the rate of evaporation is high.

Suspended Load Most streams carry the largest part of their load in *suspension*. Indeed, the muddy appearance created by suspended sediment is the most obvious portion of a stream's load (**Figure 16.17**). Usually only very fine sand, silt, and clay particles are carried this way, but during flood stage, larger particles can also be transported in suspension. During flood stage, the total quantity of material carried in suspension also increases dramatically, verifiable by people whose homes have become sites for the deposition of this material. For example, during its flood stage, the Yellow River (Hwang Ho) of China is reported to carry an amount of sediment equal in weight to the water that carries it. Rivers such as this are appropriately described as "too thick to drink but too thin to cultivate."

The type and amount of material carried in suspension are controlled by two factors: the flow velocity and the settling velocity of each sediment grain. **Settling velocity** is defined as the speed at which a particle falls through a still fluid. The larger the particle, the more rapidly it settles toward the streambed (**Figure 16.18**). In addition to size, the shape and specific gravity of

particles also influence settling velocity. Flat grains sink through water more slowly than spherical grains, and dense particles fall toward the bottom more rapidly than less-dense particles. The slower the settling velocity and higher the flow velocity, the longer a sediment particle will stay in suspension, and the farther it will be carried downstream.

Bed Load Coarse material, including coarse sands, gravels, and even boulders, typically move along the bed of a channel as bed load. The particles that make up the bed load move by rolling, sliding, and saltation. Sediment moving by **saltation** (*saltare* = to leap) appears to jump or skip along the streambed (see Figure 16.16). This

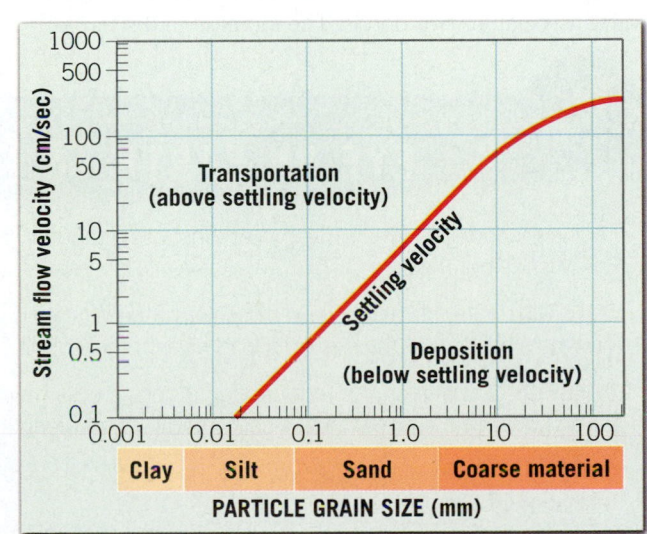

Figure 16.18
Settling velocity
The speed at which a particle falls through still water is its settling velocity. Deposition of suspended particles occurs when their settling velocities are greater than the stream's flow velocity.

occurs as particles are propelled upward by collisions or lifted by the current and then carried downstream a short distance until gravity pulls them back to the bed of the stream. Particles that are too large or heavy to move by saltation either roll or slide along the bottom, depending on their shapes.

The movement of bed load through a stream network tends to be less rapid and more localized than the movement of suspended load. A study conducted on a glacially fed river in Norway determined that suspended sediments took only a day to exit the drainage basin, while the bed load required several decades to travel the same distance. Depending on the discharge and slope of the channel, coarse gravels may only be moved during times of high flow, while boulders move only during exceptional floods. Once set in motion, large particles are usually carried short distances. Along some stretches of a stream, bed load cannot be carried at all until it is broken into smaller particles.

Capacity and Competence   A stream's ability to carry solid particles is described using two criteria: *capacity* and *competence*. **Capacity** is the maximum load of solid particles a stream can transport per unit time. The greater the discharge, the greater the stream's capacity for hauling sediment. Consequently, large rivers with high flow velocities have large capacities.

Competence is a measure of a stream's ability to transport particles based on size rather than quantity. Flow velocity is key: Swift streams have greater competencies than slow streams, regardless of channel size. A stream's competence increases proportionately to the square of its velocity. Thus, if the velocity of a stream doubles, the impact force of the water increases four times; if the velocity triples, the force increases nine times, and so forth. Hence, large boulders that are often visible in low water and seem immovable can, in fact, be transported during exceptional floods because of the stream's increased competence.

By now it should be clear why the greatest erosion and transportation of sediment occur during periods of high water associated with floods. The increase in discharge results in greater capacity; the increased velocity produces greater competency. Rising velocity makes the water more turbulent, and larger particles are set in motion. In the course of a few days, or perhaps just a few hours, a stream at flood stage can erode and transport more sediment than it does during several months of normal flow.

Deposition of Sediment by Streams

Deposition occurs whenever a stream slows, causing a reduction in competence. Put another way, particles are deposited when the flow velocity is less than the settling velocity. As a stream's flow velocity decreases, sediment begins to settle, largest particles first. In this manner, stream transport provides a mechanism by which solid particles of various sizes are separated. This process, called **sorting**, explains why particles of similar size are deposited together. For example, you may have seen stretches of a riverbed consisting mainly of gravel or boulders, while sandbars may have dominated another part of the stream.

The general term for sediment deposited by streams is **alluvium**. Many different depositional features are composed of alluvium. Some occur within stream channels, some occur on the valley floor adjacent to a channel, and some are found at the mouth of a stream. We will consider the nature of these features later in the chapter.

16.4   Concept Checks

1. Describe two processes by which streams cut channels in bedrock.

2. In what three ways does a stream transport its load? Which part of the load moves most slowly?

3. Explain the difference between capacity and competency.

4. What is settling velocity? What factors influence settling velocity? Does settling velocity affect the dissolved load?

EYE ON EARTH 16.1

The meandering White River in Arkansas is a tributary of the Mississippi River.

QUESTION 1   In this aerial view, the color of the White River is brown. What part of the stream's load gives it this color?

QUESTION 2   If a channel were created across the narrow neck of land shown by the arrow, how would the river's gradient change?

QUESTION 3   How would the flow velocity be affected by the formation of such a channel?

Michael Collier

16.5 | Stream Channels

Contrast bedrock and alluvial stream channels. Distinguish between two types of alluvial channels.

A basic characteristic that distinguishes streamflow from overland flow is that it is confined in a channel. A stream channel can be thought of as an open conduit consisting of the streambed and banks that act to confine flow except, of course, during floods.

Although this is somewhat oversimplified, we can divide stream channels into two basic types. *Bedrock channels* are channels in which the streams are actively cutting into solid rock. In contrast, when the bed and banks are composed mainly of unconsolidated sediment or alluvium, the channel is called an *alluvial channel.*

Bedrock Channels

As the name suggests, **bedrock channels** are cut into the underlying strata and typically form in the headwaters of river systems where streams have steep slopes. The energetic flow tends to transport coarse particles that actively abrade the bedrock channel. Potholes are often visible evidence of the erosional forces at work.

Steep bedrock channels often develop a sequence of *steps* and *pools*, relatively flat segments (pools) where alluvium tends to accumulate and steep segments (steps) where bedrock is exposed. The steep areas contain rapids or, occasionally, waterfalls.

The channel pattern exhibited by streams cutting into bedrock is controlled by the underlying geologic structure. Even when flowing over rather uniform bedrock, streams tend to exhibit winding or irregular patterns rather than flowing in straight channels. Anyone who has gone whitewater rafting has observed the steep, winding nature of a stream flowing in a bedrock channel.

Alluvial Channels

Alluvial channels form in sediment that was previously deposited in the valley. When the valley floor reaches

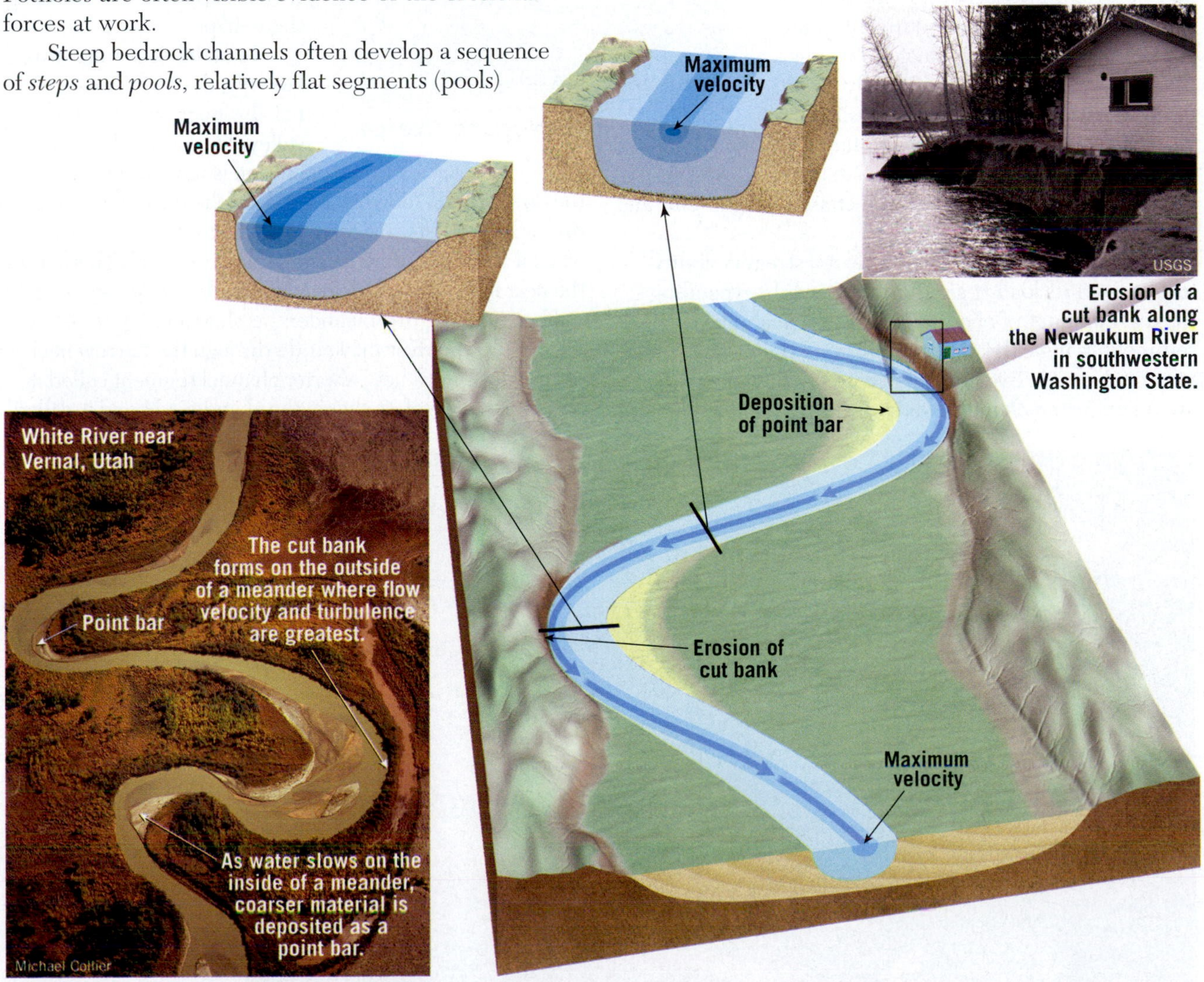

SmartFigure 16.19
Formation of cut banks and point bars By eroding its outer bank and depositing material on the inside of the bend, a stream is able to shift its channel. (https://goo.gl/4bXxsu)

Tutorial

Erosion of a cut bank along the Newaukum River in southwestern Washington State.

Maximum velocity

Maximum velocity

Deposition of point bar

Erosion of cut bank

Maximum velocity

White River near Vernal, Utah

Point bar

The cut bank forms on the outside of a meander where flow velocity and turbulence are greatest.

As water slows on the inside of a meander, coarser material is deposited as a point bar.

Michael Collier

SmartFigure 16.20

Formation of an oxbow lake Oxbow lakes occupy abandoned meanders. Aerial view of an oxbow lake created by the meandering Green River near Bronx, Wyoming. (Photo by Michael Collier) (https://goo.gl/J1KomF)

Animation

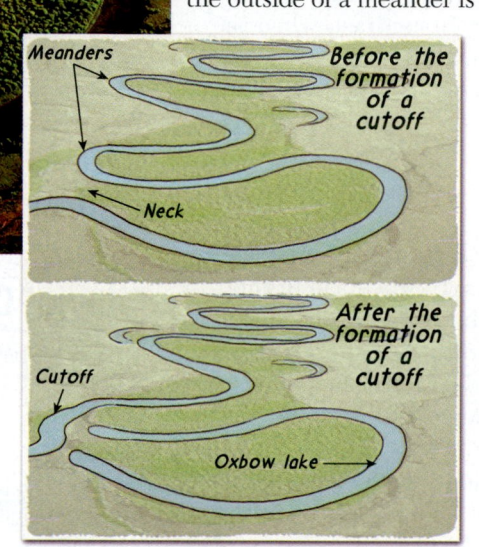

Green River, WY

Geologist's Sketch

Meandering Channels

Streams that transport much of their load in suspension generally move in sweeping bends called **meanders**. These streams flow in relatively deep, smooth channels and primarily transport mud (silt and clay), sand, and occasionally fine gravel. The lower Mississippi River exhibits this type of channel.

Meandering channels evolve over time as individual bends migrate across the floodplain. Most of the erosion is focused at the outside of the meander, where velocity and turbulence are greatest. In time, the outside bank is undermined, especially during periods of high water. Because the outside of a meander is a zone of active erosion, it is often referred to as a **cut bank** (**Figure 16.19**). Debris acquired by the stream at the cut bank moves downstream, where the coarser material is generally deposited as **point bars** on the insides of bends. In this manner, meanders migrate laterally by eroding the outside of the bends and depositing sediment on the inside, moving sideways without appreciably changing their shape.

In addition to migrating laterally, the bends in a channel also migrate down the valley. This occurs because erosion is more effective on the downstream (downslope) side of the meander. Sometimes the downstream migration of a meander is slowed when it reaches a more resistant bank material. This allows the next meander upstream to gradually erode the material between the two meanders, as shown in **Figure 16.20**. Eventually, the river may erode through the narrow neck of land, forming a new, shorter channel segment called a **cutoff**. Because of its shape, the abandoned bend is called an **oxbow lake**.

sufficient width, material deposited by the stream can form a *floodplain* that borders the channel. Because the banks and beds of alluvial channels are composed of unconsolidated sediment (alluvium), they can undergo major changes in shape as material is continually being eroded, transported, and redeposited. The major factors affecting the shapes of these channels are the average size of the sediment being transported, the channel's gradient, and discharge.

Alluvial channel patterns reflect a stream's ability to transport its load at a uniform rate while expending the least amount of energy. Thus, the size and type of sediment being carried help determine the nature of the stream channel. Two common types of alluvial channels are *meandering channels* and *braided channels*.

Braided Channels

Some streams consist of a complex network of converging and diverging channels that thread their way among numerous islands or gravel bars (**Figure 16.21**). Because these channels have an interwoven appearance, they are called **braided channels**. Braided channels form where a large portion of a

Figure 16.21

Braided stream The Knik River is a classic braided stream with multiple channels separated by migrating gravel bars. The Knik is choked with sediment from four melting glaciers in the Chugach Mountains north of Anchorage, Alaska. (Photo by Michael Collier)

stream's sediment load consists of coarse material (sand and gravel) and the stream has a highly variable discharge. Because the bank material is readily erodible, braided channels are wide and shallow.

One setting in which braided streams form is at the end of glaciers, where there is a large seasonal variation in discharge. During the summer, large amounts of ice-eroded sediment are dumped into the meltwater streams flowing away from the glacier. However, when flow is sluggish, the stream deposits the coarsest material as elongated structures called *bars*. This process causes the flow to split into several paths around the bars. During the next period of high flow, the laterally shifting channels erode and redeposit much of this coarse sediment,

thereby transforming the entire streambed. In some braided streams, the bars have built semipermanent islands anchored by vegetation.

> ### 16.5 Concept Checks
>
> 1. Are bedrock channels more likely to be found near the head or the mouth of a stream?
>
> 2. Describe or sketch the evolution of a meander, including how an oxbow lake forms.
>
> 3. Describe a situation that might cause a stream channel to become braided.

16.6 | Shaping Stream Valleys

Contrast narrow V-shaped valleys, broad valleys with floodplains, and valleys that display incised meanders or stream terraces.

A **stream valley** consists of a channel and the surrounding terrain that directs water to the stream. It includes the *valley floor*, which is the lower, flatter area that is partially or totally occupied by the stream channel, and the sloping *valley walls* that rise above the valley floor on both sides. Alluvial channels often flow in valleys that have wide valley floors consisting of sand and gravel deposited in the channel and clay and silt deposited by floods. Bedrock channels, on the other hand, tend to be located in narrow V-shaped valleys. In some arid regions, where weathering is slow and rock is particularly resistant, narrow valleys having nearly vertical walls are also found. Such features are called *slot canyons* (**Figure 16.22**). Stream valleys exist on a continuum from narrow, steep-sided valleys to valleys that are so flat and wide that the valley walls are not discernible.

Streams, with the aid of weathering and mass wasting, shape the landscape through which they flow. As a result, streams continuously modify the valleys they occupy.

Base Level and Graded Streams

In 1875 John Wesley Powell, the pioneering geologist who first explored the Grand Canyon and later headed the U.S. Geological Survey, introduced the concept of a downward limit to stream erosion, which he called **base level**. A fundamental concept in the study of stream activity, base level is defined as the lowest elevation to which a stream can erode its channel. Essentially, it is the level at which the mouth of a stream enters the ocean, a lake, or a trunk stream. Powell determined that two types of base level exist: "We may consider the level of the sea to be a grand base level, below which the dry

Figure 16.22
Halls Creek Narrows This is a classic slot canyon in Utah's Capitol Reef National Park. (Photo by Michael Collier)

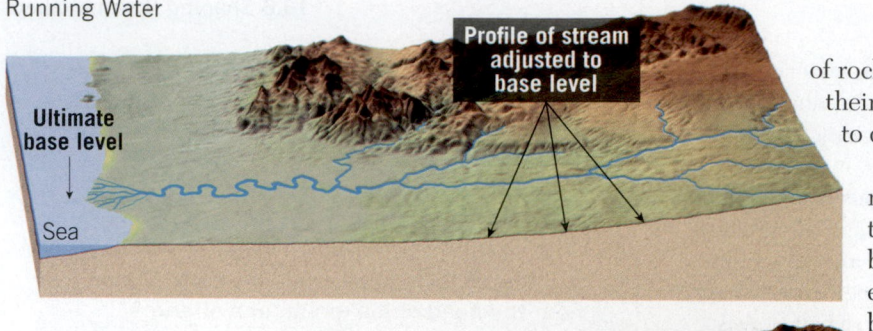

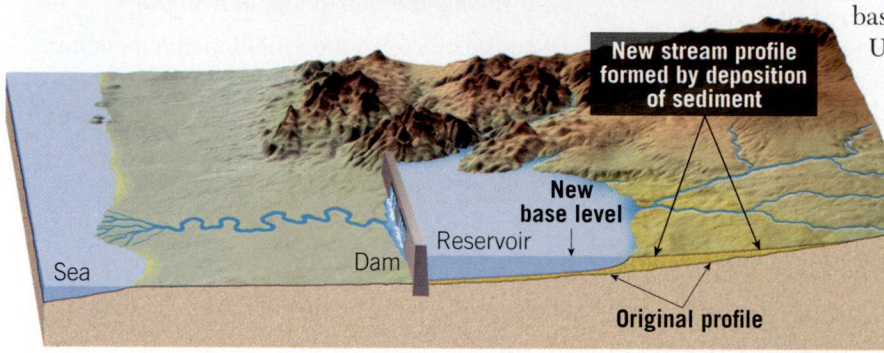

Figure 16.23
Building a dam The base level upstream from the reservoir is raised, which reduces the stream's flow velocity and leads to deposition and a reduced gradient.

of rock, and rivers that act as base levels for their tributaries. All limit a stream's ability to downcut its channel.

Changes in base level cause corresponding adjustments in the "work" that streams perform. When a dam is built along a stream course, the reservoir that forms behind it raises the base level of the stream (**Figure 16.23**). Upstream from the reservoir, the stream gradient is reduced, lowering its velocity and, hence, its sediment-transporting ability. As a result, the stream deposits material, thereby building up its channel. This process continues until the stream again has a gradient sufficient to carry its load. The profile of the new channel will be similar to the old profile but somewhat higher.

lands cannot be eroded; but we may also have, for local and temporary purposes, other base levels of erosion."*

Sea level, which Powell called "grand base level," is now referred to as **ultimate base level**. **Local** (or **temporary**) **base levels** include lakes, resistant layers

Exploration of the Colorado River of the West (Washington, DC: Smithsonian Institution, 1875), p. 203.

If, on the other hand, base level is lowered by a drop in sea level, the stream will have excess energy and downcut its channel to establish a balance with its new base level. Erosion first occurs near the mouth and then progresses upstream, creating a new stream profile.

Observing streams that adjust their profiles to changes in base level led to the concept of a graded stream. A **graded stream** has the necessary slope and other channel characteristics to maintain the minimum velocity required to transport the material supplied to it. On average, a graded system is neither eroding nor depositing material but simply transporting it. When a stream reaches equilibrium, it becomes a self-regulating system in which a change in one characteristic causes a change in the others to counteract the effect.

Figure 16.24
Changes in base level
A resistant layer of rock can act as a temporary base level. The stream concentrates its erosive energy on the resistant rock at the knickpoint. Eventually the river eliminates the knickpoint and reestablishes a smooth profile.

Consider what would happen if displacement along a fault were to raise a layer of resistant rock along the course of a graded stream. As shown in **Figure 16.24**, the resistant rock forms a waterfall and serves as a temporary base level for the stream. Because of the increased gradient, the stream concentrates its erosive energy on the resistant rock along an area called a *knickpoint*. Eventually, the river erases the knickpoint from its path and reestablishes a smooth profile.

Valley Deepening

When a stream's gradient is steep, the channel is well above base level and the dominant erosional process gradient increases significantly, a situation usually caused by variations in the erodibility of the bedrock into which a stream channel is cutting. Resistant beds create rapids by acting as a temporary base level upstream while allowing downcutting to continue downstream. Recall that, over time, erosion usually eliminates the resistant rock.

Waterfalls occur where streams make vertical drops. One type of waterfall is exemplified by Niagara Falls (**Figure 16.26**). These famous falls are supported by a resistant bed of dolostone that is underlain by less-resistant shale. As the water plunges over the lip of the falls, it erodes the less-resistant shale, undermining a section of overlying rock, which eventually breaks off. In this manner, the waterfall retains its vertical cliff while slowly, but continually, retreating upstream. Over the past 12,000 years, Niagara Falls has retreated more than 11 kilometers (7 miles) upstream.

Valley Widening

As a stream approaches a graded condition—in which it primarily works to transport sediment—downcutting becomes less dominant. At this point the stream's channel takes on a meandering pattern, and more of its energy is directed from side to side. As a result, the valley widens as the river cuts away at one bank and then the other.

Figure 16.25
Yellowstone River The V-shaped valley, rapids, and waterfalls indicate that the river is vigorously downcutting. (Photo by Charles A. Blakeslee/AGE Fotostock)

Labels on Figure 16.25: Waterfall of the Yellowstone River; Valley wall; Valley wall; Rapids; V-shaped valley formed by vigorus downcutting; *Geologist's Sketch*

is **downcutting**, the lowering of the streambed toward base level. Abrasion caused by bed load sliding and rolling along the bottom, and the hydraulic power of fast-moving water, slowly deepen the streambed. The result is usually a V-shaped valley with steep sides. A classic example of a V-shaped valley is the section of the Yellowstone River shown in **Figure 16.25**.

The most prominent features of V-shaped valleys are *rapids* and *waterfalls*. Both occur where the stream's

Figure 16.26
The retreat of Niagara Falls The force of the plunging Niagara River is causing the falls to retreat. (Photo by Michael Collier)

As the river plunges over the falls, it erodes the weaker rocks beneath the more resistant Lockport Dolostone. As a section of dolostone is undercut, it loses support and breaks off.

Labels on Figure 16.26: Lake Erie; Niagara River; Niagara Escarpment; Niagara gorge; Niagara Falls; Former position of Niagara Falls; Lake Ontario; *Geologist's Sketch*; Headward erosion of Niagara Falls; Lockport dolostone; Weak shale layers

The Niagara River flows from Lake Erie to Lake Ontario. The falls have retreated upstream (toward Lake Erie) more than 11 kilometers (7 miles) during the last 12,000 years.

SmartFigure 16.27
Development of an erosional floodplain
Continuous side-to-side erosion by shifting meanders gradually produces a broad, flat valley floor. Alluvium deposited during floods covers the valley floor.
(https://goo.gl/WNfY8s)

Condor Video

Narrow V-shaped valley

Site of erosion

Site of deposition

Well developed floodplain

TIME

another sometimes exceeds 160 kilometers (100 miles).

When a river erodes laterally and creates a floodplain as described, it is called an *erosional floodplain*. Floodplains can be depositional as well. *Depositional floodplains* are produced by major fluctuations in conditions, such as changes in base level or climate. The floodplain in California's Yosemite Valley is one such feature; it was produced when a glacier gouged the valley floor about 300 meters (1000 feet) deeper than its former level. After the glacial ice melted, running water refilled the valley with alluvium. The Merced River currently winds across a relatively flat floodplain that forms much of the floor of Yosemite Valley.

The continuous lateral erosion caused by shifting meanders gradually produces a broad, flat valley floor covered with alluvium (**Figure 16.27**). This feature, called a **floodplain**, is appropriately named because when a river overflows its banks during flood stage, it inundates the floodplain. Over time the floodplain will widen to a point where the stream is actively eroding the valley walls in only a few places. In the case of the lower Mississippi River, for example, the distance from one valley wall to

Incised Meanders and Stream Terraces

We usually find streams with highly meandering courses on floodplains in wide valleys. However, some rivers have meandering channels that flow in steep, narrow bedrock valleys. Such meanders, pictured in **Figure 16.28**, are called **incised meanders** (*incisum* = to cut into).

How do these features form? Originally the meanders probably developed on the floodplain of a stream

SmartFigure 16.28
Incised meanders Aerial view of incised meanders of the Colorado River on the Colorado Plateau. (Photo by Michael Collier)
(https://goo.gl/JWM1ib)

Tutorial

Before uplift of the Colorado Plateau, the river was meandering on a floodplain.

During uplift of the plateau, the meanders downcut because of the steepening gradient.

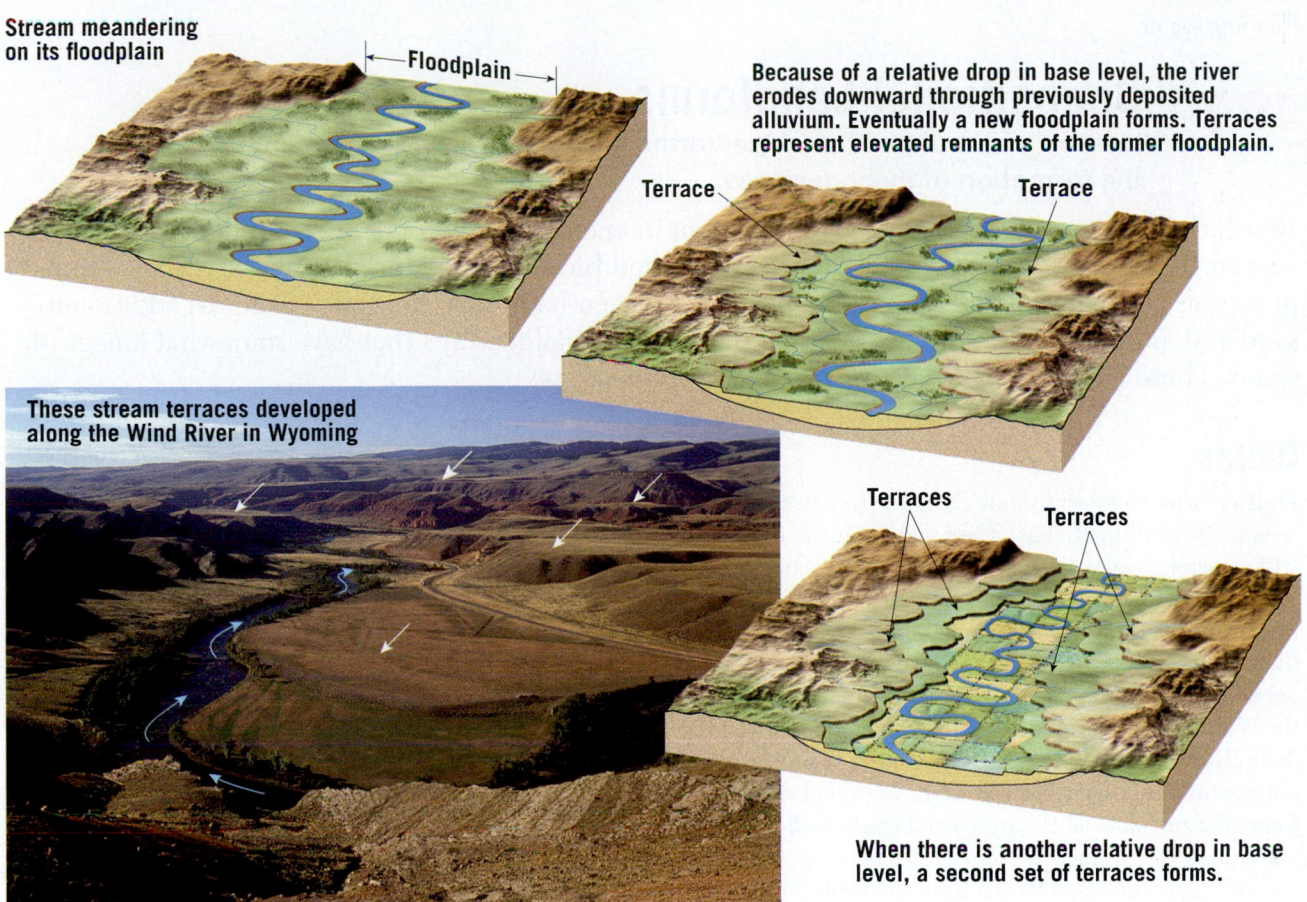

Stream meandering on its floodplain

Floodplain

Because of a relative drop in base level, the river erodes downward through previously deposited alluvium. Eventually a new floodplain forms. Terraces represent elevated remnants of the former floodplain.

Terrace

Terrace

These stream terraces developed along the Wind River in Wyoming

Terraces

Terraces

When there is another relative drop in base level, a second set of terraces forms.

SmartFigure 16.29
Stream terraces Terraces result when a stream adjusts to a relative drop in base level. (Photo by Greg Hancock) (https://goo.gl/6cDynS)

Condor Video

that was in balance with its base level. Then, a change in base level caused the stream to begin downcutting. One of two events likely occurred: Either the base level dropped, or the land on which the river was flowing was uplifted. For example, regional uplifting of the Colorado Plateau in the southwestern United States generated incised meanders on several rivers. As the plateau gradually rose, meandering rivers began downcutting because of their steepening gradient.

After a river has adjusted in this manner, it may once again produce a floodplain at a level below the old one. The remnants of a former floodplain are sometimes present as relatively flat surfaces called **terraces** (**Figure 16.29**).

16.6 Concept Checks

1. Define *base level* and distinguish between ultimate base level and local, or temporary, base level.

2. What is a graded stream?

3. Explain why V-shaped valleys often contain rapids and/or waterfalls.

4. Describe or sketch how an erosional floodplain develops.

5. List two situations that would trigger the formation of incised meanders.

EYE ON EARTH 16.2

The Middle Fork of the Salmon River flows for about 175 kilometers (110 miles) through a wilderness area in central Idaho. (Photo by Michael Collier)

QUESTION 1 Is the river flowing in an alluvial channel or a bedrock channel? Explain.

QUESTION 2 What process is dominant here: valley deepening or valley widening?

QUESTION 3 Is the area shown in this image more likely near the mouth or the head of the river?

16.7 | Depositional Landforms

List the major depositional landforms associated with streams and describe the formation of these features.

Recall that streams continually pick up sediment in one part of their channel and deposit it downstream. These small-scale channel deposits are called **bars**. Such features, however, are only temporary, as the material will be picked up again and eventually carried to the ocean. In addition to sand and gravel bars, streams also create other depositional features that have somewhat longer life spans. These include *deltas*, *natural levees*, and *alluvial fans*.

Deltas

Deltas form where sediment-charged streams reach a temporary or ultimate base level and enter the relatively still waters of a lake, an inland sea, or the ocean (**Figure 16.30**). As the stream's forward motion is slowed, sediments are deposited by the dying current, producing three types of beds. *Foreset beds* are composed of coarse particles that drop almost immediately upon entering the water to form layers that slope downcurrent from the delta front. The foreset beds are usually covered by thin, horizontal *topset beds* deposited during flood stage. The finer silts and clays settle away from the mouth in nearly horizontal layers called *bottomset beds*.

As a delta grows outward from the shoreline, the stream's gradient continually decreases. This circumstance eventually causes the channel to become choked with sediment. As a consequence, the river seeks shorter, higher-gradient routes to base level. Figure 16.30 shows the main channel dividing into several smaller ones, called **distributaries**, that carry water away from the main channel in varying paths to base level. After numerous shifts in the main flow from one distributary to the next, a delta may grow into the triangular shape of the Greek letter delta (Δ), although several other shapes exist. Differences in the configurations of shorelines and variations in the nature and strength of wave activity are responsible for the shape and structure of each delta. Many of the world's great rivers have created massive deltas, each with its own peculiarities and typically more complex than the one illustrated in Figure 16.30.

Not all rivers have deltas. A river that transports a large sediment load may lack a delta because ocean waves and powerful currents quickly redistribute the material soon after it is deposited. The Columbia River in the Pacific Northwest is an example. In other cases, rivers do not carry sufficient quantities of sediment to build a delta. The St. Lawrence River, for example, has little opportunity to pick up much sediment between its head in Lake Ontario and its mouth in the Gulf of St. Lawrence.

The Mississippi River Delta

The Mississippi River delta resulted from the accumulation of huge quantities of sediment derived from the vast region drained by the river and its tributaries. New Orleans currently rests where there was once ocean.

History and Structure The portion of the Mississippi delta that has formed during the past 6000 years is shown in **Figure 16.31**. As the figure illustrates, the delta is actually a series of seven coalescing subdeltas. Each subdelta formed when the main flow was diverted from one channel to a shorter, more direct path to the Gulf of Mexico. The individual subdeltas intertwine and partially cover each other to produce a complex structure. It is also apparent from Figure 16.31 that after each channel was abandoned, coastal erosion modified the newly formed subdelta. The present subdelta (number 7 in

Figure 16.30
Formation of a simple delta Structure and growth of a simple delta that forms in relatively quiet waters.

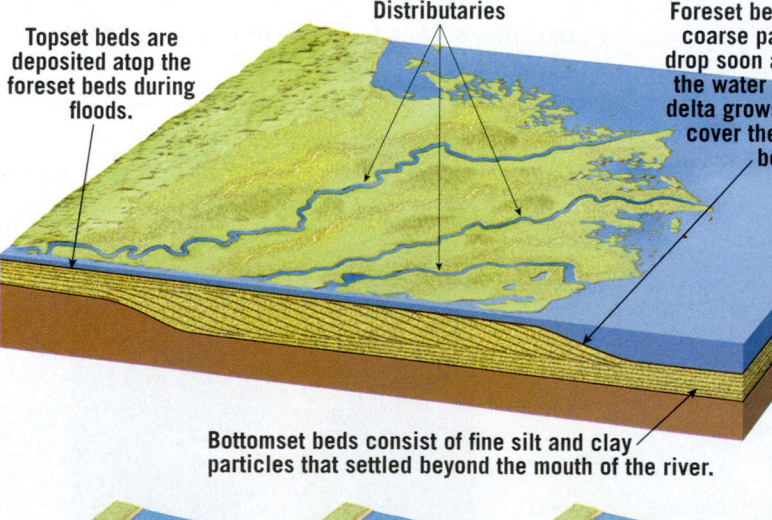

Distributaries

Topset beds are deposited atop the foreset beds during floods.

Foreset beds consist of coarse particles that drop soon after entering the water body. As the delta grows, these beds cover the bottomset beds.

Bottomset beds consist of fine silt and clay particles that settled beyond the mouth of the river.

As the stream extends its channel, the gradient is reduced. During flood stage some of the flow is diverted to a shorter, higher-gradient route forming a new distributary.

Figure 16.31), called a *bird-foot* delta because of the configuration of its distributaries, has been built by the Mississippi River over the past 500 years.

At present, this active bird-foot delta has extended seaward almost as far as natural forces will allow. In fact, for many years the river has been "struggling" to cut through a narrow neck of land and shift its course to the Atchafalaya River (see inset in Figure 16.31). If this were to happen, the Mississippi would abandon the lowermost 500 kilometers (300 miles) of its channel in favor of the Atchafalaya's much shorter 225-kilometer (140-mile) route to the Gulf.

In a concerted effort to keep the Mississippi on its present course, a dam-like structure was erected at the site where the channel was trying to break through. A flood in 1973 weakened the control structure, and the river again threatened to shift. This event caused the U.S. Army Corps of Engineers to construct a massive auxiliary dam that was completed in the mid-1980s. For the time being, at least, the inevitable has been avoided, and the Mississippi River continues to flow past Baton Rouge and New Orleans on its way to the Gulf of Mexico.

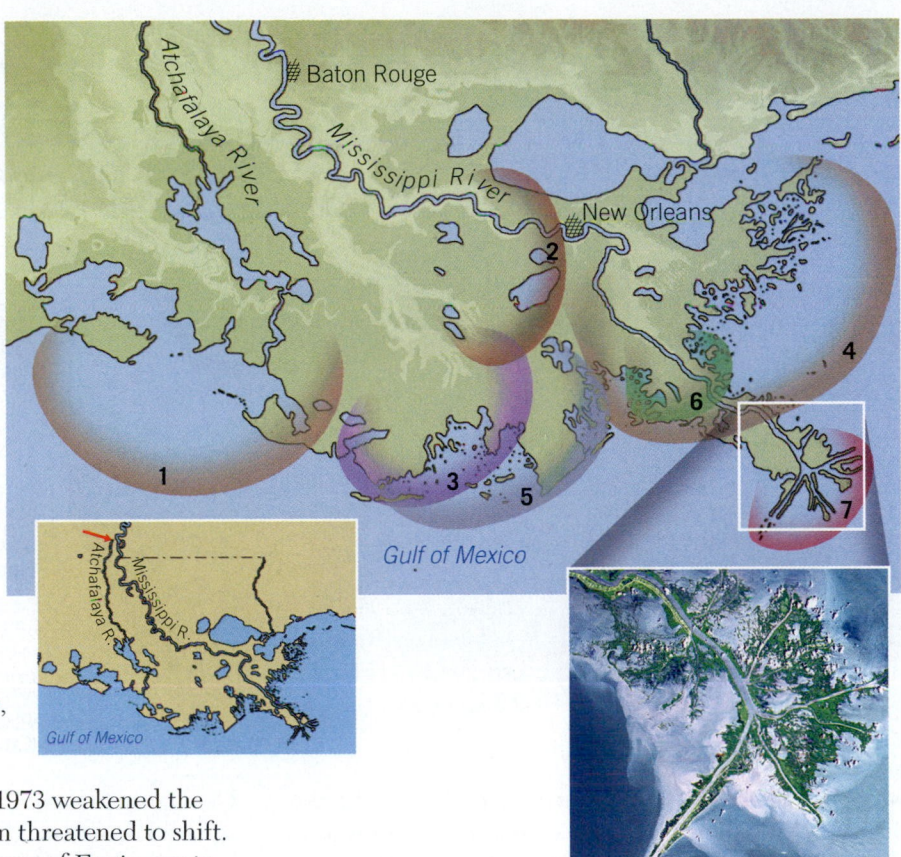

Figure 16.31
Growth of the Mississippi River delta During the past 6000 years, the river has built a series of seven coalescing subdeltas. The numbers indicate the order in which the subdeltas were deposited. The present bird-foot delta (number 7) represents the activity of the past 500 years. The left inset shows the point where the Mississippi may sometime break through (arrow) and the shorter path it would take to the Gulf of Mexico. (Image courtesy of JPL/Cal tech/NASA)

Vanishing Wetlands   The delta of the Mississippi River in Louisiana is a biologically significant region that includes about 12,000 square kilometers (3 million acres) of coastal wetlands—40 percent of all coastal wetlands in the contiguous United States (**Figure 16.32**). These flat

SmartFigure 16.32
Coastal wetlands These low-lying, sediment-starved swamps, marshes, and bayous are disappearing at an alarming rate. (Photo by Michael Collier) (https://goo.gl/YuGzP5)

Mobile Field Trip

SmartFigure 16.33
Formation of a natural levee These gently sloping structures that parallel a river channel are created by repeated floods. Because the ground next to the channel is higher than the adjacent floodplain, back swamps and yazoo tributaries may develop.
(https://goo.gl/7ZcsYJ)

Animation

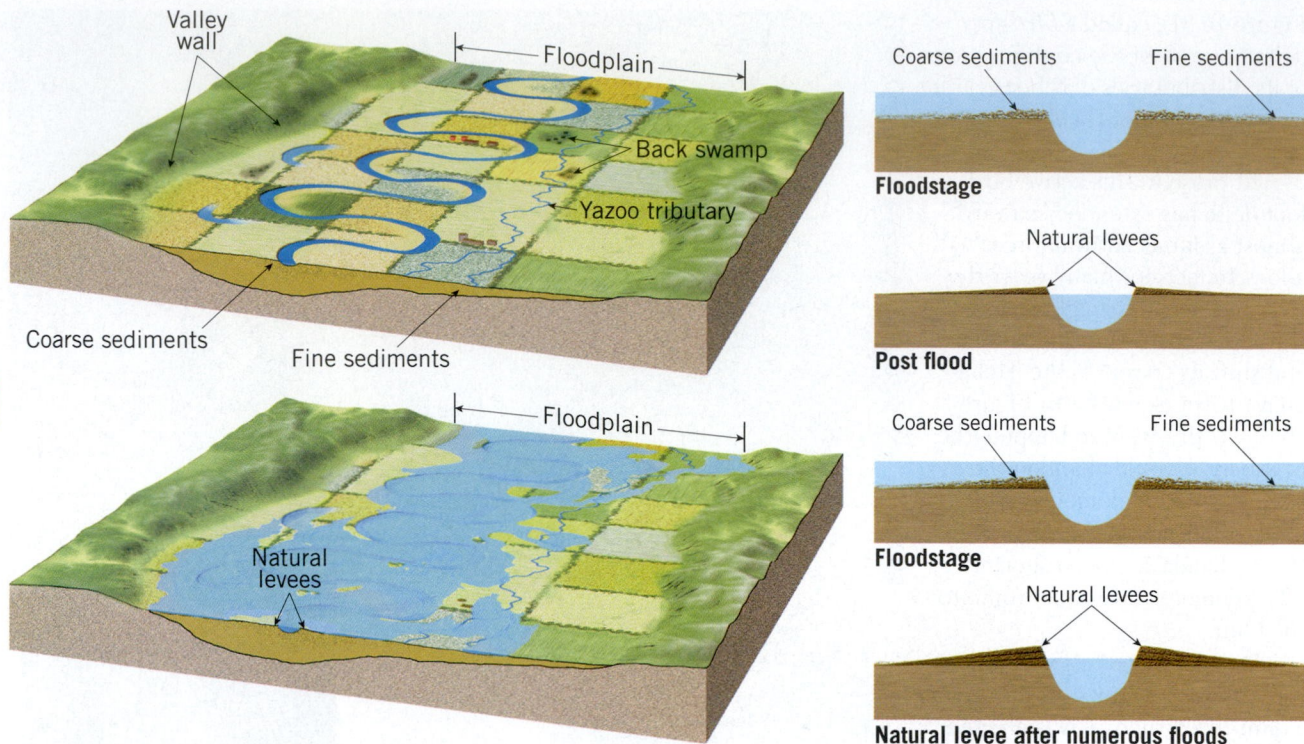

areas that are only slightly above sea level are sheltered from the wave action of hurricanes and winter storms by low-lying offshore barrier islands. Both the wetlands and offshore islands were formed and maintained by sediments carried to the Gulf of Mexico by the Mississippi River.

The coastal wetlands of Louisiana are disappearing at an alarming rate—accounting for 80 percent of the wetland loss in the lower 48 states. According to the U.S. Geological Survey, Louisiana has lost more than 5000 square kilometers (1900 square miles) of coastal land since the early 1930s. If wetland loss continues at this rate, another 3000 square kilometers (1170 square miles) will vanish by the year 2050. Why is this occurring?

Before Europeans settled the delta, the Mississippi River regularly overflowed its banks in seasonal floods. Huge quantities of sediment were deposited atop the delta, which kept the land elevated above sea level. However, with settlement came flood-control efforts and the

EYE ON EARTH 16.3

This satellite image shows the delta of the Yukon River. The river originates in northern British Columbia. It flows through the Yukon Territory and across the tundra of Alaska before entering the Bering Sea, a distance of nearly 3200 kilometers (about 2000 miles).

QUESTION 1 Explain why the river breaks into numerous channels as it crosses the delta.

QUESTION 2 What term is applied to the channels that radiate across the delta?

QUESTION 3 Notice the cloud of sediment in the water surrounding the delta. Are these sediments more likely sand and gravel or silt and clay? Explain.

QUESTION 4 After the cloud of material settles to the seafloor and the delta expands farther into the Bering Sea, which beds of the delta will these sediments become?

desire to maintain and improve navigation on the river. Artificial levees (earthen mounds built parallel to the river) were constructed to contain the rising river during flood stage. Over time the levees were extended all the way to the mouth of the Mississippi to keep the channel open to navigation.

The effects have been straightforward: The levees prevent sediment from being dispersed onto the wetlands. As a result, sediment is not added in sufficient quantities to offset compaction, subsidence, and wave erosion. Thus, the size of the delta and the extent of the wetlands are shrinking. The problem has been exacerbated by a decline in the sediment load of the Mississippi, which has decreased approximately 50 percent over the past 100 years. A substantial portion of the reduction is due to sediment being trapped upstream by dams on the river's many tributaries.

Natural Levees

Meandering rivers that occupy valleys with broad floodplains tend to build **natural levees** (*lever* = to raise) that parallel their channels on both banks (**Figure 16.33**). Natural levees are built by years of successive floods. When a stream overflows onto the floodplain, the water flows over the surface as a broad sheet. Because the flow velocity drops significantly, the coarser portion of the suspended load is immediately deposited adjacent to the channel. As the water spreads across the floodplain, a thin layer of fine sediment is laid down over the valley floor. This uneven distribution of material produces the gentle, almost imperceptible, slope of the natural levee (see Figure 16.33).

The natural levees of the lower Mississippi River rise 6 meters (20 feet) above the adjacent valley floor. The area behind a levee is characteristically poorly drained for the obvious reason that water cannot flow over the levee and into the river. Marshes called **back swamps** often result. When a tributary stream enters a river valley that has a substantial natural levee, it often flows for many kilometers through the back swamp before finding an opening where it enters the main river. Such streams are called **yazoo tributaries**, after the Yazoo River, which parallels the lower Mississippi River for more than 300 kilometers (185 miles).

Alluvial Fans

Alluvial fans are fan-shaped deposits that accumulate along steep mountain fronts (**Figure 16.34**). When a stream with a steep gradient emerges onto a relatively flat lowland, its gradient drops, and it deposits a large portion of its sediment load. Although alluvial fans are more prevalent in arid climates (see Chapter 19 for a further discussion of alluvial fans), they are occasionally found in humid regions.

Mountain streams, because of their steep gradients, carry much of their sediment load as coarse sand and

Geologist's Sketch

Dry channels
Dry stream channel
Alluvial fan
Road
Death Valley

SmartFigure 16.34
Alluvial fan in Death Valley Fans are deposited at the mouth of a valley that emerges from a mountainous or upland area onto a relatively flat lowland. Usually coarse material is dropped near the apex of the fan, while finer materials are carried toward the base of the deposit. California's Death Valley has many large fans. As adjacent fans grow larger, they may coalesce to form a steep apron of sediment called a *bajada*. (Photo by Michael Collier) (https://goo.gl/bqOZ5R)

Condor Video

gravel. Because alluvial fans are composed of these same coarse materials, water that flows across them readily soaks in. When a stream emerges from its valley onto an alluvial fan, its flow divides itself into several distributary channels. The fan shape is produced because the main flow swings back and forth between the distributaries from a fixed point where the stream exits the mountain.

Between rainy periods in deserts, little or no water flows across an alluvial fan, which is evident in the many dry channels that cross its surface. Thus, fans in dry regions grow intermittently, receiving considerable water and sediment only during wet periods. As you learned in Chapter 15, steep canyons in dry regions are prime locations for debris flows. Therefore, as would be expected, many alluvial fans have debris-flow deposits interbedded with the coarse alluvium.

16.7 Concept Checks

1. Sketch a cross section of a simple delta and distinguish among the three types of beds that compose it.

2. Explain why the Mississippi delta consists of seven coalescing subdeltas.

3. Briefly describe the formation of a natural levee. How is this feature related to back swamps and yazoo tributaries?

4. Describe the formation of an alluvial fan.

16.8 | Floods and Flood Control

Summarize the various categories of floods and the common measures of flood control.

Floods are serious and costly events that are among the most universally experienced natural hazards. **Floods** are part of the natural behavior of a stream and occur when the flow of a stream becomes so great that it exceeds the capacity of its channel and overflows its banks. Although floods are natural phenomena, the magnitude and frequency of flooding is often significantly influenced by human activities such as clearing forests, building cities, and constructing flood control structures such as dams and levees. The occurrence of floods is often linked to other natural hazards, including severe storms and mass-wasting processes such as debris flows.

Types of Floods

Most floods are caused by atmospheric processes that can vary greatly in both time and space. An hour or less of intense thunderstorm rainfall can trigger flash floods in small valleys. By contrast, major floods in large river valleys often result from an extraordinary series of precipitation events over a broad region for many days or weeks. Common flood types include *regional floods*, *flash floods*, *ice-jam floods*, and *dam-failure floods*.

Regional Floods Most regional floods are seasonal. Rapid melting of snow in spring and/or heavy spring rains often overwhelm rivers. For example, the extensive 1997 flood along the Red River of the North was preceded by an especially snowy winter and an early spring blizzard. Early April brought rapidly rising temperatures, melting the snow in a matter of days and causing a record-breaking 500-year flood. Roughly 4.5 million acres were underwater, and the losses in the Grand Forks, North Dakota, region exceeded $3.5 billion.°

*Ice jams also contribute to floods on the Red River of the North; see the section "Ice-Jam Floods."

In April 2011, unrelenting storms brought record rains to the Mississippi watershed. The Ohio Valley, which makes up the eastern portion of the Mississippi's drainage basin, received nearly 300 percent of its normal springtime precipitation. When that rainfall combined with water from the past winter's large and extensive, rapidly melting snowpack, the Mississippi River and many of its tributaries began to swell to record levels by early May. The resulting floods were among the largest and most damaging in nearly a century (**Figure 16.35**). Like most other regional floods, these were associated with weather phenomena that could be forecast with a good deal of accuracy. This allowed adequate time to warn and evacuate thousands of people who were in harm's way. Although economic losses approached $4 billion, loss of life was small.

Flash Floods Flash floods often occur with little warning and are potentially deadly because they produce rapid rises in water levels and can have devastating flow velocities (see GEOgraphics 16.2). Rainfall intensity and duration, surface conditions, and topography are among the factors that influence flash flooding. Mountainous

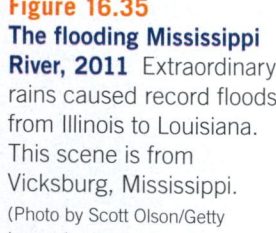

Figure 16.35
The flooding Mississippi River, 2011 Extraordinary rains caused record floods from Illinois to Louisiana. This scene is from Vicksburg, Mississippi. (Photo by Scott Olson/Getty Images)

Flash floods

Flash floods are local floods of great volume and short duration. The rapidly rising surge of water usually occurs with little advance warning and can destroy roads, bridges, homes, and other substantial structures.

USGS

The power of a flash flood is illustrated by the Big Thompson River flood of July 31, 1976, in Colorado. During a four-hour span more than 30 centimeters (12 inches) of rain fell on portions of the river's small drainage basin. This amounted to nearly three-quarters of the average yearly total. The flash flood in the narrow canyon lasted only a few hours, but cost 139 people their lives.

Michael Collier

Urban development increases runoff. As a result, peak discharge and flood frequency increase. A recent study indicated that the area of impervious surfaces in the 48 contiguous United States is roughly equal to the area of the state of Ohio (44,000 mi²).

Most people do not appreciate the power of moving water. Many automobiles will float and be swept away in a strong current that is only 2 feet deep. More than half of all U. S. flash-flood fatalities are auto related!

AP Photo/Sue Ogrocki

Question:
Briefly explain how urban development influences the peak discharge of a stream.

?

Average Annual Storm-Related Deaths in the U.S. (1984–2013)

In most years floods are responsible for the greatest number of storm-related deaths. The average number of hurricane deaths was dramatically affected by Hurricane Katrina in 2005 (more than 1000). For all other years on this graph, hurricane fatalities numbered fewer than 20.

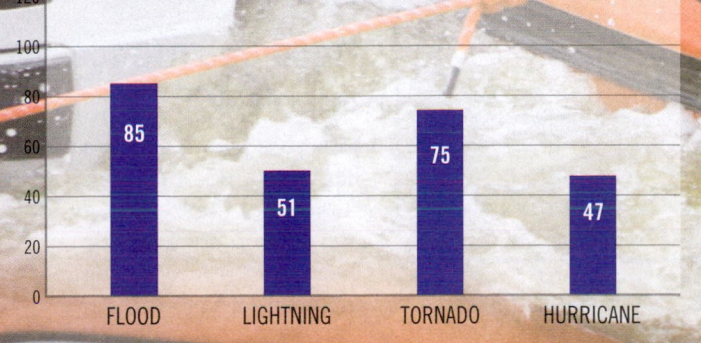

FLOOD	LIGHTNING	TORNADO	HURRICANE
85	51	75	47

Effect of Urban Development on Flooding

Streamflow in Mercer Creek, an urban stream in western Washington, increases more quickly, reaches a higher peak discharge, and has a larger volume during a one-day storm on February 1, 2000, than streamflow in Newaukum Creek, a nearby rural stream. Streamflow during the following week, however, was greater in Newaukum Creek.

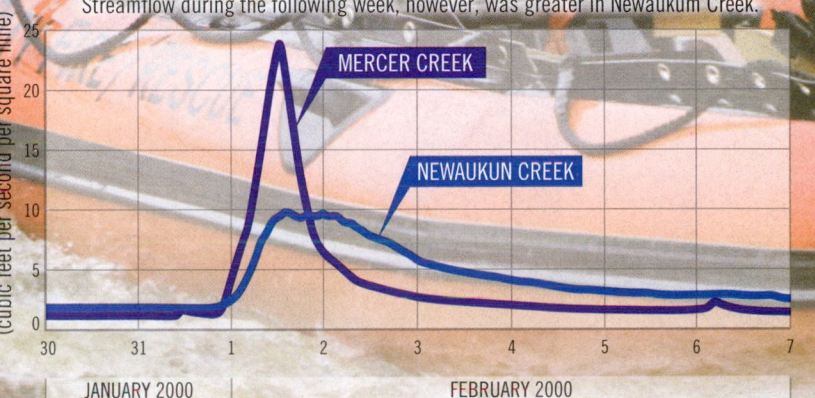

MERCER CREEK

NEWAUKUN CREEK

Hourly Unit-Area Discharge (cubic feet per second per square mile)

JANUARY 2000   FEBRUARY 2000

areas are especially susceptible because steep slopes can funnel runoff into narrow valleys with disastrous consequences.

A recent episode of flash flooding occurred in late August 2011, when heavy rains from the remnants of Hurricane Irene swamped parts of the Northeast. The hilly and mountainous terrain of upstate New York and Vermont were hit especially hard (**Figure 16.36**). Flash floods washed out dozens of roads and bridges, isolating numerous small towns and rural areas.

Urban areas are also susceptible to flash floods because a high percentage of the surface area is composed of impervious roofs, streets, and parking lots, where infiltration is minimal and runoff is rapid.

Ice-Jam Floods

Frozen rivers are especially susceptible to ice-jam floods. As the level of a stream rises, it breaks up ice and creates ice floes that can accumulate on channel obstructions. Jams of this nature create temporary ice dams across the channel. Water trapped upstream can rise rapidly and overflow the channel banks. When an ice dam fails, water behind the dam is often released with sufficient force to inflict considerable damage downstream.

These floods are often associated with northward-flowing rivers in the Northern Hemisphere. The Red

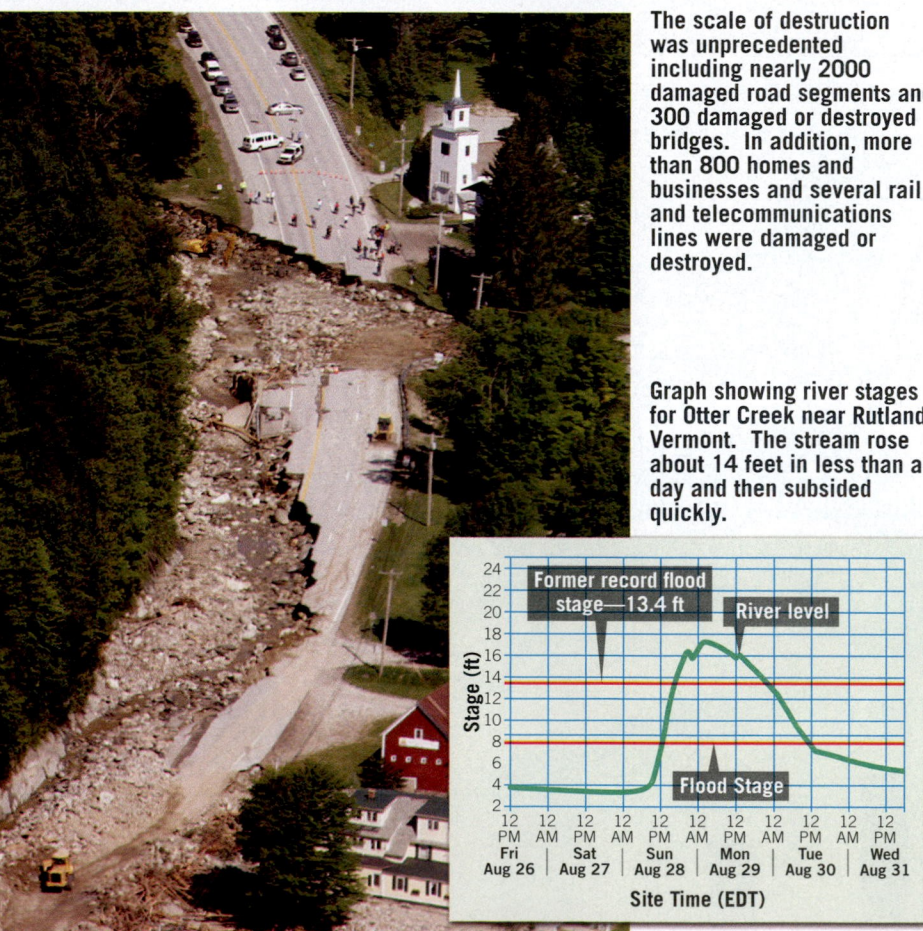

Figure 16.37
Northward-flowing Siberian rivers Ice-jam floods are relatively common in late spring and early summer in Siberia because its large rivers flow north into the Arctic Ocean.

River of the North mentioned earlier is an example. The Siberian region of Russia has several northward-flowing rivers, such as the Ob, Lena, and Yenisei, that frequently experience these floods (**Figure 16.37**). When spring arrives, ice melts in the warmer southern portions of a river and its drainage basin, while farther north the river remains frozen. Water flowing from the south "backs up" behind the frozen northern portion of the river.

Dam-Failure Floods

Human interference with stream systems can cause floods. A prime example is the failure of a dam or an artificial levee designed to contain small or moderate floods. When larger floods occur, the dam or levee may fail, resulting in the water behind it being released as a flash flood. The bursting of a dam in 1889 on the Little Conemaugh River caused the devastating Johnstown, Pennsylvania, flood that took more than 2200 lives.

Flood Recurrence Intervals

Land-use planning in river basins requires an understanding of the frequency and magnitude of floods. For every river, a relationship exists between the size of a

Figure 16.36
Flash floods in Vermont In late August 2011, the remnants of Hurricane Irene brought torrential rains that caused record-breaking flash floods in Vermont and parts of upstate New York. (Photo by AP Photo/Toby Talbot)

The scale of destruction was unprecedented including nearly 2000 damaged road segments and 300 damaged or destroyed bridges. In addition, more than 800 homes and businesses and several rail and telecommunications lines were damaged or destroyed.

Graph showing river stages for Otter Creek near Rutland, Vermont. The stream rose about 14 feet in less than a day and then subsided quickly.

flood and the frequency with which it occurs. The larger the flood, the less often it is expected to occur. You may have heard the term *100-year flood*. This describes the *recurrence interval*, which is an estimate of how often a flood of a given size can be expected to occur. A 25-year event would be much smaller but would be four times more likely to occur than a 100-year flood.

The phrase "100-year flood" does not mean that 100 years must pass between each flood of equal or greater magnitude. Floods happen irregularly! Rather, it means that there is a 1 percent (1 in 100) probability in a given year for a flood of that size. It is possible that a 100-year flood could occur 2 or 3 years in a row. It is also possible that such a flood might not occur over a span exceeding 100 years.

In order to make a reasonable calculation, stream gage data must be collected for at least 10 to 30 years. The longer the record, the better the prediction will likely be. Other factors also influence the accuracy of flood recurrence estimates. Climate cycles that involve extensive drought or rainy periods are an example. Land-use changes, such as when adjacent rural areas become urbanized, typically require reevaluation of flood recurrence intervals.

Flood Control

Several strategies have been devised to eliminate or reduce the catastrophic impact of floods on our lives and environment. Engineering efforts include the construction of artificial levees, river channelization, and the building of flood-control dams.

Artificial Levees   *Artificial levees* are earthen mounds built on riverbanks to contain a stream in its channel when the water volume increases. Levees, used since ancient times, are the most commonly used stream-containment structures. In some locations, concrete floodwalls are constructed that function as artificial levees.

Many artificial levees were not built to withstand periods of extreme flooding. For example, numerous levees failed during the summer of 1993, when the upper Mississippi and many of its tributaries experienced record flooding. During that event, floodwalls at St. Louis, Missouri, created a bottleneck for the river that led to increased flooding upstream of the city.

When exceptional floods threaten to overwhelm levees, water is sometimes intentionally diverted from a river by creating openings in artificial levees. The purpose is to spare vulnerable urban areas by allowing water to flood sparsely populated rural areas. The areas that are intentionally flooded are called *floodways*. For example, to prevent the town of Cairo, Illinois, from being inundated during the 2011 floods along the Mississippi River, a 3-kilometer- (2-mile-) wide opening was blasted in a levee. This allowed water to spill into the 130,000-acre Birds Point-New Madrid Floodway (**Figure 16.38**). Similar steps were taken downstream in Louisiana to protect the cities of Baton Rouge and New Orleans.

Figure 16.38
Birds Point-New Madrid Floodway In early May 2011, the intentional demolition of portions of an artificial levee on the west side of the Mississippi River immersed this floodway. A total of 520 square kilometers (200 square miles) of Missouri farmland were flooded. This rare action prevented the inundation of the small town of Cairo, Illinois. (Photo by David Carson/St. Louis Post Dispatch/ ZUMAPRESS.com)

Channelization   *Channelization* involves altering a stream channel in order to make the flow more efficient. This may simply involve clearing a channel of obstructions or dredging a channel (removing sediment from the channel bed) to make it wider and deeper. Another alteration involves straightening, and thus shortening, a channel by creating artificial cutoffs. Shortening a stream increases the channel's flow velocity.

Between 1929 and 1942, the U.S. Army Corps of Engineers removed 16 meander bends on the lower Mississippi for the purpose of increasing the slope of the channel and thereby reducing the threat of flooding. This river, a vital transportation corridor, was shortened about 240 kilometers (150 miles). These efforts have been somewhat successful in reducing the maximum height of the river during flood stage. However, channel shortening led to increased gradients and accelerated erosion of bank material, both of which necessitated further intervention. Following the creation of artificial cutoffs, massive bank protection was installed along several stretches of the lower Mississippi.

A similar case in which artificial cutoffs accelerated bank erosion occurred on the Blackwater River in Missouri, whose meandering course was shortened in 1910. Among the many effects of this project was a significant increase in the channel's width due to increased velocity. One particular bridge over the river collapsed in 1930 because of bank erosion. During the following 17 years, the bridge was replaced three times, each time requiring a longer span.

Flood-Control Dams   *Flood-control dams* are built to store floodwater and then release it slowly, in a controlled manner. Since the 1920s, thousands of dams have been built on nearly every major river in the United States. Many dams have significant non-flood-related functions, such as providing water for irrigated agriculture and for hydroelectric power generation. Many reservoirs are also major regional recreational facilities.

Although dams are effective in reducing flooding and provide other benefits, their construction and maintenance also have significant costs and consequences. For example, reservoirs created by dams may cover valuable farmland, forests, historic sites, and scenic valleys. Large dams can also cause significant damage to river ecosystems that have developed over thousands of years.

Furthermore, building dams is not a permanent solution to flooding. Sedimentation behind a dam gradually diminishes the volume of its reservoir, reducing the long-term effectiveness of this flood-control measure.

A Nonstructural Approach   All of the flood-control measures described so far employ structural solutions to "control" rivers. These solutions are typically expensive and often give those who reside on the floodplain a false sense of security.

Many scientists and engineers advocate a nonstructural approach to flood control. They suggest that sound floodplain management, such as limiting certain activities or land uses, is an alternative to artificial levees, dams, and channelization. In high-risk flood areas, appropriate zoning regulations can be implemented to minimize development and promote safer, more appropriate land use.

16.8 | Concept Checks

1. List and distinguish among four types of floods.

2. Describe three basic flood-control strategies. What are some drawbacks of each?

3. What is meant by a *nonstructural approach* to flood control?

16.1 Earth as a System: The Hydrologic Cycle

List the hydrosphere's major reservoirs and describe the different paths that water takes through the hydrologic cycle.

KEY TERMS hydrologic cycle, infiltration, runoff transpiration, evapotranspiration

- Water moves through the hydrosphere's many reservoirs by evaporating, condensing into clouds, and falling as precipitation. Once it reaches the ground surface, rain can either soak into the soil, evaporate, be returned to the atmosphere by plant transpiration, or run off the surface. Running water may be a small portion of the total water on Earth, but it is the most important agent in sculpting Earth's varied landscapes.

16.2 Running Water

Describe the nature of drainage basins and river systems. Sketch and briefly explain four basic drainage patterns.

KEY TERMS stream, river, drainage basin (watershed), divide, headward erosion, dendritic pattern, radial pattern, rectangular pattern, trellis pattern, water gap, antecedent stream, superposed stream

- The land area that contributes water to a stream is its drainage basin. Drainage basins are separated by imaginary lines called divides.
- As a generalization, the upstream portion of a drainage basin is a zone of sediment production where most of a stream's sediment is derived. Sediment transport characterizes the middle section, and sediment deposition is associated with the downstream end.
- A stream erodes most effectively in a headward direction, thereby lengthening its course.

- A water gap is a steep-walled notch in a ridge through which a stream flows. Such streams may either be antecedent or superposed.

Q Identify each of the drainage patterns depicted in the accompanying sketch.

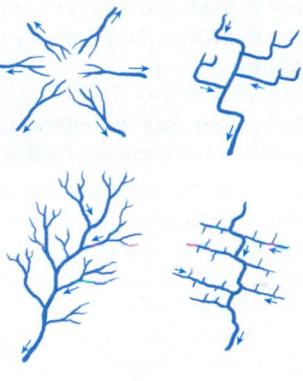

16.3 Streamflow Characteristics

Discuss streamflow and the factors that cause it to change.

KEY TERMS laminar flow, turbulent flow, gradient, wetted perimeter, discharge, longitudinal profile, head (headwaters), mouth

- The flow of water in a stream may be laminar or turbulent. A stream's flow velocity is influenced by channel gradient; size, shape, and roughness of the channel; and discharge.

- A cross-sectional view of a stream from head to mouth is a longitudinal profile. Usually the gradient and roughness of the stream channel decrease going downstream, whereas the size of the channel, stream discharge, and flow velocity increase in the downstream direction.

Q Sketch a typical longitudinal profile. Where does most erosion happen? Where is sediment transport the dominant process?

16.4 The Work of Running Water

Outline the ways in which streams erode, transport, and deposit sediment.

KEY TERMS quarrying, abrasion, pothole, corrosion, dissolved load, suspended load, bed load, settling velocity, saltation, capacity, competence, sorting, alluvium

- Streams are powerful agents of erosion that carve solid rock through quarrying, abrasion, and the focused "drilling" that results in potholes. Turbulent water also lifts loose particles from the streambed. In areas of

soluble bedrock such as limestone, stream water can also corrode the landscape by dissolving the rock.
- Sediment is transported in a stream either in solution, suspended in the water, or rolling or saltating along the bottom of the stream. Compared to slow-moving rivers, fast-moving rivers can carry a greater total amount of sediment (capacity) and larger individual particles (competence). Flooding increases both capacity and competence, which is why rivers do most of their work during short-lived times of peak flow.
- Sediment deposited by streams is called alluvium. Typically, streams are efficient agents of sorting, meaning that they deposit similarly sized grains in the same area.

16.5 Stream Channels

Contrast bedrock and alluvial stream channels. Distinguish between two types of alluvial channels.

KEY TERMS bedrock channel, alluvial channel, meander, cut bank, point bar, cutoff, oxbow lake, braided channel

- Bedrock channels are cut into solid rock. They typically exhibit steps (highlighted by waterfalls or rapids) and pools (segments that are relatively flat).
- Alluvial channels are dominated by streamflow moving through sediment previously deposited by the river. A floodplain usually covers much of a valley floor, with the river itself moving through a channel that may meander or exhibit a braided pattern.

- Meanders are enhanced through erosion at the cut bank (outside edge of the meander) and deposition of sediment on the point bar (inside edge of the meander). The shape of the meander may grow more and more exaggerated until it loops back on itself. Once a cutoff is formed, the main current abandons the old meander loop, which becomes an oxbow lake.
- Braided channels occur in streams that experience highly variable discharge. During times of low flow, the river moves in an interwoven network of channels between bars of coarse-grained alluvium.

Q The town of Carter Lake is the *only* portion of the state of Iowa that lies on the west side of the Missouri River. It is bounded on the north by its namesake, Carter Lake, on the south by the Missouri River, and on the east and west by Nebraska. After examining the map, prepare a hypothesis that explains how this unusual situation could have developed.

16.6 Shaping Stream Valleys

Contrast narrow V-shaped valleys, broad valleys with floodplains, and valleys that display incised meanders or stream terraces.

KEY TERMS stream valley, base level, ultimate base level, local (temporary) base level, graded stream, downcutting, floodplain, incised meander, terrace

- A stream valley includes the channel itself, the adjacent floodplain, and relatively steep valley walls. Stream valley width and shape vary a lot from bedrock channels (which tend to be V-shaped, with narrow floodplains) to alluvial channels, which may have broad floodplains of alluvium and relatively subdued valley walls.
- Streams erode downward until they approach base level, which is usually the level at which the stream enters another stream, a lake, or the ocean. A river flowing toward the sea (ultimate base level) may encounter several local base levels along its route. These could be lakes or resistant rock units that retard the downcutting of the stream. A graded stream has reached equilibrium with its base level and primarily works to transport sediment.

- Stream valleys are widened through the meandering action of the stream, which erodes the valley walls and widens the floodplain. If the base level drops, the stream downcuts. If it is underlain by bedrock, the stream may then develop incised meanders. Streams underlain by deep alluvium are likely to develop terraces.

Q Meanders are associated with a river that is eroding from side to side, whereas narrow canyons are associated with rivers that are vigorously downcutting. The river in this image is confined to a narrow canyon but is also meandering. Explain.

Michael Collier

16.7 Depositional Landforms

List the major depositional landforms associated with streams and describe the formation of these features.

KEY TERMS bar, delta, distributary, natural levee, back swamp, yazoo tributary, alluvial fan

- A delta may form where a river deposits sediment in another water body at its mouth. The partitioning of streamflow into multiple distributaries spreads sediment in different directions. In the United States, the Mississippi River is an example of a major river with a dynamic delta system.

- Natural levees result from sediment deposited along the margins of a river channel by many years of successive floods. Because they slope gently away from the channel, the surrounding land is poorly drained, resulting in back swamps.
- Alluvial fans are fan-shaped deposits of alluvium that form where steep mountain fronts drop down into adjacent valleys.

Q At first glance, deltas and alluvial fans may look quite similar. How are they similar, and how are they different?

16.8 Floods and Flood Control

Summarize the various categories of floods and the common measures of flood control.

KEY TERM flood

- When a stream receives more discharge than its channel can hold, a flood occurs. Four major factors that can cause a flood are large amounts of precipitation (or melting) across a region, sudden pulses of precipitation in areas of high runoff, temporary dams made of floating ice (which then break up, releasing impounded water), and failure of human-built dams, causing the sudden escape of water from a reservoir.

- Three main structural strategies exist for coping with floods: construction of artificial levees to constrain streamflow to the channel, alterations to make a stream channel's flow more efficient, and building of dams on a river's tributaries so that a sudden influx of water will be temporarily stored and released slowly to the river system. A nonstructural approach is sound floodplain management. Here, a solid scientific understanding of flood dynamics informs policy and regulation of areas that are subject to flooding.

Q Artificial levees are constructed to protect property from floods. Sometimes artificial levees in rural areas are intentionally opened up to protect a city from experiencing flooding. How does this work?

Give It Some Thought

1. A river system consists of three zones, based on the dominant process operating in each zone. On the accompanying diagram, match each process with one of the three zones: sediment production (erosion), sediment deposition, sediment transportation.

Zone #1
Zone #2
Zone #3

2. If you collect a jar of water from a stream, what part of its load will settle to the bottom of the jar? What portion will remain in the water indefinitely? What part of the stream's load would probably not be represented in your sample?

3. Streamflow is affected by several variables, including discharge, gradient, and channel roughness, size, and shape. Develop a scenario in which a mass wasting event influences a stream's flow. Explain what led up to, or triggered, the event and describe how the mass wasting process influenced the stream's flow.

4. Examine the satellite image of central Pennsylvania. Identify the feature that occurs each time the Susquehanna River crosses one of the five mountain ridges and explain how these features likely formed.

5. Several times during the past 2.5 million years, huge ice sheets (continental-size glaciers) formed and spread across large parts of Northern Hemisphere landmasses and then gradually melted away.

   a. How do you think the formation of ice sheets affected sea level?

   b. How would rivers flowing into the ocean have been affected as the ice sheets expanded?

   c. What kind of adjustments would these rivers make as the glacial ice melted?

6. Describe three ways that a stream can lengthen its course. How might a stream get shorter?

7. One day you and a friend are discussing the aftermath of a major (100-year) flood that occurred on a river in your area just a few months earlier. At the close of your conversation, your friend remarks, "At least we won't have to worry about another one of those in our lifetime." How would you respond?

8. This satellite image shows portions of the Ohio and Wabash Rivers in May 2011. What is the base level for the Wabash River? What is base level for the Ohio River (referring to Figure 16.4 will help)? Is either of the base levels you just noted considered *ultimate base level*? Explain.

9. The accompanying graphs show lag times between rainfall and peak flow (flooding) for an urban area and a rural area. Which graph (A or B) most likely represents the rural area? Explain your choice.

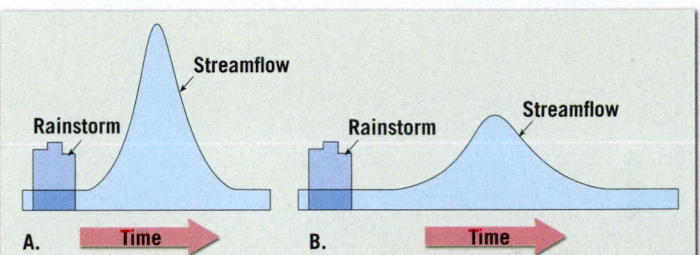

A.   Rainstorm   Streamflow   Time
B.   Rainstorm   Streamflow   Time

10. Building a dam is one method of regulating the flow of a river to control flooding. Dams and their reservoirs may also provide recreational opportunities and water for irrigation and hydroelectric power generation. This image, from near Page, Arizona, shows Glen Canyon Dam on the Colorado River upstream from the Grand Canyon and a portion of Lake Powell, the reservoir it created.

Michael Collier

   a. How did the behavior of the river likely change upstream from Lake Powell?

   b. How might the behavior of the Colorado River downstream from the dam have been affected?

   c. Given enough time, how might the reservoir change?

   d. Speculate on the possible environmental impacts of building a dam such as this one.

MasteringGeology™

17

Groundwater

Irrigation is the number-one use of groundwater in the United States. In some places, groundwater is being withdrawn much more rapidly than it is being replenished. (Photo by Debra Ferguson/Alamy)

Hidden from view, vast quantities of water exist in the cracks, crevices, and pore spaces of rock and soil. Groundwater can be found almost everywhere beneath Earth's surface and is a major source of water worldwide. Groundwater is a valuable natural resource that provides about half of our drinking water and is essential to the vitality of agriculture and industry. In addition to its utility, groundwater plays a crucial role in sustaining streamflow between precipitation events—especially during protracted dry periods. Many ecosystems depend on groundwater discharge into streams, lakes, and wetlands. In some regions, large-scale development has caused groundwater levels to decline, resulting in water shortages, streamflow depletion, land subsidence, and increased pumping costs. Groundwater pollution is also a serious issue in some places.

17.1 | The Importance of Groundwater

Describe the importance of groundwater as a source of freshwater and groundwater's roles as a geologic agent.

Groundwater is one of our most important and widely available resources, yet people's perceptions of the subsurface environment from which it comes are often unclear and incorrect. This is because the groundwater environment is largely hidden from view except in caves and mines, and the impressions people gain from these subsurface openings are misleading. Observations on the land surface give an impression that Earth is "solid." This view remains when we enter a cave and see water flowing in a channel that appears to have been cut into solid rock.

Because of such observations, many people believe that groundwater occurs only in underground "rivers." In reality, most of the subsurface environment is not "solid" at all. It includes countless tiny *pore spaces* between grains of soil and sediment, plus narrow joints and fractures in bedrock. Together, these spaces add up to an immense volume. Where these subsurface pore spaces are saturated with water, that stored water is called **groundwater**. It is in these small openings that groundwater collects and moves.

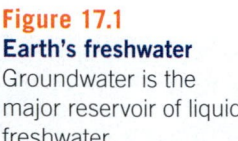

Figure 17.1
Earth's freshwater Groundwater is the major reservoir of liquid freshwater.

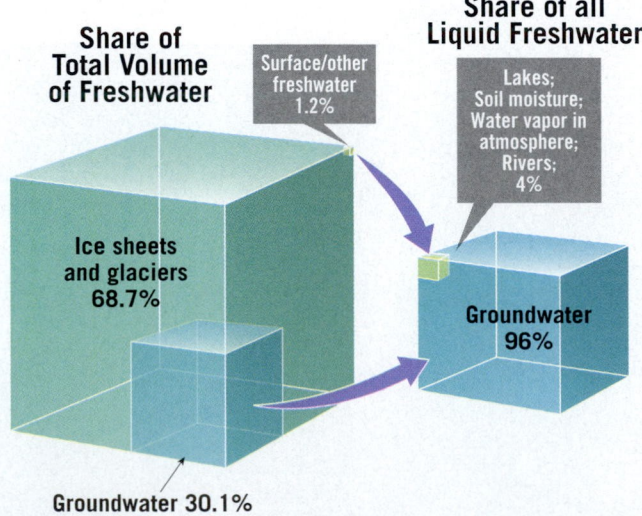

Groundwater and the Hydrosphere

When we consider the entire hydrosphere, or all of Earth's water, only about six-tenths of 1 percent is located underground. Nevertheless, this small percentage, stored in the rocks and sediments beneath Earth's surface, is a vast quantity. When the oceans are excluded and only sources of freshwater are considered, the significance of groundwater becomes more apparent.

Figure 17.1 gives estimates of the distribution of freshwater in the hydrosphere. Clearly the largest volume occurs as glacial ice. Groundwater is ranked second, with slightly more than 30 percent of the total. However, when ice is excluded and just *liquid* water is considered, 96 percent of all freshwater is groundwater. Without question, *groundwater represents the largest reservoir of freshwater that is readily available to humans.* Its value in terms of economics and human well-being is incalculable.

Geologic Importance of Groundwater

Geologically, groundwater is important as an erosional agent. The dissolving action of groundwater slowly removes soluble rock such as limestone, allowing surface depressions known as *sinkholes* to form as well as creating subterranean caverns (**Figure 17.2**). The final section of this chapter describes the landforms associated with subsurface water. Groundwater is also an equalizer of

Figure 17.2
Carlsbad Caverns, New Mexico The dissolving action of acidic groundwater created these caverns. Later, groundwater deposited the limestone decorations.
(Photo by Clint Farlinger)

streamflow. Much of the water that flows in rivers is not direct runoff from rain and snowmelt. Rather, a large percentage of precipitation soaks into the ground and then moves slowly under the surface to stream channels. Groundwater is thus a form of storage that sustains streams during periods when rain does not fall. Therefore, when we see water flowing in a river during a dry period, it is rain that fell at some earlier time and was stored underground.

Groundwater: A Basic Resource

Water is basic to life. It has been called the "bloodstream" of both the biosphere and society. Each day in the United States we use about 349 billion gallons of freshwater. According to the U.S. Geological Survey, about 77 percent comes from surface sources. Groundwater provides the remaining 23 percent (**Figure 17.3**). One of the advantages of groundwater is that it exists almost everywhere across the country and thus is often available in places that lack reliable surface sources such as lakes and rivers. Water in a groundwater system is stored in subsurface pore spaces and fractures. As water is withdrawn from a well, the connected pore spaces and fractures act as a "pipeline" that allows water to gradually move from one part of the hydrologic system to where it is being withdrawn.

What are the primary ways that we use groundwater? The U.S. Geological Survey identifies several categories, shown in Figure 17.3. More groundwater is used for irrigation than for all other uses combined. There are nearly 60 million acres (nearly 243,000 square kilometers [about 93,700 square miles]) of irrigated land in the United States. That is an area nearly the size of the state of Wyoming. The vast majority (75 percent) of the irrigated land is in the 17 conterminous western

states, where annual precipitation is typically less than 20 inches. About 42 percent of the water used for irrigation is groundwater.

Public and domestic uses include water for indoor and outdoor household purposes as well as water used for commercial purposes. Common indoor uses include drinking, cooking, bathing, washing clothes and dishes, and flushing toilets. If you are curious about how much water an average American uses each day for indoor domestic purposes, look at **Figure 17.4**. Major outdoor uses are watering lawns and gardens. Water for domestic use may come from a public supply or may be self-supplied.* Practically all (98 percent) those whose water is self-supplied rely on groundwater.

*According to the U.S. Geological Survey, *public supply* refers to water withdrawn by suppliers that furnish water to at least 25 people or have at least 15 connections.

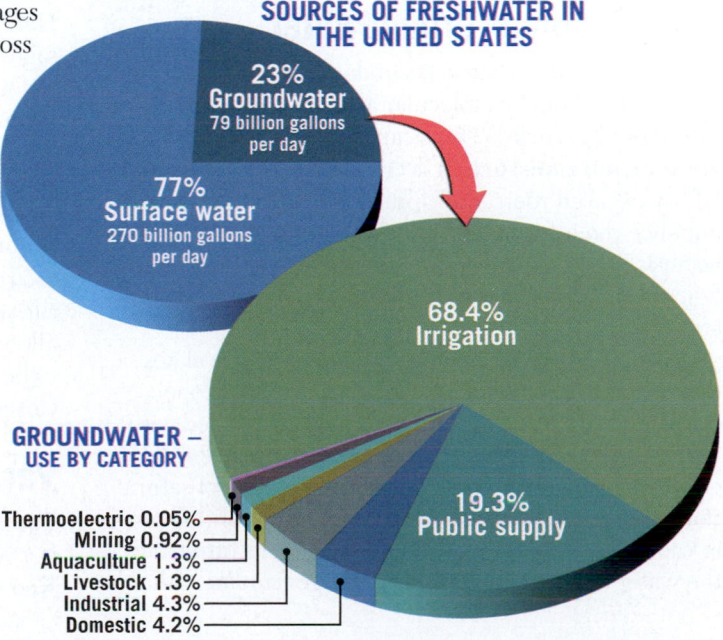

Figure 17.3

Sources and uses of freshwater Each day in the United States, we use 345 billion gallons of freshwater. Groundwater is the source of nearly one-quarter of the total. More groundwater is used for irrigation than for all other uses combined.
(Data from U.S. Geological Survey)

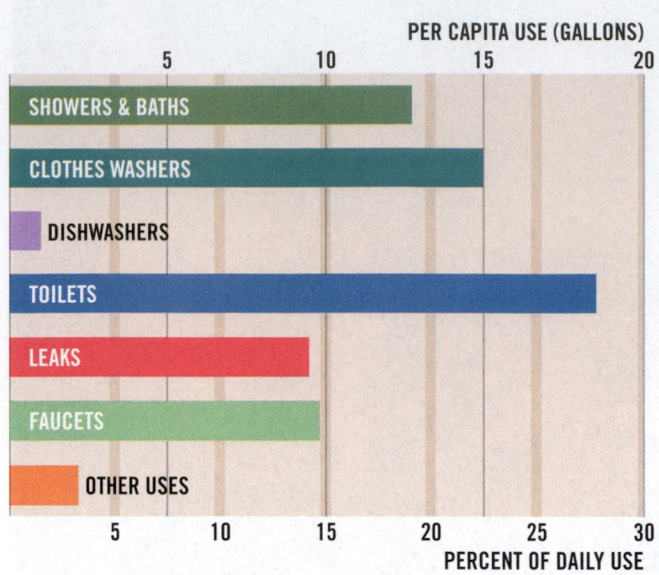

Another category, aquaculture, involves water used for fish hatcheries, fish farms, and shellfish farms. Many mining operations require significant quantities of water, as do industrial processes such as petroleum refining and the manufacture of chemicals, plastics, paper, steel, and concrete.

17.1 Concept Checks

1. What percentage of Earth's *total freshwater supply* is groundwater?

2. What share of Earth's *liquid freshwater* is groundwater?

3. List two geologic roles that groundwater plays.

4. What share of U.S. freshwater is provided by groundwater? What is most groundwater used for?

17.2 Groundwater and the Water Table

Prepare a sketch with labels that summarizes the distribution of water beneath Earth's surface. Discuss the factors that cause variations in the water table and describe the interactions between groundwater and streams.

When rain falls on Earth's land surface, some of the water runs off, some returns to the atmosphere by evaporation and transpiration, and the remainder soaks into the ground. This last path is the primary source of practically all subsurface water. The amount of water that takes each of these paths varies greatly both in time and space. Influential factors include steepness of slope, nature of the surface material, intensity of rainfall, and type and amount of vegetation. For example, heavy rains falling on steep slopes underlain by impervious materials will obviously result in a high percentage of the water running off. Conversely, rain falling steadily and gently on more gradual slopes composed of materials that are easily penetrated by the water means that a much larger percentage of water soaks into the ground.

Distribution of Groundwater

Some of the water that soaks in does not travel far because it is held by molecular attraction as a surface film on soil particles. This near-surface zone is called the **zone of soil moisture**. It is crisscrossed by roots, voids left by decayed roots, and animal and worm burrows that enhance the infiltration of rainwater into the soil. Soil water is used by plants in life functions and transpiration. Some water also evaporates directly back into the atmosphere.

Water that is not held as soil moisture percolates downward until it reaches a zone where all the open spaces in sediment and rock are completely filled with water (**Figure 17.5**). This is the **zone of saturation** (also called the **phreatic zone**). Water in the zone of saturation is called *groundwater*. The upper limit of this zone is known as the **water table**. Extending upward from the water table is the **capillary fringe** (*capillus* = hair).

Here groundwater is held by surface tension in tiny passages between grains of soil or sediment. The area above the water table that includes the capillary fringe and the zone of soil moisture is called the **unsaturated zone** (also known as the **vadose zone**). The pore spaces in this zone contain both air and water. Although a considerable amount of water can be present in the unsaturated zone, this water cannot be pumped by wells because it clings too tightly to rock and soil particles. By contrast, below the water table, the water pressure is great enough to allow water to enter wells, thus permitting groundwater to be withdrawn for use. We will examine wells more closely later in the chapter.

Variations in the Water Table

The water table, the upper limit of the zone of saturation, is a very significant feature of the groundwater system. Knowing the water table level is important in predicting the productivity of wells, explaining the changes in the flow of springs and streams, and accounting for fluctuations in the levels of lakes. This level is highly variable and can range from zero, when the zone of saturation is at the surface, to hundreds of meters below the surface in some places. An important characteristic of the water table is that its configuration varies seasonally and from year to year because the addition of water to the groundwater system is closely related to the quantity, distribution, and timing of precipitation. Except where the water table is at the surface, we cannot observe it directly. However, the elevation of the water table is mapped and studied in detail by examining water levels in wells (**Figure 17.6**). The U.S. Geological Survey and state agencies maintain and monitor an extensive network of observation wells to provide statistics about groundwater levels. Such data are the basis for maps that reveal that the water table is rarely level, as we might expect a table to be (**Figure 17.7**). Instead, its shape is usually a subdued replica of the surface topography, reaching its

highest elevations beneath hills and descending toward valleys (see Figure 17.5). Where a wetland (swamp) is encountered, the water table is right at the surface. Lakes and streams generally occupy areas low enough that the water table is above the land surface.

Several factors contribute to the irregular surface of the water table. One

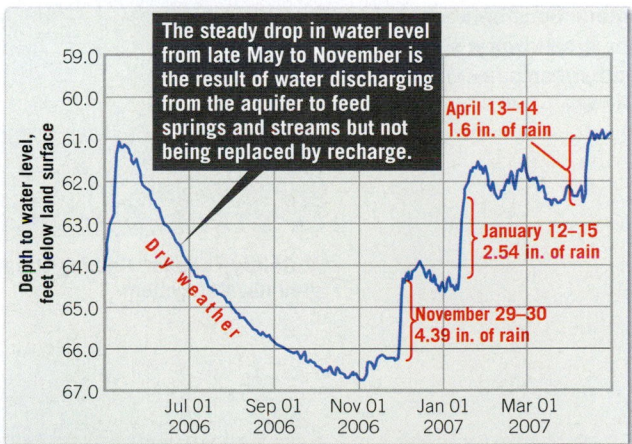

SmartFigure 17.5
Water beneath Earth's surface The shape of the water table is usually a subdued replica of the surface topography. During periods of drought, the water table falls, reducing streamflow and drying up some wells. (https://goo.gl/Z2GhM5)

Tutorial

important influence is the fact that groundwater moves very slowly and at varying rates under different conditions. Because of this, water tends to "pile up" beneath high areas between stream valleys. If rainfall were to cease completely, these water table "hills" would slowly

Figure 17.6

Monitoring the water table Water level measurements from observation wells are a basic and important source of data. The well location highlighted in red on the map is the well whose data are shown below. (Photo by Missouri Department of Natural Resources)

Water level data can be accessed remotely.

A groundwater network is an array of wells where water levels are routinely measured. The U.S. Geological Survey in conjunction with state agencies maintains an extensive network of about 20,000 observation wells. This map shows the network of wells in Missouri.

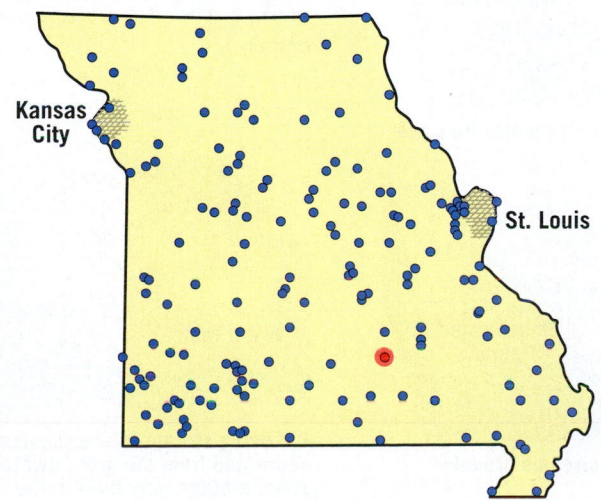

The steady drop in water level from late May to November is the result of water discharging from the aquifer to feed springs and streams but not being replaced by recharge.

Groundwater recharge and discharge at the Akers observation well in Shannon County, Missouri. The height of the water table in this well and many others in this region shows a steady decline between spring and fall. This occurs because recharge is low during these months. However, water continues to move through the aquifer to supply springs and streams in the area maintaining their flow even in dry years.

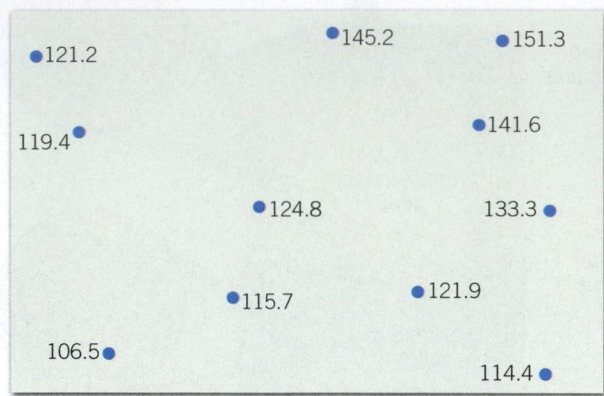

Step 1 The locations of observation wells and the elevations of the water table above sea level are plotted on the map.

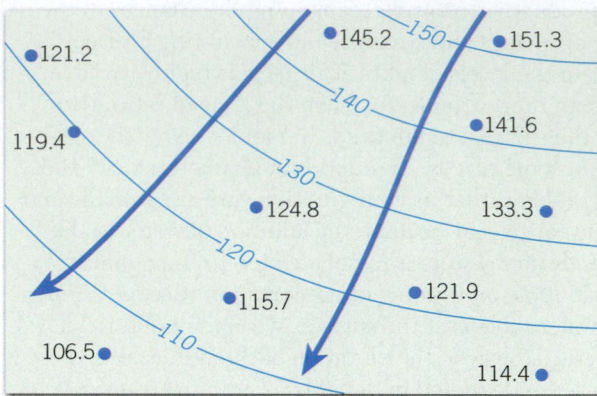

Step 2 Data points are used to guide the drawing of water-table contour lines. Groundwater flow lines can be added to show water movement in the upper portion of the zone of saturation. Groundwater moves perpendicular to the contours and down the slope of the water table.

subside and gradually approach the level of the valleys. However, new supplies of rainwater are usually added frequently enough to prevent this. Nevertheless, in times of extended drought, the water table may drop enough to dry up shallow wells (see Figure 17.5). Other causes for the uneven water table are variations in precipitation and surface permeability from place to place.

Interactions Between Groundwater and Streams

The interaction between the groundwater system and streams is a basic link in the hydrologic cycle. This interaction can take place in one of three ways. Streams may gain water from the inflow of groundwater through the streambed. Such streams are called **gaining streams** (**Figure 17.8A**). For this to occur, the elevation of the water table must be higher than the level of the surface of the stream. Streams may lose water to the groundwater system by outflow

through the streambed. The term **losing stream** is applied to this situation (**Figure 17.8B,C**). When this happens, the elevation of the water table must be lower than the surface of the stream. The third possibility is a combination of the first two: A stream gains in some sections and loses in others.

Losing streams can be connected to the groundwater system by a continuous saturated zone, or they can be disconnected from the groundwater system by an unsaturated zone. Compare parts B and C in Figure 17.8. When the stream is disconnected, the water table may have a discernible bulge beneath the stream if the rate of water movement through the streambed and unsaturated zone is greater than the rate of groundwater movement away from the bulge.

In some settings, a stream might always be a gaining stream or might always be a losing stream. However, in many situations, flow direction can vary a great deal along a stream; some sections receive groundwater, and other sections lose water to the groundwater system. Moreover, the direction of flow can change over a short time span due to storms adding water near the stream bank or when temporary flood peaks move down the channel.

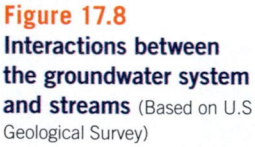

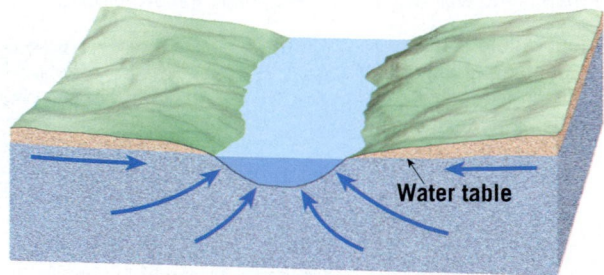

A. Gaining stream Gaining streams receive water from the groundwater system.

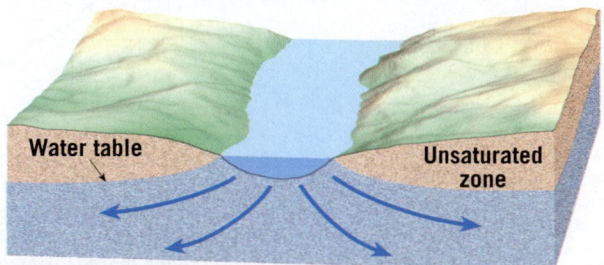

B. Losing stream (connected) Losing streams provide water to the groundwater system.

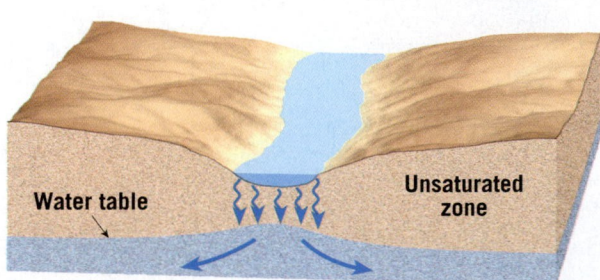

C. Losing stream (disconnected) When losing streams are separated from the groundwater system by the unsaturated zone, a bulge may form in the water table.

Marshes are characterized by saturated, poorly drained soils. This wetland is located southwest of Fort McMurray, Alberta, Canada. (Photo by Michael Collier)

QUESTION 1 Describe the position of the water table in this image.

QUESTION 2 Suggest two situations that might cause the marsh to disappear. Include one natural cause and one cause related to human activities.

Groundwater contributes to streams in most geologic and climatic settings. Even where streams are primarily losing water to the groundwater system, certain sections may receive groundwater inflow during some seasons. One study of 54 streams in all parts of the United States indicated that 52 percent of the streamflow was contributed by groundwater. The groundwater contribution ranged from a low of 14 percent to a maximum of 90 percent. Groundwater is also a major source of water for lakes and wetlands.

17.2 | Concept Checks

1. When rain falls on land, what factors influence the amount of water that soaks in?
2. Define *groundwater* and relate it to the water table.
3. A kitchen table is flat. Is this usually the case for a water table? Why?
4. Contrast a gaining stream and a losing stream.

17.3 | Factors Influencing the Storage and Movement of Groundwater

Summarize the factors that influence the storage and movement of groundwater.

The nature of subsurface materials strongly influences the rate of groundwater movement and the amount of groundwater that can be stored. Two factors are especially important: porosity and permeability.

Porosity

Water soaks into the ground because bedrock, sediment, and soil contain countless voids or openings. These openings are similar to those of a sponge and are often called *pore spaces*. The quantity of groundwater that can be stored depends on the **porosity** of the material, which is the percentage of the total volume of rock or sediment that consists of pore spaces (**Figure 17.9**). Voids most often are spaces between sedimentary particles, but also common are joints, faults, cavities formed by the dissolving of soluble rock such as limestone, and vesicles (voids left by gases escaping from lava).

Variations in porosity can be great. Sediment is commonly quite porous, and open spaces may occupy 10 to 50 percent of the sediment's total volume. Pore space depends on the size and shape of the grains, how they are packed together, the degree of sorting, and,

Figure 17.9
Porosity demonstration
Porosity is the percentage of the total volume of rock or sediment that consists of pore spaces.

The beaker on the left is filled with 1000 ml of sediment. The beaker on the right is filled with 1000 ml of water.

The sediment-filled beaker now contains 500 ml of water. Pore spaces (porosity) must represent 50 percent of the volume of the sediment.

in sedimentary rocks, the amount of cementing material. For example, clay may have a porosity as high as 50 percent, whereas some gravels may have only 20 percent voids.

Where sediments are poorly sorted, the porosity is reduced because the finer particles tend to fill the openings among the larger grains. Most igneous and metamorphic rocks, as well as some sedimentary rocks, are composed of tightly interlocking crystals such that the voids between the grains may be negligible. In these rocks, fractures must provide the porosity.

Permeability, Aquitards, and Aquifers

Porosity alone cannot measure a material's capacity to yield groundwater. Rock or sediment may be very porous yet still not allow water to move through it. The pores must be *connected* to allow water flow, and they must be *large enough* to allow flow. Thus, the **permeability** (*permeare* = to penetrate) of a material—its ability to *transmit* a fluid—is also very important.

Groundwater moves by twisting and turning through small interconnected openings. The smaller the pore spaces, the more slowly the water moves. This idea is clearly illustrated by examining the water-yielding potential of different materials in **Table 17.1**. Here groundwater is divided into two categories:

(1) the portion that will percolate downward under the influence of gravity (called *specific yield*) and (2) the part that is retained as a film on particle and rock surfaces and in tiny openings (called *specific retention*). Specific yield indicates how much water is actually available for use, whereas specific retention indicates how much water remains bound in the material. For example, the ability of a clay deposit to store water may be great, due to high porosity, but its pore spaces are so small that water is unable to move through it. Thus, the clay's porosity is high but because its permeability is poor, it has a very low specific yield.

Impermeable layers that hinder or prevent water movement are termed **aquitards** (*aqua* = water, *tard* = slow). Clay is a good example. On the other hand, larger particles, such as sand or gravel, have larger pore spaces. Therefore, the water moves through with relative ease. Permeable rock strata or sediment that transmit groundwater freely are called **aquifers** (*aqua* = water, *fer* = carry). Sands and gravels are common examples.

In summary, porosity is not always a reliable guide to the amount of surface water that can be stored as groundwater, and permeability is significant in determining the rate of groundwater movement and the quantity of water that might be pumped from a well.

17.3 | Concept Checks

1. Distinguish between porosity and permeability.
2. What is the difference between an aquifer and an aquitard?

Material	Porosity	Specific Yield	Specific Retention
Clay	50	2	48
Sand	25	22	3
Gravel	20	19	1
Limestone	20	18	2
Sandstone (semiconsolidated)	11	6	5
Granite	0.1	0.09	0.01
Basalt (fresh)	11	8	3

TABLE 17.1 Selected Values of Porosity, Specific Yield, and Specific Retention*

*Values in percent by volume.
Source: U.S. Geological Survey, Water Supply Paper 2220, 1987.

17.4 | How Groundwater Moves

Sketch and describe a simple groundwater flow system. Discuss how groundwater movement is measured and the different scales of movement.

The movement of water in the atmosphere and on the land surface is relatively easy to visualize, but the movement of groundwater is not. Near the beginning of the chapter, we mentioned the common misconception that groundwater occurs in underground rivers that resemble surface streams. Although subsurface streams do exist, they are *not* common (**Figure 17.10**). Rather, as you learned in the preceding sections, groundwater exists in the pore spaces and fractures in rock and sediment. Thus, contrary to any impressions of rapid flow that an underground river might evoke, the movement of most groundwater is exceedingly slow, from pore to pore.

A Simple Groundwater Flow System

Figure 17.11 depicts a simple example of a *groundwater flow system*—a three-dimensional body of Earth material saturated with moving groundwater. It shows groundwater moving along flow paths from **recharge areas**, where groundwater is being replenished, to a **discharge area** along a stream where groundwater is flowing back to the surface. Discharge also occurs at springs, lakes, or wetlands, and in coastal areas, as groundwater seeps into bays or the ocean. Transpiration by plants whose roots extend to near the water table is another form of groundwater discharge. Wells, where groundwater is being pumped to the surface, are artificial discharge areas.

The force that moves groundwater is gravity. In response to gravity, water moves from areas where the water table is high to zones where the water table is lower. Although some water takes the most direct path down the slope of the water table, much of the water follows long, curving paths.

Figure 17.11 shows water percolating into a stream from all possible directions. Some paths clearly turn upward, apparently against the force of gravity, and enter through the bottom of the channel. This is easily explained: The deeper you go into the zone of saturation, the greater the water pressure. Thus, the looping curves followed by water in the saturated zone may be thought of as a compromise between the downward pull of gravity and the

Figure 17.10
Underground rivers—a common misconception Although subsurface channels of flowing water occasionally occur, such as this one in France, they are rare. (Photo by Michael Collier)

Figure 17.11
Groundwater movement
Arrows show paths of groundwater movement through uniformly permeable material.

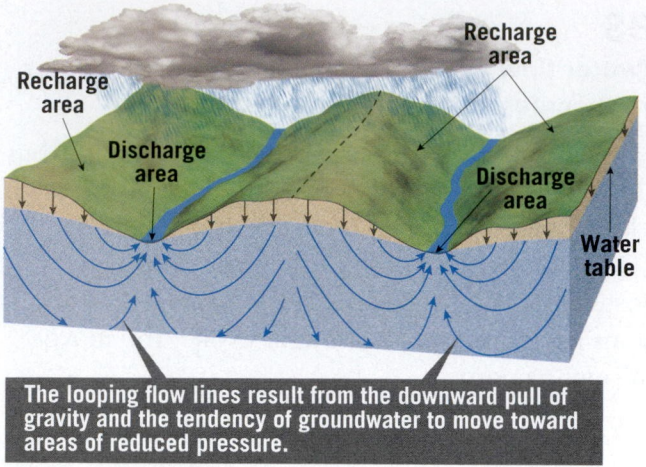

The looping flow lines result from the downward pull of gravity and the tendency of groundwater to move toward areas of reduced pressure.

Figure 17.12
Hydraulic gradient The hydraulic gradient is determined by measuring the difference in elevation between two points on the water table ($h_1 - h_2$) divided by the distance between them, d. Wells are used to determine the height of the water table.

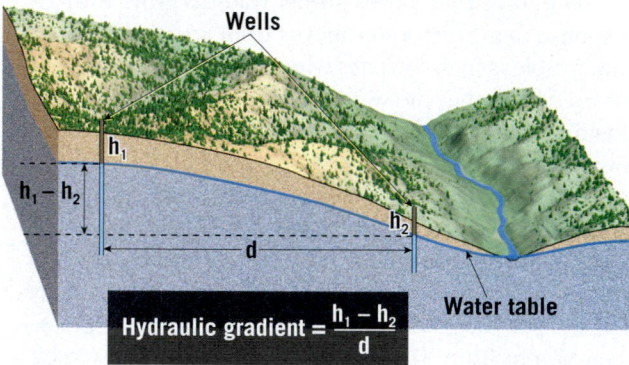

$$\text{Hydraulic gradient} = \frac{h_1 - h_2}{d}$$

tendency of water to move toward areas of reduced pressure. As a result, water at any given height is under greater pressure beneath a hill than beneath a stream channel, and the water tends to migrate toward points of lower pressure.

Measuring Groundwater Movement

The foundations of our modern understanding of groundwater movement began in the mid-nineteenth century with the work of the French scientist-engineer

Henri Darcy. One of the experiments Darcy carried out showed that the velocity of groundwater flow is proportional to the slope of the water table: The steeper the slope, the faster the water moves (because the steeper the slope, the greater the pressure difference between two points). The water table slope is known as the **hydraulic gradient** and can be expressed as follows:

$$\text{hydraulic gradient} = \frac{h_1 - h_2}{d}$$

where h_1 is the elevation of one point on the water table, h_2 is the elevation of a second point, and d is the horizontal distance between the two points (**Figure 17.12**).

Darcy also experimented with different materials, such as coarse sand and fine sand, by measuring the rate of flow through sediment-filled tubes that were tilted at varying angles. He found that flow velocity varied with the permeability of the sediment: Groundwater flows more rapidly through sediments having greater permeability than through materials having lower permeability. This factor is known as **hydraulic conductivity** and is a coefficient that takes into account the permeability of the aquifer and the viscosity of the fluid.

To determine discharge (Q)—that is, the actual volume of water that flows through an aquifer in a specified time—the following equation is used:

$$Q = \frac{KA(h_1 - h_2)}{d}$$

where $\frac{h_1 - h_2}{d}$ is the hydraulic gradient, K is the coefficient that represents hydraulic conductivity, and A is the cross-sectional area of the aquifer. This expression has come to be called **Darcy's law**, in honor of the pioneering French scientist-engineer. Using this equation, if you know an aquifer's hydraulic gradient, conductivity, and cross-sectional area, you can calculate its discharge.

SmartFigure 17.13
Hypothetical groundwater flow system The diagram includes subsystems at three different scales. Variations in surface topography and subsurface geology can produce a complex situation. The horizontal scale of the figure could range from tens to hundreds of kilometers.
(https://goo.gl/3GYAll)

▶ **Tutorial**

Explanation

High hydraulic-conductivity aquifer

Low hydraulic-conductivity aquitard

------ Water table

→ Groundwater movement in near-surface local systems

→ Groundwater movement in a subregional system

→ Groundwater movement in a deep regional system

Different Scales of Movement

The geographic extent of groundwater flow systems varies from a few square kilometers or less to tens of thousands of square kilometers. The length of flow paths ranges from a few meters to tens and sometimes hundreds of kilometers. **Figure 17.13** is a cross section of a hypothetical region in which a deep groundwater flow system is overlain by and connected to several, more shallow local flow systems. The subsurface geology exhibits a complicated arrangement

EYE ON EARTH 17.2

The drainage basin of the Republican River occupies portions of Colorado, Nebraska, and Kansas. A significant part of the basin is considered semiarid. In 1943, the three states made a legal agreement regarding sharing the river's water. In 1998, Kansas went to court to force farmers in southern Nebraska to substantially reduce the amount of groundwater used for irrigation. Nebraska officials claimed that the farmers were not taking water from the Republican River and thus were not violating the 1943 agreement. The court ruled in favor of Kansas.

QUESTION 1 Explain why the court ruled that groundwater in southern Nebraska should be considered part of the Republican River system.

QUESTION 2 How might heavy irrigation in a drainage basin influence the flow of a river?

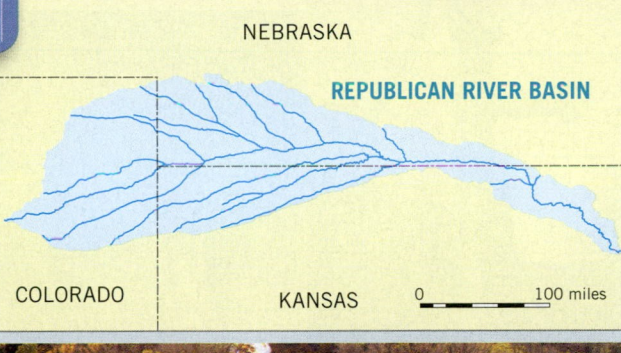

Michael Collier

of high-hydraulic-conductivity aquifer units and low-hydraulic-conductivity aquitard units.

Starting near the top of Figure 17.13, the blue arrows represent water movement in several local groundwater systems that occur in the upper water table aquifer. The groundwater systems are separated by groundwater divides at the center of the hills, and they discharge into the nearest surface water body. Beneath these most shallow systems, red arrows show water movement in a somewhat deeper system in which groundwater does not discharge into the nearest surface water body but into a more distant one. Finally, the black arrows show groundwater movement in a deep regional system that lies beneath the more shallow ones and is connected to them.

The horizontal scale of the figure could range from tens to hundreds of kilometers.

17.4 | Concept Checks

1. What factors cause water to follow the paths shown in Figure 17.11?

2. Relate groundwater movement to hydraulic gradient and hydraulic conductivity.

3. Contrast groundwater movement in a near-surface local system with that in a deep regional system.

17.5 | Wells and Artesian Systems

Discuss water wells and their relationship to the water table. Sketch and label a simple artesian system.

According to the National Groundwater Association, there are more than 16 million water wells for all purposes in the United States. Private household wells constitute the largest share—more than 13 million. About 500,000 new residential wells are drilled each year.

Wells

The most common method for removing groundwater is to use a **well**, a hole bored into the zone of saturation (**Figure 17.14**). Wells serve as small reservoirs into which groundwater migrates and from which it can be pumped to the surface. The use of wells dates back many centuries and continues to be an important method of obtaining water today. Groundwater is the principal source of

Figure 17.14
Wells Wells are the most common means by which people obtain groundwater. (Top photo by Shutterstock; bottom photo by ASP/YPP/AGE Fotostock)

drinking water for about 50 percent of the U.S. population and provides about 96 percent of the water used for rural domestic supplies.

The water table level may fluctuate considerably during the course of a year, dropping during dry periods and rising following wet periods. Therefore, to ensure a continuous supply of water, a well must penetrate below the water table. Whenever substantial water is withdrawn

from a well, the water table around the well may be lowered. This effect, termed **drawdown**, decreases with increasing distance from the well. The result is a depression in the water table, roughly conical in shape, known as a **cone of depression** (**Figure 17.15**). Because the cone of depression increases the hydraulic gradient near the well, groundwater will flow more rapidly toward the opening. For most small domestic wells, the cone of depression is negligible. However, when wells are heavily pumped for irrigation or for industrial purposes, the withdrawal of water can be great enough to create a very wide and steep cone of depression. This may substantially lower the water table in an area and cause nearby shallow wells to become dry. Figure 17.15 illustrates this situation.

Digging a successful well is a familiar challenge for people in areas where groundwater is the primary source of supply. One well may be successful at a depth of 10 meters (33 feet), whereas a neighbor may have to go twice as deep to find an adequate supply. Still others may be forced to go deeper or try a different site altogether. When subsurface materials are heterogeneous, the amount of water that a well can provide may vary a great deal over short distances. For example, when two nearby wells are drilled to the same level and only one is successful, it may be because there is a perched water table beneath one of them. As **Figure 17.16** illustrates, a **perched water table** forms where an aquitard is situated above the main water table. Massive igneous and metamorphic rocks provide a second example. These crystalline rocks are usually not very permeable, except where they are cut by many intersecting joints and fractures. Therefore, when a well drilled into such rock does not intersect an adequate network of fractures, it is likely to be unproductive.

Artesian Systems

In most wells, water cannot rise without the use of pumps. If water is first encountered at a depth of 30 meters (100 feet), it remains at that level, fluctuating perhaps 1 or 2 meters (3 to 6 feet) with seasonal wet and dry periods. However, in some wells, water rises, sometimes overflowing at the surface. Such wells are abundant in the Artois region of northern France, and so we call these self-rising wells *artesian*.

The term **artesian** is applied to *any* situation in which groundwater under pressure rises above the level of the aquifer. For an artesian system to exist, two conditions usually are met (**Figure 17.17**): (1) Water is confined to an

SmartFigure 17.15
Cone of depression For most small domestic wells, the cone of depression is negligible. When wells are heavily pumped, the cone of depression can be large and may lower the water table such that nearby shallower wells may be left dry.
(https://goo.gl/nup06m)

Animation

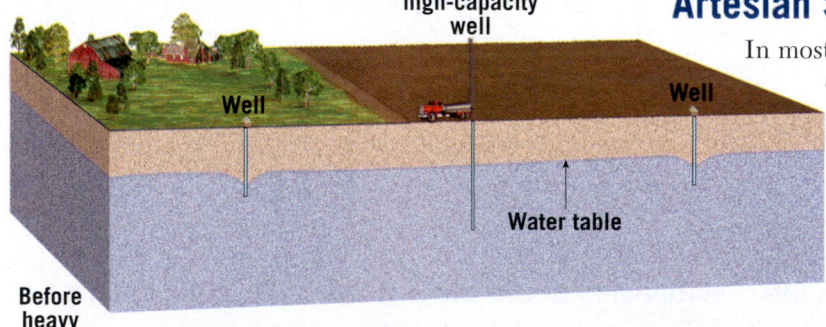

Figure 17.16
Perched water table

aquifer that is inclined so that one end can receive water, and (2) aquitards, both above and below the aquifer, must be present to prevent the water from escaping. Such an aquifer is called a **confined aquifer**. When such a layer is tapped, the pressure created by the weight of the water above will force the water to rise. If there were no friction, the water in the well would rise to the level of the water at the top of the aquifer. However, friction reduces the height of the pressure surface. The greater the distance from the recharge area (where water enters the inclined aquifer), the greater the friction and the less the rise of water.

In Figure 17.17, well 1 is a **nonflowing artesian well** because at this location, the pressure surface is below ground level. When the pressure surface is above the ground and a well is drilled into the aquifer, a **flowing artesian well** is created (well 2 in Figure 17.17). Not all artesian systems are wells. *Artesian springs* also exist. Here groundwater may reach the

surface by rising along a natural fracture such as a fault rather than through an artificially produced hole. In deserts, artesian springs are sometimes responsible for creating oases.

Artesian systems act as conduits, often transmitting water great distances from remote areas of recharge to points of discharge. A well-known artesian system in

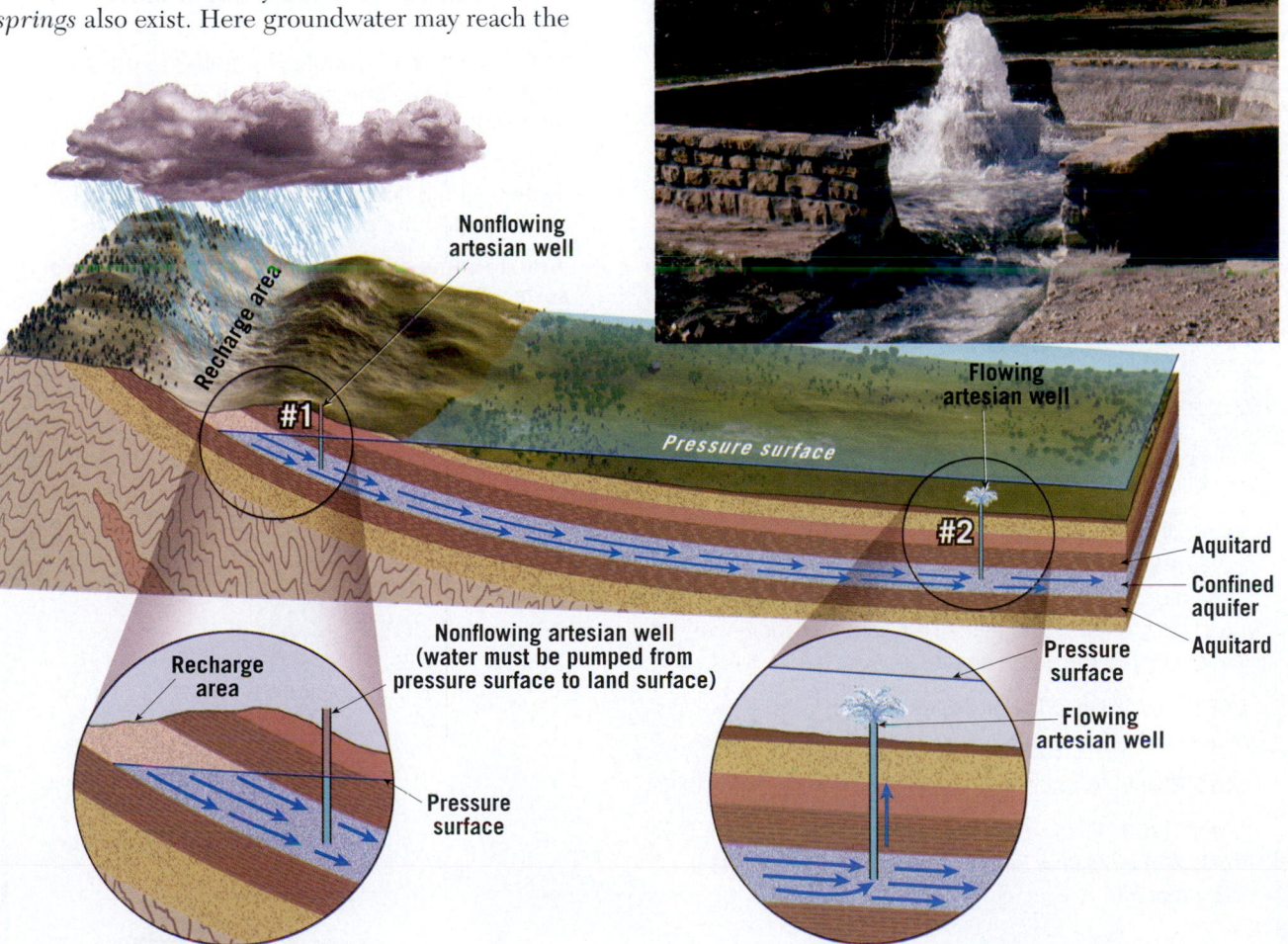

This well was unsuccessful because it missed the perched water table and was not deep enough to reach the main water table.

The perched water table allowed this well to be successful.

A natural outflow of water called a spring forms where the perched water table intersects the slope.

Aquitard

When an aquitard occurs above the main water table, it intercepts downward percolating water creating a localized zone of saturation and a perched water table.

Main water table

Nonflowing artesian well

Recharge area

#1

Pressure surface

Flowing artesian well

#2

Aquitard

Confined aquifer

Aquitard

Nonflowing artesian well (water must be pumped from pressure surface to land surface)

Recharge area

Pressure surface

Pressure surface

Flowing artesian well

SmartFigure 17.17
Artesian systems These groundwater systems occur where an inclined aquifer is surrounded by impermeable beds (aquitards). Such aquifers are called *confined aquifers*. The photo shows a flowing artesian well.
(Photo by James E. Patterson)
(https://goo.gl/yVJhLE)

Tutorial

Figure 17.18
A classic artesian system This geologic cross section across South Dakota shows the major elements of the Dakota Sandstone artesian system.

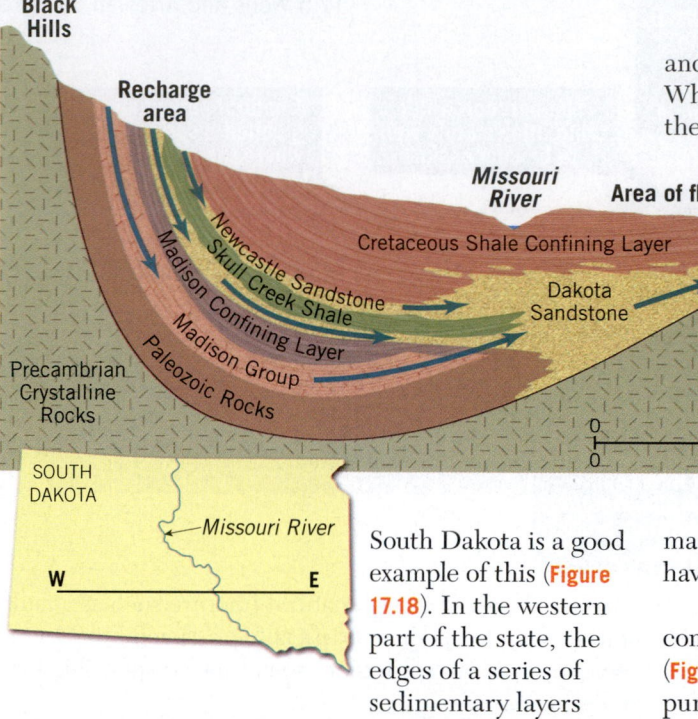

and gradually dips into the ground toward the east. When the aquifer was first tapped, water poured from the ground surface, creating fountains many meters high. In some places the force of the water was sufficient to power waterwheels. Such scenes no longer occur because thousands of additional wells now tap the same aquifer. This has depleted the reservoir and lowered the water table in the recharge area. As a consequence, the pressure has dropped to the point where many wells have stopped flowing altogether and now have to be pumped.

South Dakota is a good example of this (**Figure 17.18**). In the western part of the state, the edges of a series of sedimentary layers have been bent up to the surface along the flanks of the Black Hills. One of these beds, the permeable Dakota Sandstone, is sandwiched between impermeable strata

On a different scale, city water systems can be considered examples of artificial artesian systems (**Figure 17.19**). The water tower, into which water is pumped, would represent the area of recharge, the pipes the confined aquifer, and the faucets in homes the flowing artesian wells.

Figure 17.19
City water systems City water systems can be considered artificial artesian systems.

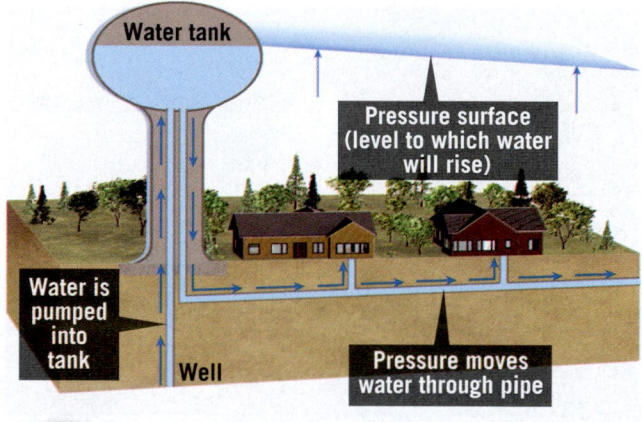

17.5 | Concept Checks

1. Define *drawdown* and relate this to *cone of depression*.

2. In Figure 17.16, two wells are drilled to the same depth. Why was one successful and the other not?

3. Sketch a simple cross section of an artesian system with a flowing well. Label aquitards, aquifers, and the pressure surface.

4. Why do some artesian wells not flow at Earth's surface?

EYE ON EARTH 17.3

In 1900, when this well was drilled near Woonsocket in eastern South Dakota, a "gusher" of water resulted. The stream of water from a 3-inch pipe reached a height of nearly 30 meters (100 feet). Thousands of additional wells now tap the same aquifer. (Photo by N.H. Darton/USGS)

QUESTION 1 Describe or sketch the subsurface geologic situation that was responsible for this fountain of water.

QUESTION 2 What term is applied to a well such as this?

QUESTION 3 Today wells that tap this aquifer do not flow freely at the surface but must be pumped. Suggest a likely reason.

17.6 | Springs, Hot Springs, and Geysers

Distinguish among springs, hot springs, and geysers.

The phenomena described in this section often arouse people's curiosity and wonder. The fact that springs, hot springs, and geysers seem rather mysterious is not difficult to understand, for here is water (sometimes very hot water) flowing or spewing from the ground in all kinds of weather, in seemingly inexhaustible supply, but with no obvious source.

Springs

Not until the middle of the seventeenth century did the French physicist Pierre Perrault invalidate the age-old assumption that precipitation could not adequately account for the amount of water emanating from springs and flowing in rivers. Over several years, Perrault computed the quantity of water that fell on France's Seine River basin. He then calculated the mean annual runoff by measuring the river's discharge. After allowing for the loss of water by evaporation, he showed that there was sufficient water remaining to feed the springs. Thanks to Perrault's pioneering efforts and the measurements by many afterward, we now know that the source of springs is water from the zone of saturation and that the ultimate source of this water is precipitation.

Whenever the water table intersects Earth's surface, a natural outflow of groundwater results, and we call this a **spring**. Springs often form when an aquitard blocks the downward movement of groundwater and causes the water to move laterally. Where the permeable bed crops out, a spring results. Another example is illustrated in Figure 17.16, which shows a perched water table intersecting a slope.

Springs, however, are not confined to places where a perched water table creates a flow at the surface. Many geologic situations lead to the formation of springs because subsurface conditions vary greatly from place to place. Even in areas underlain by impermeable crystalline rocks, permeable zones may exist in the form of fractures or solution channels. If these openings fill with water and intersect the ground surface along a slope, a spring results (**Figure 17.20**).

Figure 17.20
Thunder Spring A spring is a natural outflow of groundwater that occurs when the water table intersects the surface. Thunder Spring emerges from a deep joint and cave system in the Muav limestone along the North Rim of the Grand Canyon. (Photo by Michael Collier)

Hot Springs

There is no universally accepted definition of *hot spring*. One frequently used definition is that the water in a **hot spring** is 6°–9°C (10°–15°F) warmer than the mean annual air temperature for the locality where it occurs (**Figure 17.21**). In the United States alone, there are more than 1000 such springs.

Temperatures in deep mines and oil wells usually rise with increasing depth, an average of about 25°C (45°F) per kilometer. You learned in Chapter 4 that this is called the *geothermal gradient*. Therefore, when

Figure 17.21
Hot spring Many hot springs, including this one in Iceland, are associated with areas that have experienced recent igneous activity. (Photo by Andre Hasson/Alamy Images)

Figure 17.22
Distribution of hot springs and geysers Note the concentration in the West, where igneous activity has been most recent.

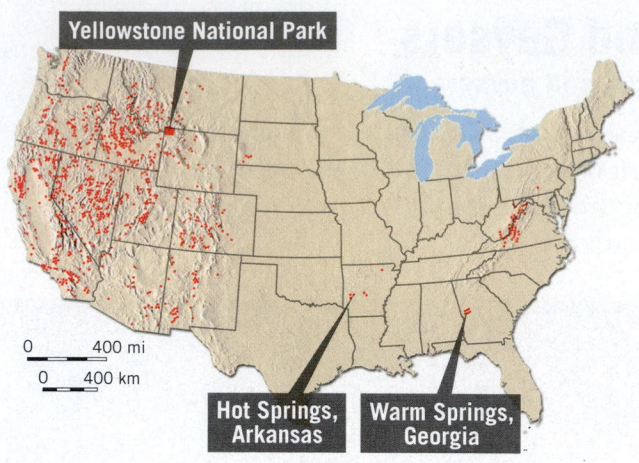

Geysers

Geysers are intermittent hot springs or fountains in which columns of water are ejected with great force at various intervals, often rising 30 to 60 meters (100 to 200 feet) into the air. After the jet of water ceases, a column of steam rushes out, usually with a thunderous roar. Perhaps the most famous geyser in the world is Old Faithful in Yellowstone National Park (**Figure 17.23**). The great abundance, diversity, and spectacular nature of geysers and other thermal features in Yellowstone undoubtedly was the primary reason for its becoming the first national park in the United States. Geysers are also found in other parts of the world, notably New Zealand and Iceland. In fact, the Icelandic word *geysa*, meaning "to gush," gives us the name *geyser*.

groundwater circulates at great depths, it becomes heated. If the hot water rises rapidly to the surface, it may emerge as a hot spring. The water of some hot springs in the eastern United States is heated in this manner. The springs at Warm Springs, Georgia, the presidential retreat of Franklin Roosevelt, are one example. The temperature of these hot springs is always near 32°C (90°F). At Hot Springs National Park, Arkansas, water temperatures average about 60°C (140°F).

The great majority (more than 95 percent) of the hot springs (and geysers) in the United States are found in the West. A glance at **Figure 17.22** reinforces this fact. This is because the heat sources for most hot springs are magma bodies and hot igneous rocks, and igneous activity has occurred more recently in the West than in the rest of the country. The hot springs and geysers of the Yellowstone region are well-known examples.

How Geysers Work
Geysers occur where extensive underground chambers exist within hot igneous rocks. How they operate is shown in **Figure 17.24**. As relatively cool groundwater enters these chambers, it is heated by the surrounding rock. At the bottom of the chambers, the water is under great pressure because of the weight of the overlying water. This great pressure prevents the water from boiling at the normal surface temperature of 100°C (212°F). For example, water at the bottom of a 300-meter (1000-foot) water-filled chamber must attain nearly 230°C (450°F) before it will boil. The heating causes the water to expand, and as a result, some is forced out at the surface. This loss of water reduces the pressure on the remaining water in the chamber, which lowers the boiling point. A portion of the water deep within the chamber quickly turns to steam, and the geyser erupts. Following eruption, cool groundwater again seeps into the chamber, and the cycle begins anew.

Geyser Deposits
When groundwater from hot springs and geysers flows out at the surface, material in solution is often precipitated, producing an accumulation of chemical sedimentary rock. The material deposited at any given place commonly reflects the chemical makeup of the rock through which the water circulated. When the water contains dissolved silica, a material called *siliceous sinter*, or *geyserite*, is deposited around the spring. When the water contains dissolved calcium carbonate, a form of limestone called *travertine*, or *calcareous tufa*, is deposited. The latter term is used if the material is spongy and porous.

The deposits at Mammoth Hot Springs in Yellowstone National Park are more spectacular than most others (**Figure 17.25**). As the hot water flows upward through a series of channels and then out at the surface,

Figure 17.23
Old Faithful This geyser in Wyoming's Yellowstone National Park is one of the most famous in the world. Contrary to popular legend, it does not erupt every hour on the hour. Time spans between eruptions vary from about 65 minutes to more than 90 minutes and have generally increased over the years due to changes in the geyser's plumbing. (Photo by Jeff Vanuga/Corbis)

A. Water near the bottom is heated to near its boiling point. The boiling point is higher at the bottom because pressure is high due to the weight of all the water above.

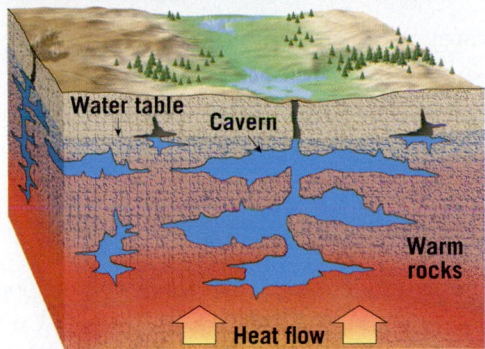

B. Higher up in the geyser the water is also heated, therefore it expands causing some to flow out at the top. This outflow reduces the pressure on the water at the bottom.

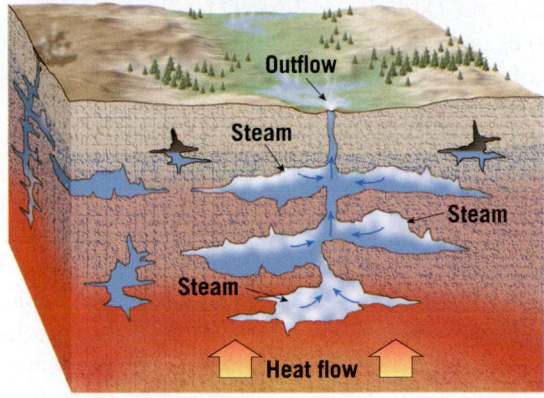

C. When pressure is reduced at the bottom, boiling occurs. Some of the bottom water flashes into steam. The expanding steam triggers an eruption. Then the water flows back in and the whole process begins anew.

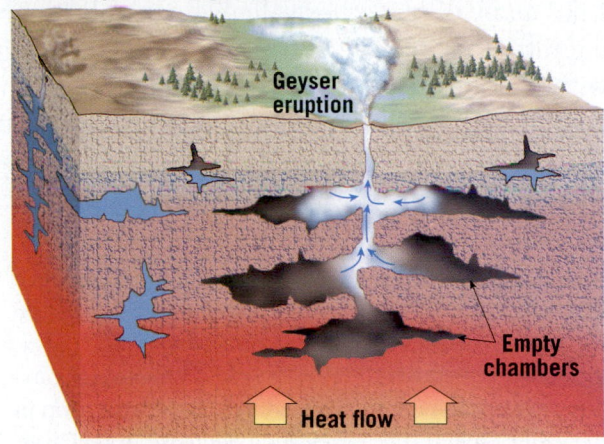

SmartFigure 17.24
How a geyser works A geyser can form if the underground plumbing does not allow heat to be readily distributed by convection. (https://goo.gl/qzYxyy)

Tutorial

the reduced pressure allows carbon dioxide to separate and escape from the water. The loss of carbon dioxide causes the water to become supersaturated with calcium carbonate, which then precipitates. In addition to containing dissolved silica and calcium carbonate, some hot springs contain sulfur, which gives water a poor taste and unpleasant odor. This is undoubtedly the case at Rotten Egg Spring, Nevada.

17.6 Concept Checks

1. Describe some circumstances that lead to the formation of a spring.

2. What warms the waters that flow at Hot Springs National Park, Arkansas, and at Warm Springs, Georgia?

3. What is the source of heat for most hot springs and geysers? How is this reflected in the distribution of these features?

4. Describe what occurs to cause a geyser to erupt.

Figure 17.25
Yellowstone's Mammoth Hot Springs Although most of the deposits associated with geysers and hot springs in Yellowstone National Park are silica-rich geyserite, the deposits here consist of a form of limestone called travertine. (Photo by Jamie and Judy Wild/ Danita Delimont/Alamy)

17.7 | Environmental Problems

List and discuss important environmental problems associated with groundwater.

Like many of our other valuable natural resources, groundwater is being exploited at an increasing rate. In some areas, overuse threatens the groundwater supply. In other places, groundwater withdrawal has caused the ground and everything resting on it to sink. Still other localities are concerned with possible contamination of the groundwater supply.

Mining Groundwater

Many natural systems tend to establish a condition of equilibrium. The groundwater system is no exception. The water table's height reflects a balance between the rate of recharge and the rate of discharge and withdrawal. Any imbalance will either raise or lower the water table. Long-term imbalances can lead to a significant drop in the water table if there is either a decrease in recharge due to prolonged drought or an increase in groundwater discharge or withdrawal (see GEOgraphics 17.1).

Many believe that groundwater is an endlessly renewable resource because it is continually replenished by rainfall and melting snow. But in some regions, groundwater has been and continues to be treated as a *nonrenewable* resource. Where this occurs, the water available to recharge the aquifer falls significantly short of the amount being withdrawn.

The High Plains aquifer provides one example (**Figure 17.26**). Underlying about 111 million acres (450,000 square kilometers [174,000 square miles]) in parts of eight western states, it is one of the largest and most agriculturally significant aquifers in the United States. It accounts for about 30 percent of all groundwater withdrawn for irrigation in the country. Mean annual precipitation on the High Plains is modest—ranging from about 40 centimeters (16 inches) in western portions to about 71 centimeters (28 inches) in eastern parts. Evaporation rates, on the other hand, are high—ranging from about 150 centimeters (60 inches) in the cooler northern parts of the region to 265 centimeters (105 inches) in the warmer southern parts. Because evaporation rates are high relative to precipitation, there is little rainwater to recharge the aquifer. Thus, in some parts of the region where intense irrigation has been practiced for an extended period, groundwater depletion has been severe. Figure 17.26 bears this out. The U.S. Geological Survey estimates that since 1950, water storage in the High Plains aquifer has declined about 267 million acre feet (about 87 trillion gallons), with 60 percent of the total decline having occurred in Texas.

Subsidence

As you will see later in this chapter, surface subsidence can result from natural processes related to groundwater. However, the ground may also sink when water is pumped from wells faster than natural recharge processes can replace it. This effect is particularly pronounced in areas underlain by thick layers of unconsolidated sediments. As the water is withdrawn, the water pressure drops, and the weight of the overburden is transferred to the sediment. The greater pressure packs the sediment grains tightly together, and the ground subsides.

Many areas illustrate land subsidence caused by excessive pumping of groundwater from relatively loose sediment. A classic U.S. example occurred in

Figure 17.26
High Plains aquifer The map shows changes in groundwater levels from predevelopment (about 1950) to 2013. Extensive pumping for irrigation has led to water level declines in excess of 30 meters (100 feet) in parts of four states. Water level rises have occurred where surface water is used for irrigation, such as along the Platte River in Nebraska. (Based on U.S. Geological Survey)

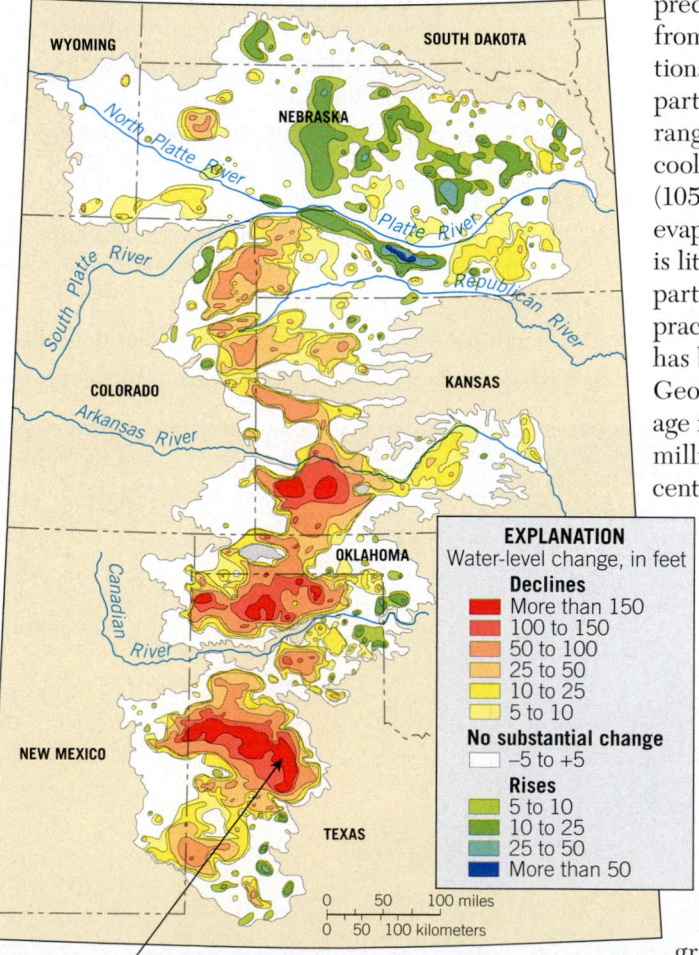

Because of its high porosity, excellent permeability, and great size, the High Plains aquifer, the largest in the United States, accumulated enough freshwater to fill Lake Huron.

EXPLANATION
Water-level change, in feet
Declines
More than 150
100 to 150
50 to 100
25 to 50
10 to 25
5 to 10
No substantial change
−5 to +5
Rises
5 to 10
10 to 25
25 to 50
More than 50

0   50   100 miles
0   50   100 kilometers

The U.S. Geological Survey estimates that during the past 60 years, water in storage in the High Plains aquifer declined by about 267 million acre feet (about 87 trillion gallons) with 60 percent of the total decline occurring in Texas.

Drought Impacts the Hydrologic System

Drought is a period of abnormally dry weather that persists long enough to produce a significant hydrologic imbalance such as crop damage or water supply shortages. Drought severity depends upon the degree of moisture deficiency, its duration, and the size of the affected area.

Drought status map for the western United States on April 7, 2015

Much of the West was experiencing drought at this time. California was suffering most. Entering its fourth year of drought, more than 93 percent of the state had at least *severe drought* conditions. *Exceptional drought* was affecting nearly 40 percent of the state. Mountain snowpack, the source that feeds California's rivers, lakes, and reservoirs, was just 19 percent of the late winter average.

Impact on groundwater

Because drought depleted surface water sources, groundwater use soared to make up the shortfall with many areas experiencing a dramatic increase in well drilling. Not only more wells, but *deeper* wells. Nearly 60 percent of the state's water needs were being met by groundwater, up from 40 percent in years when rain and snow were normal.

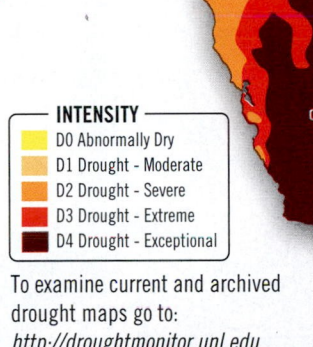

INTENSITY
- D0 Abnormally Dry
- D1 Drought - Moderate
- D2 Drought - Severe
- D3 Drought - Extreme
- D4 Drought - Exceptional

To examine current and archived drought maps go to:
http://droughtmonitor.unl.edu

Question:
Reduced rain and snow means reduced groundwater recharge and greater groundwater use. When such conditions prevail, how will the water table likely be affected?

?

Daniel Acker/Bloomberg/Getty Images

DROUGHT CATEGORIES

METEOROLOGICAL DROUGHT	AGRICULTURAL DROUGHT	HYDROLOGICAL DROUGHT
PRECIPITATION DEFICIENCY (results in reduced runoff and infiltration)	**SOIL MOISTURE DEFICIENCY** (results in low crop yields)	**REDUCED STREAMFLOW, INFLOW TO RESERVOIRS, LAKES, AND PONDS; REDUCED WETLANDS** (results in reduced domestic water supply and wildlife habitat)

After the onset of meteorological drought, agriculture is affected first, followed by reductions in streamflow and water levels in lakes, streams, and underground. When meteorological drought ends, agricultural drought ends as soil moisture is replenished. It can take much longer for hydrolgical drought to end.

COMPARING AVERAGE YEARLY COSTS—DROUGHTS, FLOODS, AND HURRICANES

HURRICANES
$1.2 to $4.8 billion

NASA

FLOODS
$2 to $3 billion

AP Photo/Jeff Roberson

DROUGHT
$6 to $8 billion

Daniel Acker/Bloomberg/Alamy

Although natural disasters such as floods and hurricanes usually generate more attention, droughts can be just as devastating and often carry a bigger price tag. Unlike other hazards which are short-lived, drought occurs in a gradual "creeping" way, making its onset and end difficult to determine.

Figure 17.27
That sinking feeling! The San Joaquin Valley, an important agricultural area, relies heavily on irrigation. Between 1925 and 1975, this part of the valley subsided almost 9 meters (30 feet) because of the withdrawal of groundwater and the resulting compaction of sediments. (Photo courtesy of U.S. Geological Survey)

of southern Arizona (**Figure 17.28**); Las Vegas, Nevada; New Orleans and Baton Rouge, Louisiana; and the Houston–Galveston area of Texas. In the low-lying coastal area between Houston and Galveston, land subsidence ranges from 1.5 to 3 meters (5 to 10 feet). The result is that an area of about 78 square kilometers (30 square miles) is permanently flooded.

Outside the United States, one of the most spectacular examples of subsidence occurred in Mexico City, a portion of which is built on a former lake bed. In the first half of the twentieth century, thousands of wells were sunk into the water-saturated sediments beneath the city. As water was withdrawn, portions of the city subsided by as much as 6 to 7 meters (20 to 23 feet). In some places buildings have sunk to such a point that access to them from the street is located at what used to be the second-floor level!

Saltwater Intrusion

In many coastal areas, the groundwater resource is being threatened by the encroachment of saltwater. To understand this problem, let us examine the relationship between fresh groundwater and salty groundwater. **Figure 17.29** shows a cross section that illustrates this relationship in a coastal area underlain by permeable homogeneous materials. Freshwater is less dense than saltwater, so it floats on the saltwater and forms a large lens-shaped body that may extend to considerable depths below sea level. In such a situation, if the water table is 1 meter (3 feet) above sea level, the base of the freshwater body will extend to a depth of about 40 meters (130 feet) below sea level. Stated another way, the depth of the freshwater below sea level is about 40 times greater than the elevation of the water table above sea level. Thus, when excessive pumping lowers the water table by a certain amount, the bottom of the freshwater zone will rise by 40 times that amount. Therefore, if groundwater withdrawal continues to exceed recharge, at some point, the elevation of the saltwater will be sufficiently high that the saltwater will be drawn into wells, thus contaminating the freshwater supply. This is called *saltwater intrusion* or *saltwater contamination*. Deep wells and wells near the shore are usually the first to be affected.

In urbanized coastal areas, the problems created by excessive pumping are compounded by a decrease in the rate of natural recharge. As more and more of the surface is covered by streets, parking lots, and buildings, surface runoff increases and infiltration into the soil is diminished.

the San Joaquin Valley of California, where subsidence approached 9 meters (30 feet) in some areas (**Figure 17.27**). Many other cases of land subsidence due to groundwater pumping exist in the United States, including in portions

Figure 17.28
Land subsidence in south-central Arizona In southern Arizona, heavy pumping has led to water table declines of up to 180 meters (600 feet). This has triggered extensive and uneven permanent compaction of sediments and the formation of large fissures (cracks) in the ground around the margins of subsiding basins. Some rural roads have signs that warn of the potential hazard. (Photo by Todd Shipman/ Arizona Geological Survey)

SUBSIDENCE AREA

Because freshwater is less dense than saltwater, it floats on the saltwater and forms a lens-shaped body that may extend to considerable depths below sea level.

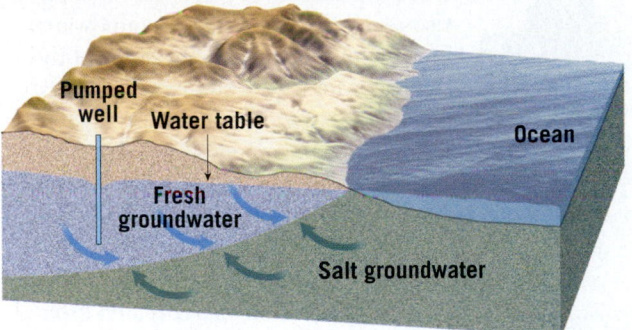

If excessive pumping lowers the water table, the base of the freshwater zone will rise 40 times that amount. The result may be saltwater contamination of wells.

Figure 17.29
Saltwater intrusion Heavy pumping in coastal areas can cause encroachment of saltwater and threaten the supply of fresh groundwater.

One way to correct the problem of saltwater intrusion of groundwater resources is to use a network of recharge wells. These wells allow wastewater to be pumped back into the groundwater system. A second method of correction is accomplished by building large recharge basins. These basins collect surface drainage and allow it to seep into the ground. On New York's Long Island, where the problem of saltwater intrusion was recognized more than 50 years ago, both of these methods have been employed with considerable success (**Figure 17.30**).

Contamination of freshwater aquifers by saltwater is primarily a problem in coastal areas, but it can also threaten noncoastal locations. Many ancient sedimentary rocks of marine origin were deposited when the ocean covered places that are now far inland. In some instances, significant amounts of seawater were trapped and still remain in the rock. These strata sometimes contain quantities of freshwater that people may pump. However, if freshwater is removed more rapidly than it is replenished, saline water may encroach and render the wells unusable. Such a situation threatened users of a deep (Cambrian age) sandstone aquifer in the Chicago area. To counteract this, water from Lake Michigan was allocated to the affected communities to offset the rate of withdrawal from the aquifer.

Groundwater Contamination

The pollution of groundwater is a serious matter, particularly in areas where aquifers provide a large part of the

water supply. One common source of groundwater pollution is sewage. Its sources include an ever-increasing number of septic tanks, as well as inadequate or broken sewer systems and farm wastes.

If sewage water that is contaminated with bacteria enters the groundwater system, it may become purified through natural processes. The harmful bacteria may be mechanically filtered by the sediment through which the water percolates, destroyed by chemical oxidation, and/or assimilated by other organisms. For purification to occur, however, the aquifer must be of the correct composition. For example, extremely permeable aquifers (such as highly fractured crystalline rock, coarse gravel, or cavernous limestone) have such large openings that contaminated groundwater may travel long distances without being filtered and cleansed. In this case, the water flows too rapidly and is not in contact with the surrounding

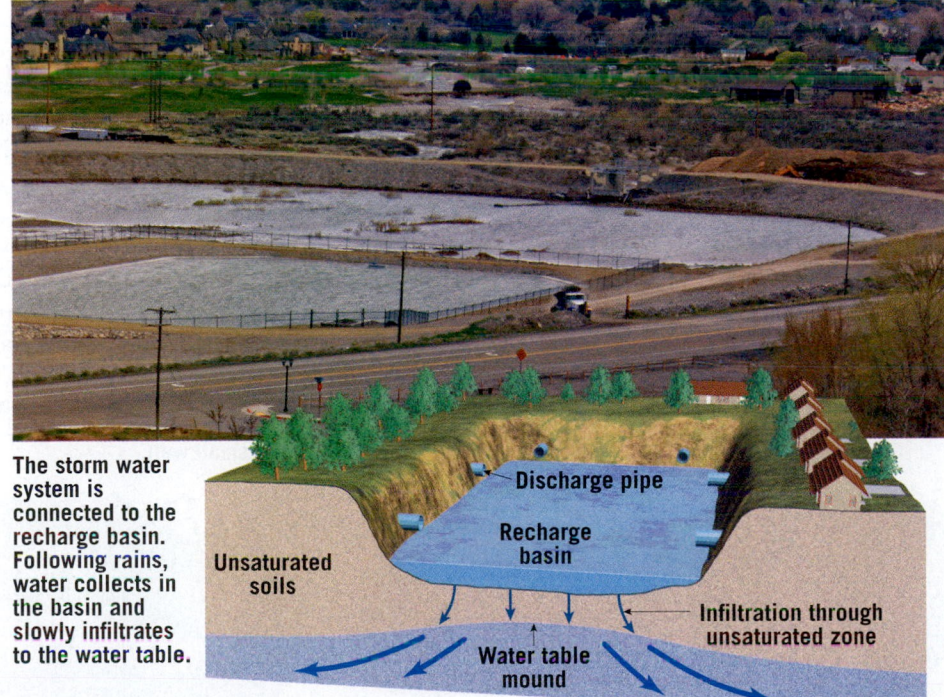

The storm water system is connected to the recharge basin. Following rains, water collects in the basin and slowly infiltrates to the water table.

Figure 17.30
Recharge basins Much of Long Island, New York, is completely dependent on groundwater. To help maintain the water table and prevent saltwater intrusion, more than 2000 recharge basins have been constructed. Recharge basins are used in many places, not just coastal areas. (Photo by Alan Cressler)

Figure 17.31
Comparing two aquifers In this example, the limestone aquifer allowed the contamination to reach a well, but the sandstone aquifer did not.

Although the contaminated water has traveled more than 100 meters before reaching Well 1, the water moves too rapidly through the cavernous limestone to be purified.

Well 1 delivering contaminated water
Contaminated water
Septic tank
Cavernous limestone
Water table
Time: Days or weeks
A.

As the discharge from the septic tank percolates through the permeable sandstone, it moves more slowly and is purified in a relatively short distance.

Permeable sandstone
Well 2 delivering clean water
Contaminated water
Septic tank
Water table
Time: Months or years
B.

slope may even be reversed. This could lead to the contamination of wells that yielded unpolluted water before heavy pumping began (**Figure 17.32**).

Also recall that the rate of groundwater movement increases as the slope of the water table gets steeper. This could produce problems because a faster rate of movement allows less time for the water to be purified in the aquifer before it is pumped to the surface.

Other sources and types of contamination also threaten groundwater supplies (**Figure 17.33**). These include widely used substances such as highway salt, fertilizers that are spread across the land surface, and pesticides. In addition, a wide array of chemicals and industrial materials may leak from pipelines, storage tanks, landfills, and holding ponds. Some of these pollutants are classified as *hazardous*, meaning that they are either flammable, corrosive, explosive, or toxic. In landfills, potential contaminants are heaped onto mounds or spread directly over the ground. As rainwater oozes through the refuse, it may dissolve a variety of organic and inorganic materials. If the leached material reaches the water table, it will mix with the groundwater and contaminate the supply. Similar problems may result from leakage of shallow excavations called *holding ponds* into which a variety of liquid wastes are disposed.

Because groundwater movement is usually slow, polluted water can go undetected for a long time. In fact, contamination is sometimes discovered only after drinking water has been affected and people become ill. By this time, the volume of polluted water may be very large, and even if the source of contamination is removed immediately, the problem is not solved. Although the sources of groundwater contamination are numerous, there are relatively few solutions.

material long enough for purification to occur. This is the problem at well 1 in **Figure 17.31A**.

On the other hand, when the aquifer is composed of sand or permeable sandstone, it can sometimes be purified after traveling only a few dozen meters through it. The openings between sand grains are large enough to permit water movement, yet the movement of the water is slow enough to allow ample time for its purification (well 2, **Figure 17.31B**).

Sometimes sinking a well can lead to groundwater pollution problems. If the well pumps a sufficient quantity of water, the cone of depression will locally increase the slope of the water table. In some instances, the original

SmartFigure 17.32
Changing direction
Drawdown at the heavily pumped well changed the slope of the water table, which led to the contamination of the small well.
(https://goo.gl/SO6Vhx)

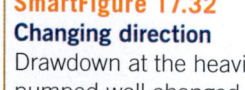

Originally the outflow from the septic tank moved away from the small well.

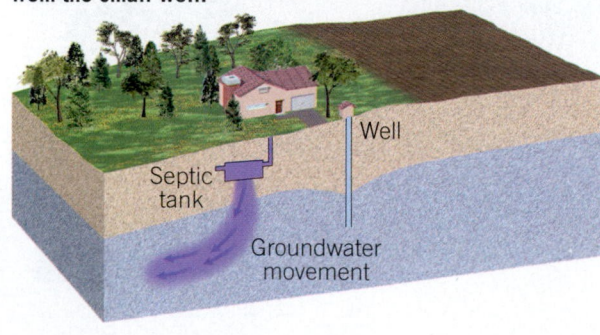

Well
Septic tank
Groundwater movement

The heavily pumped well changed the slope of the water table, causing contaminated groundwater to flow toward the small well.

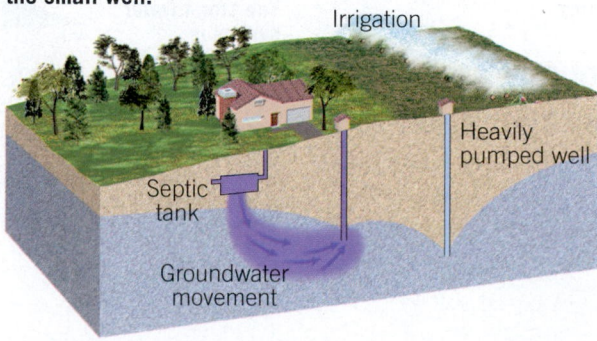

Irrigation
Heavily pumped well
Septic tank
Groundwater movement

Figure 17.33
Potential sources of contamination Sometimes materials leached from landfills and leaking gasoline storage tanks contaminate an aquifer. (Landfill photo by Deposit Photos/Glow Images; storage tank photo by Earth Gallery environment/Alamy)

Once the source of the problem has been identified and eliminated, the most common practice is simply to abandon the water supply and allow the pollutants to be flushed away gradually. This is the least costly and easiest solution, but the aquifer must remain unused for many years. To accelerate this process, polluted water is sometimes pumped out and treated. Following removal of the tainted water, the aquifer is allowed to recharge naturally, or in some cases the treated water or other freshwater is pumped back in. This process is costly, time-consuming, and may be risky because there is no way to be certain that all of the contamination has been removed. Clearly, the most effective solution to groundwater contamination is prevention.

17.7 | Concept Checks

1. Describe the problem associated with pumping groundwater for irrigation in the southern High Plains.

2. Explain why ground may subside after groundwater is pumped to the surface.

3. Which aquifer would be most effective in purifying polluted groundwater: coarse gravel, sand, or cavernous limestone?

4. Describe a significant problem that may arise when groundwater is heavily pumped at a coastal site.

17.8 | The Geologic Work of Groundwater

Explain the formation of caverns and the development of karst topography.

Groundwater dissolves rock. This fact is key to understanding how caverns and sinkholes form (**Figure 17.34**). Soluble rocks, especially limestone, underlie millions of square kilometers of Earth's surface, and it is in these rocks that groundwater carries on its important role as an erosional agent. Limestone is nearly insoluble in pure water but is quite easily dissolved by water containing small quantities of carbonic acid, and most groundwater contains this acid. It forms because rainwater readily dissolves carbon dioxide from the air and from decaying plants. When groundwater comes in contact with limestone, the carbonic acid reacts with the calcite (calcium carbonate) in the rocks to form calcium bicarbonate, a soluble material that is then carried away in solution.

SmartFigure 17.34
Kentucky's Mammoth Cave area Portions of Kentucky are underlain by limestone. Dissolution by groundwater has created a landscape characterized by caves and sinkholes. (Photo by Michael Collier) (http://goo.gl/jsqQfh)

Mobile Field Trip

Caverns

The most spectacular results of groundwater's erosional handiwork are limestone **caverns**. In the United States alone, about 17,000 caves have been discovered, and more are being found every year. Although most are relatively

A.

B.

small, some have spectacular dimensions. Mammoth Cave in Kentucky and Carlsbad Caverns in southeastern New Mexico are famous examples. The Mammoth Cave system is the most extensive in the world, with more than 540 kilometers (335 miles) of interconnected passages. The dimensions at Carlsbad Caverns are impressive in a different way. Here we find the largest and perhaps most spectacular single chamber. The Big Room at Carlsbad Caverns has an area equivalent to 14 football fields and enough height to accommodate the U.S. Capitol building.

Cavern Development   Most caverns are created at or just below the water table, in the zone of saturation. Here acidic groundwater follows lines of weakness in the rock, such as joints and bedding planes. As time passes, the dissolving process slowly creates cavities and gradually enlarges them into caverns. Material that is dissolved by the groundwater is eventually discharged into streams and carried to the ocean.

In many cases, cavern development has occurred at several levels, with the current cavern-forming activity occurring at the lowest elevation. This situation reflects the close relationship between the formation of major subterranean passages and the river valleys into which

they drain. As streams cut their valleys deeper, the water table drops as the elevation of the river drops. Consequently, during periods when surface streams are rapidly downcutting, surrounding groundwater levels drop rapidly, and cave passages are abandoned by the water while the passages are still relatively small in cross-sectional area. Conversely, when the entrenchment of streams is slow or negligible, there is time for large cave passages to form.

How Dripstone Forms   Certainly the features that arouse the greatest curiosity for most cavern visitors are the stone formations that give some caverns a wonderland appearance. These are not erosional features, like the cavern itself, but depositional features created by the seemingly endless dripping of water over great spans of time. Recall from our discussion of hot springs that the calcium carbonate left behind produces the limestone we call travertine. These cave deposits, however, are also commonly called *dripstone*, an obvious reference to their mode of origin. Although the formation of caverns takes place in the zone of saturation, the deposition of dripstone is not possible until the caverns are above the water table in the unsaturated zone. As soon as the chamber is filled with air, the stage is set for the decoration phase of cavern building to begin.

Dripstone Features—Speleothems   The various dripstone features found in caverns are collectively called **speleothems** (*spelation* = cave, *them* = put), and no two of them are exactly alike. Perhaps the most familiar speleothems are **stalactites** (*stalaktos* = trickling). These icicle-like pendants hang from the ceiling of a cavern and form where water seeps through cracks above. When the water reaches air in the cave, some of the dissolved carbon dioxide escapes from the drop, and calcite precipitates. Deposition occurs as a ring around the edge of the water drop. As drop after drop follows, each leaves an infinitesimal trace of calcite behind, and a hollow limestone tube is created. Water then moves through the tube, remains suspended momentarily at the end, contributes a tiny ring of calcite, and falls to the cavern floor. The stalactite just described is appropriately called a *soda straw* (**Figure 17.35A**). Often the hollow tube of the soda straw becomes plugged, or its supply of water increases. In either case, the water is forced to flow and hence deposit along the outside of the tube. As deposition continues, the stalactite takes on the more common conical shape.

Speleothems that form on the floor of a cavern and reach upward toward the ceiling are called **stalagmites** (*stalagmos* = dropping). The water supplying the calcite for stalagmite growth falls from the ceiling and splatters over the surface. As a result, stalagmites do not have a central tube and are usually more massive in appearance and rounded on their upper ends than stalactites. Given enough time, a downward-growing stalactite and an upward-growing stalagmite may join to form a *column* (**Figure 17.35B**).

Karst Topography

Many areas of the world have landscapes that, to a large extent, have been shaped by the dissolving power of groundwater. Such areas are said to exhibit **karst topography**, named for the Krs Plateau in Slovenia, located along the northeastern shore of the Adriatic Sea, where such topography is strikingly developed. In the United States, karst landscapes occur in many areas that are underlain by limestone, including portions of Kentucky, Tennessee, Alabama, southern Indiana, and central and northern Florida (**Figure 17.36**). Generally, arid and semiarid areas are too dry to develop karst topography. When these features exist in such regions, they are likely to be remnants of a time when rainier conditions prevailed.

Sinkholes   Karst areas typically have irregular terrain punctuated with many depressions, called **sinkholes,** or

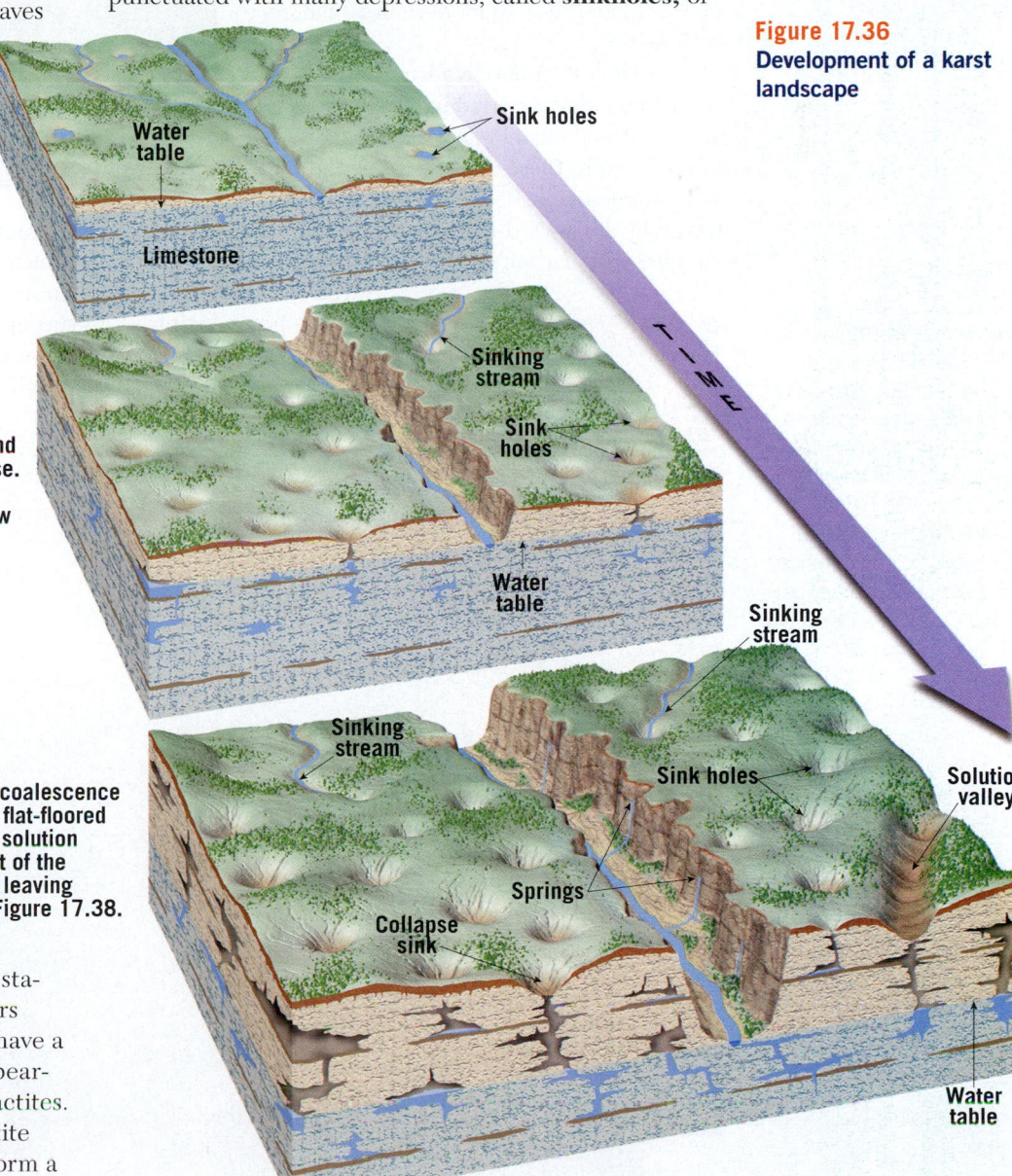

Figure 17.36 Development of a karst landscape

During early stages, groundwater percolates through limestone along joints and bedding planes. Solution activity creates and enlarges caverns at and below the water table.

With time, caverns grow larger and the number and size of sinkholes increase. Surface drainage is frequently funneled below ground.

Collapse of caverns and coalescence of sinkholes form larger, flat-floored depressions. Eventually solution activity may remove most of the limestone from the area, leaving isolated remnants as in Figure 17.38.

Figure 17.37
Sinkholes Karst land-scapes are typically punctuated with these depressions. The white spots in the top photo are grazing sheep. (Top photo by David Wall/Alamy Images; bottom photo by AP Photo/The Florida Times-Union, Jon M. Fletcher)

Groundwater was responsible for creating these sinkholes west of Timaru on New Zealand's South Island.

This small sinkhole formed suddenly when the roof of a cavern collapsed, eliminating the backyard of this house in Lake City, Florida.

sinks (**Figure 17.37**). In the limestone areas of Florida, Kentucky, and southern Indiana, there are tens of thousands of these depressions, varying in depth from just 1 or 2 meters (3 or 7 feet) to more than 50 meters (165 feet).

Sinkholes commonly form in two ways. Some develop gradually over many years, without any physical disturbance to the rock. In these situations, the limestone immediately below the soil is dissolved by downward-sweeping rainwater that is freshly charged with carbon dioxide. With time, the bedrock surface is lowered, and the fractures into which the water seeps are enlarged. As the fractures grow in size, soil subsides into the widening voids, from which it is removed by groundwater flowing in the passages below. These depressions are usually shallow and have gentle slopes.

In contrast, sinkholes can also form abruptly and without warning when the roof of a cavern collapses under its own weight. Typically, the depressions created in this manner are steep-sided and deep. When they form in populous areas, they may represent a serious geologic hazard. Such a situation is shown in the lower photo of Figure 17.37.

In addition to a surface pockmarked by sinkholes, karst regions characteristically show a striking lack of surface drainage (streams). Following rainfall, the runoff is quickly funneled belowground through sinks. It then flows through caverns until it finally reaches the water table. Where streams do exist at the surface, their paths are usually short. The names of such streams often give

Figure 17.38
Tower karst in China One of the best-known and most distinctive regions of tower karst development is along the Li River in the Guilin District of southeastern China. The painting "Peach Garden Land of Immortals" clearly depicts this distinctive landscape. (Landscape photo Philippe Michel/AGE Fotostock; painting photo by Qiu Ying)

a clue to their fate. In the Mammoth Cave area of Kentucky, for example, there is Sinking Creek, Little Sinking Creek, and Sinking Branch. Some sinkholes become plugged with clay and debris, creating small lakes or ponds.

Tower Karst   Some regions of karst development exhibit landscapes that look very different from the sinkhole-studded terrain depicted in Figure 17.36. One striking example is an extensive region in southern China that is described as exhibiting **tower karst**. As **Figure 17.38** shows, the term *tower* is appropriate because the landscape consists of a maze of isolated steep-sided hills that rise abruptly from the ground. Each is riddled with interconnected caves and passageways. This type of karst topography forms in wet tropical and subtropical regions having thick beds of highly jointed limestone. Here groundwater has dissolved large volumes of limestone, leaving only these residual towers. Karst development occurs more rapidly in tropical climates due to the abundant rainfall and the greater availability of carbon dioxide from the decay of lush tropical vegetation. The extra carbon dioxide in the soil means there is more carbonic acid for dissolving limestone. Other tropical areas of advanced karst development include portions of Puerto Rico, western Cuba, and northern Vietnam.

17.8   Concept Checks

1. How does groundwater create caverns?

2. What causes cavern formation to stop at one level (depth) but continue or begin at a lower level?

3. How do stalactites and stalagmites form?

4. Describe two ways in which sinkholes form.

17.1 The Importance of Groundwater

Describe the importance of groundwater as a source of freshwater and groundwater's roles as a geologic agent.

KEY TERM **groundwater**

- Groundwater is water stored below Earth's surface, mainly in tiny pore spaces between rock or sediment grains. Groundwater represents the largest reservoir of freshwater that is readily available to humans and is a critical resource for human civilization.

- Groundwater is important geologically because it dissolves rock to make sinkholes and caverns and supplies surface streams with additional water.
- Each day in the United States we use about 349 billion gallons of freshwater. Groundwater provides about 79 billion gallons, or 23 percent of the total. More groundwater is used for irrigation than for all other uses combined.

Q Examine Figure 17.1 to answer these questions: How much of Earth's freshwater is groundwater? How much of Earth's liquid freshwater is groundwater?

17.2 Groundwater and the Water Table

Prepare a sketch with labels that summarizes the distribution of water beneath Earth's surface. Discuss the factors that cause variations in the water table and describe the interactions between groundwater and streams.

KEY TERMS **zone of soil moisture, zone of saturation (phreatic zone), water table, capillary fringe, unsaturated zone (vadose zone), gaining stream, losing stream**

- Some of the rain that falls on land soaks into the ground. Typically, a hole dug into the ground penetrates this zone of soil moisture and then crosses the unsaturated zone where pore spaces contain both water and air. The soil grows moist again just above the water table, in the capillary fringe. Crossing the water table, the boundary between the groundwater below and the unsaturated zone above, the hole begins to fill (to the height of the water table) with water that flows in from the zone of saturation.
- Streams and groundwater interact in one of three ways: Streams gain water from the inflow of groundwater (gaining stream); they lose water through the streambed to the groundwater system (losing stream); or they do both, gaining in some sections and losing in others.

Q Examine this cross section, which shows the distribution of water in loose sediments. Provide the correct term for each lettered feature.

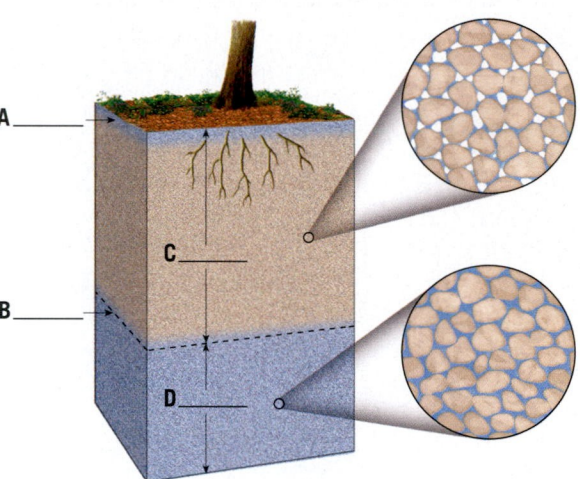

17.3 Factors Influencing the Storage and Movement of Groundwater

Summarize the factors that influence the storage and movement of groundwater.

KEY TERMS **porosity, permeability, aquitard, aquifer**

- The quantity of water that can be stored in a material depends on the material's porosity (the volume of open spaces). The permeability (the ability to transmit a fluid through interconnected pore spaces) of a material is a very important factor controlling the movement of groundwater.
- Materials with very small pore spaces (such as clay) hinder or prevent groundwater movement and are called aquitards. Aquifers consist of materials with larger pore spaces (such as sand) that are permeable and transmit groundwater freely.

Q Use Table 17.1 to select two good aquifers and two aquitards. What did you choose and why?

17.4 How Groundwater Moves

Sketch and describe a simple groundwater flow system. Discuss how groundwater movement is measured and the different scales of movement.

KEY TERMS **recharge area, discharge area, hydraulic gradient, hydraulic conductivity, Darcy's law**

- Groundwater flows slowly through underground pore spaces, moving on average only a few centimeters per day. Driven by gravity and pressure, it moves as a three-dimensional mass from areas of recharge (where water is added) to areas of discharge (where water leaves the groundwater system), such as springs, gaining streams, or even wells drilled by people.
- French scientist-engineer Henri Darcy pioneered the quantification of groundwater flow by measuring the slope of the water table (hydraulic gradient) and the permeability of the sediment or rock (hydraulic conductivity). Darcy's law combines these in an equation to estimate an aquifer's discharge.
- Groundwater flows both short and long distances at both shallow and deep levels. Closer to the surface, the flow is more local in scale, while at greater depths, the flow occurs over regional scales.

17.5 Wells and Artesian Systems

Discuss water wells and their relationship to the water table. Sketch and label a simple artesian system.

KEY TERMS well, drawdown, cone of depression, perched water table, artesian, confined aquifer, nonflowing artesian well, flowing artesian well

- For centuries, humans have been obtaining groundwater by drilling wells. As water is pumped out, the water table immediately adjacent to the well drops. This drawdown results in a "dimple" in the surface of the water table called the cone of depression. If there is sufficient drawdown, the cone of depression might encompass a large enough area that neighboring wells might go dry.
- A perched water table results from groundwater "piled up" atop an aquitard that is above the main body of groundwater. The shape of the water table is complex, which results in challenges for people trying to dig productive wells.
- Artesian wells tap into inclined aquifers bounded above and below by aquitards. For a system to qualify as artesian, the water in the well must be under sufficient pressure that it can rise above the top of the confined aquifer. Artesian wells may be flowing or nonflowing, depending on whether the pressure surface is above or below the ground surface.

Q Relate this image to an artesian system.

igone/Shutterstock

17.6 Springs, Hot Springs, and Geysers

Distinguish among springs, hot springs, and geysers.

KEY TERMS spring, hot spring, geyser

- Springs are naturally occurring spots where groundwater leaves the ground and flows out onto the surface. They may be due to the intersection of a perched water table and Earth's surface.
- Hot springs are like regular springs but hot. They transfer heat from the deeper crust to the surface. Most often, this heat comes from relatively shallow bodies of magma.
- Geysers are intermittent hot springs that "erupt" hot water periodically. They are fed by underground chambers that fill with water that warms past the boiling point. Once most of the water in the chamber has been sufficiently heated, it will flash to water vapor and rapidly expand, expelling some liquid water as it forces its way to the surface in an eruption. Geysers can precipitate silica or calcium carbonate around the geyser vent, producing the rocks siliceous sinter (geyserite) or travertine (tufa).

Q This photo from the 1930s shows Franklin Roosevelt enjoying the hot springs at the presidential retreat at Warm Springs, Georgia. The temperature of this water is always near 32°C (90°F). This area has no history of recent igneous activity. What is the likely reason these springs are so warm?

New York Daily News/Getty Images

17.7 Environmental Problems

List and discuss important environmental problems associated with groundwater.

- Groundwater can be "mined" by being extracted at a rate that is greater than the rate of replenishment. When groundwater is treated as a nonrenewable resource, as it is in parts of the High Plains aquifer, the water table drops, in some cases by more than 45 meters (150 feet).
- The extraction of groundwater can cause pore spaces to decrease in volume and the grains of loose Earth materials to pack more closely together. This overall compaction of sediment volume results in subsidence of the land surface.
- Saltwater contamination is a common environmental problem near coastal areas. Fresh groundwater "floats" on salty groundwater due to its lower density. If sufficient freshwater is pumped out to lower the water table by some amount, the base of the freshwater lens will rise about 40 times that amount. Deep wells may begin to access the deeper, salty water instead.
- Contamination of groundwater with sewage, highway salt, fertilizer, or industrial chemicals is another issue of critical concern. Once groundwater is contaminated, the problem is very difficult to solve, requiring expensive remediation or abandonment of the aquifer.

17.8 The Geologic Work of Groundwater

Explain the formation of caverns and the development of karst topography.

KEY TERMS cavern, speleothem, stalactite, stalagmite, karst topography, sinkhole (sink), tower karst

- Groundwater dissolves rock, in particular limestone, leaving behind void spaces in the rock. Caverns form at the zone of saturation, but later dropping of the water table may leave them open and dry—and available for people to explore.

- Dripstone is rock deposited by dripping of water that contains dissolved calcium carbonate inside caverns. Speleothems are features made of dripstone and include stalactites, stalagmites, and columns.

- Karst topography is a distinctive type of landscape dominated by the dissolving of limestone near Earth's surface. Collapsing caverns show up as sinkholes. Streams flowing on the surface may "sink" into the subterranean cavern system, and in other places the same water may reemerge as a spring. If enough limestone is dissolved, only isolated pinnacles of limestone will remain, towering over the landscape as tower karst.

Q Identify the three speleothems shown in this photograph. Would these speleothems have formed in the saturated zone or unsaturated zone? Why?

Miroslav/AGE Fotostock

Give It Some Thought

1. The cemetery in this photo is located in New Orleans, Louisiana. As in other cemeteries in the area, all the burial plots here are aboveground. Based on what you have learned in this chapter, suggest a reason for this rather unusual practice.

Zack Frank/Shutterstock

2. Imagine a water molecule that is part of a groundwater system in an area of gently rolling hills in the eastern United States. Describe some possible paths the molecule might take through the hydrologic cycle if:

   a. It is pumped from the ground to irrigate a farm field.

   b. There is a long period of heavy rainfall.

   c. The water table in the vicinity of the molecule develops a steep cone of depression due to heavy pumping from a nearby well.

   Combine your understanding of the hydrologic cycle with your imagination and include possible short-term and long-term destinations and information about how the molecule gets to these places via evaporation, transpiration, condensation, precipitation, infiltration, and runoff. Remember to consider possible interactions with streams, lakes, groundwater, the ocean, and the atmosphere.

3. Identify a location in the United States where you might find a gaining stream and a location where you would expect to find a losing stream. Why did you select each location? Describe a situation that would cause a gaining stream to become a losing stream. Also describe a situation in which a losing stream would become a gaining stream.

4. Imagine that you are an environmental scientist who has been hired to solve a groundwater contamination problem. Several homeowners have noticed that their well water has a funny smell and taste. Some think the contamination is coming from a landfill, but others think it might be a nearby cattle feedlot or chemical plant. Your first step is to gather data from wells in the area and prepare the map of the water table shown here.

   a. Based on your map, can any of the three potential sources of contamination be eliminated? If so, explain.

   b. What other steps would you take to determine the source of the contamination?

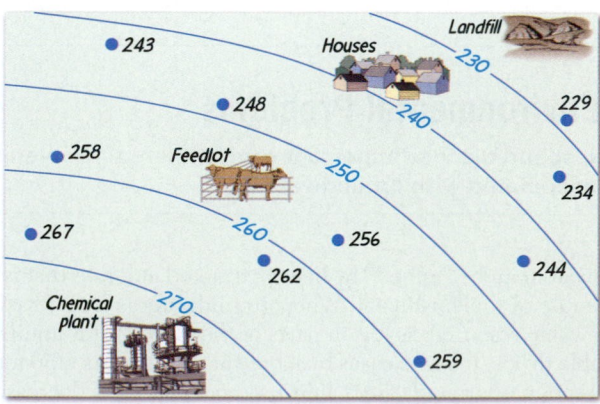

5. During a trip to a grocery store, your friend wants to buy some bottled water. Some brands promote the fact that their product is artesian. Other brands boast that their water comes from a spring. Your friend asks, "Is artesian water or spring water necessarily better than water from other sources?" How would you answer?

6. An acquaintance is considering purchasing a large tract of productive irrigated farmland in western Texas. His intention is to continue growing crops on the land for years to come. If he asked your opinion about the area he selected, what figure in this chapter would you consult before you responded? How would this figure help your friend evaluate his potential purchase?

7. Sinkholes commonly form in one of two ways. Examine the accompanying photo, which shows a sinkhole in Winter Park, Florida, and describe how it likely formed.

AP Photo

8. This satellite image shows a portion of the desert in northern Saudi Arabia, a region known for its abundant sunshine, high temperatures, and meager rainfall. The green circles are agricultural fields that are about 1 kilometer (0.62 mile) in diameter. Water for irrigation is pumped from deep aquifers and distributed around a center point within each field—a technique known as center pivot irrigation. The deep aquifers contain water that dates to the Ice Age about 20,000 years ago, a time when the climate in this region was wetter and milder.

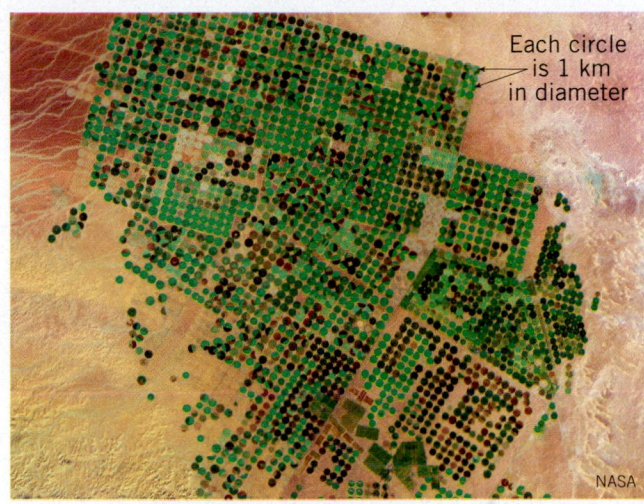

Each circle is 1 km in diameter

NASA

a. Is it likely that agricultural activity in this region is sustainable indefinitely? Explain.

b. A significant portion of the water placed on these fields is "lost" (not available to the crops). Suggest a reason for the loss of water.

c. Relate what is likely occurring to the water table in the region pictured here to an example of a similar situation in the United States.

MasteringGeology™

18

Glaciers and Glaciation

Hikers next to Exit Glacier in Alaska's Kenai Fiords National Park. (Photo by Michael Collier)

Climate has a strong influence on the nature and intensity of Earth's external processes. This fact is dramatically illustrated in this chapter because the existence and extent of glaciers is largely controlled by Earth's changing climate.

Like the running water and groundwater that were the focus of the preceding two chapters, glaciers represent a significant erosional process. These moving masses of ice are responsible for creating many unique landforms and are part of an important link in the rock cycle in which the products of weathering are transported and deposited as sediment.

Today glaciers cover nearly 10 percent of Earth's land surface; however, in the recent geologic past, ice sheets were three times more extensive, covering vast areas with ice thousands of meters thick. Many regions still bear the marks of these glaciers. The landscapes of such diverse places as the Alps, Cape Cod, and Yosemite Valley were fashioned by now-vanished masses of glacial ice. Moreover, Long Island, the Great Lakes, and the fiords of Norway and Alaska all owe their existence to glaciers. Glaciers, of course, are not just a phenomenon of the geologic past. As you will see, they are still sculpting the landscape and depositing debris in many regions today.

18.1 | Glaciers: A Part of Two Basic Cycles

Explain the role of glaciers in the hydrologic and rock cycles and describe the different types of glaciers, their characteristics, and their present-day distribution.

Glaciers are a part of two fundamental cycles in the Earth system: the hydrologic cycle and the rock cycle. The water of the hydrosphere is constantly cycled through the atmosphere, biosphere, and geosphere. Time and time again, water evaporates from the oceans into the atmosphere, precipitates on the land, and flows in rivers and underground back to the sea. However, when precipitation falls at high elevations or high latitudes, the water may not immediately make its way toward the sea. Instead, it may become part of a glacier. Although the ice will eventually melt, allowing the water to continue its path to the sea, water can be stored as glacial ice for many tens, hundreds, or even thousands of years.

A **glacier** is a thick ice mass that forms over hundreds or thousands of years. It originates on land from the accumulation, compaction, and recrystallization of snow. A glacier appears to be motionless, but it is not; glaciers move very slowly. Like running water, groundwater, wind, and waves, glaciers are dynamic erosional agents that accumulate, transport, and deposit sediment. As such, glaciers are among the agents that perform a basic function in the rock cycle. Although glaciers are found in many parts of the world today, most are located in remote areas, either near Earth's poles or in high mountains.

Valley (Alpine) Glaciers

Literally thousands of relatively small glaciers exist in lofty mountain areas, where they usually follow valleys that were originally occupied by streams. Unlike the rivers that previously flowed in these valleys, glaciers advance slowly, perhaps only a few centimeters per day. Because of their setting, these moving ice masses are termed **valley glaciers**, or **alpine glaciers** (**Figure 18.1**). Each glacier actually is a stream of ice, bounded by precipitous rock walls, that flows downvalley from an accumulation center near its head. Like rivers,

Figure 18.1
Valley glacier This tongue of ice, also called an *alpine glacier*, is still eroding the Alaskan landscape. Dark stripes of sediment within these glaciers are called medial moraines. This is Johns Hopkins Glacier in Alaska's Glacier Bay National Park. (Photo by Michael Collier)

valley glaciers can be long or short, wide or narrow, single or with branching tributaries. Generally, alpine glaciers are longer than they are wide. Some extend for just a fraction of a kilometer, whereas others go on for many tens of kilometers. The west branch of the Hubbard Glacier, for example, runs through 112 kilometers (nearly 70 miles) of mountainous terrain in Alaska and the Yukon Territory.

Ice Sheets

In contrast to valley glaciers, **ice sheets** exist on a much larger scale. The low total annual solar radiation reaching the poles makes these regions hospitable to great ice accumulations. Presently both of Earth's polar regions support ice sheets: Greenland in the Northern Hemisphere and Antarctica in the Southern Hemisphere (**Figure 18.2**).

Ice Age Ice Sheets About 18,000 years ago, glacial ice covered not only Greenland and Antarctica but also large portions of North America, Europe, and Siberia. That period in Earth history is appropriately known as the *Last Glacial Maximum*. The term implies that there were other glacial maximums, which is indeed the case. Throughout the Quaternary period, which began about 2.6 million years ago and extends to the present, ice sheets have formed, advanced over broad areas, and then

SmartFigure 18.2

Ice sheets The only present-day ice sheets are those covering Greenland and Antarctica. Their combined areas represent almost 10 percent of Earth's land area.

(https://goo.gl/JdyThn)

▶ **Video**

Greenland's ice sheet occupies 1.7 million square kilometers (663,000 square miles), about 80 percent of the island.

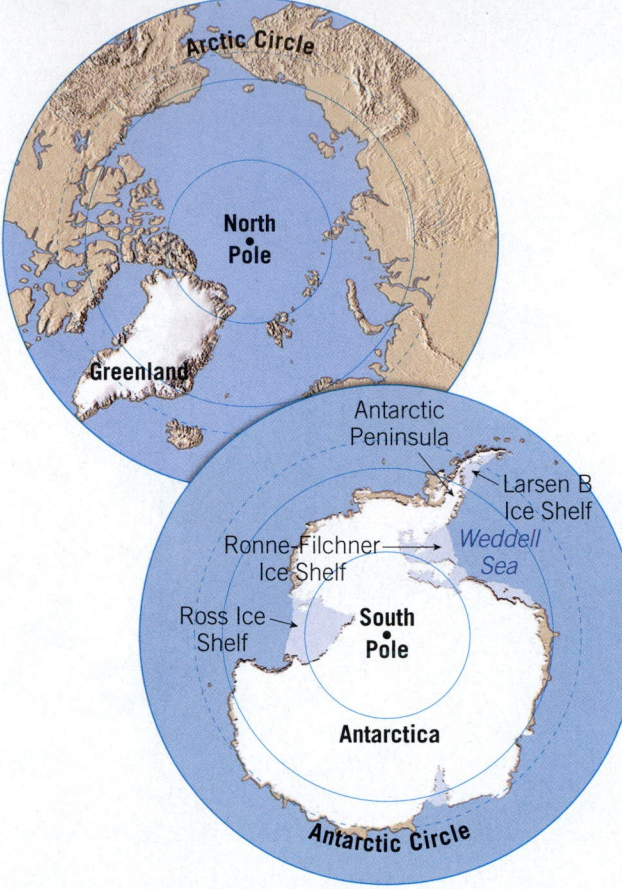

The area of the Antarctic Ice Sheet is almost 14 million square kilometers (5,460,000 square miles). Ice shelves occupy an additional 1.4 million square kilometers (546,000 square miles).

melted away. These alternating glacial and interglacial periods have occurred over and over again.

Greenland and Antarctica

Some people mistakenly think that the North Pole is covered by glacial ice,

but this is not the case. The ice that covers the Arctic Ocean is **sea ice**—frozen seawater. Sea ice floats because ice is less dense than liquid water. Although sea ice never completely disappears from the Arctic, the area covered with sea ice expands and contracts with the seasons. The thickness of sea ice ranges from a few centimeters for new ice to 4 meters (13 feet) for sea ice that has survived for years. By contrast, glaciers can be hundreds or thousands of meters thick.

Glaciers form on land, and in the Northern Hemisphere, Greenland supports an ice sheet. Greenland extends between about 60 and 80 degrees north latitude. This largest island on Earth is covered by an imposing ice sheet that occupies 1.7 million square kilometers (more than 660,000 square miles), or about 80 percent of the island. Averaging nearly 1500 meters (5000 feet) thick, the ice extends 3000 meters (10,000 feet) above the island's bedrock floor in some places.

In the Southern Hemisphere, practically all of Antarctica is covered by two huge ice sheets that extend over an area of more than 13.9 million square kilometers (5.4 million square miles). Because of the proportions of these huge features, they are often called *continental ice sheets*. There is more about Antarctica and its ice sheets in GEOgraphics 18.1. The combined areas of present-day continental ice sheets represent almost 10 percent of Earth's land area.

These enormous masses flow out in all directions from one or more snow-accumulation centers and completely obscure all but the highest areas of underlying terrain. Even sharp variations in the topography beneath a glacier usually appear as relatively subdued undulations on the surface of the ice. Such topographic differences, however, affect the behavior of the ice sheets, especially near their margins, by guiding flow in certain directions and creating zones of faster and slower movement.

Figure 18.3

Ice shelves An ice shelf forms when a glacier or an ice sheet flows into the adjacent ocean.

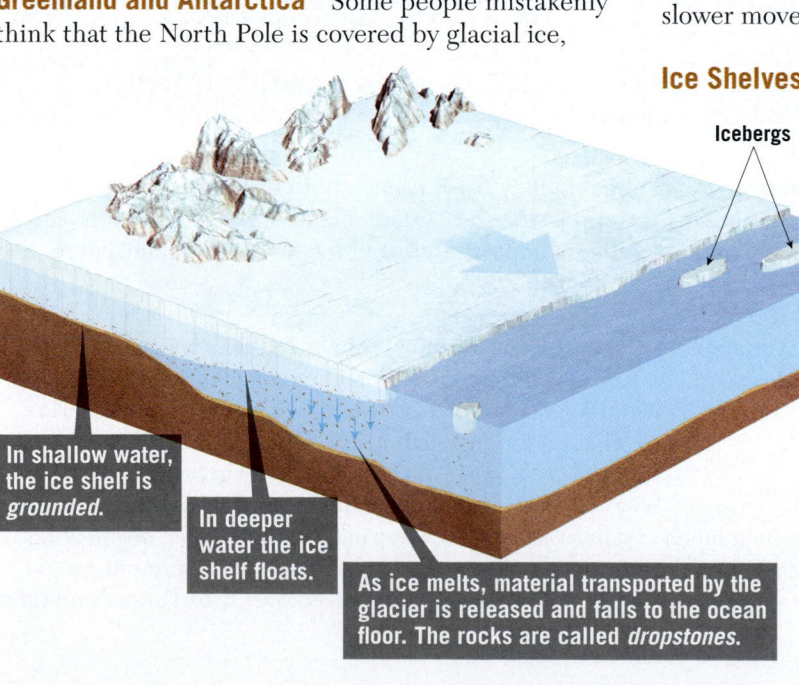

In shallow water, the ice shelf is *grounded*.

In deeper water the ice shelf floats.

As ice melts, material transported by the glacier is released and falls to the ocean floor. The rocks are called *dropstones*.

Icebergs

Ice Shelves

Along portions of the Antarctic coast, glacial ice flows into the adjacent ocean, creating features called **ice shelves**. These large, relatively flat masses of glacial ice extend seaward from the coast but remain attached to the land along one or more sides. About 80 percent of the ice lies below the surface of the ocean, so in shallow water, the ice shelf "touches bottom" and is said to be *grounded*. In deeper water, the ice shelf floats (**Figure 18.3**). There are ice shelves along more than half of the Antarctic coast, but there are relatively few in Greenland.

These shelves are thickest on their landward sides and become thinner seaward. They are sustained by ice from the

Ice shelf before breakup

January 31, 2002

50km

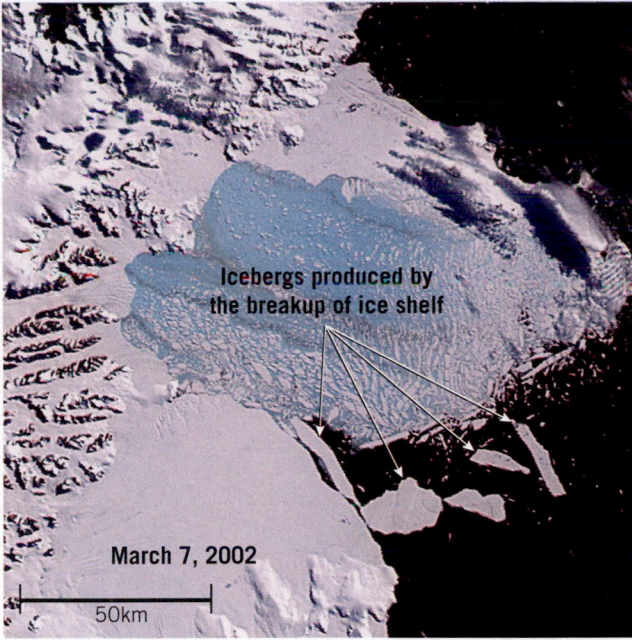

Icebergs produced by the breakup of ice shelf

March 7, 2002

50km

Figure 18.4
Collapse of an ice shelf These satellite images document the breakup of the Larsen B Ice Shelf adjacent to the Antarctic Peninsula in 2002. Thousands of icebergs were created in the process. (NASA)

adjacent ice sheet, and they are also nourished by snowfall on their surfaces and the freezing of seawater to their bases. Antarctica's ice shelves extend over approximately 1.4 million square kilometers (0.6 million square miles). The Ross and Ronne-Filchner Ice Shelves are the largest, with the Ross Ice Shelf alone covering an area approximately the size of Texas (see Figure 18.2). In recent years, satellite monitoring has shown that some ice shelves are unstable and breaking apart. For example, during a 35-day span in February and March 2002, an ice shelf on the eastern side of the Antarctic Peninsula, known as the Larsen B Ice Shelf, broke apart and separated from the continent. Thousands of icebergs were set adrift in the adjacent Weddell Sea. The event was captured in satellite imagery (Figure 18.4). This was not an isolated happening but part of a trend related to accelerated climate change. In fact, a 2015 NASA study predicts that the last remnant of the once-vast Larsen B Ice Shelf will break apart by 2020.

Other Types of Glaciers

In addition to valley glaciers and ice sheets, scientists have identified other types of glaciers. Covering some uplands and plateaus are masses of glacial ice called **ice caps**. Like ice sheets, ice caps completely bury the underlying landscape, but they are much smaller than the continental-scale features. Ice caps occur in many places, including Iceland and several of the large islands in the Arctic Ocean (Figure 18.5).

Often ice caps and ice sheets feed **outlet glaciers**. These tongues of ice flow down valleys, extending outward from the margins of these larger ice masses. The tongues are essentially valley glaciers that are avenues for ice movement from an ice cap or ice sheet through mountainous terrain to the sea. Where they encounter the ocean, some outlet glaciers spread out as floating ice shelves. Often large numbers of icebergs are produced.

ICELAND

Reykjavik

Vatnajökull ice cap

Ice caps completely bury the underlying terrain but are much smaller than ice sheets.

SmartFigure 18.5
Iceland's Vatnajökull ice cap In 1996 the Grímsvötn Volcano erupted beneath this ice cap, an event that triggered melting and floods. (NASA) (http://goo.gl/RsbHWM).

Mobile Field Trip

Antarctica Fact File

Earth's southernmost continent surrounds the South Pole (90° S. Latitude) and is almost entirely south of the Antarctic Circle (66.5° S. Latitude). This icy landmass is the fifth largest continent and is twice as large as Australia. It also has the distinction of being the coldest, driest, and windiest continent and also has the highest average elevation.

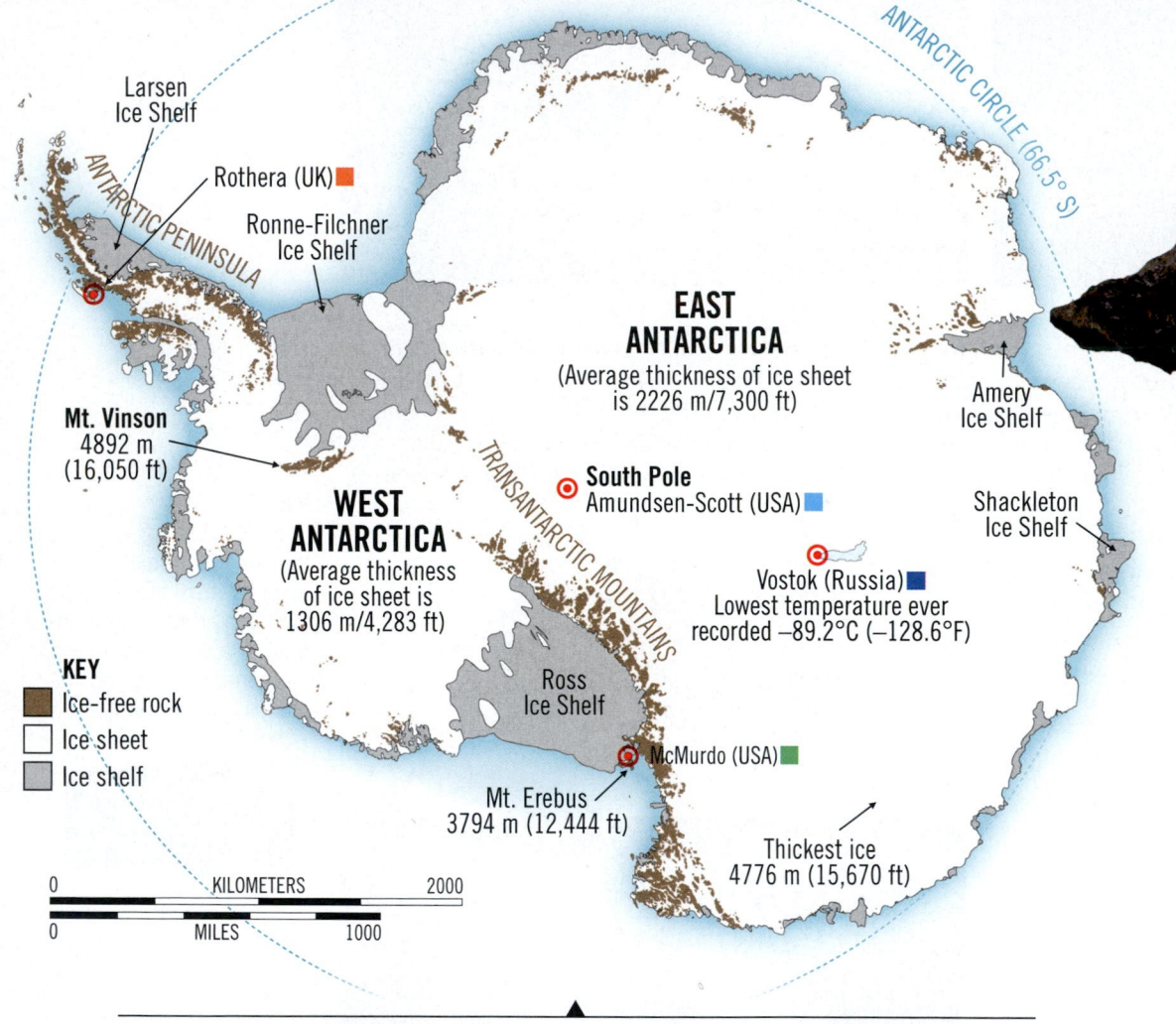

ANTARCTIC CIRCLE (66.5° S)

Larsen Ice Shelf

Rothera (UK) ■

Ronne-Filchner Ice Shelf

ANTARCTIC PENINSULA

EAST ANTARCTICA
(Average thickness of ice sheet is 2226 m/7,300 ft)

Amery Ice Shelf

Mt. Vinson 4892 m (16,050 ft)

◎ **South Pole**
Amundsen-Scott (USA) ■

Shackleton Ice Shelf

WEST ANTARCTICA
(Average thickness of ice sheet is 1306 m/4,283 ft)

TRANSANTARCTIC MOUNTAINS

◎ Vostok (Russia) ■
Lowest temperature ever recorded –89.2°C (–128.6°F)

Ross Ice Shelf

KEY
- ■ Ice-free rock
- □ Ice sheet
- ▨ Ice shelf

◎ McMurdo (USA) ■

Mt. Erebus 3794 m (12,444 ft)

Thickest ice 4776 m (15,670 ft)

| 0 | KILOMETERS | 2000 |
| 0 | MILES | 1000 |

ANSMET

1 CM

Meteorites are rocky or metallic particles from space that have fallen on Earth. These ancient fragments provide clues to the origin and history of our solar system. Antarctica is an especially good place to collect these dark masses from space because even small ones are relatively easy to spot. In addition, ice flow patterns tend to concentrate them in certain areas.

Linda Martel/ANSMET

Antarctica is 1.4 times larger than the United States and about 58 times bigger than the United Kingdom. The continent is almost completely ice covered. The ice-free area amounts to only 44,890 square kilometers (17,330 square miles) or just 0.32 percent (32/100 of 1 percent) of the continent.

Practically all of the continent belongs to the same climate classification, aptly termed ice cap climate, in which the average temperature of the warmest month is 0° C (32° F) or below. Take a look at the graph and you will see that some areas are much colder than others.

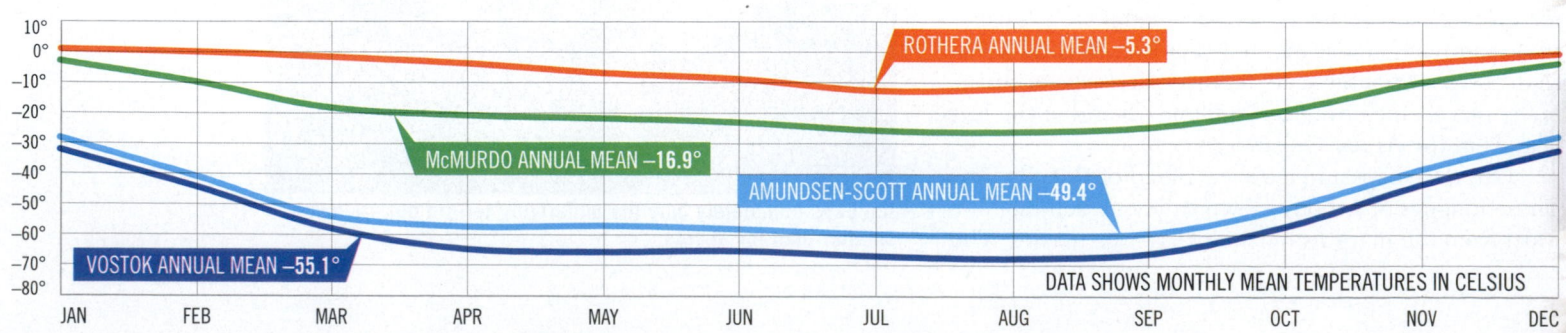

ROTHERA ANNUAL MEAN –5.3°

McMURDO ANNUAL MEAN –16.9°

AMUNDSEN-SCOTT ANNUAL MEAN –49.4°

VOSTOK ANNUAL MEAN –55.1°

DATA SHOWS MONTHLY MEAN TEMPERATURES IN CELSIUS

JAN FEB MAR APR MAY JUN JUL AUG SEP OCT NOV DEC

John Googe/NSF

Rick Price/Getty Images

What if the ice melted? The discharge at the mouth of the Mississippi River is 17,300 cubic meters (593,000 cubic feet) per second. If Antarctica's ice sheets melted at a suitable rate, they could maintain the flow of the Mississippi River for more than 50,000 years! If all of the continent's ice were to melt, sea level would rise by an estimated 56 meters (more than 180 feet). Antarctica's ice represents about 65 percent of Earth's entire supply of freshwater.

The Transantarctic Mountains are a 3300-kilometer- (2600-mile-) long range that separates the West Antarctic Ice Sheet and the East Antarctic Ice Sheet. Vinson Massif is the highest peak at 4892 meters (16,050 feet). Most of the mountains are buried beneath the continent's huge ice sheets.

About half of the continent's coastal areas are characterized by ice shelves. The Ross Ice Shelf is about the size of France, whereas the Ronne-Filchner Ice Shelf has an area similar to that of Spain.

Questions:
1. Is the glacial ice thickest in East Antarctica or West Antarctica?
2. If all of Antarctica's ice were to melt, about how much would sea level rise?

McMurdo Station is the main U.S. scientific research station. It is the largest installation on the continent, capable of supporting more than 1200 residents. The total population at all research stations is about 4000 in summer and 1000 in winter. There are no permanent (indigenous) human residents on the continent.

ZUMA Wire Service/Alamy

MCMURDO STATION ANTARCTICA

Dan Leeth/Alamy

Figure 18.6
Piedmont glacier Piedmont glaciers occur where valley glaciers exit a mountain range onto broad lowlands.

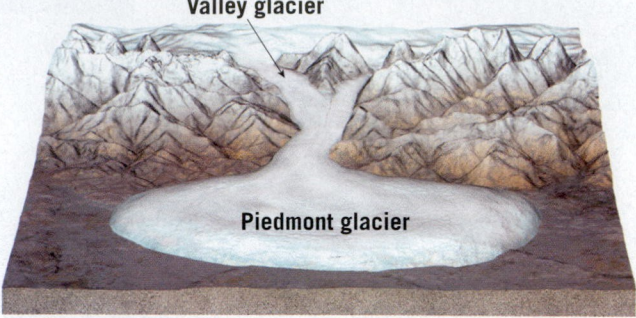

Valley glacier

Piedmont glacier

When a valley glacier is no longer confined, it spreads out to become a piedmont glacier.

Piedmont glaciers occupy broad lowlands at the bases of steep mountains and form when one or more alpine glaciers emerge from the confining walls of mountain valleys. Here the advancing ice spreads out to form a broad lobe (**Figure 18.6**). The sizes of

individual piedmont glaciers vary greatly. Among the largest is the broad Malaspina Glacier along the coast of southern Alaska. It covers thousands of square kilometers of the flat coastal plain at the foot of the lofty St. Elias Range.

18.1 Concept Checks

1. Where are glaciers found today? What percentage of Earth's land surface do they cover?

2. Describe how glaciers fit into the hydrologic cycle. What role do they play in the rock cycle?

3. List and briefly distinguish among four types of glaciers.

4. What is the difference between an ice sheet and an ice shelf? How are they related?

18.2 | Formation and Movement of Glacial Ice

Describe how glaciers move, the rates at which they move, and the significance of the glacial budget.

Snow is the raw material from which glacial ice originates; therefore, glaciers form in areas where more snow falls in winter than melts during the summer. Glaciers develop in the high-latitude polar realm because, even though annual snowfall totals are modest, temperatures are so low that little of the snow melts. Glaciers can form in mountains because temperatures drop with an increase in altitude. So even near the equator, glaciers may form at elevations above about 5000 meters (16,400 feet). For example, Tanzania's Mount Kilimanjaro, which is located practically astride the equator at an altitude of 5895 meters (19,336 feet), has glaciers at its summit. The elevation above which snow remains throughout the year varies with latitude. Near the equator, this boundary occurs high in the mountains, whereas in the vicinity of the 60th parallel, it is at or near sea level. Before a glacier is created, however, snow must be converted into glacial ice.

Glacial Ice Formation

When temperatures remain below freezing following a snowfall, the fluffy accumulation of delicate hexagonal crystals soon changes. As air infiltrates the spaces between the crystals, the extremities of the crystals evaporate, and the water vapor condenses near the centers of the crystals. In this manner, snowflakes become smaller, thicker, and more spherical, and the large pore spaces disappear. As air is forced out, what was once light, fluffy snow is recrystallized into a much denser mass of small grains having the consistency of coarse sand. This granular recrystallized snow is called **firn** and is commonly found making up old snow banks near the end of winter. As more snow is added, the pressure on the lower layers gradually increases, compacting the ice grains at depth. Once the thickness of ice and snow exceeds 50 meters (165 feet), the weight is sufficient to

fuse firn into a solid mass of interlocking ice crystals. Glacial ice has now been formed.

The rate at which this transformation occurs varies. In regions where the annual snow accumulation is great, burial is relatively rapid, and snow may turn to glacial ice in a matter of a decade or less. Where the yearly addition of snow is less abundant, burial is slow, and the transformation of snow to glacial ice may take hundreds of years.

How Glaciers Move

The movement of glacial ice is generally referred to as *flow*. The fact that glacial movement is described in this way seems paradoxical; how can a solid flow? The way in which ice moves is complex and is of two basic types. The first of these, **plastic flow**, involves movement *within* the ice. Ice behaves as a brittle solid until the pressure upon it is equivalent to the weight of about 50 meters (165 feet) of ice. Once that load is surpassed, ice behaves as a plastic material, and flow begins. Such flow occurs because of the molecular structure of ice. Glacial ice consists of layers of molecules stacked one upon the other. The bonds between layers are weaker than those within each layer. Therefore, when a stress exceeds the strength of the bonds between the layers, the layers remain intact and slide over one another.

A second and often equally important mechanism of glacial movement consists of an entire ice mass slipping along the ground. With the exception of some glaciers located in polar regions where the ice is probably frozen to the solid bedrock floor, most glaciers are thought to move by this sliding process, called **basal slip**. In this process, meltwater probably acts as a hydraulic jack and perhaps as a lubricant that helps the ice move over the rock. The source of the liquid water is related in part to the fact that the melting point of ice decreases as pressure increases. Therefore, deep within a glacier, the ice may be at the melting point even though its temperature is below 0°C (32°F).

Other factors may also contribute to the presence of meltwater deep within a glacier. Temperatures may be increased by plastic flow (an effect similar to heating due to friction), by heat added from Earth below, and by the refreezing of meltwater that has seeped down from above. This last process relies on the fact that as water changes state from liquid to solid, heat (termed *latent heat of fusion*) is released.

Figure 18.7 illustrates the effects of these two basic types of glacial motion. This vertical profile through a glacier also shows that not all the ice flows forward at the same rate. Frictional drag with the bedrock floor causes the lower portions of the glacier to move more slowly.

Ice in the zone of fracture is carried along "piggyback" style.

Below a depth of about 50 meters (160 feet), ice behaves plastically (deforms without breaking) and gradually flows.

Basal slip occurs episodically. Ice in contact with the valley floor remains fixed as stress builds to the point that the glacier lurches forward.

Total movement

Internal flow

Zone of fracture

Sliding

Bedrock

SmartFigure 18.7 Movement of a glacier This vertical cross section through a glacier shows that movement is divided into two components. Also notice that the rate of movement is slowest at the base of the glacier, where frictional drag is greatest. (https://goo.gl/adFVVp)

Tutorial

EYE ON EARTH 18.1

The central focus of this satellite image is Byrd Glacier in Antarctica. In this region, the glacier advances at a rate of about 0.8 kilometer (0.5 mile) per year. Take note of its position in relation to other labeled features. (Photo by NASA)

QUESTION 1 Is Byrd Glacier flowing toward the top or the bottom of the image? How did you figure this out?

QUESTION 2 What term describes this type of glacier?

Ice sheet

Transantarctic Mountains

Byrd Glacier

Transantarctic Mountains

Ross Ice Shelf

Figure 18.8
Crevasses As a glacier moves, internal stresses cause large cracks to develop in the brittle upper portion of the glacier, called the zone of fracture. Crevasses can extend to depths of 50 meters (165 feet) and can make travel across glaciers dangerous. (Photo by Wave/Glow Images)

in cracks called **crevasses** (**Figure 18.8**). These gaping cracks may extend to depths of 50 meters (160 feet) and can make travel across glaciers dangerous. Below this depth, plastic flow seals them off.

Observing and Measuring Movement

Unlike the movement of water in streams, the movement of glacial ice is not obvious. If we could watch a valley glacier move, we would see that, as with the water in a river, the ice moves downstream at different rates. Flow is greatest in the center of the glacier because the drag created by the walls and floor of the valley slow the base and sides.

Early in the nineteenth century, the first experiments on glacier movement were designed and carried out in the Alps. Markers were placed in a straight line across an alpine glacier, and the line's position was marked on the valley walls so that if the ice moved, the change in position could be detected. Periodically the positions of the markers were recorded, showing the movement just described. Although most glaciers move too slowly for direct visual detection, the experiments successfully demonstrated that movement nevertheless occurs. The experiment, illustrated in **Figure 18.9**, was carried out at Switzerland's Rhône Glacier later in the nineteenth century. It not only traced the movement of markers within the ice but also mapped the position of the glacier's terminus.

For many years, time-lapse photography has allowed us to observe glacial movement. Images are taken from the same vantage point on a regular basis (for example, once per day) over an extended span and then played back like a movie. More recently, satellites let us track the movement of glaciers and observe glacial behavior. This is especially useful because the remoteness and extreme weather associated with many glacial areas limit on-site study.

How rapidly does glacial ice move? Average rates vary considerably from one glacier to another. Some glaciers move so slowly that trees and other vegetation may become well established in the debris that accumulates on the glacier's surface. Others advance up to several meters each day. Recent satellite imaging provided insights into movements within the Antarctic Ice Sheet. An examination of **Figure 18.10** shows that portions of some outlet glaciers move at rates greater than 800 meters (2600 feet) per

In contrast to the lower portion of the glacier, the upper 50 meters (160 feet) or so are not under sufficient pressure to exhibit plastic flow. Rather, the ice in this uppermost zone is brittle, and this zone is appropriately referred to as the **zone of fracture**. The ice in the zone of fracture is carried along "piggyback" style by the ice below. When the glacier moves over irregular terrain, the zone of fracture is subjected to tension, resulting

SmartFigure 18.9
Measuring the movement of a glacier Ice movement and changes in the terminus of Rhône Glacier, Switzerland. In this classic study of a valley glacier, the movement of stakes clearly shows that glacial ice moves and that movement along the sides of the glacier is slower than movement in the center. Also notice that even though the ice front was retreating, the ice within the glacier was advancing. (https://goo.gl/JoKPM3)

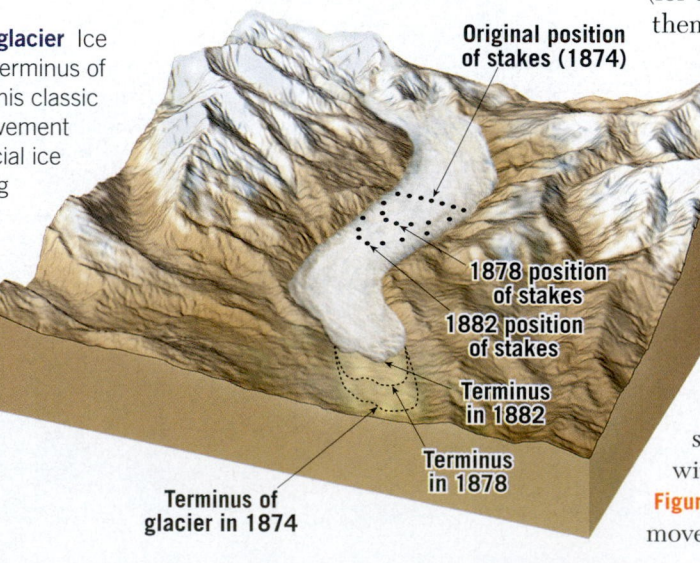

Original position of stakes (1874)

1878 position of stakes

1882 position of stakes

Terminus in 1882

Terminus in 1878

Terminus of glacier in 1874

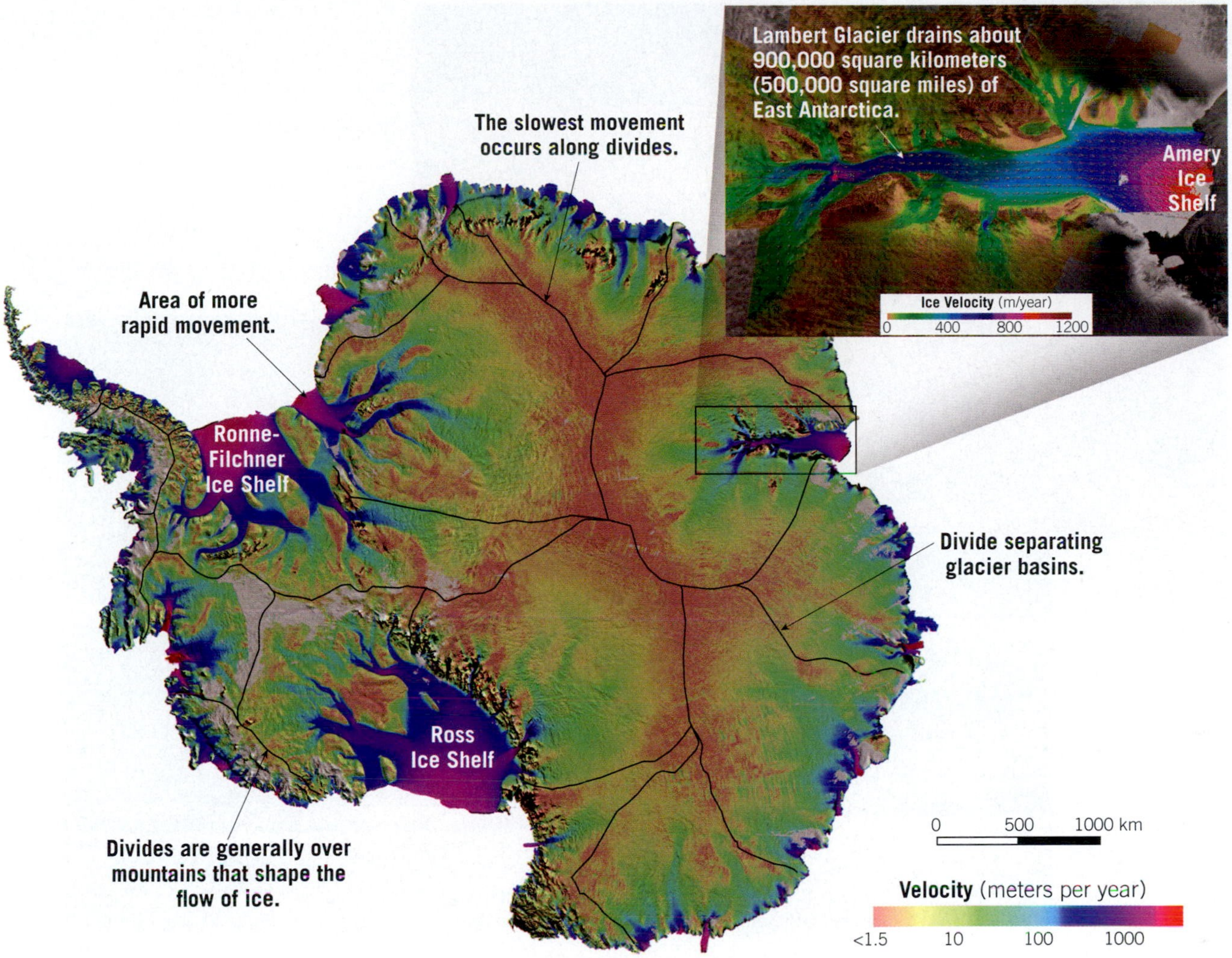

Lambert Glacier drains about 900,000 square kilometers (500,000 square miles) of East Antarctica.

Amery Ice Shelf

Ice Velocity (m/year)
0   400   800   1200

The slowest movement occurs along divides.

Area of more rapid movement.

Ronne-Filchner Ice Shelf

Divide separating glacier basins.

Ross Ice Shelf

Divides are generally over mountains that shape the flow of ice.

0   500   1000 km

Velocity (meters per year)
<1.5   10   100   1000

**Figure 18.10
Movement of Antarctic ice** These maps result from thousands of satellite measurements taken over several years. Outflow from the interior is organized into a series of drainage basins separated by ice divides. Flow is concentrated into narrow, mountain-bound glaciers or into relatively fast-moving ice streams surrounded by slower-moving ice. (NASA)

year; on the other hand, ice in some interior regions creeps along at less than 2 meters (6.5 feet) per year. Movement of some glaciers is characterized by occasional periods of extremely rapid advance called *surges*, followed by periods of much slower movement.

Budget of a Glacier: Accumulation Versus Wastage

Snow is the raw material from which glacial ice originates; therefore, glaciers form in areas where more snow falls in winter than melts during the summer. Glaciers are constantly gaining and losing ice.

Glacial Zones   Snow accumulation and ice formation occur in the **zone of accumulation** (**Figure 18.11**). Its outer limits are defined by the **snowline**, or **equilibrium line**—the elevation at which

the accumulation and wasting of glacial ice are equal. As noted earlier, the elevation of this boundary varies greatly, from sea level in polar regions to altitudes approaching 5000 meters (16,000 feet) near the equator. Above the snowline, in the zone of accumulation, the addition of snow thickens the glacier and promotes movement. Below the snowline is the **zone of wastage**. Here

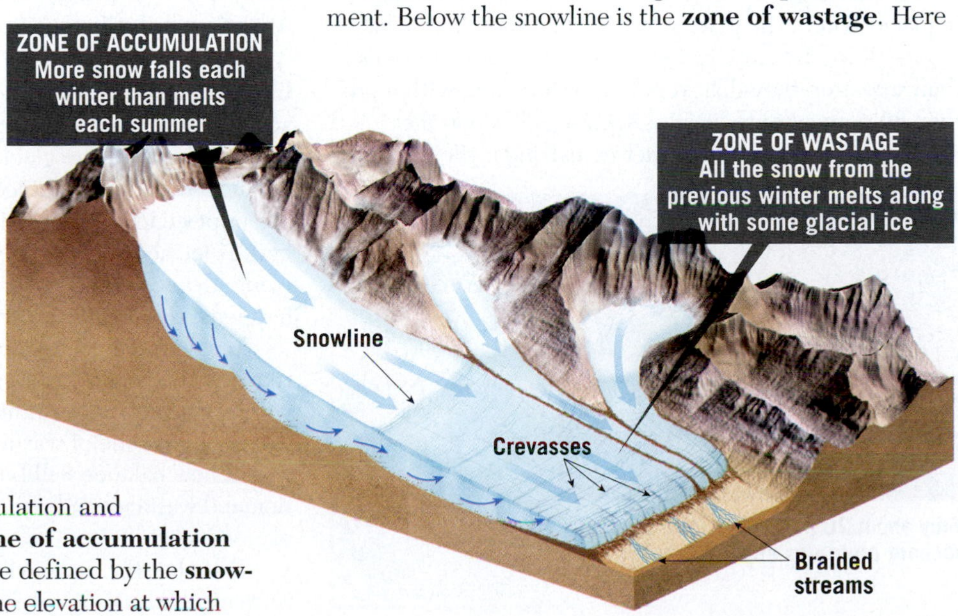

ZONE OF ACCUMULATION More snow falls each winter than melts each summer

ZONE OF WASTAGE All the snow from the previous winter melts along with some glacial ice

Snowline

Crevasses

Braided streams

**SmartFigure 18.11
Zones of a glacier** The snowline separates the zone of accumulation and the zone of wastage. Whether the ice front advances, retreats, or remains stationary depends on the balance or lack of balance between accumulation and wastage (ablation). (https://goo.gl/25XUcw)

Tutorial

Figure 18.12
Examples of ablation
A. Melting at Alaska's Root Glacier created this river atop the glacier. Notice the large rocks exposed as the ice wastes away. (Photo by Michael Collier) **B.** Ice loss by calving at Beloit Glacier. When valley glaciers or outlet glaciers terminate in the ocean, they may also be called *tidewater glaciers*. It is common for large blocks to break off the front of the glacier and form icebergs. (Photo by Michael Collier)

A.

B.

there is a net loss to the glacier as all of the snow from the previous winter melts, as does some of the glacial ice (see Figure 18.11).

The loss of ice by a glacier is termed **ablation**. In addition to melting, glaciers waste away as large pieces of ice break off the front of the glacier in a process called **calving** (**Figure 18.12**). Calving creates **icebergs** in places where the glacier has reached the sea or a lake (**Figure 18.13**). Because icebergs are just slightly less dense than seawater, they float very low in the water, with more than 80 percent of their mass submerged. Along the margins of Antarctica's ice shelves, calving is the primary

means by which these masses lose ice. The relatively flat icebergs produced here can be several kilometers across and up to about 600 meters (2000 feet) thick (see the icebergs in Figure 18.3 and 18.4). By comparison, thousands of irregularly shaped icebergs are produced by outlet glaciers flowing from the margins of the Greenland ice sheet. Many drift southward and find their way into the North Atlantic, where they can be hazardous to navigation.

Glacial Budget Whether the margin of a glacier is advancing, retreating, or remaining stationary depends on the budget of the glacier. The **glacial budget** is the balance, or lack of balance, between accumulation at the upper end of the glacier and ablation at the lower end. If ice accumulation exceeds ablation, the glacial front advances until the two factors balance. When this happens, the terminus of the glacier is stationary.

If a warming trend increases ablation and/or if reduced snowfall decreases accumulation, the ice front will retreat. As the terminus of the glacier retreats, the extent of the zone of wastage diminishes. Therefore, in time a new balance will be reached between accumulation and wastage, and the ice front will again become stationary.

Whether the margin of a glacier is advancing, retreating, or stationary, the ice within the glacier

Figure 18.13
Icebergs Icebergs form when large masses of ice break off from the front of a glacier after it reaches a water body, in a process known as calving. Other examples of icebergs appear in Figures 18.3, 18.4, and 18.12. (Photo by Radius Images/Photolibrary)

Only about 20 percent or less of an iceberg protrudes above the waterline.

Geologist's Sketch

About 200 years ago Bear Glacier extended to the end moraine at the bottom of the image.

Bear Glacier

End moraine marking the farthest advance of a glacier

A.

1935

2013

B.

Figure 18.14
Retreating glaciers A. Bear Glacier is a tidewater glacier that flows out of the Harding Icefield near Seward, Alaska. Like most other glaciers in Alaska, Bear Glacier is retreating back into the mountains. (Photo by Michael Collier) **B.** These two images were taken 78 years apart, from about the same vantage point, along the southwest coast of Greenland. Between 1935 and 2013, the outlet glacier that is the primary focus of these photos retreated about 3 kilometers (about 2 miles). (National Snow and Ice Data Center)

continues to flow forward. In the case of a receding glacier, the ice still flows forward but not rapidly enough to offset ablation. This point is illustrated well in Figure 18.9. As the line of stakes within the Rhône Glacier continued to move downvalley, the terminus of the glacier slowly retreated upvalley.

Glaciers in Retreat: Unbalanced Glacial Budgets

Because glaciers are sensitive to changes in temperature and precipitation, they provide clues about changes in climate. With few exceptions, valley glaciers around the world have been retreating at unprecedented rates over the past century. Bear Glacier is one example (**Figure 18.14A**). The photos in **Figure 18.14B** illustrate another example. Many valley glaciers have disappeared altogether. For example, 150 years ago, there were 147

glaciers in Montana's Glacier National Park. Today only 37 remain.

18.2 Concept Checks

1. Describe two components of glacial movement.

2. How rapidly does glacial ice move? Provide some examples.

3. What are crevasses, and where do they form?

4. Relate the glacial budget to the two zones of a glacier.

5. Under what circumstances will the front of a glacier advance? Retreat? Remain stationary?

This photo shows an iceberg floating in the ocean near the coast of Greenland. (Photo by Andrzej Gibasiewicz/Shutterstock)

QUESTION 1 How do icebergs form? What term applies to this process?

QUESTION 2 Using the knowledge you have gained about these features, explain the common phrase "It's only the tip of the iceberg."

QUESTION 3 Is an iceberg the same as sea ice? Explain.

QUESTION 4 If this iceberg were to melt, how would sea level be affected?

18.3 | Glacial Erosion

Discuss the processes of glacial erosion. Identify and describe major topographic features created by glacial erosion.

Glaciers are capable of great erosion. For anyone who has observed the terminus of an alpine glacier, the evidence of its erosive force is clear (**Figure 18.15**). The release of rock material of various sizes from the ice as it melts leads to the conclusion that the ice has scraped, scoured, and torn rock from the floor and walls of the valley and carried it downvalley. It should be pointed out, however, that in mountainous regions, mass-wasting processes also make substantial contributions to the sediment load of a glacier.

Once rock debris is acquired by a glacier, the enormous competence of ice will not allow the debris to settle out like the load carried by a stream or by the wind. For example, notice in Figure 18.12A the masses of rock exposed within the glacier as the ice melts. Indeed, as a medium of sediment transport, ice has no equal. Consequently, glaciers can carry large blocks that no other erosional agent could possibly budge. Although today's glaciers are of limited importance as erosional agents, many landscapes that were modified by the widespread glaciers of the Ice Age still reflect, to a high degree, the work of ice.

How Glaciers Erode

Glaciers erode the land primarily in two ways: plucking and abrasion. First, as a glacier flows over a fractured bedrock surface, it loosens and lifts blocks of rock and incorporates them into the ice. This process, known as **plucking**, occurs when meltwater penetrates the cracks and joints of bedrock beneath a glacier and freezes. As the water expands, it exerts tremendous leverage that pries the rock loose. In this manner, sediment of all sizes becomes part of the glacier's load.

Figure 18.15
Evidence of glacial erosion As the terminus of a glacier wastes away, it deposits large quantities of sediment. This close-up view shows the rock debris dropped by the melting ice. The glacier clearly acquired a significant quantity of rock debris as it advanced across the landscape. (Photo by Michael Collier)

The rock debris dropped by melting ice is a jumbled mixture of different size sediments.

Glacial abrasion created the scratches and grooves in this bedrock.

A.

Glacially polished granite in California's Yosemite National Park.

B.

Figure 18.16
Glacial abrasion Moving glacial ice, armed with sediment, acts like sandpaper, scratching and polishing rock. (Photos by Michael Collier)

The second major erosional process is **abrasion** (Figure 18.16). As the ice and its load of rock fragments slide over bedrock, they function like sandpaper, smoothing and polishing the surface below. The pulverized rock produced by the glacial "grist mill" is appropriately called **rock flour**. So much rock flour may be produced that meltwater streams flowing out of a glacier often have the cloudy appearance of skim milk and offer visible evidence of the grinding power of ice. Lakes fed by such streams frequently have a distinctive turquoise color (Figure 18.17).

When the ice at the bottom of a glacier contains large rock fragments, long scratches and grooves called **glacial striations** may even be gouged into the bedrock (see Figure 18.16A). These linear grooves provide clues to the direction of ice flow. By mapping the striations over large areas, patterns of glacial flow can often be reconstructed. On the other hand, not all abrasive action produces striations. The rock surfaces over which the glacier moves may also become highly polished by the ice and its load of finer particles. The broad expanses of smoothly polished granite in Yosemite National Park provide an excellent example (see Figure 18.16B).

As is the case with other agents of erosion, the rate of glacial erosion is highly variable. This differential erosion by ice is largely controlled by four factors: (1) rate of glacial movement; (2) thickness of the ice; (3) shape, abundance, and hardness of the rock fragments contained in the ice at the base of the glacier; and (4) erodibility of the surface beneath the glacier. Variations in any or all of these factors from time to time and/or from place to place mean that the features, effects, and

Figure 18.17
Distinctive color caused by rock flour Many lakes that are fed by glaciers have a distinctive turquoise color. The distinctive color occurs because the rock flour suspended in the lake reflects different parts of the visible spectrum more strongly than others. (Photo by Grant/Shutterstock)

SmartFigure 18.18
Erosional landforms created by alpine glaciers The unglaciated landscape (**A**) is modified by valley glaciers (**B**). After the ice recedes (**C**), the terrain looks very different than it looked before glaciation. You can see several of the landforms depicted here by viewing SmartFigure Mobile Field Trip 4.14, *Yosemite: Granite and Glaciers* (page 118). (Arête photo by James E. Patterson; cirque photo by Marli Miller; hanging valley photo by John Warden/SuperStock) (https://goo.gl/XPgbvY)

Tutorial

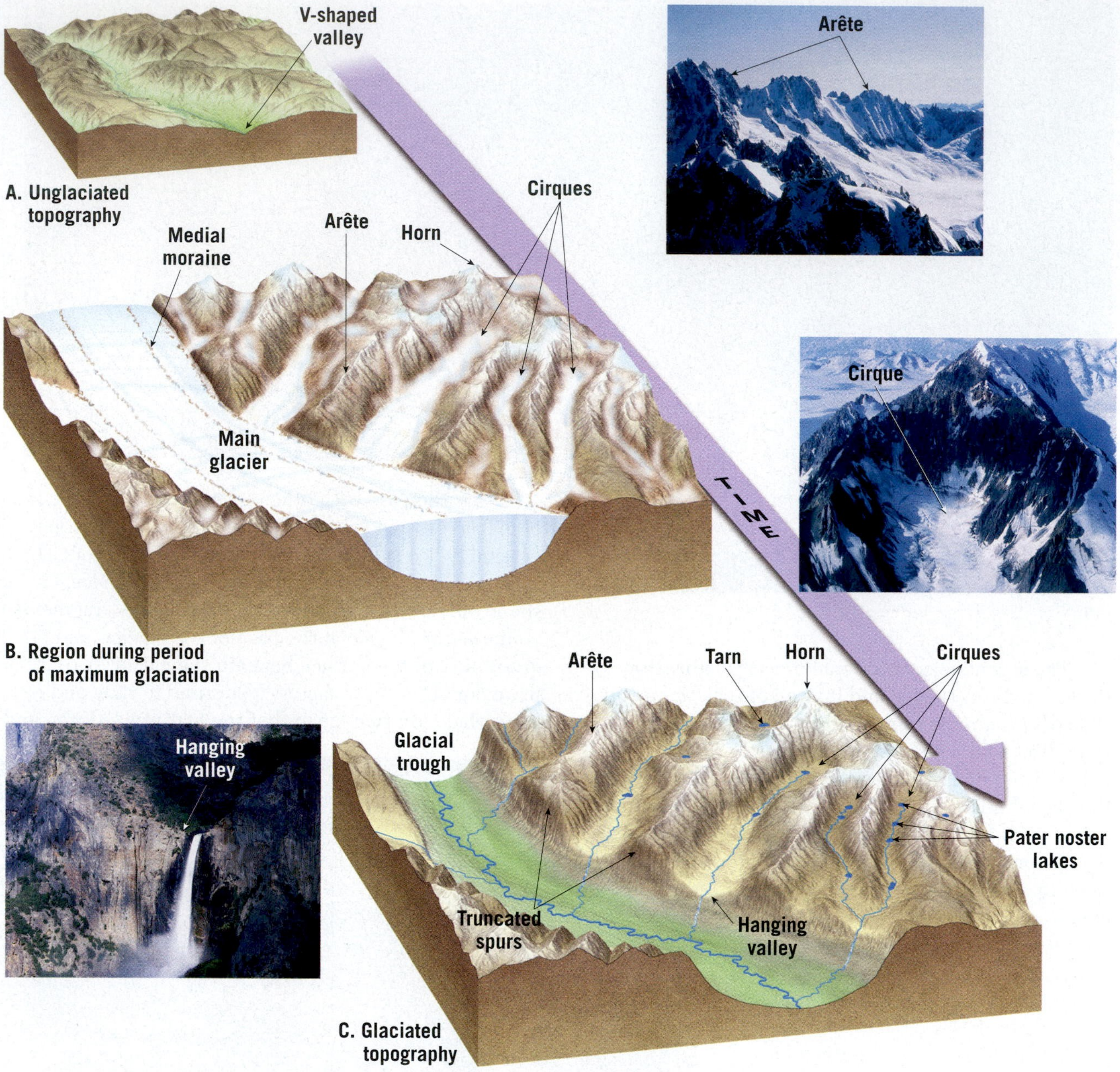

V-shaped valley

A. Unglaciated topography

Cirques

Medial moraine

Arête

Horn

Arête

Cirque

Main glacier

B. Region during period of maximum glaciation

TIME

Arête

Tarn

Horn

Cirques

Glacial trough

Hanging valley

Pater noster lakes

Truncated spurs

Hanging valley

C. Glaciated topography

degree of landscape modification in glaciated regions can vary greatly.

Landforms Created by Glacial Erosion

The erosional effects of valley glaciers and ice sheets are quite different. A visitor to a glaciated mountain region is likely to see a sharp and angular topography. This is because as the more confined alpine glaciers move downvalley, they tend to accentuate the irregularities of the mountain landscape by creating steeper canyon walls and making bold peaks even more jagged. By contrast, continental ice sheets generally override the terrain and hence subdue rather than accentuate the irregularities they encounter. Although the erosional

potential of ice sheets is enormous, landforms carved by these huge ice masses usually do not inspire the same wonderment and awe as do the erosional features created by valley glaciers. Much of the rugged mountain scenery so celebrated for its majestic beauty is produced by erosion by alpine glaciers. **Figure 18.18** shows a hypothetical mountain area before, during, and after glaciation. You will refer to this figure often in the following discussion.

Glaciated Valleys   A hike up a glaciated valley reveals a number of striking ice-created features. The valley itself is often a dramatic sight. Unlike streams, which create their own valleys, glaciers take the path of least resistance and follow existing stream valleys. Prior to

SmartFigure 18.19
A U-shaped glacial trough Prior to glaciation, a mountain valley is typically narrow and V-shaped. During glaciation, an alpine glacier widens, deepens, and straightens the valley, creating the classic U-shape shown here. (Photo by Michael Collier) (https://goo.gl/bE8lYT)

Animation

glaciation, mountain valleys are characteristically narrow and V-shaped because streams are well above base level and are therefore downcutting (see Chapter 16). However, during glaciation, these narrow valleys are transformed as the glacier widens and deepens them, creating a U-shaped **glacial trough** (see Figure 18.18 and **Figure 18.19**). In addition to producing a broader and deeper valley, the glacier also straightens the valley. As ice flows around sharp curves, its great erosional force removes the spurs of land that extend into the valley. The results of this activity are triangular-shaped cliffs called **truncated spurs**.

The amount of glacial erosion that takes place will vary in different valleys in a mountainous area. Prior to glaciation, the mouths of tributary streams join the main valley (or *trunk* valley) at the elevation of the stream in that valley. During glaciation, the amount of ice flowing through the main valley can be much greater than the amount advancing down each tributary. Consequently, the valley containing the main glacier (or *trunk* glacier) is eroded deeper than the smaller valleys that feed it. Thus, after the ice has receded, the valleys of tributary glaciers are left standing above the main glacial trough and are termed **hanging valleys** (see Figure 18.18C). Rivers flowing through hanging valleys can produce spectacular waterfalls, such as those in Yosemite National Park, California.

As hikers walk up a glacial trough, they may pass a series of bedrock depressions on the valley floor, probably created by plucking and then scouring by the abrasive force of the ice. If these depressions are filled with water, they are called **pater noster lakes** (see Figure 18.18C). The Latin name means "our Father" and is a reference to a string of rosary beads.

Cirques At the head of a glacial valley is a characteristic and often imposing feature associated with an alpine glacier, called a **cirque**. As the photo in Figure 18.18 illustrates, these bowl-shaped depressions have precipitous walls on three sides but are open on the downvalley side. The cirque is the focal point of the glacier's growth because it is the area of snow accumulation and ice formation. Cirques begin as irregularities in the mountainside that are subsequently enlarged by frost wedging and plucking along the sides and bottom of the glacier. The glacier in turn acts as a conveyor belt that carries away the debris. After the glacier has melted away, the cirque basin is sometimes occupied by a small lake called a **tarn** (see Figure 18.18C).

Sometimes, when two glaciers exist on opposite sides of a divide, each flowing away from the other, the dividing ridge between their cirques is largely eliminated as plucking and frost action enlarge each one. When this occurs, the two glacial troughs come to

Figure 18.20
The Matterhorn Horns are sharp, pyramid-like peaks that were shaped by alpine glaciers. The Matterhorn, in the Swiss Alps, is a famous example. (Photo by Andy Selinger/age fotostock)

Arêtes and Horns The Alps, Northern Rockies, and many other mountain landscapes sculpted by valley glaciers reveal more than glacial troughs and cirques. In addition, sinuous, sharp-edged ridges called **arêtes** (French for "knife-edge") and sharp, pyramid-like peaks termed **horns** project above the surroundings (see Figure 18.18C). Both features can originate from the same basic process: the enlargement of cirques produced by plucking and frost action. Several cirques around a single high mountain create the spires of rock called horns. As the cirques enlarge and converge, an isolated horn is produced. A famous example is the Matterhorn in the Swiss Alps (**Figure 18.20**).

Arêtes can form in a similar manner, except that the cirques are not clustered around a point but rather exist on opposite sides of a divide. As the cirques grow, the divide separating them is reduced to a very narrow, knifelike partition. An arête can also be created in another way. When two glaciers occupy parallel valleys, an arête can form when the land separating the moving tongues of ice is progressively narrowed as the glaciers scour and widen their valleys.

intersect, creating a gap or pass from one valley into the other. Such a feature is termed a **col**. Some important and well-known mountain passes that are cols include St. Gotthard Pass in the Swiss Alps, Tioga Pass in California's Sierra Nevada, and Berthoud Pass in the Colorado Rockies.

Roche Moutonnée In many glaciated landscapes, but most frequently where continental ice sheets have modified the terrain, the ice carves small streamlined hills from protruding bedrock knobs. Such an asymmetrical knob of bedrock is called a **roche moutonnée** (French for "sheep rock"). These features are formed when glacial abrasion smooths the gentle slope facing the oncoming ice sheet and plucking steepens the opposite side as the ice rides over the knob (**Figure 18.21**). Roches moutonnées indicate the direction of glacial flow because the gentler slope is generally on the side from which the ice advanced.

Figure 18.21
Roche moutonnée This classic example is in Yosemite National Park, California. The gentle slope was abraded, and the steep slope was plucked. The glacier moved from right to left. (Photo by E. J. Tarbuck)

Ice flow

Glacial plucking

Glacial abrasion

Bedrock

Geologist's Sketch

Fiords Sometimes spectacular steep-sided inlets of the sea called **fiords** are present at high latitudes where mountains are adjacent to the ocean (**Figure 18.22**). They are

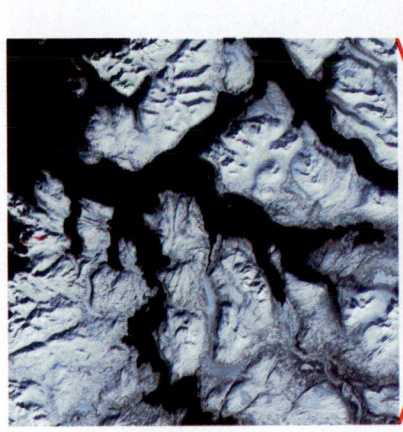
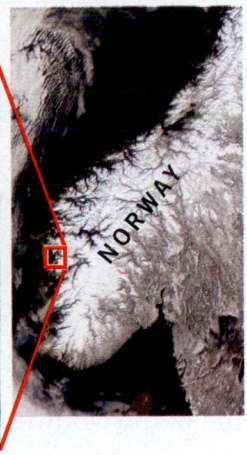

Figure 18.22
Fiords The coast of Norway is known for its many fiords. Frequently these ice-sculpted inlets of the sea are hundreds of meters deep. (Satellite images courtesy of NASA; photo by Inger Yoshio Tomii/SuperStock)

"drowned" glacial troughs that became submerged as the ice left the valleys and sea level rose following the Ice Age. The depths of fiords may exceed 1000 meters (3300 feet). However, the great depths of these flooded troughs are only partly explained by the post–Ice Age rise in sea level. Unlike the situation governing the downward erosional work of rivers, sea level does not act as base level for glaciers. As a consequence, glaciers are capable of eroding their beds far below the surface of the sea. For example, a 300-meter- (1000-foot-) thick glacier can carve its valley floor more than 250 meters (820 feet) below sea level before downward erosion ceases and the ice begins to float. Norway, British Columbia, Greenland, New Zealand, Chile, and Alaska all have coastlines characterized by fiords.

18.3 Concept Checks

1. How do glaciers acquire their load of sediment?
2. What are some visible effects of glacial erosion?
3. What factors influence a glacier's ability to erode?
4. How does a glaciated mountain valley differ in appearance from a mountain valley that was not glaciated? Describe the features created by valley glaciers.
5. Relate fiords to glacial troughs.

EYE ON EARTH 18.3

This mountain landscape is in Montana's Glacier National Park. Lake St. Mary occupies a glacial trough. (Photo by Dan Sherwood/age fotostock)

QUESTION 1 How does the current shape of the valley compare to its shape prior to glaciation? Explain.

QUESTION 2 What term is applied to the narrow ridge indicated by the arrow?

18.4 | Glacial Deposits

Distinguish between the two basic types of glacial drift. List and describe the major depositional features associated with glacial landscapes.

A glacier picks up and transports a huge load of rock debris as it slowly advances across the land. Ultimately when the ice melts, these materials are deposited. In regions where glacial sediment is deposited, it can play a significant role in forming the physical landscape. For example, in many areas once covered by the continental ice sheets of the recent Ice Age, the bedrock is rarely exposed because glacial deposits that are tens or even hundreds of meters thick completely mantle the terrain. The general effect of these deposits is to reduce the local relief and thus level the topography. Indeed, rural country scenes that are familiar to many of us—rocky pastures in New England, wheat fields in the Dakotas, rolling farmland in the Midwest—result directly from glacial deposition.

Glacial Drift

Long before the theory of an extensive Ice Age was ever proposed, much of the soil and rock debris covering portions of Europe was recognized as having come from somewhere else. At the time, these "foreign" materials were believed to have been "drifted" into their present positions by floating ice during an ancient flood. As a consequence, the term *drift* was applied to this sediment. Although rooted in an incorrect concept, this term was so well established by the time the true glacial origin of the debris became widely recognized that it remained part of the basic glacial vocabulary. Today **glacial drift** is an all-embracing term for sediments of glacial origin, no matter how, where, or in what shape they were deposited.

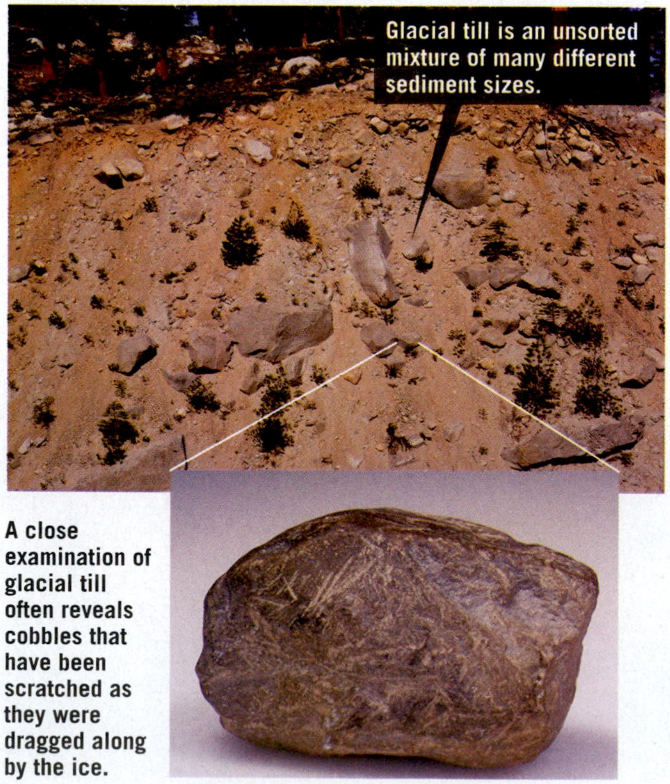

Figure 18.23
Glacial till Unlike sediment deposited by running water and wind, material deposited directly by a glacier is not sorted. Figure 18.15 provides an example of till being deposited at the terminus of a glacier. (Top photo by Michael Collier; bottom photo by E. J. Tarbuck)

Glacial till is an unsorted mixture of many different sediment sizes.

A close examination of glacial till often reveals cobbles that have been scratched as they were dragged along by the ice.

Geologists divide glacial drift into two distinct types: (1) materials deposited directly by the glacier, which are known as *till*, and (2) sediments laid down by glacial meltwater, called *stratified drift*.

Glacial Till   As glacial ice melts and drops its load of rock fragments, **till** is deposited. Unlike moving water and wind, ice cannot sort the sediment it carries; therefore, deposits of till are characteristically unsorted mixtures of many particle sizes (**Figure 18.23**). A close examination of this sediment shows that many of the pieces are scratched and polished as a result of being dragged along by the glacier. Such pieces help distinguish till from other deposits that are a mixture of different sediment sizes, such as material from a debris flow or a rockslide.

Boulders found in the till or lying free on the surface are called **glacial erratics** if they are different from the bedrock below (**Figure 18.24**). Of course, this means that they must have been derived from a source outside the area where they are found. Although the source for most erratics is unknown, the origin of some can be determined. In many cases, boulders were transported as far as 500 kilometers (300 miles) from their source area and, in a few instances, more than 1000 kilometers (600 miles). Therefore, by studying glacial erratics as well as the mineral composition of the remaining till, geologists are sometimes able to trace the path of a lobe of ice.

In portions of New England and other areas, erratics dot pastures and farm fields. In fact, in some places, these large rocks were cleared from fields and piled to make fences and walls. Keeping the fields clear, however, is an ongoing chore because each spring, newly exposed erratics appear. Wintertime frost heaving lifts them to the surface.

Stratified Drift   As the name implies, **stratified drift** is sorted according to the size and weight of the particles. Ice is not capable of sorting sediment the way running water can. Therefore, stratified drift is not deposited directly by the glacier the way till is, but instead reflects the sorting action of glacial meltwater.

Some deposits of stratified drift are made by streams issuing directly from the glacier. Other stratified deposits involve sediment that was originally laid down as till and later picked up, transported, and redeposited by meltwater beyond the margin of the ice. Accumulations of stratified drift often consist largely of sand and gravel because the meltwater is not capable of moving larger material and because the finer rock flour remains suspended and is commonly carried far from the glacier. Evidence that stratified drift consists primarily of sand and gravel can be seen in many areas where these deposits are actively mined as aggregate for road work and other construction projects.

Landforms Made of Till

Perhaps the most widespread features created by glacial deposition are *moraines*, which are simply layers or ridges of till. Several types of moraines are identified; some are common in mountain valleys, and others are associated with areas affected by either ice sheets or valley glaciers. Lateral and medial moraines fall in the first category, whereas end moraines and ground moraines are in the second.

Lateral and Medial Moraines

Alpine glaciers produce two types of moraines that occur exclusively in mountain valleys. The first of these is called a **lateral moraine**. As discussed earlier, when an alpine glacier moves down a valley, the ice erodes the sides of the valley with great efficiency. In addition, large quantities of debris are added to the glacier's surface as rubble falls or slides from higher up on the valley walls and collects on the margins of the moving ice. When the ice eventually melts, this accumulation of debris is dropped next to the valley walls. These ridges of

Figure 18.24
Glacial erratic This large glacially transported boulder, called Doane Rock, is a prominent feature near Nauset Bay on Cape Cod. Such boulders are called glacial erratics. (Photo by Michael Collier)

till paralleling the sides of the valley constitute the lateral moraines.

The second type of moraine that is unique to alpine glaciers is the **medial moraine** (Figure 18.25). Medial moraines are created when two alpine glaciers coalesce to form a single ice stream. The till that was once carried along the sides of each glacier joins to form a single dark stripe of debris within the newly enlarged glacier. The sketch in Figure 18.25 illustrates this nicely. The creation of these dark stripes within the ice stream is one obvious proof that glacial ice moves because the moraine could not form if the ice did not flow downvalley. It is common to see several medial moraines within a single large alpine glacier because a streak will

SmartFigure 18.25
Formation of a medial moraine Kennicott Glacier is a 43-kilometer- (27-mile-) long valley glacier that is sculpting the mountains in Alaska's Wrangell–St. Elias National Park. The dark stripes of sediment are medial moraines. (Photo by Michael Collier) (http://goo.gl/jkqqPA)

Mobile Field Trip

Geologist's Sketch

SmartFigure 18.26
End moraines of the Great Lakes region End moraines deposited during the most recent (Wisconsinan) stage are the most prominent.
(https://goo.gl/QNwbma)

Animation

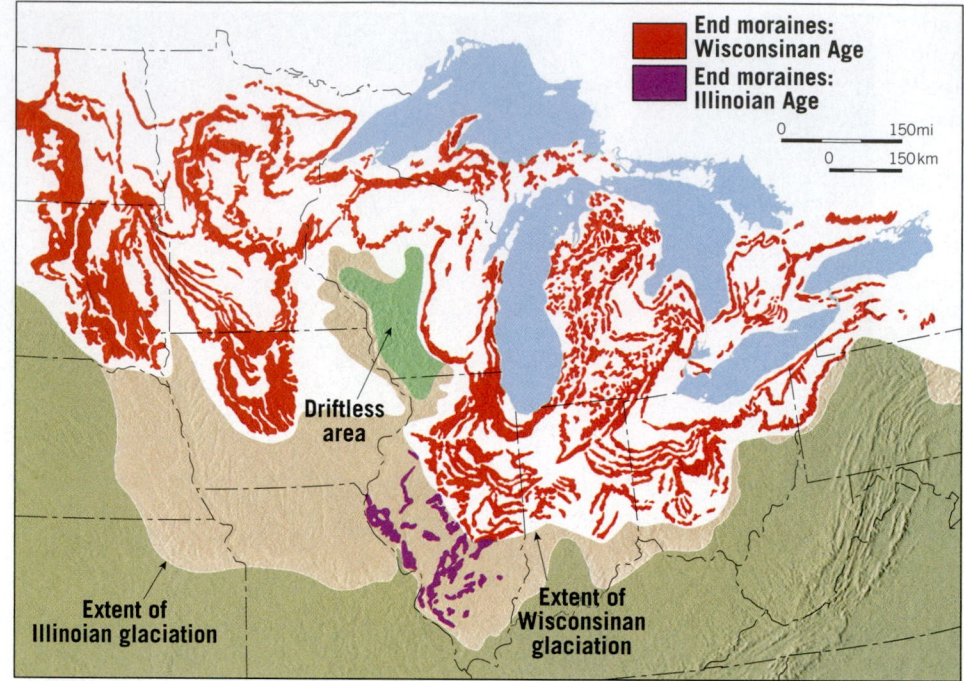

SmartFigure 18.26. End moraines of the Great Lakes region.

the end moraine grows. The longer the ice front remains stable (with ablation and accumulation in balance), the larger the ridge of till will become.

Eventually, ablation exceeds nourishment. At this point, the front of the glacier begins to recede in the direction from which it originally advanced. However, as the ice front retreats, the conveyor-belt action of the glacier continues to provide fresh supplies of sediment to the terminus. In this manner, a large quantity of till is deposited as the ice melts away, creating a rock-strewn, undulating plain. This gently rolling layer of till deposited as the ice front recedes is termed **ground moraine**. It has a leveling effect, filling in low spots and clogging old stream channels, often leading to a derangement of the existing drainage system. In areas where this layer of till is still relatively fresh, such as the northern Great Lakes region, poorly drained swampy lands are quite common.

Periodically, a glacier will retreat to a point where ablation and nourishment once again balance. When this happens, the ice front stabilizes, and a new end moraine forms. The pattern of end moraine formation and ground moraine deposition may be repeated many times before the glacier has completely vanished. Such a pattern is illustrated in **Figure 18.26**. The very first end moraine to form signifies the farthest advance of the glacier and is called the *terminal end moraine*. End moraines that form as the ice front occasionally stabilizes during retreat are termed *recessional end moraines*. Terminal and recessional moraines are essentially alike; the only difference between them is their relative positions.

End moraines deposited by the most recent stage of Ice Age glaciation are prominent features in many parts of the U.S. Midwest and Northeast. In Wisconsin, the wooded, hilly terrain of the Kettle Moraine

form whenever a tributary glacier joins the main valley glacier.

End and Ground Moraines

Sometimes a glacier is compared to a conveyor belt. No matter whether the front of a glacier or ice sheet is advancing, retreating, or stationary, it is constantly moving sediment forward and dropping it at its terminus. This is a useful analogy when considering end and ground moraines.

An **end moraine** is a ridge of till that forms at the terminus of a glacier. End moraines are characteristic of ice sheets and valley glaciers alike. These relatively common landforms are deposited when a state of equilibrium is attained between ablation and ice accumulation. That is, the end moraine forms when the ice is melting and evaporating near the end of the glacier at a rate equal to the forward advance of the glacier from its region of nourishment. Although the terminus of the glacier is stationary, the ice continues to flow forward, delivering a continuous supply of sediment in the same way a conveyor belt delivers goods to the end of a production line. As the ice melts, the till is dropped, and

SmartFigure 18.27
Two significant end moraines in the Northeast The Ronkonkoma moraine, deposited about 20,000 years ago, extends through central Long Island, Martha's Vineyard, and Nantucket. The Harbor Hill moraine formed about 14,000 years ago and extends along the north shore of Long Island, through southern Rhode Island and Cape Cod. A portion of the *Mobile Field Trip* explores the glacial origins of Cape Cod. (http://goo.gl/MfwH34)

Mobile Field Trip

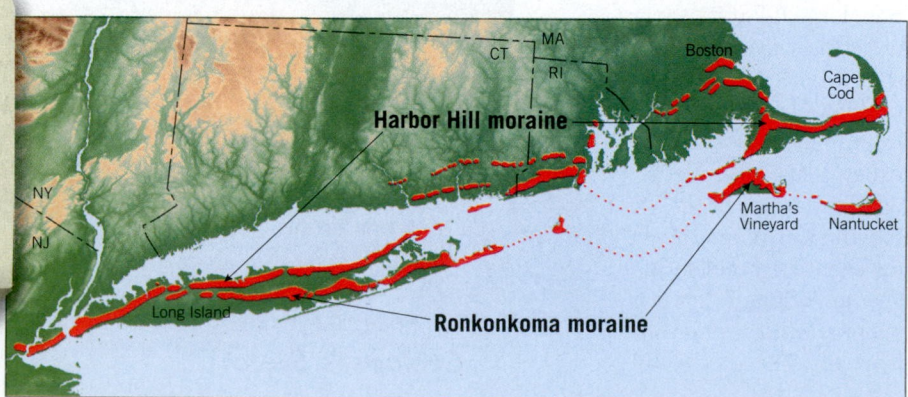

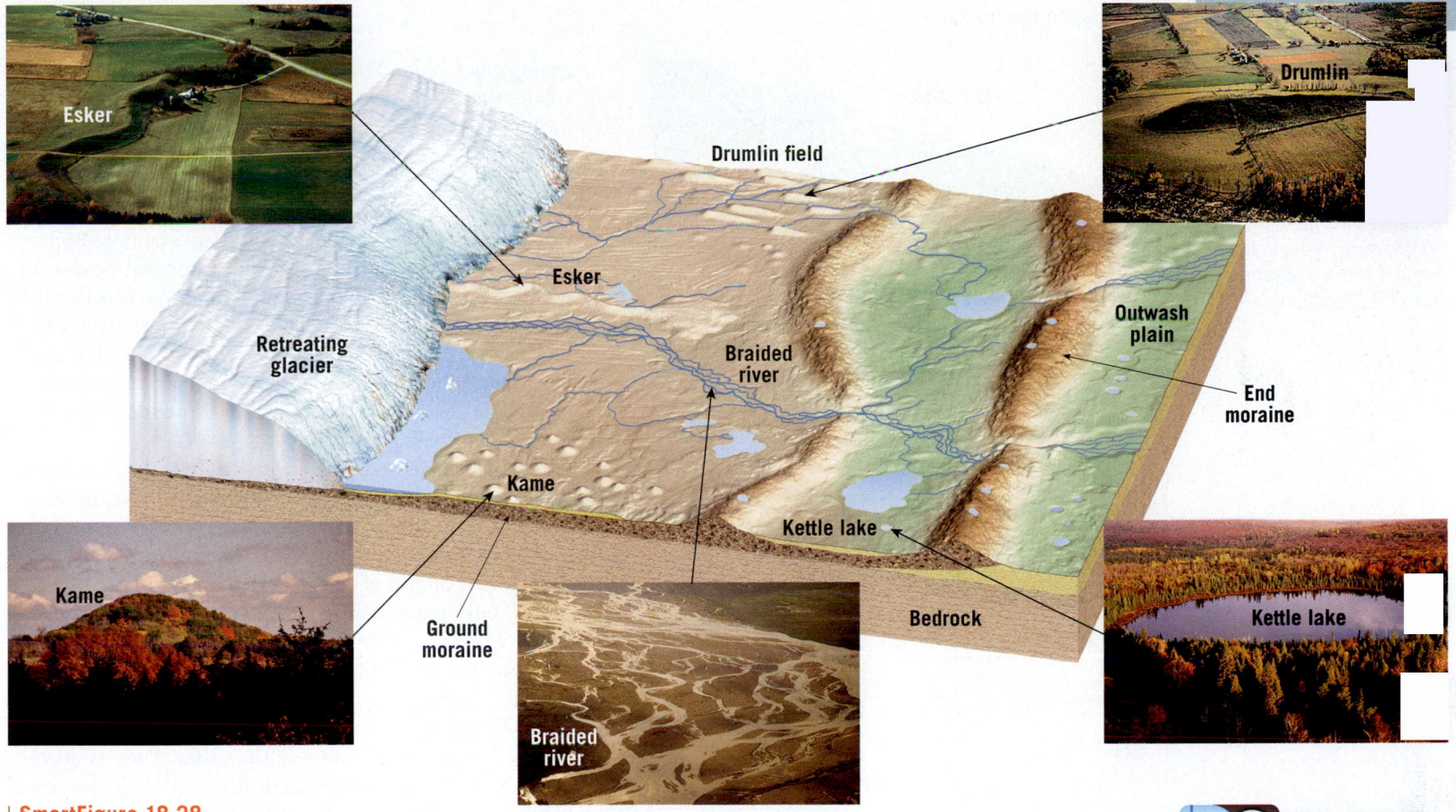

SmartFigure 18.28
Common depositional landforms This diagram depicts a hypothetical area affected by ice sheets in the recent geologic past.
(Drumlin photo courtesy of Ward's Natural Science Establishment; esker photo by Richard P. Jacobs/JLM Visuals; kame photo by John Dankwardt; kettle lake photo by Carlyn Iverson/Science Source; braided river photo by Michael Collier) (https://goo.gl/7TDPkt)

near Milwaukee is a particularly picturesque example. A well-known example in the Northeast is Long Island. This linear strip of glacial sediment extending northeastward from New York City is part of an end moraine complex that stretches from eastern Pennsylvania to Cape Cod, Massachusetts (**Figure 18.27**).

Figure 18.28 represents a hypothetical area during glaciation and after the retreat of ice sheets. This figure depicts landscape features such as the end moraines just described as well as depositional landforms, similar to what might be encountered if you were traveling in the upper Midwest or New England. You will be referred to this figure several times as you read the following paragraphs on glacial deposits.

Drumlins Moraines are not the only landforms composed of till. In some areas that were once covered by continental ice sheets, a special variety of glacial landscape exists—one characterized by smooth, elongate, parallel hills called **drumlins** (see Figure 18.28). Certainly one of the best-known drumlins is Bunker Hill in Boston, the site of the famous Revolutionary War battle in 1775.

An examination of Bunker Hill or other less famous drumlins would show that drumlins are streamlined, asymmetrical hills composed largely of till. They range in height from about 15 to 50 meters and may be up to 1 kilometer long. The steep side of the hill faces the direction from which the ice advanced, whereas the gentler, longer slope points in the direction the ice moved. Drumlins are not found as isolated landforms but rather occur in clusters called *drumlin fields* (**Figure 18.29**). One such cluster, east of Rochester, New York, is estimated to contain about 10,000 drumlins. Although drumlin formation is not fully understood, the streamlined shape of drumlins indicates that they were molded in the zone of plastic flow within an active glacier. It is believed that many drumlins originate when glaciers advance over previously deposited drift and reshape the material.

Landforms Made of Stratified Drift

Much of the material acquired and transported by a glacier is ultimately deposited by streams of glacial meltwater flowing on, within, beneath, and beyond a glacier. This sediment is termed *stratified drift*. Unlike glacial till, stratified drift shows some degree of sorting. There are two basic categories of features composed of stratified drift: *Ice-contact deposits* accumulate on, within, or immediately adjacent to a glacier. *Outwash sediment*, or simply *outwash*, is material deposited by meltwater streams beyond the terminus of a glacier.

Figure 18.29
Drumlin field A portion of the drumlin field shown on the Palmyra, New York, 7.5-minute topographic map. North is at the top. The drumlins are steepest on the north side, indicating that the ice advanced from this direction.

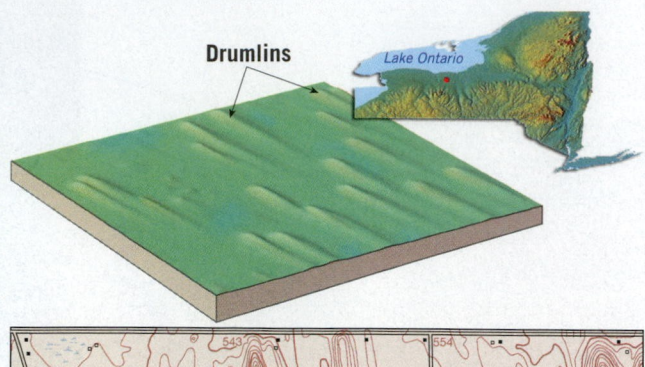

Drumlins

Lake Ontario

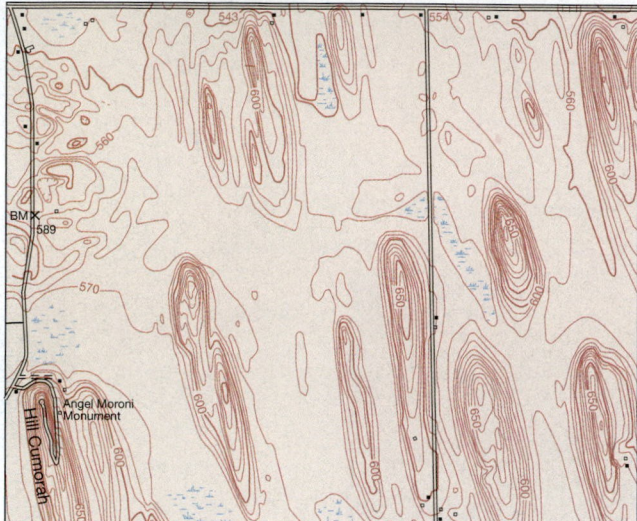

Outwash Plains and Valley Trains At the same time that an end moraine is forming, water from the melting glacier cascades over the till, sweeping some of it out in front of the growing ridge of unsorted debris. Meltwater generally emerges from the ice in rapidly moving streams that are often choked with suspended material and carry a substantial bed load as well. Water leaving the glacier moves onto the relatively flat surface beyond and rapidly loses velocity. As a consequence, much of its bed load is dropped, and the meltwater begins weaving a complex pattern of braided channels (see Figure 18.28). In this way, a broad, ramplike surface composed of stratified drift is built adjacent to the downstream edge of most end moraines. When the feature is formed in association with an ice sheet, it is termed an **outwash plain**, and when largely confined to a mountain valley, it is usually called a **valley train**.

Outwash plains and valley trains often are pock-marked with basins or depressions known as **kettles** (see Figure 18.28). Kettles also occur in deposits of till. Kettles are formed when blocks of stagnant ice become wholly or partly buried in drift and eventually melt, leaving pits in the glacial sediment. Although most kettles do not exceed 2 kilometers in diameter, some with diameters exceeding 10 kilometers occur in Minnesota. Likewise, the typical depth of most kettles is less than 10 meters, although the vertical dimensions of some approach 50 meters. In many cases water eventually fills the depression and forms a pond or lake. One well-known

example is Walden Pond near Concord, Massachusetts. It is here that Henry David Thoreau lived alone for 2 years in the 1840s and about which he wrote his famous book *Walden; or, Life in the Woods.*

Ice-Contact Deposits When the melting terminus of a glacier shrinks to a critical point, flow virtually stops, and the ice becomes stagnant. Meltwater that flows over, within, and at the base of the motionless ice lays down deposits of stratified drift. Then, as the supporting ice melts away, the stratified sediment is left behind in the form of hills, terraces, and ridges. Such accumulations are collectively termed **ice-contact deposits** and are classified according to their shapes.

When the ice-contact stratified drift is in the form of a mound or steep-sided hill, it is called a **kame** (see Figure 18.28). Some kames represent bodies of sediment deposited by meltwater in openings within or depressions on top of the ice. Others originate as deltas or fans built outward from the ice by meltwater streams. Later, when the stagnant ice melts away, these various accumulations of sediment collapse to form isolated, irregular mounds.

When glacial ice occupies a valley, **kame terraces** may be built along the sides of the valley. These features commonly are narrow masses of stratified drift laid down between the glacier and the side of the valley by meltwater streams that drop debris along the margins of the shrinking ice mass.

A third type of ice-contact deposit is a long, narrow, sinuous ridge composed largely of sand and gravel. Some are more than 100 meters high, with lengths in excess of 100 kilometers. The dimensions of many others are far less spectacular. Known as **eskers**, these ridges are deposited by meltwater rivers flowing within, on top of, and beneath a mass of motionless, stagnant glacial ice (see Figure 18.28). Many sediment sizes are carried by the torrents of meltwater in the ice-banked channels, but only the coarser material can settle out of the turbulent stream.

18.4 Concept Checks

1. What term can be applied to any glacial deposit? Distinguish between till and stratified drift.

2. How are medial moraines and lateral moraines related? In what kind of setting do these features form?

3. Contrast end moraine and ground moraine. Relate these features to the budget of a glacier.

4. Sketch a profile (side view) of a drumlin. Include an arrow to indicate the direction from which the ice advanced.

5. Distinguish between outwash deposits and ice-contact deposits.

This is a ridge of glacial drift that was deposited at the terminus of Exit Glacier in Alaska's Kenai Fiords National Park. (Photo by Michael Collier)

QUESTION 1 Is this a deposit of stratified drift or a deposit of till? Explain.

QUESTION 2 Describe the budget of the glacier (accumulation versus ablation) as this ridge of sediment accumulated.

QUESTION 3 What term is applied to features such as the one pictured here?

18.5 | Other Effects of Ice Age Glaciers

Describe and explain several important effects of Ice Age glaciers other than erosional and depositional landforms.

In addition to the massive erosional and depositional work carried out by Ice Age glaciers, the ice sheets had other effects, sometimes profound, on the landscape. For example, as the ice advanced and retreated, animals and plants were forced to migrate. This led to stresses that some organisms could not tolerate. Hence, a number of plants and animals became extinct. Other effects of Ice Age glaciers that are described in this section involve adjustments in Earth's crust due to the addition and removal of ice, and sea-level changes associated with the formation and melting of ice sheets. The advance and retreat of ice sheets also led to significant changes in the routes taken by rivers. In some regions, glaciers acted as dams that created large lakes. When these ice dams failed, the effects on the landscape were profound. In areas that today are deserts, lakes of another type, called pluvial lakes, formed.

Crustal Subsidence and Rebound

In areas that were major centers of ice accumulation, such as Scandinavia and the Canadian Shield, the land has been slowly rising over the past several thousand years. Uplifting of almost 300 meters (1000 feet) has occurred in the Hudson Bay region. This, too, is the result of the continental ice sheets. But how can glacial ice cause such vertical crustal movement? We now understand that the land is rising because the added weight of the 3-kilometer- (2-mile-) thick mass of ice caused downwarping of Earth's crust. For example, scientists have determined that Antarctica's ice sheets depress Earth's crust by an estimated 900 meters (3000 feet) or more in some places. Following the removal of this immense load, the crust adjusts by gradually rebounding upward (**Figure 18.30**).*

* For a detailed explanation of this concept, termed *isostatic adjustment*, see the discussion of isostasy in the section "Vertical Movements of the Crust" in Chapter 14.

In northern Canada and Scandinavia, where the greatest accumulation of glacial ice occurred, the added weight caused downwarping of the crust.

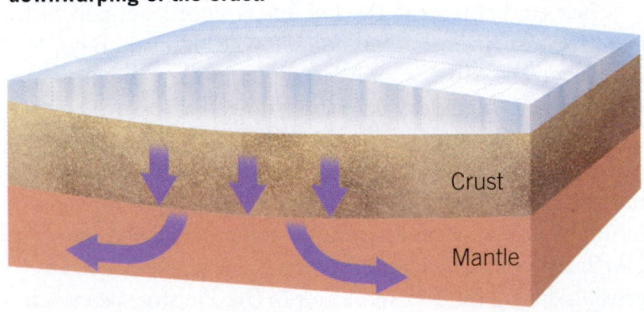

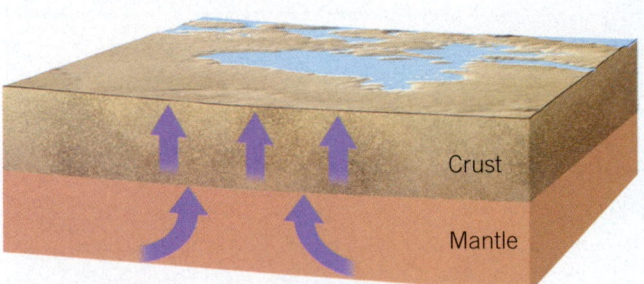

Ever since the ice melted, there has been gradual uplift or rebound of the crust.

Figure 18.30 Crustal subsidence and rebound These simplified diagrams illustrate subsidence and rebound resulting from the addition and removal of continental ice sheets.

Sea-Level Changes

One of the most interesting and perhaps dramatic effects of the Ice Age was the fall and rise of sea level that accompanied the advance and retreat of the glaciers.

Although the total volume of glacial ice today is great, exceeding 25 million cubic kilometers, during the Last Glacial Maximum the volume of glacial ice amounted to about 70 million cubic kilometers, or 45 million cubic kilometers more than at present. Because we know that the snow from which glaciers are made ultimately comes from the evaporation of ocean water, the growth of ice sheets must have caused a significant worldwide drop in sea level (**Figure 18.31**). Indeed, estimates suggest that sea level was as much as 100 meters (330 feet) lower than it is today. Thus, land that is presently flooded by the oceans was dry. The Atlantic coast of the United States lay more than 100 kilometers (60 miles) to the east of New York City, France and Britain were joined where the famous English Channel is today, Alaska and Siberia were connected across the Bering Strait, and Southeast Asia was tied by dry land to the islands of Indonesia. Conversely, if the water currently locked up in the Antarctic ice sheet were to melt completely, sea level would rise by an estimated 60 or 70 meters. Such an occurrence would flood many densely populated coastal areas.

Changes to Rivers and Valleys

Among the effects associated with the advance and retreat of North American ice sheets were changes in the routes of many rivers and the modification in the size and shape of many valleys. If we are to understand the present pattern of rivers and lakes in the central and northeastern United States (and many other places as well), we

SmartFigure 18.31
Changing sea level As ice sheets form and then melt away, sea level falls and rises, causing the shoreline to shift. (https://goo.gl/fKN1kJ)

During the Last Glacial Maximum, about 18,000 years ago, sea level was nearly 100 meters (330 feet) lower than it is today.

During the Last Glacial Maximum, the shoreline extended out onto the present-day continental shelf.

A. This map shows the Great Lakes and the familiar present-day pattern of rivers. Quaternary ice sheets played a major role in creating this pattern.

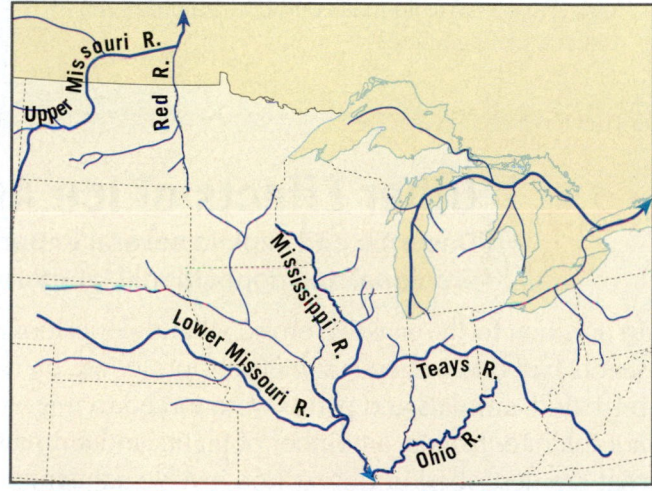

B. Reconstruction of drainage systems prior to the Ice Age. The pattern was very different from today, and the Great Lakes did not exist.

Figure 18.32
Changing rivers The advance and retreat of ice sheets caused major changes in the routes followed by rivers in the central United States.

need to be aware of glacial history. Two examples illustrate these effects.

Upper Mississippi Drainage Basin   Figure 18.32A

shows the familiar present-day pattern of rivers in the central United States, with the Missouri, Ohio, and Illinois Rivers as major tributaries to the Mississippi. **Figure 18.32B** depicts drainage systems in this region prior to the Ice Age. The pattern is *very* different from the present. This remarkable transformation of river systems resulted from the advance and retreat of the ice sheets.

Notice that prior to the Ice Age, a significant part of the Missouri River drained north toward Hudson Bay. Moreover, the Mississippi River did not follow the present Iowa–Illinois boundary but rather flowed across west-central Illinois, where the lower Illinois River flows today. The preglacial Ohio River barely reached to the present-day state of Ohio, and the rivers that today feed the Ohio in western Pennsylvania flowed north and drained into the North Atlantic. The Great Lakes were created by glacial erosion during the Ice Age. Prior to the Pleistocene epoch,

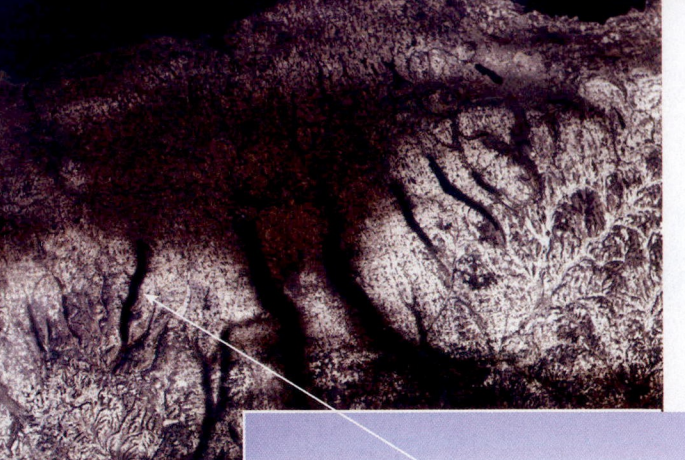

(600 feet) deep at its lowest point, and Cayuga Lake is nearly 135 meters (450 feet) deep—and the beds of both lakes lie below sea level. The depth to which the glaciers carved these basins is much greater. There are hundreds of feet of glacial sediment in the deep rock troughs below the lake beds.

Ice Dams Create Proglacial Lakes

Ice sheets and alpine glaciers can act as dams to create lakes by trapping glacial meltwater and blocking the flow of rivers. Some of these lakes are relatively small, short-lived impoundments. Others can be large and exist for hundreds or thousands of years.

Figure 18.34 is a map of Lake Agassiz—the largest lake to form during the Ice Age in North America. It came into existence about 12,000 years ago and lasted for about 4500 years. With the retreat of the ice sheet came enormous volumes of meltwater. The Great Plains generally slope upward to the west. As the terminus of the ice sheet receded northeastward, meltwater was trapped between the ice on one side and the sloping land on the other, causing Lake Agassiz to deepen and spread across the landscape. Such water bodies are termed **proglacial lakes**, referring to their position just beyond the outer limits of a glacier or an ice sheet. The history of Lake Agassiz is complicated by the dynamics of the ice sheet, which, at various times, readvanced and affected lake levels and drainage systems. Where drainage occurred depended on the water level of the lake and the position of the ice sheet.

Lake Agassiz left marks over a broad region. Former beaches, now many kilometers from any water, mark

Figure 18.33
New York's Finger Lakes The long, narrow basins occupied by these lakes were created when ice sheets scoured these river valleys into deep troughs. (Satellite image courtesy of NASA; photo by James Schwabel/Alamy Images)

the basins occupied by these huge lakes were lowlands with rivers that ran eastward to the Gulf of St. Lawrence.

The large Teays River was a significant feature prior to the Ice Age (see Figure 18.32B). The Teays flowed from West Virginia across Ohio, Indiana, and Illinois, and it discharged into the Mississippi River not far from present-day Peoria. This river valley, which would have rivaled the Mississippi in size, was completely obliterated during the Pleistocene, buried by glacial deposits hundreds of feet thick. Today the sands and gravels in the buried Teays valley make it an important aquifer.

New York's Finger Lakes The recent geologic history of west-central New York State south of Lake Ontario was dominated by ice sheets. We have already noted the drumlins in the vicinity of Rochester (see Figure 18.29). Many other depositional features and erosional effects are found in the region. Perhaps the best known are the Finger Lakes, 11 long, narrow, roughly parallel water bodies oriented north–south like fingers on a pair of outstretched hands (**Figure 18.33**). Prior to the Ice Age, the Finger Lakes area consisted of a series of river valleys that were oriented parallel to the direction of ice movement. Multiple episodes of glacial erosion transformed these valleys into deep, steep-walled lakes. Two of the lakes are very deep—Seneca Lake is more than 180 meters

Figure 18.34
Glacial Lake Agassiz This lake was an immense feature—bigger than all of the present-day Great Lakes combined. Modern-day remnants of this proglacial water body are still major landscape features.

Figure 18.35
Lake Missoula and the Channeled Scablands During a span of 1500 years, more than 40 megafloods from Lake Missoula carved the Channeled Scablands. (Photo by John S. Shelton/University of Washington Libraries)

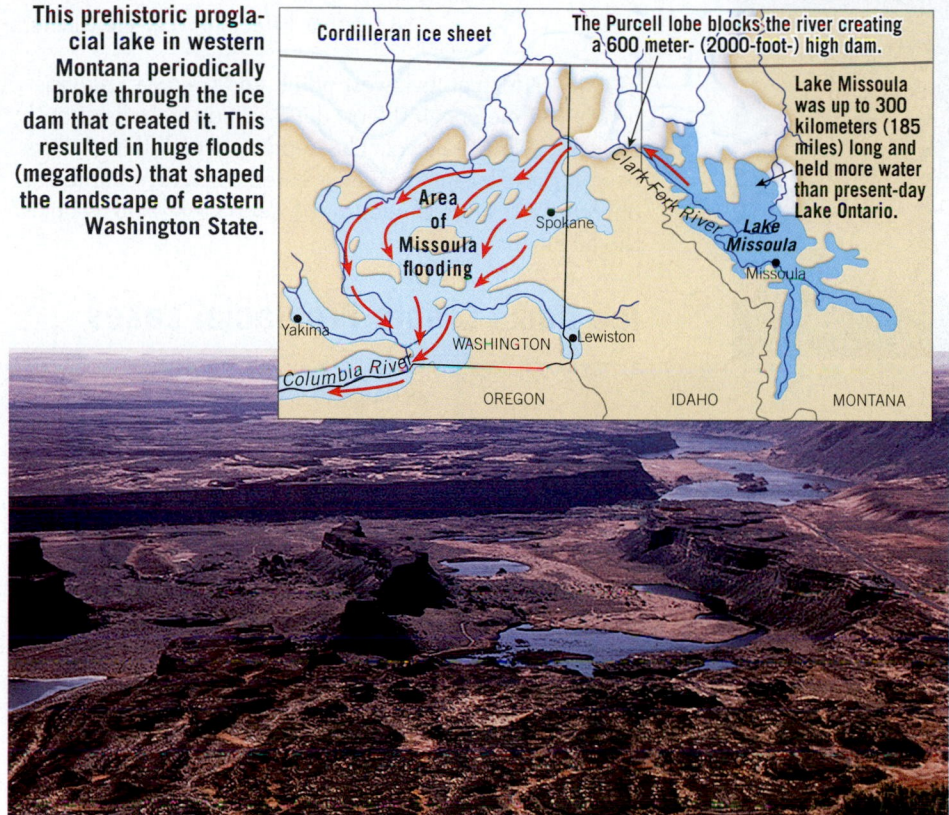

This prehistoric proglacial lake in western Montana periodically broke through the ice dam that created it. This resulted in huge floods (megafloods) that shaped the landscape of eastern Washington State.

Cordilleran ice sheet

The Purcell lobe blocks the river creating a 600 meter- (2000-foot-) high dam.

Lake Missoula was up to 300 kilometers (185 miles) long and held more water than present-day Lake Ontario.

Area of Missoula flooding

The towering mass of rushing water from each megaflood stripped away layers of sediment and soil and cut deep canyons (coulees) into the underlying layers of basalt to create the Channeled Scablands.

former shorelines. Several modern river valleys, including the Red River and the Minnesota River, were originally cut by water entering or leaving the lake. Present-day remnants of Lake Agassiz include Lake Winnipeg, Lake Manitoba, Lake Winnipegosis, and Lake of the Woods. The sediments

of the former lake basin are now fertile agricultural land.

Research shows that the shifting of glaciers and the failure of ice dams can cause the rapid release of huge volumes of water. Such events occurred during the history of Lake Agassiz. A dramatic example of such glacial outbursts occurred in the Pacific Northwest between about 15,000 and 13,000 years ago and is briefly described in **Figure 18.35**.

Pluvial Lakes

While the formation and growth of ice sheets was an obvious response to significant changes in climate, the existence of the glaciers themselves triggered important climatic changes in the regions beyond their margins. In arid and semiarid areas on all the continents, temperatures were lower and thus evaporation rates were lower, but at the same time, precipitation totals were moderate. This cooler, wetter climate formed many **pluvial lakes** (*pluvia* = rain). In North America the greatest concentration of pluvial lakes occurred in the vast Basin and Range region of Nevada and Utah (**Figure 18.36**). By far the largest of the lakes in this region was Lake Bonneville. With maximum depths exceeding 300 meters (1000 feet) and an area of 50,000 square kilometers (20,000 square miles), Lake Bonneville was nearly the same size as present-day Lake Michigan. As the ice sheets waned, the climate again grew more arid, and the lake levels lowered in response. Although most of the lakes completely disappeared, a few small remnants of Lake Bonneville remain, the Great Salt Lake being the largest and best known.

Figure 18.36
Pluvial lakes During the Ice Age, the Basin and Range region experienced a wetter climate than it has today. Many basins turned into large lakes.

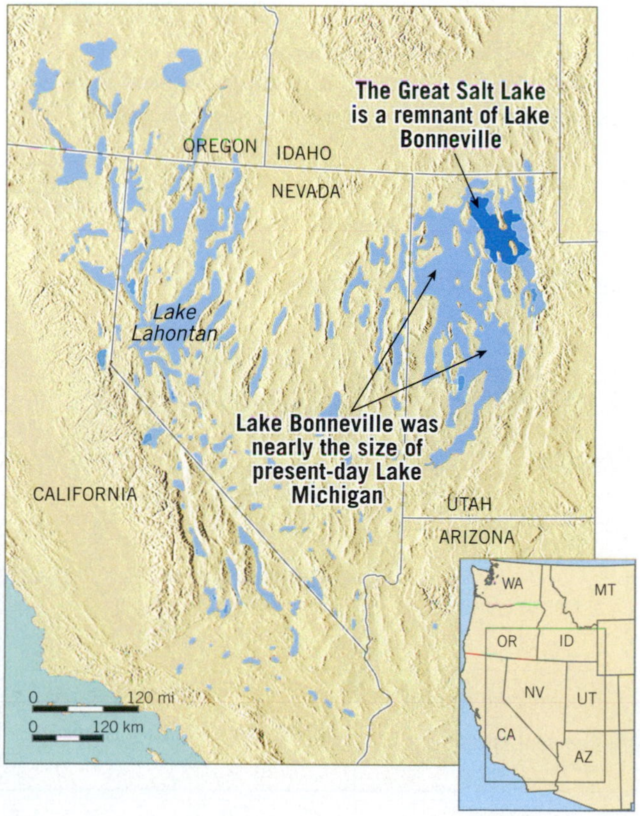

The Great Salt Lake is a remnant of Lake Bonneville

Lake Lahontan

Lake Bonneville was nearly the size of present-day Lake Michigan

0   120 mi
0   120 km

18.5   Concept Checks

1. List and briefly describe five effects of Ice Age glaciers aside from the formation of major erosional and depositional features.

2. Examine Figure 18.31 and determine how much sea level has changed since the Last Glacial Maximum.

3. Compare the two parts of Figure 18.32 and identify three major changes to the flow of rivers in the central United States during the Ice Age.

4. Contrast proglacial lakes and pluvial lakes. Give an example of each.

18.6 | The Ice Age

Briefly discuss the development of glacial theory and summarize current ideas on the causes of ice ages.

In the preceding pages, we mentioned the Ice Age, a time when ice sheets and alpine glaciers were far more extensive than they are today. As noted, there was a time when the most popular explanation for what we now know to be glacial deposits was that the materials had been drifted in by means of icebergs or perhaps simply swept across the landscape by a catastrophic flood. What convinced geologists that an extensive ice age was responsible for these deposits and many other glacial features?

Historical Development of the Glacial Theory

In 1821 a Swiss engineer, Ignaz Venetz, presented a paper suggesting that glacial landscape features occurred at considerable distances from the existing glaciers in the Alps. This implied that the glaciers had once been larger and occupied positions farther downvalley. Another Swiss scientist, Louis Agassiz, doubted the proposal of widespread glacial activity put forth by Venetz. He set out to prove that the idea was not valid. However, his 1836 fieldwork in the Alps convinced him of the merits of his colleague's hypothesis. In fact, a year later Agassiz hypothesized a great ice age that had extensive and far-reaching effects—an idea that was to give Agassiz widespread fame.

The proof of the glacial theory proposed by Agassiz and others constitutes a classic example of applying the principle of uniformitarianism. Realizing that certain features are produced by no other known process but glacial action, the scientists were able to begin reconstructing the extent of now-vanished ice sheets based on the presence of features and deposits found far beyond the margins of present-day glaciers and ice sheets. In this manner, the development and verification of the glacial theory continued during the nineteenth century, and through the efforts of many scientists, a knowledge of the nature and extent of former ice sheets became clear.

By the beginning of the twentieth century, geologists had largely determined the extent of the Ice Age glaciation. Further, during the course of their investigations, they discovered that many glaciated regions had not one but several layers of drift. Moreover, close examination of these older deposits showed well-developed zones of chemical weathering and soil formation, as well as the remains of plants that require warm temperatures. The evidence was clear: There had been not just one glacial advance but many, each separated by extended periods when climates were as warm as or even warmer than the present. The Ice Age had not simply been a time when the ice advanced over the land, lingered for a while, and then receded. Rather, the period was a very complex event, characterized by a number of advances and withdrawals of glacial ice.

By the early twentieth century, a fourfold division of the Ice Age had been established for both North America and Europe. These divisions were based largely on studies of glacial deposits. In North America each of the four major stages was named for the midwestern state where deposits of that stage were well exposed and/or were first studied. These are, in order of occurrence, the Nebraskan, Kansan, Illinoian, and Wisconsinan. These traditional divisions remained in place for many years, until it was learned that sediment cores from the ocean floor contain a much more complete record of climate change during the Ice Age. Unlike the glacial record on land, which is punctuated by many unconformities, seafloor sediments provide an uninterrupted record of climatic cycles for this period (**Figure 18.37**). Studies of these seafloor sediments showed that glacial/interglacial cycles had occurred about every 100,000 years. About 20 such cycles of cooling and warming were identified for the span we call the Ice Age.

During the Ice Age, ice left its imprint on almost 30 percent of Earth's land area, including about 10 million square kilometers of North America, 5 million square kilometers of Europe, and 4 million square kilometers of Siberia (**Figure 18.38**). The amount of glacial ice in the Northern Hemisphere was roughly twice that in the Southern Hemisphere. The primary reason is that the southern polar ice could not spread far beyond the margins of Antarctica. By contrast, North America and Eurasia provided great expanses of land for the spread of ice sheets.

Today we know that the Ice Age began between 2 million and 3 million years ago. This means that most of the major glacial stages occurred during a division of the geologic time scale called the **Quaternary period**. However, this period does not encompass all of the last glacial period. The Antarctic ice sheet, for example, probably formed at least 30 million years ago.

Figure 18.37
Evidence from the seafloor Cores of seafloor sediment provided data that led to a more complete understanding of the complexity of Ice Age climates. (Photo by Gary Braasch/ZUMA Press/Newscom)

Figure 18.38
Where was the ice? This map shows the maximum extent of ice sheets in the Northern Hemisphere during the Ice Age.

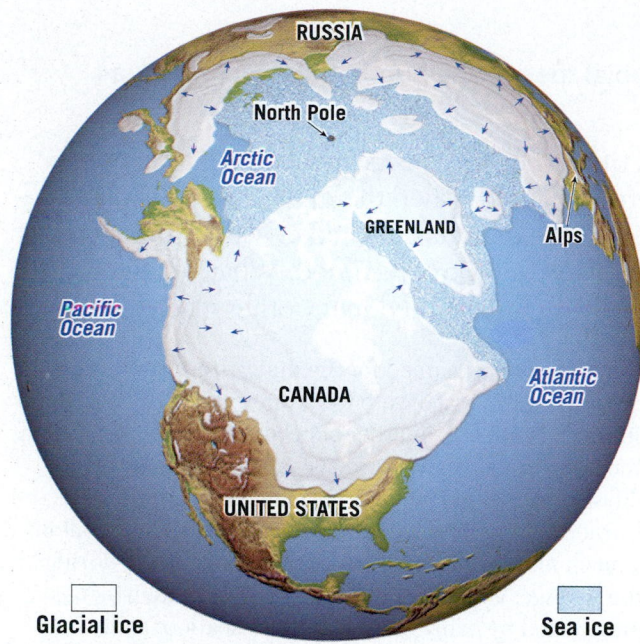

of evidence supporting the continental drift hypothesis mentioned a glacial period that occurred in late Paleozoic time (see Figure 2.7, page 42). Two Precambrian glacial episodes have been identified in the geologic record, the first approximately 2 billion years ago and the second about 600 million years ago.

Any theory that attempts to explain the causes of ice ages must successfully answer two basic questions:

- *What causes the onset of glacial conditions?* For continental ice sheets to have formed, average temperature must have been somewhat lower than at present, and perhaps substantially lower than throughout much of geologic time. Thus, a successful theory would have to account for the cooling that finally leads to glacial conditions.

- *What caused the alternating glacial and interglacial stages that have been documented for the Quaternary period?* Whereas the first question deals with long-term trends in temperature on a scale of millions of years, this question relates to much shorter-term changes.

Although the scientific literature contains many hypotheses related to possible causes of glacial periods, we will discuss only a few major ideas to summarize current thought.

Causes of Ice Ages

A great deal is known about glaciers and glaciation. Much has been learned about glacier formation and movement, the extent of glaciers past and present, and the features created by glaciers, both erosional and depositional. However, the causes of ice ages are not completely understood.

Although widespread glaciation has been rare in Earth's history, the Ice Age that encompassed most of the Quaternary period is not the only glacial period for which records exist. Earlier glaciations are indicated by deposits called **tillite**, a sedimentary rock formed when glacial till becomes lithified (**Figure 18.39**). Such deposits, found in strata of several different ages, usually contain striated rock fragments, and some overlie grooved and polished bedrock surfaces or are associated with sandstones and conglomerates that show features of outwash deposits. For example, our Chapter 2 discussion

Plate Tectonics Probably the most attractive proposal for explaining the fact that extensive glaciations have occurred only a few times in the geologic past comes from the theory of plate tectonics. Because glaciers can form only on land, we know that landmasses must exist somewhere in the higher latitudes before an ice age can commence. Many scientists suggest that ice ages have occurred only when Earth's shifting crustal plates have carried the continents from tropical latitudes to more poleward positions.

Glacial features in present-day Africa, Australia, South America, and India indicate that these regions, which are now tropical or subtropical, experienced an ice age near the end of the Paleozoic era, about 250 million years ago. However, there is no evidence that ice sheets existed during this same period in what are today the higher latitudes of North America and Eurasia. For many years this puzzled scientists. Was the climate in these relatively tropical latitudes once like it is today in Greenland and Antarctica? Why did glaciers not form in North America and Eurasia? Until the plate tectonics theory was formulated, there had been no reasonable explanation.

Today scientists understand that the areas containing these ancient glacial features were joined together as a single supercontinent called Pangaea, located at latitudes far to the south of their present positions. Later this landmass broke apart, and its pieces, each moving on a different plate, migrated toward their present locations (**Figure 18.40**). Now we know that during the geologic past, plate movements accounted for many dramatic climate changes, as landmasses shifted in relation to one another and moved to different latitudinal positions. Changes in oceanic circulation also must have occurred,

Figure 18.39
Tillite When glacial till is lithified, it becomes the sedimentary rock known as tillite. Tillite strata are evidence for ice ages that occurred prior to the Quaternary period. (Photo by Brian Roman)

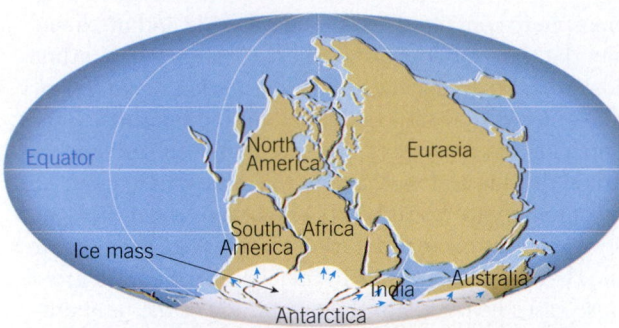

The supercontinent Pangaea showing the area covered by glacial ice near the end of the Paleozoic era.

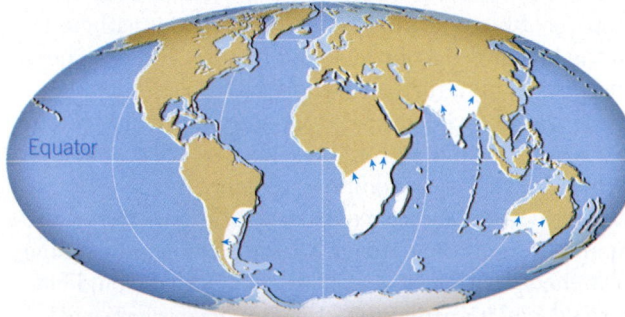

The continents as they appear today. The white areas indicate where evidence of the late Paleozoic ice sheets exists.

Figure 18.40
A late Paleozoic ice age Shifting tectonic plates sometimes move landmasses to high latitudes, where the formation of ice sheets is possible.

altering the transport of heat and moisture and consequently the climate as well. Because the rate of plate movement is very slow—a few centimeters annually—appreciable changes in the positions of the continents occur only over great spans of geologic time. Thus, climate changes triggered by shifting plates are extremely gradual and happen on a scale of millions of years.

Variations in Earth's Orbit Because climatic changes brought about by moving plates are extremely gradual, the plate tectonics theory cannot be used to explain the alternating glacial and interglacial climates that occurred during the Pleistocene epoch. Therefore, we must look to some other triggering mechanism that might cause climate change on a scale of thousands, rather than millions, of years. Today many scientists strongly suspect that the climate oscillations that characterized the Quaternary period are linked to variations in Earth's orbit. This hypothesis, first developed and strongly advocated by the Serbian astrophysicist Milutin Milankovitch, is based on the premise that variations in incoming solar radiation are a principal factor in controlling Earth's climate.

Milankovitch formulated a comprehensive mathematical model based on the following elements (**Figure 18.41**):

- Variations in the shape (*eccentricity*) of Earth's orbit about the Sun

- Changes in *obliquity*—that is, changes in the angle that Earth's axis makes with the plane of our planet's orbit
- The wobbling of Earth's axis, called *precession*

Using these factors, Milankovitch calculated variations in the receipt of solar energy and the corresponding surface temperature of Earth back into time, in an attempt to correlate these changes with the climate fluctuations of the Quaternary. In explaining climate changes that result from these three variables, note that they cause little or no variation in the *total* solar energy reaching the ground. Instead, their impact is felt because they change the degree of contrast between the seasons. Somewhat milder winters in the middle to high latitudes mean greater snowfall totals, whereas cooler summers bring a reduction in snowmelt.

Among the studies that have added credibility to the astronomical hypothesis of Milankovitch is one in which

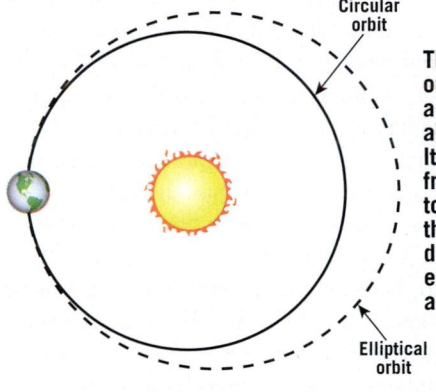

The shape of Earth's orbit changes during a cycle that spans about 100,000 years. It gradually changes from nearly circular to more elliptical and then back again. This diagram greatly exaggerates the amount of change.

SmartFigure 18.41
Orbital variations Periodic variations in Earth's orbit are linked to alternating glacial and interglacial conditions during the Ice Age.
(https://goo.gl/Kf87Ky)

 **Tutorial**

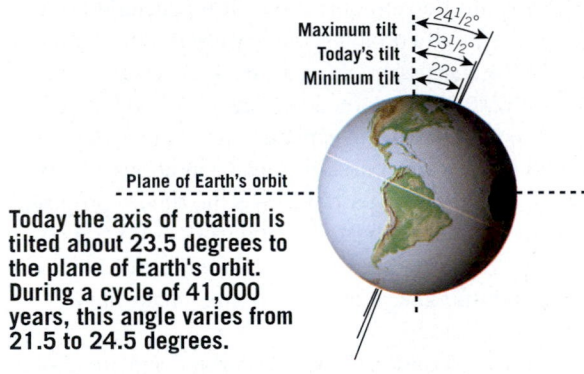

Today the axis of rotation is tilted about 23.5 degrees to the plane of Earth's orbit. During a cycle of 41,000 years, this angle varies from 21.5 to 24.5 degrees.

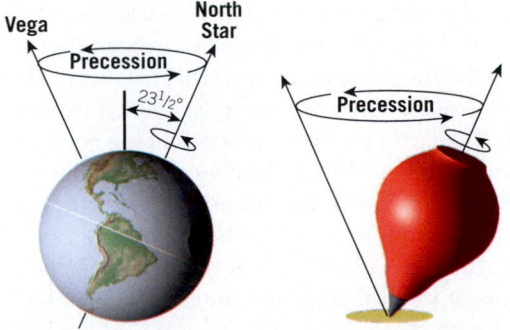

Earth's axis wobbles like a spinning top. Consequently, the axis points to different spots in the sky during a cycle of about 26,000 years.

Figure 18.42
Ice cores contain clues to shifts in climate This scientist is slicing an ice core from Antarctica for analysis. He is wearing protective clothing and a mask to minimize contamination of the sample. Chemical analyses of ice cores can provide important data about past climates. (Photo by British Antarctic Survey/Science Source)

deep-sea sediments containing certain climatically sensitive microorganisms were analyzed to establish a chronology of temperature changes going back nearly a half-million years. This time scale of climate change was then compared to astronomical calculations of eccentricity, obliquity, and precession to determine whether a correlation existed. Although the study was very involved and mathematically complex, the conclusions were straightforward: The researchers found that major variations in climate over the past several hundred thousand years were closely associated with changes in the geometry of Earth's orbit; that is, cycles of climate change were shown to correspond closely with the periods of obliquity, precession, and orbital eccentricity. More specifically, the authors stated: "It is concluded that changes in the earth's orbital geometry are the fundamental cause of the succession of Quaternary ice ages."[*]

To briefly summarize the ideas just described, plate tectonics theory provides an explanation for the widely spaced and nonperiodic onset of glacial conditions at various times in the geologic past, whereas the theory proposed by Milankovitch and supported by the work of J. D. Hays and his colleagues furnishes an explanation for the alternating glacial and interglacial episodes of the Quaternary.

Other Factors   Variations in Earth's orbit correlate closely with the timing of glacial–interglacial cycles. However, the variations in solar energy reaching Earth's surface caused by these orbital changes do not adequately explain the magnitude of the temperature changes that occurred during the most recent Ice Age. Other factors must also have contributed. One factor involves variations in the composition of the atmosphere. Other influences are related to changes in the reflectivity of Earth's surface and in ocean circulation. Let's take a brief look at these factors.

Chemical analyses of air bubbles trapped in glacial ice at the time of ice formation indicate that the Ice Age

atmosphere contained less carbon dioxide and methane than the post–Ice Age atmosphere (**Figure 18.42**). Carbon dioxide and methane are important "greenhouse" gases, which means they trap radiation emitted by Earth and contribute to atmospheric heating. When the amount of carbon dioxide and methane in the atmosphere increases, global temperatures rise, and when there is a reduction in these gases, as occurred during the Ice Age, temperatures fall. Therefore, reductions in the concentrations of greenhouse gases help explain the magnitude of the temperature drop that occurred during glacial times. Although scientists know that concentrations of carbon dioxide and methane dropped, they do not know what caused the drop. As often occurs in science, observations gathered during one investigation yield information and raise questions that require further analysis and explanation.

Obviously, whenever Earth enters an ice age, extensive areas of land that were once ice free are covered with ice and snow. In addition, a colder climate causes the area covered by sea ice (frozen surface seawater) to expand. Ice and snow reflect a large portion of incoming solar energy back to space. Thus, energy that would have warmed Earth's surface and the air above is lost, and global cooling is reinforced.

Yet another factor that influences climate during glacial times relates to ocean currents. Research has shown that ocean circulation changes during ice ages. For example, studies suggest that the warm current that transports large amounts of heat from the tropics toward higher latitudes in the North Atlantic was significantly weaker during the Ice Age. This would lead to a colder climate in Europe, amplifying the cooling attributable to orbital variations.

In conclusion, we emphasize that the ideas just discussed do not represent the only possible explanations for ice ages. Although interesting and attractive, these proposals are certainly not without critics, nor are they the only possibilities currently under study. Other factors may be, and probably are, involved.

18.6 Concept Checks

1. What was the best source of data showing Ice Age climate cycles?

2. About what percentage of Earth's land surface has been affected by glaciers during the Quaternary period?

3. Where were ice sheets more extensive during the Ice Age: the Northern Hemisphere or the Southern Hemisphere? Why?

4. How does the theory of plate tectonics help us understand the causes of ice ages? Does this theory explain alternating glacial/interglacial climates during the Ice Age?

5. Briefly summarize the climate change hypothesis that involves variations in Earth's orbit.

*J. D. Hays, John Imbrie, and N. J. Shackelton, "Variations in the Earth's Orbit: Pacemaker of the Ice Ages," *Science* 194 (1976): 1121–1132.

18.1 Glaciers: A Part of Two Basic Cycles

Explain the role of glaciers in the hydrologic and rock cycles and describe the different types of glaciers, their characteristics, and their present-day distribution.

KEY TERMS glacier, valley glacier (alpine glacier), ice sheet, sea ice, ice shelf, ice cap, outlet glacier, piedmont glacier

- A glacier is a thick mass of ice originating on land from the compaction and recrystallization of snow and shows evidence of past or present flow. Glaciers are part of both the hydrologic cycle and the rock cycle because they store and release freshwater, and they erode and transport rock material.
- Valley glaciers flow down mountain valleys, while ice sheets are very large masses of ice, such as those that cover Greenland and Antarctica. During the Last Glacial Maximum, Earth was in an Ice Age that covered large areas of the land surface with glacial ice.
- When valley glaciers leave the confining mountains, they may spread out into broad lobes called piedmont glaciers. Similarly, ice shelves form when glaciers flow into the ocean and spread out to form a wide layer of floating ice.
- Ice caps are like smaller ice sheets. Both ice sheets and ice caps may be drained by outlet glaciers.

Q What term is applied to the ice at the North Pole? What term best describes Greenland's ice? Are both considered glaciers? Explain.

NASA

18.2 Formation and Movement of Glacial Ice

Describe how glaciers move, the rates at which they move, and the significance of the glacial budget.

KEY TERMS firn, plastic flow, basal slip, zone of fracture, crevasse, zone of accumulation, snowline (equilibrium line), zone of wastage, ablation, calving, iceberg, glacial budget

- When snow piles up sufficiently, it recrystallizes to dense granules of firn, which can then pack together even more tightly to make glacial ice.
- When ice is put under pressure, it will flow very slowly. The uppermost 50 meters (165 feet) of a glacier doesn't have sufficient pressure to flow, so it breaks open in dangerous cracks called crevasses that occur in the zone of fracture. In addition, most glaciers also move by a sliding process called basal slip.
- Glaciers move at a slow but measurable rate. Fast glaciers may move 800 meters (2600 feet) per year, while slow glaciers may move only 2 meters (6.5 feet) per year. Some glaciers experience periodic surges of sudden movement.
- A glacier's rate of flow does not necessarily correlate to the position of its terminus. Instead, if the glacier has a positive "budget," in which accumulation exceeds wastage, the terminus will advance. If calving of icebergs, melting, or other forms of ablation exceed the input of new ice, then the glacier's terminus will retreat. Even a retreating glacier is still experiencing downstream flow.

Q This image shows that melting is one way that glacial ice wastes away. What is another way that ice is lost from a glacier? What is the general term for ice loss from a glacier?

Glacial meltwater

Robbie Shone/Photo Researchers, Inc

18.3 Glacial Erosion

Discuss the processes of glacial erosion. Identify and describe major topographic features created by glacial erosion.

KEY TERMS plucking, abrasion, rock flour, glacial striation, glacial trough, truncated spur, hanging valley, pater noster lake, cirque, tarn, col, arête, horn, roche moutonnée, fiord

- Glaciers are powerful agents of erosion and acquire sediment through plucking from the bedrock beneath the glacier, by abrasion of the bedrock using sediment already in the ice, and when mass-wasting processes drop debris on top of the glacier. Grinding of the bedrock produces grooves and scratches called glacial striations.
- Glacial troughs have a distinctive U-shaped profile, very different from a stream-carved mountain valley, with its typical V-shaped profile. Lining the edge of the valley may be triangle-shaped cliffs called truncated spurs. Higher up, hanging valleys mark the spots where tributary glaciers once flowed into the main glacier.
- At the head of a valley glacier is an amphitheater-shaped cirque, which may or may not hold a small lake called a tarn. Other glacial lakes in the bottom of the valley resemble beads on a string and are called pater noster lakes. The intersection of two cirques produced by glaciers flowing in opposite directions forms a col.

- If two glaciers flow parallel to each other, their troughs may intersect in a knife-edge ridge called an arête. A high point that had multiple glaciers flowing away from a point in a radial array may leave behind a pyramidal horn. Protrusions of bedrock may be glacially abraded on their upstream side and plucked on their downstream side. This produces asymmetric knobs of bedrock called roches moutonnées.
- In coastal mountain settings, a glacial trough may be eroded below sea level and subsequently flooded by the ocean to become a fiord, a narrow steep-sided inlet.

Q **Examine the illustration of a mountainous landscape after glaciation. Identify the landforms that resulted from glacial erosion.**

18.4 Glacial Deposits

Distinguish between the two basic types of glacial drift. List and describe the major depositional features associated with glacial landscapes.

KEY TERMS glacial drift, till, glacial erratic, stratified drift, lateral moraine, medial moraine, end moraine, ground moraine, drumlin, outwash plain, valley train, kettle, ice-contact deposit, kame, kame terrace, esker

- Any sediment of glacial origin is called drift. The two distinct types of glacial drift are till, which is unsorted material deposited directly by the ice, and stratified drift, which is sediment sorted and deposited by meltwater from a glacier.
- The most widespread features created by glacial deposition are layers or ridges of till, called moraines. Associated with valley glaciers are lateral moraines, formed along the sides of the valley, and medial moraines, formed between two valley glaciers that have merged. End moraines marking the former position of the front of a glacier, and ground moraines, layers of till deposited as the ice front retreats, are common to both valley glaciers and ice sheets.
- Stratified drift can be deposited immediately adjacent to a glacier or carried some distance away and laid down as outwash. Streams draining an ice sheet produce a broad outwash plain beyond the end moraine. A similar feature hemmed in by the walls of a mountain valley is called a valley train. Blocks of ice buried by the sediment may melt to produce depressions called kettles.

- Kames are mounds or steep-sided hills of stratified drift that represent the former positions of sediment-filled lakes on top of (or within) the glacier. Kame terraces are narrow masses of stratified drift deposited adjacent to a stagnant glacier. Meltwater streams flowing through tunnels in the ice may leave behind sinuous ridges of stratified drift called eskers.

Q **Examine the illustration of depositional features formed in the wake of a retreating ice sheet. Name the features and indicate which landforms are composed of till and which are composed of stratified drift.**

18.5 Other Effects of Ice Age Glaciers

Describe and explain several important effects of Ice Age glaciers other than erosional and depositional landforms.

KEY TERMS proglacial lake, pluvial lake

- Glaciers are heavy—so heavy that they can cause Earth's crust to flex downward under their tremendous load. After the glaciers melt off, that weight is released, and the crust slowly rebounds vertically upward.
- Ice sheets are nourished by water that ultimately comes from the ocean, so when ice sheets grow, sea level falls, and when they melt, sea level rises. At the Last Glacial Maximum, global sea level was about 100 meters (330 feet) lower than it is today. At that time, the coastlines of the modern continents were vastly different.

- The advance and retreat of ice sheets caused significant changes to the paths that rivers follow. In addition, glaciers deepened and widened stream valleys and lowlands to create features such as the Great Lakes.
- An ice sheet can act as a dam by trapping meltwater or blocking the flow of rivers to create proglacial lakes. Glacial Lakes Agassiz and Missoula both impounded tremendous quantities of water. In Lake Missoula's case, the water drained out in huge torrents when the ice dam periodically broke.
- Pluvial lakes existed during the height of the Ice Age but occurred far from the actual glaciers, in a climate that was cooler and wetter than today's. Lake Bonneville is a classic example that existed in the area that is now Utah and Nevada.

18.6 The Ice Age

Briefly discuss the development of glacial theory and summarize current ideas on the causes of ice ages.

KEY TERMS Quaternary period, tillite

- The idea of a geologically recent Ice Age was born in the early 1800s in Switzerland. Louis Agassiz and others established that only the former presence of tremendous quantities of glacial ice could explain the landscape of Europe (and later North America and Siberia). As additional research accumulated, especially data from the study of seafloor sediments, it was revealed that the Quaternary period was marked by numerous advances of glacial ice.
- While rare, glacial episodes have occurred in Earth history prior to the recent glaciations we call the Ice Age. Lithified till, called tillite, is a major line of evidence for these ancient ice ages. Several reasons explain why glacial ice might accumulate globally, including the position of the continents, which is driven by plate tectonics.

- The Quaternary period is marked by glacial advances alternating with episodes of glacial retreat. One way to explain these oscillations is through variations in Earth's orbit, which lead to seasonal variations in the distribution of solar radiation. The orbit's shape varies (eccentricity), the tilt of the planet's rotational axis varies (obliquity), and the axis slowly "wobbles" over time (precession). These three effects, all of which occur on different time scales, collectively account for alternating colder and warmer periods during the Quaternary.
- Additional factors that may be important for initiating or ending glaciations include rising or falling levels of greenhouse gases, changes in the reflectivity of Earth's surface, and variations in the ocean currents that redistribute heat energy from warmer to colder regions.

Q About 250 million years ago, parts of India, Africa, and Australia were covered by ice sheets, while Greenland, Siberia, and Canada were ice free. Explain why this occurred.

Give It Some Thought

1. The accompanying diagram shows the results of a classic experiment used to determine how glacial ice moves in a mountain valley. The experiment occurred over an 8-year span. Refer to the diagram and answer the following:

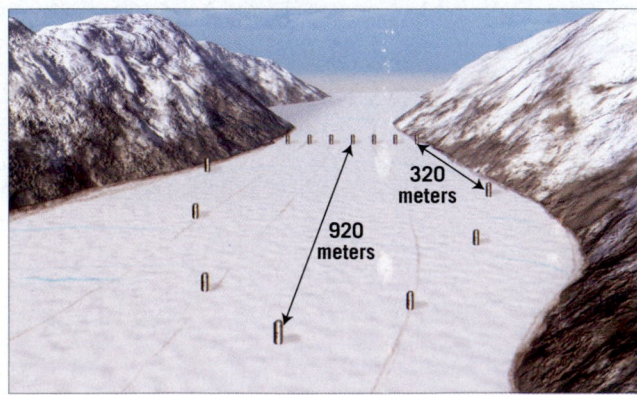

a. What was the average yearly rate at which ice in the center of the glacier advanced?

b. About how fast was the center of the glacier advancing *per day*?

c. Calculate the average rate at which ice along the sides of the glacier moved forward.

d. Why was the rate at the center different than along the sides?

2. Studies have shown that during the Ice Age, the margins of some ice sheets advanced southward from the Hudson Bay region at rates ranging from about 50 to 320 meters per year.

a. Determine the maximum amount of time required for an ice sheet to move from the southern end of Hudson Bay to the south shore of present day Lake Erie, a distance of 1600 kilometers.

b. Calculate the minimum number of years required for an ice sheet to move this distance.

3. If Earth were to experience another Ice Age, one hemisphere would have substantially more expansive ice sheets than the other. Would it be the Northern Hemisphere or the Southern Hemisphere? What is the reason for the large disparity?

4. The accompanying image shows the top of a valley glacier in which the ice is fractured.

Glow Images

a. What term is applied to fractures such as these?

b. In what vertical zone do these breaks occur?

c. Do the fractures likely extend to the base of the glacier? Explain.

5. While taking a break from a hike in the Northern Rockies with a fellow geology enthusiast, you notice that the boulder you are sitting on is part of a deposit consisting of a jumbled mixture of unsorted sediment. Since you are in an area that once had extensive valley glaciers, your colleague suggests that the deposit must be glacial till. Although you know this is certainly a good possibility, you remind your companion that other processes in mountain areas also produce unsorted deposits. What might such a process be? How might you and your friend determine whether this deposit is actually glacial till?

6. This streamlined erosional feature in New York City's Central Park was shaped by glacial ice. Did the glacier advance from the left or from the right? How did you figure this out? What term is applied to bedrock knobs such as this?

John A. Anderson/Shutterstock

7. If the budget of a valley glacier were balanced for an extended time span, what feature would you expect to find at the terminus of the glacier? Is it composed of till or stratified drift? Now assume that the glacier's budget changes so that ablation exceeds accumulation. How would the terminus of the glacier change? Describe the deposit you would expect to form under these conditions.

8. Is the glacial deposit shown here an example of till or stratified drift? Is it more likely part of an end moraine or an esker?

E. J. Tarbuck

9. Watch the Mobile Field Trip titled *Yosemite: Granite and Glaciers* (SmartFigure 4.14, page 118). List and briefly describe the erosional and depositional features associated with glaciers depicted in this video.

10. This wall, located in New England, is built of diverse stones and boulders cleared from nearby fields. In 1914, Robert Frost wrote a now-famous poem titled "Mending Wall" about a feature like this one. It begins with these lines:

Something there is that doesn't love a wall,
That sends the frozen-ground-swell under it,
And spills the upper boulders in the sun;
And makes gaps even two can pass abreast.

a. What is the likely weathering process causing the wall to swell and "spill" its boulders? (Think back to Chapter 6.)

b. Is it likely that the source of all the rocks in the wall is bedrock in the immediate vicinity? Explain.

c. What term applies to the rocks composing the wall?

Kenneth Wiedermann/Getty Images

11. Assume that you and a nongeologist friend are visiting Alaska's Hubbard Glacier. After studying the glacier for quite a long time, your friend asks, "Do these things really move?" How would you convince your friend that this glacier does indeed move, using evidence that is clearly visible in this image?

Michael Collier

MasteringGeology™   www.masteringgeology.com

Looking for additional review and test prep materials? With individualized coaching on the toughest topics of the course, MasteringGeology offers a wide variety of ways for you to move beyond memorization to begin thinking like a geologist. Visit the Study Area in www.masteringgeology.com to find practice quizzes, study tools, and multimedia that will improve your understanding of this chapter's content. Sign in today to enjoy the following features: **Self Study Quizzes, SmartFigure: Tutorials/ Animations/Condor Videos/Mobile Field Trips, Geoscience Animation Library, GEODe, RSS Feeds, Digital Study Modules,** and an optional **Pearson eText.**

19
Deserts and Wind

These huge sand dunes are among the tallest on Earth. Some approach 300 meters (nearly 1000 feet) high. They are located in the Namib Desert, a coastal desert in southern Africa. (Photo by FRIEDRICHSMEIER/Alamy)

Each statement represents the primary **LEARNING OBJECTIVE** for the corresponding major heading within the chapter. After you complete the chapter, you should be able to:

19.1 Describe the general distribution of Earth's dry lands and explain why deserts form in the subtropics and middle latitudes.

19.2 Summarize the geologic roles of weathering, running water, and wind in arid and semiarid climates.

19.3 Discuss the stages of landscape evolution in the Basin and Range region of the western United States.

19.4 List and describe the ways that wind transports sediment.

19.5 Describe the processes and features associated with wind erosion.

19.6 Discuss dune formation and movement and distinguish among different dune types. Explain how loess deposits differ from deposits of sand.

Climate has a strong influence on the nature and intensity of Earth's external processes. This was clearly demonstrated in the preceding chapter on glaciers. Desert landscapes and their development provide another excellent example of the strong link between climate and geology. The word *desert* literally means "deserted" or "unoccupied." For many dry regions this is an appropriate description, although where water is available in deserts, plants and animals thrive. Nevertheless, the world's dry regions are probably the least familiar land areas on Earth outside the polar realm.

As you will see, arid regions are not dominated by a single geologic process. Rather, the effects of tectonic forces, running water, and wind are all apparent. Because these processes combine in different ways from place to place, the appearance of desert landscapes varies a great deal as well.

19.1 | Distribution and Causes of Dry Lands

Describe the general distribution of Earth's dry lands and explain why deserts form in the subtropics and middle latitudes.

Desert landscapes frequently appear stark. Their profiles are not softened by a continuous carpet of soil and abundant plant life. Instead, barren rocky outcrops with steep, angular slopes are common. At some places the rocks are tinted orange and red. At others they are gray and brown and streaked with black. For many visitors, desert scenery exhibits a striking beauty; to others, the terrain seems bleak. No matter which feeling is elicited, it is clear that deserts are very different from the more humid places where most people live.

What Is Meant by *Dry*

We all recognize that deserts are dry places, but just what is meant by the term *dry*? That is, how much rain defines the boundary between humid and dry regions? Sometimes it is arbitrarily defined by a single rainfall figure, such as 25 centimeters (10 inches) per year of precipitation. However, the concept of dryness is a relative one that refers to any situation in which an ongoing water deficiency exists.

Dry Climates Climatologists define **dry climate** as a climate in which yearly precipitation is not as great as the potential loss of water by evaporation. Therefore, dryness is not only related to annual rainfall totals but is also a function of evaporation, which in turn is closely dependent on temperature. As temperatures climb, potential evaporation also increases. As little as 15 to 25 centimeters (6 to 10 inches) of precipitation per year may be sufficient to support coniferous forests in northern Scandinavia or Siberia, where evaporation into the cool, humid air is slight and a surplus of water remains in the soil. However, the same amount of rain falling on Nevada or Iran supports only a sparse vegetative cover because evaporation into the hot, dry air is great. So, clearly, no specific amount of precipitation can serve as a universal boundary for dry climates.

The dry regions of the world encompass about 42 million square kilometers (16 million square miles), a surprising 30 percent of Earth's land surface. No other climate category covers so large a land area. Within water-deficient regions, two climatic types are commonly recognized: **desert**, or arid, and **steppe**, or semiarid. The two share many features; their differences are primarily a matter of degree. The steppe is a marginal and more humid variant of the desert and is a transition zone that surrounds the desert and separates it from bordering humid climates. The world map of the distribution of desert and steppe regions shows that dry lands are concentrated in the subtropics and in the middle latitudes (**Figure 19.1**).

Are Deserts Expanding? Desertlike conditions are expanding worldwide. This important environmental problem, called **desertification**, refers to the persistent degradation of dry-land ecosystems primarily due to human activities. It most often, but not always, takes place on the margins of desert and steppe regions and involves a continuum of change, from slight to severe alteration of plant

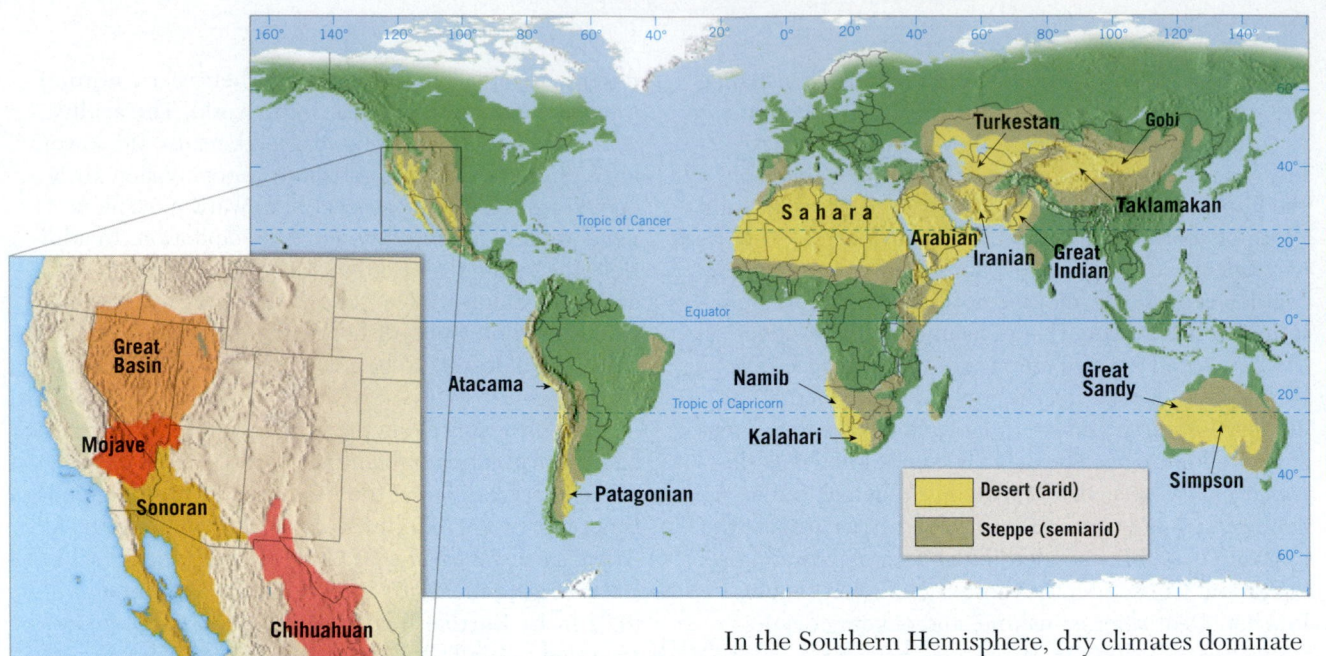

SmartFigure 19.1
Dry climates Arid and semiarid climates cover about 30 percent of Earth's land surface. The dry region of the American West is commonly divided into four deserts, two of which extend into Mexico.
(https://goo.gl/HxXqnO)

Tutorial

In the Southern Hemisphere, dry climates dominate Australia. Almost 40 percent of the continent is desert, and much of the remainder is steppe. In addition, arid and semiarid areas occur in southern Africa and make a limited appearance in coastal Chile and Peru.

Subsiding Air Masses What causes these bands of low-latitude desert? The answer is the global distribution of air pressure and winds. **Figure 19.2A**, an idealized

and soil resources. Desertification occurs when deforestation and overgrazing reduce or completely remove the tree and plant cover that anchors the soil. In some regions, intensive and unsustainable farming practices destroy the natural vegetation and deplete soil nutrients. Wind and water erosion aggravate the damage by carrying away topsoil. When drought occurs on these lands without sufficient vegetation to hold the soil against erosion, the destruction can be irreversible. Desertification is occurring in many places but is particularly serious in the region south of the Sahara Desert known as the Sahel.

Subtropical Deserts and Steppes

The heart of the subtropical dry climates lies in the vicinities of the Tropics of Cancer and Capricorn. Figure 19.1 shows a virtually unbroken desert environment stretching for more than 9300 kilometers (5800 miles) from the Atlantic coast of North Africa to the dry lands of northwestern India. In addition to this single great expanse, the Northern Hemisphere contains another, much smaller area of subtropical desert and steppe in northern Mexico and the southwestern United States.

SmartFigure 19.2
Subtropical deserts The distribution of subtropical deserts and steppes is closely related to the global distribution of air pressure.
(Photo courtesy of NASA)
(https://goo.gl/ONZDSx)

Animation

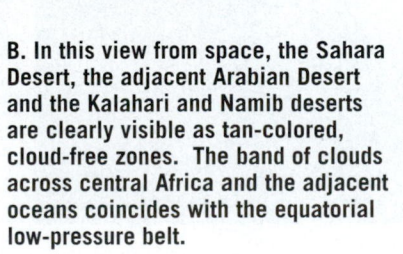

A. Subtropical deserts and steppes are centered between 20° and 30° north and south latitude in association with the subtropical high-pressure belts. Dry subsiding air inhibits cloud formation and precipitation.

B. In this view from space, the Sahara Desert, the adjacent Arabian Desert and the Kalahari and Namib deserts are clearly visible as tan-colored, cloud-free zones. The band of clouds across central Africa and the adjacent oceans coincides with the equatorial low-pressure belt.

diagram of Earth's general circulation, helps visualize the relationship. Heated air in the pressure belt known as the *equatorial low* rises to great heights (usually between 15 and 20 kilometers) and then spreads out. As the upper-level flow reaches 20 to 30 degrees latitude, north or south, it sinks toward the surface. Air that rises through the atmosphere expands and cools, a process that leads to the development of clouds and precipitation. For this reason, the areas under the influence of the equatorial low are among the rainiest on Earth. Just the opposite is true for the regions in the vicinity of 30 degrees north and south latitude, where high pressure predominates. Here, in the zones known as the *subtropical highs,* air is subsiding. When air sinks, it is compressed and warmed. Such conditions are just the opposite of what is needed to produce clouds and precipitation. Consequently, these regions are known for their clear skies, sunshine, and ongoing dryness (**Figure 19.2B**).

West Coast Subtropical Deserts   Where subtropical deserts are found along the west coasts of continents, cold ocean currents have a dramatic influence on the climate. The principal examples are the Atacama Desert in South America, which is adjacent to the cold Peru Current, and the Namib Desert in southwestern Africa, which parallels the cold Benguela Current (see Figure 19.1). These dry lands deviate considerably from the general image we have of subtropical deserts.

The most obvious effect of the cold current is reduced temperatures. Furthermore, although these deserts are adjacent to the ocean, their yearly rainfall totals are among the lowest in the world. The aridity of these coastal areas is intensified because the lower air is chilled by the cold offshore waters. When air is cooled from below, it resists the upward movement needed for cloud formation and precipitation. In addition, the cold current often chills the air enough to cause fog to form. Thus, not all subtropical deserts are sunny and hot. Cold offshore currents cause west coast subtropical deserts to be relatively cool places that sometimes are foggy.

The Atacama along the west coast of South America has the distinction of being the world's driest desert. The average rainfall at the Atacama's wettest locations is not more than 3 millimeters (0.12 inch) per year. At Arica, a coastal town near Chile's border with Peru, the average annual rainfall is a mere 0.5 millimeter (0.02 inch). Further inland, some stations have *never* recorded rainfall.

Middle-Latitude Deserts and Steppes

Unlike their low-latitude counterparts, middle-latitude deserts and steppes are not controlled by the subsiding air masses associated with zones of high pressure. Instead, many of these dry lands exist because they are sheltered in the deep interiors of large landmasses. They are far removed from the ocean, which is the ultimate source of moisture for cloud formation and precipitation. One well-known example is the Gobi Desert of central Asia (see Figure 19.1).

The presence of high mountains across the paths of prevailing winds is another factor that separates middle-latitude arid and semiarid areas from water-bearing, maritime air masses. Mountains force the air masses to lose much of their water. The mechanism is straightforward: As prevailing winds meet mountain barriers, the air is forced to ascend. When air rises, it expands and cools, a process that can produce clouds and precipitation. The windward sides of mountains, therefore, often have high precipitation. By contrast, the leeward sides of mountains are usually much drier (**Figure 19.3**). This situation exists because air reaching the leeward side has lost much of its moisture, and if the air descends, it is compressed

SmartFigure 19.3
Rainshadow deserts
Mountains frequently contribute to the aridity of middle-latitude deserts and steppes by creating a rainshadow. The Great Basin Desert is a rainshadow desert that covers nearly all of Nevada and portions of adjacent states. (Photo on left by Dean Pennala/Shutterstock; photo on right by Dennis Tasa) (https://goo.gl/xfTCht)

Animation

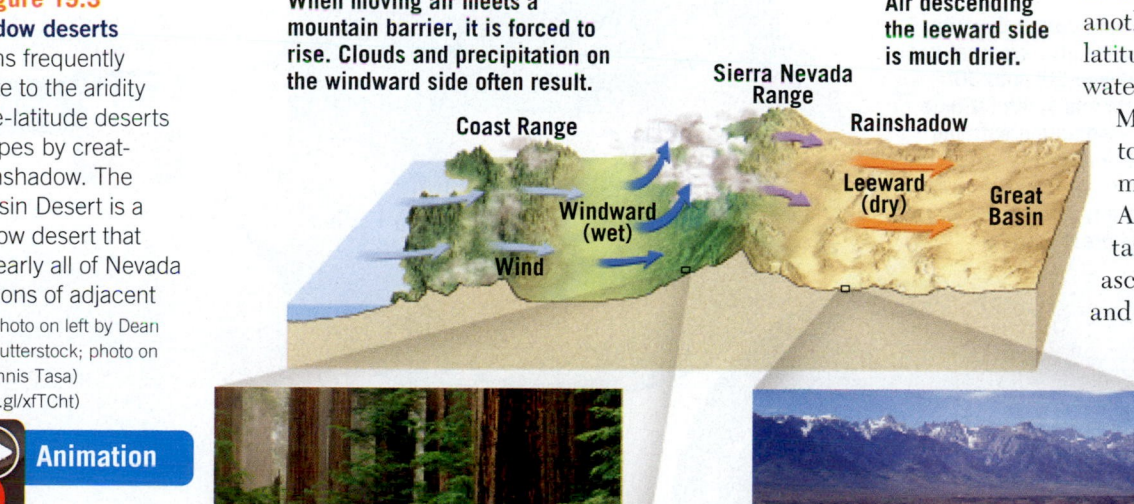

When moving air meets a mountain barrier, it is forced to rise. Clouds and precipitation on the windward side often result.

Air descending the leeward side is much drier.

Sierra Nevada Range

Coast Range

Rainshadow

Windward (wet)

Leeward (dry)

Great Basin

Wind

and warmed, making cloud formation even less likely. The dry region that results on this leeward side is often referred to as a **rainshadow**. Figure 19.4, which shows the distribution of precipitation in western Washington State, is a good example. When prevailing winds from the Pacific Ocean to the west (left) meet the mountains, rainfall totals are high. By comparison, precipitation on the leeward (eastern) side of the mountains is relatively meager. In North America, the foremost mountain barriers to moisture from the Pacific are the Coast Ranges, Sierra Nevada, and Cascades. In Asia, the great Himalayan chain prevents the summertime monsoon flow of moist Indian Ocean air from reaching the interior of the continent.

Because the Southern Hemisphere lacks extensive land areas in the middle latitudes, only small areas of desert and steppe occur in this latitude range, existing primarily near the southern tip of South America in the rainshadow of the towering Andes.

The middle-latitude deserts provide an example of how tectonic processes affect climate. Rainshadow deserts exist because of the mountains produced when plates collide. Without such mountain-building episodes, wetter climates would prevail where many dry regions exist today.

ANNUAL PRECIPITATION

cm		in.
>500		>200
405–499		160–199
250–404		100–159
150–249		60–99
50–149		20–59
25–49		10–19
<25		<10

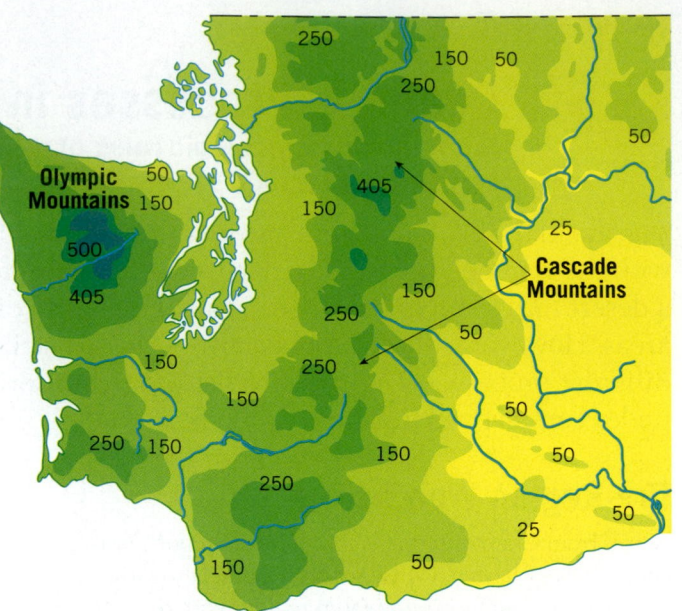

Figure 19.4
Distribution of precipitation in western Washington State The Olympic and Cascade Mountains receive abundant rainfall. The semiarid eastern portion of the area shown here is in a rainshadow.

19.1 Concept Checks

1. Explain why the boundary between humid and dry climates cannot be defined by a single rainfall amount.

2. How extensive are the desert and steppe regions of Earth?

3. What is the primary cause of subtropical deserts and steppes? How do cold ocean currents influence some subtropical deserts?

4. Why do middle-latitude dry regions exist? What role do mountains play?

5. In which hemisphere (Northern or Southern) are middle-latitude deserts most common? Explain.

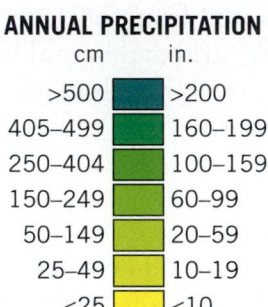

EYE ON EARTH 19.1

This view of Earth from space was acquired in April 2015. The image shows southern Africa and adjacent ocean areas. Two cloud-free desert areas are identified by the letters A and B.

QUESTION 1 Name the two deserts shown on this image.

QUESTION 2 Both deserts result from the same basic cause. What is it?

QUESTION 3 Although both deserts form from the same process, desert B exhibits some characteristics that are quite different from desert A. Describe how they differ and why.

QUESTION 4 Name another desert that has characteristics similar to desert B.

(NASA)

19.2 | Geologic Processes in Arid Climates

Summarize the geologic roles of weathering, running water, and wind in arid and semiarid climates.

The angular hills, the sheer canyon walls, and the desert surface of pebbles or sand contrast sharply with the rounded hills and curving slopes of more humid places. To a visitor from a humid region, a desert may seem to have been shaped by forces different from those operating in well-watered areas. However, although the contrasts may be striking, arid and humid landscapes do not reflect different processes. They merely disclose the differing effects of the same processes that operate under contrasting climatic conditions.

Dry-Region Weathering

Recall from Chapter 6 that water plays an important role in chemical weathering. Consequently, chemical weathering processes are not as prominent in regions with dry climates as in wetter regions. In humid regions, relatively fine-textured soils support an almost continuous cover of vegetation that mantles the surface. The slopes and rock edges are rounded, reflecting the strong influence of chemical weathering in a humid climate. By contrast, much of the weathered debris in deserts consists of unaltered rock and mineral fragments—the result of mechanical weathering processes. In dry lands, rock weathering of any type is greatly reduced because of the lack of moisture and the scarcity of organic acids from decaying plants. However, chemical weathering is not completely lacking in deserts. Over long spans of time, clays and thin soils do form, and many iron-bearing silicate minerals oxidize, producing the rust-colored stain that tints some desert landscapes.

The Role of Water

Deserts have scant precipitation and few major rivers. Nevertheless, water plays an important role in shaping landscapes in dry regions. Permanent streams are normal in humid regions, but practically all desert streambeds are dry most of the time. Deserts have intermittent streams, or **ephemeral streams** (*ephemero* = short-lived), which means they carry water only in response to specific episodes of rainfall (**Figure 19.5**).

A typical ephemeral stream might flow only a few days or perhaps just a few hours during the year, following sporadic rains. In some years the channel might carry no water at all. This fact is obvious even to the casual traveler who notices numerous bridges with no streams beneath them or numerous dips in the road where dry channels cross. When the rare heavy showers do come, however, so much rain falls in such a short time that all of it cannot soak in. Because desert vegetative cover is sparse, runoff is largely unhindered and consequently rapid, often creating flash floods along valley floors. These floods are quite unlike floods in humid regions. A flood on a river like the Mississippi may take several days to reach its crest and then subside. But desert floods arrive suddenly and subside quickly. Because much surface material in a desert is not anchored by vegetation, the amount of erosional work that occurs during a single short-lived rain event is impressive.[*]

In the dry western United States, different names are used for ephemeral streams, including *wash* and *arroyo*. In other parts of the world, a dry desert stream may be a *wadi* (Arabian Peninsula and North Africa), a *donga* (South America), or a *nullah* (India). The satellite images in **Figure 19.6** show a wadi in the Sahara Desert.

Humid regions are notable for their integrated drainage systems. But in arid regions, streams usually lack an extensive system of tributaries. In fact, a basic characteristic of desert streams is that they are small and die out before reaching the sea. Because the water table is

*See Chapter 16 for more information on flash flooding.

Figure 19.5
Ephemeral stream This example is near Arches National Park in southern Utah. (Photos by E. J. Tarbuck; inset photo by Demetrio Carrasco/Dorling Kindersley, Ltd)

Most of the time desert stream channels are dry.

A familiar sign in desert areas. Roads dip into washes which can rapidly fill with water following a heavy rain.

POTENTIAL FLASH FLOOD AREAS

NEXT 21 MILES

An ephemeral stream shortly after a heavy shower. Although such floods are short-lived, they cause large amounts of erosion.

usually far below the surface, few desert streams can draw upon it as streams do in humid regions (see Figure 17.8, page 506). Without a steady supply of water, the combination of evaporation and infiltration soon depletes the stream.

The few permanent streams that do cross arid regions, such as the Colorado and Nile Rivers, originate *outside* the desert, often in well-watered mountains. Here the water supply must be great, or the stream will lose all its water as it crosses the desert. For example, after the Nile leaves its head-waters in the lakes and mountains of central Africa, it traverses almost 3000 kilometers (1900 miles) of the Sahara without a single tributary. By contrast, in humid regions the discharge of a river grows as it flows downstream because tributaries and ground-water contribute additional water along the way.

It should be emphasized that *running water, although infrequent, nevertheless does most of the erosional work in deserts*. This is contrary to the common belief that wind is the most important erosional agent sculpturing desert landscapes. Although wind erosion is indeed more significant in dry areas than elsewhere, most desert landforms are carved by running water. As you will see in Section 19.4, the main role of wind is the transport and deposition of sediment, which creates and shapes the ridges and mounds we call dunes. Other misconceptions about deserts are presented in GEOgraphics 19.1.

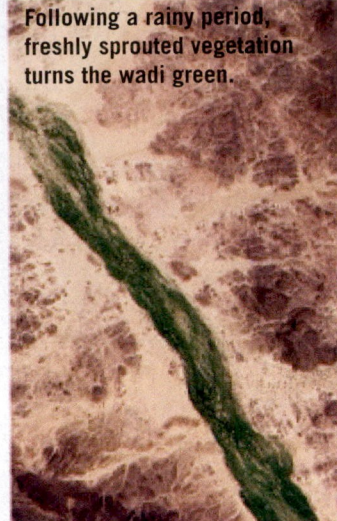

The wadi in its usual dry state.

Following a rainy period, freshly sprouted vegetation turns the wadi green.

Figure 19.6
Wadi in North Africa
These two satellite images show how rain transformed a wadi in Niger. (NASA)

19.2 Concept Checks

1. How does the rate of rock weathering in dry climates compare to the rate in humid regions?

2. What is an ephemeral stream?

3. When a permanent stream such as the Nile River crosses a desert, does the river's discharge increase or decrease? How does this compare to a river in a humid area?

3. What is the most important erosional agent in deserts?

19.3 | Basin and Range: The Evolution of a Desert Landscape

Discuss the stages of landscape evolution in the Basin and Range region of the western United States.

As discussed above, arid regions typically lack permanent streams and often have **interior drainage**. This means that they exhibit a discontinuous pattern of intermittent streams that do not flow out of the desert to the ocean. In the United States, the dry Basin and Range region provides an excellent example. The region includes southern Oregon, all of Nevada, western Utah, south-eastern California, southern Arizona, and southern New Mexico. The name *Basin and Range* is an apt description for this almost 800,000-square-kilometer (300,000-square-mile) region because it is characterized by more than 200 relatively small mountain ranges that rise 900 to 1500 meters (3000 to 5000 feet) above the basins that separate them. The origin of these fault-block mountains is examined in Chapter 14. The Mobile Field Trip associated with SmartFigure 10.18 (page 314) also provides worthwhile insight into Basin and Range geology. In this discussion we look at how surface processes change the landscape.

In this region, as in others like it around the world, erosion mostly occurs without reference to the ocean (ultimate base level) because the interior drainage never reaches the sea. Even where permanent streams flow to the ocean, few tributaries exist, and thus only a narrow strip of land adjacent to the stream has sea level as its ultimate level of land reduction.

The block models in **Figure 19.7** depict how the landscape has evolved in the Basin and Range region and illustrate the landforms described in the following

Common Misconceptions About Deserts

Deserts are hot, lifeless, sand-covered landscapes shaped largely by the forces of wind. The preceding statement summarizes the image of arid regions that many people hold, especially those living in more humid places. Is it an accurate view? The answer is no. Although there are clearly elements of reality in such an impression, it is a generalization that contains several misconceptions.

George Mongol/Alamy

In addition to record-setting heat, cold temperatures are also experienced in deserts. At Ulan Bator in Mongolia's Gobi Desert, the average high temperature in January is only -19°C (-2°F)! The average daily minimum in January at Phoenix, Arizona, is 1.7°C (35°F), just barely above freezing.

Jo Ann Snover/Shutterstock

Rick and Nora Bowers/Alamy

Rick and Nora Bowers/Alamy

Jack Goldfarb/AGE Fotostock

Although reduced in amount and different in character, both plant and animal life are usually present. All forms of desert life have developed adaptations that make them highly tolerant of drought. Plants are widely dispersed and provide little ground cover, but many kinds flourish in the desert.

Another mistaken assumption is the seemingly logical idea that wind is the most important agent of erosion in deserts. However, the greatest erosional work in deserts is done by running water. The infrequent rains often take the form of thunder-storms. Because the heavy rains cannot all soak in, rapid runoff results. Without a thick vegetative cover to protect the ground, erosion is great.

Universal Images Group/SuperStock

David Edwards/National Geographic/Getty Images

Michael Collier

A mistaken assumption is that deserts consist of mile after mile of drifting sand such as these giant dunes along the southwest coast of Africa in the Namib Desert. These huge dunes reach heights of 300 to 350 meters (1000 to 1167 feet). Although sand accumulations may be striking features in some areas, they represent only a small percentage of the total desert area. For example, in the Sahara accumulations of sand cover only one-tenth of its area. The sandiest of all deserts is the Arabian, one-third of which is sand covered.

Merwe/Shutterstock

Question:
List four common misconceptions about deserts.

?

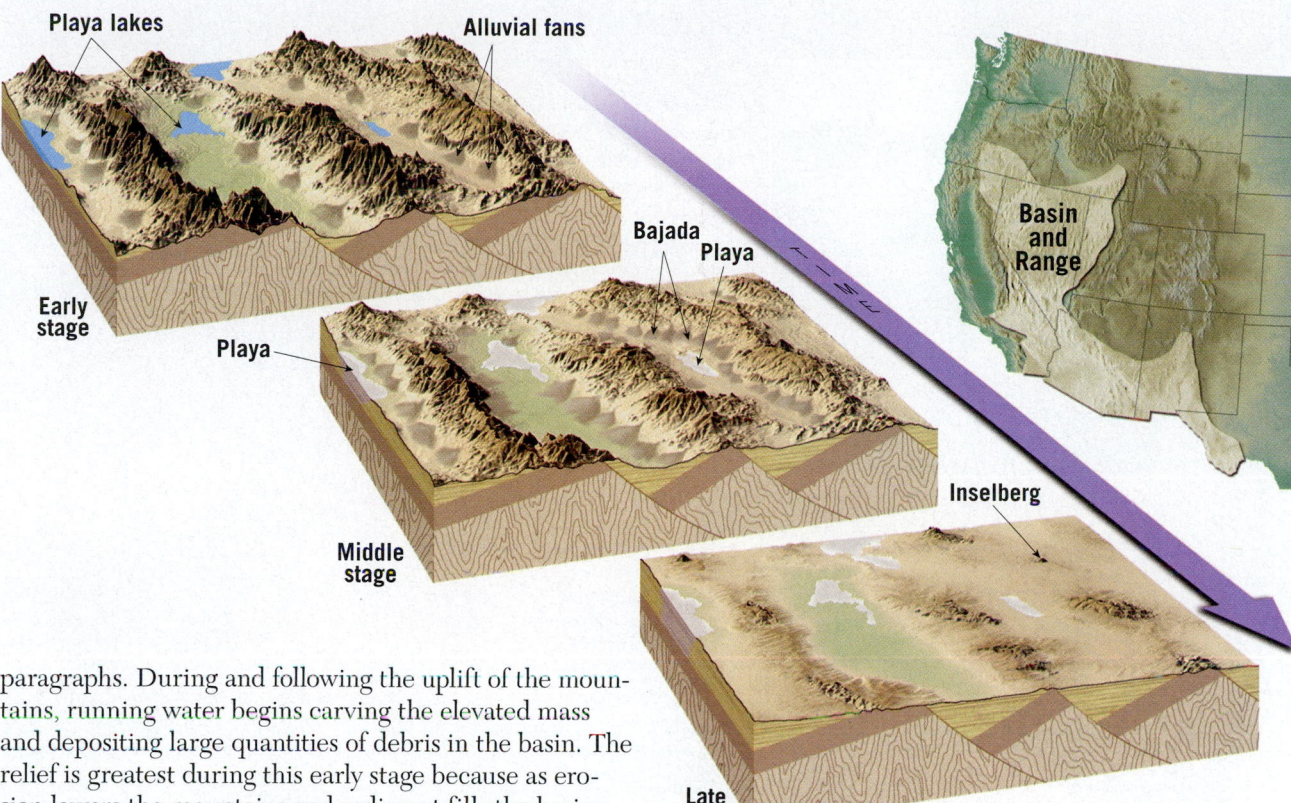

SmartFigure 19.7
Landscape evolution in the Basin and Range region As erosion of the mountains and deposition in the basins continue, relief diminishes. (https://goo.gl/dkNCpX)

Tutorial

paragraphs. During and following the uplift of the mountains, running water begins carving the elevated mass and depositing large quantities of debris in the basin. The relief is greatest during this early stage because as erosion lowers the mountains and sediment fills the basins, elevation differences gradually diminish.

When the occasional torrents of water produced by sporadic rains move down the mountain canyons, they are heavily loaded with sediment. Emerging from the confines of the canyon, the runoff spreads over the gentler slopes at the base of the mountains and quickly loses velocity. Consequently, most of its load is dumped within a short distance. The result is a cone of debris at the mouth of a canyon known as an **alluvial fan**. Because the coarsest (heaviest) material is dropped first, the head

of the fan is steepest, having a slope of perhaps 10 to 15 degrees. Moving down the fan, the size of the sediment and the steepness of the slope decrease, and the fan merges imperceptibly with the basin floor. An examination of the fan's surface would likely reveal a braided channel pattern because of the water shifting its course as successive channels became choked with sediment. Over the years a fan enlarges, eventually coalescing with fans from adjacent canyons to produce an apron of sediment called a **bajada** along the mountain front (**Figure 19.8**).

On the rare occasions of abundant rainfall, streams may flow across the bajada to the center of the basin, converting the basin floor into a shallow **playa lake**. Playa lakes are temporary features that last only a few days or at best a few weeks before evaporation and infiltration remove the water. The dry, flat lake bed that remains is called a **playa**. Playas are typically composed of fine silts and clays and are occasionally encrusted with salts precipitated during evaporation (**Figure 19.9**). These precipitated salts may be unusual. A case in point is the sodium borate (better known as

SmartFigure 19.8
Bajada As neighboring canyons keep depositing sediment, their alluvial fans grow and coalesce to form a bajada—a continuous apron across the base of the mountain. (Photo by Michael Collier) (https://goo.gl/bqOZ5R)

Condor Video

Figure 19.10
Death Valley: A classic Basin and Range landscape Shortly before this satellite image was taken in February 2005, heavy rains led to the formation of a small playa lake—the pool of greenish water on the basin floor. By May 2005, the lake had reverted to a salt-covered playa. (NASA)

Figure 19.9
Playa in Death Valley These dry, flat lake beds on the floor of the basin are typically composed of silt and clay. Sometimes, as in this example, they are encrusted with salt. (Photo by Michael Collier)

borax) mined from ancient playa lake deposits in Death Valley, California.

With the ongoing erosion of the mountain mass and the accompanying sedimentation, the local relief continues to diminish. Eventually, nearly the entire mountain mass is gone. Thus, by the late stages of erosion, the mountain areas are reduced to a few large bedrock knobs projecting above the surrounding sediment-filled basin. These isolated erosional remnants on a late-stage desert landscape are called **inselbergs**, a German word meaning "island mountains."

Each of the stages of landscape evolution in an arid climate depicted in Figure 19.7 can be observed in the Basin and Range region. Recently uplifted mountains in an early stage of erosion are found in southern Oregon and northern Nevada. Death Valley, California, and southern Nevada fit into the more advanced middle stage, whereas the late stage, with its inselbergs, can be seen in southern Arizona.

Figure 19.10 is a satellite image of a portion of Death Valley that shows many of the features just described. The image was taken in February 2005, shortly after a rare heavy rain. As has occurred many times over thousands of years, a wide, shallow playa lake formed in the lowest spot. By May 2005, only 3 months after the storm, the valley floor had returned to being a dry, salt-encrusted playa.

19.3 Concept Checks

1. What is meant by *interior drainage*?

2. Describe the features and characteristics associated with each stage in the evolution of a mountainous desert.

3. Where in the United States can each stage of desert landscape evolution be observed?

19.4 | Transportation of Sediment by Wind

List and describe the ways that wind transports sediment.

Moving air, like moving water, is turbulent and able to pick up loose debris and transport it to other locations. Just as in a stream, the wind velocity increases with height above the surface. Also as with a stream, wind transports fine particles in suspension, while heavier ones are carried as bed load. However, the transport of sediment by wind differs from that of running water in two significant ways. First, wind's lower density compared to water renders it less capable of picking up and transporting coarse materials. Second, because wind is not confined to channels, it can spread sediment over large areas, as well as high into the atmosphere.

Animation

Saltating sand grains.

resting particles. At first the sand rolls along the surface. When a moving sand grain strikes another grain, one or both of them may jump into the air. Once in the air, the grains are carried forward by the wind until gravity pulls them back toward the surface. When the sand hits the surface, it either bounces back into the air or dislodges other grains, which then jump upward. In this manner, a chain reaction is established, filling the air near the ground with saltating sand grains in a short period of time (**Figure 19.11**).

Bouncing sand grains never travel far from the surface. Even when winds are very strong, the height of the saltating sand seldom exceeds 1 meter and usually is no greater than 0.5 meter. Some sand grains are too large to be thrown into the air by impact from other particles. When this is the case, the energy provided by the impact of the smaller saltating grains drives the larger grains forward. Estimates indicate that between 20 and 25 percent of the sand transported in a sandstorm is moved in this way.

Bed Load

The **bed load** that wind carries consists of sand grains. Observations in the field and experiments using wind tunnels indicate that windblown sand moves mainly by skipping and bouncing along the surface—a process termed **saltation** (Latin for "to jump").

The movement of sand grains begins when wind reaches a velocity sufficient to overcome the inertia of the

Suspended Load

Unlike sand, finer particles of dust can be swept high into the atmosphere by the wind. Because dust is often composed of rather flat particles that have large surface areas compared to their weight, it is relatively easy for turbulent air to counterbalance the pull of gravity and keep these fine particles airborne for hours or even days.

Video

A dust storm blackens the Colorado sky in this historic image from the Dust Bowl of the 1930s.

A.

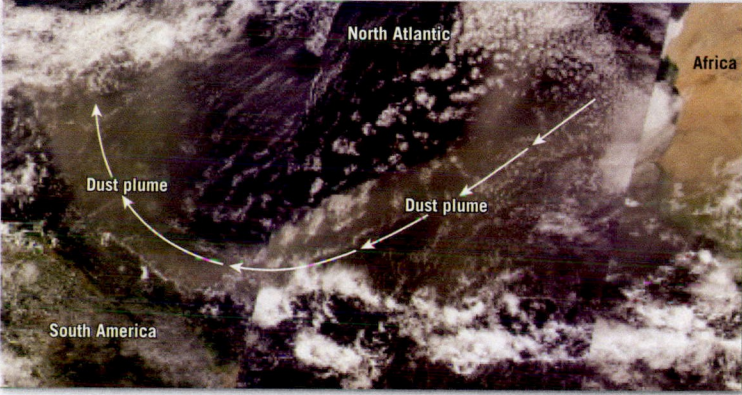
A dust storm in Africa produced a plume that reached South America. Each year such plumes transport about 40 million tons of dust from the Sahara to the Amazon Basin.

North Atlantic

Africa

Dust plume

Dust plume

South America

B.

Although both silt and clay can be carried in suspension, silt commonly makes up the bulk of the **suspended load** because the reduced level of chemical weathering in deserts provides only small amounts of clay.

Fine particles are easily carried by the wind, but they are not so easily picked up to begin with. This is because wind velocity is practically zero within a very thin layer close to the ground. Thus, the wind cannot lift the sediment by itself. Instead, the dust must be ejected or spattered into the moving air by bouncing sand grains or other disturbances. This idea is illustrated nicely by a dry, unpaved country road on a windy day. When the road is undisturbed, the wind raises little dust. However, as a car or truck moves over the road, the layer of silt is kicked up, creating a thick cloud of dust. Although the suspended load is usually deposited relatively near its

source, high winds are capable of carrying large quantities of dust great distances. In the 1930s, silt that was picked up in Kansas was transported to New England and beyond, into the North Atlantic (**Figure 19.12A**). Similarly, dust blown from the Sahara has been traced as far as South America (**Figure 19.12B**).

19.4	**Concept Checks**

1. Describe the way in which wind transports sand. When winds are strong, how high above the surface can sand be carried?

2. How does wind's suspended load differ from its bed load?

19.5 | Wind Erosion

Describe the processes and features associated with wind erosion.

Compared to running water and glaciers, wind is a relatively insignificant erosional agent. Recall that even in deserts, most erosion is performed by intermittent running water, not by wind. Wind erosion is more effective in arid lands than in humid areas because in humid places moisture binds particles together, and vegetation anchors the soil. For wind to be an effective erosional force, dryness and scant vegetation are important prerequisites. When such circumstances exist, wind may pick up, transport, and deposit great quantities of fine sediment. During the 1930s, parts of the Great Plains experienced vast dust storms (see GEOgraphics 6.3, page 204). The plowing under of the natural vegetative cover for farming, followed by severe drought, exposed the land to wind erosion, giving the area its Dust Bowl label.

Deflation and Blowouts

One way that wind erodes is by **deflation** (*de* = out, *flat* = blow), the lifting and removal of loose material. Deflation sometimes is difficult to notice because the entire surface is being lowered at the same time, but it can be significant. In portions of the 1930s Dust Bowl, vast areas of land were lowered by as much as 1 meter in only a few years.

The most noticeable results of deflation in some places are shallow depressions appropriately called **blowouts** (**Figure 19.13**). In the Great Plains region, from Texas north to Montana, thousands of blowouts are visible on the landscape. They range from small dimples less than 1 meter deep and 3 meters wide to depressions that approach 50 meters deep and several kilometers across. The factor that controls the depths of these basins (that is, acts as base level) is the local water table. When blowouts are lowered to the water table, damp ground and vegetation prevent further deflation.

Figure 19.13
Blowouts Deflation is especially effective in creating these depressions when the land is dry and largely unprotected by anchoring vegetation. (Photo courtesy of U.S.D.A./Natural Resources Conservation Service)

Blowout

In this example, deflation has removed about 1.2 meters (4 feet) of soil—the distance from the man's outstretched arm to his feet.

Figure 19.14
Desert pavement This closely packed veneer of pebbles and cobbles is only one or two stones thick. It is underlain by material containing a significant proportion of finer particles. (Photo by Bobbé Christopherson)

Lens cap

Desert Pavement

In portions of many deserts, the surface consists of a closely packed layer of coarse particles. This veneer of pebbles and cobbles, called **desert pavement**, is only one or two stones thick (**Figure 19.14**). Beneath is a layer containing a significant proportion of silt and sand. When desert pavement is present, it is an important control on wind erosion because pavement stones are too large for deflation to remove. When this armor is disturbed, wind can easily erode the exposed fine silt.

For many years, the most common explanation for the formation of desert pavement was that it develops when wind removes sand and silt from poorly sorted surface deposits. As **Figure 19.15A** illustrates, the concentration of larger particles at the surface gradually increases as the finer particles are blown away. Eventually the surface is completely covered with pebbles and cobbles too large to be moved by the wind.

Studies have shown that the process depicted in Figure 19.15A is not an adequate explanation for all environments in which desert pavement exists. For example, in many places, desert pavement is underlain by a relatively thick layer of silt that contains few if any pebbles and cobbles. In such a setting, deflation of fine sediment could not leave behind a layer of coarse particles. Geologists have also found that in some areas, the pebbles and cobbles composing desert pavement have all been exposed at the surface for about the same length of time. This would not be the case for the process shown in Figure 19.15A. Here, the coarse particles that make up the pavement reach the surface over an extended time span, as deflation gradually removes the fine material.

As a result, an alternate explanation for desert pavement was formulated (**Figure 19.15B**). This hypothesis suggests that pavement develops on a surface that initially consists of coarse particles. Over time, protruding cobbles trap fine, windblown grains that settle and sift downward through the spaces between the larger surface stones. The process is aided by infiltrating rainwater. In this model, the cobbles composing the pavement were never buried. Moreover, it successfully explains the lack of coarse particles beneath the desert pavement.

Ventifacts and Yardangs

Like glaciers and streams, wind also erodes by **abrasion**. In dry regions as well as along some beaches, windblown sand cuts and polishes exposed rock surfaces. Abrasion sometimes creates interestingly shaped stones called **ventifacts** (**Figure 19.16A**). The side of such a stone that is exposed to the prevailing wind is abraded, leaving it polished, pitted, and with sharp edges. If the wind is not consistently from one direction, or if the pebble becomes reoriented, it may have several faceted surfaces.

Unfortunately, abrasion is often given credit for accomplishments beyond its capabilities. Such features as

SmartFigure 19.15
Formation of desert pavement A. This model shows an area with poorly sorted surface deposits. Over time, deflation lowers the surface, and coarse particles become concentrated. **B.** In this model, the surface is initially covered with cobbles and pebbles. Over time, windblown dust accumulates at the surface and gradually sifts downward. (https://goo.gl/6diZff)

Tutorial

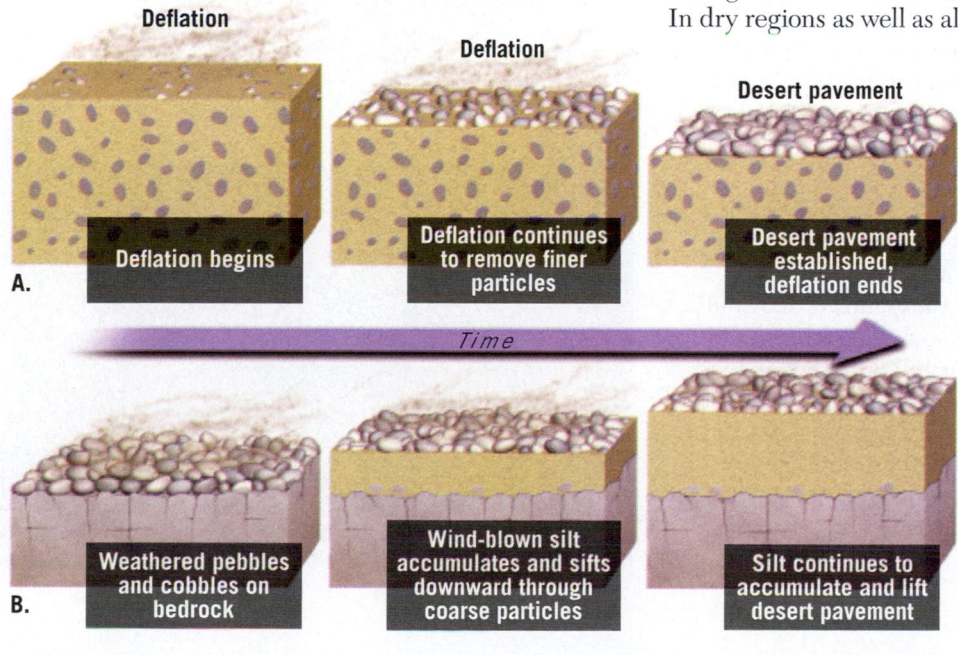

Deflation

Deflation

Desert pavement

Deflation begins

Deflation continues to remove finer particles

Desert pavement established, deflation ends

A.

Time

Weathered pebbles and cobbles on bedrock

Wind-blown silt accumulates and sifts downward through coarse particles

Silt continues to accumulate and lift desert pavement

B.

Ventifacts are stones that are shaped and polished by sandblasting.

A.

Yardangs are usually small, wind-sculpted landforms that are aligned parallel with the wind.

B.

Figure 19.16
Shaped by the wind The sandblasting effect of the wind creates ventifacts **A.** and yardangs **B.** (Photo A by Richard M. Busch; photo B by Mike P. Shepherd/Alamy Images)

balanced rocks that stand high atop narrow pedestals and intricate detailing on tall pinnacles are not the results of abrasion. Sand seldom travels more than 1 meter above the surface, so the wind's sandblasting effect is obviously limited in vertical extent.

In addition to creating ventifacts, wind erosion is responsible for creating much larger features, called yardangs (from the Turkistani word *yar*, meaning "steep bank"). A **yardang** is a streamlined, wind-sculpted landform that is oriented parallel to the prevailing wind (**Figure 19.16B**). Individual yardangs are generally small features that stand less than 5 meters (16 feet) high and no more than about 10 meters (33 feet) long. Because the sand-blasting effect of wind is greatest near the ground, these abraded bedrock remnants are usually narrower at their base. Sometimes yardangs are large features. Peru's

Ica Valley contains yardangs that approach 100 meters (330 feet) in height and several kilometers in length. Some yardangs in the desert of Iran reach 150 meters (nearly 500 feet) in height.

19.5 Concept Checks

1. Why is wind erosion relatively more important in dry regions than in humid areas?

2. What factor limits the depths of blowouts?

3. Briefly describe two hypotheses used to explain the formation of desert pavement.

4. What are yardangs and ventifacts?

EYE ON EARTH 19.2

This satellite image shows a large plume of windblown sediment covering large portions of Iran, Afghanistan, and Pakistan in March 2012. The airborne material is thick enough to completely hide the area beneath it. On either side of the plume, skies are mostly clear.

QUESTION 1 What term is applied to the erosional process responsible for producing this plume?

QUESTION 2 Is the wind-transported material in the image more likely bed load or suspended load?

QUESTION 3 People sometimes refer to events like the one pictured here as "sandstorms." Is that an appropriate description? Why or why not?

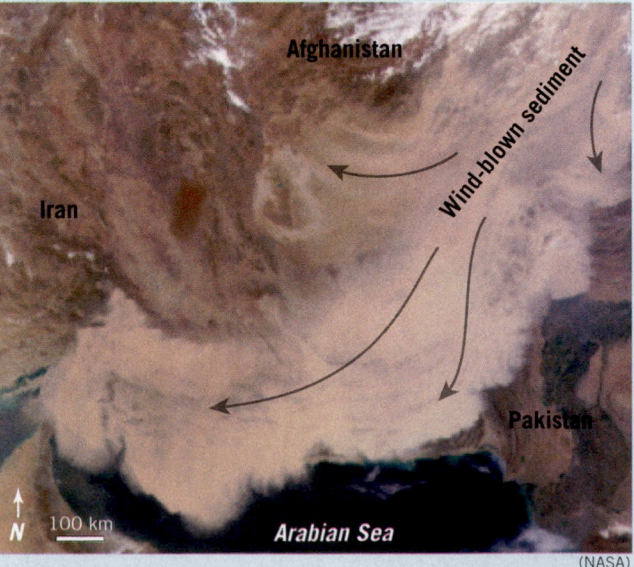

(NASA)

19.6 | Wind Deposits

Discuss dune formation and movement and distinguish among different dune types. Explain how loess deposits differ from deposits of sand.

Although wind is relatively unimportant in producing *erosional* landforms, significant *depositional* landforms are created by the wind in some regions. Accumulations of windblown sediment are particularly conspicuous in the world's dry lands and along many sandy coasts. Wind deposits are of two distinctive types: (1) mounds and ridges of sand from the wind's bed load, which we call *dunes*, and (2) extensive blankets of silt, called *loess*, that once were carried in suspension.

Sand Deposits

As is the case with running water, wind drops its load of sediment when velocity falls and the energy available for transport diminishes. Thus, sand begins to accumulate wherever an obstruction across the path of the wind slows its movement. Unlike many deposits of silt, which form blanketlike layers over large areas, winds commonly deposit sand in mounds or ridges called **dunes** (**Figure 19.17**).

Moving air encountering an object, such as a clump of vegetation or a rock, sweeps around and over the object, leaving a "shadow" of slower-moving air behind the obstacle and a smaller zone of quieter air just in front of the obstacle. Some of the saltating sand grains moving with the wind come to rest in these wind shadows. As the accumulation of sand continues, it becomes a more imposing barrier to the wind and thus a more efficient trap for even more sand. If there is a sufficient supply of sand and the wind blows steadily for a long enough time, the mound of sand grows into a dune.

Many dunes have an asymmetrical profile, with the leeward (sheltered) slope being steep and the windward slope more gently inclined. The dunes in Figure 19.17 are a good example. Sand moves up the gentler slope on the windward side by saltation. Just beyond the crest of the dune, where the wind velocity is reduced, the sand accumulates. As more sand collects, the slope steepens, and eventually some of it slides under the pull of gravity. In this way, the leeward slope of the dune, called the **slip face**, maintains an angle of about 34 degrees, the angle of repose for loose dry sand. (Recall from Chapter 15 that the angle of repose is the steepest angle at which loose material remains stable.) Continued sand accumulation, coupled with periodic slides down the slip face, results in the slow migration of the dune in the direction of air movement.

As sand is deposited on the slip face, layers form that are inclined in the direction the wind is blowing. These sloping layers are called **cross-beds** (**Figure 19.18**). When the dunes are eventually buried under other layers of sediment and become part of the sedimentary rock record, their asymmetrical shape is destroyed, but the cross-beds remain as testimony to their origin. Nowhere is cross-bedding more prominent than in the sandstone walls of Zion Canyon in southern Utah (see Figure 19.18).

For some areas, moving sand is troublesome. In **Figure 19.19**, dunes are advancing across irrigated fields in Egypt. In portions of the Middle East, valuable oil rigs must be protected from encroaching dunes. In some cases, fences are built sufficiently upwind of the dunes to stop their migration. As sand continues to collect, however, the fences must be built higher. In Kuwait protective fences extend for almost 10 kilometers (6 miles) around one important oil field. Migrating dunes can also pose a problem to the construction

SmartFigure 19.17
White Sands National Monument The dunes at this landmark in southeastern New Mexico are composed of gypsum. The dunes slowly migrate with the wind. (Photos by Michael Collier) (http://goo.gl/m8eYBH)

Mobile Field Trip

Wind

Strong winds move sand up the relatively gentle windward slope.

As sand accumulates at the dune crest, the slope steepens and some of the sand slides down the steep *slip face*.

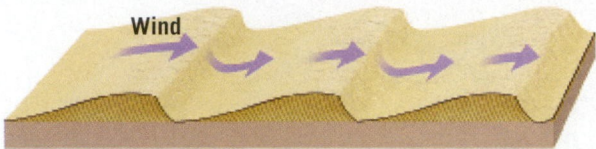

Dunes commonly have an asymmetrical shape and migrate with the wind.

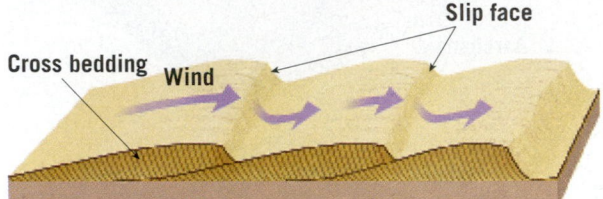

Sand grains deposited on the slip face at the angle of repose create the cross bedding of dunes.

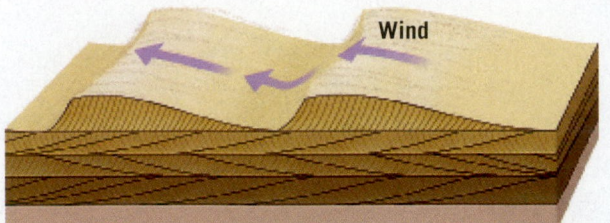

When dunes are buried and become part of the sedimentary rock record, the cross bedding is preserved.

Cross bedding is an obvious characteristic of the Navajo Sandstone in Zion National Park, Utah.

SmartFigure 19.18
Cross-bedding As sand is deposited on the slip face, layers form that are inclined in the direction the wind is blowing. With time, complex patterns develop in response to changes in wind direction. (Photo by Dennis Tasa) (https://goo.gl/Ewb76M)

Tutorial

and maintenance of highways and railroads that cross sandy desert regions. For example, to keep a portion of Highway 95 near Winnemucca, Nevada, open to traffic, sand must be taken away about three times a year. Each time, between 1500 and 4000 cubic meters of sand are removed. Attempts at stabilizing the dunes by planting different varieties of grasses have been unsuccessful because the meager rainfall cannot support the plants.

Types of Sand Dunes

Dunes are not just random heaps of windblown sediment. Rather, they are accumulations that usually assume patterns that are surprisingly consistent. A leading early investigator of dunes, the British engineer R. A. Bagnold, observed: "Instead of finding chaos and disorder, the observer never fails to be amazed at a simplicity of form, an exactitude of repetition, and a geometric order." A broad assortment of dune forms exists, generally simplified to a few major types for discussion.

Of course, gradations exist among different forms, and there are also irregularly shaped dunes that do not fit easily into any category. Several factors influence the form and size that dunes ultimately assume. These include wind direction and velocity, availability of sand, and amount of vegetation present. Six basic dune types are shown in Figure 19.20, with arrows indicating wind directions.

Barchan Dunes Solitary sand dunes shaped like crescents and with their tips pointing

downwind are called **barchan dunes** (see Figure 19.20A). These dunes form where supplies of sand are limited and the surface is relatively flat, hard, and lacking vegetation. They migrate slowly with the wind at a rate of up to 15 meters (50 feet) per year. Their size is usually modest, with the largest barchans reaching heights of about 30 meters (100 feet), while the maximum spread between their tips approaches 300 meters (1000 feet). When the wind direction is nearly constant, the crescent form of these dunes is nearly symmetrical. However, when the wind direction is not perfectly fixed, one tip becomes larger than the other.

Transverse Dunes In regions where the prevailing winds are steady, sand is plentiful, and vegetation is sparse or absent, the dunes form a series of long ridges that are separated by troughs and oriented at right angles to the prevailing wind. Because of this orientation, they

Figure 19.19
Migrating dunes These desert dunes in Egypt are encroaching on irrigated farm fields. (Photo by George Gerster/Photo Researchers, Inc.)

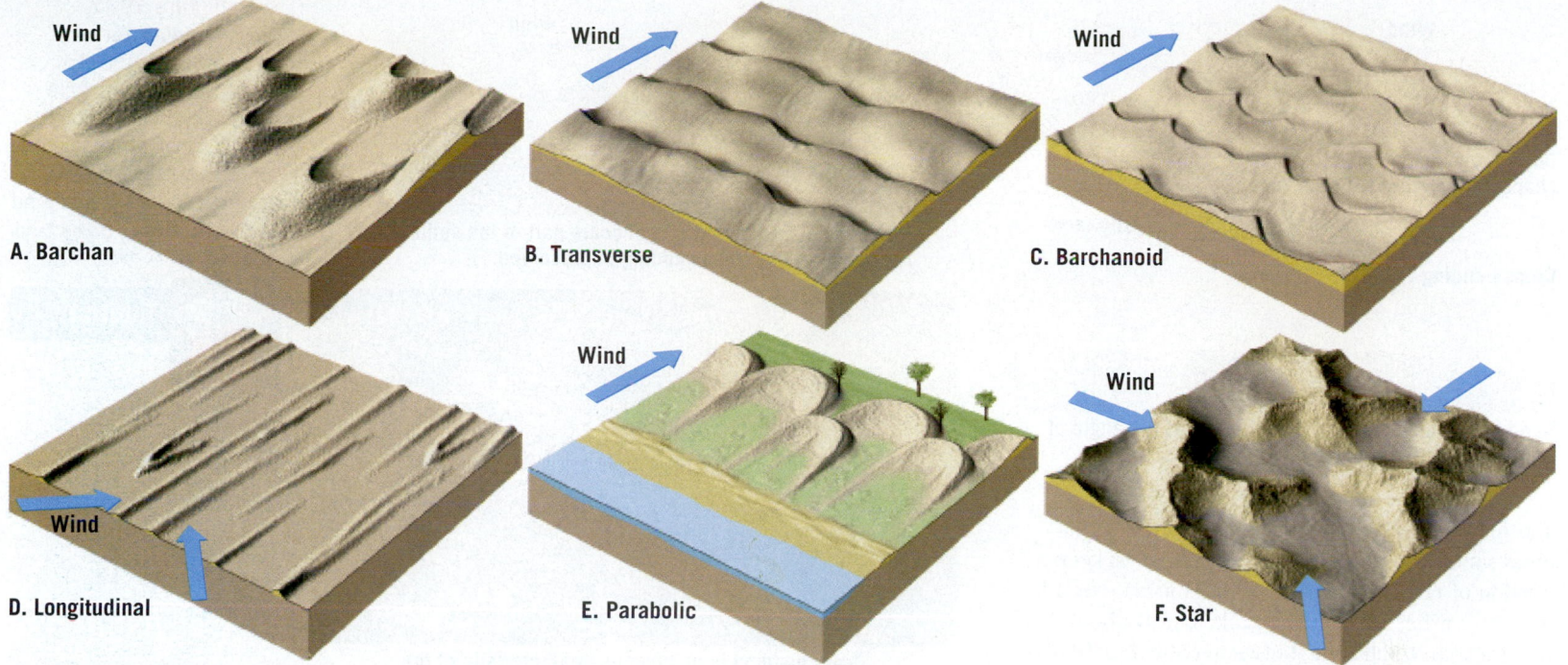

A. Barchan

Wind

B. Transverse

Wind

C. Barchanoid

Wind

D. Longitudinal

Wind   Wind

E. Parabolic

Wind

F. Star

Wind

SmartFigure 19.20
Types of sand dunes
Factors that influence the form and size of dunes include wind direction and velocity, the availability of sand, and the amount of vegetation. (https://goo.gl/fuZU6X)

Tutorial

are termed **transverse dunes** (see **Figure 19.20B**). Many coastal dunes are of this type. In addition, transverse dunes are common in many arid regions where the extensive surface of wavy sand is sometimes called a *sand sea*. In some parts of the Sahara and Arabian Deserts, transverse dunes reach heights of 200 meters (660 feet), are 1 to 3 kilometers (0.6 to 2 miles) across, and can extend for distances of 100 kilometers (60 miles) or more.

There is a relatively common dune form that is intermediate between isolated barchans and extensive waves of transverse dunes. Such dunes, called **barchanoid dunes**, form scalloped rows of sand oriented at right angles to the wind (see **Figure 19.20C**). The rows resemble a series of barchans that have been positioned side by side.

Longitudinal Dunes   Long ridges of sand that form more or less parallel to the prevailing wind where sand supplies are moderate are called **longitudinal dunes** (see **Figure 19.20D**). Apparently, the prevailing wind direction must vary somewhat but still remain in the same quadrant of the compass. Although the smaller types are only 3 or 4 meters high and several dozens of meters long, in some large deserts, longitudinal dunes can reach great size. For example, in portions of North Africa, Arabia, and central Australia, these dunes may approach a height of 100 meters (330 feet) and extend for distances of more than 100 kilometers (62 miles).

Parabolic Dunes   Unlike the other dune types described thus far, **parabolic dunes** form where vegetation partially covers the sand. The shape of these dunes resembles the shape of barchans except that their tips point into the wind rather than downwind (see **Figure 19.20E**). Parabolic

dunes often form along coasts where there are strong onshore winds and abundant sand. If the sand's sparse vegetative cover is disturbed at some spot, deflation creates a blowout. Sand is then transported out of the depression and deposited as a curved rim, which grows higher as deflation enlarges the blowout.

Star Dunes   Confined largely to parts of the Sahara and Arabian Deserts, **star dunes** are isolated hills of sand that exhibit a complex form (see **Figure 19.20F**). Their name is derived from the fact that the bases of these dunes resemble multipointed stars. Usually three or four sharp-crested ridges diverge from a central high point that in some cases may approach a height of 90 meters (300 feet). As their form suggests, star dunes develop where wind directions are variable.

Loess (Silt) Deposits

In some parts of the world, the surface topography is mantled with deposits of windblown silt, called **loess**. Over periods of perhaps thousands of years, dust storms deposited this material. When loess is breached by streams or road cuts, it tends to maintain vertical cliffs and lacks any visible layers, as you can see in **Figure 19.21**.

The distribution of loess worldwide indicates that there are two primary sources for this sediment: deserts and glacial outwash deposits. The thickest and most extensive deposits of loess on Earth occur in western and northern China. They were blown there from the extensive desert basins of central Asia. Accumulations of 30 meters (100 feet) are common, and thicknesses of

more than 100 meters (330 feet) have been measured. It is this fine, buff-colored sediment that gives the Yellow River (Huang Ho) its name.

In the United States, deposits of loess are significant in many areas, including South Dakota, Nebraska, Iowa, Missouri, and Illinois, as well as portions of the Columbia Plateau in the Pacific Northwest. The correlation between the distribution of loess and important farming regions in the Midwest and eastern Washington State is not just a coincidence because soils derived from this wind-deposited sediment are among the most fertile in the world.

Unlike the deposits in China, which originated in deserts, the loess in the United States and Europe is an indirect product of glaciation. Its source is deposits of stratified drift. During the retreat of the ice sheets, many river valleys were choked with sediment deposited by meltwater. Strong westerly winds sweeping across the barren floodplains picked up the finer sediment and dropped it as a blanket on the eastern sides of the valleys. Such an origin is confirmed by the fact that loess deposits are thickest and coarsest on the leeward side of such major glacial drainage outlets as the Mississippi and Illinois Rivers and rapidly thin with increasing distance from the valleys. Furthermore, the angular, mechanically weathered particles composing the loess are essentially the same as the rock flour produced by the grinding action of glaciers.

This vertical bluff near the Mississippi River in southern Illinois is about 3 meters (10 feet) high.

Figure 19.21
Loess In some regions, the surface is mantled with deposits of windblown silt. (Top photo by James E. Patterson; bottom photo by Ashley Cooper/Alamy)

In parts of China, loess has sufficient structural strength to permit excavation of cave-like dwellings.

19.6 Concept Checks

1. How do sand dunes migrate?

2. List and briefly distinguish among basic dune types.

3. How do loess deposits differ from sand deposits?

4. How are some loess deposits related to glaciers?

EYE ON EARTH 19.3

This is an aerial view of the Preston Mesa dunes in northern Arizona.

QUESTION 1 Which one of the basic dune types is shown here?

QUESTION 2 Sketch a simple profile (side view) of one of these dunes. Add an arrow to show the prevailing wind direction and label the dune's slip face.

QUESTION 3 These dunes gradually migrate across the surface. Describe this process.

Michael Collier

19.1 Distribution and Causes of Dry Lands

Describe the general distribution of Earth's dry lands and explain why deserts form in the subtropics and middle latitudes.

KEY TERMS dry climate, desert, steppe, desertification, rainshadow

- Dry climates cover about 30 percent of Earth's land area. These regions have yearly precipitation totals that are less than the potential loss of water through evaporation. Evaporation depends on temperature, and deserts may occur in hot or cold climates. Deserts are drier than steppes, but both climatic types are considered water deficient.
- Dry climates in subtropical latitudes are associated with the global distribution of air pressure and winds. Near the equator, warm, moist air rises (causing lots of rain) and then moves to 20 or 30 degrees latitude before sinking back toward Earth's surface. The subsiding air brings clear skies, copious sunshine, and dry conditions to these zones of subtropical high pressure.
- Deserts also occur in continental interiors of the middle latitudes. Most are subject to the rainshadow effect, in which moist air moving inland from oceans is intercepted by mountainous obstacles. As the air is forced to rise, it cools, producing clouds and precipitation on the windward slopes. By contrast, the leeward side, called the rainshadow, tends to be quite dry.

Q This map shows annual precipitation across the continent of Africa. At which latitudes is the atmosphere rising? At which latitudes is it sinking? How does this atmospheric circulation influence the continent's climates?

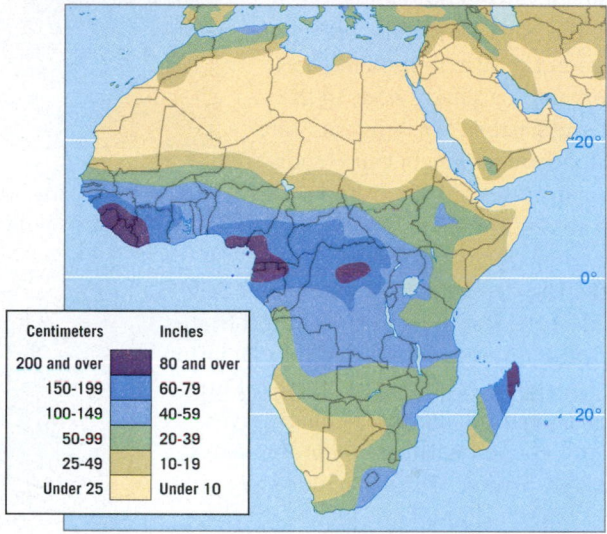

Centimeters	Inches
200 and over	80 and over
150-199	60-79
100-149	40-59
50-99	20-39
25-49	10-19
Under 25	Under 10

19.2 Geologic Processes in Arid Climates

Summarize the geologic roles of weathering, running water, and wind in arid and semiarid climates.

KEY TERMS ephemeral stream

- In dry lands rock weathering of any type is greatly reduced because of the lack of water and the scarcity of organic acids from decaying plants.
- Practically all desert streams are dry most of the time and are said to be ephemeral. Their channels are carved out largely by flash floods that occur during sporadic storm events.
- Permanent desert streams originate in wetter climates. They must carry a great volume of water to keep from losing all their water as they cross the desert.
- Running water is responsible for most of the erosional work in a desert. Although wind erosion is more significant in dry areas than in other environments, it still cannot match running water as an agent of erosion in deserts.

Q This is a typical desert stream shortly after a heavy rain. What term is applied to such a stream? How will this scene likely change in the near future?

Ben Kelmer/Alamy

19.3 Basin and Range: The Evolution of a Desert Landscape

Discuss the stages of landscape evolution in the Basin and Range region of the western United States.

KEY TERMS interior drainage, alluvial fan, bajada, playa lake, playa, inselberg

- The Basin and Range region is a distinctive area of western North America that demonstrates some key principles of mountainous desert landscapes. The mountain ranges and intervening valleys were produced through normal faulting, but then the rock mass of the mountain ranges was redistributed by weathering, erosion, transportation, and deposition. With sufficient time, the overall effect is to reduce topographic relief by grinding down the mountains and depositing the sedimentary debris in the low-lying basins.

- Alluvial fans form during the early stage of dry landscape development, when the relief is highest. Over time, these fans grow and merge, so that by the middle stage, continuous bajadas of sediment cover the line where the ranges meet the basins. On the basin floor, playa lakes may form during wet periods and then dry up to produce salty playas.
- When almost all of the mountains have been worn down and most low-lying basins have been filled in with sediment, the landscape has reached the late stage of development, marked by isolated inselbergs of bedrock poking up through a sea of sediment.

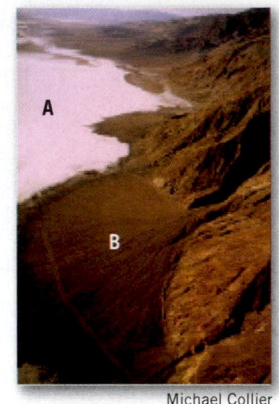

Michael Collier

Q Identify the lettered features in this photo. How did they form?

19.4 Transportation of Sediment by Wind

List and describe the ways that wind transports sediment.

KEY TERMS bed load, saltation, suspended load

- A current of air can pick up and transport sediment, though with a lower competence than a current of water or glacial ice. Wind cannot pick up the coarser particles that water can but is capable of transporting sediment over vast areas and even high into the atmosphere.
- A portion of the sediment transported by wind is bed load that bounces along Earth's surface. Generally, these saltating particles of sand never get more than 0.5 meter above the ground.
- Other sediment is fine grained enough to stay suspended in the air. Clay and silt can be carried as suspended load. Once aloft, they can be carried tremendous distances, across entire continents and ocean basins.

19.5 Wind Erosion

Describe the processes and features associated with wind erosion.

KEY TERMS deflation, blowout, desert pavement, abrasion, ventifact, yardang

- Wind is capable of erosion, though it should be emphasized that water is the most important agent of erosion in desert regions. The Dust Bowl is a classic example of a massive episode of soil erosion by wind during the 1930s. Localized deflation may produce shallow depressions known as blowouts.

- Desert pavement is a thin layer of coarse pebbles and cobbles that covers some desert surfaces. Two models have been proposed to explain the formation of desert pavement: (1) a model in which deflation strips the finer grains from a deposit of poorly sorted sediments and (2) an additive model in which pebbles and cobbles trap smaller windblown grains in the spaces between them.
- Ventifacts are produced when individual rocks become abraded by windblown sediment, giving the rocks a polished, pitted surface. Similar "sand-blasting" of the land surface may sculpt rock outcrops into streamlined yardangs that are longest parallel to the prevailing wind direction.

19.6 Wind Deposits

Discuss dune formation and movement and distinguish among different dune types. Explain how loess deposits differ from deposits of sand.

KEY TERMS dune, slip face, cross-bed, barchan dune, transverse dune, barchanoid dune, longitudinal dune, parabolic dune, star dune, loess

- Wind deposits are of two distinct types: (1) mounds and ridges of sand, called dunes, which are formed from sediment that is carried as part of the wind's bed load, and (2) extensive blankets of silt, called loess, carried by wind in suspension.
- Dunes accumulate due to the difference in wind energy on the upwind and downwind sides of some obstacle. Wind moves sand up the more gently sloping upwind side, across the crest of the dune, where it settles out in the calmer air on the downwind side, the steeply sloped slip face. When the sand on the slip face is piled up past the angle of repose, it will spontaneously collapse in small "avalanches" of sand. Over time, this will cause a dune to slowly move in the direction of the prevailing wind. Inside the dune, the buried slip faces may be preserved as cross-beds.

- There are six major kinds of dunes. Their shapes result from the pattern of prevailing winds, the amount of available sand, and the presence of vegetation.
- Loess is windblown silt, deposited over large areas sometimes in thick blankets. Most loess is derived from either (1) deserts or (2) areas that have recently been glaciated. Winds blow across stratified drift and pick up silt-size grains, carrying them in suspension to the site of deposition.

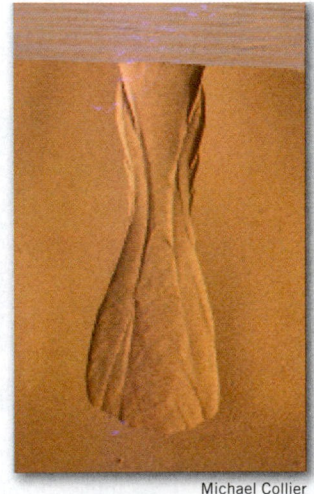

Michael Collier

Q This close-up photo shows a small portion of one side of a barchan dune. What term is applied to this side of the dune? Is the prevailing wind direction "coming out" of the photo or "going into" the photo? Explain. Why did some of the sand break away and slide?

Give It Some *Thought*

1. Albuquerque, New Mexico, receives an average of 20.7 centimeters (8.07 inches) of rainfall annually. Albuquerque is considered a desert under the commonly used Köppen climate classification. The Russian city of Verkhoyansk is located near the Arctic Circle in Siberia. Yearly precipitation at Verkhoyansk averages 15.5 centimeters (6.05 inches), about 5 centimeters (2 inches) less than Albuquerque, yet it is classified as a humid climate. Explain why this is the case.

2. Examine the precipitation map for the state of Nevada. Notice that the areas receiving the most precipitation resemble long, slender "islands" scattered across the state. Provide an explanation for this pattern.

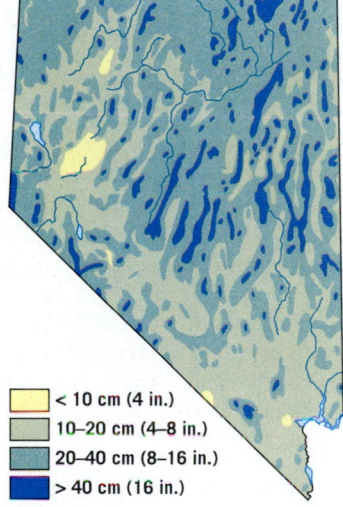

	< 10 cm (4 in.)
	10–20 cm (4–8 in.)
	20–40 cm (8–16 in.)
	> 40 cm (16 in.)

3. Name three deserts that are visible on this classic view of Earth from space. Briefly explain why these regions are so dry. Can you also explain the cause of the band of clouds in the region of the equator?

NASA

4. Compare and contrast the sediment deposited by a stream, the wind, and a glacier. Which deposit should have the most uniform grain size? Which one would exhibit the poorest sorting? Explain your choices.

5. Is either of the following statements true? Are they both true? Explain your answer.

   a. Wind is more effective as an agent of erosion in dry places than in humid places.

   b. Wind is the most important agent of erosion in deserts.

6. This satellite image shows a small portion of the Zagros Mountains in dry southern Iran. Streams in this region flow only occasionally. The green tones on the image identify productive agricultural areas.

Basin

?

Mountains

Dry river bed

2.5 km

NASA

   a. Identify the large feature that is labeled with a question mark.

   b. Explain how the feature named in Question a formed.

   c. What term is used to describe streams like the ones that occur in this region?

   d. Speculate on the likely source of water for the agricultural areas in this image.

7. Bryce Canyon National Park, shown in the photo, is in dry southern Utah. It is carved into the eastern edge of the Paunsaugunt Plateau. Erosion has sculpted the colorful limestone into bizarre shapes, including spires called "hoodoos." As you and a companion (who has not studied geology) view Bryce Canyon, your friend says, "It's amazing how wind has created this incredible scenery!" Now that you have studied arid landscapes, how would you respond to your companion's statement?

ozoptimes/Shutterstock

Mastering Geology™

Looking for additional review and test prep materials? With individualized coaching on the toughest topics of the course, MasteringGeology offers a wide variety of ways for you to move beyond memorization to begin thinking like a geologist. Visit the Study Area in www.masteringgeology.com to find practice quizzes, study tools, and multimedia that will improve your understanding of this chapter's content. Sign in today to enjoy the following features: **Self Study Quizzes, SmartFigures: Tutorials/Animations/Condor Videos/Mobile Field Trips, Geoscience Animation Library, GEODe, RSS Feeds, Digital Study Modules,** and an optional **Pearson eText.**

20
Shorelines

The lower portion of this aerial view shows a sand spit and a small portion of a beach along the shore of Nantucket, an island off the coast of Cape Cod. Offshore shoals (sandbars) are also conspicuous features. (Photo by Michael Collier)

Each statement represents the primary **LEARNING OBJECTIVE** for the corresponding major heading within the chapter. After you complete the chapter, you should be able to:

20.1 Explain why the shoreline is considered a dynamic interface and identify the basic parts of the coastal zone.

20.2 List and discuss the factors that influence the height, length, and period of a wave and describe the motion of water within a wave.

20.3 Explain how waves erode and how waves move sediment along the shore.

20.4 Describe the features typically created by wave erosion and those resulting from sediment deposited by longshore transport processes.

20.5 Distinguish between emergent and submergent coasts. Contrast the erosion problems faced on the Atlantic and Gulf coasts with those along the Pacific coast.

20.6 Describe the basic structure and characteristics of a hurricane and the three broad categories of hurricane destruction.

20.7 Summarize the ways in which people deal with shoreline erosion problems.

20.8 Explain the cause of tides, their monthly cycles, and patterns. Describe the horizontal flow of water that accompanies the rise and fall of tides.

The restless waters of the ocean are constantly in motion. Winds generate surface currents, the gravity of the Moon and Sun produces tides, and density differences create deep-ocean circulation. Further, waves carry the energy from storms to distant shores, where their impact erodes the land.

Shorelines are dynamic environments. Their topography, geologic makeup, and climate vary greatly from place to place. Continental and oceanic processes converge along the shore to create landscapes that frequently undergo rapid change. When it comes to the deposition of sediment, shore areas are transition zones between marine and continental environments.

20.1 | The Shoreline: A Dynamic Interface

Explain why the shoreline is considered a dynamic interface and identify the basic parts of the coastal zone.

Nowhere is the restless nature of the ocean's water more noticeable than along the shore—the dynamic interface among air, land, and sea. An **interface** is a common boundary where different parts of a system interact. This is certainly an appropriate designation for the coastal zone. Here we can see the rhythmic rise and fall of tides and observe waves constantly rolling in and breaking. Sometimes the waves are low and gentle. At other times they pound the shore with awesome fury.

The Coastal Zone

Although it may not be obvious, the shoreline is constantly being modified by waves. For example, along Cape Cod, Massachusetts, wave activity is eroding cliffs of poorly consolidated glacial sediment so aggressively that the cliffs are retreating inland up to 1 meter (3 feet) per year (**Figure 20.1A**). By contrast, at Point Reyes, California, the far more durable bedrock cliffs

Figure 20.1
Cape Cod and Point Reyes A. A satellite image of Cape Cod, Martha's Vineyard (left), and Nantucket (right). Although waves constantly modify this coastal landscape, they are not responsible for creating it. Rather, the original size and shape of Cape Cod resulted from the positioning of moraines and other glacial materials deposited during the Quaternary Ice Age (see Figure 18.27, page 554). (Image courtesy of Earth Satellite Corporation/Science Photo Library/ Photo Researchers, Inc.)
B. A high-altitude image of the Point Reyes area north of San Francisco. The 5.5-kilometer- (3.4-mile-) long south-facing cliffs are exposed to the full force of the waves from the Pacific Ocean. Nevertheless, this promontory retreats slowly because it is formed from highly resistant bedrock. (Image courtesy of USDA-ASCS)

A.

B.

are less susceptible to wave attack and therefore are retreating much more slowly (**Figure 20.1B**). Along both coasts, wave activity is moving sediment along the shore and building narrow sandbars that protrude into and across some bays.

Present-day Shorelines   The nature of present-day shorelines is not just the result of the relentless attack of the land by the sea. Rather, the shore has a complex character that results from multiple geologic processes. For example, practically all coastal areas were affected by the worldwide rise in sea level that accompanied the melting of glaciers following the Last Glacial Maximum (see Figure 18.31, page 558). As the sea encroached landward, the shoreline retreated, becoming superimposed upon existing landscapes that had resulted from such diverse processes as stream erosion, glaciation, volcanic activity, and the forces of mountain building.

Human Activity   The coastal zone is experiencing intensive human activity. About half of the world's human population lives on or within about 100 kilometers (60 miles) of a coast. Such large numbers of people so near the shore means that hurricanes and tsunamis place millions at risk. Unfortunately, people often treat the shoreline as if it were a stable platform on which structures can safely be built. This attitude inevitably leads to conflicts between people and nature. As you will see, many coastal landforms, especially beaches and barrier islands, are relatively fragile, short-lived features that are inappropriate sites for development. The image of the New Jersey shoreline in **Figure 20.2** is a good example. In the years to come, coastal areas will be even more vulnerable because sea level is rising due

to human-induced climate change. There is more about this in Chapter 21.

Basic Features of the Coastal Zone

In general conversation, a number of terms are used to describe the boundary between land and sea. In the preceding paragraphs, the terms *shore*, *shoreline*, *coastal zone*, and *coast* were all used. Moreover, when many think of the land–sea interface, the word *beach* comes to mind. Let's take a moment to clarify these terms and introduce some other terminology used by those who study the land–sea boundary zone. You will find it helpful to refer to **Figure 20.3**, which is an idealized profile of the coastal zone.

Shore and Coast   The **shoreline** is the line that marks the contact between land and sea. Each day, as tides rise and fall, the position of the shoreline migrates. Over longer time spans, the average position of the shoreline gradually shifts as sea level rises or falls.

The **shore** is the area that extends between the lowest tide level and the highest elevation on land that is

Figure 20.2
Hurricane Sandy A portion of the New Jersey shoreline just south of New York City shortly after this huge storm, also called Superstorm Sandy, struck in late October 2012. The extraordinary storm surge caused the damage pictured here. The fact that it struck the most populated metropolitan region in the United States clearly contributed to the storm's great financial impact. Many shoreline areas are intensively developed. Often the shifting shoreline sands and the desire of people to occupy these areas are in conflict. (Photo by Mario Tama/Getty Images)

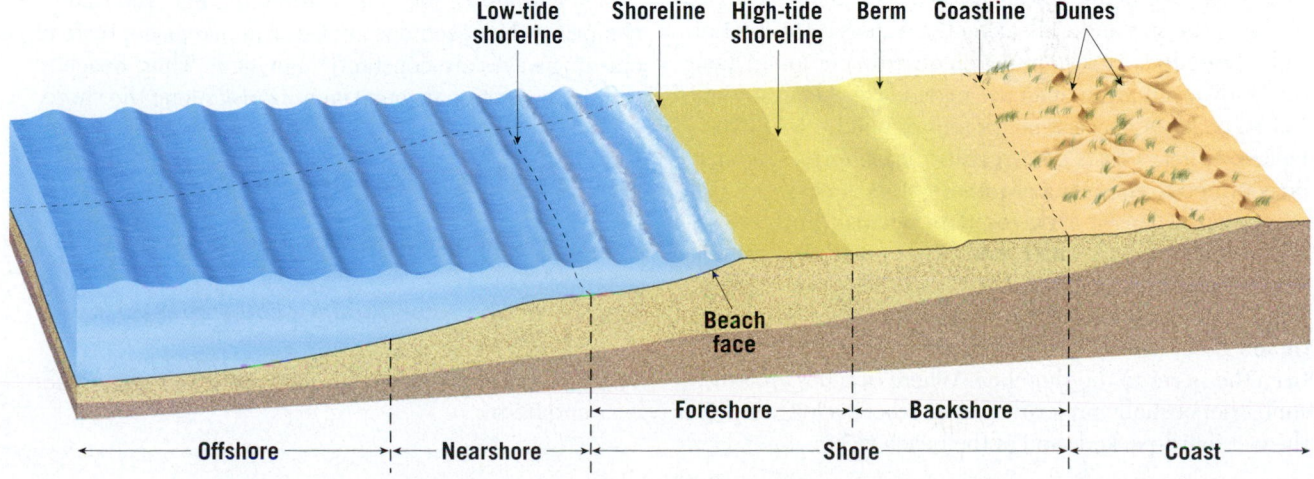

Low-tide shoreline   Shoreline   High-tide shoreline   Berm   Coastline   Dunes

Beach face

Foreshore   Backshore

Offshore — Nearshore — Shore — Coast

Figure 20.3
The coastal zone The transition zone between land and sea consists of several parts.

Figure 20.4
Beaches A beach is an accumulation of sediment on the landward margin of an ocean or a lake and can be thought of as material in transit along the shore. Beaches are composed of whatever material is locally available. (Photo A by David R. Frazier/Photo Library/Alamy Images; photo B by E. J. Tarbuck)

This beach on Florida's Sanibel Island consists of shells and shell fragments.

A.

The black sands on this beach in Hawaii were derived from the weathering of nearby basaltic lava flows.

B.

affected by storm waves. By contrast, the **coast** extends inland from the shore as far as ocean-related features can be found. The **coastline** marks the coast's seaward edge, whereas the inland boundary is not always obvious or easy to determine.

As Figure 20.3 illustrates, the shore is divided into the *foreshore* and the *backshore*. The **foreshore** is the area that is exposed when the tide is out (low tide) and submerged when the tide is in (high tide). The **backshore** is landward of the high-tide shoreline. It is usually dry, being affected by waves only during storms. Two other zones are commonly identified. The **nearshore zone** lies between the low-tide shoreline and the line where waves break at low tide. Seaward of the nearshore zone is the **offshore zone**.

Beaches   For many, a beach is the sandy area where people lie in the sun and walk along the water's edge. Technically, a **beach** is an accumulation of sediment found along the landward margin of an ocean or a lake. Along straight coasts, beaches may extend for tens or hundreds of kilometers. Where coasts are irregular, beach formation may be confined to the relatively quiet waters of bays.

Beaches consist of one or more **berms**, which are relatively flat platforms often composed of sand that are adjacent to coastal dunes or cliffs and marked by a change in slope at the seaward edge. Another part of the beach is the **beach face**, which is the wet, sloping surface that extends from the berm to the shoreline. Where beaches are sandy, sunbathers usually prefer the berm, whereas joggers prefer the wet, hard-packed sand of the beach face.

Beaches are composed of whatever material is locally abundant. The sediment for some beaches is derived from the erosion of adjacent cliffs or nearby coastal mountains. Other beaches are built from sediment delivered to the coast by rivers.

Although the mineral makeup of many beaches is dominated by durable quartz grains, other minerals may be dominant. For example, in areas such as southern Florida, where there are no mountains or other sources of rock-forming minerals nearby, most beaches are composed of shell fragments and the remains of organisms that live in coastal waters (**Figure 20.4A**). Some beaches on volcanic islands in the open ocean are composed of weathered grains of the basaltic lava that comprise the islands or of coarse debris eroded from coral reefs that develop around islands in low latitudes (**Figure 20.4B**).

Regardless of the composition, the material that comprises the beach does not stay in one place. Instead, crashing waves are constantly moving it. Thus, beaches can be thought of as material in transit along the shore.

20.1   Concept Checks

1. Why is the shoreline described as being an interface?

2. Distinguish among *shore*, *shoreline*, *coast*, and *coastline*.

3. What is a beach? Distinguish between *beach face* and *berm*.

20.2 | Ocean Waves

List and discuss the factors that influence the height, length, and period of a wave and describe the motion of water within a wave.

Ocean waves are caused by energy traveling along the interface between ocean and atmosphere, often transferring energy from a storm far out at sea over distances of several thousand kilometers. That's why even on calm days, the ocean still has waves that travel across its surface. When observing waves, always remember that you are watching *energy* travel through a medium (water). If you make waves by tossing a pebble into a pond, or by splashing in a pool, or by blowing across the surface of a cup of coffee, you are imparting *energy* to the liquid, and the waves you see are visible evidence of the energy passing through.

Wind-generated waves provide most of the energy that shapes and modifies shorelines. Where the land and sea meet, waves that may have traveled unimpeded for hundreds or thousands of kilometers suddenly encounter a barrier that will not allow them to advance farther and must absorb their energy. Stated another way, the shore is the location where a practically irresistible force confronts an almost immovable object. The conflict that results is never-ending and sometimes dramatic.

Wave Characteristics

Most ocean waves derive their energy and motion from the wind. When the velocity of a breeze is less than 3 kilometers (2 miles) per hour, only small wavelets appear. At greater wind speeds, more stable waves gradually form and advance with the wind.

Characteristics of ocean waves are illustrated in Figure 20.5, which shows a simple, nonbreaking wave form. The tops of the waves are the *crests*, which are separated by *troughs*. Halfway between the crests and troughs is the *still water level*, which is the level the water would occupy if there were no waves. The vertical distance between trough and crest is called the **wave height**, and the horizontal distance between successive crests (or troughs) is the **wavelength**. The time it takes one full wave—one wavelength—to pass a fixed position is the **wave period**.

The height, length, and period that are eventually achieved by a wave depend on three factors: (1) the wind speed, (2) the length of time the wind has blown, and (3) the **fetch**, or distance that the wind has traveled across open water. As the quantity of energy transferred from the wind to the water increases, the height and steepness of the waves increase as well. Eventually a critical point is reached where waves grow so tall

that they topple over, forming ocean breakers called *whitecaps*.

For a particular wind speed, there is a maximum fetch and duration of wind beyond which waves will no longer increase in size. This is because the waves are losing as much energy through the breaking of whitecaps as they are receiving from the wind. When the maximum fetch and duration are reached for a given wind velocity, the waves are said to be "fully developed."

When wind stops or changes direction, or if waves leave the stormy area where they were created, they continue on unrelated to local winds. The waves also undergo a gradual change to *swells*, which are lower and longer and may carry a storm's energy to distant shores. Because many independent wave systems exist at the same time, the sea surface acquires a complex, irregular pattern. Hence, the sea waves we watch from the shore are often a mixture of swells from faraway storms and waves created by local winds.

Circular Orbital Motion

Waves can travel great distances across ocean basins. In one study, waves generated near Antarctica were tracked as they traveled through the Pacific Ocean basin. After traveling more than 10,000 kilometers (over 6000 miles), the waves finally expended their energy a week later along the shoreline of Alaska's Aleutian Islands. The water itself doesn't travel this distance, but the wave form

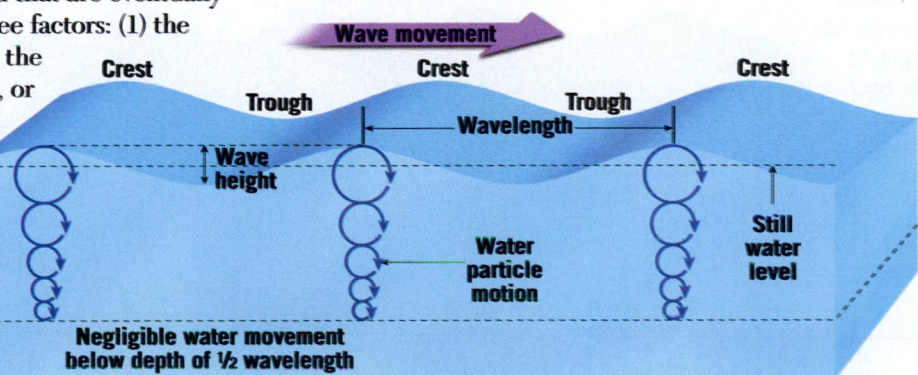

Wave movement

Crest · Crest · Crest

Trough · Trough

Wavelength

Wave height

Water particle motion

Still water level

Negligible water movement below depth of ½ wavelength

SmartFigure 20.5
Wave basics An idealized nonbreaking wave, showing its basic parts and the movement of water with increasing depth.
(https://goo.gl/57GfHl)

Animation

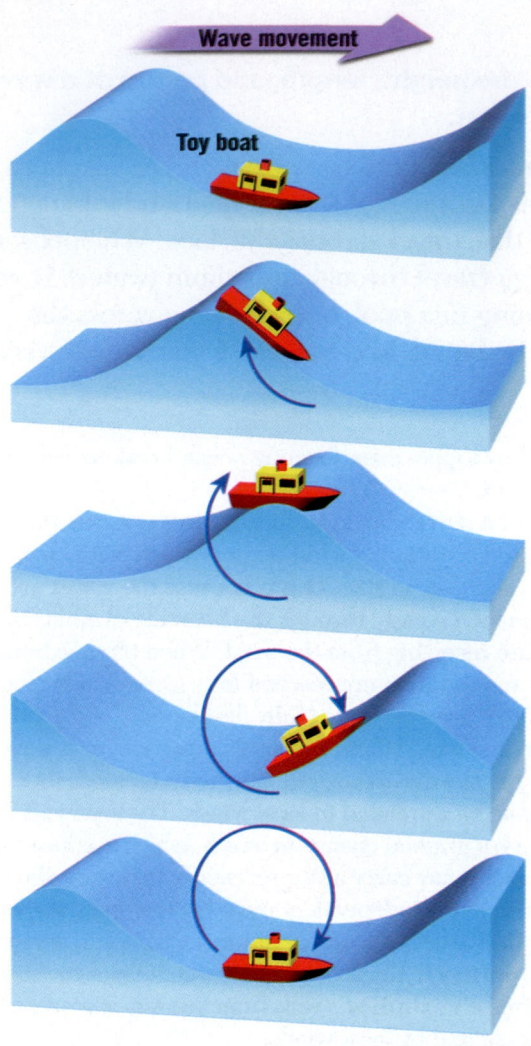

Wave movement

Toy boat

does. As the wave travels, the water passes the energy along by moving in a circle. This movement is called *circular orbital motion*.

Observation of an object floating in waves reveals that it moves not only up and down but also slightly

forward and backward with each successive wave. Figure 20.6 shows that a floating object moves up and backward as the crest approaches, up and forward as the crest passes, down and forward after the crest, and down and backward as the trough approaches; it rises and moves backward again as the next crest advances. When we trace the movement of the toy boat shown in Figure 20.6 as a wave passes, we see that the boat moves in a circle and returns to essentially the same place. Circular orbital motion allows a wave form (the wave's shape) to move forward *through the water* while the individual water particles that transmit the wave move in a circle. Wind moving across a field of wheat causes a similar phenomenon: The wheat itself doesn't travel across the field, but the waves do.

The wind energy given to the water is transmitted not only along the surface of the sea but also downward. However, beneath the surface, the circular motion rapidly diminishes until, at a depth equal to one-half the wavelength measured from the still water level, the movement of water particles becomes negligible. This depth is known as the *wave base*. The dramatic decrease of wave energy with depth is shown by the rapidly diminishing diameters of water-particle orbits in Figure 20.5.

Waves in the Surf Zone

As long as a wave is in deep water, it is unaffected by water depth (Figure 20.7, left). However, when a wave approaches the shore, the water becomes shallower and influences wave behavior. The wave begins to "feel bottom" at a water depth equal to its wave base. Such depths interfere with water movement at the base of the wave and slow its advance (see Figure 20.7, center).

As a wave advances toward the shore, the slightly faster waves farther out to sea catch up, decreasing the wavelength. As the speed and length of the wave

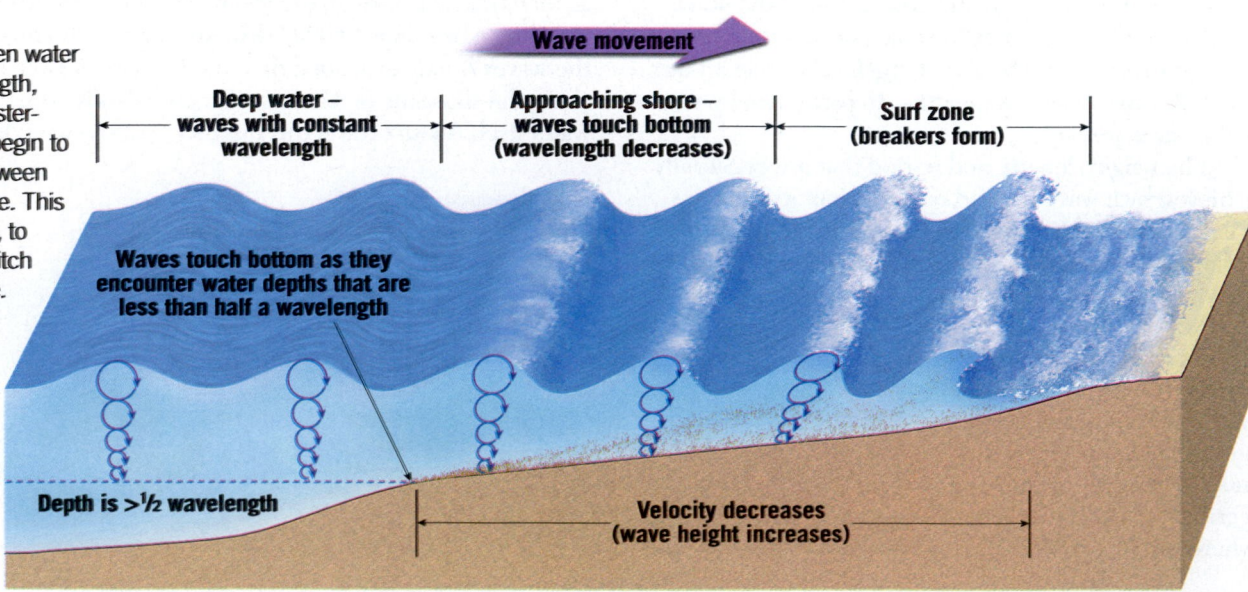

Wave movement

Deep water – waves with constant wavelength | Approaching shore – waves touch bottom (wavelength decreases) | Surf zone (breakers form)

Waves touch bottom as they encounter water depths that are less than half a wavelength

Depth is >½ wavelength

Velocity decreases (wave height increases)

diminish, the wave steadily grows higher. Finally, a critical point is reached when the wave is too steep to support itself and the wave front collapses, or *breaks* (see **Figure 20.7, right**), causing water to advance up the shore.

The turbulent water created by breaking waves is called **surf**. On the landward margin of the surf zone, the *swash*—the turbulent sheet of water from collapsing breakers—moves up the slope of the beach. When the energy of the swash has been expended, the water flows back down the beach toward the surf zone as *backwash*.

20.3 | Shoreline Processes

Explain how waves erode and how waves move sediment along the shore.

During calm weather, wave action is minimal. However, just as streams do most of their work during floods, so too do waves accomplish most of their work during storms. The impact of high, storm-induced waves against the shore can be awesome in its violence (**Figure 20.8**). The discussion of hurricanes later in the chapter serves to reinforce this fact.

Wave Erosion

Each breaking wave may hurl thousands of tons of water against the land, sometimes causing the ground to literally tremble. The forces exerted by Atlantic waves in wintertime, for example, average nearly 10,000 kilograms per square meter (more than 2000 pounds per square foot). The force during storms is even greater. It is no wonder that cracks and crevices are quickly opened in cliffs, seawalls, breakwaters, and anything else that is subjected to these enormous shocks. Water is forced into every opening, causing air in the cracks to become highly compressed by the thrust of crashing waves. When the wave subsides, the air expands rapidly, dislodging rock fragments and enlarging and extending fractures.

In addition to the erosion caused by wave impact and pressure, **abrasion**—the sawing and grinding action of the water armed with rock fragments—is also important. In fact, abrasion is probably more intense in the surf zone than in any other environment. Smooth, rounded stones and pebbles along the shore are obvious reminders of the relentless grinding action of rock against rock in

EYE ON EARTH 20.1

This surfer is enjoying a ride on a large wave along the coast of Maui. (Photo by Getty Images)

QUESTION 1 What was the source of energy that created this wave?

QUESTION 2 How was the wavelength changing just prior to the time when this photo was taken?

QUESTION 3 Why was the wavelength changing?

QUESTION 4 Many waves exhibit circular orbital motion. Is that true of the wave in this photo? Explain.

Figure 20.8
Storm waves When large waves break against the shore, the force of the water can be powerful and the erosional work that is accomplished can be great. These storm waves are breaking along the coast of Wales. (The Photo Library/Alamy Images)

swash and backwash move sand toward and away from the shoreline. Whether there is a net loss or addition of sand depends on the level of wave activity. When wave activity is relatively light (less energetic waves), much of the swash soaks into the beach, which reduces the backwash. Consequently, the swash dominates and causes a net movement of sand up the beach face toward the berm.

When high-energy waves prevail, the beach is saturated from previous waves, so much less of the swash soaks in. As a result, the berm erodes because backwash is strong and causes a net movement of sand down the beach face.

Along many beaches, light wave activity is the rule during the summer. Therefore, a wide sand berm gradually develops. During winter, when storms are frequent and more powerful, strong wave activity erodes and narrows the berm. A wide berm that may have taken months to build can be dramatically narrowed in just a few hours by the high-energy waves created by a strong winter storm.

the surf zone (**Figure 20.9A**). Further, the waves use such fragments as "tools" as they cut horizontally into the land (**Figure 20.9B**).

Sand Movement on the Beach

Energy from breaking waves can move large quantities of sand along the beach face and in the surf zone roughly parallel to the shoreline. Wave energy also causes sand to move perpendicular to (toward and away from) the shoreline.

Movement Perpendicular to the Shoreline If you stand ankle deep in water at the beach, you will see that

Wave Refraction The bending of waves, called **wave refraction**, plays an important part in shoreline processes (**Figure 20.10**). It affects the distribution of energy along the shore and thus strongly influences where and to

Figure 20.9
Abrasion: Sawing and grinding Breaking waves armed with rock debris can do a great deal of erosional work. (Photo A by Michael Collier; photo B by Fletcher and Baylis/ Science Source)

A.

Smooth, rounded rocks along the shore are an obvious reminder that abrasion can be intense in the surf zone.

B.

This sandstone cliff at Gabriola Island, British Columbia, was undercut by wave action.

As these waves approach nearly straight on, refraction causes the wave energy to be concentrated at headlands (resulting in erosion) and dispersed in bays (resulting in deposition).

Beach deposits

Headland

Waves travel at original speed in deep water

Waves "feel bottom" and slow down in surf zone

Shoreline

Result: waves bend so that they strike the shore more directly

Wave refraction at Rincon Point, California

what degree erosion, sediment transport, and deposition will take place.

Waves seldom approach the shore straight on. Rather, most waves move toward the shore at an angle. However, when they reach the shallow water of a smoothly sloping bottom, they are bent and tend to become parallel to the shore. Such bending occurs because the part of the wave nearest the shore reaches shallow water and slows first, whereas the end that is still in deep water continues forward at its full speed. The net result is a wave front that may approach nearly parallel to the shore, regardless of the original direction of the wave.

Because of refraction, wave impact is concentrated against the sides and ends of headlands that project into the water, whereas wave attack is weakened in bays. This differential wave attack along irregular coastlines is illustrated in Figure 20.10. As the waves reach the shallow water in front of the headland sooner than they do in adjacent bays, they are bent more nearly parallel to the protruding land and strike it from all three sides. By contrast, refraction in the bays causes waves to diverge and expend less energy. In these zones of weakened wave activity, sediments can accumulate and form sandy beaches. Over a long period, erosion of the headlands and deposition in the bays will straighten an irregular shoreline.

Longshore Transport Although waves are refracted, most still reach the shore at some angle, however slight. Consequently, the uprush of water from each breaking wave (the swash) is at an oblique angle to the shoreline. However, the backwash is straight down the slope of the beach. The effect of this pattern of water movement is to transport sediment in a zigzag pattern along the beach face (**Figure 20.11**). This movement is called **beach drift**, and it can transport sand and pebbles hundreds or even thousands of meters each day. However, a more typical rate is 5 to 10 meters per day.

SmartFigure 20.10
Wave refraction As waves first touch bottom in the shallows along an irregular coast, they are slowed; they then bend (refract) and align nearly parallel to the shoreline. (Photo by Rich Reid/ National Geographic/Getty Images) (https://goo.gl/3jlK4V)

Tutorial

Path of sand particles

Beach drift

Net movement of sand grains

Longshore current

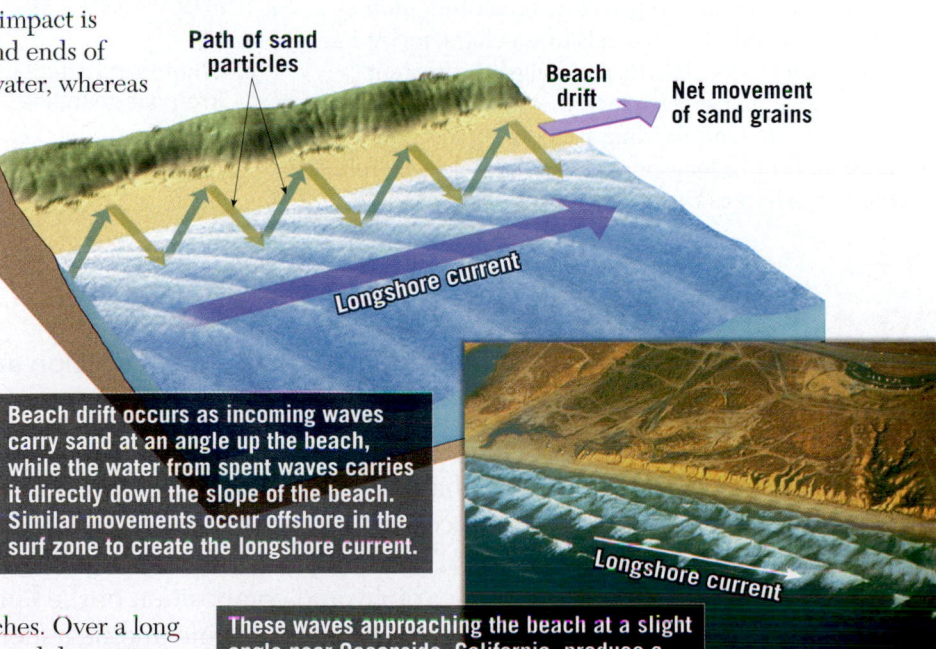

Beach drift occurs as incoming waves carry sand at an angle up the beach, while the water from spent waves carries it directly down the slope of the beach. Similar movements occur offshore in the surf zone to create the longshore current.

Longshore current

These waves approaching the beach at a slight angle near Oceanside, California, produce a longshore current moving from left to right.

SmartFigure 20.11
The longshore transport system The two components of the transport system, beach drift and longshore currents, are created by breaking waves that approach the beach at an angle. These processes transport large quantities of material along the beach and in the surf zone. (Photo by John S. Shelton/University of Washington Libraries) (https://goo.gl/Z0EFUs)

Tutorial

Rip current extends outward from shore and interferes with incoming waves.

WARNING
DANGEROUS RIP CURRENTS
NO BOARD SURFING ZONE
SURF BOARDS, SURF MATS, SURF SKIS, BODY BOARDS, HAND BOARDS, KAYAKS ARE PROHIBITED

Waves that approach the shore at an angle also produce currents within the surf zone that flow parallel to the shore and move substantially more sediment than beach drift. Because the water here is turbulent, these **longshore currents** easily move the fine suspended sand and roll larger sand and gravel along the bottom. When the sediment transported by longshore currents is added to the quantity moved by beach drift, the total amount can be very large. At Sandy Hook, New Jersey, for example, the quantity of sand transported along the shore over a 48-year period averaged almost 750,000 tons annually. For a 10-year period in Oxnard, California, more than 1.5 million tons of sediment moved along the shore each year.

Both rivers and coastal zones move water and sediment from one area (*upstream*) to another (*downstream*). As a result, the beach is often characterized as a "river of sand." Beach drift and longshore currents, however, move in a zigzag pattern, whereas rivers flow mostly in a turbulent, swirling fashion. In addition, the direction of flow of longshore currents along a shoreline can change, whereas rivers flow in the same direction

(downhill). Longshore currents change direction because the direction that waves approach the beach changes seasonally. Nevertheless, longshore currents generally flow southward along both the Atlantic and Pacific shores of the United States.

Rip Currents Rip currents are concentrated movements of water that flow in the *opposite* direction of breaking waves. (Sometimes rip currents are incorrectly called *rip tides*, although they are unrelated to tidal phenomena.) Most of the backwash from spent waves finds its way back to the open ocean as an unconfined flow across the ocean bottom called *sheet flow*. However, sometimes a portion of the returning water moves seaward in the form of surface rip currents. Rip currents do not travel far beyond the surf zone before breaking up, and they can be recognized by the way they interfere with incoming waves or by the sediment that is often suspended within (**Figure 20.12**). They can be hazardous to swimmers, who, if caught in them, can be carried out away from shore. The best strategy for exiting a rip current is to swim *parallel* to the shore for a few tens of meters.

20.3 | Concept Checks

1. Why do waves approaching the shoreline often bend?

2. What is the effect of wave refraction along an irregular coastline?

3. Describe the two processes that contribute to longshore transport.

20.4 | Shoreline Features

Describe the features typically created by wave erosion and those resulting from sediment deposited by longshore transport processes.

A fascinating assortment of shoreline features can be observed along the world's coastal regions. Although the same processes cause change along every coast, not all coasts respond in the same way. Interactions among different processes and the relative importance of each process depend on local factors. The factors include (1) the proximity of a coast to sediment-laden rivers, (2) the degree of tectonic activity, (3) the topography and composition of the land, (4) prevailing winds and weather patterns, and (5) the configuration of the coastline and nearshore areas. Features that originate primarily because of erosion are called *erosional features*, whereas accumulations of sediment produce *depositional features*.

Wave-cut platform

Marine terrace

Figure 20.13
Wave-cut platform and marine terrace This wave-cut platform is exposed at low tide along the California coast at Bolinas Point near San Francisco. A wave-cut platform was uplifted to create the marine terrace. (Photo by John S. Shelton/University of Washington Libraries)

Erosional Features

Many coastal landforms owe their origin to erosional processes. Such erosional features are common along the rugged and irregular New England coast and along the steep shorelines of the west coast of the United States.

Wave-cut Cliffs, Wave-cut Platforms, and Marine Terraces   As the name implies, **wave-cut cliffs** originate in the cutting action of the surf against the base of coastal land. As erosion progresses, rocks overhanging the notch at the base of the cliff crumble into the surf, and the cliff retreats. A relatively flat, benchlike surface, called a **wave-cut platform**, is left behind by the receding cliff (**Figure 20.13, left**). The platform broadens as wave attack continues. Some

debris produced by the breaking waves remains along the water's edge as sediment on the beach, and the remainder is transported farther seaward. If a wave-cut platform is uplifted above sea level by tectonic forces, it becomes a **marine terrace** (see **Figure 20.13, right**). Marine terraces are easily recognized by their gentle seaward-sloping shape and are often desirable sites for coastal roads, buildings, or agriculture.

Sea Arches and Sea Stacks   Waves vigorously attack headlands that extend into the sea because of refraction. The surf erodes the rock selectively, wearing away the softer or more highly fractured rock at the fastest rate. At first, sea caves may form. When two caves on opposite sides of a headland unite, a **sea arch** results (**Figure 20.14**).

Figure 20.14
Sea stack and sea arch These features at the tip of Mexico's Baja Peninsula resulted from vigorous wave attack on a headland. (Photo by Lew Robertson/Getty Images)

Sea stack

Sea arch

Baymouth bar

Spit

Tidal delta

SmartFigure 20.15
Coastal Massachusetts
A. High-altitude image of a well-developed spit and baymouth bar along the coast of Martha's Vineyard. (Image courtesy of USDA-ASCS)
B. This photograph, taken from the International Space Station, shows Provincetown Spit at the tip of Cape Cod. Can you pick out this feature on the satellite image in Figure 20.1A (NASA image) (http://goo.gl/MfwH34)

Mobile Field Trip

A.

Provincetown Spit

B.

dominant direction of the longshore current (**Figure 20.15**). The term **baymouth bar** is applied to a sandbar that completely crosses a bay, sealing it off from the open ocean (see Figure 20.15). Such a feature tends to form across a bay where currents are weak, allowing a spit to extend to the other side. A **tombolo** (*tombolo* = mound), a ridge of sand that connects an island to the mainland or to another island, forms in much the same manner as a spit.

Barrier Islands The Atlantic and Gulf coastal plains are relatively flat and slope gently seaward. The shore zone is characterized by **barrier islands**. These low ridges of land parallel the coast at distances from 3 to 30 kilometers offshore. From Cape Cod, Massachusetts, to Padre Island, Texas, nearly 300 barrier islands rim the coast (**Figure 20.16**).

Most barrier islands are 1 to 5 kilometers wide and 15 to 30 kilometers long. The tallest features are sand dunes, which usually reach heights of 5 to 10 meters; in a few areas, unvegetated dunes are more than 30 meters high. The lagoons separating these narrow islands from the

Eventually the arch falls in, leaving an isolated remnant, or **sea stack**, on the wave-cut platform (see Figure 20.14). In time, it too will be consumed by the action of the waves.

Depositional Features

Sediment eroded from the beach is transported along the shore and deposited in areas where wave energy is low. Such processes produce a variety of depositional features.

Spits, Bars, and Tombolos

Where beach drift and longshore currents are active, several features related to the movement of sediment along the shore may develop. A **spit** (*spit* = spine) is an elongated ridge of sand that projects from the land into the mouth of an adjacent bay. Often the end of a spit that is in the water hooks landward in response to the

Figure 20.16
Barrier islands Nearly 300 barrier islands rim the Gulf and Atlantic coasts. The islands along the coast of North Carolina are excellent examples. In this view, south is at the top of the photo. (Photo by Michael Collier)

VIRGINIA
NORTH CAROLINA
Albemarle Sound
NC
Pamlico Sound
Hatteras Island
Cape Lookout
ATLANTIC OCEAN

Hatteras Island
Pamlico Sound
Road
Dunes
ATLANTIC OCEAN

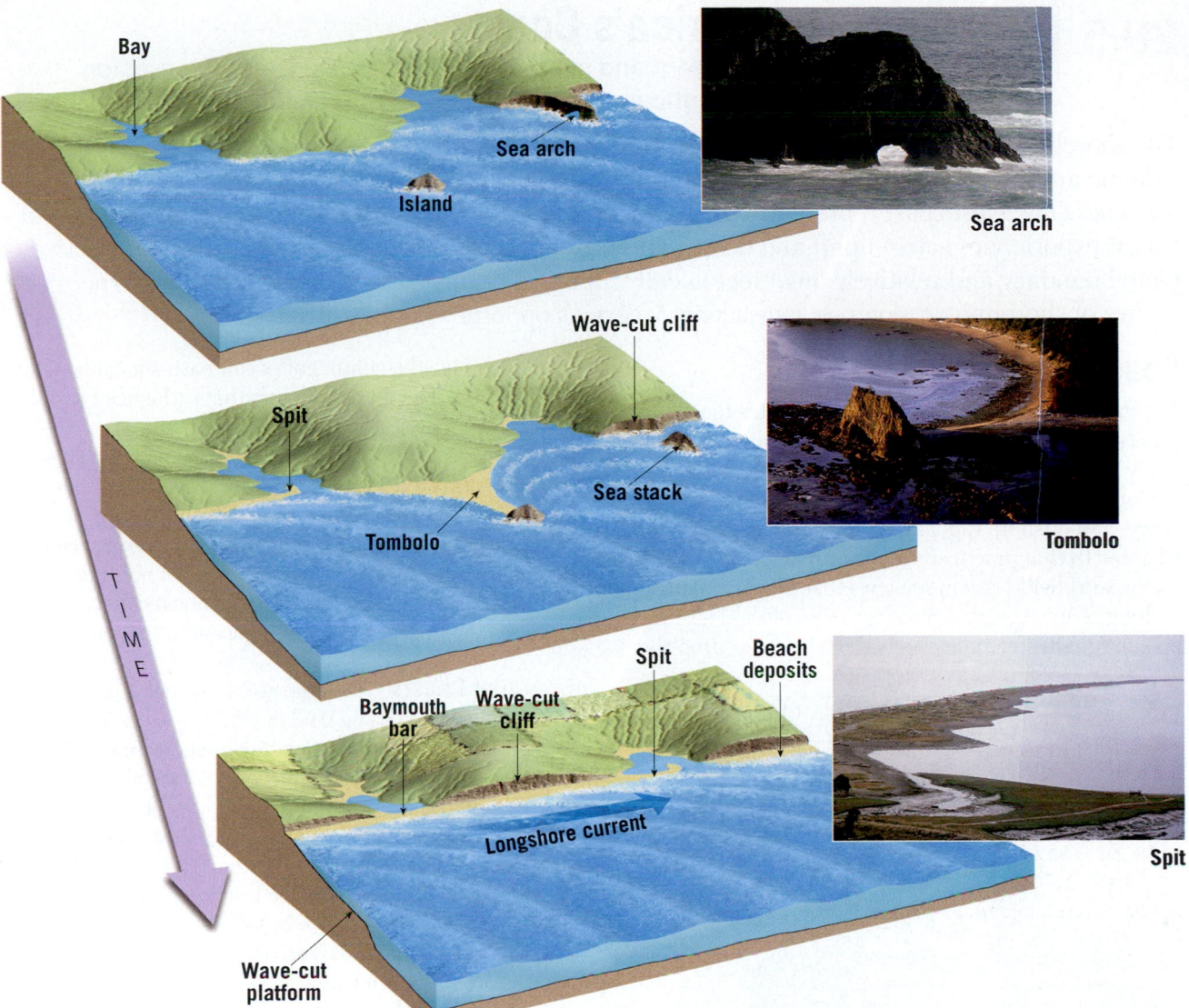

Figure 20.17
The evolving shore These diagrams illustrate changes that can take place through time along an initially irregular coastline that remains relatively stable. The diagrams also illustrate many of the features described in the section on shoreline features. (Top and bottom photos by E. J. Tarbuck; middle photo by Michael Collier)

shore represent zones of relatively quiet water that allow small craft traveling between New York and northern Florida to avoid the rough waters of the North Atlantic.

Barrier islands form in several ways. Some originated as spits that were severed from the mainland by wave erosion or by the general rise in sea level after the last episode of glaciation. Others are created when turbulent waters in the line of breakers heap up sand scoured from the ocean bottom. Because these sand barriers rise above normal sea level, the sand likely piles up as a result of the work of storm waves at high tide. Finally, some barrier islands may be former sand dune ridges that originated along the shore during the last glacial period, when sea level was lower. When the ice sheets melted, sea level rose and flooded the area behind the beach–dune complex.

The Evolving Shore

A shoreline continually undergoes modification, regardless of its initial configuration. At first, most coastlines are irregular, although the degree of and reason for the irregularity may vary considerably from place to place.

Along a coastline that is characterized by varied geology, the pounding surf may at first increase its irregularity because the waves will erode the weaker rocks more easily than the stronger ones. However, if a shoreline remains tectonically stable, marine erosion and deposition will eventually produce a straighter, more regular coast. **Figure 20.17** illustrates the evolution of an initially irregular coast. As waves erode the headlands, creating cliffs and a wave-cut platform, sediment is carried along the shore. Some material is deposited in the bays, while other debris is formed into spits and baymouth bars. At the same time, rivers fill the bays with sediment. Ultimately, a generally straight, smooth coast results.

20.4 | Concept Checks

1. How is a marine terrace related to a wave-cut platform?

2. Describe the formation of each labeled feature in Figure 20.17.

3. List three ways that a barrier island may form.

20.5 | Contrasting America's Coasts

Distinguish between emergent and submergent coasts. Contrast the erosion problems faced on the Atlantic and Gulf coasts with those along the Pacific coast.

The shoreline along the Pacific coast of the United States is strikingly different from that of the Atlantic and Gulf coast regions (see GEOgraphics 20.1). Some of the differences are related to plate tectonics. The west coast represents the leading edge of the North American plate, and because of this, it experiences active uplift and deformation. By contrast, the east coast is far from any active plate boundary and relatively quiet tectonically. Because of this basic geologic difference, the nature of shoreline erosion problems along America's opposite coasts is different.

Coastal Classification

The great variety of shorelines demonstrates their complexity. Indeed, to understand any particular coastal area, many factors must be considered, including rock types, size and direction of waves, frequency of storms, tidal range, and offshore topography. In addition, recall from Chapter 18 that practically all coastal areas were affected by the worldwide rise in sea level that accompanied the melting of Ice Age glaciers at the close of the Pleistocene epoch. Finally, tectonic events that elevate or drop the land or change the volume of ocean basins must be taken into account. The myriad factors that influence coastal areas make shoreline classification difficult.

Many geologists classify coasts based on the changes that have occurred with respect to sea level. This commonly used classification divides coasts into two general categories: emergent and submergent. **Emergent coasts** develop because an area experiences either uplift or a drop in sea level. Conversely, **submergent coasts** are created when sea level rises or the land adjacent to the sea subsides.

Emergent Coasts   In some areas, the coast is clearly emergent because rising land or a falling water level exposes wave-cut cliffs and platforms above sea level. Excellent examples include portions of coastal California, where uplift has occurred in the recent geologic past. The marine terrace shown in Figure 20.13 illustrates this situation. In the case of the Palos Verdes Hills, south of Los Angeles, seven different terrace levels exist, indicating seven episodes of uplift. The ever-persistent sea is now cutting a new platform at the base of the cliff. If uplift follows, it too will become an elevated marine terrace.

Other examples of emergent coasts include regions that were once buried beneath great ice sheets. When glaciers were present, their weight depressed the crust, and when the ice melted, the crust began gradually to spring back. Consequently, prehistoric shoreline features may now be found high above sea level. The Hudson Bay region of Canada is such an area; portions of it are still rising at a rate of more than 1 centimeter per year.

Submergent Coasts   In contrast to the preceding examples, other coastal areas show definite signs of submergence. Shorelines that have been submerged in the relatively recent past are often highly irregular because the sea typically floods the lower reaches of river valleys flowing into the ocean. The ridges separating the valleys, however, remain above sea level and project into the sea as headlands. These drowned river mouths, which are called **estuaries** (*aestus* = tide), characterize many coasts today. Along the Atlantic coastline, the Chesapeake and Delaware Bays are examples of large estuaries created by submergence (**Figure 20.18**). The picturesque coast of Maine, particularly in the vicinity of Acadia National Park, is another excellent example of an area that was

SmartFigure 20.18
East coast estuaries The lower portions of many river valleys were flooded by the rise in sea level that followed the end of the Quaternary Ice Age, creating large estuaries such as Chesapeake and Delaware Bays. (https://goo.gl/iYlc7z)

Tutorial

Various attempts to protect the lighthouse failed. They included building groins and beach nourishment. By 1999, when this photo was taken, the lighthouse was only 36 meters (120 ft.) from the water.

Former location of lighthouse

884 meters (2900 ft.)

To save the famous candy-striped landmark, the National Park Service authorized moving the structure. After the $12 million move, it is expected to be safe for 50 years or more.

Figure 20.19
Relocating the Cape Hatteras lighthouse
After the failure of a number of efforts to protect this 21-story lighthouse, the nation's tallest, from being destroyed due to a receding shoreline, the structure finally had to be moved.
(Top photo by Don Smetzer/ PhotoEdit Inc.; bottom photo by Drew C. Wilson/Virginian-Pilot/ AP Photo)

flooded by the postglacial rise in sea level and transformed into a highly irregular coastline.

Keep in mind that most coasts have complicated geologic histories. With respect to sea level, at various times many coasts have emerged and then submerged. Each time, they may retain some of the features created during the previous situation.

Atlantic and Gulf Coasts

Much of the coastal development along the Atlantic and Gulf coasts has occurred on barrier islands. Typically, a barrier island, also termed a *barrier beach* or *coastal barrier*, consists of a wide beach that is backed by dunes and separated from the mainland by marshy lagoons. The broad expanses of sand and exposure to the ocean have made barrier islands exceedingly attractive sites for development. Unfortunately, development has taken place more rapidly than our understanding of barrier island dynamics has increased.

Because barrier islands face the open ocean, they receive the full force of major storms that strike the coast. When a storm occurs, the barriers absorb the energy of the waves primarily through the movement of sand. Figure 20.19, which shows changes at Cape Hatteras National Seashore, reinforces this point. The process and problems that result were recognized years ago and accurately described as follows:

Waves may move sand from the beach to offshore areas or, conversely, into the dunes; they may erode the dunes, depositing sand onto the beach or carrying it out to sea; or they may carry sand from the beach and the dunes into the marshes behind the barrier, a process

known as overwash. The common factor is movement. Just as a flexible reed may survive a wind that destroys an oak tree, so the barriers survive hurricanes and nor'easters not through unyielding strength but by giving before the storm.

This picture changes when a barrier is developed for homes or as a resort. Storm waves that previously rushed harmlessly through gaps between the dunes now encounter buildings and roadways. Moreover, since the dynamic nature of the barriers is readily perceived only during storms, homeowners tend to attribute damage to a particular storm, rather than to the basic mobility of coastal barriers. With their homes or investments at stake, local residents are more likely to seek to hold the sand in place and the waves at bay than to admit that development was improperly placed to begin with.*

Pacific Coast

In contrast to the broad, gently sloping coastal plains of the Atlantic and Gulf coasts, much of the Pacific coast is characterized by relatively narrow beaches that are backed by steep cliffs and mountain ranges (Figure 20.20). Recall that America's western margin is a more rugged and tectonically active region than the eastern margin. Because uplift continues, a rise in sea level in the west is not so readily apparent. Nevertheless, like the shoreline erosion problems facing the Atlantic coast's barrier islands, west coast difficulties also stem largely from the alteration of a natural system by people.

*Frank Lowenstein, "Beaches or Bedrooms—The Choice as Sea Level Rises," *Oceanus* 28 (No. 3, Fall 1985): p. 22 © Woods Hole Oceanographic Institute.

A Brief Tour of America's Coasts*

1 A small portion of the Cape Cod coast shows the entrance to Nauset Bay. Depending on the whims of recent storms and the strength of coastal currents, the opening into the bay may only be a few hundred feet wide. Tidal currents have created an underwater sandbar just inside the harbor.

10 Sea ice hugs Alaska's north slope near Barrow. The Arctic shore is locked in ice for much of the year.

7 This highway clings to the California Coast south of Big Sur. Uplift is occurring in this area near the boundary separating the Pacific and North American plates.

3 North Carolina's Outer Banks. Oregon Inlet Bridge connecting Bodie Island (foreground) and Hatteras Island. These narrow wisps of sand are part of an extensive barrier island system. The beach and dunes on the left face the Atlantic Ocean. On the right are the quieter waters of Pamlico Sound.

Prepared with the assistance of Michael Collier. All photos by Michael Collier.

Coasts are among Earth's most dynamic landscapes. Waves, tides, and currents continuously shape this interface between land and sea. Coasts may also exhibit the effects of mountain building, sea level changes, rivers, glaciers, and people. Here is a very small glimpse at the diversity and beauty of America's costs.

5 The delta of the Mississippi River is a major feature in the Gulf of Mexico. This low-lying coastal zone is a maze of low soggy islands that are barely above sea level with a myriad of natural distributries and artificial channels.

9 A small glacier flows down a steep mountain slope into the Cook Inlet southwest of Anchorage, Alaska. The total length of Alaska's irregular coastline is nearly 71,000 kilometers (44,000 miles).

6 Waves have aggressively attacked the coast north of La Push Harbor, Washington. In places, remnants of the cliff remain as sea stacks.

2 Spruce and fir cover Turtle Island, part of Maine's Acadia National Park. This region was sculpted by Ice Age glaciers, then flooded when sea level rose as the ice sheets melted.

4 Mobjack Bay near Gloucester, Virginia, is part of the much larger Chesapeake Bay. These bays, called estuaries, are actually river valleys that were drowned when sea level rose at the end of the Ice Age.

8 Cliffs and waterfalls along the north coast of Hawaii's Big Island. This portion of the island receives abundant rainfall. There are still active volcanoes elsewhere on the island.

11 Padre Island is a barrier island that protects the coast of southern Texas from storms. Winds off the Gulf of Mexico create dunes, which offer a temporary footing for vegetation that is occasionally stripped away by hurricanes.

Question:
Use these images to identify at least two shoreline features described in section 20.4.

?

Figure 20.20
Pacific Coast Wave refraction along the coast of California south of Shelter Cove. These steep cliffs are a sharp contrast to typical scenes along the Atlantic and Gulf coasts. (Photo by Michael Collier)

level rises at an increasing rate in the years to come, increased shoreline erosion and sea-cliff retreat should be expected along many parts of the Pacific coast. Coastal vulnerability to sea-level rise is examined in more detail as part of a discussion of the possible consequences of global warming in Chapter 21.

20.5 Concept Checks

1. Are estuaries associated with submergent or emergent coasts? Explain.

2. What observable features would lead you to classify a coastal area as emergent?

3. Briefly describe what happens when storm waves strike an undeveloped barrier island.

4. How might building a dam on a river that flows to the sea affect a coastal beach?

A major problem facing the Pacific shoreline, and especially portions of southern California, is a significant narrowing of many beaches. The bulk of the sand on many of these beaches is supplied by rivers that transport it from the mountains to the coast. Over the years, this natural flow of material to the coast has been interrupted by dams built for irrigation and flood control. The reservoirs effectively trap the sand that would otherwise nourish the beach environment (**Figure 20.21**). When the beaches were wider, they protected the cliffs behind them from the force of storm waves. Now, however, the waves move across the narrowed beaches without losing much energy and cause more rapid erosion of the sea cliffs.

Although the retreat of the cliffs provides material to replace some of the sand impounded behind dams, it also endangers homes and roads built on the bluffs. In addition, development atop the cliffs aggravates the problem. Urbanization increases runoff, which, if not carefully controlled, can result in serious bluff erosion. Watering lawns and gardens high on the cliffs adds significant quantities of water to the slope. This water percolates downward toward the base of the cliff, where it may emerge in small seeps. This action reduces the slope's stability and facilitates mass wasting.

Shoreline erosion along the Pacific coast varies considerably from one year to the next, largely because of the sporadic occurrence of storms. As a consequence, when the infrequent but serious episodes of erosion occur, the damage is often blamed on the unusual storms and not on coastal development or the sediment-trapping dams that may be great distances away. If, as predicted, sea

Figure 20.21
Pacoima Dam and Reservoir Dams such as this one in the San Gabriel Mountains near Los Angeles trap sediment that otherwise would have nourished beaches along the nearby coast. (Photo by Michael Collier)

20.6 | Hurricanes: The Ultimate Coastal Hazard

Describe the basic structure and characteristics of a hurricane and the three broad categories of hurricane destruction.

Many view the weather in the tropics with favor—and rightfully so. Places such as islands in the South Pacific and the Caribbean are known for their lack of significant day-to-day variations. Warm breezes, steady temperatures, and rains that occur as heavy but brief tropical showers are expected. It is ironic that these relatively tranquil regions occasionally produce some of the world's most violent storms. Once formed, these storms carry severe conditions far from the tropics.

Whirling tropical cyclones—the greatest storms on Earth—occasionally have wind speeds exceeding 300 kilometers (185 miles) per hour. In the United States they are known as **hurricanes**, in the western Pacific they are called *typhoons*, and in the Indian Ocean they are simply called *cyclones*. No matter which name is used, these storms are among the most destructive of natural disasters (**Figure 20.22**).

The vast majority of hurricane-related deaths and damage are caused by relatively infrequent yet powerful storms. Of course, the deadliest and most costly storm in recent memory occurred in August 2005, when Hurricane Katrina devastated the Gulf coast of Louisiana, Mississippi, and Alabama. Although hundreds of thousands of people fled before the storm made landfall, thousands of others were caught by the storm. In addition to the human suffering and tragic loss of life left in the wake of Hurricane Katrina, the financial losses caused by the storm are practically incalculable.

Our coasts are vulnerable. People are flocking to live near the ocean. Over half of the U.S. population resides within 75 kilometers (45 miles) of a coast, placing millions at risk. Moreover, the potential costs of property damage are incredible. As sea level continues to rise in coming decades, low-lying, densely populated coastal areas will become even more vulnerable to the destructive effects of major storms.

The eye wall surrounds the eye and is the most intense part of the storm

Well-developed eye

↑ N

Figure 20.22
Super Typhoon Haiyan In the Western Pacific, hurricanes are called typhoons. This storm struck portions of the central Philippines in November 2013. With sustained winds estimated at 315 kilometers (195 miles) per hour, it was the strongest storm on record ever to make landfall. The counterclockwise spiral of the clouds indicates that it is a Northern Hemisphere storm. In the Southern Hemisphere, the spiral is clockwise. (NASA)

requirement explains why hurricane formation over the relatively cool waters of the South Atlantic and eastern South Pacific is extremely rare. For the same reason, few hurricanes form poleward of 20 degrees latitude. Although water temperatures are sufficiently high, hurricanes do not develop within about 5 degrees of the equator because the Coriolis effect (a force related to Earth's rotation that gives storms their "spin") is too weak there. **Figure 20.25** shows the regions where most hurricanes form.

Pressure Gradient   Hurricanes are intense low-pressure centers, which means that as you move toward

Profile of a Hurricane

A hurricane is a heat engine fueled by the energy liberated when huge quantities of water vapor condense. The amount of energy produced by a typical hurricane in just a single day is truly immense. To get the engine started, a large quantity of warm, moist air is required, and a continuous supply is needed to keep it going.

Hurricane Formation   As the graph in **Figure 20.23** illustrates, hurricanes most often form in late summer and early fall. It is during this span that sea-surface temperatures reach 27°C (80°F) or higher and are thus able to provide the necessary heat and moisture to the air (**Figure 20.24**). This ocean-water temperature

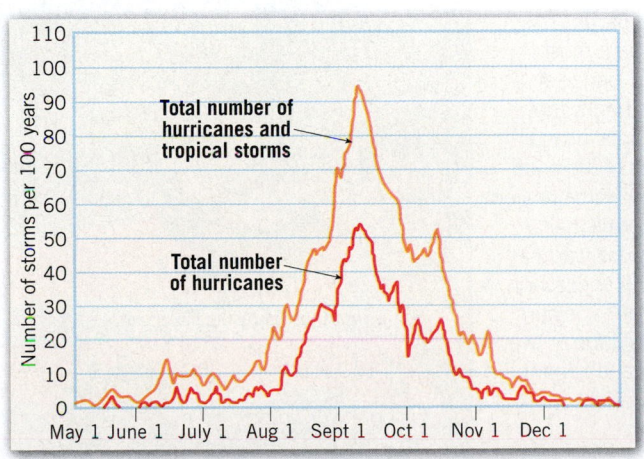

Total number of hurricanes and tropical storms

Total number of hurricanes

Figure 20.23
When do Atlantic hurricanes occur? Frequency of tropical storms and hurricanes from May 1 through December 31 in the Atlantic basin. The graph shows the number of storms to be expected over a span of 100 years. The period from late August through October is clearly the most active. (Data from National Hurricane Center/NOAA)

Figure 20.24
Sea-surface temperatures Among the necessary ingredients for a hurricane is warm ocean temperatures above 27°C (80°F). This color-coded satellite image from June 1, 2010, shows sea-surface temperatures at the beginning of hurricane season. (NASA)

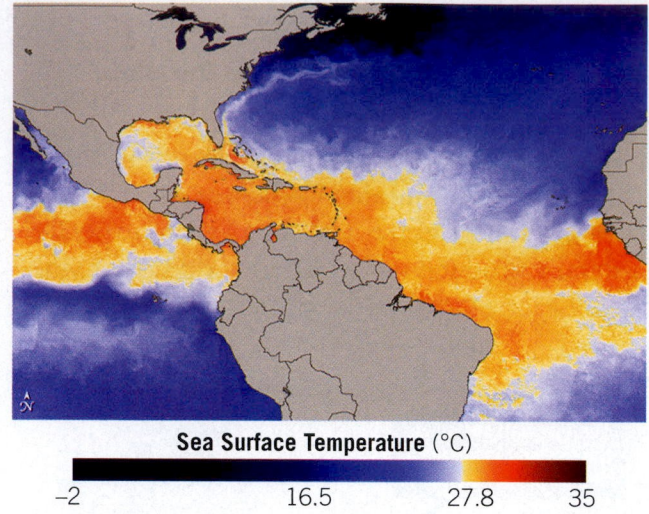

Sea Surface Temperature (°C)

−2   16.5   27.8   35

the center of the storm, air pressure gets lower and lower. Such storms are said to have a very steep pressure gradient. *Pressure gradient* refers to how rapidly the pressure changes per unit distance and is shown on a map with *isobars*, lines of equal pressure. Just as the spacing of contour lines on a topographic map indicates how steep or gentle a slope is, the spacing of isobars on a weather chart shows how rapidly air pressure is changing. Closely spaced isobars indicate a steep pressure gradient and stronger winds. A steep pressure gradient generates the rapid, inward-spiraling winds of a hurricane. As the air rushes toward the center of the storm, its velocity increases. This is similar to skaters with their arms extended spinning faster as they pull their arms in close to their bodies.

Storm Structure As the inward rush of warm, moist surface air approaches the core of the storm, it turns upward and ascends in a ring of cumulonimbus cloud towers (**Figure 20.26**). This doughnut-shaped wall of intense convective activity surrounding the center of the

storm is called the **eye wall**. It is here that the greatest wind speeds and heaviest rainfall occur. Surrounding the eye wall are curved bands of clouds that trail away in a spiral fashion. Near the top of the hurricane, the airflow is outward, carrying the rising air away from the storm center, thereby providing room for more inward flow at the surface.

At the very center of the storm is the **eye** of the hurricane (see Figure 20.26). This well-known feature is a zone about 20 kilometers (12.5 miles) in diameter where precipitation ceases and winds subside. It offers a brief but deceptive break from the extreme weather in the enormous curving wall clouds that surround it. The air within the eye gradually descends and heats by compression, making it the warmest part of the storm. Although many people believe that the eye is characterized by clear blue skies, this is usually not the case because the subsidence in the eye is seldom strong enough to produce cloudless conditions. Although the sky appears much brighter in this region, scattered clouds at various levels are common.

Hurricane Destruction

Although hurricanes are tropical or subtropical in origin, their destructive effects can be experienced far from where they originate. For example, in 2012 Hurricane Sandy (also called Superstorm Sandy) originated in the Caribbean Sea and affected the entire eastern seaboard from Florida to Maine. Destruction was especially great in New Jersey and New York, even though Sandy may have been downgraded from hurricane status by that time.

The amount of damage caused by a hurricane depends on several factors, including the size and population density of the area affected and the shape of the ocean bottom near the shore. The most significant factor, of course, is the strength of the storm. By studying past storms, a scale called the *Saffir–Simpson hurricane scale* was established to rank the relative intensities of

SmartFigure 20.25
Hurricane source regions and paths
The map shows the regions where most hurricanes form as well as their principal months of occurrence and the tracks they most commonly follow. Hurricanes do not develop within about 5 degrees of the equator because the Coriolis effect (a force related to Earth's rotation that gives storms their "spin") there is too weak. Because warm ocean-surface temperatures are necessary for hurricane formation, hurricanes seldom form poleward of 20 degrees latitude or over the cool waters of the south Atlantic and the eastern south Pacific. (https://goo.gl/M6aCWh)

Tutorial

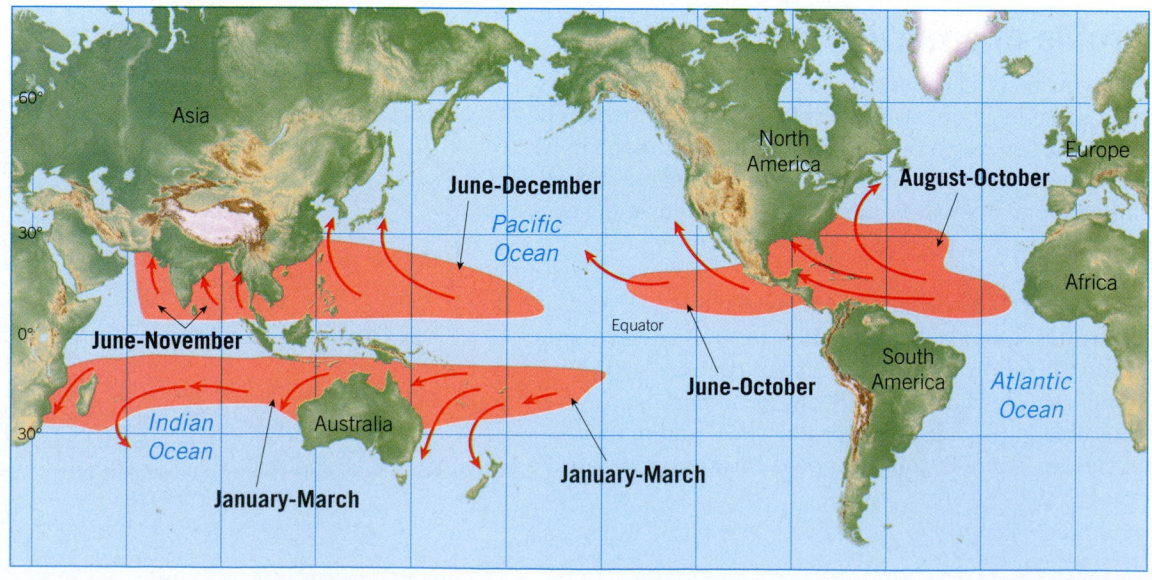

hurricanes. As **Table 20.1** indicates, a category 5 storm is the worst possible, and a category 1 hurricane is least severe.

During hurricane season, it is common to hear scientists and reporters alike use the numbers from the Saffir–Simpson scale. When Hurricane Katrina made landfall, sustained winds were 225 kilometers (140 miles) per hour, making it a strong category 4 storm. Storms that fall into category 5 are rare. Hurricane Camille, a 1969 storm that caused catastrophic damage along the coast of Mississippi, is one well-known example.

Once a hurricane makes landfall, it loses energy because it is cut off from its energy source—warm ocean water—and is usually downgraded to a lower category. However, these storms are so large and violent that their effects are felt far inland. Damage caused by hurricanes can be divided into three categories: storm surge, wind damage, and inland flooding.

Storm Surge   Without question, the most devastating damage in the coastal zone is caused by storm surge. It not only accounts for a large share of coastal property losses but is also responsible for a high percentage of all hurricane-caused deaths. A **storm surge** is a dome of water 65 to 80 kilometers (40 to 50 miles) wide that sweeps across the coast near the point where the eye makes landfall. If all wave activity were smoothed out, the storm surge would be the height of the water above normal tide level. In addition, tremendous wave activity is superimposed on the surge. This surge of water can inflict immense damage on low-lying coastal areas (**Figure 20.27**). The worst surges occur in places like the Gulf of Mexico, where

the continental shelf is very shallow and gently sloping. In addition, local features such as bays and rivers can cause the surge to double in height and increase in speed.

As a hurricane advances toward the coast in the Northern Hemisphere, storm surge is always most intense on the right side of the eye, where winds are blowing *toward* the shore. On this side of the storm, the forward movement of

SmartFigure 20.26
Cross section of a hurricane
(Data for graph from World Meteorological Organization)
(http://goo.gl/yT2d1q)

 Video

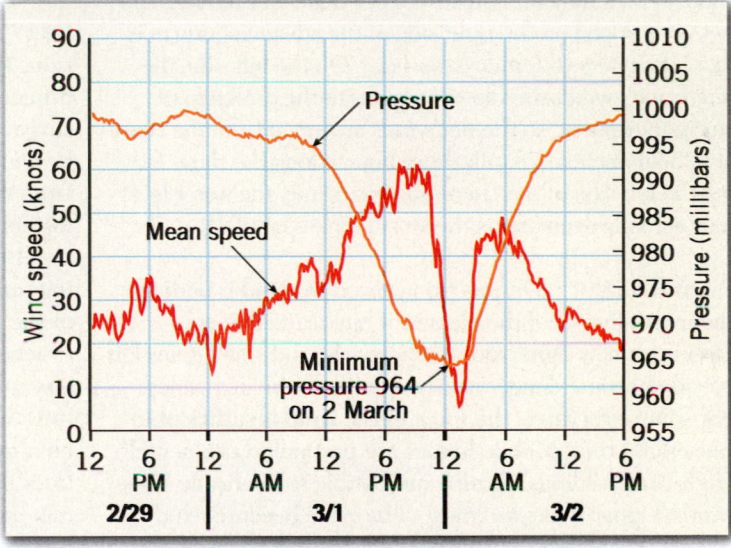

Outflow of air at the top of the hurricane is important because it prevents the convergent flow at lower levels from "filling in" the storm.

Eye

Sinking air in the eye warms by compression.

Eye wall, the zone where winds and rain are most intense.

Tropical moisture spiraling inward creates rain bands that pinwheel around the storm center.

Cross section of a hurricane. Note that the vertical dimension is greatly exaggerated. (After NOAA)

Measurements of surface pressure and wind speed during the passage of Cyclone Monty at Mardie Station, Western Australia, between February 29 and March 2, 2004. (Hurricanes are called "cyclones" in this part of the world.)

TABLE 20.1	Saffir–Simpson Hurricane Scale					
Scale Number (category)	Central Pressure (millibars)	Wind Speed (kph)	Wind Speed (mph)	Storm Surge (meters)	Storm Surge (feet)	Damage
1	≥980	119–153	74–95	1.2–1.5	4–5	Minimal
2	965–979	154–177	96–110	1.6–2.4	6–8	Moderate
3	945–964	178–209	111–130	2.5–3.6	9–12	Extensive
4	920–944	210–250	131–155	3.7–5.4	13–18	Extreme
5	<920	>250	>155	>5.4	>18	Catastrophic

Figure 20.27
Storm surge destruction
This is Crystal Beach, Texas, on September 16, 2008, 3 days after Hurricane Ike came ashore. At landfall the storm had sustained winds of 165 kilometers (105 miles) per hour. The extraordinary storm surge caused most of the damage shown here. (Photo by Smiley N. Pool/Newscom)

the hurricane also contributes to the storm surge. In Figure 20.28, assume that a hurricane with peak winds of 175 kilometers (109 miles) per hour is moving toward the shore at 50 kilometers (31 miles) per hour. In this case, the net wind speed on the right side of the advancing storm is 225 kilometers (140 miles) per hour. On the left side, the hurricane's winds are blowing opposite the direction of storm movement, so the net winds are *away* from the coast at 125 kilometers (78 miles) per hour. Along the shore facing the left side of the oncoming hurricane, the water level may actually decrease as the storm makes landfall.

Wind Damage   Destruction caused by wind is perhaps the most obvious of the classes of hurricane damage. Debris such as signs, roofing materials, and small items left outside become dangerous flying missiles in hurricanes. For some structures, the force of the wind is sufficient to cause total ruin. Mobile homes are particularly vulnerable. High-rise buildings are also susceptible to hurricane-force winds. Upper floors are most vulnerable because wind speeds usually increase with height. Recent research suggests that people should stay below the 10th floor of a building but remain above any floors at risk for flooding. In regions with good building codes, wind damage is

usually not as catastrophic as storm-surge damage. However, hurricane-force winds affect a much larger area than storm surge and can cause huge economic losses. For example, in 1992 it was largely the winds associated with Hurricane Andrew that produced more than $25 billion of damage in southern Florida and Louisiana.

Hurricanes sometimes produce tornadoes that contribute to the storm's destructive power. Studies have shown that more than half of the hurricanes that make landfall produce at least one tornado. In 2004 the number of tornadoes associated with tropical storms and hurricanes was extraordinary. Tropical Storm Bonnie and five landfalling hurricanes—Charley, Frances, Gaston, Ivan, and Jeanne—produced nearly 300 tornadoes that affected the southeastern and mid-Atlantic states.

Heavy Rains and Inland Flooding   The torrential rains that accompany most hurricanes bring a third significant threat: flooding. Whereas the effects of storm surge and strong winds are concentrated in coastal areas, heavy rains may affect places hundreds of kilometers from the coast for up to several days after the storm has lost its hurricane-force winds.

In September 1999, Hurricane Floyd brought flooding rains, high winds, and rough seas to a large portion of the Atlantic seaboard. More than 2.5 million people evacuated their homes from Florida north to the Carolinas and beyond. It was the largest peacetime evacuation in U.S. history up to that time. Torrential rains falling on already saturated ground created devastating inland flooding. Altogether Floyd dumped more than 48 centimeters (19 inches) of rain on Wilmington, North Carolina, with 33.98 centimeters (13.38 inches) falling in a single 24-hour span.

Another well-known example is Hurricane Camille (1969). Although this storm is best known for its exceptional storm surge and the devastation it brought to coastal areas, the greatest number of deaths associated with this storm occurred in the Blue Ridge Mountains of Virginia 2 days after Camille's landfall. Many places received more than 25 centimeters (10 inches) of rain.

Detecting and Tracking Hurricanes

A location only a few hundred kilometers from a hurricane—just a day's striking distance away—may experience clear skies and virtually no wind. Before the age of weather satellites, such a situation made

Figure 20.28
An approaching hurricane This hypothetical Northern Hemisphere hurricane, with peak winds of 175 kilometers (109 miles) per hour, is moving toward the coast at 50 kilometers (31 miles) per hour. On the right side of the advancing storm, the 175-kilometer-per-hour winds blow in the same direction as the movement of the storm (50 kilometers per hour). Therefore the *net* wind speed on the right side of the storm is 225 kilometers (140 miles) per hour *toward* the coast. On the left side, the hurricane's winds are blowing opposite the direction of storm movement, so the *net* winds of 125 kilometers (78 miles) per hour are *away* from the coast. Storm surge will be greatest along the part of the coast hit by the right side of the advancing hurricane.

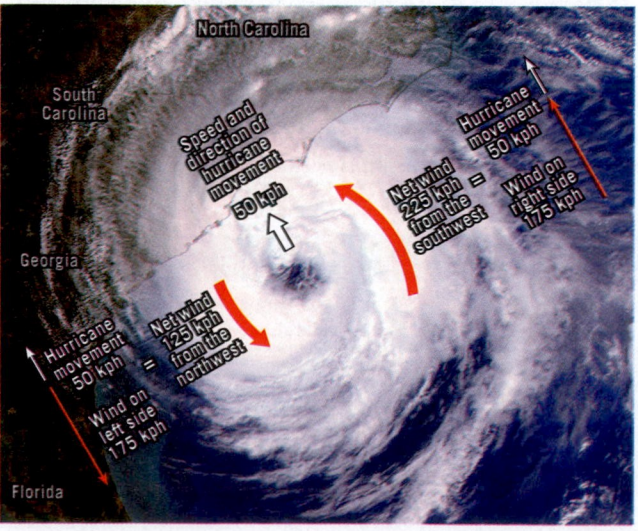

it difficult to warn people of impending storms. The worst natural disaster in U.S. history came as a result of a hurricane that struck an unprepared Galveston, Texas, on September 8, 1900. The strength of the storm, together with the lack of adequate warning, caught the population by surprise and took the lives of 8000 people (**Figure 20.29**).

In the United States, early warning systems have greatly reduced the number of deaths caused by hurricanes. At the same time, however, an astronomical rise has occurred in the amount of property damage, primarily due to the rapid population growth and accompanying development in coastal areas.

Figure 20.29
Aftermath of the historic Galveston hurricane The storm struck an unsuspecting and unprepared city on September 8, 1900. It was the worst natural disaster in U.S. history. Entire blocks were swept clean, and mountains of debris accumulated around the few remaining buildings. (AP Photo)

Satellites

The greatest single advancement in tools used for observing hurricanes has been the development of meteorological satellites. Vast areas of open ocean must be observed in order to detect a hurricane. Before satellites, this was an impossible task. Today instruments aboard satellites can detect a potential storm even before it develops its characteristic circular cloud pattern.

In recent years two methods of using satellite-acquired data to monitor hurricane intensity have been developed. One technique uses instruments aboard a satellite to estimate wind speeds within a storm. A second method uses satellites to identify areas of extraordinary cloud development, called *hot towers*, in the eye wall of an approaching hurricane (**Figure 20.30**).

Aircraft Reconnaissance

When a hurricane is within range, specially instrumented aircraft can fly directly into a threatening storm and accurately measure details of its position and current state of development. Data transmission can be made directly from an aircraft in the midst of a storm to the forecast center. The data collected are critical in analyzing the current characteristics needed to forecast the future behavior of a storm.

Radar

A third basic tool in the observation and study of hurricanes is radar. When a hurricane nears the coast, it is monitored by land-based Doppler weather radar. This tool provides detailed information on hurricane wind fields, rainfall intensity, and storm movement. As a result, local forecast centers are able to track the storm and provide short-term warnings for floods, tornadoes, and high winds for specific areas.

20.6 Concept Checks

1. What factors influence where and when hurricane formation takes place?

2. Distinguish between the eye and eye wall of a hurricane.

3. What are the three broad categories of hurricane damage? Which one is responsible for the greatest number of hurricane-related deaths?

4. Which side of an advancing hurricane in the Northern Hemisphere has the strongest winds and highest storm surge—right or left? Explain.

5. List three tools that provide data used to track and analyze hurricanes.

SmartFigure 20.30

Hot towers This *Tropical Rainfall Measuring Mission* (*TRMM*) satellite image of Hurricane Katrina was acquired early on August 28, 2005. The cutaway view of the inner portion of the storm shows cloud height on one side and rainfall rates on the other. Two hot towers (in red) are visible: one in an outer rain band and the other in the eye wall. The eye wall tower rises 16 kilometers (10 miles) above the ocean surface and is associated with an area of intense rainfall. Towers this tall near the core often indicate that a storm is intensifying. Katrina grew from a category 3 to a category 4 storm soon after this image was received. (NASA) (http://goo.gl/Dn8nOI)

Video

EYE ON EARTH 20.2

This satellite image shows Cyclone Favio as it came ashore along the coast of Mozambique, Africa, on February 22, 2007. This powerful storm was moving from east to west. Since it is a Southern Hemisphere storm, the cloud pattern shows that the winds circulate in a clockwise spiral instead of the counterclockwise pattern typical of storms in the Northern Hemisphere. Portions of the cyclone had sustained winds of 203 kilometers (126 miles) per hour as it made landfall.

QUESTION 1 Identify the eye and eye wall of the storm.

QUESTION 2 Based on wind speed, classify the storm using the Saffir–Simpson scale.

QUESTION 3 Which one of the lettered sites should experience the strongest storm surge? Explain.

20.7 | Stabilizing the Shore
Summarize the ways in which people deal with shoreline erosion problems.

Compared with natural hazards such as earthquakes, volcanic eruptions, and landslides, shoreline erosion is often perceived to be a more continuous and predictable process that may cause relatively modest damage to limited areas. In reality, the shoreline is a dynamic place that can change rapidly in response to natural forces. Exceptional storms are capable of eroding beaches and cliffs at rates that greatly exceed the long-term average. Such bursts of accelerated erosion not only significantly affect the natural evolution of a coast but also can have a profound impact on people who reside in the coastal zone. Erosion along our coasts causes significant property damage. Huge sums are spent annually not only to repair damage but also to prevent or control erosion. Already a problem at many sites, shoreline erosion is certain to become an increasingly serious problem as extensive coastal development continues.

During the past 100 years, growing affluence and increasing demands for recreation have brought unprecedented development to many coastal areas. As both the number and the value of buildings have increased, so too have efforts to protect property from storm waves by stabilizing the shore. Also, controlling the natural migration of sand is an ongoing struggle in many coastal areas. Such interference can result in unwanted changes that are difficult and expensive to correct.

Hard Stabilization

Structures built to protect a coast from erosion or to prevent the movement of sand along a beach are collectively known as **hard stabilization**. Hard stabilization can take many forms and often results in predictable yet unwanted outcomes. Hard stabilization includes jetties, groins, breakwaters, and seawalls.

Jetties Since relatively early in America's history, a principal goal in coastal areas has been the development and maintenance of harbors. In many cases, this has involved the construction of jetty systems. **Jetties** are usually built in pairs and extend into the ocean at the entrances to rivers and harbors. With the flow of water confined to a narrow zone, the ebb and flow caused by the rise and fall of the tides keep the sand in motion and prevent deposition in the channel. However, as illustrated in **Figure 20.31**, a jetty may act as a dam against which the longshore current and beach drift deposit sand. At the same time, wave activity removes sand on the other side. Because the other side is not receiving any new sand, there is soon no beach at all.

Groins To maintain or widen beaches that are losing sand, groins are sometimes constructed. A **groin** (*groin* = ground) is a barrier built at a right angle to the beach to trap sand that is moving parallel to the shore. Groins are usually constructed of large rocks but may also be composed of wood. These structures often do their job so effectively that the longshore current beyond the groin becomes sand starved. As a result, the current erodes sand from the beach on the downstream side of the groin.

To offset this effect, property owners downstream from the structure may erect a groin on their property. In this manner, the number of groins multiplies, resulting in a *groin field* (**Figure 20.32**). The New Jersey shoreline is a good example of groin proliferation, where hundreds of these structures have been built. Because it has been shown that groins often do not provide a satisfactory solution, using them is no longer the preferred method of keeping beach erosion in check.

Breakwaters and Seawalls Hard stabilization can be built parallel to the shoreline. One such structure is a **breakwater**, which protects boats from the force of large breaking waves by creating a quiet water zone near the shoreline. However, when a breakwater is constructed, the reduced wave activity along the shore behind the structure may allow sand to accumulate. If this happens, the marina will eventually fill with sand while the downstream beach erodes and retreats. At Santa Monica, California, where the building of a breakwater has created such a problem, the city uses a dredge to remove sand from the protected quiet water zone and deposit it downstream, where longshore currents and beach drift continue to move the sand down the coast (**Figure 20.33**).

Another type of hard stabilization built parallel to the shoreline is a **seawall**, which is designed to armor the coast and defend property from the force of breaking waves. Waves expend much of their energy as they move across an open beach. Seawalls cut this process short by reflecting the force of unspent waves seaward. As a consequence, the beach to the seaward side of the seawall experiences significant erosion and may in some instances be eliminated entirely (**Figure 20.34**). Once the width of the beach is reduced, the seawall is subjected

to even greater pounding by the waves. Eventually this battering will cause the wall to fail, and a larger, more expensive wall must be built to take its place.

The wisdom of building temporary protective structures along shorelines is increasingly questioned. Many coastal scientists and engineers are of the opinion that halting an eroding shoreline with protective structures benefits only a few and seriously degrades or destroys the natural beach and the value it holds for the majority. Protective structures divert the ocean's energy temporarily from private properties but usually refocus that energy on the adjacent beaches. Many structures interrupt the natural sand flow in coastal currents, robbing affected beaches of vital sand replacement.

> Jetties interrupt the movement of sand causing deposition on the upcurrent side.

> Erosion by sand-starved currents occurs downcurrent from these structures.

Jetties

Longshore current

Figure 20.31
Jetties These structures are built at the entrances to rivers and harbors and are intended to prevent deposition in the navigation channel. The photo is an aerial view at Santa Cruz Harbor, California. (Photo by U.S. Army Corps of Engineers)

Jetties

SmartFigure 20.32
Groins These wall-like structures trap sand that is moving parallel to the shore. This series of groins is along the shoreline near Chichester, Sussex, England. (Photo by Sandy Stockwell/London Aerial Photo Library/CORBIS) (https://goo.gl/WMBDw9)

Mobile Field Trip

Figure 20.33
Breakwater Aerial view of a breakwater at Santa Monica, California. The structure appears as a line in the water behind which many boats are anchored. The construction of the breakwater disrupted longshore transport and caused the seaward growth of the beach. (Photo by John S. Shelton/University of Washington Libraries)

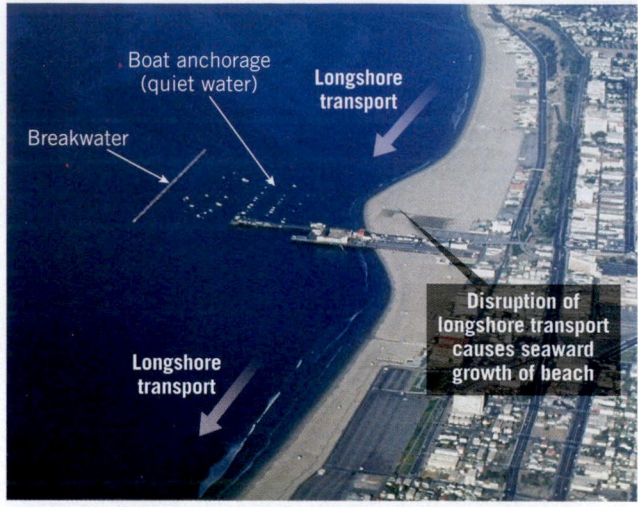

Alternatives to Hard Stabilization

Armoring the coast with hard stabilization has several potential drawbacks, including the cost of the structure and the loss of sand on the beach. Alternatives to hard stabilization include beach nourishment and relocation.

Beach Nourishment   One approach to stabilizing shoreline sands without hard stabilization is **beach nourishment**. As the term implies, this practice involves adding large quantities of sand to the beach system (**Figure 20.35**). Extending beaches seaward makes buildings along the shoreline less vulnerable to destruction by storm waves and enhances recreational uses. Without sandy beaches, tourism suffers.

The process of beach nourishment is straightforward. Sand is pumped by dredges from offshore or trucked from inland locations. The "new" beach,

however, will not be the same as the former beach. Because replenishment sand is from somewhere else, typically not another beach, it is new to the beach environment. The new sand is often different in size, shape, sorting, and composition. Such differences pose problems in terms of erodibility and the kinds of life the new sand will support.

Beach nourishment is not a permanent solution to the problem of shrinking beaches. The same processes that removed the sand in the first place will eventually remove the replacement sand as well. Nevertheless, the number of nourishment projects has increased in recent years, and many beaches, especially along the Atlantic coast, have had their sand replenished many times. Virginia Beach, Virginia, has been nourished more than 50 times.

Beach nourishment is costly. For example, a modest project might involve 38,000 cubic meters (50,000 cubic yards) of sand distributed across about 1 kilometer (0.6 mile) of shoreline. A good-sized dump truck holds about 7.6 cubic meters (10 cubic yards) of sand. So this small project would require about 5000 dump-truck loads. Many projects extend for many miles. Nourishing beaches typically costs millions of dollars per mile.

Changing Land Use   Instead of building structures such as groins and seawalls to hold the beach in place or adding sand to replenish eroding beaches, another option is available. Many coastal scientists and planners are calling for a policy shift from defending and rebuilding beaches and coastal property in high-hazard areas to relocating or abandoning storm-damaged buildings and letting nature reclaim the beach. This option is similar to an approach the federal government

Figure 20.34
Seawall Seabright in northern New Jersey once had a broad, sandy beach. A seawall 5 to 6 meters (16 to 18 feet) high and 8 kilometers (5 miles) long was built to protect the town and the railroad that brought tourists to the beach. After the wall was built, the beach narrowed dramatically. (Photo by Rafael Macia/Science Source)

adopted for river floodplains following the devastating 1993 Mississippi River floods, in which vulnerable structures are either abandoned or relocated on higher, safer ground.

A recent example of changing land use occurred on New York's Staten Island following Hurricane Sandy in 2012. The state turned some vulnerable shoreline areas of the island into waterfront parks. The parks act as buffers to protect inland homes and businesses from strong storms while providing the community with needed open space and access to recreational opportunities.

Land use changes can be controversial. People with significant nearshore investments want to rebuild and defend coastal developments from the erosional wrath of the sea. Others, however, argue that with sea level rising, the impact of coastal storms will get worse in the decades to come, and vulnerable or oft-damaged structures should be abandoned or relocated to improve personal safety and reduce costs. Such ideas will no doubt be the focus of much study and debate as states and communities evaluate and revise coastal land-use policies.

Figure 20.35
Beach nourishment
If you visit a beach along the Atlantic coast, it is more and more likely that you will walk into the surf zone atop an artificial beach. (Photo by Michael Weber/imagebroker/Alamy Images)

Dredge

Offshore sand pouring onto the beach

20.7 Concept Checks

1. List at least three examples of hard stabilization and describe what each is intended to do. How does each affect distribution of sand on a beach?

2. What are two alternatives to hard stabilization, and what are the potential problems associated with each?

EYE ON EARTH 20.3

This structure along the eastern shore of Lake Michigan was built at the entrance to Port Shelton, Michigan.

QUESTION 1 What is the purpose of the artificial structure pictured here?

QUESTION 2 What term is applied to structures such as this?

QUESTION 3 Explain why there is a greater accumulation of sand on one side of the structure than the other.

Michael Collier

20.8 | Tides

Explain the cause of tides, their monthly cycles, and patterns. Describe the horizontal flow of water that accompanies the rise and fall of tides.

Tides are daily changes in the elevation of the ocean surface caused by gravitational interactions of Earth with the Moon and Sun. Their rhythmic rise and fall along coastlines have been known since antiquity. Other than waves, they are the easiest ocean movements to observe (**Figure 20.36**).

Figure 20.36
Bay of Fundy tides High tide and low tide at Hopewell Rocks on the Bay of Fundy. Tidal flats are exposed during low tide. (High tide photo courtesy of Ray Coleman/Science Source; low tide photo by Jeffrey Greenberg/Science Source)

High tide

Low tide
Tidal flat

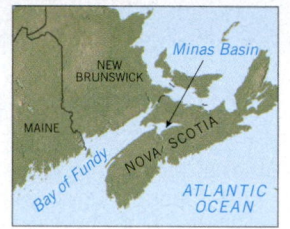

ocean, which is mobile, is deformed quite dramatically by this effect, producing the two opposing tidal bulges.

Because the position of the Moon changes only moderately in a single day, the tidal bulges remain in place while Earth rotates "through" them. For this reason, if you stand on the seashore for 24 hours, Earth will rotate you through alternating areas of deeper and shallower water. As you are carried into each tidal bulge, the tide rises, and as you are carried into the intervening troughs between the tidal bulges, the tide falls. Therefore, most places on Earth experience two high tides and two low tides each day.

Further, the tidal bulges migrate as the Moon revolves around Earth about every 29 days. As a result, the tides, like the time of moonrise, shift about 50 minutes later each day. After 29 days the cycle is complete, and a new one begins.

In many locations, there may be an inequality between the high tides during a given day. Depending on the position of the Moon, the tidal bulges may be inclined to the equator, as in Figure 20.37. This figure illustrates

Although known for centuries, tides were not explained satisfactorily until Sir Isaac Newton applied the law of gravitation to them. Newton showed that there is a mutual attractive force between two bodies and that because oceans are free to move, they are deformed by this force. Hence, ocean tides result from the gravitational attraction exerted upon Earth by the Moon and, to a lesser extent, by the Sun.

Causes of Tides

It is easy to see how the Moon's gravitational force can cause the water to bulge on the side of Earth nearest the Moon. In addition, however, an equally large tidal bulge is produced on the side of Earth directly opposite the Moon (**Figure 20.37**).

Both tidal bulges are caused, as Newton discovered, by the pull of gravity. Gravity is inversely proportional to the square of the distance between two objects, meaning simply that it quickly weakens with distance. In this case, the two objects are the Moon and Earth. Because the force of gravity decreases with distance, the Moon's gravitational pull on Earth is slightly greater on the near side of Earth than on the far side. The result of this differential pulling is to stretch (elongate) the "solid" Earth very slightly. In contrast, the world

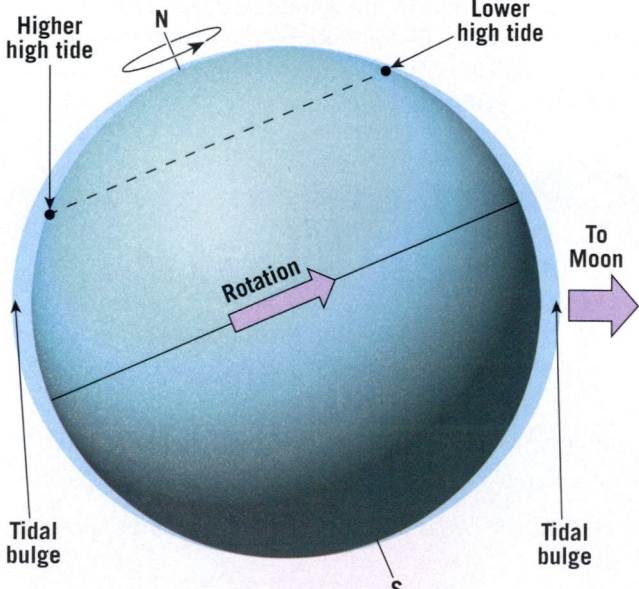

Figure 20.37
Idealized tidal bulges caused by the Moon If Earth were covered to a uniform depth with water, there would be two tidal bulges: one on the side of Earth facing the Moon (right) and the other on the opposite side of Earth (left). Depending on the Moon's position, tidal bulges may be inclined relative to Earth's equator. In this situation, Earth's rotation causes an observer to experience two unequal high tides during a day.

that one high tide experienced by an observer in the Northern Hemisphere is considerably higher than the high tide half a day later. In contrast, a Southern Hemisphere observer would experience the opposite effect.

Monthly Tidal Cycle

The primary body that influences the tides is the Moon, which makes one complete revolution around Earth every 29.5 days. The Sun, however, also influences the tides. It is far larger than the Moon, but because it is much farther away, its effect is considerably less. In fact, the Sun's tide-generating effect is only about 46 percent that of the Moon.

Near the times of new and full moons, the Sun and Moon are aligned, and their forces on tides are added together (Figure 20.38A). The combined gravity of these two tide-producing bodies causes larger tidal bulges (higher high tides) and deeper tidal troughs (lower low tides), producing a large tidal range. These are called the **spring tides** (*springen* = to rise up), which have no connection with the spring season but occur twice a month, during the time when the Earth–Moon–Sun system is aligned. Conversely, at about the time of the first and third quarters of the Moon, the gravitational forces of the Moon and Sun act on Earth at right angles, and each partially offsets the influence of the other (Figure 20.38B). As a result, the daily tidal range is less. These are called **neap tides** (*nep* = scarcely or barely touching), and they also occur twice each month. Each month, then, there are two spring tides and two neap tides, each about 1 week apart.

Tidal Patterns

Although the basic causes and types of tides have been explained, these theoretical considerations cannot be used to predict either the height or the time of actual tides at a particular place. This is because many factors—including the shape of the coastline, the configuration of ocean basins, and water depth—greatly influence the tides. Consequently, tides at various locations respond differently to tide-producing forces. This being the case, the nature of the tide at any coastal location can be determined most accurately by actual observation. The predictions in tidal tables and tidal data on nautical charts are based on such observations.

Three main tidal patterns exist worldwide (Figure 20.39). A *diurnal tidal pattern* (*diurnal* = daily) is characterized by a single high tide and a single low tide each tidal day (24-hour period). Tides of this type occur along the northern shore of the Gulf of Mexico, among other locations. A *semidiurnal tidal pattern* exhibits two high tides and two low tides each tidal day, with the two highs about the same height and the two lows about the same height. This type of tidal pattern is common along the Atlantic coast of the United States. A *mixed tidal pattern* is similar to a semidiurnal pattern except that it is characterized by a large inequality in high water heights, low

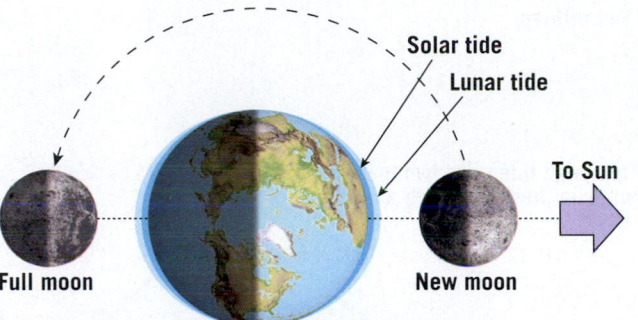

A. Spring Tide When the Moon is in the full or new position, the tidal bulges created by the Sun and Moon are aligned and there is a large tidal range.

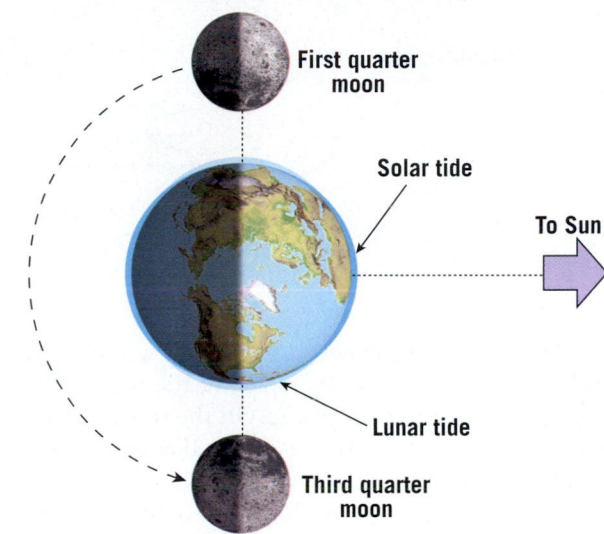

B. Neap Tide When the Moon is in the first-or third-quarter position, the tidal bulges produced by the Moon are at right angles to the bulges created by the Sun and the tidal range is smaller.

SmartFigure 20.38
Spring and neap tides
Earth–Moon–Sun positions influence the tides.
(https://goo.gl/YLHO8T)

Animation

water heights, or both. In this case, there are usually two high and two low tides each day, with high tides of different heights and low tides of different heights. Such tides are prevalent along the Pacific coast of the United States and in many other parts of the world.

Tidal Currents

Tidal current is the term used to describe the *horizontal* flow of water that accompanies the rise and fall of the tide. These water movements induced by tidal forces can be important in some coastal areas. Tidal currents flow in one direction during a portion of the tidal cycle and reverse their flow during the remainder. Tidal currents that advance into the coastal zone as the tide rises are called **flood currents**. As the tide falls, seaward-moving water generates **ebb currents**. Periods of little or no current, called *slack water*, separate flood and ebb. The areas affected by these alternating tidal currents are **tidal flats** (see Figure 20.36). Depending on the nature of the coastal zone, tidal flats vary from narrow strips seaward of the beach to extensive zones that may extend for several kilometers.

SmartFigure 20.39
Tidal patterns A diurnal tidal pattern (lower right) shows one high tide and one low tide each tidal day. A semidiurnal pattern (upper right) shows two high tides and two low tides of approximately equal heights during each tidal day. A mixed tidal pattern (left) shows two high tides and two low tides of unequal heights during each tidal day.
(https://goo.gl/NNoCGJ)

Tutorial

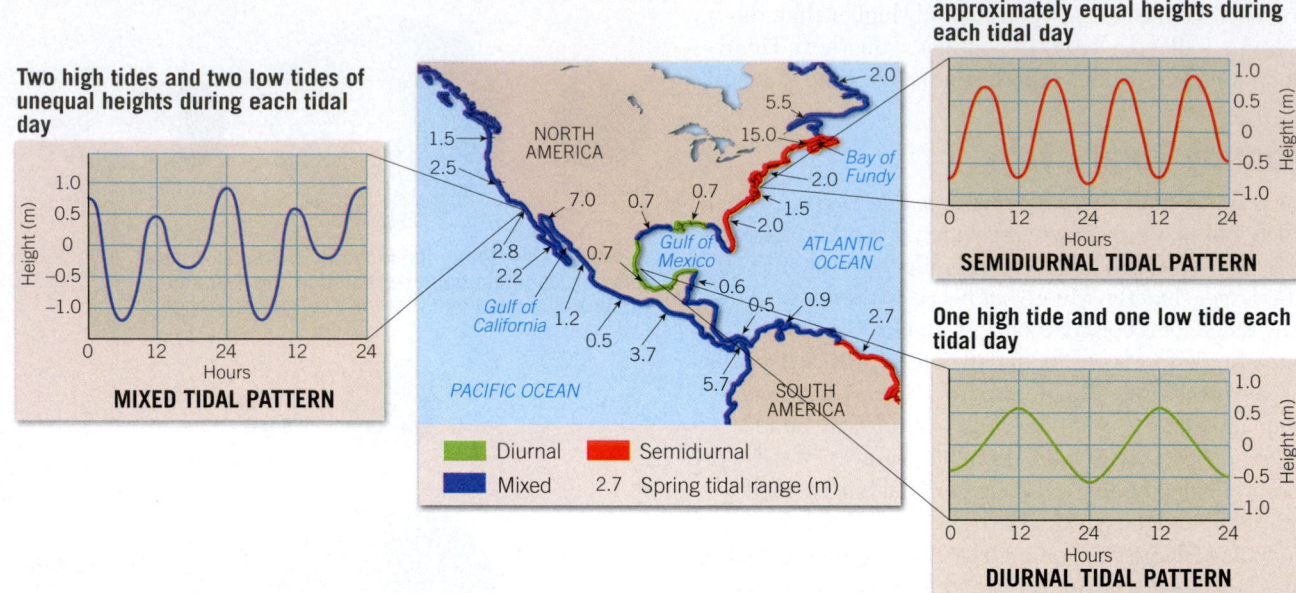

Two high tides and two low tides of unequal heights during each tidal day

MIXED TIDAL PATTERN

Two high tides and two low tides of approximately equal heights during each tidal day

SEMIDIURNAL TIDAL PATTERN

One high tide and one low tide each tidal day

DIURNAL TIDAL PATTERN

Although tidal currents are generally not important in the open sea, they can be rapid in bays, river estuaries, straits, and other narrow places. Off the coast of Brittany in France, for example, tidal currents that accompany a high tide of 12 meters (40 feet) may attain a speed of 20 kilometers (12 miles) per hour. While tidal currents are not generally major agents of erosion and sediment transport, notable exceptions occur where tides move through narrow inlets. Here they constantly scour the small entrances to many good harbors that would otherwise be blocked.

Sometimes deposits called **tidal deltas** are created by tidal currents (**Figure 20.40**). They may develop either as *flood deltas* landward of an inlet or as *ebb deltas* on the seaward side of an inlet. Because wave activity and longshore currents are reduced on the sheltered landward side, flood deltas are more common and more prominent (see Figure 20.15A). They form after the tidal current moves rapidly through an inlet. As the current emerges from the narrow passage into more open waters, it slows and deposits its load of sediment.

Figure 20.40
Tidal deltas As a rapidly moving tidal current (flood current) moves through a barrier island's inlet into the quiet waters of the lagoon, the current slows and deposits sediment, creating a tidal delta. Because this tidal delta has developed on the landward side of the inlet, it is called a *flood delta*. Such a tidal delta is shown in Figure 20.15A.

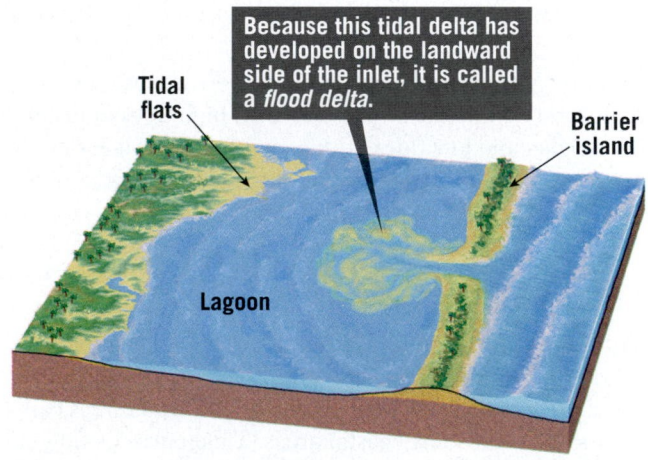

Because this tidal delta has developed on the landward side of the inlet, it is called a *flood delta*.

Tidal flats

Barrier island

Lagoon

20.8 Concept Checks

1. Explain why an observer can experience two unequal high tides during a single day.

2. Distinguish between *neap tides* and *ebb tides*.

3. Contrast *flood current* and *ebb current*.

20.1 The Shoreline: A Dynamic Interface

Explain why the shoreline is considered a dynamic interface and identify the basic parts of the coastal zone.

KEY TERMS interface, shoreline, shore, coast, coastline, foreshore, backshore, nearshore zone, offshore zone, beach, berm, beach face

- The shore is the area extending between the lowest tide level and the highest elevation on land that is affected by storm waves. The coast extends inland from the shore as far as ocean-related features can be found. The shore is divided into the foreshore and backshore. Seaward of the foreshore are the nearshore and offshore zones.
- A beach is an accumulation of sediment along the landward margin of an ocean or a lake. Among its parts are one or more berms and the beach face. Beaches are composed of whatever material is locally abundant and should be thought of as material in transit along the shore.

Q Assume that you photographed this scene and that the photo was taken at high tide. On which part of the shore are you standing?

Shutterstock

20.2 Ocean Waves

List and discuss the factors that influence the height, length, and period of a wave and describe the motion of water within a wave.

KEY TERMS wave height, wavelength, wave period, fetch, surf

- Waves are moving energy, and most ocean waves are initiated by wind. The three factors that influence the height, wavelength, and period of a wave are (1) wind speed, (2) length of time the wind has blown, and (3) fetch, the distance that the wind has traveled across open water. Once waves leave a storm area, they are termed *swells*.
- As waves travel, water particles transmit energy by circular orbital motion, which extends to a depth equal to one-half the wavelength (the wave base). When a wave enters water that is shallower than the wave base, it slows, allowing waves farther from shore to catch up. As a result, wavelength decreases and wave height increases. Eventually the wave breaks, creating turbulent surf in which water rushes toward the shore.

20.3 Shoreline Processes

Explain how waves erode and how waves move sediment along the shore.

KEY TERMS abrasion, wave refraction, beach drift, longshore current, rip current

- Wind-generated waves provide most of the energy that modifies shorelines. Each time a wave hits, it can impart tremendous force. The impact of waves, coupled with abrasion from the grinding action of rock particles, erodes material exposed along the shoreline.
- Wave refraction is a consequence of a wave encountering shallower water as it approaches shore. The shallowest part of the wave (closest to shore) slows the most, allowing the faster part (still in deeper water) to catch up. This modifies a wave's trajectory so that the wave front becomes almost parallel to the shore by the time it hits. Wave refraction concentrates impacting energy on headlands and dissipates that energy in bays, which become sites of sediment accumulation.
- Beach drift describes the movement of sediment in a zigzag pattern along a beach face. The swash of incoming waves pushes the sediment up the beach at an oblique angle, but the backwash transports it directly downhill. Net movement along the beach can be many meters per day. Longshore currents are a similar phenomenon in the surf zone, capable of transporting very large quantities of sediment parallel to a shoreline.

Q What process is causing wave energy to be concentrated on the headland? Predict how this area will appear in the future.

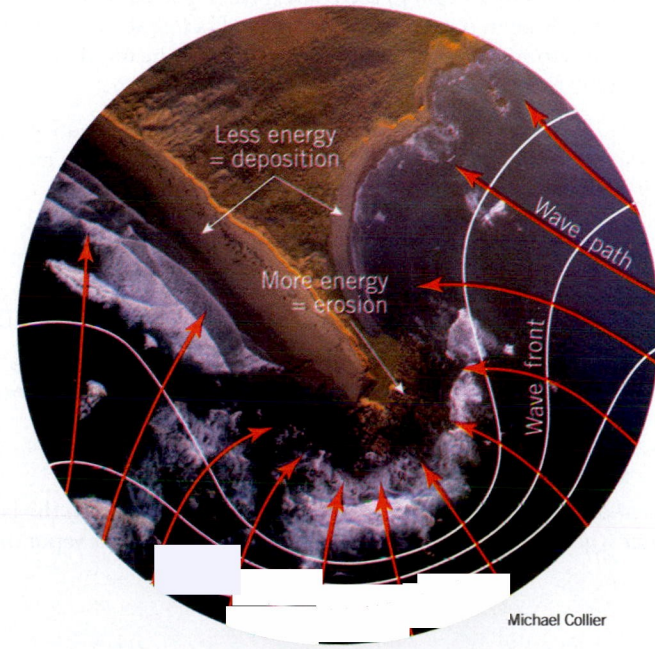

Less energy = deposition

More energy = erosion

Wave path

Wave front

Michael Collier

20.4 Shoreline Features

Describe the features typically created by wave erosion and those resulting from sediment deposited by longshore transport processes.

KEY TERMS wave-cut cliff, wave-cut platform, marine terrace, sea arch, sea stack, spit, baymouth bar, tombolo, barrier island

- Erosional features include wave-cut cliffs (created by the cutting action of the surf against the base of coastal land), wave-cut platforms (relatively flat surfaces left behind by receding cliffs), and marine terraces (uplifted wave-cut platforms). Erosional features also include sea arches (formed when a headland is eroded and two sea caves from opposite sides unite) and sea stacks (formed when the roof of a sea arch collapses).
- Depositional features that form when sediment is moved by beach drift and longshore currents include spits (elongated ridges of sand that project from the land into the mouth of a bay), baymouth bars (sandbars that completely cross a bay), and tombolos (ridges of sand connecting an island to the mainland or another island). The Atlantic and Gulf coastal region is characterized by offshore barrier islands, which are low ridges of sand that parallel the coast.
- Over time, irregular, rocky shorelines are modified by erosion and deposition to become smoother and straighter.

Q Identify the lettered features in this diagram.

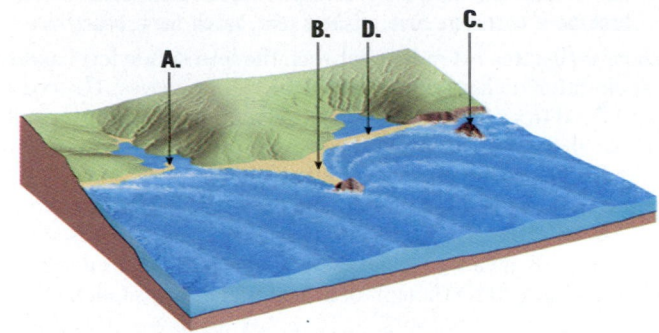

20.5 Contrasting America's Coasts

Distinguish between emergent and submergent coasts. Contrast the erosion problems faced on the Atlantic and Gulf coasts with those along the Pacific coast.

KEY TERMS emergent coast, submergent coast, estuary

- Coasts may be classified by their changes relative to sea level. Emergent coasts are sites of either land uplift or sea-level fall. Marine terraces are features of emergent coasts. Submergent coasts are sites of land subsidence or sea-level rise. One characteristic of submergent coasts is drowned river valleys called estuaries.
- The Atlantic and Gulf coasts of the United States are markedly different from the Pacific coast. The Atlantic and Gulf coasts are lined in many places by barrier islands—dynamic expanses of sand that see a lot of change during storm events. Many of these low and narrow islands have also been prime sites for real estate development.
- The Pacific coast's big issue is the narrowing of beaches due to sediment starvation. Rivers that drain to the coast (bringing it sand) have been dammed, resulting in reservoirs that trap sand before it can make it to the coast. Narrower beaches offer less resistance to incoming waves, often leading to erosion of bluffs behind the beach.

Q Is this an emergent coast or a submergent coast? Provide an easily seen line of evidence to support your answer. Is the location more likely along the coast of North Carolina or California? Explain.

Michael Collier

20.6 Hurricanes: The Ultimate Coastal Hazard

Describe the basic structure and characteristics of a hurricane and the three broad categories of hurricane destruction.

KEY TERMS hurricane, eye wall, eye, storm surge

- Hurricanes are fueled by warm, moist air and usually form in the late summer when sea-surface temperatures are highest. Water vapor in rising warm air condenses, releasing heat and triggering the formation of dense clouds and heavy rain. Because of a steep pressure gradient, air rushes into the center of the storm. The Coriolis effect and ocean-water temperatures strongly influence where hurricanes form.
- The eye at the center of a hurricane has the lowest pressure, is relatively calm, and lacks rain. The surrounding eye wall has the strongest winds and most intense rainfall. The Saffir–Simpson scale classifies storms based on their air pressure and wind speed.

- Most hurricane damage comes from one or a combination of three causes: storm surge, wind damage, or inland flooding due to heavy rains. Storm surge is ocean water that gets pushed up above the normal water level by the strong winds. In the Northern Hemisphere hurricanes rotate counterclockwise, and storm surge is greatest on the right side of an advancing hurricane. This is due to the combination of the storm's forward movement and strong winds blowing toward the shore.

Q This coastal scene shows hurricane destruction. Which one of the three basic classes of damage was most likely responsible for this destruction? What is your reasoning?

Lucas Jackson/Reuters

20.7 Stabilizing the Shore

Summarize the ways in which people deal with shoreline erosion problems.

KEY TERMS hard stabilization, jetty, groin, breakwater, seawall, beach nourishment

- Hard stabilization refers to any structures built along the coastline to prevent movement of sand. Jetties project out from the coast with the goal of keeping inlets open. Groins are also oriented perpendicular to the coast, but their goal is to slow beach erosion by longshore currents. Offshore breakwaters are constructed parallel to the coast to blunt the force of incoming ocean waves, often to protect boats. Like breakwaters, seawalls are parallel to the coast, but are built on the shoreline itself. Hard stabilization measures often result in increased erosion elsewhere.
- Beach nourishment is an expensive alternative to hard stabilization. Sand is pumped onto a beach from some other area, temporarily replenishing the sediment supply. Another option is relocating buildings away from high-risk areas and leaving the beach to be shaped by natural processes.

Q Based on their position and orientation, identify the four kinds of hard stabilization illustrated in this diagram.

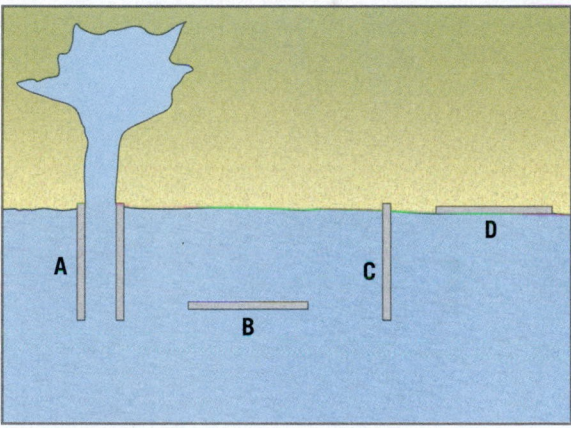

20.8 Tides

Explain the cause of tides, their monthly cycles, and patterns. Describe the horizontal flow of water that accompanies the rise and fall of tides.

KEY TERMS tide, spring tide, neap tide, tidal current, flood current, ebb current, tidal flat, tidal delta

- Tides are daily changes in ocean-surface elevation. They are caused by gravitational pull on ocean water by the Moon and, to a lesser extent, the Sun. When the Sun, Earth, and Moon all line up about every 2 weeks (full moon or new moon), the tides are most exaggerated. When

a quarter Moon is in the sky, it indicates that the Moon is pulling on Earth's water at a right angle relative to the Sun, and the daily tidal range is minimized as the two forces partially counteract one another.
- A flood current is the landward movement of water during the shift between low tide and high tide. When high tide transitions to low tide again, the movement of water away from the land is an ebb current. Ebb currents may expose tidal flats to the air. If a tide passes through an inlet, the current may carry sediment that gets deposited as a tidal delta.

Q Would spring tides and neap tides occur on an Earth-like planet that had no moon? Explain.

Give It Some Thought

1. During a visit to the beach, you and a friend get in a rubber raft and paddle out into deep water *beyond* the surf zone. Tiring, you stop and take a rest. Describe the movement of the raft during your rest. How does this movement differ, if at all, from what you would have experienced if you had stopped paddling while *in* the surf zone?

2. Examine the aerial photo that shows a portion of the New Jersey coast. What term is applied to the wall-like structures that extend into the water? What is their purpose? In what direction are beach drift and longshore currents moving sand: toward the top or toward the bottom of the photo?

John S. Shelton/University of Washington Libraries

3. You and a friend set up an umbrella and chairs at a beach. Your friend then goes into the surf zone to play Frisbee with another person. Several minutes later, your friend looks back toward the beach and is surprised to see that she is no longer near the umbrella and chairs. Although she is still in the surf zone, she is 30 yards away from where she started. How would you explain to your friend why she moved along the shore?

4. What term is applied to the masses of rock protruding from the water? How did they form? Is the location more likely along the U.S. Gulf Coast or the Pacific Coast? Explain.

Michael Collier

5. Assume that it is late September 2018, and Hurricane Gordon, a category 5 storm, is projected to follow the path shown on the accompanying map. The path of the arrow represents the path of the hurricane's eye. Answer the following questions:

Dallas–Fort Worth

Houston

Track of Hurricane Gordon

a. Should the city of Houston expect to experience Gordon's fastest winds and greatest storm surge? Explain why or why not.

b. What is the greatest threat to life and property if this storm approaches the Dallas–Fort Worth area? Explain your reasoning.

6. A friend wants to purchase a vacation home on a barrier island. If consulted, what advice would you give your friend?

7. Hurricane Rita was a major storm that struck the Gulf coast in late September 2005, less than a month after Hurricane Katrina. The accompanying graph shows changes in air pressure and wind speed from the storm's beginning as an unnamed tropical disturbance north of the Dominican Republic on September 18 until its last remnants faded away in Illinois on September 26. Use the graph to answer these questions:

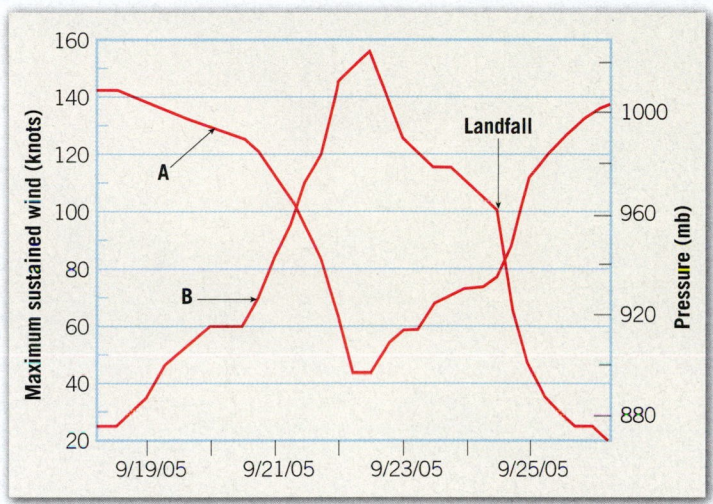

a. Which line represents air pressure, and which line represents wind speed? How did you figure this out?

b. What was the storm's maximum wind speed, in knots? Convert this answer to kilometers per hour by multiplying by 1.85.

c. What was the lowest pressure attained by Hurricane Rita?

d. Using wind speed as your guide, what was the highest category reached on the Saffir–Simpson scale? On what day was this status reached?

e. When landfall occurred, what was the category of Hurricane Rita?

8. The force of gravity plays a critical role in creating ocean tides. The more massive an object, the stronger its pull of gravity. Explain why the Sun's influence is only about half that of the Moon, even though the Sun is much more massive than the Moon.

9. This photo shows a portion of the Maine coast. The brown muddy area in the foreground is influenced by tidal currents. What term is applied to this muddy area? Name the type of tidal current this area will experience in the hours to come.

Marli Miller

Mastering Geology ™

Looking for additional review and test prep materials? With individualized coaching on the toughest topics of the course, MasteringGeology offers a wide variety of ways for you to move beyond memorization to begin thinking like a geologist. Visit the Study Area in www.masteringgeology.com to find practice quizzes, study tools, and multimedia that will improve your understanding of this chapter's content. Sign in today to enjoy the following features: **Self Study Quizzes, SmartFigures: Tutorials/Animations/Condor Videos/Mobile Field Trips, Geoscience Animation Library, GEODe, RSS Feeds, Digital Study Modules,** and an optional **Pearson eText.**

21

Global
Climate
Change

Glaciers are sensitive to changes in temperature and precipitation and therefore provide clues about changes in climate. Like most glaciers and ice sheets worldwide, Margerie Glacier in Alaska's Glacier Bay National Park, is losing mass and retreating. (Photo by Don Paulson/AGE Fotostock)

FOCUS ON CONCEPTS

Each statement represents the primary **LEARNING OBJECTIVE** for the corresponding major heading within the chapter. After you complete the chapter, you should be able to:

21.1 List the major parts of the climate system and some connections between climate and geology.

21.2 Explain why unraveling past climate changes is important and discuss several ways in which such changes are detected.

21.3 Describe the composition of the atmosphere and the atmosphere's vertical changes in pressure and temperature.

21.4 Outline the basic processes involved in heating the atmosphere.

21.5 Discuss hypotheses that relate to natural causes of climate change.

21.6 Summarize the nature and cause of the atmosphere's changing composition since about 1750. Describe the climate's response.

21.7 Contrast positive- and negative-feedback mechanisms and provide examples of each.

21.8 Discuss several likely consequences of global warming.

Climate has a significant impact on people, and we are learning that people also have a strong influence on climate. Today global climate change caused by humans is a major environmental issue. Unlike changes in the geologic past, which were natural variations, modern climate change is dominated by human influences that are sufficiently large that they exceed the bounds of natural variability. Moreover, these changes are likely to continue for many centuries. The effects of this venture into the unknown with climate could be very disruptive not only to humans but to many other life-forms as well. The latter portion of this chapter examines the ways in which humans may be changing global climate.

21.1 | Climate and Geology

List the major parts of the climate system and some connections between climate and geology.

The term **weather** refers to the state of the atmosphere at a given time and place. Changes in the weather are frequent and sometimes seemingly erratic. In contrast, **climate** is a description of aggregate weather conditions, based on observations over many decades. Climate is often defined simply as "average weather," but this definition is inadequate because variations and extremes are also important parts of a climate description.

The Climate System

Throughout this book you have frequently been reminded that Earth is a complex system that consists of many interacting parts. A change in any one part can produce changes in any or all of the other parts—often in ways that are neither obvious nor immediately apparent. Key to understanding climate change and its causes is the fact that climate is related to all parts of the Earth system. We must recognize that there is a **climate system** that derives its energy from the Sun and includes the atmosphere, hydrosphere, geosphere, biosphere, and cryosphere. The first four were discussed in Chapter 1; the **cryosphere** refers to the portion of Earth's surface where water is in solid form. This includes snow, glaciers, sea ice, freshwater ice, and frozen ground (termed *permafrost*). The climate system *involves the exchanges of energy and moisture that occur among the five spheres*. These exchanges link the atmosphere to the other spheres so that the whole functions as an extremely complex interactive unit. Changes in the climate system do not occur in isolation. Rather, when one part of the climate system changes, the other components react. The major components of the climate system are shown in **Figure 21.1**.

The climate system provides a framework for the study of climate. The interactions and exchanges among the parts of the climate system create a complex network that links the five spheres. As you will see, because the climate system involves all of Earth's spheres, data from many sources are used to study and decipher climate change.

Climate–Geology Connections

Climate has a profound impact on many geologic processes. When climate changes, these processes respond. A glance back at the rock cycle in Chapter 1 reminds us about many of the connections. Of course, rock weathering has an obvious climate connection, as do processes associated with arid, tropical, and glacial landscapes. Phenomena such as debris flows and river flooding are often triggered by atmospheric events such as periods of extraordinary rainfall. Clearly, the atmosphere is a basic link in the hydrologic cycle. Other climate–geology connections involve the impact of internal processes on the atmosphere. For example, the particles and gases emitted by volcanoes can change the composition of the atmosphere, and mountain building can have a significant impact on regional temperature, precipitation, and wind patterns.

The study of sediments, sedimentary rocks, and fossils clearly demonstrates that, during Earth's long and complex history, practically every place on our planet

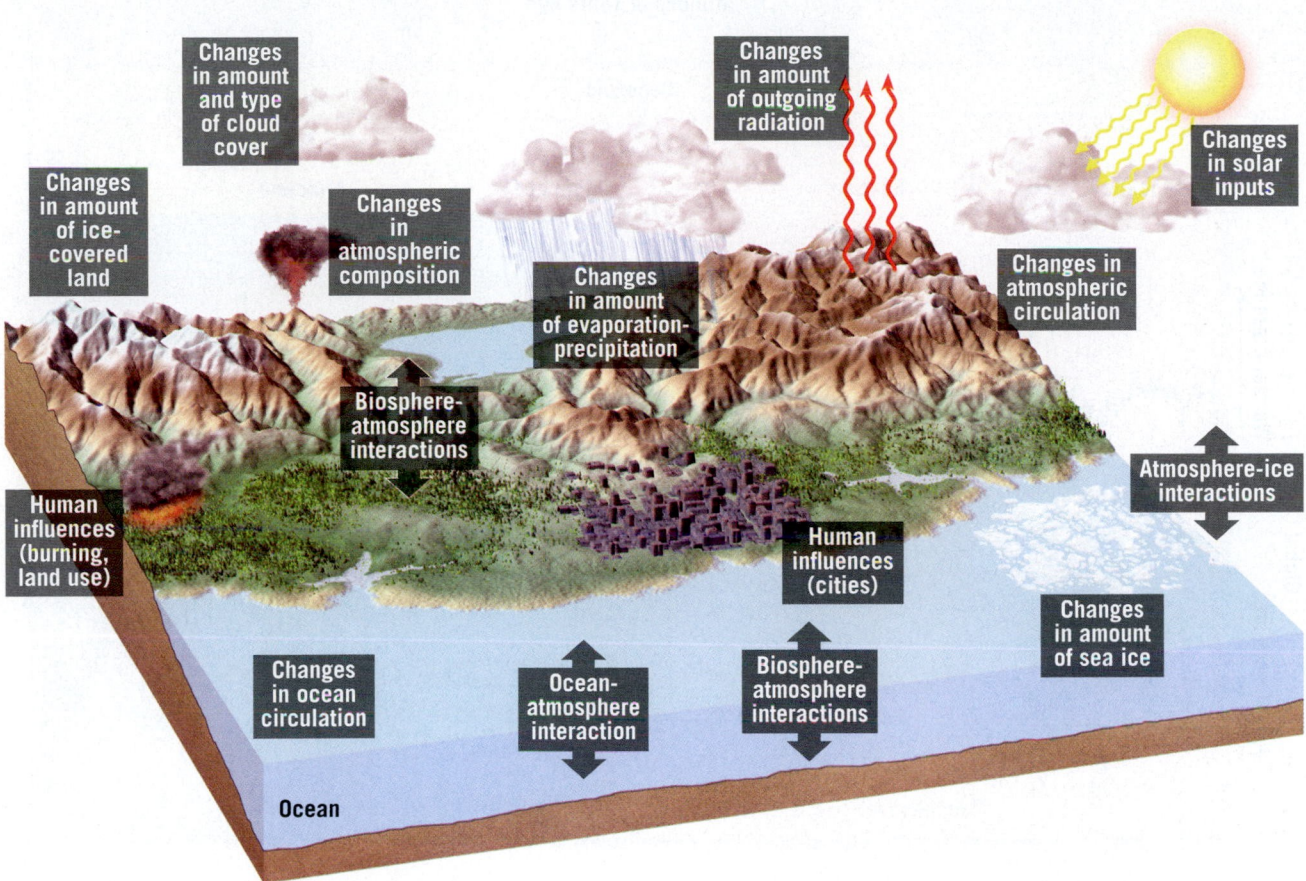

SmartFigure 21.1
Earth's climate system
Schematic view showing some important components of Earth's climate system. Many interactions occur among the various components on a wide range of space and time scales, making the system extremely complex.
(http://goo.gl/Bjd57G)

Video

---

has experienced wide swings in climate, from ice ages to conditions associated with subtropical coal swamps or desert dunes. Time scales for climate change vary from decades to millions of years. Chapter 22 "Earth's Evolution Through Geologic Time," documents many of these shifts in climate.

21.1   Concept Checks

1. Distinguish between *weather* and *climate*.

2. What are the five major parts of the climate system?

3. List at least five connections between climate and geology.

21.2 | Detecting Climate Change

Explain why unraveling past climate changes is important and discuss several ways in which such changes are detected.

Climate not only varies from place to place but is also naturally variable over time. During the great expanse of Earth history, and long before humans were roaming the planet, there were many shifts—from warm to cold and from wet to dry and back again.

Climates Change

Using fossils and many other geologic clues, scientists have reconstructed Earth's climate going back hundreds of millions of years. Chapter 22, which covers Earth's history, also discusses the geologic evidence for natural climate variability. Over long time scales (tens to hundreds of millions of years), Earth's climate can be broadly characterized as being a warm "greenhouse" or a cold "icehouse."

During greenhouse times, there is little, if any, permanent ice at either pole, and relatively warm temperate

Figure 21.2
Relative climate change during the Cenozoic era During the past 65 million years, Earth's climate shifted from being a warm "greenhouse" to being a cool "icehouse." Climate is not stable when viewed over long time spans. Earth has experienced several back-and-forth shifts between warm and cold.

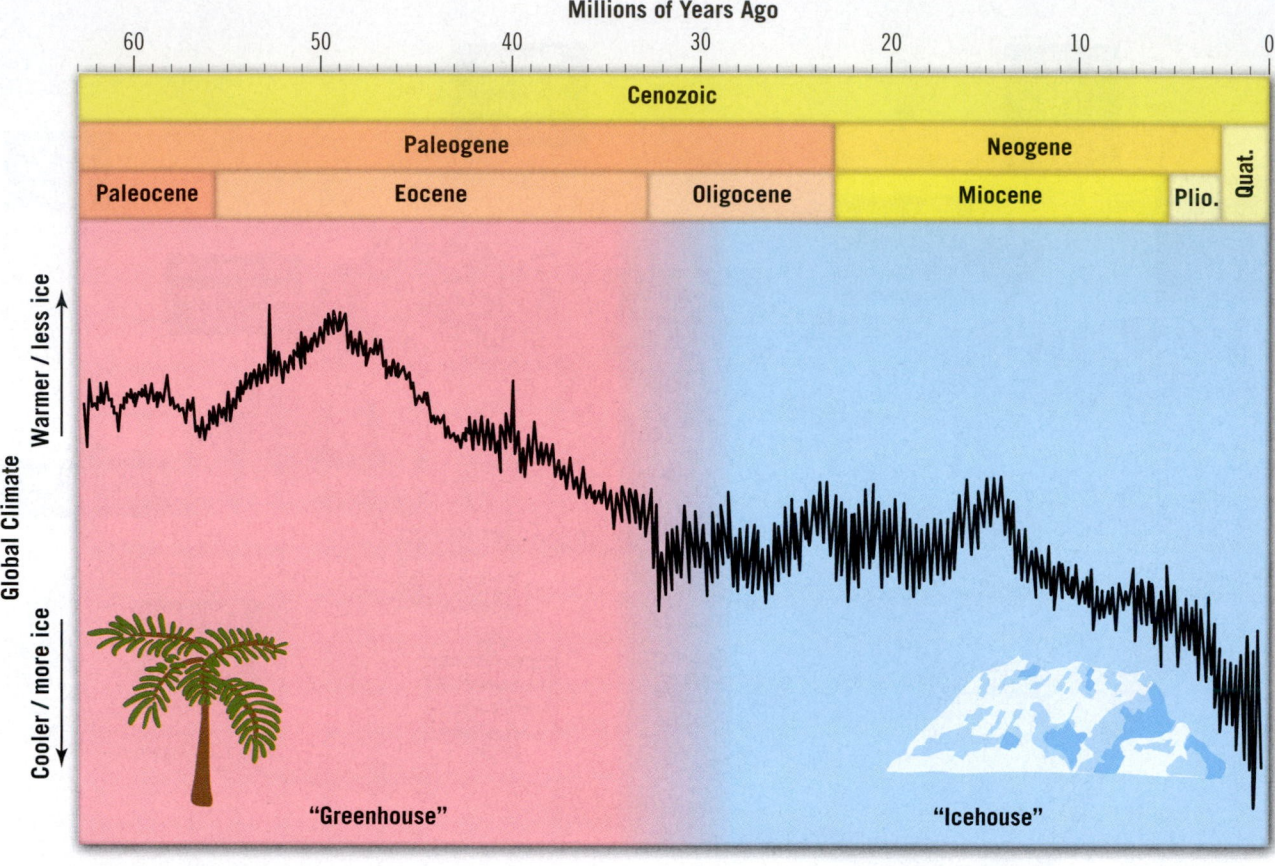

climates are found even at high latitudes. During icehouse conditions, global climate is cool enough to support ice sheets at one or both poles. Earth's climate has gradually transitioned between these two categories only a few times in the past 541 million years, the span known as the Phanerozoic ("visible life") eon. The rocks and deposits of the Phanerozoic eon contain abundant fossils that document major environmental and evolutionary trends. The most recent transition occurred during the Cenozoic era.

The early Cenozoic was a time of greenhouse climates like those the dinosaurs experienced during the preceding Mesozoic era. By about 34 million years ago, permanent ice sheets were present at the South Pole, ushering in icehouse conditions (**Figure 21.2**). Climate warmed during the Miocene epoch (about 20 million years ago) as mammal populations reached their greatest diversity. Climate then cooled. In North America, the lush "greenhouse" forests (there were palm trees in Wyoming and banana trees in Oregon) were replaced by open grasslands. Grassland ecosystems are better suited for a cooler, drier "icehouse" climate. By 2 million years ago (the start of the Quaternary epoch), Earth's climate was cold enough to support vast ice sheets at both poles. In the Northern Hemisphere ice advanced nearly as far south as the present-day Ohio River, then subsequently retreated to Greenland. For the past 800,000 years, this cycle of ice advance and retreat has occurred about every 100,000 years. The

last major ice sheet advance reached a maximum about 18,000 years ago.

How do we know about these changes? What are the causes? The next sections take a look at how scientists decipher Earth's climate history. Later we will explore some significant natural causes of climate change.

Proxy Data

High-technology and precision instrumentation are now available to study the composition and dynamics of the atmosphere. But such tools are recent inventions and therefore have been providing data for only a short time span. To understand fully the behavior of the atmosphere and to anticipate future climate change, we must somehow discover how climate has changed over broad expanses of time.

Instrumental records go back only a couple of centuries, at best, and the further back we go, the less complete and more unreliable the data become. To overcome this lack of direct measurements, scientists must decipher and reconstruct past climates by using indirect evidence. **Proxy data** come from natural recorders of climate variability, such as seafloor sediments, glacial ice, fossil pollen, and tree-growth rings, as well as from historical documents (**Figure 21.3**). Scientists who analyze proxy data and reconstruct past climates are engaged in the study of **paleoclimatology**. The main goal of such work is to understand the

Figure 21.3
Ancient bristlecone pines Some of these trees in California's White Mountains are more than 4000 years old. The study of tree-growth rings is one way that scientists reconstruct past climates. (Photo by Bill Stevenson/Alamy)

climate of the past in order to assess the current and potential future climate in the context of natural climate variability.

Seafloor Sediment: A Storehouse of Climate Data

We know that the parts of the Earth system are linked so that a change in one part can produce changes in any or all of the other parts. In this section, you will see how changes in atmospheric and oceanic temperatures are reflected in the nature of life in the sea.

Most seafloor sediments contain the remains of organisms that once lived near the sea surface (the ocean–atmosphere interface). When such near-surface organisms die, their shells slowly settle to the floor of the ocean, where they become part of the sedimentary record (**Figure 21.4**). These seafloor sediments are useful recorders of worldwide climate change because the numbers and types of organisms living near the sea surface change with the climate. For this reason, scientists are tapping the huge reservoir of data in seafloor sediments. The sediment cores gathered by drilling ships and other research vessels have provided invaluable data that have significantly expanded our knowledge and understanding of past climates.

One notable example of how seafloor sediments add to our understanding of climate change relates to unraveling the fluctuating atmospheric conditions of the Ice Age. The records of temperature changes contained in cores of sediment from the ocean floor have proven critical to our present understanding of this recent span of Earth history. There is more about this topic in the section "Causes of Ice Ages" in Chapter 18.

Oxygen Isotope Analysis

The isotopes of oxygen in water molecules or in the shells of marine organisms are an important source of proxy data on past climate conditions. **Oxygen isotope analysis** is based on precise measurement of the ratio

Figure 21.4
Foraminifera These single-celled amoeba-like organisms, also called *forams*, are extremely abundant and found throughout the world's oceans. Although the foram record in ocean sediment goes back farther, the remains of these organisms are most commonly used to study climate change during the Cenozoic era. The chemical composition of their hard parts depends on water temperature and the presence or absence of large ice sheets. Because of this relationship, scientists analyze foram shells to estimate ocean temperatures and the existence of ice sheets. (Biophoto Associates/Science Source)

SmartFigure 21.5
Ice cores: Important sources of climate data

A. The National Ice Core Laboratory is a physical plant for storing and studying cores of ice taken from glaciers around the world. These cores represent a long-term record of material deposited from the atmosphere. The lab enables scientists to conduct examinations of ice cores, and it preserves the integrity of these samples in a repository for the study of global climate change and past environmental conditions. (Photo by USGS/National Ice Core Laboratory) **B.** This graph, showing temperature variations over the past 40,000 years, is derived from oxygen isotope analysis of ice cores recovered from the Greenland ice sheet. (Based on U.S. Geological Survey) (https://goo.gl/Hx5Ggu)

Tutorial

A.

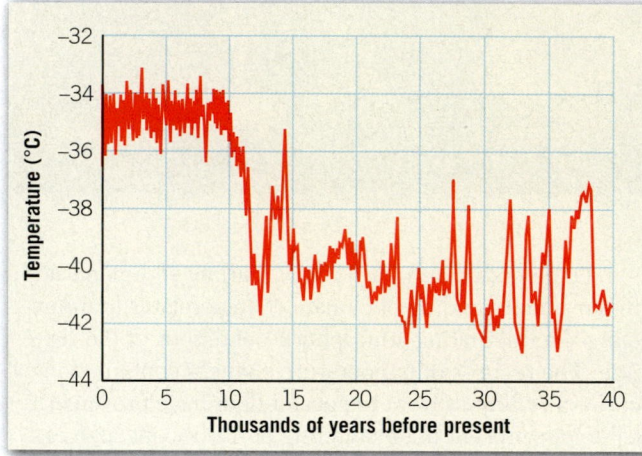

B.

in shells of certain microorganisms buried in deep-sea sediments. A higher ratio of ^{18}O to ^{16}O in shells indicates a time when ice sheets were growing larger.

The $^{18}O/^{16}O$ ratio also varies with temperature. More ^{18}O is evaporated from the oceans when temperatures are high, and less is evaporated when temperatures are low. Therefore, the heavy isotope is more abundant in the precipitation of warm eras and less abundant during colder periods. Using this principle, scientists studying the layers of ice and snow in glaciers have been able to determine past temperature changes.

Climate Change Recorded in Glacial Ice

Ice cores are an indispensable source of data for reconstructing past climates. Research based on vertical cores taken from the Greenland and Antarctic ice sheets has changed our basic understanding of how the climate system works.

Scientists collect samples by using a drilling rig, like a small version of an oil drill. A hollow shaft follows the drill head into the ice, and an ice core is extracted. In this way, cores that sometimes exceed 2000 meters (6500 feet) in length and may represent more than 200,000 years of climate history are acquired for study (**Figure 21.5A**).

Scientists are able to produce a record of changing air temperatures and snowfall by means of the oxygen isotope analysis described above. A portion of such a record is shown in **Figure 21.5B**.

Air bubbles trapped in the ice also record variations in atmospheric composition. Changes in carbon dioxide and methane are linked to fluctuating temperatures. The cores also include atmospheric fallout such as wind-blown dust, volcanic ash, pollen, and modern-day pollution.

Tree Rings: Archives of Environmental History

If you look at the end of a log, you will see that it is composed of a series of concentric rings (**Figure 21.6A**). Tree rings can be a very useful source of proxy data on past climates. Every year, in temperate regions, trees add a layer of new wood under the bark. Characteristics of each tree ring, such as thickness and density, reflect the environmental conditions (especially climate) that prevailed during the year when the ring formed. Favorable growth conditions produce a wide ring; unfavorable ones produce a narrow ring. Trees growing at the same time in the same region show similar tree-ring patterns.

Because a single growth ring is usually added each year, the age of the tree when it was cut can be determined by counting the rings. If the year of cutting is known, the age of the tree and the year in which each ring formed can be determined by counting back from

between two isotopes of oxygen: ^{16}O, which is the most common, and the heavier ^{18}O. A molecule of H_2O can form from either ^{16}O or ^{18}O, but the lighter isotope, ^{16}O, evaporates more readily from the oceans. Because of this, precipitation (and hence the glacial ice that it may form) is enriched in ^{16}O. This leaves a greater concentration of the heavier isotope, ^{18}O, in the ocean water. Thus, during periods when glaciers are extensive, more of the lighter ^{16}O is tied up in ice, so the concentration of ^{18}O in seawater increases. Conversely, during warmer interglacial periods, when the amount of glacial ice decreases dramatically, more ^{16}O is returned to the sea, so the proportion of ^{18}O relative to ^{16}O in ocean water also drops. Now, if we had some ancient recording of the changes of the $^{18}O/^{16}O$ ratio, we could determine when there were glacial periods and, therefore, when the climate grew cooler.

Fortunately, we do have such a recording. Certain marine microorganisms secrete their shells of calcium carbonate ($CaCO_3$), and the prevailing oceanic $^{18}O/^{16}O$ ratio is reflected in the composition of these hard parts. When these organisms die, their hard parts settle to the ocean floor, becoming part of the sediment layers there. Consequently, periods of glacial activity can be determined from variations in the oxygen isotope ratio found the outside ring. Scientists are not limited to working with trees that have been cut down. Small, nondestructive core samples can be taken from living trees (**Figure 21.6B**).

To make the most effective use of tree rings, extended patterns known as *ring chronologies* are established. They are produced by comparing the patterns of rings among trees in an area. If the same pattern can be identified in two samples, one of which has been dated, the second sample can be dated from the first by matching the ring patterns common to both. Tree-ring chronologies extending back thousands of years have been established for some regions. To date a timber sample of unknown age, its ring pattern is matched against the reference chronology.

Tree-ring chronologies are unique archives of environmental history and have important applications in such disciplines as climate, geology, ecology, and archaeology. For example, tree rings are used to reconstruct climate variations within a region for spans of thousands of years prior to human historical records. Knowing such long-term variations is of great value when interpreting the recent record of climate change.

Other Types of Proxy Data

In addition to the sources already discussed, other sources of proxy data that are used to gain insight into past climates include fossil pollen, corals, and historical documents.

Fossil Pollen Climate is a major factor influencing the distribution of vegetation, so the nature of the plant community occupying an area is a reflection of the climate. Pollen and spores are parts of the life cycles of many plants, and because they have very resistant walls, they are often the most abundant, easily identifiable, and best-preserved plant remains in sediments (**Figure 21.7**). By analyzing pollen from accurately dated sediments, scientists can obtain high-resolution records of vegetation changes in an area. Past climates can be reconstructed from such information.

Corals Coral reefs consist of colonies of corals, invertebrates that live in warm, shallow waters and form atop the hard material left behind by past corals (see GEOgraphics 13.1, page 398). Corals build their hard skeletons from calcium carbonate ($CaCO_3$) extracted from seawater. The carbonate

A.

B.

Figure 21.6
Tree rings A. Each year a growing tree produces a layer of new cells beneath the bark. If the tree is cut down and the trunk is examined, each year's growth can be seen as a ring. These rings are useful records of past climate because the amount of growth (the thickness of a ring) depends on precipitation and temperature. (Photo by Victor Zastolskiy/Fotolia) **B.** Scientists are not limited to working with trees that have been cut down. Small, nondestructive core samples can be taken from living trees. (Photo by Gregory K. Scott/Science Source)

contains isotopes of oxygen that can be used to determine the temperature of the water in which the coral grew. The portion of the skeleton that forms in winter has a different density than the portion that forms in summer because of variations in growth rates related to temperature and other environmental factors. Thus, corals exhibit seasonal growth bands very much like those observed in trees. The accuracy and reliability of the climate data extracted from corals has been established by comparing recent instrumental records to coral records for the same period. Oxygen

Figure 21.7
Pollen This false-color image from an electron microscope shows an assortment of pollen grains. Note how the size, shape, and surface characteristics differ from one species to another. Analysis of the types and abundance of pollen in lake sediments and peat deposits provides information about how climate has changed over time. (Photo by David AMI Images/Science Source)

Figure 21.8
Corals record sea-surface temperatures Coral colonies thrive in warm, shallow tropical waters. The tiny invertebrates extract calcium carbonate from seawater to build hard parts. They live atop the solid foundation left by past coral. Chemical analysis of the changing composition of coral reefs with depth can provide useful data on past near-surface temperatures. This graph shows a 350-year record of sea-surface temperatures obtained through oxygen isotope analysis of coral from the Galapagos Islands.

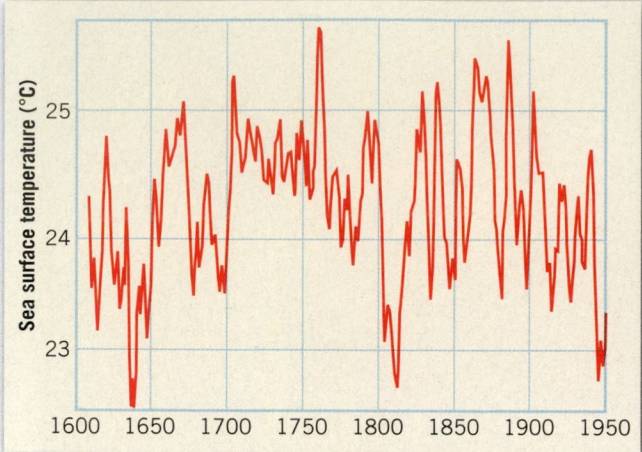

isotope analysis of coral-growth rings can also serve as a proxy measurement for precipitation, particularly in areas where large variations in annual rainfall occur.

Think of coral as a *paleothermometer* that enables us to answer important questions about climate variability in the world's oceans. The graph in **Figure 21.8** is a 350-year sea-surface temperature record based on oxygen isotope analysis of a core extracted from a reef in the Galapagos Islands.

Historical Documents   Historical documents sometimes contain helpful information. Although it may seem that such records should readily lend themselves to climate analysis, that is not the case. Most manuscripts were written for purposes other than climate description. Furthermore, writers understandably neglected periods of relatively stable atmospheric conditions and mention only droughts, severe storms, memorable blizzards, and other extremes. Nevertheless, records of crops, floods, and human migration have furnished useful evidence of the possible influences of changing climate (**Figure 21.9**).

Figure 21.9
Harvest dates as climate clues Historical records can sometimes be helpful in the analysis of past climates. The date for the beginning of the grape harvest in the fall is an integrated measure of temperature and precipitation during the growing season. These dates have been recorded for centuries in Europe and provide a useful record of year-to-year climate variations. (Photo by SGM/AGE Fotostock)

21.2 | Concept Checks

1. What are proxy data, and why are they necessary in the study of climate change?

2. Why are seafloor sediments useful in the study of past climates? Aside from seafloor sediments, list four sources of proxy climate data.

3. Explain how past temperatures are determined using oxygen isotope analysis.

21.3 | Some Atmospheric Basics

Describe the composition of the atmosphere and the atmosphere's vertical changes in pressure and temperature.

To better understand climate change, it is helpful to possess some basic knowledge about the composition and structure of the atmosphere.

Composition of the Atmosphere

Air is *not* a unique element or compound. Rather, air is a *mixture* of many discrete gases, each with its own physical properties, in which varying quantities of tiny solid and liquid particles are suspended.

Clean, Dry Air   As you can see in **Figure 21.10**, clean, dry air is composed almost entirely of two gases—78 percent nitrogen and 21 percent oxygen. Although these gases are the most plentiful components of air and are of great significance to life on Earth, they are of little or no importance in affecting weather phenomena. The remaining 1 percent of dry air is mostly the inert gas argon (0.93 percent) plus tiny quantities of a number of other gases. Carbon dioxide, although present in only minute amounts (0.0400 percent, or 400 parts per million), is nevertheless an important constituent of air because it has the ability to absorb heat energy radiated by Earth and thus influences the heating of the atmosphere.

Air includes many gases and particles that vary significantly from time to time and from place to place. Important examples of these variable gases include water vapor, ozone, and tiny solid and liquid particles.

Water Vapor   The amount of *water vapor* in the air varies considerably, from practically none at all up to about 4 percent by volume. Why is such a small fraction of the atmosphere so significant? Certainly the fact that water vapor is the source of all clouds and precipitation is enough to explain its importance. However, water vapor has other roles. Like carbon dioxide, it has the ability to absorb heat energy given off by Earth as well as some solar energy. It is, therefore, important when we examine the heating of the atmosphere.

Ozone   Another important component of the atmosphere is *ozone*. It is a form of oxygen that combines three oxygen atoms into each molecule (O_3). Ozone is not the same as the oxygen we breathe, which has two atoms per molecule (O_2). There is very little ozone in the atmosphere, and its distribution is not uniform. It is concentrated well above Earth's surface in a layer called the *stratosphere*, at an altitude of between 10 and 50 kilometers (6 and 31 miles). The presence of the ozone layer in our atmosphere is crucial to those who dwell on Earth. The reason is that ozone absorbs the potentially harmful ultraviolet (UV) radiation from the Sun. If ozone did not filter a great deal of the ultraviolet radiation, and if the Sun's UV rays reached the surface of Earth undiminished, our planet would be uninhabitable for most life as we know it.

Aerosols   The movements of the atmosphere are sufficient to keep a large quantity of solid and liquid particles suspended within it. Although visible dust sometimes clouds the sky, these relatively large particles are too heavy to stay in the air for very long. Still, many particles are microscopic and remain suspended for considerable periods of time. They may originate from many sources, both natural and human made, and include sea salts from breaking waves, fine soil blown into the air, smoke and soot from fires, pollen and microorganisms lifted by the wind, ash and dust from volcanic eruptions, and more. Collectively, these tiny solid and liquid particles are called **aerosols**.

From a meteorological standpoint, these tiny, often invisible particles can be significant. First, many act as surfaces on which water vapor can condense, an important function in the formation of clouds and fog. Second, aerosols can absorb or reflect incoming solar radiation. Thus, when an air pollution episode is occurring, or when ash fills the sky following a volcanic eruption, the amount of sunlight reaching Earth's surface can be measurably reduced (**Figure 21.11**).

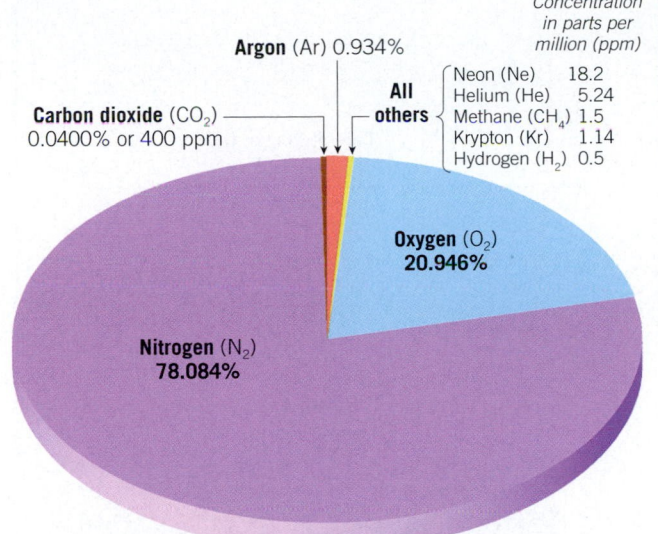

SmartFigure 21.10
Composition of the atmosphere Proportional volume of gases composing dry air. Nitrogen and oxygen obviously dominate.
(https://goo.gl/LIOYJI)

Tutorial

Extent and Structure of the Atmosphere

To say that the atmosphere begins at Earth's surface and extends upward is obvious. But where does the atmosphere end, and where does outer space begin? There is no sharp boundary; the atmosphere rapidly thins as you travel away from Earth, until there are too few gas molecules to detect.

SmartFigure 21.11
Aerosols This satellite image shows two examples of aerosols. First, a dust storm is blowing across northeastern China toward the Korean Peninsula. Second, a dense haze toward the south (bottom center) is human-generated air pollution.
(NASA) (http://goo.gl/P4peqp)

Video

Figure 21.12
Vertical changes in air pressure Pressure decreases rapidly near Earth's surface and more gradually at greater heights.

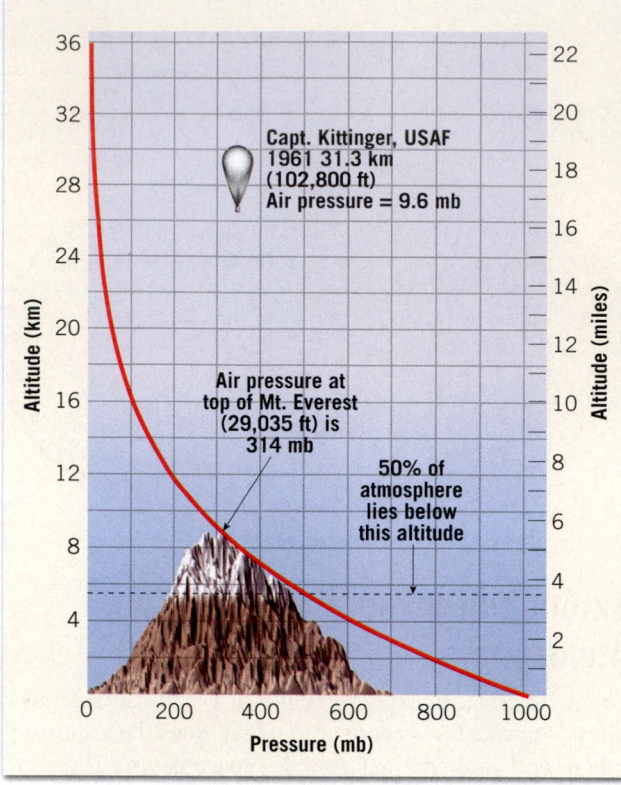

Pressure Changes with Height To understand the vertical extent of the atmosphere, let us examine changes in atmospheric pressure with height. Atmospheric pressure is simply the weight of the air above. At sea level, the average pressure is slightly more than 1000 millibars.

Figure 21.13
Thermal structure of the atmosphere

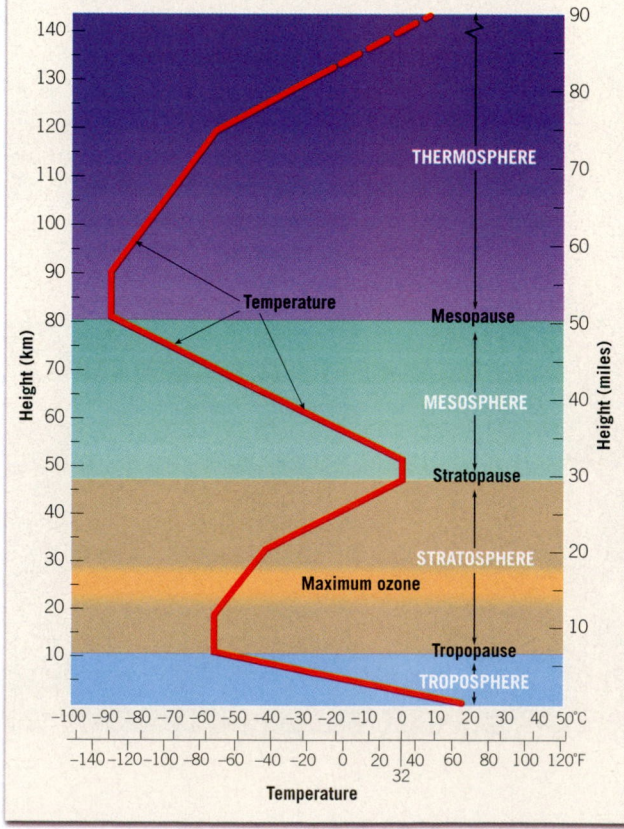

This corresponds to a weight of slightly more than 1 kilogram per square centimeter (14.7 pounds per square inch). Obviously, the pressure at higher altitudes is less (**Figure 21.12**).

One-half of the atmosphere lies below an altitude of 5.6 kilometers (3.5 miles). At about 16 kilometers (10 miles), 90 percent of the atmosphere has been traversed, and above 100 kilometers (62 miles), only 0.00003 percent of all the gases making up the atmosphere remains. Even so, traces of our atmosphere extend far beyond this altitude, gradually merging with the emptiness of space.

Temperature Changes In addition to vertical changes in air pressure, there are also changes in air temperature as we ascend through the atmosphere. Earth's atmosphere is divided vertically into four layers, on the basis of temperature (**Figure 21.13**):

- **Troposphere.** We live in the bottom layer, which is characterized by a decrease in temperature with increasing altitude and is called the **troposphere**. The term literally means the region where air "turns over," a reference to the appreciable vertical mixing of air in this lowermost zone. The troposphere is the chief focus of meteorologists because it is in this layer that essentially all important weather phenomena occur. The temperature decrease in the troposphere is called the *environmental lapse rate*. Its average value is 6.5°C per kilometer (3.5°F per 1000 feet), a figure known as the *normal lapse rate*. It should be emphasized, however, that the environmental lapse rate is not a constant but rather can be highly variable and must be regularly measured. To determine the actual environmental lapse rate as well as to gather information about vertical changes in pressure, wind, and humidity, radiosondes are used. A **radiosonde** is an instrument package that is attached to a weather balloon and transmits data by radio as it ascends through the atmosphere (**Figure 21.14**). The thickness of the troposphere is not the same everywhere; it varies with latitude and season. On the average, the temperature drop continues to a height of about 12 kilometers (7.4 miles). The outer boundary of the troposphere is the *tropopause*.

- **Stratosphere.** Beyond the tropopause is the **stratosphere**. In the stratosphere, the temperature remains constant to a height of about 20 kilometers (12 miles) and then begins a gradual increase that continues until the *stratopause*, at a height of nearly 50 kilometers (30 miles) above Earth's surface. Below the tropopause, atmospheric properties such as temperature and humidity are readily transferred by large-scale turbulence and mixing. Above the tropopause, in the stratosphere, they are not. Temperatures increase in the stratosphere because it is in this layer that the atmosphere's ozone is concentrated. Recall that ozone absorbs ultraviolet radiation from the Sun. As a consequence, the stratosphere is heated.

Figure 21.14
Radiosonde A radiosonde is a lightweight package of instruments that is carried aloft by a small weather balloon. It transmits data on vertical changes in temperature, pressure, and humidity in the troposphere. The troposphere is where practically all weather phenomena occur; therefore, it is very important to have frequent measurements. (Photo by David R. Frazier/Newscom)

- **Mesosphere.** In the third layer, the **mesosphere**, temperatures again decrease with height until at the *mesopause*, more than 80 kilometers (50 miles) above Earth's surface, the temperature approaches −90°C (−130°F). The coldest temperatures anywhere in the atmosphere occur at the mesopause.

- **Thermosphere.** The fourth layer extends outward from the mesopause and has no well-defined upper limit. This is the **thermosphere**, a layer that contains only a *tiny fraction* of the atmosphere's mass. In the extremely rarefied air of this outermost layer, temperatures again increase, due to the absorption of very short-wave, high-energy solar radiation by atoms of oxygen and nitrogen. Temperatures rise to extremely high values of more than 1000°C (1800°F) in the thermosphere. But such temperatures are not comparable to those experienced near Earth's surface. Temperature is defined in terms of the average speed at which molecules move. Because the gases of the thermosphere are moving at very high speeds, the temperature is very high. But the gases are so sparse that, collectively, they process only an insignificant amount of heat.

21.3 Concept Checks

1. What are the major components of clean, dry air? List two significant variable components.

2. Describe how air pressure changes with an increase in altitude. Does it change at a constant rate?

3. The atmosphere is divided vertically into four layers, on the basis of temperature. Name the layers from bottom to top and indicate how temperatures change in each.

EYE ON EARTH 21.1

When this weather balloon was launched, the surface temperature was 17°C. The balloon is now at an altitude of 1 kilometer. (Photo by David R. Frazier/Science Source)

QUESTION 1 What term is applied to the instrument package being carried aloft by the balloon?

QUESTION 2 In what layer of the atmosphere is the balloon?

QUESTION 3 If average conditions prevail, what is the air temperature at this altitude? How did you figure this out?

QUESTION 4 How will the size of the balloon change, if at all, as it rises through the atmosphere? Explain.

21.4 | Heating the Atmosphere

Outline the basic processes involved in heating the atmosphere.

Nearly all the energy that drives Earth's variable weather and climate comes from the Sun. Before we can adequately describe how Earth's atmosphere is heated, it is helpful to know something about solar energy and what happens to this energy once it is intercepted by Earth.

Energy from the Sun

From our everyday experience, we know that the Sun emits light and heat as well as the ultraviolet rays that cause suntan. Although these forms of energy comprise a major portion of the total energy that radiates from the Sun, they are only part of a large array of energy called *radiation*, or *electromagnetic radiation*. This array, or spectrum, of electromagnetic energy is shown in **Figure 21.15**. All radiation—whether x-rays, microwaves, or radio waves—transmits energy through the vacuum of space at 300,000 kilometers (186,000 miles) per second and

only slightly slower through our atmosphere. When an object absorbs any form of radiant energy, the result is an increase in molecular motion, which causes a corresponding increase in temperature.

To better understand how the atmosphere is heated, it is useful to have a general understanding of the basic laws governing radiation:

- **All objects, at whatever temperature, emit radiant energy.** Not only hot objects like the Sun but also Earth, including its polar ice caps, continually emit energy.

SmartFigure 21.15

The electromagnetic spectrum This diagram illustrates the wavelengths and names of various types of radiation. Visible light consists of an array of colors we commonly call the "colors of the rainbow." (Photo courtesy of Dennis Tasa) (http://goo.gl/QBJdNJ)

Video

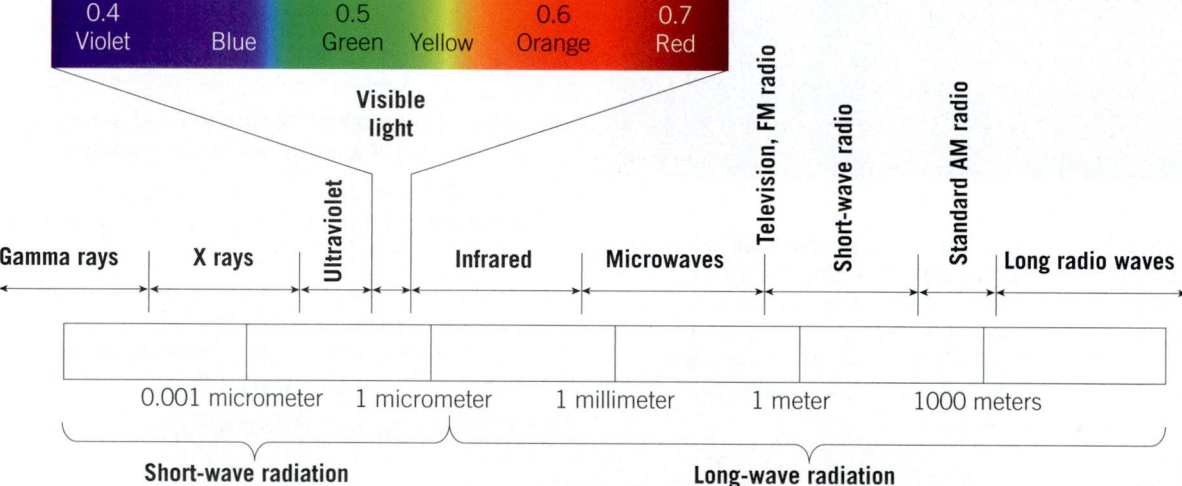

- **Hotter objects radiate more total energy per unit area than do colder objects.**
- **The hotter the radiating body, the shorter the wavelength of maximum radiation.** The Sun, with a surface temperature of about 5700°C, radiates maximum energy at 0.5 micrometer, which is in the visible range. The maximum radiation for Earth occurs at a wavelength of 10 micrometers, well within the infrared (heat) range. Because the wavelength for maximum Earth radiation is roughly 20 times longer than the maximum solar radiation, Earth radiation is often called *long-wave radiation*, and solar radiation is called *short-wave radiation*.
- **Objects that are good absorbers of radiation are good emitters as well.** Earth's surface and the Sun approach being perfect radiators because they absorb and radiate with nearly 100 percent efficiency for their respective temperatures. On the other hand, *gases are selective absorbers and emitters of radiation.* For some wavelengths, the atmosphere is nearly transparent (that is, little radiation is absorbed). For other wavelengths, however, the atmosphere is nearly opaque (that is, it is a good absorber). Experience tells us that the atmosphere is transparent to visible light; hence, these wavelengths readily reach Earth's surface. This is not the case for the longer-wavelength radiation emitted by Earth.

The Paths of Incoming Solar Energy

Figure 21.16 shows the paths taken by incoming solar radiation averaged for the entire globe. Notice that the atmosphere is quite transparent to incoming solar radiation. On average, about 50 percent of the solar energy reaching the top of the atmosphere passes through the atmosphere and is absorbed at Earth's surface. Another 20 percent is absorbed directly by clouds and certain atmospheric gases (including oxygen and ozone) before reaching the surface. The remaining 30 percent is reflected back to space by the atmosphere, clouds, and reflective surfaces such as snow and ice. The fraction of the total radiation that is reflected by a surface is called its **albedo** (**Figure 21.17**). Thus, the albedo for Earth as a whole (the *planetary albedo*) is 30 percent.

What determines whether solar radiation will be transmitted to the surface, scattered, or reflected outward? It depends greatly on the wavelength of the energy being transmitted, as well as on the nature of the intervening material.

The numbers shown in Figure 21.16 represent global averages. The actual percentages can vary greatly, primarily due to changes in the percentage of light reflected and scattered back to space. For example, if the sky is overcast, a higher percentage of light is reflected back to space than when the sky is clear.

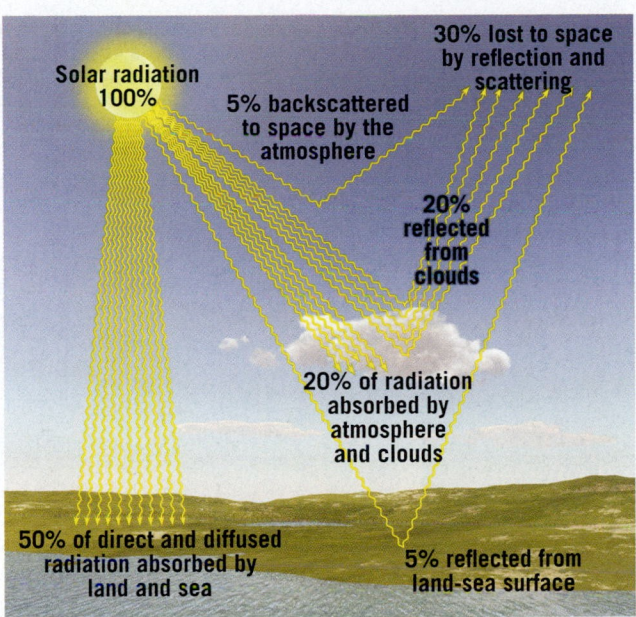

SmartFigure 21.16
Paths taken by solar radiation This diagram shows the average distribution of incoming solar radiation by percentage. More solar radiation is absorbed by Earth's surface than by the atmosphere.
(https://goo.gl/REGnn9)

Tutorial

Heating the Atmosphere: The Greenhouse Effect

If Earth had no atmosphere, it would experience an average surface temperature far below freezing. But the atmosphere warms the planet and makes Earth livable. The extremely important role the atmosphere plays in heating Earth's surface has been named the **greenhouse effect**.

As discussed earlier, cloudless air is largely transparent to incoming short-wave solar radiation and, hence, transmits it to Earth's surface. By contrast, a significant fraction of the long-wave radiation emitted by Earth's land–sea surface is absorbed by water vapor, carbon dioxide, and other trace gases in the atmosphere. This energy heats the air and increases the rate at which it radiates energy, both out to space

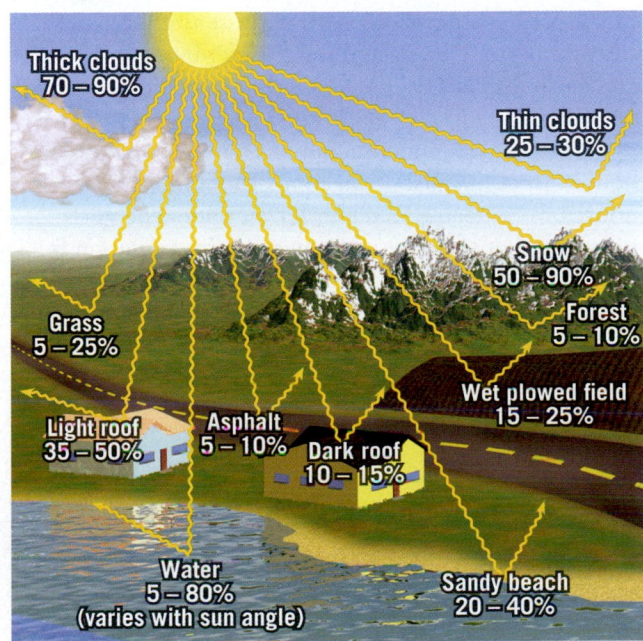

Figure 21.17
Albedo (reflectivity) of various surfaces In general, light-colored surfaces tend to be more reflective than dark-colored surfaces and thus have higher albedos.

Airless bodies like the Moon All incoming solar radiation reaches the surface. Some is reflected back to space. The rest is absorbed by the surface and radiated directly back to space. As a result the lunar surface has a much lower average surface temperature than Earth.

Bodies with modest amounts of greenhouse gases like Earth The atmosphere absorbs some of the longwave radiation emitted by the surface. A portion of this energy is radiated back to the surface and is responsible for keeping Earth's surface 33°C (59°F) warmer than it would otherwise be.

Bodies with abundant greenhouse gases like Venus Venus experiences extraordinary greenhouse warming, which is estimated to raise its surface temperature by 523°C (941°F).

All outgoing longwave energy is reradiated directly back to space

Incoming shortwave solar radiation

David Cole/Alamy

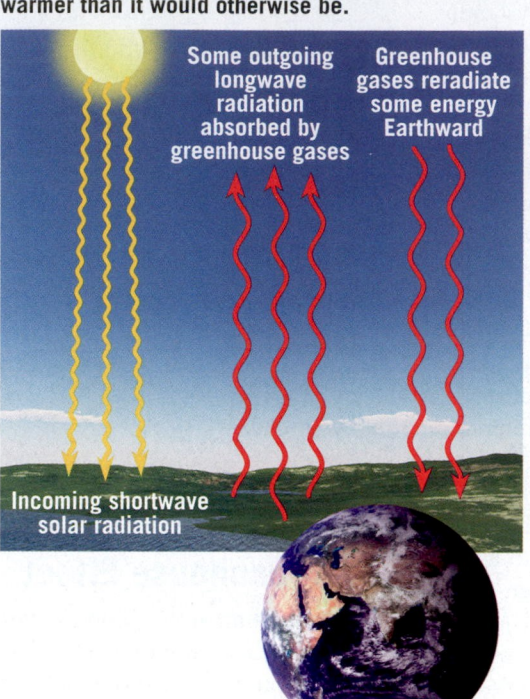

Some outgoing longwave radiation absorbed by greenhouse gases

Greenhouse gases reradiate some energy Earthward

Incoming shortwave solar radiation

NASA

Most outgoing longwave radiation absorbed by greenhouse gases

Greenhouse gases reradiate considerable energy toward the Venusian surface

Incoming shortwave solar radiation

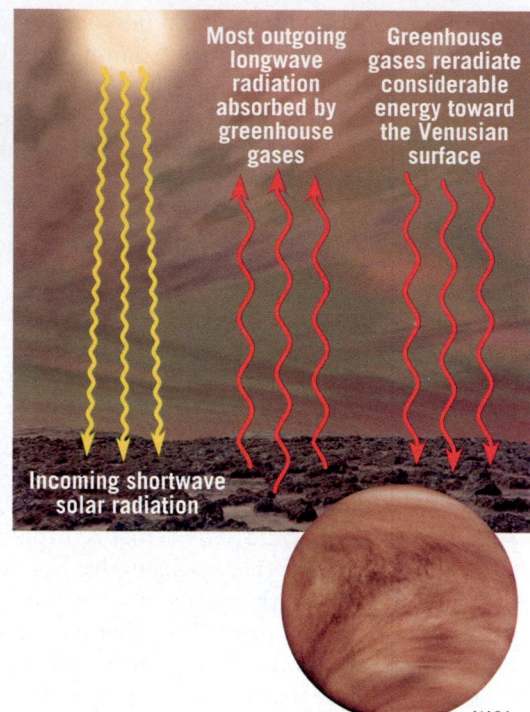

NASA

SmartFigure 21.18
The greenhouse effect The greenhouse effect of Earth compared with two of our close solar system neighbors.
(https://goo.gl/e150CY)

Tutorial

and back toward Earth's surface. Without this complicated game of "pass the hot potato," Earth's average temperature would be –18°C (–0.4°F) rather than the current average temperature of 15°C (59°F) (**Figure 21.18**). These absorptive gases in our atmosphere make Earth habitable for humans and other life-forms. As you will see in the sections that follow, changes in the air's composition, both natural and human caused, impact the greenhouse effect in ways that can cause the atmosphere to either become warmer or cooler.

This natural phenomenon was named the greenhouse effect because it was once thought that greenhouses were heated in a similar manner. The glass in a greenhouse allows short-wave solar radiation to enter and be absorbed by the objects inside. These objects, in turn, radiate energy but at longer wavelengths, to which glass is nearly opaque. The heat,

therefore, is "trapped" in the greenhouse. It has been shown, however, that air inside greenhouses becomes warmer than outside air mainly because greenhouses restrict the exchange of air between the inside and outside. Nevertheless, the term *greenhouse effect* remains.

21.4   Concept Checks

1. What are the three paths taken by incoming solar radiation? What might cause the percentage taking each path to vary?

2. Explain why the atmosphere is heated chiefly by radiation from Earth's surface.

3. Prepare a sketch with labels that explains the greenhouse effect.

21.5 | Natural Causes of Climate Change

Discuss hypotheses that relate to natural causes of climate change.

A great variety of hypotheses have been proposed to explain climate change. Several have gained wide support, only to lose it and then sometimes to regain it. Some explanations are controversial. This is to be expected because planetary atmospheric processes are so large-scale and complex that they cannot be reproduced physically in laboratory experiments. Rather, climate and its changes must be simulated mathematically (modeled), using powerful computers.

In this section we examine several current hypotheses that have earned serious consideration from the scientific community. They describe "natural" mechanisms of climatic change, causes that are unrelated to human activities. A later section examines human-induced climate changes, including the effect of rising carbon dioxide levels caused primarily by our burning of fossil fuels.

As you read this section, you will find that more than one hypothesis may explain the same change in climate. In fact, several mechanisms may interact to shift climate. Also, no single hypothesis can explain climate change on all time scales. A proposal that explains variations over millions of years generally cannot explain fluctuations over hundreds of years. If our atmosphere and its changes ever become fully understood, we will probably see that climate change is caused by many of the mechanisms discussed here, plus new ones yet to be proposed.

Plate Movements and Orbital Variations

In Chapter 18, the section "Causes of Ice Ages" describes two natural mechanisms of climate change. Recall that the movement of lithospheric plates gradually moves Earth's continents closer to or farther from the equator. Although these shifts in plates are very slow, they can have a dramatic impact on climate over spans of millions of years. Moving landmasses can also lead to significant shifts in ocean circulation, which influences heat transport around the globe.*

A second natural mechanism of climate change related to the causes of ice ages involves variations in Earth's orbit. Changes in the shape of the orbit (*eccentricity*), variations in the angle that Earth's axis makes with the plane of its orbit (*obliquity*), and the wobbling of the axis (*precession*) cause fluctuations in the seasonal and latitudinal distribution of solar radiation (see Figure 18.41, page 563). These variations, in turn, contributed to the alternating glacial–interglacial episodes of the Ice Age.

Volcanic Activity and Climate Change

The idea that explosive volcanic eruptions might alter Earth's climate was first proposed many years ago. It is still regarded as a plausible explanation for some aspects of climatic variability. Explosive eruptions emit huge quantities of gases and fine-grained debris into the atmosphere (**Figure 21.19**). The greatest eruptions are sufficiently

*For more on this, see the section "Supercontinents, Mountain Building, and Climate" in Chapter 22.

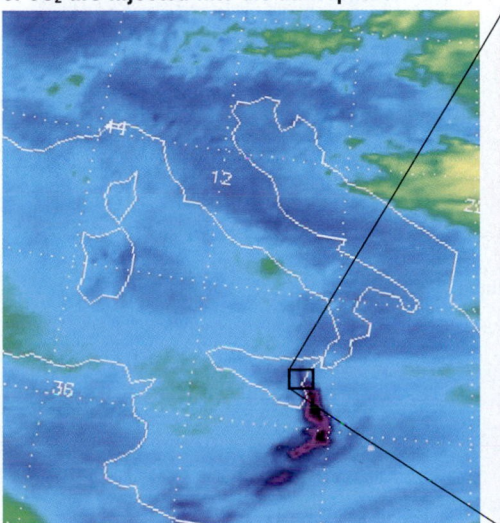

This satellite image shows the sulfur dioxide (SO₂) plume in shades of purple and black. Climate may be affected when large quantities of SO₂ are injected into the atmosphere.

This image was taken from the International Space Station and shows a plume of volcanic ash streaming southeastward from the volcano.

Figure 21.19
Mount Etna erupting in October 2002 This volcano on the island of Sicily is Europe's largest and most active volcano. (NASA images)

powerful to inject material high into the atmosphere; strong upper-level winds then spread it around the globe, where it remains for many months or even years.

The Effect of Volcanic Aerosols on Climate

Suspended volcanic material filters out a portion of the incoming solar radiation, which in turn lowers temperatures in the troposphere. More than 200 years ago, Benjamin Franklin used this idea to argue that material from the eruption of a large Icelandic volcano could have reflected sunlight back to space and therefore might have been responsible for the unusually cold winter of 1783–1784.

Perhaps the most notable cool period linked to a volcanic event is the "year without a summer" that followed the 1815 eruption of Mount Tambora in Indonesia. Tambora's is the largest eruption of modern times. During April 7–12, 1815, this nearly 4000-meter- (13,000-foot-) high volcano violently expelled an estimated 100 cubic kilometers (24 cubic miles) of volcanic debris. The impact of the volcanic aerosols on climate is believed to have been widespread in the Northern Hemisphere. From May through September 1816, an unprecedented series of cold spells affected the northeastern United States and adjacent portions of Canada. There was heavy snow in June and frost in July and August. Abnormal cold was also experienced in much of Western Europe. Similar,

although apparently less dramatic, effects have been associated with other great explosive volcanoes, including Indonesia's Krakatoa in 1883.

Three more recent volcanic events have provided considerable data and insight regarding the impact of volcanoes on global temperatures. The eruptions of Washington State's Mount St. Helens in 1980, the Mexican volcano El Chichón in 1982, and Mount Pinatubo in the Philippines in 1991 have given scientists opportunities to study the atmospheric effects of volcanic eruptions with the aid of more sophisticated technology than was previously available. Satellite images and remote-sensing instruments allowed scientists to closely monitor the effects of the clouds of gases and ash that these volcanoes emitted.

Volcanic Ash and Dust When Mount St. Helens erupted, there was immediate speculation about the possible effects on climate. Could such an eruption cause our climate to change? There is no doubt that the large quantity of volcanic ash emitted by the explosive eruption had significant local and regional effects for a short period. Still, studies indicated that any longer-term lowering of hemispheric temperatures was negligible. The cooling was so slight—probably less than 0.1°C (0.2°F)—that it could not be distinguished from other natural temperature fluctuations.

Sulfuric Acid Droplets Two years of monitoring and studies following the 1982 El Chichón eruption indicated that it had a greater cooling effect on global mean temperature than Mount St. Helens—on the order of 0.3° to 0.5°C (0.5° to 0.9°F). El Chichón's eruption was *less explosive* than the Mount St. Helens blast, so why did it have a greater impact on global temperatures? The reason is that the material emitted by Mount St. Helens was largely fine ash that settled out in a relatively short time. El Chichón, on the other hand, emitted far greater quantities of sulfur dioxide gas (an estimated 40 times more) than Mount St. Helens. This gas combines with water vapor in the stratosphere to produce a dense cloud of tiny sulfuric acid particles (**Figure 21.20A**). These particles take several years to settle out completely. They lower the troposphere's mean

Figure 21.20
Volcanic haze reducing sunlight at Earth's surface The reflective haze produced by some volcanic eruptions is not volcanic ash but tiny sulfuric acid aerosols. (NASA)

A plume of white haze from Anatahan Volcano blankets a portion of the Philippine Sea in April 2005. The haze consisted of tiny droplets of sulfuric acid formed when sulfur dioxide from the volcano combined with water in the atmosphere. The plume is bright and reflects sunlight back to space.

A.

Net solar radiation at Hawaii's Mauna Loa Observatory relative to 1970 (zero on the graph). The eruptions of El Chichón and Mt. Pinatubo clearly caused temporary drops in solar radiation reaching the surface.

B.

temperature because they reflect solar radiation back to space (**Figure 21.20B**).

We now understand that volcanic clouds that remain in the stratosphere for a year or more are composed largely of sulfuric-acid droplets and not of dust, as was once thought. Thus, the volume of fine debris emitted during an explosive event is not an accurate criterion for predicting the global atmospheric effects of an eruption.

Mount Pinatubo in the Philippines erupted explosively in June 1991, injecting 25 to 30 million tons of sulfur dioxide into the stratosphere. The event provided scientists with an opportunity to study the climatic impact of a major explosive volcanic eruption using NASA's spaceborne Earth Radiation Budget Experiment. During the next year, the haze of tiny aerosols increased reflectivity and lowered global temperatures by 0.5°C (0.9°F).

The impact on global temperature of eruptions like El Chichón and Mount Pinatubo is relatively minor, but many scientists agree that the cooling produced could alter the general pattern of atmospheric circulation for a limited period. Such a change, in turn, could influence the weather in some regions. Predicting, or even identifying, specific regional effects still presents a considerable challenge to atmospheric scientists.

The impact on climate of any single volcanic eruption, no matter how great, is relatively small and short-lived. The graph in Figure 21.20B reinforces this point. Therefore, many great eruptions, closely spaced in time, need to occur if volcanic processes are to have a pronounced impact on climate for an extended period. Because no extensive episode of explosive volcanism is known to have occurred in historic times, such an occurrence is most often mentioned as a possible contributor to prehistoric climatic shifts.

Volcanism and Global Warming   The Cretaceous period is the last period of the Mesozoic era, the era of *middle life* that is often called the "age of dinosaurs." It began about 145.5 million years ago and ended about 65.5 million years ago, with the extinction of the dinosaurs (and many other life-forms as well).[*]

The Cretaceous climate was among the warmest in Earth's long history. Dinosaurs, which are associated with mild temperatures, ranged north of the Arctic Circle. Tropical forests existed in Greenland and Antarctica, and coral reefs grew as much as 15 degrees latitude closer to the poles than at present. Deposits of peat that would eventually form widespread coal beds accumulated at high latitudes. Sea level was as much as 200 meters (650 feet) higher than it is today, consistent with a lack of polar ice sheets.

What caused the unusually warm climates of the Cretaceous period? Among the significant factors that may have contributed was an enhanced greenhouse effect due to an increase in the amount of carbon dioxide in the atmosphere. But where did the additional CO_2 come from?

Many geologists suggest that the probable source was volcanic activity. Carbon dioxide is one of the gases emitted during volcanism, and there is now considerable geologic evidence that the Middle Cretaceous was a time when there was an unusually high rate of volcanic activity. Several huge oceanic lava plateaus were produced on the floor of the western Pacific during this span. These vast features were associated with hot spots that may have been produced by large mantle plumes. Massive outpourings of lava over millions of years would have been accompanied by the release of huge quantities of CO_2, which in turn would have enhanced the

[*]For more about the end of the Cretaceous, see Chapter 22.

EYE ON EARTH 21.2

This satellite image shows an extensive plume of ash from an explosive volcanic eruption of Indonesia's Sinabung volcano on January 16, 2014. (NASA)

QUESTION 1 How might the volcanic ash from this eruption influence air temperatures?

QUESTION 2 Would this effect likely be long-lasting—perhaps for years? Explain.

QUESTION 3 What "invisible" volcanic emission might have a greater effect than the volcanic ash?

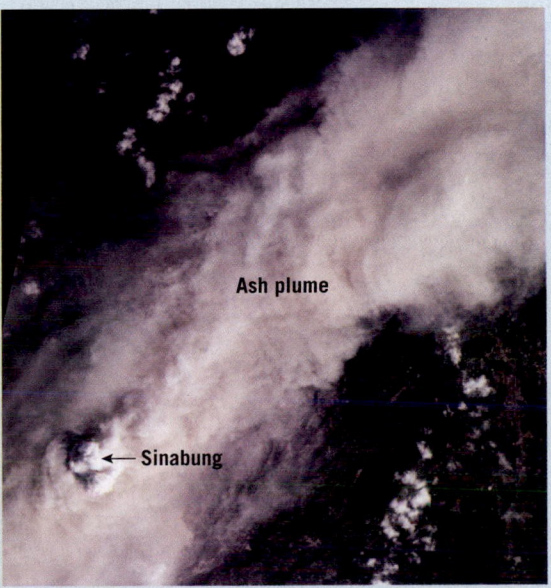

Ash plume

← Sinabung

This view shows an approximation of the Sun's surface. The black spots surrounded by deep orange is a sunspot region where magnetic activity is extremely intense.

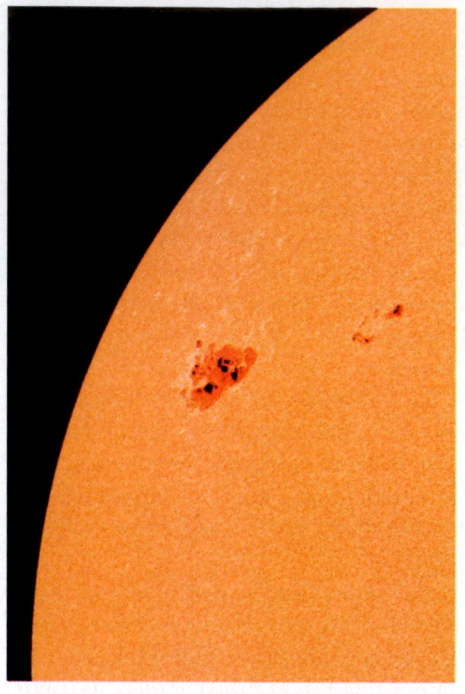

The instrument that produced this image used ultraviolet, radio, and other parts of the electromagnetic spectrum. Looping lines show solar plasma following magnetic field lines.

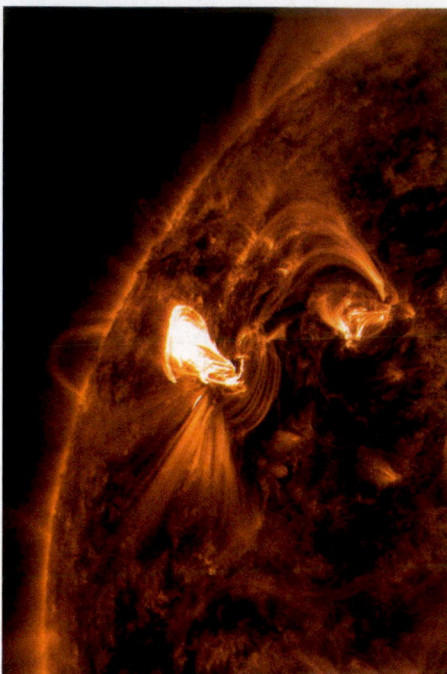

However, no major *long-term* variations in the total intensity of solar radiation have yet been measured outside the atmosphere. Such measurements were not even possible until satellite technology became available. We can now measure solar output but still need many decades of records before we will begin to sense how variable (or invariable) energy from the Sun really is.

Some hypotheses for climate change have related to sunspot cycles. The most conspicuous and best-known features on the surface of the Sun are the dark blemishes called **sunspots** (**Figure 21.21**). Sunspots are huge magnetic storms that extend from the Sun's surface deep into the interior. Moreover, these spots are associated with the Sun's ejection of huge masses of particles that, on reaching Earth's upper atmosphere, interact with gases to produce displays known as the *aurora borealis*, or Northern Lights, in the Northern Hemisphere.

atmospheric greenhouse effect. *Thus, the warmth that characterized the Cretaceous may have had its origins deep in Earth's mantle.*

This example illustrates the interrelationships among parts of the Earth system. Seemingly unrelated materials and processes turn out to be linked. Here you have seen how processes originating deep in Earth's interior are connected directly or indirectly to the atmosphere, the oceans, and the biosphere.

Solar Variability and Climate

Among the most persistent hypotheses of climate change have been those based on the idea that the Sun is a variable star and that its output of energy varies through time. The effect of such changes would seem direct and easily understood: Increases in solar output would cause the atmosphere to warm, and reductions would result in cooling. This notion is appealing because it can be used to explain climate change of any length or intensity.

Sunspots occur in cycles, with the number of sunspots reaching a maximum about every 11 years (**Figure 21.22**). During periods of maximum sunspot activity, the Sun emits slightly more energy than during sunspot minimums. Based on measurements from space that began in 1978, the variation during an 11-year cycle is about 0.1 percent. Although sunspots are dark, they are surrounded by brighter areas, which apparently offset the effect of the dark spots. It appears that this change in solar output is too small and the cycles are too short to have any appreciable effect on global temperatures. However, there is a possibility that longer-term variations in solar output may affect climates on Earth.

For example, the span between 1645 and 1715 is a period known as the *Maunder minimum*, during which

Figure 21.22
Mean annual sunspot numbers The number of sunspots reaches a maximum about every 11 years.

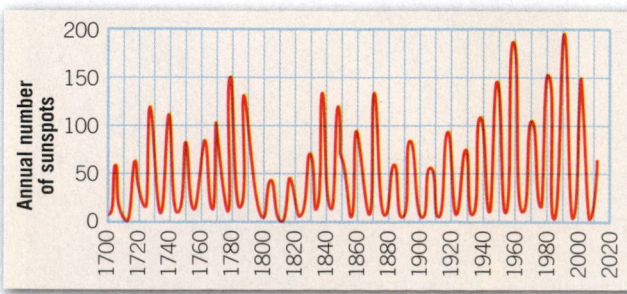

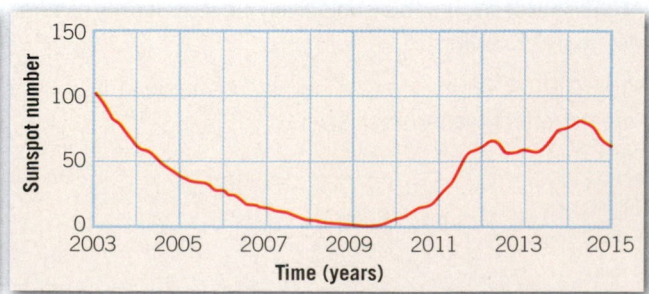

sunspots were largely absent. This period of missing sunspots closely corresponds with a period in climate history known as the *Little Ice Age*, an especially cold period in Europe. For some scientists, this correlation suggests that a reduction in the Sun's output was likely responsible at least in part for this cold episode. Other scientists seriously question this notion. Their hesitation stems in part from subsequent investigations using different climate records from around the world that failed to find a significant correlation between sunspot activity and climate.

21.5 | **Concept Checks**

1. Describe and briefly explain the effect of the El Chichón and Mount Pinatubo eruptions on global temperatures.

2. How might volcanism lead to global warming?

3. What are sunspots? How does solar output change as sunspot numbers change? Is there a solid connection between sunspot numbers and climate change on Earth?

21.6 | Human Impact on Global Climate

Summarize the nature and cause of the atmosphere's changing composition since about 1750. Describe the climate's response.

So far we have examined potential causes of climate change that are natural. In this section, we examine how humans contribute to global climate change. One impact largely results from the addition of carbon dioxide and other greenhouse gases to the atmosphere. A second impact stems from the addition of human-generated aerosols to the atmosphere.

Human influence on regional and global climate did not just begin with the onset of the modern industrial period. There is good evidence that people have been modifying the environment over extensive areas for thousands of years. The use of fire and the overgrazing of marginal lands by domesticated animals have both reduced the abundance and distribution of vegetation. By altering ground cover, humans have modified such important climate factors as surface albedo, evaporation rates, and surface winds.

Rising CO_2 Levels

Earlier you learned that carbon dioxide (CO_2) represents only about 0.0400 percent (400 parts per million) of the gases that make up clean, dry air. Nevertheless, it is a very significant component meteorologically. Carbon dioxide is influential because it is transparent to incoming short-wavelength solar radiation, but it is not transparent to some of the longer-wavelength, outgoing Earth radiation. A portion of the energy leaving Earth's surface is absorbed by atmospheric CO_2. This heat energy is subsequently re-emitted, part of it back toward the surface, thereby keeping the air near the ground warmer than it would be without CO_2. Thus, along with water vapor, carbon dioxide is largely responsible for the atmosphere's greenhouse effect.

The tremendous industrialization of the past two centuries has been fueled—and still is fueled—by burning fossil fuels: coal, natural gas, and petroleum (**Figure 21.23**). Combustion

of these fuels has added great quantities of carbon dioxide to the atmosphere. **Figure 21.24** shows changes in CO_2 concentrations at Hawaii's Mauna Loa Observatory, where measurements have been made since 1958. The graph shows an annual seasonal cycle and a steady upward trend over the years. The up-and-down of the seasonal cycle is due to the vast land area of the Northern Hemisphere, which contains the majority of land-based vegetation. During spring and summer in the Northern Hemisphere, when plants are absorbing CO_2 as part of photosynthesis, concentrations decrease. The annual increase in atmospheric CO_2 during the cold months occurs as vegetation dies and leaves fall and decompose, releasing CO_2 back into the air.

The use of coal and other fuels is the most prominent means by which humans add CO_2 to the atmosphere, but it is not the only way. The clearing of forests also contributes substantially because CO_2 is released as vegetation is burned or decays (**Figure 21.25**). Deforestation is particularly pronounced in the tropics, where vast tracts

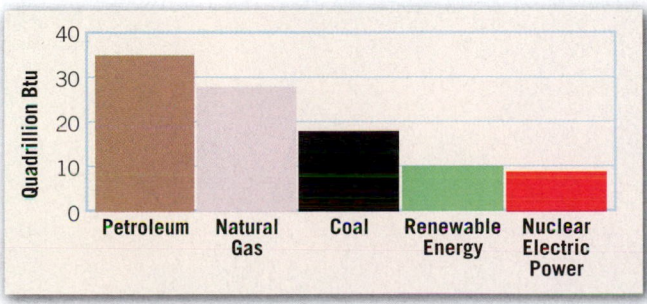

Figure 21.23
U.S. energy consumption The graph shows energy consumption in 2014. The total was 98.3 quadrillion Btu. A quadrillion is 10 raised to the 12th power, or a million million. The burning of fossil fuels represents about 81 percent of the total. (Based on data from U.S. Energy Information Administration)

SmartFigure 21.24
Monthly CO₂ concentrations Atmospheric CO_2 has been measured at Mauna Loa Observatory, Hawaii, since 1958. There has been a consistent increase since monitoring began. This graphic portrayal is known as the *Keeling Curve*, in honor of the scientist who originated the measurements. (Based on NOAA)
(https://goo.gl/5HSuIS)

▶ **Tutorial**

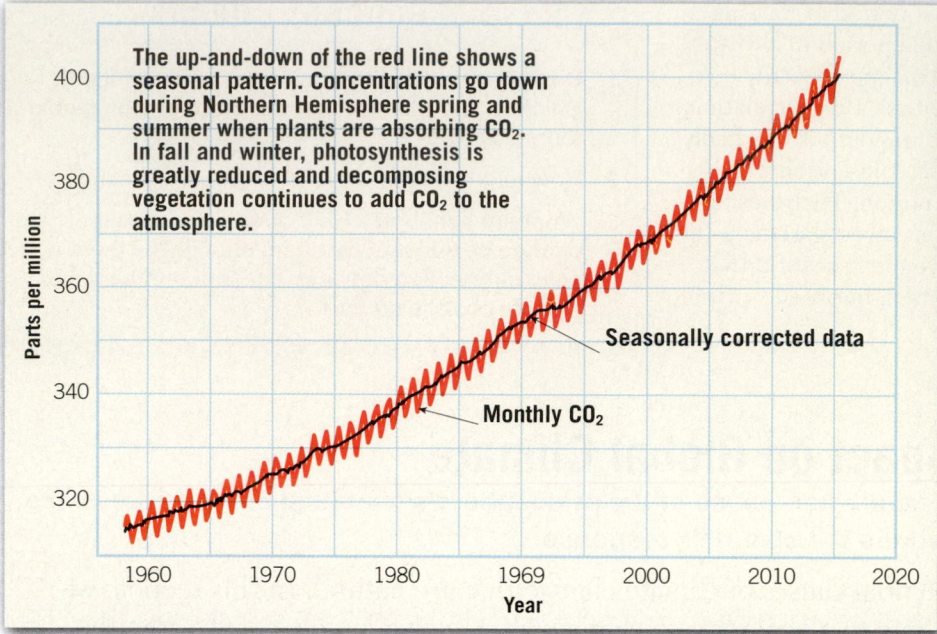

The up-and-down of the red line shows a seasonal pattern. Concentrations go down during Northern Hemisphere spring and summer when plants are absorbing CO_2. In fall and winter, photosynthesis is greatly reduced and decomposing vegetation continues to add CO_2 to the atmosphere.

Seasonally corrected data

Monthly CO₂

have varied from about 180 to 300 ppm. As a result of human activities, the present CO_2 level is about 30 percent higher than its highest level over at least the past 600,000 years. The rapid increase in CO_2 concentrations since the onset of industrialization is obvious. The annual rate at which atmospheric CO_2 concentration is growing has been increasing over the past several decades.

The Atmosphere's Response

Given the increase in the atmosphere's CO_2 content, have global temperatures actually increased? The answer is yes. According to a 2013 report by the Intergovernmental Panel on Climate Change (IPCC), "Warming of the climate system is unequivocal, as is now evident from observations of increases in global average air and ocean temperatures, widespread melting of snow and ice, and rising global sea level."* Most of the observed increase in global average temperatures since the mid-twentieth century is

are cleared for ranching and agriculture or subjected to inefficient commercial logging operations. All major tropical forests—including those in South America, Africa, Southeast Asia, and Indonesia—are disappearing. According to United Nations estimates, more than 10 million hectares (25 million acres) of tropical forest were permanently destroyed *each year* during the decades of the 1990s and 2000s; this rate has slowed in recent years.

Some of the excess CO_2 is taken up by plants or is dissolved in the ocean, but an estimated 45 percent remains in the atmosphere. **Figure 21.26** is a graphic record of changes in atmospheric CO_2 extending back more than 400,000 years. Over this long span, natural fluctuations

* IPCC, "Summary for Policymakers," in *Climate Change 2013: The Physical Science Basis*. The Intergovernmental Panel on Climate Change is an authoritative group of scientists that provides advice to the world community through periodic reports that assess the state of knowledge of the causes and effects of climate change.

Figure 21.25
Tropical deforestation Clearing the tropical rain forest is a serious environmental issue. In addition to causing a loss of biodiversity, deforestation is a significant source of carbon dioxide. Fires are frequently used to clear the land. This scene is in Brazil's Amazon basin. (Photo by Nigel Dickinson/Alamy)

extremely likely due to the observed increase in human-generated greenhouse gas concentrations. As used by the IPCC, *extremely likely* indicates a probability of 95–100 percent. Global warming since the mid-1970s is now about 0.6°C (1°F), and total warming in the past century is about 0.8°C (1.4°F). The upward trend in surface temperatures is shown in **Figure 21.27**. With the exception of 1998, the 10 warmest years in the 135-year record have all occurred since 2000, with 2014 ranking as the warmest year on record.

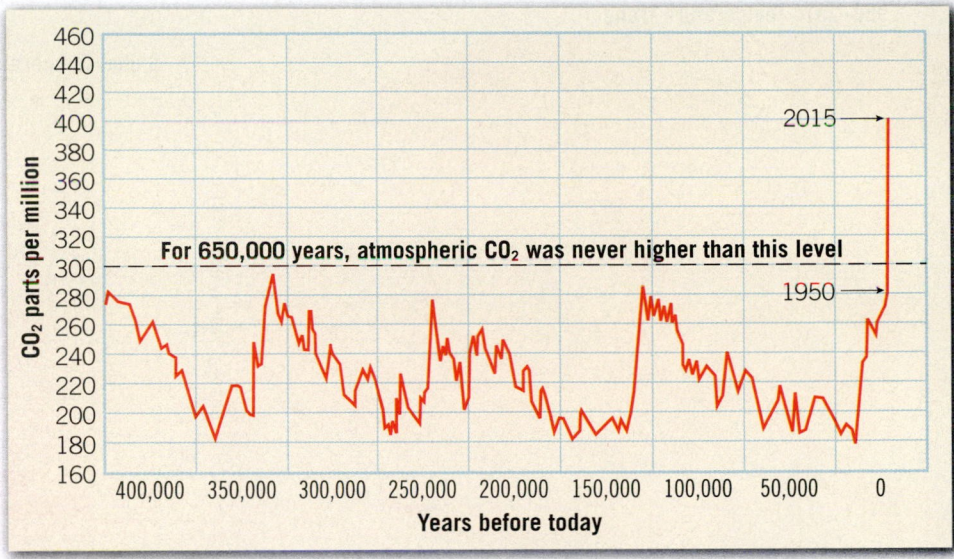

Figure 21.26
CO$_2$ concentrations over the past 400,000 years Most of these data come from the analysis of air bubbles trapped in ice cores. The record since 1958 comes from direct measurements at Mauna Loa Observatory, Hawaii. The rapid increase in CO$_2$ concentrations since the onset of the Industrial Revolution is obvious. (Based on NOAA)

Weather patterns and other natural cycles cause fluctuations in average temperatures from year to year. This is especially true on regional and local levels. For example, while the globe experienced notably warm temperatures in 2014, parts of the continental United States were cooler than normal. By contrast, 2014 was the warmest year on record for much of Europe and parts of Russia, and ocean temperatures were at a record high. Regardless of regional differences in any year, increases in greenhouse gas levels are causing a long-term rise in global temperatures. Each calendar year will not necessarily be warmer than the one before, but scientists expect each decade to be warmer than the previous one. An examination of the decade-by-decade temperature trend in **Figure 21.28** bears this out.

What about the future? Projections for the years ahead depend in part on the quantities of greenhouse gases that are emitted. **Figure 21.29** shows the best estimates of global warming for several different scenarios. The 2013 IPCC report states that if there is a doubling of the pre-industrial level of carbon dioxide (280 ppm) to 560 ppm, the *likely* temperature increase will be in the range of 2° to 4.5°C (3.5° to 8.1°F). The increase is *very unlikely* (1 to 10 percent probability) to be less than 1.5°C (2.7°F), and values higher than 4.5°C (8.1°F) are possible.

The Role of Trace Gases

Carbon dioxide is not the only gas contributing to a possible global increase in temperature. In recent years atmospheric scientists have come to realize that human industrial and agricultural activities are causing a buildup of several trace gases that also play significant roles. The substances are called **trace gases** because their concentrations are much lower than the concentration of carbon dioxide. The most important trace gases are methane

(CH$_4$), nitrous oxide (N$_2$O), and chlorofluorocarbons (CFCs). These gases absorb wavelengths of outgoing radiation from Earth that would otherwise escape into space. Although individually their impact is modest, taken together these trace gases play a significant role in warming the troposphere.

Methane　Although methane is present in much smaller amounts than CO$_2$, its significance is greater than its relatively small concentration would indicate (**Figure 21.30**). This is because methane is about 20 times more effective than CO$_2$ at absorbing infrared radiation emitted by Earth.

Methane is produced by *anaerobic* bacteria in wet places where oxygen is scarce. (*Anaerobic* means "without air," specifically oxygen.) Such places include swamps, bogs, wetlands, and the guts of termites and grazing animals such as cattle and sheep. Methane is also generated in flooded paddy fields ("artificial swamps") used for growing rice. Mining of coal and drilling for oil and natural gas are other sources, because methane is a product of their formation.

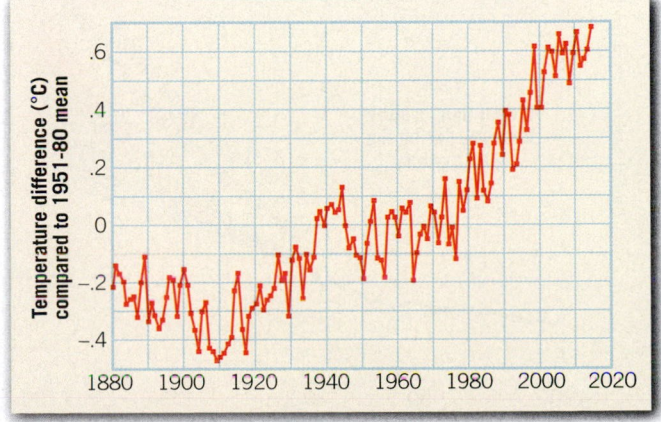

SmartFigure 21.27
Global temperatures, 1880–2014 With the exception of 1998, 10 of the warmest years in this 135-year temperature record have occurred since 2000. (https://goo.gl/V60gFp)

　**Video**

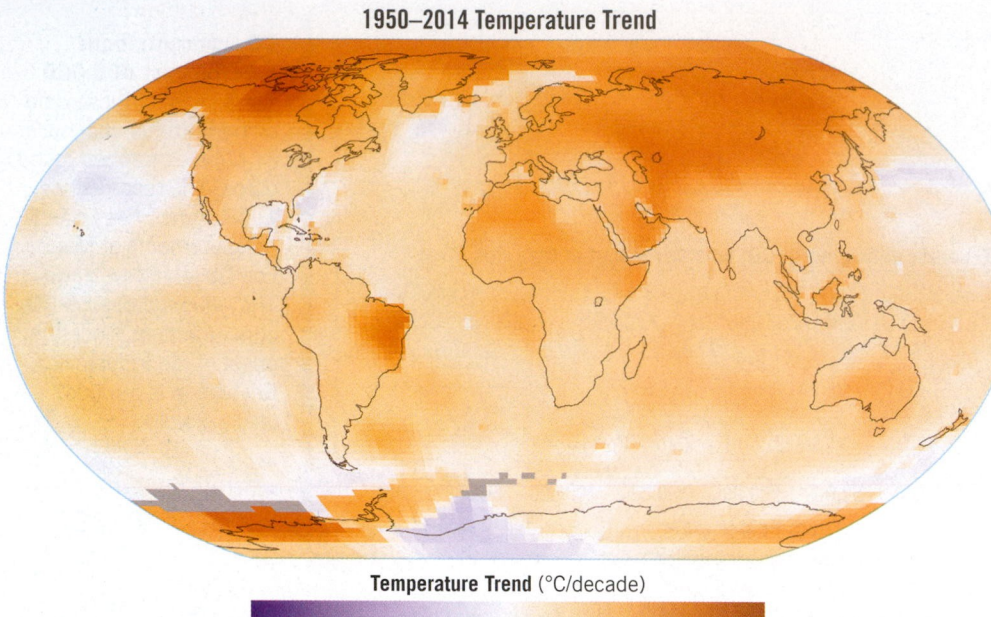

1950–2014 Temperature Trend

Temperature Trend (°C/decade)

–0.5   –0.25   0   +0.25   +0.5

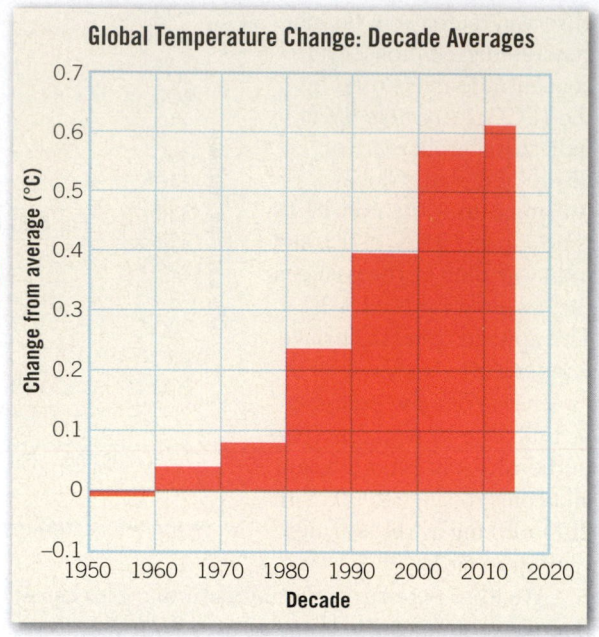

Global Temperature Change: Decade Averages

Figure 21.28

Decade-by-decade temperature trend Continued increases in the atmosphere's greenhouse gas levels are driving a long-term increase in global temperatures. Each calendar year is not necessarily warmer than the year before, but, since 1950, each decade has been warmer than the previous one. The graph clearly shows this. The world map shows the regional differences in the rate of global warming.

The increase in the concentration of methane in the atmosphere has been in step with the growth in human population. This relationship reflects the close link between methane formation and agriculture. As population increases, so do the numbers of cattle and rice paddies.

Nitrous Oxide Sometimes called "laughing gas," nitrous oxide is also building in the atmosphere, although not as rapidly as methane (see Figure 21.30). The increase results primarily from agricultural activity. When farmers use nitrogen fertilizers to boost crop yield, some of the nitrogen enters the air as nitrous oxide. This gas is also produced by high-temperature combustion of fossil fuels. Although the annual release into the atmosphere is small, the lifetime of a nitrous oxide molecule is about 150 years! If nitrogen fertilizer and fossil fuel use grow at projected rates, nitrous oxide's contribution to greenhouse warming may approach half that of methane.

CFCs Unlike methane and nitrous oxide, chlorofluorocarbons (CFCs) are not naturally present in the atmosphere. CFCs are manufactured chemicals with many uses that have gained notoriety because they are responsible for ozone depletion in the stratosphere. The role of CFCs in global warming is less well known. CFCs are very effective greenhouse gases. They were not developed until the 1920s and were not used in great quantities until the 1950s. Although corrective action has been taken, CFC levels will not drop rapidly. CFCs remain in the atmosphere for decades, so even if all CFC emissions were to stop immediately, the atmosphere would not be free of them for many years.

A Combined Effect Carbon dioxide is clearly the most important single cause of the projected global greenhouse warming. However, it is not the only contributor.

SmartFigure 21.29
Temperature projections to 2100 The right half of the graph shows projected global warming based on different emissions scenarios. The shaded zone adjacent to each colored line shows the uncertainty range for each scenario. The basis for comparison (0.0 on the vertical axis) is the global average for the period 1980 to 1999. The orange line represents the scenario in which CO_2 concentrations were held constant at values for the year 2000.
(http://goo.gl/B3UbWC)

 **Video**

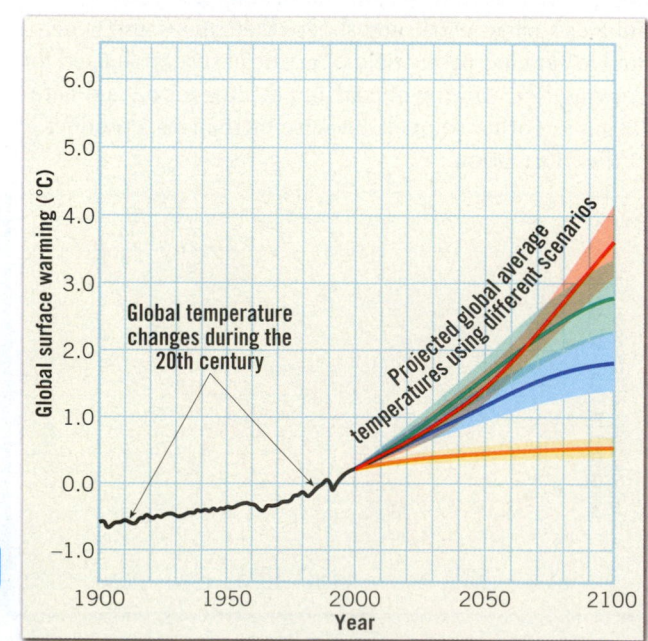

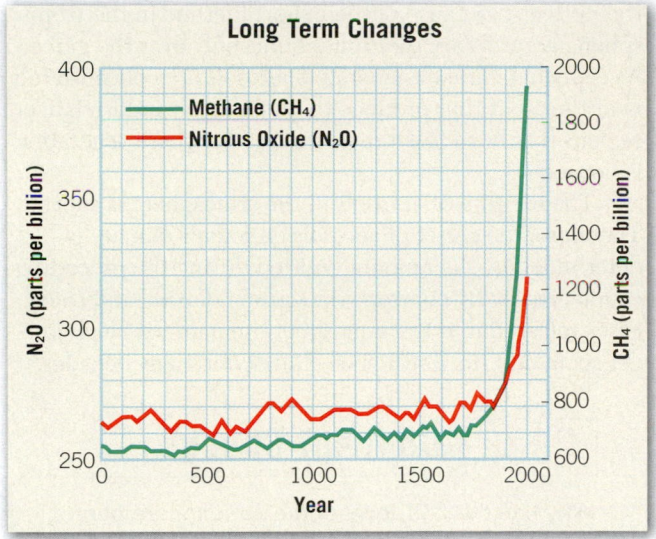

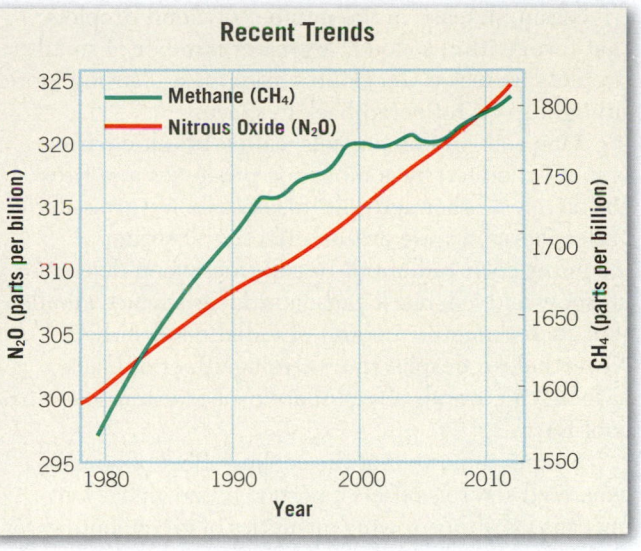

Figure 21.30
Methane and nitrous oxide Although CO_2 is most important, these trace gases also contribute to global warming. Over the 2000-year span shown here, there were relatively minor fluctuations until the industrial era. The graph on the right shows recent trends.

When the effects of all human-generated greenhouse gases other than CO_2 are added together and projected into the future, their collective impact significantly increases the impact of CO_2 alone.

Sophisticated computer models show that the warming of the lower atmosphere caused by CO_2 and trace gases will not be the same everywhere. Rather, the temperature response in polar regions could be two to three times greater than the global average. Because the polar troposphere is very stable, vertical mixing is suppressed, which limits the amount of surface heat that is transferred upward. In addition, an expected reduction in sea ice would contribute to the greater temperature increase. This topic will be explored more fully in the next section.

How Aerosols Influence Climate

Increasing the levels of carbon dioxide and other greenhouse gases in the atmosphere is the most direct human influence on global climate. But it is not the only impact. Global climate is also affected by human activities that contribute to the atmosphere's aerosol content. Recall that aerosols are the tiny, often microscopic, liquid and solid particles that are suspended in the air. Unlike cloud droplets, aerosols are present even in relatively dry air. Atmospheric aerosols are composed of many different materials, including soil, smoke, sea salt, and sulfuric acid. Natural sources are numerous and include such phenomena as dust storms and volcanoes.

Most human-generated aerosols come from the sulfur dioxide emitted during the combustion of fossil fuels and result from burning vegetation to clear agricultural land. Chemical reactions in the atmosphere convert the sulfur dioxide into sulfate aerosols, the same material that produces acid precipitation. The satellite images in **Figure 21.31** provide an example.

How do aerosols affect climate? Aerosols act directly by reflecting sunlight back to space and indirectly by making clouds "brighter" reflectors. The second effect relates to the fact that many aerosols (such as those composed of salt or sulfuric acid) attract water and thus are especially effective as cloud condensation nuclei. The large quantity of aerosols produced by human activities (especially industrial emissions)

This satellite image shows the extremely high levels of aerosols associated with this air pollution episode. At an index value of 4, aerosols are so dense that you would have difficulty seeing the midday sun.

Figure 21.31
Human-generated aerosols These satellite images show a serious air pollution episode in China on October 8, 2010. (NASA)

trigger an increase in the number of cloud droplets that form within a cloud. A greater number of small droplets increases the cloud's brightness, causing more sunlight to be reflected back to space.

One category of aerosols, called **black carbon**, is soot generated by combustion processes and fires. Unlike most other aerosols, black carbon warms the atmosphere because it is an effective absorber of incoming solar radiation. In addition, when deposited on snow and ice, black carbon reduces surface albedo, thus increasing the amount of radiation absorbed. Nevertheless, despite the warming effect of black carbon, the overall effect of atmospheric aerosols is to cool Earth.

Studies indicate that the cooling effect of human-generated aerosols offsets a portion of the global warming caused by the growing quantities of greenhouse gases in the atmosphere. The magnitude and extent of the cooling effect of aerosols is uncertain. This uncertainty is a significant hurdle in advancing our understanding of how humans alter Earth's climate.

It is important to point out some significant differences between global warming by greenhouse gases and aerosol cooling. After being emitted, carbon dioxide and trace gases remain in the atmosphere for many decades. By contrast, aerosols released into the troposphere remain there for only a few days or, at most, a few weeks before they are "washed out" by precipitation, limiting their effects. Because of their short lifetime in the troposphere, aerosols are distributed unevenly over the globe. As expected, human-generated aerosols are concentrated near the areas that produce them—namely industrialized regions that burn fossil fuels and places where vegetation is burned.

The lifetime of aerosols in the atmosphere is short. Therefore, the effect of aerosols on today's climate is determined by the amount emitted during the preceding couple weeks. By contrast, the carbon dioxide and trace gases released into the atmosphere remain for much longer spans and thus influence climate for many decades.

21.6 Concept Checks

1. Why has the CO_2 level of the atmosphere been increasing over the past 200 years?

2. How has the atmosphere responded to the growing CO_2 levels? How are temperatures in the lower atmosphere likely to change as CO_2 levels continue to increase?

3. Aside from CO_2, what trace gases are contributing to global temperature change?

4. List the main sources of human-generated aerosols and describe their net effect on atmospheric temperatures.

EYE ON EARTH 21.3

The Amundsen–Scott South Pole Station is a U.S. research facility. Among the phenomena monitored here are variations in atmospheric composition. The accompanying graph shows changes in the air's CO_2 content at South Pole Station (90 degrees south latitude) and at a similar facility at Barrow, Alaska (71 degrees north latitude). (Photo by Scot Jackson, National Science Foundation/ZUMA/Newscom)

QUESTION 1 Describe how the two lines on the graph differ.

QUESTION 2 Which line on the graph represents the South Pole, and which represents Barrow, Alaska?

QUESTION 3 Explain how you were able to determine which line is which.

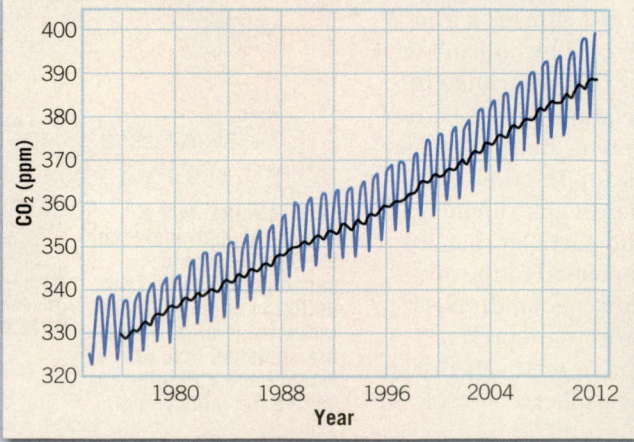

21.7 | Climate-Feedback Mechanisms

Contrast positive- and negative-feedback mechanisms and provide examples of each.

Climate is a very complex interactive physical system. Thus, when any component of the climate system is altered, scientists must consider many possible outcomes. These possible outcomes are called **climate-feedback mechanisms**. They complicate climate-modeling efforts and add greater uncertainty to climate predictions.

Types of Feedback Mechanisms

What climate-feedback mechanisms are related to carbon dioxide and other greenhouse gases? One important mechanism is that warmer surface temperatures increase evaporation rates. This, in turn, increases the water vapor in the atmosphere. Remember that water vapor is an even more powerful absorber of radiation emitted by Earth than is carbon dioxide. Therefore, with more water vapor in the air, the temperature increase caused by carbon dioxide and trace gases is reinforced.

Scientists who model global climate change indicate that the temperature increase at high latitudes may be two to three times greater than the global average. This assumption is based in part on the likelihood that the area covered by sea ice will decrease as surface temperatures rise. Because ice reflects a much larger percentage of incoming solar radiation than does open water, the melting of sea ice replaces a highly reflective surface with a relatively dark surface (**Figure 21.32**). The result is a substantial increase in the solar energy absorbed at the surface. This in turn feeds back to the atmosphere and magnifies the initial temperature increase created by higher levels of greenhouse gases.

The climate-feedback mechanisms discussed thus far magnify the temperature rise caused by the buildup of greenhouse gases. Because these effects reinforce the initial change, they are called **positive-feedback mechanisms**. On the other hand, **negative-feedback mechanisms** are those that produce results that are just the opposite of the initial change and tend to offset it.

One probable result of a global temperature rise would be an accompanying increase in cloud cover due to the higher moisture content of the atmosphere. Most clouds are good reflectors of solar radiation. At the same time, however, they are also good absorbers and emitters of radiation emitted by Earth. Consequently, clouds produce two opposite effects. They are a negative-feedback mechanism because they increase Earth's albedo and thus reflect some of the solar energy available to heat the atmosphere. On the other hand, clouds act as a positive-feedback mechanism by absorbing and emitting radiation that would otherwise be lost from the troposphere. Which effect, if either, is stronger? Scientists still are not sure whether clouds will produce a net positive or negative feedback. Although recent studies have not settled the question, they seem to lean

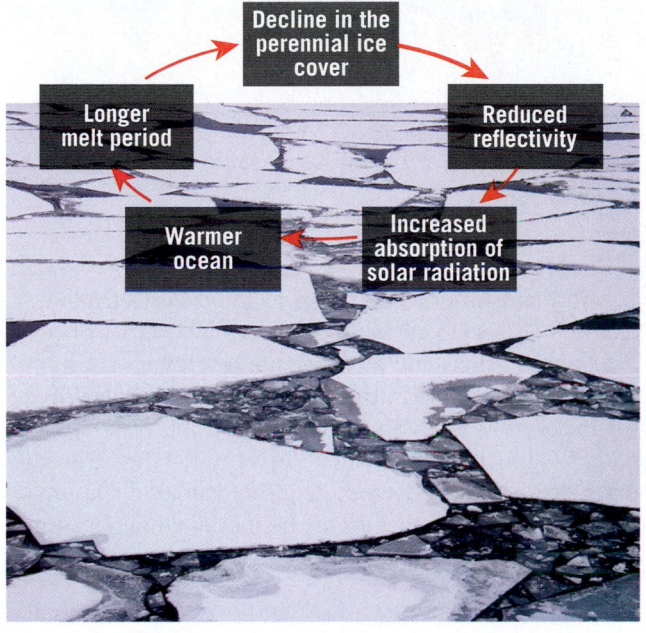

SmartFigure 21.32
Sea ice as a feedback mechanism The image shows the springtime breakup of sea ice near Antarctica. The diagram shows a likely feedback loop. A reduction in sea ice acts as a positive-feedback mechanism because surface albedo decreases, and the amount of energy absorbed at the surface increases. (Radius Images/ Alamy) (https://goo.gl/FrSgZK)

Video

toward the idea that clouds do not dampen global warming but rather produce a small positive feedback overall.[*]

The problem of global warming caused by human-induced changes in atmospheric composition continues to be one of the most studied aspects of climate change. Although no models yet incorporate the full range of potential factors and feedbacks, the scientific consensus is that the increasing levels of atmospheric carbon dioxide and trace gases will lead to a warmer planet with a different distribution of climate regimes.

Computer Models of Climate: Important yet Imperfect Tools

Earth's climate system is amazingly complex. Comprehensive state-of-the-science climate simulation models are among the basic tools used to develop possible climate-change scenarios. Called *general circulation models* (*GCMs*), they are based on fundamental laws of physics and chemistry and incorporate human and biological interactions. GCMs are used to simulate many variables, including temperature, rainfall, snow cover, soil moisture, winds, clouds, sea ice, and ocean circulation

*A. E. Dessler, "A Determination of the Cloud Feedback from Climate Variations over the Past Decade," *Science* 330 (December 10, 2010), 1523–1526.

Figure 21.33
Separating human and natural influences on climate The blue band shows how global average temperatures would have changed due to natural forces only, as simulated by climate models. The pink band shows model projections of the effects of human and natural forces combined. The black line shows actual observed global average temperatures. As the blue band indicates, without human influences, temperatures over the past century would actually have first warmed and then cooled slightly over recent decades. Bands of color are used to express the range of uncertainty.

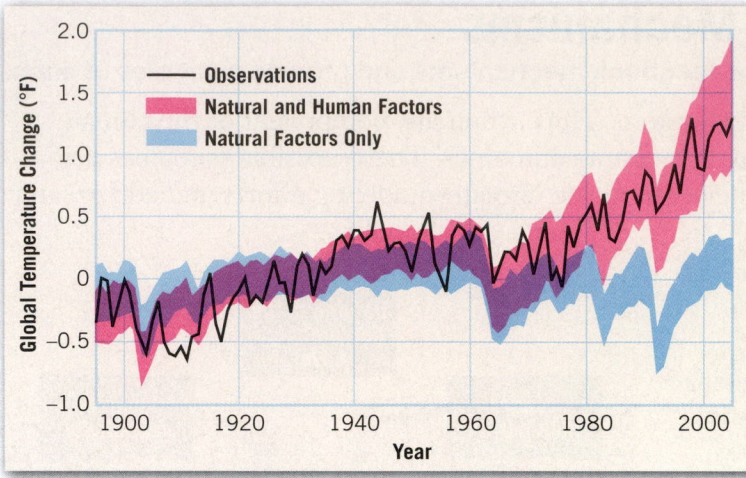

over the entire globe through the seasons and over spans of decades.

In many other fields of study, hypotheses can be tested by direct experimentation in a laboratory or by field observations and measurements. However, this is often not possible in the study of climate. Rather, scientists must construct computer models of how our planet's climate system works. If we understand the climate system correctly and construct the model appropriately, then the behavior of the model climate system should mimic the behavior of Earth's climate system (**Figure 21.33**).

What factors influence the accuracy of climate models? Clearly, mathematical models are *simplified* versions of the real Earth and cannot capture its full complexity, especially at smaller geographic scales. Moreover, computer models used to simulate future climate change must make many assumptions that significantly influence predictions. They must consider a wide range of possible changes in population, economic growth, fossil fuel consumption, technological development, improvements in energy efficiency, and more.

Despite many obstacles, our ability to use supercomputers to simulate climate is very good and continues to improve. Although today's models are far from infallible, they are powerful tools for understanding what Earth's future climate might be like.

21.7   Concept Checks

1. Distinguish between positive- and negative-feedback mechanisms.

2. Provide at least one example of each type of feedback mechanism.

3. What factors influence the accuracy of computer models of climate?

EYE ON EARTH 21.4

This satellite image from August 2007 shows the effects of tropical deforestation in a portion of the Amazon basin in western Brazil. Intact forest is dark green, whereas cleared areas are tan (bare ground) or light green (crops and pasture). Notice the relatively dense smoke in the left center of the image. (NASA)

QUESTION 1 How does the destruction of tropical forests change the composition of the atmosphere?

QUESTION 2 Describe the effect that tropical deforestation has on global warming.

21.8 | Some Consequences of Global Warming

Discuss several likely consequences of global warming.

What consequences can be expected as the carbon dioxide content of the atmosphere reaches a level that is twice what it was early in the twentieth century? Because the climate system is complex, predicting the occurrence of specific effects in particular places is speculative. It is not yet possible to pinpoint such changes. Nevertheless, there are plausible scenarios for larger scales of space and time.

As noted, the magnitude of the temperature increase will not be the same everywhere. The temperature rise will probably be smallest in the tropics and increase toward the poles. As for precipitation, the models indicate that some regions will experience significantly more precipitation and runoff. However, others will experience a decrease in runoff due to reduced precipitation or greater evaporation caused by higher temperatures.

Table 21.1 lists possible effects of global warming based on the IPCC's projections for the late twenty-first century, ranked in decreasing order of certainty. Probabilities are based on the quality, volume, and consistency of the evidence and the extent of agreement among scientists. The risk associated with any of the projections in Table 21.1 is a combination of the probability of occurrence and the severity of the damage if it were to occur. This means that even projections labeled "unlikely" should not be ignored, because if any of those events were to happen, the consequences would be extremely serious.

Sea-Level Rise

A significant impact of human-induced global warming is a rise in sea level. As this occurs, coastal cities, wetlands, and low-lying islands could be threatened with more frequent flooding, increased shoreline erosion, and saltwater encroachment into coastal rivers and aquifers.

How is a warmer atmosphere related to a rise in sea level? One significant factor is thermal expansion. Higher air temperatures warm the adjacent upper layers of the ocean, which in turn causes the water to expand and sea level to rise.

A second factor contributing to global sea-level rise is melting glaciers. With few exceptions, glaciers around

TABLE 21.1 IPCC Projections for the Late 21st Century	
• Cold days and nights will be warmer and less frequent over most land areas • Hot days and nights will be warmer and more frequent over most land areas • The extent of permafrost will decline • Ocean acidification will increase as the atmosphere accumulates CO_2 • Northern Hemisphere glaciation will not initiate before the year 3000 • Global mean sea level will rise and continue to do so for many centuries	**Virtually certain (99–100%)**
• Arctic sea ice cover will continue to shrink and thin, and Northern Hemisphere spring snow cover will decrease • The dissolved oxygen content of the ocean will decrease by a few percent • The rate of increase in atmospheric CO_2, methane, and nitrous oxide will reach levels unprecedented in the last 10,000 years • The frequency of warm spells and heat waves will increase • The frequency of heavy precipitation events will increase • Precipitation amounts will increase in high latitudes • The ocean's conveyor-belt circulation will weaken • The rate of sea-level rise will exceed that of the late 20th century • Extreme high sea-level events will increase, as will ocean wave heights of midlatitude storms	**Very likely (90–100%)**
• If the atmospheric CO_2 level stabilizes at double the present level, global temperatures will rise by between 1.5°C (2.7°F) and 4.5°C (8.1°F) • Areas affected by drought will increase • Precipitation amounts will decline in the subtropics • The loss of glaciers will accelerate in the next few decades	**Likely (66–100%)**
• Intense tropical cyclone activity will increase • The West Antarctic ice sheet will pass the melting point if global warming exceeds 5°C (9°F)—this is relative not absolute	**About as likely as not (33–66%)**
• Antarctic and Greenland ice sheets will collapse due to surface warming	← **Not likely (0–33%)**

(Michael E. Mann and Lee R. Kump, Dire Predictions: Understanding Climate Change, 2nd edition, © 2015, p. 71.)

0 10 20 30 40 50 60 70 80 90
Probability (%)

SmartFigure 21.34
Changing sea level This
graph shows changes in
sea level between 1900
and 2012 and projections
to 2100 using four different scenarios. Currently
the highest and lowest
projections are considered
to be extremely unlikely.
The greatest uncertainty
surrounding estimates is
the rate and magnitude
of ice sheet loss from
Greenland and Antarctica.
Zero on the graph repre-
sents mean sea level in
1992. (https://goo.gl/IgyxOb)

Video

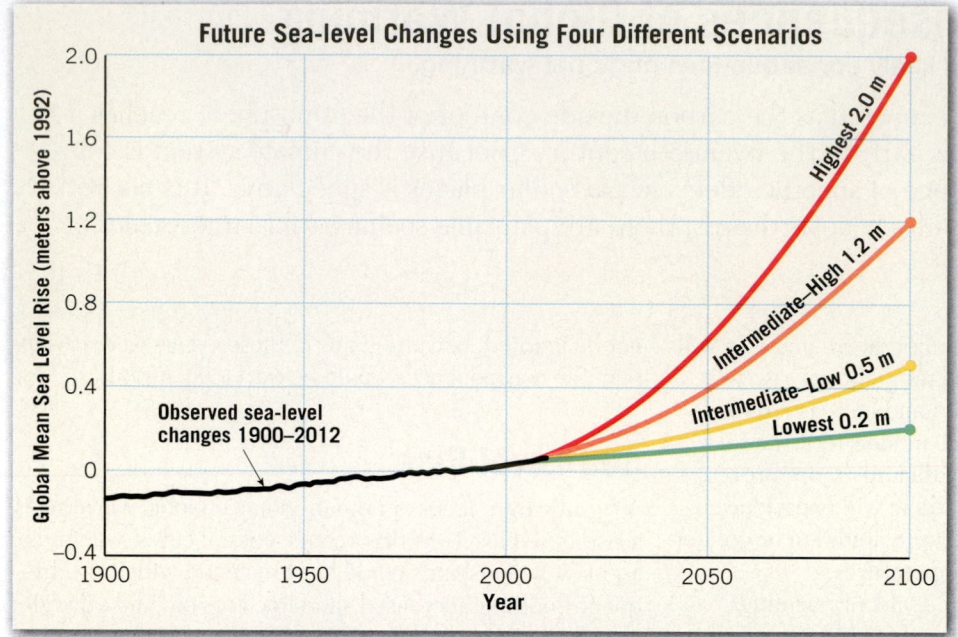

estimates of future sea-level rise are uncertain. The four scenarios depicted on the graph represent estimates based on different degrees of ocean warming and ice sheet loss and range from 0.2 meter (8 inches) to 2 meters (6.6 feet). The lowest sce- nario is an extrapolation of the annual rate of sea-level rise that occurred between 1870 and 2000 (1.7 milli- meters per year). However, when the rate of sea-level rise for the period 1993 to 2012 is examined, the annual change is 3.17 millimeters per year. Such data show that there is a reasonable chance that sea level will rise considerably more than the lowest scenario indicates.

the world have been retreating at unprecedented rates over the past century. Some mountain glaciers have disap- peared altogether (see Figure 18.14, page 545). A satel- lite study spanning 20 years showed that the mass of the Greenland and Antarctic ice sheets dropped an average of 475 gigatons per year. (A gigaton is 1 billion metric tons.) That is enough water to raise sea level 1.5 millimeters (0.05 inch) per year. The loss of ice was not steady but was occurring at an accelerating rate during the study period. During the same span, mountain glaciers and ice caps lost an average of slightly more than 400 gigatons per year.

Research indicates that sea level has risen about 25 centimeters (9.75 inches) since 1870, with the rate of sea-level rise accelerating in recent years. What about future changes in sea level? As **Figure 21.34** indicates, the

Scientists realize that even modest rises in sea level along a *gently* sloping shoreline, such as the Atlantic and Gulf coasts of the United States, will lead to significant erosion and severe permanent inland flooding (**Figure 21.35**). If this happens, many beaches and wetlands will disappear, and coastal civilization will be severely dis- rupted. Low-lying and densely populated places such as Bangladesh and the small island nation of the Maldives are especially vulnerable. The average elevation in the Maldives is 1.5 meters (less than 5 feet), and its highest point is just 2.4 meters (less than 8 feet) above sea level.

Because rising sea level is a gradual phenomenon, coastal residents may overlook it as an important contrib- utor to shoreline flooding and erosion problems. Rather, the blame may be assigned to other forces, especially storm activity. Although a given storm may be the immediate cause, the mag- nitude of its destruction may result from the relatively small sea-level rise that allowed the storm's power to cross a much greater land area.

The Changing Arctic

The effects of global warming are most pronounced in the high latitudes of the North- ern Hemisphere. For more than 30 years, the extent and thickness of sea ice have been rapidly declining. In addi- tion, permafrost temperatures have been rapidly rising, and the area affected by permafrost has been decreasing. Meanwhile,

SmartFigure 21.35
Slope of the shoreline The
slope of the shoreline is
critical to determining the
degree to which sea-level
changes will affect it. As
sea level gradually rises,
the shoreline retreats, and
structures that were once
thought to be safe from wave
attack become vulnerable.
(https://goo.gl/ww4WSi)

Tutorial

alpine glaciers and the Greenland ice sheet have been shrinking. Another sign that the Arctic is rapidly warming is related to plant growth (**Figure 21.36**). A 2013 study showed that vegetation growth at northern latitudes now resembles that which characterized areas 4 to 6 degrees of latitude farther south as recently as 1982. That is a distance of 400 to 700 kilometers (250 to 430 miles). One researcher characterized the finding this way: "It's like Winnipeg, Manitoba, moving to Minneapolis-St. Paul in only 30 years."

Arctic Sea Ice Climate models are in general agreement that one of the strongest signals of global warming should be a loss of sea ice in the Arctic. This is indeed occurring. The map in **Figure 21.37A** compares the average sea ice extent for September 2014 to the long-term average for the period 1981–2010. On that date the extent was about 5 million square kilometers (1.94 square miles)—the sixth lowest extent of the satellite era which began in 1979. (September represents the end of the melt period, when the area covered by sea ice is at a minimum.) **Figure 21.37B**, which shows year-to-year changes, clearly depicts the trend. Not only is the area covered by sea ice declining, but the remaining sea ice has become thinner, making it more vulnerable to further melting.

Models that best match historical trends project that Arctic waters may be virtually ice-free in the late summer by the 2030s. As was noted earlier in this chapter, a reduction in sea ice is a positive-feedback mechanism that reinforces global warming.

Permafrost Chapter 15 included a brief discussion of permafrost landscapes. The map in Figure 15.23 (page 461) shows that permafrost occurs in large portions of the high latitudes of the Northern Hemisphere. Mounting evidence indicates that the extent of permafrost in the Northern Hemisphere has decreased over the past

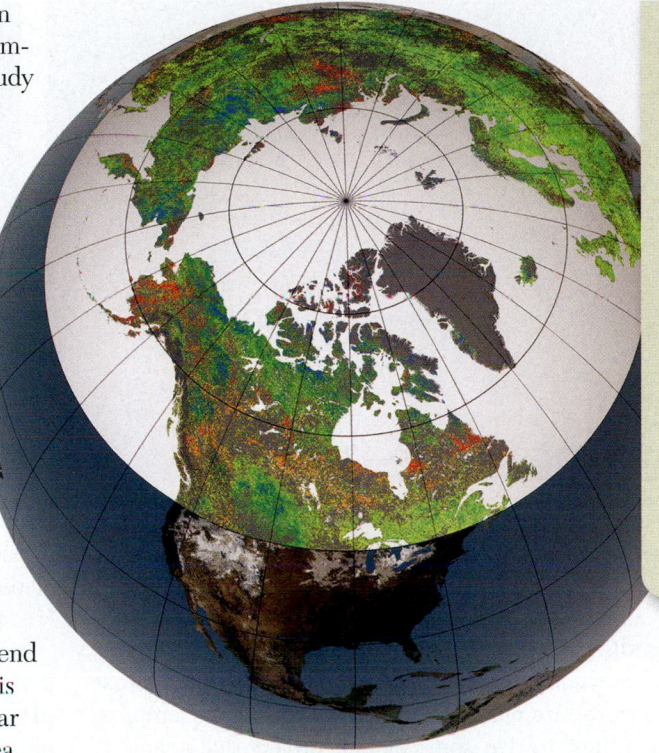

Plant Growth Change

−5   0   5   10

Percent per decade

SmartFigure 21.36
Climate change spurs plant growth beyond 45 degrees north latitude Of the 26 million square kilometers (10 million square miles) of northern vegetated lands, about 40 percent showed increases in plant growth during the 30-year period ending in 2012 (green and blue on the satellite image). The Mobile Field Trip explores several aspects of climate change in the Arctic. (Data from National Snow and Ice Data Center) (https://goo.gl/wL2oAg)

decade, as would be expected under long-term warming conditions. **Figure 21.38** presents an example of this decline. In the Arctic, short summers thaw only the top layer of frozen ground. The permafrost beneath this *active layer* is like the cement bottom of a swimming pool. In summer, water cannot percolate downward, so it saturates the soil above the permafrost and collects on the surface in thousands of lakes. However, as Arctic temperatures climb, the bottom of the "pool" seems to be "cracking." Satellite imagery shows that a significant

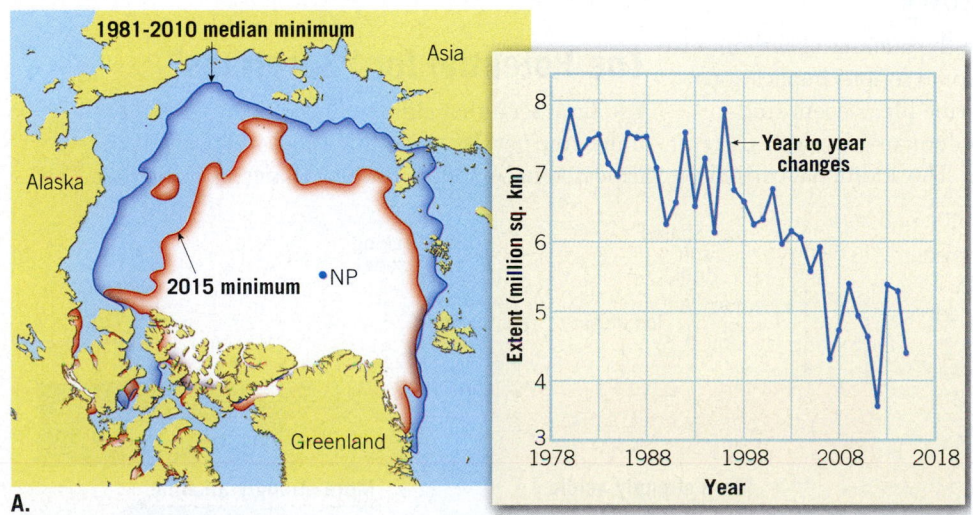

A.

B.

SmartFigure 21.37
Tracking sea ice changes Sea ice is frozen seawater. In winter the Arctic Ocean is completely ice covered. In summer, a portion of the ice melts. **A.** This map shows the extent of sea ice in early September 2015 compared to the average extent for the period 1981 to 2010. The sea ice that does not melt in summer is getting thinner. **B.** The graph clearly depicts the trend in the area covered by sea ice at the end of the summer melt period. (Data from National Snow and Ice Data Center) (https://goo.gl/Cxmt5k)

Figure 21.38
Siberian lakes This false-color image pair shows lakes dotting the tundra in 1973 and 2002. The tundra vegetation is colored a faded red, whereas lakes appear blue or blue-green. Many lakes disappeared or shrunk considerably between 1973 and 2002. After studying satellite imagery of about 10,000 large lakes in a 500,000-square-kilometer (195,000- square-mile) area in northern Siberia, scientists documented an 11 percent decline in the number of lakes. (Based on NASA)

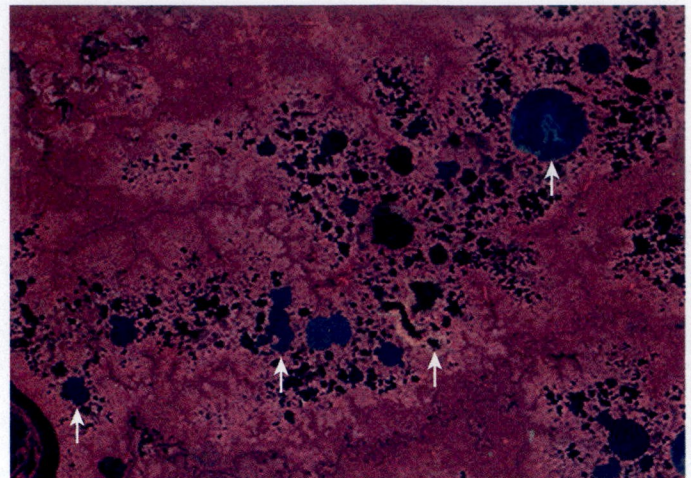

A. June 27, 1973

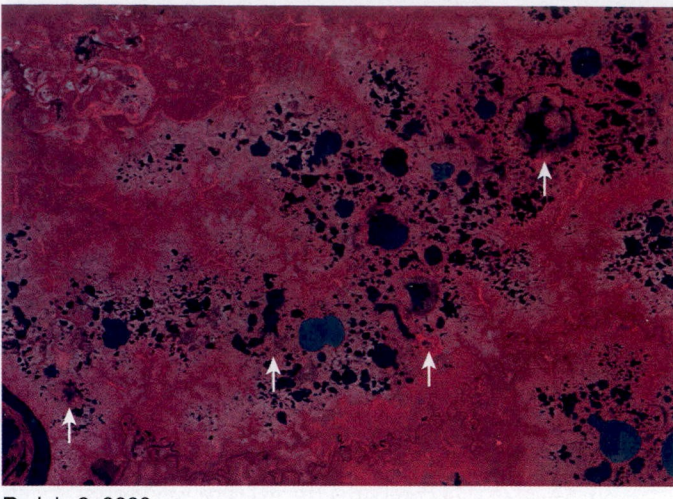

B. July 2, 2002

number of lakes have shrunk or disappeared altogether. As the permafrost thaws, lake water drains deeper into the ground.

Studies in Alaska show that thawing is occurring in interior and southern parts of the state where permafrost temperatures are near the thaw point. As Arctic temperatures continue to rise, some models project that near-surface permafrost may be lost entirely from large parts of Alaska by the end of the century.

Thawing permafrost represents a potentially significant positive-feedback mechanism that may reinforce global warming. When vegetation dies in the Arctic, cold temperatures inhibit its decomposition. As a consequence, over thousands of years, a great deal of organic matter has become stored in the permafrost. When the permafrost thaws, organic matter that may have been frozen for millennia comes out of "cold storage" and decomposes. The result is the release of carbon dioxide and methane—greenhouse gases that contribute to global warming. Thus, like decreasing sea ice, thawing permafrost is a positive-feedback mechanism.

Increasing Ocean Acidity

The human-induced increase in the amount of carbon dioxide in the atmosphere has some serious implications for ocean chemistry and for marine life. Recent studies show that about one-third of human-generated CO_2 currently ends up in the oceans. The additional carbon

dioxide lowers the ocean's pH, making seawater more acidic. The pH scale is shown and briefly described in **Figure 21.39**.

When atmospheric CO_2 dissolves in seawater (H_2O), it forms carbonic acid (H_2CO_3). This lowers the ocean's pH and changes the balance of certain chemicals found naturally in seawater. In fact, the oceans have already absorbed enough carbon dioxide for surface waters to have experienced a decrease of 0.1 pH units since preindustrial times, with an additional pH decrease likely in the future (**Figure 21.40**). Moreover, if the current trend in carbon dioxide emissions continues, the ocean will experience a pH decrease of at least 0.2 pH units by 2100, which represents a change in ocean chemistry that has not occurred for millions of years. This shift toward acidity and the resulting changes in ocean chemistry make it more difficult for certain marine creatures to build hard parts out of calcium carbonate. The decline in pH thus threatens a variety of calcite-secreting organisms as diverse as microbes and corals, which concerns marine scientists because of the potential consequences for other sea life that depend on the health and availability of these organisms.

The Potential for "Surprises"

You have seen that climate in the twenty-first century, unlike during the preceding thousand years, is not expected to be stable. Rather, change is occurring. The

Figure 21.39
The pH scale This is the common measure of the degree of acidity or alkalinity of a solution. The scale ranges from 0 to 14, with a value of 7 indicating a solution that is neutral. Values below 7 indicate greater acidity, whereas numbers above 7 indicate greater alkalinity. It is important to note that the pH scale is logarithmic; that is, each whole number increment indicates a tenfold difference. Thus, pH 4 is 10 times more acidic than pH 5 and 100 times (10 × 10) more acidic than pH 6.

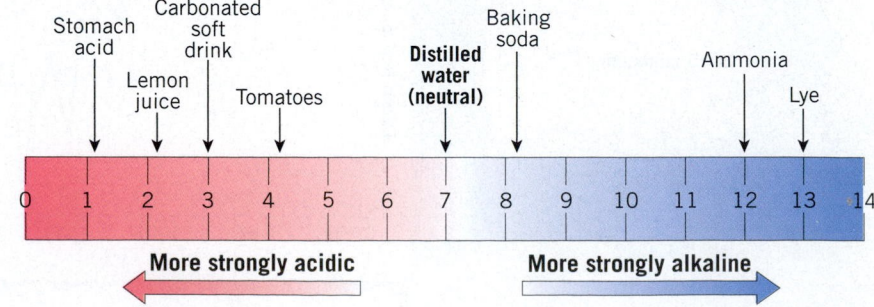

amount and rate of future climate shifts depends primarily on current and future human-caused emissions of heat-trapping gases and airborne particles. Many of the changes will probably be gradual, nearly imperceptible from year to year. Nevertheless the effects, accumulated over decades, will have powerful economic, social, and political consequences.

Despite our best efforts to understand future climate shifts, there is also the potential for "surprises." This simply means that due to the complexity of Earth's climate system, we might experience relatively sudden, unexpected changes or see some aspects of climate shift in unanticipated ways. Many future climate scenarios predict steadily changing conditions, giving the impression that humanity will have time to adapt. However, the scientific community has indicated that at least some changes will be abrupt, perhaps crossing a threshold, or "tipping point," so quickly that there will be little time to react. This is a reasonable concern because abrupt changes occurring over periods as short as decades or even years have been a natural part of the climate system throughout Earth history. The paleoclimate record described earlier in the chapter contains ample evidence of such abrupt changes. One such abrupt change occurred at the end of a time span known as the *Younger Dryas*, a time of abnormal cold and drought in the Northern Hemisphere that occurred about 12,000 years ago. Following this 1000-year-long cold period, the Younger Dryas abruptly ended in a few decades or less and is associated with the extinction of more than 70 percent of large-bodied mammals in North America.

There are many examples of potential surprises, each of which would have large consequences. We simply do not know how far the climate system or other systems it affects can be pushed before they respond in unexpected ways. Even if the chance of any particular surprise happening is small, the chance that at least one such surprise will occur is much greater. In other words, although we may not know which of these events will occur, it is likely that one or more will eventually occur.

The impact on climate of an increase in atmospheric CO_2 and trace gases is obscured by some uncertainties. Yet climate scientists continue to improve our understanding of the climate system and the potential impacts and effects of global climate change. Policymakers are confronted with responding to the risks posed by greenhouse gas emissions, knowing that our understanding

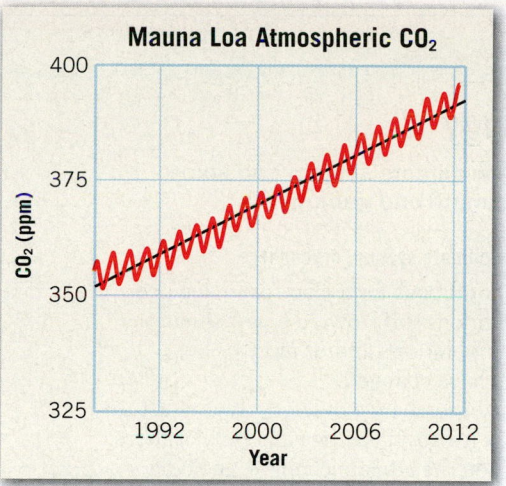

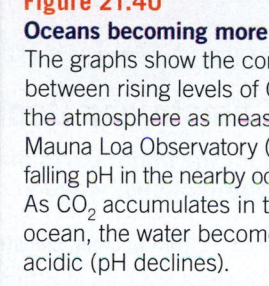

Figure 21.40
Oceans becoming more acidic The graphs show the correlation between rising levels of CO_2 in the atmosphere as measured at Mauna Loa Observatory (**A**) and falling pH in the nearby ocean (**B**). As CO_2 accumulates in the ocean, the water becomes more acidic (pH declines).

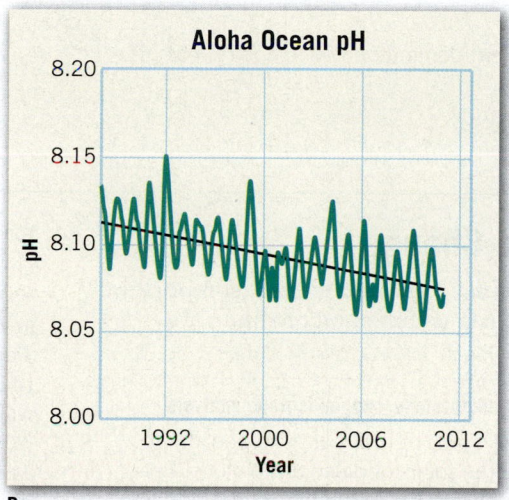

is imperfect. They must also face the fact that climate-induced environmental changes cannot be reversed quickly, if at all, due to the lengthy time scales associated with the climate system.

21.8 | Concept Checks

1. Describe the factors that are causing sea level to rise.

2. Is global warming greater near the equator or near the poles? Explain.

3. Based on Table 21.1, what projected changes relate to something other than temperature?

21.1 Climate and Geology

List the major parts of the climate system and some connections between climate and geology.

KEY TERMS weather, climate, climate system, cryosphere

- Climate is the aggregate weather conditions for a place or region over a long period of time. If those conditions shift toward a new situation over time, with hotter or cooler temperatures and/or more or less precipitation, the climate is said to have changed.
- Earth's climate system is a complex interchange of energy and moisture that occurs among the atmosphere, hydrosphere, geosphere, biosphere, and cryosphere (ice and snow). When the climate changes, geologic processes such as weathering, mass wasting, and erosion may change as well.

Q Which sphere of the climate system dominates this image? What other sphere or spheres are present?

James Balog/Getty Images

21.2 Detecting Climate Change

Explain why unraveling past climate changes is important and discuss several ways in which such changes are detected.

KEY TERMS proxy data, paleoclimatology, oxygen isotope analysis

- The geologic record yields multiple kinds of indirect evidence about past climate. These proxy data are the focus of paleoclimatology. Proxy data can come from seafloor sediment, oxygen isotopes, cores of glacial ice, tree rings, coral-growth bands, fossil pollen, and even historical documents.
- Trees grow thicker rings in warmer, wetter years and thinner rings in colder, drier years. The pattern of ring thickness can be matched up between trees of overlapping ages for a long-term record of a region's climate.
- Oxygen isotope analysis is based on the difference between heavier ^{18}O and lighter ^{16}O and their relative amounts in water molecules (H_2O). The $^{18}O/^{16}O$ ratio of ocean water rises during cold times because water that contains lightweight oxygen evaporates more readily. During warmer times, more energy is available to evaporate water containing ^{18}O, plus glacial ice melts and sends some of its ^{16}O back to the sea. This makes the marine $^{18}O/^{16}O$ ratio drop. Oxygen isotopes can also be measured in the shells of fossil marine organisms or the water molecules that make up glacial ice. Glacial ice also traps small samples of the atmosphere in air bubbles.

21.3 Some Atmospheric Basics

Describe the composition of the atmosphere and the atmosphere's vertical changes in pressure and temperature.

KEY TERMS aerosol, troposphere, radiosonde, stratosphere, mesosphere, thermosphere

- Air is a mixture of many discrete gases, and its composition varies from time to time and place to place. Two gases, nitrogen and oxygen, make up 99 percent of the volume of the remaining clean, dry air. Carbon dioxide, although present in only minute amounts (0.0397 percent, or 397 parts per million), is an efficient absorber of energy emitted by Earth and thus influences the heating of the atmosphere.
- Two important variable components of air are water vapor and aerosols. Like carbon dioxide, water vapor can absorb heat given off by Earth. Aerosols are important because these often invisible particles act as surfaces on which water vapor can condense and are also good absorbers and reflectors (depending on the particles) of incoming solar radiation.
- The atmosphere is densest closest to the surface of Earth. It thins rapidly with increasing altitude and gradually fades off into space. Temperature varies through a vertical section of the atmosphere. Generally, temperatures drop in the troposphere, warm in the stratosphere, cool in the mesosphere, and increase in the thermosphere.

Q This graph shows changes in an atmospheric element from Earth's surface to a height of about 140 kilometers (90 miles). Which element is being depicted: air pressure, humidity, or temperature? Show how this graph is used to divide the atmosphere into layers.

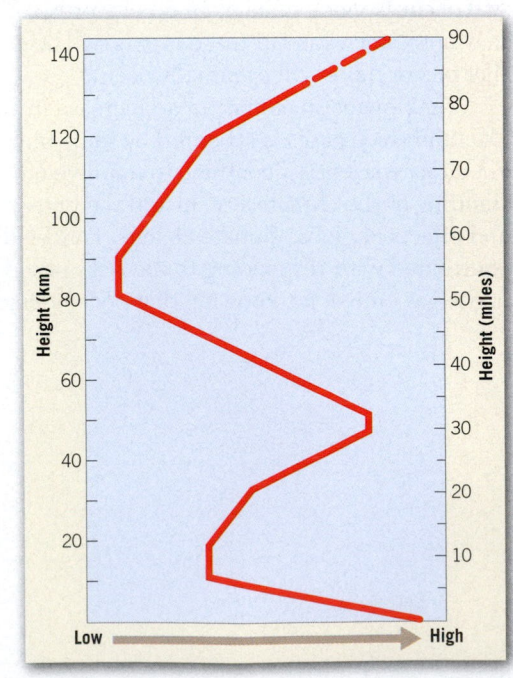

21.4 Heating the Atmosphere

Outline the basic processes involved in heating the atmosphere.

KEY TERMS albedo, greenhouse effect

- Electromagnetic radiation is energy emitted in the form of rays, or waves, called electromagnetic waves. All radiation can transmit energy through the vacuum of space. Electromagnetic waves have different wavelengths. Visible light is the only portion of the electromagnetic spectrum we can see. Some basic laws that govern radiation as it heats the atmosphere are (1) all objects emit radiant energy, (2) hotter objects radiate more total energy than do colder objects, (3) the hotter the radiating body, the shorter the wavelengths of maximum radiation, and (4) objects that are good absorbers of radiation are good emitters as well. Gases are selective absorbers, meaning that they absorb and emit certain wavelengths but not others.

- Approximately 50 percent of the solar energy striking the top of the atmosphere reaches Earth's surface. About 30 percent is reflected back to space. The remaining 20 percent is absorbed by clouds and the atmosphere's gases. The wavelengths of the energy being transmitted, as well as the size and nature of the absorbing or reflecting substance, determine whether radiation will be scattered and reflected back to space or absorbed.

- Radiant energy that is absorbed heats Earth and eventually is reradiated skyward. Because Earth has a much lower surface temperature than the Sun, its radiation is in the form of long-wave infrared radiation. Because atmospheric gases, primarily water vapor and carbon dioxide, are more efficient absorbers of long-wave radiation, the atmosphere is heated from the ground up. The transmission of short-wave solar radiation by the atmosphere, coupled with the selective absorption of Earth radiation by atmospheric gases, results in the warming of the atmosphere and is called the greenhouse effect.

21.5 Natural Causes of Climate Change

Discuss hypotheses that relate to natural causes of climate change.

KEY TERM sunspot

- The natural functions of the Earth system produce climate change. The position of lithospheric plates can influence the climate of the continents as well as oceanic circulation. Variations in the shape of Earth's orbit, angle of axial tilt, and orientation of the axis all cause changes in the distribution of solar energy.

- Volcanic aerosols act like a Sun shade, screening out a portion of incoming solar radiation. Volcanic sulfur dioxide emissions that reach the stratosphere are particularly important. Combined with water to form tiny droplets of sulfuric acid, these aerosols can remain aloft for several years.

- Volcanoes emit carbon dioxide. During times of especially large eruptions, such as those that produced oceanic lava plateaus during the Cretaceous period, volcanic CO_2 emissions may contribute to the greenhouse effect sufficiently to cause global warming.

- Since Earth's climate is fueled by solar energy, variations in the Sun's energy output affect Earth temperatures. Sunspots are dark features on the surface of the Sun associated with periods of increased solar energy output. The number of sunspots rises and drops throughout an 11-year cycle. During the peak of the cycle, the Sun puts out about 0.1 percent more energy than during the lowest part of the cycle, but this small cyclical effect is not correlated with the current episode of global warming.

21.6 Human Impact on Global Climate

Summarize the nature and cause of the atmosphere's changing composition since about 1750. Describe the climate's response.

KEY TERMS trace gases, black carbon

- Humans have been modifying the environment for thousands of years. Clearing or burning ground cover and overgrazing land have changed important climatic factors as surface albedo, evaporation rates, and surface winds.

- Human activities produce climate change by releasing carbon dioxide (CO_2) and trace gases. Humans release CO_2 when they cut down forests or burn fossil fuels such as coal, oil, and natural gas. A steady rise in atmospheric CO_2 levels has been documented at Mauna Loa, Hawaii, and other locations around the world.

- More than half of the carbon released by humans is absorbed by new plant matter or dissolved in the oceans. About 45 percent remains in the atmosphere, where it can influence climate for decades. Air bubbles trapped in glacial ice reveal that there is currently about 30 percent more CO_2 than the atmosphere has contained in the past 650,000 years.

- As a result of extra heat retained by added CO_2, Earth's atmosphere has warmed by about 0.8°C (1.4°F) in the past 100 years, most of it since the 1970s. Temperatures are projected to increase by another 2° to 4.5°C (3.5° to 8.1°F) in the future.

- Trace gases such as methane, nitrous oxide, and CFCs also play significant roles in increasing global temperature.

- When emitted due to human activities, aerosols can reflect a portion of incoming solar radiation back to space and therefore have a cooling effect. Yet some aerosols, called black carbon, absorb incoming solar radiation and act to warm the atmosphere. When black carbon is deposited on snow and ice, it reduces surface albedo and increases the amount of light absorbed at the surface.

Q Do aerosols spend more or less time in the atmosphere than greenhouse gases such as carbon dioxide? What is the significance of this difference in residence time? Explain.

21.7 Climate-Feedback Mechanisms

Contrast positive- and negative-feedback mechanisms and provide examples of each.

KEY TERMS climate-feedback mechanism, positive-feedback mechanism, negative-feedback mechanism

- A change in one part of the climate system may trigger changes in other parts of the climate system that amplify or diminish the initial effect. These climate-feedback mechanisms are called positive-feedback mechanisms if they reinforce the initial change and negative-feedback mechanisms if they counteract the initial effect.
- The melting of sea ice due to global warming (decreasing albedo and increasing the initial effect of warming) is one example of a positive-feedback mechanism. The production of more clouds (blotting out incoming solar radiation, leading to cooling) is an example of a negative-feedback mechanism.
- Computer models of climate give scientists a tool for testing hypotheses about climate change. Although these models are simpler than the real climate system, they are useful tools for predicting the future climate.

Q Changes in precipitation and temperature due to climate change can increase the risk of forest fires. Describe two ways that the event shown in this photo could contribute to global warming.

Michael Collier

21.8 Some Consequences of Global Warming

Discuss several likely consequences of global warming.

- In the future, Earth's surface temperature is likely to continue to rise. The temperature increase will likely be greatest in the polar regions and least in the tropics. Some areas will get drier, and other areas will get wetter.
- Sea level is predicted to rise for several reasons, including the melting of glacial ice and thermal expansion. Low-lying, gently sloped, highly populated coastal areas are most at risk.
- Sea ice cover and thickness in the Arctic have been declining since satellite observations began in 1979.
- Because of the warming of the Arctic, permafrost is melting, releasing CO_2 and methane to the atmosphere in a positive-feedback mechanism.
- Because the climate system is complicated, dynamic, and imperfectly understood, it could produce sudden, unexpected changes with little warning.

Q This ice breaker is plowing through sea ice in the Arctic Ocean. What spheres of the climate system are represented in this photo? How has the area covered by summer sea ice been changing since satellite monitoring began in 1979? How does this change influence temperatures in the Arctic?

Wolfgang Bechtold/imagebroker/Alamy

Give It Some *Thought*

1. Refer to Figure 21.1, which illustrates various components of Earth's climate system. Boxes represent interactions or changes that occur in the climate system. Select three boxes and provide an example of an interaction or change associated with each. Explain how these interactions may influence temperature.

2. Describe one way in which changes in the biosphere can cause changes in the climate system. Next, suggest one way in which the biosphere is affected by changes in some other part of the climate system. Finally, indicate one way in which the biosphere records changes in the climate system.

3. Figure 21.16 shows that about 30 percent of the Sun's energy intercepted by Earth is reflected or scattered back to space. If Earth's albedo were to increase to 50 percent, how would you expect Earth's average surface temperature to change? Explain.

4. Volcanic events, such as the eruptions of El Chichón and Mount Pinatubo, have been associated with drops in global temperatures. During the Cretaceous period, volcanic activity was associated with global warming. Explain the apparent paradox.

5. The accompanying photo is a 2005 view of Athabasca Glacier in the Canadian Rockies. A line of boulders in the foreground marks the outer limit of the glacier in 1992. Is the behavior of Athabasca Glacier shown in this image typical of other glaciers around the world? Describe a significant impact of such behavior.

Hughrocks

6. Motor vehicles are a significant source of CO_2. Using electric cars, such as the one pictured here, is one way to reduce emissions from this source. Even though these vehicles emit little or no CO_2 or other pollutants directly into the air, can they still be connected to such emissions? Explain.

David Pearson/Alamy

7. During a conversation, an acquaintance indicates that he is skeptical about global warming. When you ask him why, he says, "The past couple of years in this area have been among the coolest I can remember." While you assure this person that it is useful to question scientific findings, you suggest to him that his reasoning in this case may be flawed. Use your understanding of the definition of *climate* along with one or more graphs in the chapter to persuade this person to reevaluate his reasoning.

8. A 2015 report by the National Research Council on climate intervention recommends conducting research on strategies that would help offset global warming. One strategy would involve injecting aerosols into the stratosphere.

a. Explain how such an aerosol strategy might influence global warming.

b. What *natural* cause of climate change operates on the same principle as the one you describe in your answer to part a?

9. This large cattle feedlot is in the Texas Panhandle. How might consuming less beef influence global climate change?

Glow Images

MasteringGeology™

www.masteringgeology.com

Looking for additional review and test prep materials? With individualized coaching on the toughest topics of the course, MasteringGeology offers a wide variety of ways for you to move beyond memorization to begin thinking like a geologist. Visit the Study Area in www.masteringgeology. com to find practice quizzes, study tools, and multimedia that will improve your understanding of this chapter's content. Sign in today to enjoy the following features: **Self Study Quizzes, SmartFigure: Tutorials/ Animations/Condor Videos/Mobile Field Trips, Geoscience Animation Library, GEODe, RSS Feeds, Digital Study Modules,** and an optional **Pearson eText.**

22

Earth's Evolution Through Geologic Time

Grand Prismatic Pool in Yellowstone National Park. This hot-water pool gets its blue color from several species of heat-tolerant cyanobacteria. Microscopic fossils of organisms similar to modern cyanobacteria are among Earth's oldest fossils. (Photo by Don Johnston/ Glow Images)

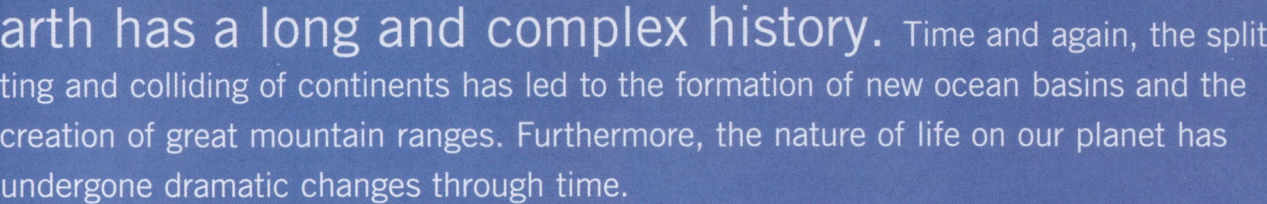

E arth has a long and complex history. Time and again, the splitting and colliding of continents has led to the formation of new ocean basins and the creation of great mountain ranges. Furthermore, the nature of life on our planet has undergone dramatic changes through time.

22.1 | Is Earth Unique?

List the principle characteristics that make Earth unique among the planets.

There is only one place in the universe, as far as we know, that can support life—a modest-sized planet called Earth that orbits an average-sized star, the Sun. Life on Earth is ubiquitous; it is found in boiling mudpots and hot springs, in the deep abyss of the ocean, and even under the Antarctic ice sheet. Living space on our planet, however, is significantly limited when we consider the needs of individual organisms, particularly humans. The global ocean covers 71 percent of Earth's surface, but only a few hundred meters below the water's surface, pressures are so intense that humans cannot survive without an atmospheric diving suit. In addition, many continental areas are too steep, too high, or too cold for us to inhabit (Figure 22.1). Nevertheless, based on what we know about other bodies in the solar system—and the hundreds of planets recently discovered orbiting around other stars—Earth is still by far the most accommodating.

Figure 22.1
Climbers near the top of Mount Everest Much of Earth's surface is uninhabitable by humans. At this altitude, the level of oxygen is only one-third the amount available at sea level. (Photo by STR/AFP/Getty Images)

What fortuitous events produced a planet so hospitable to life? Earth was not always as we find it today. During its formative years, our planet became hot enough to support a magma ocean. It also survived a several-hundred-million-year period of extreme bombardment by asteroids, to which the heavily cratered surfaces of Mars and the Moon testify. The oxygen-rich atmosphere that makes higher life-forms possible developed relatively recently. Serendipitously, Earth seems to be the right planet, in the right location, at the right time.

The Right Planet

What are some of the characteristics that make Earth unique among the planets? Consider the following:

- If Earth were considerably larger (more massive), its force of gravity would be proportionately greater. Like the giant planets, Earth would have retained a thick, hostile atmosphere consisting of ammonia and methane, and possibly hydrogen and helium.
- If Earth were much smaller, oxygen, water vapor, and other volatiles would escape into space and be lost forever. Thus, like the Moon and Mercury, both of which lack atmospheres, Earth would be devoid of life.
- If Earth did not have a rigid lithosphere overlaying a weak asthenosphere, plate tectonics would not operate. The continental crust (Earth's "highlands") would not have formed without the recycling of

plates. Consequently, the entire planet would likely be covered by an ocean a few kilometers deep. As author Bill Bryson so aptly stated, "There might be life in that lonesome ocean, but there certainly wouldn't be baseball."*

- Most surprisingly, perhaps, is the fact that if our planet did not have a molten metallic outer core, most of the life-forms on Earth would not exist. Fundamentally, without the flow of iron in the core, Earth could not support a magnetic field. It is the magnetic field that prevents lethal cosmic rays from showering Earth's surface and from stripping away our atmosphere.

The Right Location

One of the primary factors that determine whether a planet is suitable for higher life-forms is its location in the solar system. The following scenarios substantiate Earth's advantageous position:

- If Earth were about 10 percent closer to the Sun, our atmosphere would be more like that of Venus and consist mainly of the greenhouse gas carbon dioxide. Earth's surface temperature would then be too hot to support higher life-forms.
- If Earth were about 10 percent farther from the Sun, the problem would be reversed—it would be too cold. The oceans would freeze over, and Earth's active water cycle would not exist. Without liquid water, all life would perish.
- Earth is near a star of modest size. Stars like the Sun have a life span of roughly 10 billion years and emit radiant energy at a fairly constant level during most of this time. Giant stars, on the other hand, consume their nuclear fuel at very high rates and "burn out" in a few hundred million years. Therefore, Earth's proximity to a modest-sized star allowed enough time for the evolution of humans, who first appeared on this planet only a few million years ago.

The Right Time

The last, but certainly not the least, fortuitous factor for Earth is timing. The first organisms to inhabit Earth were extremely primitive and came into existence roughly 3.8 billion years ago. From

that point in Earth's history, innumerable changes occurred: Life-forms came and went, and there were many changes in the physical environment of our planet. Consider two of the many timely Earth-altering events:

- Earth's atmosphere has developed over time. Earth's primitive atmosphere is thought to have been composed mostly of methane, water vapor, ammonia, and carbon dioxide—but no free oxygen. Fortunately, microorganisms evolved that released oxygen into the atmosphere through the process of *photosynthesis*. About 2.5 billion years ago, an atmosphere with free oxygen came into existence. The result was the evolution of the ancestors of the vast array of organisms that occupy Earth today.
- About 66 million years ago, our planet was struck by an asteroid 10 kilometers (6 miles) in diameter. This impact likely caused a mass extinction during which nearly three-quarters of all plant and animal species were obliterated—including dinosaurs. Although this may not seem fortuitous, the extinction of dinosaurs opened new habitats for small mammals that survived the impact. These habitats, along with evolutionary forces, led to the development of many large mammals that occupy our modern world (**Figure 22.2**). Without this event, mammals may not have evolved beyond small rodent-like creatures that live in burrows.

As various observers have noted, Earth developed under "just right" conditions to support higher life-forms. Astronomers refer to this as the *Goldilocks scenario.* Like the classic "Goldilocks and the Three Bears" fable, Venus is too hot (Papa Bear's porridge), Mars is too cold (Mama Bear's porridge), but Earth is just right (Baby Bear's porridge).

Figure 22.2
Paleontologists uncover the remains of a 10-million-year-old rhinoceros at a dig site near Orchard, Nebraska. (Photo by Annie Griffiths Belt/Corbis)

*A Short History of Nearly Everything (Broadway Books, 2003).

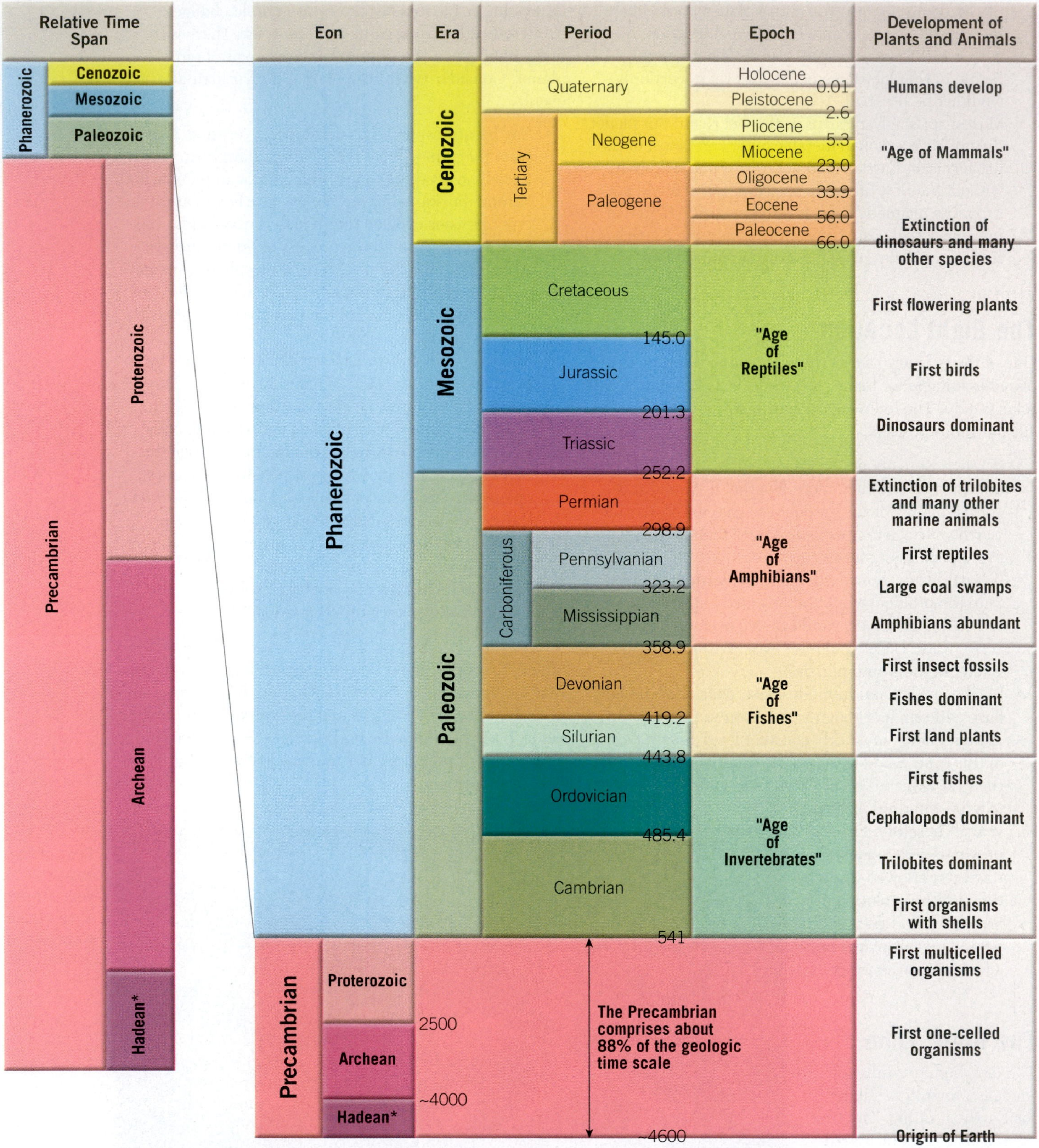

Relative Time Span	Eon	Era	Period	Epoch		Development of Plants and Animals
Phanerozoic — Cenozoic / Mesozoic / Paleozoic	Phanerozoic	Cenozoic	Quaternary	Holocene	0.01	Humans develop
				Pleistocene	2.6	
			Tertiary — Neogene	Pliocene	5.3	"Age of Mammals"
				Miocene	23.0	
			Tertiary — Paleogene	Oligocene	33.9	
				Eocene	56.0	
				Paleocene	66.0	Extinction of dinosaurs and many other species
Precambrian — Proterozoic / Archean / Hadean*		Mesozoic	Cretaceous	"Age of Reptiles"		First flowering plants
			145.0			First birds
			Jurassic			
			201.3			Dinosaurs dominant
			Triassic			
			252.2			
		Paleozoic	Permian	"Age of Amphibians"		Extinction of trilobites and many other marine animals
			298.9			
			Carboniferous — Pennsylvanian			First reptiles
			323.2			Large coal swamps
			Carboniferous — Mississippian			Amphibians abundant
			358.9			
			Devonian	"Age of Fishes"		First insect fossils
						Fishes dominant
			419.2			
			Silurian			First land plants
			443.8			
			Ordovician	"Age of Invertebrates"		First fishes
						Cephalopods dominant
			485.4			
			Cambrian			Trilobites dominant
						First organisms with shells
			541			
	Precambrian	Proterozoic	The Precambrian comprises about 88% of the geologic time scale			First multicelled organisms
		2500				
		Archean				First one-celled organisms
		~4000				
		Hadean*				
		~4600				Origin of Earth

* Hadean is the informal name for the span that begins at Earth's formation and ends with Earth's earliest-known rocks.

Figure 22.3

The geologic time scale Numbers represent time in millions of years before the present. The Precambrian accounts for about 88 percent of geologic time.

Viewing Earth's History

The remainder of this chapter focuses on the origin and evolution of planet Earth—the one place in the universe we know fosters life. As you learned in Chapter 9, researchers utilize many tools to interpret clues about Earth's past. Using these tools, as well as clues contained in the rock record, scientists continue to unravel many complex events of the geologic past. The goal of this chapter is to provide a brief overview of the history of our planet and its life-forms—a journey that takes us back about 4.6 billion years, to the formation of Earth. Later, we will consider how our physical world assumed its present state and how Earth's inhabitants changed through time. We suggest that you reacquaint yourself with the *geologic time scale* presented in **Figure 22.3** and refer to it throughout the chapter.

22.1 Concept Checks

1. In what way is Earth unique among the planets of our solar system?

2. Explain why Earth is just the right size.

3. Why is Earth's molten, metallic core important to humans living today?

4. Why is Earth's location in the solar system ideal for the development of higher life-forms?

22.2 | Birth of a Planet

Outline the major stages in Earth's evolution, from the Big Bang to the formation of our planet's layered internal structure.

The universe had been evolving for several billion years before our solar system and Earth began to form. The universe began about 13.7 billion years ago with the *Big Bang*, when all matter and space came into existence. Shortly thereafter the two simplest elements, hydrogen and helium, formed. These basic elements were the ingredients for the first star systems. Several billion years later, our home galaxy, the Milky Way, came into existence. It was within a band of stars and nebular debris in an arm of this spiral galaxy that the Sun and planets took form nearly 4.6 billion years ago.

From the Big Bang to Heavy Elements

According to the Big Bang theory, the formation of our planet began about 13.7 billion years ago with a cataclysmic explosion that created all matter and space (**Figure 22.4**). Initially, subatomic particles (protons, neutrons, and electrons) formed. Later, as this debris cooled, atoms of hydrogen and helium, the two lightest elements, began to form. Within a few hundred million years, clouds of these gases condensed and coalesced into stars that compose the galactic systems we now observe.

As these gases contracted to become the first stars, heating triggered the process of *nuclear fusion*. Within the interiors of stars, hydrogen nuclei convert to helium nuclei, releasing enormous amounts of radiant energy (heat, light, cosmic rays). Astronomers have determined that in stars more massive than our Sun, other thermonuclear reactions occur that generate all the elements on the periodic table up to number 26, iron. The heaviest elements (beyond number 26) are created only at extreme temperatures during the explosive death of a star eight or more times as massive as the Sun. During these cataclysmic **supernova** events, exploding stars produce all the elements heavier than iron and spew them into interstellar space. It is from such debris that our Sun and solar system formed. Based on the Big Bang scenario, atoms in your body were produced billions of years ago, in the hot interior of now-defunct stars, and the gold in your jewelry was produced during a supernova explosion that occurred in some distant place.

From Planetesimals to Protoplanets

Recall that the solar system, including Earth, formed about 4.6 billion years ago from the **solar nebula**, a large rotating cloud of interstellar dust and gas (see Figure 22.4E). As the solar nebula contracted, most of the matter collected in the center to create the hot *protosun*. The remaining materials formed a thick, flattened, rotating disk, within which matter gradually cooled and condensed into grains and clumps of icy, rocky, and metallic material. Repeated collisions resulted in most of the material eventually collecting into asteroid-sized objects called **planetesimals**.

The composition of planetesimals was largely determined by their proximity to the protosun. As you might expect, temperatures were highest in the inner solar system and decreased toward the outer edge of the disk. Therefore, between the present orbits of Mercury and Mars, the planetesimals were composed mainly of materials with high melting temperatures—metals and rocky substances. The planetesimals that formed beyond the orbit of Mars, where temperatures are low, contained high percentages of ices—water, carbon dioxide, ammonia, and methane—as well as smaller amounts of rocky and metallic debris.

Through repeated collisions and accretion (sticking together), these planetesimals grew into eight **protoplanets** and some larger moons (see Figure 22.4G). During this process, the same amount of matter was concentrated into fewer and fewer bodies, each having greater and greater masses.

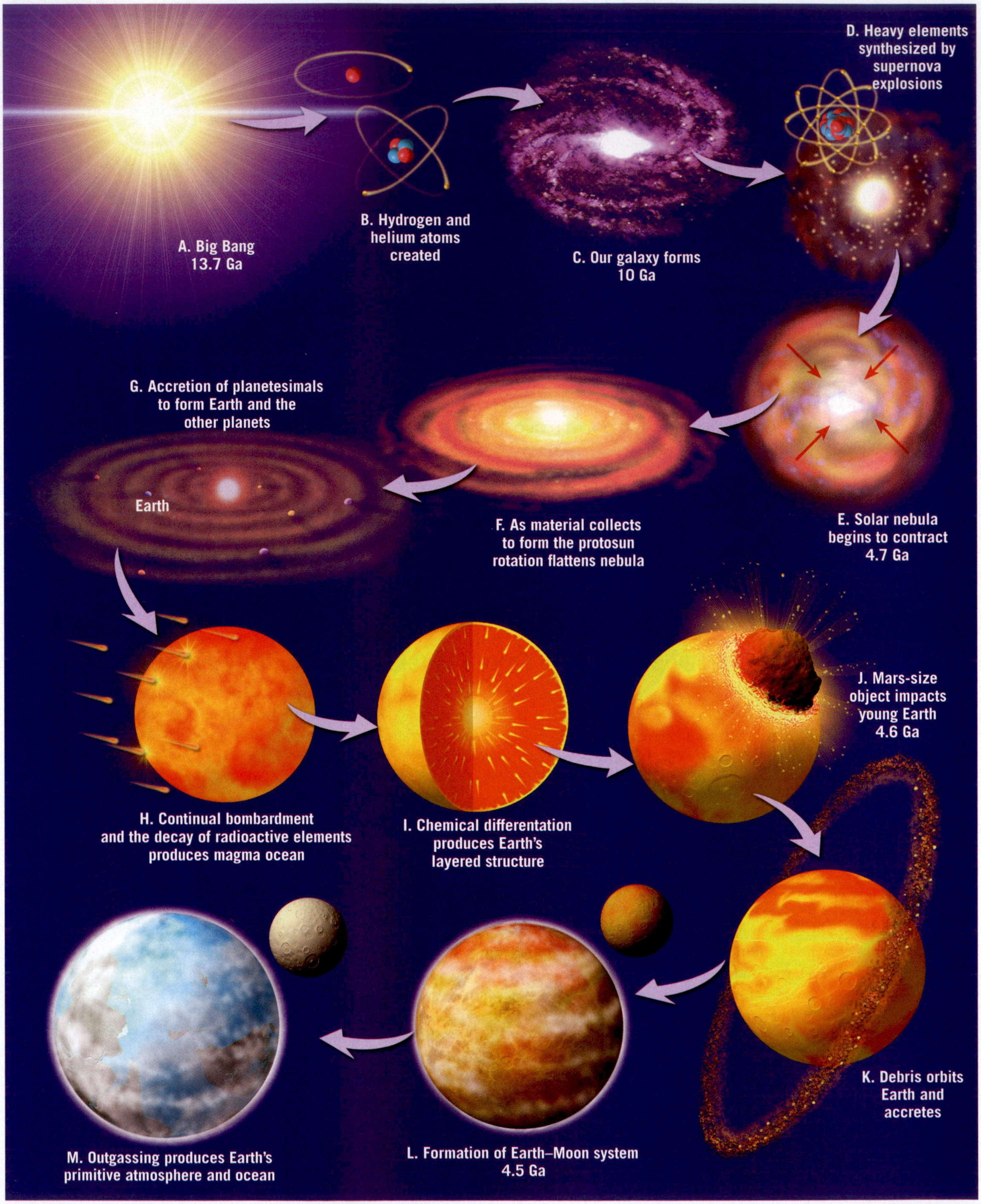

D. Heavy elements synthesized by supernova explosions

A. Big Bang 13.7 Ga

B. Hydrogen and helium atoms created

C. Our galaxy forms 10 Ga

G. Accretion of planetesimals to form Earth and the other planets

Earth

F. As material collects to form the protosun rotation flattens nebula

E. Solar nebula begins to contract 4.7 Ga

H. Continual bombardment and the decay of radioactive elements produces magma ocean

I. Chemical differentation produces Earth's layered structure

J. Mars-size object impacts young Earth 4.6 Ga

K. Debris orbits Earth and accretes

M. Outgassing produces Earth's primitive atmosphere and ocean

L. Formation of Earth–Moon system 4.5 Ga

SmartFigure 22.4
Major events that led to the formation of early Earth
(https://goo.gl/b11eLc)

 Tutorial

At some point in Earth's early evolution, a giant impact occurred between a Mars-sized object and a young, semimolten Earth. This collision ejected huge amounts of debris into space, some of which coalesced to form the Moon (see Figure 22.4J,K,L).

Earth's Early Evolution

As material continued to accumulate, the high-velocity impact of interplanetary debris (planetesimals) and the decay of radioactive elements caused the temperature of our planet to steadily increase. This early period of heating resulted in a magma ocean that was perhaps several hundred kilometers deep. Within the magma ocean, buoyant masses of molten rock rose toward the surface and eventually solidified to produce thin rafts of crustal rocks. Geologists call this early period of Earth's history the **Hadean**, and it began with Earth's formation about 4.6 billion years ago and ended roughly 4 billion years ago (**Figure 22.5**). The name *Hadean* is derived from the Greek word *Hades*, meaning "the underworld," referring to the "hellish" conditions on Earth at the time.

During this period of intense heating, Earth became hot enough that iron and nickel began to melt. Melting produced liquid blobs of heavy metal that sank under their own weight. This process occurred rapidly on the scale of geologic time and produced Earth's dense iron-rich core. As you learned in Chapter 12, the formation of a molten iron core was the first of many stages of chemical differentiation in which Earth converted from a homogeneous body, with roughly the same matter at all depths, to a layered planet with material sorted by density (see Figure 22.4I).

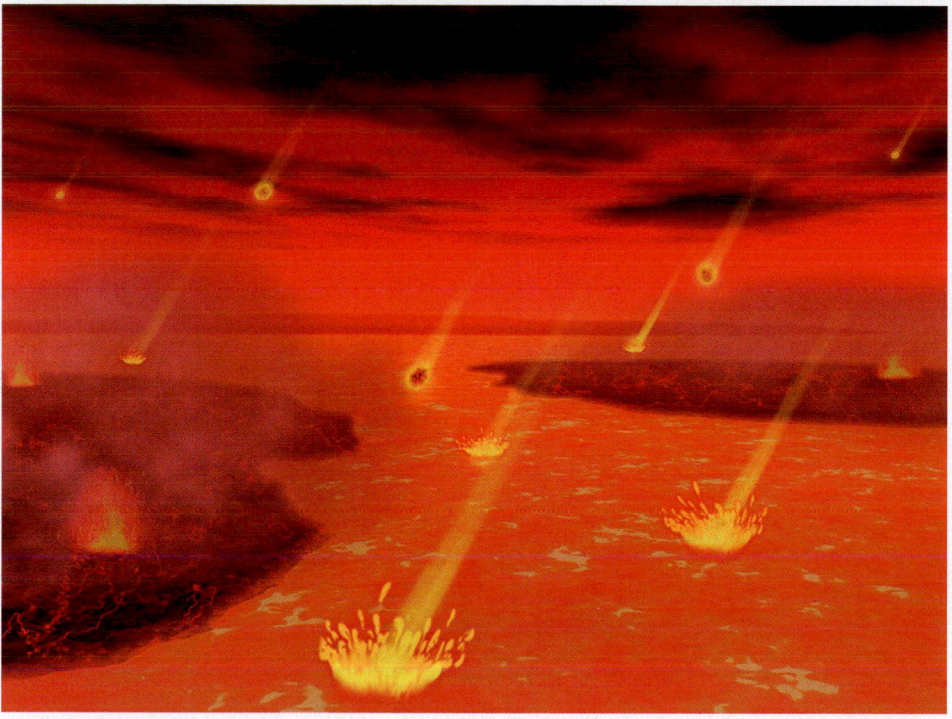

Figure 22.5
Artistic depiction of Earth during the Hadean The Hadean is an unofficial eon of geologic time that occurred before the Archean. Its name refers to the "hellish" conditions on Earth. During part of the Hadean, Earth had a magma ocean and experienced intense bombardment by nebular debris.

This period of chemical differentiation established the three major divisions of Earth's interior—the iron-rich *core*; the thin *primitive crust*; and Earth's thickest layer, the *mantle*, located between the core and the crust. In addition, the lightest materials—including water vapor, carbon dioxide, and other gases—escaped to form a primitive atmosphere and, shortly thereafter, the oceans.

22.2 | Concept Checks

1. What two elements made up most of the very early universe?

2. What is the name of the cataclysmic event in which an exploding star produces all the elements heavier than iron?

3. Briefly describe the formation of the planets from the solar nebula.

4. Describe the conditions on Earth during the Hadean.

22.3 | Origin and Evolution of the Atmosphere and Oceans
Describe how Earth's atmosphere and oceans formed and evolved through time.

We can be thankful for our atmosphere; without it, there would be no greenhouse effect, and Earth would be nearly 60°F colder. Earth's water bodies would be frozen over, making the hydrologic cycle nonexistent.

The air we breathe is a stable mixture of 78 percent nitrogen, 21 percent oxygen, about 1 percent argon (an inert gas), and small amounts of gases such as carbon dioxide and water vapor. However, our planet's original atmosphere was substantially different.

Earth's Primitive Atmosphere

Early in Earth's formation, its atmosphere likely consisted of gases most common in the early solar system: hydrogen, helium, methane, ammonia, carbon dioxide, and water

**Figure 22.6
Outgassing produced Earth's first enduring atmosphere** Outgassing continues today from hundreds of active volcanoes worldwide. (Photo by Lee Frost/Robert Harding)

submarine volcanism and associated hydrothermal vents. Iron has tremendous affinity for oxygen. When these two elements join, they become iron oxide (rust). These early iron oxide accumulations on the seafloor created alternating layers of iron-rich rocks and chert, called **banded iron formations**. Most banded iron deposits accumulated in the Precambrian, between 3.5 and 2 billion years ago, and represent the world's most important reservoir of iron ore.

As the number of oxygen-generating organisms increased, oxygen began to build in the atmosphere. Chemical analysis of rock suggests that oxygen first appeared in significant amounts in the atmosphere around 2.5 billion years ago, a phenomenon termed the **Great Oxygenation Event**. Thereafter, oxygen levels in the atmosphere gradually climbed.

For the next billion years, oxygen levels in the atmosphere probably fluctuated but remained below 10 percent of current levels. Prior to the start of the Cambrian period 541 million years ago, which coincided with the evolution of organisms with skeletal hard parts, the level of free oxygen in the atmosphere began to increase. The availability of abundant oxygen in the atmosphere contributed to the proliferation of aerobic life-forms (oxygen-consuming organisms). On the other hand, it likely wiped out huge portions of Earth's anaerobic organisms (organisms that do not require oxygen), for which oxygen is poisonous. One apparent spike in oxygen levels occurred during the Pennsylvanian period (300 million years ago), when oxygen made up about 35 percent of the atmosphere, compared to today's level of 21 percent. This abundance of oxygen has been attributed to the large size of insects and amphibians during the Pennsylvanian period.

Another positive benefit of the Great Oxygenation Event is that, for the first time, oxygen molecules (O_2) became a major component of our atmosphere. When struck by sunlight, oxygen molecules form a compound called *ozone* (O_3), a type of oxygen molecule composed of three oxygen atoms. Ozone, which absorbs much of the harmful ultraviolet radiation before it reaches Earth's surface, is concentrated between 10 and 50 kilometers (6 to 30 miles) above Earth's surface, in a layer called the *stratosphere*. Thus, as a result of the Great Oxygenation Event, Earth's landmasses were protected from ultraviolet radiation, which is particularly harmful to DNA—the

vapor. The lightest of these—hydrogen and helium—most likely escaped into space because Earth's gravity was too weak to hold them. The remaining gases—methane, ammonia, carbon dioxide, and water vapor—contain the basic ingredients of life: carbon, hydrogen, oxygen, and nitrogen. This early atmosphere was enhanced by a process called **outgassing**, through which gases trapped in the planet's interior are released. Outgassing from hundreds of active volcanoes still remains an important planetary function worldwide (**Figure 22.6**). However, early in Earth's history, when massive heating and fluidlike motion occurred in the mantle, the gas output would likely have been immense. These early eruptions probably released mainly water vapor, carbon dioxide, and sulfur dioxide, with minor amounts of other gases. Most importantly, free oxygen was not present in Earth's primitive atmosphere.

Oxygen in the Atmosphere

As Earth cooled, water vapor condensed to form clouds, and torrential rains began to fill low-lying areas, eventually becoming the oceans. In those oceans, nearly 3.5 billion years ago, photosynthesizing bacteria began to release oxygen into the water. During *photosynthesis*, organisms use the Sun's energy to produce organic material (energetic molecules of sugar containing hydrogen and carbon) from carbon dioxide (CO_2) and water (H_2O). The first bacteria probably used hydrogen sulfide (H_2S) rather than water as a hydrogen source. One of the earliest known bacteria, *cyanobacteria* (once called blue-green algae), began to produce oxygen as a by-product of photosynthesis.

Initially, the newly released free oxygen was readily captured by chemical reactions with organic matter and dissolved iron in the ocean. It seems that large quantities of iron were released into the early ocean through

These prominent chalk cliffs are composed largely of tiny shells of marine organisms, such as foraminifera.

Figure 22.7
White chalk cliffs, England
Similar deposits are also found in northern France.
(Photo by Imagesources/Glow Images)

genetic blueprints for living organisms. Marine organisms had always been shielded from harmful ultraviolet radiation by seawater, but the development of the atmosphere's protective ozone layer allowed the continents to become more hospitable to plants and animals as well.

Evolution of the Oceans

When Earth cooled sufficiently to allow water vapor to condense, rainwater fell and collected in low-lying areas. By 4 billion years ago, scientists estimate that as much as 90 percent of the current volume of seawater was contained in the developing ocean basins. Because volcanic eruptions released into the atmosphere large quantities of sulfur dioxide, which readily combines with water to form sulfuric acid, the earliest rainwater was highly acidic. The level of acidity was even greater than in the acid rain that damaged lakes and streams in eastern North America during the latter part of the twentieth century. Consequently, Earth's rocky surface weathered at an accelerated rate. The products released by chemical weathering included atoms and molecules of various substances—including sodium, calcium, potassium, and silica—that were carried into the newly formed oceans. Some of these dissolved substances precipitated to become chemical sediment that mantled the ocean floor. Other substances formed soluble salts, which increased the salinity of seawater. Research suggests that the salinity of the oceans initially increased rapidly but has remained relatively constant over the past 2 billion years.

Earth's oceans also serve as a repository for tremendous volumes of carbon dioxide, a major constituent of the primitive atmosphere. This is significant because carbon dioxide is a greenhouse gas that strongly influences the heating of the atmosphere. Venus, once thought to be very similar to Earth, has an atmosphere composed of 97 percent carbon dioxide

that produced an extreme greenhouse effect. As a result, its surface temperature is 475°C (nearly 900°F).

Carbon dioxide is readily soluble in seawater, where it often joins other atoms or molecules to produce various chemical precipitates. The most common compound generated by precipitation is calcium carbonate ($CaCO_3$), which makes up limestone, the most abundant chemical sedimentary rock. About 542 million years ago, marine organisms began to extract calcium carbonate from seawater to make their shells and other hard parts. Included were trillions of tiny marine organisms, such as foraminifera, whose shells were deposited on the seafloor at the end of their life cycle. Some of these deposits can be observed today in the chalk beds exposed along the White Cliffs of Dover, England, shown in **Figure 22.7**. By "locking up" carbon dioxide, these limestone deposits store this greenhouse gas so it cannot easily reenter the atmosphere.

22.3 Concept Checks

1. What is meant by *outgassing*, and what modern phenomenon serves that role today?

2. Identify the most abundant gases that were added to Earth's early atmosphere through the process of outgassing.

3. Why was the evolution of photosynthesizing bacteria important to most modern organisms?

4. Why was rainwater highly acidic early in Earth's history?

5. How does the ocean remove carbon dioxide from Earth's atmosphere? What role do tiny marine organisms, such as foraminifera, have in the removal of carbon dioxide?

22.4 | Precambrian History: The Formation of Earth's Continents

Explain the formation of continental crust, how continental crust becomes assembled into continents, and the role that the supercontinent cycle has played in this process.

Earth's first 4 billion years are encompassed in the time span called the *Precambrian*. Representing nearly 90 percent of Earth's history, the Precambrian is divided into the *Archean eon* ("ancient age") and the *Proterozoic eon* ("early life age"). Our knowledge of this ancient time is limited because much of the early rock record has been obscured by the very Earth processes you have been studying, especially plate tectonics, erosion, and deposition. Most Precambrian rocks lack fossils, which hinders correlation of rock units. In addition, rocks this old are often metamorphosed and deformed, extensively eroded, and frequently concealed by younger strata. Indeed, Precambrian history is written in scattered, speculative episodes, like a long book with many missing chapters.

Earth's First Continents

Geologists have discovered tiny crystals of the mineral zircon in continental rocks that formed 4.4 billion years ago—evidence that the continents began to form early in Earth's history. By contrast, the oldest rocks found in the ocean basins are generally less than 200 million years old.

What differentiates continental crust from oceanic crust? Recall that oceanic crust is a relatively dense (3.0 g/cm^3) homogeneous layer of basaltic rocks derived from partial melting of the rocky upper mantle. In addition, oceanic crust is thin, averaging only 7 kilometers (4.3 miles) thick. Continental crust, on the other hand, is composed of a variety of rock types, has an average thickness of nearly 40 kilometers (25 miles), and contains a large percentage of low-density (2.7 g/cm^3), silica-rich rocks such as granite.

The significance of the differences between continental crust and oceanic crust cannot be overstated in a review of Earth's geologic evolution. Oceanic crust, because it is relatively thin and dense, is found several kilometers below sea level—unless of course it has been pushed onto a landmass by tectonic forces. Continental crust, because of its great thickness and lower density, may extend well above sea level. Also, recall that oceanic crust of normal thickness readily subducts, whereas thick, buoyant blocks of continental crust resist being recycled into the mantle.

Making Continental Crust   The formation of continental crust is a continuation of the gravitational segregation of Earth materials that began during the final stage of our planet's formation. Dense metallic material, mainly iron and nickel, sank to form Earth's core, leaving behind the less dense rocky material that forms the mantle. It is from Earth's rocky mantle that low-density, silica-rich minerals were gradually distilled to form continental crust. This process is analogous to making sour mash whiskeys. In the production of whiskeys, various grains such as corn are fermented to generate alcohol, with sour mash being the by-product. This mixture is then heated or distilled, which drives off the lighter material (alcohol) and leaves behind the sour mash. In a similar manner, partial melting of mantle rocks generates low-density, silica-rich materials that buoyantly rise to the surface to form Earth's crust, leaving behind the dense mantle rocks (see Chapter 4). However, little is known about the details of the mechanisms that generated these silica-rich melts during the Archean.

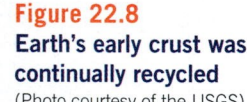

Figure 22.8
Earth's early crust was continually recycled
(Photo courtesy of the USGS)

The crust covering this lava lake is continually being replaced with fresh lava from below, much like the way Earth's crust was recycled early in its history.

Earth's first crust was probably ultramafic in composition, but because physical evidence no longer exists, we are not certain. The hot, turbulent mantle that most likely existed during the Archean eon recycled most of this material back into the mantle. In fact, it may have been continuously recycled, in much the same way that the "crust" that forms on a lava lake is repeatedly replaced with fresh lava from below (**Figure 22.8**).

The oldest preserved continental rocks occur as small, highly deformed *terranes*, which are incorporated within somewhat younger blocks of continental crust (**Figure 22.9**). One of these is a 3.8-billion-year-old terrane located near Isua, Greenland. Slightly older crustal rocks, called the Acasta Gneiss, have been discovered in Canada's Northwest Territories (see Figure 22.11).

Some geologists think that some type of plate-like motion operated early in Earth's history. In addition, hot-spot volcanism was likely active during this time. However, because the mantle was hotter in the Archean than it is today, both of these phenomena would have progressed at higher rates than their modern counterparts. Hot-spot volcanism is thought to have created immense

These rocks at Isua, Greenland, some of the world's oldest, have been dated at 3.8 billion years.

Figure 22.9
Earth's oldest preserved continental rocks are more than 3.8 billion years old
(Photo courtesy of James L. Amos/ CORBIS)

shield volcanoes as well as oceanic plateaus. Simultaneously, subduction of oceanic crust generated volcanic island arcs. Collectively, these relatively small crustal fragments represent the first phase in creating stable, continent-size landmasses.

From Continental Crust to Continents   The growth of larger continental masses was accomplished through collision and accretion of various types of crustal fragments, as illustrated in **Figure 22.10.** This type of collision tectonics

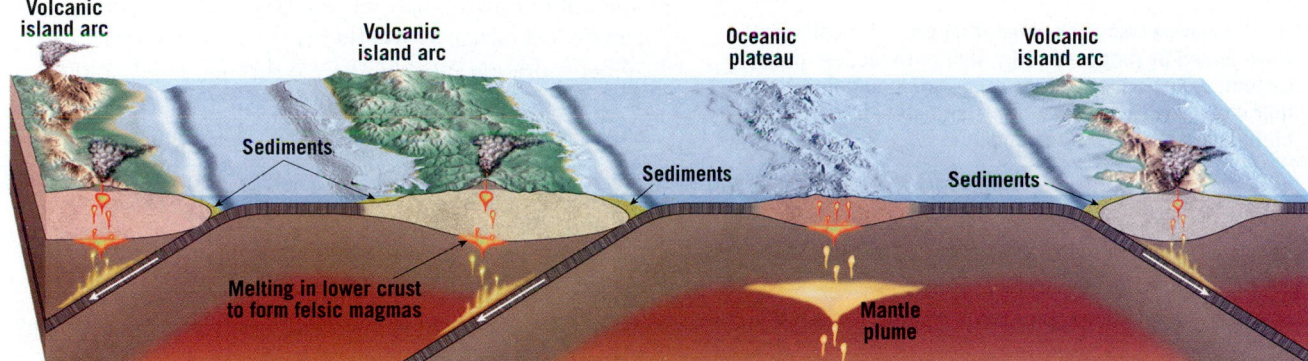

A. Scattered crustal fragments separated by ocean basins

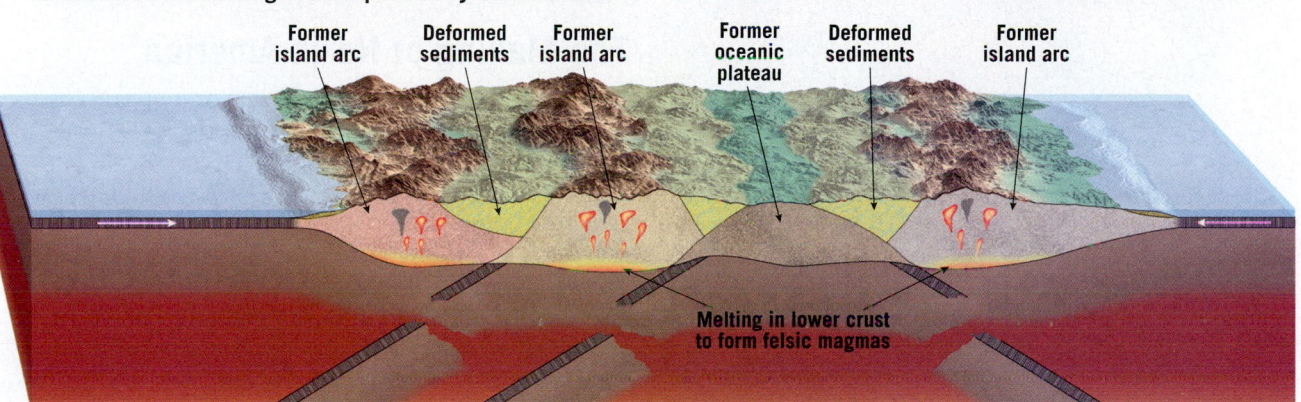

B. Collision of volcanic island arcs and oceanic plateau to form a larger crustal block

SmartFigure 22.10
The formation of continents The growth of large continental masses occurs through the collision and accretion of smaller crustal fragments.
(https://goo.gl/uVIk9X)

Tutorial

Figure 22.11
Distribution of crustal material remaining from the Archean and Proterozoic eons

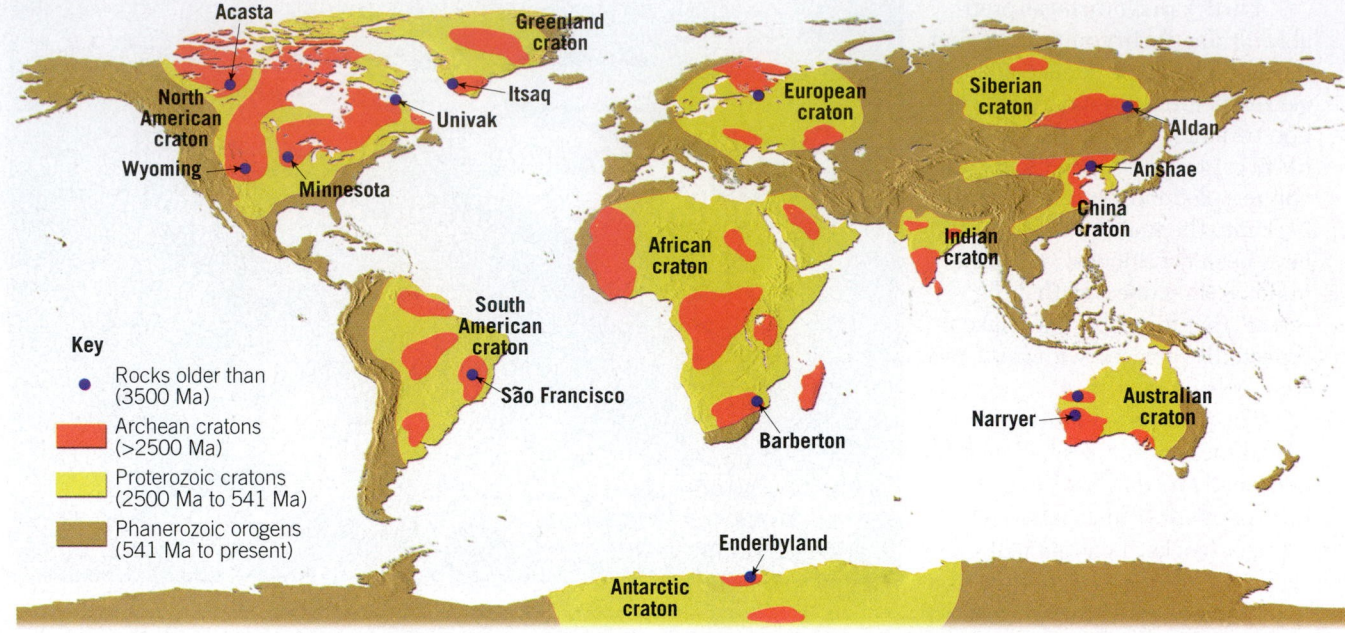

Key
- Rocks older than (3500 Ma)
- Archean cratons (>2500 Ma)
- Proterozoic cratons (2500 Ma to 541 Ma)
- Phanerozoic orogens (541 Ma to present)

deformed and metamorphosed sediments caught between converging crustal fragments, thereby shortening and thickening the developing crust. In the deepest regions of these collision zones, partial melting of the thickened crust generated silica-rich magmas that ascended and intruded the rocks above. The result was the formation of large crustal provinces that, in turn, accreted with others to form even larger crustal blocks called **cratons**.

The portion of a modern craton that is exposed at the surface is called a **shield**. The assembly of a large

SmartFigure 22.12
The major geologic provinces of North America
Age of each province is in billions of years (Ga).
(https://goo.gl/RDXNEA)

North America was assembled from crustal blocks that were joined by processes very similar to modern plate tectonics. Ancient collisions produced mountain belts that include remnant volcanic island arcs, trapped by colliding continental fragments.

Age (Ga)
- <1.0
- 1.0–1.2
- 1.6–1.7
- 1.7–1.8
- 1.8–2.0
- 2.5–3.0
- >3.5

craton involves the accretion of several crustal blocks that cause major mountain-building episodes similar to India's collision with Asia. **Figure 22.11** shows the extent of crustal material that was produced during the Archean and Proterozoic eons. This was accomplished through the collision and accretion of many thin, highly mobile terranes into nearly recognizable continental masses.

Although the Precambrian was a time when much of Earth's continental crust was generated, a substantial amount of crustal material was destroyed as well. Crust can be lost either by weathering and erosion or by direct reincorporation into the mantle through subduction. Evidence suggests that during much of the Archean, thin slabs of continental crust were eliminated, mainly by subduction into the mantle. However, by about 3 billion years ago, cratons grew sufficiently large and thick to resist subduction. After that time, weathering and erosion became the primary processes of crustal destruction. By the close of the Precambrian, an estimated 85 percent of the modern continental crust had formed.

The Making of North America

North America provides an excellent example of the development of continental crust and its piecemeal assembly into a continent. Notice in **Figure 22.12** that very little continental crust older than 3.5 billion years remains. In the late Archean, between 3 and 2.5 billion years ago, there was a period of major continental growth. During this span, the accretion of numerous island arcs and other fragments generated several large crustal provinces. North America contains some of these crustal units, including the Superior and Hearne-Rae cratons shown in Figure 22.12. It remains unknown where these ancient continental blocks formed.

About 1.9 billion years ago, these crustal provinces collided to produce the Trans-Hudson mountain belt (see Figure 22.12). (Such mountain-building episodes were not restricted to North America; ancient deformed strata of similar age are also found on other continents.) This event built the North American craton, around which several large and numerous small crustal fragments were later added. One of these late arrivals is the Piedmont province of the Appalachians. In addition, several terranes were added to the western margin of North America during the Mesozoic and Cenozoic eras to generate the mountainous North American Cordillera.

Supercontinents of the Precambrian

At different times, parts of what is now North America have combined with other continental landmasses to form a supercontinent. **Supercontinents** are large landmasses that contain all, or nearly all, the existing continents. Pangaea was the most recent, but certainly not the only, supercontinent to exist in the geologic past. The earliest well-documented supercontinent, *Rodinia*, formed during the Proterozoic eon, about 1.1 billion years ago **(Figure 22.13)**. Although geologists are still studying its construction, it is clear that Rodinia's configuration was quite different from Pangaea's. One obvious distinction is that North America was located near the center of this ancient landmass.

Between 800 and 600 million years ago, Rodinia gradually split apart. By the end of the Precambrian, many of the fragments reassembled, producing a large landmass in the Southern Hemisphere called *Gondwana*, comprised mainly of present-day South America, Africa, India, Australia, and Antarctica **(Figure 22.14)**. Other continental fragments also developed—North America, Siberia, and Northern Europe. We consider the fate of these Precambrian landmasses in the next section.

Supercontinent Cycle   The idea that rifting and dispersal of one supercontinent is followed by a long period during which the fragments are gradually reassembled into a new supercontinent with a different configuration is called the **supercontinent cycle**. The assembly and dispersal of supercontinents had a profound impact on the evolution of Earth's continents. In addition, this phenomenon greatly influenced global climates and contributed to periodic episodes of rising and falling sea level.

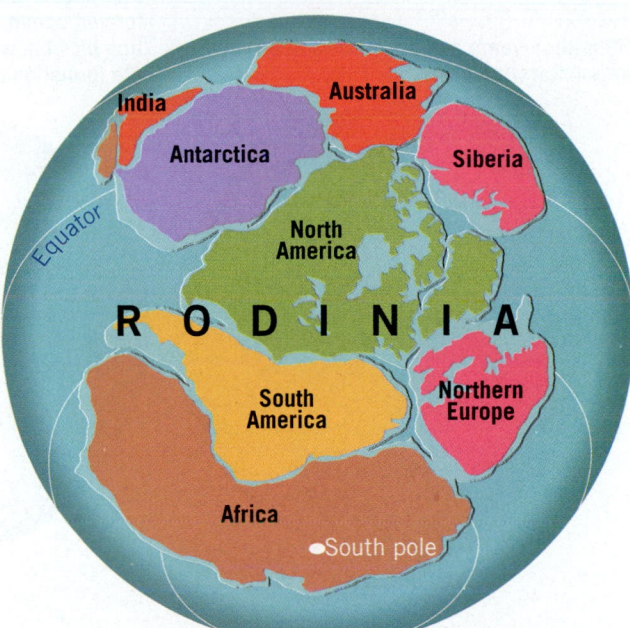

Figure 22.13
Possible configuration of the supercontinent Rodinia For clarity, the continents are drawn with somewhat modern shapes, not their actual shapes from 1 billion years ago. (After P. Hoffman, J. Rogers, and others)

Supercontinents, Mountain Building, and Climate

As continents move, the patterns of ocean currents and global winds change, influencing the global distribution of temperature and precipitation. One example of how a supercontinent's dispersal influenced climate is the formation of the Antarctic ice sheet. Although eastern Antarctica remained over the South Pole for more than 100 million years, it was not glaciated until about 25 million years ago. Prior to this period of glaciation, South America was connected to the Antarctic Peninsula. This arrangement of landmasses helped maintain a circulation pattern in which warm ocean currents reached the coast of Antarctica, as shown in **Figure 22.15A**. This is similar to the way in which the modern Gulf Stream keeps Iceland mostly ice free, despite its name. However, as South America separated from Antarctica, it moved northward, permitting ocean circulation to flow from west to east around the entire continent of Antarctica **(Figure 22.15B)**. This cold current, called the West Wind Drift, effectively isolated the entire

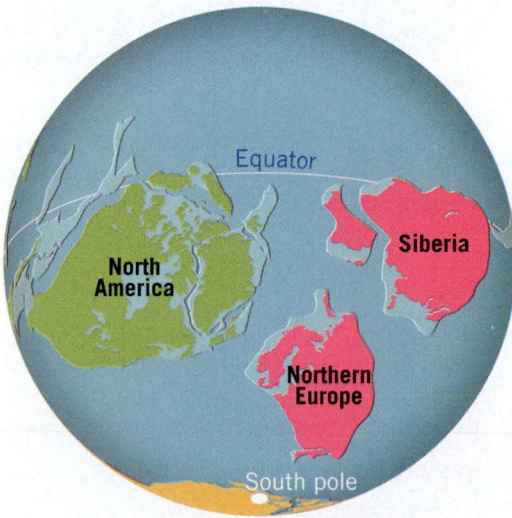

A. Continent of Gondwana   B. Continents not a part of Gondwana

Figure 22.14
Reconstruction of Earth as it may have appeared in late Precambrian time The southern continents were joined into a single landmass called Gondwana. Other landmasses that were not part of Gondwana include North America, northwestern Europe, and northern Asia. (After P. Hoffman, J. Rogers, and others)

SmartFigure 22.15
Connection between ocean circulation and the climate in Antarctica
(https://goo.gl/7nRRRF)

▶ **Tutorial**

50 million years ago warm ocean currents kept Antarctica nearly ice free.

As South America separated from Antarctica, the West Wind Drift developed. This newly formed ocean current effectively cut Antarctica off from warm currents and contributed to the formation of its vast ice sheets.

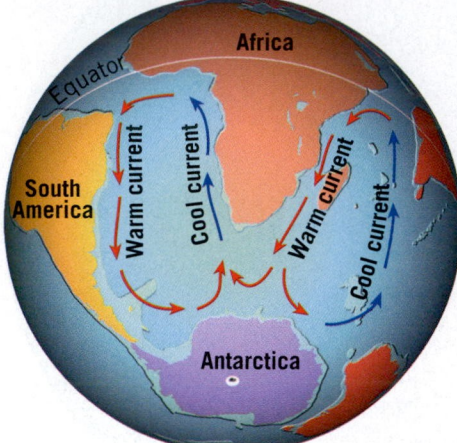

A. Not glaciated

B. Glaciated

Antarctic coast from the warm, poleward-directed currents in the southern oceans. As a result, most of the Antarctic landmass became covered with glacial ice.

Local and regional climates have also been impacted by large mountain systems created by the collision of large cratons. Because of their high elevations, mountains exhibit markedly lower average temperatures than surrounding lowlands. In addition, air rising over these lofty structures promotes condensation and precipitation, leaving the region downwind relatively dry. A modern analogy is the wet, heavily forested western slopes of the Sierra Nevada compared to the dry climate of the Great Basin Desert that lies directly to the east.

Supercontinents and Sea-Level Changes
Significant and numerous sea-level changes have been documented in geologic history, and many of them appear to be related to the assembly and dispersal of supercontinents. If sea level rises, shallow seas advance onto the

continents. Evidence for periods when the seas advanced onto the continents include thick sequences of ancient marine sedimentary rocks that blanket large areas of modern landmasses—including much of the eastern two-thirds of the United States.

The supercontinent cycle and sea-level changes are directly related to rates of *seafloor spreading*. When the rate of spreading is rapid, as it is along the East Pacific Rise today, the production of warm oceanic crust is also high. Because warm oceanic crust is less dense (takes up more space) than cold crust, fast-spreading ridges occupy more volume in the ocean basins than do slow-spreading centers. (Think of getting into a tub filled with water.) As a result, when rates of seafloor spreading increase, more seawater is displaced, which results in the sea level rising. This, in turn, causes shallow seas to advance onto the low-lying portions of the continents.

22.4 | Concept Checks

1. Briefly explain how low-density continental crust was produced from Earth's rocky mantle.

2. Describe how cratons came into being.

3. What is the supercontinent cycle? What supercontinent proceeded Pangaea?

4. Give an example of how the movement of a continent can trigger climate change.

5. Explain how seafloor spreading rates are related to sea-level changes.

EYE ON EARTH 22.1

The oldest-known sample of Earth is a 4.4-billion-year-old zircon crystal found in a metaconglomerate in the Jack Hills area of western Australia. Zircon is a silicate mineral that commonly occurs in trace amounts in most granitic rocks. (Photo by John W. Valley/NSF)

QUESTION 1 What is the parent rock of a metaconglomerate?

QUESTION 2 Assuming that this zircon crystal originated as part of a granite intrusion, briefly describe its journey from the time of its formation until it was discovered at Jack Hill.

QUESTION 3 Is this zircon crystal younger or older than the metaconglomerate in which it was found? Explain.

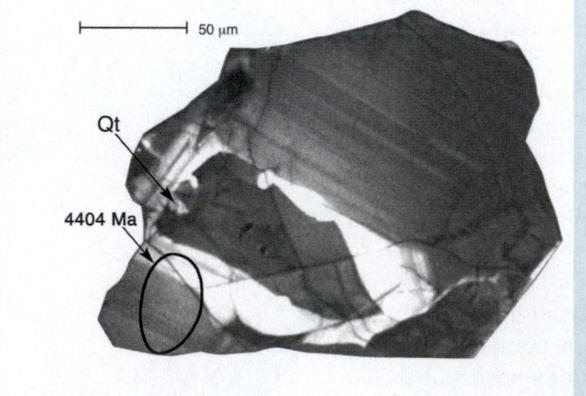

22.5 | Geologic History of the Phanerozoic: The Formation of Earth's Modern Continents

List and discuss the major geologic events in the Paleozoic, Mesozoic, and Cenozoic eras.

The time span since the close of the Precambrian, called the *Phanerozoic eon*, encompasses 541 million years and is divided into three eras: *Paleozoic, Mesozoic,* and *Cenozoic.* The beginning of the Phanerozoic is marked by the appearance of the first life-forms with hard parts such as shells, scales, bones, or teeth—all of which greatly enhance the chances for an organism to be preserved in the fossil record. Thus, the study of Phanerozoic crustal history was aided by the availability of fossils, which improved our ability to date and correlate geologic events. Moreover, because every organism is associated with its own particular niche, the greatly improved fossil record provided invaluable information for deciphering ancient environments.

Paleozoic History

As the Paleozoic era opened, what is now North America hosted no plants or animals large enough to be seen—just tiny microorganisms such as bacteria. There were no Appalachian or Rocky Mountains; the continent was largely a barren lowland. Several times during the early Paleozoic, shallow seas moved inland and then receded from the continental interior and left behind the thick deposits of limestone, shale, and clean sandstone that mark the shorelines of these previously midcontinent shallow seas.

Formation of Pangaea One of the major events of the Paleozoic era was the formation of the supercontinent **Pangaea**. This event began with a series of collisions that over millions of years joined North America, Europe, Siberia, and other smaller crustal fragments to form a large continent called **Laurasia** (**Figure 22.16**). Located south of Laurasia was the vast southern continent called **Gondwana**, which encompassed five modern landmasses— South America, Africa, Australia, Antarctica, and India—and perhaps portions of China. Evidence of extensive continental glaciation places this landmass near the South Pole. By the late Paleozoic, Gondwana had migrated northward and collided with Laurasia to begin the final stage in Pangaea's assembly.

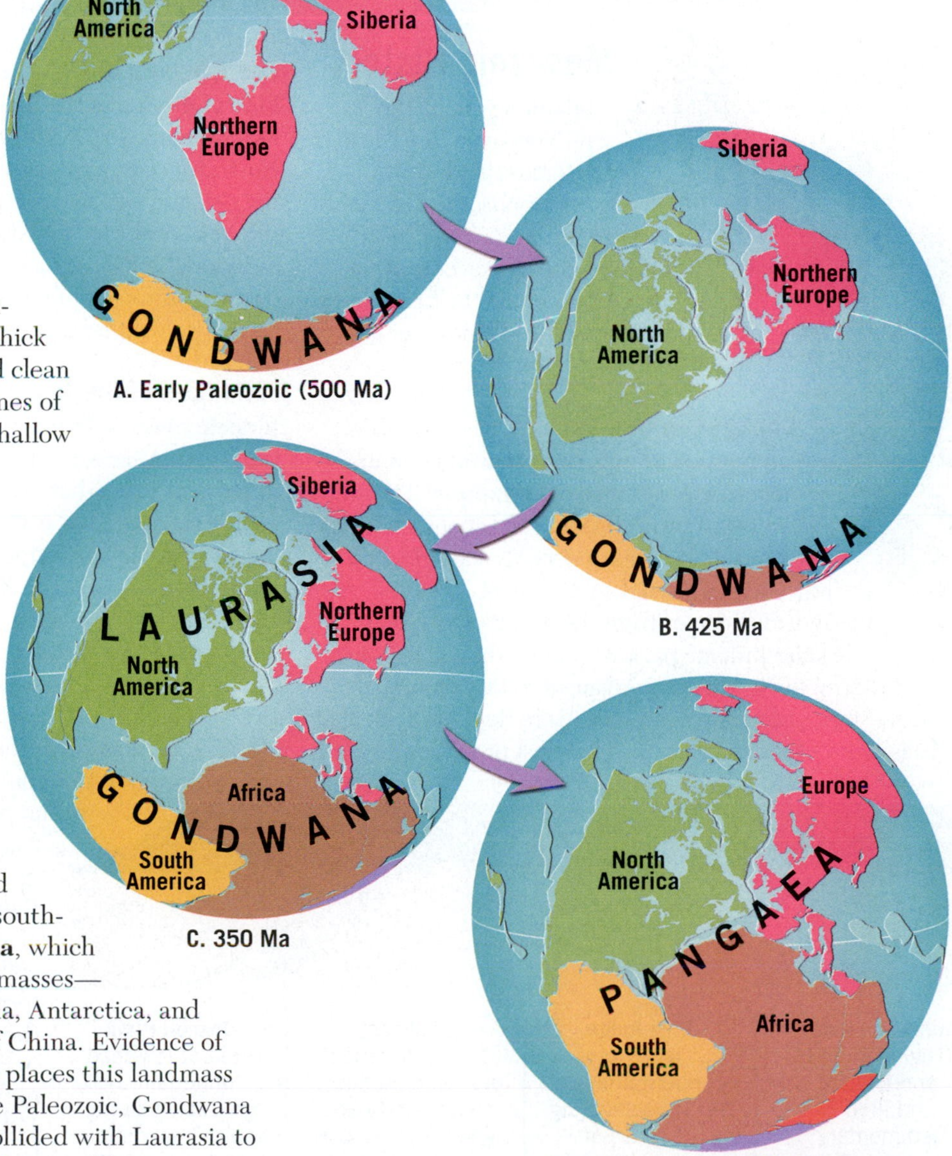

Figure 22.16
Formation of Pangaea During the late Paleozoic, Earth's major landmasses joined to produce the supercontinent of Pangaea. (After P. Hoffman, J. Rogers, and others)

A. Early Paleozoic (500 Ma)

B. 425 Ma

C. 350 Ma

D. Late Paleozoic (300–250 Ma)

The accretion of all of Earth's major landmasses to form Pangaea spans more than 300 million years and resulted in the formation of several mountain belts. The collision of northern Europe (mainly Norway) with Greenland produced the Caledonian Mountains, whereas the joining of northern Asia (Siberia) and Europe created the Ural Mountains. Northern China is also thought to have accreted to Asia by the end of the Paleozoic, whereas southern China may not have become part of Asia until after Pangaea had begun to rift. (Recall that India did not begin to accrete to Asia until about 50 million years ago.)

Pangaea reached its maximum size between 300 and 250 million years ago, as Africa collided with North America (see Figure 22.16D). This event marked the final and most intense period of mountain building in the long history of the Appalachian Mountains. This mountain-building event produced the Central Appalachians of the Atlantic states, as well as New England's northern Appalachians and mountainous structures that extend into Canada (**Figure 22.17**).

SmartFigure 22.17
Major provinces of the Appalachian Mountains
(https://goo.gl/b2Deug)

▶ **Tutorial**

Mesozoic History

Spanning about 186 million years, the Mesozoic era is divided into three periods: the *Triassic*, *Jurassic*, and *Cretaceous*. Major geologic events of the Mesozoic include the breakup of Pangaea and the evolution of our modern ocean basins.

Changes in Sea Levels

The Mesozoic era began with much of the world's continents above sea level. The exposed Triassic strata are primarily red sandstones and mudstones that lack marine fossils, features that indicate a terrestrial environment. (The red color in sandstone comes from the oxidation of iron.)

As the Jurassic period opened, the sea invaded western North America. Adjacent to this shallow sea, extensive continental sediments were deposited on what is now the Colorado Plateau. The most prominent is the

Navajo Sandstone, a cross-bedded, quartz-rich layer that in some places approaches 300 meters (1000 feet) thick. These remnants of massive dunes indicate that an enormous desert occupied much of the American Southwest during early Jurassic times (**Figure 22.18**). Another well-known Jurassic deposit is the Morrison Formation—the world's richest storehouse of dinosaur fossils. Included are the fossilized bones of massive dinosaurs such as *Apatosaurus* (formerly *Brontosaurus*), *Brachiosaurus*, and *Stegosaurus*.

Coal Formation in Western North America   As the Jurassic period gave way to the Cretaceous, shallow seas again encroached upon much of western North America, as well as the Atlantic and Gulf coastal regions. This led to the formation of "coal swamps" (see Chapter 7) similar to those of the Paleozoic era. Today, the Cretaceous coal deposits in the western United States and Canada are economically important. For example, on the Crow Native American reservation in Montana, there are nearly 20 billion tons of high-quality, Cretaceous-age coal.

The Breakup of Pangaea   Another major event of the Mesozoic era was the breakup of Pangaea. About 185 million years ago, a rift developed between what is now North America and western Africa, marking the birth of the Atlantic Ocean. As Pangaea gradually broke apart, the westward-moving North American plate began to override the Pacific basin. This tectonic event triggered a continuous wave of deformation that moved inland along the entire western margin of North America.

Formation of the North American Cordillera   By Jurassic times, subduction of the Pacific basin under the North American plate began to produce the chaotic mixture of rocks that exist today in the Coast Ranges of California (see Figure 14.5, page 424). Further inland, igneous activity was widespread, and for more than 100 million years volcanism was rampant as huge masses of magma rose within a few miles of Earth's surface. The remnants of this activity include the granitic plutons of the Sierra Nevada, as well as the Idaho batholith and British Columbia's Coast Range batholith.

The subduction of the Pacific basin under the western margin of North America also resulted in the piecemeal addition of crustal fragments to the entire Pacific margin of the continent—from Mexico's Baja Peninsula to northern Alaska (see Figure 14.9, page 428). Each collision displaced earlier accreted terranes further inland, adding to the zone of deformation as well as to the thickness and lateral extent of the continental margin.

Compressional forces moved huge rock units in a shingle-like fashion toward the east. Across

A.

Map labels: Central Appalachians, Northern Appalachians, Appalachian Plateau, Valley and Ridge, Blue Ridge Mountains, Piedmont, Coastal Plain

B.

Appalachian Plateau	Valley and Ridge	Blue Ridge	Piedmont	Coastal Plain
(Underlain by nearly flat-lying sedimentary strata of Paleozoic age.)	(Highly folded and thrust-faulted sedimentary rocks of Paleozoic age.)	(Hilly to mountainous terrain consisting of slices of basement rock of Precambrian age.)	(Crustal fragments of metamorphosed sedimentary and igneous rocks that were added to North America.)	(Area of low relief underlain by gradualy sloping sedimentary strata and unlithified sediments.)

much of North America's western margin, older rocks were thrust eastward over younger strata, for distances exceeding 150 kilometers (90 miles). Ultimately, this activity was responsible for generating a vast portion of the North American Cordillera that extends from Wyoming to Alaska.

Toward the end of the Mesozoic, the southern portions of the Rocky Mountains developed. This mountain-building event, called the *Laramide Orogeny*, occurred when large blocks of deeply buried Precambrian rocks were lifted nearly vertically along steeply dipping faults, upwarping the overlying younger sedimentary strata. The mountain ranges produced by the Laramide Orogeny include Colorado's Front Range, the Sangre de Cristo of New Mexico and Colorado, and the Bighorns of Wyoming (see GEOgraphics 14.1, page 436).

Cenozoic History

The Cenozoic era, or "era of recent life," encompasses the past 66 million years of Earth history. It was during this span that the physical landscapes and life-forms of our modern world came into existence. The Cenozoic era represents a considerably smaller fraction of geologic time than either the Paleozoic or the Mesozoic, but we know much more about this time span because the rock formations are more widespread and less disturbed than those of any preceding era.

Most of North America was above sea level during the Cenozoic era. However, the eastern and western margins of the continent experienced markedly contrasting events because of their different plate boundary relationships. The Atlantic and Gulf coastal regions, far removed from an active plate boundary, were tectonically stable. By contrast, western North America was the leading edge of the North American plate, and the plate interactions during the Cenozoic account for many events of mountain building, volcanism, and earthquakes.

Eastern North America
The stable continental margin of eastern North America was the site of abundant marine sedimentation. The most extensive deposition surrounded the Gulf of Mexico, from the Yucatan Peninsula to Florida, where a massive buildup of sediment caused the crust to downwarp. In many instances, faulting created structures in which oil and natural gas accumulated. Today, these and other petroleum traps (see Chapter 23) are the Gulf Coast's most economically important resource, evidenced by numerous offshore drilling platforms.

Early in the Cenozoic, the Appalachians had eroded to create a low plain. Later, isostatic adjustments again raised the region and rejuvenated its rivers. Streams eroded with renewed vigor, gradually sculpting the surface into its present-day topography. Sediments from this erosion were deposited along the eastern continental margin, where they accumulated to a thickness of many kilometers. Today, portions of the strata deposited during the Cenozoic are exposed as the gently sloping Atlantic and Gulf coastal plains, where a large percentage of the eastern and southeastern United States population resides (see Figure 22.17).

Western North America
In the West, the Laramide Orogeny responsible for building the southern Rocky Mountains was coming to an end. As erosional forces lowered the mountains, the basins between uplifted ranges began to fill with sediment. East of the Rockies, a large wedge of sediment from the eroding mountains created the gently sloping Great Plains.

Beginning in the Miocene epoch, about 20 million years ago, a broad region from northern Nevada into Mexico experienced crustal extension that created more than 100 fault-block mountain ranges. Today, they rise abruptly above the adjacent basins, forming the Basin and Range province (see Chapter 14).

During the development of the Basin and Range province, the entire western interior of the continent gradually uplifted. This event elevated the Rockies and rejuvenated many of the West's major rivers. As the rivers became incised, many spectacular gorges were created, including the Grand Canyon of the Colorado River, the Grand Canyon of the Snake River, and the Black Canyon of the Gunnison River.

Volcanic activity was also common in the West during much of the Cenozoic. Beginning in the Miocene epoch, great volumes of fluid basaltic lava flowed from

Figure 22.18
Massive, cross-bedded sandstone cliffs in Zion National Park These sandstone cliffs are the remnants of ancient sand dunes that were part of an enormous desert during the Jurassic period. (Photo by Michael Collier; inset photo by Dennis Tasa)

Close-up view of cross bedding in the Navajo Sandstone, Zion National Park

Figure 22.19
Mount Shasta, California This volcano is one of several large composite cones that comprise the Cascade Range. (Photo by Michael Collier)

fissures in portions of present-day Washington, Oregon, and Idaho. These eruptions built the 3.4-million-square-kilometer (1.3-million-square–mile) Columbia Plateau. Immediately west of the vast Columbia Plateau, volcanic activity was different in character. Here, more viscous magmas with higher silica content erupted explosively, creating the Cascades, a chain of stratovolcanoes extending from northern California into Canada, some of which are still active (**Figure 22.19**).

As the Cenozoic was drawing to a close, the effects of mountain building, volcanic activity, isostatic adjustments, and extensive erosion and sedimentation created the physical landscape we know today. All that remained of Cenozoic time was the final 2.6-million-year episode called the Quaternary period. During this most recent, and ongoing, phase of Earth's history,

humans evolved and the action of glacial ice, wind, and running water added to our planet's long, complex geologic history.

22.5 | Concept Checks

1. During which period of geologic history did the supercontinent Pangaea come into existence? During which period did it begin to break apart?

2. Describe the climate of the present-day American Southwest during early Jurassic time.

3. Where is most Cretaceous age coal found today in the United States?

4. Compare and contrast eastern and western North America during the Cenozoic era.

22.6 | Earth's First Life

Describe some of the hypotheses on the origin of life and the characteristics of early prokaryotes, eukaryotes, and multicelled organisms.

The oldest fossils provide evidence that life on Earth was established at least 3.5 billion years ago (**Figure 22.20**). Microscopic fossils similar to modern cyanobacteria have been found in silica-rich chert deposits worldwide. Notable examples include southern Africa, where rocks date to more than 3.1 billion years, and the Lake Superior region of western Ontario and northern Minnesota, where the Gunflint Chert contains some fossils older than 2 billion years. Chemical traces of organic matter in even older rocks have led paleontologists to conclude that life may have existed as early as 3.8 billion years ago.

Origin of Life

How did life begin? This question sparks considerable debate, and hypotheses abound. Requirements for life, assuming the presence of a hospitable environment,

include the chemical raw materials that are found in essential molecules such as proteins. Proteins are made from organic compounds called *amino acids*. The first amino acids may have been synthesized from methane

Evolution of Life Through Geologic Time

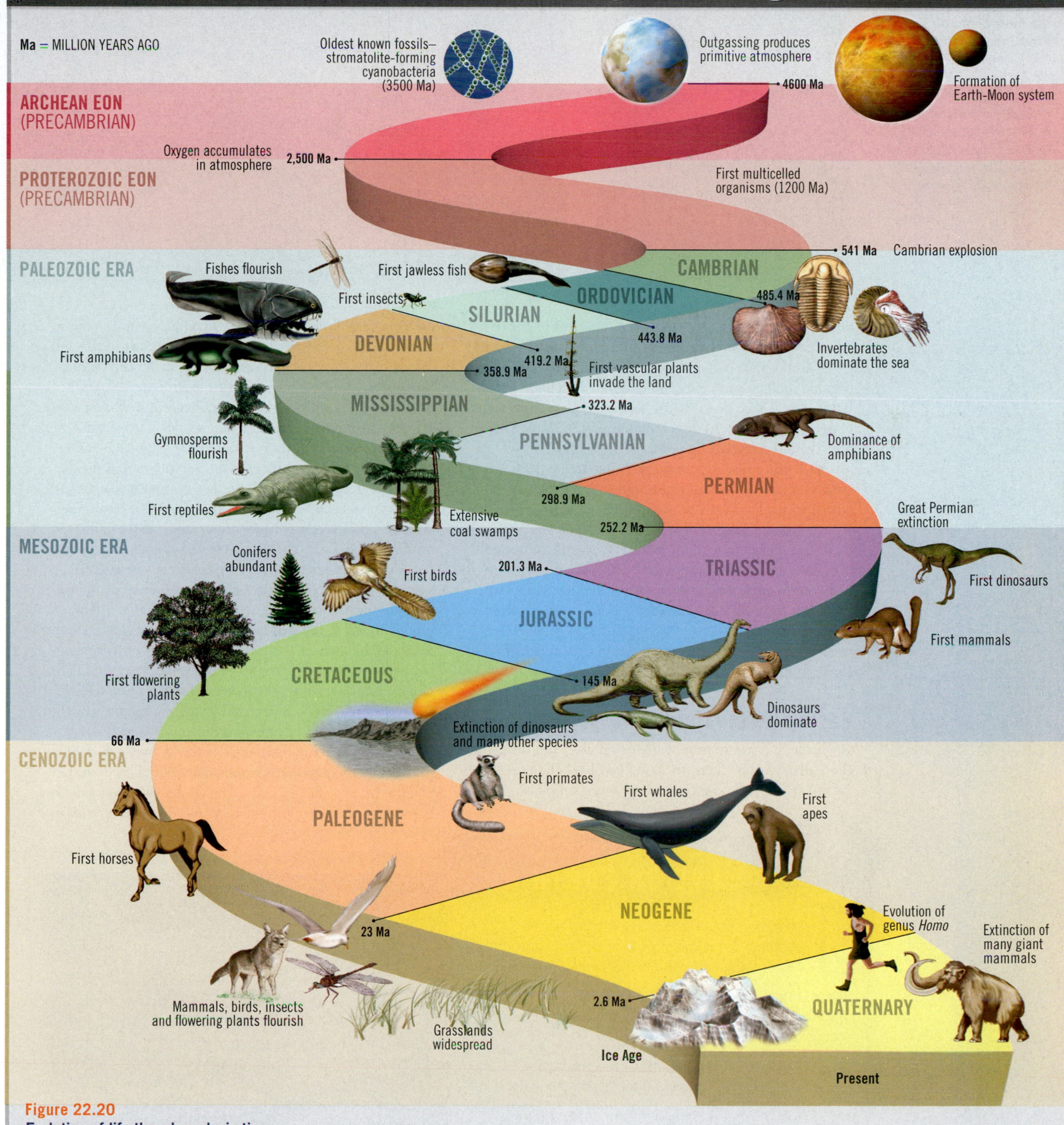

Ma = MILLION YEARS AGO

Oldest known fossils— stromatolite-forming cyanobacteria (3500 Ma)

Outgassing produces primitive atmosphere

4600 Ma

Formation of Earth-Moon system

ARCHEAN EON (PRECAMBRIAN)

Oxygen accumulates in atmosphere — 2,500 Ma

PROTEROZOIC EON (PRECAMBRIAN)

First multicelled organisms (1200 Ma)

541 Ma — Cambrian explosion

PALEOZOIC ERA

Fishes flourish

First jawless fish

CAMBRIAN

First insects

ORDOVICIAN

485.4 Ma

SILURIAN

443.8 Ma

Invertebrates dominate the sea

DEVONIAN

419.2 Ma

First amphibians

358.9 Ma

First vascular plants invade the land

MISSISSIPPIAN

323.2 Ma

Gymnosperms flourish

PENNSYLVANIAN

Dominance of amphibians

First reptiles

298.9 Ma

PERMIAN

Extensive coal swamps

252.2 Ma

Great Permian extinction

MESOZOIC ERA

Conifers abundant

First birds

201.3 Ma

TRIASSIC

First dinosaurs

JURASSIC

First mammals

First flowering plants

CRETACEOUS

145 Ma

Dinosaurs dominate

66 Ma

Extinction of dinosaurs and many other species

CENOZOIC ERA

First primates

First whales

First apes

PALEOGENE

First horses

NEOGENE

Evolution of genus *Homo*

Extinction of many giant mammals

23 Ma

2.6 Ma

QUATERNARY

Mammals, birds, insects and flowering plants flourish

Grasslands widespread

Ice Age

Present

Figure 22.20
Evolution of life though geologic time

Figure 22.21
Stromatolites are among the most common Precambrian fossils A. Cross-section through fossil stromatolites deposited by cyanobacteria. (Photo by Sinclair Stammers/Science Source) **B.** Modern stromatolites exposed at low tide in western Australia. (Photo by Bill Bachman/Science Source)

A.   B.

and ammonia, both of which were plentiful in Earth's primitive atmosphere. Some scientists suggest that these gases could have been easily reorganized into useful organic molecules by ultraviolet light. Others consider lightning to have been the impetus, as the well-known experiments conducted by biochemists Stanley Miller and Harold Urey attempted to demonstrate.

Still other researchers suggest that amino acids arrived "ready-made," delivered by asteroids or comets that collided with a young Earth. Evidence for this hypothesis comes from a group of meteorites, called *carbonaceous chondrites*, which contain amino acid–like organic compounds.

Yet another hypothesis proposes that the organic material needed for life came from the methane and hydrogen sulfide that spews from deep-sea hydrothermal vents (black smokers) (see Figure 8.22, page 257). It is also possible that life originated in hot springs similar to those found in Yellowstone National Park.

Earth's First Life: Prokaryotes

Regardless of where or how life originated, it is clear that the journey from "then" to "now" involved change (Figure 22.20). The first known organisms were simple single-cell

bacteria called **prokaryotes**, which means their genetic material (DNA) is *not separated* from the rest of the cell by a nucleus. Because oxygen was largely absent from Earth's early atmosphere and oceans, the first organisms employed anaerobic (without oxygen) metabolism to extract energy from "food." Their food source was likely organic molecules in their surroundings, but that supply was very limited. Later, bacteria evolved that used solar energy to synthesize organic compounds (sugars). This event was an important turning point in biological evolution: For the first time, organisms had the ability to produce food for themselves as well as for other life-forms.

Recall that photosynthesis by ancient cyanobacteria, a type of prokaryote, contributed to the gradual rise in the level of oxygen, first in the ocean and later in the atmosphere. It was these early organisms, which began to inhabit Earth 3.5 billion years ago, that dramatically transformed our planet. Fossil evidence for the existence of these microscopic bacteria includes distinctively layered mats, called **stromatolites**, composed of slimy material secreted by these organisms, along with trapped sediments (**Figure 22.21A**). What is known about these ancient fossils comes mainly from the study of modern stromatolites like those found in Shark Bay, Australia (**Figure 22.21B**). Today's stromatolites look like stubby pillars built as microbes slowly move upward to

EYE ON EARTH 22.2

The rocks shown here are Cambrian-age stromatolites of the Hoyt Limestone, exposed at Lester Park, near Saratoga Springs, New York.

QUESTION 1 Using Figure 22.3, determine approximately how many years ago these rocks were deposited.

QUESTION 2 What is the name of the group of organisms that likely produced these limestone deposits?

QUESTION 3 What was the environment like in this part of New York when these rocks were deposited?

M. C. Rygel

Figure 22.22
Ediacaran fossil The Ediacarans are a group of sea-dwelling animals that may have come into existence about 600 million years ago. These soft-bodied organisms were up to 1 meter (3 feet) in length and are the oldest animal fossils so far discovered. (Photo Sinclair Stammers/Science Source)

During much of the Precambrian, life consisted exclusively of single-celled organisms. It wasn't until perhaps 1.2 billion years ago that multicelled eukaryotes evolved. Green algae, one of the first multicelled organisms, contained chloroplasts (used in photosynthesis) and were the likely ancestors of modern plants. The first primitive marine animals did not appear until somewhat later, perhaps 600 million years ago (**Figure 22.22**).

Fossil evidence suggests that organic evolution progressed at an excruciatingly slow pace until the end of the Precambrian. At that time, Earth's continents were largely barren and the oceans were populated mainly with tiny organisms, many too small to be seen with the naked eye. Nevertheless, the stage was set for the evolution of larger and more complex plants and animals.

avoid being buried by the sediment that is continually deposited on them.

Evolution of Eukaryotes The oldest fossils of more advanced organisms, called **eukaryotes**, are about 2.1 billion years old. The first eukaryotes were microscopic, single-cell organisms, but unlike prokaryotes, eukaryotes contain nuclei. This distinctive cellular structure is what all multicellular organisms that now inhabit our planet—trees, birds, fish, reptiles, and humans—have in common.

22.6 Concept Checks

1. What group of organic compounds is essential for the formation of DNA and RNA, and therefore necessary for life as we know it?

2. What are stromatolites? What group of organisms is thought to have produced them?

3. Compare *prokaryotes* with *eukaryotes*. To which group do all multicelled organisms belong?

22.7 | Paleozoic Era: Life Explodes

List the major developments in the history of life during the Paleozoic era.

The Cambrian period marks the beginning of the Paleozoic era, a time span that saw the emergence of a spectacular variety of new life-forms. All major invertebrate (animals lacking backbones) groups made their appearance, including jellyfish, sponges, worms, mollusks (clams and snails), and arthropods (insects and crabs). This huge expansion in biodiversity is often referred to as the **Cambrian explosion**.

Early Paleozoic Life-Forms

The Cambrian period was the golden age of *trilobites* (**Figure 22.23**). Trilobites developed a flexible exoskeleton of a protein called chitin (similar to a lobster shell), which enabled them to be mobile and search for food by burrowing through soft sediment. More than 600 genera of these mud-burrowing scavengers flourished worldwide.

The Ordovician marked the appearance of abundant cephalopods—mobile, highly developed mollusks that became the major predators of their time (**Figure 22.24**). Descendants of these cephalopods include the squid, octopus, and chambered nautilus that inhabit our modern oceans. Cephalopods were the first truly large organisms on Earth, including one species that reached a length of nearly 10 meters (30 feet).

The early diversification of animals was driven, in part, by the emergence of predatory lifestyles. The

Figure 22.23
Fossil of a trilobite
Trilobites dominated the early Paleozoic ocean, scavenging food from the bottom. (Photo by Ed Reschke/ Peter Arnold, Inc.)

larger mobile cephalopods preyed on trilobites that were typically smaller than a child's hand. The evolution of efficient movement was often associated with the development of greater sensory capabilities and more complex

Figure 22.24
Artistic depiction of a shallow Ordovician sea During the Ordovician period (488–444 million years ago), the shallow waters of an inland sea over central North America contained an abundance of marine invertebrates. Shown in this reconstruction are (1) *corals*, (2) *trilobite*, (3) *snails*, (4) *brachiopods*, and (5) *straight-shelled cephalopod*. (The Field Museum/Getty Images)

Figure 22.25
Land plants of the Paleozoic The Silurian saw the first upright-growing (vascular) plants. Plant fossils became increasingly common from the Devonian onward.

Small upright–growing, vascular plants begin to invade the land

SILURIAN PERIOD

First tree-size plants become common

DEVONIAN PERIOD

TIME

Extensive forests cover vast areas of the continents

MISSISSIPPIAN PERIOD

nervous systems. These early animals developed sensory devices for detecting light, odor, and touch.

Approximately 400 million years ago, green algae that had adapted to survive at the water's edge gave rise to the first multicellular land plants. The primary difficulty in sustaining plant life on land was obtaining water and staying upright, despite gravity and winds. These earliest land plants were leafless vertical spikes about the size of a human index finger, but the fossil record indicates that by the beginning of the Mississippian period, there were forests with trees tens of meters tall (**Figure 22.25**).

In the ocean, fish perfected an internal skeleton as a new form of support, and they were the first creatures to have jaws. Armor-plated fish that evolved during the Ordovician continued to adapt. Their armor plates thinned to lightweight scales that increased their speed and mobility. Other fish evolved during the Devonian, including primitive sharks with cartilage skeletons and bony fish—the groups in which many modern fish are classified. Fish, the first large vertebrates, proved to be faster swimmers than invertebrates and possessed more acute senses and larger brains. They became the dominant predators of the sea, which is why the Devonian period is often referred to as the "Age of the Fishes."

Vertebrates Move to Land

During the Devonian, a group of fish called the *lobe-finned fish* began to adapt to terrestrial environments (**Figure 22.26A**). Lobe-finned fish had internal sacks that could be filled with air to supplement their "breathing" through gills. One group of lobe-finned fish probably occupied freshwater tidal flats or small ponds. Some began to use their fins to move from one pond to another in search of food or to evacuate deteriorating ponds. This favored the evolution of a group of animals able to stay out of water longer and move on land more efficiently. By the late Devonian, lobe-finned fish had evolved into air-breathing amphibians with strong legs yet retaining a fishlike head and tail (**Figure 22.26B**).

Modern amphibians, such as frogs, toads, and salamanders, are small and occupy limited biological niches. However, conditions during the late Paleozoic were ideal for these newcomers to land. Large tropical swamps teeming with large insects and millipedes extended across North America, Europe, and Siberia (**Figure 22.27**). With virtually no predatory risks, amphibians diversified rapidly. Some even took on lifestyles and forms similar to modern reptiles such as crocodiles.

Despite their success, amphibians were not fully adapted to life out of water. In fact, *amphibian* means "double life" because these animals need both the water from where they came and the land to where they moved. Amphibians are born in water, as exemplified by tadpoles, complete with gills and tails. These features disappear during the maturation process, resulting in air-breathing adults with legs.

Reptiles: The First True Terrestrial Vertebrates

Vertebrates advanced from water-dwelling lobefin fish, to amphibians, to reptiles—the first true terrestrial vertebrates (**Figure 22.28**), with improved lungs for active lifestyles and "waterproof" skin that prevented the loss of

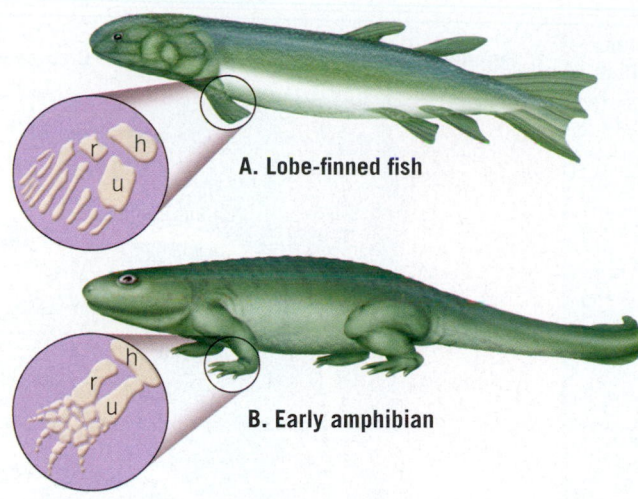

A. Lobe-finned fish

B. Early amphibian

body fluids. Most importantly, reptiles developed shell-covered eggs laid on land. Eliminating the water-dwelling stage (like the tadpole stage in frogs) was an important evolutionary step. Of interest is the fact that the watery fluid within the reptilian egg closely resembles seawater in chemical composition. Because the reptile embryo develops in this watery environment, the shelled egg has been characterized as a "private aquarium" in which the embryos of these land vertebrates spend their water-dwelling stage of life. With this "sturdy egg," the remaining ties to the water were broken, and reptiles moved inland.

The Great Permian Extinction

By the close of the Permian period a **mass extinction** occurred, in which a large number of Earth's species became extinct. During this mass extinction, 70 percent

Figure 22.26
Comparison of the anatomical features of the lobe-finned fish and early amphibians A. The fins on the lobe-finned fish contained the same basic elements (*h*, humerus, or upper arm; *r*, radius; and *u*, ulna, or lower arm) as those of the amphibians. **B.** This amphibian is shown with the standard five toes, but early amphibians had as many as eight toes. Eventually the amphibians evolved to have a standard toe count of five.

Figure 22.27
Artistic depiction of a Pennsylvanian-age coal swamp Shown are scale trees (left), seed ferns (lower left), and scouring rushes (right). Also note the large dragonfly. (The Field Museum/Getty Images)

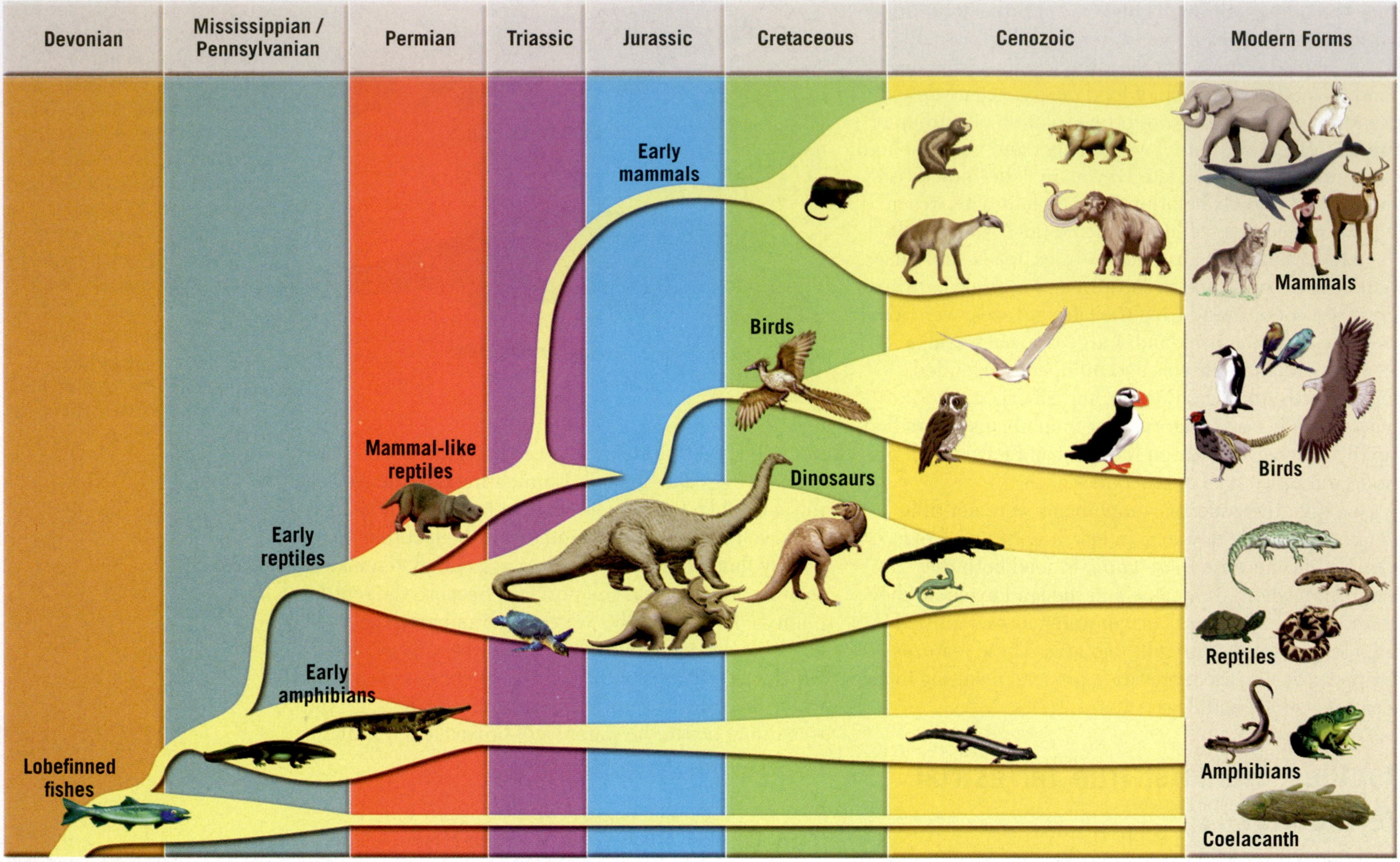

Devonian	Mississippian / Pennsylvanian	Permian	Triassic	Jurassic	Cretaceous	Cenozoic	Modern Forms

Early mammals

Birds

Mammal-like reptiles

Early reptiles

Dinosaurs

Early amphibians

Lobefinned fishes

Mammals

Birds

Reptiles

Amphibians

Coelacanth

SmartFigure 22.28

Relationships of vertebrate groups and their divergence from lobefin fish

(https://goo.gl/Ly53xz)

Tutorial

of all land-dwelling vertebrate species and perhaps 90 percent of all marine organisms were obliterated; it was the most significant of five mass extinctions that occurred over the past 500 million years. Each extinction wreaked havoc with the existing biosphere, wiping out large numbers of species. In each case, however, survivors entered new biological communities that were ultimately more diverse. Therefore, mass extinctions actually invigorated life on Earth, as the few hardy survivors eventually filled more environmental niches than those left behind by the victims.

Several mechanisms have been proposed to explain these ancient mass extinctions. Initially, paleontologists believed they were gradual events caused by a combination of climate change and biological forces, such as predation and competition. Other research groups have attempted to link certain mass extinctions to the explosive impact of a large asteroid striking Earth's surface.

The most widely held view is that the Permian mass extinction was driven mainly by volcanic activity because it coincided with a period of voluminous eruptions of flood basalts that blanketed about 1.6 million square kilometers (624,000 square miles), an area nearly

the size of Alaska. This series of eruptions, which lasted roughly 1 million years, occurred in northern Russia, in an area called the Siberian Traps. It was the largest volcanic eruption in the past 500 million years. The release of huge amounts of carbon dioxide likely generated a period of accelerated greenhouse warming, while the emission of sulfur dioxide is credited with producing copious amounts of acid rain. These drastic environmental changes likely put excessive stress on many of Earth's life-forms.

22.7 Concept Checks

1. What is the Cambrian explosion?

2. Describe the obstacles plants had to overcome in order to inhabit the continents.

3. What group of animals is thought to have left the ocean to become the first amphibians?

4. Why are amphibians not considered "true" land animals?

5. What major development allowed reptiles to move inland?

22.8 | Mesozoic Era: Age of the Dinosaurs

Briefly explain the major developments in the history of life during the Mesozoic era.

As the Mesozoic era dawned, its life-forms were the survivors of the great Permian extinction. These organisms diversified in many ways to fill the biological voids created at the close of the Paleozoic. While life on land underwent a radical transformation with the rise of the dinosaurs, life in the sea also entered a dramatic phase of transformation that produced many of the animal groups that prevail in the oceans today, including modern groups of predatory fish, crustaceans, mollusks, and sand dollars.

Gymnosperms: The Dominant Mesozoic Trees

On land, conditions favored organisms that could adapt to drier climates. One such group of plants, **gymnosperms**, produced "naked" seeds that are exposed on modified leaves that usually form cones. The seeds are not enclosed in fruits, as are apple seeds, for example. Unlike the first plants to invade the land, seed-bearing gymnosperms did not depend on free-standing water for fertilization. Consequently, these plants were not restricted to a life near the water's edge.

The gymnosperms quickly became the dominant trees of the Mesozoic. Examples of this group include cycads that resembled large pineapple plants (**Figure 22.29**); ginkgo plants having fan-shaped leaves, much like their modern relatives; and the largest plants, the conifers, whose modern descendants include the pines, firs, and junipers. The best-known fossil occurrence of these ancient trees is in northern Arizona's Petrified Forest National Park. Here, huge petrified logs lie exposed at the surface, having been weathered from rocks of the Triassic Chinle Formation (**Figure 22.30**).

Figure 22.29
Cycads, a type of gymnosperm that was very common in the Mesozoic These plants have palm-like leaves and large cones. (Photo by Jiri Loun/Science Source)

Reptiles Take Over the Land, Sea, and Sky

Among the animals, reptiles readily adapted to the drier Mesozoic environment, thereby relegating amphibians to the swamps and wetlands, where most remain today. The first reptiles were small, but larger forms evolved rapidly, particularly the dinosaurs. One of the largest was *Ultrasaurus*, which weighed more than 80 tons and measured over 30 meters (100 feet) from head to tail. Some of the largest dinosaurs were carnivorous (for example, *Tyrannosaurus*), whereas others were herbivorous (like ponderous *Apatosaurus*).

Some reptiles evolved specialized characteristics that allowed them to occupy drastically different environments. One group, the *pterosaurs*, became airborne. These "dragons of the sky" possessed huge membranous wings that allowed them rudimentary flight. How the largest pterosaurs (some had wing spans of 8 meters [26 feet] and weighed 90 kilograms [200 pounds]) took flight is

still unknown. Another group, exemplified by the fossil *Archaeopteryx*, led to more successful flyers—birds (**Figure 22.31**). This ancestor of modern birds had feathered wings but retained reptilian characteristics, such

Figure 22.30
Petrified logs of Triassic age, Arizona's Petrified Forest National Park (Photo by Bernd Siering/AGE Fotostock)

Figure 22.31

***Archaeopteryx*, ancestors of modern birds** Fossil evidence indicates that *Archaeopteryx* had feathers like modern birds but retained many characteristics of reptiles. The sketch shows an artist's reconstruction of *Archaeopteryx*. (Photo by Michael Collier)

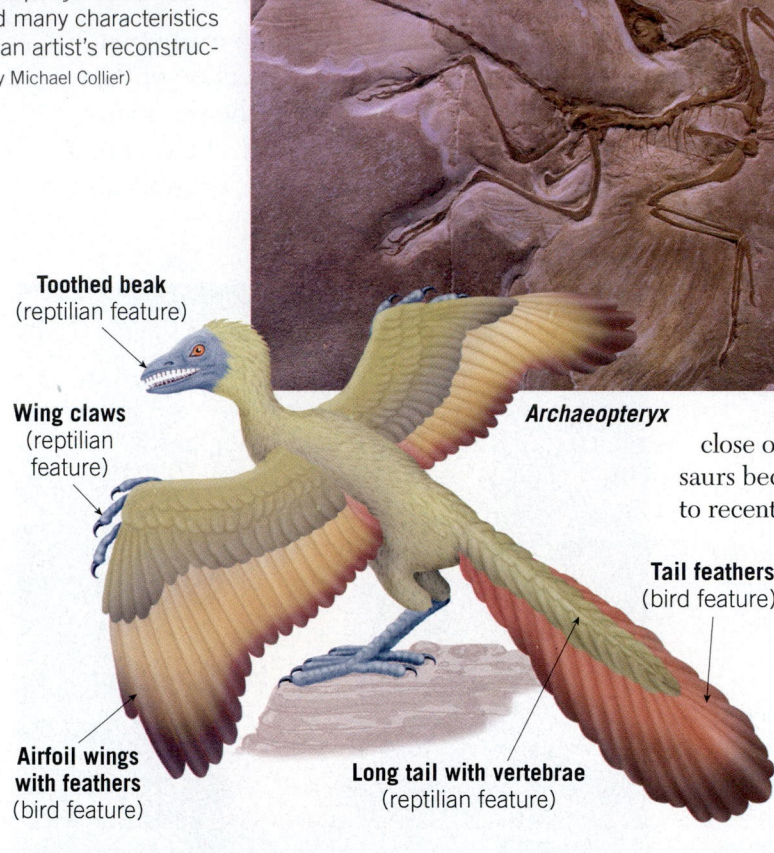

Toothed beak
(reptilian feature)

Wing claws
(reptilian feature)

Archaeopteryx

Tail feathers
(bird feature)

Airfoil wings with feathers
(bird feature)

Long tail with vertebrae
(reptilian feature)

birds took to the air from the ground *up* or from the trees *down* is a question scientists continue to debate.

Other reptiles returned to the sea, including fish-eating *plesiosaurs* and *ichthyosaurs* (**Figure 22.32**). These reptiles became proficient swimmers but retained their reptilian teeth and breathed by means of lungs rather than gills.

For nearly 160 million years, dinosaurs reigned supreme. However, by the close of the Mesozoic, like many other reptiles, dinosaurs became extinct. Select reptile groups have survived to recent times, including turtles, snakes, crocodiles, and lizards. The huge, land-dwelling dinosaurs, the marine plesiosaurs, and the flying pterosaurs are known only through the fossil record. What caused this great extinction?

Demise of the Dinosaurs

The boundaries between divisions on the geologic time scale represent times of significant geological and/or biological change. Of special interest is the boundary between the Mesozoic era ("middle life") and the Cenozoic era ("recent life"), about 66 million years ago. During this transition, roughly three-quarters of all plant and animal species died out in another mass extinction. This boundary marks the end of the era in which dinosaurs and other large reptiles dominated the landscape (**Figure 22.33**) and the beginning of the era when mammals assumed that role.

as sharp teeth, clawed digits in its wings, and a long tail with many vertebrae. A recent study concluded that *Archaeopteryx* were unable to use flapping flight. Rather, by running and leaping into the air, these bird-like reptiles escaped predators with glides and downstrokes. Other researchers disagree and see them as climbing animals that glided down to the ground, following the idea that birds evolved from tree-dwelling gliders. Whether

Figure 22.32

Reptiles returned to the sea Reptiles, including *Ichthyosaur*, became the dominant marine animals during the Mesozoic. (Photo by Chip Clark)

Figure 22.33
Artist's rendering of Allosaurus Allosaurus was a large carnivorous dinosaur that lived in the late Jurassic period (155 to 145 million years ago). (Image by Roger Harris/Science Source)

What could have triggered the extinction of one of the most successful groups of land animals? An increasing number of researchers support the view that the dinosaurs fell victim to a "one–two punch." The first blow occurred during the last few million years of the Mesozoic era when climate data indicates that average temperature over the land increased by more than 20°C (40°F) over a few tens of thousand years—a blink of the eye in geologic time. This period of global warming coincided with massive basaltic eruptions that produced the Deccan Plateau, a volcanic area in present-day India. The Deccan eruptions presumably released massive amounts of carbon dioxide, which caused a period of greenhouse warming that resulted in a dramatic rise in temperatures. This period of global warming is thought to have snuffed out some species and hobbled others.

The final blow came about 66 million years ago, when our planet was struck by a stony meteorite, a relic from the formation of the solar system. The errant mass of rock was approximately 10 kilometers (6 miles) in diameter and was traveling at about 90,000 kilometers per hour at the time of impact. It collided with the southern portion of North America in a shallow tropical sea—now Mexico's Yucatan Peninsula (**Figure 22.34**).

Following the impact, suspended dust greatly reduced the amount of sunlight reaching Earth's

Figure 22.34
Chicxulub crater The Chicxulub crater is a giant impact crater that formed about 65.5 million years ago and has since filled with sediment. It is 180 kilometers (110 miles) in diameter and is the likely impact site that resulted in the demise of the dinosaurs.

UNITED STATES

Gulf of Mexico

CHICXULUB CRATER

MEXICO

Yucatán

Figure 22.35
Iridium layer: Evidence for a catastrophic impact 66 million years ago A thin layer of sediment has been discovered worldwide at Earth's physical boundary separating the Mesozoic and Cenozoic eras. This sediment contains a high level of iridium, an element that is rare in Earth's crust but found in much higher concentrations in stony meteorites. (National Parks)

surface, which resulted in global cooling ("impact winter") and inhibited photosynthesis, disrupting food production. Long after the dust settled, the sulfur oxides that were added to the atmosphere by the blast remained. Sulfate aerosols, because of their high reflectivity, perpetuated the unusually cold temperatures.

One piece of evidence that points to a catastrophic collision 66 million years ago is a thin layer of sediment, less than 1 centimeter thick, discovered in several places around the globe. This sediment contains a high level of the element *iridium*, rare in Earth's crust but found in high proportions in stony meteorites (**Figure 22.35**).

This layer is presumably the scattered remains of the meteorite responsible for the environmental changes that provided the second and final blow that led to the demise of many reptile groups.

Regardless of what caused this massive extinction, its outcome provides a valuable lesson in understanding the role that catastrophic events play in shaping our planet's physical landscape and sustainability of life-forms. The extinction of the dinosaurs opened habitats for the small mammals that survived. These new habitats, along with evolutionary forces, led to the development of the large mammals that occupy our modern world.

22.8 | Concept Checks

1. What group of plants became the dominant trees during the Mesozoic era? Name a modern descendant of this group.

2. What group of reptiles led to the evolution of modern birds?

3. What was the dominant reptile group on land during the Mesozoic?

4. Identify two reptiles that returned to life in the sea.

22.9 | Cenozoic Era: Age of Mammals

Discuss the major developments in the history of life during the Cenozoic era.

During the Cenozoic, mammals replaced reptiles as the dominant land animals. At nearly the same time, **angiosperms** (flowering plants with covered seeds) replaced gymnosperms as the dominant plants. The Cenozoic is often called the "Age of Mammals" but can also be considered the "Age of Flowering Plants" because, in the plant world, angiosperms enjoy a status similar to that of mammals in the animal world.

The development of flowering plants strongly influenced the evolution of both birds and mammals that feed on seeds and fruits, as well as many insect groups. During the middle of the Cenozoic, another type of angiosperm, grasses, developed and spread rapidly over the plains (**Figure 22.36**). This fostered the emergence of herbivorous (plant-eating) mammals, which, in turn, provided the evolutionary foundation for large predatory mammals.

During the Cenozoic, the ocean teemed with modern fish such as tuna, swordfish, and barracuda. In addition, some mammals, including seals, whales, and walruses, took up life in the sea.

From Reptiles to Mammals

The earliest mammals coexisted with dinosaurs for nearly 100 million years but were small rodent-like

A.

B.

Figure 22.36
Angiosperms became the dominant plants during the Cenozoic Angiosperms, commonly known as flowering plants, consist of a group of seed-plants that have reproductive structures called flowers and fruits. **A.** The most diverse and widespread of modern plants, many angiosperms display easily recognizable flowers. **B.** Some angiosperms, including grasses, have very tiny flowers. The expansion of the grasslands during the Cenozoic era greatly increased the diversity of grazing mammals and the predators that feed on them. (Photo A by WDG Photo/Shutterstock; Photo B by Cusp/SuperStock)

creatures that gathered food at night, when dinosaurs were probably less active. Then, about 66 million years ago, fate intervened when the large meteorite collided with Earth and dealt the final crashing blow to the reign of the dinosaurs. This transition from one dominant group to another is clearly visible in the fossil record.

Mammals are distinct from reptiles in that they are warm blooded and give birth to live young that suckle on milk. Being warm blooded allowed mammals to lead more active lives and to occupy more diverse habitats than reptiles because they could survive in cold regions. (Modern reptiles are cold blooded and dormant during cold weather. However, recent studies suggest that dinosaurs may have been warm blooded.) Other mammalian adaptations included the development of insulating body hair and more efficient hearts, lungs, and other organs.

With the demise of the large Mesozoic reptiles, Cenozoic mammals diversified rapidly. The many forms that exist today evolved from small primitive mammals that were characterized by short legs; flat, five-toed feet; and small brains. Their development and specialization took four principal directions: increase in size, increase in brain capacity, specialization of teeth to better accommodate their diet, and specialization of limbs to be better equipped for a particular lifestyle or environment.

Marsupial and Placental Mammals

Two groups of mammals, the marsupials and the placentals, evolved and diversified during the Cenozoic. The groups differ principally in their modes of reproduction. Young marsupials are born live at a very early

stage of development. At birth, the tiny and immature young enter the mother's pouch to suckle and complete their development. Today, marsupials are found primarily in Australia, where they took a separate evolutionary path largely isolated from placental mammals. Modern marsupials include kangaroos, opossums, and koalas (**Figure 22.37**).

Figure 22.37
Kangaroos, examples of marsupial mammals After the breakup of Pangaea, the Australian marsupials evolved differently than their relatives in the Americas. (Photo by Martin Harvey/Peter Arnold Inc.)

Humans: Mammals with Large Brains and Bipedal Locomotion

Both fossil and genetic evidence suggest that around 7 or 8 million years ago in Africa, several populations of anthropoids (informally called apes) diverged. One line would eventually produce modern apes such as gorillas, orangutans, and chimpanzees, while the other would produce several varieties of human ancestors. We have a good record of this evolution in fossils found in several sedimentary basins in Africa, in particular the rift valley system in East Africa.

The genus *Australopithecus*, which came into existence about 4.2 million years ago, showed skeletal characteristics that were intermediate between our apelike ancestors and modern humans. Over time, these human ancestors evolved features that suggest an upright posture and therefore a habit of walking around on two legs rather than four. Evidence for this bipedal stride includes footprints preserved in 3.2-million-year-old ash deposits at Laetoli, Tanzania (**Figure 22.38**). This new way of moving around is correlated with our human ancestors leaving forest habitat in Africa and moving to open grasslands for hunting and gathering food.

The earliest fossils of our genus, *Homo*, include *Homo habilis*, nicknamed "handy man" because their remains were often found with sharp stone tools in sedimentary deposits from 2.4 to 1.5 million years ago. *Homo habilis* had a shorter jaw and a larger brain than its ancestors. The development of a larger brain size is thought to be correlated with an increase in tool use.

During the next 1.3 million years, our ancestors developed substantially larger brains and long slender legs with hip joints adapted for long-distance walking. These species (including *Homo erectus*) ultimately gave

Placental mammals (eutherians), conversely, develop within the mother's body for a much longer period, so birth occurs when the young are comparatively mature. Members of this group include wolves, elephants, bats, manatees, and monkeys. Most modern mammals, including humans, are placental.

Figure 22.40
Mammoths These relatives of modern elephants were among the large mammals that became extinct at the close of the Ice Age. (Image courtesy of INTERFOTO/Alamy)

rise to our species, *Homo sapiens*, as well as to some extinct related species, including the Neanderthals (*Homo neanderthalis*). Despite having the same-sized brain as present-day humans and being able to fashion hunting tools from wood and stone, Neanderthals became extinct about 28,000 years ago. At one time, Neanderthals were considered a stage in the evolution of *Homo sapiens*, but that view has largely been abandoned.

Based on our current understanding, *Homo sapiens* originated in Africa about 200,000 years ago and began to spread around the globe. The oldest-known *Homo sapiens* fossils outside of Africa were found in the Middle East and date back to 115,000 years ago. Humans are known to have coexisted with Neanderthals and other prehistoric populations, with remains found in Siberia, China, and Indonesia. Further, there is mounting genetic evidence that our ancestors may have interbred with members of some of these groups.

By 36,000 years ago, humans were producing spectacular cave paintings in Europe (**Figure 22.39**). About 11,500 years ago, all prehistoric human populations except for *Homo sapiens* died out.

Large Mammals and Extinction

During the rapid mammal diversification of the Cenozoic era, some species became very large. For example, by the Oligocene epoch, a hornless rhinoceros evolved that stood nearly 5 meters (16 feet) high. It is the largest land mammal known to have existed. As time passed, many other mammals evolved to larger forms—more, in fact, than now exist. Many of these large forms were common

as recently as 11,000 years ago. However, a wave of late Pleistocene extinctions rapidly eliminated these animals from the landscape.

North America experienced the extinction of mastodons and mammoths, both huge relatives of the modern elephant (**Figure 22.40**). In addition, saber-toothed cats, giant beavers, large ground sloths, horses, giant bison, and others died out. In Europe, late Pleistocene extinctions included woolly rhinos, large cave bears, and Irish elk. Scientists remain puzzled about the reasons for this recent wave of extinctions of large animals. Because these large animals survived several major glacial advances and interglacial periods, it is difficult to ascribe extinctions of these animals to climate change. Some scientists hypothesize that early humans hastened the decline of these mammals by selectively hunting large forms.

22.9 Concept Checks

1. What animal group became the dominant land animals of the Cenozoic era?

2. Explain how the demise of the large Mesozoic reptiles impacted the development of mammals.

3. Where has most of the evidence for the early evolution of our ancestors been discovered?

4. What two characteristics best separate humans from other mammals?

5. Describe one hypothesis that explains the extinction of large mammals in the late Pleistocene.

22.1 Is Earth Unique?

List the principle characteristics that make Earth unique among the planets.

- As far as we know, Earth is unique among planets in the fact that it hosts life. The planet's size, composition, and location all contribute to conditions that support life (or at least our kind of life).

22.2 Birth of a Planet

Outline the major stages in Earth's evolution, from the Big Bang to the formation of our planet's layered internal structure.

KEY TERMS supernova, solar nebula, planetesimal, protoplanet, Hadean

- The universe is thought to have formed around 13.7 billion years ago, with the Big Bang, which generated space, time, energy, and matter. Early stars grew from the lightest elements, hydrogen and helium, and the process of nuclear fusion produced the other low-mass elements. Some large stars exploded in supernovae, generating heavier atoms and spewing them into space.
- Earth and the solar system formed around 4.6 billion years ago, with the contraction of a solar nebula. Collisions between clumps of matter in this spinning disc resulted in the growth of planetesimals and then protoplanets. Over time, the matter of the solar nebula was concentrated into a smaller number of larger bodies: the Sun, the rocky inner planets, the icy outer planets, moons, comets, and asteroids.
- Heat production within Earth during its formative years was much higher than it is today, thanks to the kinetic energy of impacting asteroids and planetesimals, as well as the decay of short-lived radioactive isotopes. These high temperatures caused rock and iron to melt. This allowed iron to sink to form Earth's core and rocky material to rise to form the mantle and crust.

Comet Shoemaker-Levy 9 impacted Jupiter in 1994.

NASA

Q The accompanying image shows Comet Shoemaker-Levy 9 impacting Jupiter in 1994. After this event, what happened to Jupiter's total mass? How was the number of objects in the solar system affected?

22.3 Origin and Evolution of the Atmosphere and Oceans

Describe how Earth's atmosphere and oceans formed and evolved through time.

KEY TERMS outgassing, banded iron formation, Great Oxygenation Event

- Earth's atmosphere is essential for life. It evolved as volcanic outgassing added mainly water vapor and carbon dioxide to the primordial atmosphere of gases common in the early solar system: methane and ammonia.

- Free oxygen began to accumulate partly through photosynthesis by cyanobacteria, which released oxygen as a waste product. Much of this early oxygen immediately reacted with iron dissolved in seawater and settled to the ocean floor as chemical sediments called banded iron formations. The Great Oxygenation Event of 2.5 billion years ago marks the first evidence of significant amounts of free oxygen in the atmosphere.
- Earth's oceans formed after the planet's surface had cooled. Soluble ions weathered from the crust were carried to the ocean, making it salty. The oceans also absorbed tremendous amounts of carbon dioxide from the atmosphere.

22.4 Precambrian History: The Formation of Earth's Continents

Explain the formation of continental crust, how continental crust becomes assembled into continents, and the role that the supercontinent cycle has played in this process.

KEY TERMS craton, shield, supercontinent, supercontinent cycle

- The Precambrian includes the Archean and Proterozoic eons. However, our knowledge of what occurred during these eons of geologic time is rather limited because erosion destroyed much of the rock record.
- Continental crust was produced over time through the recycling of ultramafic and mafic crust in an early version of plate tectonics. Small crustal fragments formed and amalgamated into large crustal provinces called cratons. Over time, North America and other continents grew through the accretion of new terranes around the edges of this central "nucleus" of crust.
- Early cratons not only merged but sometimes rifted apart. The supercontinent Rodinia formed around 1.1 billion years ago and then rifted apart, opening new ocean basins. Over time, these ocean basins also closed and formed a new supercontinent called Pangaea around 250 million years ago. Like Rodinia before it, Pangaea broke up as part of the ongoing supercontinent cycle.
- The formation of elevated oceanic ridges upon the breakup of a supercontinent displaced enough water that sea level rose, and shallow seas flooded the continents. The breakup of continents can also influence the direction of ocean currents, with important consequences for climate.

Q Consult Figure 22.12 and briefly summarize the history of the assembly of North America over the past 3.5 billion years.

22.5 Geologic History of the Phanerozoic: The Formation of Earth's Modern Continents

List and discuss the major geologic events in the Paleozoic, Mesozoic, and Cenozoic eras.

KEY TERMS Pangaea, Laurasia, Gondwana

- The Phanerozoic eon encompasses the Paleozoic, Mesozoic, and Cenozoic eras, which cover the past 542 million years of geologic time.
- In the Paleozoic era, North America experienced a series of collisions that resulted in the rise of the young Appalachian mountain belt and the assembly of Pangaea. High sea levels caused the ocean to cover vast areas of the continent and resulted in a thick sequence of sedimentary strata.
- During the Mesozoic, Pangaea broke up, and the Atlantic Ocean began to form. As the continent moved westward, the Cordillera began to rise due to subduction and the accretion of terranes along the west coast. In the Southwest, vast deserts deposited thick layers of dune sand, while environments in the East were conducive to the formation and subsequent burial of coal swamps.
- In the Cenozoic era, a thick sequence of sediments was deposited along the Atlantic margin and the Gulf of Mexico. Meanwhile, western North America experienced an extraordinary episode of crustal extension; the Basin and Range province resulted.

Q Contrast the tectonics of eastern and western North America during the Mesozoic era.

22.6 Earth's First Life

Describe some of the hypotheses on the origin of life and the characteristics of early prokaryotes, eukaryotes, and multicelled organisms.

KEY TERMS prokaryote, stromatolite, eukaryote

- Life began from nonlife. Amino acids are a necessary building block for proteins. They may have been assembled with energy from ultraviolet light or lightning, or in a hot spring, or on another planet, only to be delivered later to Earth on meteorites.
- The first organisms were relatively simple single-celled prokaryotes that thrived in low-oxygen environments. They formed around 3.8 billion years ago. The advent of photosynthesis allowed microbial mats to build up and form stromatolites.
- Eukaryotes have larger, more complex cells than prokaryotes. The oldest eukaryotic cells formed around 2.1 billion years ago. Eventually, some eukaryotic cells linked together and differentiated their structures and functions, producing the earliest multicellular organisms.

Q The accompanying image shows fossil stromatolites. How did these ancient cyanobacteria transform our planet?

Biophoto Associates/Science Source

22.7 Paleozoic Era: Life Explodes

List the major developments in the history of life during the Paleozoic era.

KEY TERMS Cambrian explosion, mass extinction

- At the beginning of the Cambrian period, abundant fossil hard parts appear in sedimentary rocks. The source of these shells and other skeletal material were a profusion of new animals, including trilobites and cephalopods.
- Plants colonized the land around 400 million years ago and soon diversified into forests.
- In the Devonian, some lobe-finned fish began to spend time out of water and gradually evolved into the first amphibians. A subset of the amphibian population evolved waterproof skin and shelled eggs and split off to become the reptile line.
- The Paleozoic era ended with the largest mass extinction in the geologic record. This deadly event may have been related to the eruption of the Siberian Traps flood basalts.

Q What advantages do reptiles have over amphibians? What advantages do amphibians have over fish?

22.8 Mesozoic Era: Age of the Dinosaurs

Briefly explain the major developments in the history of life during the Mesozoic era.

KEY TERM gymnosperm

- Plants diversified during the Mesozoic. The flora of that time was dominated by gymnosperms, the first plants with seeds that allowed them to migrate beyond the edges of water bodies.
- Reptiles diversified, too. The dinosaurs came to dominate the land, pterosaurs dominated the air, and a suite of several different marine reptiles swam the seas. The first birds evolved during the Mesozoic, exemplified by *Archaeopteryx*, a transitional fossil.
- As with the Paleozoic, the Mesozoic ended with a mass extinction, probably due to a massive meteorite impact in what is now Chicxulub, Mexico.

22.9 Cenozoic Era: Age of Mammals

Discuss the major developments in the history of life during the Cenozoic era.

KEY TERM angiosperm

- Once the giant Mesozoic reptiles were extinct, mammals were able to diversify on the land, in the air, and in the oceans. Mammals are warm blooded, have hair on their bodies, and nurse their young with milk. Marsupial mammals are born very young and then move to a pouch on the mother, while placental mammals spend a longer time *in utero* and are born in a relatively mature state compared to marsupials.
- Flowering plants, called angiosperms, diversified and spread around the world through the Cenozoic era.
- Humans evolved from primate ancestors in Africa over a period of 8 million years. They are distinguished from their ape ancestors by an upright, bipedal posture, large brains, and tool use. The oldest anatomically modern human fossils are 200,000 years old. Some of these humans migrated out of Africa and coexisted with Neanderthals and other human-like populations.

Give It Some *Thought*

1. Refer to the geologic time scale in Figure 22.3. The Precambrian accounts for nearly 90 percent of geologic time. Why do you think it has fewer divisions than the rest of the time scale?

2. Referring to Figure 22.4, write a brief summary of the events that led to the formation of Earth.

3. The accompanying photograph shows layered iron-rich rocks called banded iron formations. What does the existence of these 2.5-billion-year-old rocks tell us about the evolution of Earth's atmosphere?

Blue Gum Pictures/Alamy

4. Describe two ways in which the sudden appearance of oxygen in the atmosphere about 2.5 billion years ago influenced the development of modern life-forms.

5. Currently, oceans cover about 71 percent of Earth's surface. However, early in Earth history the oceans covered a greater percentage of Earth's surface. Explain.

6. Contrast the eastern and western margins of North America during the Cenozoic era in terms of their relationships to plate boundaries.

7. Suggest at least one reason why plants moved onto land before animals.

8. Five mass extinctions, in which 50 percent or more of Earth's marine species became extinct, are documented in the fossil record. Use the accompanying graph, which depicts the time and extent of each mass extinction, to answer the following:

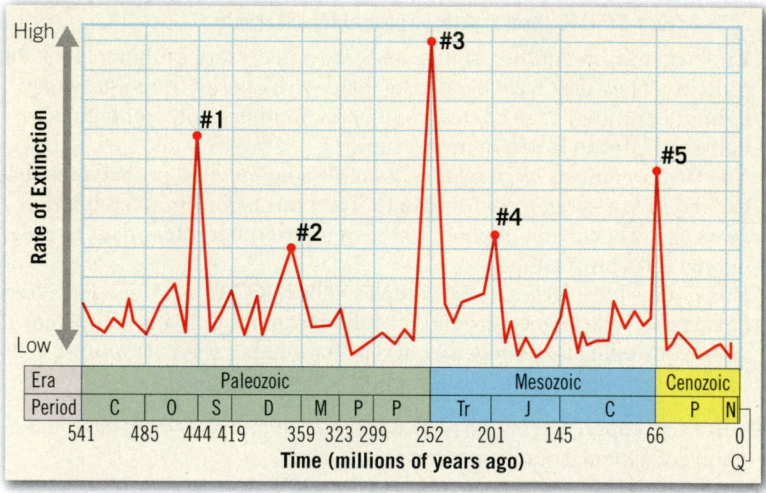

a. Which of the five mass extinctions was the *most extreme*? Identify this extinction by name and when it occurred.

b. What group of animals was most affected by the extinction referred to in Question a?

c. When did the *most recent* mass extinction occur?

d. During the most recent mass extinction, what prominent animal group was eliminated?

e. What animal group experienced a major period of diversification following the most recent mass extinction?

9. Some scientists have proposed that the environments around black smokers may be similar to the extreme conditions that existed early in Earth history. Therefore, these scientists look to the unusual life that exists around black smokers for clues about how the earliest life may have survived. Compare and contrast the environment of a black smoker to the environment on Earth approximately 3 to 4 billion years ago. Do you think there are parallels between the two, and if so, do you think black smokers are good examples of the environment that earliest life may have experienced? Explain.

10. Between 300 and 250 million years ago, plate movement assembled all the previously separated landmasses together to form the supercontinent Pangaea. The formation of Pangaea resulted in deeper ocean basins and a drop in sea level, causing shallow coastal areas to dry up. Thus, in addition to rearranging the geography of our planet, continental drift had a major impact on life on Earth. Use the accompanying diagrams and the information above to answer the following:

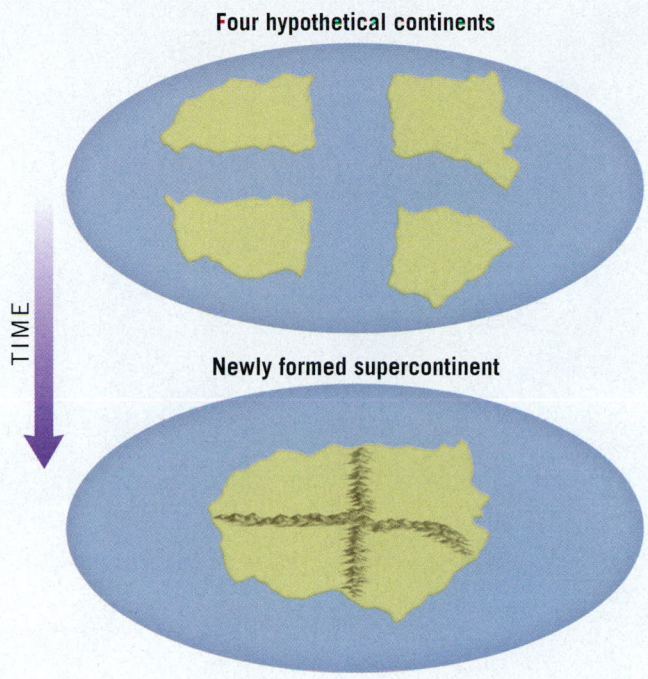

Four hypothetical continents

TIME

Newly formed supercontinent

a. Which of the following types of habitats would likely diminish in size during the formation of a supercontinent: deep-ocean habitats, wetlands, shallow marine environments, or terrestrial (land) habitats? Explain.

b. During the breakup of a supercontinent, what would happen to sea level—would it remain the same, rise, or fall?

c. Explain how and why the development of an extensive oceanic ridge system that forms during the breakup of a supercontinent affects sea level.

11. Suggest a geologic reason why the rift valley system of East Africa is so rich in human ancestor fossils.

12. Most of the vast North American coal resources located from Pennsylvania to Illinois formed during the Pennsylvanian and Mississippian periods of Earth history. (This time period is also known as the Carboniferous period.) Examine the accompanying diagram, which illustrates the geographic position of North America during the period of coal formation.

a. Where, relative to the equator, was North America located during the time of coal formation?

b. Why is it unlikely that a similar coal-forming environment will repeat itself in North America in the near future?

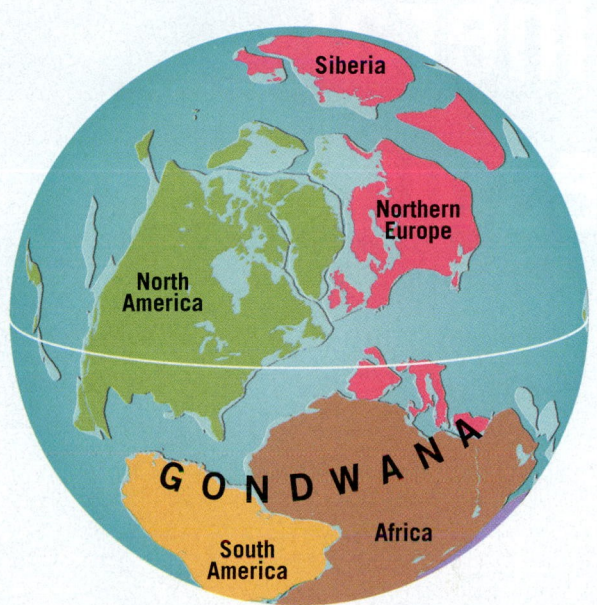

Siberia

Northern Europe

North America

G O N D W A N A

Africa

South America

MasteringGeology™

23

Energy and Mineral Resources

In 2014, renewable energy consumption represented about 10 percent of the U.S. total—a record high. Wind energy is a fast-growing part of this category. These wind turbines are in California's Mojave Desert. (Photo by John W. Warden)

Each statement represents the primary **LEARNING OBJECTIVE** for the corresponding major heading within the chapter. After you complete the chapter, you should be able to:

23.1 Distinguish between renewable and nonrenewable resources.

23.2 Compare and contrast fossil fuel types and describe how each satisfies U.S. energy consumption.

23.3 Describe the importance of nuclear energy and discuss its pros and cons.

23.4 List and discuss the major sources of renewable energy. Describe the contribution of renewable energy to the overall U.S. energy supply.

23.5 Distinguish among resource, reserve, and ore.

23.6 Explain how different igneous and metamorphic processes produce economically significant mineral deposits.

23.7 Discuss ways in which surface processes produce ore deposits.

23.8 Distinguish between two broad categories of nonmetallic mineral resources and list examples of each.

Materials that we extract from Earth are the basis of modern civilization. A high percentage of the products and substances we use every day were mined from Earth's crust and/or manufactured using energy extracted from the crust. We humans consume huge quantities of energy and mineral resources, and demand continues to grow. In 1960, the world population was 3 billion. In 2015, Earth was home to nearly 7.4 billion people. By 2025, the planet's population is expected to approach 8 billion. Improving living standards in many parts of the world add to resource demand as well.

How long can our remaining resources sustain the rising standard of living in today's industrialized countries and still provide for the growing needs of developing regions? How much environmental deterioration are we willing to accept in pursuit of resources? Can alternatives be found? If we are to satisfy an increasing per capita demand and a growing world population, we must understand our resources and their limits.

23.1 | Renewable and Nonrenewable Resources

Distinguish between renewable and nonrenewable resources.

Resources are commonly divided into two broad categories: renewable and nonrenewable. **Renewable resources** can be replenished over relatively short time spans such as months, years, or decades. Common examples are plants and animals for food, natural fibers for clothing, and trees for lumber and paper. Energy from flowing water, wind, and the Sun are also considered renewable.

By contrast, **nonrenewable resources** continue to be formed in Earth, but the processes that create them are so slow that significant deposits take millions of years to accumulate. For human purposes, Earth contains fixed quantities of these substances. When the present supplies are mined or pumped from the ground, there will be no more. Examples are fossil fuels (for example, coal, oil, natural gas) and many important metals (for example, iron, copper, uranium, gold). Some of these nonrenewable resources, such as aluminum, can be used over and over again; others, such as oil, cannot be recycled.

Occasionally some resources can be placed in either category, depending on how they are used. Groundwater is one such example. Where it is pumped from the ground at a rate that can be replenished, groundwater can be classified as a renewable resource. However, in places where groundwater is withdrawn faster than it is replenished, the water table drops steadily. In this case, the groundwater is being "mined" just like other nonrenewable resources. The problem of declining water-table levels is discussed in Chapter 17.

Most of the energy and mineral resources we use are nonrenewable. How much of these does the average

Figure 23.1
How much do each of us use? The annual per capita consumption of nonmetallic and metallic resources for the United States is about 11,000 kilograms (12 tons). About 97 percent of the materials used are nonmetallic. The per capita use of oil, coal, and natural gas exceeds 11,000 kilograms. (U.S. Geological Survey)

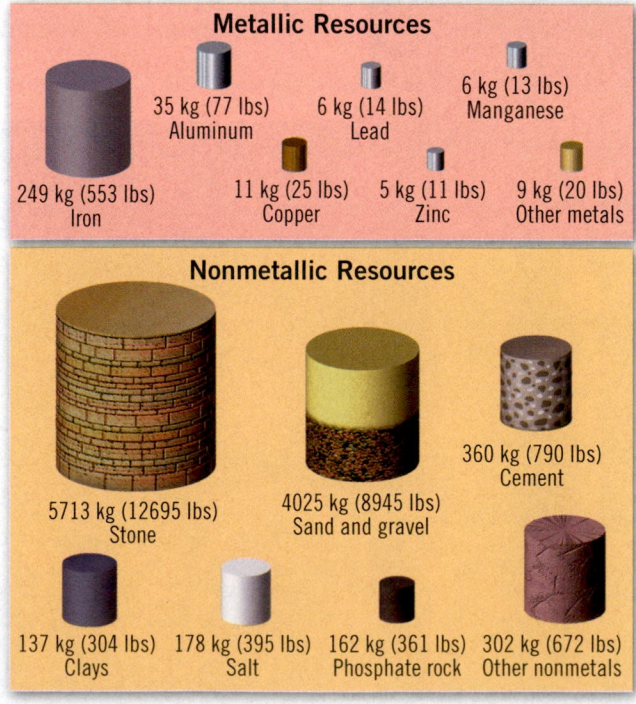

Metallic Resources

249 kg (553 lbs) Iron
35 kg (77 lbs) Aluminum
11 kg (25 lbs) Copper
6 kg (14 lbs) Lead
5 kg (11 lbs) Zinc
6 kg (13 lbs) Manganese
9 kg (20 lbs) Other metals

Nonmetallic Resources

5713 kg (12695 lbs) Stone
4025 kg (8945 lbs) Sand and gravel
360 kg (790 lbs) Cement
137 kg (304 lbs) Clays
178 kg (395 lbs) Salt
162 kg (361 lbs) Phosphate rock
302 kg (672 lbs) Other nonmetals

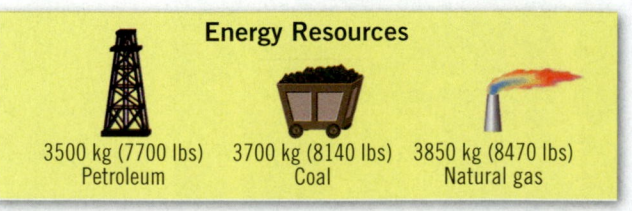

Energy Resources

3500 kg (7700 lbs) Petroleum
3700 kg (8140 lbs) Coal
3850 kg (8470 lbs) Natural gas

person in the United States use each year? **Figure 23.1** shows the estimated annual per capita consumption of several important mineral and energy resources. This is each person's prorated share of the materials required by industry to provide the vast array of homes, cars, electronics, cosmetics, packaging, and so on that modern society requires. Figures for other highly industrialized countries, such as Canada, Australia, and several nations in Western Europe are comparable.

23.2 | Energy Resources: Fossil Fuels
Compare and contrast fossil fuel types and describe how each satisfies U.S. energy consumption.

Earth's tremendous industrialization over the past two centuries has been fueled—and is still fueled—by burning coal, petroleum, and natural gas. About 81 percent of the energy consumed in the United States today comes from these basic sources. **Figure 23.2** shows where our energy comes from and what it is used for. Our reliance on fossil fuels is obvious.

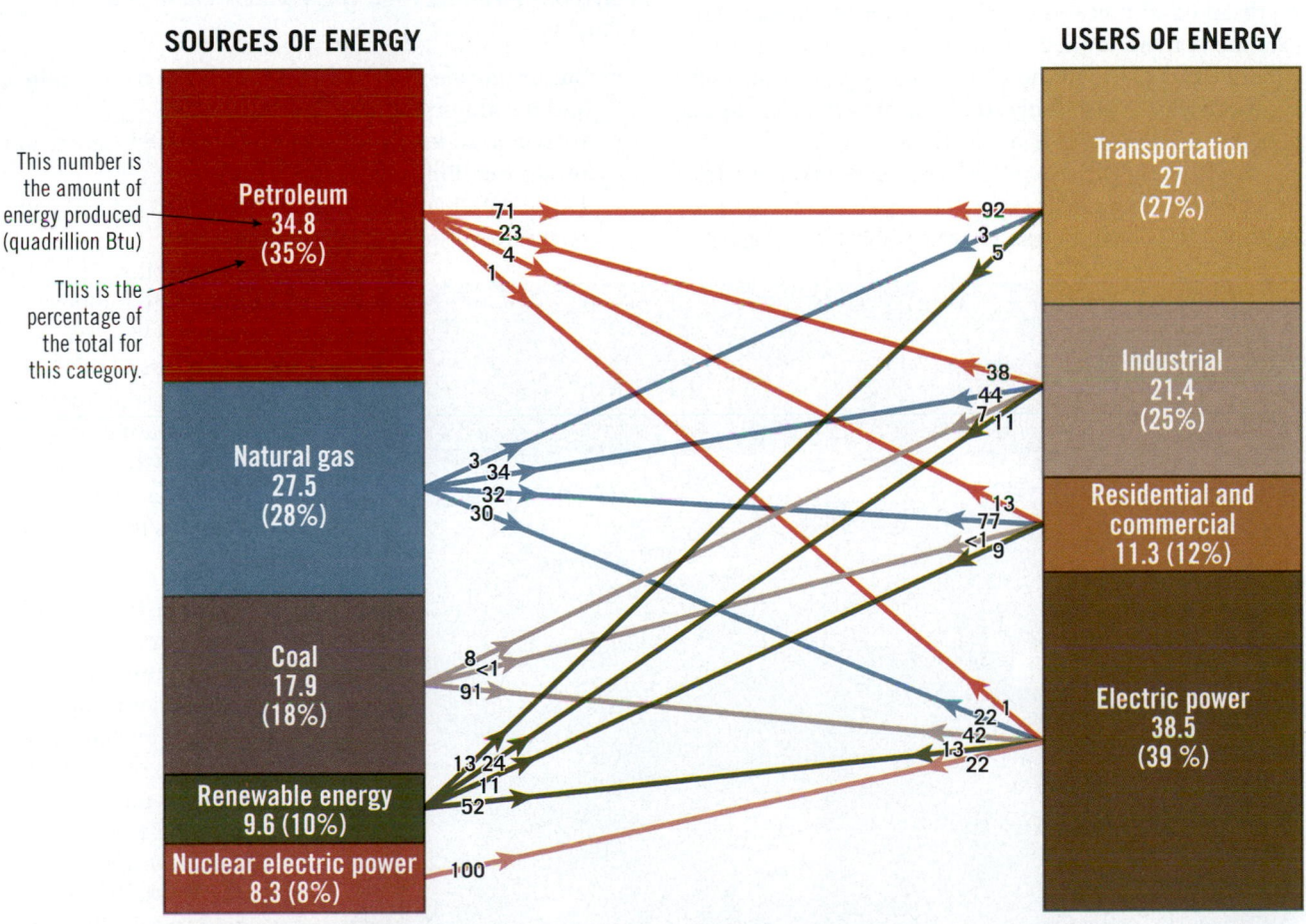

SmartFigure 23.2
U.S. energy consumption, 2014 The total was 98.3 quadrillion Btu (British thermal units). A quadrillion is 10 raised to the 12th power, or a million million. A quadrillion Btu is a convenient unit for referring to U.S. energy use as a whole. (U.S. Department of Energy/Energy Information Agency) (https://goo.gl/6gUJ3J)

Reading this double graph:
The left side indicates what energy sources we use. The right side shows where we use the energy.
The lines with numbers that connect the graphs provide more details. Use the top line as an example. It shows that 71% of the petroleum is used by the transportation sector. It also indicates that 92% of the energy used by the transportation sector is petroleum.

Coal

Along with oil and natural gas, coal is commonly called a **fossil fuel**. This designation is appropriate because each time we burn coal, we are using energy from the Sun that was stored by plants many millions of years ago—so we are indeed burning a "fossil." A discussion of coal formation can be found in Chapter 7, pages 225–226.

Coal has been an important fuel for centuries. In the nineteenth and early twentieth centuries, cheap and plentiful coal powered the Industrial Revolution. By 1900 coal was providing 90 percent of the energy used in the United States. Although still important, coal currently accounts for a much smaller proportion of the nation's energy needs: about 18 percent.

Until the 1950s, coal was an important domestic heating fuel as well as a power source for industry. However, its direct use in the home has been largely replaced by oil, natural gas, and electricity. These fuels are preferred because they are more readily available (delivered via pipes, tanks, or wiring) and cleaner to use.

Nevertheless, coal remains the major fuel used in power plants to generate electricity and is therefore indirectly an important source of energy for our homes. An examination of Figure 23.2 shows that in 2014, 91 percent of the coal use in the United States was for generating electricity. It also shows that coal was the fuel used to generate 42 percent of U.S. electrical power.

In the decades to come, the percentage of our electricity generated from coal is expected to decrease because of greater use of natural gas and renewable technologies.

However, this does not mean that we will use less coal. Because overall demand for electricity will increase, the actual amount of coal used may increase as well. Expanded coal production is possible because the world has enormous reserves and the technology to mine coal efficiently. In the United States, coal fields are widespread and contain supplies that should last for hundreds of years. GEOgraphics 23.1 focuses on coal production in the United States.

Coal is plentiful, but its recovery and use present a number of challenges. Surface mining can turn the countryside into a scarred wasteland if careful (and costly) reclamation is not carried out to restore the land. Today all U.S. surface mines are required to reclaim the land. Although underground mining does not scar the landscape to the same degree, it has been costly in terms of human life and health. Strong federal safety regulations have made U.S. mining quite safe. However, collapsing roofs, gas explosions, and the required heavy equipment remain hazards. Over the years, the share of coal produced from surface mines has increased significantly, from 51 percent in 1949 to about 65 percent in 2014 (**Figure 23.3**).

Burning coal produces emissions that adversely influence the environment and human health. Principal emissions resulting from coal combustion include the following:

- Sulfur dioxide (SO_2), which contributes to acid rain and respiratory illnesses
- Nitrogen oxides (NO_x), which contribute to smog and respiratory illnesses
- Particulate matter, which contributes to smog, haze, respiratory illnesses, and lung disease
- Carbon dioxide (CO_2), the primary greenhouse gas produced from the burning of fossil fuels, which plays a significant role in the heating of our atmosphere (Chapter 21, "Global Climate Change," examines this issue in some detail.)

The coal industry has found several ways to reduce sulfur, nitrogen oxides, and other impurities from coal, and it has developed more effective ways of cleaning coal after it is mined. Coal consumers have shifted toward greater use of less-polluting low-sulfur coal. Yet significant challenges remain.

Figure 23.3
Surface coal mine About 65 percent of U.S. coal production is from surface mines. Wyoming is by far the largest producer, accounting for nearly 40 percent of the coal mined in the United States in 2013. This mine is in Campbell County, Wyoming. (Photo by David R. Frazier photography, Inc./Alamy Images)

Coal
A Major Energy Source

In 2014 the United States produced nearly 1 billion tons of coal. It will likely remain a major energy source for many years to come.

Where is coal mined in the United States?

MT · ND · WY

WYOMING

Wyoming is far and away the largest producer of coal in the United States. With more than 39% of U. S. production, Wyoming is home to the eight largest mines in the country.

IL · IN · OH · PA · WV · KY · TX

Glow Images

The top 10 coal-producing states in 2014
(thousands of tons/percentage of total)

Wyoming	387,924	39.4%
West Virginia	112,786	11.5%
Kentucky	80,380	8.2%
Pennsylvania	54,009	5.5%
Illinois	52,147	5.3%
Texas	42,851	4.4%
Montana	42,231	4.3%
Indiana	39,102	4.0%
North Dakota	27,639	2.8%
Ohio	25,113	2.5%

What type of coal is mined?

Subbituminous coal is 35 to 45% carbon. Large quantities occur in thick beds near the surface, resulting in low mining costs. Wyoming is the major source of this type.

Bituminous coal Coal classified as bituminous has the widest range of carbon content—45 to 86%. West Virginia leads production, followed by Kentucky and Pennsylvania.

47% · 45% · 0.2% · 7%

Anthracite coal has the highest carbon content—86 to 97%. It is relatively rare in the United States and is more difficult to mine. All of the mines are in northeastern Pennsylvania.

Lignite coal has a low carbon content of 25 to 35% and therefore the lowest energy content of the four types. Texas and North Dakota are the leading producers.

Coal fields of the United States
Production by region in millions of tons
(percentage of total)

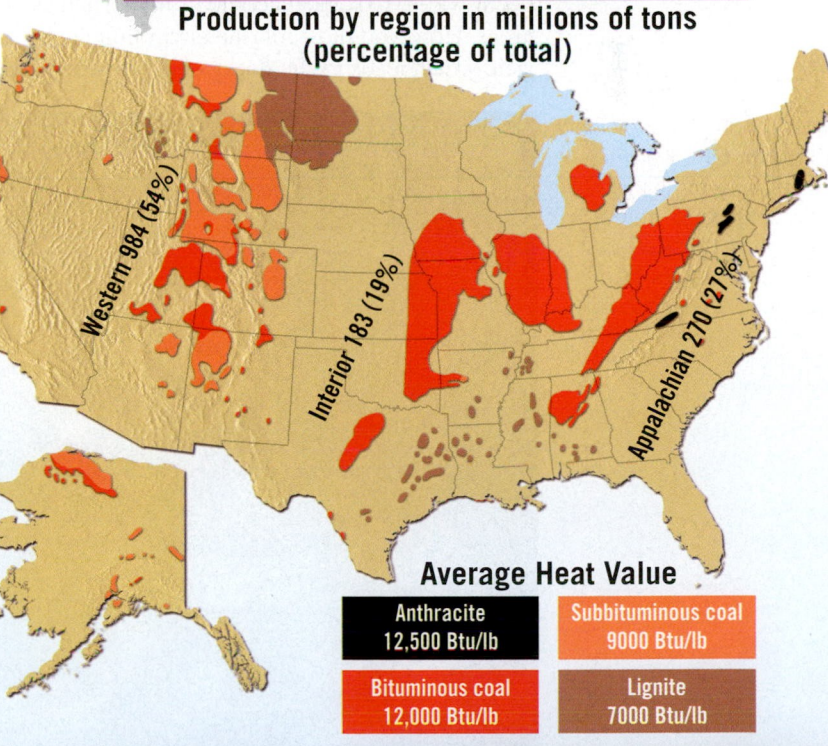

Western 984 (54%) · Interior 183 (19%) · Appalachian 270 (27%)

Average Heat Value

Anthracite 12,500 Btu/lb	**Subbituminous coal** 9000 Btu/lb
Bituminous coal 12,000 Btu/lb	**Lignite** 7000 Btu/lb

Questions:
1. In what state is the most coal mined?
2. What is the primary type of coal mined in that state?

Oil and Natural Gas

Together, oil and natural gas provided more than 60 percent of the energy consumed in the United States in 2014. By examining Figure 23.2, you can see that the transportation sector of the U.S. economy relies almost totally on petroleum as an energy source. In 2011, natural gas surpassed coal for the first time in more than 30 years as a source of energy in the United States. An important reason for this is the use of new technologies that have increased production from shale formations. Users of natural gas are almost evenly divided among the three categories other than transportation in Figure 23.2.

Petroleum Formation   Petroleum and natural gas are found in similar environments and frequently occur together. Both consist of various hydrocarbon compounds (compounds consisting of hydrogen and carbon mixed together). They may also contain relatively small quantities of other elements, such as sulfur, nitrogen, and oxygen. Like coal, petroleum and natural gas are biological products derived from the remains of organisms. However, the environments in which they form are very different, as are the organisms. Coal is formed mostly from plant material that accumulated in a swampy environment above sea level (see Figure 7.20, page 225), while oil and gas are derived from the remains of both plants and animals having a marine origin.

The formation of oil and natural gas begins with the accumulation of sediment in ocean areas that are rich in plant and animal remains. These accumulations must occur where biological activity is high, such as nearshore areas. However, most marine environments are oxygen rich, which leads to the decay of organic remains before they can be buried by other sediments. Therefore, accumulations of oil and gas are not as widespread as are the marine environments that support abundant biological activity. This limiting factor notwithstanding, large quantities of organic matter are buried and protected from oxidation in many offshore sedimentary basins. With ever-deeper burial over millions of years, chemical reactions gradually transform some of the original organic matter into the liquid and gaseous hydrocarbons we call petroleum and natural gas.

Unlike the organic matter from which they formed, the newly created petroleum and natural gas are mobile. These fluids are gradually squeezed from the compacting, mud-rich layers where they originate into adjacent permeable beds such as sandstone, where openings between sediment grains are larger. Because this occurs underwater, the rock layers containing the oil and gas are saturated with water. Because oil and gas are less dense than water, they migrate upward through the water-filled pore spaces of the enclosing rocks. Unless something halts this upward migration, the fluids will eventually reach the surface, at which point the volatile components will evaporate.

Traps for Oil and Gas   Sometimes the upward migration of oil and natural gas is halted. A geologic environment that allows for economically significant amounts of oil and gas to accumulate underground is termed an **oil trap**. Several geologic structures may act as oil traps, but all have two basic conditions in common: a porous, permeable **reservoir rock** that will yield petroleum and natural gas in sufficient quantities to make drilling worthwhile, and a **cap rock**, such as shale, that is virtually impermeable to oil and gas. The cap rock halts the upwardly

Figure 23.4
Drilling for oil Oil accumulates in oil traps that consist of porous, permeable *reservoir rock* overlain by an impermeable *cap rock*. (North Sea photo by Peter Bowater/Photo Researchers, Inc.; historic photo by Hulton Archive/Getty Images)

Modern offshore oil production platform in the North Sea.

The first successful oil well was completed by Edwin Drake (right) on August 27, 1859, near Titusville, PA. The oil-bearing reservoir rock was encountered at a Depth of 21 meters (69 feet).

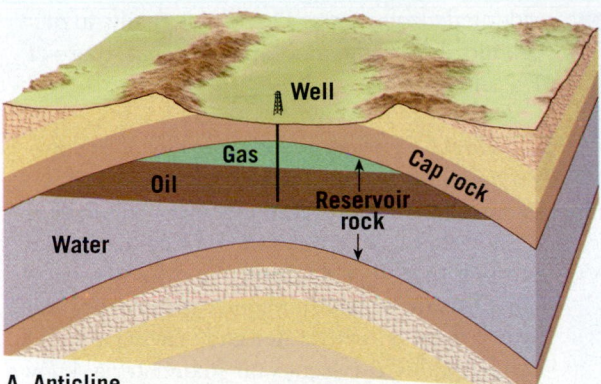

A. Anticline

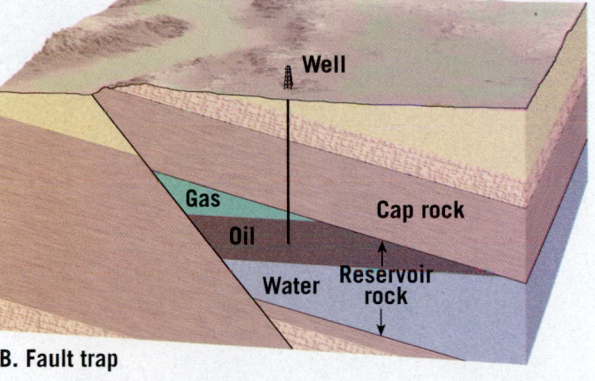

B. Fault trap

SmartFigure 23.5
Common oil traps
(https://goo.gl/39yDY8)

Tutorial

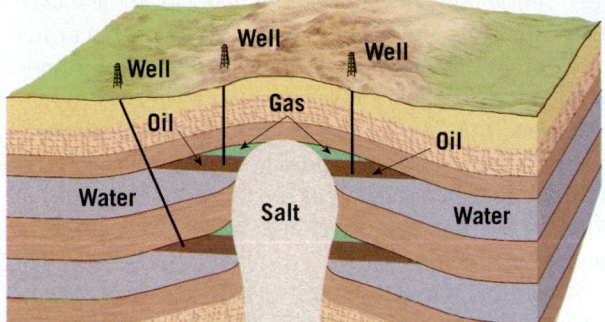

C. Salt dome

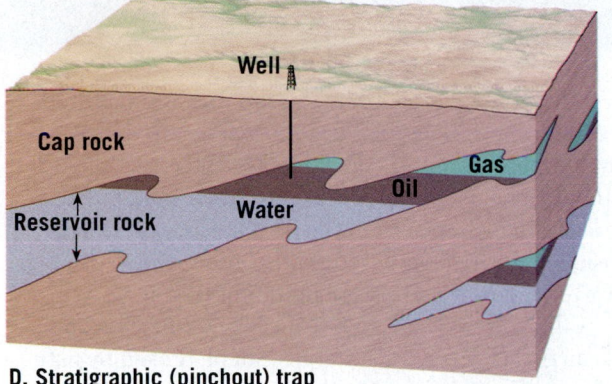

D. Stratigraphic (pinchout) trap

mobile oil and gas and keeps both from escaping at the surface (**Figure 23.4**). **Figure 23.5** illustrates some common oil and natural gas traps, described in the following list:

- **Anticline.** One of the simplest traps is an *anticline*, an uparched series of sedimentary strata (Figure 23.5A). As the strata are bent, the rising oil and gas collect at the apex (top) of the fold. Because of its lower density, the natural gas collects above the oil. Both rest upon the denser water that saturates the reservoir rock. One of the world's largest oil fields, El Nala in Saudi Arabia, is the result of an anticlinal trap, as is the famous Teapot Dome in Wyoming.
- **Fault trap.** When strata are displaced in such a manner as to bring a dipping reservoir rock into position opposite an impermeable bed, a *fault trap* forms, as shown in Figure 23.5B. In this case, the upward migration of the oil and gas is halted where it encounters the fault.
- **Salt dome.** In the Gulf coastal plain region of the United States, important accumulations of oil occur in association with *salt domes*. Such areas have thick accumulations of sedimentary strata, including layers of rock salt. Salt occurring at great depths was forced to rise in columns by the pressure of overlying beds. These rising salt columns gradually deform the overlying strata. Because oil and gas migrate to the highest level possible, they accumulate in the upturned sandstone beds adjacent to the salt column (Figure 23.5C).
- **Stratigraphic (pinchout) trap.** Yet another important geologic circumstance that may lead to significant accumulations of oil and gas is termed

a *stratigraphic trap*. These oil-bearing structures result primarily from the original pattern of sedimentation rather than structural deformation. The stratigraphic trap illustrated in Figure 23.5D exists because a sloping bed of sandstone thins to the point of disappearance.

How do geologists locate oil traps? Recall from Chapter 12 that changes in the composition or structure of rock cause seismic waves to reflect off boundaries between layers. This characteristic of waves is especially useful in the exploration for oil and natural gas, as artificially generated seismic waves can be used to probe the crust (**Figure 23.6**). Oil and natural gas would be much more difficult and expensive to find without seismic imaging because a huge number of wells would have to be randomly drilled to locate oil traps.

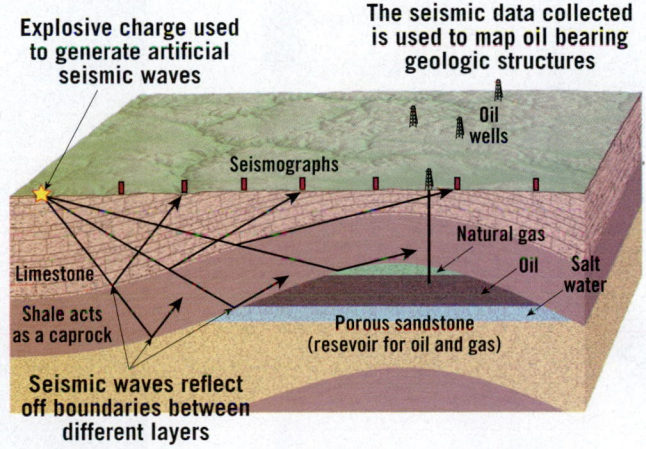

Figure 23.6
Seismic search for oil and natural gas Reflected seismic waves are used to search for underground reservoirs of oil and natural gas. The seismic waves from explosions reflect off the boundaries between layers of different composition. Using computer programs, the data show the geometry of the strata, including folds and faults. Using this information, geologists map potential petroleum reservoirs in Earth's crust.

Figure 23.7
A close-up of oil sand Notice how solid the material is. Sometimes the term *tar sand* is used to describe bitumen deposits, but this is inaccurate because tar is a human-made substance produced by distilling organic matter. Bitumen looks like tar but is a naturally occurring substance. (Petter Essick/ Aurora Photos/Alamy Images)

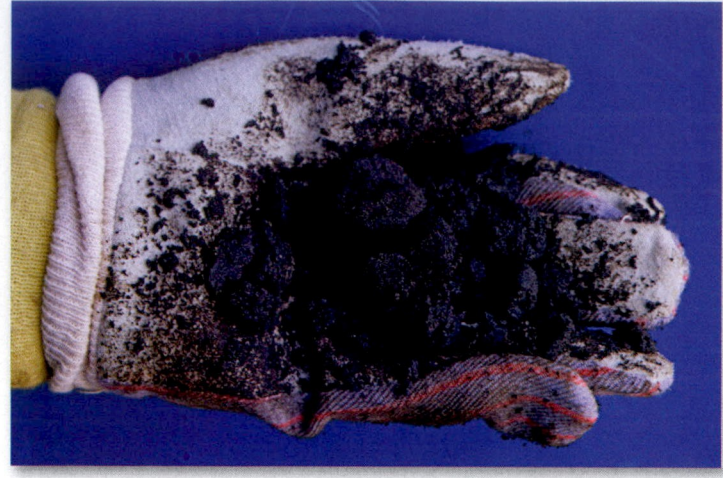

When the lid created by the cap rock is punctured by drilling, the oil and natural gas, which are under pressure, migrate from the pore spaces of the reservoir rock to the drill hole. On rare occasions, when fluid pressure is great, it may force oil up the drill hole to the surface, causing a "gusher" or oil fountain at the surface. Usually, however, a pump is required to lift out the oil.

A drill hole is not the only means by which oil and gas can escape from a trap. Traps can be broken by natural forces. For example, Earth movements may create fractures that allow the hydrocarbon fluids to escape. Surface erosion may breach a trap with similar results. The older the rock strata, the greater the chance that deformation or erosion has affected a trap. Indeed, not all ages of rock yield oil and gas in the same proportions. The greatest production comes from the youngest rocks, those of Cenozoic age. Older Mesozoic rocks produce considerably less, followed by even smaller yields from the still older Paleozoic strata. There is virtually no oil produced from the most ancient rocks, those of Precambrian age.

Oil Sands Oil sands are a somewhat unconventional yet significant source of oil that is likely to become increasingly important in the decades to come. Unlike the oil that accumulates in traps, this oil is obtained in a different manner. **Oil sands** are usually mixtures of clay and sand combined with water and varying amounts of a black, highly viscous tar-like material known as *bitumen*. The use of the term *sand* can be misleading because not all deposits are associated with sands and sandstones. Some occur in other materials, including shales and limestones. The oil in these deposits is very similar to heavy crude oils pumped from wells. The major difference between conventional oil reservoirs and oil sand deposits is in the viscosity (resistance to flow) of the oil they contain. In oil sands, the oil is much more viscous and cannot simply be pumped out (**Figure 23.7**).

Substantial oil sand deposits occur in several locations around the world. By far the largest are in the Canadian province of Alberta (**Figure 23.8**). Some oil sands are mined at the surface in a manner similar to the strip mining of coal. The excavated material is then heated with pressurized steam until the bitumen softens and rises. Once collected, the oily material is treated to remove impurities, and then hydrogen is added. This last step upgrades the material to a synthetic crude, which can then be refined. Extracting and refining oil sands requires a great deal of energy—nearly half as much as the end product yields! Nevertheless, oil sands from Alberta's vast deposits are the source of about 50 percent of Canada's oil production. In 2013, nearly 2 million barrels of crude bitumen were produced per day. It is estimated that this amount will reach 3.7 million barrels per day by 2020 and 5.2 million barrels per day by 2030.

SmartFigure 23.8
Alberta's oil sands Huge oil reserves occur across more than 140,000 square kilometers (54,000 square miles) of northern Alberta. Some bitumen-rich material can be mined at the surface, as in the photo, but most will be pumped to the surface after injecting the material with steam. The Mobile Field Trip explores this unconventional source of oil. (Photo by Michael Collier) (https://goo.gl/55ZRJX)

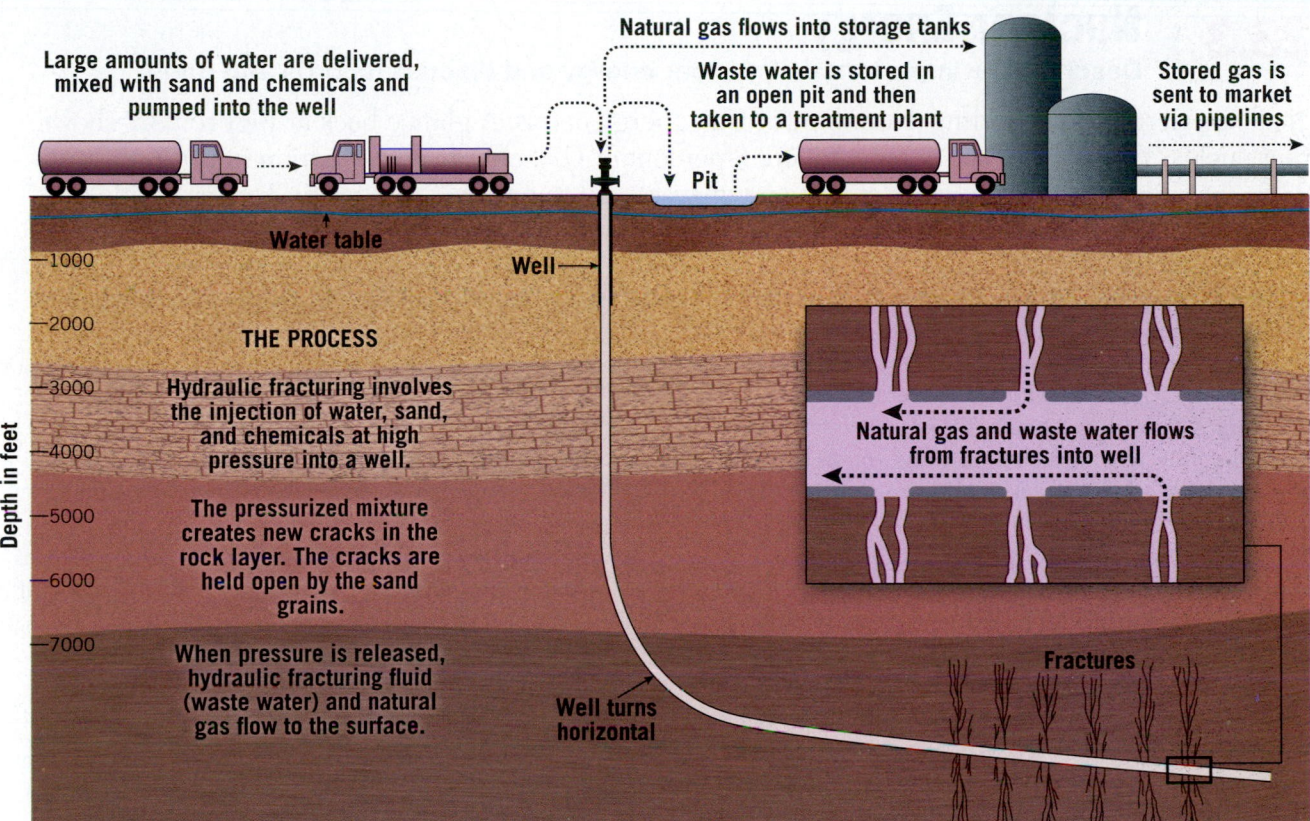

Natural gas flows into storage tanks

Large amounts of water are delivered, mixed with sand and chemicals and pumped into the well

Waste water is stored in an open pit and then taken to a treatment plant

Stored gas is sent to market via pipelines

Pit

Water table

Well

Depth in feet

−1000

−2000

THE PROCESS

−3000

Hydraulic fracturing involves the injection of water, sand, and chemicals at high pressure into a well.

−4000

−5000

The pressurized mixture creates new cracks in the rock layer. The cracks are held open by the sand grains.

−6000

−7000

When pressure is released, hydraulic fracturing fluid (waste water) and natural gas flow to the surface.

Well turns horizontal

Natural gas and waste water flows from fractures into well

Fractures

Figure 23.9
Hydraulic fracturing ("fracking") This well-stimulation process is commonly used in low-permeability rocks such as shale to increase oil and/or natural gas flow.

Obtaining oil from oil sand has environmental drawbacks. Substantial land disturbance is associated with mining huge quantities of rock and sediment. Moreover, large quantities of water are required for processing, and when processing is completed, contaminated water and sediments accumulate in toxic disposal ponds.

About 80 percent of the oil sands in Alberta are buried too deep for surface mining. Oil from these deep deposits must be recovered by using *in situ* (Latin for "in place") techniques. Using drilling technology, steam is injected into the deposit to heat the oil sand, which reduces the viscosity of the bitumen. The hot, mobile bitumen migrates toward producing wells, which pump it to the surface, while the sand is left in place. Production using *in situ* techniques already rivals open-pit mining and in the future will replace mining as the main source of bitumen production from the oil sands.

Challenges facing *in situ* processes include increasing the efficiency of oil recovery, managing the water used to make steam, and reducing the costs of energy required for the process.

Hydraulic Fracturing In some shale deposits, there are significant reserves of natural gas that cannot naturally leave because of the rock's low permeability. The practice of **hydraulic fracturing** (often called *fracking*) shatters the shale, opening up cracks through which the natural gas can flow into wells and then be brought to the surface. **Figure 23.9** illustrates the process. The fracturing of the shale is initiated by pumping fluids into the rock at very high pressures. The fluid is mostly water but also includes other chemicals that aid in the fracturing process. Some of these chemicals may be toxic, and there are concerns about fracking fluids leaking into freshwater aquifers that supply people with drinking water. The injection fluid also includes sand, so once fractures open up in the shale, the sand grains can keep them propped open and permit the gas to continue to flow. Once the fracturing has been accomplished, the fracking fluid is brought back to the surface. This wastewater is then injected into deep disposal wells. In some locations, these injections appear to trigger numerous minor earthquakes. Due to concerns about potential groundwater contamination and induced seismicity, hydraulic fracturing remains a controversial practice. Its environmental effects remain a focus of continuing research.

23.2 Concept Checks

1. Why are coal, oil, and natural gas called *fossil fuels*? Do all three form under the same circumstances?

2. What part of U.S. energy consumption does each of the fossil fuels represent? Where is most coal used? How about petroleum?

3. What is an oil trap? Sketch two examples. What do all oil traps have in common?

4. What are oil sands, and where are they most plentiful?

5. Describe the circumstances in which hydraulic fracturing is used.

23.3 | Nuclear Energy

Describe the importance of nuclear energy and discuss its pros and cons.

Nuclear energy meets an important part of U.S. energy needs. A glance back at Figure 23.2 shows that nuclear power was the source of about 8 percent of U.S. energy consumption in 2014. Figure 23.2 also shows that 100 percent of that energy was used to produce electricity. The right side of Figure 23.2 also shows that 22 percent of U.S. electricity is from nuclear energy.

The fuel for these nuclear power plants comes from radioactive materials that release energy through the process of **nuclear fission**. Fission is accomplished by bombarding the nuclei of heavy atoms, commonly uranium-235, with neutrons. This causes the uranium nuclei to split into smaller nuclei and to emit neutrons and heat energy. The ejected neutrons, in turn, bombard the nuclei of adjacent uranium atoms, producing a *chain reaction*. If the supply of fissionable material is sufficient and if the reaction is allowed to proceed in an uncontrolled manner, an enormous amount of energy is released, in the form of an atomic explosion.

In a nuclear power plant, the fission reaction is controlled by moving neutron-absorbing rods into or out of the nuclear reactor. The result is a controlled nuclear chain reaction that releases great amounts of heat. The heat energy produced is transported from the reactor and used to drive steam turbines that turn electrical generators, which is similar to what occurs in most conventional power plants.

Uranium

Uranium-235 is the only naturally occurring isotope that is readily fissionable, and it is therefore the primary fuel used in nuclear power plants.° Although large quantities of uranium ore have been discovered, most contain less than 0.05 percent uranium. Of this small amount, 99.3 percent is the nonfissionable isotope uranium-238, and just 0.7 percent consists of the fissionable isotope uranium-235. Because most nuclear reactors operate with fuels that are at least 3 percent uranium-235, the two isotopes must be separated in order to concentrate the fissionable uranium-235. The process of separating the uranium isotopes is difficult and substantially increases the cost of nuclear power generation.

Although uranium is a rare element in Earth's crust, it does occur in enriched deposits. Some of the most important occurrences are associated with what are believed to be ancient placer deposits in streambeds.°° For example, in Witwatersrand, South Africa, grains of uranium ore (as well as rich gold deposits) were concentrated by virtue of their high density in rocks made largely of quartz pebbles. In the United States, the richest uranium deposits are found in Jurassic and Triassic sandstones in the Colorado Plateau and in younger rocks in Wyoming. Most of these deposits have formed through the precipitation of uranium compounds from groundwater. Here, precipitation of uranium occurs as a result of a chemical reaction with organic matter, as evidenced by the concentration of uranium in fossil logs and organic-rich black shales.

Concerns Regarding Nuclear Development

At one time, nuclear power was heralded as the clean, cheap source of energy that would replace fossil fuels. However, several obstacles have emerged to hinder the development of nuclear power as a major energy source. One is the skyrocketing costs of building nuclear facilities that must contain numerous safety features. There is also concern about the possibility of a serious accidents.

The March 2011 emergency at Japan's Fukushima nuclear power plant is a recent example (**Figure 23.10**). A powerful earthquake generated a tsunami that devastated the coastal zone where the plant was situated. A series of equipment failures, nuclear meltdowns, and releases of radioactive materials followed.

*Thorium, although not capable by itself of sustaining a chain reaction, can be used with uranium-235 as a nuclear fuel.

**Placer deposits are discussed later in the chapter.

Figure 23.10

Tsunami destroys nuclear power plant In March 2011, an earthquake and massive tsunami destroyed Japan's Fukushima nuclear power plant. The event triggered a reevaluation of nuclear power in many countries. (Photo by SSEI KATO/AFP/Getty Images)

It should be emphasized that the concentrations of fissionable uranium-235 and the design of reactors are such that nuclear power plants cannot explode like atomic bombs. The dangers arise from the possible escape of radioactive debris during a meltdown of the core or other malfunction. In addition, hazards such as the disposal of nuclear waste and the relationship that exists between nuclear energy programs and the proliferation of nuclear weapons must be considered when evaluating the pros and cons of nuclear power.

Worldwide, 11 percent of electricity is generated by nuclear power plants, and the amount is increasing. Although nuclear power capacity has remained relatively unchanged in the United States for many years, it continues to grow worldwide. In 2015 more than 430 nuclear plants were operating worldwide, and another 60 reactors were under construction, the majority in Asian countries.

Among the benefits, or "pros," for nuclear energy is the fact that nuclear power plants do not emit carbon dioxide—the greenhouse gas that contributes significantly to global warming (see Chapter 21). By contrast, the generation of electricity from fossil fuels produces large quantities of carbon dioxide. Thus, substituting nuclear power for power generated by fossil fuels represents one option for reducing carbon emissions.

23.3 Concept Checks

1. What portion of U.S. energy consumption is provided by nuclear power?
2. What is the primary fuel used in nuclear power plants?
3. List some pros and cons of nuclear energy.

23.4 Renewable Energy

List and discuss the major sources of renewable energy. Describe the contribution of renewable energy to the overall U.S. energy supply.

Unlike fossils fuels, which are exhaustible, renewable energy sources regenerate and can be sustained indefinitely. The use of renewable energy is not new. More than 150 years ago, wood supplied a high percentage of our energy needs. Today coal, oil, and natural gas dominate, but the use of renewable forms of energy is on the rise. In 2014 U.S. consumption of renewable energy represented about 10 percent of all energy used (**Figure 23.11**). About 13 percent of U.S. electricity is generated from renewable sources. The use of renewable fuels has been increasing during the past 40 years and is expected to continue growing in the decades to come. Nevertheless, the U.S. Department of Energy projects that we will still rely on nonrenewable fuels to meet a significant portion of our needs.

Solar Energy

The term *solar energy* generally refers to the direct use of the Sun's rays to supply energy for the needs of people. The simplest and perhaps most widely used *passive solar collectors* are south-facing windows. As sunlight passes through the glass, its energy is absorbed by objects in the room. These objects in turn radiate heat that warms the air.

More elaborate systems used for home heating involve *active solar collectors*. These roof-mounted devices are usually large, blackened boxes that are covered with a transparent material. The heat they collect can be transferred to where it is needed by circulating air or fluids through pipes. In Israel, for example, about 80 percent of all homes are equipped with solar collectors used to provide hot water.

Research is currently under way to improve the technologies for concentrating sunlight. One method uses parabolic troughs as solar energy collectors. Each

SmartFigure 23.11 Renewable energy In 2014 renewable energy represented about 10 percent of U.S. energy consumption. (U.S. Energy Information Administration) (https://goo.gl/Le6OGn)

TOTAL: 9.6 quadrillion Btu — Solar 4%, Geothermal 2%, Wind 18%, Biomass waste 5%, Biofuels 22%, Wood 23%, Hydropower 26%, Biomass 50%

TOTAL: 98.3 quadrillion Btu — Petroleum 35%, Coal 18%, Natural gas 28%, Nuclear electric power 8%, Renewable energy 10%

Figure 23.12
Parabolic troughs concentrate sunlight These solar collectors focus sunlight onto collection pipes filled with a fluid. The heat is used to make steam that drives turbines used to generate electricity. (Photo by Jim West/Alamy Images)

collector resembles a large tube that has been cut in half. Their highly polished surfaces reflect sunlight onto a collection pipe (**Figure 23.12**). A fluid (usually oil) runs through the pipe and absorbs the concentrated sunlight. The fluid is heated to over 200°C (400°F) and is typically used to make steam that drives a turbine to produce electricity. Parabolic troughs gradually rotate to track the Sun.

Another type of collector uses photovoltaic (solar) cells that convert the Sun's energy directly into electricity (**Figure 23.13**). Photovoltaic cells are usually connected together to create solar panels, in which sunlight knocks electrons into higher energy states to produce electricity. For many years solar cells were used mainly to power calculators and novelty devices. Today, however, large photovoltaic power stations are connected to electrical grids to supplement other power-generating facilities. The relatively high cost of solar cells makes generating electricity from this source more expensive than conventional sources. However, as the cost of fossil fuels increases, advances in photovoltaic technology should narrow the price difference.

Another technology that is being developed, called the *Stirling dish*, converts thermal energy to electricity by using a mirror array to focus the Sun's rays on the receiver end of a Stirling engine. The internal side of a receiver then heats hydrogen gas, causing it to expand. The pressure created by the expanding gas drives a piston, which turns a small electric generator.

Wind Energy

Air has mass, and when it moves (that is, when the wind blows), it contains the energy of that motion—kinetic energy. A portion of that energy can be converted into other forms—mechanical force or electricity—that we can use to perform work.

Mechanical energy from wind is commonly used for pumping water in rural or remote places. The "farm windmill," still a familiar sight in many rural areas, is an example. Mechanical energy converted from wind can also be used for other purposes, such as sawing logs, grinding grain, and propelling sailboats. In addition, wind-powered electric turbines generate electricity for homes and businesses and for sale to utilities (see chapter-opening photo).

Today, modern wind turbines are being installed at breakneck speed. In fact, worldwide, the installed wind power capacity was nearly 370,000 megawatts (million watts; MW) in 2014, an increase of more than 85 percent over 2010 (**Figure 23.14**).° China has the greatest installed capacity (more than 114,000 megawatts in 2014), followed by the United States (about 66,000 megawatts in 2014). According to the World Wind Energy Association, wind turbines were supplying about 3 percent of worldwide electricity demand at the beginning of 2015.

Wind speed is crucial in determining whether a place is a suitable site for installing a wind-energy facility. Generally a minimum average wind speed of 21 kilometers (13 miles) per hour (or 6 meters per second) is necessary for a large-scale wind-power plant to be profitable. A small difference in wind speed results in a large difference in energy production and, therefore, a large difference in the cost of the electricity generated. For example, a turbine operating on a site with an average wind speed of 19.3 kilometers (12 miles) per hour would generate about 33 percent more electricity than one operating at 17.7 kilometers (11 miles) per hour. Also, there is little energy to be harvested at low wind speeds; air moving

Figure 23.13
Photovoltaic (solar) cells Photovoltaic cells turn solar radiation directly into electricity. The desert Southwest, with its generally cloud-free skies, has the greatest potential for solar energy development. (Map data from National Renewable Energy Laboratory; photo by Andrei Orlov/Shutterstock)

Photovoltaic Solar Resource Potential of the United States

kWh/m²/Day
<6.8 6.0 5.0 4.0 3.0 >2.2

*One megawatt is enough electricity to supply 250–300 average American households.

at 9.6 kilometers (6 miles) per hour contains less than one-eighth the energy of air moving at twice that speed.

Although the modern U.S. wind industry began in California, many states have greater wind potential. **Figure 23.15** shows the estimated average wind speeds at a height of 80 meters (260 feet) above the surface—the level where most commercial wind turbines rotate. Areas with average wind speeds greater than about 6 meters per second (13 miles per hour) are considered to have potential for development. At the end of 2014, Texas (14,208 MW) had the most installed wind capacity, followed by California (5,914 MW), Iowa (5,708 MW), Oklahoma (3,782 MW), and Illinois (3,568 MW).

Compared to burning fossil fuels to generate electricity, wind power produces little impact in terms of pollution. However, there are other problems related to wind energy, most of which are local. One of the most critical is bird kills, because birds occasionally collide with wind turbines. Land erosion caused by improper installation, noise, and visual impact are also cited as potential environmental issues.

The U.S. Department of Energy currently provides funds to develop offshore wind technologies. Offshore winds are stronger and blow more consistently than winds over land. Data suggest that more than 4 million megawatts of capacity is available on public land along the coasts of the United States. This is four times more than the country's total electrical-generating capacity. Although only a small fraction of U.S. electricity currently generated comes from wind energy, the Department of Energy's goal is to obtain 20 percent of U.S. electricity from wind by 2030, with 4 percent of that amount anticipated from offshore wind farms. This target seems consistent with the current growth rate of wind energy nationwide. Thus, wind-generated electricity appears to be shifting from being an alternative to being a mainstream energy source.

Hydroelectric Power

Falling water has been an energy source for centuries. Through much of human history, the mechanical energy produced by waterwheels has been used to power mills and other machinery. Today, the power generated by falling water is used to drive turbines that produce electricity, hence the term **hydroelectric power**. In the United States, hydroelectric power plants satisfied about 7 percent of the country's demand in 2014. Most of this energy is produced at large dams, which allow for a controlled flow of water (**Figure 23.16**). The water impounded in a reservoir is a form of stored energy that can be released at any time to produce electricity.

Although water power is considered a renewable resource, the dams built to provide hydroelectricity have finite lifetimes. Recall from Chapter 16 that all rivers carry suspended sediment, which is deposited behind the dam as soon as it is built. Eventually sediment may completely fill the reservoir. This takes 50 to 300 years, depending on the quantity of suspended material transported by the river. An example is Egypt's huge Aswan High Dam, which was completed in the 1960s. It is estimated that half of the reservoir will be filled with sediment from the Nile River by 2025.

The availability of appropriate sites is an important limiting factor in the development of large-scale

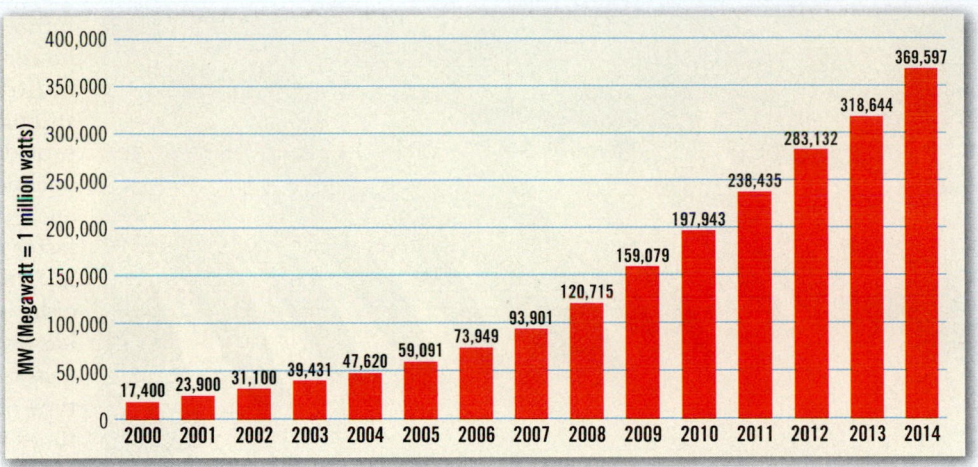

Figure 23.14
Global cumulative installed wind capacity, 2000–2014

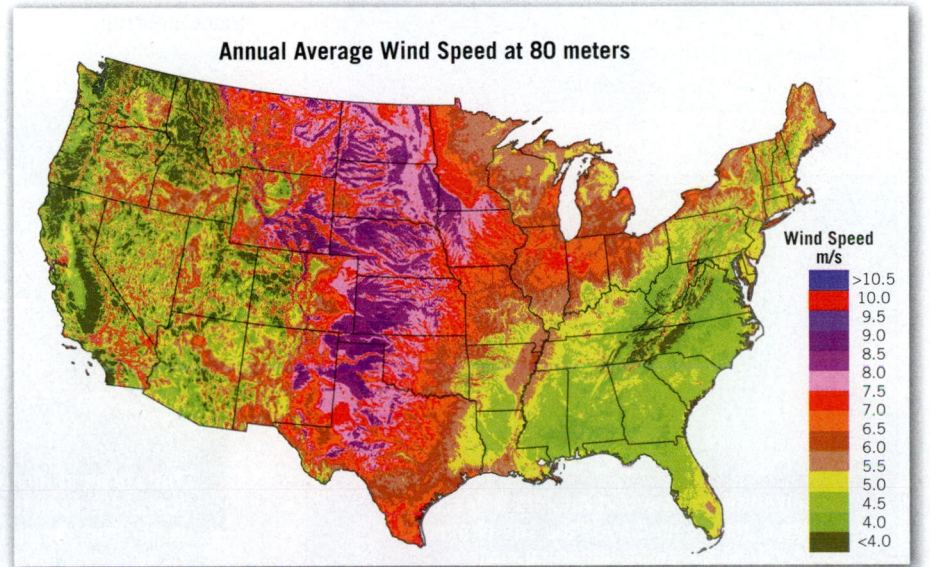

Figure 23.15
Wind energy potential for the United States Large wind systems require average wind speeds of about 6 meters per second (13 miles per hour). (Map data from National Renewable Energy Laboratory)

Figure 23.16
Grand Coulee Dam More than one-half of U.S. hydroelectric capacity is concentrated in Washington, Oregon, and California. In 2014 about 29 percent of the total hydropower was generated in Washington, the location of the nation's largest hydroelectric facility—Grand Coulee Dam. (Photo by Science Source)

hydroelectric power plants. A good site provides a significant height for the water to fall and a high rate of flow. Hydroelectric dams exist in many parts of the United States, with the greatest concentrations occurring in the Southeast and the Pacific Northwest. Most of the best U.S. sites have already been developed, limiting the future expansion of hydroelectric power. The total power produced by hydroelectric sources might still increase, but the relative share provided by this source will likely decline because other alternate energy sources will increase at a faster rate.

In recent years, a different type of hydroelectric power production has come into use. Called a *pumped water-storage system*, it is actually a type of energy management. During times when demand for electricity is low, unneeded power produced by nonhydroelectric sources is used to pump water from a lower reservoir to a storage area at a higher elevation. Then, when demand for electricity is great, the water stored in the higher reservoir is available to drive turbines and produce electricity to supplement the power supply.

Geothermal Energy

Geothermal energy is harnessed by tapping natural underground reservoirs of steam and hot water. These occur where subsurface temperatures are high, due to relatively recent volcanic activity or nearby magma chambers. Geothermal energy is put to use in two ways: The steam and hot water are used for heating and to generate electricity.

Iceland is a large volcanic island with current volcanic activity (**Figure 23.17**). In Iceland's capital, Reykjavik, underground steam and hot water are pumped into buildings throughout the city for space heating. They also warm greenhouses, where fruits and vegetables are grown all year. In the United States, localities in several western states use hot water from geothermal sources for space heating.

Figure 23.17
Geothermal development in Iceland Geothermal sources account for 66 percent of Iceland's primary energy use. Much is used directly for space heating—9 of 10 homes are heated this way. About 25 percent of Iceland's electricity is generated by geothermal sources. (Photo by Simon Fraser/Science Photo Library/Photo Researchers, Inc.)

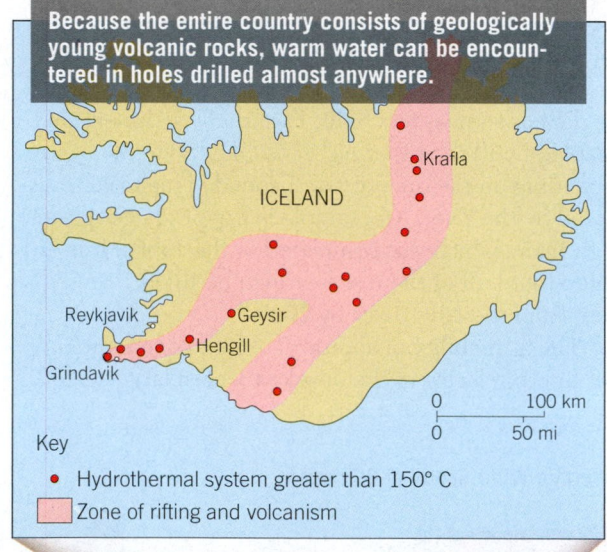

Because the entire country consists of geologically young volcanic rocks, warm water can be encountered in holes drilled almost anywhere.

ICELAND

Krafla

Reykjavik • Geysir
Hengill
Grindavik

0   100 km
0   50 mi

Key
• Hydrothermal system greater than 150° C
▮ Zone of rifting and volcanism

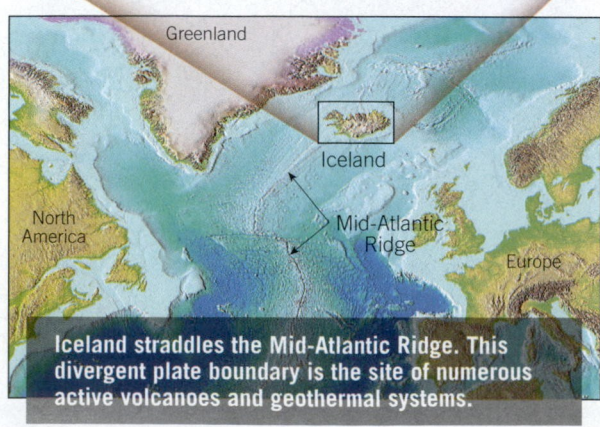

Greenland

Iceland

North America

Mid-Atlantic Ridge

Europe

Iceland straddles the Mid-Atlantic Ridge. This divergent plate boundary is the site of numerous active volcanoes and geothermal systems.

The steam at this power station in southwestern Iceland is used to generate electricity. Hot (83°C) water from the plant is sent via an insulated pipeline to Reykjavik for space heating.

The Italians began generating electricity geothermally in 1904, so the idea is not new. In 2014, 24 countries produced more than 12,800 megawatts of power this way. The United States was the leading producer, responsible for more than 25 percent of the total. Nevertheless, geothermal power represents just 0.4 percent (four-tenths of 1 percent) of U.S. electric power.

The first commercial geothermal power plant in the United States was built in 1960 at The Geysers, north of San Francisco (**Figure 23.18**). The Geysers remains the world's largest geothermal power plant, generating nearly 1000 megawatts annually. In addition to The Geysers, geothermal development is occurring elsewhere in the western United States, including Nevada, Utah, and the Imperial Valley in southern California. The U.S. geothermal generating capacity in 2014 of 3525 megawatts is enough to supply more than 3 million homes. This is comparable to burning about 70 million barrels of oil each year.

The following geologic factors favor a geothermal reservoir of commercial value:

- *A potent source of heat*, such as a large magma chamber deep enough to ensure adequate pressure and slow cooling but not so deep that the natural water circulation is inhibited. Such magma chambers are most likely in regions of recent volcanic activity.
- *Large and porous reservoirs with channels connected to the heat source*, near which water can circulate and then be stored in the reservoir.
- *A cap of low-permeability rocks* that inhibits the flow of water and heat to the surface. A deep, well-insulated reservoir contains much more stored energy than a similar but uninsulated reservoir.

As with other alternative methods of power production, geothermal sources are not expected to provide a high percentage of the world's growing energy needs.

Figure 23.18
The Geysers This facility, near the city of Santa Rosa in northern California, is the world's largest electricity-generating geothermal development. Most of the steam wells are about 3000 meters (10,000 feet) deep. California's 35 geothermal power plants were responsible for 80 percent of U.S. geothermal power production in 2014. (Kim Steele/Getty Images)

Nevertheless, in regions where its potential can be developed, the use of geothermal energy will continue to grow.

Biomass: Renewable Energy from Plants and Animals

Biomass is organic material made from plants and animals. It is a renewable energy source because we can always grow more trees and crops, and waste will always exist. Some examples of biomass fuels are wood, crops, manure, and some garbage. When burned, the chemical energy in biomass is released as heat. A common example is burning a log in a wood stove or fireplace. Wood and garbage can be burned to produce steam for generating electricity or to provide heat to industries and homes.

EYE ON EARTH 23.1

Lake Powell is the reservoir that was created when Glen Canyon Dam was built across the Colorado River. As water in the lake is released, it drives turbines and produces electricity. (Photo by Michael Collier)

QUESTION 1 What term is applied to the energy generated at Glen Canyon Dam?

QUESTION 2 Is this energy considered renewable or nonrenewable?

QUESTION 3 Explain why most dams and reservoirs do not last indefinitely.

Figure 23.19
Tidal power **A.** Simplified diagram showing the principle of the tidal dam. Electricity is generated only when a sufficient water-height difference exists between the bay and the ocean. **B.** Tidal power plant at Annapolis Royal, Nova Scotia, on the Bay of Fundy. (Photo by James P. Blair/National Geographic Image Collection)

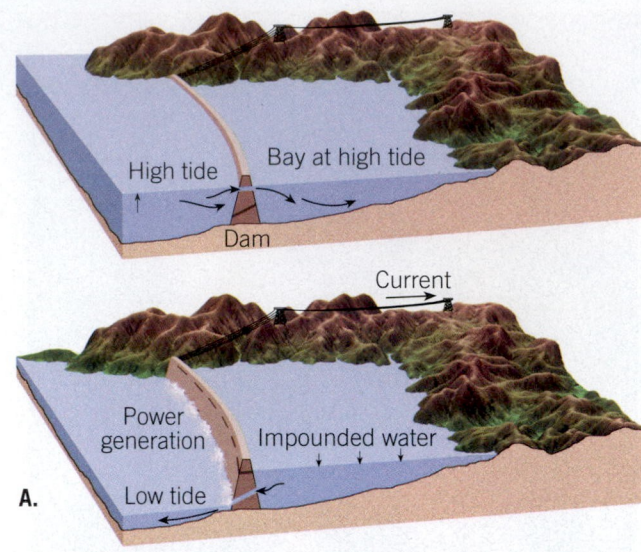

B.

Burning biomass is not the only way to release its energy. Biomass can be converted to other usable forms of energy, such as methane gas, or transportation fuels, such as ethanol and biodiesel. In 2014, biomass fuels provided about 4.8 percent of the energy used in the United States. Primary sources of biomass energy are as follows:

- Wood biomass includes wood chips from forestry operations, residues from lumber, pulp/paper, and furniture mills, and fuel wood for space heating. The largest single source of wood energy is "black liquor," a residue of pulp, paper, and paperboard production.
- Biofuels include alcohol fuels such as ethanol, and "biodiesel," a fuel from grain oils and animal fats. Most biofuel used in the United States is ethanol produced from corn.
- Municipal waste contains biomass such as paper, cardboard, food scraps, grass clippings, and leaves. Waste can be recycled, composted, sent to landfills, or used in waste-to-energy plants. There are hundreds of landfills in the United States that recover *biogas*, which is methane that forms when waste matter decomposes in low-oxygen (anaerobic) conditions. The methane is burned to produce electricity and heat.

Tidal Power

Several methods of generating electrical energy from the oceans have been proposed, but the ocean's energy potential remains largely untapped. The development of tidal power is the principal example of energy production from the ocean.

Tides have been used as a source of power for centuries. Beginning in the twelfth century, waterwheels driven by the tides were used to power gristmills and sawmills. During the seventeenth and eighteenth centuries, much of Boston's flour was produced at a tidal mill. Today far greater energy demands must be satisfied, and more sophisticated ways of using the force created by the perpetual rise and fall of the ocean must be employed.

Tidal power is harnessed by constructing a dam across the mouth of a bay or an estuary in a coastal area having a large tidal range (**Figure 23.19A**). The narrow opening between the bay and the open ocean magnifies the variations in water level that occur as the tides rise and fall. The strong in-and-out flow that results at such a site is then used to drive turbines and electrical generators.

Tidal energy utilization is exemplified by the tidal power plant at the mouth of the Rance River in France. By far the largest yet constructed, this plant went into operation in 1966 and produces enough power to satisfy the needs of Brittany and also meet the demands of other regions. Much smaller experimental facilities have been built near Murmansk in Russia, near Taliang in China, and on the Annapolis River estuary, an arm of the Bay of Fundy in the Canadian province of Nova Scotia (**Figure 23.19B**).

Along most of the world's coasts, it is not possible to harness tidal energy. If the tidal range is less than 8 meters (25 feet), or if narrow, enclosed bays are absent, tidal power development is not economically feasible. For this reason, the tides will never provide a very high portion of our ever-increasing electrical energy requirements. Nevertheless, the development of tidal power may be worth pursuing at feasible sites because electricity produced by the tides consumes no exhaustible fuels and creates no noxious wastes.

23.4 Concept Checks

1. How important is renewable energy to the overall U.S. energy supply?

2. What is the greatest use of renewable energy?

3. Describe two ways in which solar energy is used to produce electricity.

4. Which has had the more rapid rate of growth rate in the United States in recent years: wind power or hydroelectric power?

5. What is biomass? List four examples.

6. Where in the United States is hydroelectric power development most concentrated? How about geothermal power development?

EYE ON EARTH 23.2

This image shows an active landfill where tons of trash and garbage are dumped every day. Eventually this site will be reclaimed to resemble the area shown on the right and will become a source of energy. (Left Photo by NHPA/SuperStock; Right Photo by Jim West/Alamy Images)

QUESTION 1 Explain how an area filled with trash and waste could become a source of energy. What form of energy will it be? How might it be used?

QUESTION 2 Will this energy be considered renewable or nonrenewable? Explain.

23.5 | Mineral Resources
Distinguish among resource, reserve, and ore.

Earth's crust is the source of a wide variety of useful and essential substances. In fact, practically every manufactured product contains substances derived from minerals. Table 23.1 lists some important examples.

TABLE 23.1   Occurrence of Metallic Minerals

Metal	Principal Ores	Geologic Occurrences
Aluminum	Bauxite	Residual product of weathering
Chromium	Chromite	Magmatic differentiation
Copper	Chalcopyrite	Hydrothermal deposits; contact metamorphism; enrichment by weathering processes
	Bornite	
	Chalcocite	
Gold	Native gold	Hydrothermal deposits; placers
Iron	Hematite	Banded sedimentary formations; magmatic differentiation
	Magnetite	
	Limonite	
Lead	Galena	Hydrothermal deposits
Magnesium	Magnesite	Hydrothermal deposits
	Dolomite	
Manganese	Pyrolusite	Residual product of weathering
Mercury	Cinnabar	Hydrothermal deposits
Molybdenum	Molybdenite	Hydrothermal deposits
Nickel	Pentlandite	Magmatic differentiation
Platinum	Native platinum	Magmatic differentiation; placers
Silver	Native silver	Hydrothermal deposits; enrichment by weathering processes
	Argentite	
Tin	Cassiterite	Hydrothermal deposits; placers
Titanium	Ilmenite	Magmatic differentiation; placers
	Rutile	
Tungsten	Wolframite	Pegmatites; contact metamorphic deposits; placers
	Scheelite	
Uranium	Uraninite (pitchblende)	Pegmatites; sedimentary deposits
Zinc	Sphalerite	Hydrothermal deposits

Figure 23.20
Most abundant elements in the continental crust More than 98 percent of the continental crust is composed of just eight elements. The concentration of many elements that are important mineral resources is part of the "others" category on the graph.

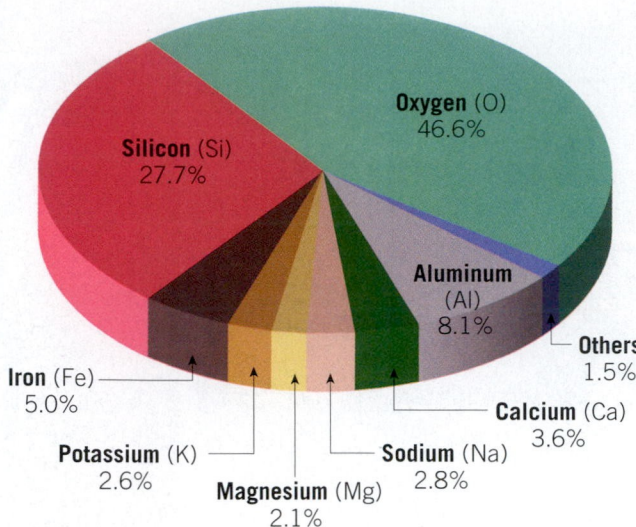

Oxygen (O)
46.6%

Silicon (Si)
27.7%

Aluminum (Al)
8.1%

Others
1.5%

Iron (Fe)
5.0%

Calcium (Ca)
3.6%

Potassium (K)
2.6%

Sodium (Na)
2.8%

Magnesium (Mg)
2.1%

Mineral resources are the endowment of useful minerals ultimately available commercially. Resources include already identified deposits from which minerals can be extracted profitably, called **reserves**, as well as known deposits that are not yet economically or technologically recoverable. Deposits inferred to exist but not yet discovered are also considered mineral resources. In addition, the term **ore** is used to denote useful metallic minerals that can be mined at a profit. In common usage, the term *ore* is also applied to some nonmetallic minerals, such as fluorite and sulfur. However, materials used for such purposes as building stone, road aggregate, abrasives, ceramics, and fertilizers are not usually called ores; rather, they are classified as industrial rocks and minerals.

Recall from Chapter 3 that more than 98 percent of the continental crust is composed of only eight elements.

Except for oxygen and silicon, all other elements make up a relatively small fraction of common crustal rocks (**Figure 23.20**). Indeed, the natural concentrations of many elements are exceedingly small. A deposit containing the average percentage of a valuable element is worthless if the cost of extracting it exceeds the value of the material recovered. To be considered valuable, an element must be concentrated above the level of its average crustal abundance. Generally, the lower the crustal abundance, the greater the concentration.

For example, copper makes up about 0.0135 percent of the crust. However, for a material to be considered a copper ore, it must contain a concentration that is about 50 times this amount. Aluminum, in contrast, represents 8.13 percent of the crust and must be concentrated to only about four times its average crustal percentage before it can be extracted profitably.

It is important to realize that a deposit may become profitable to extract or lose its profitability because of economic changes. If demand for a metal increases and prices rise, the status of a previously unprofitable deposit changes, and it becomes an ore. The status of unprofitable deposits may also change if a technological advance allows the useful element to be extracted at a lower cost than before. This occurred at the copper-mining operation located at Bingham Canyon, Utah, one of the largest open-pit mines on Earth (**Figure 23.21**). Mining was halted here in 1985 because outmoded equipment had driven up the cost of extracting the copper beyond the selling price. The owners responded by replacing an antiquated 1000-car railroad with conveyor belts and pipelines for transporting the ore and waste. These devices achieved a cost reduction

Figure 23.21
Utah's Bingham Canyon copper mine This is one of the largest open-pit mines in the world. This huge hole is nearly 4 kilometers (2.5 miles) across and 900 meters (3000 feet) deep. Although the amount of copper in the rock is less than 0.5 percent, the huge volume of material removed each day yields enough metal to be profitable. In addition to copper, the mine produces gold, silver, and molybdenum. (Photo by Michael Collier)

of nearly 30 percent and returned this mining operation to profitability.

Over the years, geologists have been keenly interested in learning how natural processes produce localized concentrations of essential metallic minerals. One well-established fact is that occurrences of valuable mineral resources are closely related to the rock cycle. That is, the mechanisms that generate igneous, sedimentary, and metamorphic rocks, including the processes of weathering and erosion, play a major role in producing concentrated accumulations of useful elements. Moreover, with the development of plate tectonics theory, geologists added yet another tool for understanding the processes by which one rock is transformed into another.

23.6 | Igneous and Metamorphic Processes

Explain how different igneous and metamorphic processes produce economically significant mineral deposits.

Some of the most important accumulations of metals, such as gold, silver, copper, mercury, lead, platinum, and nickel, are produced by igneous and metamorphic processes (see Table 23.1). These mineral resources, like most others, result from processes that concentrate desirable materials to the extent that extraction is economically feasible.

Magmatic Differentiation and Ore Deposits

The igneous processes that generate some metal deposits are quite straightforward. For example, as a large magma body cools, the heavy minerals that crystallize early tend to settle to the lower portion of the magma chamber (*crystal settling*). This type of magmatic differentiation (see Figure 4.22, page 126) is particularly active in large basaltic magmas where chromite (ore of chromium), magnetite, and platinum are occasionally generated. Layers of chromite, interbedded with other heavy minerals, are mined from such deposits in the Stillwater Complex of Montana. Another example is the Bushveld Complex in South Africa, which contains over 70 percent of the world's known reserves of platinum.

Pegmatite Deposits Magmatic differentiation is important in the late stages of the magmatic process. This is particularly true of granitic magmas, in which the residual melt can become enriched in rare elements and heavy metals. Further, because water and other volatile substances do not crystallize along with the bulk of the magma body, these fluids make up a high percentage of the melt during the final phase of solidification. Crystallization in a fluid-rich environment, where ion migration is enhanced, results in the formation of crystals several centimeters, or even a few meters, in length. The resulting rocks, called **pegmatites**, are composed of these unusually large crystals.

Most pegmatites are granitic in composition and consist of unusually large crystals of quartz, feldspar, and muscovite. Feldspar is used in the production of ceramics, and muscovite is used for electrical insulation and glitter. Further, pegmatites often contain some of the least abundant elements. Thus, in addition to the common silicates, some pegmatites include semiprecious gems such as beryl, topaz, and tourmaline. Moreover, minerals containing the elements lithium, cesium, uranium, and the rare earths° are occasionally found (**Figure 23.22**). Most pegmatites are located within large igneous masses or as dikes or veins that cut into the host rock surrounding the magma chamber (**Figure 23.23**).

*The rare earths are a group of 15 elements (atomic numbers 57 through 71) that possess similar properties. They are useful catalysts in petroleum refining and are used to manufacture strong magnets for turbines and rechargeable batteries for cell phones and computers.

Figure 23.22
Pegmatites This pegmatite in the Black Hills of South Dakota was mined for its large crystals of spodumene, an important source of lithium. Arrows are pointing to impressions left by crystals. Note the person in the upper center of the photo for scale. (Photo by James G. Kirchner)

SmartFigure 23.23
Pegmatites and hydrothermal deposits Illustration of the relationship between an igneous body and associated pegmatites and hydrothermal mineral deposits.
(Photo by Greenshoots Communications/Alamy)
(https://goo.gl/AUDiZv)

Tutorial

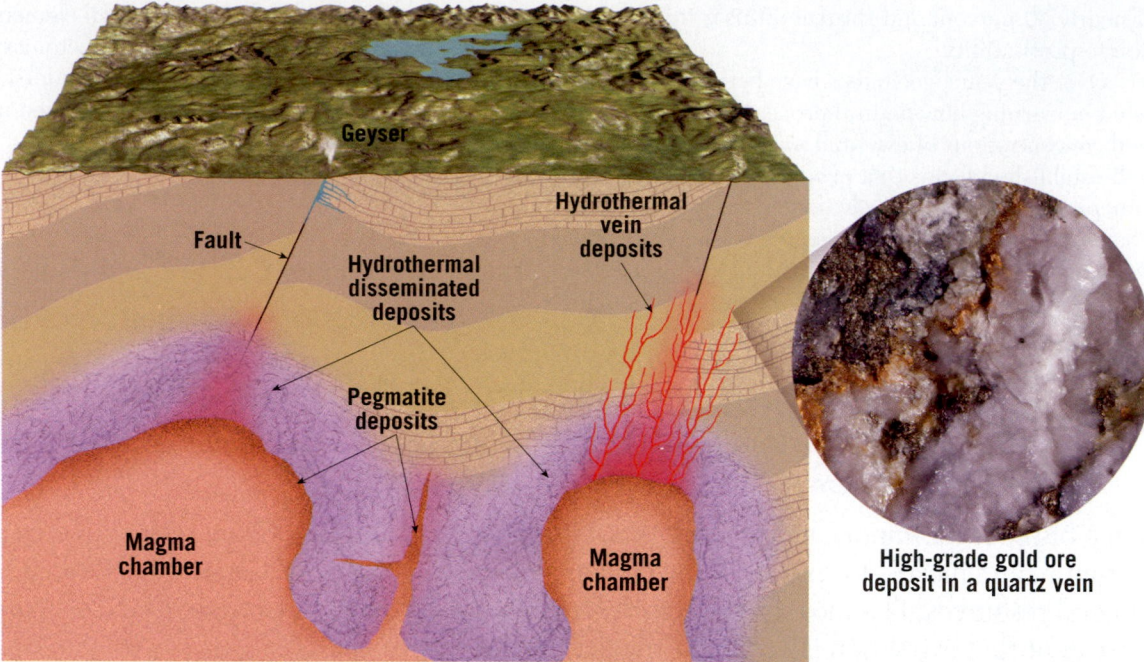

High-grade gold ore deposit in a quartz vein

Not all late-stage magmas produce pegmatites, nor do all have a granitic composition. Rather, some magmas become enriched in iron or occasionally copper. For example, at Kiruna, Sweden, magma composed of more than 60 percent magnetite solidified to produce one of the largest iron deposits in the world.

Hydrothermal Deposits

Among the best-known and most important ore deposits are those generated from hot, ion-rich fluids called **hydrothermal** (hot-water) **solutions**. Included in this group are the gold deposits of the Homestake Mine in South Dakota; the lead, zinc, and silver ores near Coeur d'Alene, Idaho; the silver deposits of the Comstock Lode in Nevada; and the copper ores of the Keweenaw Peninsula in Michigan (**Figure 23.24**).

Hydrothermal Vein Deposits The majority of hydrothermal deposits originate from hot, metal-rich fluids that are remnants of late-stage magmatic processes. During solidification, liquids plus various metallic ions accumulate near the top of the magma chamber. Because of their mobility, these ion-rich solutions can migrate great distances through the surrounding rock before they are eventually deposited, usually as sulfides of various metals. Some of this fluid moves along openings such as fractures or bedding planes, where it cools and precipitates the metallic ions to produce **vein deposits** (see photo in Figure 23.23). Many of the most productive deposits of gold, silver, and mercury occur as hydrothermal vein deposits.

Disseminated Deposits Another important type of accumulation generated by hydrothermal activity is called a **disseminated deposit**. Rather than being concentrated in narrow veins and dikes, these ores are distributed as minute masses throughout the entire rock mass. Much of the world's copper is extracted from disseminated deposits, including those at Chuquicamata, Chile, and the huge Bingham Canyon copper mine in Utah (see Figure 23.21). Because these accumulations contain only 0.4 to 0.8 percent copper, between 125 and 250 kilograms of ore must be mined for every 1 kilogram of metal recovered. The environmental impact of these large excavations, including the problem of waste disposal, is significant.

Some hydrothermal deposits have been generated by the circulation of ordinary groundwater in regions where magma was emplaced near the surface. The Yellowstone National Park area is a modern example of

Figure 23.24
Native copper This nearly pure metal from northern Michigan's Keweenaw Peninsula is an excellent example of a hydrothermal deposit. At one time this area was an important source of copper, but it is now largely depleted. (Photo by E. J. Tarbuck)

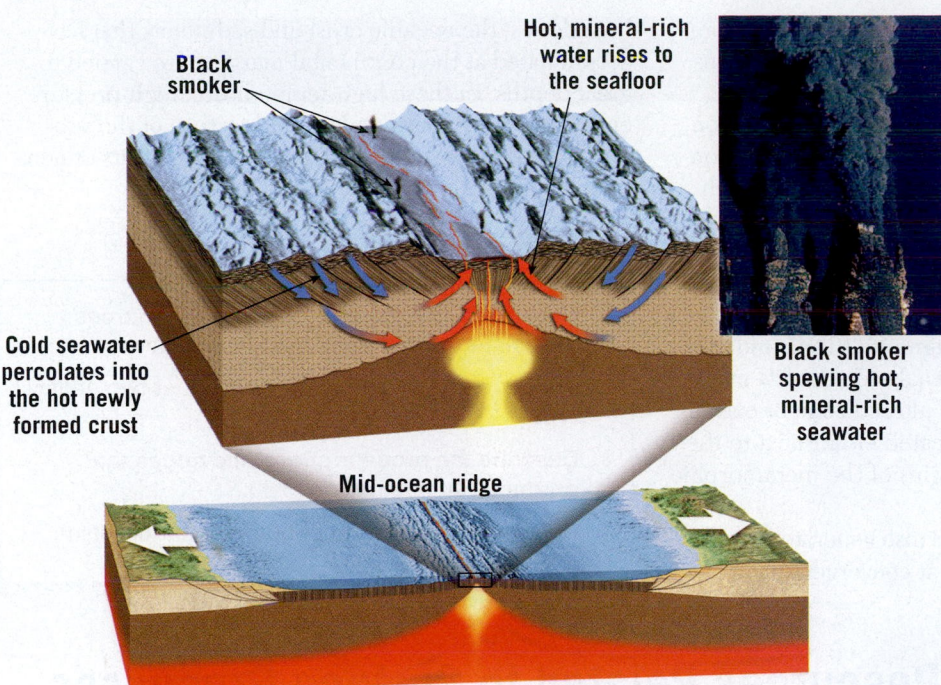

Figure 23.25
Sulfide deposits along a mid-ocean ridge Massive sulfide deposits can result from the circulation of seawater through the oceanic crust along active spreading centers. As seawater infiltrates the hot basaltic crust, it leaches sulfur, iron, copper, and other metals. The hot, enriched fluid returns to the seafloor near the ridge axis along faults and fractures. Some metal sulfides may be precipitated in these channels as the rising fluid begins to cool. When the hot liquid emerges from the seafloor and mixes with cold seawater, the sulfides precipitate to form massive deposits. Photo shows a close-up view of a black smoker spewing hot, mineral-rich seawater along the East Pacific Rise. (Photo by Fisheries and Oceans Canada/Uvic-Verena Tunnicliffe/Newscom)

such a situation. When groundwater invades a zone of recent igneous activity, its temperature rises, greatly enhancing its ability to dissolve minerals. These migrating hot waters remove metallic ions from intrusive igneous rocks and carry them upward, where they may be deposited as an ore body. Depending on the conditions, the resulting accumulations may occur as vein deposits, as disseminated deposits, or, where hydrothermal solutions reach the surface in the form of hot springs or geysers, as surface deposits.

Hydrothermal Activity at Oceanic Ridges With the development of plate tectonics theory, it became clear that some hydrothermal deposits originated along ancient oceanic ridges. A well-known example is found on the island of Cyprus, where copper has been mined for more than 4000 years. Apparently, these deposits represent ores that formed on the seafloor at an ancient oceanic spreading center.

Since the mid-1970s, active hot springs and metal-rich sulfide deposits have been detected at several sites, including study areas along the East Pacific Rise and the Juan de Fuca Ridge. The deposits form where heated seawater, rich in dissolved metals and sulfur, gushes from the seafloor as particle-filled clouds called *black smokers*. As shown in **Figure 23.25**, seawater infiltrates the hot oceanic crust along the flanks of the ridge. As the water moves through the newly formed material, it is heated and chemically interacts with the basalt, extracting and transporting sulfur, iron, copper, and other metals. Near the ridge axis, the hot, metal-rich fluid rises along faults. Upon reaching the seafloor, the spewing liquid mixes with the cold seawater, and the sulfides precipitate to form massive sulfide deposits.

Origin of Diamonds

Another economically important mineral with an igneous origin is diamond. Although best known as gems, diamonds are used extensively as abrasives. Diamonds originate at depths of nearly 200 kilometers (125 miles), where the confining pressure is great enough to generate this high-pressure form of carbon. Once crystallized, the diamonds are carried upward through pipe-shaped conduits that increase in diameter toward the surface. In diamond-bearing pipes, nearly the entire pipe contains diamond crystals that are disseminated throughout an ultramafic rock called *kimberlite*. The most productive kimberlite pipes are found in South Africa. The only equivalent source of diamonds in the United States is located near Murfreesboro, Arkansas, but this deposit is exhausted and today serves merely as a tourist attraction.

Metamorphic Processes

The role of metamorphism in producing mineral deposits is frequently tied to igneous processes. For example, many of the most important metamorphic ore deposits are produced by contact metamorphism. Here the host rock is recrystallized and chemically altered by heat, pressure, and hydrothermal solutions emanating from an intruding igneous body. The extent to which the host rock is altered depends on the nature of the intruding igneous mass as well as the nature of the host rock.

Some resistant materials, such as quartz sandstone, may show very little alteration, whereas others, including limestone, might exhibit the effects of metamorphism for several kilometers from the igneous pluton. As hot, ion-rich fluids move through limestone, chemical reactions take place, producing useful minerals such as garnet and corundum. Further, these reactions release carbon dioxide, which greatly facilitates the outward migration of metallic ions. Thus, extensive aureoles of metal-rich deposits commonly surround igneous plutons that have invaded limestone strata.

The most common metallic minerals associated with contact metamorphism are sphalerite (zinc), galena (lead), chalcopyrite (copper), magnetite (iron), and bornite (copper). The hydrothermal ore deposits may be disseminated throughout the altered zone or exist as concentrated masses that are located either next to the intrusive body or along the margins of the metamorphic zone.

Regional metamorphism can also generate useful mineral deposits. Recall that at convergent plate boundaries, the oceanic crust and sediments that have accumulated at the continental margins are carried to great depths. In these high-temperature, high-pressure environments, the mineralogy and texture of the subducted materials are altered, producing deposits of non-metallic minerals such as talc and graphite.

23.6 | Concept Checks

1. Describe two examples of mineral resources associated with magmatic differentiation.

2. What are hydrothermal deposits? List two general types.

3. Describe the process at oceanic ridges that produces sulfide ore deposits.

4. Relate ore deposits to contact metamorphism.

23.7 | Mineral Resources Related to Surface Processes

Discuss ways in which surface processes produce ore deposits.

The preceding discussion focused on mineral resources that are closely linked to igneous activity and metamorphism. In this section, we look at examples of metallic mineral resources that accumulate as a result of surface processes.

Weathering and Ore Deposits

Weathering creates many important mineral deposits by concentrating minor amounts of metals that are scattered through unweathered rock into economically valuable concentrations. Such a transformation is often termed **secondary enrichment**, and it takes place in one of two ways. In one situation, chemical weathering coupled with downward-percolating water removes undesirable materials from decomposing rock, leaving the desirable elements enriched in the upper zones of the soil. The second way is basically the reverse of the first. That is, the desirable elements that are found in low concentrations near the surface are removed and carried to lower zones, where they are redeposited and become more concentrated.

Bauxite *Bauxite*, the principal ore of aluminum, is one important example of an ore created as a result of enrichment by weathering processes (**Figure 23.26**). Although aluminum is the third-most-abundant element in Earth's crust, economically valuable concentrations of this important metal are not common because most aluminum is tied up in silicate minerals, from which it is extremely difficult to extract.

Bauxite forms in rainy tropical climates, in association with the deeply weathered soils known as laterites. (In fact, bauxite is sometimes called *aluminum laterite*.) When aluminum-rich source rocks are subjected to the intense and prolonged chemical weathering of the tropics, most of the common elements, including calcium, sodium, and silicon, are removed by leaching. Because aluminum is extremely insoluble, it becomes concentrated in the soil as bauxite, a hydrated aluminum oxide. Thus, the formation of bauxite depends both on climatic conditions in which chemical weathering and leaching are pronounced and on the presence of an aluminum-rich source rock. Important deposits of nickel and cobalt

Figure 23.26
Bauxite This ore of aluminum forms as a result of weathering processes under tropical conditions. Its color varies from red or brown to nearly white.
(Photo by E. J. Tarbuck)

are also found in laterite soils that develop from igneous rocks rich in ferromagnesian silicate minerals.

Other Deposits Many copper and silver deposits result when weathering processes concentrate metals that are deposited through a low-grade primary ore. Usually such enrichment occurs in deposits containing pyrite (FeS_2), the most common and widespread sulfide mineral. Pyrite is important because when it chemically weathers, sulfuric acid forms, which enables percolating waters to dissolve the ore metals. Once dissolved, the metals gradually migrate downward through the primary ore body until they are precipitated. Deposition takes place because of changes that occur in the chemistry of the solution when it reaches the saturated zone (the zone beneath the surface where all pore spaces are filled with groundwater). In this manner, the small percentage of dispersed metal can be removed from a large volume of rock and redeposited as a higher-grade ore in a smaller volume of rock.

This enrichment process is responsible for the economic success of many copper deposits, including one located in Miami, Arizona. Here the ore was upgraded from less than 1 percent copper in the primary deposit to as much as 5 percent copper in some localized zones of enrichment. When pyrite weathers (oxidizes) near the surface, residues of iron oxide remain. The presence of these rusty-colored caps at the surface indicates the possibility of an enriched copper ore below and is a visible sign for prospectors.

Placer Deposits

The sorting of sediments by wind or water usually results in like-sized grains being deposited together. However, sorting according to the specific gravity of particles also occurs. This latter type of sorting is responsible for the creation of **placers**, which are deposits formed when heavy minerals are mechanically concentrated by currents. Placers associated with streams are among the most common and best known, but the sorting action of waves can also create placers along the shore. Placer deposits usually involve minerals that are not only heavy but also durable (to withstand physical destruction during transportation) and chemically resistant (to endure weathering processes). Placers form because heavy minerals settle quickly from a current, whereas less-dense particles remain suspended and are carried onward. Common sites of accumulation include point bars on the insides of meanders as well as cracks, depressions, and other irregularities on streambeds.

Many economically important placer deposits exist, with accumulations of gold the best known. Indeed, it was the placer deposits discovered in 1848 that led to the famous California Gold Rush. Years later, similar deposits created a gold rush to Alaska as well (**Figure 23.27**). Panning for gold by washing sand and gravel from a flat pan to concentrate the fine "dust" at the bottom was a common method early prospectors used to recover the precious metal, and it is a process similar to that which created the placer in the first place.

In addition to gold, other heavy and durable minerals form placers. These include platinum, diamonds, and tin. The Ural Mountains contain placers rich in platinum, and placers are important sources of diamonds in southern Africa. Significant portions of the world's supply of cassiterite, the principal ore of tin, have come from placer deposits in Malaysia and Indonesia. Cassiterite is often widely disseminated through granitic igneous rocks. In this state, the mineral is not sufficiently concentrated to be extracted profitably. However, as the enclosing rock dissolves and disintegrates, the heavy and durable cassiterite grains are set free. Eventually the freed particles are washed to a stream, where they are deposited in placers that are significantly more

Figure 23.27
Placers These deposits form when heavy minerals are mechanically concentrated by currents. (Photo on left by Underwood Archives/Getty Images; photo on right from the Library of Congress)

concentrated than the original deposit. Similar circumstances and events are common to many minerals mined from placers.

In some cases, if the source rock for a placer deposit can be located, it too may become an important ore body. By following placer deposits upstream, one can sometimes locate the original ore body. This is how the gold-bearing veins of the mother lode in California's Sierra Nevada batholith were found, as well as the famous Kimberley diamond mine of South Africa. The placers were discovered first, and their source was found at a later time.

23.7 | **Concept Checks**

1. **What is secondary enrichment?**

2. **Name the primary ore of aluminum and describe its formation.**

3. **How might the mineral pyrite play a role in creating an ore deposit?**

4. **Describe the way in which minerals accumulate in placers. List four minerals that are mined from such deposits.**

23.8 | Nonmetallic Mineral Resources

Distinguish between two broad categories of nonmetallic mineral resources and list examples of each.

Earth materials that are not used as fuels or processed for the metals they contain are referred to as **nonmetallic mineral resources**. Realize that use of the word *mineral* is very broad in this economic context and is quite different from the geologist's strict definition of *mineral* found in Chapter 3. Nonmetallic mineral resources are extracted and processed either for the nonmetallic elements they contain or for the physical and chemical properties they possess.

People often do not realize the importance of nonmetallic minerals because they see only the products that result from their use and not the minerals themselves. That is, many nonmetallics are used up in the process of creating other products. Examples include the fluorite and limestone that are part of the steelmaking process, the abrasives required to make a piece of machinery, and the fertilizers needed to grow a food crop (**Table 23.2**).

The quantities of nonmetallic minerals used each year are enormous. A glance back at Figure 23.1 reminds us that the per capita consumption of nonfuel resources in the United States totals more than 11 metric tons, of which about 94 percent are nonmetallics. Nonmetallic mineral resources are commonly divided into two broad groups: *building materials* and *industrial minerals*. Because some substances have many different uses, they are found in both categories.

TABLE 23.2   Occurrences and Uses of Nonmetallic Minerals

Mineral	Uses	Geologic Occurrences
Apatite	Phosphorus fertilizers	Sedimentary deposits
Asbestos	Incombustible fibers	Metamorphic alteration (chrysotile)
Calcite	Aggregate; steelmaking; soil conditioning; chemicals; cement; building stone	Sedimentary deposits
Clay minerals	Ceramics; china	Residual product of weathering (kaolinite)
Corundum	Gemstones; abrasives	Metamorphic deposits
Diamond	Gemstones; abrasives	Kimberlite pipes; placers
Fluorite	Steelmaking; aluminum refining; glass; chemicals	Hydrothermal deposits
Garnet	Abrasives; gemstones	Metamorphic deposits
Graphite	Pencil lead; lubricant; refractories	Metamorphic deposits
Gypsum	Plaster of Paris	Evaporite deposits
Halite	Table salt; chemicals; ice control	Evaporite deposits; salt domes
Muscovite	Insulator in electrical applications	Pegmatites
Quartz	Primary ingredient in glass	Igneous intrusions; sedimentary deposits
Sulfur	Chemicals; fertilizer manufacture	Sedimentary deposits; hydrothermal deposits
Sylvite	Potassium fertilizers	Evaporite deposits
Talc	Powder used in paints, cosmetics, etc.	Metamorphic deposits

Limestone, perhaps the most versatile and widely used rock of all, is the best example (**Figure 23.28**). As a building material, it is used not only as crushed rock and building stone but also in the making of cement. Moreover, as an industrial mineral, limestone is an ingredient in the manufacture of steel and is used in agriculture to reduce the acidity of soils (see GEOgraphics 7.1, page 220).

Figure 23.28
Limestone quarry
Limestone is considered both a building material and an industrial mineral. This quarry is near Amsterdam, Indiana. (Photo by Daniel Dempster/Alamy Images)

Building Materials

Natural aggregate consists of crushed stone, sand, and gravel. From the standpoint of quantity and value, aggregate is a very important building material. The United States produces nearly 2 billion tons of aggregate per year, which represents about one-half of the country's entire nonenergy mining volume. Aggregate is produced commercially in every state and is used in nearly all building construction and in most public works projects.

Besides aggregate, other important building materials include gypsum for plaster and wallboard, clay for tile and bricks, and limestone and shale, which combine to form cement. Cement and aggregate go into the making of concrete, a material that is essential to practically all construction. Aggregate gives concrete its strength and volume, and cement binds the mixture into a rock-hard substance. Building just 2 kilometers of four-lane highway requires more than 85 thousand tons of aggregate. Building an average six-room house requires 90 tons of aggregate.

Because most building materials are widely distributed and present in almost unlimited quantities, they have little intrinsic value. Their economic worth comes only after the materials are removed from the ground and processed. Since their per-ton value compared with the values of metals and industrial minerals is low, mining and quarrying operations are usually undertaken to satisfy local needs. Except for special types of cut stone used for building and monuments, transportation costs greatly limit the distance most building materials can be profitably moved.

Industrial Minerals

Many nonmetallic resources are classified as industrial minerals. In some instances, these materials are important because they are sources of specific chemical elements or compounds. Such minerals are used to manufacture chemicals and produce fertilizers. In other cases, their importance is related to the physical properties they exhibit. Examples include minerals such as corundum and garnet, which are used as abrasives. Although supplies of industrial minerals are generally plentiful, most of these minerals are not nearly as abundant as building materials. Moreover, deposits are far more restricted in distribution and extent. As a result, many of these nonmetallic resources must be transported considerable distances, which of course adds to their cost. Unlike most building materials, which need a minimum of processing before they are ready to use, many industrial minerals require considerable processing to extract the desired substance at the proper degree of purity for its ultimate use.

Fertilizers The growth in world population beyond 7 billion requires that the production of basic food crops continue to expand. Therefore, fertilizers—primarily nitrate, phosphate, and potassium compounds—are extremely important to agriculture. The synthetic nitrate industry, which derives nitrogen from the atmosphere, is the source of practically all the world's nitrogen fertilizers. The primary source of phosphorus and potassium, however, remains Earth's crust. The mineral apatite is the primary source of phosphate. In the United States, most production comes from marine sedimentary deposits in Florida and North Carolina. Although potassium is an abundant element in many minerals, the primary commercial sources are evaporite deposits containing the mineral sylvite. In the United States, deposits near Carlsbad, New Mexico, have been especially important. **Figure 23.29** shows a mining operation in southern Utah that produces sylvite (commonly called potash).

Figure 23.29
Potash mine evaporation ponds west of Moab, Utah Potash (potassium chloride) at this site is obtained using a system that combines solution mining and solar evaporation. Water from the nearby Colorado River is injected into deposits that are 900 meters (3000 feet) below the surface. The mineral-rich water is brought to the surface and piped to 400 acres of shallow ponds. The water evaporates, leaving behind deposits of potash and common salt (sodium chloride). A blue dye is added to assist the evaporation process. The crystals are harvested and sent to a mill, where the potash is separated from the salt by using a flotation process. (Photo by Michael Collier)

Sulfur Because it has many uses, sulfur is an important nonmetallic resource. In fact, the quantity of sulfur used is considered one indicator of a country's level of industrialization. More than 80 percent is used to produce sulfuric acid. Although its principal use is in the manufacture of phosphate fertilizer, sulfuric acid has many other applications as well. Sources include deposits of native sulfur associated with salt domes and volcanic areas, as well as common iron sulfides such as pyrite. In recent years an increasingly important source has been the sulfur removed from coal, oil, and natural gas in order to make these fuels less polluting.

Salt Common salt, known by the mineral name *halite*, is another important and versatile resource. It is among the most prominent nonmetallic minerals used as a raw material in the chemical industry. In addition, large quantities are used to "soften" water and to keep streets and highways free of ice. Of course, most people are aware that it is also a basic nutrient and a part of many food products.

Salt is a common evaporite, and thick deposits are exploited using conventional underground mining techniques. Subsurface deposits are also tapped, using brine wells in which a pipe is introduced into a salt deposit and water is pumped down the pipe. The salt dissolved by the water is brought to the surface through a second pipe. This is also the process used to obtain the potassium chloride (potash) in Figure 23.29. In addition, seawater continues to serve as a source of salt, as it has for centuries. The salt is harvested after the Sun evaporates the water.

23.8 Concept Checks

1. Which is greater, the per capita consumption of metallic resources or of nonmetallic mineral resources?

2. Distinguish between *building materials* and *industrial minerals*.

3. List two examples of building materials. What are three examples of industrial minerals?

4. What is an example of a substance that may be either a building material or an industrial mineral?

23.1 Renewable and Nonrenewable Resources

Distinguish between renewable and nonrenewable resources.

KEY TERMS renewable resources, nonrenewable resources

- Renewable resources can be replenished over relatively short time spans. An example is trees used for lumber and paper pulp. Nonrenewable resources form so slowly that, from a human standpoint, Earth contains fixed supplies. Examples include fuels such as oil and coal and metals such as copper and gold.

Q Some resources can be recycled. Does that mean they can be classified as renewable? Explain.

23.2 Energy Resources: Traditional Fossil Fuels

Compare and contrast fossil fuel types and describe how each satisfies U.S. energy consumption.

KEY TERMS fossil fuel, oil trap, reservoir rock, cap rock, oil sands, hydraulic fracturing

- Coal, oil, and natural gas are all fossil fuels. In each, the energy of ancient sunlight, captured by photosynthesis, is stored in the hydrocarbons that made up plants or other living things and were then buried by sediments.
- Coal is formed from compressed plant fragments, originally deposited in ancient swamps. Burning it supplies about 18 percent of U.S. energy use. Coal mining can be risky and environmentally damaging, and burning coal generates several kinds of pollution.
- Oil and natural gas are formed from the heated remains of marine plankton. Together, they account for about 63 percent of U.S. energy use. Both oil and natural gas leave their source rock (typically shale) and migrate to an oil trap made up of other, more porous rocks, called reservoir rocks, covered by a suitable impermeable cap rock.
- Oil sands are sedimentary deposits that contain bitumen in their pore spaces. Because of its high viscosity, bitumen cannot be pumped out of the rock and must be processed with steam (a significant energy input).
- Hydraulic fracturing (or "fracking") is a method of opening up pore space in otherwise impermeable rocks, permitting natural gas to flow out into wells.

Q As you can see here, the Sinclair Oil Company's logo speaks directly to the "fossil" nature of the fuel it sells. However, is it likely that any *dinosaur* carbon ended up in Sinclair's oil? Explain.

Vespasian/Alamy

23.3 Nuclear Energy

Describe the importance of nuclear energy and discuss its pros and cons.

KEY TERM nuclear fission

- Nuclear power plants use a controlled nuclear fission chain reaction, in which heavy atoms such as uranium are bombarded with neutrons to cause atoms to split. This releases heat that boils water and drives steam turbines that generate electricity.
- Uranium ore is mined from old placer deposits or sites of groundwater precipitation. The proportion of uranium-235 must be concentrated before it is capable of supporting a fission reaction.
- Nuclear power plants carry significant risks, and building a safe plant is a very expensive proposition.

Q Current U.S. uranium reserves have been estimated at about 1227 million pounds of uraninite. At current rates of usage, this represents about 23 years' worth of demand. If the United States were to raise its nuclear power generation capacity to 16 percent (compare to Figure 23.2), how long would the U.S. uranium reserves last?

23.4 Renewable Energy

List and discuss the major sources of renewable energy. Describe the contribution of renewable energy to the overall U.S. energy supply.

KEY TERMS hydroelectric power, geothermal energy, biomass

- Renewable energy resources are replenished rapidly, and their use can be sustained for an indefinite period of time. Free solar energy, which will last as long as the Sun shines, can be captured through passive solar design, active solar collectors (to heat liquids), and photovoltaic cells (for electricity).
- Wind energy is harvested with windmills or turbines. Wind speed is an important variable when considering a site to erect a turbine.
- When running water is dammed, its energy may be harvested as hydroelectric power. At suitable natural sites for damming, pumped water-storage systems are sometimes used to produce "artificial" hydroelectric power.
- Geothermal energy uses heat in Earth's subsurface to produce hot water or electricity. A geothermal site requires (1) a potent source of heat, (2) large subsurface reservoirs, and (3) a cap of low-permeability rock.
- Biomass is animal or plant matter that may be burned directly or converted into a fuel. Campfires, corn ethanol gasoline additives, and methane capture at landfills are all examples of biomass energy.
- Tidal energy works by impounding the waters of the high tide and then releasing them at low tide through a dam, much like with hydroelectric power. It requires a suitable coastline and tidal range, so it is not applicable in most areas.

23.5 Mineral Resources

Distinguish among resource, reserve, and ore.

KEY TERMS mineral resource, reserve, ore

- Ores are metallic mineral resources that can be economically mined. We refer to the "stockpile" of as-yet-unmined but economically recoverable mineral resources as *reserves*. To be considered valuable, a mineral resource must have the element of interest concentrated above the level of its average crustal abundance.

Q Give two examples of events that would cause estimates of U.S. platinum reserves to be revised.

23.6 Igneous and Metamorphic Processes

Explain how different igneous and metamorphic processes produce economically significant mineral deposits.

KEY TERMS pegmatite, hydrothermal solution, vein deposit, disseminated deposit

- Igneous processes concentrate some elements through magmatic differentiation, emplacement of pegmatites, and eruption of kimberlites.
- Magmas may give off hydrothermal (hot-water) solutions that penetrate surrounding rock, carrying dissolved metals in them. The metal ores may be precipitated as fracture-filling deposits (veins) or may "soak" into surrounding strata, producing a great many tiny deposits disseminated throughout the host rock. Black smokers at deep-sea vents are another example of metal-rich hydrothermal deposits.
- Contact and regional metamorphism may both produce concentrations of nonmetallic and metallic mineral resources, particularly where igneous plutons intrude limestone.

Q This sample consists mainly of quartz, potassium feldspar, and tourmaline. The large tourmaline crystal at the top right is more than 4 centimeters (1.5 inches) long. What term is applied to rocks such as this that are composed of unusually large crystals? With what process is this rock associated?

Harry Taylor/DK Images

23.7 Mineral Resources Related to Surface Processes

Discuss ways in which surface processes produce ore deposits.

KEY TERMS secondary enrichment, placer

- Weathering creates ore deposits by concentrating minor amounts of metals into economically valuable deposits. The process, called secondary enrichment, is accomplished either by (1) removing undesirable materials and leaving the desired elements enriched in the upper zones of the soil or (2) removing and carrying the desirable elements to lower zones, where they are redeposited and become more concentrated. Bauxite, the principle ore of aluminum, is associated with the first of these processes.
- Placer deposits form when tough, dense minerals such as gold are sorted by water currents and separated from lower-density sediments.

Q Would Jamaica or Sweden be more likely to have bauxite deposits forming at the present time? Explain.

23.8 Nonmetallic Mineral Resources

Distinguish between two broad categories of nonmetallic mineral resources and list examples of each.

KEY TERM nonmetallic mineral resources

- Earth materials that are not used as fuels or processed for the metals they contain are referred to as nonmetallic resources. Many are sediments or sedimentary rocks. The two broad groups of nonmetallic resources are building materials and industrial minerals.

Q Is limestone considered an industrial mineral or a building material? Include examples as part of your explanation.

Give It Some Thought

1. At one time, most of the energy used in the United States was renewable. Subsequently, things changed, and most of our energy began to come from nonrenewable sources. What was the renewable energy source that was once dominant? What replaced it?

2. While you're in the car with a friend, a radio news story mentions a coal-mine accident in which some miners have been injured. After hearing this, your friend says, "I thought coal was sort of an old-fashioned fuel that wasn't used much anymore." How would you convince your friend that he's mistaken?

3. The scene in this photo seems an unlikely source of renewable energy, but it is. Describe how this could be the case.

Radharc Images/Alamy

4. In the years to come, coal will likely represent a smaller percentage of U.S. energy consumption than it does at present. However, it is also possible that coal production may increase. Explain this apparent paradox.

5. As you and a companion are taking a walk, you mention that you have been reading about renewable and nonrenewable resources. Shortly after mentioning this fact, the two of you pass a refuse container with a sign that says, "Aluminum cans only." Upon seeing this, your companion says, "They sure recycle lots of aluminum these days. That makes it a renewable resource, right?" How would you reply?

Texas Stock Photo/Alamy

6. This chapter includes a discussion of industrial minerals. Are these substances actually minerals? That is, do they meet the definition of *mineral* outlined in Chapter 3? Explain.

7. Assume that you just read a magazine article about a copper mine that is closing. In the article, a mining geologist states that there is plenty of copper-bearing rock remaining and that the concentration of copper is uniform and of equal quality to what has been mined in the past. Later in the article, a spokesperson for the mine owner states, "No ore remains." Assume that both the geologist and the spokesperson are correct and suggest an explanation for the apparently contradicting statements.

MasteringGeology™

24

Touring Our Solar System[1]

Selfie of *Curiosity* rover as it moves up the slopes of Mount Sharp on Mars. This is a composite of several images captured by a camera located at the end of the rover's robotic arm. (Photo courtesy of NASA/JPL-Caltech/MSSS)

[1]This chapter was revised with the assistance of Professors Teresa Tarbuck and Mark Watry.

Planetary geology is the study of the formation and evolution of the bodies in our solar system—including the eight planets and myriad smaller objects: moons, dwarf planets, asteroids, comets, and meteoroids. Studying these objects provides valuable insights into the dynamic processes that operate on Earth. Understanding how other atmospheres evolve helps scientists build better models for predicting climate change. Studying tectonic processes on other planets helps us appreciate how these complex interactions alter Earth. In addition, seeing how erosional forces work on other bodies allows us to observe the many ways landscapes are created. Finally, the uniqueness of Earth, a body that harbors life, is revealed through the exploration of other planetary bodies.

24.1 | Our Solar System: An Overview

Describe the formation of the solar system according to the nebular theory. Compare and contrast the terrestrial and Jovian planets.

The Sun is at the center of a revolving system, trillions of miles wide, consisting of eight planets, their satellites, and numerous smaller asteroids, comets, and meteoroids. An estimated 99.85 percent of the mass of our solar system is contained within the Sun. Collectively, the planets account for most of the remaining 0.15 percent. Starting from the Sun, the planets are Mercury, Venus, Earth, Mars, Jupiter, Saturn, Uranus, and Neptune (**Figure 24.1**). Pluto was reclassified in 2003 as a member of a new class of solar system bodies called *dwarf planets*.

Tethered to the Sun by gravity, all the planets travel in the same direction, on slightly elliptical orbits (**Table 24.1**). Gravity causes objects nearest the Sun to travel fastest. Therefore, Mercury has the highest orbital velocity, 48 kilometers (30 miles) per second, and the shortest period of revolution around the Sun, 88 Earth days. By contrast, the distant dwarf planet Pluto has an orbital speed of just 5 kilometers (3 miles) per second and requires 248 Earth-years to complete one revolution. Most large bodies orbit the Sun approximately in the same plane. The planets' inclination with respect to the Earth–Sun orbital plane, known as the *ecliptic*, is shown in Table 24.1.

Nebular Theory: Formation of the Solar System

The **nebular theory**, which explains the formation of the solar system, proposes that the Sun and planets formed from a rotating cloud of interstellar gases (mainly hydrogen and helium) and dust called the **solar nebula**. As the solar nebula contracted due to gravity, most of the material collected in the center to form the hot *protosun*.

The remaining materials formed a thick, flattened, rotating disk, within which matter gradually cooled and condensed into grains and clumps of icy, rocky material. Repeated collisions resulted in most of the material clumping together into larger and larger chunks that eventually became asteroid-sized objects called **planetesimals**.

The composition of planetesimals was largely determined by their proximity to the protosun. As you might expect, temperatures were highest in the inner solar system and decreased toward the outer edge of the disk. Therefore, between the present orbits of Mercury and Mars, the planetesimals were composed of materials with high melting temperatures—metals and rocky substances. Then, through repeated collisions and accretion, these asteroid-sized rocky bodies combined to form the four **protoplanets** that eventually became Mercury, Venus, Earth, and Mars.

The planetesimals that formed beyond the orbit of Mars, where temperatures were low, contained high percentages of ices—water, carbon dioxide, ammonia, and methane—as well as small amounts of rocky and metallic debris. It was mainly from these planetesimals that the four outer planets eventually formed. The accumulation

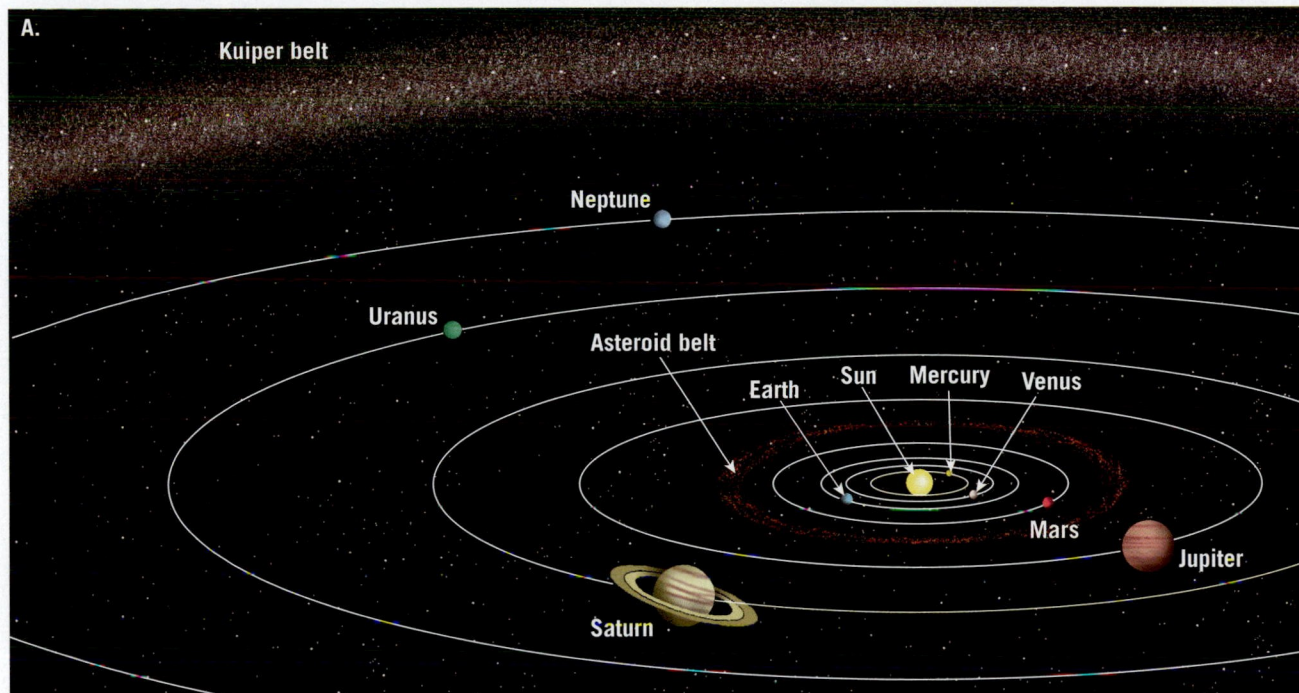

Kuiper belt

Neptune

Uranus

Asteroid belt

Earth   Sun   Mercury   Venus

Mars

Jupiter

Saturn

SmartFigure 24.1
Orbits of the planets
A. Artistic view of the solar system, in which planets are not drawn to scale.
B. Positions of the planets shown to scale using astronomical units (AU), where 1 AU is equal to the average distance from Earth to the Sun—150 million kilometers (93 million miles).
(https://goo.gl/ZFtHNL)

Tutorial

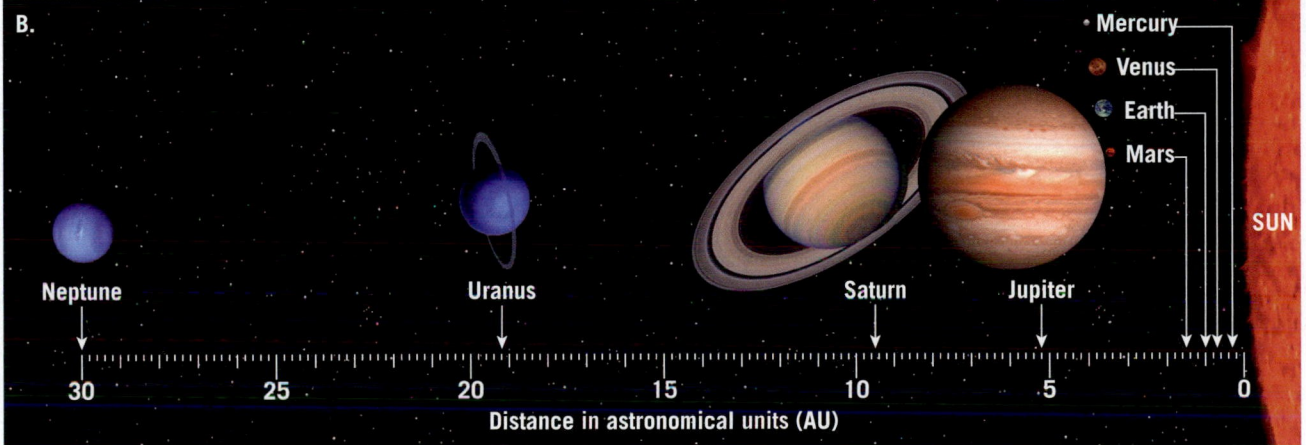

Mercury
Venus
Earth
Mars
SUN

Neptune   Uranus   Saturn   Jupiter

30   25   20   15   10   5   0

Distance in astronomical units (AU)

of ices accounts, in part, for the large sizes and low densities of these outer planets. The two most massive planets, Jupiter and Saturn, had surface gravities sufficient to attract and retain large quantities of hydrogen and helium, the lightest elements.

It took roughly 1 billion years after the protoplanets formed for the planets to gravitationally accumulate most of the interplanetary debris. This was a period of intense bombardment as the planets cleared their orbits by collecting much of the leftover material. The "scars" of this period are still evident on the Moon's surface. Because of the gravitational effect of the planets, particularly Jupiter, small bodies were flung into planet-crossing orbits or into interstellar space. The small fraction of interplanetary matter that escaped this violent period became either asteroids or comets. By comparison, the present-day solar system is a much quieter place, although many of these processes continue today at a reduced pace.

The Planets: Internal Structures and Atmospheres

The planets fall into two groups, based on location, size, and density: the **terrestrial (Earth-like) planets** (Mercury, Venus, Earth, and Mars) and the **Jovian (Jupiter-like) planets** (Jupiter, Saturn, Uranus, and Neptune). Because of their locations relative to the Sun, the four terrestrial planets are also known as *inner planets*, and the four Jovian planets are known as *outer planets*. A correlation exists between planetary locations and sizes: The inner planets are substantially smaller than the outer planets, also known as *gas giants*. For example, the diameter of Neptune (the smallest Jovian planet) is nearly 4 times larger than the diameter of Earth. Furthermore, Neptune's mass is 17 times greater than that of Earth or Venus.

Other properties that differ among the planets include densities, chemical compositions, orbital periods,

TABLE 24.1 Planetary Data

Planet	Symbol	AU*	Mean Distance from Sun: Millions of Miles	Mean Distance from Sun: Millions of Kilometers	Period of Revolution	Inclination of Orbit	Orbital Velocity: mi/s	Orbital Velocity: km/s
Mercury	☿	0.39	36	58	88^d	7°00'	29.5	47.5
Venus	♀	0.72	67	108	225^d	3°24'	21.8	35.0
Earth	⊕	1.00	93	150	365.25^d	0°00'	18.5	29.8
Mars	♂	1.52	142	248	687^d	1°51'	14.9	24.1
Jupiter	♃	5.20	483	778	12^{yr}	1°18'	8.1	13.1
Saturn	♄	9.54	886	1427	30^{yr}	2°29'	6.0	9.6
Uranus	⛢	19.18	1783	2870	84^{yr}	0°46'	4.2	6.8
Neptune	♆	30.06	2794	4497	165^{yr}	1°46'	3.3	5.3

Planet	Period of Rotation	Diameter: Miles	Diameter: Kilometers	Relative Mass (Earth = 1)	Average Density (g/cm^3)	Polar Flattening (%)	Eccentricity†	Number of Known Satellites‡
Mercury	59^d	3015	4878	0.06	5.4	0.0	0.206	0
Venus	243^d	7526	12,104	0.82	5.2	0.0	0.007	0
Earth	$23^h56^m04^s$	7920	12,756	1.00	5.5	0.3	0.017	1
Mars	$24^h37^m23^s$	4216	6794	0.11	3.9	0.5	0.093	2
Jupiter	9^h56^m	88,700	143,884	317.87	1.3	6.7	0.048	67
Saturn	10^h30^m	75,000	120,536	95.14	0.7	10.4	0.056	62
Uranus	17^h14^m	29,000	51,118	14.56	1.2	2.3	0.047	27
Neptune	16^h07^m	28,900	50,530	17.21	1.7	1.8	0.009	14

* AU = astronomical unit, Earth's mean distance from the Sun.

†Eccentricity is a measure of the amount an orbit deviates from a circular shape. The larger the number, the less circular the orbit.

‡Includes all satellites discovered as of July 2015. Satellites are celestial bodies that orbit a planet rather than orbiting a star like the Sun.

and numbers of satellites (see Table 24.1). Variations in the chemical composition of planets are largely responsible for their density differences. Specifically, the average density of the terrestrial planets is about 5 times the density of water, whereas the average density of the Jovian planets is only 1.5 times that of water. In fact, Saturn has a density only 0.7 times that of water, which means that it would float in a sufficiently large tank of water. The outer planets are also characterized by long orbital periods and numerous satellites.

Internal Structures Recall from Chapter 12 that shortly after Earth formed, segregation of material resulted in the formation of three major layers, defined by their chemical composition: the crust, mantle, and core. This type of chemical separation occurred in the other planets as well. However, because the terrestrial planets are compositionally different from the Jovian planets, the nature of these layers is different as well (Figure 24.2).

The terrestrial planets are dense, having relatively large cores of iron and nickel. The outer cores of Earth and Mercury are liquid, whereas the cores of Venus and Mars are thought to be only partially molten. This difference is attributable to Venus and Mars having lower internal temperatures than Earth and Mercury. Silicate minerals and other lighter compounds make up the mantles of the terrestrial planets. Finally, the silicate crusts of terrestrial planets are relatively thin compared to their mantles.

The two largest Jovian planets, Jupiter and Saturn, likely have small, solid cores consisting of iron compounds, like the cores of the terrestrial planets, and rocky material similar to Earth's mantle. Progressing outward, the layer above the core consists of liquid hydrogen that is under extremely high temperatures and pressures. There is substantial evidence that under these conditions, hydrogen behaves like a metal in that its electrons move freely about and are efficient conductors of both heat and electricity. Jupiter's intense magnetic field is thought to be the result of electric currents flowing within a spinning layer of liquid metallic hydrogen. Saturn's magnetic field is much weaker than Jupiter's, due to its smaller shell of liquid metallic hydrogen. Above this metallic layer, both Jupiter and Saturn are thought to be composed of molecular liquid hydrogen that is intermixed with helium. The outermost layers are gases of hydrogen and helium, as well as ices of water, ammonia, and methane—which mainly account for the low densities of these giants.

Uranus and Neptune also have small iron-rich, rocky cores, but their mantles are likely hot, dense water and ammonia. Above their mantles, the amount of hydrogen and helium increases, but these gases exist in much lower concentrations than in Jupiter and Saturn.

All planets except Venus and Mars have significant magnetic fields generated by flow of metallic materials in their liquid cores, or mantles. Magnetic fields play an important role in protecting a planet's surface from bombardment by the solar wind, a stream of charged particles ejected from the Sun. This protection is necessary for the survival of any life-forms. Venus has a weak field due to the interaction between the solar wind and its uppermost atmosphere (ionosphere), while the weak Martian magnetic field is thought to be a remnant from when its interior was hotter.

The Atmospheres of the Planets

The Jovian planets have very thick atmospheres composed mainly of hydrogen and helium, with lesser amounts of water, methane, ammonia, and other hydrocarbons. By contrast, the terrestrial planets, including Earth, have relatively meager atmospheres composed of carbon dioxide, nitrogen, and oxygen. Two factors explain these significant differences: solar heating (temperature) and gravity (**Figure 24.3**). These variables determine what planetary gases, if any, were captured by planets during the formation of the solar system and which of them were ultimately retained.

During planetary formation, the inner regions of the developing solar system were too hot for ices and gases to condense. By contrast, the Jovian planets formed where temperatures were low and solar heating of planetesimals was minimal. This allowed water vapor, ammonia, and methane to condense into ices. Hence, the gas giants contain large amounts of these volatiles. As the planets grew, the largest Jovian planets, Jupiter and Saturn, also attracted large quantities of the lightest gases, hydrogen and helium.

How did Earth acquire water and other volatile gases? It seems that early in the history of the solar system, gravitational tugs by the developing protoplanets sent planetesimals into very eccentric orbits. As a result, Earth was bombarded with

Figure 24.2
Comparing internal structures of the planets

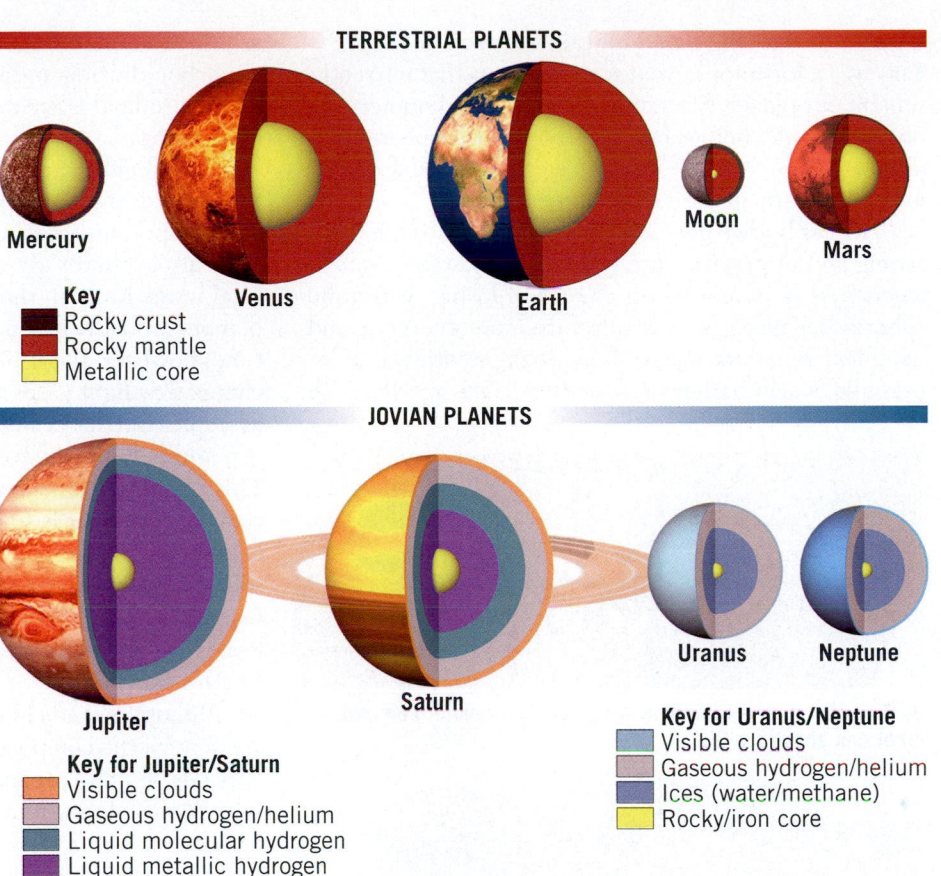

TERRESTRIAL PLANETS

Mercury   Venus   Earth   Moon   Mars

Key
Rocky crust
Rocky mantle
Metallic core

JOVIAN PLANETS

Jupiter   Saturn   Uranus   Neptune

Key for Jupiter/Saturn
Visible clouds
Gaseous hydrogen/helium
Liquid molecular hydrogen
Liquid metallic hydrogen
Rocky/iron core

Key for Uranus/Neptune
Visible clouds
Gaseous hydrogen/helium
Ices (water/methane)
Rocky/iron core

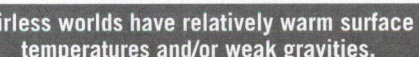
Airless worlds have relatively warm surface temperatures and/or weak gravities.

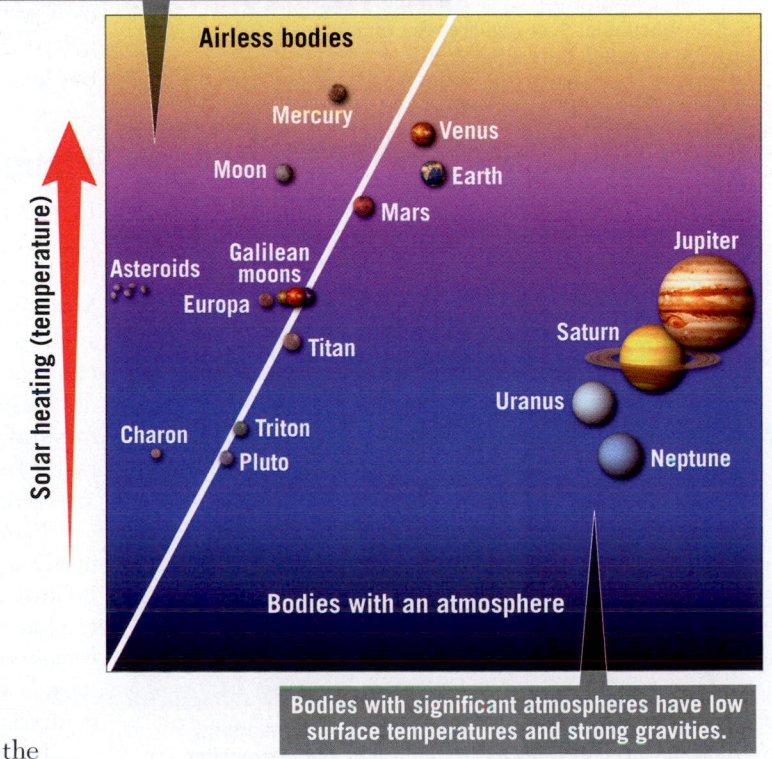

Airless bodies

Solar heating (temperature)

Mercury   Venus
Moon   Earth
Mars
Asteroids   Galilean moons   Jupiter
Europa
Titan   Saturn
Uranus
Charon   Triton
Pluto   Neptune

Bodies with an atmosphere

Bodies with significant atmospheres have low surface temperatures and strong gravities.

Gravity

SmartFigure 24.3
Bodies with atmospheres versus airless bodies Two factors largely explain why some solar system bodies have thick atmospheres, whereas others are airless. Airless worlds have relatively warm surface temperatures and/or weak gravities. Bodies with significant atmospheres have low surface temperatures and strong gravities. (https://goo.gl/0H6CPo)

Tutorial

icy objects that originated beyond the orbit of Mars. This was a fortuitous event for organisms that currently inhabit our planet. Mercury, our Moon, and numerous other small bodies lack significant atmospheres, although they certainly would have been bombarded by icy bodies early in their development.

Airless bodies develop where solar heating is strong and/or gravities are weak. Simply stated, *small warm bodies* have a better chance of losing their atmospheres because gas molecules are more energetic and need less speed to escape their weak gravities. For example, warm bodies with small surface gravity, such

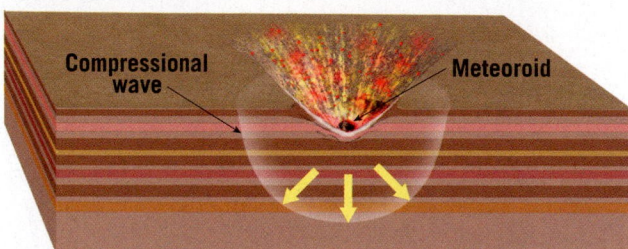

A. The energy of a rapidly moving body is transformed into heat and shock waves.

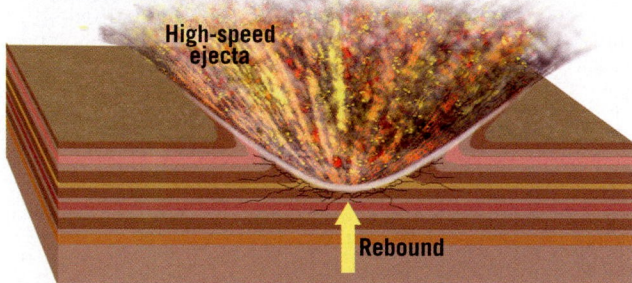

B. The rebound of over-compressed rock causes debris to be explosively ejected from the crater.

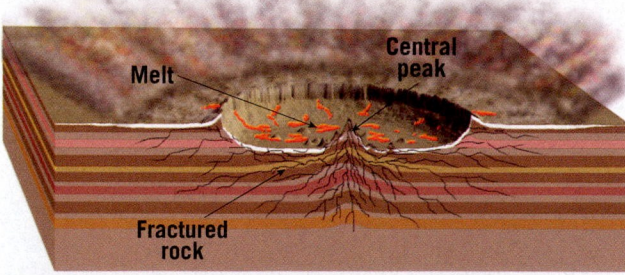

C. Heating melts some material that may be ejected from the crater as glass beads.

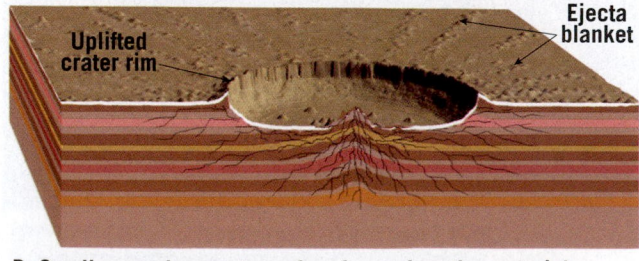

D. Small secondary craters often form when the material "splashed" from the impact crater strikes the surrounding landscape.

Figure 24.4
Formation of an impact crater

as our Moon, are unable to hold even heavy gases such as carbon dioxide and nitrogen. Mercury is massive enough to hold trace amounts of hydrogen, helium, and oxygen gases.

The slightly larger terrestrial planets, Earth, Venus, and Mars, retain some heavy gases, including water vapor, nitrogen, and carbon dioxide. However, their atmospheres are miniscule compared to their total mass. Early in their development, the terrestrial planets probably had much thicker atmospheres. Over time, however, these primitive atmospheres gradually changed as light gases trickled away into space. For example, Earth's atmosphere continues to leak hydrogen and helium (the two lightest gases) into space. This phenomenon occurs near the top of Earth's atmosphere, where air is so tenuous that nothing stops the fastest-moving particles from flying off into space. The speed required to escape a planet's gravity is called **escape velocity**. Because hydrogen is the lightest gas, it most easily reaches the speed needed to overcome Earth's gravity.

Billions of years in the future, the loss of hydrogen (one of the components of water) will eventually "dry out" Earth's oceans, ending its hydrologic cycle. Life, however, may remain sustainable in Earth's polar regions.

The massive Jovian planets have strong gravitational fields and thick atmospheres. Furthermore, because of their great distance from the Sun, solar heating is minimal. This explains why Saturn's moon Titan, which is small compared to Earth but much farther from the Sun, retains an atmosphere. Because the molecular motion of a gas is temperature dependent, even hydrogen and helium move too slowly to escape the gravitational pull of the Jovian planets.

Planetary Impacts

Planetary impacts between solar system bodies have occurred throughout the history of the solar system. On bodies that have little or no atmosphere (like the Moon) and, therefore, no air resistance, even the smallest pieces of interplanetary debris (meteorites) can reach the surface. At high enough velocities, this debris can produce microscopic cavities on individual mineral grains. By contrast, large **impact craters** result from collisions with massive bodies, such as asteroids and comets (**Figure 24.4**).

Planetary impacts were considerably more common in the early history of the solar system than they are today. Following that early period of intense bombardment, the rate of cratering diminished dramatically, and it now remains essentially constant. Because weathering and erosion are almost nonexistent on the Moon and Mercury, evidence of their cratered past is clearly evident.

On larger bodies, the presence of an atmosphere may cause the impacting objects to break up and/or decelerate. For example, Earth's atmosphere causes meteoroids with masses of less than 10 kilograms (22 pounds) to lose up to 90 percent of their speed as they penetrate the atmosphere. Therefore, impacts of low-mass bodies produce only small craters on Earth. Our atmosphere is much less effective in slowing large bodies; fortunately, they make very rare appearances.

The formation of a large impact crater is illustrated in Figure 24.4. The meteoroid's high-speed impact compresses the material it strikes, causing an almost instantaneous rebound, which ejects material from the surface. On Earth, impacts can occur at speeds that exceed 50 kilometers (30 miles) per second. Impacts at such high speeds produce shock waves that compress both the impactor and the material being impacted. Almost instantaneously, the over-compressed material rebounds and explosively ejects material out of the newly formed crater. This process is analogous to the detonation of an explosive device that has been buried underground.

Craters excavated by objects that are several kilometers across often exhibit a central peak, such as the one in the large crater in **Figure 24.5**. Much of the material expelled, called *ejecta*, lands in or near the crater, where it accumulates to form a rim. Large meteoroids may generate sufficient heat to melt and eject some of the impacted rock as glass beads. Specimens of glass beads produced in this manner, as well as melt breccia consisting of broken fragments welded by the heat of impact, have been collected on Earth, as well as the Moon, allowing planetary geologists to more fully understand such events.

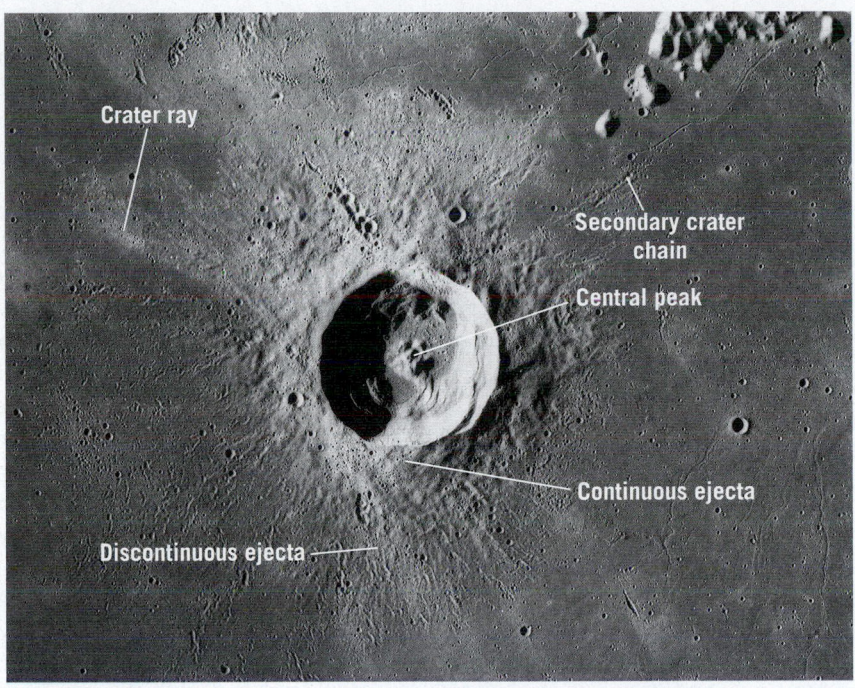

Figure 24.5
Lunar crater Euler
This 20-kilometer-wide (12-mile-wide) crater is located in the southwestern part of Mare Imbrium. Clearly visible are the bright rays, central peak, secondary craters, and large accumulation of ejecta near the crater rim. (Courtesy of NASA)

Labels on image: Crater ray; Secondary crater chain; Central peak; Continuous ejecta; Discontinuous ejecta

---

24.1 Concept Checks

1. Briefly outline the steps in the formation of our solar system, according to the nebular theory.

2. By what criteria are planets considered either terrestrial or Jovian?

3. What accounts for the large density differences between the terrestrial and Jovian planets?

4. Explain why the terrestrial planets have meager atmospheres compared to the Jovian planets.

24.2 | Earth's Moon: A Chip off the Old Block

List and describe the major features of Earth's Moon and explain how maria basins were formed.

The Earth–Moon system is unique partially because of the Moons large size relative to other bodies in the inner solar system. Mars is the only other terrestrial planet that has moons, but its tiny satellites are likely captured asteroids. Most of the 150 or so satellites of the Jovian planets are composed of low-density rock–ice mixtures, none of which resemble the Moon. As we will see later, our unique planet–satellite system is closely related to the mechanism that created it.

The diameter of the Moon is 3475 kilometers (2160 miles), about one-fourth of Earth's 12,756 kilometers (7926 miles). The Moon's surface temperature averages about 107°C (225°F) during daylight hours and −153°C (−243°F) at night. Because its period of rotation on its axis equals its period of revolution around Earth, the same lunar hemisphere always faces Earth. All of the landings of staffed *Apollo* missions were confined to the side of the Moon that faces Earth.

The Moon's density is 3.3 times that of water, comparable to that of *mantle* rocks on Earth but considerably less than Earth's average density (5.5 times that of water). The Moon's relatively small iron core is thought to account for much of this difference.

The Moon's low mass relative to Earth results in a lunar gravitational attraction that is one-sixth that of Earth. A person who weighs 150 pounds on Earth weighs only 25 pounds on the Moon, although the person's mass remains the same. This difference allows an astronaut to carry a heavy life-support system with relative ease. If not burdened with such a load, an astronaut could jump six times higher on the Moon than on Earth. The Moon's small mass (and low gravity) is the primary reason it was not able to retain an atmosphere.

How Did the Moon Form?

Until recently, the origin of the Moon—our nearest planetary neighbor—was a topic of considerable debate among scientists. Current models show that Earth is too small to have formed with a moon, particularly one so large. Furthermore, a captured moon would likely have an eccentric orbit similar to the captured moons that orbit the Jovian planets.

The current consensus is that the Moon formed as a result of a collision between a Mars-sized body and a youthful, semimolten Earth about 4.5 billion years ago. During this collision, some of the ejected debris was thrown into orbit around Earth and gradually coalesced to form the Moon. Computer simulations show that most of the ejected material would have come from the rocky mantle of the impactor, while its core was assimilated into the growing Earth. This *impact model* is consistent with the Moon having a proportionately smaller core than Earth's and, hence, a lower density.

The Lunar Surface When Galileo first pointed his telescope toward the Moon, he observed two different types of terrain: dark lowlands and brighter, highly cratered highlands (**Figure 24.6**). Because the dark regions appeared to be smooth, resembling seas on Earth, they were called **maria** (*mar* = sea, singular *mare*). The *Apollo 11* mission showed conclusively that the maria are exceedingly smooth plains composed of basaltic lavas. These vast plains are strongly concentrated on the side of the Moon facing Earth and cover about 16 percent of the lunar surface. The lack of large volcanic cones on these surfaces is evidence of high eruption rates of very fluid basaltic lavas similar to the Columbia Plateau flood basalts on Earth.

By contrast, the Moon's light-colored areas resemble Earth's continents, so the first observers dubbed them **terrae** (Latin for "lands"). These areas are now generally referred to as the **lunar highlands** because they are elevated several kilometers above the maria. Rocks retrieved from the highlands are mainly breccias, pulverized by massive bombardment early in the Moon's history. The arrangement of terrae and maria has resulted in the legendary "face" of the "man in the Moon."

Some of the most obvious lunar features are impact craters. A meteoroid 3 meters (10 feet) in diameter can blast out a crater 50 times larger, or about 150 meters (500 feet) in diameter. The larger craters shown in Figure 24.6, such as Kepler and Copernicus (32 and 93 kilometers [20 and 57 miles] in diameter, respectively), were created from bombardment by bodies 1 kilometer (0.62 mile) or more in diameter. These two craters are thought to be relatively young because of the bright *rays* (light-colored ejected material) that radiate from them for hundreds of kilometers.

History of the Lunar Surface The evidence used to unravel the history of the lunar surface comes primarily from radiometric dating of rocks returned from *Apollo* missions and studies of crater densities—counting the number of craters per unit area. The greater the crater density, the older the feature is inferred to be. Such evidence suggests that, after the Moon coalesced, it passed through the following four phases: (1) formation of the original

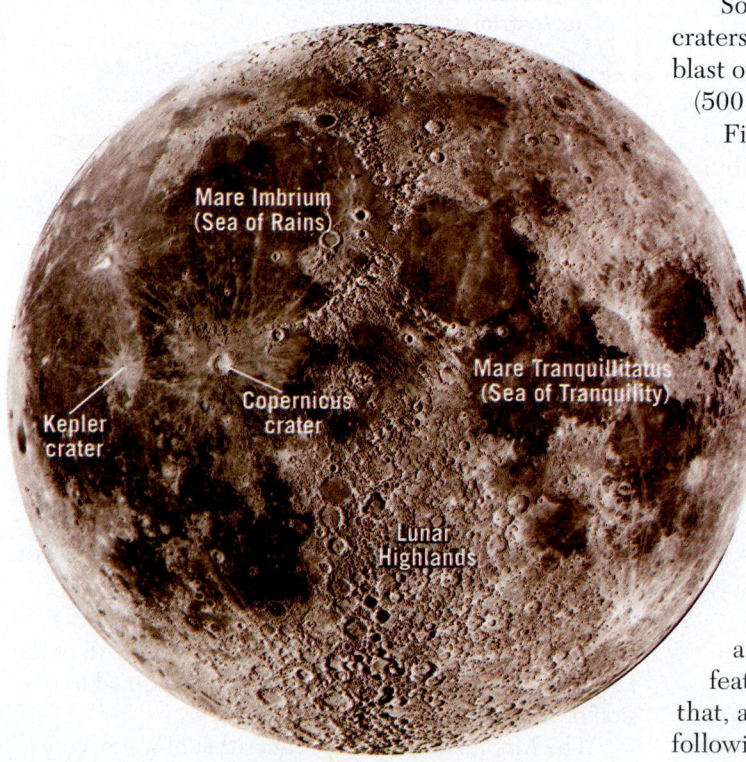

Figure 24.6
Telescopic view of the lunar surface The major features are the dark maria and the light, highly cratered highlands. (Lick Observatory Publications Office)

Mare Imbrium (Sea of Rains)

Mare Tranquillitatus (Sea of Tranquility)

Copernicus crater

Kepler crater

Lunar Highlands

crust, (2) excavation of the large impact basins, (3) filling of maria basins, and (4) formation of rayed craters.

During the late stages of its accretion, the Moon's outer shell was most likely completely melted—literally a magma ocean. Then, about 4.4 billion years ago, the magma ocean began to cool and underwent magmatic differentiation (see Chapter 4). Most of the dense minerals, olivine and pyroxene, sank, while less-dense silicate minerals floated to form the Moon's crust. The highlands are made of these igneous rocks, which rose buoyantly like "scum" from the crystallizing magma. The most common highland rock type is *anorthosite*, which is composed mainly of calcium-rich plagioclase feldspar.

Once formed, the lunar crust was continually impacted as the Moon swept up debris from the solar nebula. During this time, several large impact basins were created. Then, about 3.8 billion years ago, the Moon, as well as the rest of the solar system, experienced a sudden drop in the rate of meteoritic bombardment.

The Moon's next major event was the filling of the large impact basins, which had been created at least 300 million years earlier (**Figure 24.7**). Radiometric dating of the maria basalts puts their age between 3.0 billion and 3.5 billion years, considerably younger than the initial lunar crust.

The maria basalts are thought to have originated at depths between 200 and 400 kilometers (125 and 250 miles). They were likely generated by a slow rise in temperature attributed to the decay of radioactive elements. Partial melting probably occurred in several isolated pockets, as indicated by the diverse chemical makeup of the rocks retrieved during the *Apollo* missions. Recent evidence suggests that some mare-forming eruptions may have occurred as recently as 1 billion years ago.

Other lunar surface features related to this period of volcanism include small shield volcanoes (8–12 kilometers [5–7.5 miles] in diameter), evidence of pyroclastic eruptions, rilles (narrow winding valleys thought to be lava channels), and grabens (down-faulted valleys).

The last prominent features to form were rayed craters, as exemplified by the 93-kilometer-wide (56-mile-wide) Copernicus crater shown in Figure 24.7. Material ejected from these craters blankets the maria surfaces and many older, rayless craters. The relatively young Copernicus crater is thought to be about 1 billion years old. Had it formed on Earth, weathering and erosion would have long since obliterated it.

Today's Lunar Surface
The Moon's small mass and low gravity account for its lack of atmosphere and flowing water; therefore, the processes of weathering and erosion

that continually modify Earth's surface are absent. In addition, tectonic forces are no longer active on the Moon, so quakes and volcanic eruptions have ceased. Because the Moon is unprotected by an atmosphere, erosion is dominated by the impact of tiny particles from space (*micrometeorites*) that continually bombard its surface and gradually smooth the landscape. This activity has crushed and repeatedly mixed the upper portions of the lunar crust.

SmartFigure 24.7
Formation and filling of large impact basins
(Photo by NASA)
(https://goo.gl/9g1WUy)

Tutorial

Impact of an asteroid-size body produced a huge crater hundreds of kilometers in diameter and disturbed the lunar crust far beyond the crater.

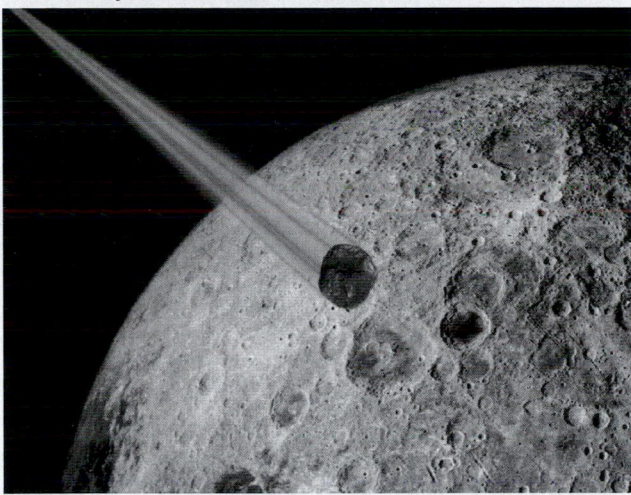

Filling of the impact crater with fluid basalts, perhaps derived from partial melting deep within the lunar mantle.

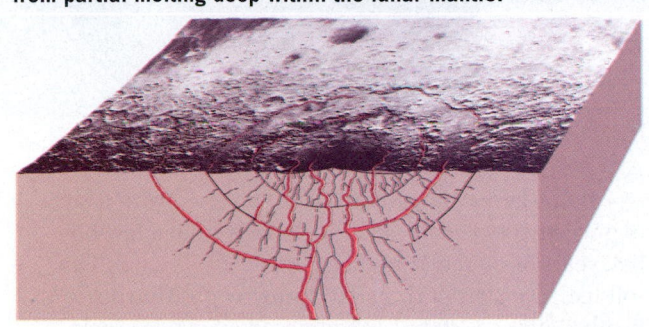

Today these basins make up the lunar maria and a few similar large structures on Mercury.

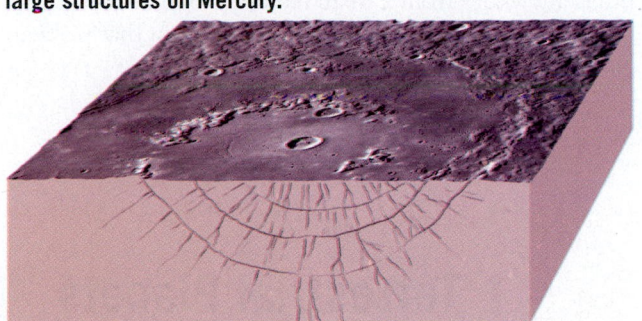

Figure 24.8
Astronaut Harrison Schmitt, sampling the lunar surface Notice the footprint (inset) in the lunar "soil," called regolith, which lacks organic material and is therefore not a true soil. Harrison Schmitt was the only geologist to go to the Moon. (Courtesy of NASA)

Both the maria and terrae are mantled with a layer of gray, unconsolidated debris derived from a few billion years of meteoric bombardment (**Figure 24.8**). This soil-like layer, properly called **lunar regolith** (*rhegos* = blanket, *lithos* = stone), is composed of igneous rocks, breccia, glass beads, and fine *lunar dust*. The lunar regolith is anywhere from 2 to 20 meters (6.6 to 66 feet) thick, depending on the age of the lunar surface in that particular region.

24.2   Concept Checks

1. Briefly describe the origin of the Moon.

2. Compare and contrast the Moon's maria and highlands.

3. How are maria on the Moon similar to the Columbia Plateau in the Pacific Northwest?

4. How is crater density used in the relative dating of the Moon's surface features?

5. Summarize the major stages in the development of the modern lunar surface.

24.3 | Terrestrial Planets

Outline the principal characteristics of Mercury, Venus, and Mars. Describe their similarities to and differences from Earth.

The terrestrial planets, in order from the Sun, are Mercury, Venus, Earth, and Mars. Here we consider the three other Earth-like planets and compare their features to those of Earth.

Mercury: The Innermost Planet

Mercury, the innermost and smallest planet, revolves around the Sun quickly (88 days) but rotates slowly on its axis. Mercury's day–night cycle, which lasts 176 Earth days, is very long compared to Earth's 24-hour cycle. One "night" on Mercury is roughly equivalent to 3 months on Earth and is followed by the same duration of daylight.

Mercury has the greatest temperature extremes, from as low as −173°C (−280°F) at night to noontime temperatures exceeding 427°C (800°F), hot enough to melt tin and lead. These extreme temperatures make life as we know it impossible on Mercury.

Mercury absorbs most of the sunlight that strikes it, reflecting only 6 percent into space, a characteristic of terrestrial bodies with little or no atmosphere. The minuscule amount of gas that is present on Mercury may have originated from several sources, including ionized gas from the Sun, ices that vaporized during a relatively recent comet impact, and/or outgassing of the planet's interior.

Although Mercury is small and scientists expected the planet's interior to have already cooled, NASA's *Messenger* spacecraft measured Mercury's magnetic field in 2012 and found it to be about 100 times less than Earth's. This suggests that Mercury has a large core that remains hot and fluid—a requirement for generating a magnetic field.

Mercury resembles Earth's Moon in that it has very low reflectivity, no sustained atmosphere, numerous volcanic features, and a heavily cratered terrain (**Figure 24.9**). The largest-known impact crater (1300 kilometers [800 miles] in diameter) on Mercury is Caloris Basin. Images and other data gathered by *Mariner 10* show evidence of volcanism in and around Caloris Basin and a few other smaller basins. Also like our Moon, Mercury has smooth plains that cover nearly 40 percent of the area imaged by *Mariner 10*. Most of these smooth areas are associated with large impact basins, including Caloris Basin, where lava partially filled the basins and surrounding lowlands. Consequently, these smooth plains appear to be similar in origin to lunar maria. Recently, *Messenger* found evidence of volcanism by revealing thick volcanic deposits similar to those in the Columbia Basin on Earth. Researchers were also surprised by the recent detection of probable ice caps on Mercury.

Venus: The Veiled Planet

Venus, second only to the Moon in brilliance in the night sky, is named for the Roman goddess of love and beauty. It orbits the Sun in a nearly perfect circle once every 225 Earth days. However, Venus rotates in the opposite direction of the other planets (*retrograde motion*) at an agonizingly slow pace: 1 Venus day is equivalent to about 243 Earth days. Venus has the densest atmosphere of the terrestrial planets, consisting mostly of carbon dioxide (97 percent)—the prototype for an extreme *greenhouse effect*. As a consequence, the surface temperature of Venus averages more than 450°C (900°F) day and night. Temperature variations at the surface are generally minimal because of the intense mixing within the planet's dense atmosphere. Investigations of the planet's extreme and uniform

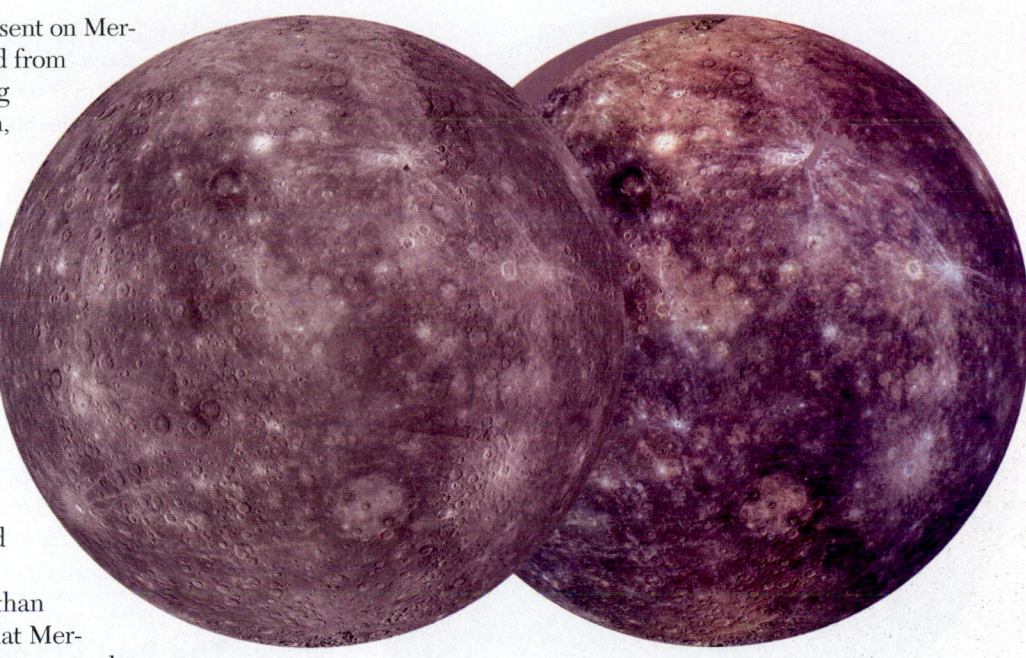

Figure 24.9
Two views of Mercury On the left is a monochromatic image, while the image on the right is color enhanced. These are high-resolution mosaics constructed from thousands of images obtained by the *Messenger* orbiter. (Courtesy of NASA)

surface temperature led scientists to more fully understand how the greenhouse effect operates on Earth.

The composition of the Venusian interior is probably similar to Earth's. However, Venus's weak magnetic field means its internal dynamics must be very different from Earth's. Scientists think that mantle convection operates on Venus, but the processes of plate tectonics do not appear to have contributed to the present Venusian topography.

The surface of Venus is completely hidden from view by a thick cloud layer composed mainly of tiny sulfuric acid droplets. In the 1970s, despite extreme temperatures and pressures, four Russian spacecraft landed successfully and obtained surface images. As expected, however, all the probes were crushed by the planet's immense atmospheric pressure, approximately 90 times that on Earth, within an hour of landing. Using radar imaging, the unstaffed spacecraft *Magellan* mapped Venus's surface in stunning detail (**Figure 24.10**).

A few thousand impact craters have been identified on Venus—far fewer than on Mercury and Mars but more than on Earth. Researchers expected that Venus would show evidence of extensive cratering from the heavy bombardment period but found instead that a period of extensive volcanism was responsible for resurfacing Venus. The planet's thick atmosphere also limits the number of impacts by breaking up large incoming meteoroids and incinerating most of the small debris.

About 80 percent of the Venusian surface consists of low-lying plains covered by lava flows, some of which traveled along lava channels that extend hundreds of kilometers (**Figure 24.11**). Venus's Baltis Vallis, the longest-known lava channel in the solar system, meanders 6800 kilometers (4225 miles) across the planet. More than 1000 volcanoes with diameters greater than 20 kilometers (12 miles) have been identified on Venus. However, high surface pressures keep the gaseous components in lava

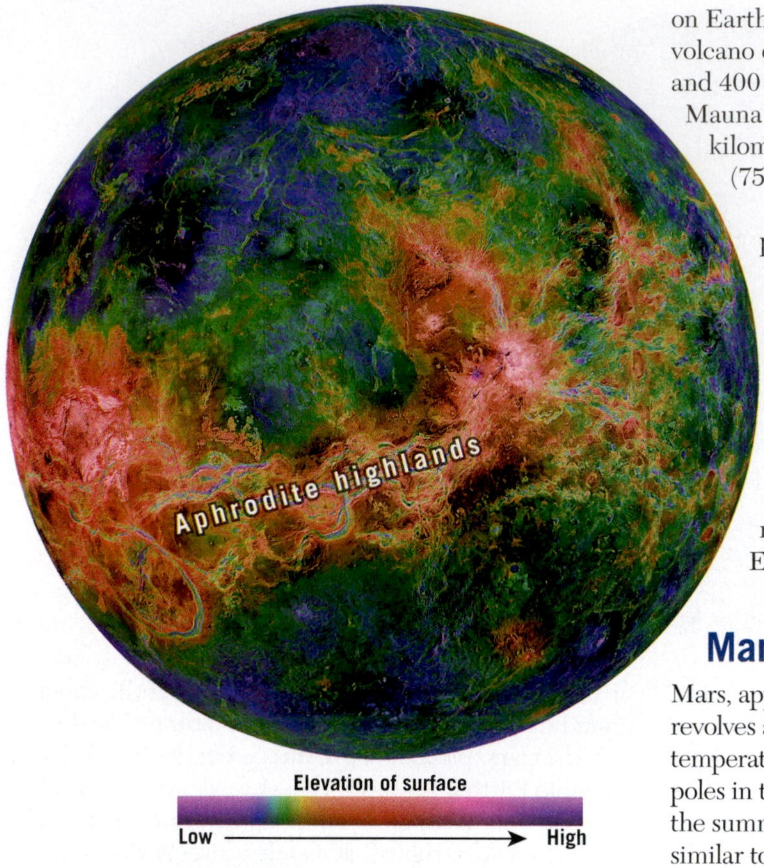

Figure 24.10
Global view of the surface of Venus This computer-generated image of Venus was constructed from years of investigations, culminating with the *Magellan* mission. The twisting bright features that cross the globe are highly fractured ridges and canyons of the eastern Aphrodite highland. (Courtesy of NASA)

Aphrodite highlands

Elevation of surface
Low ⟶ High

on Earth or Mars (**Figure 24.12**). Maat Mons, the largest volcano on Venus, is about 8.5 kilometers (5 miles) high and 400 kilometers (250 miles) wide. By comparison, Mauna Loa, the largest volcano on Earth, is about 9 kilometers high (5.5 miles) and only 120 kilometers (75 miles) wide.

Venus also has major highlands consisting of plateaus, ridges, and topographic rises that stand above the plains. The rises are thought to have formed where hot mantle plumes encountered the base of the planet's crust, causing uplift. Much like with mantle plumes on Earth, abundant volcanism is associated with mantle upwelling on Venus. Recent data collected by the European Space Agency's *Venus Express* suggest that Venus's highlands contain silica-rich granitic rock. As such, these elevated landmasses resemble Earth's continents, albeit on a much smaller scale.

Mars: The Red Planet

Mars, approximately one-half the diameter of Earth, revolves around the Sun in 687 Earth days. Mean surface temperatures range from lows of −140°C (−220°F) at the poles in the winter to highs of 20°C (68°F) at the equator in the summer. Although seasonal temperature variations are similar to Earth's, daily temperature variations are greater due to the very thin atmosphere of Mars (only 1 percent as dense as Earth's). The tenuous Martian atmosphere consists primarily of carbon dioxide (95 percent), with small amounts of nitrogen, oxygen, and water vapor.

from escaping. This retards the production of pyroclastic material and lava fountaining, phenomena that tend to steepen volcanic cones. In addition, Venus's high temperature allows lava to remain mobile longer and thus flow far from the vent. Both of these factors result in volcanoes that tend to be flatter and wider than those

Martian Topography Mars, like the Moon, is pitted with impact craters. The smaller craters are usually filled with wind-blown dust—confirming that Mars is a dry, desert world. The reddish color of the Martian landscape is due to iron oxide (rust). Large impact craters provide information about the nature of the Martian surface. For example, where the Martian surface is composed of dry rocky debris, ejecta similar in size and shape to that surrounding lunar craters is found. However, some Martian craters feature ejecta that looks like a muddy slurry was splashed from the crater. Planetary geologists infer that a layer of permafrost (frozen, icy soil) lies below portions of the Martian surface and that the heat of impacts melted the ice to produce the fluid-like appearance of these ejecta.

About two-thirds of the surface of Mars consists of heavily cratered highlands, concentrated mostly in its

Figure 24.11
Extensive lava flows on Venus This *Magellan* radar image shows a system of lava flows that originated from a volcano named Ammavaru, approximately 300 kilometers (186 miles) west of the scene. The lava, which appears bright in this radar image, has rough surfaces, whereas the darker flows are smooth. Upon breaking through the ridge belt (left of center), the lava collected in a 100,000-square-kilometer pool. (Courtesy of NASA)

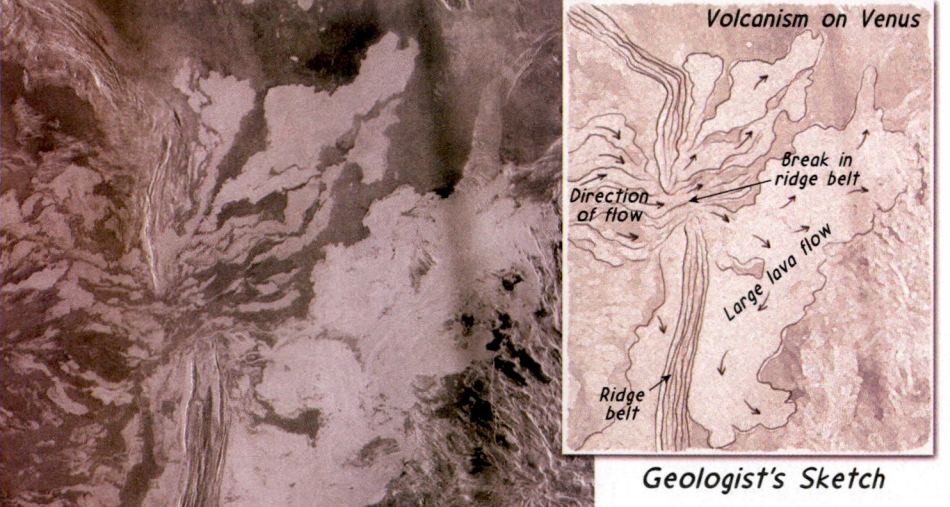

Volcanism on Venus

Break in ridge belt

Direction of flow

Large lava flow

Ridge belt

Geologist's Sketch

southern hemisphere (**Figure 24.13**). The period of extreme cratering occurred early in the planet's history and ended about 3.8 billion years ago, as it did in the rest of the solar system. Thus, Martian highlands are similar in age to the lunar highlands.

Based on relatively low crater counts, the northern plains, which account for the remaining one-third of the planet, are younger than the highlands. If Mars once had abundant water, it would have flowed to the north, which is lower in elevation, forming an expansive ocean (shown in blue in Figure 24.13, indicating lower elevation). The relatively flat topography of the northern plains, possibly the smoothest surface in the solar system, is consistent with vast outpourings of fluid basaltic lavas. Visible on these plains are volcanic cones, some with summit pits (craters) and lava flows with wrinkled edges.

Located along the Martian equator is an enormous elevated region about the size of North America, called the *Tharsis bulge*. This feature, about 10 kilometers (6 miles) high, appears to have been uplifted and capped with a massive accumulation of volcanic rock that includes the solar system's largest volcanoes.

The tectonic forces that created the Tharsis region also produced fractures that radiate from its center, like spokes on a bicycle wheel. Along the eastern flanks of the bulge, a series of vast canyons called *Valles Marineris* (Mariner Valleys) developed. Valles Marineris is so vast that it can be seen in the image of Mars in Figure 24.13. This canyon network was largely created by down-faulting rather than the stream erosion that carved Arizona's

Grand Canyon. Thus, it consists of graben-like valleys similar to the East African Rift valleys. Once formed, Valles Marineris grew by water erosion and collapse of the rift walls. The main canyon is more than 5000 kilometers (3000 miles) long, 7 kilometers (4 miles) deep, and 100 kilometers (60 miles) wide.

Other prominent features on the Martian landscape are large impact basins. Hellas, the largest identifiable basin, is about 2300 kilometers (1400 miles) in diameter and has the planet's lowest elevation. Debris ejected from this basin contributed to the elevation of the adjacent highlands. Other buried crater basins that are even larger than Hellas probably exist.

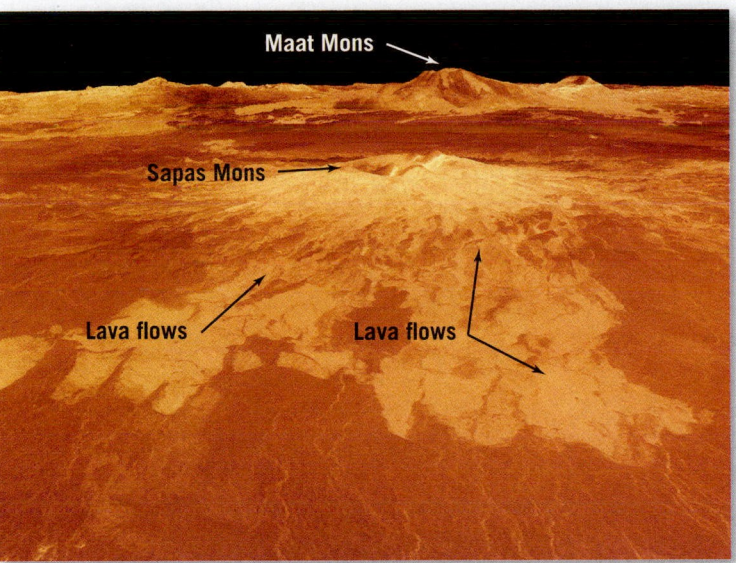

Figure 24.12
Volcanoes on Venus Sapas Mons is a broad volcano, 400 kilometers (250 miles) wide. The bright areas in the foreground are lava flows. Another large volcano, Maat Mons, is in the background. (Courtesy of NASA)

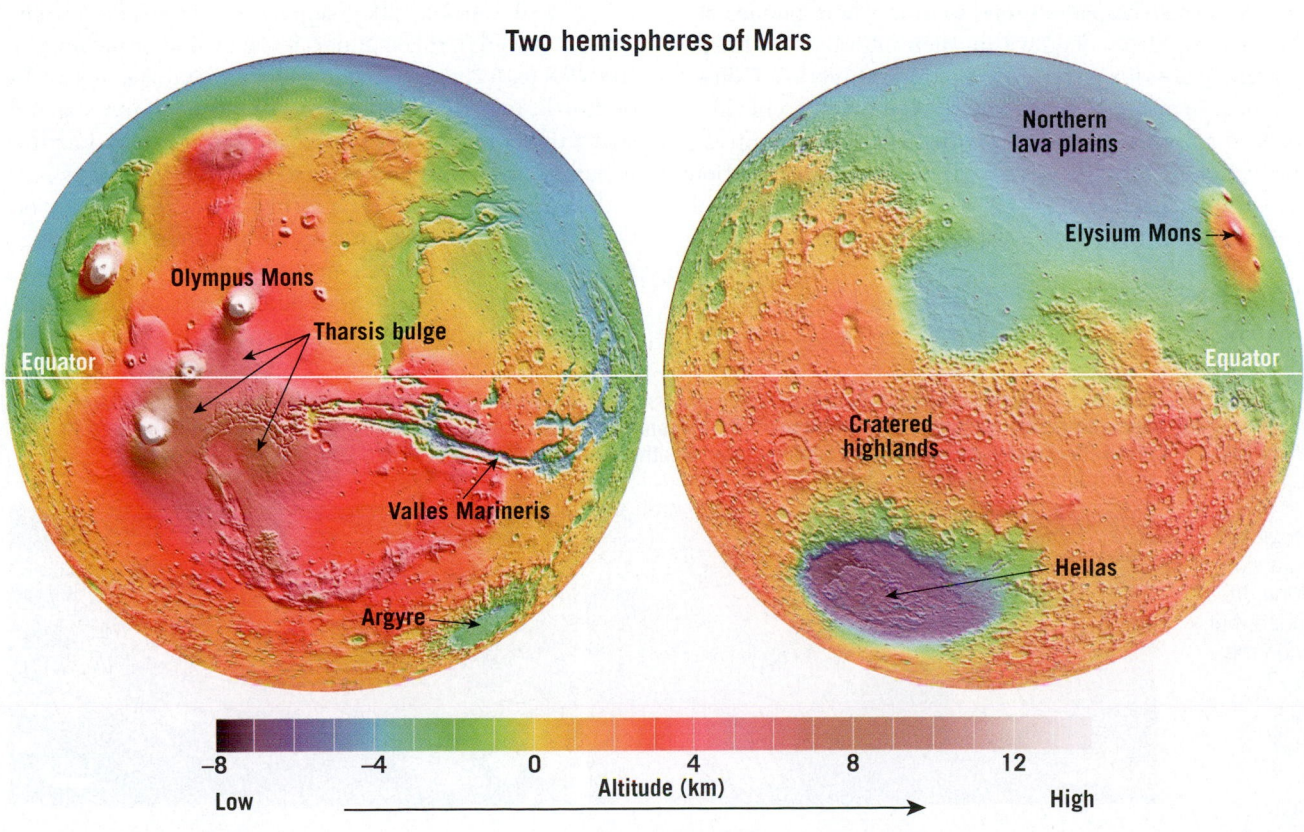

Two hemispheres of Mars

Figure 24.13
Two Hemispheres of Mars Color represents height above (or below) the mean planetary radius: White is about 12 kilometers above average, and dark blue is 8 kilometers below average. (Courtesy of NASA)

SmartFigure 24.14
Olympus Mons This massive inactive shield volcano on Mars covers an area about the size of the state of Arizona. (Courtesy of NASA) (https://goo.gl/RlbimQ)

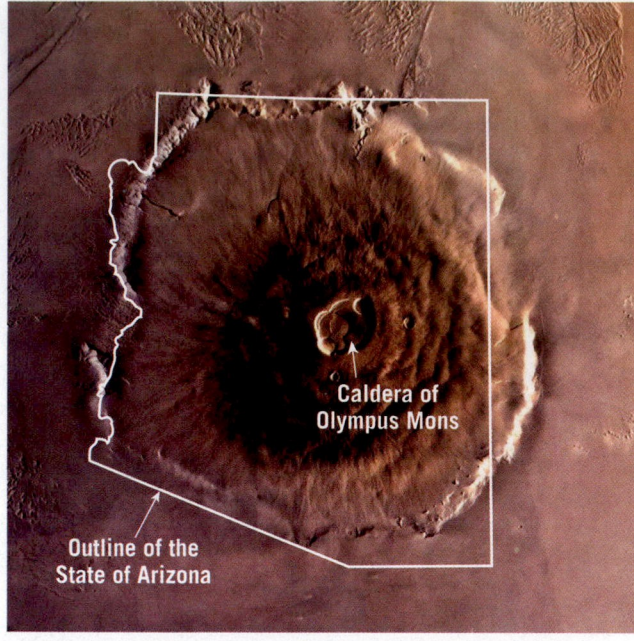

Caldera of Olympus Mons

Outline of the State of Arizona

dust storms with winds up to 270 kilometers (170 miles) per hour can persist for weeks. Dust devils have also been photographed. Most of the Martian landscape resembles Earth's rocky deserts, with abundant dunes and low areas partially filled with dust.

Water Ice on Mars Liquid water does not appear to exist anywhere on the Martian surface. However, poleward of about 30 degrees latitude, ice can be found within 1 meter (3 feet) of the surface. In the polar regions, it forms small permanent ice caps. Current estimates place the maximum amount of water ice held by the Martian polar ice caps at about 1.5 times the amount covering Greenland.

Considerable evidence indicates that in the first 1 billion years of the planet's history, liquid water flowed on the surface, creating stream valleys and related features (**Figure 24.15**). One location where running water was involved in carving valleys can be seen in the *Mars Reconnaissance Orbiter* image in **Figure 24.16**. Notice the stream-like banks that contain numerous teardrop-shaped islands. These valleys appear to have been cut by catastrophic floods with discharge rates that were more than 1000 times greater than those of the Mississippi River. Most of these large flood channels emerge from areas of chaotic topography that appear to have formed when the surface collapsed. The most likely source of water for these flood-created valleys was the melting of subsurface ice. If the meltwater was trapped beneath a thick layer of permafrost, pressure could mount until a catastrophic release occurred. As the water escaped, the overlying surface would collapse, creating the chaotic terrain.

Not all Martian valleys appear to have resulted from water released in this manner. Some exhibit branching, tree-like patterns resembling dendritic drainage networks on Earth. In addition, the *Opportunity* rover investigated structures similar to features created by water on Earth—including layered sedimentary rocks, playas (salt flats), and lake beds. Minerals that form only in the presence of water, such as hydrated sulfates, were also detected. Small spherical concretions of hematite, dubbed "blueberries," were found that probably precipitated from water to form lake sediments. Nevertheless, except in the polar regions,

Volcanoes on Mars Volcanism has been prevalent on Mars throughout most of its history. The scarcity of impact craters on some volcanic surfaces suggests that the planet is still active. Mars has several of the solar system's largest known volcanoes, including the largest, Olympus Mons, which is about the size of Arizona and stands nearly three times higher than Mount Everest. This gigantic volcano was last active about 100 million years ago and resembles Earth's Hawaiian shield volcanoes (**Figure 24.14**).

How did the volcanoes on Mars grow so much larger than similar structures on Earth? The largest volcanoes on the terrestrial planets tend to form where plumes of hot rock rise from deep within their interiors. On Earth, moving plates keep the crust in constant motion. Consequently, mantle plumes tend to produce a chain of volcanic structures, like the Hawaiian Islands. By contrast, plate tectonics on Mars is absent, so successive eruptions accumulate in the same location, creating enormous volcanoes rather than a string of smaller ones.

Wind Erosion on Mars The dominant force currently shaping the Martian surface is wind erosion. Extensive

Figure 24.15
Similar rock outcrops on Mars and Earth This set of images compares a rock outcrop on Mars (left) with similar rocks on Earth. The rock outcrop on Earth formed in a streambed, which suggests that the Martian rocks formed in a similar environment. Based on this finding, the scientist John Grotzinger concluded that there was once "a vigorous flow on the surface of Mars." (Courtesy of NASA)

This NASA image obtained by *Curiosity* rover shows rounded gravel fragments within a rock outcrop consistent with the sedimentary rock conglomerate. Weathered rock fragments can be seen below.

A typical sample of the sedimentary rock conglomerate that contains rounded gravel fragments deposited in a stream bed on Earth.

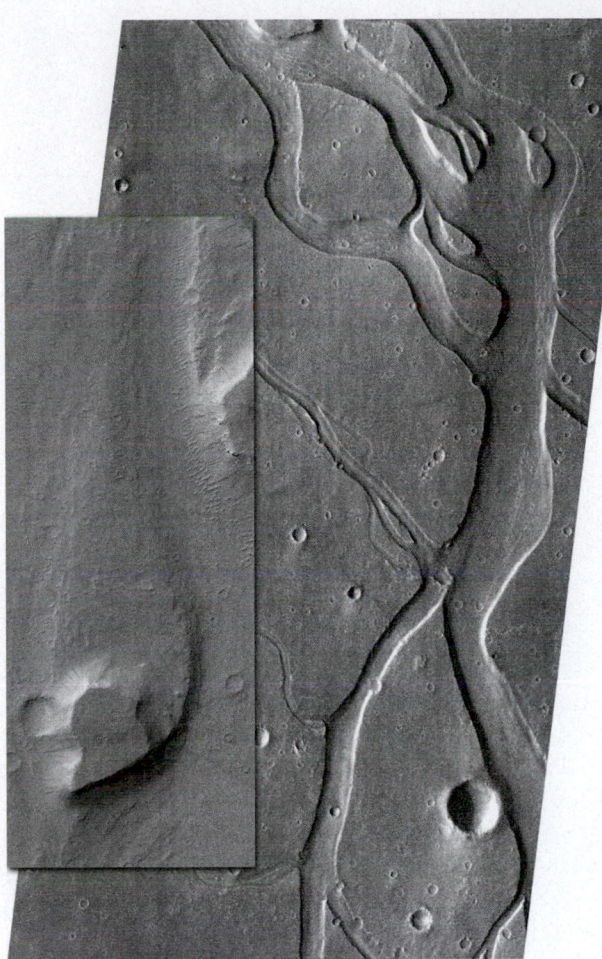

Figure 24.16
Earth-like stream channels are strong evidence that Mars once had flowing water Inset shows a close-up of a stream-lined island where running water encountered resistant material along its channel. (Courtesy of NASA)

water does not appear to have significantly altered the topography of Mars for more than 1 billion years.

On August 6, 2012, the Mars rover *Curiosity* landed in Gale Crater, near what NASA calls Mount Sharp. As of April 2015, *Curiosity* had traveled over 10 kilometers (6 miles), examining the lower slopes of this layered mountain in an effort to investigate how the region's ancient environment evolved from having lakes and rivers to the much drier conditions observed today. At a target zone NASA calls "Big Arm," the Mars Hand Lens Imager (MAHLI) took images of sandstone bedrock containing some rounded grains, indicating that the grains traveled long distances before becoming part of the sediment that later hardened into this sandstone (see GEOgraphics 24.1). The fact that these grains appeared to have been carried by running water is strong evidence that climatic conditions dramatically changed in the distant past. We can only imagine what *Curiosity* will uncover in its quest to study the climate and geology of Mars.

24.3 Concept Checks

1. What body in our solar system is most like Mercury?

2. Venus was once referred to as "Earth's twin." How are these two planets similar? How do they differ from one another?

3. What surface features do Mars and Earth have in common?

4. Why are the largest volcanoes on Earth so much smaller than the largest ones on Mars?

5. What evidence suggests that Mars had an active hydrologic cycle in the past?

EYE ON EARTH 24.1

Mariner 9 obtained this image of Phobos, one of two tiny moons of Mars. Phobos has a diameter of only 24 kilometers (15 miles). The two moons of Mars were not discovered until the late 1800s because their size made them nearly impossible to view telescopically. (Photo by NASA)

QUESTION 1 In what way is Phobos similar to Earth's moon?

QUESTION 2 List characteristics of Phobos that make it different from Earth's moon.

QUESTION 3 Search the Internet to learn how Phobos and its companion moon, Deimos, got their names.

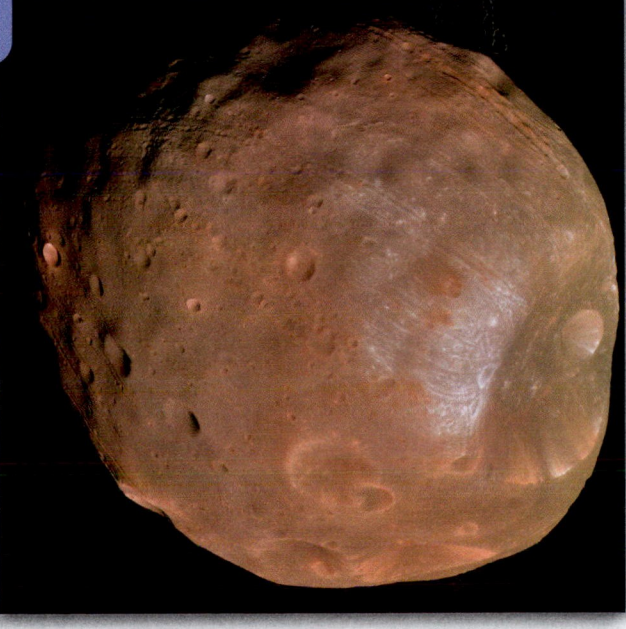

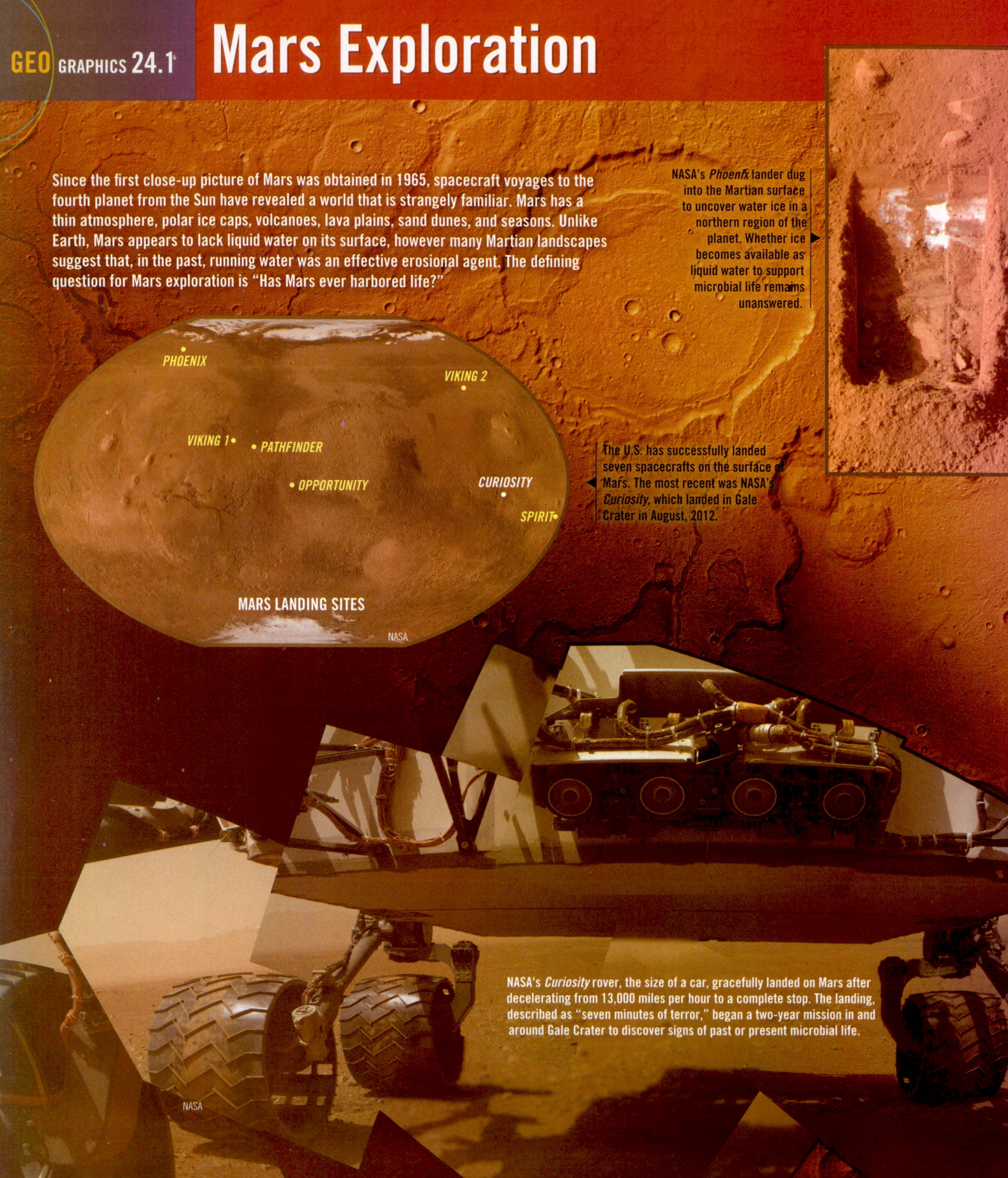

Mars Exploration

Since the first close-up picture of Mars was obtained in 1965, spacecraft voyages to the fourth planet from the Sun have revealed a world that is strangely familiar. Mars has a thin atmosphere, polar ice caps, volcanoes, lava plains, sand dunes, and seasons. Unlike Earth, Mars appears to lack liquid water on its surface, however many Martian landscapes suggest that, in the past, running water was an effective erosional agent. The defining question for Mars exploration is "Has Mars ever harbored life?"

NASA's *Phoenix* lander dug into the Martian surface to uncover water ice in a northern region of the planet. Whether ice becomes available as liquid water to support microbial life remains unanswered.

PHOENIX

VIKING 2

VIKING 1 • • PATHFINDER

• OPPORTUNITY

CURIOSITY

SPIRIT

MARS LANDING SITES

NASA

The U.S. has successfully landed seven spacecrafts on the surface of Mars. The most recent was NASA's *Curiosity*, which landed in Gale Crater in August, 2012.

NASA's *Curiosity* rover, the size of a car, gracefully landed on Mars after decelerating from 13,000 miles per hour to a complete stop. The landing, described as "seven minutes of terror," began a two-year mission in and around Gale Crater to discover signs of past or present microbial life.

NASA

NASA

NASA

The layers in the background, located at the base of Mount Sharp, are thought to be surviving remnants of extensive deposits laid down in a lake long ago, or possibly wind-delivered sediments subsequently cemented together by ground water. *Curiosity* uses 10 instruments to investigate whether Mars had ever provided a water-rich environment.

This image captured by *Curiosity* shows rounded sand grains, which suggests the grains were carried long distances, possibly by running water.

NASA

CAPE ST. VINCENT

VICTORIA CRATER

NASA

This false-color image obtained by Rover *Opportunity* shows Cape St. Vincent, one of many promontories that jut out from the walls of Victoria Crater. Below the loose, jumbled rocks, layering in the crater walls shows evidence of ancient wind-blown dunes.

	Spacecraft	Type	Landed or entered orbit	Years Active**
1	*Curiosity*	Rover	August 2012	Remains in operation
2	*Phoenix*	Lander	May 2008	Ran out of power during its first Martian winter
3	*Mars Reconnaissance*	Orbiter	March 2006	Planned 2-year mission, remains in operation
4	*Spirit*	Rover	January 2004	Planned 4-year mission, operated for more than 6 years
5	*Opportunity*	Rover	January 2004	Planned 90-day mission, remains in operation
6	*Odyssey*	Orbiter	October 2001	Remains active, longest active spacecraft in orbit around another planet
7	*Viking I*	Lander*	July 1976	Operational for more than 6 years
8	*Viking II*	Lander*	September 1976	Operational for more than 3 years
9	*Maven*	Orbiter	September 2014	Remains in operation

* Also had an orbiter; ** As of 2015

Question:
What evidence indicates that Mars may have been habitable in the distant past?

?

NASA

24.4 | Jovian Planets

Summarize and compare the features of Jupiter, Saturn, Uranus, and Neptune, including their ring systems.

The four Jovian planets, in order from the Sun, are Jupiter, Saturn, Uranus, and Neptune. Because of their location within the solar system and their size and composition, they are also commonly called the *outer planets* and the *gas giants*.

Jupiter: Lord of the Heavens

The giant among planets, Jupiter has a mass 2.5 times greater than the combined mass of all other planets, satellites, and asteroids in the solar system. However, it pales in comparison to the Sun, with only 1/800 of the Sun's mass.

Jupiter orbits the Sun once every 12 Earth years, and it rotates more rapidly than any other planet, completing one rotation in slightly less than 10 hours. When viewed telescopically, the effect of this fast spin is noticeable. The bulge of the equatorial region and the slight flattening at the poles are evident (see the "Polar Flattening" column in Table 24.1).

Figure 24.17

The structure of Jupiter's atmosphere The areas of light clouds (*zones*) are regions where gases are ascending and cooling. Sinking and warming dominate the flow in the darker cloud layers (*belts*). This convective circulation, along with the planet's rapid rotation, generates the high-speed winds observed between the belts and zones. (NASA)

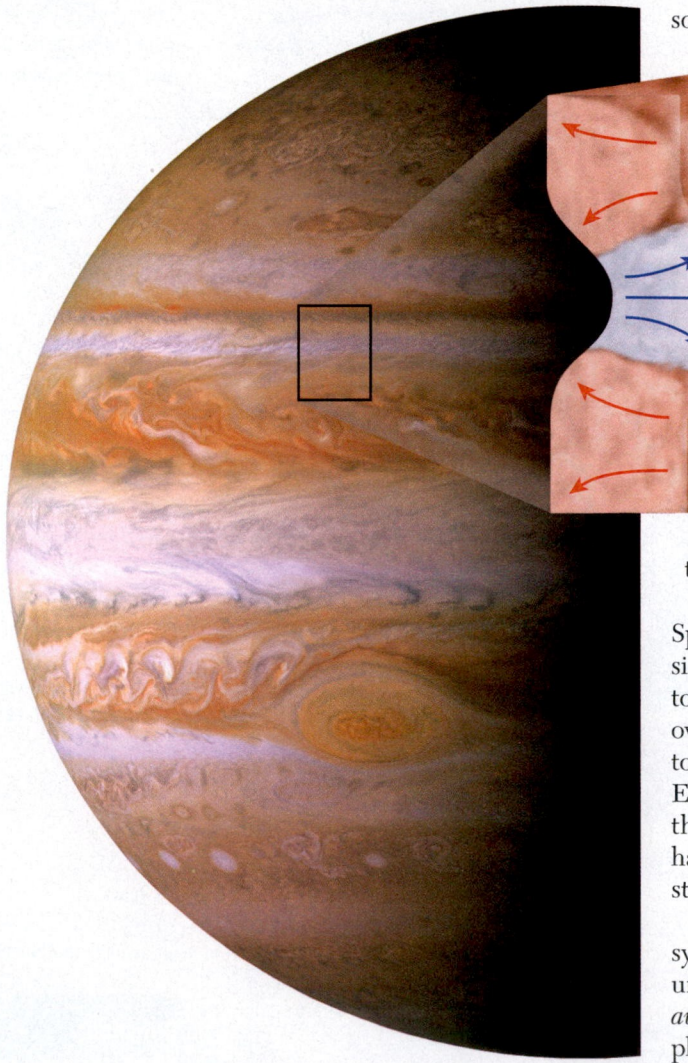

Belts
(dark clouds)

Strong winds

Zones
(bright clouds)

Strong winds

Belts
(dark clouds)

Jupiter's appearance is mainly attributable to the colors of light reflected from its three main cloud layers (**Figure 24.17**). The warmest, and lowest, layer is composed mainly of water ice and appears blue-gray; it is generally not seen in visible-light images. The middle layer, where temperatures are lower, consists of brown to orange-brown clouds of ammonium hydrosulfide droplets. These colors are thought to be by-products of chemical reactions occurring in Jupiter's atmosphere. Near the top of its atmosphere lie white wispy clouds of ammonia ice.

Because of its immense gravity, Jupiter is shrinking a few centimeters each year. This contraction generates most of the heat that drives Jupiter's atmospheric circulation. Thus, unlike winds on Earth, which are driven by solar energy, the huge convection currents observed in its atmosphere are produced by heat emanating from Jupiter's interior.

Jupiter's convective flow produces alternating dark-colored *belts* and light-colored *zones*, as shown in Figure 24.17. The light clouds (*zones*) are regions where warm material is ascending and cooling, whereas the dark belts indicate cool material that is sinking and warming. This convective circulation, along with Jupiter's rapid rotation, generates the high-speed, east–west flow observed between the belts and zones.

The largest storm on the planet is the Great Red Spot. This enormous anticyclonic storm that is twice the size of Earth has been known for 300 years. In addition to the Great Red Spot, there are various white and brown oval-shaped storms. The white ovals are the cold cloud tops of huge storms many times larger than hurricanes on Earth. The brown storm clouds reside at lower levels in the atmosphere. Lightning in various white oval storms has been photographed by the *Cassini* spacecraft, but the strikes appear to be less frequent than on Earth.

Jupiter's magnetic field, the strongest in the solar system, is probably generated by a rapidly rotating, liquid metallic hydrogen layer surrounding its core. Bright *auroras* associated with the magnetic field have been photographed over Jupiter's poles (**Figure 24.18**). Unlike

Earth's auroras, which occur only in conjunction with heightened solar activity, Jupiter's auroras are continuous.

Jupiter's Moons   Jupiter's satellite system, consisting of 67 moons discovered thus far, resembles a miniature solar system. Galileo discovered the 4 largest satellites, referred to as Galilean satellites, in 1610 (**Figure 24.19**). The two largest, Ganymede and Callisto, are roughly the size of Mercury, whereas the two smaller ones, Europa and Io, are about the size of Earth's Moon. The eight largest moons appear to have formed around Jupiter as the solar system condensed.

Jupiter also has many very small satellites (about 20 kilometers [12 miles] in diameter) that revolve in the opposite direction (retrograde motion) of the largest moons and have eccentric (elongated) orbits steeply inclined to the Jovian equator. These satellites appear to be asteroids or comets that passed near enough to be either gravitationally captured by Jupiter or remnants of the collisions of larger bodies.

The Galilean moons can be observed with binoculars or a small telescope and are interesting in their own right. Images from *Voyagers 1* and *2* revealed, to the surprise of most geoscientists, that each of the four Galilean satellites is a unique world (see Figure 24.19). The *Galileo* mission also unexpectedly revealed that the composition of each satellite is strikingly different, implying a different evolution for each. For example, Ganymede has a dynamic core that generates a strong magnetic field not observed in other satellites.

The innermost of the Galilean moons, Io, is perhaps the most volcanically active body in our solar system. More than 80 active, sulfurous volcanic centers have been discovered. Umbrella-shaped plumes have been observed rising from Io's surface to heights approaching 200 kilometers (125 miles) (**Figure 24.20A**). The heat source for volcanic activity is tidal energy generated by a relentless "tug of war" between Jupiter and the other Galilean satellites—with Io as the rope. The gravitational field of Jupiter and the other nearby satellites pull and push on Io's tidal bulge as its slightly eccentric orbit takes it alternately closer to and farther from Jupiter. This gravitational flexing of Io is transformed into heat (similar to the back-and-forth bending of a piece of sheet metal) and results in Io's spectacular sulfurous volcanic eruptions. Lava, thought to be mainly composed of silicate minerals, regularly erupts on its surface (**Figure 24.20B**).

The planets closer to the Sun than Earth are considered too warm to contain liquid water, and those farther from the Sun are generally too cold (although recall that Mars probably had abundant liquid water at some point in its history). The best prospects of finding liquid water within our

Figure 24.18
View of Jupiter's aurora, taken by the Hubble Space Telescope This phenomenon is produced by high-energy electrons racing along Jupiter's magnetic field. The electrons excite atmospheric gases and make them glow. (Courtesy of NASA/John Clark)

solar system lie beneath the icy surfaces of some of Jupiter's moons. For example, detailed images from *Galileo* have revealed that Europa's icy surface is quite young and exhibits cracks apparently filled with dark fluid from below. This suggests that under its icy shell, Europa must have a warm, mobile interior—perhaps an ocean. Because liquid water is a necessity for life as we know it, there is considerable interest in sending an orbiter to Europa—and, eventually, a lander capable of launching a robotic submarine—to determine whether it harbors life.

Jupiter's Rings   One of the surprising aspects of the *Voyager 1* mission was the discovery of Jupiter's ring

Figure 24.19
Jupiter's four largest moons These moons are often referred to as the Galilean moons because Galileo discovered them. (Courtesy of NASA)

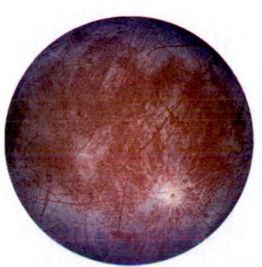

A. Io is one of only three volcanically active bodies other than Earth known to exist in the solar system.

B. Europa, the smallest of the Galilean moons, has an icy surface that is crisscrossed by many linear features.

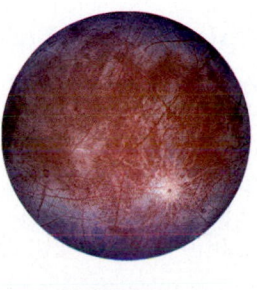
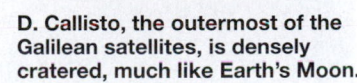

C. Ganymede, the largest Jovian satellite, exhibits cratered areas, smooth regions, and areas covered by numerous parallel grooves.

D. Callisto, the outermost of the Galilean satellites, is densely cratered, much like Earth's Moon.

Figure 24.20
A volcanic eruption on Jupiter's moon Io
(Courtesy of NASA; Jet Propulsion Laboratory/University of Arizona/NASA)

A.

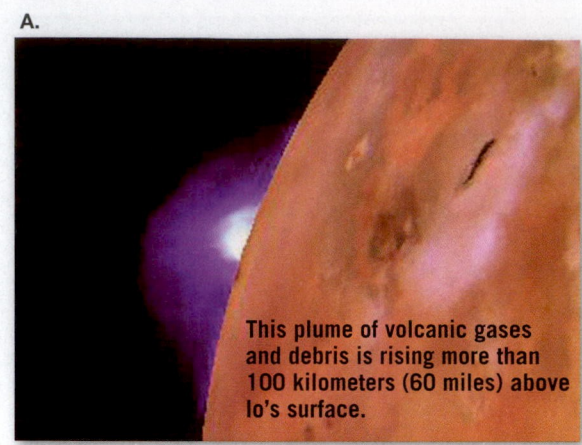

This plume of volcanic gases and debris is rising more than 100 kilometers (60 miles) above Io's surface.

B.

The bright red area on the left side of the image (see arrow) is newly erupted lava.

system. More recently, the ring system was thoroughly investigated by the *Galileo* mission. By analyzing how these rings scatter light, researchers determined that the rings are composed of fine, dark particles similar in size to smoke particles. Furthermore, the faint nature of the rings indicates that these minute particles are widely dispersed. The main ring is composed of particles believed to be fragments blasted from the surfaces of Metis and Adrastea, two small moons of Jupiter. Impacts on Jupiter's moons Amalthea and Thebe are believed to be the source of the debris from which the outer gossamer ring formed.

Saturn: The Elegant Planet

Requiring more than 29 Earth years to make one revolution, Saturn is almost twice as far from the Sun as Jupiter, yet their atmospheres, compositions, and internal structures are remarkably similar. The most striking feature of Saturn is its system of rings, first observed by Galileo in 1610 (**Figure 24.21**). Through his primitive telescope,

the rings appeared as two small bodies adjacent to the planet. Their ring nature was determined 50 years later by Dutch astronomer Christian Huygens.

Saturn's atmosphere, like Jupiter's, is dynamic. Although the bands of clouds are fainter and wider near the equator, rotating "storms" similar to Jupiter's Great Red Spot occur in Saturn's atmosphere, as does intense lightning. Although the atmosphere is about 93 percent hydrogen and 3 percent helium by volume, the clouds (or condensed gases) are composed of ammonia, ammonia hydrosulfide, and water, each segregated by temperature. As on Jupiter, the dynamics of Saturn's atmosphere are driven by the heat released by gravitational compression.

Saturn's Moons   The Saturnian satellite system consists of 62 known moons, of which 53 have been named. The moons vary significantly in size, shape, surface age, and origin. Twenty-three of the moons are "original" satellites that formed in tandem with their parent planet.

Figure 24.21
Saturn's dynamic ring system The two bright rings, called the A ring (outer) and B ring (inner), are separated by the Cassini division. A second small gap (Encke gap) is also visible as a thin line in the outer portion of the A ring. (Courtesy of NASA)

Encke gap

Cassini division

Saturn

D C B A

At least 3 (Rhea, Dione, and Tethys) show evidence of tectonic activity, where internal forces have ripped apart their icy surfaces. Others, like Hyperion, are so porous that impacts punch into their surfaces (**Figure 24.22**). Many of Saturn's smallest moons have irregular shapes and are only a few tens of kilometers in diameter.

Saturn's largest moon, Titan, is larger than Mercury and is the second-largest satellite in the solar system. Titan and Neptune's Triton are the only satellites in the solar system known to have substantial atmospheres. Titan was visited and photographed by the *Cassini-Huygens* probe in 2005. The atmospheric pressure at Titan's surface is about 1.5 times that at Earth's surface, and the atmospheric composition is about 98 percent nitrogen and 2 percent methane, with trace organic compounds. Titan has Earth-like landforms and geologic processes, such as dune formation and streamlike erosion caused by methane "rain." In addition, the northern latitudes appear to have lakes of liquid methane.

Enceladus is another unique satellite of Saturn—one of the few where active eruptions have been observed (**Figure 24.23**). The outgassing, comprised mostly of water, is thought to be the source that replenishes the material in Saturn's E ring. The volcanic-like activity occurs in areas called "tiger stripes" that consist of four large fractures with ridges on either side.

Saturn's Ring System In the early 1980s, the nuclear-powered *Voyagers 1* and *2* explored Saturn within 160,000 kilometers (100,000 miles) of its surface. More information was collected about Saturn in that short time than had been acquired since Galileo first viewed this "elegant planet" in the early 1600s. More recently, observations from ground-based telescopes, the Hubble Space Telescope, and the *Cassini-Huygens* spacecraft, have added to our knowledge of Saturn's ring system. In 1995 and 1996, when the positions of Earth and Saturn allowed the rings to be viewed edge-on, Saturn's faintest rings and satellites became visible. (The rings were visible edge-on again in 2009.)

Saturn's ring system is more like a large rotating disk of varying density and brightness than a series of independent ringlets. Each ring is composed of individual particles—mainly water ice, with lesser amounts of rocky debris—that circle the planet while regularly impacting one another. There are only a few gaps; most of the areas that look like empty space either contain fine dust particles or coated ice particles that are inefficient light reflectors.

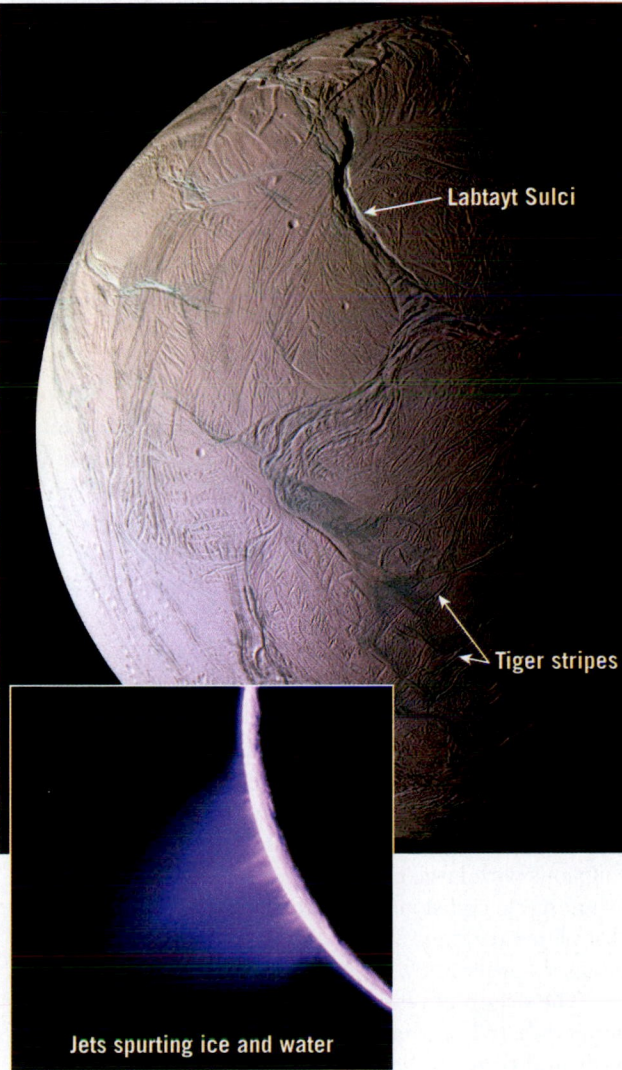

Labtayt Sulci

Tiger stripes

Jets spurting ice and water

Figure 24.24
Two of Saturn's ring moons (Courtesy of NASA)

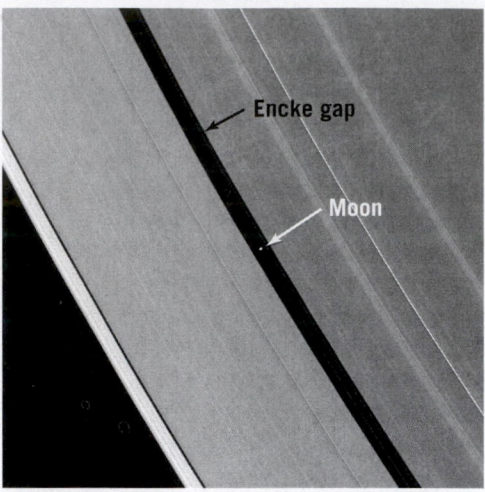

A. Pan is a small moon about 30 kilometers in diameter that orbits in the Encke gap, located in the A ring. It is responsible for keeping the Encke gap open by sweeping up any stray material that may enter.

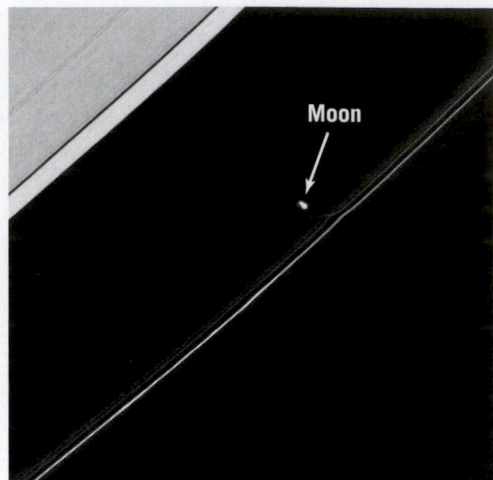

B. Prometheus, a potato-shaped moon, acts as a ring shepherd. Its gravity helps confine the moonlets in Saturn's thin Fring.

Most of Saturn's rings fall into one of two categories, based on density. Saturn's main (bright) rings, designated A and B, are tightly packed and contain particles ranging in size from a few centimeters (pebble-size) to tens of meters (house-size), with most of the particles being roughly the size of a large snowball (see Figure 24.21). In these dense rings, particles collide frequently as they orbit the planet. Although Saturn's main rings (A and B) are 40,000 kilometers (25,000 miles) wide, they are very thin, only 10–30 meters (30–100 feet) from top to bottom.

At the other extreme are Saturn's faint rings. Saturn's outermost ring (E ring), not visible in Figure 24.21, is composed of widely dispersed, tiny particles. Recall that volcanic-like activity on Saturn's satellite Enceladus is thought to be the source of material for the E ring.

Studies have shown that the gravitational tugs of nearby moons tend to "shepherd" the ring particles by gravitationally altering their orbits (**Figure 24.24**). For example, the F ring, which is very narrow, appears to be the work of satellites located on either side that confine the ring by pulling back particles that try to escape. On the other hand, the Cassini division, a clearly visible gap in Figure 24.21, arises from the gravitational pull of Mimas, one of Saturn's moons.

Some of the ring particles are believed to be debris ejected from the moons embedded in them. It is also possible that material is continually recycled between the rings and the ring moons. The ring moons gradually sweep up particles, which are subsequently ejected by collisions with large chunks of ring material or perhaps by energetic collisions with other moons. It seems, then, that planetary rings are not timeless features as we once thought; rather, they are continually recycled.

The origin of planetary ring systems is still being debated. Perhaps the rings formed simultaneously and from the same material as the planets and moons—condensing from a flattened cloud of dust and gases that encircled the parent planet. Or perhaps the rings formed later, when a moon or large asteroid was gravitationally pulled apart after straying too close to a planet. Yet another hypothesis suggests that a foreign body collided catastrophically with one of the planet's moons, the fragments of which would tend to jostle one another and form a flat, thin ring. Researchers expect more light to be shed on the origin of planetary rings as the *Cassini* spacecraft continues its tour of Saturn.

Uranus and Neptune: Twins

Although Earth and Venus have many similar traits, Uranus and Neptune are perhaps more deserving of being called "twins." They are nearly equal in diameter (both about four times the size of Earth), and they are both bluish in appearance, as a result of methane in their atmospheres. Their days are nearly the same length, and their cores are made of rocky silicates and iron—similar to the other gas giants. Their mantles, made mainly of water, ammonia, and methane, are thought to be very different from those of Jupiter and Saturn. One of the most pronounced differences between Uranus and Neptune is the time they take to complete one revolution around the Sun—84 and 165 Earth years, respectively.

Uranus: The Sideways Planet   Unique to Uranus is the orientation of its axis of rotation. Whereas the other planets resemble spinning toy tops as they circle the Sun, Uranus is like a top that has been knocked on its side but remains spinning (**Figure 24.25**). This unusual characteristic of Uranus is likely due to one or more impacts that essentially knocked the planet sideways from its original orientation early in its evolution.

Uranus shows evidence of huge storm systems equivalent in size to those in the United States. Recent photographs from the Hubble Space Telescope also reveal banded clouds composed mainly of ammonia and methane ice—similar to the cloud systems of the other gas giants.

Uranus's Moons   Spectacular views from *Voyager 2* showed that Uranus's five largest moons have varied terrains. Some have long, deep canyons and linear scars, whereas others possess large, smooth areas on otherwise crater-riddled surfaces. Studies conducted at NASA's Jet Propulsion Laboratory suggest that Miranda, the innermost of the five largest moons, was recently geologically active—most likely driven by gravitational heating, as occurs on Io.

Uranus's Rings

A surprise discovery in 1977 showed that Uranus has a ring system. The discovery was made as Uranus passed in front of a distant star and blocked its view, a process called *occultation* (*occult* = hidden). Observers saw the star "wink" briefly five times (meaning five rings) before the primary occultation and again five times afterward. More recent ground- and space-based observations indicate that Uranus has at least 10 sharp-edged, distinct rings orbiting its equatorial region. Interspersed among these distinct structures are broad sheets of dust.

Neptune: The Windy Planet

Because of Neptune's great distance from Earth, astronomers knew very little about this planet until 1989. Twelve years and nearly 3 billion miles of *Voyager 2* travel provided investigators an amazing opportunity to view the outermost planet.

Neptune has a dynamic atmosphere, much like that of the other Jovian planets (**Figure 24.26**). Record wind speeds exceeding 2400 kilometers (1500 miles) per hour encircle the planet, making Neptune one of the windiest places in the solar system. Neptune also exhibits large dark spots, thought to be rotating storms similar to Jupiter's Great Red Spot. However, Neptune's storms appear to have comparatively short life spans—usually only a few years. Another feature that Neptune has in common with the other Jovian planets is layers of white, cirrus-like clouds (probably frozen methane) about 50 kilometers (30 miles) above the main cloud deck.

Neptune's Moons

Neptune has 14 known satellites, the largest of which is Triton; the other 13 are small, irregularly shaped bodies. Triton is the only large moon in the solar system that exhibits retrograde motion, indicating that it most likely formed independently and was later gravitationally captured by Neptune (**Figure 24.27**).

Triton and a few other icy moons erupt "fluid" ices—an amazing manifestation of volcanism. **Cryovolcanism** (from the Greek *kryos*, meaning "frost") describes the eruption of magmas derived from the partial melting of ice instead of silicate rocks. Triton's icy magma is a mixture of water ice, methane, and probably ammonia. When partially melted, this mixture behaves as molten rock does on Earth. In fact, upon reaching the surface, these magmas can generate quiet outpourings of ice lavas that can flow great distances from their source—similar to the fluid basaltic flows on Hawaii. They also occasionally produce explosive eruptions that can generate the ice equivalent of volcanic ash. In 1989, *Voyager 2* detected active plumes on Triton that rose 8 kilometers (5 miles)

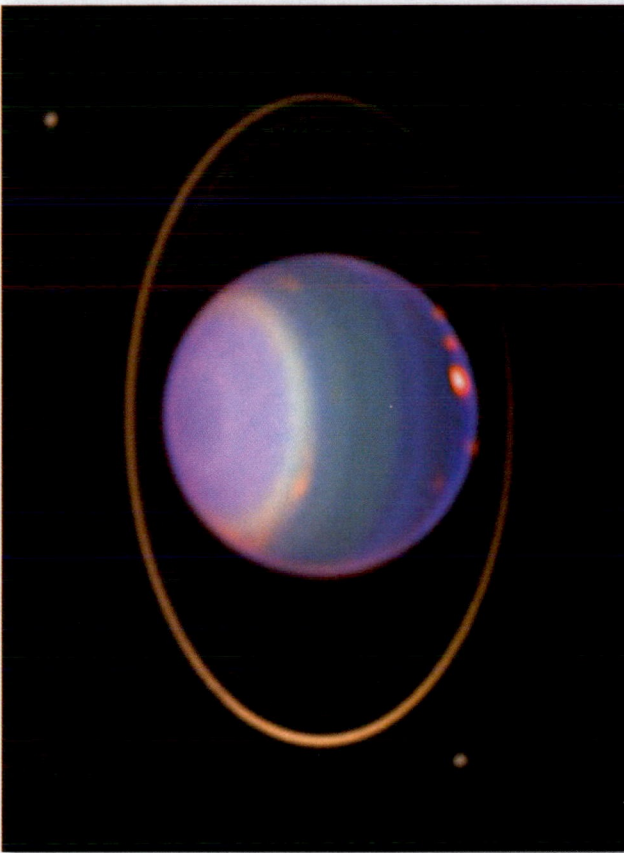

Figure 24.25
Uranus, surrounded by its major rings and a few of its known moons Also visible in this image are cloud patterns and several oval storm systems. This false-color image was generated from data obtained by Hubble's Near Infrared Camera. (Image by Hubble Space Telescope, courtesy of NASA)

above the surface and were blown downwind for more than 100 kilometers (60 miles).

Neptune's Rings

Neptune has five named rings; two of them are broad, and three are narrow, perhaps no more than 100 kilometers (60 miles) wide. The

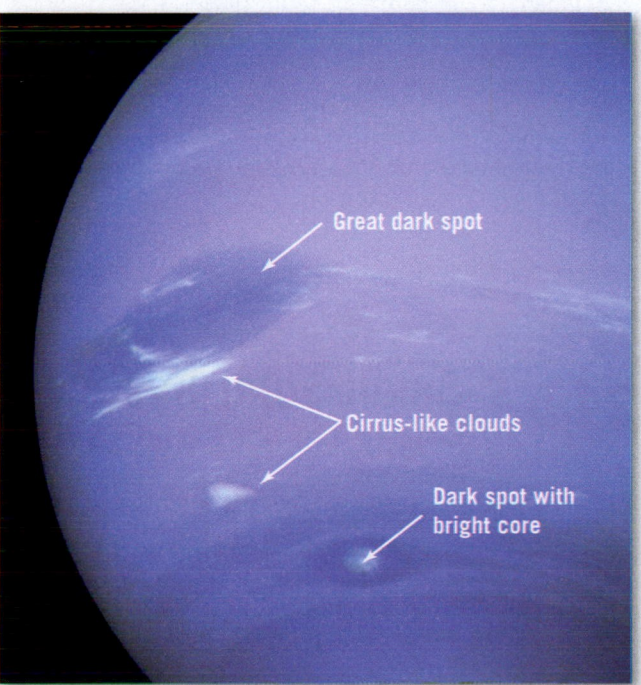

Figure 24.26
Neptune's dynamic atmosphere
(Courtesy of NASA)

Great dark spot

Cirrus-like clouds

Dark spot with bright core

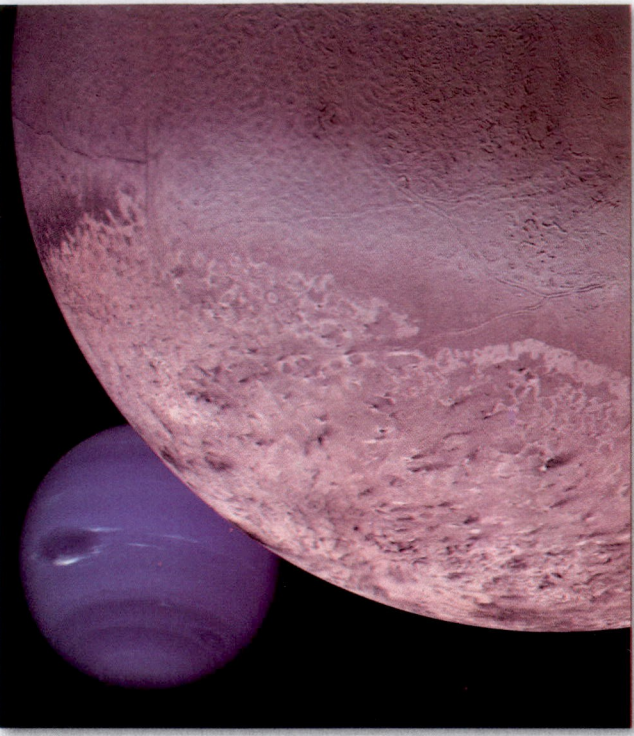

Figure 24.27
Triton, Neptune's largest moon The bottom of the image shows Triton's wind- and sublimation-eroded south polar cap. Sublimation is the process whereby a solid (ice) changes directly to a gas. (Courtesy of NASA)

outermost ring appears to be partially confined by the satellite Galatea. Neptune's rings, like Jupiter's, appear faint, which suggests that they are composed mostly of dust-size particles. Neptune's rings also display red colors, indicating that the dust is composed of organic compounds.

24.4   Concept Checks

1. What is the nature of Jupiter's Great Red Spot?

2. What is distinctive about Jupiter's satellite Io?

3. How are Jupiter and Saturn similar to one another?

4. What two roles do ring moons play in the nature of planetary ring systems?

5. How are Saturn's satellite Titan and Neptune's satellite Triton similar to one another?

6. Name three bodies in the solar system that exhibit active volcanism.

EYE ON EARTH 24.2

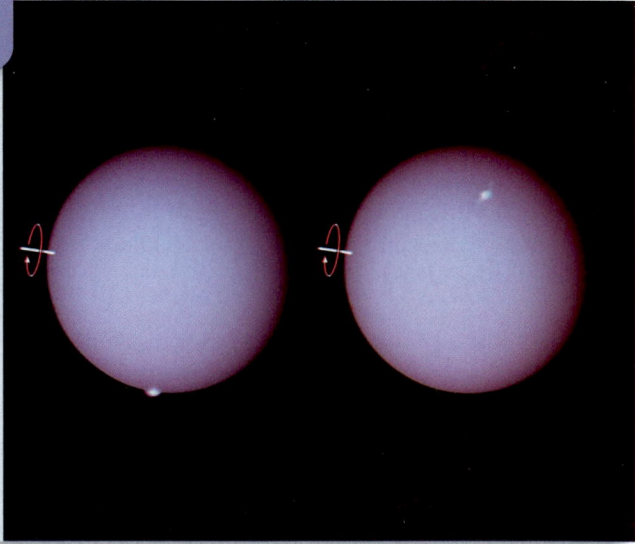

In 2012, the Hubble Space Telescope captured these images of auroras above the giant planet Uranus. These light shows on Uranus appear to last for only a few minutes and consist of faint glowing dots. These are unlike auroras on Earth, which can color the sky shades of green, red, or purple for hours. (Photo by NASA)

QUESTION 1   What is unusual about the location of the auroras on Uranus?

QUESTION 2   What does this indicate about the locations of Uranus's "north" and "south" magnetic poles?

24.5   Small Solar System Bodies

List and describe the principal characteristics of the small bodies that inhabit the solar system.

There are countless chunks of debris in the vast spaces separating the eight planets and in the outer reaches of the solar system. In 2006, the International Astronomical Union organized solar system objects not classified as planets or moons into two broad categories: (1) **small solar system bodies**, including *asteroids*, *comets*, and *meteoroids*, and (2) **dwarf planets**. The newest grouping, dwarf planets, includes Ceres, a body about 1000 kilometers (600 miles) in diameter and the largest known object in the asteroid belt, and Pluto, a former planet.

Asteroids and meteoroids are composed of rocky and/or metallic material with compositions somewhat like the terrestrial planets. They are distinguished according to size: Asteroids are much larger than meteoroids, which range in size from minute grains to 1 meter-wide objects. Comets, on the other hand, are collections of ices, dust, and small rocky particles that originate in the outer reaches of the solar system.

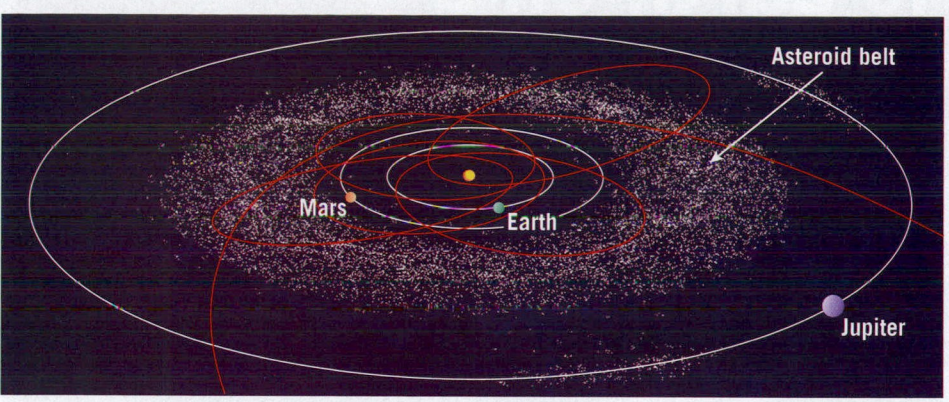

Figure 24.28
The asteroid belt The orbits of most asteroids lie between Mars and Jupiter. Also shown in red are the orbits of a few known near-Earth asteroids.

Asteroids: Leftover Planetesimals

Asteroids are small bodies (planetesimals) that remain from the formation of the solar system, which means they are about 4.6 billion years old. Most asteroids orbit the Sun between Mars and Jupiter, in the region known as the **asteroid belt** (**Figure 24.28**). There are only four asteroids with diameters greater than 400 kilometers (250 miles). However, the solar system hosts an estimated 1 to 2 million asteroids larger than 1 kilometer (0.6 mile) and many millions that are smaller. Some travel along eccentric orbits that take them very near the Sun, and others regularly pass close to Earth and the Moon (Earth-crossing asteroids). Many of the recent large-impact craters on the Moon and Earth probably resulted from collisions with asteroids. An estimated 1000 to 2000 Earth-crossing asteroids are more than 0.6 kilometer (0.37 mile) in diameter. Inevitably, Earth–asteroid collisions will occur again (see GEOgraphics 24.2).

Because most asteroids have irregular shapes, planetary geologists initially speculated that they might be fragments of a broken planet that once orbited between Mars and Jupiter. However, the combined mass of all asteroids is now estimated to be less than 1/1000 of the modest-sized Earth. Today, most researchers agree that asteroids are leftover debris from the solar nebula. Asteroids have lower densities than scientists originally thought, suggesting that they are porous bodies, like "piles of rubble," loosely bound together (**Figure 24.29**).

In February 2001, an American spacecraft became the first visitor to an asteroid. Although it was not designed for landing, *NEAR Shoemaker* landed successfully on Eros and collected information that has planetary geologists both intrigued and perplexed. Images obtained as the spacecraft drifted toward the surface of Eros revealed a barren, rocky surface composed of particles ranging in size from fine dust to boulders up to 10 meters (30 feet) across. Researchers unexpectedly discovered that fine debris tends to concentrate in the low areas, where it forms flat deposits resembling ponds. Surrounding the low areas, the landscape is marked by an abundance of large boulders.

One of several hypotheses to explain the boulder-strewn topography is seismic shaking, which would cause the boulders to move upward as the finer materials sink. This is analogous to what happens when a jar of sand and various-sized pebbles is shaken: The larger pebbles rise to the top, while the smaller sand grains settle to the bottom (sometimes referred to as the Brazil nut effect).

Indirect evidence from meteorites suggests that some asteroids might have been heated by a large impact event. A few large asteroids may have completely melted, causing them to differentiate into a dense iron core and a rocky mantle. In November 2005, the Japanese probe *Hayabusa* landed on a small near-Earth asteroid named 25143 Itokawa; it returned to Earth in June 2010. Analyzed samples suggest that the surface of the asteroid was identical in composition to meteorites and was once part of a larger asteroid. *Hayabusa 2*, launched in 2014, is expected to land on its target, asteroid 1999 JU3, in July 2018. After exploring this asteroid by digging into its surface to extract fresh samples, it is scheduled to return to Earth in 2020.

Figure 24.29
Giant asteroid Vesta
(Photo courtesy of NASA)

Is Earth on a Collision Course?

The solar system is cluttered with asteroids and comets that travel at great speeds and can strike Earth with an explosive force many times greater than a nuclear weapon.

MAJOR IMPACT STRUCTURES

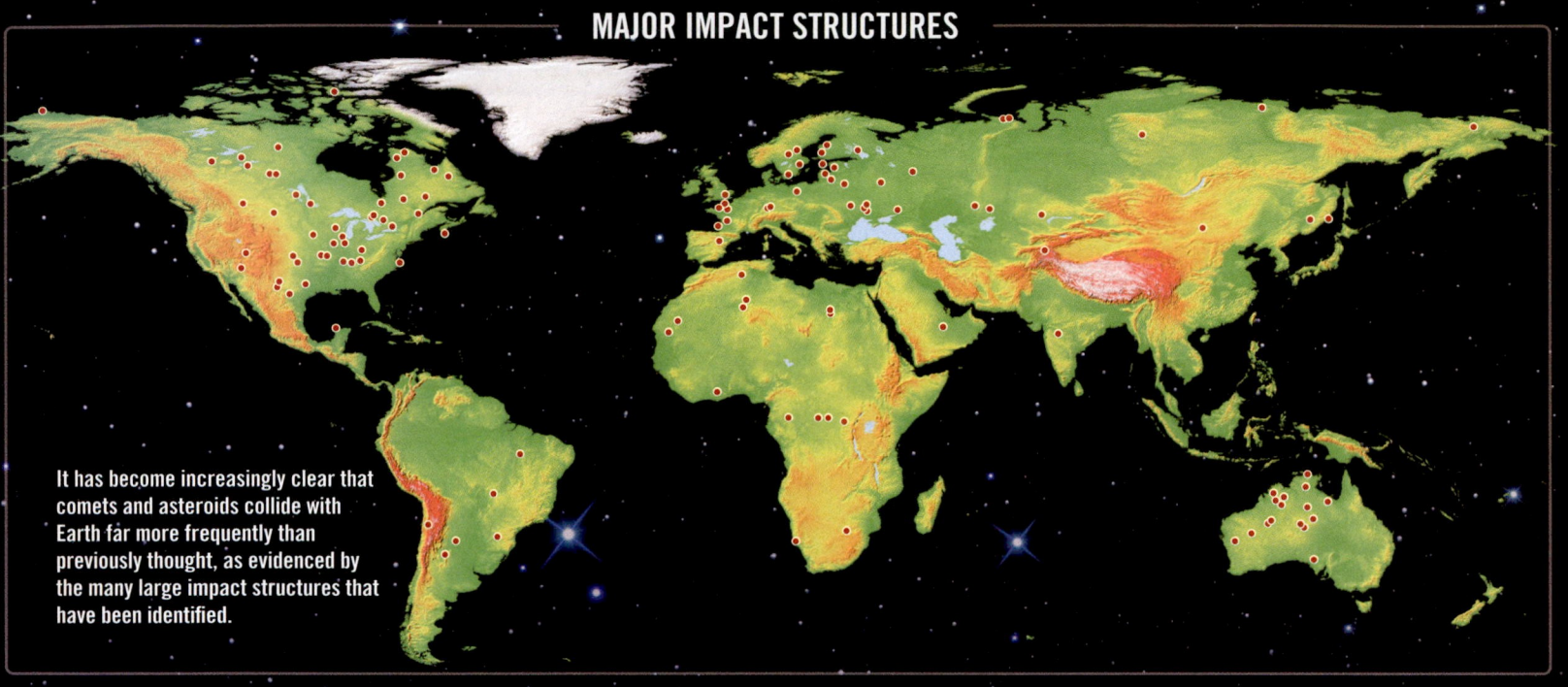

It has become increasingly clear that comets and asteroids collide with Earth far more frequently than previously thought, as evidenced by the many large impact structures that have been identified.

A Near-Earth Asteroid Census

During a one-year period, NASA's Wide-field Infrared Survey Explorer, or WISE, scanned the entire sky, continuously recording pictures. Based on information obtained by WISE, NASA researchers estimate that there are about 980 near-Earth asteroids larger than 1000 meters in diameter, of which 911 have already been located. These objects, roughly the size of small mountains would have global consequences if any of them were to strike Earth.

● known asteroids   ● predicted total   (each image represents 100 objects)

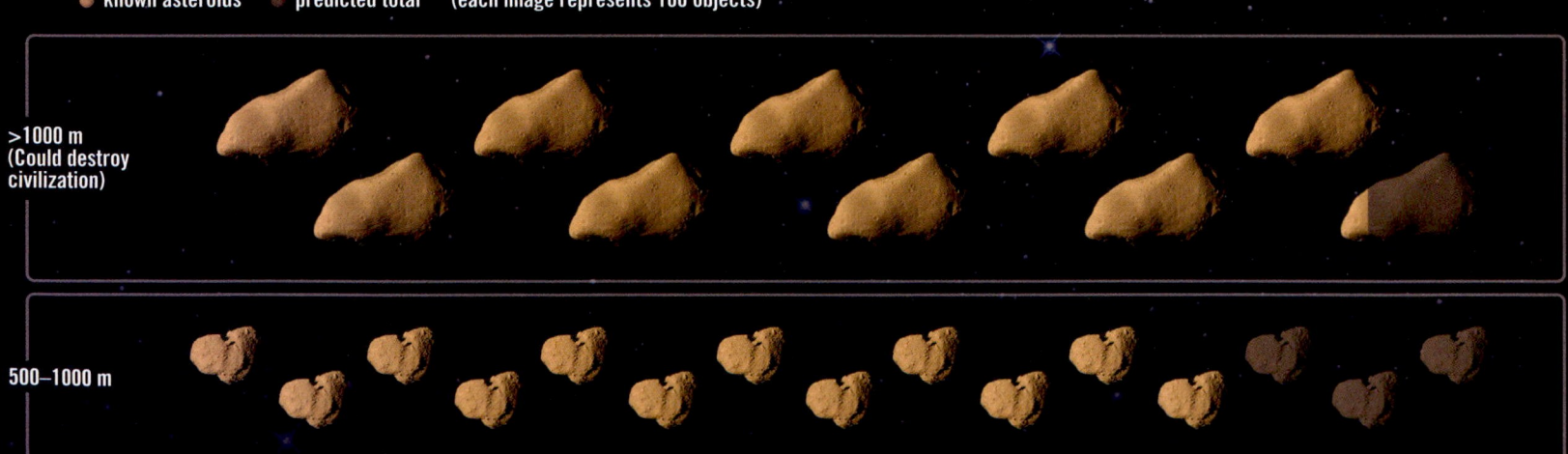

>1000 m
(Could destroy
civilization)

500–1000 m

300–500 m

100–300 m
(Could destroy
a small country)

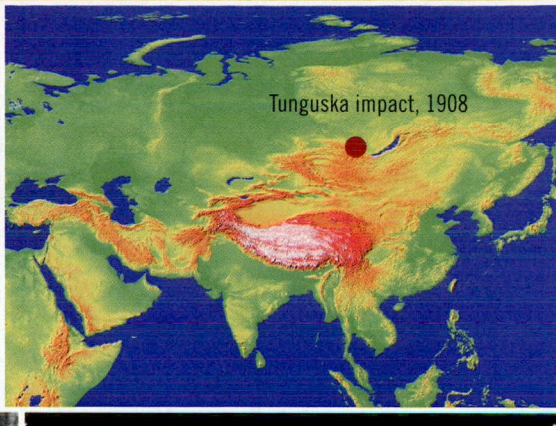
Tunguska impact, 1908

In 1908, in a remote region of Siberia, a "fireball" that appeared more brilliant than the Sun exploded violently. The shock waves rattled windows and triggered reverberations heard up to 1000 kilometers away. Called the Tunguska event, it scorched, delimbed, and flattened 80 million trees over an area of 2000 square kilometers. This spectacular explosion has been attributed to an asteroid or comet entering Earth's atmosphere at speeds exceeding 50,000-kilometers per hour. Surprisingly, expeditions to the area found no evidence of a crater or any rocky or metallic fragments that indicated a body impacted Earth. The explosion, which released the energy equivalent to about 185 Hiroshima atomic bombs, is thought to have occurred nearly 9 kilometers above the surface. Had it occurred over a major population center, the destruction would have been devastating.

NASA

? **Question:**
Conduct an Internet search to learn about and describe the damage caused by Chelyabinsk meteor when it exploded over western Russia in 2013.

The dangers of living with these small, but deadly objects from space came to public attention again in 1989 when an asteroid nearly 1 kilometer across shot past Earth in a "near miss." Traveling at 70,000 kilometers (44,000 miles) per hour, it could have produced a crater 10 kilometers (6 miles) wide, and perhaps 2 kilometers (1.2 miles) deep. As an observer noted, "Sooner or later it will be back." Statistics show that collisions with bodies larger than 1 kilometer should be expected every few hundred thousand years. Collisions with bodies larger than 6 kilometers, resulting in mass extinctions, are anticipated every 100 million years.

MARS
VENUS
MERCURY
SUN
EARTH

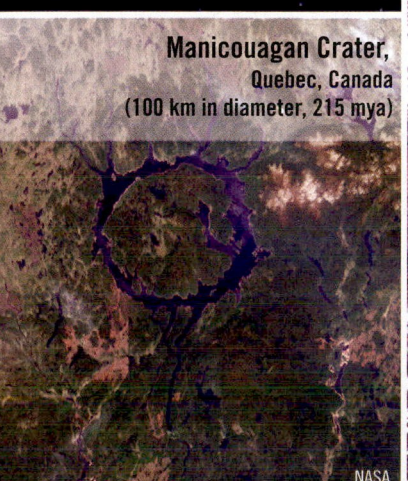
Manicouagan Crater,
Quebec, Canada
(100 km in diameter, 215 mya)
NASA

Kara Crater,
Nenetsia, Russia
(65 km in diameter, 70 mya)
NASA

Tenoumer Crater,
Mauritania, Africa
(2 km in diameter, < 30,000 years ago)
NASA

Goat Paddock Crater,
Northwest Australia
(5 km in diameter, < 50 mya)
NASA

Most impact structures on Earth are so old and highly eroded that they were not discovered until satellite images became available.

Figure 24.30
Changing orientation of a comet's tail as it orbits the Sun (Photo by Dan Schechter/Science Source)

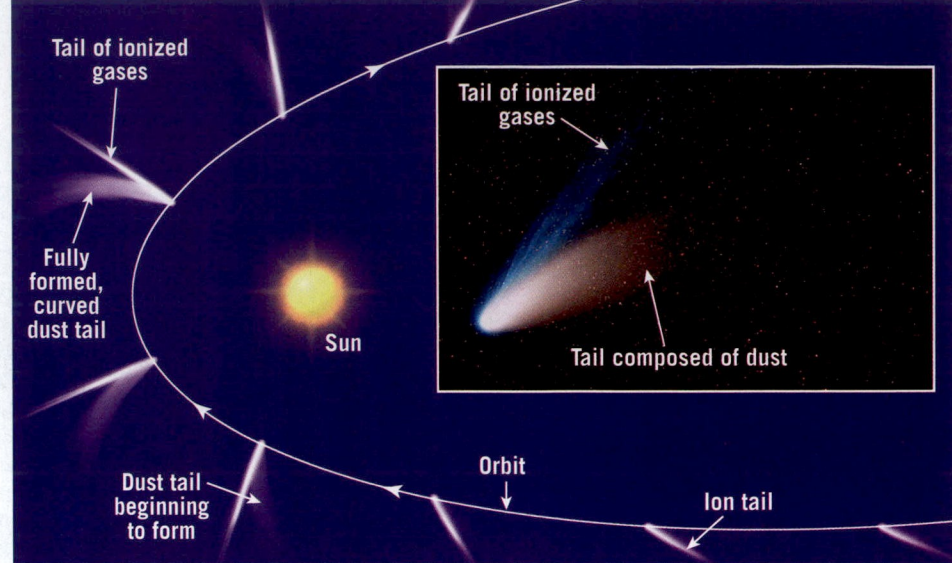

Tail of ionized gases

Tail of ionized gases

Fully formed, curved dust tail

Sun

Tail composed of dust

Dust tail beginning to form

Orbit

Ion tail

Comets: Dirty Snowballs

Comets, like asteroids, are leftover material from the formation of the solar system. They are loose collections of rocky material, dust, water ice, and frozen gases (ammonia, methane, and carbon dioxide) and thus are nicknamed "dirty snowballs." Recent space missions to comets have shown their surfaces to be dry and dusty, which indicates that their ices are hidden beneath a rocky layer.

Most comets reside in the outer reaches of the solar system and take hundreds of thousands of years to complete a single orbit around the Sun. However, a smaller

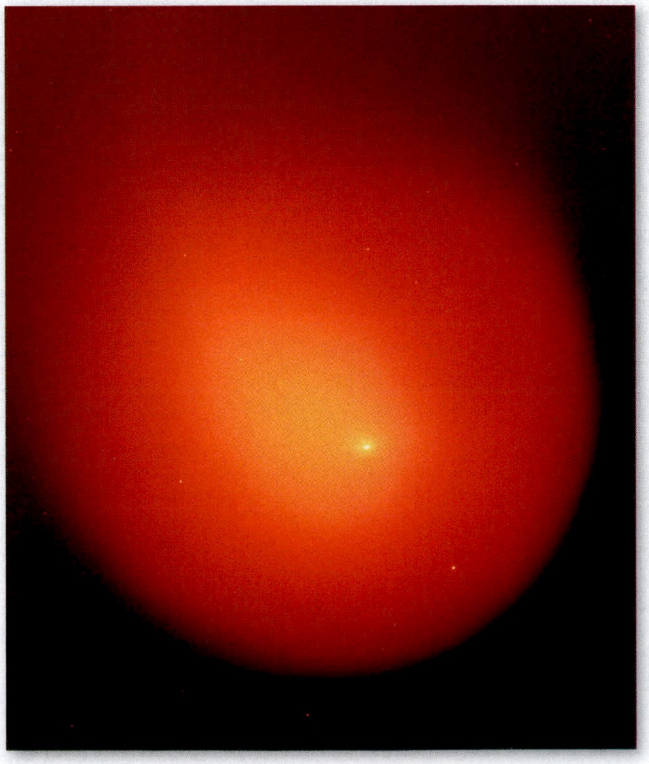

Figure 24.31
Coma of Comet Holmes The nucleus of the comet is within the bright spot in the center. Comet Holmes, which orbits the Sun every 6 years, was uncharacteristically active during its most recent entry into the inner solar system. (Courtesy of NASA)

number of *short-period comets* (those having orbital periods of less than 200 years), such as the famous Halley's Comet, make regular encounters with the inner solar system (**Figure 24.30**). The shortest-period comet (Encke's Comet) orbits the Sun once every 3 years.

Structure and Composition of Comets

All the phenomena associated with comets come from a small central body called the **nucleus**. These structures are typically 1 to 10 kilometers in diameter, but nuclei 40 kilometers across have been observed. When comets reach the inner solar system, solar energy begins to vaporize their ices. The escaping gases carry dust from the comet's surface, producing a highly reflective halo called a **coma** (**Figure 24.31**). Within the coma, the small glowing nucleus with a diameter of only a few kilometers can sometimes be detected.

As comets approach the Sun, most develop tails that can extend for millions of kilometers. The tail of a comet points away from the Sun in a slightly curved manner (see Figure 24.30), which led early astronomers to believe that the Sun has a repulsive force that pushes away particles of the coma to form the tail. Scientists have identified two solar forces known to contribute to tail formation: *radiation pressure* caused by radiant energy (light) emitted by the Sun and the charged particles of the *solar wind*. Sometimes a single tail composed of both dust and ionized gases is produced, but two tails are often observed (see Figure 24.30). The heavier dust particles produce a slightly curved tail that follows the comet's orbit, whereas the extremely light ionized gases are "pushed" directly away from the Sun, forming the second tail.

As a comet's orbit carries it away from the Sun, the gases forming the coma recondense, the tail disappears, and the comet returns to cold storage. Material that was blown from the coma to form the tail is lost forever. When all the gases are expelled, the inactive comet, which closely resembles an asteroid, continues its orbit without a coma or tail. It is believed that few comets remain active for more than a few hundred close orbits of the Sun.

The very first samples from a comet's coma (Comet Wild 2) were returned to Earth in January 2006 by NASA's *Stardust* spacecraft (**Figure 24.32**). Images from *Stardust* show that the comet's surface was riddled with flat-bottomed depressions and appeared dry, although at least 10 gas jets were active. Laboratory studies revealed that the coma contained a wide range of organic compounds and substantial amounts of silicate crystals.

The Realm of Comets: The Kuiper Belt and Oort Cloud

Most comets originate in one of two regions: the *Kuiper belt* or the *Oort cloud*. Named in honor of astronomer Gerald Kuiper, who predicted its existence, the **Kuiper belt** hosts comets that orbit in the outer solar system, beyond Neptune (see Figure 24.1). This disc-shaped structure is thought to contain about a billion objects over 1 kilometer (0.62 mile) in size. However, most comets are too small and too distant to be observed from Earth, even using the Hubble Space Telescope. Like the asteroids in the inner solar system, most Kuiper belt comets move in slightly elliptical orbits that lie roughly in the same plane as the planets. A chance collision between two Kuiper belt comets or the gravitational influence of one of the Jovian planets occasionally alters their orbits sufficiently to send them into our view.

Halley's Comet originated in the Kuiper belt. Its orbital period averages 76 years, and every one of its 29 appearances since 240 B.C.E. has been recorded, thanks to ancient Chinese astronomers—testimony to their dedication as astronomical observers and the endurance of Chinese culture. In 1910, Halley's Comet made a very close approach to Earth, making for a spectacular display.

Named for Dutch astronomer Jan Oort, the **Oort cloud** consists of comets that are distributed in all directions from the Sun, forming a spherical shell around the solar system. Most Oort cloud comets orbit the Sun at distances greater than 10,000 times the Earth–Sun distance. The gravitational effect of a distant passing star may send an occasional Oort cloud comet into a highly eccentric orbit that carries it toward the Sun. However, only a tiny fraction of Oort cloud comets have orbits that bring them into the inner solar system.

Meteoroids: Visitors to Earth

Nearly everyone has seen **meteors**, commonly (but inaccurately) called "shooting stars." These streaks of light can be observed in as little as the blink of an eye or can last as "long" as a few seconds. They occur when a small solid particle, a **meteoroid**, enters Earth's atmosphere from interplanetary space. Heat, created by friction between the meteoroid and the air, produces the light we see trailing across the sky. Most meteoroids originate from one of three sources: (1) interplanetary debris missed by the gravitational sweep of the planets

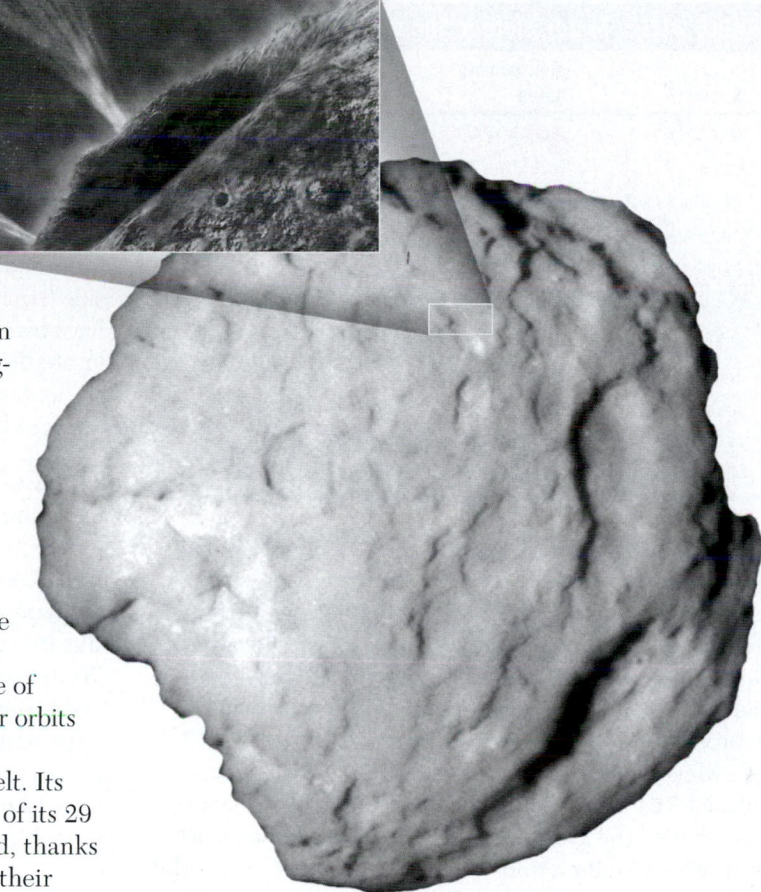

Figure 24.32
Comet Wild 2 This image shows Comet Wild 2, as seen by NASA's *Stardust* spacecraft. The inset shows an artist's depiction of jets of gas and dust erupting from Comet Wild 2. (Courtesy of NASA)

during formation of the solar system, (2) material that is continually being ejected from the asteroid belt, or (3) the rocky and/or metallic remains of comets that once passed through Earth's orbit. A few meteoroids are probably fragments of the Moon, Mars, or possibly Mercury, ejected by a violent asteroid impact. Before *Apollo* astronauts brought Moon rocks back to Earth, meteorites were the only extraterrestrial materials that could be studied in the laboratory.

Meteoroids less than about 1 meter (3 feet) in diameter generally vaporize before reaching Earth's surface. Some, called *micrometeorites*, are so tiny and their rate of fall so slow that they drift to Earth continually as space dust. Researchers estimate that thousands of meteoroids enter Earth's atmosphere every day. After sunset on a clear, dark night, many are bright enough to be seen with the naked eye.

Meteor Showers

Occasionally, meteor sightings increase dramatically to 60 or more per hour. These displays, called **meteor showers**, result when Earth encounters a swarm of meteoroids traveling in the same direction at nearly the same speed as Earth. The close association of these swarms to the orbits of some short-term comets strongly suggests that they represent

TABLE 24.2 Major Meteor Showers

Shower	Approximate Dates	Associated Comet
Quadrantids	January 4–6	
Lyrids	April 20–23	Comet 1861 I
Eta Aquarids	May 3–5	Halley's Comet
Delta Aquarids	July 30	
Perseids	August 12	Comet Swift-Tuttle
Draconids	October 7–10	Comet Giacobini-Zinner
Orionids	October 20	Halley's Comet
Taurids	November 3–13	Encke's Comet
Andromedids	November 14	Comet Biela
Leonids	November 18	Comet 1866 I
Geminids	December 4–16	

material lost by these comets (**Table 24.2**). Some swarms, not associated with the orbits of known comets, are probably the scattered remains of the nucleus of a long-defunct comet. The notable *Perseid meteor shower* that occurs each year around August 12 is likely material ejected from the comet Swift–Tuttle on previous approaches to the Sun.

Most meteoroids large enough to survive passage through the atmosphere to impact Earth probably originate among the asteroids, where chance collisions or gravitational interactions with Jupiter modify their orbits and send them toward Earth. Earth's gravity does the rest.

A few very large meteoroids have blasted craters on Earth's surface that strongly resemble those on our Moon. At least 40 terrestrial craters exhibit features that could be produced only by an explosive impact of a large asteroid or perhaps even a comet nucleus. More than 250 others may be of impact origin. Notable among them is Arizona's Meteor Crater, a huge cavity more than 1 kilometer (0.6 mile) wide and 170 meters (560 feet) deep, with an upturned rim that rises above the surrounding countryside (**Figure 24.33**). More than 30 tons of iron fragments have been found in the immediate area, but attempts to locate the main body have been unsuccessful. Based on the amount of erosion observed on the crater rim, the impact likely occurred within the past 50,000 years.

Types of Meteorites The remains of meteoroids, when found on Earth, are referred to as **meteorites** (**Figure 24.34**). Classified by their composition, meteorites are either (1) *irons*, mostly aggregates of iron with 5–20 percent nickel; (2) *stony* (also called *chondrites*), silicate minerals with inclusions of other minerals; or (3) *stony–irons*, mixtures of the two. Although stony meteorites are the most common, irons are found in large numbers because metallic meteorites withstand impacts better, weather more slowly, and are easily distinguished from terrestrial rocks. Iron meteorites are probably fragments of once-molten cores of large asteroids or small planets.

One type of stony meteorite, called a *carbonaceous chondrite*, contains organic compounds and occasionally

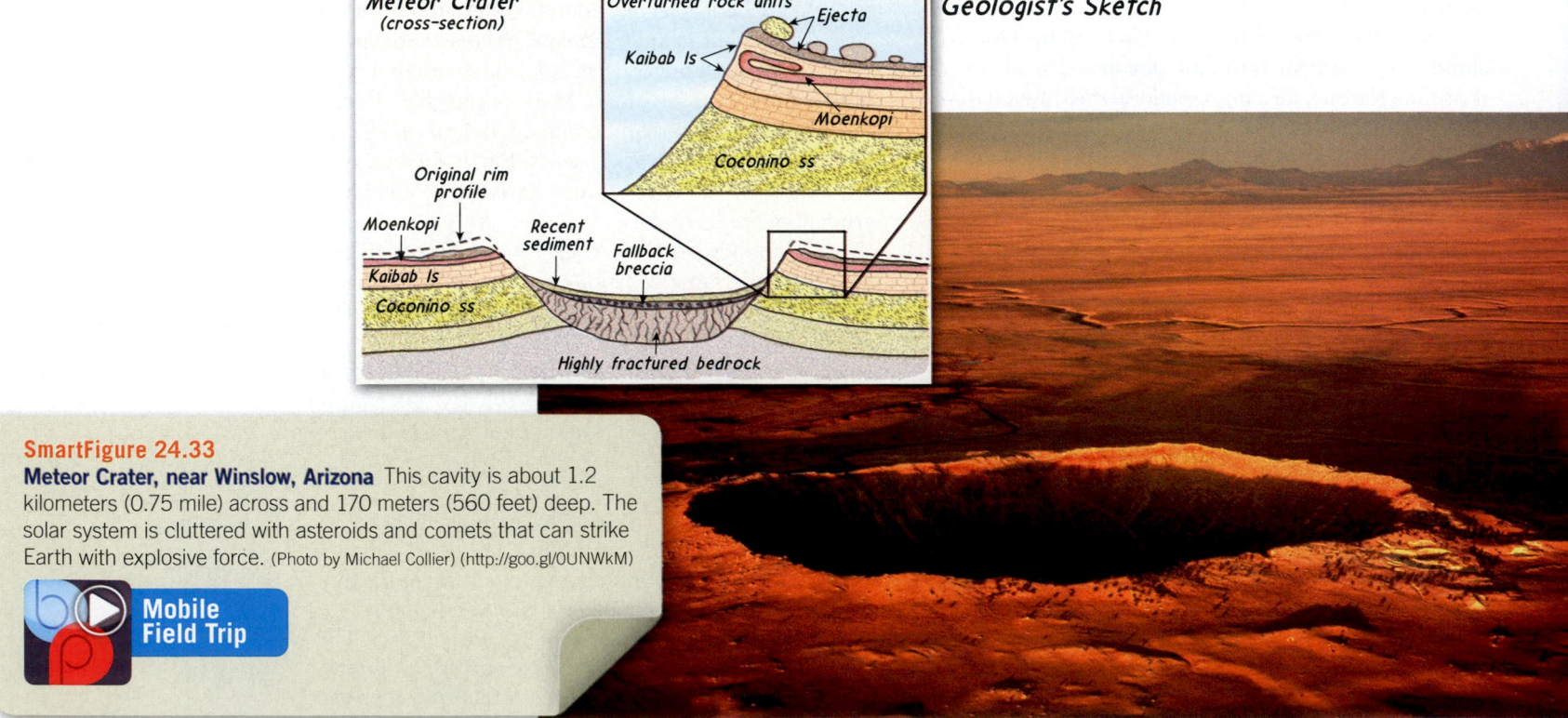

SmartFigure 24.33
Meteor Crater, near Winslow, Arizona This cavity is about 1.2 kilometers (0.75 mile) across and 170 meters (560 feet) deep. The solar system is cluttered with asteroids and comets that can strike Earth with explosive force. (Photo by Michael Collier) (http://goo.gl/0UNWkM)

Mobile Field Trip

simple amino acids, which are some of the basic building blocks of life. This discovery confirms similar findings in observational astronomy, which indicate that numerous organic compounds exist in interstellar space.

Data from meteorites have been used to ascertain the internal structure of Earth and the age of the solar system. If meteorites represent the composition of the terrestrial planets, as some planetary geologists suggest, our planet must contain a much larger percentage of iron than is indicated by surface rocks. This is one reason that geologists think Earth's core is mostly iron and nickel. In addition, radiometric dating of meteorites indicates that the age of our solar system is about 4.6 billion years. This "old age" has been confirmed by data obtained from lunar samples.

Dwarf Planets

Since its discovery in 1930, Pluto has been a mystery to astronomers who were searching for another planet in order to explain irregularities in Neptune's orbit. At the time of its discovery, Pluto was thought to be the size of Earth—too small to significantly alter Neptune's orbit. Recently, it was determined that Pluto's diameter is about 2370 kilometers (1470 miles), about one-fifth the diameter of Earth and less than half that of Mercury (long considered the solar system's "runt").

More attention was given to Pluto's status as a planet when astronomers discovered another large icy body in orbit beyond Neptune. Soon, more than 1000 of these *Kuiper belt objects* were discovered forming a band of objects—a second "asteroid belt," but located at the outskirts of the solar system. The Kuiper belt objects are rich in ices and have physical properties similar to those of comets. Many other planetary objects, some perhaps larger than Pluto, are thought to exist in this belt of icy worlds beyond Neptune's orbit. Researchers soon recognized that Pluto was unique among the planets—completely different from the four rocky, innermost planets, as well as the four gaseous giants.

In 2006, the International Astronomical Union, the group responsible for naming and classifying celestial objects, voted to designate a new class of solar system objects called *dwarf planets*. These celestial bodies orbit the Sun and are essentially spherical due to their own gravity but are not large enough to sweep their orbits

clear of other debris. By this definition, Pluto is recognized as a dwarf planet and was the prototype for this new category of planetary objects. Other dwarf planets include Eris, a Kuiper belt object, and Ceres, the largest-known asteroid.

In July 2015, when this chapter was being finalized, NASA's *New Horizons* spacecraft shot past Pluto, after a 9-year journey that brought it within 12,500 kilometers (7800 miles) of Pluto's surface. Images transmitted from *New Horizons* showed Pluto to be a complex body with several distinct terrains including mountainous areas, ice plains, and rugged areas showing a history of impacts (**Figure 24.35**). One of the most interesting regions, named Sputnik Planum (*planum* = plain or flat area), is found in Pluto's southern hemisphere. Sputnik Planum

SmartFigure 24.34
Iron meteorite found near Meteor Crater, Arizona
(Courtesy of M2 Photography/Alamy)
(https://goo.gl/bbT16J)

Tutorial

Figure 24.35
This enhanced color image is used to detect differences in the composition and texture of Pluto's surface The bright area in the lower-central region, sometimes referred to as "the heart of the heart" is formally named Sputnik Planum and is thought to be the source of exotic ices that flowed to produce the two bluish lobes near the bottom of the image.
(Courtesy of NASA)

Sputnik Planum

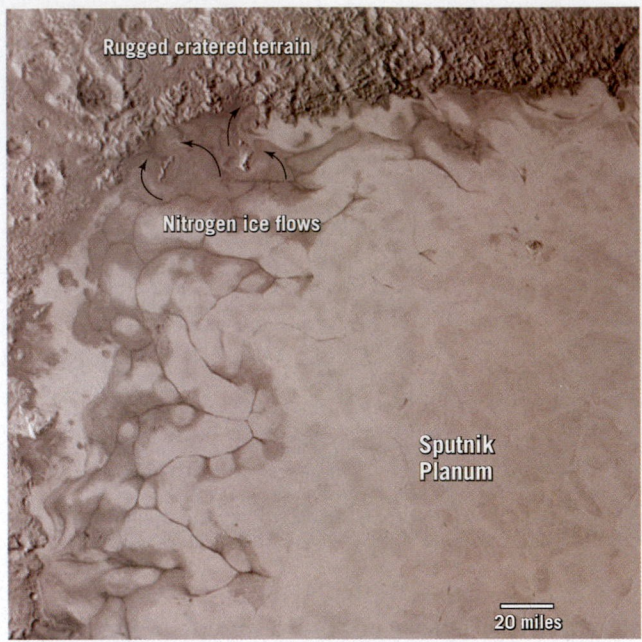

is most likely frozen nitrogen, which will flow at these frigid temperatures.

Because of Pluto's distance from Earth (about 32 times the distance between Earth and the Sun), a complete set of data cannot be retrieved until autumn 2016. After its flyby of Pluto, *New Horizons* will change course and is targeted at an object about 30-45 kilometers in diameter located in the Kuiper belt—a trove of comets and asteroids—some large enough to be classified as dwarf planets.

is presumed to be a large ice field that has tongues of ice flowing from its edges, similar to glaciers found on Earth (**Figure 24.36**). With surface temperatures of about −235°C (just shy of −400°F), Pluto is too cold for these ice flows to be made of water. Instead, Pluto's surface ice

24.5 Concept Checks

1. Compare and contrast asteroids and comets. Where are most asteroids found?

2. Where are most comets thought to reside? What eventually becomes of comets that orbit close to the Sun?

3. Differentiate among the following solar system bodies: meteoroid, meteor, and meteorite.

4. What are the three main sources of meteoroids?

5. Why was Pluto reclassified as a dwarf planet?

An artist's drawing showing the relative sizes of the best known dwarf planets compared to Earth and its moon. Eris, the largest known dwarf planet, has a very eccentric orbit that takes it as far as 100 AU from the Sun. Both Eris an Pluto are composed mainly of ice and water, methane, and ammonia. Ceres is the only identified dwarf planet in the asteroid belt. (NASA)

24.1 Our Solar System: An Overview

Describe the formation of the solar system according to the nebular theory. Compare and contrast the terrestrial and Jovian planets.

KEY TERMS nebular theory, solar nebula, planetesimal, protoplanet, terrestrial (Earth-like) planet, Jovian (Jupiter-like) planet, escape velocity, impact crater

- Our Sun is the most massive body in our solar system, which includes planets, dwarf planets, moons, and other small bodies. The planets orbit in the same direction and at speeds proportional to their distance from the Sun, with inner planets moving faster and outer planets moving more slowly.

- The solar system began as a solar nebula before condensing due to gravity. While most of the matter ended up in the Sun, some material formed a thick disc around the early Sun and later clumped together into larger and larger bodies. Planetesimals collided to form protoplanets, and protoplanets grew into planets.
- The four terrestrial planets are enriched in rocky materials, whereas the Jovian planets have a higher proportion of ice and gas. The terrestrial planets are relatively dense, with thin atmospheres, while the Jovian planets are less dense and have thick atmospheres.
- Smaller planets have less gravity to retain gases in their atmosphere. Lightweight gases such as hydrogen and helium more easily reach escape velocity, so the atmospheres of the terrestrial planets tend to be enriched in heavier gases, such as water vapor, carbon dioxide, and nitrogen.

24.2 Earth's Moon: A Chip off the Old Block

List and describe the major features of Earth's Moon and explain how maria basins were formed.

KEY TERMS maria, lunar highlands (terrae), lunar regolith

- Earth's Moon is the largest moon relative to its planet, and its composition is approximately the same as that of Earth's mantle. The

Moon likely formed from a collision between a Mars-sized protoplanet and the early Earth.
- The lunar surface is dominated by light-colored lunar highlands (or terrae) and darker lowlands called maria, which are dominated by younger flood basalts. Both terrae and maria are partially covered by lunar regolith produced by micrometeorite bombardment.

Q Briefly describe the formation of our Moon and how its formation accounts for its low density compared to that of Earth.

24.3 Terrestrial Planets

Outline the principal characteristics of Mercury, Venus, and Mars. Describe their similarities to and differences from Earth.

- Mercury has a very thin atmosphere and a weak magnetic field. Like Earth's moon, Mercury has a heavily cratered terrain and smooth plains similar to maria.
- Venus has a very dense atmosphere dominated by carbon dioxide. The resulting extreme greenhouse effect produces surface temperatures around 450°C (900°F). The topography of Venus has been resurfaced by active volcanism.
- Mars has about 1 percent as much atmosphere as Earth, so it is relatively cold (−140°C to 20°C [−220°F to 68°F]). Mars appears to be the closest planetary analogue to Earth, showing surface evidence of rifting, volcanism, and modification by flowing water. Volcanoes on Mars are much bigger than volcanoes on Earth because of the lack of plate motion on Mars.

Q As you can see from this graph, Mercury's temperature varies a lot from "day" to "night," but Venus's temperature is relatively constant "around the clock." Suggest a reason for this difference.

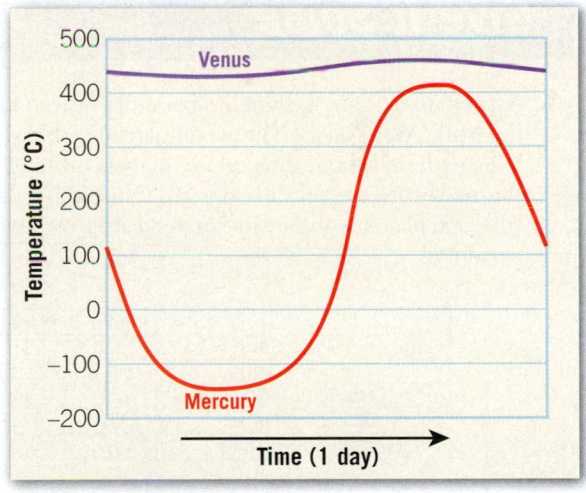

Idealized graph comparing the daily temperature variations on Venus and Mercury.

24.4 Jovian Planets

Summarize and compare the features of Jupiter, Saturn, Uranus, and Neptune, including their ring systems.

KEY TERM: cryovolcanism

- Jupiter is several times larger than the combined mass of everything else in the solar system except for the Sun. Convective flow among its three cloud layers produces its characteristic banded appearance. Persistent, giant rotating storms exist between these bands. Many moons orbit Jupiter, including Io, which shows active volcanism, and Europa, which has an icy shell.
- Saturn, like Jupiter, is big, gaseous, and endowed with dozens of moons. Some of these moons show evidence of tectonics, while Titan has its own

atmosphere. Saturn's well-developed rings are made of many particles of water ice and rocky debris.
- Uranus, like its "twin" Neptune, has a blue atmosphere dominated by methane, and its diameter is about four times greater than Earth's. Uranus rotates sideways relative to the plane of the solar system. It has a relatively thin ring system and at least five moons.
- Neptune has an active atmosphere, with fierce wind speeds and giant storms. It has 1 large moon, Triton, which shows evidence of cryovolcanism, as well as 13 smaller moons and a ring system.

Q Prepare and label a sketch comparing the typical characteristics of terrestrial planets and those of Jovian planets.

24.5 Small Solar System Bodies

List and describe the principal characteristics of the small bodies that inhabit the solar system.

KEY TERMS small solar system body, dwarf planet, asteroid, asteroid belt, comet, nucleus, coma, Kuiper belt, Oort cloud, meteor, meteoroid, meteor shower, meteorite

- Small solar system bodies include rocky asteroids and icy comets. Both are basically scraps left over from the formation of the solar system or fragments from later impacts.
- Most asteroids are concentrated in a wide belt between the orbits of Mars and Jupiter. Some are rocky, some are metallic, and some are basically "piles of rubble" loosely held together by their own weak gravity.
- Comets are dominated by ices that are "dirtied" by rocky material and dust. They originate in either the Kuiper belt beyond Neptune or the Oort cloud. When a comet's orbit brings it through the inner solar system, solar radiation causes its ices to begin to vaporize, generating the coma and its characteristic "tail."
- A meteoroid is debris that enters Earth's atmosphere, flaring briefly as a meteor before either burning up or striking Earth's surface to become a meteorite. Asteroids and material lost from comets as they travel through the inner solar system are the most common sources of meteoroids.
- Dwarf planets include Ceres (located in the asteroid belt), Pluto, and Eris, a Kuiper belt object. They are spherical bodies that orbit the Sun but are not massive enough to have cleared their orbits of debris.

Q Shown here are four small solar system bodies. Identify each and explain the differences among them.

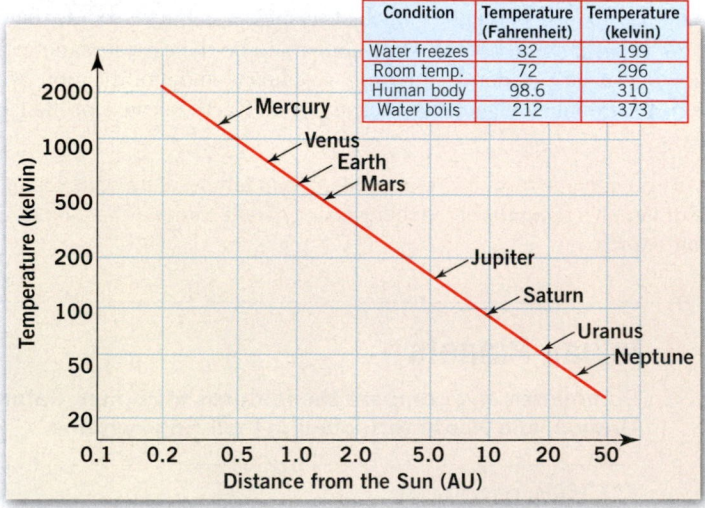

Give It Some *Thought*

1. Assume that a solar system has been discovered in a nearby region of the Milky Way Galaxy. The accompanying table shows data that have been gathered about three of the planets orbiting the central star of this newly discovered solar system. Using Table 24.1 as a guide, classify each planet as either Jovian, terrestrial, or neither. Explain your reasoning.

	Planet 1	Planet 2	Planet 3
Relative Mass (Earth = 1)	1.2	15	0.1
Diameter (km)	15,000	52,000	5000
Mean Distance from Star (AU)	1.4	17	35
Density (g/cm3)	4.8	1.22	5.3
Orbital Eccentricity	0.01	0.05	0.23

2. In order to conceptualize the size and scale of Earth and Moon as they relate to the solar system, complete the following:

   a. Approximately how many Moons (diameter 3475 kilometers [2160 miles]) would fit side-by-side across the diameter of Earth (diameter 12,756 kilometers [7926 miles])?

   b. Given that the Moon's orbital radius is 384,798 kilometers, approximately how many Earths would fit side-by-side between Earth and the Moon?

   c. Approximately how many Earths would fit side-by-side across the Sun, whose diameter is about 1,390,000 kilometers?

   d. Approximately how many Suns would fit side-by-side between Earth and the Sun, a distance of about 150,000,000 kilometers?

3. The accompanying graph shows the temperatures at various distances from the Sun during the formation of our solar system. Use it to complete the following:

Condition	Temperature (Fahrenheit)	Temperature (kelvin)
Water freezes	32	199
Room temp.	72	296
Human body	98.6	310
Water boils	212	373

   a. Which planets formed at locations in the solar system where the temperature was hotter than the boiling point of water?

   b. Which planets formed at locations where the temperature was cooler than the freezing point of water?

4. This sketch shows four primary craters (A, B, C, and D). The impact that produced crater A produced two secondary craters (labeled "a") and three rays. Crater D has one secondary crater (labeled "d"). Rank the four primary craters from oldest to youngest and explain your ranking.

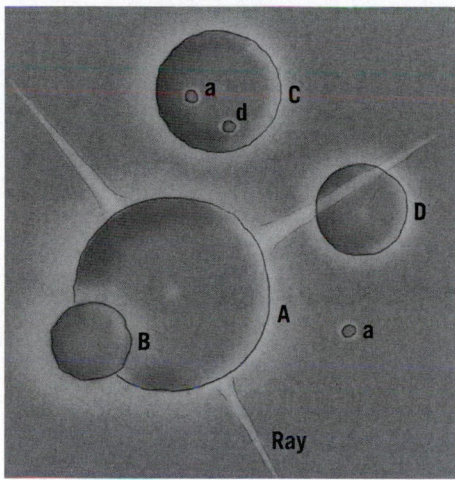

5. The accompanying diagram shows two of Uranus's moons, Ophelia and Cordelia, which act as shepherd moons for the Epsilon ring. Explain what would happen to the Epsilon ring if a large asteroid struck Ophelia, knocking it out of the Uranian system.

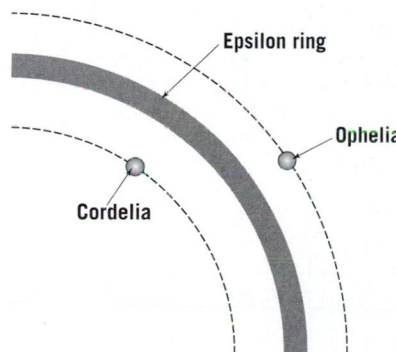

6. Halley's Comet has a mass estimated at 100 billion tons. Furthermore, it is estimated to lose about 100 million tons of material when its orbit brings it close to the Sun. If its orbital period is 76 years, calculate the maximum remaining life span of Halley's Comet.

7. Assume that three irregularly shaped planet-like objects, each smaller than our Moon, have just been discovered orbiting the Sun at a distance of 35 AU. One of your friends argues that the objects should be classified as planets because they are large and orbit the Sun. Another friend argues that the objects should be classified as dwarf planets, such as Pluto. State whether you agree or disagree with either or both of your friends. Explain your reasoning.

8. This diagram shows a comet traveling toward the Sun at the first position where it has both an ion tail and a dust tail. Refer to this diagram to answer the following:

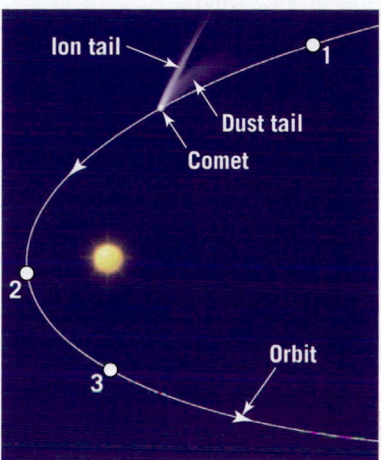

   a. For each of the three numbered sites, indicate whether the comet will have no tails, one tail, or two tails. If it will have one tail or two tails, in what direction will the tail(s) point?

   b. Would your answers to Question a change if the Sun's energy output were to increase significantly? If so, how would they change?

   c. If the solar wind suddenly ceased, how would this affect the comet and its tails?

9. Among the planets in our solar system, Earth is unique because water exists in all three states (solid, liquid, and gas) on and near its surface. In what state(s) of matter is water found on Mercury, Venus, and Mars?

   a. How would Earth's hydrologic cycle be different if Earth's orbit were inside the orbit of Venus?

   b. How would Earth's hydrologic cycle be different if Earth's orbit were outside the orbit of Mars?

10. If a large meteorite were to strike Earth in the near future, what effect might this event have on the atmosphere (in particular, on average temperatures and climate)? If these conditions persisted for several years, how might the changes influence the biosphere?

Appendix A
Metric and English Units Compared

Units

1 kilometer (km) = 1000 meters (m)
1 meter (m) = 100 centimeters (cm)
1 centimeter (cm) = 0.39 inch (in.)
1 mile (mi) = 5280 feet (ft)
1 foot (ft) = 12 inches (in.)
1 inch (in.) = 2.54 centimeters (cm)
1 square mile (mi^2) = 640 acres (a)
1 kilogram (kg) = 1000 grams (g)
1 pound (lb) = 16 ounces (oz)
1 fathom = 6 feet (ft)

Conversions

When you want

to convert:	multiply by:	to find:

Length

inches	2.54	centimeters
centimeters	0.39	inches
feet	0.30	meters
meters	3.28	feet
yards	0.91	meters
meters	1.09	yards
miles	1.61	kilometers
kilometers	0.62	miles

Area

square inches	6.45	square centimeters
square centimeters	0.15	square inches
square feet	0.09	square meters
square meters	10.76	square feet
square miles	2.59	square kilometers
square kilometers	0.39	square miles

Volume

cubic inches	16.38	cubic centimeters
cubic centimeters	0.06	cubic inches
cubic feet	0.028	cubic meters
cubic meters	35.3	cubic feet
cubic miles	4.17	cubic kilometers
cubic kilometers	0.24	cubic miles
liters	1.06	quarts
liters	0.26	gallons
gallons	3.78	liters

Masses and Weights

ounces	28.35	grams
grams	0.035	ounces
pounds	0.45	kilograms
kilograms	2.205	pounds

Temperature

When you want to convert degrees Fahrenheit (°F) to degrees Celsius (°C), subtract 32 degrees and divide by 1.8.

When you want to convert degrees Celsius (°C) to degrees Fahrenheit (°F), multiply by 1.8 and add 32 degrees.

When you want to convert degrees Celsius (°C) to Kelvins (K), delete the degree symbol and add 273. When you want to convert Kelvins (K) to degrees Celsius (°C), add the degree symbol and subtract 273.

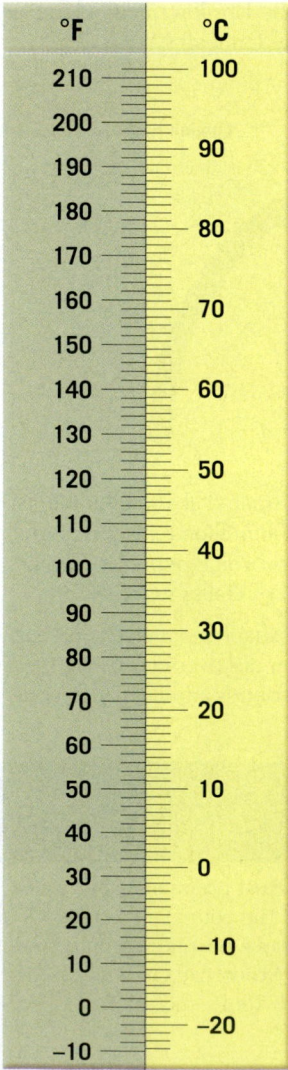

Figure A.1
A comparison of Fahrenheit and Celsius temperature scales.

Glossary

A

Aa flow A type of lava flow that has a jagged, blocky surface.

Ablation A general term for the loss of ice and snow from a glacier.

Abrasion The grinding and scraping of a rock surface by the friction and impact of rock particles carried by water, wind, and ice.

Abyssal plain A very level area of the deep-ocean floor, usually lying at the foot of the continental rise.

Accretionary wedge A large wedge-shaped mass of sediment that accumulates in subduction zones. Here sediment is scraped from the subducting oceanic plate and accreted to the overriding crustal block.

Active continental margin A margin that is usually narrow and consists of highly deformed sediments. Such margins occur where oceanic lithosphere is being subducted beneath the margin of a continent.

Active layer The zone above the permafrost that thaws in summer and refreezes in winter.

Aerosols Tiny solid and liquid particles suspended in the atmosphere.

Aftershock A smaller earthquake that follows the main earthquake.

Albedo The reflectivity of a substance, usually expressed as a percentage of the incident radiation reflected.

Alluvial channel A stream channel in which the bed and banks are composed largely of unconsolidated sediment (alluvium) that was previously deposited in the valley.

Alluvial fan A fan-shaped deposit of sediment formed when a stream's slope is abruptly reduced.

Alluvium Unconsolidated sediment deposited by a stream.

Alpine glacier A glacier confined to a mountain valley, which in most instances had previously been a stream valley.

Ambiguous properties Properties of minerals that may vary among different samples of the same mineral, such as color.

Andesite A gray, fine-grained igneous rock, primarily of volcanic origin and commonly exhibiting a porphyritic texture.

Andesitic composition See *Intermediate composition.*

Angiosperm A flowering plant in which fruits contain the seeds.

Angle of repose The steepest angle at which loose material remains stationary without sliding downslope.

Angular unconformity An unconformity in which the older strata dip at an angle different from that of the younger beds.

Antecedent stream A stream that continued to downcut and maintain its original course as an area along its course was uplifted by faulting or folding.

Anthracite A hard metamorphic form of coal that burns cleanly and hot.

Anticline A fold in sedimentary strata that resembles an arch.

Aphanitic texture A texture of igneous rocks in which the crystals are too small for individual minerals to be distinguished without the aid of a microscope.

Aquifer Rock or sediment through which groundwater moves easily.

Aquitard An impermeable bed that hinders or prevents groundwater movement.

Archean eon The first eon of Precambrian time. The eon preceding the Proterozoic. It extends between 4.5 and 2.5 billion years ago.

Arête A narrow, knifelike ridge separating two adjacent glaciated valleys.

Arkose A feldspar-rich sandstone.

Artesian well A well in which the water rises above the level where it was initially encountered.

Assimilation In igneous activity, the process of incorporating country rock into a magma body.

Asteroid One of thousands of small planetlike bodies, ranging in size from a few hundred kilometers to less than 1 kilometer across. Most asteroids' orbits lie between those of Mars and Jupiter.

Asthenosphere A subdivision of the mantle situated below the lithosphere. This zone of weak material exists below a depth of about 100 kilometers (60 miles) and in some regions extends as deep as 700 kilometers (430 miles). The rock within this zone is easily deformed.

Atmosphere The gaseous portion of a planet, the planet's envelope of air. One of the traditional subdivisions of Earth's physical environment.

Atoll A coral island that consists of a nearly continuous ring of coral reef surrounding a central lagoon.

Atom The smallest particle that exists as an element.

Atomic mass unit A mass unit equal to exactly one-twelfth the mass of a carbon-12 atom.

Atomic number The number of protons in the nucleus of an atom.

Atomic weight The average of the atomic masses of isotopes for a given element.

Augite A black, opaque silicate mineral of the pyroxene group that is a dominant component of basalt.

Aureole A zone or halo of contact metamorphism found in the country rock surrounding an igneous intrusion.

B

Backarc basin A basin that forms on the side of a volcanic arc away from the trench.

Backshore The inner portion of the shore, lying landward of the high-tide shoreline. It is usually dry, being affected by waves only during storms.

Backswamp A poorly drained area on a floodplain resulting when natural levees are present.

Bajada An apron of sediment along a mountain front created by the coalescence of alluvial fans.

Banded iron formations A finely layered iron and silica-rich (chert) layer deposited mainly during the Precambrian.

Bar Common term for sand and gravel deposits in a stream channel.

Barchan dune A solitary sand dune shaped like a crescent, with its tips pointing downwind.

Barchanoid dune A type of dune that forms scalloped rows of sand oriented at right angles to the wind. This form is intermediate between isolated barchans and extensive waves of transverse dunes.

Barrier island A low, elongate ridge of sand that parallels the coast.

Basal slip A mechanism of glacial movement in which the ice mass slides over the surface below.

Basalt A fine-grained igneous rock of mafic composition.

Basaltic composition A compositional group of igneous rocks indicating that the rock contains substantial dark silicate minerals and calcium-rich plagioclase feldspar.

Base level The level below which a stream cannot erode.

Basin A circular downfolded structure.

Batholith A large mass of igneous rock that formed when magma was emplaced at depth, crystallized, and subsequently exposed by erosion.

Bathymetry The measurement of ocean depths and the charting of the topography of the ocean floor.

Baymouth bar A sandbar that completely crosses a bay, sealing it off from the main body of water.

Beach An accumulation of sediment found along the landward margin of the ocean or a lake.

Beach drift The transport of sediment in a zigzag pattern along a beach, caused by the uprush of water from obliquely breaking waves.

Beach face The wet, sloping surface that extends from the berm to the shoreline.

Beach nourishment A process in which large quantities of sand are added to the beach system to offset losses caused by wave erosion. Building beaches seaward improves beach quality and storm protection.

Bed See *Strata.*

Bed load Sediment moved along the bottom of a stream by moving water, or particles moved along the ground surface by wind.

Bedding plane A nearly flat surface that separates two beds of sedimentary rock. Each bedding plane marks the end of one deposit and the beginning of another having different characteristics.

Bedrock channel A channel in which a stream is cutting into solid rock. Such channels typically form in the headwaters or river systems where gradients are high.

Benioff zone See *Wadati–Benioff zone.*

Berm The dry, gently sloping zone on the backshore of a beach at the foot of coastal cliffs or dunes.

Biochemical A type of chemical sediment that forms when material dissolved in water is precipitated by water-dwelling organisms. Shells are common examples.

Biogenous sediment Seafloor sediments consisting of material of marine-organic origin.

Biomass Organic material that is renewable energy derived from trees, crops, and waste. Examples include biofuels such as ethanol and biodiesel, as well as biogas, which is methane recovered from landfills.

Biosphere The totality of life-forms on Earth.

Biotite A dark, iron-rich mineral and a member of the mica family with excellent cleavage.

Bituminous coal The most common form of coal, often called soft, black coal.

Black carbon Soot generated by combustion processes and fires.

Black smoker A hydrothermal vent on the ocean floor that emits a black cloud of hot, metal-rich water.

Block lava Lava that has a surface of angular blocks associated with material having andesitic and rhyolitic compositions.

Blowout A depression excavated by wind in easily eroded materials.

Body wave A seismic wave that travels through Earth's interior.

Bottomset bed A layer of fine sediment deposited beyond the advancing edge of a delta and then buried by continued delta growth.

Bowen's reaction series A concept proposed by N. L. Bowen that illustrates the relationships between magma and the minerals crystallizing from it during the formation of igneous rocks.

Braided stream A stream that consists of numerous intertwining channels.

Breakwater A structure that protects a near-shore area from breaking waves.

Breccia A sedimentary rock composed of angular fragments that were lithified.

Brittle deformation Deformation that involves the fracturing of rock. Associated with rocks near the surface.

Burial metamorphism Low-grade metamorphism that occurs in the lowest layers of very thick accumulations of sedimentary strata.

C

Caldera A large depression typically caused by collapse or ejection of the summit area of a volcano.

Caliche A hard layer, rich in calcium carbonate, that forms beneath the *B* horizon in soils of arid regions.

Calving Wastage of a glacier that occurs when large pieces of ice break into the water.

Cambrian explosion The huge expansion in biodiversity that occurred at the beginning of the Paleozoic era.

Cap rock A necessary part of an oil trap. The cap rock is impermeable and hence keeps upwardly mobile oil and gas from escaping at the surface.

Capacity The total amount of sediment that a stream is able to transport.

Capillary fringe A relatively narrow zone at the base of the zone of aeration. Here water rises from the water table in tiny, threadlike openings between grains of soil or sediment.

Carbon cycle An Earth system in which carbon moves through the atmosphere, hydrosphere, biosphere, and geosphere, in different directions.

Cassini gap A wide gap in the ring system of Saturn between the A ring and the B ring.

Catastrophism The concept that Earth was shaped by catastrophic events of a short-term nature.

Cavern A naturally formed underground chamber or series of chambers most commonly produced by solution activity in limestone.

Cementation One way in which sedimentary rocks are lithified. As material precipitates from water that percolates through the sediment, open spaces are filled and particles are joined into a solid mass.

Cenozoic era A time span on the geologic time scale beginning about 65.5 million years ago, following the Mesozoic era.

Chemical bond A strong attractive force that exists between atoms in a substance. It involves the transfer or sharing of electrons that allows each atom to attain a full valence shell.

Chemical compound See *Compound*.

Chemical sedimentary rock Sedimentary rock consisting of material that was precipitated from water by either inorganic or organic means.

Chemical weathering The processes by which the internal structure of a mineral is altered by the removal and/or addition of elements.

Chert A durable sedimentary rock formed of microcrystalline quartz.

Cinder cone A rather small volcano built primarily of ejected lava fragments that consist mostly of pea- to walnut-size lapilli.

Cirque An amphitheater-shaped basin at the head of a glaciated valley produced by frost wedging and plucking.

Clastic texture A sedimentary rock texture consisting of broken fragments of preexisting rock.

Cleavage The tendency of a mineral to break along planes of weak bonding.

Climate A description of aggregate weather conditions; the sum of all statistical weather information that helps describe a place or region.

Climate feedback mechanism Various outcomes that may result when one of this complex interactive physical system's elements is altered.

Climate system Exchanges of energy and moisture occurring among the atmosphere, hydrosphere, lithosphere, biosphere, and cryosphere.

Closed system A system that is self-contained with regard to matter—that is, no matter enters or leaves.

Coarse-grained See *Phaneritic texture*.

Coast A strip of land that extends inland from the coastline as far as ocean-related features can be found.

Coastline The coast's seaward edge. The landward limit of the effect of the highest storm waves on the shore.

Col A pass between mountain valleys where the headwalls of two cirques intersect.

Color A phenomenon of light by which otherwise identical objects may be differentiated.

Column A feature found in caves that is formed when a stalactite and stalagmite join.

Columnar joints A pattern of cracks that forms during cooling of molten rock to generate columns.

Coma The fuzzy, gaseous component of a comet's head.

Comet A small body that generally revolves about the Sun in an elongated orbit.

Compaction A type of lithification in which the weight of overlying material compresses more deeply buried sediment. It is most important in the fine-grained sedimentary rocks such as shale.

Competence A measure of the largest particle a stream can transport; a factor dependent on velocity.

Composite cone A volcano composed of both lava flows and pyroclastic material.

Composite volcano See *Composite cone*.

Compound A substance formed by the chemical combination of two or more elements in definite proportions and usually having properties different from those of its constituent elements.

Compressional mountains Mountains in which great horizontal forces have shortened and thickened the crust. Most major mountain belts are of this type.

Compressional stress Differential stress that shortens a rock body.

Concordant A term used to describe intrusive igneous masses that form parallel to the bedding of the surrounding rock.

Conduction The transfer of heat through matter by molecular activity.

Conduit A pipelike opening through which magma moves toward Earth's surface. It terminates at a surface opening called a vent.

Cone of depression A cone-shaped depression in the water table immediately surrounding a well.

Confined aquifer An aquifer that has impermeable layers (aquitards) both above and below.

Confining pressure Stress that is applied uniformly in all directions.

Conformable layers Rock layers that were deposited without interruption.

Conglomerate A sedimentary rock composed of rounded, gravel-size particles.

Contact metamorphism Changes in rock caused by the heat from a nearby magma body.

Continental drift A hypothesis, credited largely to Alfred Wegener, which suggested that all present continents once existed as a single supercontinent. Further, beginning about 200 million years ago, the supercontinent began breaking into smaller continents, which then "drifted" to their present positions.

Continental margin The portion of the seafloor that is adjacent to the continents. It may include the continental shelf, continental slope, and continental rise.

Continental rift A linear zone along which continental lithosphere stretches and pulls apart. Its creation may mark the beginning of a new ocean basin.

Continental rise The gently sloping surface at the base of the continental slope.

Continental shelf The gently sloping submerged portion of the continental margin, extending from the shoreline to the continental slope.

Continental slope The steep gradient that leads to the deep-ocean floor and marks the seaward edge of the continental shelf.

Continental volcanic arc Mountains formed in part by igneous activity associated with the subduction of oceanic lithosphere beneath a continent. Examples include the Andes and the Cascades.

Convection The transfer of heat by the mass movement or circulation of a substance.

Convergent plate boundary A boundary in which two plates move together, resulting in oceanic lithosphere being thrust beneath an overriding plate, eventually to be reabsorbed into the mantle. It can also involve the collision of two continental plates to create a mountain system.

Coral reef A structure formed in a warm, shallow, sunlit ocean environment that consists primarily of the calcite-rich remains of corals as well as the limy secretions of algae and the hard parts of many other small organisms.

Core The innermost layer of Earth. It is thought to be largely an iron–nickel alloy, with minor amounts of oxygen, silicon, and sulfur.

Correlation The process of establishing the equivalence of rocks of similar age in different areas.

Corrosion The process by which soluble rock is gradually dissolved by flowing water.

Country rock See *Host rock.*

Covalent bond A chemical bond produced by the sharing of electrons.

Crater The depression at the summit of a volcano or a depression that is produced by a meteorite impact.

Craton The part of the continental crust that has attained stability; that is, it has not been affected by significant tectonic activity during the Phanerozoic eon. It consists of the shield and the stable platform.

Creep The slow downhill movement of soil and regolith.

Crevasse A deep crack in the brittle surface of a glacier.

Cross-bedding A structure in which relatively thin layers are inclined at an angle to the main bedding. Cross-bedding is formed by currents of wind or water.

Cross-cutting A principle of relative dating. A rock or fault is younger than any rock (or fault) through which it cuts.

Crust The very thin, outermost layer of Earth.

Cryosphere The portion of Earth's surface where water is in solid form, including snow, glaciers, sea ice, freshwater ice, and frozen ground. It is one of the spheres of the climate system.

Cryovolcanism A type of volcanism that results from the eruption of magmas derived from the partial melting of ice.

Crystal Any natural solid with an ordered, repetitive atomic structure.

Crystal settling A process that occurs during the crystallization of magma, in which the earlier-formed minerals are denser than the liquid portion and settle to the bottom of the magma chamber.

Crystal shape See *Habit.*

Crystalline See *Crystal.*

Crystalline texture See *Nonclastic texture.*

Crystallization The formation and growth of a crystalline solid from a liquid or gas.

Curie point The temperature above which a material loses its magnetization.

Cut bank The area of active erosion on the outside of a meander.

Cutoff A short channel segment created when a river erodes through the narrow neck of land between meanders.

D

D" layer A region in roughly the lowermost 200 kilometers (125 miles) of the mantle where P waves experience a sharp decrease in velocity.

Darcy's law An equation which states that groundwater discharge depends on the hydraulic gradient, hydraulic conductivity, and cross-sectional area of an aquifer.

Dark silicate A silicate mineral that contains ions of iron and/or magnesium in its structure. Dark silicates are dark in color and have a higher specific gravity than nonferromagnesian silicates.

Daughter product An isotope that results from radioactive decay.

Debris flow A flow of soil and regolith that contains a large amount of water. Most common in semiarid mountainous regions and on the slopes of some volcanoes.

Debris slide See *Rockslide.*

Decompression melting Melting that occurs as rock ascends due to a drop in confining pressure.

Deep-ocean basin The portion of seafloor that lies between the continental margin and the oceanic ridge system. This region comprises almost 30 percent of Earth's surface.

Deep-ocean trench A narrow, elongated depression of the seafloor.

Deep-sea fan A cone-shaped deposit at the base of the continental slope. The sediment is transported to the fan by turbidity currents that follow submarine canyons.

Deflation The lifting and removal of loose material by wind.

Deformation General term for the processes of folding, faulting, shearing, compression, or extension of rocks as the result of various natural forces.

Delta An accumulation of sediment formed where a stream enters a lake or an ocean.

Dendritic pattern A stream system that resembles the pattern of a branching tree.

Density A property of matter defined as mass per unit volume.

Desalination The removal of salts and other chemicals from seawater.

Desert One of the two types of dry climate; the driest of the dry climates.

Desert pavement A layer of closely spaced coarse pebbles and gravel that cover barren, rocky deserts to form a relatively smooth surface.

Desertification The degradation of dryland ecosystems on desert margins, primarily due to human activities such as deforestation and overgrazing that remove tree and plant cover anchoring the soil.

Detachment fault A nearly horizontal fault that may extend for hundreds of kilometers below the surface. Such a fault represents a boundary between rocks that exhibit ductile deformation and rocks that exhibit brittle deformation.

Detrital sedimentary rocks Rocks that form from the accumulation of materials that originate and are transported as solid particles derived from both mechanical and chemical weathering.

Diagenesis A collective term for all the chemical, physical, and biological changes that take place after sediments are deposited and during and after lithification.

Diagnostic properties Properties of minerals that aid in mineral identification. Taste or feel, crystal shape, and streak are examples of diagnostic properties.

Differential stress Forces that are unequal in different directions.

Differential weathering The variation in the rate and degree of weathering caused by such factors as mineral makeup, degree of jointing, and climate.

Diffraction The bending of waves as they pass by a curved surface between two compositionally different layers. For example, seismic waves diffract at the boundary between Earth's mantle and outer core.

Dike A tabular-shaped intrusive igneous feature that cuts through the surrounding rock.

Diorite A coarse-grained, intrusive igneous rock primarily composed of plagioclase feldspar and amphibole minerals.

Dip The angle at which a rock layer or fault is inclined from the horizontal. The direction of dip is at a right angle to the strike.

Dip-slip fault A fault in which the movement is parallel to the dip of the fault.

Discharge The quantity of water in a stream that passes a given point in a period of time.

Discharge area A location, such as a spring or a stream, where groundwater flows back to the surface.

Disconformity A type of unconformity in which the beds above and below are parallel.

Discontinuity A sudden change of depth in one or more of the physical properties of the material making up Earth's interior. The boundary between two dissimilar materials in Earth's interior as determined by the behavior of seismic waves.

Discordant A term used to describe plutons that cut across existing rock structures, such as bedding planes.

Disseminated deposit Any economic mineral deposit in which the desired mineral occurs as scattered particles in the rock but in sufficient quantity to make the deposit an ore.

Dissolution A common form of chemical weathering, it is the process of dissolving into a homogeneous solution, as when an acidic solution dissolves limestone.

Dissolved load The portion of a stream's load that is carried in solution.

Distributary A section of a stream that leaves the main flow.

Divergent plate boundary A boundary in which two plates move apart, resulting in upwelling of material from the mantle to create new seafloor.

Divide An imaginary line that separates the drainage of two streams, often found along a ridge.

Dolostone A chemical sedimentary rock formed from dolomite, a calcium-magnesium carbonate mineral.

Dome A roughly circular upfolded structure.

Downcutting The lowering of a streambed toward base level as turbulent water lifts unconsolidated material or when bedrock channels are lowered by means of quarrying, abrasion, and corrosion.

Drainage basin The land area that contributes water to a stream.

Drawdown The difference in height between the bottom of a cone of depression and the original height of the water table.

Drift See *Glacial drift.*

Drumlin A streamlined symmetrical hill composed of glacial till. The steep side of the hill faces the direction from which the ice advanced.

Dry climate A climate in which yearly precipitation is less than the potential loss of water by evaporation.

Ductile deformation A type of solid-state flow that produces a change in the size and shape of a rock body without fracturing. Occurs at depths where temperatures and confining pressures are high.

Dune A hill or ridge of wind-deposited sand.

Dwarf planet Celestial bodies that orbit stars and are massive enough to be spherical but have not cleared their neighboring regions of planetesimals.

E

Earth system science An interdisciplinary study that seeks to examine Earth as a system composed of numerous interacting parts or subsystems.

Earthflow The downslope movement of water-saturated, clay-rich sediment. Most characteristic of humid regions.

Earthquake Vibration of Earth produced by the rapid release of energy.

Ebb current The movement of tidal current away from the shore.

Echo sounder An instrument used to determine the depth of water by measuring the time interval between emission of a sound signal and the return of its echo from the bottom.

Economic mineral A concentration of a mineral resource or reserve that can be profitably extracted from Earth.

Elastic deformation Rock deformation in which the rock will return to nearly its original size and shape when the stress is removed.

Elastic rebound The sudden release of stored strain in rocks that results in movement along a fault.

Electron A negatively charged subatomic particle that has a negligible mass and is found outside an atom's nucleus.

Element A substance that cannot be decomposed into simpler substances by ordinary chemical or physical means.

Eluviation The washing out of fine soil components from the A horizon by downward-percolating water.

Emergent coast A coast where land formerly below sea level has been exposed by crustal uplift or a drop in sea level or both.

End moraine A ridge of till marking a former position of the front of a glacier.

Energy levels, or shells Spherically shaped, negatively charged zones that surround the nucleus of an atom.

Environment of deposition A geographic setting where sediment accumulates. Each site is characterized by a particular combination of geologic processes and environmental conditions.

Eon The largest time unit on the geologic time scale, next in order of magnitude above era.

Ephemeral stream A stream that is usually dry because it carries water only in response to specific episodes of rainfall. Most desert streams are of this type.

Epicenter The location on Earth's surface that lies directly above the focus of an earthquake.

Epoch A unit of the geologic time scale that is a subdivision of a period.

Era A major division on the geologic time scale; eras are divided into shorter units called periods.

Erosion The incorporation and transportation of material by a mobile agent, such as water, wind, or ice.

Eruption column Buoyant plumes of hot, ash-laden gases that can extend thousands of meters into the atmosphere.

Escape velocity The initial velocity an object needs to escape from the surface of a celestial body.

Esker A sinuous ridge composed largely of sand and gravel deposited by a stream flowing in a tunnel beneath a glacier near its terminus.

Estuary A funnel-shaped inlet of the sea that formed when a rise in sea level or subsidence of land caused the mouth of a river to be flooded.

Eukaryotes An organism whose genetic material is enclosed in a nucleus; plants, animals, and fungi are eukaryotes.

Evaporite A sedimentary rock formed of material deposited from solution by evaporation of the water.

Evapotranspiration The combined effect of evaporation and transpiration.

Exfoliation dome A large, dome-shaped structure, usually composed of granite, that is formed by sheeting.

Exotic stream A permanent stream that traverses a desert and has its source in well-watered areas outside the desert.

External process A process such as weathering, mass wasting, or erosion that is powered by the Sun and contributes to the transformation of solid rock into sediment.

Extrusive Igneous activity that occurs at Earth's surface.

Extrusive igneous rock Igneous rock formed when magma solidifies at Earth's surface.

Eye wall The doughnut-shaped area of intense cumulonimbus development and very strong winds that surrounds the eye of a hurricane.

F

Facies A portion of a rock unit that possesses a distinctive set of characteristics that distinguishes it from other parts of the same unit.

Fall A type of movement that is common to mass-wasting processes that refers to the free falling of detached individual pieces of any size.

Fault A break in a rock mass along which movement has occurred.

Fault creep Gradual displacement along a fault. Such activity occurs relatively smoothly and with little noticeable seismic activity.

Fault scarp A cliff created by movement along a fault. It represents the exposed surface of the fault prior to modification by weathering and erosion.

Fault-block mountain A mountain that is formed by the displacement of rock along a fault.

Felsic composition See *Granitic composition*.

Ferromagnesian silicate See *Dark silicate*.

Fetch The distance that the wind has traveled across the open water.

Fine-grained See *Aphanitic texture*.

Fiord A steep-sided inlet of the sea formed when a glacial trough was partially submerged.

Firn Granular, recrystallized snow. A transitional stage between snow and glacial ice.

Fissility The property of splitting easily into thin layers along closely spaced, parallel surfaces, such as bedding planes in shale.

Fission (nuclear) The splitting of a heavy nucleus into two or more lighter nuclei, caused by the collision with a neutron. During this process, a large amount of energy is released.

Fissure A crack in rock along which there is a distinct separation.

Fissure eruption An eruption in which lava is extruded from narrow fractures or cracks in the crust.

Flood The overflow of a stream channel that occurs when discharge exceeds the channel's capacity. The most common and destructive geologic hazard.

Flood basalts Flows of basaltic lava that issue from numerous cracks or fissures and commonly cover extensive areas to thicknesses of hundreds of meters.

Flood current The tidal current associated with the increase in the height of the tide.

Floodplain The flat, low-lying portion of a stream valley subject to periodic inundation.

Flow A type of movement common to mass-wasting processes in which water-saturated material moves downslope as a viscous fluid.

Flowing artesian well An artesian well in which water flows freely at Earth's surface because the pressure surface is above ground level.

Fluorescence The absorption of ultraviolet light, which is reemitted as visible light.

Focus (earthquake) The zone within Earth where rock displacement produces an earthquake.

Fold A bent layer or series of layers that were originally horizontal and subsequently deformed.

Fold-and-thrust belts Regions within compressional mountain systems where large areas have been shortened and thickened by the processes of folding and thrust faulting, as exemplified by the Valley and Ridge province of the Appalachians.

Foliated texture A texture of metamorphic rocks that gives the rock a layered appearance.

Foliation A term for a linear arrangement of textural features often exhibited by metamorphic rocks.

Footwall block The rock surface below a fault.

Forced subduction A process that occurs at Peru–Chile–type subduction zones in which lithosphere is too buoyant to subduct spontaneously but is forced beneath the overriding plate.

Forearc basin The region located between a volcanic arc and an accretionary wedge where shallow-water marine sediments typically accumulate.

Foreset bed An inclined bed deposited along the front of a delta.

Foreshocks Small earthquakes that often precede a major earthquake.

Foreshore The portion of the shore that lies between the normal high and low water marks; the intertidal zone.

Fossil The remains or traces of organisms preserved from the geologic past.

Fossil assemblage The overlapping ranges of a group of fossils (assemblage) collected from a layer.

By examining such an assemblage, the age of the sedimentary layer can be established.

Fossil fuel General term for any hydrocarbon that may be used as a fuel, including coal, oil, natural gas, bitumen from tar sands, and shale oil.

Fossil magnetism See *Paleomagnetism*.

Fossil succession The definite and determinable order in which fossil organisms occur. Fossil succession enables us to identify many time periods by their fossil content.

Fracture Any break or rupture in rock along which no appreciable movement has taken place.

Fracture zone A linear zone of irregular topography on the deep-ocean floor that follows transform faults and their inactive extensions.

Fragmental texture See *Pyroclastic texture*.

Frost wedging The mechanical breakup of rock caused by the expansion of freezing water in cracks and crevices.

Fumarole A vent in a volcanic area from which fumes or gases escape.

G

Gabbro A dark-green to black intrusive igneous rock composed of dark silicate minerals. Gabbro makes up a significant percentage of oceanic crust.

Gaining stream Streams that gain water from the inflow of groundwater through the streambed.

Gas hydrates Compact chemical structures made of water and natural gas (usually methane) that occur in permafrost and under the ocean floor at depths greater than 525 meters (1720 feet).

Geodynamo The generation and maintenance of Earth's magnetic field by the rising iron-rich fluid in the outer core.

Geologic map Graphic depiction of an area of geologic study, with labels and annotations.

Geologic structure See *Rock structure*.

Geologic time scale The division of Earth history into blocks of time—eons, eras, periods, and epochs. The time scale was created using relative dating principles.

Geologic time The span of time since the formation of Earth, about 4.6 billion years.

Geology The science that examines Earth, its form and composition, and the changes that it has undergone and is undergoing.

Geosphere The solid Earth; one of Earth's four basic spheres.

Geothermal energy Natural steam used for power generation.

Geothermal gradient The gradual increase in temperature with depth in the crust. The average is 30°C per kilometer in the upper crust.

Geyser A fountain of hot water ejected periodically from the ground.

Glacial budget The balance, or lack of balance, between ice formation at the upper end of a glacier and ice loss in the zone of wastage.

Glacial drift An all-embracing term for sediments of glacial origin, no matter how, where, or in what shape they were deposited.

Glacial erratic An ice-transported boulder that was not derived from the bedrock near its present site.

Glacial striations Scratches and grooves on bedrock caused by glacial abrasion.

Glacial trough A mountain valley that has been widened, deepened, and straightened by a glacier.

Glacier A thick mass of ice originating on land from the compaction and recrystallization of snow that shows evidence of past or present flow.

Glass (volcanic) Natural glass that is produced when molten lava cools too rapidly to permit recrystallization. Volcanic glass is a solid composed of unordered atoms.

Glassy A term used to describe the texture of certain igneous rocks, such as obsidian, that contain no crystals.

Gneiss Medium- to coarse-grained banded metamorphic rocks in which granular and elongated minerals dominate.

Gneissic texture A texture of metamorphic rocks in which dark and light silicate minerals are separated, giving the rock a banded appearance.

Gondwanaland The southern portion of Pangaea consisting of South America, Africa, Australia, India, and Antarctica.

Graben A valley formed by the downward displacement of a fault-bounded block.

Graded bed A sediment layer characterized by a decrease in sediment size from bottom to top.

Graded stream A stream that has the correct channel characteristics to maintain exactly the velocity required to transport the material supplied to it.

Gradient The slope of a stream, generally expressed as the vertical drop over a fixed distance.

Granite An abundant, coarse-grained igneous rock composed of about 10–20 percent quartz and 50 percent potassium feldspar. Granite is used as a building material.

Granitic composition A compositional group of igneous rocks indicating the rock is composed almost entirely of light-colored silicates.

Gravitational collapse The gradual subsidence of mountains caused by lateral spreading of weak material located deep within these structures.

Great Oxygenation Event A time about 2.5 billion years ago, when a significant amount of oxygen appeared in the atmosphere.

Greenhouse effect The transmission of short-wave solar radiation by the atmosphere coupled with the selective absorption of longer-wavelength terrestrial radiation, especially by water vapor and carbon dioxide, resulting in warming of the atmosphere.

Groin A short wall built at a right angle to the seashore to trap moving sand.

Ground moraine An undulating layer of till deposited as an ice front retreats.

Groundmass The matrix of smaller crystals within an igneous rock that has porphyritic texture.

Groundwater Water in the zone of saturation.

Guyot A submerged, flat-topped seamount.

Gymnosperm A group of seed-bearing plants that includes conifers and Ginkgo. The term means "naked seed," a reference to the unenclosed condition of the seeds.

H

Habit Refers to the common or characteristic shape of a crystal or an aggregate of crystals.

Hadean The earliest time interval (eon) of Earth history. The time before the planet's first rocks.

Half graben A tilted fault block in which the higher side is associated with mountainous topography and the lower side is a basin that fills with sediment.

Half-life The time required for one-half of the atoms of a radioactive substance to decay.

Hanging valley A tributary valley that enters a glacial trough at a considerable height above the floor of the trough.

Hanging wall block The rock surface immediately above a fault.

Hardness A mineral's resistance to scratching and abrasion.

Head (stream) The beginning or source area for a stream. Also called the headwaters.

Headward erosion The extension upslope of the head of a valley due to erosion.

Historical geology A major division of geology that deals with the origin of Earth and its development through time. Usually involves the study of fossils and their sequence in rock beds.

Hogback A narrow, sharp-crested ridge formed by the upturned edge of a steeply dipping bed of resistant rock.

Horizon A layer in a soil profile.

Horn A pyramid-like peak formed by glacial action in three or more cirques surrounding a mountain summit.

Hornblende A dark green to black mineral of the amphibole group, often found in igneous rocks.

Hornfels A fine-grained nonfoliated metamorphic rock formed from various minerals.

Horst An elongate, uplifted block of crust bounded by faults.

Host rock Pre-existing crustal rocks intruded by magma. Host rock may be displaced or assimilated by magmas.

Hot spot A concentration of heat in the mantle, capable of producing magma that, in turn, extrudes onto Earth's surface. The intraplate volcanism that produced the Hawaiian Islands is one example.

Hot spot tracks A chain of volcanic structures produced as a lithospheric plate moves over a mantle plume.

Hot spring A spring in which the water is 6–9°C (10–15°F) warmer than the mean annual air temperature of its locality.

Humus Organic matter in soil that is produced by the decomposition of plants and animals.

Hurricane A tropical cyclonic storm having winds in excess of 119 kilometers (74 miles) per hour.

Hydraulic conductivity A factor relating to groundwater flow; it is a coefficient that takes into account the permeability of the aquifer and the viscosity of the fluid.

Hydraulic fracturing A method of opening up pore space in otherwise impermeable rocks, permitting natural gas to flow out into wells.

Hydraulic gradient The slope of the water table. It is determined by finding the height difference between two points on the water table and dividing by the horizontal distance between the two points.

Hydroelectric power Electricity generated by falling water that is used to drive turbines.

Hydrogenous sediment Seafloor sediment consisting of minerals that crystallize from seawater. An important example is manganese nodules.

Hydrologic cycle The unending circulation of Earth's water supply. The cycle is powered by energy from the Sun and is characterized by continuous exchanges of water among the oceans, the atmosphere, and the continents.

Hydrolysis A chemical weathering process in which minerals are altered by chemically reacting with water and acids.

Hydrosphere The water portion of our planet; one of the traditional subdivisions of Earth's physical environment.

Hydrothermal metamorphism Chemical alterations that occur as hot, ion-rich water circulates through fractures in rock.

Hydrothermal solution The hot, watery solution that escapes from a mass of magma during the latter stages of crystallization. Such solutions may alter the surrounding country rock and are frequently the source of significant ore deposits.

Hypocenter See *Focus (earthquake)*.

Hypothesis A tentative explanation that is then tested to determine if it is valid.

I

Ice cap A mass of glacial ice covering a high upland or plateau and spreading out radially.

Ice sheet A very large, thick mass of glacial ice flowing outward in all directions from one or more accumulation centers.

Ice shelf A large, relatively flat mass of floating ice that forms where glacial ice flows into bays and that extends seaward from the coast but remains attached to the land along one or more sides.

Iceberg A mass of floating ice produced by a calving glacier. Usually 20 percent or less of the iceberg protrudes above the waterline.

Ice-contact deposit An accumulation of stratified drift deposited in contact with a supporting mass of ice.

Igneous rock Rock formed from the crystallization of magma.

Immature soil A soil that lacks horizons.

Impact crater A depression that results from collisions with bodies such as asteroids and comets.

Impact metamorphism Metamorphism that occurs when meteorites strike Earth's surface.

Incised meander A meandering channel that flows in a steep, narrow valley. It forms either when an area is uplifted or when the base level drops.

Inclusion A piece of one rock unit that is contained within another. Inclusions are used in relative dating. The rock mass adjacent to the one containing the inclusion must have been there first in order to provide the fragment.

Index fossil A fossil that is associated with a particular span of geologic time.

Index mineral A mineral that is a good indicator of the metamorphic environment in which it formed. Used to distinguish different zones of regional metamorphism.

Inertia A property by which objects at rest tend to remain at rest, and objects in motion tend to stay in motion unless either is acted upon by an outside force.

Infiltration The movement of surface water into rock or soil through cracks and pore spaces.

Infiltration capacity The maximum rate at which soil can absorb water.

Inner core The solid innermost layer of Earth, about 1216 kilometers (754 miles) in radius.

Inner planets The innermost planets of our solar system, which include Mercury, Venus, Earth, and Mars. Also known as the terrestrial planets because of their Earth-like internal structure and composition.

Inselberg An isolated mountain remnant characteristic of the late stage of erosion in a mountainous arid region.

Intensity (earthquake) A measure of the degree of earthquake shaking at a given locale, based on the amount of damage.

Interface A common boundary where different parts of a system interact.

Interior drainage A discontinuous pattern of intermittent streams that do not flow to the ocean.

Intermediate composition A compositional group of igneous rocks that contains at least 25 percent dark silicate minerals. The other dominant mineral is plagioclase feldspar.

Internal process A process such as mountain building or volcanism that derives its energy from Earth's interior and elevates Earth's surface.

Intraplate volcanism Igneous activity that occurs within a tectonic plate, away from plate boundaries.

Intrusion See *Pluton*.

Intrusive rock Igneous rock that formed below Earth's surface.

Ion An atom or a molecule that possesses an electrical charge.

Ionic bond A chemical bond between two oppositely charged ions that is formed by the transfer of valence electrons from one atom to the other.

Iron meteorite One of the three main categories of meteorites. This group is composed largely of iron, with varying amounts of nickel (5–20 percent). Most meteorite finds are irons.

Island arc See *Volcanic island arc*.

Isostasy The concept that Earth's crust is "floating" in gravitational balance upon the material of the mantle.

Isostatic adjustment Compensation of the lithosphere when weight is added or removed. When weight is added, the lithosphere responds by subsiding, and when weight is removed, there is uplift.

Isotopes Varieties of the same element that have different mass numbers; their nuclei contain the same number of protons but different numbers of neutrons.

J

Jetties A pair of structures extending into the ocean at the entrance to a harbor or river that are built for the purpose of protecting against storm waves and sediment deposition.

Joint A fracture in rock along which there has been no movement.

Jovian planet One of the Jupiter-like planets, Jupiter, Saturn, Uranus, and Neptune. These planets have relatively low densities.

K

Kame A steep-sided hill composed of sand and gravel, originating when sediment collected in openings in stagnant glacial ice.

Kame terrace A narrow, terracelike mass of stratified drift deposited between a glacier and an adjacent valley wall.

Karst A type of topography formed on soluble rock (especially limestone) primarily by dissolution. It is characterized by sinkholes, caves, and underground drainage.

Kettle holes Depressions created when blocks of ice become lodged in glacial deposits and subsequently melt.

Klippe A remnant or an outlier of a thrust sheet that was isolated by erosion.

Kuiper belt A region outside the orbit of Neptune where most short-period comets are thought to originate.

L

Laccolith A massive igneous body intruded between preexisting strata.

Lag time The amount of time between a rainstorm and the occurrence of flooding.

Lahar A debris flow on the slopes of a volcano that results when unstable layers of ash and debris become saturated and flow downslope, usually following stream channels.

Laminar flow The movement of water particles in straight-line paths that are parallel to the channel. The water particles move downstream without mixing.

Large igneous province Voluminous accumulations of lava extruded along fissures that produce broad, flat features that are also referred to as basalt plateaus.

Lateral continuity (principle of) A principle which states that sedimentary beds originate as continuous layers that extend in all directions until they grade into a different type of sediment or thin out at the edge of a sedimentary basin.

Lateral moraine A ridge of till along the sides of a valley glacier composed primarily of debris that fell to the glacier from the valley walls.

Laterite A red, highly leached soil type found in the tropics that is rich in oxides of iron and aluminum.

Laurasia The northern portion of Pangaea, consisting of North America and Eurasia.

Lava Magma that reaches Earth's surface.

Lava dome A bulbous mass associated with an old-age volcano, produced when thick lava is slowly squeezed from the vent. Lava domes may act as plugs to deflect subsequent gaseous eruptions.

Lava tube A tunnel in hardened lava that acts as a horizontal conduit for lava flowing from a volcanic vent. Lava tubes allow fluid lavas to advance great distances.

Law A formal statement of the regular manner in which a natural phenomenon occurs under given conditions.

Law of Constancy of Interfacial Angles A law which states that the angle between equivalent faces of the same mineral is always the same.

Leaching The depletion of soluble materials from the upper soil by downward-percolating water.

Light silicate A silicate mineral that lacks iron and/ or magnesium. Light silicates are generally lighter in color and have lower specific gravities than dark silicates.

Limestone A chemical sedimentary rock composed chiefly of calcite. Limestone can form by inorganic means or from biochemical processes.

Liquefaction The transformation of a stable soil into a fluid that is often unable to support buildings or other structures.

Lithification The process, generally involving cementation and/or compaction, of converting sediments to solid rock.

Lithosphere The rigid outer layer of Earth, including the crust and upper mantle.

Lithospheric mantle The uppermost portion of the mantle, below Earth's crust, ranging from a few kilometers thick to about 200 kilometers thick under continental crust.

Lithospheric plate A coherent unit of Earth's rigid outer layer that includes the crust and upper unit.

Local base level See *Temporary (local) base level*.

Loess Deposits of windblown silt, lacking visible layers, generally buff-colored, and capable of maintaining a nearly vertical cliff.

Long (L) waves Earthquake-generated waves that travel along the outer layer of Earth and are responsible for most of the surface damage. L waves have longer periods than other seismic waves.

Longitudinal dunes Long ridges of sand oriented parallel to the prevailing wind; these dunes form where sand supplies are limited.

Longitudinal profile A cross section of a stream channel along its descending course from the head to the mouth.

Longshore current A near-shore current that flows parallel to the shore.

Losing stream A stream that loses water to the groundwater system by outflow through the streambed.

Lower mantle See *Mesosphere*.

Low-velocity zone A subdivision of the mantle located between 100 and 250 kilometers (60 and 150 miles) and discernible by a marked decrease in the velocity of seismic waves. This zone does not encircle Earth.

Lunar breccia A lunar rock formed when angular fragments and dust are welded together by the heat generated by the impact of a meteoroid.

Lunar highlands See *Terrae*.

Lunar regolith A thin, gray layer on the surface of the Moon, consisting of loosely compacted, fragmented material believed to have been formed by repeated meteoritic impacts.

Luster The appearance or quality of light reflected from the surface of a mineral.

M

Mafic composition See *Basaltic composition*.

Magma A body of molten rock found at depth, including any dissolved gases and crystals.

Magma mixing The process of altering the composition of a magma through the mixing of material from another magma body.

Magmatic differentiation The process of generating more than one rock type from a single magma.

Magnetic field A phenomenon occurring around a magnet or an electric charge, characterized by a magnetic force at every point in the region. Earth's magnetic field is dipolar and extends from the core out to the solar wind.

Magnetic reversal A change in Earth's magnetic field from normal to reverse or vice versa.

Magnetic time scale A scale that shows the ages of magnetic reversals and is based on the polarity of lava flows of various ages.

Magnetometer A sensitive instrument used to measure the intensity of Earth's magnetic field at various points.

Magnitude (earthquake) An estimate of the total amount of energy released during an earthquake, based on seismic records.

Manganese nodules A type of hydrogenous sediment scattered on the ocean floor, consisting mainly of manganese and iron and usually containing small amounts of copper, nickel, and cobalt.

Mantle One of Earth's compositional layers. The solid rocky shell that extends from the base of the crust to a depth of 2900 kilometers (1800 miles).

Mantle plume A mass of hotter-than-typical mantle material that ascends toward the surface, where it may lead to igneous activity. These plumes of solid yet mobile material may originate as deep as the core–mantle boundary.

Marble A soft metamorphic rock formed from limestone or dolostone. Marble of various colors is used for building stones and monuments.

Maria The smooth areas on the Moon's surface that were incorrectly thought to be seas.

Marine terrace A wave-cut platform that has been exposed above sea level.

Mass extinction An event in which a large percentage of species become extinct.

Mass number The sum of the number of neutrons and protons in the nucleus of an atom.

Mass wasting The downslope movement of rock, regolith, and soil under the direct influence of gravity.

Massive An igneous pluton that is not tabular in shape.

Meander A looplike bend in the course of a stream.

Mechanical weathering The physical disintegration of rock, resulting in smaller fragments.

Medial moraine A ridge of till formed when lateral moraines from two coalescing alpine glaciers join.

Megathrust fault The plate boundary separating a subducting slab of oceanic lithosphere and the overlying plate.

Melt The liquid portion of magma excluding the solid crystals.

Mercalli intensity scale See *Modified Mercalli Intensity scale*.

Mesosphere The part of the mantle that extends from the core–mantle boundary to a depth of 660 kilometers (410 miles). Also known as the lower mantle.

Mesozoic era A time span on the geologic time scale between the Paleozoic and Cenozoic eras—from about 248 to 65.5 million years ago.

Metallic bond A chemical bond that is present in all metals that may be characterized as an extreme type of electron sharing in which the electrons move freely from atom to atom.

Metamorphic facies A group of associated minerals that are used to establish the pressures and temperatures at which rocks undergo metamorphism.

Metamorphic grade The degree to which a parent rock changes during metamorphism. It varies from low grade (low temperatures and pressures) to high grade (high temperatures and pressures).

Metamorphic rock Rock formed by the alteration of preexisting rock deep within Earth (but still in the solid state) by heat, pressure, and/or chemically active fluids.

Metamorphism The changes in mineral composition and texture of a rock subjected to high temperatures and pressures within Earth.

Meteor The luminous phenomenon observed when a meteoroid enters Earth's atmosphere and burns up; popularly called a "shooting star."

Meteor shower Numerous meteoroids traveling in the same direction and at nearly the same speed. They are thought to be material lost by comets.

Meteorite Any portion of a meteoroid that survives its traverse through Earth's atmosphere and strikes the surface.

Meteoroid Any small, solid particle that has an orbit in the solar system.

Microcontinents Relatively small fragments of continental crust that may lie above sea level, such as the island of Madagascar, or that may be submerged, as exemplified by the Campbell Plateau located near New Zealand.

Micrometeorite A very small meteorite that does not create sufficient friction to burn up in the atmosphere but slowly drifts down to Earth.

Mid-ocean ridge A continuous mountainous ridge on the floor of all the major ocean basins and varying in width from 500 to 5000 kilometers (300 to 3000 miles). The rifts at the crests of these ridges represent divergent plate boundaries.

Migmatite A rock exhibiting both igneous and metamorphic rock characteristics. Such rocks may form when light-colored silicate minerals melt and then crystallize, while the dark silicate minerals remain solid.

Mineral A naturally occurring, inorganic crystalline material with a unique chemical structure.

Mineral phase change A change that occurs when a mineral is subjected to intense pressure; in this change, the structure of a mineral may become unstable, causing its atoms to rearrange into a denser, more stable structure.

Mineral resource All discovered and undiscovered deposits of a useful mineral that can be extracted now or at some time in the future.

Mineralogy The study of minerals.

Modified Mercalli Intensity scale A 12-point scale developed to evaluate earthquake intensity, based on the amount of damage to various structures.

Mohorovičić discontinuity (Moho) The boundary separating the crust and the mantle, discernible by an increase in seismic velocity.

Mohs scale A series of 10 minerals used as a standard in determining hardness.

Moment magnitude A more precise measure of earthquake magnitude than the Richter scale that is derived from the amount of displacement that occurs along a fault zone.

Monocline A one-limbed flexure in strata. The strata are usually flat-lying or very gently dipping on both sides of the monocline.

Mountain belt A geographic area of roughly parallel and geologically connected mountain ranges developed as a result of plate tectonics.

Mouth The point downstream where a river empties into another stream or water body.

Mud crack A feature in some sedimentary rocks that forms when wet mud dries out, shrinks, and cracks.

Mudflow See *Debris flow*.

Muscovite A common member of the mica family of minerals, with excellent cleavage.

N

Natural levee An elevated landform composed of alluvium that parallels some streams and acts to confine their waters, except during floodstage.

Neap tide The lowest tidal range, occurring near the times of the first and third quarters of the Moon.

Near-shore The zone of a beach that extends from the low-tide shoreline seaward to where waves break at low tide.

Nebular theory A model for the origin of the solar system that supposes a rotating nebula of dust and gases that contracted to form the Sun and planets.

Negative-feedback mechanism As used in climatic change, any effect that is opposite the initial change and tends to offset it.

Neutron A subatomic particle found in the nucleus of an atom. The neutron is electrically neutral, with a mass approximately equal to that of a proton.

Nonclastic texture A term for the texture of sedimentary rocks in which the minerals form a pattern of interlocking crystals.

Nonconformity An unconformity in which older metamorphic or intrusive igneous rocks are overlain by younger sedimentary strata.

Nonferromagnesian silicate See *Light silicate*.

Nonflowing artesian well An artesian well in which water does not rise to the surface because the pressure surface is below ground level.

Nonfoliated rocks Metamorphic rocks that do not exhibit foliation.

Nonmetallic mineral resource A mineral resource that is not a fuel or processed for the metals it contains.

Nonrenewable resource A resource that forms or accumulates over such long time spans that it must be considered as fixed in total quantity.

Nonsilicates Mineral groups that lack silicas in their structures and account for less than 10 percent of Earth's crust.

Normal fault A fault in which the rock above the fault plane has moved down relative to the rock below.

Normal polarity A magnetic field the same as that which presently exists.

Nuclear fission The splitting of atomic nuclei into smaller nuclei, causing neutrons to be emitted and heat energy to be released.

Nucleus The small, heavy core of an atom that contains all of its positive charge and most of its mass.

Nuée ardente Incandescent volcanic debris buoyed up by hot gases that moves downslope in an avalanche fashion.

Numerical date The number of years that have passed since an event occurred.

O

Oblique slip fault A fault that exhibits both dip-slip and strike-slip movement.

Obsidian A volcanic glass of felsic composition.

Occultation The disappearance of light that results when one object passes behind an apparently larger one. For example, the passage of Uranus in front of a distant star.

Ocean basin A deep submarine region that lies beyond the continental margins.

Oceanic plateau An extensive region on the ocean floor that is composed of thick accumulations of pillow basalts and other mafic rocks that, in some cases, exceed 30 kilometers (20 miles) in thickness.

Oceanic ridge See *Mid-ocean ridge*.

Oceanic ridge system A continuous elevated zone on the floor of all the major ocean basins and varying in width from 500 to 5000 kilometers (300–3000 miles). The rifts at the crests of ridges represent divergent plate boundaries.

Octet rule A rule which states that atoms combine in order that each may have the electron arrangement of a noble gas (that is, the outer energy level contains eight neutrons).

Offshore The relatively flat submerged zone that extends from the breaker line to the edge of the continental shelf.

Oil sands Mixtures of clay, sand, water, and a black viscous form of petroleum known as bitumen.

Oil shale A fine-grained sedimentary rock that contains a solid mixture of organic compounds from which liquid hydrocarbons called shale oil can be produced.

Oil trap A geologic structure that allows for significant amounts of oil and gas to accumulate.

Olivine A high temperature, dark silicate mineral typically found in basalt.

Oort cloud A spherical shell composed of comets that orbit the Sun at distances generally greater than 10,000 times the Earth–Sun distance.

Open system A system in which both matter and energy flow into and out of the system. Most natural systems are of this type.

Ophiolite complex The sequence of rocks that make up the oceanic crust. The three-layer sequence includes an upper layer of pillow basalts, a middle zone of sheeted dikes, and a lower layer of gabbro.

Ore Usually a useful metallic mineral that can be mined at a profit. The term is also applied to certain nonmetallic minerals such as fluorite and sulfur.

Organic sedimentary rock Sedimentary rock composed of organic carbon from the remains of plants that died and accumulated on the floor of a swamp. Coal is the primary example.

Original horizontality Layers of sediment that are generally deposited in a horizontal or nearly horizontal position.

Orogenesis The processes that collectively result in the formation of mountains.

Outcrop Sites where bedrock is exposed at the surface.

Outer core A layer beneath the mantle about 2270 kilometers (1410 miles) thick, which has the properties of a liquid.

Outer planets The outermost planets of our solar system, which include Jupiter, Saturn, Uranus, and Neptune. They are also known as the Jovian planets.

Outgassing The escape of dissolved gases from molten rocks.

Outlet glacier A tongue of ice normally flowing rapidly outward from an ice cap or ice sheet, usually through mountainous terrain to the sea.

Outwash plain A relatively flat, gently sloping plain consisting of materials deposited by melt-water streams in front of the margin of an ice sheet.

Oxbow lake A curved lake that is created when a stream cuts off a meander.

Oxidation The removal of one or more electrons from an atom or ion. So named because elements commonly combine with oxygen.

Oxygen isotope analysis A method of deciphering past temperatures based on the precise measurement of the ratio between two isotopes of oxygen, ^{16}O and ^{18}O. Analysis is commonly made of seafloor sediments and cores from ice sheets.

P

P wave The fastest earthquake wave, which travels by compression and expansion of the medium.

Pahoehoe flow A lava flow with a smooth to ropy surface.

Paleoclimatology The study of ancient climates; the study of climate and climate change prior to the period of instrumental records using proxy data.

Paleomagnetism The natural remnant magnetism in rock bodies. The permanent magnetization acquired by rock that can be used to determine the location of the magnetic poles and the latitude of the rock at the time it became magnetized.

Paleontology The systematic study of fossils and the history of life on Earth.

Paleoseismology The study of the timing, location, and size of prehistoric earthquakes.

Paleozoic era A time span on the geologic time scale between the Precambrian and Mesozoic eras—from about 542 million to 251 million years ago.

Pangaea The proposed supercontinent that 200 million years ago began to break apart and form the present landmasses.

Parabolic dune A sand dune that is similar in shape to a barchan dune except that its tips point into the wind. These dunes often form along coasts that have strong onshore winds, abundant sand, and vegetation that partly covers the sand.

Parasitic cone A volcanic cone that forms on the flank of a larger volcano.

Parent material The material on which a soil develops.

Parent rock The rock from which a metamorphic rock formed.

Partial melting The process by which most igneous rocks melt. Since individual minerals have different melting points, most igneous rocks melt over a temperature range of a few hundred degrees. If the liquid is squeezed out after some melting has occurred, a melt with a higher silica content results.

Passive continental margin A margin that consists of a continental shelf, continental slope, and continental rise. They are not associated with plate boundaries and therefore experience little volcanism and few earthquakes.

Pater noster lakes A chain of small lakes in a glacial trough that occupies basins created by glacial erosion.

Pegmatite A very coarse-grained igneous rock (typically granite) commonly found as a dike associated with a large mass of plutonic rock that has smaller crystals. Crystallization in a water-rich environment is believed to be responsible for the very large crystals.

Pegmatitic texture A texture of igneous rocks in which the interlocking crystals are all larger than one centimeter in diameter.

Perched water table A localized zone of saturation above the main water table, created by an impermeable layer (aquiclude).

Peridotite An igneous rock of ultramafic composition thought to be abundant in the upper mantle.

Period A basic unit of the geologic time scale that is a subdivision of an era. Periods may be divided into smaller units called epochs.

Periodic table An arrangement of the elements in which atomic number increases from the left to right and elements with similar properties appear in columns called families or groups.

Permafrost Any permanently frozen subsoil. Usually found in the subarctic and arctic regions.

Permeability A measure of a material's ability to transmit water.

Phaneritic texture An igneous rock texture in which the crystals are roughly equal in size and large enough so the individual minerals can be identified without the aid of a microscope.

Phanerozoic eon The part of geologic time that is represented by rocks containing abundant fossil evidence. The eon extending from the end of the Proterozoic eon (540 million years ago) to the present.

Phase change In geology, the process by which the atomic structure of a mineral changes although its composition remains the same.

Phenocryst A conspicuously large crystal embedded in a matrix of finer-grained crystals.

Phreatic zone See *Saturated zone*.

Phyllite A metamorphic rock composed mainly of fine crystals of muscovite, chlorite, or both.

Physical geology A major division of geology that examines the materials of Earth and seeks to understand the processes and forces acting beneath and upon Earth's surface.

Piedmont glacier A glacier that forms when one or more alpine glaciers emerge from the confining walls of mountain valleys and spread out to create a broad sheet in the lowlands at the base of the mountains.

Pillow basalts Basaltic lava that solidifies in an underwater environment and develops a structure that resembles a pile of pillows.

Pipe A vertical conduit through which magmatic materials have passed.

Placer A deposit formed when heavy minerals are mechanically concentrated by currents, most commonly streams and waves. Placers are sources of gold, tin, platinum, diamonds, and other valuable minerals.

Plagioclase feldspar A relatively hard light silicate mineral containing both sodium and calcium ions that freely substitute for one another depending on the crystallization environment.

Planetesimal A solid celestial body that accumulated during the first stages of planetary formation.

Planetesimals aggregated into increasingly larger bodies, ultimately forming the planets.

Plastic flow A type of glacial movement that occurs within a glacier, below a depth of approximately 50 meters (165 feet), in which the ice is not fractured.

Plate See *Lithospheric plate*.

Plate resistance A force that counteracts plate motion as a subducting plate scrapes against an overriding plate.

Plate tectonics A theory which proposes that Earth's outer shell consists of individual plates that interact in various ways and thereby produce earthquakes, volcanoes, mountains, and the crust itself.

Playa The flat central area of an undrained desert basin.

Playa lake A temporary lake in a playa.

Pleistocene epoch An epoch of the Quaternary period that began about 2.6 million years ago and ended about 10,000 years ago. Best known as a time of extensive continental glaciation.

Plucking A process by which pieces of bedrock are lifted out of place by a glacier.

Plug See *Volcanic neck*.

Pluton A structure that results from the emplacement and crystallization of magma beneath the surface of Earth.

Plutonic rock Igneous rocks that form at depth. Named after Pluto, the god of the lower world in classical mythology.

Pluvial lake A lake formed during a period of increased rainfall. For example, this occurred in many nonglaciated areas during periods of ice advance elsewhere.

Point bar A crescent-shaped accumulation of sand and gravel deposited on the inside of a meander.

Polymerization The ability of silicate tetrahedra to link to one another in a variety of configurations, including chains, sheets, and three-dimensional structures.

Polymorphs Two or more minerals having the same chemical composition but different crystalline structures. Exemplified by the diamond and graphite forms of carbon.

Porosity The volume of open spaces in rock or soil.

Porphyritic texture An igneous rock texture characterized by two distinctively different crystal sizes. The larger crystals are called phenocrysts, whereas the matrix of smaller crystals is termed the groundmass.

Porphyroblastic texture A texture of metamorphic rocks in which particularly large grains (porphyroblasts) are surrounded by a fine-grained matrix of other minerals.

Porphyry An igneous rock that has a porphyritic texture.

Positive feedback mechanism As used in climatic change, any effect that acts to reinforce the initial change.

Potassium feldspar An abundant, relatively hard light silicate mineral containing potassium ions in its structure.

Pothole A depression formed in a stream channel by the abrasive action of the water's sediment load.

Precambrian All geologic time prior to the Phanerozoic eon. A term encompassing both the Archean and Proterozoic eons.

Precursor Events or changes that precede an earthquake and may provide a warning.

Primary (P) wave A type of seismic wave that involves alternating compression and expansion of the material through which it passes.

Principal shell The shell or energy level an electron occupies.

Principle of cross-cutting relationships The geologic principle stating that geologic features that cut across rocks must form after the rocks they cut through.

Principle of fossil succession A principle by which fossil organisms succeed one another in a definite and determinable order, and any time period can be recognized by its fossil content.

Principle of inclusions The principle stating that a rock mass adjacent to one containing inclusions must have been there first in order to provide the rock fragments, and is therefore the older rock mass.

Principle of original horizontality A principle by which layers of sediment are generally deposited in a horizontal or nearly horizontal position.

Principle of superposition A principle which states that in any undeformed sequence of sedimentary rocks, each bed is older than the one above and younger than the one below.

Proglacial lake A lake created when a glacier acts as a dam blocking the flow of a river or trapping glacial meltwater. The term refers to the position of such lakes just beyond the outer limits of a glacier.

Prokaryotes Refers to the cells or organisms such as bacteria whose genetic material is not enclosed in a nucleus.

Proterozoic eon The eon following the Archean and preceding the Phanerozoic. It extends between 2500 and 542 million years ago.

Proton A positively charged subatomic particle found in the nucleus of an atom.

Protoplanets A developing planetary body that grows by the accumulation of planetesimals.

Proxy data Data gathered from natural recorders of climate variability such as tree rings, ice cores, and ocean-floor sediments.

Pumice A light-colored, glassy vesicular rock commonly having a granitic composition.

Pyroclastic flow A highly heated mixture, largely of ash and pumice fragments, that travels down the flanks of a volcano or along the surface of the ground.

Pyroclastic material The volcanic rock ejected during an eruption. Pyroclastics include ash, bombs, and blocks.

Pyroclastic texture An igneous rock texture resulting from the consolidation of individual rock fragments that are ejected during a violent volcanic eruption.

Q

Quarrying Removing loosened blocks from the bed of a channel during times of high flow rates.

Quartz A common silicate mineral consisting entirely of silicon and oxygen that resists weathering.

Quartzite A hard metamorphic rock formed from quartz sandstone.

Quaternary period The most recent period on the geologic time scale. It began about 2.6 million years ago and extends to the present.

R

Radial pattern A system of streams running in all directions, away from a central elevated structure, such as a volcano.

Radioactive decay The spontaneous decay of certain unstable atomic nuclei.

Radioactivity See *Radioactive decay*.

Radiocarbon (carbon-14) dating Dating of events from the very recent geologic past (the past few tens of thousands of years) based on the fact that the radioactive isotope of carbon is produced continuously in the atmosphere.

Radiometric dating The procedure of calculating the absolute ages of rocks and minerals that contain certain radioactive isotopes.

Rainshadow desert A dry area on the lee side of a mountain range. Many middle-latitude deserts are of this type.

Rapids A part of a stream channel in which the water suddenly begins flowing more swiftly and turbulently because of an abrupt steepening of the gradient.

Rays Bright streaks that appear to radiate from certain craters on the lunar surface. The rays consist of fine debris ejected from the primary crater.

Recessional moraine An end moraine formed as the ice front stagnated during glacial retreat.

Recharge area An area where groundwater is replenished.

Recrystallization The formation of new mineral crystals in a rock that tend to be larger than the original crystals.

Rectangular pattern A drainage pattern characterized by numerous right angle bends that develops on jointed or fractured bedrock.

Recurrence interval The average time interval between occurrences of hydrologic events such as floods of a given or greater magnitude.

Reflection (seismic) The redirection of some waves back to the surface when seismic waves hit a boundary between different Earth materials.

Refraction See *Wave refraction*.

Regional metamorphism Metamorphism associated with large-scale mountain building.

Regolith The layer of rock and mineral fragments that nearly everywhere covers Earth's land surface.

Rejuvenation A change in relation to base level, often caused by regional uplift, which causes the forces of erosion to intensify.

Relative dating A process of determining the chronological order of events by placing rocks and structures in their proper sequence or order.

Renewable resource A resource that is virtually inexhaustible or that can be replenished over relatively short time spans.

Reserve Already identified deposits from which minerals can be extracted profitably.

Reservoir rock The porous, permeable portion of an oil trap that yields oil and gas.

Residual soil Soil developed directly from the weathering of the bedrock below.

Return period See *Recurrence interval*.

Reverse fault A fault in which the material above the fault plane moves up in relation to the material below.

Reverse polarity A magnetic field opposite that which presently exists.

Rhyolite The fine-grained equivalent of the igneous rock granite, composed primarily of the light-colored silicates.

Richter scale A scale of earthquake magnitude based on the amplitude of the largest seismic wave.

Ridge push A mechanism that may contribute to plate motion. It involves the oceanic lithosphere sliding down the oceanic ridge under the pull of gravity.

Rift valley A long, narrow trough bounded by normal faults. It represents a region where divergence is taking place.

Rills Tiny channels that develop as unconfined flow begins producing threads of current.

Ring of Fire The zone of active volcanoes surrounding the Pacific Ocean.

Rip current A strong, narrow surface or near-surface current of short duration and high speed that moves seaward through the breaker zone at nearly a right angle to the shore.

Ripple marks Small waves of sand that develop on the surface of a sediment layer by the action of moving water or air.

River A general term for a stream that carries a substantial amount of water and has numerous tributaries.

Roche moutonnée An asymmetrical knob of bedrock that is formed when glacial abrasion smoothes the gentle slope facing the advancing ice sheet and plucking steepens the opposite side as the ice overrides the knob.

Rock A consolidated mixture of minerals.

Rock avalanche Very rapid downslope movement of rock and debris. These rapid movements may be aided by a layer of air trapped beneath the debris, and they have been known to reach speeds of over 200 kilometers (125 miles) per hour.

Rock cleavage The tendency of rocks to split along parallel, closely spaced surfaces. These surfaces are often highly inclined to the bedding planes in the rock.

Rock cycle A model that illustrates the origin of the three basic rock types and the interrelatedness of Earth materials and processes.

Rock flour Ground-up rock produced by the grinding effect of a glacier.

Rock structure All features created by the processes of deformation from minor fractures in bedrock to a major mountain chain.

Rock-forming minerals The relatively few minerals that make up most of the rocks in Earth's crust.

Rockslide The rapid slide of a mass of rock downslope, along planes of weakness.

Runoff Water that flows over land rather than infiltrating into the ground.

S

S wave An earthquake wave, slower than a P wave, that travels only in solids.

Salinity The proportion of dissolved salts to pure water, usually expressed in parts per thousand (‰).

Salt flat A white crust on the ground that is produced when water evaporates and leaves behind its dissolved materials.

Saltation Transportation of sediment through a series of leaps or bounces.

Sandstone An abundant, durable sedimentary rock primarily composed of sand-size grains.

Schist Medium- to coarse-grained metamorphic rocks having a foliated texture, in which platy minerals dominate.

Schistosity A type of foliation that is characteristic of coarser-grained metamorphic rocks. Such rocks have a parallel arrangement of platy minerals such as the micas.

Scientific method The process by which researchers raise questions, gather data, and formulate and test scientific hypotheses.

Scoria Vesicular ejecta that is the product of basaltic magma.

Scoria cone See *Cinder cone*.

Sea arch An arch formed by wave erosion when caves on opposite sides of a headland unite.

Sea ice Frozen seawater that is associated with polar regions. The area covered by sea ice expands in winter and shrinks in summer.

Sea stack An isolated mass of rock standing just offshore, produced by wave erosion of a headland.

Seafloor spreading A hypothesis, first proposed in the 1960s by Harry Hess, which suggested that new oceanic crust is produced at the crests of mid-ocean ridges, which are the sites of divergence.

Seamount An isolated volcanic peak that rises at least 1000 meters (3300 feet) above the deep-ocean floor.

Seawall A barrier constructed to prevent waves from reaching the area behind the wall. Its purpose is to defend property from the force of breaking waves.

Secondary (S) wave A seismic wave that involves oscillation perpendicular to the direction of propagation.

Secondary enrichment The concentration of minor amounts of metals that are scattered through unweathered rock into economically valuable concentrations by weathering processes.

Sediment Unconsolidated particles created by the weathering and erosion of rock by chemical precipitation from solution in water, or from the secretions of organisms, and transported by water, wind, or glaciers.

Sedimentary environment See *Environment of deposition*.

Sedimentary rock Rock formed from the weathered products of preexisting rocks that have been transported, deposited, and lithified.

Seiche The sloshing of water in an enclosed basin, generated by seismic waves.

Seismic gap A segment of an active fault zone that has not experienced a major earthquake over a span when most other segments have. Such segments are probable sites for future major earthquakes.

Seismic reflection profile A method of viewing the rock structure beneath a blanket of sediment by using strong, low-frequency sound waves that penetrate the sediments and reflect off the contacts between rock layers and fault zones.

Seismic sea wave A rapidly moving ocean wave, generated by earthquake activity, that is capable of inflicting heavy damage in coastal regions.

Seismic tomography A technique in which seismic signals collected from many earthquakes are used to make images that map locations within Earth where seismic waves travel slower or faster.

Seismic wave A rapidly moving ocean wave generated by earthquake activity capable of inflicting heavy damage in coastal regions.

Seismogram A record made by a seismograph.

Seismograph An instrument that records earthquake waves.

Seismology The study of earthquakes and seismic waves.

Settling velocity The speed at which a particle falls through a still fluid. The size, shape, and specific gravity of particles influence settling velocity.

Shadow zone The zone between 105 and 140 degrees from an earthquake epicenter. Direct waves do not penetrate the shadow zone because of refraction by Earth's core.

Shale The most common sedimentary rock, consisting of silt- and clay-size particles.

Shear Stress that causes two adjacent parts of a body to slide past one another.

Sheet flow Runoff moving in unconfined thin sheets.

Sheeted dike complex A large group of nearly parallel dikes.

Sheeting A mechanical weathering process that is characterized by the splitting off of slablike sheets of rock.

Shelf break The point at which a rapid steepening of the gradient occurs, marking the outer edge of the continental shelf and the beginning of the continental slope.

Shield A large, relatively flat expanse of ancient igneous and metamorphic rocks within the craton.

Shield volcano A broad, gently sloping volcano built from fluid basaltic lavas.

Shore Seaward of the coast, a zone that extends from the highest level of wave action during storms to the lowest tide level.

Shoreline The line that marks the contact between land and sea. It migrates up and down as the tide rises and falls.

Silicate mineral Any one of numerous minerals that have the silicon-oxygen tetrahedron as their basic structure.

Silicon-oxygen tetrahedron A structure composed of four oxygen atoms surrounding a silicon atom that constitutes the basic building block of silicate minerals.

Sill A tabular igneous body that was intruded parallel to the layering of preexisting rock.

Sinkhole A depression produced in a region where soluble rock has been removed by groundwater.

Slab pull A mechanism that contributes to plate motion in which cool, dense oceanic crust sinks into the mantle and "pulls" the trailing lithosphere along.

Slab suction One of the driving forces of plate motion, which arises from the drag of the subducting plate on the adjacent mantle. It is an induced mantle circulation that pulls both the subducting and overriding plates toward the trench.

Slate A very fine-grained metamorphic rock containing platy minerals and having excellent rock cleavage.

Slaty cleavage A type of foliation that is characteristic of slates, in which there is a parallel arrangement of fine-grained metamorphic minerals.

Slickenslide Polished and grooved rock surfaces etched as crustal rocks slide past one another.

Slide A movement common to mass-wasting processes in which the material moving downslope remains fairly coherent and moves along a well-defined surface.

Slip face The steep, leeward surface of a sand dune that maintains a slope of about 34 degrees.

Slump The downward slipping of a mass of rock or unconsolidated material moving as a unit along a curved surface.

Small solar system body Solar system objects such as asteroids, comets, and meteoroids.

Snowball Earth A hypothesis that relates a period of global glaciation to the Great Oxygenation Event.

Snowfield An area where snow persists throughout the year.

Snowline The lower limit of perennial snow.

Soil A combination of mineral and organic matter, water, and air; the portion of the regolith that supports plant growth.

Soil horizon A layer of soil that has identifiable characteristics produced by chemical weathering and other soil-forming processes.

Soil profile A vertical section through a soil, showing its succession of horizons and the underlying parent material.

Soil taxonomy A soil classification system that consists of six hierarchical categories, based on observable soil characteristics. The system recognizes 12 soil orders.

Soil texture The relative proportions of clay, silt, and sand in a soil. A soil's texture strongly influences its ability to retain and transmit water and air.

Solar nebula The cloud of interstellar gas and/or dust from which the bodies of our solar system formed.

Solifluction The slow, downslope flow of water-saturated materials common to permafrost areas.

Solum The O, A, and B horizons in a soil profile. Living roots and other plant and animal life are largely confined to this zone.

Sonar An instrument that uses acoustic signals (sound energy) to measure water depths. Sonar is an acronym for *so*und *na*vigation and *r*anging.

Sorting The degree of similarity in particle size in sediment or sedimentary rock.

Specific gravity The ratio of a substance's weight to the weight of an equal volume of water.

Speleothem A collective term for the dripstone features found in caverns.

Spheroidal weathering Any weathering process that tends to produce a spherical shape from an initially blocky shape.

Spit An elongate ridge of sand that projects from the land into the mouth of an adjacent bay.

Spontaneous subduction A process that occurs at Mariana-type subduction zones in which old, dense lithosphere sinks into the mantle at a steep angle by its own weight creating a deep trench.

Spreading center See *Divergent plate boundary*.

Spring A flow of groundwater that emerges naturally at the ground surface.

Spring tide The highest tidal range. Occurs near the times of the new and full moons.

Stable platform That part of a craton that is mantled by relatively undeformed sedimentary rocks and underlain by a basement complex of igneous and metamorphic rocks.

Stalactite An icicle-like structure that hangs from the ceiling of a cavern.

Stalagmite A columnlike form that grows upward from the floor of a cavern.

Star dune An isolated hill of sand that exhibits a complex form and develops where wind directions are variable.

Steno's Law See *Law of Constancy of Interfacial Angles*.

Steppe One of the two types of dry climate. A marginal and more humid variant of the desert that separates it from bordering humid climates.

Stock A pluton similar to but smaller than a batholith.

Stony meteorite One of the three main categories of meteorites. Such meteorites are composed largely of silicate minerals with inclusions of other minerals.

Stony-iron meteorite One of the three main categories of meteorites. This group, as the name implies, is a mixture of iron and silicate minerals.

Storm surge The abnormal rise of the sea along a shore as a result of strong winds.

Strain An irreversible change in the shape and size of a rock body caused by stress.

Strata Parallel layers of sedimentary rock.

Stratified drift Sediments deposited by glacial meltwater.

Stratosphere The layer of the atmosphere immediately above the troposphere, characterized by increasing temperatures with height, due to the concentration of ozone.

Stratovolcano See *Composite cone*.

Streak The color of a mineral in powdered form.

Stream A general term to denote the flow of water within any natural channel. Thus, a small creek and a large river are both streams.

Stream piracy Diversion of the drainage of one stream that results from the headward erosion of another stream.

Stream valley The channel, valley floor, and sloping valley walls of a stream.

Stress The force per unit area acting on any surface within a solid.

Striations (glacial) Scratches or grooves in a bedrock surface caused by the grinding action of a glacier and its load of sediment.

Strike The compass direction of the line of intersection created by a dipping bed or fault and a horizontal surface. A strike is always perpendicular to the direction of dip.

Strike-slip fault A fault along which movement occurs horizontally.

Stromatolites Distinctively layered mounds of calcium carbonate, which are fossil evidence for the existence of ancient microscopic bacteria.

Subduction The process by which oceanic lithosphere plunges into the mantle along a convergent zone.

Subduction erosion A process in subduction zones in which sediment and rock are scraped off the bottom of the overriding plate and transported into the mantle.

Subduction zone A long, narrow zone where one lithospheric plate descends beneath another.

Subduction zone metamorphism High-pressure, low-temperature metamorphism that occurs where sediments are carried to great depths by a subducting plate.

Submarine canyon A seaward extension of a valley that was cut on the continental shelf during a time when sea level was lower, or a canyon carved into the outer continental shelf, slope, and rise by turbidity currents.

Submergent coast A coast whose form is largely a result of the partial drowning of a former land surface due to a rise of sea level or subsidence of the crust, or both.

Subsoil A term applied to the *B* horizon of a soil profile.

Sulfur dioxide A gas with the chemical formula SO_2, that is associated naturally with volcanic activity, and as a waste gas (air pollutant) with the burning of fossil fuels and various industrial processes.

Sunspot A dark area on the Sun that is associated with powerful magnetic storms that extend from the Sun's surface deep into the interior.

Supercontinent A large landmass that contains all, or nearly all, of the existing continents.

Supercontinent cycle The idea that the rifting and dispersal of one supercontinent is followed by a long period during which the fragments gradually reassemble into a new supercontinent.

Supernova An exploding star that increases its brightness many thousands of times.

Superposed stream A stream that cuts through a ridge lying across its path. The stream established its course on uniform layers at a higher level without regard to underlying structures and subsequently downcut.

Surf A collective term for breakers; also the wave activity in the area between the shoreline and the outer limit of breakers.

Surface waves Seismic waves that travel along the outer layer of Earth.

Surge A period of rapid glacial advance. Surges are typically sporadic and short-lived.

Suspended load Fine sediment carried within the body of flowing water or air.

Suture A zone along which two crustal fragments are jointed together. For example, following a continental collision, the two continental blocks are sutured together.

Swells Wind-generated waves that have moved into an area of weaker winds or calm.

Syncline A linear downfold in sedimentary strata; the opposite of anticline.

System A group of interacting or interdependent parts that form a complex whole.

T

Tablemount See *Guyot.*

Tabular Describing a feature such as an igneous pluton that has two dimensions that are much longer than the third.

Talus An accumulation of rock debris at the base of a cliff.

Tarn A small lake in a cirque.

Tectonic plate See *Lithospheric plate.*

Tectonic structure A basic geologic feature, such as a fold, fault, or rock foliation, that results from forces associated with the interaction of tectonic plates.

Tectonics The study of the large-scale processes that collectively deform Earth's crust.

Temporary (local) base level The level of a lake, resistant rock layer, or any other base level that stands above sea level.

Tenacity Describes a mineral's toughness or resistance to breaking or deforming.

Tensional stress The type of stress that tends to pull a body apart.

Tephra See *Pyroclastic materials.*

Terminal moraine The end moraine that marks the farthest advance of a glacier.

Terrace A flat, benchlike structure produced by a stream, which was left elevated as the stream cut downward.

Terrae The extensively cratered highland areas of the Moon. Also known as *lunar highlands.*

Terrane A crustal block bounded by faults, whose geologic history is distinct from the histories of adjoining crustal blocks.

Terrestrial planet One of the Earthlike planets: Mercury, Venus, Earth, and Mars. These planets have similar densities.

Terrigenous sediment Seafloor sediments derived from terrestrial weathering and erosion.

Texture The size, shape, and distribution of the particles that collectively constitute a rock.

Theory A well-tested and widely accepted view that explains certain observable facts.

Thermal metamorphism See *Contact metamorphism.*

Thermosphere The region of the atmosphere immediately above the mesosphere and characterized by increasing temperatures due to absorption of very shortwave solar energy by oxygen.

Thrust fault A low-angle reverse fault.

Tidal current The alternating horizontal movement of water associated with the rise and fall of the tide.

Tidal delta A deltalike feature created when a rapidly moving tidal current emerges from a narrow inlet and slows, depositing its load of sediment.

Tidal flat A marshy or muddy area that is alternately covered and uncovered by the rise and fall of the tide.

Tide The periodic change in the elevation of the ocean surface.

Till Unsorted sediment deposited directly by a glacier.

Tillite A rock formed when glacial till is lithified.

Tombolo A ridge of sand that connects an island to the mainland or to another island.

Topset bed An essentially horizontal sedimentary layer deposited on top of a delta during flood-stage.

Tower karst Steep-sided hills formed in wet tropical and subtropical regions with thick beds of highly jointed limestone. The limestone is dissolved by groundwater, leaving residual towers.

Trace gases Gases present in Earth's atmosphere at concentrations much lower than that of carbon dioxide. Methane and nitrous oxide are important trace gases that absorb outgoing radiation and help warm the atmosphere.

Transform fault A major strike-slip fault that cuts through the lithosphere and accommodates motion between two plates.

Transform fault boundary A boundary in which two plates slide past one another without creating or destroying lithosphere.

Transition zone The lowest portion of the upper mantle.

Transpiration The release of water vapor to the atmosphere by plants.

Transported soil Soil that forms on unconsolidated deposits.

Transverse dunes A series of long ridges oriented at right angles to the prevailing wind; these dunes form where vegetation is sparse and sand is very plentiful.

Travertine A form of limestone ($CaCO_3$) that is deposited by hot springs or as a cave deposit.

Trellis drainage pattern A system of streams in which nearly parallel tributaries occupy valleys cut in folded strata.

Trench See *Deep-ocean trench.*

Trigger A factor or event, such as soil saturation, oversteepened slopes, removal of vegetation, or ground shaking, that initiates downslope movement of rock material.

Triple junction A point where three lithospheric plates meet.

Troposphere The lowermost layer of the atmosphere. It is generally characterized by a decrease in temperature with height.

Truncated spurs Triangular-shaped cliffs produced when spurs of land that extend into a valley are removed by the great erosional force of a valley glacier.

Tsunami The Japanese word for a seismic sea wave.

Turbidite A turbidity current deposit characterized by graded bedding.

Turbidity current A downslope movement of dense, sediment-laden water created when sand and mud on the continental shelf and slope are dislodged and thrown into suspension.

Turbulent flow Erratic movement of water often characterized by swirling, whirlpool-like eddies. Most streamflow is of this type.

U

Ultimate base level Sea level; the lowest level to which stream erosion could lower the land.

Ultramafic composition A compositional group of igneous rocks containing mostly olivine and pyroxene.

Unconformity A surface that represents a break in the rock record, caused by erosion and nondeposition.

Uniformitarianism The concept that the processes that have shaped Earth in the geologic past are essentially the same as those operating today.

Unit cell The smallest group of atoms, ions, or molecules that form the building block of a crystal.

Unsaturated zone The area above the water table where openings in soil, sediment, and rock are not saturated but filled mainly with air.

Upper mantle The top portion of the mantle extending from the Moho to a depth of about 660 km and comprising the lithospheric mantle, asthenosphere, and transition zone.

V

Vadose zone See *Unsaturated zone*.

Valence electron The electrons involved in the bonding process; the electrons occupying the highest principal energy level of an atom.

Valley glacier See *Alpine glacier*.

Valley train A relatively narrow body of stratified drift deposited on a valley floor by meltwater streams that issue from the terminus of an alpine glacier.

Vein deposit A mineral that fills a fracture or fault in a host rock. Such deposits have a sheetlike, or tabular, form.

Vent The surface opening of a conduit or pipe.

Ventifact A cobble or pebble polished and shaped by the sandblasting effect of wind.

Vesicles Spherical or elongated openings on the outer portion of a lava flow that were created by escaping gases.

Vesicular texture A term applied to aphanitic igneous rocks that contain many small cavities called vesicles.

Viscosity A measure of a fluid's resistance to flow.

Volatiles Gaseous components of magma dissolved in the melt. Volatiles will readily vaporize (form a gas) at surface pressures.

Volcanic Pertaining to the activities, structures, or rock types of a volcano.

Volcanic bomb A streamlined pyroclastic fragment ejected from a volcano while still semimolten.

Volcanic cone A cone-shaped structure built by successive eruptions of lava and/or pyroclastic materials.

Volcanic island A seamount that has grown large enough to rise above sea level.

Volcanic island arc A chain of volcanic islands generally located a few hundred kilometers from a trench where there is active subduction of one oceanic plate beneath another.

Volcanic neck An isolated, steep-sided, erosional remnant consisting of lava that once occupied the vent of a volcano.

Volcanic rock See *Extrusive igneous rock*.

Volcano A mountain formed from lava and/or pyroclastics.

W

Wadati–Benioff zone The narrow zone of inclined seismic activity that extends from a trench downward into the asthenosphere.

Water gap A pass through a ridge or mountain in which a stream flows.

Water table The upper level of the saturated zone of groundwater.

Waterfall A precipitous drop in a stream channel that causes water to fall to a lower level.

Wave height The vertical distance between the trough and crest of a wave.

Wave of oscillation A water wave in which the wave form advances as the water particles move in circular orbits.

Wave of translation The turbulent advance of water created by breaking waves.

Wave period The time interval between the passage of successive crests at a stationary point.

Wave refraction A change in direction of waves as they enter shallow water. The portion of the wave in shallow water is slowed, which causes the waves to bend and align with the underwater contours.

Wave-cut cliff A seaward-facing cliff along a steep shoreline formed by wave erosion at its base and mass wasting.

Wave-cut platform A bench or shelf along a shore at sea level, cut by wave erosion.

Wavelength The horizontal distance separating successive crests or troughs.

Weather The state of the atmosphere at any given time.

Weathering The disintegration and decomposition of rock at or near the surface of Earth.

Welded tuff A pyroclastic deposit composed of particles fused together by the combination of heat still contained in the deposit after it has come to rest and the weight of overlying material.

Well An opening bored into the zone of saturation.

Wetted perimeter The total distance in a linear cross-section of a stream that is in contact with water.

Wilson Cycle See *Supercontinent cycle*.

Wind gap An abandoned water gap. These gorges typically result from stream piracy.

X

Xenolith An inclusion of unmelted country rock in an igneous pluton.

Xerophyte A plant that is highly tolerant of drought.

Y

Yardang A streamlined, wind-sculpted ridge that has the appearance of an inverted ship's hull that is oriented parallel to the prevailing wind.

Yazoo tributary A tributary that flows parallel to the main stream because a natural levee is present.

Z

Zone of accumulation The part of a glacier that is characterized by snow accumulation and ice formation. The outer limit of this zone is the snowline.

Zone of fracture The upper portion of a glacier consisting of brittle ice.

Zone of saturation The zone where all open spaces in sediment and rock are completely filled with water.

Zone of soil moisture A zone in which water is held as a film on the surface of soil particles and may be used by plants or withdrawn by evaporation. The uppermost subdivision of the unsaturated zone.

Zone of wastage The part of a glacier beyond the snowline, where annually there is a net loss of ice.

Index